Proceedings of the 1994 Snowmass Summer Study

PARTICLE AND NUCLEAR ASTROPHYSICS AND COSMOLOGY IN THE NEXT MILLENNIUM

Proceedings of the 1994 Snowmass Summer Study

PARTICLE AND NUCLEAR ASTROPHYSICS AND COSMOLOGY IN THE NEXT MILLENNIUM

Snowmass, Colorado June 29 – July 14, 1994

Editors

E W Kolb
Fermi National Laboratory

R D Peccei
Department of Physics, University of California at Los Angeles

Organized jointly by:
the Astrophysics Division of the American Physical Society,
the Division of Nuclear Physics of the American Physical Society,
and the Division of Particles and Fields of the American Physical Society

World Scientific
Singapore • New Jersey • London • Hong Kong

Published by

World Scientific Publishing Co Pte Ltd
P O Box 128, Farrer Road, Singapore 912805
USA office: Suite 1B, 1060 Main Street, River Edge, NJ 07661
UK office: 57 Shelton Street, Covent Garden, London WC2H 9HE

British Library Cataloguing-in-Publication Data
A catalogue record for this book is available from the British Library.

Proceedings of the 1994 Snowmass Summer Study
PARTICLE AND NUCLEAR ASTROPHYSICS AND COSMOLOGY IN THE NEXT MILLENNIUM

ISBN 981-02-2488-5

Printed in Singapore by Uto-Print

Snowmass '94

PARTICLE AND NUCLEAR ASTROPHYSICS AND COSMOLOGY IN THE NEXT MILLENNIUM

Organizing Committee:
E. Bertcshinger (MIT), E. Bloom (SLAC), R. Cahn (LBL), C. Fichtel (NASA), G. Fuller (UCSD), F. Halzen (Wisconsin), W. Haxton (U. Washington), E. Kolb (FNAL/Chicago), N. Koller (Rutgers), A. Olinto (Chicago), R. Peccei (UCLA), H. Robertson (LANL), B. Sadoulet (UCB), C. Sazama (FNAL)

Neutrinos *Wick Haxton (Univ. of Washington)*

Neutrino Masses
Boris Kayser (NSF) & Robert Lanou (Brown)
Neutrino Oscillations, Long Baseline and Short Baseline
Bill Reay (Kansas State) & Lincoln Wolfenstein (Carnegie-Mellon)
Solar Neutrinos
John Bahcall (IAS), Ken Lande (Penn) & S. Peter Rosen (UT Arlington)
Neutrino Astrophysics
David Cline (UCLA), George Fuller (UCSD) & Francis Halzen (Wisconsin)

Cosmic Rays *Jim Cronin (Chicago) & Jonathan Ormes (NASA/Goddard)*

Space-Based Gamma Ray Astronomy
Josh Grindlay (Harvard/CfA)
Ground-Based Gamma Ray Astronomy
Richard Lamb (Iowa State) & Rene Ong (Chicago)
Over the Knee and Cosmic Rays in Space
Todor Stanev (Bartol) & Simon Swordy (Chicago)
Highest Energy Cosmic Rays
Pierre Sokolsky (Utah) & Alan Watson (Leeds)

Gravitational Phenomena *Kip Thorne (Caltech)*

Quantum Aspects -- Black Holes and Cosmology
Larus Thorlacius (UCSB) & Valery Frolov (Alberta)
Black Hole Astrophysics
Elliott Bloom (SLAC) & Igor Novikov (NORDITA)
Gravity Waves
Leonid Grishchuk (Washington U.) & Rainer Weiss (MIT)

Low Background Experiments *Bernard Sadoulet (Berkeley)*

Proton Decay
Todd Haines (LANL)
Dark Matter
Charles Alcock (LLNL), Brian Dougherty (Stanford) & Leslie Rosenberg (MIT)
Other Underground Experiments
Frank Avignone (South Carolina)

Cosmology *Paul Steinhardt (Penn)*

Cosmological Models, Theory and Phenomenology
Josh Frieman (Fermilab)
Baryogenesis
Andy Cohen (Boston Univ.) & Larry McLerran (Minnesota)
Nucleosynthesis
Grant Mathews (LLNL) & David Schramm (Chicago)
Cosmic Background Radiation
Stephan Meyer (Chicago), George Smoot (LBL) & Edward Wright (UCLA)
Structure Formation (Simulation and Phenomenology)
Joel Primack (UCSC) & Alex Szalay (Johns Hopkins)
Observational Cosmology, Large-Scale Structure and Cosmological Parameters
Marc Davis (Berkeley) & Rich Kron (Fermilab/Chicago)

PREFACE

This book contains the Proceedings of the Summer Study entitled *Particle and Nuclear Astrophysics and Cosmology in the Next Millennium*, held in July, 1994 in Snowmass, Colorado. This intensive two-week workshop was jointly sponsored by three separate Divisions of the American Physical Society: the Division of Astrophysics, the Division of Nuclear Physics, and the Division of Particles and Fields. The principal goal of the Summer Study was to examine the opportunities emerging in this cross-disciplinary field, and help provide a critical evaluation of the future course of this developing discipline. The Summer Study's title aptly conveyed the necessary joining of forces in the different disciplines needed for future progress. *Snowmass '94* represented the first time that members of the high-energy and nuclear-physics communities met with the community of astronomers, astrophysicists, and cosmologists with the explicit purpose of exploring jointly areas of overlap and plotting a common future. This experiment proved to be both popular and successful, with nearly 450 participants attending the Summer Study.

The Summer Study started with a day and a half of plenary lectures, reviewing critical areas in particle and nuclear astrophysics and cosmology. The participants then split up into smaller working groups, organized under five main rubrics: neutrinos; cosmic rays; gravitational phenomena; low-background experiments; and cosmology. Within each of these main areas there were typically four or five working groups, which met together for a period of three to five days. Because different working groups met during mornings or afternoons, most Snowmass participants had the opportunity to be involved in three or four separate working groups. This really helped foster the interdisciplinary nature of the Summer Study. At the end of the Summer Study the conveners of the various working groups summarized the findings for all of the participants.

Crucial to the success of the Summer Study were the contributions of the five "Superconveners" (Wick Haxton, Jim Cronin, Bernard Sadoulet, Kip Thorne, and Paul Steinhardt) and of the over twenty working group conveners who provided the intellectual leadership at Snowmass.

The 1994 Snowmass Summer Study would not have been possible without the generous support of both the funding agencies and the national laboratories. We are grateful for the support provided to the Summer Study by the Department of Energy (High Energy Physics and Nuclear Physics Divisions); the International Science Foundation; the National Science Foundation (cross-divisional support from many divisions); the Center for Particle Astrophysics at Berkeley; Fermi National Accelerator Laboratory; the Lawrence Berkeley Laboratory; and the Stanford Linear Accelerator Center. The success of the Summer Study is due in great measure to the excellent Secretariat and Computing Support, which under the able leadership of Cynthia Sazama were able to resolve even the most vexing problems with magical ease. All participants owe a vote of thanks to both the Secretariat Staff (Mercy Bicad, Valerie MacLean, Sherry Melby, Joyce Rossi and Suzanne Weber) and the Computing Staff (Gina Carranza, Dave Fagan, Lisa Giacchetti, Donna Lamore, David Tang) for their efficiency and cheerful help.

Rocky Kolb
Roberto Peccei

CONTENTS

Neutrinos

Cosmic Rays

Gravitational Phenomena

Cosmology

Overview

Particle and Nuclear Astrophysics and Cosmology in the Next Millennium — Snowmass 1994

J. W. Cronin, Wick C. Haxton, Edward W. Kolb,
Roberto D. Peccei, Bernard Sadoulet, Paul J. Steinhardt, and Kip S. Thorne

Organizers and conveners of the 1994 Snowmass Summer Study

From June 29 to July 14, 1994, nearly 450 astronomers, astrophysicists, cosmologists, high-energy physicists, nuclear physicists, and gravitation physicists met in the mountains of Colorado to define and develop a vision for the future of an emerging interdisciplinary field. This book is the permanent record of their deliberations. The purpose of this introduction is to express some unifying themes, and provide an overview of the Snowmass summer study.

§1 Introduction

Snowmass 1994 brought together for the first time a very disparate, yet interconnected, group of astrophysicists, cosmologists, particle physicists, nuclear physicists, gravitation physicists, and astronomers for an intensive two-week Summer Study to discuss the gamut of problems that link them intellectually. The range of topics discussed was vast, but clear connections could be easily discerned. Thus, even though the Summer Study was organized in terms of five topical "supergroups" (Neutrinos, Cosmic Rays, Low-Background Experiments, Gravitational Phenomena, and Cosmology), there were clear overlaps. For instance, short-baseline neutrino oscillation experiments probe neutrino masses which are cosmological significant. Results in this area may well impact other searches for dark matter in low-background experiments. Similarly, searches for gravitational waves produced by violent events in the universe may illuminate other mysterious phenomena, like intense rapidly varying gamma-ray bursts.

The remarkable high attendance at Snowmass 1994 (nearly 450 participants) signalled a field coming of age. Although it is still difficult to draw a clear boundary around this emerging discipline of Particle and Nuclear Astrophysics and Cosmology, two facts are contributing to its vitality. First, there is a strong symbiotic synergy resulting from the joining of disparate disciplines. Thus, for example, the organizational and computational talents of particle physicists have emboldened astronomers to scan millions of stars for the Macho searches. Conversely, remarkable astrophysical phenomena like super energetic air showers have enticed physicists away from their laboratories to construct extensive open-air detector arrays to better understand these phenomena. Second, the development of new instruments, telescopes, and detectors has been most crucial. Indeed, it is this fact that drives the field. From COBE to Gallex and Kamiokande, to the new 8m telescopes and the instruments on CGRO and the Hubble observatory, data in profuse quantities has been flowing. These data is the lifeblood of this new field. What is particularly exciting is that this flow of new information is just beginning, with a number of second generation projects underway and many new initiatives already well advanced.

In this overview, we want to describe briefly some of the accomplishments of this field and the intellectual goals that guide it. At the same time we want to delineate some of the areas where one can expect progress in the future, outlining some of the proposed new initiatives. In keeping with the organization of the Summer Study this overview is structured in a similar fashion, except that we have incorporated the discussion of low-background experiments into that of the other areas.

§2 Neutrinos

Neutrinos are playing an increasingly crucial role in tests of the standard model of electroweak interactions, in cosmology, and in probing the nuclear and particle astrophysics of stars and supernovae. In particle physics neutrino properties may provide us with a window on new physics phenomena at energies scales well beyond the direct reach of accelerators. Measurements of neutrino masses and mixing could provide the experimental foundations for building a new and more general standard model. In cosmology, neutrinos are a leading candidate for the "missing mass" that appears to govern the clustering of galaxies on very large scales. Measurements by COBE and other groups of the angular variations of the temperature of cosmic microwave photons suggest that such "hot dark matter" is present. In nuclear and particle astrophysics, neutrinos provide a probe of the interiors of our sun and supernovae; higher energy neutrinos could allow us to look into the centers of active galactic nuclei.

There is also no small measure of serendipity in the relationship between these neutrino astrophysics subfields. In the past decade it was discovered that the effects of neutrino mass and flavor mixing are marvelously enhanced as solar neutrinos traverse the sun. It also appears that the tau neutrino masses favored by cosmologists could help explain how supernovae explode, and thus how our galaxy has become enriched in the heavy elements responsible for life. Such connections have greatly stimulated the field, allowing us to look at neutrino physics from new angles. It also has meant that we have added to our "toolbox" of laboratory neutrino experiments powerful new cosmological and astrophysical probes of neutrino properties.

As we enter the new millennium, neutrino physics stands at an important threshold, one driven by the remarkable technological revolution taking place in astrophysics. In the past decade detectors have been developed to see solar neutrinos event-by-event. Neutrinos from a supernova were detected for the first time, beginning an era where we can monitor our galaxy, and perhaps beyond, for stars in gravitational collapse. Detectors using the oceans, lakes, and antarctic ice have begun a new era of high-energy neutrino astronomy. Indirect probes, such as COBE and gravitational lensing measurements, have provided evidence of neutrino mass. In parallel with these astrophysical advances, laboratory experimenters have strived for and achieved unprecedented precision in measurements of the beta decay spectrum, double beta decay, and neutrino oscillations.

One of the driving forces behind this field is the expectation that new physics lies very close to our present experimental horizons. Almost all extensions of the standard model predict neutrino masses. In many of these the smallness of the neutrino mass - many orders of magnitude below the masses of other fermions - has a natural explanation in terms of the seesaw mechanism: $m_\nu = m_D^2/m_R$ where m_D is a typical quark or charged lepton mass and m_R is the scale of new physics. Values of $m_R \sim 10^{12}$ GeV produce neutrino masses relevant to the dark matter and solar neutrino problems, thus demonstrating the power of such experiments to probe beyond the scales of present colliders. The seesaw mechanism beautifully explains why neutrino masses are so different from the natural mass scale m_D of other standard model fermions: the small parameter m_D/m_R arises as a ratio of a familiar Dirac mass of the standard model and a new mass scale. Demonstration that neutrino masses follow a seesaw pattern might be the first step in understanding the puzzling pattern of masses throughout the standard model.

This possibility has motivated heroic efforts to detect neutrino masses. In the past decade direct searches for mass effects in the shape of the β decay spectrum of tritium have reduced the limits on the electron antineutrino mass to about 5 eV. Double beta decay experiments using isotopically enriched detectors have set limits on the Majorana mass of the electron neutrino of about 1eV. Majorana masses are associated with the identity of the neutrino and its antiparticle, a possibility that does not arise for other standard model fermions. This question is intimately connected with the seesaw mechanism and with the conservation of total lepton number.

While no evidence for neutrino masses has arisen from these experiments, others do suggest nonzero masses. The cosmological evidence for neutrino mass has often led to the conjecture that the third-generation neutrino, ν_τ, has a mass near 10 eV. Masses of this scale can be probed in short-baseline neutrino oscillation experiments. Indeed, one current experiment claims evidence for neutrino masses of a few eV.

But perhaps the most tantalizing suggestion for neutrino masses comes from efforts to detect solar neutrinos. The combined results of the ^{37}Cl, Kamiokande II/III, and SAGE/GALLEX experiments reveal a pattern of fluxes that is very difficult to reconcile with plausible variations in the standard solar model, but quite compatible with neutrino oscillations occurring in matter. The preferred solution corresponds to a mass difference between neutrino mass eigenstates of $\delta m^2 = m_2^2 - m_1^2 \sim 10^{-5}$ eV^2. If the seesaw mechanism is correct, this suggests that the heavier neutrino involved in the oscillation has a mass $\sim$ few $\times 10^{-3}$ eV. This would be a natural value for the mass of the ν_μ if the ν_τ is the source of hot dark matter. Flavor oscillations in matter (the MSW mechanism) produce distinctive experimental signals: there is a characteristic distortion of the ν_e spectrum and an appearance of ν_μ (or ν_τ) neutrinos. Thus there is great excitement in the community about new experiments, such as the Sudbury Neutrino Observatory (SNO) and Super Kamiokande, that can be mounted to detect such signals.

Several underground neutrino detectors indicate that there may be a deficit in the ratio of ν_μ to ν_e in neutrino interactions following cosmic ray interactions. If neutrino oscillations are responsible for the anomalies, the deficit is consistent with an m_2 of about 0.1 eV and requires large mixing angles. As these parameters can be explored in accelerator neutrino oscillation experiments employing long baselines, they have stimulated efforts to mount new experiments at Fermilab, CERN, BNL, and KEK.

It is likely that some of these hints for neutrino mass are false: all of the evidence is difficult to incorporate into a simple theory, given the Z width constraint of three light active neutrinos and the big-bang nucleosynthesis constraint on sterile neutrinos. But many believe that certain results, such as the solar neutrino problem, may hold up: four experiments are consistent and indicate a neutrino mass in a range favored by theory. Thus great hope exists that neutrino physics may lead us beyond the standard electroweak model.

Because neutrinos are so weakly interacting, they play a special role in energy transport within dense stars. The central helium cores of red giants are cooled by neutrino emission. Because the onset of helium burning depends delicately on this cooling, red giant observations can be used to constrain neutrino properties such as magnetic moments. The astrophysical limits exceed the sensitivity of laboratory limits by about two orders of magnitude. Similarly, almost the entire binding energy of core-collapse supernovae (10^{53} ergs!) is carried off by neutrinos. One of the great challenges for the coming years is to measure the temperatures of the ν_e, $\bar{\nu}_e$, and heavy-flavor supernova neutrinos. Because of the MSW mechanism, a distinctive temperature inversion between ν_e and ν_τ should signal the existence of a cosmological interesting ν_τ mass. This phenomenon makes supernova neutrino physics enormously important to cosmology. If such an MSW crossing occurs, it would also lead to a more robust explosion by enhancing the energy neutrinos deposit in the star's mantle.

The nuclei synthesized in the big bang and during the evolution of our galaxy provide fossil evidence of past neutrino reactions. Big bang constraints on the production of helium suggested there were only 3-4 light neutrino species well before measurements of the Z width. Supernovae are the factories that produce, and then eject into the interstellar medium, most of the heavy nuclei found in second-generation stars like our sun. Many of these are made by rapid neutron capture in the hot plasma blown off the star by a neutrino wind. The MSW mechanism mentioned above frequently destroys the conditions necessary to this synthesis. Thus the issue of a massive ν_τ ties together cosmology, the supernova explosion mechanism, and the associated nucleosynthesis. An improvement in our understanding of one of these problems will limit the possibilities for the others.

This short introduction to the presentations at Snowmass is meant to provide a snapshot of a field in great flux. Driven by new experiments and by theories, such as supersymmetric grand unification, that predict new phenomenon within the reach of these experiments, interest in this field has exploded. The path for the next decade, not to mention the next millennium, is both exciting and unclear. It is possible, with SNO and Super Kamiokande nearing completion, that definitive proof of massive neutrinos and neutrino mixing could be in hand well before the 1990's end. If this occurs, neutrino physics will have provided the experimental foundations for the next standard model of particle physics, one having profound implications for cosmology and astrophysics. If no such "smoking gun" is found, we will continue to seek the pattern that accounts for dark matter, the absence of solar neutrinos, and the origin of the elements.

§3 Cosmic Rays

The array of astrophysical subjects gathered in the category of cosmic rays was deliberately chosen to be very broad, ranging from hard X-rays to protons and atomic nuclei with energies beyond 10^{20} eV. This grouping is related to the fact that there is no natural break in this vast energy range which covers sixteen orders of magnitude. There may be natural divisions marked by differences of instrumentation but in most cases these are bridged by natural scientific links. Examples abound. The space based gamma-ray astronomy as exemplified by the results of the Compton Gamma-Ray Observatory (CGRO) is linked with ground based observations which, when fully developed, can extend the energy range beyond the upper limit of the CGRO (20 GeV). The ground based techniques for gamma-ray astronomy merge into techniques of more traditional cosmic ray physics when the energies surpass 10 TeV. There are strong scientific links as well between phenomena at greatly different scales of energy. The synchrotron emission of low energy gamma-rays from the Crab nebula can be related to higher energy gamma-rays produced by inverse Compton scattering of the high energy electrons responsible for that synchrotron emission. Gamma-rays, unlike the charged cosmic ray particles, are unaffected by magnetic fields and can reveal sources of the charged cosmic ray particles. Thus, the study of all kinds of radiation incident on the earth provide insight into a broad range of astrophysical problems and mysteries.

3.1 Space-based gamma-ray astronomy

The existence of the earth's atmosphere dictates the choice of detection techniques to observe astrophysical objects in various ranges of the electromagnetic spectrum. From the ultra-violet to gamma-rays of energy $\sim$ 20 GeV the atmosphere is opaque. Thus, observations must be made from instruments on satellites or on high altitude balloons. It is a general rule that the intensity of electromagnetic radiation from an astrophysical source decreases with increasing photon energy. Fortunately the typical intensities are sufficient so that the acceptance of detectors placed above the atmosphere is reasonably matched to the aforementioned range. Interest at Snowmass was primarily devoted to the future of observations in three ranges: 0.01-1 MeV, 1-30 MeV, and 30 Mev to 10 GeV; with improvements suggested in both spatial and energy resolution. Most physicists and astronomers are aware of discoveries made by the CGRO, in particular of the gamma-ray bursts (GRB) made by the BATSE detector. These intense bursts of gamma-rays are isotropically distributed in the sky and do not correlate with bursts in other parts of the electromagnetic spectrum. The EGRET detector aboard the CGRO has discovered many active galactic nuclei (AGN's), which are very bright as seen in GeV gamma-rays. This is a discovery which was completely unexpected.

A number of new concepts for a space-based instrument to follow the CGRO are under consideration. These include new survey instruments to explore the entire sky with a wide field, good spatial and energy resolution, and increased sensitivity in the the hard X-ray region. These instruments require the large field of view because most of the activity in this area is strongly time varying. They will be dedicated to the exploration of a range of phenomena involving pulsars, AGN's, and black holes—all objects that are poorly understood.

3.2 Ground-based gamma-ray astronomy

For gamma-ray energies above 20 GeV, the atmosphere shifts from being an opaque shield to being an essential part of the detector. At these energies the gamma-rays produce air showers in the atmosphere that can be observed either by the Čerenkov radiation produced by the shower particles or, at high altitudes, the shower particles themselves. At present, positive results have been obtained only with the Čerenkov technique. Mirrors image the Čerenkov light on to a cluster of photomultipliers. The effective area of this detector is given by the size of the pool of Čerenkov light on the ground which is $\sim 10^4$ km^2. Nature cooperates by allowing the gamma-ray signal to be detected on the ground with an effective area much larger than what could have been possible for a space based detector.

In recent years two pulsars (the Crab and PSR 1706-44) have been convincingly observed at TeV energies. Even more remarkable has been the observation of an AGN, Markarian

421. AGN's are extra galactic objects thought to be powered by accretion of material onto massive black holes. These same objects are observed in the 10 GeV range by the EGRET detector on the CGRO. Other AGN's, seen by the CGRO much brighter than Markarian 421, are not seen at TeV energies. The explanation of these remarkable observations may involve the softening of the spectra at the higher energies or the absorption of the TeV gammas by infrared photons. Only the extension of gamma-ray astronomy between 20 GeV and 200 Gev will be able to answer these questions. To lower the threshold for ground-based gamma-ray astronomy is an imperative and is technically feasible. This was discussed at Snowmass, along with other projects, including a large water detector to extend the sensitivity to the higher energy components of the GRB spectra.

3.3 Over the Knee

The cosmic rays which strike the earth isotropically comprise nuclei with abundances which are very similar to solar system abundances, except that elements with high ionization potential are systematically suppressed. It is as if the particles are injected at low energy into an accelerator. It is believed that this scenario is correct with supernova shock waves serving as the accelerator. Shock acceleration naturally produces a power law spectrum as observed. Detailed examination of this acceleration mechanism suggests that the upper limit of this process occurs at about E/Z= 10^{15} eV. Curiously the cosmic ray spectrum steepens at about 3×10^{15} eV which may be related to the upper limit of acceleration.

The cosmic rays are contained in the galaxy by its magnetic field. As the cosmic rays pass through the dust and material in the galaxy, nuclear species, far in excess of the solar abundance, are produced by spallation reactions and give a measure of the average amount of galactic material traversed by the cosmic rays and hence their lifetime. The spallation products show that the mean life of the cosmic rays in the galaxy is about 10^7 years and that the particles with higher magnetic rigidity escape from the galaxy more easily. These facts imply that the mean atomic weight of the cosmic rays should increase with their total energy in the region of the knee. Since the flux of cosmic rays falls rapidly with energy, it has been difficult to directly measure the abundances above 10^{14} eV. New techniques and new instruments will permit the direct measurement of the relative abundances up to 10^{15} eV. Indirect measurements of the mean atomic number will be possible by means of the simultaneous measurement of a number of shower parameters. This technique will be effective above 10^{14} eV, so that an overlap of the two techniques is possible.

Beyond the knee little is known about the cosmic rays other than the energy spectrum. One knows neither the acceleration mechanism, nor the source (galactic or extra galactic), nor the mean atomic number. In the next decade technical means will be developed to answer these questions. Cosmic rays do not exist independently of other powerful astrophysical phenomena; the effort to understand the origin of cosmic rays bears on a much broader domain of astrophysics.

Among the charged cosmic rays there are also electrons, positrons, and antiprotons which are much less abundant. Their presence in the cosmic rays is expected at a predictable level due to interactions of the primary cosmic rays with galactic material. Excesses of the antiparticle components are signals of exotic phenomena such as annihilating dark matter. Improvements in technology will permit much more sensitive investigation of these components.

3.4 Highest Energy Cosmic Rays

Over thirty years ago a cosmic ray was observed that had an energy $\sim 10^{20}$ eV. Since that time some ten events have been observed with energies in excess of 10^{20} eV. With these extraordinary events, attention is brought to the upper end of the cosmic ray spectrum. Cosmic rays with energy above 10^{19} eV defy easy explanations for their acceleration. These cosmic rays have been observed by four independent detectors which agree on the shape and flux. There is structure at the end of the spectrum suggesting a complex set of sources. The fluxes are low, about one cosmic ray above 10^{19} eV per km^2 per year and about one above 10^{20} eV per century.

Recently two events with energies of 2.0 and 3.2×10^{20} eV have been observed. These events must have originated at distances cosmologically close to earth. Cosmic rays have a strong energy loss in their transport through the 2.7° cosmic background radiation and these two cosmic rays must have traveled ≤ 30 Mpc. Because of their high energy they travel in the galactic and extragalactic magnetic field with little deflection and hence should point close to their source. Neither of these cosmic rays points to any plausible source within the proscribed distance. Furthermore, there is no clear understanding how these energies can be reached using the properties of known astrophysical objects. This puzzle can only be answered with new massive detectors.

At present, a detector with acceptance of 200 km^2-ster is operating in Japan and a detector with 500 km^2-ster is under construction in Utah. While these detectors will begin to work on this puzzle, the final resolution will require massive new detectors of $\sim 5,000$ km^2 area placed in the northern and southern hemispheres.

§4 Gravitational Phenomena

Gravitation physicists seek to understand the nature of gravity, both classical and quantum. Is general relativity really the correct classical theory, not only in the solar system where gravity is weak, but also in compact objects and on cosmological scales where it is nonlinearly strong? What are the (as yet ill-understood) laws of quantum gravity that govern the birth of the universe and the cores of black holes and black-hole evaporation? What kinds of new phenomena are predicted to exist by general relativity (black holes, singularities, soliton stars, cosmic strings, ...) and by quantum gravity (black-hole evaporation, the birth of the Universe, the creation and destruction of classical spacetime in "singularites", ...).

Many of these issues are beyond the reach of 20th century human technology and remain the sole province of theorists—both gravitation theorists and particle theorists. However, other gravitational issues have become experimentally accessible or will be so in the near future, most notably black holes, and gravitational waves as a tool for probing black holes, neutron stars, soliton stars, cosmic strings, and the early universe. These experimentally accessible phenomena have much intellectual contact with Particle and Nuclear Astrophysics and Cosmology, and thus were included in the Snowmass Study. Also included were two quantum-gravity phenomena (quantum cosmology and black-hole evaporation) which, though not directly experimentally accessible today, nevertheless may have long-term import for Particle and Nuclear Astrophysics and Cosmology.

4.1 Quantum Aspects of Gravity

General relativity theory insists that the Universe began in a big-bang singularity where the density of matter and the curvature of spacetime were both infinite. Relativity also insists that singularities of infinite density and curvature reside inside black holes. Simple quantum mechanical considerations predict, however, that when the density and curvature exceed "Planckian" values constructed from Planck's constant $\hbar$, Newton's gravitation constant G and the speed of light c (density $c^5/\hbar G^2 \sim 10^{94}\text{g/cm}^3$, radius of curvature $\sqrt{\hbar G/c^3} \sim 10^{-33}\text{cm}$), general relativity must break down, space and time as we know them must cease to exist, and a new set of physical laws and phenomena called "quantum gravity" must take over. Thus, the cores of black holes and the origin of the universe should be the domain of quantum gravity.

There is hope, from cosmological observations of what came out of the big bang, to get a handle on the quantum-gravity origin of our Universe, i.e., on its "Planck era". One can come at this from two different directions. The first is that of quantum cosmologists who use candidate, partial formulations of the laws of quantum gravity to try to deduce what emerged from the big bang and how the Universe made its transition from an initially quantum mechanical object to the largely classical object in which we now live. This is the direction taken by participants in the Snowmass G1 working group. The second direction is that of astrophysicists and physical cosmologists who begin with observations of the universe today and try to extrapolate back toward the Planck era. This direction, which was taken by Snowmass supergroup C and is explored in Sec. 5, is now awash with wonderful, new observational data and new ideas that might lead it into firm contact with the quantum-gravity direction. Such contact might seriously constrain candidate theories of quantum gravity, or it might simply reveal that (by virtue of inflation of the early universe; cf. Sec. 5) the Universe today is not very sensitive to the details of quantum gravity and the Planck era.

A successful partial step toward quantizing gravity was achieved in the 1970s, when several theorists, coming from different directions, converged on an apparently unique way to treat quantum fields that reside in the classical, curved spacetime of general relativity. Much to everyone's amazement, the resulting ***quantum field theory in curved spacetime*** predicted that a black hole must emit radiation ("Hawking radiation") and thereby must evaporate, if one waits long enough (far far longer than the Universe's age for stellar mass black holes, but much less for holes less massive than $\sim 10^{-19}M_\odot$).

The prediction of black-hole evaporation has led to a theoretical conundrum, the resolution of which may teach us much about the full laws of quantum gravity: One can imagine forming a black hole by the implosion of matter that is in a quantum mechanically pure state. Since the Hawking radiation, by which the hole ultimately evaporates, is (or appears to be) in a thermally randomized, mixed state, the hole's formation and subsequent evaporation seems to transform a pure state into a mixed state. In other words, information about quantum mechanical correlations is lost not just in practice, but even in principle. Such an information loss and pure-to-mixed transition is forbidden by the standard Hamiltonian formulation of quantum mechanics, but is permitted by certain generalizations of quantum theory based on Feynman's path integral methods.

Since the endpoint of the evaporation is governed by the (ill-understood) full laws of quantum gravity, it may be this information loss is trying to tell us that quantum gravity cannot be formulated in a Hamiltonian way. However, this is just one of several possible implications of the apparent information loss—and the one that most theorists find least palatable. While there is great disagreement about the real message, there is general agreement that theorists are likely to learn much about the interface between general relativity, quantum theory, and particle theory by struggling to decipher the endpoint of black-hole evaporation and other quantum aspects of small black holes. That struggle was the principle focus of Snowmass Group G1.

4.2 Black Hole Astrophysics

Black holes are predicted to exist by general relativity, and there is compelling *circumstantial* evidence that they do exist in relative profusion in the Universe in two varieties: stellar-mass black holes ($M \sim 3$ to $50M_\odot$) that are remnants of the evolution of massive but normal stars, and supermassive black holes ($M \sim 10^5$ to $10^9 M_\odot$) that reside in the nuclei of galaxies and quasars. A third variety, primordial black holes formed in the very early universe with masses as small as $\sim 10^{-19}M_\odot$ and quantum mechanical evaporation times as short as the Universe's age, might well exist but there is no compelling observational evidence for them.

The general relativistic, classical theory of black holes (which is relevant to all stellar-mass and supermassive holes) is in fairly complete shape: Thanks to the "no-hair" theorem, we know that all the properties of such a hole should be fully determined by its mass and spin; and thanks to many years of analysis by many gravitation theorists, we now fully understand those predicted properties, with one major exception. We do not yet understand in any detail the behaviors of highly dynamical black holes (e.g., colliding and coalescing black holes). That dynamical understanding may come within the next decade, as a result of combined numerical solutions of Einstein's equations on supercomputers and gravitational-wave observations of black-hole collisions (Sec. 4.3).

With classical black-hole theory mostly in hand, black-hole research has become largely observational. Current and future observational studies have four main goals: (i) to test, observationally, the predicted properties of black holes (for example, to measure the details of the curvature of spacetime around a black hole and see whether they are in accord with the no-hair theorem), (ii) to prove unequivocally that one or more of the observed black-hole candidates is indeed a black hole, so we no longer have to make do with circumstantial evidence, (iii) to determine the distributions of black holes in the universe (their numbers, spatial distributions, and distributions of mass and spin), and (iv) to explore the roles of black holes in astrophysical phenomena (their births, and their interaction with stellar companions and with accretion disks and the interstellar medium).

The latter goals (black-hole distributions and astrophysical roles) are part of main-stream astronomy and astrophysics, and are being pursued with moderate success (assuming the objects studied really are black holes), using a variety of astronomical instruments and analyses. The former goals (uniquivocal proof that an observed candidate is truly a black hole, and quantitative tests of black-hole theory) have been more problematic, and were a primary focus of the Snowmass black-hole group G2.

The keys to these elusive goals are observational studies of a black hole's immediate vicinity, from its horizon (its "surface", inside which one can never see) out to roughly ten horizon radii. There are two promising vehicles for such studies: X and gamma rays emitted by hot gas in an accretion disk swirling inward toward the horizon of a stellar-mass black hole, and gravitational waves produced when two black holes coalesce or when a neutron star or white dwarf spirals into a more massive black hole (Sec. 4.3).

The X and gamma rays, emitted by gas spiraling into black-hole candidates, are observed to fluctuate on a wide variety of timescales, from years down to milliseconds. The shortest timescales are thought to be associated with gas near the hole's horizon: It seems reasonable to expect the radiation from a blob of near-horizon gas to fluctuate, due to moderate beaming and gravitational lensing effects, at the blob's orbital period (which is a few milliseconds); and as the blob spirals inward, that period should decrease. The result should be an X-ray or gamma-ray "chirp" that cuts off at the period of the last stable orbit or a bit shorter. A number of such chirps may occur at once, but by statistical studies of the millisecond fluctuations one may hope to determine whether such chirping is indeed occuring, and if it is, one may hope to learn details of the near-horizon environment and confirm firmly that the central body is a black hole.

In the 1970s there was much hope that such studies would be performed by HEAO-A, the first of the high-energy astronomical observatories, which carried a large-area, fast-timing X-ray detector. Unfortunately, a malfunction prevented HEAO-A from taking extensive data of the required sort. It has been nearly 20 years, but at last two new X-ray missions with the required timing capabilities are being readied: the X-Ray Timing Explorer (XTE), and the Unconventional Stellar Aspect (USA) X-Ray Telescope. The XTE and USA data, when analyzed with new techniques such as wavelets, give promise of much improved understanding of the inner regions of accretion disks, and perhaps they will bring the long-sought unequivocal proof of black holes and the first evidence of their detailed properties.

4.3 Gravitational Wave Astrophysics

Gravitational waves are ripples of warpage in the fabric of spacetime. General relativity predicts their existence and predicts that they should be produced strongly by highly compact, massive bodies (e.g., black holes and neutron stars) that orbit, collide, or vibrate in a highly nonspherical manner. Although gravitational waves have not yet been detected experimentally, a massive international effort is likely to capture them within the coming decade, and harness them for research in fundamental physics, astrophysics, and cosmology.

There are enormous differences between these gravitational waves, and the electromagnetic waves on which our present knowledge of the Universe is based: (i) Astronomical electromagnetic waves have frequencies that begin at $f \sim 10^7$ Hz and extend on *upward* by roughly 20 orders of magnitude; by contrast, astronomical gravitational waves should begin at $\sim 10^4$ Hz (1000-fold lower than the lowest-frequency astronomical electromagnetic waves), and should extend on *downward* from there by roughly 20 orders of magnitude. (ii) Astronomical electromagnetic waves are almost always incoherent superpositions of emission from individual electrons, atoms, or molecules; by contrast, cosmic gravitational waves are produced by coherent, bulk motions of huge amounts of mass-energy—either material mass, or the electromagnetically-dark energy of oscillating, nonlinear spacetime curvature as in the collision of two black holes. (iii) Electromagnetic waves are easily absorbed, scattered, and dispersed by matter and thus can only be seen coming from optically thin regions such as diffuse gas clouds and stellar atmospheres; by contrast, gravitational waves travel nearly unscathed through all forms and amounts of intervening matter and thus, for example, can emerge from the core of a supernova or the Planck era of our Universe unscathed by material absorption or scattering.

These enormous differences make it likely that gravitational waves not only will bring us valuable new information about phenomena such as black holes, for which we already have electromagnetic evidence; they may also bring us great surprises. In the past, when a radically new window has been opened onto the Universe, the resulting surprises have had a profound, indeed revolutionary, impact. For example, the radio universe, as discovered in the 1940s, 50s and 60s, turned out to be far more violent than the optical universe; radio waves brought us quasars, pulsars, and the cosmic microwave radiation, and with them our first observational evidence for black holes, neutron stars, and the heat of the big bang It is reasonable to hope that gravitational waves will bring a similar "revolution".

The technology for gravitational wave detection has been under development for 35 years, and is now nearing fruition in several different frequency bands:

In the *extremely low frequency band*, $\sim 10^{-18}$ to 10^{-15} Hz (wavelengths of order the size of the observable universe), observations of the cosmic microwave anisotropy are constraining a predicted stochastic background of primordial gravitational waves. It may even be that a portion of the quadrupolar microwave anisotropy is due to such primordial gravitational waves with strengths (energy density in units of the energy to close the universe) $\Omega_g \sim 10^{-10}$.

In the *very low frequency band*, $\sim 10^{-9}$ to 10^{-7} Hz (periods of a few years), the observed steadiness of millisecond-pulsar periods is placing limits $\Omega_g \sim 10^{-7}$ on any stochastic gravitational wave background due to early-universe processes (primordial waves, waves from phase transitions, waves from cosmic strings).

Black holes, neutron stars, and other compact, strongly gravitating bodies that exist in the Universe today are expected to radiate in the *low-frequency band*, $\sim 10^{-4}$ to 1 Hz, and the *high-frequency band*, ~ 1 to 10^4 Hz.

In the high-frequency band, the "LIGO-VIRGO" international network of laser interferometer gravitational wave detectors is now under construction, and spherical, resonant-mass detectors ("bars") are being designed to operate at the upper end of this band, where the interferometers lose sensitivity. This network of interferometers and bars is likely to see and study the waves from the inspiral and final coalescence of stellar-mass black-hole black-hole binaries, black-hole neutron-star binaries, and neutron-star neutron-star binaries out to near-cosmological distances; and from spinning, slightly asymmetric neutron stars in our own galaxy, and perhaps nonaxisymmetric supernovae well beyond the VIRGO cluster of galaxies.

When the first LIGO/VIRGO interferometers turn on, most and perhaps all these sources will be beyond their reach; but the interferometers will then be improved step by step by more than an order of magnitude, with a resulting event rate enhancement of more than 1000, thereby probably bringing these sources into view.

The development of this enhanced interferometer technology — which involves monitoring distances between kilometer-separated test masses to a precision $\sim 1/1000$ the diameter of the nucleus of an atom — is a major effort involving a number of research groups world wide, as is the development of the resonant-mass detectors. These technologies, which were a central topic of discussion by Snowmass Group G3, are likely to find many applications outside the gravitational-wave field.

The low-frequency band, $\sim 10^{-4}$ to 1 Hz, is to the high-frequency band what radio astronomy is to optical astronomy. Each band will bring us different kinds of information about different kinds of phenomena.

Low frequencies are the domain of massive and supermassive black holes ($M \sim 100$ to $10^8 M_\odot$) — their births and collisions, and the inspiral of smaller objects into them —, and also of known binary star systems such as 44 i Boo, and our galaxy's shortest-period compact-body binaries (white-dwarf, neutron-star, and black-hole).

The premier instrument for this low-frequency band will be a space-based, several-million-kilometer-long variant of the LIGO/VIRGO earth-based interferometers. This *Laser Interferometer Space Anteanna* (LISA) has been recommended by the European Space Agency's Survey Committee as the third of three Cornerstone Missions in ESA's Horizon 2000 Plus Program, with a flight in $\sim$ 2014. However, its implementation may require an augmentation of the Program's budget level by 5%, and a final decision remains to be made. Members of the American gravitation community and of the LISA team hope that NASA will join together with ESA in this endeavor, and that working jointly, ESA and NASA will be able to fly this exciting mission considerably sooner than 2014.

The scientific payoffs of LISA and the LIGO/VIRGO network arise from their broad-band nature: they can monitor the two waveforms of a gravitational wave in the time domain, over frequency bands of several orders of magnitude, and they can determine the directions to the waves' sources with accuracies of $\sim$ 1 degree or better.

Waveform studies, most especially with LISA, are likely to bring us maps of the spacetime warpage around black holes and tests of the black-hole no-hair theorem, and with these maps and tests, unequivocal proof that black holes do exist in our Universe. If the LIGO/VIRGO network were now in a mature stage of operation, it would tell us whether the enigmatic, observed gamma-ray bursts are coming from the coalescence of binary neutron stars at near cosmological distances, since those coalescences should produce observable gravitational waves. Moreover, the gravitational observations would identify, to within about one millisecond, the time at which the final neutron-star collision occurs, thereby (depending on one's viewpoint) determining how long after the collision the gamma ray burst begins, or determining to within a fraction of a second (out of roughly a billion years) the relative propagation times and thus speeds of gravitational and electromagnetic waves. If LIGO/VIRGO had been in operation at the time of supernova 1987A, it should have seen gravitational waves produced by the boiling supernova core whose bubbling neutrino-sphere (analog of photosphere) is thought to have produced the observed neutrinos; and by cross correlating the observed gravitational and neutrino signals, we could have gained valuable additional information about the supernova mechanism.

These are just a few examples of the payoffs that may come from the gravitational window onto the universe. But as with radio astronomy, the biggest payoffs of all are likely to be wholly unexpected discoveries.

§5 Cosmology

As the new millennium approaches, cosmology is entering an historic epoch in which some of the fundamental questions concerning the origin and evolution of the Universe may be answered. Many of the questions date back before recorded history: how big is the Universe? how old is the Universe? How did it begin and how did it evolve? Many solutions have been proposed over time by scientists, philosophers, and religious leaders. What makes this period in history distinctive is the advent of new technologies which give us the capability of peering far into space and gathering data which can test our answers to these questions. The technology has developed over the course of this century, beginning with the advent of the giant optical telescopes, and is now progressing at an incredible pace. Numerous new technologies are coming together at the same time and breaking new ground in cosmology, including: the use of novel bolometer and solid state detectors to measure the anisotropy in cosmic microwave background radiation at the microkelvin level; the application of CCDs to measure red shifts of distant galaxies and to construct three-dimensional maps of the universe; the development of satellites to detect potential cosmological sources of gamma-ray bursts, x-ray glow, and infrared radiation; computer-coordinated surveys of gravitational lensing by intragalactic and extragalactic sources; and experiments for direct detection of axions and supersymmetric weakly interacting massive particles (WIMPs) utilizing novel detectors. It is fair to say that, for the first time in human history, cosmology is undergoing a period in which the subject is as much observation-driven as theory-driven, just what is desired for a true, healthy science. Historical precedents suggest that, when a discipline reaches this balance for the first time, a heroic age of major breakthroughs ensues. There is every reason to suspect that cosmology will follow this precedent.

The grand ambition of cosmology is to explain the evolution of the universe in terms of a simple, predictive model. At this point in time, there is no single complete picture. The hot big bang model explains the expansion and thermal history of the universe, and is extremely well tested. But the big bang picture is incomplete: it does not explain why the universe is so homogeneous, or why the universe is made almost entirely of matter with little antimatter or how large-scale structure formed. A number of concepts, including inflation, cosmic defects, dark matter, and baryogenesis, have been put forward as additions to the big bang model that may provide explanations. Entering the new millennium, the primary focus will be in precision tests of the big bang predictions, measurement of cosmic parameters, and resolution of the various competing ideas that enhance the big bang picture.

5.1 The Big Bang Model: Present Status and Future Tests

We enter the new millennium with a highly successful paradigm in hand: the hot big bang model. According to this model, the universe began as an infinitesimal patch of space filled with hot, dense gas which suddenly began to expand and cool. The universe we see today is the result of fifteen billion years of expansion and cooling. This simple picture successfully explains: 1) the Hubble red shift of distant galaxies, a result of the continued expansion of the universe; 2) the abundance of light nuclei, a consequence of fusion processes that took place in the hot universe during the first seconds after the big bang; and, 3) the existence of the cosmic microwave background, a remnant of radiation that filled the universe and decoupled from matter when the first atoms began to form.

The successful predictions of the big bang model verify it as a valid description of the universe from the present back to the first hundredths of a second after the big bang. The Hubble-distance relation, which relates the red shift of distant galaxies to the expansion of the universe, is verified out to several hundred megaparsecs (Mpc=3×10^{14} cm), using galactic markers that date back to the first billion years or so after the big bang. In the past few years, the COBE Far Infrared Absolute Spectrophotometer (FIRAS) experiment and the the rocket experiments of the University of British Columbia group have verified the Planckian spectrum of the cosmic microwave background in exquisite detail. This result is an extraordinarily compelling verification that the universe was once hot and has been expanding and cooling dating back to the period when the cosmic microwave background photons decoupled from matter, some 100,000 years or so after the big bang. The successful predictions of nucleosynthesis of the light elements verify the big bang picture back in time to when the universe was hot enough to fuse protons and neutrons into nuclei, 0.01 sec or so after the big bang.

Important new tests of the big bang model are anticipated in the next millennium. Although the linearity of the Hubble-distance relation is well-verified, the magnitude of the proportionality factor, the Hubble constant, H_0, remains uncertain to within a factor of two. Since this constant determines the age of the universe, and thereby substantially constrains competing models for large-scale structure formation, eliminating the uncertainty in the value of H_0 is one of the primary goals of cosmology. Although the issue has been the subject of considerable controversy for decades, there is real hope for resolution in the new millennium thanks to several new technologies and observational approaches. The Hubble Space Telescope (HST) will provide a major improvement in the traditional astrometric methods of measuring H_0, using Cepheid variable stars in other galaxies as standard candles. The first results from this work based on 13 Cepheids in one galaxy in the Virgo cluster have produced the result $H_0 = 80 \pm 17$ km/s/Mpc. It seems likely that further observations will significantly reduce the uncertainty. Various alternative approaches for measuring H_0 are also being pursued, including studies of supernova photospheres, supernova luminosities, gravitational lenses, and the Sunyaev-Zel'dovich effect (the rescattering of microwave background radiation by the hot gas within galaxy clusters). Progress can best be made in this field by developing all of these techniques and seeking convergence among them.

The age of the universe has been traditionally estimated using nuclear cosmochronology, cooling rates of white dwarfs, and globular clusters. Globular clusters provide the most stringent age constraint, $t_0 > 11.5$ Gyrs. The age is determined by plotting the distribution of stars in globular clusters on a color-magnitude diagram. One can identify stars that are turning off from the main sequence and measure their apparent luminosity. Knowing the distance to the cluster then determines the absolute luminosity. Then from stellar evolution, it is well-known how long it takes for stars to turn off the main sequence branch as a function of their luminosity. The limitation in this approach is in capturing stars close to the main sequence turn-off. Recent reviews based on taking limits of stars on the main sequence suggest an age of $t_0 > 13.7$ Gyrs, whereas the limit cited above is based on studies along the giant sequence. Hubble Space Telescope studies may improve resolution of stars near the turnoff and significantly improve the limits.

The product of the Hubble constant and the age, H_0t_0, are directly linked to Ω, the ratio of the energy density to the critical density of the universe, and to the energy content. For the simplest model, a universe with $\Omega = 1$ comprised of non-relativistic matter, $H_0t_0 = 2/3$. Yet, the current best-estimates correspond to H_0t_0 near one. This conflict comprises the "age crisis," which either indicates a problem with the measurements or a different energy density or energy content. For example, if there is a significant vacuum density (or, equivalently, cosmological constant) contribution to the total energy density, a value of H_0t_0 can be obtained which is consistent with present measurements. At present, the error bars on the measurements are too large to definitively determine if there is an age crisis, but the anticipated progress in the new millennium will settle the issue.

The Planckian form of the cosmic microwave background spectrum has been precisely verified near the peak of the spectrum (tens to hundreds of GHz). The spectrum obeys a thermal distribution with a temperature of $T = 2.326 \pm .010$ degrees, corresponding to a photon density of 420 cm^{-3}. However, verification of the Planckian shape at long-wavelengths (< 1 GHz) is much less precise. Improvements in these measurements would confirm simple big bang evolution and further rule out exotic models, such as late-decaying particles.

Primordial nucleosynthesis has been established as a primary test of the big bang picture and as one of the best means of measuring the abundance of baryons in the universe. Current best-estimates on Ω_B, the ratio of baryon density to the critical density needed to close the universe, lie between 0.01 and 0.1. To improve upon these limits, more quantitative measurements of primordial abundances are needed along with improved understanding of chemical evolution, stellar processing, stellar atmospheres, and recombination/collisional excitation rates. The development of multi-dimensional hydrodynamic codes to model stellar atmospheres and supernova will be important developments. Also, recent attempts to measure the abundances of light nuclear elements using pre-galactic hydrogen clouds (Lyman-α clouds) appears to be a very promising, independent approach perhaps less subject to evolutionary assumptions. Continued improvement in lithium abundances is especially critical both as corroborating evidence and as the most precise method for narrowing the uncertainty in Ω_B.

The big bang models is left incompletely specified unless one also determines Ω, the ratio of the energy density to the critical density separating an open from closed universe, and Λ, the cosmological constant or vacuum density. The best approaches, in principle, are global approaches which measure the universe over very large distances and/or very long times. The cosmic microwave background anisotropy entails radiation from the furthest observational reaches of the universe. In many cosmological models, including inflation, the temperature autocorrelation spectrum includes sharp features which can be used to determine Ω (see below) using instruments with current sensitivities. Resolving other cosmological parameters, including Λ, through the cosmic wave background anisotropy alone will require substantial improvements in detector sensitivities and new satellite and/or long-duration balloon projects which can scan nearly the full sky. Measurements of cosmic deceleration, q_0, have been notoriously difficult. The method relies on comparing the geometrical sizes of similar objects near and far, but one must beware that the far objects are also much older and that evolutionary effects may alter their perceived size or luminosity. With improved modeling of evolution and a major effort to obtain spectra from a large sample of faint galaxies (see below), progress may be made in the coming decades. Surveys of gravitational lenses is another global approach. Compared to $\Omega = 1$ and $\Lambda = 0$ models, flat universes with non-zero Λ predict significantly more lensing of quasars by galaxies. Detection of "standard events" over a range of cosmological distances is another evolving approach. An example is the effort to detect Type Ia supernovae in high redshift galaxies.

5.2 *Beyond the Big Bang*

The most exciting developments in cosmology for the next millennium entail ideas that stretch beyond the standard hot big bang model. It is clear that, in spite of its phenomenal successes, the big bang picture is an incomplete model of the evolution of the universe. The big bang model does not provide any answer to a number of key questions: Why is the universe so homogeneous or anisotropic? Why is the energy density of the universe today so close to the critical density (Ω near one)? How did the inhomogeneities in the universe arise that are observed in the cosmic microwave background anisotropy and/or are responsible for the formation of large-scale structure? How did galaxies form? Why is the universe composed predominantly of matter with insignificant fractions of antimatter? What is the dark matter that comprises the halos of galaxies and perhaps the missing matter of the universe?

Most likely, the answers to these questions entail events that took place in the first instants after the big bang. Whereas the concepts and tests of the big bang model entail nuclear physics, atomic physics, and astrophysics, new ideas that go beyond the big bang picture entail early times in the universe when the temperature throughout the cosmos was sufficient to excite interactions among elementary particles ($\gg$ 1 GeV). Hence, the answers to the problems of cosmology may be directly linked to our understanding of the fundamental forces and constituents of nature. Theory and observation in cosmology in the next millennium will be focused on exploring the cosmic connection between the very large and the very small.

A New Generation of Cosmological Models

To address the questions left unanswered by the big bang model, new cosmological models are needed. The inflationary model of the universe is the leading candidate for an explanatory and predictive theory that extends beyond the standard big bang picture. The inflationary model proposes that the universe underwent a brief period of extraordinary, superluminal expansion, "inflation," during the first instants after the big bang. The remarkable stretching smooths the distribution of matter and energy, explaining why the universe is so homogeneous and isotropic. The stretching flattens any spatial curvature, explaining why space appears to be Euclidean. According to Einstein's theory of general relativity, a spatially flat universe must have energy density equal to the critical value that divides an open from a closed universe. Hence, the flattening induced by inflation also explains why the observed energy density, ρ, is close the critical density today. A key, testable prediction is that the ratio of the energy density to the critical value, $\Omega \equiv \rho/\rho_{\mathrm{crit}}$, is indistinguishable from unity today. The inflationary stretching dilutes the density of magnetic monopoles and other particle monsters created near the Planck temperature, thus explaining their absence in the universe today. Finally, inflation predicts a spectrum of inhomogeneities were produced during inflation that might leave an imprint on the cosmic microwave background and act as seeds for large scale structure formation. The spectrum of fluctuations is predicted to be nearly scale-invariant, which is consistent with the observations of the COBE Differential Microwave Radiometer (DMR) experiment. Inflation is the only viable cosmological model which explains so many, diverse aspects of the universe.

One important theoretical challenge in inflationary cosmology is to understand that process that may have caused the brief period of superluminal expansion. If the microwave background anisotropy ($\Delta T/T \approx 10^{-5}$) is a consequence of inflation, then most inflationary models would predict that the magnitude of the observed fluctuations in $\Delta T/T \approx 10^{-5}$ is equal to M^2/M_P^2, where M_P is the Planck mass, 1.2×10^{19} GeV, and $M \approx 10^{16}$ GeV is the characteristic energy of whatever physics drove inflation. Hence, inflation is linked by the microwave background to unification scales. Relating inflation to unification models of particle interactions, such as superstrings, is an important element needed to complete the inflationary picture.

Another important theoretical challenge, which applies both to inflation and more general theoretical models, is to understand the source and the value of the cosmological constant. Inflation relies on the notion that particle physics interactions produce a positive contribution to the vacuum density of the universe, adding a non-negligible contribution to the cosmological constant. It is normally presumed, based on observations, that the cosmological constant is zero or near-zero today. But, the inflation picture (or any other cosmological model) is not complete until it is understood why the cosmological constant is small today. Particle theorists believe that the answer lies in the unification of gravity and particle interactions into a unified quantum theory, such as superstrings.

At present, there are no models competing with inflation to explain the homogeneity, isotropy, flatness, and mass density of the universe. The only alternative has been to assume these unusual and unstable properties as part of the initial conditions of the universe. However, there are numerous competing models for explaining the source of inhomogeneities that are observed in the cosmic microwave background and that may be the seeds for large-scale structure formation.

According to inflation, the inhomogeneities are the result of quantum fluctuations in the energy density and space-time metric that ran rampant when the universe occupied a subnuclear volume prior to inflation. As space inflated, the fluctuations, ripples in the fabric of space, stretched also, ultimately spanning a cosmological range of wavelengths. The predicted fluctuations are gaussian, adiabatic (equal fluctuations in all species of energy), and nearly scale-invariant.

Cosmic defect models presume that the universe underwent a phase transition which resulted in the formation of topological defects, which act as the seeds for large-scale structure formation. A topological defect results if, following a phase transition, there are many possible vacuum states corresponding to different values of quantum fields. Different, causally-disconnected regions of space fall randomly into one or another vacuum state. After time passes, the formerly disconnected regions make contact and, where there are mismatches in vacuum states, stable knots in the quantum fields form. Depending on the nature of the vacuum degeneracy, these knots may take the form of points ("monopoles"), curves ("cosmic strings"), surfaces ("domain walls"), or textures. The notion of defect models is that these concentrated knots of energy might be the origin of inhomogeneity in the universe. One key difference from inflation is that these inhomogeneities are strongly non-gaussian. General arguments suggest that the distribution of defects is scale-invariant. A consequence is that there are always defects within our Hubble horizon. For each type of defect, there is a characteristic signal to be found if one should pass through the field of view. For example, a cosmic string would leave a line-like discontinuity in a high-resolution map of the cosmic microwave background.

A theoretical challenge for cosmic defect models continues to be finding reliable methods for computing their predictions. Whereas inflation predicts a simple spectrum of fluctuations which can be understood analytically, defect models require very large-scale numerical simulations. The defects enter the horizon with cosmological size, but decay and interact on microscopic scales. The implications for cosmology are sensitive to the entire range of dimensions. New theoretical methods are needed to reliably circumvent this problem and obtain trustworthy predictions of cosmic microwave background anisotropy and large-scale structure formation.

Both inflation and cosmic defect models invoke dynamical processes based on particle physics to explain the origin of inhomogeneity in the universe. Some cosmologists prefer a more phenomenological approach in which one uses present observations to infer an initial spectrum of inhomogeneities without explaining their origin. One such model is the primeval isocurvature baryon (PIB) model. The model is intended to be conservative with the virtue that it does not require dark matter and any other unverified physical processes. So, it presumes only baryons comprise the matter of the universe and, given nucleosynthesis constraints, this means that the universe is open. Also, one presumes an ad hoc initial spectrum of perturbations in the baryons relative to the photons. The obvious disadvantage of such models is that they are not truly predictive. By presuming different initial spectra, one can get arbitrarily different answers. However, the development of such phenomenological fitting models is an important tool for guiding the development of alternatives to our present, rather restrictive set of predictive models.

Why is There an Excess of Matter over Antimatter in the Universe?

A striking feature of the observable universe is that it is composed primarily of matter, with a negligible proportion of antimatter. The observed baryon excess is ten orders of magnitude higher than would be obtained if the universe began with equal numbers of baryons and antibaryons and simply had them annihilate as the universe cooled and expanded. One explanation may be that the universe began with precisely the observed baryon excess, and that the excess has simply maintained itself over time. Not only is this unsatisfying, but, if inflation is correct, then any initial excess would have been wiped out during the inflationary stretching of the universe. Hence, current research has focused on the notion that the matter excess was generated by dynamical processes as the universe cooled from Planckian temperatures, e.g., after inflation.

In the late 1960's, Sakharov realized that dynamical baryogenesis would require three conditions: (1) deviation from thermal equilibrium; (2) violation of baryon conservation; and, (3) violation of CP conservation. The advent of grand unified theories in the 1970's provided a theoretical framework for achieving all three conditions at grand unification energy scales, 10^{14} GeV or so, using decays of long-lived relics to achieve the deviation from equilibrium. In the past decade, the focus has switched to lower energy-scale (100 GeV) baryogenesis associated with the electroweak phase transition. It has been noted that the required baryon-violation could arise from non-perturbative effects in the standard model.

The biggest uncertainty is the origin of CP violation. It now seems likely that the CP violation associated with the standard model is insufficient, so that new CP violation sources are needed. There are numerous workable suggestions, but none that is compelling. Future experiments to improve constraints on the electron and neutron electric dipole moments could strongly influence developments in the field since most proposals suggest a substantial non-zero value.

On the theoretical side, there remain open issues about the detailed mechanism by which baryogenesis takes place in the case of the electroweak phase transition. One issue is whether the phase transition is a sufficiently strong first-order phase transition to provide the needed deviation from thermal equilibrium. Investigations thus far suggest that the experimentally allowed mass range for the Higgs in the minimal standard model precludes a first order transition, although the issue remains controversial. The minimal model also appears inadequate for providing sufficient CP violation. Consequently, the focus of the field is likely to be on other models of weak symmetry breaking. If a sufficiently strong first-order transition is achieved, it will proceed through the nucleation of bubbles of true vacuum which grow and coalesce to complete the transition. The interaction of quarks and gluons with the bubbles is believed to be a critical contributor to baryogenesis, but the progress is needed in developing a quantitative and predictive understanding of the process. More generally, progress on electroweak baryogenesis will benefit from measurements at the LHC and future accelerators which will explore electroweak symmetry breaking and quark and lepton Yukawa couplings.

5.3 Dark Matter

Another area of overlap and complementarity between cosmology and particle physics may be provided by dark matter. There are increasingly compelling pieces of evidence that at least 90 percent of the mass in the universe is dark, by which we mean that it does not emit or absorb any form of electromagnetic radiation. Once a subject of controversy among astronomers, its existence is now acknowledged by a large majority as its presence is inferred on a variety of scales. Not only does the non- Keplerian character of rotation curves around spiral galaxies indicate that they are at least ten times more massive than the sum of the stars that we can see, but velocity dispersion of stars and x-ray emission in elliptical galaxies show that these objects are also dominated by a dark component. These galactic measurements show that dark matter represents at least a few percent of the critical density. At the scale of clusters of galaxies, three independent sets of measurements point at an even greater amount of the dark matter. The large velocity dispersion of galaxies inside the cluster, the temperature of the x-rays emitted by gas falling into the center of the clusters, the arclets resulting from gravitational lensing of galaxies located behind the clusters, all lead to very similar estimate of the depth of the potential well. This pushes estimates of the mean density in the universe to at least 20 percent of the critical density. Velocity correlation and flows on large scale indicate even greater values.

The combination of all these observations makes it rather convincing that dark matter does indeed exist. The only other (even more earthshaking) possibility is that the laws of gravity are violated on the large scale. Uncovering the nature of this "dark matter" has become one of the more central problems in astronomy and cosmology: it is certainly embarrassing not to understand the dominant component of the universe.

What could it be? The average density may be an important clue. If $\Omega < 0.1$, dark matter may be made of ordinary baryonic matter, as this density is quite consistent with the measurement of primordial abundances of light elements and the standard nucleosynthesis scenario in a homogeneous universe. However, this baryonic component must neither radiate nor absorb light. This basically excludes gas (unless it is in a very exotic state) and dust, and leads to the conclusion that this form of dark matter is viable only in the form of condensed objects: stars too small to start thermonuclear reactions or black holes. Both types can be combined under the name Massive Compact Halo Objects (MACHOs).If, as may be indicated by cluster observations and large scale velocity flows, Ω is significantly higher than 0.1, it would be incompatible with Ω_B determined from the primordial nucleosynthesis. We may then be forced into the hypothesis that at least some dark matter is nonbaryonic. So far attempts to modify the standard nucleosynthesis scenario (e.g., through inhomogeneities generated by the quark-hadron phase transition) have been unsuccessful in extending significantly the upper bound on Ω_B. If indeed we could show that dark matter is not made of ordinary matter, we would have to deeply modify our vision of the universe and of our place in it: a few clumped baryons floating in a sea of foreign particles. Paradoxically, as it dominates gravity, this most inert component of the universe may be responsible for the formation of structure by gravitational collapse and therefore of galaxies, stars, planets and ultimately life. Elucidating the nature of dark matter is therefore a high priority endeavor which can be approached in two complementary ways. Cosmology gives us information on this nature through the value of the cosmological parameters, the detailed shape of the temperature fluctuations of the cosmic microwave background, and the evolution of the large scale structure. And we could make progress through attempts to detect this dark matter directly. A number of such direct searches are already in progress.

Microlensing provides us with a method of detecting the only natural form of baryonic dark matter still compatible with observations, Massive Compact Halo Objects. If one of these MACHOs happens to cross the line of sight to a star, say in the Large Magellanic Cloud, a temporary increase of the star's intensity will be observed. This increase would be symmetric in time, achromatic and non-repetitive, contrary to sporadic phenomena in stars. At least five collaborations are now actively searching for such events, an effort which requires the regular observation of some ten million stars. At this writing, microlensing has clearly been established but the results are puzzling: We observe too many events towards the galactic bulge, and apparently too few (within meager statistics) towards the Large Magellanic Cloud, indicating that our previous ideas of the structure of the galaxies were too naive. Before we can conclude, we need both to increase the statistics and to better pin down the halo structure.

As explained above, if Ω is significantly greater than 0.1, some dark matter has to be nonbaryonic. If we discard exotica such as a shadow universe or primordial black holes, the most attractive hypothesis then is that dark matter is made of particles that were created in the hot early universe and managed to stay around.

One of the well-motivated candidates is the axion. Such a particle has been proposed in order to suppress the strong CP violation implied by the otherwise very successful Quantum Chromodynamics. It has not been observed, but it is interesting to note that the combination of laboratory and astrophysics experiments has constrained its mass in such a way that if it does exist, it must be cosmologically significant, accounting for a large fraction of the critical density. These "invisible" axions from the halo of our galaxy could in fact be detected through their conversion into monochromatic microwave photons inside a tunable microwave cavity in a large magnetic field. In the past ten years, two pilot efforts explored the technology, but lacked about three orders of magnitude in sensitivity to reach a cosmologically interesting limit. A second generation experiment is currently in preparation at Livermore, which should have the needed sensitivity range over one decade in mass. If no signal is unraveled, a significant experimental challenge will be to cover the other two decades which will still be allowed.

Without further information from a specific model, it is quite natural to assume that these dark matter particles were once in thermodynamic equilibrium with the quarks and leptons. In this case, their current density depends on whether they were relativistic or not at the time they decoupled from the rest of the universe. If they are light enough to be relativistic at that time, their density is just related to the decoupling temperature and is basically equal to that of the photons in the universe. This is, for instance, what is expected to have happened for light neutrinos, and a neutrino of 25 eV would give a value of Ω of the order of unity. Unfortunately such a neutrino is extremely difficult to detect in the astrophysical environment. It should, however, be possible to test this hypothesis in the laboratory through neutrino oscillation experiments which are described in this volume. Note that these efforts are complementary to attempts to solve the solar neutrino puzzle, and to confirm atmospheric neutrino oscillations. Although the neutrino mass range covered by these experiments is much lower than necessary to account for the dark matter, confirmation that neutrinos have indeed finite masses will be invaluable in reconstructing the general framework.

For particles which happen to have decoupled when they were non- relativistic, their density today is inversely proportional to their annihilation cross section. A density close to the critical density leads to a cross-section of the order of the Weak Interaction, indicating that the physics at the W/Z^0 intermediate vector boson scale (e.g., supersymmetry) may be responsible for the dark matter in the universe. This generic class of particles is usually called Weakly Interacting Massive Particles (WIMPs). A first generation of experiments looking for elastic scattering of such WIMPs in the laboratory with solid state detectors proved that heavy Dirac neutrinos cannot be a major component of our galactic halo, and almost eliminated a class of WIMPs designed both to be dark matter and to account for the paucity of solar neutrinos. A second generation of laboratory experiments is being brought into operation; depending on the groups, they use larger masses of germanium detectors, large scintillating crystals of NaI, or novel "cryogenic detectors" working at millikelvin temperatures. While the first methods promise sensitivity gains of a factor of a few, the cryogenic detectors allow an active rejection of the radioactive background (e.g., through the simultaneous measurement of the phonons and ionization produced in particle interactions) and may give gains of one hundred or more. These experiments will begin to probe the rate region expected for the theoretically favored neutralinos, the lightest particles in supersymmetry. It will also be possible to use the large high energy neutrino detectors to indirectly search for these particles.

To conclude, the dark matter problem occupies a central place in the current cosmological debate, and elucidating its nature is closely linked to a number of other questions and observations in particle astrophysics. We are poised to make significant experimental progress in the coming years. The beginning of the next millennium may well see the solution of this fascinating puzzle!

5.4 The Large-scale Structure of the Universe

One of the greatest challenges of cosmology is to explain the large-scale structure of the universe. The present notion is that tiny inhomogeneities in the distribution of energy were generated in the early universe, and then these were amplified through the action of gravity over time into the structure we observe today. To transform this notion into a predictive theory, three key questions need to be answered. What are the values of cosmological parameters: the mass density, the cosmological constant, and the Hubble constant? What is the quantity and composition of matter/energy in the universe? And, what is the origin of the initial inhomogeneities? Current and future progress on these issues have been discussed under the prior sections on Big Bang cosmology, on Dark Matter and on Cosmological Models, respectively.

In addition, for a truly detailed understanding of the formation of large-scale structure, substantial advances in both theory and observation are needed. On the theoretical front, the detailed tests of theory require numerical simulation of the formation of large-scale structure, including hydrodynamics, star formation, supernova, chemical evolution, etc. At present, numerical simulations only span a dynamical range of three or four orders of magnitude. One key challenge is to improve numerical techniques to expand the simulation range. Another is to improve physical understanding of the process of galaxy and star formation, which is crucial input into the simulations. Better statistical or analytical methods of comparing observational data with models are needed which take account of the real selection biases in experiment.

The advent of large-area red shift surveys is radically transforming observational studies of large-scale structure. Current surveys probe only a relatively small volume of the local cosmic neighborhood spanning of order 10,000 galaxies. The next generation of surveys, such as the Sloan Digital Sky Survey and the Anglo-Australian 2dF, will measure the red shifts of over one million galaxies and extend the cosmic range by more than an order of magnitude. In addition, the surveys will capture images of tens of millions of galaxies. The surveys will measure the galactic power spectrum over a range of wavelengths that will overlap measurements of the cosmic microwave background, providing critically important, redundant tests of the power-spectrum at long-wavelengths, measurements of Ω, and determinations of the bias parameter as a function of wavelength. The surveys will also provide information about the distribution of peculiar velocities at large scales, which is an additional test of cosmological models of large-scale structure formation. While the red shift surveys are direct probes of luminous matter, another much-anticipated development are studies of weak lensing, which probes the distribution of foreground dark matter. The diverse array of new measurements being initiated as we enter the new millennium will provide our first detailed look of the universe at large scales and profoundly affect our ideas about the origin of the universe.

5.5 The Cosmic Microwave Background

The cosmic microwave background is, in many ways, our best cosmological probe. Assuming the standard interpretation of its origin, the cosmic microwave background provides direct information about the largest distances and the earliest times accessible (until some means is found for probing the cosmological graviton or neutrino backgrounds). In addition, cosmic microwave background measurements are much less sensitive to subjective or model-dependent assumptions compared to most cosmological measurements.

Measurements of the energy spectrum of the cosmic microwave background, beginning with the initial discovery of the 3 K background by Penzias and Wilson through the recent COBE FIRAS and UBC rocket precision measurements, have provided exquisite tests of the standard hot big bang explanation of how the universe evolved from the 100,000 year mark to the present. Although current measurements are quite impressive, extending the spectral measurements to longer wavelengths will provide further support for the simple, adiabatic expansion picture or could provide evidence of some exotic thermal processes (e.g., late decays of elementary particles). Significant improvements probably require a space-borne mission using a small-sized satellite.

The revolution that is occurring as we enter the new millennium is the first detections of anisotropy, spatial variations in the temperature of the microwave background. These variations provide a detailed, quantitative fingerprint which can be used to decisively discriminate competing models for the evolution of the universe. Precise measurements of anisotropy also provide novel ways of determining the values of cosmological parameters, such as Ω, the Hubble constant and the cosmological constant.

The revolution began in 1992 with the announcement by the COBE Differential Microwave Radiometer (DMR) experimental that they had detected anisotropies of $\Delta T/T \sim 10^{-5}$. Since COBE DMR, more than a dozen new detections have been reported on angular scales ranging down to one-half degree, and significant upper bounds have been reported on yet smaller scales. The field is still in its infancy, though: Improvements in instrumentation and sky coverage could dramatically improve the precision of the measurements within the next decade.

Future cosmic microwave background anisotropy experiments can be categorized into three regimes, large-, intermediate-, and small-angular scale measurements, each of which reveals a different, key aspect of our cosmology. Experiments measuring inhomogeneities stretching over large angles in the sky (> 2 degrees) probe the largest structures in the universe. When the microwave background radiation last scattered from matter and began its trek across the universe, there had not yet been time for these large structures to evolve since whatever events created them, e.g., inflation. Hence, large-angle experiments reveal the directly primal universe dating back to the first instants after creation. In terms of our efforts to understand large-scale structure formation, large-angle experiments measure the magnitude of the initial inhomogeneities before gravity had a chance to amplify them. The most important features to be determined are the amplitude and the spectral index of the power spectrum, the key input parameters for any theory of large-scale structure formation. COBE has provided a rough measure, but greater precision is needed to discriminate and refine models.

Experiments measuring inhomogeneities spanning intermediate scales (half-degree to several degrees) probe features in the sharp features in the spectrum which discriminate qualitatively different models of large-scale structure formation (e.g., inflation vs. cosmic defects). For any given type of model, these features can also be used as a novel means of determining Ω, and for determining the ionization history of the universe.

What we learn from large- and intermediate-scale measurements will play a leading and profound role in determining our vision of the evolution of our universe. The challenge for both regimes is that a very large fraction of the sky must be measured in order to obtain good statistical measures. This dictates an experiment that flies at high altitude for the very long periods of time need to scan the sky with fine resolution. One set of proposal entails long-duration balloons which circumnavigate Antarctica, for example, for weeks or months. The balloon projects will evolve quickly, obtain good results soon, and be critical in developing advanced technologies. The most precise results, however, are likely to come from a future satellite mission which avoids atmospheric and sidelobe problems of earth- and balloon-borne missions and is able to measure the full-sky in a controlled, redundant patterns. Such a mission would be a monumental and historic contribution to our understanding of cosmology.

Experiments at small-angle scales (less than half-degree) are important because features observed in the spectrum at these scales can be used to determine the values of cosmological parameters, such as the cosmological constant, the Hubble constant, and the baryon density. They can also be used to distinguish the nature of dark matter, e.g., the proportion of hot or cold dark matter. Measurements at these angular scales are also optimal for detecting the polarization of the microwave background, non-gaussian contributions to primordial fluctuations, the Sunyaev-Zel'dovich effect, and secondary anisotropies association with reionization of the microwave background. Land-based and balloon-borne experiments will be the dominant contributors to our understanding of this regime, since larger instruments are needed to obtain the fine resolution and there is less demand for full-sky coverage.

In sum, a program of high-resolution measurements of the cosmic microwave background anisotropy is the highest priority for microwave background studies and probably for cosmology in general. Each of the three angular-scale regimes reveals different, fundamental facets of the universe. Improving long-wavelength spectral measurements, e.g., by a small satellite mission, is a second priority.

A balanced program of land, air and space missions is needed to extract the extraordinary wealth of information which the cosmic microwave background has to offer. Combined with measurements of large-scale structure and peculiar velocities, the cosmic microwave background will provide a new understanding of the origin and evolution of the universe, a truly profound breakthrough that will be one of the historic achievements of the new millennium.

§6 Structure of the Field

By its very nature science is a continually evolving endeavor, with exciting new fields arising at the interface between well established disciplines. In fostering and nurturing new fields, three important issues of science policy arise:

1. In a severely constrained budget climate, how can one support a developing field without an established "budget line?"

2. How can cross-disciplinary priorities be established so that new endeavors may be judged relative to well established activities?

3. As new fields develop and their financial needs grow, how can the community organize to set long-term strategies that can serve as a basis by which new proposals can be evaluated?

Particle and nuclear astrophysics and cosmology is a case in point. Most agree that a fascinating field is now emerging at the border between particle and nuclear physics, cosmology, stellar astrophysics, high-energy astrophysics, and gravitation. This area of research is undergoing a dramatic expansion that occurs only rarely in the history of science, akin perhaps to the intellectual and technological impetus that gave rise to particle and nuclear physics in the 30's, 40's, and 50's. This meeting, organized by three separate divisions of the American Physical Society and attended by over 450 physicists, testifies to the breath and depth of a field that barely existed 15 years ago.

The number of scientists and the total funding of the field are already quite substantial. Even restricting the definition of the field to astrophysical activities explicitly linked to particle and nuclear physics, we estimate that more than 300 experimentalists are involved at a substantial level in these activities. With an equal number of theorists, the total number of people in the field is quite impressive. The Department of Energy supports such activities in universities and national labs at a total level of roughly \$26M/yr, the NSF at a level of \$13M, and the NASA at a level of \$2M (restricting the field to cosmic rays above 1 GeV). The "gravitational field" represents another \$11M (of which perhaps half is astrophysics/cosmology related), plus current construction money for LIGO.

Irrespective of funding, there is a strong feeling in the community that this emerging field is under-represented, and tends to "fall between the cracks." This is perhaps unavoidable in a field that has so many distinct roots and where new generation of experiments incorporate methodology and people across many disciplines. Nevertheless, as a result of its somewhat helter-skelter growth, this burgeoning field finds itself in an agency management structure not quite reflective of natural intellectual relationships. It is obvious that it does not fit readily within the traditional framework of NSF and NASA astronomy funding. DOE officials historically have worried that such activities may conflict with their interpretation of the agency mission. There is a general perception within the community that the review process could be improved: mail reviewers do not have access to the overall picture and *ad hoc* subcommittees (*e.g.*, of HEPAP) have been sparse and have lacked the continuity required to develop a long-term vision for the field. There is widespread consensus that there should be more intra-agency and inter-agency communication. Finally, there is the worry that in any triage generated by a financial crisis, entrenched fields of science will receive attention while emerging fields will be declared dead on arrival.

To be sure, over the last few years significant progress has been made. For instance, the NSF has put into effect an explicit mechanism of collaboration in particle astrophysics encompassing the Physics, Astrophysics, and Polar Programs Divisions. Solar neutrinos have been adopted by the DOE Nuclear Physics Division, and DOE high-energy physics laboratories have become deeply involved in the field. More importantly, there is a widening recognition at DOE that particle and nuclear astrophysics is an integral part of its basic science mission of fundamental physics.

However, it is clear that we do not yet have in place the machinery necessary to address in a coherent fashion large international projects on the drawing boards. There is no shortage of proposed projects with price tags between \$15M and \$100M: proposals for second-generation cosmic microwave background satellites, dedicated cosmology telescopes, a new generation of solar neutrino detectors, an air-Cerenkov farm to extend the measurement of the gamma-ray spectra of AGN and black hole candidates, giant air shower arrays to explore the highest energy cosmic rays, cubic-kilometer neutrino detectors to look for high energy neutrino sources, a space-based gravitational wave interferometer, and so on.

While NASA may have in place the necessary mechanisms to compare satellite proposals, the other funding agencies (DOE and NSF) lack the reviewing and prioritization tools normally employed for sizable projects (e.g., program advisory committees at accelerators). None of the existing advisory committees to the three agencies (HEPAP, NSAC, SAC) are fully suitable for the advocacy role, and setting up a specific standing committee may be difficult in the current political climate. Regular summer studies sponsored by relevant divisions of the American Physical Society are important, but they are only a part of the process of developing a long-term vision. The APS may be ill equipped for a difficult prioritization. One might envision one or more of the national laboratories stepping in as the main support structure for this new science and setting up a program advisory committee which may *de facto* develop into a national advisory role. But a committee so constituted may lack the proper balance. Finally, one could think of extending the role that the National Research Council, through the various reports and strategic analysis from the Board on Physics and Astronomy, the Space Studies Board, and their joint Committee on Astronomy and Astrophysics, plays in the process.

There is much we could do to decrease the potential barrier encountered by excellent proposals, to welcome young investigators into a more nurturing environment, to optimize the scientific output in the framework of a very limited budget, and to pursue and develop the necessary international partnerships in large projects. An innovative mix of some of the above suggestions may go a long way towards building the framework required to realize the potential of the exciting scientific issues.

If there was any spirit that characterized the two weeks of the Snowmass Summer Study, it was a shared feeling across all disciplines that we are in the midst of a unique conjunction of theoretical ideas, experimental realities, and technological possibilities, which allow us for the first time to address many of the most fundamental questions about our Universe. The most repeated phrase during the many forward-looking talks of the Summer Study was "... now for the first time we have the ability to..."

It is a sad testament to the times that the potential for new discoveries does not seem to be limited by the lack of ideas, technology, or proposals, but by fiscal realities and the artificially constructed barriers of the existing science policy framework.

At the end of this millennium, both as scientists and members of society, we inherit the cultural legacy and reap the benefits of a proud scientific tradition. The 1994 Snowmass Summer Study was witness to the fact that we do not lack the intellectual boldness or technological imagination to address questions once thought to be beyond the realm of human comprehension. For two weeks in the summer of 1994, 450 physicists in the mountains of Colorado united in the conviction that we must enter the next millennium with the same intellectual fervor and hope for the future that led to the great scientific achievements of the 20th century. From the top of the mountains of Colorado we saw unlimited horizons in the field, and we are dedicated to the establishment of a framework to realize the vision of Snowmass.

Opening Talks

The Particle- and Astro-Physics of Dark Matter

Kim Griest

Physics Department, University of California, San Diego, La Jolla CA 92093

We review some recent determinations of the amount of dark matter on galactic, cluster, and large scales, noting some puzzles and their possible resolutions. We discuss the interpretation of big bang nucleosynthesis for dark matter, and then review the motivation for and basic physics of several dark matter candidates, including Machos, Wimps, axions, and neutrinos. Finally, we discuss how the uncertainty in the models of the Milky Way dark halo will affect the dark matter detection experiments.

1 Introduction

It is remarkable that here, at the end of the 20th Century, when science has produced the top quark, and measured tiny fluctuations in the microwave background, we still don't know what the primary constituent of the Universe is. This is the stuff that dominates gravity on galactic scales, and determines the ultimate fate of the Universe, but we have almost no idea what it is. It doesn't emit or absorb electromagnetic radiation at any known wavelength, but it is "seen" through its gravitational effects on scales from tiny dwarf galaxies, to large spirals like the Milky Way, to the largest scales yet observed. The next speaker (Caldwell, these proceedings) will describe exciting new developments in the possibility of detecting this dark matter; a topic which has progressed dramatically in the past few years. I will set the stage for his talk by reviewing recent developments in the amount and location of dark matter, mention some puzzles, and describe some of the more popular dark matter candidates.

2 Dark Matter Inventory

Evidence for dark matter (DM) exists on many scales, and it is important to keep in mind that the dark matter on different scales may be different – the dark matter in dwarf spirals may not be the dark matter which contributes $\Omega = 1$; in fact, the $\Omega = 1$ dark matter may not exist. This consideration is especially important when discussing dark matter detection, since detection is done in the Milky Way, and evidence for dark matter outside the Milky Way may not be relevant.

So, let me start with an inventory of dark matter in the Universe. The quantity of dark matter on different scales is quoted using $\Omega = \rho/\rho_{crit}$, where ρ is the density of some material averaged over the Universe, and ρ_{crit} is the critical density. Most determinations of Omega are made by measuring the mass-to-light ratio Υ of some system and then multiplying this by the average luminosity density of the Universe: $j_0 = 1.7 \pm .6 \times 10^8 h^{-1} L_\odot/M_\odot$ (Binney and Tremaine 1987; Davis & Huchra 1982; Kirshner *et al.* 1983). Here $h = 0.4 - 1$ parameterizes our uncertainty of the Hubble constant. Note the factor of two uncertainty in this number, which implies that all determinations of Ω which use this method will be uncertain by at least this amount. In fact, almost all determinations of Ω do use this method; the exceptions being determination of Ω_{baryon} from big bang nucleosynthesis, and the large scale determinations from bulk flows. For example, the mass-to-light ratio in the solar neighborhood is $\Upsilon \approx 5$, giving $\Omega_{lum} = 0.003h^{-1} = 0.003 - 0.007$. If the solar neighborhood is typical, the amount of material in stars, dust and gas is far below the critical value.

2.1 Spiral Galaxies

The most robust evidence for dark matter comes from the rotation curves of spiral galaxies. Using 21 cm emission, the velocities of clouds of neutral hydrogen can be measured as a function of r, the distance from the center of the galaxy. In almost all cases, after a rise near $r = 0$, the velocities remain constant out as far as can be measured. By Newton's law for circular motion $GM(r)/r^2 = v^2/r$, this implies that the density drops like r^{-2} at large radius and that the mass $M(r) \propto r$ at large radii. Once r becomes greater than the extent of the mass, one expects the velocities to drop $\propto r^{-1/2}$, but this is never seen, implying that we do not know how large the extended dark halos around spirals are. For example, the rotation curve of NGC3198 (Binney & Tremaine 1987) implies $\Upsilon > 30h$, or $\Omega_{halo} > 0.017$. The large discrepancy between this number and Ω_{lum} is seen in many external galaxies and is the most robust evidence for dark matter. This is fortunate since searches for dark matter can be made only in spiral galaxies; in fact only in our spiral, the Milky Way. Unfortunately, the rotation curve of the Milky Way is poorly constrained, which leads to uncertainty in the amount of dark matter in our Galaxy. However, there is less secure evidence for substantial dark matter in our Galaxy. By studying the motion of dwarf galaxies (especially Leo I at a distance of 230 kpc) Zaritsky, *et al.* (1989) find a mass of the Milky Way of $M_{MW} = 1.25^{+0.8}_{-0.3} \times 10^{12} M_{\odot}$, for $\Upsilon_{MW} \approx 90$, and $\Omega_{MW} \approx 0.054h^{-1}$ (assuming the Universe is like the Milky Way). There are a limited number of small satellite galaxies around the Milky Way, so the uncertainty in this measurement is large. However, for external galaxies Zaritsky (1992) used a sample of 69 small satellite galaxies around 45 spirals similar to the Milky Way to estimate the total mass for a "typical" spiral. He found that $M \approx 10^{12} M_{\odot}$ at 200 kpc from the center, implying $\Omega_{spirals} \approx 0.087h^{-1}$ out to this radius. Even in this case, there is not strong evidence that the rotation speeds drop, so there is no good upper limit to Ω_{spiral}. This number is similar to the number found by the Local Group Timing method (e.g. Binney & Tremaine 1987).

2.2 Clusters of Galaxies

Moving to larger scales, the methods of determining Ω become less secure, but give larger values. There is a great deal of new evidence on dark matter in clusters of galaxies, coming from gravitational lensing which Tyson (these proceedings) will discuss, from X-ray gas temperatures, and from the motions of cluster member galaxies. For example, consider the Coma cluster of around a thousand galaxies; the cluster which Zwicky (1933) used to first hypothesize that dark matter existed. White *et al.* (1993) recently collated some of the data on the Coma cluster, reporting separate measurements of the amount of mass in stars, hot gas, and in total. Within a radius of $1.5h^{-1}$ Mpc, they give

$$\begin{aligned} M_{star} &= 1.0 \pm 0.2 \times 10^{13} h^{-1} M_{\odot} \\ M_{gas} &= 5.4 \pm 1 \times 10^{13} h^{-5/4} M_{\odot} \\ M_{total} &= 5.7 - 11 \times 10^{14} h^{-1} M_{\odot}, \end{aligned} \tag{1}$$

where the total mass is estimated in two completely different ways. The first method is a refinement of Zwicky's method of using the radial velocities of the member galaxies, and the assumption of virialization to gauge the depth of the gravitational potential well. The second method makes use of the ROSAT X-ray maps and the assumption of a constant temperature equilibrium to get the same information. Remarkably the two methods give the same mass within errors. Thus with a mass-to-light ratio of $\Upsilon = 330 - 620 M_{\odot}/L_{\odot}$, one finds $\Omega = 0.2 - 0.4$, if the inner 1.5 Mpc of Coma is representative of the Universe as a whole.

There is, however, a very disconcerting fact about the above numbers. As pointed out by White, *et al.* (1993),

$$\frac{M_{baryon}}{M_{total}} > 0.009 + 0.05h^{-3/2}.$$

Now the Coma cluster is large enough that one might expect its baryon to dark matter ratio to be the Universal value, ($\Omega_{baryon}/\Omega_{total} = M_{baryon}/M_{total}$), and in fact White, *et al.* argue that this is the case. Then the inequality above should apply to the entire Universe. But, as we discuss later, big bang nucleosynthesis limits $\Omega_{baryon} < 0.015h^{-2}$. If $\Omega_{total} = 1$, as many would like, the two inequalities are in quite strong disagreement for any value of h! So this is a big puzzle. The conclusions of White, *et al.*, are that either Ω is not unity, or that big bang nucleosynthesis is not working. As I think Joel Primack (these proceedings) and Tony Tyson (these proceedings) will discuss, there are other possible explanations. Notably that measurements of the the total mass in clusters by weak or strong gravitational lensing tend to give twice as much total mass as the X-ray and virial methods, and that mass and velocity bias may mean that clusters are not so representative of the Universe as a whole.

2.3 Large Scale Flows

Next, we turn to some of the most interesting new results on the amount of dark matter. One would really like to measure the amount of dark matter on the largest possible scales so that one can be sure the sample is representative of the entire Universe. Within the past several years a host of related methods have been tried, and while they initially gave uncertain results, the most recent determinations do impressively well. These methods have the advantage stated above, but the disadvantage that they depend upon assumptions about galaxy formation; which means they depend upon gravitational instability theory, biasing, etc. Also, the errors in these results are still large and the calculations are complicated, but they do have the promise to answer the fundamental question of how large Ω is. They also tend to give values of Ω_{total} near unity!

The simplest example comes from the observation that the local group of galaxies moves at 627 ± 22 km/sec with respect to the Cosmic Microwave Background (CMB) (measured from the amplitude of the CMB dipole). If this motion comes from gravity, then the direction of the motion should line up with the direction where there is an excess of mass, and the velocity should be determined by the size of this excess. Thus, taking into account the expansion of the Universe, one has

$$v \propto \Omega^{0.6} \frac{\delta\rho}{\rho} = \frac{\Omega^{0.6}}{b} \frac{\delta n}{n},$$

where the bias factor b has slipped in because observers use the excess in galaxy number counts $\delta n/n$ to estimate the excess in mass density $\delta\rho/\rho$. So, using galaxy counts from the IRAS satellite survey, Yahil *et al.* (1986) find that the direction of the $\delta n/n$ excess agrees with the direction of the velocity vector to within $\sim 20^0$, and that

$$\frac{\Omega^{0.6}}{b} = 0.9 \pm 0.2.$$

Thus with the very conservative limit $b > 0.5$, one has $\Omega > 0.2$, and with the reasonable limit $b > 1$, one finds $\Omega > 0.5$. So there is now observational evidence that Ω is near the theoretically favored value of unity. However, for the method above to be reliable, one must measure $\delta n/n$ on very large scales to ensure that convergence has been reached. I don't think this has yet been convincingly demonstrated.

The above technique is only one of a host of methods used to determine Ω on large scales. Especially promising is the detailed comparison of the peculiar velocities of many galaxies to the detailed maps of $\delta n/n$. This should not only determine Ω, but serve as a stringent test on the theory that large scale structure is formed by gravitational instability. The trick is to get the peculiar velocities, which means subtracting the much larger Hubble flow velocity. Since the redshift measurements gives only the radial component of velocity it seemed difficult to get complete information, but Bertshinger and Dekel (1989) proposed an ingenious method, in which one assumes that the velocity field is curl free, and proceeded to reconstruct the entire three dimensional field. They use

$$\nabla \cdot \mathbf{v} = -\frac{\Omega^{0.6}}{b}\frac{\delta n}{n},$$

and solve for $\Omega^{0.6}/b$. They find $0.3 < \Omega < 1$ for reasonable choices of b. Especially notable is that the detailed $\delta n/n$ maps agree remarkably well with the reconstructed velocity fields, and I interpret this as evidence that gravitational instability is the most likely cause of the structure. In fact, this is now the story for most of the large scale flow methods. They find reasonable to excellent agreement with the theory, and predict a large value of Ω. A very nice table of values and a review of many of these methods can be found in Dekel (1994).

2.4 Big Bang Nucleosynthesis

Next, I turn to a very important and recently controversial ingredient in the dark matter story; namely the predictions of big bang nucleosynthesis. The standard story has been (Smith, Kawano, Malaney 1993; Walker, *et al.* 1991) that to get agreement with the measured abundances of helium, deuterium, and lithium, the baryonic content of the Universe had to be between $0.01 \leq \Omega_{baryon}h^2 \leq 0.015$. Given the large uncertainty in h this meant $0.01 \leq \Omega_{baryon} \leq 0.1$. This value is inconsistent with having a total $\Omega = 1$ in baryons, and so, together with the predilection for $\Omega_{total} = 1$, is the main motivation for postulating non-baryonic (elementary particle) dark matter. The majority of dark matter searches are targeted towards these non-baryonic particles. The lower limit of this range is actually *above* the abundance of known stars, gas, etc., and so there also seems to be evidence for substantial *baryonic* dark matter as well.

Much recent excitement was caused in this field when Songaila, *et al.* (1994) reported a possible detection of deuterium in a Lyman limit cloud at a redshift of 2.9 (via absorption of light emitted by an even more distant quasar). The controversy was caused by the extremely large value $D/H \approx 2.4 \times 10^{-4}$, when measurements from the interstellar medium are in the range $1.9 \times 10^{-5} < D/H < 6.8 \times 10^{-5}$ (Steigman & Tosi 1992). If the new measurement is true, the limits from big bang nucleosynthesis would move to roughly to $0.004 < \Omega h^2 < 0.0068$, or $0.006 \leq \Omega \leq 0.035$, substantially lower than the current limits. In this case, most of the baryonic material may well be known, and models of structure formation which require substantial baryonic content may be in trouble. However, it is too soon to tell whether the measurement will stand up. Several groups are making these measurements and hopefully soon there will be either confirmation or refutation of this result.

2.5 Distribution of Dark Matter in the Milky Way

While we don't have a clue as to what the dark matter (DM) is, we have a reasonable idea as to how much of it there is in the Galaxy, how it is distributed, and how fast it is moving. This information comes from the rotation curve of the Milky Way, and is crucial to all the direct searches for dark matter. If we say that the rotation curve of the Milky Way is constant at about $v_c = 220$ km/sec out to as far as it is measured, then we know that the density must drop as r^{-2} at large distances. This velocity also sets the scale for the depth of the potential well and says that the dark matter must also move with velocities in this range. Assuming a spherical and isotropic velocity distribution is common, and a usual parameterization is

$$\rho(\mathbf{r}) = \rho_0 \frac{a^2 + r_0^2}{a^2 + r^2},$$

where $r_0 \approx 8.5$ kpc is the distance of the Sun from the galactic center, a is the core radius of the halo, and $\rho_0 \approx 0.3$ GeV cm^{-3} is the density of dark matter near the Sun. Also, a typical velocity distribution is

$$f(v)d^3v = \frac{e^{-v^2/v_c^2}}{\pi^{3/2}v_c^3}d^3v.$$

It should be noted that the specifics of the above models are not very secure. For example, it is quite possible that the halo of our Galaxy is flattened into an ellipsoid, and there may be a component of the halo velocity which is rotational and not isotropic. Also, some (or even much) of the rotation curve of the Milky Way at the solar radius could be due to the stellar disk. Canonical models of the disk have the disk contributing about half the rotation velocity, but larger disks have been envisioned. Recent microlensing results may be indicative of a larger disk as well. I will talk about this later if there is time.

Finally, other important points about our Galaxy's geography include the fact that the nearest two galaxies are the LMC and SMC, located at a distance of 50 kpc and 60kpc respectively, that the halo of the Milky Way is thought to extend out at least this far, and that the bulge of the Milky Way is a concentration of stars in the center of our Galaxy (8.5 kpc away) with a size of about 1 kpc.

3 Dark Matter Candidates

There is no shortage of ideas as to what the dark matter could be. In fact, the problem is the opposite. Serious candidates have been proposed with masses ranging from 10^{-5} eV $= 1.8 \times 10^{-41}$ kg $= 9 \times 10^{-72} M_\odot$ (axions) up to $10^4 M_\odot$ black holes. That's a range of masses of over 75 orders of magnitude! It should be clear that no one search technique could be used for all dark matter candidates.

Even finding a consistent categorization scheme is difficult, so we will try a few. First, as discussed above, is the baryonic vs non-baryonic distinction. The main baryonic candidates are the Massive Compact Halo Object (Macho) class of candidates. These include brown dwarf stars, jupiters, and 100 $M_\odot$ black holes. Brown dwarfs are balls of H and He with masses below 0.08 $M_\odot$, so they never begin nuclear fusion of hydrogen. Jupiters are similar but with masses near 0.001 $M_\odot$. Black holes with masses near 100 $M_\odot$ could be the remnants of an early generation of stars which were massive enough so that not many heavy elements were dispersed when they went supernova. Other, less popular, baryonic possibilities include fractal or specially conditioned clouds of neutral hydrogen. The non-baryonic candidates are basically elementary particles which are either not yet discovered or have non-standard properties. Outside the baryonic/non-baryonic categories are two other possibilities which don't get much attention, but which I think should be kept in mind until the nature of the dark matter is discovered. The first is non-Newtonian gravity. See Begeman *et al.* (1991) for a provocative discussion of this possibility; but watch for results from gravitational lensing which may place very strong constraints. I won't talk about this possibility here. Second, we shouldn't ignore the "none-of-the-above" possibility which has several times surprised the physics/astronomy community in the past.

Among the non-baryonic candidates there are several classes of particles which are distinguished by how they came to exist in large quantity during the Early Universe, and also how they are most easily detected. The axion (Section 3.5) is motivated as a possible solution to the strong CP problem and is in a class by itself. The largest class is the Weakly Interacting Massive Particle (Wimp) class (Section 3.3), which consists of literally hundreds of suggested particles. The most popular of these Wimps is the neutralino from supersymmetry (Section 3.4). Finally, if the tau or muon neutrino had a mass in the 5 eV to 100 eV range, it could make up all or much of the dark matter (Section 3.2).

Another important categorization scheme is the "hot" vs "cold" classification. A dark matter candidate is called "hot" if it was moving at relativistic speeds at the time galaxies could just start to form (when the horizon first contained about $10^{12} M_\odot$). It is called "cold" if it was moving non-relativistically at that time. This categorization has important ramifications for structure formation, and there is a chance of determining whether the dark matter is hot or cold from studies of galaxy formation. Hot dark matter cannot cluster on galaxy scales until it has cooled to non-relativistic speeds, and so gives rise to a considerably different primordial fluctuation spectrum. Of the above candidates only the light neutrinos would be hot; all the others would be cold.

3.1 Machos

Probably the most exciting development in the dark matter story is the detection of Machos by three separate groups (Alcock, *et al.* 1993; Aubourg, *et al.* 1993; Udalski, *et al.* 1993). All three groups monitored millions of stars, either in the LMC or in the galactic bulge, for signs of gravitational microlensing, and all three groups seem to have found it. It is still not clear whether the objects which are causing the microlensing are numerous enough to make up the Milky Way dark matter, or even whether they are in the galactic halo or disk. Whether they are light enough to qualify as bona fide brown dwarfs, or whether they are just faint stars is also not yet known. See David Caldwell's talk (these proceedings) for a review of the current status of these important experiments.

However, from the more general point of view, I'd like to note that these experiments very well may have the capability to give a definitive answer to the question of whether the dark matter in our Galaxy is baryonic. The microlensing searches are probably sensitive to any objects in the range $\sim 10^{-8} M_\odot < m < 10^3 M_\odot$, just the range in which such objects are theoretically allowed to exist. Objects made purely of H and He with masses less than $\sim 10^{-9} - 10^{-7} M_\odot$ are expected to evaporate due to the microwave background in less than a Hubble time, while objects with masses greater than $\sim 10^3 M_\odot$ would have disrupted known globular clusters.

3.2 Light Neutrinos

Hot dark matter was out of favor for several years, but has come back into style in a big way. Neutrino dark matter was unpopular for two main reasons. First, by Jeans theorem, the current phase space density of any dissipationless particle should be less than or equal to its phase space density at decoupling. Tremaine & Gunn (1979), showed that this meant very light neutrinos could not be packed endlessly into dwarf galaxies. Gerhard & Spergel (1992) applied this result to the dwarf galaxy Ursa Minor, which is known to have a great deal of dark matter, and very conservatively found that if all the dark matter in Ursa Minor were neutrinos, the neutrinos would have to have masses greater than 81 eV. Since $\Omega_\nu h^2 = m_\nu/90$ eV, and $\Omega_\nu h^2 < 1$, the maximum mass which neutrino dark matter could have is around 90 eV, with values of around 30 eV being favored. So the dark matter in dwarf galaxies such as Ursa Minor and Draco probably cannot be light neutrinos. Thus, one would need at least two types of dark matter to have hot dark matter play an important role. By simplicity, this led to neutrino dark matter being disfavored.

The other reason neutrino dark matter became less popular was that galaxy formation with pure hot dark matter and a Zeldovich perturbation spectrum just doesn't match the observations.

While the arguments above are still valid, several recent developments have changed the attitude towards them. First, it is now known that galaxy formation with pure cold dark matter and a Zeldovich perturbation spectrum also doesn't match the observations. So some additional ingredient is needed. Second, one of the few models which does match the observations is a mixed dark matter Universe with $\Omega_{cold} = .7$, $\Omega_\nu = 0.25$, and $\Omega_{baryon} = 0.05$. Thus, since it seems we may need two types of dark matter anyway, the objection that the dark matter in dwarf galaxies cannot all be neutrinos becomes less important. We may well have a tau or muon neutrino with a mass near 6 eV. In these mixed models, the cold dark matter dominates, and so the ongoing searches for the other dark matter candidates are hardly affected.

See talks by Joel Primack and David Caldwell (these proceedings) for more discussion of these models and attempts at detection of neutrino dark matter by measuring the neutrino mass.

3.3 Thermal Relics as Dark Matter (Wimps)

Among the particle dark matter candidates an important distinction is whether the particles were created thermally in the Early Universe, or whether they were created non-thermally in a phase transition. Thermal and non-thermal relics have a different relationship between their relic abundance Ω and their properties such as mass and couplings, so the distinction is especially important for dark matter detection efforts. For example, the Wimp class of particles can be defined as those particles which are created thermally, while dark matter axions come mostly from non-thermal processes.

In thermal creation one imagines that early on, when the Universe was at very high temperature, thermal equilibrium obtained, and the number density of Wimps (or any other particle species) was roughly equal to the number density of photons. As the Universe cooled the number of Wimps and photons would decrease together as long as the temperature remained higher than the Wimp mass. When the temperature finally dropped below the Wimp mass, creation of Wimps would require being on the tail of the thermal distribution, so in equilibrium, the number density of Wimps would drop exponentially $\propto \exp(-m_{Wimp}/T)$. If equilibrium were maintained until today there would be very few Wimps left, but at some point the Wimp density would drop low enough that the probability of one Wimp finding another to annihilate would become small. (Remember we must assume that an individual Wimp is stable if it is to become the dark matter.) The Wimp number density would "freeze-out" at this point and we would be left with a substantial number of Wimps today. Detailed evolution of the Boltzmann equation can be done for an accurate prediction, but roughly

$$\Omega_{Wimp} \approx \frac{10^{-26}\mathrm{cm}^3\mathrm{s}^{-1}}{\langle\sigma v\rangle},$$

where $\langle\sigma v\rangle$ is the thermally averaged cross section for two Wimps to annihilate into ordinary particles. The remarkable fact is that for $\Omega \approx 1$, as required by the dark matter problem, the annihilation cross section $\langle\sigma v\rangle$ for any thermally created particle turns out to be just what would be predicted for particles with electroweak scale interactions. Thus the name "Wimp". There are several theoretical problems with the Standard Model of particle physics which are solved by new electroweak scale physics such as supersymmetry. Thus these theoretical problems may be clues that the dark matter does indeed consist of Wimps. Said another way, any stable particle which annihilates with an electroweak scale cross section is bound to contribute to the dark matter of the Universe. It is interesting that theories such as supersymmetry, invented for entirely different reasons, typically predict just such a particle.

The fact that thermally created dark matter has weak scale interactions also means that it may be within reach of accelerators such as LEP at CERN, and CDF at Fermilab. After all these accelerators were built precisely to probe the electroweak scale. Thus many accelerator searches for exotic particles are also searches for the dark matter of the Universe. Also, due to the weak scale interactions, Wimp-nuclear interaction rates are within reach of many direct and indirect detection methods. Caldwell (these proceedings) will review these.

3.4 Example of the Neutralino

The most popular of the Wimp candidates is the neutralino from supersymmetry. Supersymmetry seems to be a necessary ingredient of any theory which which consistently combines general relativity with the Standard Model of particle physics.

When Dirac attempted to combine special relativity with quantum mechanics, he came up with the Dirac equation and discovered (to his chagrin) that it contained a new CPT symmetry which required the existence of a CPT partner, or anti-particle for every known particle. Dirac's initial hopes that the electron might be the anti-particle of the proton were soon dashed, but the discovery of the positron vindicated Dirac's theory. Today this "doubling" of the number of particles by CPT symmetry is taken for granted to such an extent that anti-particles are not listed in the particle data book.

It is interesting that attempts to combine general relativity with quantum field theory seem to require supersymmetry in a similar way. Here the symmetry relates integral spin particles to half integral spin superpartners and vice versa. Analogous initial hopes that the photon might be the superpartner of the neutrino were soon dashed, and now, if supersymmetry exists, it is known that there would be a doubling of the number of known particles. While there has been no quickly following discovery of supersymmetric particles, the hypothetical particles have been named and thoroughly studied. Intense searches for these particles are taking place at all the major accelerators. Examples of superpartners are the spin 1/2 photino, Higgsino, Z-ino, and W-ino (superpartners of the photon, Higgs, Z, and W), and the spin 0 squark and selectron (partners of the quark and electron).

There are actually several other strong motivations for the existence of supersymmetry, among which, the stabilization of the electroweak scale (hierarchy problem), and the remarkable coupling constant unification which occurs in supersymmetric grand unified models, stand out. The interest for dark matter, however, arises because typically there is a conserved multiplicative R-parity, which means that the lightest supersymmetric particle (LSP) is stable. The most likely LSP is the neutralino, which is a linear combination of the photino, Z-ino, and Higgsino. The precise components of the combination are determined by the parameters of the underlying supersymmetric model, and there are many of these parameters. Typically, the minimal supersymmetric model (MSSM) is considered, which is the supersymmetric extension of the Standard Model with the minimal number of new particles. While supersymmetry relates many couplings and masses, there are still dozens of free parameters in the most general MSSM, so many times a restricted set of these parameters is used which derives from assuming a simple grand unification scheme. The particle accelerators are searching this parameter space for supersymmetry, but have turned up no new particles. It is interesting is that in much of this parameter space calculation of the relic abundance predicts $\Omega \approx 1$ in neutralinos. For example a recent calculation by Drees and Nojiri (1993) (as programmed by Jungman, *et al.* 1994) shows this clearly.

These supersymmetric models are also within reach of the new generation of direct and indirect Wimp detection experiments as Caldwell (these proceedings) will review.

3.5 Axions

The best example of a non-thermal particle dark matter candidate is the axion. Actually thermal axions are produced in the standard way, but if such axions existed in such numbers as to make up the dark matter, they would have lifetimes too short to still be around in quantity. However, there is another, more important, production mechanism for axions in the early Universe.

The axion arises because the QCD Lagrangian contains a term

$$L \supset \frac{\bar{\theta} g^2}{32\pi^2} G\tilde{G},$$

where G is the gluon field strength. This term predicts an electric dipole moment of the neutron of $d_n \approx 5\times10^{-16}\bar{\theta}$. Experimentally, however, the neutron dipole moment $d_n < 10^{-25}$, which means $\bar{\theta} < 10^{-10}$. The question becomes why does this $\bar{\theta}$ parameter have such a small value, when it naturally would have a value near unity? This is the strong CP problem, and one way to resolve this problem is to introduce a new Peccei-Quinn symmetry which predicts a new particle named the axion. The P-Q symmetry forces $\bar{\theta} = 0$ at low temperatures today, but in the early Universe, the axion field was free to roll around the bottom of its Mexican hat potential. The axion field motion in the angular direction is called θ, and since the curvature of the potential in this direction is zero, the axion at high temperatures was massless. However, when the temperature of the Universe cooled below a few hundred MeV (QCD energy scale), the axion potential "tilts" due to QCD instanton effects, and the axion begins to oscillate around the minimum, like a marble in the rim of a tilted Mexican hat. The minimum of the potential forces the average $\bar{\theta}$ to zero, solving the strong CP problem, and the curvature of the potential means the axion now has a mass. There is no damping mechanism for the axion oscillations, so the energy density which goes into oscillation remains until today as a coherent axion field condensate filling the Universe. This is a zero momentum condensate and so constitutes cold dark matter. One can identify this energy density as a bunch of axion particles, which later can become the dark matter in halos of galaxies. The relic energy density Ω is thus related to the tilt of the potential, which in turn is related to the axion mass, a free parameter of the model. If the axion mass $m_a \approx 10^{-5}$ eV, then $\Omega_a \approx 1$. This rather unusual story is still probably the most elegant solution to the strong CP problem, and several groups are mounting laboratory searches for the coherent axions which may make up the major component of mass in the Galaxy. Again Caldwell (these proceedings) will review these experiments.

4 Uncertainties in the Milky Way Dark Matter

Finally, let me turn to a subject which has not received a lot of attention, but which I think will get increasing attention now that one of the dark matter searches (Macho) is returning positive results. That is, the uncertainties in the results of dark matter searches due to uncertainties in the model of Milky Way dark matter halo, where by model I mean the density and velocity distributions of the dark matter. This uncertainty affects Wimp, axion, and Macho searches, and so should be given some attention. For example, axion detection rates are proportional to the local density of dark matter, which can vary greatly in different galactic models, while Wimp rates are proportional to this as well as to the average Wimp velocity. Macho microlensing rates, being averages over the line-of-sight to the stellar source are more complicated functions of all the parameters in the halo model.

Much of the model dependence comes because the parameters of Milky Way (and especially of the dark halo) are not well known. For example, the local circular velocity has been measured at values in the range 190 km/s $< v_c <$ 250 km/s. Estimates of the distance of the Sun from the galactic center range between $7 < r_0 < 9$ kpc. And the halo core radius has been estimated at between $2 < a < 10$ kpc. In addition, the halo may not be spherical, but may be flattened into an ellipsoidal configuration, and the rotation curve may be gently rising or falling (by about 15%).

To investigate this model dependence one can use some nice self consistent models of galactic halos due to Evans (1994). For example, one can calculate the microlensing rate towards the LMC, SMC, and bulge using these models and see how the predicted rates change as the parameters vary within their observationally allowed ranges. Allowing the halo to be flattened up to a very reasonable axis ratio of three-to-one, one finds a variation in the predicted microlensing rate of more than a factor of ten. Since the purpose of the microlensing experiments is to measure the fraction of the dark halo made of Machos, and the connection between the measurement and fraction requires a model of the halo, one sees that this large model uncertainty will translate into an uncertainty in the dark matter fraction. Thus it may be difficult for the Macho searches to place strong constraints on the amount of non-baryonic dark matter in the halo.

However, there is an additional, and perhaps even larger, uncertainty which comes about because the mass of the stellar disk is not well known. In the above derivations of the allowed halo parameters a "canonical" exponential stellar disk was used. This says that about half the rotation curve of the Galaxy comes from the halo, and about half from the stellar disk (Evans & Jijina 1994). However, if, as has been suggested for many years, the disk is actually twice as massive (Oort 1960; Bahcall 1984), then very little need would exist for a dark halo component at the solar radius of 8.5 kpc. (At large r there would still be a need for halo dark matter). This might mean that the Wimp and axion searches would not have much dark matter locally to look for. The Macho searches are hoping to measure the amount of material in the disk and so resolve this part of the problem (Griest, *et al.* 1991; Paczyński, 1991; Alcock, *et al.* 1994; Udalski, *et al.* 1994).

The hope is that the microlensing searches can also resolve the halo model uncertainties (if the halo consists of machos!). The idea is to look in several different lines-of-sight (LMC, SMC, M31, and bulge) and find quantities which differ enough among the different models to distinguish them.

5 Conclusions

Rapid progress is being made in the dark matter question, but as usual with new results, more questions are being opened than closed.

Acknowledgements

We thank Andrew Gould, George Fuller, Joel Primack, and members of the MACHO collaboration for valuable help. K.G. acknowledges support from a DoE OJI Award, the NSF Center for Particle Astrophysics, and the Alfred P. Sloan foundation.

References

Alcock, C. *et al.* 1993, Nature, 365, 621
Alcock, C. *et al.* 1994, submitted to ApJ
Ashman, K. 1992, PASP, 104, 1109
Aubourg, E. *et al.* 1993, Nature, 365, 623
Bahcall, J.N. 1984, ApJ, 287, 926
Begeman, K.G, Broeils, A.H, & Sander, R.H. 1991, MNRAS, 249, 523
Binney, J. & Tremaine, S. 1987, Galactic Dynamics (Princeton University Press, Princeton)
Bertshinger & Dekel 1989, ApJ Lett, 336, L5

Davis, M. & Huchra, J. 1982, ApJ, 254, 437
Dekel, A. 1994, to appear in ARAA 32.
Drees, M. & Nojiri, M.M., 1993, Phys. Rev. D48, 3483
Evans, N.W. 1994 MNRAS, 267, 333
Evans, N.W., & Jijina, J. 1994, MNRAS 267, L21
Gerhard, O.E. & Spergel, D.N. 1992, ApJ. 389, L9
Griest, K. 1991, ApJ, 366, 412
Griest, K. *et al.* 1991, ApJ Lett, 372, L79
Jungman, J., Griest, K, & Kamionkowski, M. 1994, in progress
Kirshner, R.D., *et al.* 1983, AJ, 88, 1285
Oort, J.H. 1960, Bull. Astr. Inst. Netherlands, 6, 249
Paczyński, B. 1986, ApJ, 304, 1
Paczyński, B. 1991, ApJ Lett, 371, L63
Primack, J. R., Seckel, D., & Sadoulet, B. 1988, Ann. Rev. Nucl. Part. Sci., 38, 751
Smith, M.S., Kawano, L.H., & Malaney, R.A. 1993, ApJ Supp, 85,219
Songaila, A., *et al.* 1993, Nature, 368, 599
Steigman, G & Tosi, M. 1992, ApJ, 401, 150
Tremaine, S. & Gunn, J. 1979. Phys. Rev. Lett., 42, 407
Udalski *et al.* 1993, Acta Astron, 43, 289
Udalski *et al.* 1994, preprint
Walker, T.P. *et al.* 1991, ApJ, 376, 51
White, S.D.M., *et al.* 1993, Nature, 366, 429
Yahil, A., Walker, X., & Rowan-Robinson, M. 1986, ApJ Lett, 301, L1
Zaritsky, D. 1992, PASP, 394, 1
Zaritsky, D. *et al.*, 1989, ApJ, 345, 759
Zwicky, F., 1933. Helvetica Physica Acta, 11, 110

MAPPING COSMIC DARK MATTER

J. A. Tyson

AT&T Bell Laboratories, Murray Hill, NJ 07974

June 30, 1995

Gravitational lensing, while still a young field, is beginning to transform our understanding of dark matter and the process of structure formation in the universe. So-called "weak gravitational lensing" provides a direct measure of mass overdensity on scales of kpc to several Mpc. Through comparison with large N-body simulations for various cosmogonies, this new window on mass in the universe will provide constraints on the nature of dark matter.

§1 INTRODUCTION

Over the past decade the discovery of a high surface number density of faint distant galaxies has led to their use as a tool for statistical cosmology; because most of these faint galaxies are resolved and are distributed up to high redshifts they may be employed in statistical gravitational lens studies of foreground mass distributions. Foreground mass concentrations can be detected and mapped directly by the way they distort images of background galaxies.

There are several source requirements for mass mapping by statistical gravitational lens inversion. To be a useful tool for mass mapping the sources must (1) have redshifts large compared with the lens, (2) have a number density on the sky sufficient to sample the lens on relevant scales, (3) have an intrinsic angular diameter larger than the ratio of seeing FWHM to the magnification of the lens, and (4) have other properties (blue color and very low surface brightness) enabling efficient separation of the sources from lens and other foreground objects.

At faint optical surface brightness one finds over 300,000 blue galaxies per square degree. We may be seeing many of these 25–28 magnitude galaxies at an epoch of star formation. The resulting UV-bright spectrum, when redshifted to redshifts of 1–3, could produce the observed blue spectral shape. Their average angular size is large and their correlations are small by comparison to theories. The seeing-deconvolved half light radii of these galaxies approach 0.6 arcsecond at the faint end.

§2 FAINT GALAXY SURVEYS

Even with mild luminosity evolution the surface brightness of galaxies at $z > 1$ is expected to be considerably fainter than nearby L* galaxies. Photographic surveys to a surface brightness of 26 magnitude arcsec^{-2} showed roughly 17,000 faint galaxies per square degree at 24th magnitude (Jarvis & Tyson 1981; Kron 1982), and led to the first search for statistical gravitational lens distortions from dark matter clumping on large scales (Valdes, et al 1983). However, photographic surveys of galaxies suffer from a selection bias due to the achievable limiting surface brightness for small objects occupying less than a few square arcseconds. The number of galaxies found at $z > 0.7$ is severely reduced by the surface brightness cut typical of photographic limit plates; about an 90% reduction, in mild evolution models.

2.1 CCD Surveys

Spot CCD imaging surveys have been made to 29 U, B_j, 28 R, and 26 I mag arcsec^{-2} limiting surface brightness, and ultra-deep CCD imaging to 30 mag arcsec^{-2} has been obtained in several fields (Guhathakurta *et al* 1990, Tyson 1988). [The blue B_j passband (close to the photographic IIIaJ+GG385 passband) is effectively 3700–5100 Å (Gullixson *et al* 1993)]. The deepest survey involves as many as 100 disregistered exposures of one field in a given filter. This 3-d stack of disregistered exposures contains all the information about the systematic errors in the CCD, telescope and sky, and the true object luminosity distribution in separable form (see Tyson 1990), producing a catalog of multi-band photometric data.

Figure 1 summarizes faint galaxy count data for five wavelength bands. The U counts are from Guhathakurta *et al* 1990, Majewski 1988, 1993, and Koo 1986; the B_j, R, and I counts are from our most recent data on 15 fields; and the K counts are from Gardner *et al* 1993. Number-magnitude counts which rise with magnitude like dex (0.4 mag) or faster continue to add to the integrated light at the faintest magnitudes, and if this slope never falls below 0.4 the surface brightness of the extragalactic sky would approach that of a typical galaxy — a type of Olbers' paradox. The finite luminous lifetime of stellar populations offers a way out of both problems. The N(m) counts in the B_j band have average slope 0.45, while the N(m) slope in the U-band is even steeper. The slope of the N(m) counts for longer wavelengths is sub-critical, going to a "no-evolution" slope of 0.3 at 0.9 μm wavelength. The integrated light of the FBGs is influenced strongly by galaxy evolution (stellar lifetimes and population evolution), and to a much lesser extent by cosmology (Harrison 1964).

The enhancement of the blue and U band counts is appar-

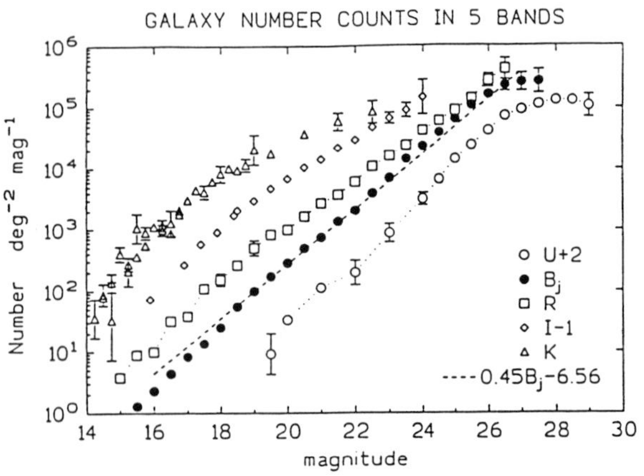

Figure 1. *Faint galaxy differential number counts dN(m) as a function of magnitude in five color bands are plotted to their completeness limits. Data are from several surveys (see text). Note the decrease in slope* $d\log N/dm$ *of the faint part of the number counts with wavelength.*

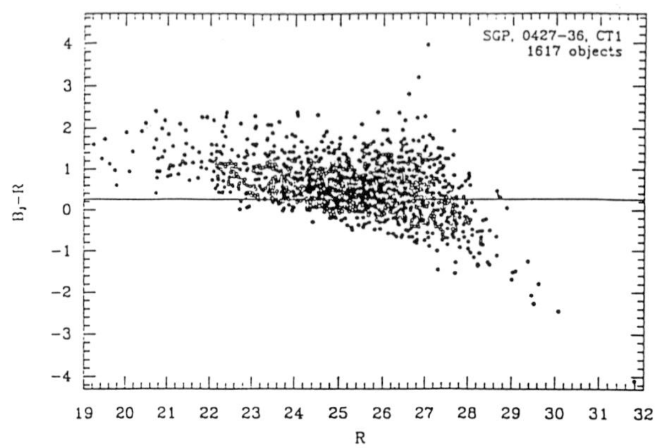

Figure 2. *Galaxy* B$_j$-*R color vs total R magnitude (Guhathakurta* et al. *1990). Identical isophotes, determined from the* B$_j$+*R sum image, were used in the two bands in order to avoid systematics. The bluing trend continues to the completeness limit.*

ently coming from objects which are UV-bright in their rest frame. Tinsley (1980) was the first to point out that a high star formation rate at a given epoch would produce a bump in the number-magnitude counts. The small but highly significant bump seen in the blue counts around 25 B_j mag, and more prominently in the U counts, is possibly an effect of this kind. The average blue counts over the range 15–27 B_j mag follow the relation: $\log N = 1.43\,B_j - 0.14\,B_j^{3/2} - 13.59$, where N has units of deg^{-2} mag^{-1}. Counts at blue wavelengths exceed simple no-evolution models using a locally normalized Schechter luminosity function for a $(\Omega, \Lambda) = (1{,}0)$ cosmology by as much as a factor of ten at the faint end. The longer look-back times and larger volumes in open cosmologies make it easier to build models with larger numbers of faint galaxies.

2.2 Colors of faint galaxies

While the mean B_j-R color of zero redshift galaxies is about 1.5–2 mag at 21 B_j mag, many FBGs at 26–28 B_j mag have $B_j - R < -0.2$. The 1.5 mag of bluing in B-R of these FBGs to 27 B magnitude is shown in Figure 2. These extreme blue colors are evidence that one is seeing starburst galaxies at redshifts of 1–2, so that the UV excess is redshifted into the B_j passband. Unfortunately, there are insufficient data on local starburst galaxy UV spectral energy distributions shortward of 1500 Å to construct semi-empirical models for high z.

2.3 The nature of the faint blue galaxies

What are the FBGs? A population of extreme blue galaxies is expected from metallicity production arguments: an estimate by Cowie, *et al* (1988) suggested that star-forming galaxies would produce as much as 40% of the observed metals. Babul & Rees (1992) argued that the majority of the FBGs may be dwarf ellipticals undergoing starburst at redshift about 1, most fading below detectability at the present epoch via supernova wind dispersal. Further studies to ultra faint flux levels at wavelengths from radio to x-ray will be required for their characterization (see Treyer & Silk 1993).

Based on estimates of the local galaxy luminosity function Loveday, *et al* (1992) conclude that the surface density of galaxies exceeds a no-evolution estimate by a factor of two by B_j = 20–21 mag; Colless *et al* (1993) claim a factor of two excess by 22.5 B_j magnitude. A review of the galaxy luminosity function problem is given by Eales (1993).

What should we expect for the redshift of a 24th to 26th magnitude galaxy? Figure 3 shows differential number-redshift counts at 24–26 B magnitude in a variety of models. Model number – log redshift dN/dLog z distributions (ΔLog z = 0.04) for the $\Omega = 0.1$ C model of Guiderdoni & Rocca Volmerange (1989), which does not include a lost-light correction, are shown, together with one of our mild evolution models for $\Omega = 0.2$ and $H_0 = 50$; and LGRS93 (Lacey et al. 1993 The $(\Omega, \Lambda) = (.2,.8)$, $(.2,0)$ and $(1,0)$ models of Yoshii *et al* (1993); and the no-evolution heuristic model KGB93 (Koo et al. 1993). These models use somewhat different luminosity functions. Models corrected for observational selection should be compared with raw data, while the others should be compared with corrected data. Redshifts of 23 mag

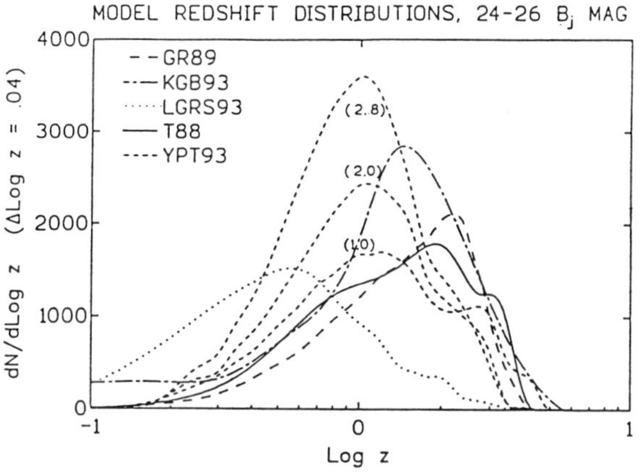

Figure 3. *Theoretical differential number-redshift relation dN(z)/dLog z at magnitudes 24–26* B_j*, for galaxies in several models which are roughly consistent with the observed* N(z, m < 22)*. The models cover a wide range of evolution and cosmology.*

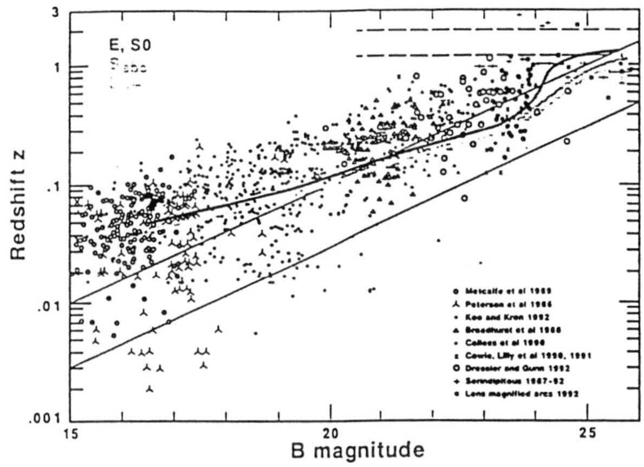

Figure 4. *A plot of field galaxy redshift vs* B_j *magnitude, from all the redshift surveys to date (adapted from Koo & Kron 1992). New survey data have been added.*

field galaxies in most models peak around $z = 0.4$, consistent with existing galaxy spectroscopic data. On the other hand, the bulk of the FBG population is over 16 times fainter ($m > 26$ mag) and should span redshifts up to 3.

2.4 Observed Redshift Distribution

Redshift data for samples of galaxies complete to very faint limits will eventually provide the only hope of disentangling the trends of galaxy mix, luminosity functions, evolution scenarios, and cosmology. Figure 4 collects all redshift surveys of field galaxies, excluding AGNs and QSOs, to date. The recent LDSS-2 survey at 23-24 mag is shown as filled circles. Overplotted are mean redshift lines from a mild evolution model, for three different galaxy types.

The trend to redshift ≈ 1 at 25th B_j magnitude is clear. Note that a typical galaxy seen at z=1 may be a $0.1\ L^*$ galaxy and that at 25th magnitude there may be a wide range in redshifts extending from 1–3. There is an excluded band (dashed lines) between redshifts 1.2 and 1.8 for ground based surveys due to inaccessibility of Ly-α or oxygen 3727 $\AA$ emission lines. The Keck Deimos spectrograph will enable wholesale exploration of galaxy redshifts to 25th magnitude and fainter.

Although redshifts of individual galaxies fainter than 24th magnitude are generally not known, rough statistical limits have been set via gravitational lens image distortion in the case of several foreground massive clusters with independently determined masses. Thus, a lower redshift limit for these faint galaxies may be derived from their response to a gravitational lens placed in the foreground. A good test is provided by the dark matter lens in the well studied cluster 0024+16 (z=0.39). Roughly the same number of FBGs per square arcminute are seen behind these red clusters, but their images are stretched by the gravitational lens and are aligned orthogonal to the vector to the cluster center. We take this as evidence that most of the background galaxies are at redshifts larger than 0.7.

An upper redshift limit for these FBGs may be determined from their spectral energy distribution over the wavelength range 0.3–1 μm, and the discontinuity in the spectrum of stars and galaxies at the Lyman break of hydrogen at 912 Å (Guhathakurta *et al* 1990). Most of these faint galaxies appear to have redshift less than three. If a significant fraction of these galaxies were at redshift greater than 2.8, the Lyman break would be shifted through the 0.32–0.36 μm "U" passband, causing these galaxies to either drop out in this U band or to have a drop in flux between B_j (0.37–0.5 μm) and U wavelengths. This is not seen.

2.5 Sizes of Faint Galaxies

To be useful as sources for mass mapping by statistical gravitational lens distortion galaxies must have apparent angular diameters larger than the point spread divided by the lens magnification. For 1 arcsec FWHM and the range of lens mass profiles typical of either CDM overdensities at z=0-0.2 or cluster mass profiles this minimum useful source diameter ranges from 0.1 to 0.5 arcsec. Are the FGBs large enough? When surface brightness effects are taken into account, detected galaxies at high redshift are expected to be sub-L^* in all models. When combined with the additional expected magnitude-size correlation, predicted scale lengths decrease monotonically with apparent magnitude. Statistical average

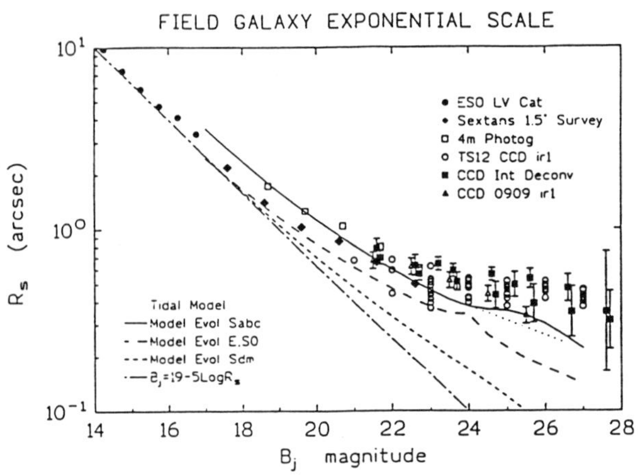

Figure 5. *Data and mild evolution models for the seeing-deconvolved angular exponential scale radius as a function of apparent* $\mathrm{B_j}$ *magnitude. These are averages; the spread in sizes at each magnitude is at least a factor of three.*

radial profiles of the FBGs, at a given magnitude, can be obtained: after seeing deconvolution galaxies fainter than 26 B_j mag are found to have average exponential scale length 0.2–0.5 arcsec and typical half light diameters of 1 arcsec.

Galaxy angular scale data from a number of sources are shown in Figure 5, in which the best fit exponential scale radii are plotted vs total B_j magnitude. The lines plotted in Figure 5 show schematically the behavior of the scale length in simple single population mild luminosity evolution models for several types of galaxies. Comoving sizes are multiplied by 1+z in the model to account roughly for collapse of the star forming region. A tidally triggered galaxy formation model of Lacey *et al* 1993 is also shown. The bump in R_s for the mild evolution model at 24–26 mag is due to higher SFR at z=2, which also produces the bump in the counts similar to that seen in Figure 1.

Broadhurst *et al* (1988), and more recently Cowie *et al* (1991) and Babul & Rees (1992), suggested that the FBGs may be dwarfs undergoing starburst which have faded below detection in recent times. Starbursting dwarf galaxies at high redshift may have exponential scale radii less than 0.1 arcsec (see Carlberg & Charlot 1992), so that the FBGs may not be these unless star formation in dwarf galaxies at z=1 takes place over a very much larger comoving volume than in nearby dwarfs. Baron & White 1987 showed that in CDM theory the properties of the FBGs are consistent with their being primeval galaxies at z = 2–3, with formation continuing to z = 1–2. In any case these FBGs, if imaged at > 29 $mag\ arcsec^{-2}$ surface brightness, have apparent angular diameters larger than the minimum required for statistical lensing studies of foreground mass. Since angular sizes of galaxies at each magnitude show a big range, the Hubble Space Telescope has been useful in resolving the small size end of this distribution (see Im, et al. 1995 for a review and updated model comparisons). This complements the lower surface brightness studies using large ground-based telescopes.

2.6 *Are evolution models unique?*

While it is possible to fit the observed N(m,λ) with low Ω and the traditional models, it is also possible to get a good fit by varying the faint tail of the luminosity function vs look-back time or introducing galaxy number non-conservation: merging models have been proposed by White 1989, Cowie *et al* 1991, and Broadhurst *et al* 1992. There is a problem getting rid of enough galaxies by merging: Dalcanton (1993) has shown that only 5–15% of the mass in galaxies at z=0.35 can have merged into the local population of ellipticals and spirals. Finally, we should perhaps be cautious about over-interpreting morphology, for it is risky to argue that irregular 2-dimensional morphology implies merging. At this time, we know even less about the nature of galaxies at z=1 than we do at z=0.3, where model calculations already extrapolate galaxy luminosity functions eight magnitudes below $L^{\star}$. Galaxy evolution models have more adjustable parameters than necessary for a good fit to the N(m,λ) data.

The faint galaxy counts in 15 randomly chosen high galactic latitude fields and 33 other fields are isotropic on the angular scales between fields (10–180 degrees) to about 10% (2σ). Since these galaxies are at redshifts of order 1 and above, the galaxies in different fields are out of causal communication (reminiscent of the cosmic microwave background). The FBGs are also weakly clustered (Efstathiou, *et al* 1991).

§3 MAPPING DARK MATTER WITH FBGs

Traditional methods of estimating total mass are indirect; mass estimators based on dynamics or X-ray flux maps involve models of orbits or the state of the hot gas, leading to systematics in the derived gravitational potential. While the dynamical evidence for dark matter is strong on scales from galaxies to superclusters of galaxies, it is worthwhile exploring independent observational techniques which do not rest on assumptions about the orbits of test particles.

3.1 *Gravitational Light-bending: Displacements and Distortions*

An excellent review of gravitational lenses was given by Blandford and Narayan (1992). The path of a photon from a distant source is bent as it passes by the foreground mass (gravitational lens), making the source appear at an altered position (Figure 6). This light-bending is accompanied by another first-order effect: systematic image distortions. If the background source image is resolved then this image stretching is observable. As shown in Figure 6 a galaxy of angular size one arcsecond may appear to be moved by an angle

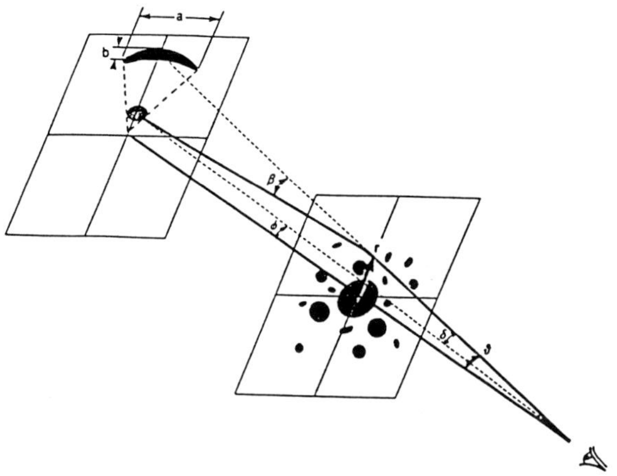

Figure 6. *The source-lens-observer geometry and the background galaxy image distortion which would be observed. A foreground cluster of mass bends light rays from a distant galaxy, distorting its image.*

$\beta = 4GM(< r)/rc^2$ and distorted into an arc many arcseconds long by sufficient foreground mass $M(< r)$ interior to projected radius r. The scattering time is small compared with the period of orbiting test particles, thus avoiding orbit assumptions. This technique is independent of luminous matter in the scatterer (lens), so that it may be used to discover mass concentrations which are totally dark.

For clusters of galaxies the measurement of cluster mass distribution via the gravitational lens technique will be a more direct and accurate measure of the projected total mass density and will provide a valuable check on the virial and orbit assumptions underlying dynamical mass estimates. Since clusters of galaxies are dominated by dark matter, the detailed distribution of mass in clusters is a useful constraint on the nature of dark matter. More generally, its nature and dynamical history will be constrained by the distance scales and time scales on which it clumps and its correlation with luminous matter.

Zwicky (1937) first argued that galaxies would act as gravitational lenses, distorting and amplifying background objects. The light deflection is proportional to the mass in the lens and is about 2 arcsec for a typical galaxy (Turner *et al* 1984). With respect to the foreground lens centroid, the source appears at angle θ, rather than the smaller angle ϕ (the intrinsic source position in the absence of the lens):

$$\vec{\phi} = \vec{\theta} - \vec{\delta}(\vec{\theta}), \qquad (1)$$

$$\vec{\delta}(\vec{\theta}) = (D_{LS}/D_S)\ \vec{\beta}(\vec{r}),$$

where r is the impact parameter for the ray at the lens, and D_S and D_{LS} are the observer-source and lens-source angular diameter distances. For a thin lens the light bending angle $\vec{\beta}$ is given in terms of the projected two dimensional mass density Σ:

$$\vec{\beta} = (4G/c^2) \int d^2u\ \Sigma(\vec{u})\ (\vec{r} - \vec{u})\ /\ |\,\vec{r} - \vec{u}\,|^2\ . \qquad (2)$$

Since we do not know how the sources were distributed on the sky before the lensing, this deflection itself is not observable. But its gradient, the shear, is observable. In the simple case of a pont mass, $\beta = 4GM/rc^2$, and a source exactly behind the mass appears as an "Einstein ring" image of radius $\theta_E = (M\ /\ 10^{11}\ M_\odot)^{1/2}\ (D\ /\ 10^6 kpc)^{-1/2}$ arcseconds, where D = $D_L\ D_S\ /\ D_{LS}$, and $M_\odot = 1$ solar mass. If the lens mass is elliptical in shape or has multiple components this circular ring symmetry is broken, so that in nature circular rings are rare. If the source angle ϕ is less than θ_E then two images of the source with separation approximately $2\theta_E$ are formed.

How big are these effects? Galaxy lenses ($M \approx 10^{12}\ M_\odot$) will produce multiple images with separations $\approx$ 3 arcseconds, while galaxy cluster lenses ($M \approx 10^{14}\ M_\odot$) can produce image separations of one arcminute. For a gravitationally bound collection of luminous objects the total interior mass is related to the observed velocities. The virial theorem relates the kinetic and potential energies (2T + U = 0), hence the velocity dispersion and the mass interior to radius r, M/r: $\Delta v^2 \geq GM(< r)/r$. In the case of an isothermal distribution of gravitationally bound masses of one-dimensional velocity dispersion σ_v,

$$\beta(r) = 4\,\pi\ (\sigma_v\ /\ c)^2, \qquad (3)$$

$$\theta_E = 29\ (\sigma_v\ /\ 10^3\ km\ sec^{-1})^2\ (D_{LS}\ /\ D_S) \qquad arcseconds.$$

Many clusters have measured velocity dispersions $\approx 10^3$ km sec^{-1}. More realistic mass distributions, such as a soft core isothermal sphere, may be calculated by substituting the corresponding density $\Sigma(r)$ (Hinshaw and Krauss, 1987).

Press and Gunn (1973) showed how a cosmologically significant density of compact objects would create small arcs and split images of background objects. In 1979 the first gravitational lens was discovered: a distant QSO split into at least two images by the mass of a foreground galaxy (Walsh *et al* 1979). After searching four thousand QSOs for multiple images, nine unambiguous gravitationally lensed QSOs have been found. Several of the lensed QSOs found to date are caused by some isolated foreground galaxy dominated by its dark matter, and the data are consistent with galaxy masses of about $10^{12}\ M_\odot$, the number density of foreground galaxies, and the small numbers of background QSOs. The cleanest cases yield values of M/r for the lensing galaxy around 3×10^{10} $M_\odot\ kpc^{-1}$, typical of a heavy galaxy dominated by dark matter (Kochanek 1991). So the lensed QSOs, despite their small number, have provided an independent confirmation of the dark matter halos of ordinary galaxies.

In the case of a lensed QSO, there are at most a few images of the QSO, so that there is little information on the mass distribution in the lens, due to the undersampling of the lens plane. The small number of strongly lensed background QSOs are due to the relative scarcity of QSOs compared with galaxies at similar redshifts.

3.2 *Tomographic reconstruction of mass in galaxy clusters*

Bright blue arcs in two clusters of galaxies were reported by Lynds and Petrosian (1986, 1989) and by Soucail, *et al* (1988), and the possibility that they were gravitationally lensed background galaxies was seriously advanced. A large blue arc occurs if a distant evolving galaxy falls near a point of infinite magnification (caustic) in the foreground lens (Grossman & Narayan 1988). Some of these highly magnified background galaxies are bright enough for spectroscopic study, and it is interesting that the nine redshifts of bright arcs, corresponding to unmagnified apparent magnitudes of about 25 mag, obtained to date are in the range 0.5-2.2. However, these caustic projections of background objects are rare, occurring only when a background object coincides with a point of infinite magnification in the lens.

If the source angle relative to the lens ϕ is larger than θ_E, multiple images of that source will not be formed, but the image of the source will be elongated as shown in Figure 6. The large mass associated with rich clusters of galaxies distorts all the faint background galaxies within several arcminutes of the cluster. Foreground galaxy clusters at redshifts 0.2-0.5 with radial velocity dispersions above 700 km s^{-1} have sufficient mass density to significantly distort background galaxies of redshift greater than 0.4-1. Lensing preserves the surface brightness and spectrum of the source, so that most arcs have the very faint surface brightness and blue color of the FBGs.

The background galaxy population is selected by its extreme blue color and low surface brightness compared with the red cluster galaxies. The high-redshift background galaxies projected within the central square arcminute of a foreground compact cluster of galaxies will be strongly lensed by the cluster mass distribution, distorting background galaxy images into faint arcs 1-30 arcsecond long. These are not the relatively rare caustic alignments: every background galaxy will be seen as stretched along a circle centered on the lens. By obtaining images which reveal the high density of faint sources, the improved statistics of these distortions permits the tomographic mapping of the projected dark matter distribution (Tyson *et al.* 1990).

3.3 *The Inverse Problem: Mass from Image Distortion*

This image stretching is quantified using the intensity weighted second moment of the galaxy image orthogonal and along the radius relative to the lens center (labeled a and b in Figure 6). A dimensionless scalar alignment T, calculated from these principle axis transformed source ellipticities, is related to the projected mass density clumping and is defined at each point (x,y) in the image plane by:

$$T_g(x,y) \;=\; (a^2-b^2)/(a^2+b^2), \tag{4}$$

where a(x,y) and b(x,y) are the (r,θ) principal-axis transformed second moments of the background galaxy image g. Referred to position (x,y) the radial moment is b(x,y) and the tangential moment is a(x,y): $a_g \;=\; M_{20}\ sin^2\phi\ +\ M_{02}\ cos^2\phi$ - 2 $M_{11}\ sin\phi\ cos\phi$, and $b_g = M_{20}\ cos^2\phi\ +\ M_{02}\ sin^2\phi$ + 2 $M_{11}\ sin\phi\ cos\phi$, where ϕ is the position angle of the vector from the point (x,y) to the background galaxy, relative to the x-axis. The intensity-weighted second moment $M_{lm,i}$ of background galaxy g is defined by

$$M_{lm,g} \;=\; M_{0,g}^{-1} \int (\delta x)^l\ (\delta y)^m\ W(\delta x,\delta y)\ [I_g(\delta x,\delta y) - I_0]\ dx\ dy, \tag{5}$$

$$M_{0,g} \;=\; \int W(\delta x,\delta y)\ [I_g(\delta x,\delta y)-I_0]\ dx\ dy,$$

where $\delta x = (x - < x >_g)$ and similarly for δy, the sky intensity near this background galaxy is given by I_0, and the optimal normalized weights $W(\delta x, \delta y, I(x,y))$ are calculated from the half luminosity radius. A randomly placed unlensed population of galaxies randomly oriented will give a net distortion $T(\vec{r})$ of zero at every point in the image plane, while a population of lensed galaxies will give a positive value at the point corresponding to the lens center. To construct an approximate image of the gravitational lens projected mass distribution, the distortion statistic $T(\vec{r})$ is computed over a grid of positions as candidate lens centers. In the weak lensing limit (image produced outside the critical radius), it can be shown that the tangential alignment T is a measure of the mass contrast (Tyson, *et al* 1984; Miralda-Escudé 1991a):

$$T(\vec{r}) \;=\; 2\,[\Sigma_{av}(< r) - \Sigma(r)]\ \Sigma_c^{-1}, \tag{6}$$

where the critical density Σ_c is given by

$$\Sigma_c \;=\; c^2\ D_S\ /\ (4\ \pi\ G\ D_L D_{LS}). \tag{7}$$

At any point $\vec{r}$ in the image plane we can sum over the tangential alignment of all source images about that point, creating a continuous scalar distortion statistic D:

$$D(\vec{r}) = \int K(\vec{u})\ T(\vec{r}-\vec{u})\ d\vec{u}, \tag{8}$$

where the apodization kernel $K(\vec{u})$ weights source images at large radius less, and is generally of the form $K(\vec{u}) \;=\; (u^2\ +\ u0^2)^{-1}$. Kaiser and Squires (1993) showed that K(s) must asymptotically approach the power law s^{-2} at large s, for D to be simply related to the mass. Since light bending angles from different mass components add, the distortion D at any point should be related to the mass density contrast, if the mass can be represented as a sum of cylindrically symmetric distributions. The solution to this inverse problem for the contrast of the projected lens mass density Σ is then given by a simple integral of the shear over radius from the lens center. The average projected mass density interior to radius r is given by

$$\bar{\Sigma}(r) = \Sigma_c C(r) B(r) \int_r^{r_{max}} T(r)$$

where C(r) is the seeing correction found in Monte Carlo tests of the data, $B(r) = (1 - r^2/r_{max}^2)^{-1}$, and $\bar{\Sigma}(r, r_{max})$ is the average density in the annulus between r and r_{max}. For a sufficiently large field $\bar{\Sigma}(r, r_{max})$ is small compared with with peak density. While complicated mass distributions can be recovered from their distortion images in this way, the technique fails for mass concentrations corresponding to velocity dispersions less than 500 km sec^{-1} because the random ellipticities of the background galaxies are larger than the shear produced by the lens. The distortion image D(x,y) uniquely locates the lens mass, obtains $M(< r)/r$, and gives its morphological shape on the sky. Equation 9 gives a high signal-to-noise solution to the radial projected mass profile. A useful check of this procedure uses the mass map (Equ 8) as input to a Bayesian search in model mass distribution space, solving for the dark matter mass and core size. For the known source ellipticity distribution, a given source redshift distribution, and a test radial mass distribution model, a maximum likelihood calculation yields the lens M/r (or equivalent velocity dispersion) and core radius.

We are obligated to test this software for solution of the inverse problem by feeding it realistic simulated data. We simulate cluster lens distortions of background galaxies by ray-tracing realistic sources as a function of cluster and dark matter parameters and to compare them with the observations. We begin this process with an investigation of the intrinsic statistical properties of the source galaxy images before lensing and before atmospheric bluring. Some knowledge of the pre-seeing ellipticity distribution is required for accurate lens Monte Carlo simulations. In addition, the redshifts of individual background faint blue galaxies are generally not known, although rough statistical limits have been set in the case of several foreground massive clusters with independently determined masses. The observed ellipticity distribution of this population (after seeing deconvolution) is similar to that of nearby field galaxies. The background galaxy fields are synthesized to have the same properties (number-magnitude counts, ellipticity distribution, galaxy intensity profiles) as found in the deep survey images when corrected for atmospheric blur.

These background galaxies are placed at redshifts ranging between z = 0.5 - 2. The main dark matter component in the cluster, a soft core isothermal mass distribution, is superposed on the relatively small mass cluster galaxies. The velocity dispersion and core radius of the dark matter are varied in different Monte Carlo simulations. After the distortion from all lens components is calculated, the image is convolved with the atmospheric blur profile, binned down to the CCD resolution, and noise is added. The distorted images of background galaxies become mini-arcs, short segments of circles centered on the foreground cluster. In the weak lensing limit (arclets outside the critical radius θ_E) a significant excess of faint blue background galaxies are aligned tangent to circles centered on the foreground cluster lens by the distortion of the gravitational lens.

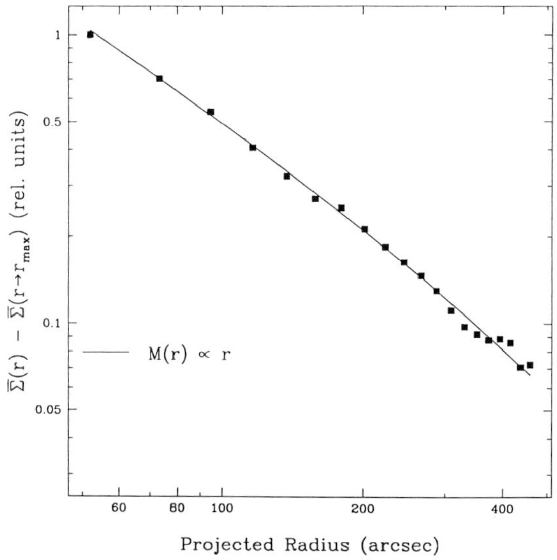

Figure 7. *A radial plot of the projected mass density obtained by inversion of the arclets in a 29 B mag arcsec*$^{-2}$ *noisy simulation with* $\sigma_v = 1000$ *km sec*$^{-1}$. *The solid line is the input mass density to the simulation.*

Automated detection, sub-image splitting and photometry reduce the CCD images to a catalog of galaxy image properties. Faint blue galaxies have their positions, total magnitudes, and second moments passed to the lens arclet inversion software. Finally, the lens projected mass overdensity (Equ 9) as a function of radius from the centroid of the mass distribution may be compared to the lens cluster light (baryonic) distribution. A radial plot of the projected surface mass density found in this way is shown in Figure 7 for $z_l = 0.3$, $\sigma_v = 1000$ km sec^{-1}. In this noisy simulation note that the recovered mass density signal-to-noise ratio is high, and nearly independent of radius, as expected since the number of arclets grows as the square.

A null result is found in fifteen control fields devoid of any foreground cluster of galaxies, each with a surveyed area similar to that of the cluster data fields. Statistical significance of the lens alignments (7-10σ per cluster) and the integral distortion statistic is particularly high for the optically compact clusters we have examined.

The mass core radius is significantly smaller than some observed X-ray core radii in nearby clusters, suggesting that the X-ray gas may be less relaxed dynamically than the dark matter. While a large X-ray luminosity from hot gas trapped in a deep gravitational potential well is a suggestive tracer of mass overdensity at late times, there are a variety of reasons why one might not expect a detailed similarity between the mass and X-ray surface brightness. At least for rich clusters these lens studies appear to confirm the large mass which was implied by virial calculations using velocity dispersion. The distribution of dark matter in rich galaxy clusters is found to be similar to the distribution of total red luminosity (although on a smoother scale of about 100 h^{-1} *kpc*). Some of the less compact clusters show no strong lens distortion effects and, thus, have velocity dispersions smaller than 700 km sec^{-1} and/or dark matter core radii larger than 200 kpc. The central projected surface density (or the cluster velocity dispersion), being an integral over the distortion function, is the best determined parameter.

To date we have surveyed 33 clusters of galaxies for dark matter to a uniform surface brightness limit of 29 B and 28 R mag arcsec^{-2}. In the best studied clusters we find that mass traces the cluster red light. Generally, we find cluster mass-to-light ratios of a few hundred, roughly a factor of five smaller than required in a mass-traces-light $\Omega = 1$ universe. For the optically compact clusters we find the dark matter peak densities are about 10^5 times the critical density for closure, with core radii up to 70 kpc. The scale of the sub-clumping of the dark matter within clusters of galaxies is also of interest. The dark matter cannot all be clumped on galaxy mass scales, otherwise the long arcs seen in several relaxed cluster lenses would be broken up on several arcsecond scales and arcs would form around cluster galaxies. Large CCD mosaics now permit data on the dark matter surrounding clusters. Individual cluster galaxies appear to be no more massive than field galaxies, so that the dark matter in clusters is primarily in the diffuse component.

3.4 Prospecting for Dark Lenses

If $\Omega \approx 1$, dark matter will fill the Universe and will likely exist in places where there is no current star formation activity. Not surprisingly, we have found dark foreground lenses in the wide separation double QSO fields 0957+561 and 2345+007. Larger scale applications of this dark matter mapping may eventually find clumped dark matter unrelated to galaxies or clusters of galaxies. If we wish to extract only information on the statistics of the foreground mass overdensities, rather than map them, there is a tensor statistic analogous to the scalar two-point correlation function (Valdes, *et al* 1983; Miralda-Escudé 1991b; Blandford, *et al* 1991). All pairs of galaxies separated by some angle on the sky are summed separately in bins of mutual orientation. This integral tensor statistic called the orientation correlation function (OCF) is then built from these summed orientation data. The OCF has both parallel and orthogonal orientation components, each a function of angle on the sky. The OCF detects an excess over random for background galaxies to be oriented parallel or orthogonal to one another. In this way much larger areas of the sky may be covered and a smaller amplitude statistical mass fluctuation spectrum may be detected.

Faint galaxy orientation correlations induced by clumped dark matter in various theoretical scenarios can in principle be detected by ultra-deep CCD imaging over sufficiently large fields. CDM theory predicts an observable correlation at about the 2% level, so contiguous samples of 100,000 faint galaxies are a useful test of CDM. As with any statistical effect at the 2% level, systematic errors in the experiment must be understood. Mosaics of CCDs make such a large scale search for coherent alignment in the distant faint galaxies particularly attractive, and dark matter on angular scales up to degrees can in principle be studied in this way. The primary limitation in this work is access to large telescopes. While mosaics of CCDs make effective use of the focal plane it is necessary to spend several nights of large telescope time per 0.1 square degree field.

The observational challenge in OCF measurement is systematic errors due to variations across the detector of the point spread function, optics distortion, and atmospheric refraction. These systematics can be overcome by chopping and trailing techniques, and calibration in rich star fields. We check the OCF mass calibration on smaller fields centered on rich clusters, and also with realistic simulations. We find a OCF detection in several fields, all with known lenses nearby on the sky. Preliminary measurements of the OCF in random fields are at or below the CDM prediction.

3.5 A dedicated dark matter telescope

How big an effort would be required for a definitive measure of the overdensity spectrum of the dark matter on 1 - 10 Mpc scales? Deep imaging of hundreds of thousands of distant galaxies at several wavelengths (to get statistical redshifts from their colors), repeated in many different directions, would have the precision to detect 10 percent of the statistical orientation correlation predicted by the standard cold dark matter theory. An accurate OCF measurement in ten (to get the cosmic variance) 2-degree areas of the sky would provide a direct measure of Ω and the mass spectral index. With a dedicated 4-meter "Dark Matter Telescope", an advanced 8000 *times* 8000 pixel mosaic of CCDs covering half a degree, and deep imaging in 4 wavelength bands, this project would take five years to complete. The payoff in our understanding of dark matter and cosmology would be well worth the expense.

3.6 Acknowledgements

My collaborators in this research are Gary Bernstein, Phil Fischer, Raja Guhathakurta, Greg Kochanski, Jordi Miralda-Escudé, and Rick Wenk.

§REFERENCES

Babul, A. & Rees, M.J. 1992, *MNRAS*, **255**, 346.

Baron, E., & White, S.D.M. 1987, *Ap. J.*, **322**, 585.

Blandford, R.D., Saust, A.B., Brainerd, T., and Villumsen, J.V. 1991, *MNRAS*, **251**, 600.

Blandford, R.D. and Narayan, R. 1992, *Ann. Rev. Astron. Astrophys.*, 311.

Broadhurst, T.J., Ellis, R.S., & Shanks, T. 1988, *MNRAS*, **235**, 827.

Broadhurst, T.J., Ellis, R.S., & Glazebrook, K. 1992, *Nature*, **355**, 55.

Carlberg, R.G., & Charlot, S. 1992, *Ap. J.*, **397**, 5.

Colless, M.M., Ellis, R.S., Broadhurst, T.J., Taylor, K., & Peterson, B.A. 1993, *MNRAS*, **261** 19.

Cowie, L.L., Lilly, S.J., & Gardner, J. 1988, *Ap. J.*, **332**, L29.

Cowie, L.L., Songaila, A., & Hu, E.M. 1991, Nature, **354**, 460.

Dalcanton, J.J. 1993, *Ap. J.*, **415**, L87.

Eales, S. 1993, *Ap. J.*, **404** 51.

Efstathiou, G., Bernstein, G., Katz, N., Tyson, J.A. & Guhathakurta, P. 1991, *Ap. J. Lett.*, **380** L47.

Gardner, J.P., Cowie, L.L., & Wainscoat, R.J. 1993, *Ap. J.*, **415**, L9.

Grossman, S.A. and Narayan, R. 1988, *Ap. J. Lett.*, **324**, L37.

Gullixson, C.A., Boeshaar, P.C., Tyson, J.A., & Seitzer, P. 1995, ApJS, in press.

Guhathakurta, P., Tyson, J.A., & Majewski, S. 1990, *Ap. J.*, **357**, L19.

Guiderdoni, B., & Rocca-Volmerange, B. 1989, *Astron. Astrophys.*, **227**, 362.

Harrison, E.R. 1964, *Nature*, **204**, 271.

Hinshaw, G. and Krauss L. 1987, *Ap. J.*, **320**, 468.

Im, M., Casertano, S., Griffiths, R.E., Ratnatunga, K.U. & Tyson, J.A. 1995 *Ap. J.*, **441**, 494.

Jarvis, J.F., & Tyson, J.A. 1981, *AJ*, **86**, 476.

Kaiser, N. and Squires, G. 1993, *Ap. J.*, **404**, 441.

Kochanek, C.S. 1991, *Ap. J.*, **373**, 354.

Koo, D. & Kron, R. 1992, *Ann. Rev. Astron. Astrophys.*, **30**, 613.

Koo, D.C. 1986, *Ap. J.*, **311**, 651.

Koo, D.C., Gronwall, C. & Bruzual, G.A. 1993, *Ap. J. Lett.*, **415** L21.

Kron, R.G. 1982, *Vistas in Astronomy*, **26**, 37.

Lacey, C., Guiderdoni, B., Rocca-Volmerange, B., & Silk, J. 1993, *Ap. J.*, **402**, 15.

Loveday, J., Peterson, B.A., Efstathiou, G. & Maddox, S.J. 1992, *Ap. J.*, **390**, 338.

Lynds, R. and Petrosian, V. 1986, *Bull. AAS*, **18**, 1014; 1989, *Ap. J.*, **336**, 1.

Miralda-Escudé, J. 1991a, *Ap. J.*, **370**, 1.

Miralda-Escudé, J. 1991b, *Ap. J.*, **380**, 1.

Miralda-Escudé, J. 1992, *Ap. J. Lett.*, **390**, L65.

Press, W. and Gunn, J.E. 1973, *Ap. J.*, **185**, 397.

Soucail, G., Mellier, Y., Fort, B., Mathez, G., and Cailloux, M., 1988, *Astron. Astrophys.*, **191**, L19.

Tinsley, B. M. 1980, *Ap. J.*, **241**, 41.

Treyer, M-A. & Silk, J. 1993, *Ap. J. Lett.*, **408**, L1.

Turner, E.L., Ostriker, J.P. and Gott, J.R. 1984, *Ap. J.*, **284**, 1.

Tyson, J. A., 1988, *AJ*, **96**, 1.

Tyson, J.A. 1990, in *CCDs in Astronomy*, G. Jacoby, Ed. (Astron. Soc. Pacific), p 1.

Tyson, J.A., Valdes, F., Jarvis, J.F., and Mills, A.P. 1984, *Ap. J.*, **281**, L59.

Tyson, J.A., Valdes, F., and Wenk, R.A. 1990, *Ap. J.*, **349**, L1.

Valdes, F., Tyson, J.A. and Jarvis J.F. 1983, *Ap. J.*, **271**, 431.

White, S. D. M. 1989, in The Epoch of Galaxy Formation, ed. C. S. Frenk, R. S. Ellis, T. Shanks, A. F. Heavens, & J. A. Peacock (Dordrecht: Kluwer), 15.

Yoshii, Y. & Takahara, F. 1988, *Ap. J.*, **326**, 1.

Yoshii, Y., Peterson, B. A., & Takahara, F. 1993, *Ap. J.*, **414**, 431.

Zwicky, F. 1937, *Phys. Rev.*, **51**, 290.

Detecting Dark Matter[0]

David O. Caldwell

Physics Department, University of California, Santa Barbara, CA 93106-9530, USA

It is increasingly likely that the universe is flat, having critical density ($\Omega = 1$), which is the only time-stable value. Of the required non-visible matter, <10% is baryonic, perhaps most of which is small, non-luminous stars, such as have started to be observed by gravitational microlensing. Of the rest $\gtrsim$70% of this mass is cold dark matter and ~20% is hot, in the form of neutrinos with total mass summed over all active species of $\sim$ 5 eV. If the solar ν_e and atmospheric ν_μ deficits also require neutrino mass, then there are only two viable patterns for those masses: (1) the ν_e, ν_μ, and ν_τ are all $\sim$ 1.6 eV, or (2) the ν_e and a sterile neutrino, ν_s, can be light, and the ν_μ and ν_τ are $\sim$ 2.5 eV. Terrestrial experiments are needed to determine the neutrino masses, and there are already hints of effects from the LSND neutrino oscillation experiment which would be consistent with the $\sim$ 2.5 eV ν_μ mass. Splitting the 5 eV between the ν_μ and ν_τ gives a model which fits the structure of the universe on all scales. Determining the nature of the cold dark matter can be attempted using accelerator experiments, searches for dark matter annihilation products, and direct detection of the dark matter particles. This third approach has already eliminated a span of 13 orders of magnitude in mass for Dirac particles and 20 orders of magnitude in their interaction cross section. New direct detection experiments include a search for axions through their conversion to microwave photons in a magnetic field and numerous searches using a variety of scintillators and new types of cryogenic devices to detect nuclei struck by dark matter particles. The latter searches and those attempting to find annihilation products in balloon-borne and large underground detectors now especially seek the supersymmetric neutralino.

§ 1 Baryonic Dark Matter

There is no doubt at this time that the majority of the mass of the universe is far greater than one can detect by other than gravitational means. Some of that non-luminous mass is in familiar baryonic form, but much more of it must be particles which are not in the Standard Model of particle physics. Thus the search for this dark matter is of great importance to particle physics, astrophysics, and cosmology. Before discussing more exotic types of dark matter, we deal with the baryonic contribution.

The success of nucleosynthesis theory in predicting the abundance in the universe of ^{4}He, ^{3}H, ^{2}H, and ^{7}Li, as well as three flavors of neutrinos, gives confidence that the ratio of the mass density in ordinary baryonic matter to the density required to just close the universe is $0.01 \leq \Omega_b \leq 0.09$.[1] Observed baryons give only $\Omega_b \approx 0.007$, so there must be some unseen baryonic matter. This is most likely in the form of Jupiters (stars too small to have initiated nuclear burning), very small black holes ($< 10^{-6}$ solar masses), old white dwarfs, cold hydrogen gas, or neutron stars.

The one surely successful search for a dark matter component so far is the use of gravitational microlensing to look for objects which can be between 10^{-7} and 10^{-1} solar masses. If lighter than the lower limit, they would have evaporated away in a galactic time scale, and if heavier than the upper limit, they could have become visible by nuclear ignition. These MAssive Compact Halo Objects, or MACHOs, are detectable if the MACHO passes across the line of sight to some distant star, so that the gravitational field of the MACHO bends the star's light around the dark object, giving a time-symmetric, achromatic brightening of the star's image. Two groups have been monitoring the brightness in two colors (blue and red) of $\sim 10^7$ stars in the Large Magellanic Cloud (LMC) and have reported[2] six MACHO candidates. The U.S.-Australian group has two 4096×4096 charge coupled device (CCD) cameras at the two foci of a 1.22 m telescope, and they have seen four candidates in one year's data. The French group's CCD cameras on a 40 cm telescope have seen no candidates in two year's data, with exposures sensitive to 10^{-7} to 10^{-3} solar-mass objects. They also exposed $5^\circ \times 5^\circ$ Schmidt photographic plates for three

[0]Supported in part by the U.S. Department of Energy

years with sensitivity to 10^{-4} to 1 solar-mass objects, and have seen two candidates. In contrast to this slow rate of detection, in a much shorter time a Polish-American collaboration[3] using CCD detectors on a 1 m telescope and the U.S.-Australian group have seen 13 microlensing events by monitoring stars in the galactic bulge. This result was quite unexpected.

The interpretation of these events is unclear at this time. Only the French group have stated their efficiency, on the basis of which they suggest that MACHOs do not constitute more than 30% of the galactic dark matter. This issue is dependent on the distribution of matter in the galaxy, which is quite uncertain, especially in view of the rate of MACHO events looking toward the bulge. These observations will eventually provide significant information about our galaxy. It is also not clear what masses of objects are acting as lenses, although the majority may be ~ 0.1 of a solar mass. The event duration depends on the MACHO mass, its relative velocity transverse to the line of sight, and the distance to the MACHO, so again a reliable model of halo dark matter is needed, although planned more complex observations could provide the information directly.

§ 2 Apparent Need for Nonbaryonic Mixed Dark Matter

While baryonic dark matter could be important in our galaxy, the need for nonbaryonic mass increases with increasing distance scale. The tangential velocity, v, of a mass, m, in a galaxy remains constant out to a radius, r, beyond the luminous mass, M, whereas from $GmM/r^2 = mv^2/r$, $v \propto r^{-1/2}$ would be expected. Unobserved mass increasing linearly with r must exist, giving a mass density relative to critical density of at least $\Omega = 0.05$–0.10, but the fall-off in v is seldom observed. On the scale of galactic clusters, need for $\Omega = 0.1$–0.3 is required both by the motion of galaxies with respect to the center of luminous mass of the cluster and by gravitational lensing by the cluster of a distant light source. Very large scale velocity fields from the flow of galaxies when related to galactic densities require $\Omega > 0.3$ at the 4–6 standard deviation level,[4] and these data favor $\Omega = 1$. The flat universe of $\Omega = 1$ is very likely because it is the only time-stable value for a zero cosmological constant. Unless there is severe fine tuning in the early universe, the present density should have been driven far out of the present range, toward 0 or ∞, unless $\Omega = 1$. Inflation theory, which provides an explanation for several otherwise inexplicable puzzles, also gives a justification for a universe at critical density.

It has been widely assumed in the past that structure grew by gravitational collapse from initial Gaussian fluctuations in the density of an $\Omega = 1$ universe made up, aside from the small admixture of baryons, of slowly moving "cold" dark matter, which preserves the fluctuations during the era of radiation domination. As observations have improved, it has become clear that a cold dark matter model cannot fit the amount of structure on all distance scales. If normalized to the COBE measurements[5] of the anisotropy in the cosmic microwave background radiation, a cold dark matter model gives too much structure on small scales, since baryons readily clump around the cold dark matter.

Because the cold dark matter (CDM) model seems to be mainly correct, several variations of it have been tried for patching it up, such as introducing "bias" (changing the normalization). The variants of CDM which agree best with observations add either a cosmological constant (ΛCDM), so that $\Lambda + \Omega = 1$, or a little hot (neutrino) dark matter (CHDM). The neutrinos, being relativistic in the early universe (hence "hot"), tended to wash out density fluctuations, so the CDM and HDM can be played off against each other. If there is 30% HDM the structure on almost all scales can be fit very well.[6] The problem is that such a large number of neutrinos causes structure to form too late. If the neutrinos in the form of one neutrino species of $94h^2 F_H \Omega \approx 5$ eV (where $h = 0.5$ is the Hubble constant in units of 100 km $\cdot$ s^{-1} $\cdot$ Mpc^{-1}, F_H is the fraction of HDM, and Ω is taken to be unity) are reduced to 20%, then there is no problem with early enough structure formation,[7] but now the number density of galactic clusters is too large. If instead the 5 eV is shared about equally between two species of neutrinos, then something rather remarkable occurs.[8] While 5 eV in one species or two makes essentially no difference at very large or very small scales, at $\sim 10h^{-1}$ Mpc the larger free-streaming length of the 2.5 eV neutrinos lowers the abundance of clusters. Thus the model (Cν^2DM) with two, ~ 2.5 eV neutrinos fits structure information on all scales. A comparison of these cosmological models is shown in Fig. 1, where all except biased CDM are normalized to COBE, and five other comparisons are made to data at various distance scales in terms of very generous errors in those observations. Expressing those errors as "standard deviations" should not be taken too literally in some cases. The lines have no significance, except to aid the eye in connecting the 6 points for each type of observation.

The Cν^2DM model works only if $\Omega = 1$ and $h = 0.5$. Although there are determinations of h near that value with very small stated errors, there are also values around 0.7 to 0.8, also with errors given which are very small, but such values give trouble with $\Omega = 1$. Should the Cν^2DM model be proved correct by terrestrial determinations of neutrino mass, the long-sought answers concerning h and Ω would be settled. That terrestrial experiments can settle the issues results from the fact that there are

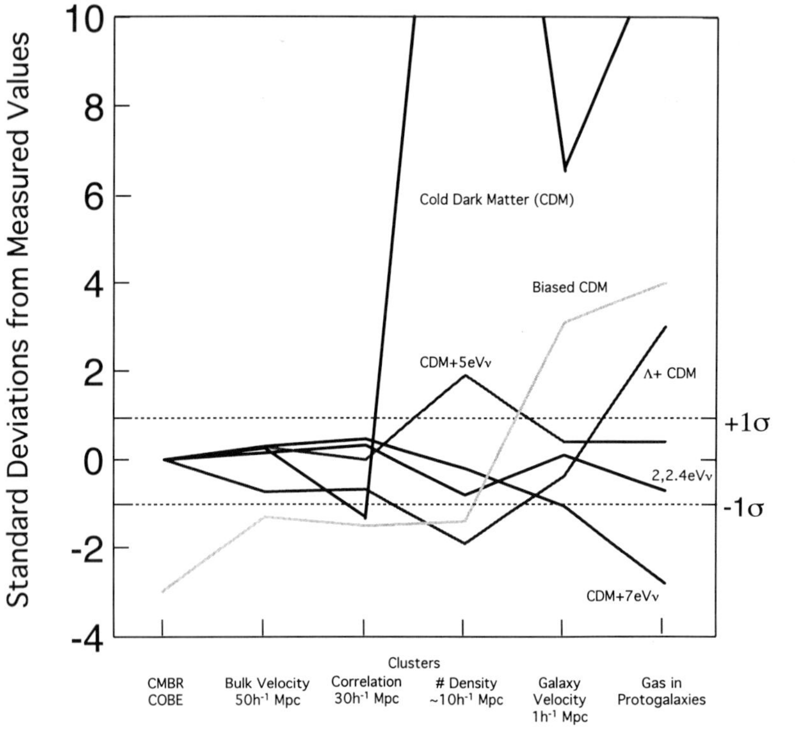

Figure 1: Various cosmological models compared with observations over a wide range of distance scales. That fitting the data best has cold plus hot dark matter, with the latter supplied by two species of 2.4 eV neutrinos.

$\sim 10^2/\text{cm}^3$ of each species of the three active neutrinos (ν_e, ν_μ, μ_τ) everywhere. Unfortunately, direct detection of these relic neutrinos, whatever their mass may be, appears to be impossible at this time, since they are now near the temperature of the cosmic microwave background radiation and hence have too little energy.

§ 3 Possible Constituents of the Hot Dark Matter

In addition to the need for some hot dark matter, there are two rather long standing and increasingly plausible indications of neutrino mass. One is the deficit of ν_es from the sun observed by four experiments of three types. Of those three types, two have to be making incorrect observations in order for an astrophysical explanation of the deficit to work.[9] A solution in terms of neutrino oscillations requires that the mass squared difference between the ν_e and whatever it turns into be no more than $\Delta m^2_{ei} \sim 10^{-5}$ eV^2. The second indication for neutrino mass results from evidence for a deficit of ν_μs relative to ν_es in atmospheric secondary cosmic rays. There are compatible results from three experiments[10] and new information from Kamiokande.[11] The latter includes accelerator confirmation of the ability to separate ν_e and ν_μ events, as well as an independent higher energy data set giving not only a ν_μ/ν_e ratio agreeing with the lower energy data, but also a zenith-angle (hence source-to-detector) dependence compatible with oscillations requiring $\Delta m^2_{\mu i} \sim 10^{-2}$ eV^2. The oscillation $\nu_\mu \to \nu_e$ is almost excluded by data from the Bugey and Krasnoyarsk reactor oscillation experiments, but in addition the calculated ν_e and ν_μ fluxes – backed by measurements of μ fluxes – agree with the ν_e data but show a ν_μ deficit.[12] Thus $\nu_\mu \to \nu_\tau$ oscillations are favored as an explanation of this atmospheric ν_μ deficit.

If the solar ν_e and atmospheric ν_μ deficits and the need for some hot dark matter arise from the existence of neutrino mass, then there are only two viable patterns of those masses.[13] If there are no sterile neutrinos, the only possibility allowed by the small Δm^2 values required by $\nu_e \to \nu_\mu$ for the solar deficit and $\nu_\mu \to \nu_\tau$ for the atmospheric one is $m_{\nu_e} \approx m_{\nu_\mu} \approx m_{\nu_\tau} \approx 1.6$ eV to give the needed ~ 5 eV for hot dark matter. If neutrinos are Majorana particles, for which there is strong theoretical prejudice, then neutrinoless double beta decay should soon eliminate this possibility or confirm its existence, since this is in the range of current experimental limits, given the uncertainty in nuclear matrix elements.

A sterile neutrino may be introduced, provided it does not exceed the limit of 3.5 neutrino species at the time of nucleosynthesis. This limitation can be avoided only for the small mixing angle solution for solar $\nu_e \to \nu_s$, since any other use of a light ν_s would bring it into equilibrium in the early universe.[14] The ν_μ and ν_τ then provide the atmospheric ν_μ deficit and share the dark matter, making $m_{\nu_\mu} \approx m_{\nu_\tau} \approx 2.5$ eV. While a combination of the SNO and SuperKamiokande experiments will be able to demonstrate $\nu_e \rightarrow \nu_s$ to check this mass pattern, the LSND oscillation experiment at Los Alamos will check for a mass difference between the light ν_e-ν_s sector and the heavier ν_μ-ν_τ pair via $\nu_\mu \rightarrow \nu_e$, for which $\Delta m^2_{e\mu} \approx 6$ eV^2 (where LSND is most sensitive).

It is intriguing that there is a hint of oscillations from the 1993 initial run of the LSND experiment.[15] There was an excess of 6 beam-on events of a type which could be interpreted as $\bar{\nu}_\mu \rightarrow \bar{\nu}_e$, wheras a background of $\lesssim 1$ event mimicking a $\bar{\nu}_e$ was expected. The experiment also has a higher energy ν_μ beam to search for $\nu_\mu \rightarrow \nu_e$, but the collaboration has not yet shown those data, nor is it claiming to have observed oscillations, preferring to await the results of a three-times longer run concluding at the end of November 1994. Nevertheless, the LSND positron energy distribution (if assumed to be from $\bar{\nu}_\mu \rightarrow \bar{\nu}_e$, $\bar{\nu}_e + p \rightarrow e^+ + n$, where $n + p \rightarrow d + \gamma$ gives an observed 2.2 MeV gamma ray) appears[16] to be compatible with $\Delta m^2_{e_\mu} \sim 6$ eV^2. This value is not in conflict with the KARMEN[17] or BNL E776[18] experiments, which are least sensitive at that Δm^2 where LSND is most sensitive. If the ν_e mass is relatively small ($\lesssim 1$ eV, as indicated for Majorana ν mass from neutrinoless double beta decay experiments), then the ν_μ mass is ~ 2.5 eV. Even one such neutrino added to low-Ω models (such as ΛCDM) causes clusters to be seriously underproduced and structure to form much too late.[8] Combined with the $\nu_\mu \rightarrow \nu_\tau$ explanation of the atmospheric ν_μ deficit, these results would force the very successful Cν^2DM model.

Even if the LSND results become quite convincing, other experimental checks will be required, since the implications are so far reaching. Not only would the existence of neutrino mass be a violation of the Standard Model, but also the implied pattern of neutrino masses and mixings would provide important guidance to a more comprehensive theory. The cosmological implications would be even more profound. First, the mixed dark matter scenario for structure formation would be validated, although it would be desirable to check that the ν_τ contributes equally with the ν_μ, which could be done by a long-baseline oscillation experiment. Confirming evidence would be provided if the CHORUS and NOMAD experiments now getting underway at CERN gave negative results, since they are not sensitive to $\Delta m^2_{\mu\tau} \sim 10^{-2}$ eV^2. Second, as indicated above, a value of $m_\nu = m_{\nu_\mu} + m_{\nu_\tau} = 5$ eV would imply that $h = 0.5$ and $\Omega = 1$, favoring the lower measurements of h and a critical density for the universe.

§ 4 Cold Dark Matter

While hot dark matter in the form of neutrinos must be found using terrestrial sources, the cold form of dark matter which presumably dominates the mass of the universe can be ascertained in numerous ways, including the use of accelerators, direct detection, and the search for annihilation products. An example is the search for the axion, some mass ranges of which have been limited by accelerator experiments and by astronomical searches for end products, but for which the direct search is most promising. The axion, which would result from a possible solution to the strong CP problem, is one of the few dark matter candidates which are motivated by needs to exist for reasons other than to be dark matter. Despite requiring a mass $\sim 10^{-5}$ eV to give $\Omega = 1$, axions are a cold dark matter candidate. The lack of evidence for energy transport by axions in stellar cooling and from Supernova 1987A, plus the searches mentioned above, limits axion mass to $\lesssim 10^{-3}$ eV. Two experiments[19] have been developed to detect axions by looking for the conversion of axions to photons in a magnetic field, but lack a factor of $\sim 10^2$ in sensitivity for doing so. A new experiment[20] is currently being constructed with a much higher field superconducting magnet which could have the required sensitivity. That sensitivity is proportional to the square of the magnetic field's value and to the first power of its volume. The axion is looked for by the signal the resulting photon would produce in a tuned microwave cavity, and sensitivity will be improved also by reducing the noise temperature. Data taking, which should begin in 1995, is a lengthy process, since each possible value of axion mass requires a specific cavity tuning. The existing results and expected sensitivity of the new experiment are shown in Fig. 2 in comparison with axion models.

Just as there is a reason for the axion's existence beyond being dark matter, so also there used to be considerable motivation for a heavy fourth-generation neutrino as dark matter. The SLC/LEP limitation to three generations of weak isodoublet neutrinos of mass $\lesssim 48$ GeV eliminated a fourth Majorana neutrino, since to be dark matter its mass would have to be $\sim$ 6–8 GeV, which is determined by its annihilation rate when such neutrinos could have been in thermal equilibrium in the early universe. Such a Majorana neutrino may have been ruled out previously by indirect dark matter searches performed at the proton decay detectors, Fréjus, Kamioka, and IMB, as will be discussed in the next section. In contrast to the Majorana case, a Dirac neutrino could have had a wide mass range, since an initial particle-antiparticle asymmetry would allow the annihilation rate to be adjusted suitably. The SLC/LEP result conclusively ruled out the 4–10 GeV Dirac mass range. Masses larger then 10 GeV had already been eliminated by underground Ge semiconductor detectors

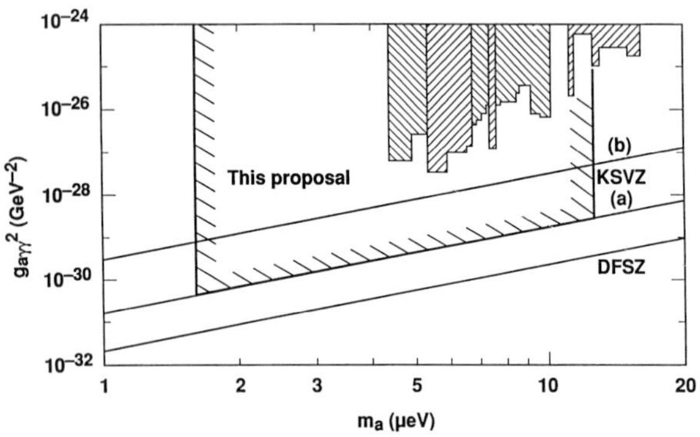

Figure 2: Axion coupling to photons vs. axion mass. Previous experimental results are shown cross hatched, and the region to be covered by the new experiment is labeled "This proposal." The range of possible axion models is indicated by the diagonal lines.

sensitive to nuclear recoils struck by such a particle, a technique which will be discussed in Section 6.

§ 5 Indirect Detection: Annihilation Products

Although fourth-generation neutrinos are ruled out as dark matter, other heavy Majorana or Dirac particles are viable candidates. One way to ascertain their existence would be by observing the products of the annihilation of pairs of such Weakly Interacting Massive Particles (WIMPs) in the halo of our galaxy.[21] These products could be antiprotons, positrons, and photons which could be detected by cosmic ray experiments. In general, these end products provide a discovery potential, but astrophysical uncertainties ($\sim 10^2$), especially the time the WIMP has been confined in the halo, prevent measurements from being utilized to set limits. This also means, of course, that the strength of a possible signal cannot be predicted accurately.

Photons could produce two types of signals, a diffuse γ-ray flux or a line. The former would probably be less than one-tenth of the background, but it would have a distinctive angular signature, and if the dark matter halo were clumpy, a γ telescope with good angular resolution might be able to pick out hot spots at high galactic latitudes. Some models suggest a concentration of WIMPs at the galactic center, again giving the possibility of using the angular distribution to pick out the γ signal, which would be broad in energy but centered roughly at about one-tenth the WIMP mass. A more readily identifiable signal could be obtained from monochromatic γ-ray lines which could result from $\chi\bar{\chi} \to \gamma\gamma$ or $\chi\bar{\chi} \to V\gamma$, where χ is the WIMP and V is a $c\bar{c}$ or $b\bar{b}$ bound state. A successful search would require that the annihilation cross section be large and that the backgrounds, presently not that well known, be favorable.

A less distinctive but perhaps more likely detection signal could be provided by antiprotons, especially those of very low energy, since the background of secondary antiprotons from interactions of cosmic rays in the interstellar gas is suppressed by kinematics, falling rapidly as the energy decreases. A likely dark matter candidate is the lightest supersymmetric particle, presumably the neutralino, which will be discussed below. Constraints from experiments and the needed universe density would make the neutralino mainly a bino, which can have an appreciable annihilation to two gluons. Such a final state would enhance the low-energy antiproton signal and is being sought in space and balloon-borne experiments, such as HEAT, which has recently taken data.

These experiments also will measure positrons, for which there is an interesting excess rising to $\sim 25\%$ at 30 GeV in past measurements. It has been proposed that WIMP annihilation to a line signal from $\chi\bar{\chi} \to e^+e^-$ may be responsible for this rise, with the energy loss of the positrons as they propagate broadening the line. The WIMP annihilation signature would be a sharp drop in the number of e^+ for $E_{e^+} \gtrsim m_\chi$. If the WIMP mass $m_\chi > m_W$, high energy e^+ can be made via $\chi\bar{\chi} \to W^+W^-$ or Z^0Z^0, giving a signal at $E_{e^+} \sim m_\chi/2$. This latter possibility is important because the leading WIMP candidates (e.g., neutralinos) are Majorana particles, which do not decay to e^+e^- pairs, but which can give positrons from W^+ or Z^0 decays.

WIMPs can interact with nuclei in the Sun or Earth and be captured, sinking to the core gravitationally. There they can annihilate in pairs to heavy quarks, and among the final annihilation products will be ordinary neutrinos which can be detected by the large proton-decay or cosmic-ray detectors nearer the surface of the earth. The neutrino signals are of two types: 1) contained events in which the ν_e or ν_μ interacts with a nucleus in the detector; or 2) throughgoing events in which a ν_μ interacts with rock outside the detector, making a muon which is identified as it passes through the detector. This means of detecting WIMPs has several advantages over looking for halo annihilation products: (1) the halo annihilation rate $\propto n_\chi^2$, where $n_\chi = \rho_\chi/m_\chi$ is the number density of WIMPs (and since the density, ρ_χ, is fixed, $n_\chi \propto 1/m_\chi$), whereas the annihilation rate in the Sun or Earth $\propto n_\chi$, making it easier to find large m_χ; (2) the neutrino signal, especially at higher energies, is more easily distinguished from background ($\sim E_\nu^{-3}$); and (3) event rates are much more predictable, so that null results can be used to set meaningful limits.

The indirect detection method has also placed some limits on Dirac mass particles, and indeed the Kamiokande II result[22] has closed a possible small loop-hole near 10 GeV left by LEP and the direct detection data. One must be careful in interpreting the indirect detection exclusion plots for Dirac fermions, as in Ref. 22, since they apply for particle-antiparticle symmetry, so $\Omega = 1$ for a mass of $\sim$ 3–4 GeV only. Other masses, to give $\Omega = 1$ and appear on the plot, correspond to some sub-Z^0 coupling, and hence not to a fourth-generation neutrino. Putting the direct and indirect detection limits on the same plot is then misleading, since they could apply to totally different particles.

The main usefulness of indirect detection is for Majorana fermions. These could be weak isospin singlet neutrinos coupled to a Z', or the lightest supersymmetric particle (LSP), which is of particular interest. The LSP is a well motivated dark matter candidate, since it was invented for a purpose other than to be dark matter, and an $\Omega = 1$ density is achieved quite naturally for a range of LSP parameters. If R-parity (+1 for particles, −1 for their supersymmetric partners) is conserved, the LSP is stable. The LSP is generally considered to be a neutralino, some mass state which is a mixture of the interaction states of the superpartners of the hypercharge gauge boson, the neutral SU(2) gauge boson, and the neutral components of Higgs doublets. The nature of that mixture determines how readily the particle can be detected, but already accelerator and nonaccelerator experiments have considerably restricted the parameter space, so that the bino component is dominant, and this favors annihilation to antiprotons, rather than final state neutrinos. The large number of free supersymmetric parameters are reduced in the Minimal Supersymmetric Standard Model (MSSM), particularly when grand unification (GU) is assumed, typified by the association with SU(5). Under these conditions LEP and CDF results restrict the neutralino to be more massive than $\sim$ 20 GeV. If one relaxes the assumption of GU,[23] neutralino masses giving $\Omega = 1$ can be $\sim$ 10 GeV, and outside the MSSM there are not well-defined limits on neutralino mass. However, the accelerator results and the requirement of $\Omega = 1$ still restrict the parameters considerably. In general it is likely that if the neutralino is dark matter, its mass is between 20 and 150 GeV.[24]

Already the Kamiokande experiment has ruled out[25] some previously viable portions of the vast parameter space available to the LSP. Their exposure was 770 m^2·yr for LSP's captured in the earth and 215 m^2·yr for those captured in the sun, and an appreciable part of their sensitivity came from the spin-independent interaction from the exchange of a light Higgs boson. Observing spin-dependent interactions, which are more generic to Majorana particles, is much more difficult, and a calculation[26] of detection probability for the possible range of supersymmetric parameters shows that $\sim 10^{10}$ m^2·yr is needed to cover it all, even for the Minimal Supersymmetric Model! MACRO and Super-Kamiokande are to have $\sim 10^3$ m^2, while DUMAND and NESTOR, featuring strings of photomultipliers in the ocean, and NT-36 having such strings in Lake Baikal might become $\sim 10^4$ m^2, as will AMANDA with its photomultiplier strings in the ice. DUMAND or AMANDA may eventually go to $\sim 10^6$ m^2, and NESTOR has similar prospects. Already a few photomultiplier strings in AMANDA and Lake Baikal have taken some data. Especially if the LSP mass is low, these detectors have some chance of finding it.

A different sort of possible dark matter annihilation signal could come from a dark matter candidate which is difficult to eliminate experimentally, the primordial black hole (PBH). PBH's could have formed from initial density fluctuations in the early universe on very small scales ($\sim 10^{15}$ grams), whereas COBE measured fluctuations[5] on scales 40 orders of magnitude larger. The temperature of the PBH is $T = 10^{13}$/M GeV, where M is in grams, and the luminosity of its radiation $\propto 1/M^2$. As it loses mass toward the end of its life, T increases abruptly, and it can radiate all fundamental particles of rest mass less than a few times T. The final stages of evaporation are sensitive to physics up to the Planck scale and could hold important clues to the coupling of gravity and quantum theory. The CYGNUS air shower array, which is sensitive to γ-ray bursts, has recently improved the upper limit on the local rate-density of evaporating PBH's by two orders of magnitude, and the recently authorized MILAGRO experiment (air shower array plus water Čerenkov detector) could improve that limit by another three orders of magnitude, raising the possibility of detections.

The MILAGRO detector or air-Čerenkov telescopes may have the sensitivity to provide vital information on another aspect of dark matter. A smoking gun for the mixed dark matter scenario is the redshift at which the initial large galaxies formed, and that information is contained in the spectrum of redshifted starlight. Active galactic nuclei (AGNs) produce a spectrum of γ-rays which can interact with that starlight, producing e^+e^- pairs (if the center of mass energy exceeds 1 MeV) and obscuring the AGN. By seeing at what γ energies various AGNs can be detected between 1 TeV and 10 GeV, the spectrum of the starlight could be determined, if there are enough nearby AGNs.

§ 6 Direct Detection

While devices for the indirect detection of WIMPs are generally very large, direct detection experiments are on quite a different scale of size, so far using mainly a few semiconductor ionization detectors of Ge of $\sim$ 1 kg each.

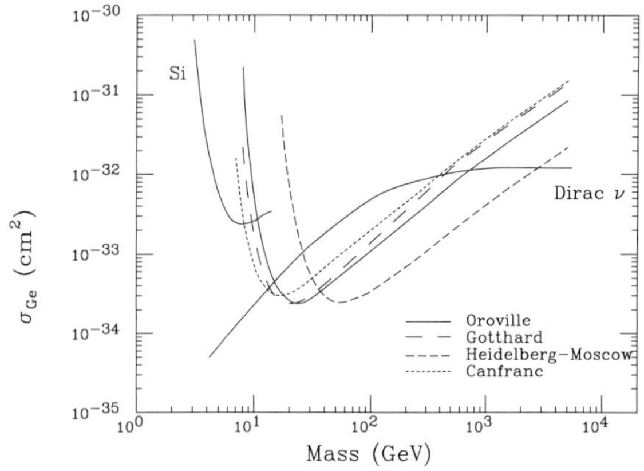

Figure 3: Exclusion plot for the mass and elastic cross section on Ge for dark matter particles. The weak interaction cross section is indicated by "*Dirac ν*". The solid-line result is from Ref. 28, the one with long dashes from Ref. 29, the one with short dashes from Ref. 30, and the very short dashes is the one from Ref. 31. The "Si" curve is the Si detector limit (converted to Ge) from the Oroville cosmion search.

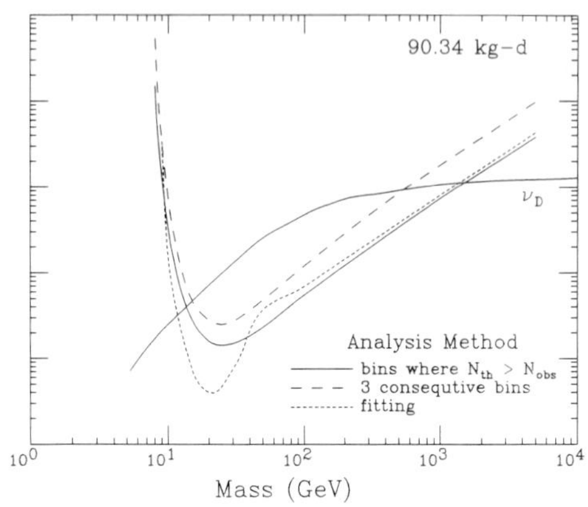

Figure 4: The data of Ref. 28 analyzed by three methods: (1) all data points utilized in a χ^2-like comparison of theory and observation; (2) three consecutive bins have 1.2σ excess of theory over observation; (3) information about backgrounds used in the fitting.

This is because one is looking for the recoil of a detector nucleus struck by a dark matter particle, and the expected flux of cold dark matter particles is surprisingly large. The density of dark matter particles required gravitationally in our galaxy has a local conservative value of 0.3 GeV/cm^3 $= 5 \times 10^{-25}$ g/cm^3. Since the Maxwellian dark matter velocity distribution has $v_{\rm rms} \approx 300$ km/s, the flux of particles of mass M GeV is $\sim 10^7$ cm$^{-2} \cdot$ s^{-1}/M. Thus recoil event rates could then be the order of 0.1–1/keV·kg·day at the level of the weak interaction. This rate is very small compared to typical radioactivity and cosmic-ray backgrounds, however. Underground double beta decay experiments had achieved unprecedentedly low background rates in the MeV region, but when they were converted to search for dark matter, new problems arose in the 10–100 keV region of the nuclear recoils. The worst backgrounds were those induced by cosmic rays when the material associated with the detector had been above ground. Because only a fraction (typically 1/5–1/3) of the recoil energy goes into ionization measured by the semiconductor (Ge or Si) detectors, achieving much lower energy thresholds was also difficult.

Five such experiments[27–31] have searched for dark matter, and a compilation of the results which contribute to the boundaries are shown in the exclusion plot of Fig. 3. The plot is made for each particle mass in turn to be the source of all galactic dark matter, and the results are for the case of spin-independent interactions, for which there is nuclear coherence (cross sections proportional to the square of the number of neutrons). Particles having the normal weak interaction would lie along the line in Fig. 3 labelled "Dirac ν" and are excluded between 10 GeV and $\sim$ 4 TeV. Since $\sim$ 30 eV–4 GeV would give $\Omega > 1$, and LEP results and those of Ref. 22 close the 4–10 GeV range, all isodoublet Dirac neutrinos $\gtrsim$ 30 eV are eliminated as dark matter, if the upper mass limit is really[32] $\sim$ 1 TeV. Also ruled out as dark matter are the lightest technibaryon with weak interactions and any particle above and to the right of the solid line.

The results shown in Fig. 3 are not the published results for the data of Refs. 27–31, but rather they have been analyzed by a common procedure. To understand the issue better, Fig. 4 shows the same data[28] analyzed by three different procedures, all of which involve adding a calculated signal for that particular mass of particle until it can no longer be consistent with the data at some confidence level. In one case (long dashes) the signal must exceed the limit at three adjacent data points by 1.2 standard deviations each, corresponding to a 90% confidence level; in a second (solid line), an effective χ^2 is calculated using all data points, and this requires a Monte Carlo to determine the confidence level; in the third (short dashes), whatever information one has about the backgrounds is included in a fit to the data. As one can see, this third method gives the lowest limits, but it is riskier, since one may be fooled by incorrect knowledge of the backgrounds. The first procedure was used in Fig. 3 because it is the easiest, but we believe the second method to be the least biased. Since Ref. 28 used the third method, the analysis of Fig. 3 gives poorer limits. Ref. 29 used the first method, but there

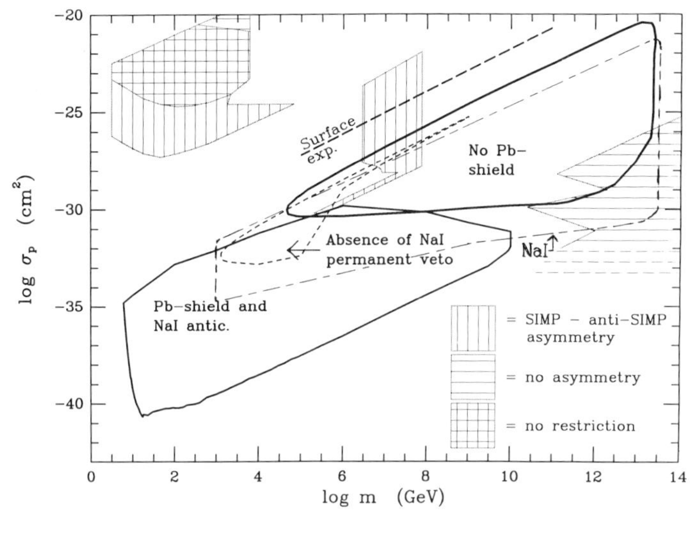

Figure 5: Exclusion plots for the mass and cross section on protons for dark matter particles, showing also other regions Ref. 33 found were not excluded by other information.

was a significant error in their analysis, so the limits are again poorer than in the published value. The results of Ref. 30 change only a little, while those of Ref. 31 become somewhat poorer, but only because of using this first method; their published results are quite satisfactory.

The exclusion region does not end at the boundaries of Fig. 3, as seen in Fig. 5. Note that the cross section of Fig. 3 is on Ge, while that of Fig. 5 is for protons. Interest in Strongly Interacting Massive Particles (SIMPs) as dark matter prompted Starkman et al.[33] to find regions of mass-cross section space viable for dark matter candidates. The most likely remaining region was the mass range $10^6 - 10^8$ GeV with a cross section on protons of $10^{-28} - 10^{-22}$ cm^2 and a narrow "finger" reaching down to 10^5 GeV and 10^{-30} cm^2. To search this region the Pb shielding and NaI veto above the Ge detectors in the UCSB/LBL/UCB apparatus were removed, giving the "no-lid, NaI" exclusion region shown. To extend the upper boundary, some data were taken with a Ge detector at the surface, giving the dashed line. Another exclusion region is dashed in and labeled "absence of a NaI permanent veto", since in this region there would always be a SIMP in the NaI within the electronics dead time. This region extends into the exclusion area of Fig. 3, which is shown completely here for the UCSB/LBL/UCB data only. The region bounded by long and short dashes and labeled "NaI" is from recent data[34] of the BPRS group and will be discussed in a later section.

Although now rather unlikely to be dark matter, SIMPs have been searched for in balloon flights.[35] Because of their interactions, the greatest sensitivity to SIMPs is available high in the atmosphere. Some of these experiments are still being analyzed, and perhaps some others will be implemented, but nearly all the possible interaction cross section vs. SIMP mass space has been ruled out. Thus it is improbable that any particle with stronger than the standard weak interaction can be the main component of dark matter. There is also a limit[32,36] that any stable elementary particle which was once in thermal equilibrium in the early universe has to have a mass less than $\sim 10^3$ TeV.

§ 7 Future Direct Detection Experiments

Future prospects for indirect detection experiments have already been discussed and are generally better and bigger applications of existing techniques. For direct detection, however, a wide variety of new techniques is being tried, and hence these require a more extensive discussion. The Ge semiconductor detectors used previously essentially have reached their limits, with enriched ^{76}Ge recently reducing backgrounds,[30] and application of diurnal effects through suitable geographic location[37] promising to make some further improvement. The large additional background reductions and sensitivity increases needed now to go to cross sections well below the weak level for Dirac particles and especially to search for Majorana particles, particularly the neutralino, call for totally new approaches.

The latter requirement gives some emphasis to utilizing detector nuclei with spin, since Majorana particles have axial, spin-dependent scattering. Initial limits on the neutralino, however, will come from the spin-independent interactions, such as arise from Higgs particle exchange, as these provide nuclear coherence and hence dominate experimentally for the most accessible part of the supersymmetric parameter space. If the neutralino, or another cold dark matter particle, is not found in this larger cross section region, nuclei with spin and even lower backgrounds will be needed to continue the search.

Because masses $\gtrsim 20$ GeV are likely for both Dirac and Majorana particles, the emphasis now is not so much on low energy detection thresholds, where the most sensitivity for small masses is found, but rather on further reduction of backgrounds. Because of the kinematics, heavier WIMP masses also make heavier detector nuclei more desirable, although form factors can eliminate this advantage.

Emphasis on nuclei with spin has increased interest in the use of scintillators, such as NaI and CaF_2. The NaI is ^{23}Na and ^{127}I, both 100% isotopes and both with spin. The ^{127}I would be by far the more effective component on two grounds: its matrix element is much more favorable for scattering, and its greater mass would be kinematically preferred for the expected heavy WIMP mass. Unfortunately, that mass also means that at these very small recoil energies the ^{127}I is very ineffective at producing ionization and hence light. At a typical recoil energy the scintillation efficiency relative to that for electrons is 0.25 for Na and 0.07 for I. For coherent, rather than spin-dependent, scattering, the mass of the I is a detriment for very heavy nuclei because that coherence is lost more quickly as the mass of the WIMP increases.

So far backgrounds and thresholds are not quite as good as those with Ge detectors, but these are being improved, and large quantities (easily 10 kg and eventually $\sim 10^2$ kg) of NaI are feasible, so that better statistics aid the search for dark matter above the fluctuation in background. Results have already been reported[38,39] with NaI which are better than the limits for Majorana particles on the basis of scaling the Ge data by 7.8%, the abundance of ^{73}Ge, which is the only isotope in normal Ge having spin. Of course, if the particle had some coherent interactions (as the neutralino would), such a scaling is not an appropriate comparison. Another result from the same BPRS group[34] using delayed coincidences between two NaI counters has produced some extension of the exclusion plot of Fig. 5 for SIMPs, as is shown in the figure and labelled "NaI".

Two other groups[40,41] are using NaI, and interesting attempts are being made to obtain discrimination against radioactivity background electrons, which come from β decays, Compton scattering, and photoelectrons. The United Kingdom dark matter group,[39] which has an excellent facility with a large highly purified water shield in the Boulby salt mine, is using NaI with low doping levels operated at 100° K, because uv and visible light are emitted in proportions which differ for nuclear and electron recoils. Discrimination can be achieved using either optical filtering or by the timing of the pulse.

In addition to NaI, both the BPRS (with Lyon)[42] and Osaka[40] groups have also been working with CaF_2 (Eu), since ^{19}F has a particularly large matrix element for spin-dependent scattering, although a small scintillation efficiency relative to electrons (7% for F and 5% for Ca). The former group,[42] despite backgrounds much worse than their NaI detector, has limits on spin-dependent scattering which are better for masses > 40 GeV than with NaI (and also ^{73}Ge, if the 7.8% scaling is used) because of the ^{19}F matrix element. Some caution is needed in such comparisons because of uncertainties in those matrix elements. This work was done with a 0.37 kg crystal, whereas the Osaka group[40] is planning to use ~ 10 kg, so future measurements could provide excellent sensitivity to spin-dependent scattering.

Liquid Xe as a scintillator appears especially interesting because it has 48% isotopic content with nuclear spin, has high light output, can be scaled up to large mass, and particularly has possibilities for nuclear/electron recoil discrimination. The United Kingdom group[43] has been investigating pulse shape differences resulting from the two mechanisms for light production which are stimulated to different extents by nuclear and electron recoils. Perhaps even more promising is the approach being investigated by a UCLA/CERN/Aquila/Padova/Paris/Torino collaboration in which an electric field is applied to prevent recombination, and then two signals are measured: (1) the primary scintillation, and (2) the ionization component, which is drifted and produces a secondary scintillation. Calculations indicate that this latter method is capable of a discrimination against electrons approaching 10^3. In addition to these two efforts, a Rome group[44] has already operated a 3.5-liter prototype at Gran Sasso. By subtracting the background measured with a gas filling from the signal with a liquid filling, they achieved a spin-dependent exclusion plot which is not very much less restrictive than those obtained with NaI. A new, lower activity detector is being installed with an improved shield and ^{85}Kr-free Xe.

The ionization and scintillation detectors suffer from the inherent problem that at the very low recoil energies ($\sim$ 10–100 keV) most of that energy is turned into heat, resulting in the small efficiencies quoted above in a few cases. This fact, plus the potential of very good energy resolution, has promoted widespread work on cryogenic detectors. Such a detector, when used as a bolometer, measures a temperature rise, ΔT, due to the recoil energy deposit, ΔE. Now $\Delta T = \Delta E/C$, where C is the heat capacity of the material, and since at low temperatures $C \propto MT^3$, in order to have a large mass, M, of the detector it is necessary to go to a very low temperature, T, in order to get an appreciable ΔT from a certain ΔE. In practice the issue is more complicated than this, but these relations show why dilution refrigerator temperatures (0.01–0.1° K, or even less) are used.

The use of heat detection also opens up the possibility of employing a wide variety of materials, since there is no restriction to materials that scintillate or give measurable ionization. For example, a Tokyo group[45] has tried a 2.8 g LiF crystal, which is an ideal material for spin-dependent interactions, and obtained 3.8 keV resolution at 60 keV. Similarly, sapphire (Al_2O_3) has been used, since the Al is a 100% isotope with spin. A French group[46] has operated a 24 g sapphire crystal at 0.055° K underground, achieving a good background, a threshold of 3 keV, and an energy resolution of 2 keV for 6 keV X-rays. These groups used Ge thermistors to measure the temperature rise, but the Munich group[47] has employed on a 31-g sapphire crystal a superconducting phase transition thermometer at 0.044° K with a SQUID readout to obtain a 210 eV resolution for 6 keV X-rays. While even better resolutions have been achieved with very small crystals, this is the best reported resolution per unit detector mass.

Another 100% isotope with spin is Nb, which becomes a superconductor at low temperatures. The Oxford group[48] has utilized differences in the phonon/quasi-particle ratio in Nb to provide some nuclear/electron discrimination. Also promising for providing this important reduction in background is the Stanford[49] approach using Si and phase transition (or transition edge) sensors to detect differences electron and nuclear recoils give in proportions of ballistic and quasi-diffusive phonons.

Recently the Stanford group has achieved simultaneous measurements of phonon and ionization energies in Si, an excellent discriminant which has been pioneered by the Berkeley group[50] using Ge. This technique requires very low temperatures for another reason. To detect ionization a semiconductor detector requires an electric field of $\sim 10^3$ V/cm to remove free charge carriers and fully deplete the crystal, but such a field moving the ionization charge creates phonons which swamp the original phonon component of the particle interaction. However, the phonon signal can be detected using very low temperatures ($\sim$ 0.03° K), since the impurity charges are then frozen out from the bands, and only about 0.5 V/cm is needed to collect ionization electrons and holes. This technique was demonstrated[50] using a 60 g. crystal of Ge and neutron-transmutation-doped Ge thermistors. Figure 6 shows the ionization and phonon response to X-rays from ^{241}Am, while Fig. 7 displays a measurement with that X-ray source and a ^{252}Cf neutron source. Figures 6 and 7 are for data from a detector which had resolutions (FWHM of 1.5 keV for ionization and 1.7 keV for phonons at 60 keV) from which an additional rejection of background electrons of $\sim 10^2$ is calculated. A later 60 g. detector has shown a FWHM resolution of 0.7 keV for phonons, so background rejection in even larger crystals should be very good.

This technique will be applied in an experiment presently being implemented at a shallow (17 m.w.e.) site on the Stanford campus by the UCSB/LBL/UCB/Stanford/INR Baksan group. Since the technique is new and dilution refrigerators require attention, a location was chosen on the basis of ready access. The Oroville site[28] at 600 m.w.e. was by far the shallowest heretofore used for a dark matter search. Cosmic ray muons are a problem in that they produce neutrons which can give nuclear recoils. A highly efficient muon veto counter is needed and will provide a deadtime after the passage of the muon until the neutron is at least of too low an energy to give a measurable recoil. The few unvetoed neutrons, produced mainly in the Pb shield needed to reduce external radioactivity, are moderated by polyethylene. According to calculations, the principal remaining source of neutrons are the Cu fixtures necessary to contain the detectors at low temperature. The shielding can be quite complete because the dilution refrigerator has been constructed with an unusual side exit so that cooling can be achieved in a separate cold box.

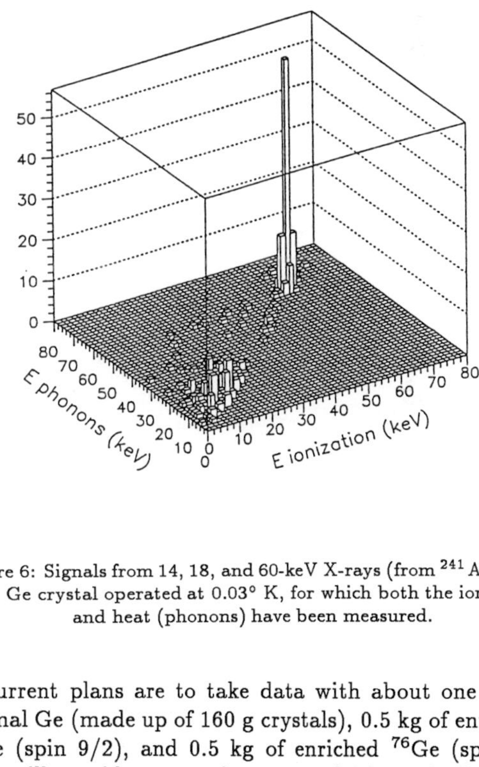

Figure 6: Signals from 14, 18, and 60-keV X-rays (from ^{241}Am) in a 60 g. Ge crystal operated at 0.03° K, for which both the ionization and heat (phonons) have been measured.

Current plans are to take data with about one kg of normal Ge (made up of 160 g crystals), 0.5 kg of enriched ^{73}Ge (spin 9/2), and 0.5 kg of enriched ^{76}Ge (spin 0). This will provide appropriate cross checks and allow the simultaneous search for Dirac and Majorana dark matter. Some Stanford Si detectors will also be installed. If we can achieve electron background levels comparable to those we have had in the semiconductor experiment at Oroville,[28] this should begin the exploration of the region in which the lightest supersymmetric particle could exist, as shown roughly in Fig. 8.

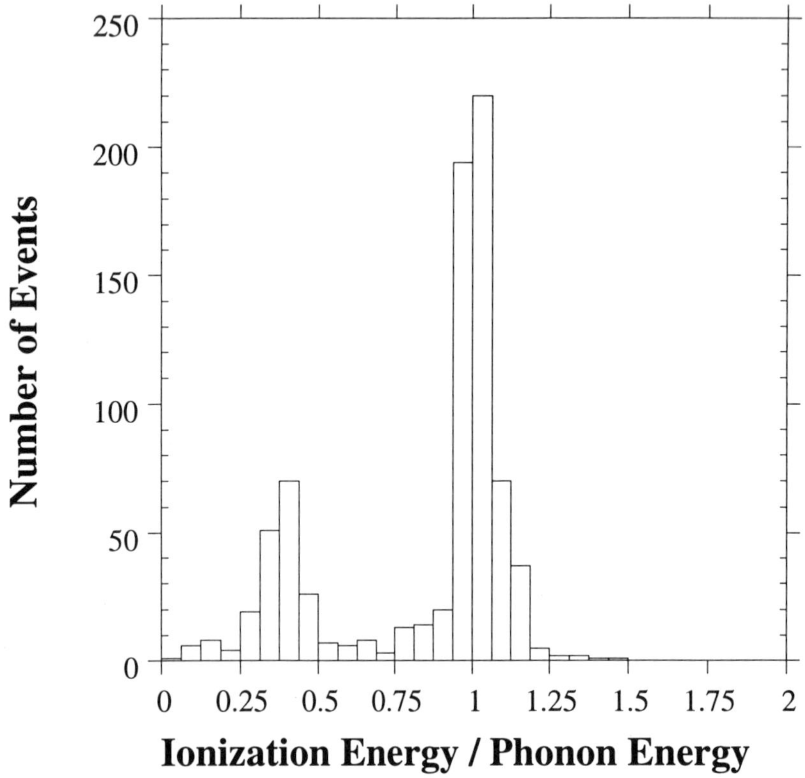

Figure 7: Response of the 60 g. Ge crystal to a ^{252}Cf neutron source, shielded by 11" of Pb, and X-rays from ^{241}Am. The separation of the two peaks indicates that 99% of photons and electrons can be rejected when looking for nuclear recoils.

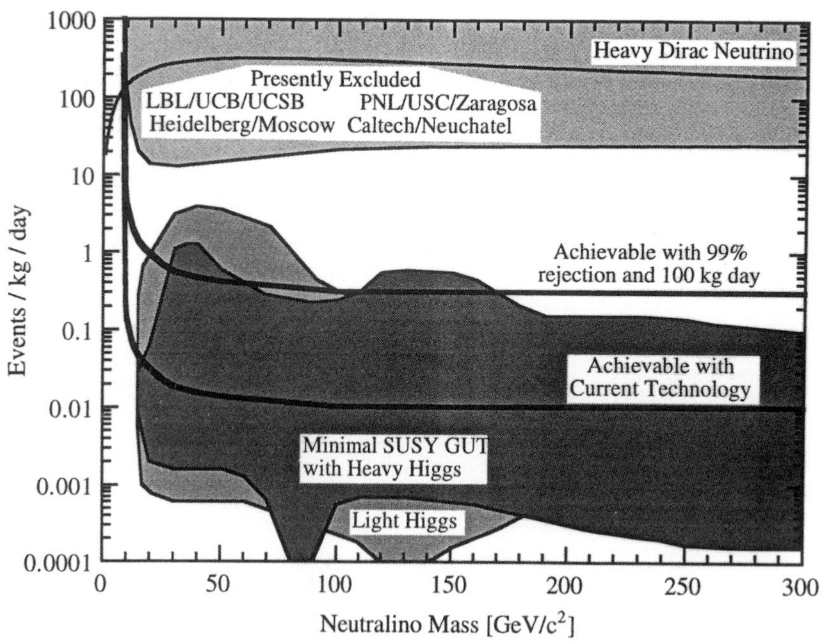

Figure 8: Exclusion plot of event rate vs. particle mass showing the already excluded region of Fig. 3 and the neutralino region with various theoretical assumptions. The lines show limits to be achieved after 100 kg days using a Ge crystal and having 1% and 0.1% additional rejection from simultaneous ionization and phonon measurements.

§ 8 Conclusions

There is increasing observational evidence for a flat universe ($\Omega = 1$) and hence for the need for $> 90\%$ of the mass of the universe to be in the form of particles outside the Standard Model, on the basis of limits on baryonic matter from nucleosynthesis. The structure of the universe over a very large range of distance scales favors a mix of cold and hot nonbaryonic dark matter. If the dark matter is indeed $\sim 7\%$ baryonic, $\sim 73\%$ cold, and $\sim 20\%$ hot, the nature of this last component will have to be determined by terrestrial experiments, guidance for which comes from present indications for neutrino mass. If the solar ν_e and atmospheric ν_μ deficits are manifestations of neutrino mass, one of two patterns of neutrino masses are viable, and there are currently hints for one of these from the LSND neutrino oscillation experiment. That favored scheme, which has a ν_μ and ν_τ of ~ 2.5 eV, provides a fit to the structure of the universe on all scales. The cold component is amenable to indirect and direct searches, with both accelerator and underground experiments. Underground direct searches have already eliminated Dirac candidates covering 13 orders of magnitude in mass and 20 orders of magnitude in cross section. While some Majorana particles, such as a fourth-generation neutrino, have also been ruled out, the emphasis of current searches is on other Majorana particles, especially the lightest supersymmetric particle. Larger detectors will look for annihilation products from cold dark matter, and detectors sensitive to nuclear recoils from dark matter are being deployed using scintillators of NaI, CaF_2, and liquid xenon, and cryogenic detectors of sapphire, LiF, Nb, Si, and Ge. Especially promising is the use of more than one type of measurement to discriminate between the recoils of background electrons and signal nuclei. Very soon there will be much enhanced sensitivity for detecting not only heavy cold dark matter, but also the axion. The era of discovery is starting.

§ References

[1] M.S. Smith, L.H. Kawano, and R.A. Malaney. *The Astrophysical Journal (Supplement)*, 85:219, 1993; T.P. Walker, "Relativistic Astrophysics and Cosmology" (eds. C.W. Akerlof and M.A. Srednicki) *Ann. of N.Y. Academy of Sciences*, 688:745, 1993.

[2] C. Alcock et al. *Nature*, 365:621, 1993; E. Auborg et al. *ibid.* 623.

[3] A. Udalski et al. *Acta Astron.*, 43:289, 1993.

[4] A. Nusser and A. Dekel. *The Astrophysical Journal*, 405:437, 1993.

[5] G.F. Smoot et al. *The Astrophysical Journal*, 396:L1, 1992; C.L. Bennett et al. *The Astrophysical Journal*, in press, 1994.

[6] A. Klypin, J. Holtzman, J.R. Primack, and E. Regös. *The Astrophysical Journal*, 416:1, 1993.

[7] A. Klypin, S. Borgani, J. Holtzman, and J.R. Primack. *The Astrophysical Journal*, submitted. (preprint astro-ph./9410022).

[8] J.R. Primack, J. Holtzman, A. Klypin, and D.O. Caldwell. *Physical Review Letters*, submitted. (preprint astro-ph/9411020).

[9] J.N. Bahcall. Princeton IAS Preprint IASSNS-AST-94-37, 1994; V. Berezinzky, G. Forentini, and M. Lissia. Gran Sasso Preprint LNGS-94-104, 1994; P.I. Krastev and A.Yu. Smirnov. Preprint INT94-13-02, 1994; S.P. Rosen. "27th International Conference on High Energy Physics", Glasgow, 1994; N. Hata and P. Langacker. Preprint UPR-0625T, 1994.

[10] K.S. Hirata et al. *Physics Letters*, B280:146, 1992; R. Becker-Szendy et al. *Physical Review*, D46:3720, 1992; P.J. Litchfield. "Proc. Int. Europhys. Conf. on High Energy Physics", (eds. J. Carr and M. Perrottel), Gif-sur-Yvette, France: Editions Frontières, p.557, 1993.

[11] Y. Fukuda et al. *Physics Letters B*, 335:237, 1994.

[12] G. Barr, T.K. Gaisser, and T. Stanev. *Physical Review D*, 39:3532, 1989; W. Frati et al. *Physical Review D*, 48:1140, 1993; D.H. Perkins. *Astroparticle Physics*, 2:249, 1994.

[13] These mass patterns were first discussed at the January 1993 Moriond Workshop [D.O. Caldwell in "Perspectives in Neutrinos, Atomic Physics and Gravitation", (ed. J. Trân Thanh Vân et al.), Gif-sur-Yvette, France: Editions Frontières, p.187, 1993 and a theoretical structure for them was provided in D.O. Caldwell and R.N. Mohapatra *Physical Review D*, 48:3259, 1993; *ibid.*, 50:3477, 1994; the four-neutrino scheme has also been discussed by J.T. Peltoniemi and J.W.F. Valle *Nuclear Physics B*, 406:409, 1993 and by Z. Shi, D.N. Schramm, and B.D. Fields *Physical Review D*, 48:2563, 1993.

[14] The ν_s could be utilized for the vacuum oscillation solution to the solar ν_e deficit, since the Δm^2 is so

small, but this explanation is disfavored by information from Supernova 1987A. It does not work for the atmospheric $\nu_\mu \to \nu_s$ or as the dark matter (see Ref. 13), because of the nucleosynthesis limitations.

[15] W.C. Louis. "XVI Conference on Neutrino Physics & Astrophysics", Eilat, Israel, 1994.

[16] D.O. Caldwell. *ibid.*

[17] G. Drexlin. *Progress in Particle and Nuclear Physics*, 32:in press, 1994.

[18] L. Borodovsky et al. *Physical Review Letters*, 68:274, 1992.

[19] W. Wuensch et al. *Physical Review D*, 40:3153, 1989; C. Hagmann et al. *Physical Review D*, 42:1297, 1990.

[20] LLNL, MIT, Florida, LBL, Fermilab, INR collaboration; K. van Bibber et al. Preprint UCRL-JC-118357, unpublished, 1994.

[21] For a recent review and references, see M. Kamionkowski. IASSNS-HEP-94-8, 1994, to appear in "Particle Astrophysics, Atomic Physics, and Gravitation".

[22] M. Mori et al. *Physics Letters* B289:463, 1992.

[23] K. Griest and L. Roszkowski. *Physical Review D*, 46:3309, 1992.

[24] L. Roszkowski. *Physics Letters*, B262:59, 1991.

[25] M. Mori et al. *Physical Review*, D48:5505, 1993.

[26] F. Halzen, T. Stelzer, and M. Kamionkowski. *Physical Review D*, 45:4439, 1992.

[27] S.P. Ahlen et al. *Physics Letters*, B195:603, 1987.

[28] D.O. Caldwell et al. *Physical Review Letters*, 61:510, 1988.

[29] D. Reusser et al. *Physics Letters*, B255:143, 1991.

[30] M. Beck et al. *Physics Letters*, B336:141, 1994.

[31] M. Sarsa et al. *Nuclear Physics B (Proc. Suppl.)*, 35:154, 1994.

[32] K. Griest and M. Kamionkowski. *Physical Review Letters*, 64:615, 1990.

[33] G.D. Starkman et al. *Physical Review D*, 41:3594, 1990.

[34] C. Bacci et al. *Astroparticle Physics*, 3:13, 1994.

[35] P. McGuire et al. "Proceedings of the International Cosmic Ray Conference", Calgary, Canada, 1993, to be published.

[36] A.-C. Davis and R.H. Brandenberger. *Physics Letters*, B284:81, 1992.

[37] J.I. Collar and F.T. Avignone III. *Physics Review D*, 47:5238, 1993.

[38] C. Bacci et al. *Physics Letters*, B293:460, 1992; A. Bottino et al. *Physics Letters*, B295:330, 1992; C. Bacci et al. LNGS preprint 94/94, unpublished, 1994.

[39] J.J. Quenby et al. *Physics Letters B*, submitted, 1994.

[40] K. Fushimi et al. *Physical Review*, C47:R425, 1993.

[41] Pacific Northwest Laboratory, South Carolina, Zaragoza Collaboration; Osaka group.

[42] C. Bacci et al. *Astroparticle Physics*, 2:117, 1994.

[43] G.J. Davies et al. *Physics Letters*, B320:395, 1994.

[44] P. Belli et al. *Nuclear Instruments and Methods*, A299:191, 1990; A310:150, 1991; A316:55, 1992; A327:207, 1992; A336:336, 1993; *Nuclear Physics B (Proc. Suppl.)*, 35:165, 1994.

[45] M. Minowa et al. *Nuclear Instruments and Methods*, A327:612, 1993.

[46] N. Coron et al. *Astronomy and Astrophysics*, 278:L31, 1993.

[47] P. Ferger et al. *Physics Letters*, B323:95, 1994.

[48] R.J. Gaitskell, N.E. Booth, and G.L. Salmon, "Workshop on Low Temperature Detectors", (eds. N.E. Booth and G.L. Salmon), Oxford: Editions Frontières, p.435, 1992; N.E. Booth et al. *Nuclear Instruments and Methods*, A315:201, 1992; R.J. Gaitskell, Oxford University preprint OUNP-93-22, unpublished, 1993.

[49] B.A. Young et al. *Nuclear Instruments and Methods*, A311:195, 1992.

[50] T. Shutt et al. *Physical Review Letters*, 69:3425,3531, 1992.

Cosmology at the Crossroads

Paul J. Steinhardt

Department of Physics and Astronomy, University of Pennsylvania, Philadelphia PA 19104, USA

Observational tests during the next decade may determine if the evolution of the Universe can be understood from fundamental physical principles, or if special initial conditions, coincidences, and new, untestable physical laws must be invoked. The inflationary model of the Universe is an important example of a predictive cosmological theory based on physical principles. In this talk, we discuss the distinctive fingerprint that inflation leaves on the cosmic microwave background anisotropy. We then suggest a series of five milestone experimental tests of the microwave background which could determine the validity of the inflationary hypothesis within the next decade.

§1 Introduction

This talk focuses on how measurements of the cosmic microwave background anisotropy can be used to test the inflationary hypothesis. Compared to the other plenary presentations, the scope may seem rather narrow. This choice has been made intentionally, though, as a means of illustrating the dramatic transformation which cosmology is undergoing and of highlighting why the coming decade is especially critical to the future of this science.

Cosmology is the one of the oldest subjects of human inquiry and, at the same time, one of the newest sciences. Questioning the origin and evolution of the Universe has been characteristic of human endeavor since before recorded history. However, until the 20th century, cosmology lay in the domain of metaphysics, a subject of pure speculation. With little observational evidence to confirm or deny proposals, cosmology could not develop as a true science. In large part, its practitioners were philosophers, religious leaders, teachers and writers, rather than scientists.

The science of cosmology emerged in the 20th Century when it first became possible to probe great distances through the Universe. The great optical telescopes and, later, radio telescopes and satellites provided images and data that could begin to discriminate competing theories. At first, the observational breakthroughs occurred infrequently. Hubble discovered the expansion of the Universe in the 1920's; Penzias and Wilson discovered the cosmic microwave background in the 1960's; and the compelling evidence for primordial nucleosynthesis of the elements was amassed during the 1970's and 1980's.

As the 20th century comes to a close and the new millennium begins, the pace of discovery has accelerated markedly. In the next decade or so, major projects will measure the distribution and velocities of galaxies, dark matter, and radiation at cosmological distances. The new data will completely dwarf all previous observations in quality and quantity. The results will tightly constrain all present theories of the evolution of the Universe and may point to fundamentally new paradigms.

Measurements of the cosmic microwave background anisotropy will be among the most decisive cosmological tests because the microwave background probes the oldest and farthest features of the Universe. Anisotropy measurements will provide a spectrum of precise, quantitative information that, by itself, can confirm or rule out present models of the origin of large scale structure, reveal the ionization history of the intergalactic medium, and significantly improve the determination of cosmological parameters. More generally, the bounds on human capability to explain the Universe are likely to be decided by what is discovered in the microwave background during the coming decade.

§2 At the Crossroads

Cosmology has reached a crossroads which may set its course as a scientific endeavor for the next millennium. By its very nature, the field entails explaining a single series of irreproducible events. Our ability to explore the Universe is physically limited to those regions which are within causal contact. (The causal limit, the maximum distance from which we can receive light or other information, is the Hubble distance $H^{-1} \lesssim 15$ billion light-years.) Given these considerations, it is natural to question whether the evolution of the Universe is completely comprehensible scientifically. Or, more explicitly, which of the following two paths lie ahead of us?

> *Path I:* The basic features of the Universe are explainable as a consequence of symmetry and general physical laws that can be learned and tested near the Earth.

Path II: Some key features of the Universe are largely determined by special initial conditions, extraordinary coincidences and/or physical laws that are untestable locally (*e.g.*, in the most extreme case, accessible only by exploring beyond our causal domain).

At any given time, there may not be complete agreement as to which *Path* we are taking. An observation that seems to suggest special initial conditions (*e.g.*, the flatness of the Universe) may later be explained by new, dynamical concepts (*e.g.*, inflation). This kind of evolution in thinking is common to all sciences. The issue being raised here, though, is whether there is an *ultimate*, fundamental, insuperable limitation to the explanatory and predictive power of cosmology.

Certainly, it is the hope of most cosmologists, at least theorists, that cosmology belongs to *Path I.* We are an ambitious lot, and we aspire to explain all that is observed. However, nature may not be so kind to human cosmologists. Considering that we are trying to explain a single sequence of events and the range over which we can measure is bounded, it seems plausible that cosmology could ultimately belong to *Path II.* In either case, the exploration of the Universe remains a captivating and important enterprise. But, without doubt, cosmological science is different along the two *Paths.* Along *Path I*, cosmology has the character of physical science, where the field ultimately evolves towards a unified, simple explanation of what is observed. Along *Path II*, cosmology has the character of archaeology or paleontology, where we can classify and quantify phenomena and explain some features, but where many general aspects seem to be accidents of environment or history.

Remarkably, it seems possible that the next decade will determine which *Path* cosmology is to take. Measurements of the microwave background anisotropy and large-scale structure may indicate that the Universe can be explained in terms of a few parameters, known physical laws, and simple initial conditions, all of which suggest *Path I.* Or, we may find that many specially-chosen parameters and complex initial conditions must be invoked, which excludes *Path I* as a possibility.

This review concentrates on testing inflationary cosmology because it is an example of an explanatory model in the sense of *Path I.* Even if inflation is proved wrong, the same tests might indicate if *Path I* might survive under the guise of some other model. The same tests are also relevant for evaluating cold dark matter (CDM) and mixed dark matter (MDM) of large-scale structure formation, which are built upon the conditions created in an inflationary Universe. In addition to testing inflation, the tests might distinguish the CDM vs. MDM possibilities. The discussion is confined to measurements of the microwave background anisotropy because these are precision, quantitative, discriminating tests whose interpretation is least dependent on unverified assumptions.

§3 Inflationary Cosmology

The standard big bang model is extraordinarily successful in explaining many features of our Universe: the Hubble expansion, the abundances of light elements, and the cosmic microwave background radiation. However, it does not address some important questions: Why is the Universe so homogeneous? Why is the Universe spatially flat? Why are there no magnetic monopoles or other remnants from phase transitions that took place early in the Universe? What produced the inhomogeneities in the distribution of matter that seeded the evolution of galaxies? Prior to inflationary theory, the only explanations assumed special initial conditions (suggesting that cosmology is condemned to *Path II*).

Inflationary cosmology [1]–[4] has been proposed as a modification of the standard big bang picture that could explain these mysteries in terms of a well-defined sequence of dynamical processes occurring in the first instants (10^{-35} seconds or so) after the big bang. The central feature is a brief epoch in which the expansion of the Universe accelerates ("inflation"), resulting in an extraordinarily rapid expansion rate. The hyperexpansion of space flattens and smooths the Universe and dilutes the density of monopoles and other remnants to negligible values. Quantum fluctuations produced during the accelerating phase are stretched into a spectrum of energy density perturbations that can seed galaxy formation [5].

Inflation is induced by a change in the equation-of-state. The stretching of a homogeneous and isotropic Universe is described by the Robertson-Walker scale factor, $R(t)$. Inflation or accelerated expansion means that $\ddot{R} > 0$. The time-dependence of $R(t)$ is given by Einstein's equation of motion:

$$\ddot{R} = -\frac{4\pi G}{3}(\rho + 3p)R = -\frac{1}{2}H^2(1+3\gamma)R, \qquad (1)$$

where G is Newton's constant, ρ is the energy density, p is the pressure, and H is the Hubble parameter where $H^2 \equiv 8\pi G\rho/3$. The ratio $\gamma = p/\rho$ defines the equation-of-state. Hence, the expansion rate inflates ($\ddot{R} > 0$) if the equation-of-state satisfies $\gamma < -1/3$. Since ρ is always a positive quantity, a large ***negative*** pressure is required. Any physics which leads to a large negative pressure averaged over cosmological distances (at least a Hubble volume, H^{-3}) can induce inflation.

The standard example is a Universe with energy density dominated by a single, scalar "inflaton" field, ϕ. The equation-of-state is then:

$$\gamma = \frac{p}{\rho} = \frac{\frac{1}{2}\dot{\phi}^2 - V}{\frac{1}{2}\dot{\phi}^2 + V}, \tag{2}$$

where V is the effective potential for the inflaton. Here we see that $\gamma < -1/3$ can be achieved if the potential energy density dominates the kinetic energy density. Inflation continues until ϕ "rolls" to a state of negligible potential energy density. The progress of ϕ can be tracked by solving the Einstein equation, Eq. (1), and the slow-roll equation:

$$\ddot{\phi} + 3H\dot{\phi} = -V'(\phi), \tag{3}$$

where $H \equiv d\ \ln R/dt$ is the Hubble parameter and the prime denotes the derivative with respect to ϕ. During inflation, V' must be sufficiently small that the evolution of ϕ is slow, *i.e.*, $\ddot{\phi}$ is negligible. This condition must be maintained long enough for the Universe to have expanded by $N \sim 60$ e-foldings $(R(t_{end})/R(t_{begin}) = e^{60})$ in order to resolve the cosmological problems of the standard big bang model. As ϕ proceeds towards a steeper part of the potential, ϕ rapidly accelerates and inflation ends. The potential energy, $V(\phi)$, is converted to kinetic energy and, then, is ultimately converted into radiation and matter which reheats the Universe [6, 7].

The most important feature of inflation so far as this talk is concerned is that inflation smoothes out any initial non-uniformity while producing a new spectrum of inhomogeneities [5]. The inflaton and any other light fields all experience quantum de Sitter fluctuations on subatomic scales which inflation stretches to cosmological dimensions. It is convenient to discuss the spectrum of fluctuations in terms of its Fourier components, *i.e.*, a linear combination of plane wave modes. The wavelength of the modes grows as the Universe expands. As the accelerating expansion stretches the wavelength of a given mode beyond the Hubble distance, H_I^{-1} (where H_I is the Hubble parameter during inflation), the amplitude of the quantum de Sitter fluctuations in ϕ (or any other light field) is $\sim H_I/2\pi$. The additional stretching of the wavelength beyond the Hubble length does not change this amplitude since causal physical processes are unable to act over distances greater than the Hubble length. Over the course of inflation, many waves are stretched in this way, ultimately leading to a broad-band spectrum of macroscopic fluctuations with (nearly) the same amplitude.

After inflation ends, the stretching of the Universe decelerates ($\ddot{R} < 0$). The Hubble distance H^{-1} begins to increase at a rate that exceeds the expansion rate, $R(t)$. Hence, even though the waves continue to be stretched, the Hubble distance grows faster, catching up to and ultimately exceeding the wavelengths of some modes. At the point where the Hubble distance equals the wavelength of a given Fourier mode, it is sometimes said that the "fluctuation re-enters the horizon," referring to the fact that the wavelength was initially less than H_I^{-1} during inflation and has become less than H^{-1} in the post-inflationary epoch. (It would perhaps be more accurate to say "the horizon catches up to the fluctuation.")

The primordial spectrum is determined by the amplitudes of the waves as they re-enter the horizon in the matter- or radiation-dominated epoch. In inflationary models, these amplitudes are precisely the amplitudes as the fluctuations were stretched beyond the horizon during the de Sitter epoch.

The fluctuations of the inflaton, which dominates the energy density of the Universe during inflation, induce a spectrum of energy density perturbations with amplitude:

$$\left.\frac{\delta\rho}{\rho}\right|_{\lambda=H_I^{-1}} \propto \left.\frac{H_I^2}{\dot{\phi}}\right|_{\lambda=H_I^{-1}} \propto \left.\frac{\frac{1}{2}\dot{\phi}^2 + V}{\dot{\phi}}\right|_{\lambda=H_I^{-1}}, \tag{4}$$

where H and $\dot{\phi}$ are evaluated as a given wavelength λ is stretched beyond the horizon ($\lambda = H_I^{-1}$) during inflation. Since all microphysical parameters ($V(\phi)$, H, $\dot{\phi}$, etc.) change slowly during inflation compared to the stretching rate ($R(t)$), the ratio in Eq. (4) is nearly constant for all waves. That is, the fluctuations are produced with the (nearly) the same amplitude on average, a nearly scale-invariant (Harrison-Zel'dovich [8]) spectrum of energy-density perturbations.

Inflation also generates similar fluctuations in other light fields. For most fields, these fluctuations are irrelevant because they are insignificant contributors to the total energy density, and the fluctuations leave no distinctive signature. An important exception is quantum fluctuations of massless gravitons, which result in a nearly scale-invariant spectrum of gravitational waves [9]–[11]. Because the gravitational waves are weakly coupled, the spectrum is not erased by reheating or other interactions. Because of their tensor symmetry, their signature on the cosmic microwave background anisotropy is quite different from that of the scalar fluctuations [12, 13]. The predicted gravitational wave amplitude for a mode re-entering the horizon is:

$$|h_k|\ |_{\lambda=H_I^{-1}} \propto \left.\frac{H_I^2}{m_p^2}\right|_{\lambda=H_I^{-1}} \propto \left.\frac{\frac{1}{2}\dot{\phi}^2 + V}{m_p^2}\right|_{\lambda=H_I^{-1}}, \tag{5}$$

where m_p is the Planck mass and the expression is to be evaluated as the wavelength is stretched beyond the Hubble length during inflation. As with the scalar fluctuations, the parameters in the right-hand expression change slowly compared to the stretching so that the gravitational wave spectrum is also nearly scale-invariant.

A critical test for inflation is whether the observed cosmic microwave background anisotropy can be explained in terms of the predicted spectrum of scalar and tensor fluctuations.

§4 What Does Inflation Predict?

- *Spatial Flatness*: Inflation flattens the Universe [1], or, more explicitly, exponentially suppresses the spatial curvature contribution to the Hubble expansion relative to the matter and radiation density and relative to any cosmological constant (Λ). Hence, if Ω_{total} is defined as including matter, radiation and vacuum energy density contributions, then inflation predicts that $\Omega_{total} = 1 \pm \epsilon$, where ϵ is exponentially small.

- *Gaussian Primordial Perturbations*: The quantum fluctuations generated in inflation are Gaussian [5]. For a gaussian distribution, the total fluctuation spectrum can be determined from the temperature autocorrelation function (see Section 5).

- *Scale-free Spectrum of Scalar and Tensor Perturbations*: Energy density and gravitational wave fluctuations are generated during inflation with a scale-free spectrum; *i.e.*, a spectrum with no characteristic scale, such as a power-law. The scalar spectrum at time t is conventionally parameterized in terms of its Fourier components by a power-law,

$$\left|\left(\delta\rho/\rho|_{\lambda=H_I^{-1}}\right)^2\right| \propto k^{n_s-1}$$
$$\text{or} \quad k^3\left\langle|\tfrac{\delta\rho}{\rho}(k,t)|^2\right\rangle \propto k^{n_s+3}, \tag{6}$$

where n_s is the called the *scalar spectral index*. The first expression is in terms of $\delta\rho/\rho$ evaluated at different times, as each mode is stretched beyond the horizon during inflation; the second expression is in terms of $\delta\rho/\rho$ at fixed time. In this convention, $n_s = 1$ corresponds to strict scale-invariance (Harrison-Zel'dovich [8]). The analogous parameterization for the gravitational wave spectrum is

$$\left|\left(h_k\ |_{\lambda=H_I^{-1}}\right)^2\right| \propto k^{n_t} \quad \text{or} \quad k^3\left\langle|h_{+,\times}(k,t)|^2\right\rangle \propto k^{n_t}, \tag{7}$$

where $h_{+,\times}$ are the amplitudes of the tensor metric fluctuations (for two polarizations), n_t is the *tensor spectral index* and $n_t = 0$ corresponds to strict scale-invariance.[1]

[1] I apologize in behalf of the CMB community for the disgusting convention that defines the indices such that $n_s = 1$ and $n_t = 0$ both correspond to scale-invariant; however, I will maintain the convention in order for readers of this review to be able to comprehend the rest of the literature.

In most inflationary models, n_s and n_t actually have "weak" k-dependence. However, microwave background experiments and large-scale structure measurements probe only a narrow range of wavenumbers: Consider a mode with physical wavelength λ today. The physical wavelength at earlier times is $\lambda' = R\lambda$, where we choose the convention that $R = 1$ today. The physical wavelength of the mode at the end of inflation is $\lambda'_{end} = (R_{end}/R_{RH})R_{RH}\lambda$, where R_{RH} is the value of the scale factor after the Universe reheats following inflation. The number of e-folds of inflation between the time that the given mode was stretched beyond the Hubble distance (H_I^{-1}) and the time that inflation ends is

$$\begin{aligned} N(\lambda) &\equiv \ln\left[\tfrac{R_{end}}{R_{RH}} R_{RH}\right] \\ &= 57 + \ln\left(\tfrac{\lambda}{6000\,\text{Mpc}}\right) \\ &\quad + \tfrac{1}{3}\ln\left(\tfrac{V(\phi_{end})^{1/2}T_{RH}}{(10^{14}\,GeV)^3}\right), \end{aligned} \tag{8}$$

where $V(\phi_{end})$ is the potential energy density at the end of inflation, $T_{RH} \leq V(\phi_{end})^{1/4}$ is the temperature at reheating, and 6000 Mpc is the present Hubble distance (for $h = .5$). Microwave background and large scale structure observations span distances between $\sim$ 1 Mpc and $\sim$ 6000 Mpc. These observations cover modes generated during the 10 e-folds between $N(1\text{ Mpc}) \sim 50$ and $N(6000\text{ Mpc}) \sim 60$. Over this narrow range, it is an excellent approximation to treat n_s and n_t as k-independent [14]. In the remaining discussion, n_s and n_t always refer to the values averaged between e-folds 50 and 60.

The total fluctuation spectrum consists of two components, scalar and tensor, each of which is characterized by an amplitude and a spectral index. One convention is to define the amplitudes in terms of the scalar and tensor contributions to the quadrupole moment $C_2^{(S,T)}$ of the CMB temperature autocorrelation function. The scalar and tensor fluctuations are predicted to be statistically independent, so the total quadrupole is the simply the sum of the two contributions. The scalar and tensor quadrupole moments are related to the values of parameters $N_H \equiv N(\sim 6000\,\text{Mpc}) \sim 60$ e-folds before the end of inflation:

$$C_2^S \approx \frac{1}{240\pi^2}\frac{H_I^4}{\dot{\phi}^2}\bigg|_{N_H} \tag{9}$$

and

$$C_2^T \approx 0.073\frac{H_I^2}{m_p^2}\bigg|_{N_H} \quad . \tag{10}$$

- *Nearly Scale-Invariant Primordial Spectra*: The spectral indices are determined by the equation-of-state, γ, at $50-60$ e-folds before the end of inflation. The derivation of the relations is straightforward and important, so we digress to provide a detailed derivation below. The reader anxious to skip to the answers should proceed to Eqs. (17) and (19) and the discussion below Eq. (19).

The equation-of-state can be re-expressed in terms of an inflaton field using the relation:

$$\gamma = \frac{p}{\rho} = \frac{\frac{1}{2}\dot{\phi}^2 - V}{\frac{1}{2}\dot{\phi}^2 + V} = \frac{8\pi}{3m_p^2}\left(\frac{\dot{\phi}^2}{H_I^2}\right) - 1, \qquad (11)$$

where $H_I^2 \equiv (8\pi/3m_p^2)[\frac{1}{2}\dot{\phi}^2 + V(\phi)]$. Instead of γ, it is useful in this derivation to introduce a related parameter:

$$\alpha^2 \equiv 24\pi\frac{\dot{\phi}^2}{\frac{1}{2}\dot{\phi}^2 + V} == 24\pi(1+\gamma) = \left(\frac{8\pi\dot{\phi}}{m_p H_I}\right)^2. \qquad (12)$$

A mode that is stretched so that its wavenumber is $k = H^{-1}$ at time when there are $N(\phi)$ e-foldings remaining has wavenumber $k = H^{-1}\exp N(\phi)$ when inflation ends, where

$$N(\phi) \equiv \int H dt = \int_{\phi}^{\phi_{end}} \frac{H}{\dot{\phi}} d\phi. \qquad (13)$$

$N(\phi)$ is the number of e-folds that remain before the end of inflation when the inflaton field has value ϕ. Then, we have

$$\frac{d\ln k}{d\phi} = \frac{d\,N(\phi)}{d\phi} = \frac{H}{\dot{\phi}}. \qquad (14)$$

The tensor fluctuation amplitude as modes are stretched beyond the Hubble distance inflation is, according to Eqs. (5) and (7), $\propto H_I^2/m_p^2 \propto k^{n_t}$, Hence, the tensor spectral index can be computed according to:

$$\begin{aligned} n_t &= \frac{d\ln(H^2/m_p^2)}{d\ln k} \\ &= \frac{d\phi}{d\ln k}\frac{d\ln(H^2/m_p^2)}{d\phi} \\ &= \frac{\dot{\phi}}{H^3}(H^2)', \end{aligned} \qquad (15)$$

where the prime will be used to denote derivatives with respect to ϕ. The inflaton satisfies the equation-of-motion

$$\ddot{\phi} + 3H\dot{\phi} = -V'(\phi), \qquad (16)$$

but the $\ddot{\phi}$ term is negligibly small during inflation. This also implies that the kinetic energy density is small compared to the potential energy density; or, $(H^2)' \approx 8\pi V'(\phi)/3m_p^2$. Using the slow-roll equation and the expression for $(H^2)'$, we find that:

$$\begin{aligned} n_t &= -\frac{8\pi}{m_p^2}\left(\frac{\dot{\phi}}{H}\right)^2 = -\frac{\alpha^2}{8\pi} \\ &= -3(1+\gamma). \end{aligned} \qquad (17)$$

In the end, there is a very simple relation between n_t and the equation-of-state.

The analogous derivation for the scalar spectral index is somewhat more tedious. The amplitude is proportional to $\sim H^4/\dot{\phi}^2 \propto k^{n_s-1}$. Then, we have

$$\begin{aligned} n_s - 1 &= \frac{d\ln(H^4/\dot{\phi}^2)}{d\ln k} \\ &= \frac{d\phi}{d\ln k}\frac{d\ln(H^4/\dot{\phi}^2)}{d\phi} \\ &= 2\frac{\dot{\phi}^2}{H^3}\left[\frac{2HH'}{\dot{\phi}} - \frac{H^2}{\dot{\phi}^2}\dot{\phi}'\right] \\ &= \frac{\dot{\phi}}{H^3}(H^2)' + 2\left(\frac{\dot{\phi}H' - \dot{\phi}'H}{H^2}\right). \end{aligned} \qquad (18)$$

The first term, precisely the same as the intermediate expression we obtained for n_t in Eq. (15), is $-3(1+\gamma)$. By the use of Eq. (12) and a bit of algebra, the second term can be expressed as

$$-\frac{m_p}{4\pi}\frac{d\alpha}{d\phi} = \frac{d\ln\alpha^2}{d\ln k} = \frac{d\ln(1+\gamma)}{d\ln k}.$$

Hence, the total expression is

$$n_s = 1 - 3(1+\gamma) + [d\ln(1+\gamma)/d\ln k]; \qquad (19)$$

Strict exponential (de Sitter) expansion, $R(t) \propto \exp(H_I t)$, corresponds to $\gamma = -1$, in which limit one obtains precise scale-invariance, $n_s = 1$ and $n_t = 0$, according to Eqs. (19) and (17). However, in any realistic inflation model, the expansion rate must slow down near the end of inflation in order to return to Friedmann-Robertson-Walker expansion. If $R(t)$ is inflating but not exponentially, then $-1/3 > \gamma > -1$, γ' may not be zero, and $n_s \neq 1$ and $n_t < 0$. Inflationary models fall in the range $0.7 \lesssim n_s \lesssim 1.2$ and $0.3 \lesssim n_t \leq 0$; pushing n_s and n_t beyond this range entails exceptional models with special choices of parameters and/or initial conditions [14, 15]. (See comments at the end of this section.)

- *Relations between* ($C_2^{(S)}$, $C_2^{(T)}$, n_t, n_t): COBE and other large-angular scale experiments can determine the total quadrupole moment, $C_2 = C_2^{(S)} + C_2^{(T)}$, placing one constraint on the four parameters that define the inflationary spectrum. The three remaining degrees of freedom are: $r \equiv C_2^{(T)}/C_2^{(S)}$, n_s and n_t. These three parameters are all expressible in terms of the equation-of-state, γ, and, hence, can be related to one another [12, 13]. The tensor-to-scalar quadrupole ratio, r, is obtained by taking the ratio of Eq. (10) to Eq. (9): $r \approx 173\dot{\phi}^2/(H_I^2 m_p^2)$. Using Eq. (11), we can convert this to:

$$r \equiv C_2^{(T)}/C_2^{(S)} \approx 21(1+\gamma). \tag{20}$$

Comparing the last relation to Eq. (17), we find that

$$r \approx -7n_t. \tag{21}$$

or, we can compare it to Eq. (19) and obtain

$$n_t = n_s - 1 - [d \ln (1+\gamma)/d \ln k], \tag{22}$$

and

$$r \approx 7(1-n_s) + [d \ln (1+\gamma)/d \ln k] \tag{23}$$

(N.B. r is non-negative, by definition. It is possible to construct models in which the right-hand-side of Eq. (23) is formally negative; this result should be interpreted as indicating negligible tensor contribution, $r \approx 0$. In particular, models with $n_s > 1$, such as some hybrid inflation [16] models, have negligible tensor fluctuations.) These three relations constitute a set of testable signatures of inflation. If observations establish that the primordial spectrum of perturbations is scale-free, observational support for these additional relations would be evidence that the perturbations were generated by inflation.

The predictions are partially illustrated in Fig. 1, which shows a range of parameter-space in the r-n_s plane. Whereas the entire plane describes tensor and scalar perturbations which are scale-free, the range allowed by inflation is confined to the range $0.7 \lesssim n_s \lesssim 1.2$, which is nearly scale-invariant. Then, Eq. (23) places a constraint on r. Hence, over the entire r-n_s plane, inflationary predictions are confined to a small box. (The boundaries of the box are not precisely defined; one can expand the box 10 per cent or so at the cost of additional fine-tuning of parameters.)

Inflation is falsifiable if observations show that the CMB spectrum lies far outside the box. For example, early reports from COBE analysis suggested that $n_s > 1.5$ [17], which would be inconsistent. These results have since been revised to $n_s \approx 1.0 \pm 0.3$, which is consistent with inflation [18, 19].

Fig. 1 further illustrates that the predictions of inflation do not uniformly cover the box. For most models, $d \ln (1+\gamma)/d \ln k$ is negligible during inflation (which is a way of saying that the equation-of-state changes very slowly during inflation), and, hence,

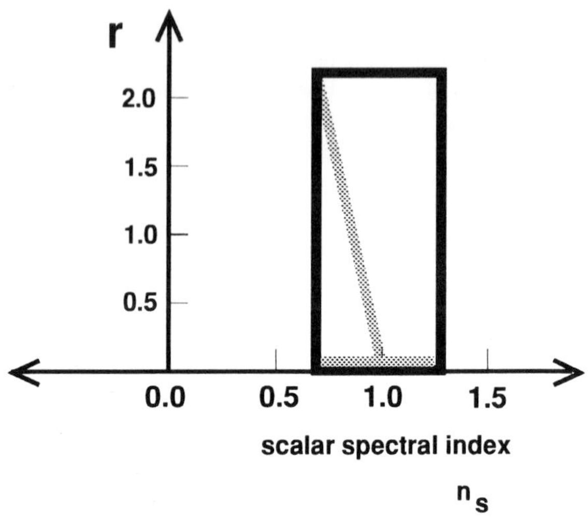

Figure 1: The range of the r-n_s plane consistent with generic inflationary models is enclosed by the box. Most models are constrained to lie along the grey diagonal curve; models in which the inflaton encounters an extremum in the inflaton potential near the last 60 e-folds have negligible tensor contribution, $r \approx 0$, along the abscissa inside the box.

$$r \approx 7(1-n_s). \tag{24}$$

These models lie along the grey curve of negative slope shown within the box. A subclass of models has the property that the inflaton encounters an extremum of the inflation potential 60 or so e-folds before the end of inflation. In these models, $d \ln (1+\gamma)/d \ln k \propto V''(\phi)$ changes significantly during inflation; in particular, $\dot{\phi} \propto V'(\phi)$ shrinks significantly near the extremum, which amplifies the scalar perturbations relative to the tensor (see Eqs. (4) and (5)). Consequently, these models predict that $r \approx 0$, along the abscissa of Fig. 1. Evidence that the CMB spectrum lies in the box but far from the abscissa or the ($r \approx 7(1-n_s)$) diagonal would be problematic for inflation.

The "generic" predictions of inflation outlined above presume no theoretical prejudice about the "brand" of inflation. They apply to new, chaotic, supersymmetric, extended, hyperextended, hybrid and natural inflation. What determines the prediction is the equation-of-state (*e.g.*, the shape of the inflaton potential $V(\phi)$) during the last 60 e-folds of inflation. Table 1 summarizes the predictions of inflation for some particular forms of the inflaton potential, $V(\phi)$. Since the examples in the Table run the gamut from potentials which are steep to those which are flat, it may be used to estimate the predictions

$V(\phi)$	$n_s - 1$	n_t	r
$V_0 \exp(-\frac{c\phi}{m_p})$	$-\frac{c^2}{8\pi}$	$-\frac{c^2}{8\pi}$	$.28c^2$
$A\phi^n$	$-.02 - \frac{n}{100}$	$-\frac{n}{100}$	$.08n$
$V_0 + \lambda\phi^4(\ln\frac{\phi^2}{\sigma^2} - \frac{1}{2})$	$-4 \times 10^{-6}\left(\frac{\sigma}{m_p}\right)^4$	$-.06 - 4 \times 10^{-6}\left(\frac{\sigma}{m_p}\right)^4$	$3 \times 10^{-5}\left(\frac{\sigma}{m_p}\right)^4$
$V_0\left(1 - \frac{\phi^2}{f^2}\right)$	$-\frac{m_p^2}{2\pi f^2}$	$-\frac{\pi m_p^2}{8f^2} e^{-Nm_p^2/2\pi f^2}$	$2.8\frac{m_p^2}{f^2} e^{-Nm_p^2/2\pi f^2}$

Table 1: Predictions of inflationary models for some common potentials.

for more general potentials. Conversely, it is possible to reconstruct a section of the inflationary potential over the range of ϕ covered during e-folds 50 to 60 from measurements of the spectral index and the ratio of tensor-to-scalar quadrupole moments [20]. This reconstructed section is extremely narrow since ϕ evolves only a tiny distance down the potential during e-folds 50 to 60. Furthermore, the reconstructed section typically lies far from the false or true vacuum. Hence, the reconstructed section is of limited value in determining the full potential or underlying physics driving inflation.

In the Table, $m_p \equiv 1.2 \times 10^{19}$ GeV is the Planck mass, $N = 60$ is e-folds corresponding to the present horizon (*i.e.*, the natural log of the ratio of the present horizon to the horizon during inflation in comoving coordinates). The first two examples correspond to cases where Eq 24 is a good approximation. The approximation correctly predicts a negligibly small value of r for the third example, but the numerical value is not well-estimated. The predictions for these first three models lie close to the diagonal line in Fig. 1. Eq 24 is a poor approximation for the fourth example, in which the inflaton rolls from the top of a quadratic potential and $d\ln(1+\gamma)/d\ln k$ is large. The prediction for this case is a spectrum with tilt ($n_s < 1$) but insignificant gravitational wave contribution, corresponding to points along the abscissa of Fig. 1.

Exceptional inflationary models can be constructed which violate any or all of the conditions described above. In fact, because theorists enjoy dwelling on such matters, there are nearly as many papers written on exceptional models as on generic ones. This causes some experimentalists, observers, and non-experts to give these exceptional predictions undue weight. Therefore, it is important to emphasize that these exceptional models, such as those which predict an open or closed Universe or primordial spectra which are not scale-free, are extremely unattractive. First, they require extraordinary fine-tuning beyond what is required to have sufficient inflation and solve the conundra of the big bang model. If one maps out the range of parameter-space which gives sufficient inflation, the exceptional models occupy an exponentially tiny corner. Second, the predictions are not robust. Moving from one point to another within the tiny corner of parameter-space significantly changes the predictions. For example, if one choice of parameters produces a Universe with $\Omega = 0.1$, a slightly different choice increases or decreases the number of e-foldings by one, which results in a change in Ω by a factor of ten. Consequently, there is little or no real predictive power to the exceptional models.

Focusing on the generic tests of inflation is well-motivated for broader reasons: The same tests might determine whether the Universe can be explained on the basis of physical laws testable in the laboratory (*Path I*, as defined in Section 2). The microphysics which we presently understand or can hope to test in the laboratory involves time-scales infinitesimally smaller than the age of the Universe and length-scales infinitesimally smaller than the Hubble length (or the sizes of galaxies). If our Universe is to be comprehended from these physical laws alone without special choices of initial conditions or parameters (*i.e.*, *Path I*), we should not find that there is something special about the present epoch (compared, say, to 10 Hubble times from now or 10 Hubble times earlier) or that there are special features (bumps, dips, etc.) in the primordial spectrum of fluctuations of cosmological wavelength. If we find evidence for new time- or length-scales of cosmological dimensions which cannot be probed in the laboratory, cosmology is thrust into *Path II*. Important features of our Universe must be attributed to initial conditions or physical laws which probably can never be independently tested. (N.B. Testing whether the spectrum is scale-free is less specific than testing inflation. One can imagine finding evidence which supports a flat Universe with scale-free primordial perturbations, but which conflicts with the inflationary relations between r and n_s.)

§5 Translating Inflationary Predictions into Precision Tests

The predictions of inflation described in the previous Section can be translated into precise tests of the CMB anisotropy. The implications for the CMB anisotropy can be obtained by numerical integration of the general relativistic Boltzmann, Einstein, and hydrodynamic equations [21]. Included in the dynamical evolution are all the relevant matter-energy components: baryons, photons, dark matter, and massless neutrinos. The temperature anisotropy, $\Delta T/T(\mathbf{x}) = \sum_{\ell m} a_{\ell m} Y_{\ell m}(\theta,\phi)$, is computed in terms of scalar [22, 23] and tensor [21] multipole components, $a_{\ell m}^{(S)}$ and $a_{\ell m}^{(T)}$, respectively.

The common method of characterizing the CMB fluctuation spectrum is in terms of multipole moments. Suppose that one measures the temperature distribution on the sky, $\Delta T/T(\mathbf{x})$. The temperature autocorrelation function (which compares points in the sky separated by angle α) is defined as:

$$\begin{aligned} C(\alpha) &= \langle \tfrac{\Delta T}{T}(\mathbf{x}) \tfrac{\Delta T}{T}(\mathbf{x}') \rangle \\ &= \tfrac{1}{4\pi} \textstyle\sum_\ell (2\ell+1) C_\ell P_\ell(\cos\alpha), \end{aligned} \tag{25}$$

where $\langle\rangle$ represents an average over the sky and $\mathbf{x}\cdot\mathbf{x}' = \cos\alpha$. The coefficients, C_ℓ, are the *multipole moments* (for example, C_2 is the quadrupole, C_3 is the octopole, *etc.*). Roughly speaking, the value of C_ℓ is determined by fluctuations on angular scales $\theta \sim \pi/\ell$. A plot of $\ell(\ell+1)C_\ell$ is referred to as the *power spectrum*. This definition is chosen so that an exactly scale-invariant spectrum, assuming no evolution when the fluctuations pass inside the Hubble horizon, produces a flat power spectrum (*i.e.*, $\ell(\ell+1)C_\ell$ is independent of ℓ). For inflation, the contributions of scalar and tensor fluctuations to the $a_{\ell m}$'s are predicted to be statistically independent. Consequently, the total multipole moment C_ℓ is the sum of the scalar and tensor contributions, $C_\ell^{(S)}$ and $C_\ell^{(T)}$, respectively. The $a_{\ell m}$'s are also predicted to be Gaussian-distributed. The cosmic mean value predicted by inflation is $C_\ell^{(S)} = \left\langle |a_{\ell m}^{(S)}|^2 \right\rangle_E$ and $C_\ell^{(T)} = \left\langle |a_{\ell m}^{(T)}|^2 \right\rangle_E$ where these are averaged over an ensemble of universes $\{E\}$ and over m. The C_ℓ's, which are an average over $2\ell+1$ Gaussian-distributed variables, have a χ^2-distribution.

There is additional valuable information in higher-point temperature correlation functions; *e.g.*, tests for non-gaussianity. However, statistical and systematic errors increase for higher-point correlations; for the short run, the most reliable information will be the angular power spectrum (C_ℓ vs. ℓ).

The predicted spectrum depends not only on the inflationary parameters (r, n_s, n_t) but also on other cosmological parameters. Inflation produces a flat Universe, $\Omega_{total} \approx 1$. In the examples shown in this article, we assume no hot dark matter, $\Omega_{HDM} = 0$, but note that, for angular scales $\gtrsim 10'$, the anisotropy for mixed hot and cold dark matter models with $\Omega_{CDM} + \Omega_{HDM} \approx 1$ is quite similar to the anisotropy if all of the dark matter is cold. For a given value of the Hubble parameter, $H = 100h$ km/sec/Mpc, we impose the nucleosynthesis estimate, $\Omega_B h^2 = 0.0125$, to determine Ω_B. We also satisfy globular cluster and other age bounds [24], and gravitational lens limits [25]: we range from $h \lesssim 0.65$ for $\Omega_\Lambda = 0$ to $h \lesssim 0.88$ for $\Omega_\Lambda \lesssim 0.6$. We also consider a range of reionization scenarios in which the intergalactic medium is fully reionized at some red shift z_R after recombination.

Fig. 2 shows the predicted power spectrum for the central range of inflationary models consistent with the generic predictions outlined in Section 4. Since the value of C_ℓ is determined by fluctuations on angular scales $\theta \sim \pi/\ell$, moving left-to-right in the Figure corresponds to moving from large-angular scales to small-angular scales. Since large-angular scale fluctuations entered the Hubble horizon recently compared to small-angular scale fluctuations, moving left-to-right also corresponds to moving from unevolved primordial fluctuations to fluctuations which have evolved inside the Hubble horizon for a significant time. For these examples, we have chosen $h = 0.5$, $\Omega_\Lambda = 0$, and $\Omega_B = 0.05$.

The predictions of theoretical models, including those for inflation shown in Fig. 2, are expressed in terms of the cosmic mean value of the C_ℓ's averaged over an ensemble of universes or, equivalently, over an ensemble of Hubble horizon-sized patches. In reality, experiments can measure, at best, over a single Hubble horizon distance. Since the experiments are limited in this way, it is important to know not only the theoretical prediction for the cosmic mean, but also the the theoretical variance about that prediction for experiments confined to a sin-

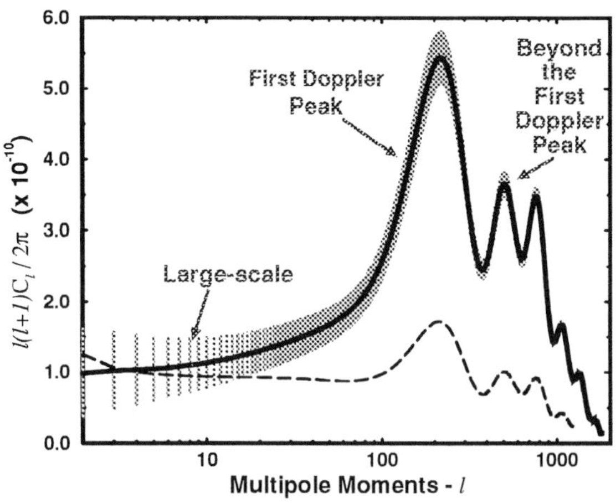

Figure 2: Predicted power spectrum predicted for inflationary cosmology. The spectra are for h=0.5, $\Lambda = 0$, $\Omega_B h^2 = .0125$ and cold dark matter (CDM). The upper (solid) curve has spectral index $n = 1$ (Harrison-Zel'dovich) and pure scalar fluctuations, $r = 0$; the vertical hashmarks represent the one-sigma, full-sky cosmic variance. The lower (dashed) curve has $n = 0.85$ and 50-50 mixture of scalar and tensor quadrupole perturbations, $r = 1$.

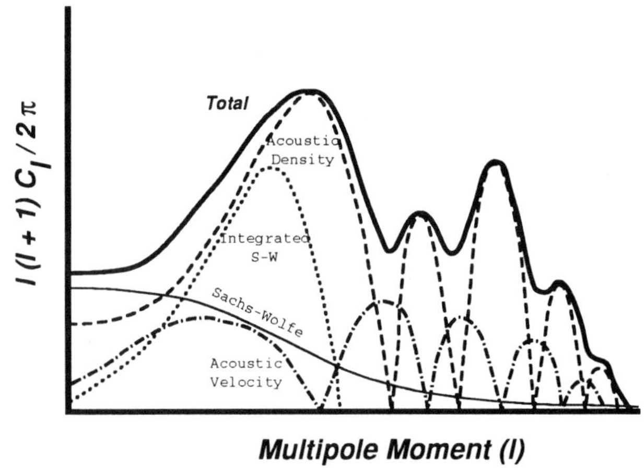

Figure 3: Schematic of the power spectrum predicted by a typical inflationary model and the major contributions to it: the Sachs-Wolfe effect (scalar and tensor modes combined), acoustic oscillations of the baryon-photon fluid (density and velocity contributions), and the integrated Sachs-Wolfe effect (significant in models where recombination occurs before matter-domination). [Adapted from Hu & Sugiyama 1994.]

gle Hubble horizon distance. This uncertainty, known as the full-sky "cosmic variance," is equal to $C_\ell/\sqrt{2\ell+1}$ for Gaussian-distributed fluctuations, such as those predicted by inflation. Note that the variance decreases with increasing ℓ. Many current small-angular scale experiments cover only a tiny fraction of Hubble horizon-sized patch. If the area fraction of full-sky coverage is A, the theoretical uncertainty scales as $A^{-\frac{1}{2}}$. (In cases where there is much less than full-sky coverage, the theoretical uncertainty is often referred to as "sample variance.")

A more realistic situation is where there are errors due to both sample variance and detector noise [26]. Consider a detection obtained from measurements $(\Delta T/T)_i \pm \sigma_D$ (σ_D represents detector noise) at $i = 1, \ldots, N_D$ experimental patches sufficiently isolated from each other to be largely uncorrelated. For large N_D, the likelihood function falls by $e^{-\nu^2/2}$ from a maximum at $(\Delta T/T)_{max}$ when

$$\left(\frac{\Delta T}{T}\right)^2 = \left(\frac{\Delta T}{T}\right)^2_{max} \pm \sqrt{\frac{2}{N_D}}\,\nu[\left(\frac{\Delta T}{T}\right)^2_{max} + \sigma_D^2]. \tag{26}$$

An experimental noise σ_D below 10^{-5} is standard now, and a few times 10^{-6} will soon be achievable; hence, if systematic errors and unwanted signals can be eliminated, the one-sigma ($\nu = 1$) relative uncertainty in $\Delta T/T$ will be from cosmic (or sample) variance alone, $1/\sqrt{2N_D}$, falling below 10% for $N_D > 50$. The hashing in Fig. 2 corresponds to full-sky cosmic variance, roughly equivalent to filling the sky with patches separated by $2\theta_{fwhm}$.

§6 The Fingerprint of Inflation

The CMB power-spectrum (Fig. 2) is a fingerprint which is evidence both for inflation and for certain values of cosmological parameters. C_ℓ's for $\ell \lesssim 200$ are dominantly determined by fluctuations outside the Hubble horizon at recombination. These multipoles measure the primordial spectrum of fluctuations. C_ℓ's for $\ell \gtrsim 200$ are sensitive to fluctuations inside the Hubble horizon at recombination. These fluctuations, which had time to evolve prior to last scattering, are sensitive to evolutionary effects which depend on the matter density, the expansion rate, and the density, type, and distribution of dark matter. Fig. 3 shows a schematic of the spectrum and the various contributions to it. The distinctive features of the fingerprint are (reading Fig. 2 from left to right):

Plateau at large angular scales ($\ell \lesssim 100$) is due to fluctuations in the gravitational potential on the last-scattering surface (the Sachs-Wolfe effect[27]). Fluctuations in the potential induce red shifts and blue shifts in the CMB photon distribution which create apparent temperature fluctuations. For a precisely scale-invariant ($n_s = 1$) spectrum of scalar fluctuations, the Sachs-Wolfe contribution to C_ℓ is proportional to $1/(\ell(\ell+1))$, and so the contribution to the power spectrum, $\ell(\ell+1)C_\ell/2\pi$ (the ordinate in Fig. 2), is flat. The full computation reveals a slightly upward slope at small ℓ due to other, higher-order contributions, the same effects which are responsible for the Doppler peaks described below. For $n_s < 1$, there is a downward tilt to the Sachs-Wolfe contribution relative to $n_s = 1$ (less power on smaller scales). The $n_s < 1$ curves in Fig. 2 include the tensor contribution predicted by generic inflation models, as expressed by Eq. (23). From observations at small ℓ only, it is difficult to distinguish the tensor contribution because the spectral slope due to the Sachs-Wolfe effect is nearly the same for tensor and scalar contributions for $\ell > 3$.[2] The slope is not very sensitive to the value of h, Ω_B, Ω_Λ or other cosmological parameters.

First Doppler Peak at $\ell \approx 200$ probes wavelengths smaller than the horizon at last-scattering ($\lesssim 1^\circ$). Gravitational wave perturbations begin to oscillate and red shift away once their wavelength falls within the horizon. Hence, for $\ell \gtrsim 200$, the tensor perturbations do not contribute significantly, even if they were dominant over the scalar fluctuations at larger angular scales ($\ell \ll 200$).

The prominent features, known as the Doppler peaks, are due to scalar fluctuations. Fig. 3 illustrates both the acoustic density and acoustic velocity contributions. The peaks are the remnant of adiabatic oscillations in the baryon-photon fluid density. The oscillations in a given mode begin when the wavelength falls below the Jeans length (*i.e.*, pressure dominates over gravity).[3] The Jean's length near recombination is roughly $2\pi c_s H^{-1}$, where $c_s \approx 1/\sqrt{3}$ is the sound speed.

The value of ℓ at the maximum [28] of the first Doppler peak is $\ell_{peak} \approx 220/\sqrt{\Omega_{total}}$. Since the location is insensitive to the value of h or Ω_B, measuring ℓ_{peak} is a novel means of measuring Ω_{total}. The height of the Doppler peak depends on the primordial spectral amplitude, the scalar spectral index (n_s), the tensor-to-scalar quadrupole ratio (r), the expansion rate, and the pressure [26]. If the spectrum is fixed at large angular scales by COBE DMR, smaller values of n_s imply decreasing primordial amplitudes on smaller angular scales and, consequently, a smaller Doppler peak. Gravitational waves add to the plateau at $\ell < 200$, but their contribution to the Doppler peak is red shifted to insignificant values. Consequently, increasing r decreases the height of the Doppler peak relative to the plateau. Increasing the expansion rate (h) pushes back matter-radiation equality relative to recombination, thereby increasing the adiabatic growth of perturbations. Photons escaping from the deeper gravitational potential are red shifted more. Hence, increasing h suppresses the Doppler peak. Increasing the pressure (by decreasing $\Omega_B h^2$) also decreases the anisotropy since the fluctuations stop growing once pressure dominates the gravitational infall. (According to the last two remarks, increasing h produces opposite effects. For $\Omega \lesssim 0.1$, the net effect is that increasing h decreases the anisotropy [29].) The height of the first Doppler peak is relatively insensitive to whether the dark matter is cold or a mixture of hot and cold dark matter.

Second and Subsequent Doppler peaks are due to modes that have undergone further adiabatic oscillations. The peaks are roughly periodically-spaced. The deviation from periodicity is due to time-variation in c_s. The amplitudes decrease as n_s decreases and r increases.

Anisotropies are caused by inhomogeneities in the baryon-photon fluid density and velocity. The acoustic density and velocity oscillations are 180 degrees out-of-phase with one another, as shown in Fig. 3. The acoustic density contribution is larger. The Doppler peak maxima and minima correspond to maxima and minima of the acoustic density oscillations; the minima do not extend to zero because they are filled in by the maxima in the acoustic velocity contribution [23, 29, 30]. The first and other odd-numbered peaks correspond to compressions and the even-numbered peaks correspond to rarefactions. Gravity tends to enhance the compressions and suppress the rarefactions. The effect is especially noticeable at low pressure (high $\Omega_B h^2$), for which the even-numbered peaks are greatly suppressed or absent altogether. The peaks are also sensitive to whether the dark matter is cold or a mixture of hot and cold.

Damping at $\ell \gtrsim 1000$: CMB fluctuations are suppressed by photon diffusion (Silk Damping [31]). The baryons and photons are imperfectly coupled. The photons tend to diffuse out of the fluctuations and smooth their distribution. Through their collisions with the baryons, the baryons distribution is smoothed as well, thereby suppressing the anisotropy. A second damping effect is due to the destructive interference of modes with wavelengths smaller than the thickness of the last-scattering surface [22, 23]. Doppler peaks from five or so adiabatic oscillations can be distinguished before the damping overwhelms.

§7 Five Milestones for Testing Inflation

The fingerprint imprinted by inflation on the CMB anisotropy suggests a series of five milestone tests. Below, the proposed tests are compared with current experimental results. Included are [32]–[42]: COBE DMR (COsmic Background Explorer Differential Microwave Radiometer), FIRS (Far InfraRed Survey), TENerife, SP91 and SP94 (South Pole 1991 and 1994), BP (Big Plate), PYTHON, ARGO, MAX (Millimeter Anisotropy Exper-

[2] There is an intriguing, notable difference in the ratio of the mean quadrupole-to-octopole moment for the models with tensor contribution. However, the effect is not a useful discriminant because there is a large cosmic variance for the small-ℓ multipoles.

[3] The term, "Doppler peak," is a misnomer since, with standard recombination, the electrons and photons oscillate together and there is little difference in their velocities. The Doppler effect does not become significant until $\ell \gtrsim 1000$.

iment), MSAM (Medium Scale Anisotropy Experiment), White Dish, and OVRO7 (Owens Valley Radio Observatory, 7 degree experiment).

Milestone 1: Observation of Large-Scale Fluctuations with $\Delta T/T \approx 10^{-5}$

The nearly scale-invariant spectrum of fluctuations generated by inflation includes modes whose wavelength is much greater than the Hubble horizon at recombination ($\gg 1°$). If inflation is responsible for the formation of large-scale structure, the magnitude of the perturbations should be at the level of $\Delta T/T \approx 10^{-5}$. If $\Delta T/T$ were a factor of five or more smaller, the amplitude would be too small to account for galaxy formation. A value of $\Delta T/T$ a factor of five or more greater would lead to unacceptable clumping of large-scale structure.

In starting a long journey, it is encouraging to begin from a point where the first milestone has already been passed. In this case, COBE DMR [32], with some corroboration from the FIRS [33] and Tenerife [34] observations, finds $\Delta T/T = 1.1 \pm 0.1 \times 10^{-5}$ scales (two-year result for 53 GHz scan smoothed over 10°) [17], just within the range consistent with inflation and dark matter models. Other models, such as cosmic defects and isocurvature baryon (PIB) models, are also consistent with these observations.

Milestone 2: Observation of Scale-free and Nearly Scale-Invariant Spectrum at Intermediate Angular Scales ($20° \gtrsim \theta \gtrsim 1°$)

Inflation predicts a primordial spectrum that is scale-free and nearly scale-invariant. Fig. 4 shows the predictions for the central range of parameters consistent with inflation compared with current measurements of the CMB anisotropy. The curves have been normalized to the COBE DMR two-year result.

A notable effect is that the spectrum for the $n_s = 1$, scale-invariant (Harrison-Zel'dovich) spectrum is not precisely flat. The slight, upward tilt is due to the small contributions of short wavelength ($\lesssim 1°$) modes, which include effects other than simple Sachs-Wolfe. The same contributions become dominant at $\ell \geq 100$ and are responsible for the Doppler peaks addressed by Milestones 3 and 4. Consequently, the *apparent* spectral index — the slope of C_ℓ vs. ℓ determined directly from the CMB anisotropy — differs from the *primordial* spectral index which generated the spectrum. For example, the upper curve has been computed for a primordial index of $n_s = 1$, but that apparent index (the upward tilt) corresponds to $n_s^{app} = 1.15$, a 15 per cent correction. An important feature of the added contributions is that, over the range $\ell \lesssim 100$, they are relatively insensitive to the value of cosmological parameters (h, Ω_B, and Ω_Λ). Fig. 6 illustrates the range of predictions for fixed spectral index ($n_s = 1$) when all other parameters are varied by the maximal amounts consistent with astrophysical observations. The spectra form a narrow sheath around the original spectrum, clearly separated from the prediction for $n_s = 0.85$. Hence, the spectrum of C_ℓ's for $\ell \lesssim 100$ can test whether the spectrum is scale-free and measure the spectral index without any additional assumptions about other cosmological parameters.

Fig. 4 also shows current detections. The error bars represent the one-sigma limits. (The method of flat band power estimation [44, 45], an important tool for converting experimental results into model-independent bounds on the C_ℓ's, is discussed in the Appendix.) The theoretical curves are normalized to match COBE DMR (two-year). The experiments on this figure illustrate the diversity in CMB experiment. COBE DMR is aboard a space-borne platform; FIRS is a high-altitude balloon experiment; Tenerife is mounted on a mountain in the Canary Islands; South Pole is in Antarctica, of course; and Big Plate is set on the ground in Saskatoon.

At this point, experiments are consistent with a scale-free form for the power spectrum with spectral-index near $n_s = 1$ and $n_t = 0$, but the error bars are large. After four years of data, COBE DMR may be able to reduce its error by two. Much more dramatic improvements are expected for the other experiments. The chief limitation is small sample area (and, hence, overwhelming sample variance) which should be overcome with continued measurements.

By itself, COBE DMR is not a powerful discriminant among theoretical models. For example, Fig. 7 shows the cross-correlation between the 53 GHz and 90 GHz frequency channels. The cross-correlation for a full-sky map is $K(\alpha) \equiv \langle \Delta T_{53}(\mathbf{x}) \Delta T_{90}(\mathbf{x}') \rangle$ where $\mathbf{x}\cdot\mathbf{x}' = \cos\alpha$; the coefficients of the multipole expansion of $K(\alpha)$ analogous to Eq. (25) are K_ℓ. If COBE DMR provided a full-sky map and there was no detector noise or foreground contamination, the K_ℓ should equal C_ℓ. In reality, COBE DMR analysis is based on a full-sky from which the galaxy has been cut. Plotted in Fig. 7 are the coefficients of orthonormal moments on the cut sky, using methods developed by Gorski [18]. The fact that there is noise in the COBE experiment accounts for the anticorrelation at large ℓ. At $\ell \lesssim 20$, though, the two channels are highly correlated (open circles) and there is strong evidence for the CMB signal. Superimposed on the plot are predictions for the cross-correlation for diverse models. The dotted lines correspond to one-sigma cosmic variance. The figure shows that the models cannot be distinguished to better than cosmic variance.

Fig. 4 and Fig. 7 illustrate important lessons concerning COBE DMR: First, the initial emphasis on extracting the COBE quadrupole moment and determining the spectral index has perhaps been misplaced. The quadrupole is difficult to extract empirically because of the galactic background. It has limited use theoretically because of

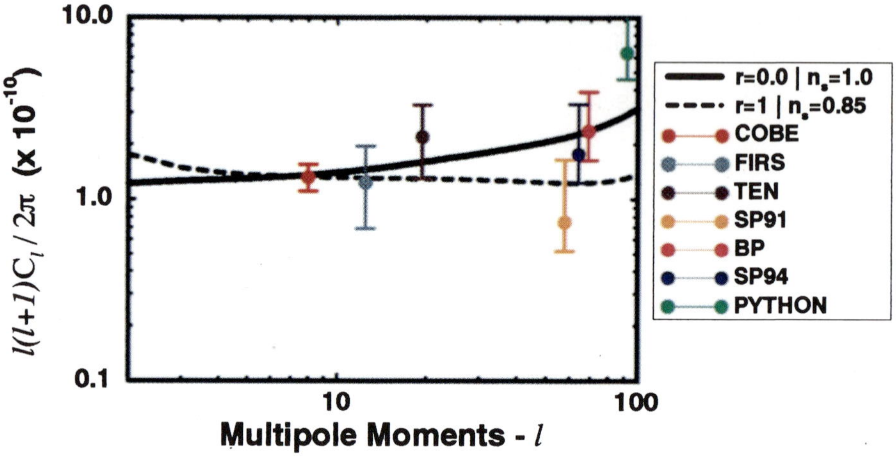

Figure 4: Blow-up of Fig. 2 showing the power spectrum for intermediate angular scales, $2 < \ell < 100$. Note that the upper curve, which corresponds to $n_s = 1$ (scale-invariant), has an upward tilt corresponding to an apparent spectral index of ~ 1.15 at large angular scales. Superimposed are the experimental flat band power detections with one-sigma error bars (see Appendix) for: (a) COBE; (b) FIRS; (c) Tenerife; (d) South Pole 1991; (e) South Pole 1994; (f) Big Plate 1993-4; and (g) PYTHON.

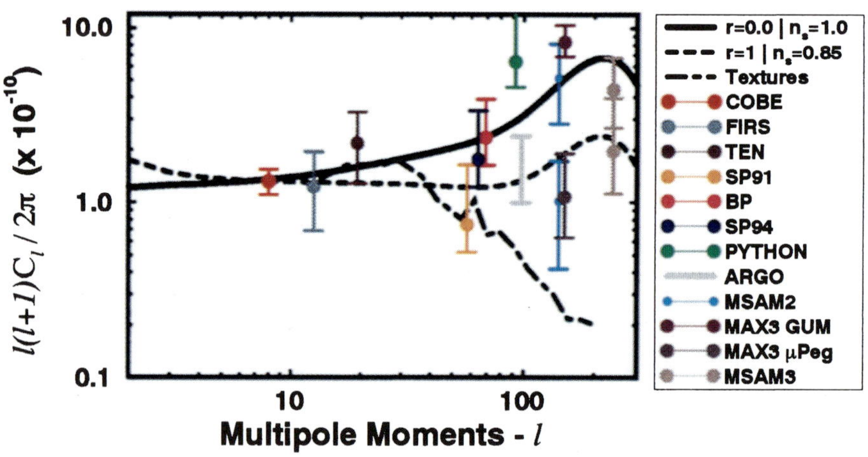

Figure 5: CMB spectra through the first Doppler peak. The solid and dashed black curves correspond to the inflationary/CDM predictions for spectral index ($r = 0$, $n_s = 1$) (Harrison-Zel'dovich) and ($r = 1$, $n = 0.85$), respectively. The dot-dashed curve is the prediction for cosmic texture models. For the data, error bars represent one-sigma. The experiments correspond to: (a) COBE; (b) FIRS; (c) Tenerife; (d) South Pole 1991; (e) South Pole 1994; (f) Big Plate 1993-4; (g) PYTHON; (h) ARGO; (i) MSAM (2-beam) - upper point uses entire data set, the lower point has unidentified point sources removed; (j) MAX3 (GUM region); (k) MAX3 (μPegasus region) showing here unidentified residual after dust subtraction; and (l) MSAM (3-beam) - the upper point uses the entire data set, lower point has unidentified point sources removed.

the large cosmic variance for this multipole (as shown in Fig. 7). Most important, the quadrupole contains no information that cannot be obtained by measuring the higher multipole moments, which can be fixed empirically and theoretically to greater accuracy. The second lesson is that measuring the spectral index using COBE DMR alone is limited by the fact that it is based on the detection of the first 20-30 multipole moments only. The range of ℓ is rather short for determining accurately the tilt in the power spectrum with increasing ℓ. Fig. 4 suggests that experiments at somewhat larger ℓ ($\gtrsim 50$), when combined with COBE DMR, can produce a more accurate measure of the spectral index because of the larger lever-arm in ℓ. There is the additional advantage that the larger ℓ multipole moments have smaller cosmic variance.

Milestone 3: Observations of the First Doppler Peak

The existence of the first Doppler peak is a generic feature of inflationary models (and CDM or HDM models of large-scale structure formation) assuming standard recombination or reionization at $z \lesssim 100$. The position of the Doppler peak, at $\ell \approx 220/\sqrt{\Omega_{total}}$, tests whether the Universe is spatially flat, as inflation predicts [28]-[30]. The height of the first Doppler peak depends sensitively on h, Ω_B, and Ω_Λ, and on the reionization history. Discriminating the various effects is difficult since rather different choices of cosmological parameters can lead to virtually indistinguishable CMB anisotropy spectra. We refer to this problem as *cosmic confusion*. Its implications are discussed in the next section.

Fig. 5 illustrates the theoretical predictions and the present experimental limits. Error bars represent one-sigma bounds. It is difficult to draw any firm conclusions from this range of observational results. However, it is interesting to note that a sequence of new MAX results re-examining the GUM (Gamma Ursa Major) and two other regions are consistent with the large amplitude found by the earlier MAX3 GUM experiment (see Fig. 8). Present results suggest a rather large Doppler peak.

Observation of the first Doppler peak is an especially important discriminant between cosmic defect (strings, textures, etc.) and inflationary/CDM models for large-scale structure formation. Fig. 5 also illustrates the predictions for cosmic texture models [46]. The predictions for other cosmic defect models are similar. The cosmic texture predictions shown here assume reionization at $z \gtrsim 200$, which accounts partially for the suppression of fluctuations just where inflation predicts a Doppler peak. However, even if no reionization is assumed, cosmic defect models normalized to COBE DMR predict smaller amplitude fluctuations at 1° scales than inflationary/CDM models (which is related to why they require high bias parameters) [47]. For both reasons, cosmic defect mod-

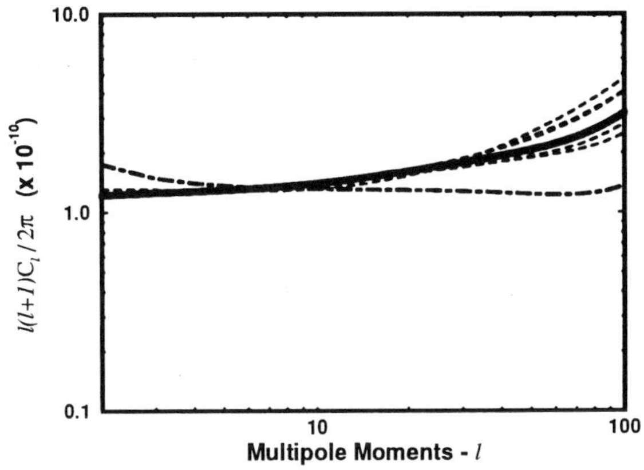

Figure 6: Effect of varying cosmological parameters on the power spectrum over intermediate ($2 < \ell \leq 100$) scales. The solid curve is the baseline spectrum, $n_s = 1$ (Harrison-Zel'dovich), $r = 0$, h=0.5, $\Omega_\Lambda = 0$ and $\Omega_B h^2 = .0125$. The enveloping dashed curves also have $n_s = 1$, but other cosmological parameters are varied within their astrophysical bounds. Dashed curves from top to bottom: (1) Ω_Λ increased to 0.6; (2) h decreased to 0.4; (3) $\Omega_B h^2$ induced to 0.025; (4) $\Omega_B h^2$ reduced to 0.005; (5) h increased to 0.75. Curves (2) and (3) are difficult to distinguish because they overlap considerably. Note how this whole family of curves is well-separated from the lower dot-dashed curve which corresponds to $n = .85$ and $r = 1$.

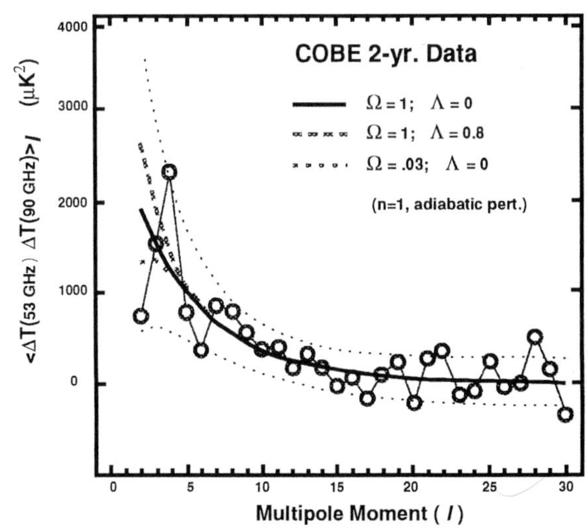

Figure 7: Cross-correlation between 53 GHz and 90 GHz frequency channels in the COBE DMR 2-year map compared with three theoretical models: (a) flat CDM universe (solid curve); (b) Λ-dominated flat universe with $\Omega_\Lambda = 0.8$ (upper dashed curve); (c) open baryon-dominated universe with $\Omega = 0.03$ (lower dashed curve). The cross-correlation has been expanded into orthogonal functions on the cut sky labeled by the index ℓ. (Courtesy of K. Gorski.)

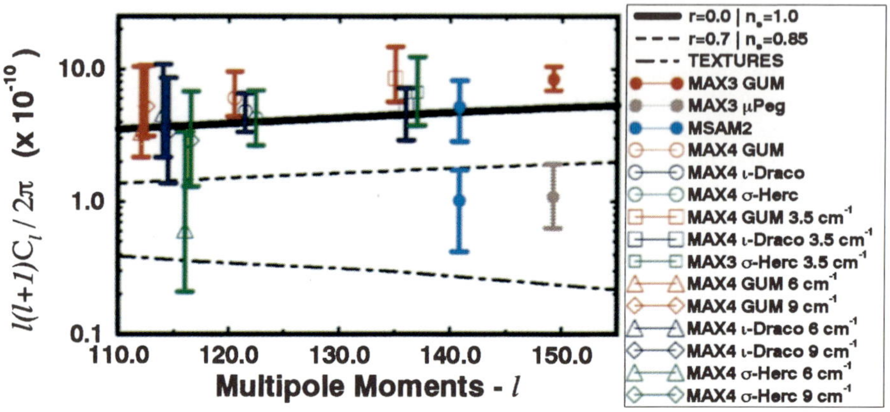

Figure 8: Blow-up of the power spectrum over a narrow range of ℓ near the left (low ℓ) slope of the first Doppler peak showing recent MAX4 results from the ι-Draconis, GUM (Gamma Ursa Major), and σ-Hercules regions. The two theoretical curves are ($r = 0$, $n_s = 1$, solid) and ($r = 1$, $n_s = 0.085$, dashed) inflationary/CDM predictions, respectively. The dot-dashed curve shows the prediction for cosmic textures (with reionization). The data are: (a) MAX4 GUM 6 cm^{-1} (triangle) and 9 cm^{-1} (diamond) channels; (b) MAX4 ι-Draconis 6 cm^{-1} (triangle) and 9 cm^{-1} (diamond) channels; (c) MAX4 σ-Hercules 6 cm^{-1} (triangle) and 9 cm^{-1} (diamond) channels; (d) MAX4 average (open circles) over all channels for (left-to-right) GUM, ι-Draconis and σ-Hercules; (e) MAX4 GUM 3.5 cm^{-1} for (left-to-right) GUM, ι-Draconis and σ-Hercules; (f) MSAM2 (2-beam) - the upper point is based on the entire data set, the lower point has unidentified point sources removed; (g) MAX3 GUM; (h) MAX3 μ Pegasus (showing here unidentified residual after dust subtraction).

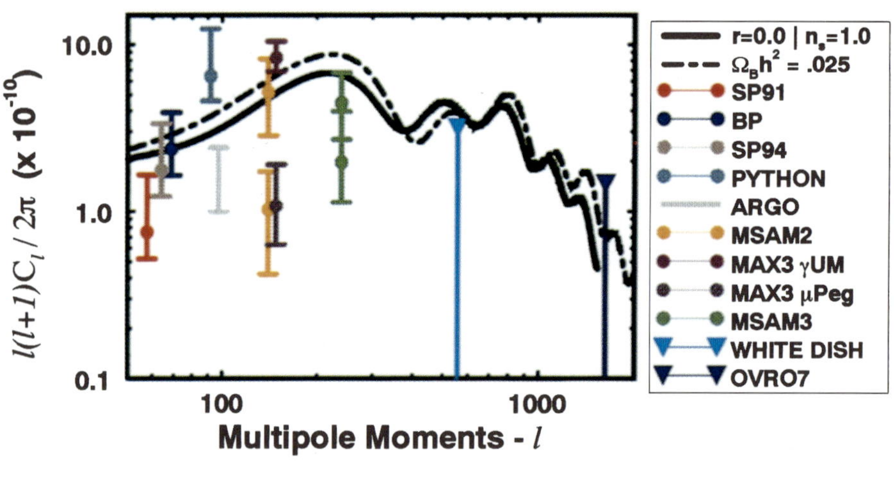

Figure 9: Power-spectrum beginning from $\ell = 50$ spanning all Doppler peaks and the Silk damping region with experimental detections superimposed. Error bars represent one-sigma detections and triangles represent 95% confidence upper limits. The two curves both correspond to pure scalar ($r = 0$), $n_s = 1$ spectra with $h = 0.5$. The solid curve corresponds to the standard value $\Omega_B h^2 = 0.0125$ and the dot-dashed curve corresponds to $\Omega_B h^2 = 0.025$. Note that the second Doppler peak is suppressed relative to the first and second as $\Omega_B h^2$ increases. The experiments correspond to: (a) South Pole 1991; (b) South Pole 1994; (c) Big Plate 1993-4; (d) PYTHON; (e) ARGO; (f) MSAM (2-beam) - the upper point uses the entire data set, the lower point has unidentified point sources removed; (g) MAX3 (GUM region); (h) MAX3 (μPegasus region) showing here unidentified residual after dust subtraction; and (i) MSAM (3-beam) - the upper point uses the entire data set, the lower point has unidentified point sources removed;. (j) White Dish; and (k) OVRO7.

els generally predict a signal considerably less than the Doppler peak predicted for $n_s = 1$ inflationary models, although precise calculations of the anisotropy are not yet available for cosmic defect models with standard recombination.

Milestone 4: Observations of Second and Higher Order Doppler Peaks
If the second Doppler peak can be resolved to near cosmic variance uncertainty, additional constraints on $\Omega_B h^2$ can be extracted. For the standard value $\Omega_B h^2 = 0.0125$, there is a sizable second peak, but this peak is increasingly suppressed as $\Omega_B h^2$ increases (see Fig. 9 and discussion in Section 6). Also, the second and higher order Doppler peaks probe sufficiently small scales to test whether the dark matter is cold or a mixture of hot and cold.

An increasing challenge for observers as the range proceeds to smaller angular scales is foreground subtraction from sources and from the Sunyaev-Zel'dovich effect. Fig. 9 illustrates the few experimental results that presently span this range.

Milestone 5: Observations of Damping
If experiments successfully trace the predicted power spectrum through the second Doppler peak, there is already overwhelming evidence in favor of inflation and dark matter models of large-scale structure formation. Observation of damping at very small angular scales ($\lesssim 5'$) due to photon diffusion (Silk damping [31]) and interference through the thickness of the last-scattering surface is corroboration of more subtle effects on the cosmic microwave background [22]. If there was significant reionization (which can already be determined by experiments at larger angular scales), then experiments probing this range might find evidence for secondary perturbations. Fig. 9 illustrates this range and the present limit from OVRO7.

§8 Cosmic Confusion??

The CMB anisotropy power spectrum entails thousands of C_ℓ's and depends only upon a handful of parameters, (n_s, n_t, r, h, Ω_B, Ω_Λ, ...). One may have hoped that measurements of the power spectrum would be able to test inflation and resolve independently each of the parameters. Unfortunately, it is possible to continuously vary certain combinations of the cosmological parameters without significantly changing the power spectrum. We refer to this "degeneracy" as *cosmic confusion*.[48, 26] It means that CMB anisotropy experiments can localize a hypersurface in cosmic parameter space but the likelihood along that hypersurface will hardly vary. This confounds efforts to separately determine the values of cosmological parameters.

The degree of cosmic confusion depends on the range in ℓ over which the anisotropy spectrum is measured and the precision of the measurement. It seems possible that measurements over the next decade will determine the spectrum from small ℓ through the first Doppler peak $\ell \sim 300$ with an error comparable to cosmic variance. In this section, we will assume that these observations can be made and show that cosmic confusion is a significant problem. If precise observations are restricted to this limited range of multipoles, other cosmic observations must be combined with CMB anisotropy measurements to break the degeneracy in fitting cosmological parameters. The degree of cosmic confusion can be reduced if even more of the power spectrum can be determined precisely through the second Doppler peak ($l \sim 500$ or $\sim$10 arcminute scales). However, mapping the full sky at such high resolution is an extraordinary challenge and, even if one succeeds, it is possible that foregrounds will obstruct measurements at these small angular scales. Hence, it seems reasonable, at least for the next decade, to consider the more conservative case in which precise measurements are made only through the first Doppler peak.

It should be emphasized at the outset that extraordinarily valuable information can be gained from microwave background measurements in spite of cosmic confusion. For example, as discussed under Milestone 2, it is possible to test with high precision whether the large-scale spectrum is scale-free and measure the spectral slope. As shown in Fig. 6, there is negligible interference due to uncertainties in other cosmological parameters. Hence, it is certainly possible to test unambiguously some of the key predictions of inflation.

However, one may have hoped for more. Some may argue that Harrison and Zel'dovich made the case for a scale-invariant spectrum without invoking inflation [8]. In this case, verifying the additional relations Eqs. (21-23), which have no natural explanation from Harrison and Zel'dovich's point-of-view, would be significant, added support for inflation (and there is the obvious converse). Also, it would have been a tremendous boon if other cosmological parameters could be unambiguously determined. This is where cosmic confusion disappoints. All is not lost, though. In some cases (see, for example, the comments at the end of this section), the degree of confusion is minor. Even in the worst cases, powerful relations result when CMB measurements are combined with other astrophysical measurements.

As an illustration of cosmic confusion, consider a baseline (solid line) spectrum ($r = 0|n_s = 1, \mathrm{h} = 0.5, \Omega_\Lambda = 0$). Increasing Ω_Λ (or decreasing h) enhances small-angular scale anisotropy by reducing the red shift $z_{\rm eq}$ at which radiation-matter equality occurs. CMB anisotropy experiments can determine either $r|n_s$, Ω_Λ, or h quite accurately if the other two parameters are known [26]. How-

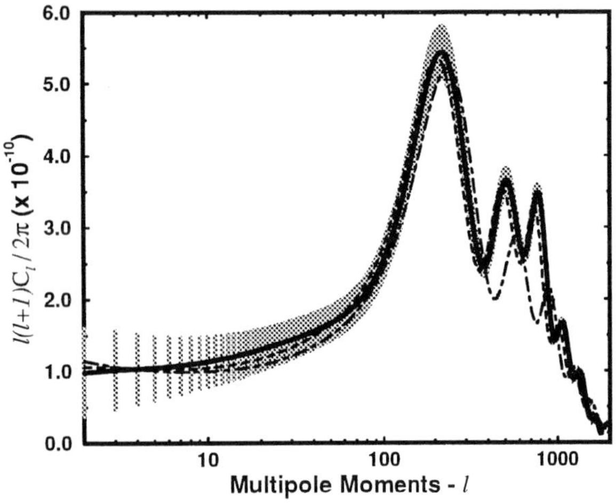

Figure 10: Examples of different cosmologies with nearly identical power spectra and band powers. The solid curve is ($r = 0|n_s = 1, h = 0.5, \Omega_\Lambda = 0$). The other two curves explore degeneracies in the ($r = 0|n_s = 1, \mathrm{h}, \Omega_\Lambda$) and ($r|n_s, h = 0.5, \Omega_\Lambda$) planes. In the dashed curve, increasing Ω_Λ is almost exactly compensated by increasing h. For the dot-dashed curve, the effect of changing to $r = 0.42|n_s = 0.94$ is nearly compensated by increasing Ω_Λ to 0.6. Restricting attention to the range from $\ell = 2$ through the first Doppler peak, one finds that the curves are difficult to distinguish, especially given cosmic variance (vertical hashing is one-sigma variance). If precision measurements of the second Doppler peak or beyond can be made, it should be possible to distinguish some of the models.

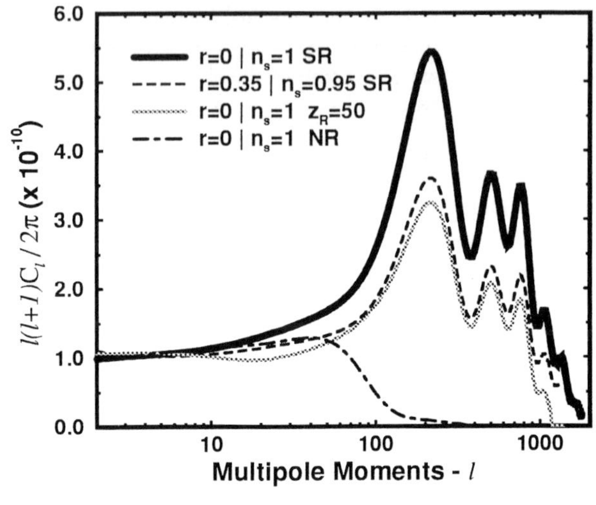

Figure 11: Power spectra for models with standard recombination (SR), no recombination (NR), and 'late' reionization (LR) at $z = 50$. In all models, $h = 0.5$ and $\Omega_\Lambda = 0$. NR or reionization at $z \geq 150$ results in substantial suppression at $\ell \geq 100$. Models with reionization at $20 \leq z \leq 150$ give moderate suppression that can mimic decreasing n_s or increasing h; for example, compare the $n_s = 0.95$ spectrum with SR (black, dashed) to the $n_s = 1$ spectrum with reionization at $z = 50$ (grey solid).

ever, cosmic confusion arises if $r|n_s$, Ω_Λ and h vary simultaneously. Fig. 10 shows spectra for models lying in a two-dimensional surface of $(r|n_s, \mathrm{h}, \Omega_\Lambda)$ which produce nearly identical spectra. In one case, $r|n_s$ is fixed, and increasing Ω_Λ is nearly compensated by increasing h. In the second case, h is fixed, but increasing Ω_Λ is nearly compensated by decreasing n_s.

Further cosmic confusion arises if we consider ionization history. Let z_R be the red shift at which we suppose sudden, total reionization of the intergalactic medium. Fig. 11 compares spectra with standard recombination (SR), no recombination (NR) and late reionization (LR) at $z_R = 50$, where $h = 0.5$ and $\Omega_\Lambda = 0$. Reionization or no recombination suppresses the small angular scale anisotropy, which can be confused with a decrease in n_s (see figure). Inflation-inspired models, *e.g.*, cold dark matter models, are likely to have negligibly small z_R.

The results can be epitomized by some simple rules-of-thumb: Over the $30' - 2^\circ$ range, the C_ℓ's for fixed ℓ are roughly proportional to the maximum of C_ℓ at the top of the first Doppler peak. The maximum (corresponding to $\sim .5^\circ$ scales) is exponentially sensitive to n_s. Since scalar fluctuations account for the Doppler peak, the maximum decreases as the fraction of tensor fluctuations (or r) increases. The maximum is also sensitive to the red shift at matter-radiation equality (or, equivalently, $(1-\Omega_\Lambda)h^2$), and to the optical depth at last scattering for late-reionization models, $\sim z_R^{3/2}$. These observations are the basis of an empirical formula (accurate to $\lesssim 15\%$)

$$\left.\frac{C_\ell}{\langle C_\ell \rangle_{dmr}}\right|_{max} \approx A\, e^{B\, \tilde{n}_s}\,, \tag{27}$$

with $A = 0.15$, $B = 3.56$, and

$$\begin{aligned}\tilde{n}_s \quad \approx \quad & n_s - 0.28 \log(1 + 0.8r) \\ & -0.52[(1-\Omega_\Lambda)h^2]^{\frac{1}{2}} - 0.00036\, z_R^{3/2} + .26\,,\end{aligned} \tag{28}$$

where r and n_s are related by Eq. (23) for generic inflation models. ($\tilde{n}_s$ has been defined such that $\tilde{n}_s = n_s$ for $r = 0$, $h = 0.5$, $\Omega_\Lambda = 0$, and $z_R = 0$.) Hence, the predicted anisotropy for any experiment in the range $10'$ and larger is not separately dependent on n_s, r, Ω_Λ, etc.; rather, it is function of the combination $\tilde{n}_s$.

Eq. (27) implies that the CMB anisotropy measurements are exponentially sensitive to $\tilde{n}_s$. Hence, we envisage that $\tilde{n}_s$ will be accurately determined in the foreseeable future. Then, Eq. (28) implies that the values of the cosmological parameters are constrained to a surface in parameter-space. Cosmological models corresponding to any point on this surface yield indistinguishable CMB anisotropy power spectra (thru the first Doppler peak). To determine which point on the surface corresponds to our Universe requires other astrophysical measurements. For example, limits on the age of the Universe from globular clusters, on h from Tully-Fisher techniques, on n_s from galaxy and cluster counts, and on Λ from gravitational lenses all reduce the range of viable parameter space to a considerable degree. It is by this combination of measurements that the CMB power spectrum can develop into a high precision test of cosmological models [21, 49].

The difficulty posed by cosmic confusion depends on which way the experimental results break. We have already stated that cosmic confusion can be lifted if precise measurements can be made through the second Doppler peak and beyond. But it should also be noted that, even if we are limited to the first Doppler peak only, there are situations in which cosmic confusion is significantly reduced. Consider our baseline spectrum, $(r = 0|n_s = 1, \mathrm{h} = 0.5, \Omega_\Lambda = 0)$. If measurements indicate a first Doppler peak which is significantly below the baseline value, then there are numerous effects which might be responsible: tilt ($n_s < 1$), increased gravitational wave contribution (r), increased Hubble parameter ($h > 0.5$), decreased $\Omega_B < 0.05$, reionization, or perhaps cosmic defects. On the other hand, if measurements indicate a first Doppler peak that is significantly higher than the baseline value, the causative effects might be: upward tilt ($n_s > 1$), cosmological constant $\Omega_\Lambda > 0$, decreased Hubble parameter ($h < 0.5$), or increased $\Omega_B > 0.05$. Of these four possibilities, the first three are extremely unattractive due to theoretical and observational constraints; the data strongly suggest $\Omega_B > 0.05$ with all other parameters held at their original values. The degree of real cosmic confusion is considerably less in the second case. Since several of the present experiments (including MAX) suggest anisotropy somewhat greater than the baseline value, these comments are especially timely.

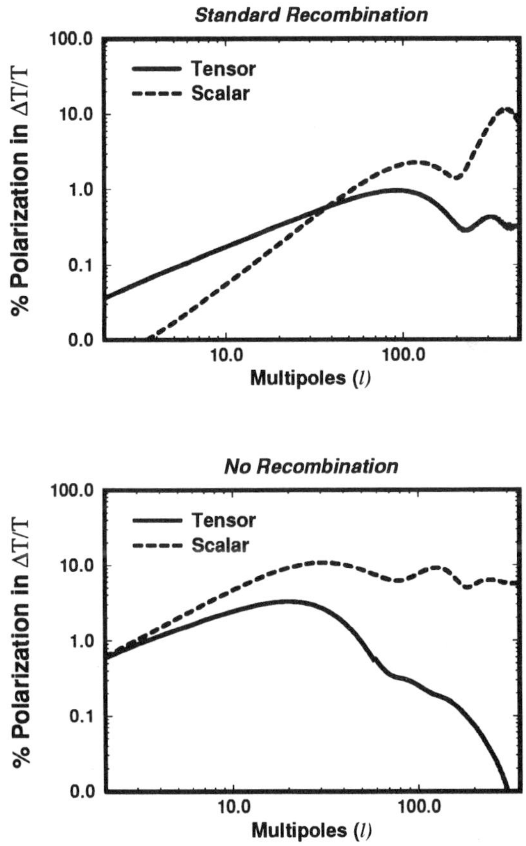

Figure 12: The percentage polarization in $\Delta T/T$ versus multipole moment ℓ predicted for an inflationary model with $n_s = 0.85$, $h = 0.5$, cold dark matter, and standard recombination. (For this value of n_s, inflation predicts equal scalar and tensor contributions to the unpolarized quadrupole.) The upper panel represents the prediction for standard recombination and the lower panel is for a model with no recombination.

§9 Polarization

In the discussion thus far, we have focused on what can be learned from the CMB anisotropy measurements based on the power spectrum only. The power spectrum represents only the two-point temperature correlation function. From a CMB anisotropy map, one can hope to measure three- and higher-point correlation functions, for example, to test for non-gaussianity of the primordial spectrum. Another conceivable test is the CMB polarization. Measurements of the polarized temperature autocorrelation function [50] and of the cross-correlation between polarized and unpolarized anisotropy [51] are further tests of inflation.

Calculations for realistic parameters, though, suggest that the polarization is unlikely to be detected or to provide particularly useful tests of cosmological parameters [52]. For example, it had been hoped that large-scale (small ℓ) polarization measurements would be useful for discriminating scalar and tensor contributions to the CMB anisotropy, thereby measuring r. In the upper panel of Fig. 12, we show the percentage polarization (in $\Delta T/T$) for scalar and tensor modes for a model with $r = 1$ and $n_s = .85$, an example where there are equal tensor and scalar contributions to the quadrupole moment. The figure shows that, indeed, there is a dramatically different polarization expected for scalar versus tensor modes for small ℓ. However, the magnitude of the polarization is less than 0.1%, probably too small to be detected in the foreseeable future. On scales less than one degree ($\ell > 100$), the total polarization rises and approaches 10%, a more plausible target for detection. However, the tensor contribution on these angular scales is negligible, so detection does not permit us to distinguish tensor and scalar modes. In fact, the predictions are relatively insensitive to the cosmological model, a notable exception being the reionization history. The lower panel of Fig. 12 illustrates the prediction for a model with no recombination. The overall level of polarization is increased. The tensor contribution is suppressed relative to scalar, so polarization remains a poor method of measuring r. However, the polarization at angular scales of a few degrees ($\ell \approx 50$) rises to nearly 5%, perhaps sufficient for detection. An observation of polarization at these angular scales would be consistent with a non-standard reionization history.

Not only is the predicted polarization small in all cases, but little is known about what the foreground contamination will be. At a minimum, the foreground from dust, synchrotron, etc., interferes with measurements at the 1% level [53]. Polarization measurements can provide useful corroborating evidence or surprises. For example, a sizable polarization (20%, say) would be unexpected in all models. However, if inflation is correct, then (unpolarized) anisotropy, rather than polarization, appears to be the most useful discriminant for the foreseeable future.

§10 Conclusions

The next decade is sure to be a historic period in the endeavor to understand the origin and evolution of the Universe. In this article, we have focused only on measurements of the cosmic microwave background. We have shown how these measurements alone will allow us to test the inflationary hypothesis and place new constraints on almost all cosmological parameters. Table 2 below summarizes the specific sequence of five milestones in CMB anisotropy experiments that need to be achieved to accomplish these goals, the range of multipole moments (C_ℓ) that need to be probed, and which aspects of inflationary predictions is tested at each milestone. Passing all milestones is overwhelming support for inflation; failing to pass one or more milestones is either invalidation or, at least, indication of some significant, additional surprise.

The ability to separately resolve inflationary and other cosmological parameters from microwave background measurements alone is limited by cosmic confusion, the phenomenon that different choices of cosmological parameters result in virtually identical CMB anisotropy spectra. Confusion is not a significant problem if the observed anisotropy at the first Doppler peak ($\sim 0.5°$) is found to be large or if precise measurements can be extended to the second Doppler peak and beyond (see remarks at the end of Section 8). However, it is not clear whether either of these conditions will be met. Even in that case, it is still possible to use the CMB anisotropy measurements as a powerful tool for determining cosmological parameters, but only after combining CMB results with other known astrophysical constraints.

Another extraordinary series of efforts during this same decade entails measurements of large-scale distribution and velocities of galaxies. These observations probe the distribution of matter on scales generally smaller than but overlapping the range probed by microwave background experiment. The ultimate goal is to find a simple, intuitive theory which joins together the microwave background and large-scale structure observations.

The present situation is summarized in Fig. 13, in which the predictions of dark matter models (in linear approximation only) are superimposed [54]. (Non-linear corrections have been computed for some models via numerical simulation.) At present, the cold dark matter models are the simplest from a theoretical point-of-view, but predict too much power on small scales when normalized to COBE DMR. Adding new dark matter components, as in mixed dark matter models, leads to a better fit, but at the cost of introducing the special condition of nearly equal hot and cold matter densities that is difficult to explain from microphysics. Tilting the spectrum to reduce power at small scales is naturally incorporated into inflation, but the numerical simulations do not fit large-scale structure as successfully. An interesting puzzle appears to be brewing.

Microwave background and large-scale structure observations are dramatic examples of the transformation of cosmology from metaphysics to hard science. These, combined with the more classical cosmological measurements of h, Ω, and Λ, are rapidly evolving into highly precise tools for testing theories and measuring fundamental parameters. As the new millennium commences, cosmological science stands at a critical crossroads: we do not know whether the Universe can be explained solely on the ba-

Milestone	Range of ℓ	What It Tests
1. Large Scale Fluctuations	$2 \lesssim \ell \lesssim 30$	Spectral Amplitude
2. Plateau at Intermediate Scales	$10 \lesssim \ell \lesssim 100$	Spectral shape/slope
3. First Doppler Peak	$100 \lesssim \ell \lesssim 300$	
a. Value of ℓ at the maximum		Flatness
b. Height		Constraints on h, Ω_B, Ω_Λ and reionization?
4. Second & Higher Doppler Peaks	$300 \lesssim \ell \lesssim 800$	Constraints on $\Omega_B h^2$, CDM vs. MDM
5. Damping	$\ell \gtrsim 1000$	Silk Effect, Cosmo. parameters?

Table 2: Five Milestone Tests of Inflation and Dark Matter Models of Large-scale Structure

sis of physical laws that can be tested locally, or whether knowledge of initial conditions or new physics beyond our grasp is required. Evidence that the Universe is scale-free (*e.g.*, spatially flat, scale-free primordial spectrum of fluctuations) is consistent with the notion that the physical laws governing the Universe involve microphysical scales only, as suggested by present understanding. Evidence that the Universe is measurably open (or closed) or that there are large-scale features in the primordial spectrum suggests initial conditions or coincidences or new physical laws that may not be probed by other means. The series of measurements anticipated in the next decade may determine the future path of cosmology, and, thereby, the ultimate limitations on human comprehension of the Universe.

§Acknowledgements

Many results in this paper are drawn from collaborations with R. Crittenden, J. R. Bond, R. Davis, G. Efstathiou, H. Hodges, G. Smoot, and M. S. Turner. I thank P. Lubin, P. Meinhold and J. Gundersen for showing the South Pole 1994 results prior to publication and discussions on MAX data, and K. Gorski for providing Fig. 7. I thank

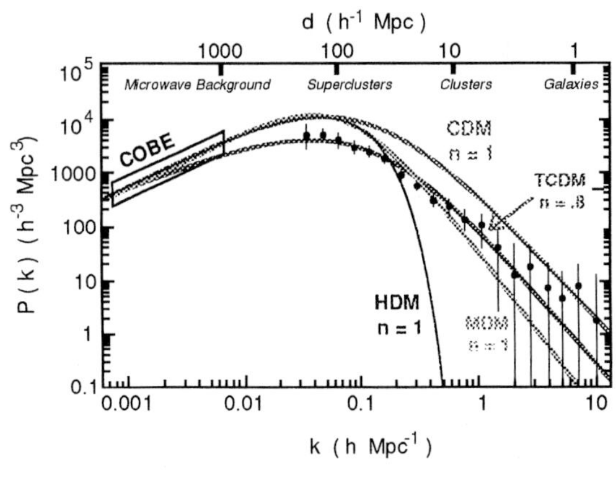

Figure 13: Comparison of matter density power spectra for cold dark matter (CDM) tilted cold dark matter (TCDM), hot dark matter (HDM) and mixed cold and hot dark matter models (MDM) of largescale structure formation. All theoretical curves are normalized to COBE DMR and include only linear approximation. (Non-linear corrections become important at small angular scales ($\lesssim$ 10 Mpc).

the Snowmass Workshop organizers, Roberto Peccei and Rocky Kolb, and the participants in the Cosmology working group for their interest, encouragement and support. I also thank Cynthia Sazama and her staff for their outstanding assistance (and the miraculous weather at the final, mountaintop banquet) . This research was supported by the DOE at Penn (DOE-EY-76-C-02-3071).

Appendix: Flat Band Power Estimates

The observational results reported in Figs. 4, 5, 8 and 9 represent "flat band power estimates" derived for each experiment. In this appendix, we explain the reasoning behind band power estimation; we briefly discuss the particular cases of GACF (gaussian auto-correlation function) and flat band power estimation, and approximate methods for converting from one estimate to another.

Each experiment, depending on the geometry, scanning strategy and detectors, has different sensitivity to the C_ℓ's which can be expressed through a filter function [43], W_ℓ. A number which can be directly extracted from experiment is the rms variation in temperature,

$$(\Delta T/T)^2_{rms} \equiv \sum_\ell \frac{\ell+\frac{1}{2}}{\ell(\ell+1)}\left[\frac{1}{2\pi}\ell(\ell+1)C_\ell^{sky}\right] W_\ell; \quad (29)$$

the square brackets enclose the expression for the power spectrum of the real sky described multipole moments C_ℓ^{sky}. It is difficult to compare $(\Delta T/T)_{rms}$ directly with theoretical predictions for three reasons: (1) the value of $(\Delta T/T)_{rms}$ is dependent upon the normalization and shape of the filter function, W_ℓ; (2) a significant fraction of $(\Delta T/T)_{rms}$ is noise; and, (3) the value of $(\Delta T/T)_{rms}$ itself gives no information about the functional form of the real sky power spectrum (that is, how $\ell(\ell+1)C_\ell^{sky}$ varies with ℓ).

A first improvement is to compute the rms band power, defined by

$$\frac{1}{2\pi}\bar{\ell}(\bar{\ell}+1)C_{\bar{\ell}} \equiv \frac{(\Delta T/T)^2_{rms}}{\sum_\ell \frac{\ell+\frac{1}{2}}{\ell(\ell+1)} W_\ell}, \quad (30)$$

where

$$\bar{\ell} \equiv \frac{\sum_\ell \frac{\ell+\frac{1}{2}}{\ell+1} W_\ell}{\sum_\ell \frac{\ell+\frac{1}{2}}{\ell(\ell+1)} W_\ell} \quad (31)$$

is the mean value of ℓ over the filter function band. (Some authors also delineate span of ℓ corresponding to the half-width of the filter function by a horizontal error bar;[44, 45] we have not do so in the interests of clarity.) Band power is independent of the normalization of W_ℓ, and only weakly dependent on its shape. However, problems (2) and (3) remain.

A second step is to introduce "$\langle$*functionalform*$\rangle$ band power estimation," a fit to the data assuming some particular functional form for the power spectrum. The assumed functional form could be the prediction of some specific theoretical model, such as the spectra shown in Fig. 2; however, there are too many weakly constrained parameters in these models at present. Hence, a less biased approach for now is to assume a simpler form, such as gaussian (GACF) or flat, which depends on only a few free parameters. A maximum likelihood fit of the data to the functional form fixes the best-fit choices of the parameters. The $\langle$*functionalform*$\rangle$ band power estimate is then:

$$\frac{1}{2\pi}\bar{\ell}(\bar{\ell}+1)C_{\bar{\ell}} \equiv \sum_\ell \frac{\ell+\frac{1}{2}}{\ell(\ell+1)}\left[\frac{1}{2\pi}\ell(\ell+1)C_\ell^{bf}\right] W_\ell, \quad (32)$$

where $\ell(\ell+1)C_\ell^{bf}/2\pi$ has the assumed functional form with free parameters fixed at the best-fit values. Best-fitting the data to a functional form which resembles the real sky has the advantage that it filters out noise. By including parameters which change the shape of the assumed power spectrum, some shape information can be extracted.

A "flat band power estimate" assumes $\ell(\ell+1)C_\ell^{bf}/2\pi$ is ℓ-independent; the one free parameter is the amplitude. A gaussian autocorrelation function (GACF) estimate assumes that the correlation function is gaussian:

$$\begin{aligned} C(\theta) &= \langle \tfrac{\Delta T}{T}(\mathbf{x})\tfrac{\Delta T}{T}(\mathbf{x}')\rangle \\ &= C(0)\,\tfrac{\theta_c^2}{\theta_c^2+2\sigma^2}\exp\{-\theta^2/[2(\theta_c^2+2\sigma^2)\} \end{aligned}$$

where σ is the FWHM beamwidth of the experiment divided by $\sqrt{8\ln 2}$. This corresponds to a gaussian power spectrum $\ell(\ell+1)C_\ell^{bf}/2\pi$. The spectrum has two free parameters, the correlation angle θ_c and the best-fit amplitude $C(0)$ for the given θ_c. (Some experiments report two-parameter best-fits. Others report single-parameter fits to $C(0)$ for fixed θ_c.)

Ultimately, a likelihood fit over a spectrum of theoretical models is the proper way to determine the best-fit model. However, it is difficult to show widely disparate models and many different parameters within a single likelihood plot. It would be highly desirable if one could somehow find unbiased band power estimates extracted from the experiment that can be compared directly to the theoretical power spectra.

After testing various forms, Bond *et al.* [44, 45] have demonstrated that flat band power estimates are a good choice for direct comparison with theory, and this is what has been applied in Figs. 4, 5, 8 and 9. If the real sky power spectrum is similar to one of the theoretical curves, the flat band power approximation is clearly a better than a gaussian (GACF) one for the plateau region (shown in Fig. 4); flat band power estimatation works surprisingly well for the large ℓ's as well. For example, if one computes flat band power estimates assuming the theoretical predictions, the estimates closely hug the theoretical curves even through the Doppler peaks [45].

As described above, the proper method of obtaining the flat band power estimate is through a full, likelihood fit to the data assuming the flat spectrum. At this point, though, many experimental groups only publish a GACF analysis. The following five easy steps are an approximate conversion formula works well for most experiments. From a reported θ_c and $\Delta T/T_{GACF}$:

1. Compute $\ell_c \equiv 1/[2\sin(\theta_c/2)]$.

2. Compute the filter function W_ℓ for the given experiment.

3. Compute

$$N \equiv \sum_\ell \frac{\ell+\frac{1}{2}}{\ell(\ell+1)}\left(\frac{\ell+\frac{1}{2}}{\ell_c+\frac{1}{2}}\right)^2 \exp\left[-\frac{1}{2}\left(\frac{\ell+\frac{1}{2}}{\ell_c+\frac{1}{2}}\right)^2\right] W_\ell$$

4. Compute

$$D = \sum_\ell \frac{\ell+\frac{1}{2}}{\ell(\ell+1)} W_\ell$$

5. Then the flat band power estimate for $\ell(\ell+1)C_\ell$ is

$$\frac{\bar{\ell}(\bar{\ell}+1)C_{\bar{\ell}}}{2\pi} = \left(\frac{\Delta T}{T}\right)^2_{GACF} \frac{N}{D}$$

§References

[1] A. H. Guth, *Phys. Rev.* *D*23, 347, (1981).

[2] A. D. Linde, *Phys. Lett.* **108B**, 389 (1982); A. Albrecht and P. J. Steinhardt, *Phys. Rev. Lett.* **48**, 1220 (1982).

[3] A. H. Guth and P. J. Steinhardt, "The Inflationary Universe" in **The New Physics**, ed. by P. Davies, (Cambridge U. Press, Cambridge, 1989) pp. 34-60.

[4] A. D. Linde, **Particle physics and inflationary cosmology**, (New York : Harwood Academic Publishers, 1990).

[5] J. Bardeen, P. J. Steinhardt and M. S. Turner, *Phys. Rev.* *D*28, 679 (1983); A. H. Guth and S.-Y. Pi, *Phys. Rev. Lett.* **49**, 1110 (1982); A. A. Starobinskii, *Phys. Lett.* *B*117, 175 (1982); S. W. Hawking, *Phys. Lett.* *B*115, 295 (1982).

[6] A. Albrecht, P. J. Steinhardt, M.S. Turner, and F. Wilczek, *Phys. Rev. Lett.* 48, 1437 (1982).

[7] L. Kofman, A. Linde, A. A. Starobinsky, submitted to *Phys. Rev. Lett.*, preprint UH-IfA-94/35 (1994).

[8] E.R. Harrison, *Phys. Rev.* *D*1, 2726 (1970); Ya.B. Zel'dovich, *Mon. Not. R. astr. Soc.* **160**, 1p (1972).

[9] V.A. Rubakov, M.V. Sazhin, A.V. Veryaskin, *Phys. Lett.* *B*115, 189 (1982).

[10] A.I. Starobinskii, *Sov. Astron. Lett.* **11**, 133 (1985).

[11] L.F. Abbott M. Wise, *Nucl. Phys. B* **244**, 541 (1984).

[12] R.L. Davis, H.M. Hodges, G.F. Smoot, P.J. Steinhardt, and M.S. Turner, *Phys. Rev. Lett.* **69**, 1856 (1992).

[13] F. Lucchin, S. Matarrese, and S. Mollerach, *Ap. J. Lett.* **401**, 49 (1992); D. Salopek *Phys. Rev. Lett.* **69**, 3602 (1992); A. Liddle and D. Lyth, *Phys.Lett.B* **291**, 391 (1992); V. Sahni and T. Souradeep, *Mod. Phys. Lett.* **A7**, 3541i (1992); J.E. Lidsey and P. Coles, *Mon. Not. Roy. astr. Soc.* **258**, 57P (1992); L. Krauss and M. White, *Phys. Rev. Lett.***69**, 869 (1992).

[14] R. Crittenden and P.J. Steinhardt, *Phys. Lett.* *B*293, 32 (1992).

[15] H.M. Hodges, G.R. Blumenthal, L.A. Kofman and J.R. Primack, *Nucl. Phys.* *B*335, 197 (1990).

[16] A. D. Linde, *Phys. Rev.* *D*49, 748 (1994).

[17] C.L. Bennett, *et al.*, to appear (1994).

[18] K. M. Gorski, *et al.*, *Ap. J.* **430**, L89 (1994). (1994).

[19] J.R. Bond, submitted to *Phys. Rev. Lett.* (1994).

[20] M.S. Turner, *Phys. Rev.* *D*48, 5539 (1993); E.J. Copeland, E.W. Kolb, A.R. Liddle, J.E. Lidsey, *Phys. Rev.* *D*48, 2529 (1993); *Phys. Rev.* *D*49, 1840 (1994); A.R. Liddle and M.S. Turner, *Phys. Rev.* *D*50, 758 (1994).

[21] R. Crittenden, J.R. Bond, R.L. Davis., G. Efstathiou, and P. J. Steinhardt, *Phys. Rev. Lett.***71**, 324 (1993).

[22] J.R. Bond, in *The Early Universe*, ed. W. Unruh and G. Semenoff, Dordrecht: Reidel (1988); Efstathiou, G., in *Physics of the Early Universe*ed. Heavens, A., Davies, A., and Peacock, A., Scottish Universities Summer School Publication, pp. 361-463 (1990).

[23] J.R. Bond and G. Efstathiou, *Ap. J.***285** L45 (1984); *Mon. Not. R. astr. Soc.* **226**, 655 (1987).

[24] E.W. Kolb and M. S. Turner, *The Early Universe*, (Redwood City: Addison-Wesley, 1990).

[25] See, for example, D. Maoz and H.-W. Rix, *Ap. J* **415**, L18 (1993).

[26] J.R. Bond, R. Crittenden, J.R. Bond, R.L. Davis., G. Efstathiou, and P. J. Steinhardt, *Phys. Rev. Lett.* **72**, 13 (1994).

[27] R. Sachs and A.M. Wolfe, *Ap. J.* **147**, 73 (1967).

[28] M. Kamionkowski, D.N. Spergel, N. Sugiyama, submitted to *Ap. J. Lett.*, IAS preprint, IASSNS-HEP-93/91 (1993).

[29] W. Hu and N. Sugiyama, preprint CfPA-TH-94-55, submitted to *Phys. Rev.***D** (1994).

[30] U. Seljak, preprint MIT-CSR-94-13 (1994), submitted to *Ap. J. Lett.*.

[31] J. Silk, *Ap. J.* **151**, 459 (1968).

[32] G.F. Smoot *et al*, *Ap. J. Lett.***396**, L1 (1992); C. L. Bennett *et al*, submitted to *Ap. J. Lett.* (1994).

[33] K. Ganga *et al Ap. J.* **410**, L57 (1993).

[34] R.A. Watson *et al*, *Nature* **357**, 660 (1992).

[35] T. Gaier *et al*, *Ap. J. Lett.* **398**, L1 (1992); J. Schuster *et al*, *Ap. J. Lett* **412**, L47 (1993); J. Gundersen *et al*, preprint (1994).

[36] E.J. Wollack, *et al.*, *Ap. J.* **419**, L49 (1993); C.B. Netterfield *et al.*, preprint (1994).

[37] P. de Bernardis *et al.*, *Ap. J.* **422**, L33 (1994).

[38] M. Dragovan *et al*, *Ap. J.* **355**, L67 (1994).

[39] E. S. Cheng *et al*, *Ap. J.* **422**, L37 (1994).

[40] P. Meinhold *et al*, *Ap. J. Lett.* **409**, L1 (1993); J. Gunderson *et al*, *Ap. J. Lett.* **413**, L1 (1993). M. Devlin, *et al.*, *Ap. J.* **430** L1 (1994); A.C. Clapp, *et al.*, submitted to *Ap. J.* (1994).

[41] G.S. Tucker, G.S. Griffin, H. Nguyen, and J.B. Peterson, *Astrophys. J. Lett.* **419**, L45 (1993).

[42] A.C.S. Readhead *et al,Ap. J.* **346** 556, (1989).

[43] J.R. Bond, G. Efstathiou, P. M. Lubin, P. Meinhold, . *Phys. Rev. Lett.* **66**, 2179 (1991).

[44] J.R. Bond, *Ap. Lett. & Comm.* (1994).

[45] P.J. Steinhardt, J.R. Bond, R.L. Davis, *Ap. Lett. & Comm.* (1994).

[46] D. Coulson, P. Ferreira, P. Graham, N. Turok, *Nature*, **368** 27 (1994).

[47] N. Turok, private communication.

[48] P. J. Steinhardt, in **Evolution of the Universe and its Observational Quest**, ed. by K. Sato, (University Academy Press, Inc., Tokyo, 1994) pp. 159-180.

[49] L. Knox and M.S. Turner, FERMILAB-PUB-94-175-A (1994).

[50] A.G. Polnarev, *Sov. Astron.* **29**, 607 (1985).

[51] D. Coulson, R. Crittenden, N. Turok, *Phys.Rev.Lett.* **73**, 2390 (1994); also, preprint PUPT-1514 (1994).

[52] R. Crittenden, R.L. Davis, and P. Steinhardt, *Ap. J.*, L13 (1993).

[53] P. Timbie, private communication.

[54] Adapted from E.W. Kolb and M. S. Turner, *The Early Universe*, (Redwood City: Addison-Wesley, 1990), preface to new edition.

Cosmic Background Radiation

E. L. Wright

UCLA Astronomy, Los Angeles, CA 90024-1562, USA

The Cosmic Background Radiation provides observational evidence about the earliest history of the Universe. The COsmic Background Explorer (*COBE*) has measured the spectrum of this radiation, which is very close to a 2.726 K blackbody, showing that less than 0.01-0.02% of the energy of the CBR has been added at times later than 1 year after the Big Bang. *COBE* also measured the anisotropy of the CBR, which consists of a dipole anisotropy indicating the motion of the Solar System relative to the "cosmic standard of rest", galactic emission, and finally a small (± 30 μK RMS) but very significant intrinsic anisotropy that indicates magnitude of the gravitational forces on very large scales that existed 3×10^5 years after the Big Bang. Since the structures seen by COBE were at least 7×10^6 light years across at this time only 3×10^5 years after the Big Bang, they are larger than *ct*, and are thus either the result of an inflationary epoch at $t < 10^{-12}$ seconds after the Big Bang, or initial conditions set up at $t = 0$.

§1 Introduction

Modern cosmology is based on many different observations of the Universe: the ages of stars and clusters; the Hubble expansion seen in the redshifts of distant galaxies; the spatial distribution and peculiar velocities of galaxies; the density and composition of the Universe; the spectrum and anisotropy of the cosmic microwave background radiation; and finally backgrounds in the infrared, ultraviolet, x-ray and gamma-ray spectral regions. The *COBE*[1] mission was designed to make definitive measurements of the spectral and anisotropy of the microwave background, and to search for a cosmic infrared background. The three scientific instruments on *COBE* are the Far Infrared Absolute Spectrophotometer (FIRAS), the Differential Microwave Radiometers (DMR), and the Diffuse Infrared Background Experiment (DIRBE). The purpose of FIRAS is to make a precision measurement of the spectrum of the CMB from 1 cm to 100 μm. The purpose of DMR is to search for variations in the temperature of the CMB on angular scales larger than 7° at frequencies of 31.5, 53, and 90 GHz. The purpose of DIRBE is to search for a CIB by making absolute brightness measurements of the diffuse infrared radiation in 10 photometric bands from 1 to 240 μm and polarimetric measurements from 1 to 3.5 μm. The FIRAS and DIRBE instruments are located inside a 650 liter superfluid liquid helium dewar. See Boggess *et al.*[1] for a full description of the *COBE* mission.

The proposals that led to the *COBE* mission were submitted in 1974 s a response to AO 6 & 7. *COBE* was launched on 18 November 1989 and operated until 23 December 1993.

The success of the *COBE* mission has been followed by a number of experiments that have extended anisotropy measurements to smaller angular scales. These experiments from balloons, Arctic and Antarctic observing sites suggest that an extensive effort to accurately measure the anisotropy of the cosmic microwave background can provide information about geometry of the Universe and the nature of the dark matter that dominates the mass of the Universe.

While the *COBE* mission was very successful, making the "discovery of the century"[2], one must remember that this work is based on the earlier work (in the 20th century!) of Hubble [3] and Penzias and Wilson[4] who discovered the expansion of the Universe and the microwave background itself. As a consequence of these two discoveries, one knows that the early Universe was hot and dense. When the density and temperature are high, the photon creation and destruction rates are high,

[1] The National Aeronautics and Space Administration/Goddard Space Flight Center (NASA/GSFC) is responsible for the design, development, and operation of the Cosmic Background Explorer (*COBE*). Scientific guidance is provided by the *COBE* Science Working Group. GSFC is also responsible for the development of the analysis software and for the production of the mission data sets.

and are sufficient to guarantee the formation of a good blackbody spectrum. Later, as the Universe expands and cools, the photon creation and destruction rates become much slower than the expansion rate of the Universe, which allow distortions of the spectrum to survive. In the standard model the time from which distortions could survive is 1 year after the Big Bang at a redshift $z \approx 10^{6.4}$.

However, the expansion of the Universe will not produce a distortion by itself, since adiabatically expanding a blackbody radiation field results in another blackbody with a lower temperature. Thus the existence of a distorted spectrum in the hot Big Bang model requires the existence at time later than 1 year after the Big Bang of both an energy source and an emission mechanism that can produce photons that are now in the millimeter spectral range. Conversely, a lack of distortions can be used to place limits on any such energy source, such as decaying neutrinos, dissipation of turbulence, etc.

At a time $10^{5.4}$ years[2] after the Big Bang, at $z \approx 1360$, the temperature has fallen to the point where helium and then hydrogen have 50% (re)combined into transparent gases[5]. The surface of last scattering ($\partial\tau/\partial \ln(1+z) = 1$) occurs later, at $z \approx 1160$ or $10^{5.7}$ years after the Big Bang. The electron scattering which had impeded the free motion of the CMB photons until this epoch is removed, and the photons stream across the Universe. Before recombination, the radiation field at any point was constrained to be nearly isotropic because the rapid scattering scrambled the directions of photons. The radiation field was not required to be homogeneous, because the photons remained approximately fixed in comoving co-ordinates. After recombination, the free streaming of the photons has the effect of averaging the intensity of the microwave background over a region with a size equal to the horizon size. Thus after recombination any inhomogeneity in the microwave background spectrum is smoothed out. Note that this inhomogeneity is not lost: instead, it is converted into anisotropy. When we study the isotropy of the microwave background, we are looking back to the surface of last scattering $\approx 10^{5.5}$ years after the Big Bang. But the hot spots and cold spots we are studying existed as inhomogeneities in the Universe before recombination. Since the 7° beam used by the DMR instrument on *COBE* is larger than the horizon size at recombination, these inhomogeneities cannot be constructed in a causal fashion during the epoch before recombination in the standard Big Bang model. Instead, they must be installed "just so" in the initial conditions. In the inflationary scenario[5] these large scale structures were once smaller than the horizon size during the inflationary epoch, but grew to be much larger than the horizon. Causal physics acting 10^{-35} seconds after the Big Bang can produce the large-scale inhomogeneities studied by the DMR.

[2] With $H_0 = 50$, $\Omega = 1$.

§2 The Spectrum of the Cosmic Microwave Background

The

Far InfraRed Absolute Spectrophotometer instrument on *COBE* is a polarizing Michelson[6] interferometer. The optical layout is very symmetrical, and it has two inputs and two outputs. If the two inputs are denoted SKY and ICAL, then the two outputs, which are denoted LEFT and RIGHT, are given symbolically as

$$LEFT = SKY - ICAL \quad (1)$$
$$RIGHT = ICAL - SKY \quad (2)$$

The FIRAS has achieved its incredible sensitivity to small deviations from a blackbody spectrum by connecting the ICAL input to an internal calibrator, a reference blackbody that can be set to a temperature very close to the temperature T_0 of the sky. Thus this "absolute" spectrophotometer is so successful because it is differential. In addition, each output is further divided by a dichroic beamsplitter into a low frequency channel (2-21 cm^{-1}) and a high frequency channel (23-95 cm^{-1}). Thus there are four overall outputs. These are labeled LL (left low) through RH (right high).

Each output has a large composite bolometer that senses the output power by absorbing the radiation, converting this power into heat, and then detecting the temperature rise of a substrate using a very small and sensitive silicon resistance thermometer. In order to minimize the specific heat of the bolometer, which maximizes the temperature rise for a given input power, the substrate is made of the material with the highest known Debye temperature: diamond. Diamond is quite transparent to millimeter waves, however, so an absorbing layer is needed. Traditionally a layer of bismuth has been used as the absorber on composite bolometers, but the FIRAS detectors use a very thin layer of gold alloyed with chromium. The absorbing layer needs to have a surface impedance about one-half of the 377 Ohms/□ impedance of free space in order to absorb efficiently, and this surface impedance can be obtained with a much thinner layer of gold rather than bismuth. Again, this serves to minimize the specific heat of the detector. In order to detect the longest waves observed by FIRAS, the diameter of the octagonal diamond substrate is quite large: about 8 mm. Each detector sits behind a compound parabolic concentrator, or Winston cone, which provides a full π steradians in the incoming beam. As a result, the FIRAS instrument has a very large étendue of 1.5 cm^2 sr.

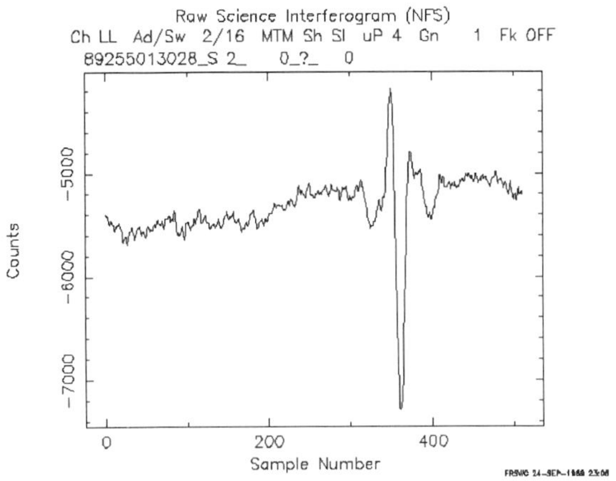

Figure 1: A FIRAS interferogram from the Left Low channel taken during ground test with the XCAL at 2.2 K and the ICAL at 2.7 K

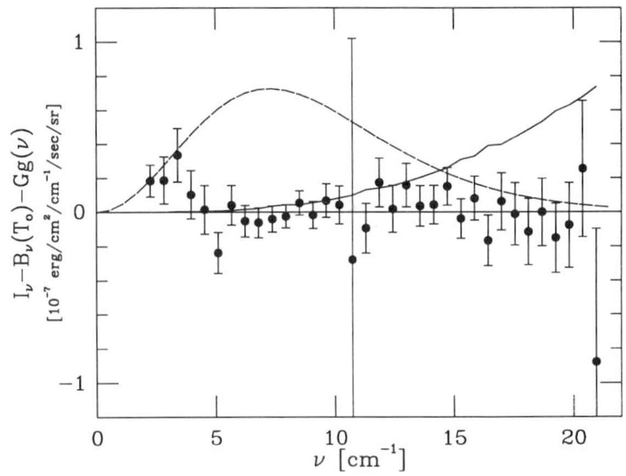

Figure 2: FIRAS residuals (points) after fitting for a blackbody and galactic emission, with an 0.5 mK ΔT and a $\Delta(\csc|b|) = 0.25$ galaxy curve shown for comparison. Full-scale is 0.1% of the peak of the CMB.

Since the FIRAS is a Michelson interferometer, the spectral data is obtained in the form of interferograms. Thus the LEFT output is approximately

$$I_L(x) = \int_0^\infty \cos(2\pi\nu x)G(\nu) \times \left(I_\nu + \sum_i \epsilon_i(\nu)B_\nu(T_i) + U_\nu\right) d\nu \quad (3)$$

The index i above runs over all the components in the FIRAS that had thermometers to measure T_i. These include the ICAL, the reference horn that connects to the ICAL, the sky horn, the bolometer housing, the optical structure of the FIRAS, and the dihedral mirrors that move to provide the variation in path length difference x. The ϵ_i's are the effective emissivities of the various components. The ICAL itself has $\epsilon \approx -1$, while the sky and reference horns have ϵ's of $\pm$ a few percent. The offset term U_ν was observed during flight to be approximately $10^{-5}\exp(-t/\tau)B_\nu(T_U)$ with a time constant τ of two months and a temperature $T_U \approx 15$ K. The SKY input I_ν above can be either the sky or an external calibrator. The XCAL is a movable re-entrant absorber that can be inserted at the top of the sky horn. The combination of sky horn plus XCAL forms a cavity with an absorbtivity > 0.999, and we believe $\gtrsim 0.99999$. With the XCAL inserted during periodic calibration runs, the SKY input is known to be $B_\nu(T_X)$. By varying T_X and the other T_i's, the calibration coefficients have been determined. See Fixsen *et al.*[7] for a more complete description of the calibration procedure. Figure 1 shows an interferogram from ground testing with the ICAL at 2.7 K and the XCAL at 2.2 K.

Once the calibration coefficients are known, the sky data can be analyzed to determine $I_\nu(l,b)$ the intensity of the sky as a function of frequency, galactic longitude and galactic latitude. This is a "data cube". The data from each direction on the sky can be written as a combination of cosmic plus galactic signals:

$$I_\nu(l,b) = e^{-\tau_\nu(l,b,\infty)}\left(B_\nu\left(T_\circ + \Delta T(l,b)\right) + \Delta I_\nu\right) + \int e^{-\tau_\nu(l,b,s)} j_\nu(l,b,s)ds \quad (4)$$

where $\tau_\nu(l,b,s)$ is the optical depth between the Solar system and the point at distance s in the direction (l,b) at frequency ν, ΔI_ν is an isotropic cosmic distortion, and $\Delta T(l,b)$ is the variation of the background temperature around its mean value $T_\circ$. This equation can be simplified because the optical depth of the galactic dust emission is always very small in the millimeter and sub-millimeter bands covered by FIRAS.

$$I_\nu(l,b) \approx B_\nu\left(T_\circ + \Delta T(l,b)\right) + \Delta I_\nu + \int j_\nu(l,b,s)ds \quad (5)$$

Even in the optically thin limit, some restrictive assumptions about the galactic emissivity j_ν are needed, since the galactic intensity $\int j_\nu(l,b,s)ds$ is a function of three variables, just like the observed data. The simplest reasonable model for the galactic emission is[8]: let

$$\int j_\nu(l,b,s)ds = G(l,b)g(\nu). \quad (6)$$

This model assumes that the shape of the galactic spectrum is independent of direction on the sky. It is reasonably successful except that the galactic center region is clearly hotter than the rest of the galaxy. The application of this model proceeds in two steps. The first step assumes that the cosmic distortions vanish, and that an approximation $g_0(\nu)$ to the galactic spectrum is known. A least squares fit over the spectrum in each pixel then gives the maps $\Delta T(l,b)$ and $G(l,b)$. The high frequency channel of FIRAS is used to derive $G(l,b)$ because the galactic emission is strongest there. An alternative way to derive $G(l,b)$ is to smooth the DIRBE map at 240 μm to the FIRAS 7° beam. This DIRBE method has been used in the latest FIRAS spectral results[9, 10, 11]. The second step in the galactic fitting then derives spectra associated with the main components of the millimeter wave sky: the isotropic cosmic background, the dipole anisotropy, and the galactic emission. This fit is done by fitting all the pixels (except for the galactic center region with $|b| < 20$ and $|l| < 40$) at each frequency to the form

$$I_\nu(l,b) = I_0(\nu) + D(\nu)\cos\theta + G(l,b)g(\nu). \qquad (7)$$

The spectra of the anisotropic components derived from this fit applied to the entire mission data set are shown in Figures 2 and 3. Figure 3 shows the residual after the predicted dipole spectrum $\Delta T \partial B_\nu/\partial T$ evaluated at T_0 is subtracted from $D(\nu)$.

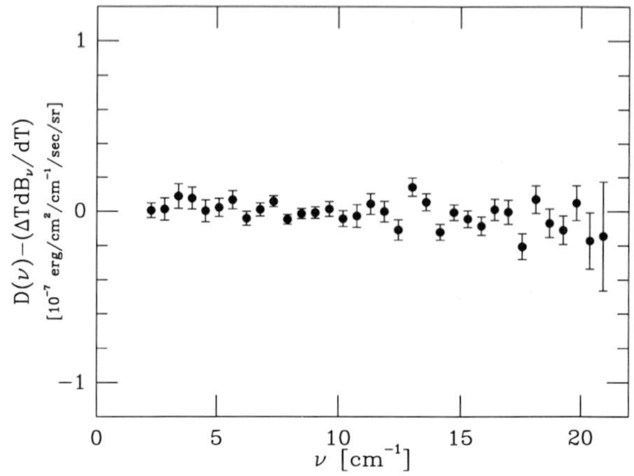

Figure 3: FIRAS Dipole spectrum residuals (points). Full-scale is 0.1% of the peak of the CMB.

The absolute temperature of the cosmic background, T_0, can be determined two ways using FIRAS. The first way is to use the readings of the germanium resistance thermometers in the XCAL when the XCAL temperature is set to match the sky. This gives $T_0 = 2.730$ K. The second way is to measure the frequency of the peak of $\partial B_\nu/\partial T$ by varying T_X by small amounts around the temperature which matches the sky, and then apply the Wien displacement law to convert this frequency into a temperature. This calculation is done automatically by the calibration software, and it gives a value of 2.722 K for T_0. The final adopted value is 2.726 ± 0.010 K (95% confidence), which just splits the difference between the two methods.

When interpreting the implications of the FIRAS spectrum on effects that could distort the microwave background, it is important to remember some basic order-of-magnitude facts. Theoretical analyses of light element abundances gives limits on the current density of baryons through models of Big Bang NucleoSynthesis (BBNS). The ratio of baryons to photons is $10^{10}\eta = (3.5 \pm 0.7)$[12] or $10^{10}\eta = (3.95 \pm 0.25)$[13]. Since the number density of photons is well determined by the FIRAS spectrum

$$N = 8\pi\zeta(3)\Gamma(3)\left(\frac{kT_0}{hc}\right)^3 = 410 \pm 5\ cm^{-3} \qquad (8)$$

the baryon density is well determined: $1.62 \times 10^{-7}\ cm^{-3}$. This gives $\Omega_B h^2 = 0.0144$, where $h = H_0/(100\ km/sec/Mpc)$. Not all of these baryons are protons, however. For a primordial helium abundance of 23.5% by mass, one has 13 protons and 1 α particle in 17 baryons. This gives an electron density of $n_{e,0} = 1.43 \times 10^{-7}\ cm^{-3}$. For a fully ionized Universe, the optical depth for electron scattering over a path length equal to the Hubble radius c/H_0 is $0.00088/h$.

The expansion of the Universe H varies with time and redshift in a way that depends on ratios of the current matter, radiation and vacuum densities to the critical density $\rho_c = 3H^2/8\pi G$. Let these ratios be Ω_m, Ω_r, and Ω_v. Then

$$\begin{aligned} H(z) &= (1+z)H_0[(1-\Omega_m-\Omega_r-\Omega_v) \\ &\quad +\Omega_m(1+z)+\Omega_r(1+z)^2 \\ &\quad +\Omega_v(1+z)^{-2}]^{0.5} \end{aligned} \qquad (9)$$

This equation is not exact for massive neutrinos which shift from being part of Ω_m into part of Ω_r as z increases. The value of $\Omega_r h^2 = (4.16 \pm 0.06) \times 10^{-5}$ is fairly well-determined for the case of three massless neutrinos. Thus if $\Omega_m \approx 1$ now, the Universe becomes radiation dominated for $z > z_{eq} = 2.4h^2 \times 10^4$ when the background temperature was $kT_{eq} = 5.65h^2\ eV$. In the radiation dominated epoch, $H = 0.645(1+z)^2\ km/sec/Mpc$ or $H = 2.09(1+z)^2 \times 10^{-20}\ s^{-1}$.

For early epochs when the Universe was ionized, electron scattering is the dominant mechanism for transferring energy between the radiation field and the matter. As long as the electron temperature is less than about 10^8 K, the effect of electron scattering on the spectrum can be calculated using the Kompaneets equation[14]:

$$\frac{\partial n}{\partial y} = x^{-2}\frac{\partial}{\partial x}\left[x^4\left(n + n^2 + \frac{\partial n}{\partial x}\right)\right] \qquad (10)$$

where n is the number of photons per mode ($n = 1/(e^x - 1)$ for a blackbody), $x = h\nu/kT_e$, and the Kompaneets y is defined by

$$dy = \frac{kT_e}{m_e c^2} n_e \sigma_T c dt. \qquad (11)$$

Thus y is the electron scattering optical depth times the electron temperature in units of the electron rest mass. Note that the electrons are assumed to follow a Maxwellian distribution, but that the photon spectrum is completely arbitrary.

Since the Kompaneets equation is describing electron scattering, which preserves the number of photons, one finds that the y derivative of the photon density N vanishes:

$$\begin{aligned}\frac{\partial N}{\partial y} &\propto \int x^2 \frac{\partial n}{\partial y} dx \\ &= \int \frac{\partial}{\partial x}\left[x^4\left(n + n^2 + \frac{\partial n}{\partial x}\right)\right] dx \\ &= 0 \end{aligned} \qquad (12)$$

The stationary solutions of the equation $\partial n/\partial y = 0$ are the photon distributions in thermal equilibrium with the electrons. Since photons are conserved, the photon number density does not have to agree with the photon number density in a blackbody at the electron temperature. Thus a more general Bose-Einstein thermal distribution is allowed: $n = 1/(\exp(x+\mu) - 1)$. This gives $\partial n/\partial y = 0$ for all μ. Since the Bose-Einstein spectrum is a stationary point of the Kompaneets equation, it is the expected form for distortions produced at epochs when

$$(1+z)\frac{\partial y}{\partial z} = \sigma_T n_{e,\circ} \frac{kT_\circ}{m_e c^2}\frac{c}{H}(1+z)^4 > 1 \qquad (13)$$

For the $\Omega_B h^2$ given by BBNS, this redshift z_y where this inequality is crossed is well within the radiation dominated era, with a value $z_y = 10^{5.1}/\sqrt{70\Omega_B h^2}$.

There are two simple solutions to the Kompaneets equation with non-zero $\partial n/\partial y$ which both give the same spectral distortion. The first simple case occurs when the electron temperature is much higher than the typical photon energy. This is the case for clusters of galaxies today with $T_e \approx 10^8$ K, or for a hot intergalactic medium. Since the variable x in the Kompaneets equation is the photon energy measured in units of the electron thermal energy, x and hence Δx are both very small in this case. Therefore the $\partial n/\partial x$ term in the Kompaneets equation is much larger than the n or n^2 terms which may be dropped. This gives a simpler equation

$$\frac{\partial n}{\partial y} = x^{-2}\frac{\partial}{\partial x}\left[x^4\left(\frac{\partial n}{\partial x}\right)\right] \qquad (14)$$

which was used by Sunyaev and Zeldovich[15]. Note that the powers of x on the right hand side cancel out, so the RHS can be evaluated using $x = h\nu/kT_\gamma$ if the initial photon field is a blackbody with temperature T_γ. The solution is easily evaluated and usually expressed in terms of a frequency dependent brightness temperature:

$$\frac{\partial T}{\partial y} = T_\gamma\left(\frac{x(e^x+1)}{e^x-1} - 4\right). \qquad (15)$$

The second simple case occurs when the initial photon field is a blackbody with a temperature T_γ which is only slightly below the electron temperature. Letting $f = T_e/T_\gamma$, we find that the initial photon field is given by $n = 1/(\exp(fx) - 1)$. Therefore

$$\left(n + n^2 + \frac{\partial n}{\partial x}\right) = \frac{(1-f)\exp(fx)}{(\exp(fx)-1)^2} = (1-f^{-1})\frac{\partial n}{\partial x}. \qquad (16)$$

Thus the distortion has a Sunyaev-Zeldovich shape but is reduced in magnitude by a factor $(1 - T_\gamma/T_e)$. Defining the "distorting" y as

$$dy_D = \frac{k(T_e - T_\gamma)}{m_e c^2} n_e \sigma_T c dt \qquad (17)$$

we find that the final spectrum is given by a frequency-dependent temperature given by

$$T_\nu = T_\circ\left[1 + y_D\left(\frac{x(e^x+1)}{e^x-1} - 4\right) + \ldots\right]. \qquad (18)$$

where $x = h\nu/kT_\circ$. The FIRAS spectrum in Figure 2 shows that $|y_D| < 2.5 \times 10^{-5}$.

The energy density transferred from the hotter electrons to the cooler photons in the y distortion is easily computed. The energy density is given by $U \propto \int x^3 n dx$ so

$$\frac{\partial U}{\partial y_D} = \int x\frac{\partial}{\partial x}\left(x^4\frac{\partial n}{\partial x}\right) dx \qquad (19)$$

which when integrated by parts twice gives

$$\begin{aligned}\frac{\partial U}{\partial y_D} &= -\int\left(x^4\frac{\partial n}{\partial x}\right) dx \\ &= 4\int x^3 n dx = 4U. \end{aligned} \qquad (20)$$

Thus the limit on y_D gives a corresponding limit on energy transfer: $\Delta U/U < 10^{-4}$. Any energy which is transferred into the electrons at redshifts $z > 7$ where the Compton cooling time is less than the Hubble time will be transferred into the photon field and produce a y

distortion. Since there are 10^9 times more photons that any other particles except for the neutrinos, the specific heat of the photon gas is overwhelmingly dominant, and the electrons rapidly cool (in a Compton cooling time) back into equilibrium with the photons. The energy gained by the photons is $\Delta U = 4yaT_\gamma^4$ which must be equal to the energy lost by the electrons and ions: $1.5(n_e+n_i)k\Delta T_e$. Since $y = \sigma_T n_e (kT_e/m_e c^2)c\Delta t$ we find the Compton cooling time

$$t_C = \frac{1.5(1+n_i/n_e)m_e c^2}{4\sigma_T c U_{rad}} = \frac{7.4 \times 10^{19}\ sec}{(1+z)^4} \tag{21}$$

for an ion to electron ratio of 14/15. This becomes equal to the Hubble time $(3.08568 \times 10^{17} sec)/(h(1+z)^{1.5})$ at $(1+z) = 9h^{0.4}$.

At redshifts $z > z_y = 10^5$, there will be enough electron scattering to force the photons into a thermal distribution with a μ distortion instead of a y distortion. However, the normal form for writing a μ distortion does not preserve the photon number density. Thus we should combine the μ distortion with a temperature change to give an effect that preserves photon number. The photon number density change with μ is given by

$$\begin{aligned} N &\propto \int \frac{x^2 dx}{\exp(x+\mu)-1} \\ &= \sum_{k=1}^{\infty} e^{-k\mu} \int x^2 e^{-kx} dx \\ &= 2\sum_{k=1}^{\infty} \frac{e^{-k\mu}}{k^3} \\ &= 2\left(\zeta(3) - \mu\zeta(2) + \ldots\right). \end{aligned} \tag{22}$$

A similar calculation for the energy density shows that

$$U \propto 6\left(\zeta(4) - \mu\zeta(3) + \ldots\right). \tag{23}$$

In order to maintain $N = const$, the temperature of the photon field changes by an amount $\Delta T/T = \mu\zeta(2)/(3\zeta(3))$. Therefore, the energy density change at constant N is

$$\frac{\Delta U}{U} = \left(\frac{4\zeta(2)}{3\zeta(3)} - \frac{\zeta(3)}{\zeta(4)}\right)\mu = 0.714\mu. \tag{24}$$

Thus the FIRAS limit $|\mu| < 3.3 \times 10^{-4}$ implies $\Delta U/U < 2.4 \times 10^{-4}$. The "improved" form of the μ distortion, with a ΔT added to keep N constant, can be given as a frequency dependent brightness temperature:

$$T_\nu = T_\circ \left(1 + \mu\left[\frac{\zeta(2)}{3\zeta(3)} - x^{-1}\right] + \ldots\right). \tag{25}$$

In this form it is clear that a μ distortion has a deficit of low energy photons and a surplus of high energy photons with respect to a blackbody. In this it is like the y distortion, but the crossover frequency is lower.

Finally, at high enough redshift the process of double photon Compton scattering becomes fast enough to produce the extra photons needed to convert a distorted spectrum into a blackbody. Whenever a photon with frequency ν scatters off an electron, there is an impulse $\propto h\nu/c$ transferred to the electron. This corresponds to an acceleration $a \propto h\nu^2/(m_e c)$ for a time interval $\Delta t \propto 1/\nu$. The energy radiated in new photons is thus $\propto e^2 h^2 \nu^3/(m_e^2 c^5)$ which is $\propto \alpha h\nu(h\nu/(m_e c^2))^2$. Since the rate of scatterings per photon is $\propto \Omega_B h^2 (1+z)^3$, the overall rate of new photon creation is $\propto \Omega_B h^2 (1+z)^5$ while the Hubble time is $\propto (1+z)^{-2}$. Proper consideration of the interaction of this photon creation process with the Kompaneets equation shows[16] that the redshift from which $1/e$ of an initial distortion can survive is

$$z_{th} = \frac{4.24 \times 10^5}{[\Omega_B h^2]^{0.4}} \tag{26}$$

which is $z_{th} = 2.3 \times 10^6$ for the BBNS value of $\Omega_B h^2$.

§3 The Anisotropy of the Cosmic Microwave Background

The Differential Microwave Radiometers (DMR) experiment on *COBE* is designed to measure small temperature differences from place to place on the sky. The DMR consists of three separate units, one for each of the three frequencies of 31.5, 53 and 90 GHz. The field of view of each unit consists of two beams that are separated by a 60° angle that is bisected by the spin axis. Each beam has a 7° FWHM. The DMR is only sensitive to the brightness difference between these two beams. This differencing is performed by a ferrite waveguide switch that connects the receiver input to one horn and then the other at a rate of 100 Hz. The signal then goes through a mixer, an IF amplifier and a video detector. The output of the video detector is demodulated by a lock-in amplifier synchronized to the input switch. The difference signal that results is telemetered to the ground every 0.5 seconds. Each radiometer has two channels: A and B. In the case of the 31.5 GHz radiometer, the two channels use a single pair of horns in opposite senses of circular polarization. In the 53 and 90 GHz radiometers there are 4 horns, and all observe the same sense of linear polarization.

The basic problem in the DMR data analysis is to construct a map of the sky using the 6×10^7 differences that are collected each year in each of the six channels. There are only about 10^3 beam areas on the sky, so the problem is highly over-determined. We have chosen to analyze the data using 6,144 pixels to cover the sky. These are approximately equal area, and arranged in a square grid of 32 × 32 pixels on each of the 6 faces of a cube. Within each pixel we assume that the temperature is constant, so we are modeling the sky with a staircase function. Therefore the basic problem can be represented as a least-squares problem with 12×10^7 equations in 6,144 variables. One also has to solve for other terms, such as the magnetic sensitivity of the ferrite switch used in the DMR instrument. One can represent the predicted output of the DMR instrument as

$$S(t) = V(t)X^T \tag{27}$$

where $V(t) = [0, \ldots, 0, +1, 0, \ldots, 0, -1, 0, \ldots, B_x(t), B_y(t), B_z(t)]$ with the +1 in the pixel containing the plus horn line-of-sight at time t, and the -1 in the pixel containing the minus horn LOS, and $B_x(t)$ being the x component of the magnetic field in spacecraft coordinates. I have assumed that a slowly varying baseline has been removed, and that the signals have been converted into milliKelvins. The parameter vector $X = [T_0, \ldots, T_{6143}, M_x, M_y, M_z]$, where T_i is the temperature of the ith pixel in mK, and M is the magnetic sensitivity of the radiometer in mK/gauss. For a constant noise per observation the least squares solution for the map is

$$\begin{aligned} A &= \sum_t V(t)^T V(t) \\ B &= \sum_t S(t) V(t)^T \\ X &= lim_{\epsilon \to 0^+} (A + \epsilon I)^{-1} B \end{aligned} \tag{28}$$

Note that A is *sparse*, *symmetric*, and *singular*. While A has 38×10^6 elements, all but 1.8×10^6 of these elements are zero, and $A_{ij} = A_{ji}$. Thus only 9×10^5 elements need to be kept. The limit $\epsilon \to 0^+$ takes care of the singularity of A, and gives a map with a sum of zero.

Note that the component of M that is parallel to the difference between the plus and minus LOS's is highly correlated with a dipole pattern on the sky oriented toward the celestial poles. Fortunately the Earth's magnetic field is not an aligned, centered dipole field. The tilt and offset of the Earth's magnetic dipole produces a 24 hour period modulation in the magnetic effect that is not present in the sky signal. This reduces the correlation to a value that can easily be handled. The statistical noise on the dipole and to lesser extent the quadrupole is increased by simultaneously solving for the systematic error terms, but these low order spherical harmonics are dominated by cosmic variance in any case.

The basic DMR result is the discovery of an intrinsic anisotropy[17] of the microwave background, beyond the dipole anisotropy[18, 19]. This anisotropy, when a monopole and dipole fit to $|b| > 20°$ is removed

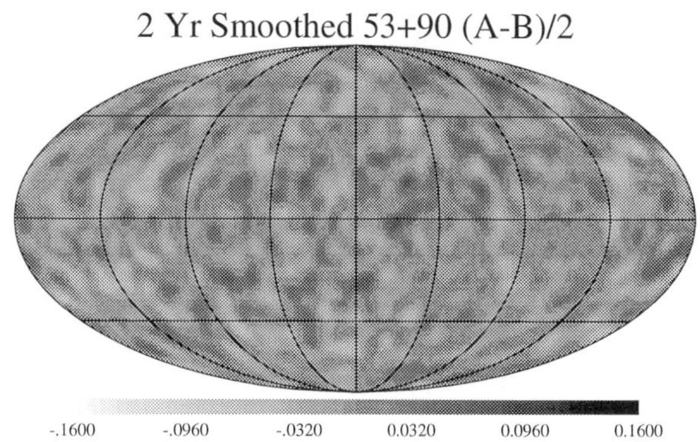

Figure 4: Noise map from (A-B)/2 for a smoothed weighted sum of the 53 GHz and 90 GHz channels for 2 years of DMR data.

from the map, and the map is then smoothed to a resolution of $\approx 10°$, is 30 μK. The correlation function of this anisotropy is well fit by the expected correlation function for the Harrison-Zeldovich[20, 21, 22] spectrum of primordial density perturbations predicted by the inflationary scenario, and the amplitude is consistent with many models of structure formation[23]. Galactic emission[24], systematic errors[25] and known extra-galactic sources[26] cannot explain the observed signal.

Even though the noise per pixel is much larger than the cosmic anisotropy, the detection of anisotropy has a high statistical confidence level because of the large number of pixels in the DMR maps. To illustrate this point, Figure 4 shows the difference of the smoothed A and B side (53+90) GHz maps from the two year datasets, while Figure 5 shows the smoothed sum map. The DMR maps show statistically significant correlated structure. Note that the good agreement between the DMR correlation function and the Harrison-Zeldovich model[17], while supporting inflation, does not prove inflation, especially since the precision with which the spectral index n can be determined is poor due to the small range of angular scales probed by *COBE*.

The power spectrum of the DMR maps has been computed[27] using the Hauser-Peebles method to allow for incomplete sky coverage caused by masking out the galactic plane region. Figure 6 shows the power in each multipole normalized to a pure $n = 1$ (Harrison-Zeldovich) power law with $\langle Q^2_{RMS} \rangle^{0.5} = 17$ μK. Also shown are results from various small and medium scale ΔT experiments: FIRS[28], Tenerife[29, 30], Ulisse[31], SP91[32], SK93[33], ARGO[34], PYTHON[35], MAX[36, 37, 38], MSAM[39], White Dish[40], OVRO[41], RING[42], and the Australia Telescope[43]. For perturbations at scales smaller than the horizon at the end of radiation dominance, dynamical effects boost the

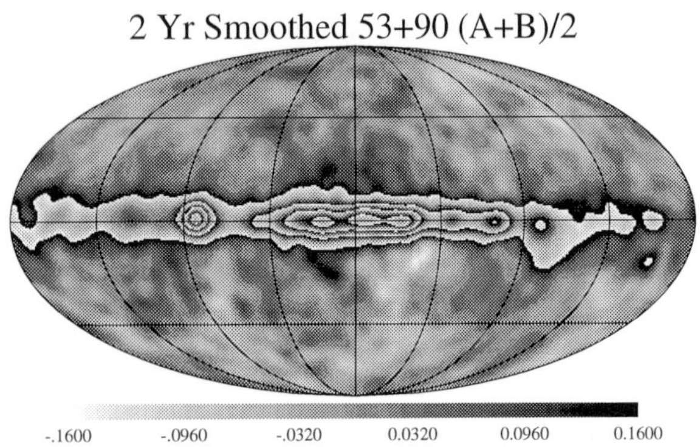

Figure 5: Signal map from (A+B)/2 for a smoothed weighted sum of the 53 GHz and 90 GHz channels for 2 years of DMR data.

expected anisotropy. At very small scales, the finite thickness of the recombination surface filters out most of the anisotropy. The models shown in Figure 6 are an $n = 0.85$ tilted CDM model[44] (lower curve, with $H_\circ = 50\ km\ s^{-1}\ Mpc^{-1}$ and $\Omega_B h^2 = 0.0125$), and the same model with the $n = 0.96$ expected from chaotic ϕ^4 inflation. The *COBE* data is consistent with the n_{app} slightly greater than 1 that is predicted by inflation when the "toe" of the Doppler peak is included. Analysis with linear[45] statistics instead of the quadratic Hauser-Peebles statistics confirms this conclusion. Note that the RING experiment[42] is known to be contaminated by discrete sources, but the power spectrum of uncorrelated point sources is $\propto \ell^2$ in Figure 6, so this population has a negligible effect on the data at lower ℓ's for sources which emit a spectrum similar to the spectrum of the Milky Way. When comparing the *COBE* DMR data to theoretical power spectra the corrections for the galactic plane cut and the non-Gaussian shape of the DMR beam should be used[46].

Note that the good agreement between the DMR correlation function or power spectrum and the Harrison-Zeldovich model, while supporting inflation, does not prove inflation, especially since the precision with which the spectral index n can be determined is poor due to the small range of angular scales probed by *COBE*.

The inflationary scenario also predicts that the temperature fluctuations should have a Gaussian distribution. This can be tested by looking at higher-order moments of the maps such as three point correlation functions[47], or by directly studying the probability distributions. The random measurement noise from the radiometers is still sufficient to interfere with these studies to a great extent, but a preliminary analysis finds that the Gaussian model is consistent with the observations. The three-point correlation function of the *COBE* DMR maps is consistent with the level of three-point correlation expected in a Harrison-Zeldovich model with Gaussian fluctuations. While the expected value of the three-point correlation vanishes for Gaussian models, the variance does not, so any individual realization of the Gaussian process will have a non-zero three-point correlation function.

Figure 7 shows histograms constructed from the smoothed DMR maps. To minimize the radiometer noise, the thermodynamic ΔT's derived from the 53 and 90 GHz channels have been averaged together with weights of 0.6 and 0.4 respectively. The histogram of the $(A - B)/2$ difference map is not expected to be Gaussian, because of the 5:1 variation in integration time per pixel between the rings 30° from the ecliptic poles and the ecliptic. However, a Gaussian convolved with the $(A - B)/2$ histogram should match the histogram of the $(A + B)/2$ map if the sky has a Gaussian distribution. As Figure 7 shows, the "Gaussian" curve, which is this convolution, matches the observed histogram well. Detailed calculations show that models which predict non-Gaussian temperature fluctuations, such as cosmic strings[48], or global monopoles[49], can also match the observed histogram. This convergence toward Gaussian probability distributions is caused by the 10° effective resolution of the smoothed maps, which averages many small patches together. Thus cosmic strings, which predict dramatically non-Gaussian behavior (sharp edges) on small scales, give almost Gaussian results after smoothing.

The ratio of the gravitational potential fluctuations seen by the DMR through the Sachs-Wolfe[50] effect to the gravitational potentials inferred from the bulk flows of galaxies using the POTENT method[51], shows that the Harrison-Zeldovich spectrum has the correct slope to connect the DMR data at 3×10^5 km/sec scales to the bulk flow data at 6000 km/sec scale. This comparison of the horizon-scale perturbations seen by *COBE* and the "large-scale" ($60/h$ Mpc, where $h = H_\circ/(100\ km\ s^{-1}\ Mpc^{-1})$) structure data gives the best determination of the slope of the power spectrum of the primordial perturbations, even though it is model-dependent. The POTENT bulk flow, if translated into a ΔT_ℓ^2 in Figure 6, would correspond to a point on the higher model curve at $\ell \approx 100$. The Lauer & Postman[52] bulk flow would correspond to a ΔT_ℓ^2 that is ≈ 40 times higher than the higher model at $\ell \approx 40$. Clearly the tilt of the primordial power spectrum cannot be determined without much better measurements of small scale ΔT and bulk flow velocities.

Wright *et al.*[23] discussed the implications of the *COBE* DMR data for models of structure formation, and selected 4 models from a large collection[53] for detailed discussion[23]: a "CDM" model, with $H_\circ = 50\ km\ s^{-1}\ Mpc^{-1}$, $\Omega_{CDM} = 0.9$, and $\Omega_B = 0.1$; a mixed

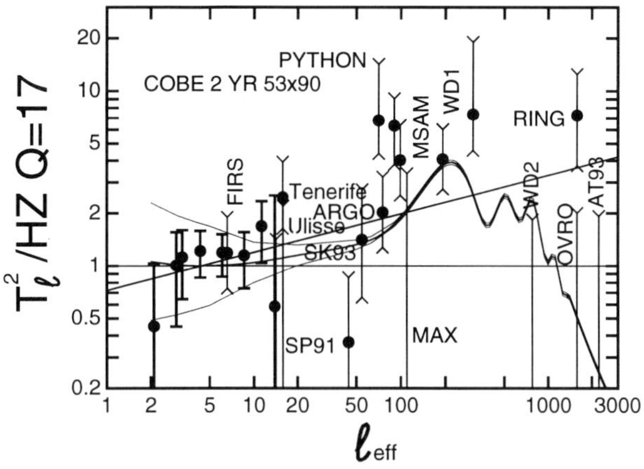

Figure 6: The angular power spectrum of the CMBR, normalized to an $\langle Q^2_{RMS} \rangle^{0.5} = 17\ \mu K$ Harrison-Zeldovich model. Data points with straight error bars are from COBE. Other data and upper limits have error bars with bent ends.

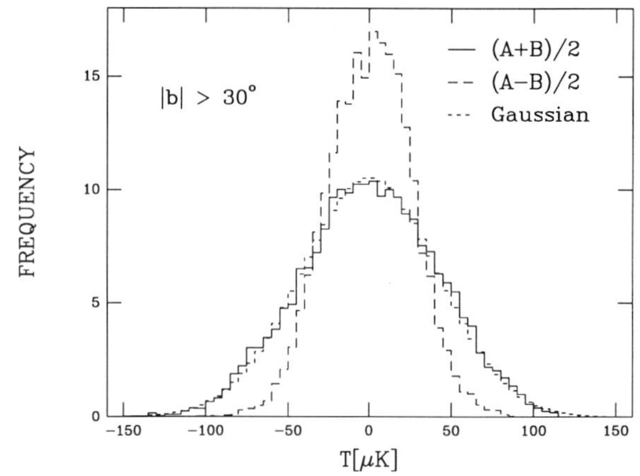

Figure 7: The histogram of the smoothed (A-B)/2 noise map, the noise convolved with a 30 μK Gaussian, and the smoothed (A+B)/2 map for the two 53+90 GHz COBE data.

"CDM+HDM" model, with $H_o = 50\ km\, s^{-1}\, Mpc^{-1}$, $\Omega_{CDM} = 0.6$, $\Omega_{HDM} = 0.3$ (a 7 eV neutrino), and $\Omega_B = 0.1$; an open model, with $H_o = 100\ km\, s^{-1}\, Mpc^{-1}$, $\Omega_{CDM} = 0.18$, and $\Omega_B = 0.02$; and a vacuum dominated model, with $H_o = 100\ km\, s^{-1}\, Mpc^{-1}$, $\Omega_{CDM} = 0.18$, $\Omega_B = 0.02$, and $\Omega_{vac} = 0.8$. The vacuum dominated model and especially the open model have potential perturbations now that are too small to explain the POTENT bulk flows[51].

The degree-scale and sub-degree scale results shown in Figure 6 indicate the presence of the Doppler peak. This has been interpreted[54] as indicating $0.01 < \Omega_B < 0.03$.

The CMBR field is moving so rapidly that any review article is quickly out-of-date, but [55] is a review of the literature on CMBR anisotropies and their interpretations up to early 1994. Remember that all of the detections on Figure 6 are less than three years old.

§4 Future CMB Work

Improving the FIRAS limits on spectral distortions will require a new space mission, and I am not aware of current plans for such a mission. The long wavelength end of the spectrum can be measured from balloons and mountaintops, and several groups are working to improve the long wavelength limits on spectral distortion.

Theoretical studies[56] show that many cosmological parameters have an effect on the power spectrum of the CMB. For example, the value of Ω_o determines the location of the Doppler peak at $\ell \approx 200$ in Figure 6, while the baryon density affects its height. The nature of the dark matter has an effect on the secondary Doppler peaks, since hot dark matter will stream out of the smaller scale structures whose "bouncing" causes the peaks. Current measurements of the small angular scale anisotropy of the CMB are limited by their small sky coverage to relatively poor precision. But a new generation of proposed experiments can improve this situation dramatically. With full sky coverage (actually only the usable 8 sr away from the galactic plane) and a small instrument beam one can achieve the accuracy shown by the error band in Figure 6. This band is limited by cosmic variance in bins with $\Delta\ell/\ell = 0.2$ for large angular scales, and a 3% calibration uncertainty for small angular scales. The 3% uncertainty in $\Delta\ell/\ell = 0.2$ requires mapping at least 10,000 beam areas, which requires for scales $\geq 2°$. Mapping the whole sky with this resolution will require a new satellite. For smaller beams, long duration balloons and observations from the South Pole will play a role along with satellites. New space missions being planned include the ESA mission COBRAS/SAMBA, and the US mission MAP and PSI. These new satellites will have beam sizes in the $0.2 - 0.7°$ range, which is 10 or more times smaller than the *COBE* DMR beam. If succesful, these satellites will provide data with the quality indicated on Figure 6 out to $\ell =$ 200 to 700. New balloon experiments being planned include the US-Italian BOOMERANG mission and the US TOPHAT mission. When these precise new anisotropy measurements are obtained, they will establish several new constraints that cosmological models must satisfy.

§References

[1] Nancy W. Boggess et al. The COBE mission, its design and performance two years after launch.

Astrophysical Journal, 397:420–429, 1992.

[2] Nigel Hawkes. Hunt on for dark secret of the universe. *Times of London*, page 1, April 25 1992.

[3] Edwin Hubble. A relation between distance and radial velocity among extragalactic nebulae. *Proceedings of the National Academy of Sciences*, 15:168–173, 1929.

[4] Arno A. Penzias and Robert W. Wilson. A measurement of excess antenna temperature at 4080 Mc/s. *Astrophysical Journal*, 142:419, 1965.

[5] Alan Guth. The inflationary universe: A possible solution to the horizon and flatness problems. *Physical Review D*, 23:347, 1981.

[6] D. H. Martin and E. Puplett. Polarised interferometric spectrometry for the millimeter and sub-millimeter spectrum. *Infrared Physics*, 10:105–109, 1970.

[7] Dale J. Fixsen et al. Calibration of the COBE FIRAS instrument. *Astrophysical Journal*, 420:457–473, 1994.

[8] E. L. Wright et al. Preliminary spectral observations of the galaxy with a 7° beam by the COsmic Background Explorer (COBE). *Astrophysical Journal*, 381:200–209, 1991.

[9] John C. Mather et al. Measurement of the cosmic microwave background spectrum by the COBE FIRAS instrument. *Astrophysical Journal*, 420:439–444, 1994.

[10] Dale J. Fixsen et al. Cosmic microwave background dipole spectrum measured by the COBE FIRAS instrument. *Astrophysical Journal*, 420:445–449, 1994.

[11] E. L. Wright et al. Interpretation of the COBE FIRAS spectrum. *Astrophysical Journal*, 420:450–456, 1994.

[12] G. Steigman and M. Tosi. Galactic evolution of D and He-3. *Astrophysical Journal*, 401:150–156, 1992.

[13] T. P. Walker, G. Steigman, D. N. Schramm, K. A. Olive, and H-S. Kang. Primordial nucleosynthesis redux. *Astrophysical Journal*, 376:51–69, 1991.

[14] A. S. Kompaneets. The establishment of thermal equilibrium between quanta and electrons. *Soviet Physics JETP*, 4:370, 1957.

[15] Ya. B. Zeldovich and R. A. Sunyaev. The interaction of matter and radiation in the hot model Universe. *Astrophysics and Space Science*, 4:301–316, 1969.

[16] C. Burigana, G. F. De Zotti, and L. Danese. Constraints on thermal history of the Universe from the CMB spectrum. *Astrophysical Journal*, 379:1–5, 1991.

[17] George F. Smoot et al. Structure in the COBE DMR first-year maps. *Astrophysical Journal (Letters)*, 396:L1–L6, 1992.

[18] E. K. Conklin. Velocity of the earth with respect to the cosmic background radiation. *Nature*, 222:971–972, 1969.

[19] A. Kogut et al. Dipole anisotropy in the COBE-DMR first-year sky maps. *Astrophysical Journal*, 419:1, 1993.

[20] Edward R. Harrison. Fluctuations at the threshold of classical cosmology. *Physical Review D*, 1:2726–2730, 1970.

[21] Peebles P. J. E. and J. T. Yu. Primeval adiabatic perturbations in an expanding universe. *Astrophysical Journal*, 162:815–636, 1970.

[22] Ya. B. Zeldovich. A hypothesis unifying the structure and entropy of the Universe. *Monthly Notices of the RAS*, 160:1p, 1972.

[23] Edward L. Wright et al. Interpretation of the cmb anisotropy detected by the COBE DMR. *Astrophysical Journal (Letters)*, 396:L13–L18, 1992.

[24] Charles L. Bennett et al. Preliminary separation of galactic and cosmic microwave emission for the COBE DMR. *Astrophysical Journal (Letters)*, 396:L7–L12, 1992.

[25] Al Kogut et al. COBE differential microwave radiometers (DMR): Preliminary systematic error analysis. *Astrophysical Journal*, 401:1–18, 1992.

[26] Charles L. Bennett, Gary Hinshaw, Anthony Banday, Al Kogut, Edward L. Wright, Karen Loewenstein, and Edward S. Cheng. Non-cosmological signal contributions to the COBE DMR anisotropy maps. *Astrophysical Journal (Letters)*, 414:L77, 1993.

[27] Edward L. Wright, George F. Smoot, Charles L. Bennett, and Phillip M. Lubin. Angular power spectrum of the microwave background anisotropy seen by the COBE DMR. *Astrophysical Journal*, TBD:TBD, 1994.

[28] K. Ganga, E. Cheng, S. Meyer, and L. Page. Cross-correlation between the 170 GHz survey map and the COBE DMR 1-yr maps. *Astrophysical Journal (Letters)*, 410:L57, 1993.

[29] R. A. Watson, C. M. Gutierrez de la Cruz, R. D. Davies, A. N. Lasenby, R. Rebolo, J. E. Beckman, and S. Hancock. Anisotropy measurements of the MBR at intermediate angular scales. *Nature*, 357:660–665, 1992.

[30] S. Hancock, R. D. Davies, A. N. Lasenby, C. M. Gutierrez de la Cruz, R. A. Watson, R. Rebolo, and J. E. Beckman. Direct observations of structure in the CMB. *Nature*, 367:333–337, 1994.

[31] P. de Bernardis, S. Masi, F. Melchiorri, B. Melchiorri, and N. Vittorio. Testing the isotropy of the CMB: The Ulisse experiment. *Astrophysical Journal (Letters)*, 396:L57–L60, 1992.

[32] J. Schuster, T. Gaier, J. Gundersen, P. Meinhold, T. Koch, M. Seiffert, C. Wuensche, and P. Lubin. CBR anisotropy at degree angular scales: Further results from the South Pole. *Astrophysical Journal (Letters)*, 412:L47–L50, 1993.

[33] E. J. Wollack, N. C. Jarosik, C. B. Netterfield, L. A. Page, and D. T. Wilkinson. A measurement of the anisotropy in the cosmic microwave background radiation at degree angular scales. *Astrophysical Journal (Letters)*, 419:L49, 1993.

[34] P. de Bernardis, E. Aquilini, A. Boscaleri, M. De Petris, G. D'Andreta, M. Gervasi, E. Kreysa, L. Martinis, S. Masi, P. Palumbo, and F. Scaramuzzi. Degree-scale observations of CMB anisotropies. *Astrophysical Journal (Letters)*, 422:L33–L36, 1994.

[35] M. Dragovan, J. Ruhl, G. Novak, S. R. Platt, B. Crone, J. B. Peterson, and R. Pernic. Anisotropy in the microwave sky at intermediate angular scales. *Astrophysical Journal (Letters)*, 427:L67–L70, 1994.

[36] J. Gundersen, A. Clapp, M. Devlin, W. Holmes, M. Fischer, P. Meinhold, A. Lange, P. Lubin, P. Richards, and G. Smoot. A degree scale anisotropy measurement of the cosmic microwave background near the star gamma Ursa Minoris. *Astrophysical Journal (Letters)*, 413:L1, 1993.

[37] A. C. Clapp, M. J. Devlin, J. O. Gundersen, C. A. Hagmann, V. V. Hristov, A. E. Lange, M. Lim, P. M. Lubin, P. D. Mauskopf, P. R. Meinhold, P. L. Richards, G. F. Smoot, S. T. Tanaka, P. T. Timbie, and C. A. Wuensche. Measurements of anisotropy in the cosmic microwave background radiation at degree angular scales near the stars sigma Hercules and iota Draconis. *Astrophysical Journal (Letters)*, 433:L57–L60, 1994.

[38] M. J. Devlin, A. C. Clapp, J. O. Gundersen, C. A. Hagmann, V. V. Hristov, A. E. Lange, M. Lim, P. M. Lubin, P. D. Mauskopf, P. R. Meinhold, P. L. Richards, G. F. Smoot, S. T. Tanaka, P. T. Timbie, and C. A. Wuensche. Measurements of anisotropy in the cosmic microwave background radiation at 0.5 degree angular scales near the star gamma Ursae Minoris. *Astrophysical Journal (Letters)*, 430:L1–L4, 1994.

[39] E. S. Cheng, D. A. Cottingham, D. J. Fixsen, C. A. Inman, M. S. Kowitt, S. S. Meyer, L. A. Page, J. L. Puchalla, and R. F. Silverberg. A measurement of the medium-scale anisotropy in the CMBR. *Astrophysical Journal (Letters)*, 422:L37–L40, 1994.

[40] G. S. Tucker, G. S. Griffin, H. T. Nguyen, and J. B. Peterson. A search for small scale anisotropy in the cosmic microwave background. *Astrophysical Journal (Letters)*, 419:L45, 1993.

[41] A. C. S. Readhead, C. R. Lawrence, S. T. Myers, W. L. W. Sargent, H. E. Hardebeck, and A. T. Moffet. A limit on the anisotropy of the MBR at arcminute scales. *Astrophysical Journal*, 346:566, 1989.

[42] S. T. Myers, A. C. S. Readhead, and C. R. Lawrence. Limits on the anisotropy of the mbr on arc minute scales. II. the RING experiment. *Astrophysical Journal*, 405:8–29, 1993.

[43] R. Subrahmayan, R. D. Ekers, M. Sinclair, and J. Silk. A search for arc-minute scale anisotropy in the CMB. *Monthly Notices of the RAS*, 263:416–424, 1993.

[44] R. Crittenden, J. R. Bond, R. L. Davis, G. Efstathiou, and P. J. Steinhardt. The imprint of gravitational waves on the CMB. *Physical Review Letters*, 71:324–327, 1993.

[45] K. M. Górski, G. Hinshaw, A. J. Banday, C. L. Bennett, E. L. Wright, A. Kogut, G. F. Smoot, and P. Lubin. On determining the spectrum of primordial inhomogeneity from the COBE DMR sky maps: Ii. results of two year data analysis. *Astrophysical Journal (Letters)*, 430:L89, 1994.

[46] E. L. Wright, G. F. Smoot, A. Kogut, G. Hinshaw, L. Tenorio, C. Lineweaver, C. L. Bennett, and P. M. Lubin. Comments on the statistical analysis of excess

variance in the COBE DMR maps. *Astrophysical Journal*, 420:1, 1994.

[47] G. Hinshaw, A. Kogut, K. Górski, A. Banday, C. L. Bennett, C. Lineweaver, P. Lubin, G. F. Smoot, and E. L. Wright. Limits on three-point correlations in the COBE DMR first year anisotropy maps. *Astrophysical Journal*, 431:1–5, 1993.

[48] D. P. Bennett, A. Stebbins, and F. R. Bouchet. The implications of the COBE DMR results for cosmic strings. *Astrophysical Journal (Letters)*, 399:L5–L8, 1992.

[49] D. P. Bennett and S. H. Rhie. Cobe's constraints on the global monopole and texture theories of cosmic structure formation. *Astrophysical Journal (Letters)*, 406:L7, 1993.

[50] R. K. Sachs and A. M. Wolfe. Perturbations of a cosmological model and angular variations of the microwave background. *Astrophysical Journal*, 147:73, 1967.

[51] Edmund Bertschinger, Avishai Dekel, Sandra M. Faber, Alan Dressler, and David Burstein. Potential, velocity and density fields from redshift-distance samples: Application: Cosmography within 6000 km/sec. *Astrophysical Journal*, 364:370–395, 1990.

[52] Todd Lauer and Marc Postman. The motion of the local group with respect to the 15,000 km/sec Abell cluster frame. *BAAS*, 24:1264, 1992.

[53] Jon Holtzman. Microwave background anisotropies and large scale structure in Universes with CDM, baryons, radiation and massive and massless neutrinos. *Astrophysical Journal Supplement*, 71:1–24, 1989.

[54] Douglas Scott and Martin White. The existence of baryons at z=1000. *TBD*, TBD:TBD, 1994.

[55] Martin White, Douglas Scott, and Joseph Silk. Anisotropies in the cosmic microwave background. *Annual Review of Astronomy and Astrophysics*, 32:319–370, 1994.

[56] Marc Kamionkowski, David Spergel, and Naoshi Sugiyama. Small-scale CMB Δt's as a probe of the geometry of the universe. *Astrophysical Journal (Letters)*, 426:L57–L60, 1994.

Status of Cosmological Parameters: Can $\Omega = 1$?

Joel R. Primack

Santa Cruz Institute for Particle Physics, University of California, Santa Cruz, CA 95064

The cosmological parameters that I will discuss are the traditional ones: the Hubble parameter $H_0 \equiv 100h$ km s^{-1} Mpc^{-1}, the age of the universe t_0, the average density $\Omega_0 \equiv \bar{\rho}/\rho_c$ in units of critical density ρ_c, and the cosmological constant Λ. To focus the discussion, I will concentrate on the issue of the value of the density Ω_0 in currently popular cosmological models in which most of the dark matter is cold, especially Cold + Hot Dark Matter (CHDM) and flat low-Ω CDM with a Cosmological Constant (ΛCDM). The evidence would favor a small $\Omega_0 \approx 0.3$ if (1) the Hubble parameter actually has the high value $h \approx 0.8$ favored by many observers, and the age of the universe $t_0 \geq 13$ Gy; or (2) the baryonic/total mass ratio in clusters is actually $\sim 20\%$, about 3-4 times larger than expected for standard Big Bang Nucleosynthesis in an $\Omega = 1$ universe, and standard BBN is actually right in predicting that the density of ordinary matter Ω_b lies in the range $0.009 \leq \Omega_b h^2 \leq 0.02$. The evidence would favor $\Omega = 1$ if (1) the POTENT analysis of galaxy peculiar velocity data is right, in particular regarding outflows from voids or the inability to obtain the present-epoch non-Gaussian density distribution from Gaussian initial fluctuations in a low-Ω universe; or (2) the preliminary report from LSND indicating a neutrino mass ≥ 2.4 eV is right, since that would be too much hot dark matter to allow significant structure formation in a low-Ω ΛCDM model. Statistics on gravitational lensing of quasars provide a strong upper limit on Λ. The era of structure formation is another important discriminant between these alternatives, low Ω favoring earlier structure formation, and $\Omega = 1$ favoring later formation with many clusters and larger-scale structures still forming today. Reliable data on all of these issues is becoming available so rapidly today that there is reason to hope that a clear decision between these alternatives will be possible within the next few years.

§1 Introduction

As I write this in early 1995, shortly after publication of the first article [1] using HST observations of Cepheid variable stars to determine a distance to a relatively distant galaxy (17.1 ± 1.8 Mpc for M100), articles in the popular news media are full of talk about a crisis in cosmology: "Big Bang Threatened..." The reason is of course that, with the additional assumptions that M100 lies in the core of the Virgo cluster and that the recession velocity of Virgo corrected for infall is about 1400 km s^{-1}, the value obtained for the Hubble parameter is at the high end of recent estimates: $H_0 = 80 \pm 17\,\mathrm{km\,s^{-1}\,Mpc}$. Using $h = 0.8$ gives, for $\Omega = 1$ and a vanishing cosmological constant $\Lambda = 0$, a very short age for the universe $t_0 = 8.15$ Gy, almost certainly younger than the ages of Milky Way globular clusters and even some nearby white dwarfs. Even with $\Omega_0 = 0.3$, about as low as permitted by observations, and with $\Omega_\Lambda \equiv \Lambda/(3H_0^2) = 0.7$, as high as permitted by observations, $t_0 = 11.8$ Gy for $h = 0.8$, which is also uncomfortably short. Is this a crisis? Does it undermine the strong evidence for the standard Big Bang? I don't think so. Given the considerable uncertainties reflected in the large quoted error on H_0, I think even $\Omega = 1$ models are not excluded. But this Cepheid measurement of the distance to M100 bodes well for the success of the HST Key Project on the Extragalactic Distance Scale, which seeks to measure H_0 to 10% within a few years. The expectation that accurate measurements of the key cosmological parameters will soon be available is great news for theorists trying to construct a fundamental theory of cosmology, and helps motivate the present summary.

In addition to the Hubble parameter $H_0 \equiv 100h$ km s^{-1} Mpc^{-1}, I will discuss the age of the universe t_0, the average density Ω_0, and the cosmological constant Λ. But there are several additional cosmological parameters whose values are critical for modern theories: the densities of ordinary matter Ω_b, cold dark matter Ω_c, and hot dark matter Ω_ν, and, for primordial fluctuation spectra

$P(k) = Ak^{n_p}$, the index n_p and the amplitude A, or equivalently (for a given model) the bias parameter $b \equiv 1/\sigma_8$, where $\sigma_8 \equiv (\delta M/M)_{rms}$ on a scale of $8\,h^{-1}$ Mpc. A full treatment of these parameters would take a much longer article than this one, so to focus the discussion I will concentrate on the issue of the value of the density Ω_0 in currently popular cosmological models in which most of the dark matter is cold. Although much of the following discussion will be quite general, it will be helpful to focus on two specific cosmological models which are perhaps the most popular today of the potentially realistic models: low-Ω Cold Dark Matter with a Cosmological Constant (ΛCDM, discussed as an alternative to $\Omega = 1$ CDM since the beginning of CDM [2, 3], and worked out in greatest detail in [4]), and $\Omega = 1$ Cold + Hot Dark Matter (CHDM, proposed in 1984 [5], and first worked out in detail in 1992-3 [6, 7]). I will begin by summarizing the rationale for these models.

§2 Models with Mostly Cold Dark Matter

Let me begin here by recalling the definitions of "hot" and "cold" dark matter. These terms describe the astrophysically relevant aspects of candidate dark matter particles. The fact that the observational lower bound on Ω_0 — namely $0.3 \lesssim \Omega_0$ — exceeds the upper limit on baryonic mass $\Omega_b \lesssim 0.02h^{-2}$ from Big Bang Nucleosynthesis [8] is the main evidence that there must be such nonbaryonic dark matter particles.

About a year after the big bang, the horizon surrounding any point encompassed a mass of about $10^{12}M_\odot$, the mass now in the dark matter halo of a large galaxy like the Milky Way. The temperature then was about a kilovolt. We define *cold* dark matter as particles that were moving sluggishly, and *hot* dark matter as particles that were still relativistic, at that time. As Kim Griest discussed in his Snowmass plenary talk, the lightest superpartner particle (LSP neutralino) and the axion remain the best motivated cold dark matter candidates, although of course many other possibilities have been suggested.

The three known neutrino species ν_e, ν_μ, and ν_τ are the standard hot dark matter candidates. Their contribution to the cosmological density today is

$$\Omega_\nu = \frac{\sum_i m(\nu_i)}{94h^2\,\mathrm{eV}}$$

Since $\Omega_\nu < \Omega_0 \lesssim 2$, each neutrino's mass must be much less than a keV, so they were certainly moving at relativistic speeds a year after the big bang. Any of these neutrinos that has a cosmologically significant mass ($\gtrsim 1$ eV) is therefore a hot dark matter particle.

If a horizon-sized region has slightly higher than average density at this time, cold dark matter — moving sluggishly — will preserve such a fluctuation. But neutrinos — moving at nearly the speed of light — will damp such fluctuations by "free streaming." For example, two years after the big bang, the extra neutrinos will have spread out over the now-larger horizon. The smallest fluctuations that will not suffer this fate are those that come into the horizon when the neutrinos become nonrelativistic, i.e. when the temperature drops to the neutrino mass. In a universe in which most of the dark matter is hot, primordial fluctuations will damp on all scales up to superclusters (with mass $\sim 10^{16}M_\odot$), leading to a sequence of cosmogony (cosmological structure formation) in which galaxies form only after superclusters. But this is contrary to observations, which show galaxies to be old but superclusters still forming. Indeed, with fluctuations on large scales consistent with COBE, pure HDM models (i.e. with the dark matter being mostly neutrinos, and a Zel'dovich spectrum of Gaussian adiabatic fluctuations) cannot form any significant number of galaxies by the present. Thus most current comparisons of cosmological models with observations have focused on models in which most of the dark matter is cold.

The standard CDM model [2] assumed a Zel'dovich (i.e. $n_p = 1$) spectrum of primordial Gaussian adiabatic fluctuations with $\Omega = 1$. It had the great virtues of simplicity and predictive power, since it had only one free parameter, the amplitude or bias b. Moreover, for a while it even looked like it agreed with all available data, with $b \approx 2.5$. One early warning that all was not well for CDM was the cosmic background dipole anisotropy, indicating a large velocity of the local group with respect to the cosmic background radiation rest frame, about 600 km s^{-1}. I must say, I and many other theorists did not immediately appreciate its possibly devastating impact. However, as evidence began to accumulate, starting in 1986, that such velocities were common on large scales — indeed, that there were large-scale flows of galaxies with such velocities [9] — it became clear that standard CDM could fit these large-scale galaxy peculiar velocities (i.e. motions in addition to the general Hubble expansion) only for $b \approx 1$. Standard CDM had various problems for any value of b; for example, the CDM matter correlation function, and hence also the galaxy and cluster correlations, are negative on scales larger than about $30\,h^{-1}$ Mpc, while observations on these large scales show that the cluster correlations are at least $\sim 3\sigma$ positive. A low value of the bias parameter subsequently also turned out to be required by the COBE DMR data, which was first announced in April 1992. But for such a small $b \lesssim 1$, CDM produces far too many clusters and predicts small-scale galaxy velocities that are much too large [10]. Thus standard CDM does not look like a very good match to the now-abundant observational data. But it did not miss by much: if the bias parameter b is adjusted to fit the COBE data, the fluctuation amplitude is too large on small scales by perhaps a factor of $\sim 2-3$.

In the wake of the discovery of the existence of large-scale galaxy peculiar velocities, I suggested that Jon Holtzman (then a UCSC graduate student whose planned Ph.D. research based on HST observations had been indefinitely postponed by the Challenger explosion) improve the program that George Blumenthal and I had written to do linear CDM calculations, and use it to investigate a variety of models in which the dark matter was mostly cold. He ultimately worked out a total of 94 such models, about half of them including some hot dark matter, and (since this was the largest such suite of interesting models all worked out the same way) his thesis [11] provided the basis for the COBE-DMR interpretation paper [12]. Meanwhile, in a follow-up paper [13], we showed that of all these CDM-like models the ones that best fit the available data — especially the cluster correlations — were $\Omega = 1$ Cold + Hot Dark Matter (CHDM), and low-Ω Cold Dark Matter with a Cosmological Constant (ΛCDM). Since both of these models turned out to fit all available data rather well when their fluctuation amplitudes were normalized to COBE observations, they remain perhaps the most popular models for galaxy formation and large scale structure. Moreover, since CHDM works best for $h \approx 0.5$ while ΛCDM works best for higher h, they will serve nicely for this review as representatives of these two opposing alternatives.

§3 Age of the Universe t_0

The strongest lower limits for t_0 come from studies of the stellar populations of globular clusters (GCs). Standard estimates of the ages of the oldest GCs are 14-18 Gy, and a conservative lower limit on the age of GCs is 13 ± 2 Gy, which is then a lower limit on t_0. The main uncertainty in the GC age estimates comes from the uncertain distance to the GCs: a 0.25 magnitude error in the distance modulus translates to a 22% error in the derived cluster age [14]. Stellar mass loss is the latest idea for lowering the GC t_0[15], but observations constrain the reduction in t_0 to be less than ~ 1 Gy. Allowing $\sim 1-2$ Gy for galaxy and GC formation, we conclude that $t_0 \gtrsim 11$ Gy from GCs, with $t_0 \approx 13$ Gy a "likely" lower limit on t_0, obtained by pushing most but not all the parameters to their limits.

The GC age estimates are of course based on standard stellar evolution calculations. New calculations using new stellar opacities now underway are not expected to change the estimates by more than a few %. But the solar neutrino problem reminds us that we are not really sure that we understand how even our nearest star operates; and the sun plays an important role in calibrating stellar evolution, since it is the only star whose age we know independently (from radioactive dating of early solar system material). What if the GC age estimates are wrong for some unknown reason?

The only independent estimates of the age of the universe come from cosmochronometry — the chemical evolution of the Galaxy — and white dwarf cooling. Cosmochronometry age estimates are sensitive to a number of uncertain effects such as the formation history of the disk and its stars, and possible actinide destruction in stars [16]. Age estimates also come from the cooling of white dwarfs in the neighborhood of the sun. The key observation is that there is a lower limit to the temperature of nearby white dwarfs; although cooler ones could have been seen, none have been found. The only plausible explanation is that the white dwarfs have not had sufficient time to cool to lower temperatures, which initially led to an estimate of 9.3 ± 2 Gy for the age of the Galactic disk [17]. Since there is evidence that the stellar disk of our Galaxy is about 2 Gy younger than the oldest GCs [18], this in turn gave an estimate of the age of the universe of $t_0 \sim 11 \pm 2$ Gy. However, more recent analyses [19] conclude that sensitivity to disk star formation history, and to effects on the white dwarf cooling rates due to C/O separation at crystallization and possible presence of trace elements such as ^{22}Ne, allow a rather wide range of ages for the disk of about 10 ± 4 Gy.

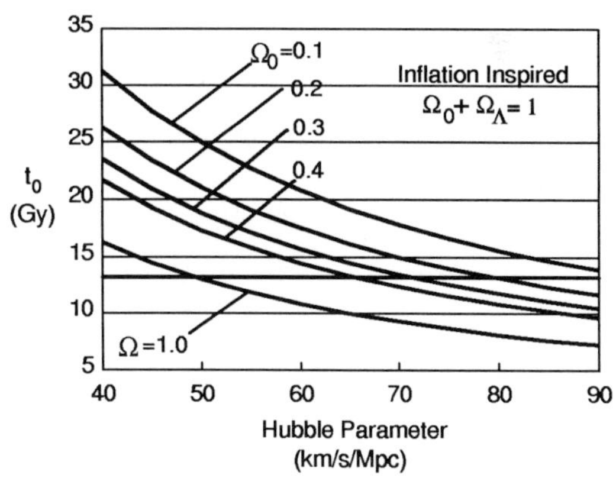

Figure 1: Age of the universe t_0 as a function of Hubble parameter H_0 in inflation inspired models with $\Omega_0 + \Omega_\Lambda = 1$, for several values of the present-epoch cosmological density parameter Ω_0.

Suppose that the GC stellar age estimates that $t_0 \gtrsim 13$ Gy are right. Fig. 1 shows that $t_0 > 13$ Gy implies that $H_0 \le 50$ km s^{-1} Mpc^{-1} for $\Omega = 1$, and that $H_0 \le 81$ km s^{-1} Mpc^{-1} even for Ω_0 as small as 0.2 (in flat cosmologies with $\Omega_0 + \Omega_\Lambda = 1$).

§4 Hubble Parameter H_0

The Hubble parameter $H_0 \equiv 100h$ km s^{-1} Mpc^{-1} remains uncertain by about a factor of two: $0.4 \lesssim h \lesssim 1$. Sandage has long contended that $h \approx 0.5$, and he still concludes [20] that the latest data are consistent with this. de Vaucouleurs long contended that $h \approx 1$. A majority of observers currently favor a value intermediate between these two extremes (recent reviews include [21, 22, 23]).

The Hubble parameter has been measured in two basic ways: (A) Measuring the distance to some nearby galaxies, typically by measuring the periods and luminosities of Cepheid variables in them; and then using these "calibrator galaxies" to set the zero point in any of the several methods of measuring the relative distances to galaxies. (B) Using fundamental physics to measure the distance to some distant object directly, thereby avoiding at least some of the uncertainties of the cosmic distance ladder [24]. The difficulty with method (A) is that there are so far only a handful of calibrator galaxies close enough for Cepheids to be resolved in them. However, the success of the HST Cepheid measurement of the distance to M100 [1] shows that the HST Key Project on the Extragalactic Distance Scale can significantly increase the set of calibrator galaxies within a few years. Adaptive optics from the ground may also be able to contribute to this effort, although I am not very impressed by the first published result of this approach [25]. The difficulty with method (B) is that in every case studied so far, some aspect of the observed system or the underlying physics remains somewhat uncertain. It is nevertheless remarkable that the results of several different methods of type (B) are rather similar, and indeed not very far from those of method (A). This gives reason to hope for convergence.

4.1 (A) Relative Distance Methods

One piece of good news is that the several methods of measuring the relative distances to galaxies now mostly seem to be consistent with each other [22, 23]. These methods use either (1) "standard candles" or (2) empirical relations between two measurable properties of a galaxy, one distance-independent and the other distance-dependent. The old favorite standard candle is Type Ia supernovae; a new one is the apparent maximum luminosity of planetary nebulae [22]. Sandage and others still get low values of $h \approx 0.4 - 0.5$ from HST Cepheid distances to SN Ia host galaxies [26]. There are claims that taking account of an empirical relationship between the SN Ia light curve shape and maximum luminosity leads to higher h [27], but Sandage and Tammann counter that any such effect is small [28]. The old favorite empirical relation used as a relative distance indicator is the Tully-Fisher relation between the rotation velocity and luminosity of spiral galaxies (and the related Faber-Jackson or $D_n - \sigma$ relation); a new one is based on the decrease in the fluctuations in elliptical galaxy surface brightness on a given angular scale as galaxies are seen at greater distances [29].

4.2 (B) Fundamental Physics Approaches

The fundamental physics approaches involve either Type Ia or Type II supernovae, the Sunyaev-Zel'dovich (S-Z) effect, or gravitational lensing. This Type Ia SN method for determining H_0 avoids the uncertainties of the distance ladder by calculating the absolute luminosity of Type Ia supernovae from first principles using a plausible but as yet unproved physical model. The result obtained is that $h = 61 \pm 10$ [30]; however, another study [31] finds that uncertainties in extinction (i.e., light absorption) toward each supernova increases the range of allowed h. The Type II SN method compares the expansion rate of the SN envelope measured by redshift with its size increase inferred from its temperature and magnitude; the 1992 result $h = 0.6 \pm 0.1$ [32] has since been revised upward by the same authors to $h = 0.73 \pm 0.06 \pm 0.07$ [33]. However, there are various complications with the physics of the expanding envelope [34].

The S-Z effect is the Compton scattering of microwave background photons from the hot electrons in a foreground galaxy cluster. This can be used to measure H_0 since properties of the cluster gas measured via the S-Z effect and from X-ray observations have different dependences on H_0. The result from the first cluster for which sufficiently detailed data was available, A665 (at $z = 0.182$), was $h = (0.4 - 0.5) \pm 0.12$ [35]; combining this with data on A2218 ($z = 0.171$) raises this somewhat to $h = 0.55 \pm 0.17$ [36]. Early results from the ASCA X-ray satellite gave $h = 0.47 \pm 0.17$ for A665 ($z = 0.182$) and $h = 0.41^{+0.15}_{-0.12}$ for CL0016+16 ($z = 0.545$) [37]. A few S-Z results have been obtained using millimeter-wave observations, and this promising method should allow many more such measurements soon [38]. Corrections for the near-relativistic electron motions will raise these estimates for H_0 a little [39], but it seems clear that the S-Z results favor a smaller value than many optical astronomers obtain. However, since the S-Z measurement of H_0 is affected by the orientation of the cluster ellipticity with respect to the line of sight, this will only become convincing if it agrees with results from observations of a significant number of additional clusters. Fortunately, this now appears to be possible within the next several years.

Several quasars have been observed to have multiple images separated by a few arc seconds; this phenomenon is interpreted as arising from gravitational lensing of the source quasar by a galaxy along the line of sight. In the first such system discovered, QSO 0957+561 ($z = 1.41$), the time delay Δt between arrival at the earth of variations in the quasar's luminosity in the two images has been measured to be 409 ± 23 days [41] (although other authors found a value of 540 ± 12 days [40]). Since $\Delta t \approx \theta^2 H_0^{-1}$, this observation allows an estimate of the Hubble parameter, with the results $h = 0.50 \pm 0.17$ [42], or $h = 0.63 \pm 0.21$ ($h = 0.42 \pm 0.14$) including (neglecting) dark matter in the lensing galaxy [43], with additional uncertainties associated with possible microlensing and unknown matter distribution in the lensing galaxy. However, recent deep images have allowed mapping of the gravitational potential of the lensing cluster (at $z = 0.36$) using weak gravitational lensing, which leads to the conclusion that $h \leq 0.70$ if $\Delta t \geq 1.1$ y [44]. Although the allowed range for H_0 remains rather large, it is reassuring that this method gives results consistent with the other determinations. The time-delay method is promising, and when delays are reliably measured in several other multiple-image quasar systems, that should lead to a reliable value for h.

4.3 Correcting for Virgocentric Infall

What about the recent HST Cepheid measurement of H_0, giving $h \approx 0.8$ [1]? This calculated value is based on neither of the two methods (A) or (B) above, and I do not regard it as being very reliable. Instead this result is obtained by assuming that M100 is at the core of the Virgo cluster, and dividing the sum of the recession velocity of Virgo, about 1100 km s^{-1}, plus the calculated "infall velocity" of the local group toward Virgo, about 300 km s^{-1}, by the measured distance to M100 of 17.1 Mpc. (These recession and infall velocities are both a little on the high side, compared to other values one finds in the literature.) Adding the "infall velocity" is necessary in this method in order to correct the Virgo recession velocity to what it would be if it were not for the gravitational attraction of Virgo for the Local Group of galaxies, but the problem with this is that the net motion of the Local Group with respect to Virgo is undoubtedly affected by much besides the Virgo cluster — e.g., the "Great Attractor." For example, in our CHDM supercomputer simulations (which appear to be a rather realistic match to observations) Anatoly Klypin and I have found that galaxies and groups at about 20 Mpc from a Virgo-sized cluster often have net outflowing rather than infalling velocities. Note that if there were no net "infall," or if M100 were in the foreground of the Virgo cluster (in which case the actual distance to Virgo would be larger than 17.1 Mpc), then the indicated H_0 would be smaller.

The authors of Ref. [1] gave an alternative argument that avoids the "infall velocity" uncertainty: the relative galaxy luminosities indicate that the Coma cluster is about six times farther away than the Virgo cluster, and peculiar motions of the Local Group and the Coma cluster are much smaller corrections to the much larger recession velocity of Coma; dividing the recession velocity of the Coma cluster by six times the distance to M100 again gives $H_0 \approx 80$. However, this approach still assumes that M100 is in the core rather than the foreground of the Virgo cluster; and in deducing the relative distance of the Coma and Virgo clusters it assumes that the galaxy luminosity functions in each are comparable, which is dubious in view of the very different environments.

To summarize, many observers, using mainly method (A), favor a value $h \approx 0.6 - 0.8$ although Sandage and collaborators continue to get $h \approx 0.4 - 0.6$, while the methods I have grouped together as (B) typically lead to $h \approx 0.4 - 0.7$. The fact that the latter measurements are mostly of more distant objects has suggested [45] that the local universe may actually be underdense and therefore be expanding faster than is typical. But in reasonable models where structure forms from Gaussian fluctuations via gravitational instability, it is extremely unlikely that a sufficiently large region has a density sufficiently smaller than average to make more than a rather small difference in the measured value of h [46].

There has been recent observational progress in both methods (A) and (B), and I think it likely that the Hubble parameter will be known reliably to 10% within a few years. But until then, we must keep an open mind.

§5 Cosmological Constant Λ, and t_0 Again

Inflation is the only known solution to the horizon and flatness problems and the avoidance of too many GUT monopoles. And inflation has the added bonus that with no extra charge (except the perhaps implausibly fine-tuned adjustment of the self-coupling of the inflaton field to be adequately small), simple inflationary models predict a near-Zel'dovich spectrum (i.e., with $n_p \approx 1$) of adiabatic Gaussian primordial fluctuations — which seems to be consistent with observations. All simple inflationary models predict that the curvature constant k is vanishingly small, although inflationary models that are extremely contrived (at least, to my mind) can be constructed with negative curvature and therefore $\Omega_0 \lesssim 1$ without a cosmological constant [47]. Thus most authors who consider inflationary models impose the condition $k = 0$, or $\Omega_0 + \Omega_\Lambda = 1$ where $\Omega_\Lambda \equiv \Lambda/(3H_0^2)$. This is what is assumed in ΛCDM models, and it is what was assumed in Fig. 1. (I hope it has been clear from the foregoing that I use Ω to refer only to the density of matter and energy, not including the cosmological constant, whose contribution in the Ω units is Ω_Λ.)

I know of no one who actually finds the idea of a non-vanishing Λ intrinsically attractive. There is no known physical reason why Λ should be so small (from the viewpoint of particle physics), though there is also no known reason why it should vanish. The most unattractive features of $\Lambda \neq 0$ cosmologies are the fact that Λ must become important only at relatively low redshift — why not much earlier or much later? — and also that $\Omega_\Lambda \gtrsim \Omega_0$ implies that the universe has recently entered an inflationary epoch (with a de Sitter horizon comparable to the present horizon). The main motivations for $\Lambda > 0$ cosmologies are (1) reconciling inflation with observations that seem to imply $\Omega \lesssim 1$, and (2) avoiding a contradiction between the lower limit $t_0 \gtrsim 13$ Gy from globular clusters and $t_0 = (2/3)H_0^{-1} = 6.52h^{-1}$ Gy for the standard $\Omega = 1$, $\Lambda = 0$ Einstein-de Sitter cosmology, if it is really true that $h > 0.5$.

The cosmological effects of a cosmological constant are not difficult to understand [48, 49]. With a positive Λ, there is a repulsion of space by space. In the early universe, the density of energy and matter is far more important than Λ on the r.h.s. of the Friedmann equation. But the average matter density decreases as the universe expands, and at a rather low redshift $z \sim 1$ the Λ term finally becomes significant. If it has been adjusted just right, Λ can almost balance the attraction of the matter, and the expansion nearly stops: for a long time, the scale factor $a \equiv (1+z)^{-1}$ increases very slowly, although it ultimately starts increasing exponentially as the universe starts inflating under the influence of the increasingly dominant Λ term. The existence of a period during which expansion slows while the clock runs explains why t_0 can be greater than for $\Lambda = 0$, but this also shows that there is a increased likelihood of finding galaxies at the redshift interval when the expansion slowed, and a correspondingly increased opportunity for lensing of quasars at higher redshift $z \gtrsim 2$ by these galaxies.

The frequency of such lensed quasars is about what would be expected in a standard $\Omega = 1$, $\Lambda = 0$ cosmology, so this data sets fairly stringent upper limits: $\Omega_\Lambda \leq 0.70$ at 90% C.L. [50, 51], with more recent data likely to give even tighter constraints [52].

A weaker but independent constraint comes from the cosmic background radiation data. In standard $\Omega = 1$ models, the quantity $\ell(\ell+1)C_\ell$ (where $C_\ell = < a^2_{\ell,m} >_m$ is the average of squared coefficients of the spherical harmonic expansion of the CMB data) is predicted to be roughly constant for $2 \leq \ell \lesssim 10$ (with an increase for higher multiples toward the Doppler peak at $\ell \sim 200$), while in models with $\Lambda > 0$ $\ell(\ell+1)C_\ell$ is predicted to dip before rising toward the Doppler peak. Comparison with the two-year COBE data, in which such a dip is not seen, implies that $\Omega_\Lambda \leq 0.78$ at the 90% C.L. [53].

Fig. 1 shows that with $\Omega_\Lambda \leq 0.7$, the cosmological constant does not lead to a very large increase in t_0 compared to the Einstein-de Sitter case, although it may still be enough to be significant. For example, the constraint that $t_0 \geq 13$ Gy requires $h \leq 0.5$ for $\Omega = 1$ and $\Lambda = 0$, but this becomes $h \leq 0.73$ for $\Omega_\Lambda \leq 0.7$.

§6 Measuring Ω_0

Although it would be desirable to measure Ω_0 and Λ through their effects on the large-scale geometry of space-time, this has proved difficult in practice since it requires comparing objects at higher and lower redshift, and it is hard to separate the effects of the evolution of the objects from those of the evolution of the universe. For example, in "redshift-volume" tests involving number counts of galaxies per redshift interval, how can we tell whether the galaxies at redshift $z \sim 1$ correspond to those at $z \sim 0$? Several galaxies at higher redshift might have merged, and galaxies might have formed or changed luminosity at lower redshift. Eventually, with extensive surveys of galaxy properties as a function of redshift using the largest telescopes such as Keck, it should be possible to perform these classical cosmological tests at least on a particular class of galaxies — that is one of the goals of the Keck DEEP project. At present, perhaps the most promising technique involves searching for Type Ia supernovae at high-redshift, since these are the brightest supernovae and the spread in their intrinsic brightness appears to be relatively small. Gerson Goldhaber, Saul Perlmutter, and collaborators have recently demonstrated the feasibility of finding significant numbers of such supernovae [54], but a dedicated campaign of follow-up observations of each one will be required in order to measure Ω_0 by determining how the apparent brightness of the supernovae depends on their redshift. This is therefore a project that will take at least several years.

6.1 Large-scale Measurements

The largest scales on which Ω_0 has been measured with some precision today are about $\sim 50\,h^{-1}$ Mpc, using the data on peculiar velocities of galaxies, and on a somewhat larger scale using redshift surveys based on the IRAS galaxy catalog. Since the results of all such measurements to date have recently been summarized in an excellent review article [55], I will only comment briefly on them. The analyses such as "POTENT" that try to recover the scalar velocity potential from the galaxy peculiar velocities are looking increasingly reliable, since they reproduce the observed large scale distribution of galaxies – that is, many galaxies are found where the converging velocities indicate that there is a lot of matter, and there are voids in the galaxy distribution where the diverging velocities indicate that the density is lower than average. The comparison of the IRAS redshift surveys with POTENT and related analyses typically give fairly large values for the parameter $\beta_I \equiv \Omega_0^{0.6}/b_I$ (where b_I is the biasing parameter for IRAS galaxies), corresponding to $0.3 \lesssim \Omega_0 \lesssim 3$ (for an assumed $b_I = 1.15$). It is not clear whether it will be possible to reduce the spread in these values significantly in the near future — probably both additional data and a better understanding of systematic and statistical effects will be required.

A particularly simple way to deduce a lower limit on Ω_0 from the POTENT peculiar velocity data has recently been proposed [56], based on the fact that high-velocity outflows from voids are not expected in low-Ω models. Data on just one void indicates that $\Omega_0 \geq 0.3$ at the 97% C.L. This argument is independent of assumptions about Λ, the initial fluctuations, or galaxy formation, but of course it does depend on the success of POTENT in recovering the peculiar velocities of galaxies.

However, for the particular cosmological models that I am focussing on in this review — CHDM and ΛCDM — stronger constraints are available. This is because these models, in common with almost all CDM variants, assume that the probability distribution function (PDF) of the primordial fluctuations was Gaussian. The PDF deduced by POTENT from observed velocities (i.e., the PDF of the mass, not that of the galaxies) is far from Gaussian today. It agrees with a Gaussian initial PDF if and only if Ω is about unity or larger: $\Omega_0 < 1$ is rejected at the 2σ level, and $\Omega_0 \leq 0.3$ is ruled out at $\geq 4\sigma$ [57]. Evolution from a Gaussian initial PDF to the non-Gaussian mass distribution observed today requires considerable gravitational nonlinearity, i.e. large Ω.

6.2 Measurements on Scales of a Few Mpc

On smaller length scales, there are many measurements that are consistent with a smaller value of Ω_0 [58]. For example, the cosmic virial theorem gives $\Omega(\sim 1h^{-1}\,\mathrm{Mpc}) \approx 0.15[\sigma(1h^{-1}\,\mathrm{Mpc})/(300\,\mathrm{km\,s^{-1}})]^2$, where $\sigma(1h^{-1}\,\mathrm{Mpc})$ here represents the relative velocity dispersion of galaxy pairs at a separation of $1h^{-1}\,\mathrm{Mpc}$. Although the classic paper [59] which first measured $\sigma(1h^{-1}\,\mathrm{Mpc})$ using a large redshift survey (CfA1) got a value of 340 km s^{-1}, this result is now known to be in error since the entire core of the Virgo cluster was inadvertently omitted [60]; if Virgo is included, the result is $\sim 500 - 600$ km s^{-1} [61, 60], corresponding to $\Omega(\sim 1h^{-1}\,\mathrm{Mpc}) \approx 0.4 - 0.6$. Various redshift surveys give a wide range of values for $\sigma(1h^{-1}\,\mathrm{Mpc}) \sim 300 - 750$ km s^{-1}, with the most salient feature being the presence or absence of rich clusters of galaxies; for example, the IRAS galaxies, which are not found in clusters, have $\sigma(1h^{-1}\,\mathrm{Mpc}) \approx 320$ km s^{-1} [62], while the northern CfA2 sample, with several rich clusters, has much larger σ than the SSRS2 sample, with only a few relatively poor clusters. It is evident that the $\sigma(1h^{-1}\,\mathrm{Mpc})$ statistic is not a very robust one.

A standard method for estimating Ω on scales of a few Mpc is based on applying virial estimates to groups and clusters of galaxies to try to deduce the total mass of the galaxies including their dark matter halos from the velocities and radii of the groups; roughly, $GM \sim rv^2$. (What one actually does is to assume that all galaxies have the same mass-to-light ratio M/L, given by the median M/L of the groups, and integrate over the luminosity function to get the mass density [63, 64, 65].) The typical result is that $\Omega(\sim 1\,h^{-1}\,\mathrm{Mpc}) \sim 0.1{-}0.2$. However, such estimates are at best lower limits, since they can only include the mass within the region where the galaxies in each group can act as test particles. In CHDM simulations, my colleagues and I [66] have found that the effective radius of the dark matter distribution associated with galaxy groups is typically 2-3 times larger than that of the galaxy distribution. Moreover, we find a velocity biasing [67] factor in CHDM groups $b_v^{grp} \equiv v_{\mathrm{gal,rms}}/v_{\mathrm{DM,rms}} \approx 0.75$, whose inverse squared enters in the Ω estimate. Finally, we find that groups and clusters are typically elongated, so only part of the mass is included in spherical estimators. These factors explain how it can be that our $\Omega = 1$ CHDM simulations produce group velocities that are fully consistent with those of observed groups, even with sophisticated robust and discriminatory statistical tests such as the median rms group velocity vs. the fraction of galaxies grouped [68, 66]. This emphasizes the point that local estimates of Ω are at best lower limits on its true value.

Another approach to estimating Ω from information on relatively small scales has been pioneered by Peebles [69]. It is based on using the least action principle (LAP) to reconstruct the trajectories of the Local Group galaxies, and the assumption that the mass is concentrated around the galaxies. This is a reasonable assumption in a low-Ω universe, but it is not at all what must occur in an $\Omega = 1$ universe where most of the mass must lie between the galaxies. Although comparison with $\Omega = 1$ N-body simulations showed that the LAP often succeeds in qualitatively reconstructing the trajectories, the mass is systematically underestimated by a large factor by the LAP method [70]. Unexpectedly, a different study [71] found that the LAP method underestimates Ω by a factor of 4-5 even in an $\Omega_0 = 0.2$ simulation; the authors say that this discrepancy is due to the LAP neglecting the effect of "orphans" — dark matter particles that are not members of any halo.

6.3 Estimates on Galaxy Halo Scales

Recent work by Zaritsky and White [72] and collaborators has shown that spiral galaxies have massive halos. A classic paper by Little and Tremaine [73] argued that the available data on the Milky Way satellite galaxies required that the Galaxy's halo terminate at about 50 kpc, with a total mass of only about $2.5 \times 10^{11} M_\odot$. But by 1991, new data on local satellite galaxies, especially Leo I, became available, and the Little-Tremaine estimator increased to $1.25 \times 10^{12} M_\odot$. Zaritsky and collaborators have collected data on satellites of other spiral galaxies, and conclude that the fact that the relative velocities do not fall off out to a separation of at least 200 kpc shows that massive halos are the norm. The typical rotation velocity of $\sim 200 - 250$ km s^{-1} implies a mass within 200 kpc of $2 \times 10^{12} M_\odot$. A careful analysis taking into account selection effects and satellite orbit uncertainties concluded that the indicated value of Ω_0 exceeds 0.13 at 90% confidence, with preferred values exceeding 0.3 [72].

§7 Clusters

7.1 Cluster Baryons vs Big Bang Nucleosynthesis

A recent review [8] of Big Bang Nucleosynthesis (BBN) and observations indicating primordial abundances of the light isotopes concludes that $0.009h^{-2} \leq \Omega_b \leq 0.02h^{-2}$ for concordance with all the abundances, and $0.006h^{-2} \leq \Omega_b \leq 0.03h^{-2}$ if only deuterium is used. For $h = 0.5$, the corresponding upper limits on Ω_b are 0.08 and 0.12, respectively. The recent observations [74] of a possible deuterium line in a hydrogen cloud at redshift $z = 3.32$ indicating a deuterium abundance of $\sim 2 \times 10^{-4}$ (and therefore $\Omega_b \leq 0.006h^{-2}$) are contradicted by a similar observation [75] in a system at $z = 3.58$ but with a deuterium abundance about ten times lower, consistent with solar system measurements of D and ^{3}He and the higher upper limit on Ω_b. (The earlier observations [74] were most probably of a Lyα forest line.)

White et al. [76] have emphasized that recent X-ray observations of clusters, especially Coma, show that the abundance of baryons, mostly in the form of gas (which typically amounts to several times the total mass of the cluster galaxies), is as much as 20% if h is as low as 0.5. For the Coma cluster they find that the baryon fraction within the Abell radius is

$$f_b \equiv \frac{M_b}{M_{tot}} \geq 0.009 + 0.050h^{-3/2},$$

where the first term comes from the galaxies and the second from gas. If clusters are a fair sample of both baryons and dark matter, as they are expected to be based on simulations, then this is 2-3 times the amount of baryonic mass expected on the basis of BBN in an $\Omega = 1$, $h \approx 0.5$ universe, though it is just what one would expect in a universe with $\Omega_0 \approx 0.3$. The fair sample hypothesis implies that

$$\Omega_0 = \frac{\Omega_b}{f_b} = 0.3 \left(\frac{\Omega_b}{0.06}\right)\left(\frac{0.2}{f_b}\right).$$

A recent review of gas in a sample of clusters [77] finds that the baryon mass fraction within about 1 Mpc lies between 10 and 22%, and argues that it is unlikely that (a) the gas could be clumped enough to lead to significant overestimates of the total gas mass — the main escape route considered in [76]. If $\Omega = 1$, the alternatives are then either (b) that clusters have more mass than virial estimates based on the cluster galaxy velocities or estimates based on hydrostatic equilibrium of the gas at the measured X-ray temperature (which is surprising since they agree [78]), or (c) that the BBN upper limit on Ω_b is wrong. It is interesting that there are indications from weak lensing [79] and galaxy velocities [80] that at least some clusters may actually have extended halos of dark matter — something that is expected to a greater extent if the dark matter is a mixture of cold and hot components, since the hot component clusters less than the cold [66, 81]. If so, the number density of clusters as a function of mass is higher than usually estimated, which has interesting cosmological implications (e.g. σ_8 is higher than usually estimated). It is of course possible that the solution is some combination of alternatives (a), (b), and (c). If none of the alternatives is right, then the only conclusion left is that $\Omega_0 \approx 0.3$. The cluster baryon problem is clearly an issue that deserves very careful examination.

7.2 Cluster Morphology

Richstone, Loeb, and Turner [82] showed that clusters are expected to be evolved — i.e. rather spherical and featureless — in low-Ω cosmologies, in which structures form at relatively high redshift, and that clusters should be more irregular in $\Omega = 1$ cosmologies, where they have formed relatively recently and are still undergoing significant merger activity. There are very few known clusters that seem to be highly evolved and relaxed, and many which are irregular — some of which are obviously undergoing mergers now or have recently done so (see e.g. [83]). This disfavors low-Ω models, but it remains to be seen just how low. Recent papers have addressed this. In one [84] a total of 24 CDM simulations with $\Omega = 1$ or 0.2, the latter with $\Omega_\Lambda = 0$ or 0.8, were compared with data on a sample of 57 clusters. The conclusion was that clusters with the observed range of X-ray morphologies are very unlikely in the low-Ω cosmologies. However, these simulations have been criticized because the $\Omega_0 = 0.2$ ones included rather a large amount of ordinary matter: $\Omega_b = 0.1$. (This is unrealistic both because $h \approx 0.8$ provides the best fit for $\Omega_0 = 0.2$, but then the standard BBN upper limit is $\Omega_b < 0.02h^{-2} = 0.03$; and also because observed clusters have a gas fraction of $\sim 0.15(h/0.5)^{-3/2}$.)

Another study [85] using dissipationless simulations and not comparing directly to observational data found that ΛCDM with $\Omega_0 = 0.3$ and $h = 0.75$ produced clusters with some substructure, perhaps enough to be observationally acceptable. Clearly, this important issue deserves study with higher resolution hydrodynamic simulations, with a range of assumed Ω_b, and possibly including at least some of the additional physics associated with the galaxies which must produce the metallicity observed in clusters, and perhaps some of the heat as well. Better statistics for comparing simulations to data may also be useful [86].

7.3 Cluster Evolution

There is evidence for strong evolution of clusters at relatively low redshift, both in their X-ray properties [87] and in the properties of their galaxies. In particular, there is a strong increase in the fraction of blue galaxies with increasing redshift (the "Butcher-Oemler effect"), which may be difficult to explain in a low-density universe [89]. Field galaxies do not appear to show such strong evolution; indeed, a recent study concludes that over the redshift range $0.2 \leq z \leq 1.0$ there is no significant evolution in the number density of "normal" galaxies [88]. This is compatible with the predictions of CHDM with two neutrinos sharing a total mass of about 5 eV [90] (see below).

§8 Early Structure Formation

In linear theory, adiabatic density fluctuations grow linearly with the scale factor in an $\Omega = 1$ universe, but more slowly if $\Omega < 1$ with or without a cosmological constant [58]. As a result, if fluctuations of a certain size in an $\Omega = 1$ and an $\Omega_0 = 0.3$ theory are equal in amplitude at the present epoch ($z = 0$), then at higher redshift the fluctuations in the low-Ω model had higher amplitude. Thus, structures typically form earlier in low-Ω models than in $\Omega = 1$ models.

Since quasars are seen at the highest redshifts, they have been used to try to constrain $\Omega = 1$ theories, especially CHDM which because of the hot component has additional suppression of small-scale fluctuations that are presumably required to make early structure (e.g., [91]). The difficulty is that dissipationless simulations predict the number density of halos of a given mass as a function of redshift, but not enough is known about the nature of quasars — for example, the mass of the host galaxy — to allow a simple prediction of the number of quasars as a function of redshift in any given cosmological model. A recent study [92] concludes that very efficient cooling of the gas in early structures, and angular momentum transfer from it to the dark halo, allows for formation of *at least* the observed number of quasars even in models where most galaxy formation occurs late.

Another sort of high redshift object which holds more promise for constraining theories is damped Lyman α systems (DLAS). DLAS are dense clouds of neutral hydrogen, generally thought to be protogalactic disks, which are observed as wide absorption features in quasar spectra [93]. They are relatively common, seen in roughly a third of all quasar spectra, so statistical inferences about DLAS are possible. At the highest redshift for which data is published, $z = 3-3.4$, the density of neutral gas in such systems in units of critical density is $\Omega_{gas} \approx 0.6\%$, comparable to the total density of visible matter in the universe today [94]. Several recent papers [95] pointed out that the CHDM model with $\Omega_\nu = 0.3$ could not produce such a high Ω_{gas}. However, my colleagues and I showed that CHDM with $\Omega_\nu = 0.2$ could do so [96], since the power spectrum on small scales is a very sensitive function of the total neutrino mass in CHDM models. This theory makes two crucial predictions [96]: Ω_{gas} must fall off at higher redshifts z, and the DLAS at $z \gtrsim 3$ correspond to systems of internal rotation velocity or velocity dispersion less than about 100 km s^{-1} (this can be measured from the Doppler widths of the metal line systems associated with the DLAS). Preliminary reports regarding the latest data at redshifts above 3.5 appear to be consistent with these predictions [97].

One of the best ways of probing early structure formation would be to look at the main light output of the stars of the earliest galaxies, which is redshifted by the expansion of the universe to wavelengths beyond about 5 microns today. Unfortunately, it is not possible to make such observations with existing telescopes; since the atmosphere blocks almost all such infrared radiation, what is required is a large infrared telescope in space. The Space Infrared Telescope Facility (SIRTF) has long been a high priority, and it would be great to have access to the data such an instrument would produce. But even if NASA started such a mission immediately, it would not be available until the next millennium. In the meantime, an alternative method is look for the starlight from the earliest stars as extragalactic background infrared light (EBL). Although it is difficult to see this background light directly because our Galaxy is so bright in the near infrared, it may be possible to detect it indirectly through its absorption of TeV gamma rays (via the process $\gamma\gamma \rightarrow e^+e^-$). Of the more than twenty AGNs that have been seen at ~ 10 GeV by the EGRET detector on the Compton Gamma Ray Observatory, only one, the nearest, Mk421, has also been clearly detected in TeV gamma rays by the Whipple Atmospheric Cerenkov Telescope. Absorption of $\sim$ *TeV* gamma rays from active galactic nuclei (AGNs) at redshifts $z \sim 0.2$ has been shown to be a sensitive probe of the era of galaxy formation [98].

§9 Neutrino mass

There are several experiments which suggest that neutrinos have mass. In particular, the recent announcement of the observation of $\bar{\nu}_\mu \rightarrow \bar{\nu}_e$ oscillations at the Liquid Scintillator Neutrino Detector (LSND) experiment at Los Alamos suggests that $\delta m^2 \equiv |m(\nu_\mu)^2 - m(\nu_e)^2| \approx 6$ eV^2 [99] , and the observation of the angular dependence of the atmospheric muon neutrino deficit at Kamiokande [100] suggests $\nu_\mu \rightarrow \nu_\tau$ oscillations are occurring with an oscillation length comparable to the depth of the atmosphere, which requires that the muon and tau neutrinos have approximately the same mass. If, for example, $m(\nu_e) \ll m(\nu_\mu)$, then this means that $m(\nu_\mu) \approx m(\nu_\tau) \approx 2.4$ eV [101]. Clearly, discovery of neutrino mass in the few eV range favors CHDM; and, as I mentioned above, this total neutrino mass of about 5 eV is just what seems to be necessary to fit the large scale structure observations [96]. Dividing the mass between two neutrinos results in somewhat lower fluctuation amplitude on the scale of clusters of galaxies because of the longer neutrino free streaming length, which improves agreement between CHDM normalized to COBE and observations of cluster abundance [101].

Of course, one cannot prove a theory since contrary evidence may always turn up. But one can certainly disprove theories. The minimum neutrino mass required by the preliminary LSND result [99] $\delta m^2 = 6$ eV^2 is 2.4 eV. This is too much hot dark matter to permit significant structure formation in a low-Ω universe; for example, in a ΛCDM model with $\Omega_0 = 0.3$, the cluster number density is more than two orders of magnitude lower than observations indicate [101]. Thus if this preliminary LSND result is correct, it implies a strong lower limit on Ω_0, and a corresponding upper bound on Λ, in ΛCDM models.

§10 Conclusions

The main issue that I have tried to address is the value of the cosmological density parameter Ω. Strong arguments can be made for $\Omega_0 \approx 0.3$ (and models such as ΛCDM) or for $\Omega = 1$ (for which the best class of models that I know about is CHDM), but it is too early to tell for sure which is right.

The evidence would favor a small $\Omega_0 \approx 0.3$ if (1) the Hubble parameter actually has the high value $H_0 \approx 80$ favored by many observers, and the age of the universe $t_0 \geq 13$ Gy; or (2) the baryonic fraction $f_b = M_b/M_{tot}$ in clusters is actually $\sim 20\%$, about 3-4 times larger than expected for standard Big Bang Nucleosynthesis in an $\Omega = 1$ universe. This assumes that standard BBN is actually right in predicting that the density of ordinary matter Ω_b lies in the range $0.009 \leq \Omega_b h^2 \leq 0.02$; if the systematic errors in the ^{4}He data are larger than currently estimated, using the deuterium upper limit $\Omega_b h^2 \leq 0.03$ lessens the discrepancy between f_b and Ω_b somewhat. High-resolution high-redshift spectra are now providing important new data on primordial abundances of the light isotopes that should clarify the reliability of the BBN limits on Ω_b. Another important constraint on Ω_b will come from the new data on small angle CMB anisotropies — in particular, the height of the Doppler peaks [102].

The evidence would favor $\Omega = 1$ if (1) the POTENT analysis of galaxy peculiar velocity data is right, in particular regarding outflows from voids or the inability to obtain the present-epoch non-Gaussian density distribution from Gaussian initial fluctuations in a low-Ω universe; or (2) the preliminary report from LSND indicating a neutrino mass ≥ 2.4 eV is right, since that would be too much hot dark matter to allow significant structure formation in a low-Ω ΛCDM model.

The statistics of gravitational lensing of quasars is incompatible with large cosmological constant Λ and low cosmological density Ω_0. Discrimination between models will improve fairly rapidly as additional examples of lensed quasars are searched for.

The era of structure formation is another important discriminant between these alternatives, low Ω favoring earlier structure formation, and $\Omega = 1$ favoring later formation with many clusters and larger-scale structures still forming today. A particularly critical test for models like CHDM is the evolution as a function of redshift of Ω_{gas} in damped Lyα systems.

Reliable data on all of these issues is becoming available so rapidly today that there is reason to hope that a clear decision between these alternatives will be possible within the next few years.

What if the data ends up supporting what appear to be contradictory possibilities, e.g. large Ω_0 *and* large H_0? Exotic initial conditions (e.g. "designer" primordial fluctuation spectra) or exotic dark matter particles beyond the simple "cold" vs. "hot" alternatives (e.g. decaying intermediate mass neutrinos) could increase the space of possible inflationary theories somewhat. But it may ultimately be necessary to go outside the framework of inflationary cosmological models and consider models with large scale spatial curvature, with a fairly large Λ as well as large Ω_0. This seems particularly unattractive, since in addition to implying that the universe is now entering a final inflationary period, it means that inflation did not happen at the beginning of the universe, when it would solve the flatness, horizon, monopole, and structure generation problems. Therefore, along with most cosmologists, I am rooting for the success of inflation-inspired cosmologies, with $\Omega_0 + \Omega_\Lambda = 1$. But the universe is under no obligation to live up to our expectations.

ACKNOWLEDGEMENTS. I have benefited from conversations or correspondence with S. Bonometto, S. Borgani, E. Branchini, D. Caldwell, L. Da Costa, M. Davis, A. Dekel, S. Faber, K. Griest, J. Holtzman, A. Klypin, C. Kochanek, K. Lanzetta, C.-P. Ma, R. Nolthenius, J. Ostriker, P.J.E. Peebles, D. Richstone, A. Sandage, L. Storrie-Lombardie, A. Wolfe, and E. Wright, and fruitful interactions with UCSC graduate students R. Davé, M. Gross, D. Hellinger, and R. Somerville. This research was supported by NSF and faculty research grants at UCSC. Simulations were done on the Convex C-3880 at NCSA.

§References

[1] W.L. Freedman et al., Nature 371, 757 (1994).

[2] G.R. Blumenthal, S.M. Faber, J.R. Primack, and M.J. Rees, Nature 311, 517 (1984); Erratum: 313, 72 (1985). M. Davis, G. Efstathiou, C.S. Frenk, and S.D.M. White, Astrophys. J. 292, 371 (1985).

[3] P.J.E. Peebles, Astrophys. J. 284, 439 (1984).

[4] R. Cen, N.Y. Gnedin, and J.P. Ostriker, Astrophys. J. 417, 387 (1993); R. Cen and J.P. Ostriker, Astrophys. J. 429, 4 (1994).

[5] S.A. Bonometto and R. Valdarnini, Phys. Lett. 103A, 369 (1984); L.Z. Fang, S.X. Li, S.P. Xiang, Astron. Astrophys. 140, 77 (1984); A. Dekel and S.J. Aarseth, Astroph. J. 283, 1 (1984); Q. Shafi and F.W. Stecker, Phys. Rev. Lett. 53, 1292 (1984).

[6] M. Davis, F. Summers, and D. Schlegel, Nature 359, 393 (1992).

[7] A. Klypin, J. Holtzman, J.R. Primack, and E. Regős, Astrophys. J. 416, 1 (1993).

[8] C.J. Copi, D.N. Schramm, M.S. Turner, Science 267, 192 (1995); R.A. Malaney and G.J. Mathews, Phys. Rep. 229, 145 (1993).

[9] A. Dressler et al., Astrophys. J. 313, L37 (1987); D. Lynden-Bell et al., Astrophys. J. 326, 19 (1988).

[10] S.D.M. White, G. Efstathiou, and C.S. Frenk, Mon. Not. R. Astron. Soc. 262, 1023 (1993).

[11] J. Holtzman, Astrophys. J., Supp. 71, 1 (1989).

[12] E.L. Wright, Astrophys. J. 396, L13 (1992).

[13] J.A. Holtzman and J.R. Primack, Astrophys. J. 405, 428 (1993). Cf. T. van Dalen and R.K. Schaefer, Astrophys. J. 398, 33 (1992); R.K. Schaefer and Q. Shafi, Phys. Rev. D 47, 1333 (1993).

[14] A. Renzini, in *Observational Tests of Cosological Inflation*, eds. T. Shanks et al. (Dordrecht: Kluwer, 1991). B. Chaboyer, Astrophys. J., Lett., submitted (1994); the latter reference includes a table showing the effect on the GC ages of more than 20 possible changes in input physics and abundances.

[15] X. Shi, Astrophys. J., submitted (1994).

[16] R.A. Malaney, G.J. Mathews, and D.S.P. Dearborn, Astrophys. J. 345, 169 (1989); G.J. Mathews and D.N. Schramm, Astrophys. J. 404, 468 (1993).

[17] D.E. Winget et al., Astrophys. J. 315, L77 (1987).

[18] D.A. VandenBerg et al., Astron. J. 100, 445 (1990); ibid., 102, 1043 (1991).

[19] J.W. Yuan, Astron. Astrophys. 224, 108 (1989); 261, 105 (1992). V. Weidemann, Ann. Rev. Astron. Astrophys. 28, 103 (1990). M.A. Wood, Astrophys. J. 386, 539 (1992). M. Hernanz et al. Astrophys. J. 434, 652 (1994).

[20] A. Sandage, in *Practical Cosmology: Inventing the Past*, 23rd Sass Fee lectures, ed. B. Binggeli and R. Buser (Berlin: Springer, 1995). A. Sandage and G.A. Tammann, in *Proc. 3rd "D. Chalonge School,"* Erice, Sept. 4-16, 1994, Carnegie Observatories preprint, in press (1995).

[21] J.P. Huchra, Science, 256, 321 (1992). S. Van den Bergh, Science 258, 421 (1992).

[22] G.H. Jacoby et al., Publ. Astron. Soc. Pac., 104, 599 (1992). R. Ciradullo, G.H. Jacoby, and J.L. Tonry, Astrophys. J. 419, 479 (1993).

[23] M. Fukugita, C.J. Hogan, and P.J.E. Peebles, Nature 366, 309 (1993).

[24] M. Rowan-Robinson, *The Cosmic Distance Ladder* (1985).

[25] M.J. Pierce et al., Nature 371, 385 (1994).

[26] A. Saha et al., Astrophys. J. 425, 14 (1994); A. Sandage et al., Astrophys. J. 423, L13 (1994); ibid., 401, L7 (1992); B.D. Schaefer, Astrophys. J., 426, 493 (1994).

[27] M.M. Phillips, Astrophys. J. 413, L105 (1993); M. Hamuy et al., Astron. J. 109, 1 (1995). A.G. Riess, W.H. Press, and R.P. Kirshner, Astrophys. J. 438, L17 (1995); they get $h = 0.67 \pm 0.07$.

[28] G.A. Tamann and S. Sandage, Astrophys. J. submitted (1995).

[29] J. Tonry, Astrophys. J. 373, L1 (1991).

[30] D. Branch, Astroph. J. 392, 35 (1992).

[31] B. Leibundgut and P.A. Pinto, Astroph. J. 401, 49 (1992). But cf. T.E. Vaughan et al., Astrophys. J. 439, 558 (1995).

[32] B.P. Schmidt, R.P. Kirschner, and R.G. Eastman, Astrophys. J. 395, 366 (1992).

[33] B.P. Schmidt et al., Astrophys. J. 432, 42 (1994).

[34] D. Branch, in prep. (1995). M. Best and R. Wehrse, Astron. Astroph. 284, 507 (1994). P. Ruiz-Lapuente et al., Astrophys. J. 439, 60 (1995).

[35] M. Birkinshaw, J.P. Hughes, and K.A. Arnoud, Astrophys. J. 379, 466 (1991).

[36] M. Birkinshaw and J.P. Hughes, Astrophys. J. 420, 33 (1994).

[37] K. Yamashita, in *New Horizon of X-ray Astronomy — First Reslts from ASCA*, eds. F. Makino and T. Ohashi (Universal Academy Press, Tokyo, 1994), p. 279.

[38] T.M. Wilbanks et al., Astrophys. J. 427, L75 (1994), and parallel session talk at Snowmass 94.

[39] Y. Rephaeli, private communication, December 1994.

[40] W.H. Press, G.B. Rybicki, and J.N. Hewitt, Astrophys. J. 385, 416 (1992).

[41] J. Pelt et al., Astron. Astroph. 286, 775 (1994).

[42] G.F.R.N. Rhee, Nature 350, 211 (1991).

[43] D.H. Roberts et al., Nature 352, 43 (1991).

[44] H. Danle, S.J. Maddox, and P.B. Lilje, Astrophys. J. 435, L79 (1994).

[45] E.L. Turner, R. Cen, and J.P. Ostriker, Astron. J. 103, 1427 (1992); X.-P. Wu et al., preprint (1995).

[46] Y. Suto, T. Suginohara, and Y. Inagaki, preprint astro-ph/9412090, submitted to Prog. Theor. Phys. Lett. (1994).

[47] B. Ratra and P.J.E. Peebles, Astrophys. J. 432, L5 (1994). M. Kamionkowski et al., Astrophys. J. 434, L1 (1994). M. Bucher, A.S. Goldhaber, and N. Turok, preprint hep-ph/9411206 (1994). K. Yamamoto, M. Sasaki, and T. Tanaka, preprint astro-ph/9501109 (1995).

[48] O. Lahav, P. Lilje, J.R. Primack, and M.J. Rees, Mon. Not. R. Astron. Soc. 251, 128 (1991).

[49] S.M. Carroll, W.H. Press, and E.L. Turner, Ann. Rev. Astron. Astrophys, 30, 499 (1992),

[50] D. Maoz and H.W. Rix, Astrophys. J. 416, 425 (1993).

[51] C. Kochanek, Astrophys. J. 419, 12 (1993).

[52] C. Kochanek, CfA preprint 3859 (1994); the 90% C.L. limit quoted there is $\Omega_\Lambda < 0.5$. Taking into account selection effects (cf. P. Helbig and R. Kayser, preprint astro-ph/9501013, 1995; P. Helbig, preprint astro-ph/9502041, 1995) raises this upper limit only slightly (C. Kochanek, private communication, Feb. 1995).

[53] E. Bunn and N. Sugiyama, preprint astro-ph/9407069 (1994).

[54] G. Goldhaber et al., Nucl. Phys. B, S38, 435 (1995).

[55] A. Dekel, Ann. Rev. Astron. Astroph. 32, 371 (1994). Cf. also M.A. Strauss and J.A. Willick, Phys. Rep. in press (1995).

[56] A. Dekel and M.J. Rees, Astrophys. J. 422, L1 (1994).

[57] A. Nusser and A. Dekel, Astrophys. J. 405, 437 (1993). Similar constraints have been derived by a different technique by Bernardeau, Juskiewicz, and Dekel.

[58] P.J.E. Peebles, *Principles of Physical Cosmology* (Princeton Univ. Press, 1992), esp. §20.

[59] M. Davis and P.J.E. Peebles, Astrophys. J. 267, 465 (1983).

[60] R. Somerville, R. Nolthenius, M. Davis, and J.R. Primack, in prep. (1995).

[61] H.J. Mo, Y.P. Jing, and G. Börner, Astron. Astrophys. (1993). W. Zurek *et al.*, Astrophys. J. (1994). Also ref. [60].

[62] K.B. Fisher, et al., Mon. Not. R. Astron. Soc. 267, 927.

[63] R. Kirschner, A. Oemler, and P. Schechter, AJ 84, 951 (1979).

[64] J.P. Huchra and M.J. Geller, Astrophys. J. 257, 423 (1982).

[65] M. Ramella, M.J. Geller, and J.P. Huchra, Astrophys. J. 344, 57 (1989).

[66] R. Nolthenius, A. Klypin, and J.R. Primack, preprint astro-ph/9410095, Astrophys. J., submitted (1994).

[67] R.G, Carlberg and H.M.P. Couchman, Astrophys. J. 340, 47 (1989).

[68] R. Nolthenius, A. Klypin, and J.R. Primack, Astrophys. J. 422, L45 (1994).

[69] P.J.E. Peebles, Astrophys. J. 344, 53 (1989); 362, 1 (1990); 429, 43 (1994).

[70] E. Branchini and R.G. Carlberg, Astrophys. J. 434, 37 (1994).

[71] A.M. Dunn and R. Laflamme, preprint LA-UR 94-2231 (1994).

[72] D. Zaritsky et al., Astrophys. J. 405, 464 (1993); D. Zaritsky and S.D.M. White, Astrophys. J. 435, 599 (1994).

[73] B. Little and S. Tremaine, Astrophys. J. 320, 493 (1987).

[74] A. Songaila et al., Nature 368, 599 (1994); R.F. Carswell et al. Mon. Not. R. Astron. Soc. 268, L1 (1994).

[75] D. Tytler and X. Fan, Bull. Am. Astron. Soc. 26, 1424 (1994).

[76] S.D.M. White and C.S. Frenk, Astrophys. J. 379, 52 (1991). S.D.M. White et al. Nature 366, 429 (1993).

[77] D.A. White and A.C. Fabian, Mon. Not. R. Astron. Soc., in press (1995).

[78] N.A. Bahcall and L.M. Lubin, Astrophys. J. 426, 513 (1994). M. Bartelmann and R. Narayan, preprint astro-ph/9411033, (1994).

[79] G. Falhman et al. Astrophys. J. 437, 56 (1994). N. Kaiser et al., preprint astro-ph/9407004 (1994).

[80] R.G. Carlberg et al., preprint (1994).

[81] This is shown visually in my group's CHDM simulations in D. Brodbeck et al., Astrophys. J. submitted (with accompanying video; high-resolution frames and low-resolution mpeg excerpts from the video can be viewed or downloaded from our WWW page http://physics.ucsc.edu/groups/cosmology.html); cf. also A. Klypin, R. Nolthenius, and J.R. Primack, preprint astro-ph/9502062 (1995). Observed cluster temperatures and even X-ray luminosities appear to be well fit, according to the first CHDM hydrodynamic cluster simulation, G. Bryan et al., Astrophys. J. 437, L5 (1994).

[82] D. Richstone, A. Loeb, and E.L. Turner, Astrophys. J. 393, 477 (1992).

[83] J.O. Burns et al., Astrophys. J. 427, L87 (1994).

[84] J.J. Mohr, A.E. Evrard, D.G. Fabricant, and M.J. Geller, Astrophys. J. in press (1995).

[85] Y.P. Jing, H.J. Mo, G. Börner, and L.Z. Fang, Proc. Potsdam Cosmology Workshop (Sept. 1994), preprint astro-ph/9411045, in press (1995); preprint 9412072, submitted to Mon. Not. R. Astron. Soc. 1995).

[86] D.A. Buote and J.C. Tsai, Astrophys. J., submitted (1994).

[87] J.P. Henry et al., Astrophys. J. 386, 408 (1992).

[88] C.C. Steidel, M. Dickinson, and S.E. Persson, Astrophys. J. 437, L75.

[89] G. Kauffman, preprint astro-ph/9409075 (1994).

[90] A. Klypin et al., in prep. (1995).

[91] M.G. Haehnelt, Mon. Not. R. Astron. Soc. 265, 727 (1993).

[92] N. Katz, T. Quinn, E. Bertschinger, and J.M. Gelb, Mon. Not. R. Astron. Soc. 270, L71.

[93] A.M. Wolfe, in *Relativistic Astrophysics and Particle Cosmology*, eds. C.W. Ackerlof, M.A. Srednicki (New York, New York Academy of Science, 1993), p. 281.

[94] K.M. Lanzetta, Publ/ Astron. Soc. Pac., 105, 1063, (1993); K.M. Lanzetta, A.M. Wolfe and D.A. Turnshek, Astrophys. J., in press (1995).

[95] H.J. Mo and J. Miralda-Escude, Astrophys. J. 430, L25 (1994); G. Kauffmann and S. Charlot, Astrophys. J. 430, L97 (1994); C.-P. Ma and E. Bertschinger, Astrophys. J. 434, L5 (1994).

[96] A. Klypin, S. Borgani, J. Holtzman, and J.R. Primack, Astrophys. J. in press (1995).

[97] L.J. Storrie-Lombardie, R.G. McMahon, M.J. Irwin, and C. Hazard, Astrophys. J. 427, L13-16 (1994); —-, *Proc. ESO Workshop on QSO Absorption Lines*, in press (1995).

[98] D. MacMinn and J.R. Primack, preprint astro-ph/9408076, submitted to Astroph. J. Lett.; D. MacMinn and J.R. Primack, in *TeV Gamma Ray Astrophysics*, ed. Heinz Völk and F. Aharonian, *Space Science Reviews*, in press (1995). Cf. Report of Working Group R2: Ground Based Gamma-Ray Astronomy, these Proceedings.

[99] D.O. Caldwell, in *Trends in Astroparticle Physics*, Stockholm, Sweden 22-25 September 1994, eds. L. Bergstrom, P. Carlson, P.O. Hulth and N. Snellman, Nucl. Phys. B, Proc. Supplements (in press, 1995). J.N. Wilford, *New York Times* (Jan. 31, 1995), p. 1. See also Caldwell's paper in these Proceedings.

[100] Y. Fukuda, Phys. Lett. B 335, 237 (1994).

[101] J.R. Primack, J. Holtzman, A. Klypin, and D.O. Caldwell, Phys. Rev. Lett. in press (1995). An inverted neutrino mass hierarchy with ν_e the most massive appears to be required if all the current experimental evidence for neutrino mass (solar and atmospheric neutrino deficits, and LSND) are valid, and r-process nucleosynthesis (responsible for production of the heavy elements) takes place in the hot ν bubble a few hundred km above the neutron star in Type II supernovae; see G.M. Fuller, J.R. Primack, and Y. Qian, preprint astro-ph/9502081, submitted to Phys. Rev.

[102] See P.J. Steinhardt's plenary talk, in these Proceedings.

MeV and GeV Astronomy: Highlights from the Compton Observatory

Brenda L. Dingus

Universities Space Research Association, NASA/Goddard Space Flight Center, Code 662, Greenbelt, MD 20771

The Compton Gamma-Ray Observatory, one of NASA's two great observatories, was launched on April 5, 1991. The four instruments on the observatory have unprecedented capabilities that have opened the window of gamma-ray astronomy. A wide variety of astrophysical objects have been detected by these instruments. Active galactic nuclei at distances of billions of light years are observed to vary in less that a week. The number of known gamma-ray sources within our own galaxy has also increased. Six gamma-ray pulsars are now known, and several x-ray novae have been disovered. Evidence of nucleosynthesis is seen in gamma-ray lines from radioactive isotopes. While the number of gamma-ray bursts detected has increased greatly, the origin still remains a mystery. Finally, diffuse gamma-ray emission is used to probe cosmic rays and matter at distances far from the neighborhood of our Sun. Some of the highlights of these varied observations will be mentioned in this paper; however, only a subset can be given here. A more complete treatment is found in the proceedings published by AIP of three recent meetings - The Second Compton Symposium[1], Gamma-Ray Bursts[2], High Energy Phenomena in Solar Flares[3].

§1 Compton Observatory Mission

On April 5, 1991, the Compton Gamma-Ray Observatory (*Compton*) was launched on the space shuttle *Atlantis*, and gamma-ray astronomy received a major boost in sensitivity, angular resolution, and energy range. The spacecraft and the four scientific intruments of board have been operating almost flawlessy since the launch. (The failure of the tape recorders in March 1992 reduced the data coverage to $\sim$ 65%, but the addition of a tracking station in Australia in December 1993 improved the coverage to over 80%.) The mission duration is planned to be 5 to 10 years.

The four scientific instruments - EGRET, COMPTEL, OSSE, and BATSE - on *Compton* combine to cover the energy range from 15 keV to 30 GeV. All instruments operate simultaneously and have overlapping fields of view. Figure 1 is an illustration of the four instruments on the spacecraft. Figure 2 delineates the major characteristics of each of the instruments.

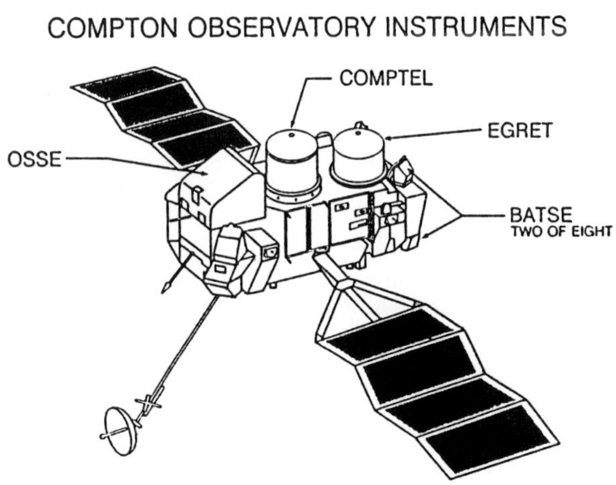

Figure 1: Compton Gamma-Ray Observatory with the 4 scientific instruments labelled. The spacecraft weighs nearly 16000 kg with 6000 kg for the scientific payload, uses less than 2.5 kW power, and is 3-axis stabilized.

1.1 EGRET

The Energetic Gamma Ray Experiment Telescope (EGRET) detects gamma rays which pair produce in its central tracking detector. Therefore, this telescope is sensitive to highest energy gamma-rays detectable by *Compton*, and has significant sensitive area over the broad energy range of 30 MeV to 30 GeV. The wide field of view of $\sim$ 0.6 sr is possible by measuring the direction of the electron-positron pair. Point sources can be localized with an accuracy of 5-30 arcmin, and the energy is measured to $\sim$ 20% by a block of NaI(Tl) located below the tracking

	OSSE	COMPTEL	EGRET	BATSE LARGE AREA	BATSE SPECTROSCOPY
ENERGY RANGE (MeV)	0.06 to 10.0	0.8 to 30.0	20 to 3 x 10^4	0.03 to 1.9	0.015 to 110
ENERGY RESOLUTION (FWHM)	12.5% at 0.2 MeV 6.8% at 1.0 MeV 4.0% at 5.0 MeV	8.8% at 1.27 MeV 6.5% at 2.75 MeV 6.3% at 4.43 MeV	~20% 100 to 2000 MeV	32% at 0.06 MeV 27% at 0.09 MeV 20% at 0.66 MeV	8.2% at 0.09 MeV 7.2% at 0.66 MeV 5.8% at 1.17 MeV
EFFECTIVE AREA (cm^2)	2013 at 0.2 MeV 1480 at 1.0 MeV 569 at 5.0 MeV	25.8 at 1.27 MeV 29.3 at 2.75 MeV 29.4 at 4.43 MeV	1200 at 100 MeV 1600 at 500 MeV 1400 at 3000 MeV	1000 ea. at 0.03 MeV 1800 ea. at 0.1 MeV 550 ea. at 0.66 MeV	100 ea. at 0.3 MeV 127 ea. at 0.2 MeV 52 ea. at 3 MeV
POSITION LOCALIZATION (STRONG SOURCE)	10 arc min square error box (special mode; 0.1 x Crab spectrum)	0.5 - 1.0 deg (90% confidence 0.2 x Crab spectrum)	5 to 10 arc min (1σ radius; 0.2 x Crab spectrum)	3° (strong burst)	-----
FIELD OF VIEW	3.8° x 11.4°	~ 64°	~ 0.6 sr	4 x sr	4 x sr
MAXIMUM EFFECTIVE GEOMETRIC FACTOR (cm^2 sr)	13	30	1050 (~ 500 MeV)	15000	5000
ESTIMATED SOURCE SENSITIVITY (5 x10^5 sec; on source, off Galactic Plane) — LINE	(3-8) x 10^{-5} cm^{-2} s^{-1}	1.5 x 10^{-5} to 6 x 10^{-5} cm^{-2} s^{-1}			0.4% equivalent width (5 sec integration)
CONTINUUM	3 x 10^{-7} cm^{-2} s^{-1} keV^{-1} (@1 MeV)	1.6 x 10^{-4} cm^{-2} s^{-1} (3 σ detection, 1-30 MeV)	7 x 10^{-8} cm^{-2} s^{-1} (> 100 MeV) 2 x 10^{-8} cm^{-2} s^{-1} (> 1000 MeV)	3 x 10^{-8} erg cm^{-2} (1 sec–burst)	

Figure 2: Scientific capabilities of the 4 instruments on the Compton Gamma-Ray Observatory

detector. The imaging of individual gamma rays as well as the active anticoincidence shield and directional, time-of-flight trigger results in negligible background contamination from the abundant charged particles in low earth orbit. The Co-Principal Investigators of EGRET are Dr. C. E. Fichtel of the NASA Goddard Space Flight Center, Dr. K. Pinkau of the Max-Planck Institute for Plasma Physics, and Dr. R. Hofstadter (deceased) of Stanford University.

1.2 COMPTEL

The Compton Telescope (COMPTEL) detects gamma rays which Compton scatter in an upper layer of liquid scintillator followed by a second interaction in a lower layer of NaI(Tl). The measurement of the energy and position of both of these interactions gives the incident gamma ray direction to within a ring on the sky. The field of view is very large (~ 1 sr), and point sources can be located to within about 1 deg. The energy range is 0.8 MeV to 30 MeV and the energy resolution is about 9% at 1 MeV. Active anticoincidence shields, time-of-flight, and pulse shape discrimination are all used to reject background. The Principal Investigator of COMPTEL is Dr. V. Schoenfelder of the Max-Planck Institute for Extraterrestrial Physics.

1.3 OSSE

The Oriented Scintillation Spectroscopy Experiment (OSSE) consists of four collimated, NaI(Tl)-CsI(Tl) phoswich detectors for gamma-rays from 0.1 to 10 MeV. The four detectors operate as pairs alternating between measuring background and on source flux. The field of view is 3.8 deg $\times$11.4 deg and the energy resolution is ~ 8% at 0.661 MeV. In the normal mode of operation, energy spectra are collected every 4 sec, but this interval can be made as short as 1/8 of a msec. Because the detectors can slew 192 deg independent of the spacecraft, a secondary target is observed when the primary target is occulted by the earth. The Principal Investigator of OSSE is Dr. J. D. Kurfess of the U.S. Naval Research Laboratory.

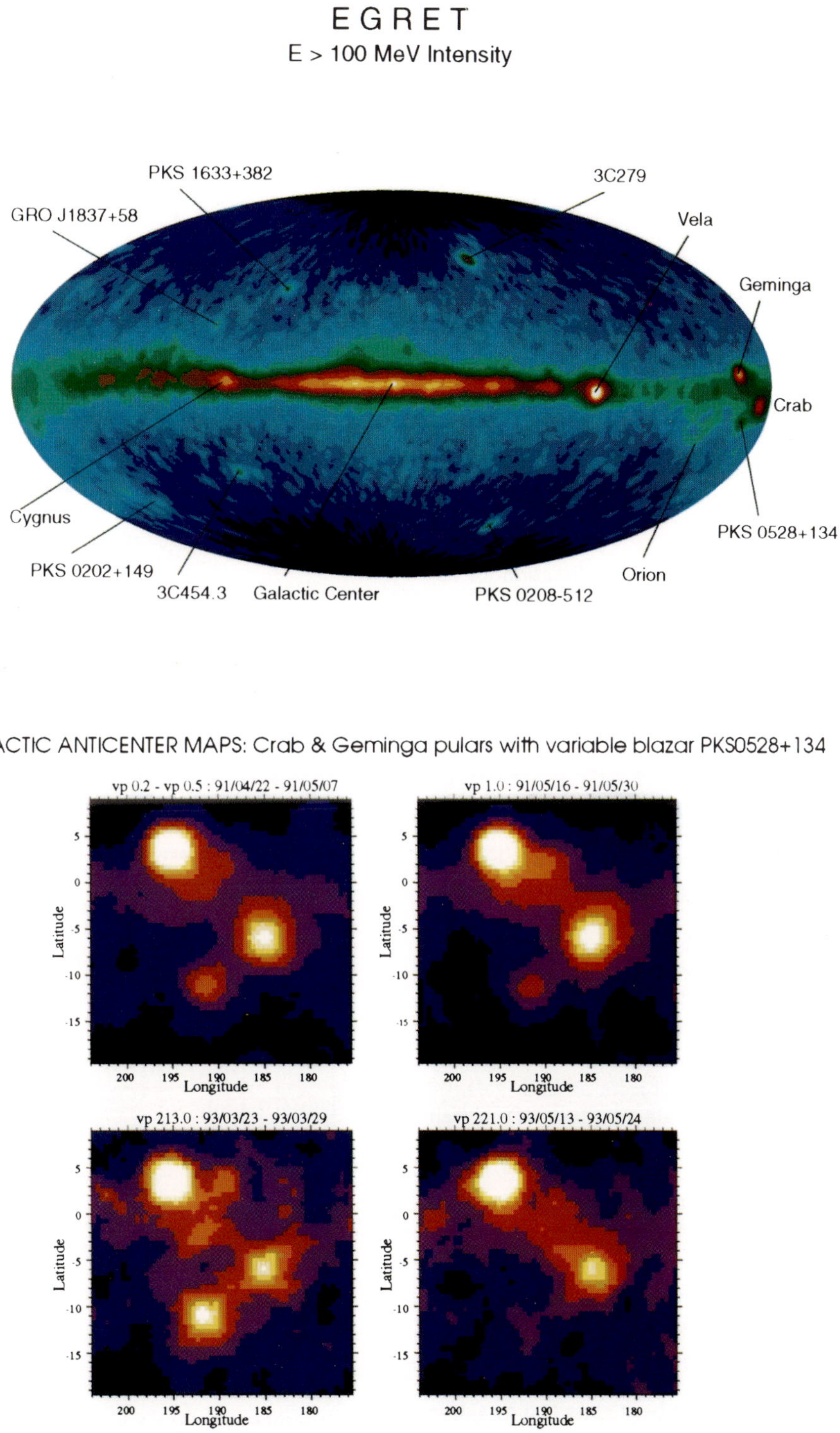

Figure 3: Gamma-Ray (> 100 MeV) Intensity Maps from EGRET. The top map is in Galactic coordinates and is the sum of over 40 overlapping observations. The four bottom maps are different observations of the Galactic anticenter region. The two bright, steady sources are the Crab and Geminga pulsars, and the third variable source is the active galactic nuclei PKS0528+134 which is one of the most distant objects detected by EGRET at $z = 2.06$.

1.4 BATSE

The Burst and Transient Source Experiment (BATSE) observes nearly the full sky simultaneously with 8 identical modules at each corner of the spacecraft. Each module consists of a Large Area Detector (LAD) of NaI(Tl), 50 cm diameter by 1.3 cm thick, and a Spectroscopy Detector (SD) also of NaI(Tl) but 13 cm diameter by 8 cm thick. The onboard electronics identifies transient events and stores the data from these events for transmission to the ground. Continuous data is also sampled at ~1 sec intervals and allows studies of steady sources by epoch-folding and earth occultation techniques. BATSE also provides a gamma-ray burst trigger for the OSSE, COMPTEL, and EGRET so that higher energy data is collected for these events. The Principal Investigator of BATSE is Dr. G. J. Fishman of NASA Marshall Space Flight Center.

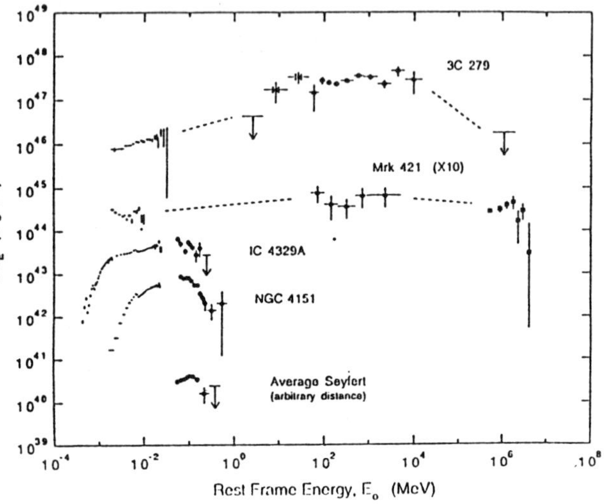

Figure 4: Typical X-ray and gamma-ray energy spectra of two classes of AGN.[5]

1.5 Observing Plan

The satellite typically points in the same direction for approximately 2 weeks; however, the spacecraft has been reoriented several times to allow deeper observations of flaring objects. The sun and several x-ray binaries are examples of these targets of opportunity. While the initial observation phases had guaranteed time for the instrument teams observations, currently and in the future all observing time is awarded purely by peer reviewed proposals. Proposal cycles typically last one year, and the current cycle proposals were due last April. Further information on past and future observations can be obtained over internet from grossc.gsfc.nasa.gov and using the login name gronews. The Project Scientist is Neil Gehrels at NASA Goddard Space Flight Center.

The first year and a half of the mission was devoted to a survey of the whole sky. EGRET has completed the first all sky survey above 100 MeV and the intensity map is shown in Figure 3. The diffuse emission from the galaxy is obvious as are many discrete objects at high galactic latitudes and a few within the galactic plane. Over half of the point sources have been identified with active galactic nuclei or pulsars. A catalog of the EGRET detected sources is published.[4] Also, obvious from the first observations is the variability of the high energy emission which can be seen from Figure 3 in the 4 different time intervals that the galactic anticenter was observed.

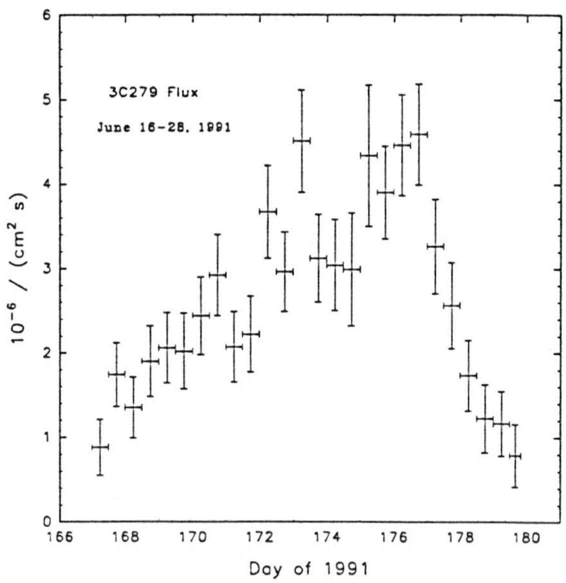

Figure 5: Light curve of the > 100 MeV flux from the blazar 3C 279 which is located at a redshift of 0.538.[9]

§2 Active Galactic Nuclei (AGN)

A few percent of all galaxies have been classified as active galactic nuclei (AGN) due to their bright central nucleus.

Two subsets of AGN have been detected by *Compton*. The soft gamma-ray emitters, observed by OSSE and known as Seyferts, have weak radio emission, and the hard gamma-ray emitters, observed by EGRET and known as blazars, have strong and hard radio spectra. Typical x-ray and gamma-ray spectra for these two subclasses are shown in Figure 4.[5] The spectra from Seyferts can be fit by thermal Comptonization models[6] which break at a few 100 keV, yet the blazar spectra show no evidence of a high energy cut off. The nearest and one of the weakest EGRET detections is Mrk 421, and this AGN has been detected by the Whipple Observatory to TeV energies.[7]

Theoretical models of AGN have geometrical configurations similar to the drawing in Figure 6 and 7 [8] where different features are observed from different regions and therefore depend on the viewing angle. The main difference between the radio quiet and radio bright AGN is the presence of a jet. The association of the EGRET detections with radio bright objects, and therefore jets, is consistent with the variability observed by EGRET on the timescale of a few days for the blazar 3C279.[9] Figure 5 is the light curve of the bright flare from 3C 279.

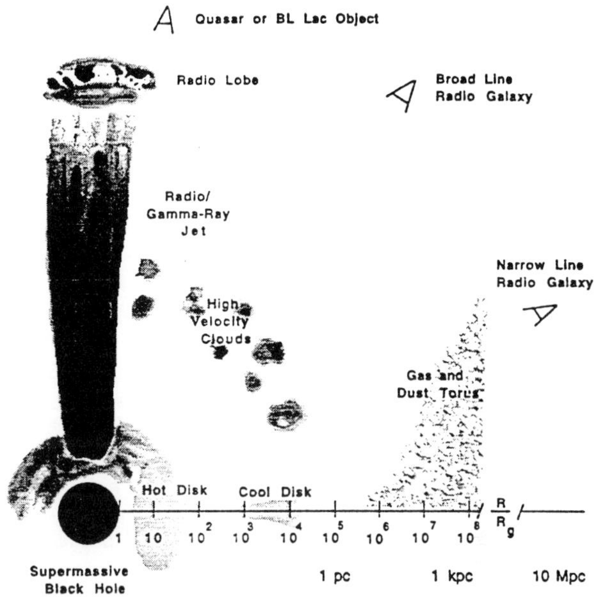

Figure 6: Schematic of Radio Bright AGN[8]. The AGN with > 100 MeV emission probably have the jet nearly aligned with our line of sight.

§3 Isolated Pulsars

Prior to the launch of *Compton*, only two isolated pulsars Crab and Vela were known to emit gamma-rays. Four more pulsars are now known. The light curves at different wavelengths are shown in Figure 8. Except for the Crab, pulsar light curves vary with wavelength, indicating a complex geometry with different energies emitted from different regions. Even the spectra above 30 MeV are observed to vary with phase as seen in the brightest gamma-ray pulsar, Vela.[10] The shape of the light curves and the energy spectra are used to constrain the emission site. Possible sites range from the polar cap of the neutron star[11] out to nearly the speed of light cylinder.[12]

The gamma-ray spectra are relatively flat, and the observed power per decade of energy is dominated by the gamma-ray emission. A large fraction of the rotational spin down energy is being emitted in gamma-rays. A general trend has been observed that the older the pulsar, the larger the fraction of the rotational energy is emitted in gamma rays, which can be nearly 100% depending on the solid angle of the gamma-ray beam.[13]

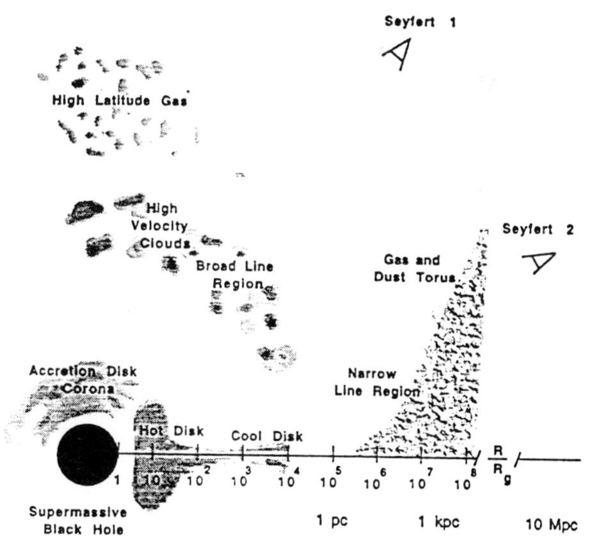

Figure 7: Schematic of Radio Quiet AGN[8]. These AGN have no radio jet and the gamma-ray emission cuts off rapidly below a few 100 keV.

§4 X-ray Binaries

Both BATSE and OSSE have detected several x-ray binaries. Althought BATSE is not an imaging detector, point sources are detected by epoch folding and by measuring the change in gamma-ray flux before and after a source is occulted by the earth. In this manner, BATSE has

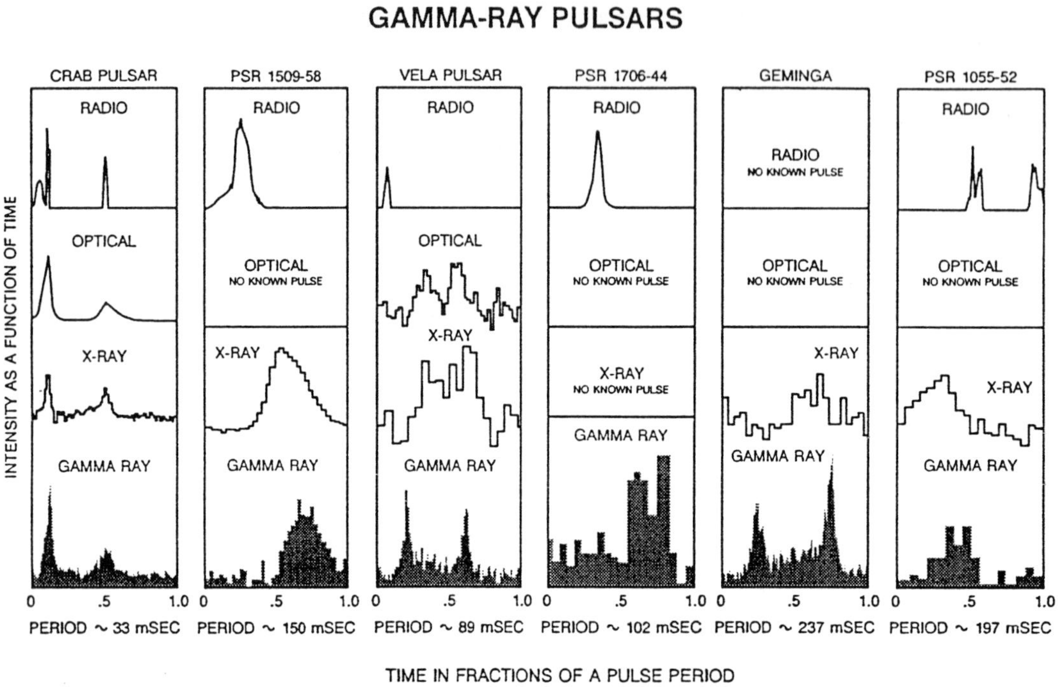

Figure 8: Light curves of the 6 rotation-driven pulsars with detected gamma-ray emission.

observed over a dozen accretion-driven pulsars and has provided an essentially continuous monitor of their period and flux.[14] Many of these sources are variable. For example, the January 1992 outburst of Be/x-ray binary system A1118-616 was detected by BATSE, and lead to a major multiwavelength study.[15]

Another subclass of binaries are x-ray novae. Many of these sources are believed to contain black hole primaries which undergo extreme accretion-induced transient episodes. The brightest source ever seen at 100 keV was GRO J0422+32 which was discovered by BATSE using earth occultation measurements and detected by COMPTEL up to 2 MeV.[16]

§5 Solar Flares

BATSE has detected over 700 solar flares with emission above 60 keV. During 1991 June the Sun was particularly active with active region 6659 producing 6 X-class flares. A target of opportunity was declared and the spacecraft, and all the *Compton*instruments pointed at the Sun. The longest duration, highest energy flare observed to date was on 1991 June 11. EGRET detected emission up to 1 GeV and the > 50 MeV emission lasted 8 hours.[17]

§6 Nucleosynthesis

Radioactive isotopes are produced in supernovae and several isotopes are observable as gamma-ray lines. ^{57}Co was detected by OSSE in SN1987A and constrains isotopic abundances.[18] These abundances are significantly less than those interpreted from the inflection in the bolometric light curve.[19]

^{44}Ti has a 78 year half-life and should be a useful tracer of recent supernovae. COMPTEL has detected the 300 year old supernova remnant, Cas A in the 1.16 MeV line of ^{44}Ti.[20] The abundance of ^{44}Ti interpreted from the flux is consistent with nucleosynthesis calculations within the uncertainties of the data and the model. OSSE has also observed Cas A without detecting ^{44}Ti emission, and the upper limit is just below the COMPTEL detection.[21]

^{26}Al has a decay time of one million years, and can be detected by the 1.809 MeV gamma-ray line. Production of this isotope is likely in core-collapse supernovae or Wolf-Rayet stars. COMPTEL detects emission from this line from the Galactic plane as shown in Figure 9.[22] While the emission is highly structured, no known sources – with the possible acception of the Vela supernova remnant – are apparent.

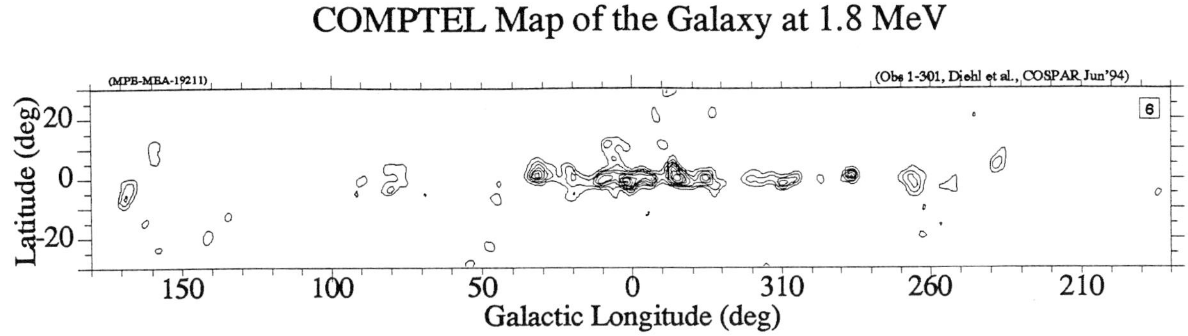

Figure 9: Map of the gamma-ray line emission from ^{26}Al measured by COMPTEL.

§7 Gamma-Ray Bursts

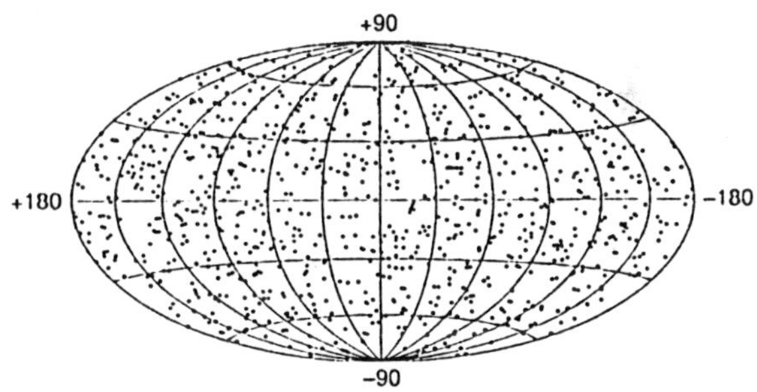

Figure 10: Positions in Galactic coordinates of the first 1000 bursts detected by BATSE. The distribution is consistent with isotropy.

Approximately once a day, a flash of gamma rays lasting a fraction of a second to a few minutes is detected by BATSE. The direction to the source is calculated from the relative intensities of the emission in each of the 8 detector modules which have different but overlapping fields of view. The systematic uncertainty in this technique is $\sim$ 4 degrees. The locations of the first 1000 bursts are shown in Figure 10. The distribution of these directions is consistent with isotropy. However, if the sources are associated with the normal stellar population in our galaxy, then more gamma-ray bursts should come from the galactic plane. If the sources are in a galactic halo centered on our galaxy, then, because we are not at the center of the galaxy, there should be a broad enhancement towards the galactic center. However, if the halo is large enough, the enhancement would not be detectable with only 1000 bursts. If other galaxies have large halos, then we should be able to detect nearby galaxies, such as M31. These constraints combine to place a limit on the range of distances of a galactic population of bursts.[23]

If gamma-ray bursts are extragalactic, then one expects an isotropic distribution of sources. Also, the deficit of weak bursts seen in Figure 11 is explained by cosmological effects if these bursts are at a distances with redshift of near one.[24] In addition, the time structure of weak bursts is longer than bright bursts which is also consistent with cosmological effects.[25] However, these effects also could be consistent with a galactic population. An extragalactic origin would be difficult if sources repeat, since the luminosity of an extragalactic burst is so large that it must be a catastrophic event. Evidence of repetition has been reported[26], but the statistical significance is weak. Recent observations of high-velocity neutron stars could provide a galactic halo population of large size.[27]

Therefore, the distance to gamma-ray bursts is uncertain by many orders of magnitude leading to larger uncertainty in the luminosity. Identification of a gamma-ray burst source at lower energies could be very useful to determine the source distance, and higher energies could place an upper limit on the distance to the source because of the opacity due to photon-photon pair production with the infrared and microwave backgrounds. Optical, radio and ultra-high energy gamma-ray searches have been done, but sensitive and simultaneous observations are difficult. BATSE now has a rapid notification network which announces burst locations within a few seconds of the burst; however, the position uncertainties are at best a few degrees. Better positions can be determined by triangulation if several spacecraft observe the burst. Also, both EGRET and COMPTEL can image the individual gamma rays to measure the burst direction, but only the brightest bursts are observed at the highest energies. A

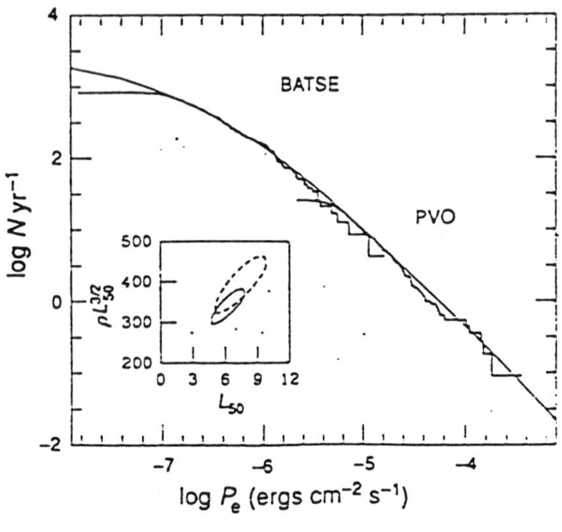

Figure 11: Cumulative frequency distribution of burst flux. For a homogeneous population of sources, this curve would be fit by a -3/2 power law. However, the distribution clearly shows a deficit of weak bursts which can be explained as seeing to the far edge of the source population. A fit to a cosmological model is shown with the derived intrisic burst luminosity described by the inset graph.[24]

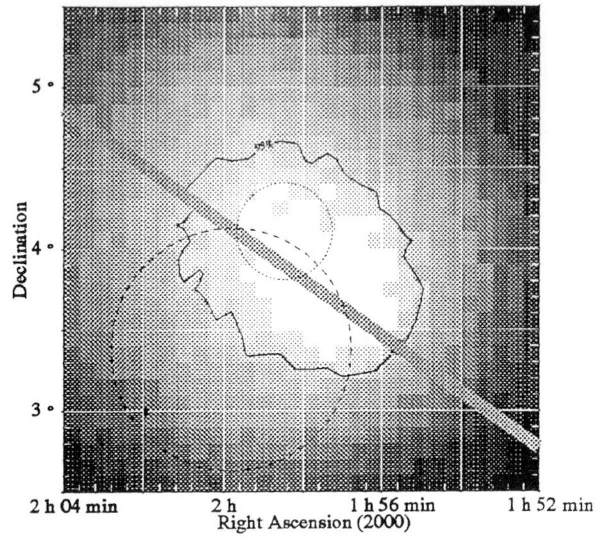

Figure 12: Localizations of the 1994 Feb 17 burst. The EGRET and COMPTEL 95% confidence intervals are given by the solid and dashed line, respectively. The broad line is the position estimated by triangulation from the Ulysses and BATSE lightcurves. The small dashed circle is the position of an 18 GeV gamma-ray which was detected by EGRET over an hour after the burst.

bright burst localization is shown in Figure 12. [28]

Soft Gamma Repeaters (SGR) are an example of a small subset of gamma-ray bursts which are identified with a known source. As suggested by their name, these sources have a softer spectrum than classical gamma-ray bursts, and several episodes of emission have been detected from each of the three known objects. Because they repeat, it has been possible to image one of the sources with a positional uncertainty of one arcmin using ASCA.[29] Radio observations of this position indicate a supernovae remnant.[30] The first SGR ever detected was also coincident with a supernovae remnant, N49 in the Large Magellanic Cloud;[31] however, the coincidence was believed by many to be accidental because of the large luminosity implied by the distance.

§8 Diffuse Emission

Gamma rays are obviously produced in the interactions of cosmic ray nuclei and electrons with matter and to a lesser extent with photons. The gamma-ray spectrum from these interactions has been predicted[32, 33] and is shown in Figure xx along with the EGRET measurement of the spectrum from a broad region near the Galactic center.[34] The spatial distribution of the diffuse emission above 100 MeV has been modeled using radio observations of atomic hydrogen and CO.[35] The model assumes that the ratio of CO to molecular hydrogen is constant in the Galaxy and that the cosmic ray density is coupled to the matter density. The model is compared to the EGRET data in order to search for point source excesses. A similar model has been made for the energy range of COMPTEL.[36]

The gamma-ray emission from molecular clouds has also been detected by EGRET,[37, 38] and some surprising results have come from the COMPTEL observations of Orion.[39] COMPTEL has observed gamma-ray line emission due to carbon and oxygen from the Orion region, which has been interpreted as an enhancement of low energy cosmic rays.

Electron-positron annihilation emits 511 keV gamma radiation, and OSSE has studied this emission from the Galactic center region. OSSE measurements are best fit with bulge of 10 degrees FWHM size centered on the galactic center and a disk extending along the galactic plane ± 20 degrees in longitude. No evidence of variability is seen by OSSE.[40]

The diffuse emission from other galaxies is generally too weak, due to the large distances, to be detected: however, the Large Magellanic Cloud has been detected by EGRET with a flux consistent with a quasi-stable equilibrium calculation.[41] The nondetection of the Small Magellanic Cloud by EGRET has yielded an interesting upper limit which proves that cosmic rays are not uniformly distributed but are confined to galaxies.[42]

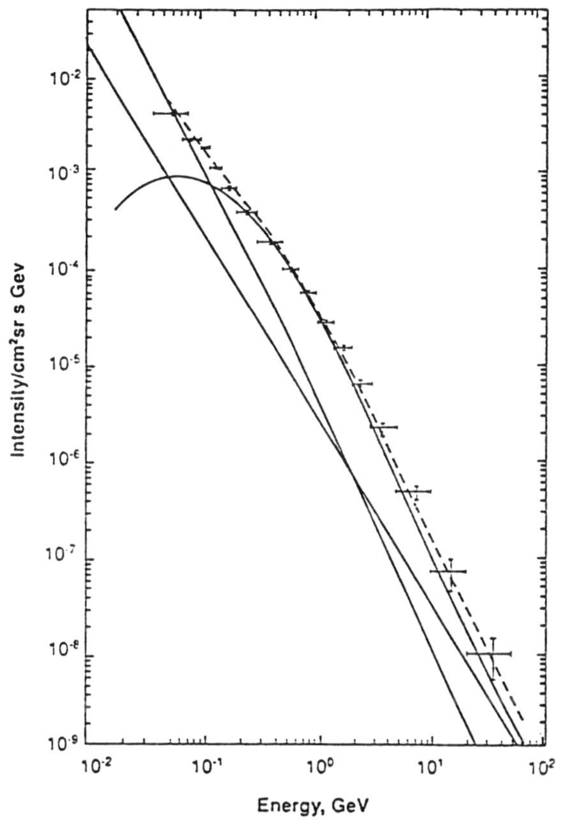

Figure 13: Energy spectrum of the gamma-ray emission from within 10 degrees in longitude and 5 degrees in latitude of the Galactic center. The expected spectrum is indicated by the dotted line with the various components as solid lines. The broad bump is due to neutral pion decay from nucleon-nucleon interactions. The steep power law is from electron bremstrahlung, and the flatter power law is from inverse Compton.

§9 The Future

The Compton Gamma-Ray Observatory will operate for several more years; however, more sensitive instruments are needed to answer the many questions raised by the exciting results of the first few years of data. The European Space Agency will launch INTEGRAL, an imaging and spectroscopy telescope for the range of xxxx to xxx MeV. Several groups are also working on prototypes for a successor to EGRET with improved angular resolution and sensitivity above 100 MeV; however, no funding is yet available for a mission.

§10 Acknowledgments

I would like to acknowledge the many people who worked on the construction of the Compton Observatory and the analysis of the data. I specifically recognize Neil Gehrels, Hans Bloemen, Chuck Dermer, and Dave Thompson for their assistance. Any opinions, misrepresentations, or omissions are my responsibility and should not be attributed to others. This work has been supported by a NASA grant.

§References

[1] C. E. Fichtel, N. Gehrels, and J.P. Norris, editors. *The Second Compton Symposium.* AIP Vol 304, 1994.

[2] K. Hurley G.J. Fishman, J.J. Brainerd, editor. *Gamma-Ray Bursts Second Workshop.* AIP Vol 307, 1994.

[3] J.M. Ryan and W.T. Vestrand, editors. *High Energy Solar Phenomena.* AIP Vol 294, 1994.

[4] C.E. Fichtel et al. The first Energetic Gamma-Ray Experiment Telescope EGRET source catalog. *Ap. J. Supp.*, 94:551–581, 1994.

[5] C.D. Dermer and Neil Gehrels. Two classes of gamma-ray emitting active galactic nuclei. *Ap.J.*, submitted, 1995.

[6] R.A. Sunyaev and L.G. Titarchuk. Comptonization of x-rays in plasma clouds. typical radiation spectra. *Astron. Astrophys.*, 86:121, 1980.

[7] M. Punch et al. Detection of TeV photons from the active galaxy markarian 421. *Nature*, 160:477, 1992.

[8] C.D. Dermer. Private communication, 1994.

[9] D.A. Kniffen et al. Time variability in the gamma ray emmision of 3c 279. *Ap. J.*, 411:133–136, 1993.

[10] G. Kanbach et al. EGRET observations of the Vela pulsar. *Astron. Astrophys*, 289:885–867, 1994.

[11] J.K. Daugherty and A.K. Harding. Polar cap models of gamma-ray pulsars: Emission from single poles of nearly aligned rotators. *Ap.J.*, 429:325–330, 1994.

[12] K. Chen and M. Ruderman. Pulsar death lines and death valley. *Ap. J.*, 402:264, 1993.

[13] D.J. Thompson et al. EGRET high-energy gamma-ray pulsar studies, part 1: Young spin-powered pulsars. *Ap. J.*, 436:229, 1994.

[14] G.J. Fishman et al. Overview of observations from BATSE on the compton observatory. *Astron. Astrophys. Supp.*, 97:17, 1993.

[15] M.J. Coe et al Multiwaveband Study of a Major X-ray outburst from the Be/X-ray Transient system A1118-616 C. E. Fichtel, N. Gehrels, and J.P. Norris, editors. *The Second Compton Symposium.* AIP Vol 304:187-191, 1994.

[16] R. van Dijk et al MeV Emission from the black-hole candidate GRO J0422+32 measured with COMPTEL C. E. Fichtel, N. Gehrels, and J.P. Norris, editors. *The Second Compton Symposium.* AIP Vol 304:197-201, 1994.

[17] G. Kanbach et al. Detection of long-duration solar gamma-ray flare on June 11, 1991 with EGRET on compton GRO. *Astron. Astrophys. Supp.*, 97:349–354, 1993.

[18] J.D. Kurfess et al. Oriented scintillation spectrometer experment observation of ^{57}co in SN1987A. *Ap. J. (Lett.)*, 399:L137, 1992.

[19] N.B. Suntzeff et al. The energy sources powering the late-time bolometric evolution of SN1987A. *Ap. J. (Lett.)*, 384:L33, 1992.

[20] A.F. Iyudin et al COMPTEL observations of ^{44}Ti Gamma-ray line emission from Cas A C. E. Fichtel, N. Gehrels, and J.P. Norris, editors. *The Second Compton Symposium.* AIP Vol 304:156-160, 1994.

[21] L.S. The et al OSSE observations of the Cassiopeia A Supernova Remnant C. E. Fichtel, N. Gehrels, and J.P. Norris, editors. *The Second Compton Symposium.* AIP Vol 304:180-186, 1994.

[22] R. Diehl et al. *Astron. Astrophys.*, in press, 1994.

[23] J. Hakkila et al. Constraints of Galactic distributions of gamma-ray burst sources from BATSE observations. *Ap.J.*, 422:659–670, 1994.

[24] E.E. Fenimore et al. The intrinsic luminosity of gamma-ray bursts and their host galaxies. *Nature*, 366:40–42, 1993.

[25] J.P. Norris. Detection of signature consistent with cosmological time dilation in gamma-ray bursts. *Ap.J.*, 424:540–545, 1994.

[26] J.M. Quashnock and D.Q. Lamb. Evidence that gamma-ray burst sources repeat. *Mon. Not. R. Astron. Soc.*, 265:L59–L64, 1993.

[27] A.G. Lyne and D.R. Lorimer. *Nature*, 369:127, 1994.

[28] K. Hurley et al. Exceptionally long duration, high energy gamma ray burst. *Nature*, in press, 1994.

[29] T. Murakami et al X-ray identification of SGR1806-20 G.J. Fishman, J.J. Brainerd, and K. Hurley, editors. *Gamma-Ray Bursts Second Workshop.* AIP Vol 307:489-494, 1994.

[30] S.R. Kulkarni and D.A. Frail. Identification of a supernova remnant coincident with the soft gamma-ray repeater SGR1806-20. *Nature*, 365:33–35, 1993.

[31] T.L. Cline et al. *Ap.J.*, 348:485, 1990.

[32] F.W. Stecker. *Cosmic Gamma Rays.* Mono Books Corp., Baltimore, 1971.

[33] C. Dermer. Secondary production of neutral pi-mesons and the diffuse galactic gamma radiation. *Astron. Astrophys.*, 157:223, 1986.

[34] C.E. Fichtel. Overview of the first results from EGRET. *Astron. Astrophys. Supp.*, 97:13–16, 1993.

[35] D.L. Bertsch et al. Diffuse gamma-ray emission in the galactic plane from cosmic-ray, matter, and photon interactions. *Ap. J.*, 416:587–600, 1993.

[36] A.W. Strong et al. Diffuse continuum gamma rays from the Galaxy observed by COMPTEL. *Astron. Astrophys*, in press, 1994.

[37] S.W. Digel et al. EGRET observations of gamma-ray emission from the interstellar gas in Orion. *Ap.J.*, in press, 1995.

[38] S.D. Hunter et al. *Ap.J.*, in press, 1995.

[39] H. Bloemen et al. COMPTEL observations of the Orion complex: evidence for cosmic-ray induced gamma-ray lines. *Astron. Astrophys.*, 281:L5–8, 1994.

[40] W.R. Purcell et al. OSSE observation of Galactic 511 keV positron annihilation radiation: Initial phase 1 results. *Ap.J.*, 413:L85, 1993.

[41] P. Sreekumar et al. Observations of the Large Magellanic Cloud in high-energy gamma rays. *Ap. J. (Lett.)*, 400:L67–70, 1992.

[42] P. Sreekumar et al. Constraints on the cosmic rays in the Small Magellanic Cloud. *Physical Review Letters*, 70:127–129, 1993.

High Energy Cosmic Ray Physics and Gamma-Ray Astronomy

René A. Ong

Enrico Fermi Institute, The University of Chicago, 5640 S. Ellis Ave. Chicago, IL 60637, USA

The cosmic radiation, first discovered early in this century, continues to provide mysteries as we approach the next century. From experiments using direct direction methods, we have a reasonable description of the cosmic rays below 100 TeV (10^{14} eV). Our understanding becomes less certain above that energy, and in particular the elemental composition is poorly known. As a result, at the present time we are unable to answer the basic question of where the high energy cosmic rays come from. In principle, astronomy using neutral radiation (photons or neutrinos) may help answer this question. Remarkably, because of exciting new results from the Compton Gamma Ray Observatory and from ground-based Cherenkov telescopes, high energy gamma-ray astronomy has clearly emerged as a field of its own, with its own set of questions. This talk summarizes the current status of the fields of high energy cosmic ray physics and gamma-ray astronomy and discusses the outlook for future progress in these fields.

§1 Overview

Cosmic rays are remarkable for a variety of reasons. Firstly, they reach tremendous energies (E> 10^{20} eV) and therefore provide evidence for powerful astrophysical accelerators. In addition, cosmic rays are abundant, with a galactic energy density of ~1 eV/cm^3. This energy density is comparable to that contained in starlight, the galactic magnetic field, or in the microwave background radiation. The energy sources needed to power the cosmic radiation must have a total luminosity exceeding 10^{40} ergs/sec. We have evidence that cosmic rays have existed for as long as the galaxy itself (~ 10^{10} years). Yet, we can estimate the galactic containment time to be on the order of ten million (10^7) years. Therefore, these energy sources must be renewed during the life of the galaxy in order to continually replenish the cosmic rays. Cosmic ray particles represent the largest source of material reaching earth from outside of our solar system. Therefore, although modulated by propagation effects, the chemical composition of these particles reflects the nucleosynthetic processes occurring at their origin.

A general talk on the subject of cosmic rays has a great deal of ground to cover. For one thing the energy range involved is enormous, from the MeV and GeV radiation produced by our Sun to extremely high energy events beyond 10^{20} eV. Furthermore, cosmic rays come in many flavors, including charged particles (electrons, protons, helium, and heavier nuclei), their antiparticles, and neutrals (photons, neutrons, and neutrinos). For the purposes of this talk I concentrate on two topics of particular interest:

- Nuclei in the energy range between 10^{13} and 10^{17} eV, centered around the region of the "knee" in the cosmic ray spectrum.
- Gamma-rays above 20 GeV, which is the upper end of the energy range covered by current space-based gamma-ray instruments.

These topics are the subject matter for working groups R3 and R2 of this workshop, respectively.

This talk is one of three overview talks at Snowmass on cosmic rays and gamma-ray astronomy. The subject of gamma-rays below 20 GeV are covered in the talk of Brenda Dingus. The study of the highest energy cosmic rays (above 10^{17} eV) are discussed by Alan Watson.

§2 Cosmic Rays Between 10^{13} and 10^{17} eV

2.1 Current Status

Spectrum: A compilation of data on the high energy cosmic ray spectrum is shown in Figure 1 [1]. This figure shows the "all-particle" spectrum, i.e. the differential flux of the total number of cosmic ray particles as a function of energy where no identification by particle type has been made. On the vertical scale the flux has been multiplied by a factor of $E^{2.5}$ which serves to flatten out the spectrum. Although accentuated by the flattening factor, the primary feature is the break in the spectrum (the "knee") around 3×10^{15} eV. Other than this break, the spectrum exhibits power-law behaviour which is in itself important

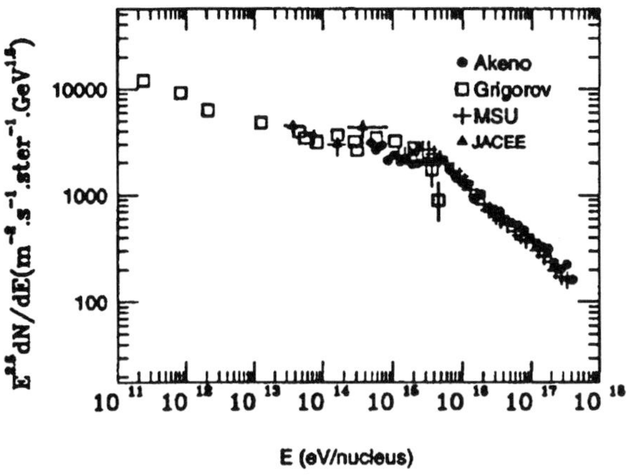

Figure 1: A compilation of data on the all-particle cosmic ray spectrum [1].

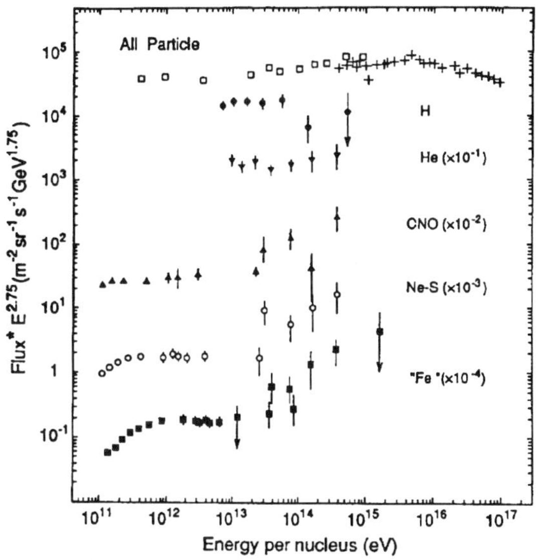

Figure 2: Compilation of data on the composition of high energy cosmic rays [6]. The data come from the HEAO spacecraft, the CRN detector on Spacelab-2, and the JACEE balloon packages. The all-particle spectrum is shown at the top of the figure.

because it shows that the production of the high energy cosmic rays cannot be due to thermal processes like those successful at explaining diffuse radiation at lower energies. The data in Figure 1 come from experiments on satellites and balloons at lower energies [2][3][4], and from the Akeno air shower experiment at higher energies [5].

Composition: From Figure 1 it is clear that the total particle spectra is reasonably well determined over the energy range $10^{13}-10^{17}$ eV. What about the particle type or composition? Figure 2 shows a compilation of measurements of the elemental composition. In this figure, the differential flux has been multiplied by a flattening factor of $E^{2.75}$ and the spectra for the different charge groups are plotted as a function of the total particle energy. This latter feature permits an easy comparison with the all-particle spectrum even though the experiments contributing data to this figure typically measure the Lorentz factor of particles (i.e. the energy per nucleon).

Figure 2 shows that the cosmic ray composition is reasonably well known at energies below 5×10^{14} eV. These data come from experiments on balloons or spacecraft that detect cosmic ray particles above the 25 radiation lengths of the earth's atmosphere. The experiments make *direct measurements* of the particle properties– in other words, they are sensitive to the first interaction of the cosmic ray particle. Above 5×10^{14} eV the cosmic ray flux is so greatly reduced that measurements using direct techniques have so far been impossible due to the limited detector area that can be carried aloft by a balloon or spacecraft. Ironically, our ignorance in the elemental composition of the cosmic rays occurs near the knee of the all-particle spectrum. The possibility that the knee signals a change in the origin of the cosmic rays is an important driving force to better determine the particle type near 10^{15} eV and beyond. At these high energies cosmic rays can be detected by ground-based experiments that make use of the extensive air showers created when a cosmic ray interacts in the atmosphere. We discuss such *indirect measurements* of cosmic rays by air shower experiments in Section 2.3.

Anisotropy: Since cosmic rays are predominantly charged, their trajectories are altered by the magnetic fields they encounter on their journey from their production to earth. At low energies cosmic rays are affected by solar modulation and by the earth's magnetic field. At energies above 10^{10} eV, however, the only important deflection is caused by the galactic magnetic field of a few μG. For a particle of energy 10^{15} eV, this field strength results in a radius of curvature of ~ 0.3 pc. Such a short radius (on galactic length scales), coupled with the fact that the magnetic field itself is quite inhomogeneous, leads one to expect that any directional information would be completely lost by the time the cosmic rays reach Earth. Indeed, experimental studies of the arrival directions of cosmic rays indicate that they are highly isotropic at these energies. Below 10^{16} eV, the amplitude of the first term in a simple harmonic analysis is found to be less than 0.1%

[7]. The high degree of isotropy means that we cannot expect to learn anything about the origin of cosmic rays by studying their arrival directions. We can deduce something about the nature of the acceleration site by accurate measurements of the composition and spectrum. In addition, we can try to directly pinpoint powerful astrophysical accelerators by the detection of *neutral* radiation from them. The possibility of detecting high energy gamma-rays from astrophysical sources is discussed later in this paper and in the report from Group R2 of this workshop [8].

Conventional Understanding: Even though we cannot pinpoint with certainty the acceleration sites of high energy cosmic rays, a general picture has been developed to discuss their probable origins. We believe that cosmic rays below 10^{14} eV are produced in our galaxy. The most likely power sources are supernova remnants (SNR) and the most likely acceleration method is that of diffusive shocks. Shock acceleration involves the scattering of particles off magnetic irregularities as the particles propagate back and forth across the shock boundary. The particle observes converging scattering centers that result in an energy gain with each cycle from upstream to downstream [9][10]. Shock acceleration is attractive because it naturally generates power-law spectra. For strong shocks, the power law index, α, is expected to be $\sim$2.

At low energies cosmic rays are comprised of almost all of the elements in the periodic table. Their composition reflects the material available at the source. We classify as *primary* cosmic rays those elements that are stable nucleosynthetic products. The primary cosmic rays (H, He, C, O, Ne, Mg, Si, Fe) are abundant. Elements that are less abundant in the cosmic rays can be produced by collisions of the primary particles with interstellar gas and dust. These *secondary* cosmic rays give information about the amount of material ($\sim 10\,\mathrm{g/cm^2}$) encountered by the primary particles on their trip from their origin to the Earth. We expect this picture of mixed composition to continue to high energies (Figure 2).

Shock acceleration naturally leads to power-law spectra of $\alpha \sim 2$, but the measured cosmic ray spectrum has an index $\alpha \sim 2.65$. The difference in these indices is explained by a model of propagation in which the cosmic rays are loosely confined in our galaxy by its magnetic field. This "leaky-box" model incorporates a rate of escape from our galaxy that is energy dependent with a spectral index of ~ 0.6. This energy dependence is supported by measurements of the ratio of the secondary cosmic ray flux to the primary flux which goes as $E^{-0.6}$. The equilibrium between the acceleration rate and the escape rate results in the overall spectrum observed at Earth.

The picture of a galactic origin for cosmic rays is plausible and is well supported by data for energies less than 10^{14} eV. To extrapolate to higher energies is a bit problematic. On the one hand, it is natural to assume that the same SNR acceleration mechanism continues beyond 10^{15} and that the knee in the spectrum results from the gradual breakdown of this mechanism. It may be difficult to explain the apparent sharpness of the knee in this picture, but this problem may be mitigated by better measurements of the spectrum. On the other hand, the acceleration limit for single strong shocks is believed to be $\sim Z \times 10^{14}$ eV, where Z is the particle charge [11]. Therefore, our conventional picture breaks down at energies well below the knee.

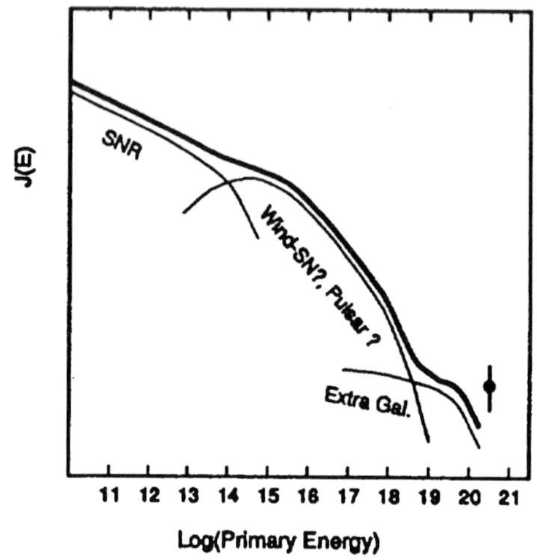

Figure 3: Schematic figure of the cosmic ray spectrum and current ideas relating to cosmic ray origins [12]. At low energies ($< 10^{14}$ eV, it is generally assumed that cosmic rays are galactic in origin and are produced via shock acceleration at supernova remnants (SNR). In the intermediate energy regime ($10^{15} - 10^{17}$ eV), there are models which extend the energy reach of SNR. There are also scenarios in which the cosmic rays are extra-galactic in origin (e.g. produced by active galactic nuclei). At high energies ($> 10^{17}$ eV), models assuming an extra-galactic origin predominate.

There are a variety of ideas that are proposed to explain the existence of cosmic rays above the cutoff of simple shock acceleration. These ideas can be separated into two broad categories, either those that postulate an extension of the supernova mechanism or those that postulate a new component that is unrelated to lower energy cosmic rays. Examples of the first possibility include models in which the shock front expands into the wind of its progenitor star or in which the acceleration is done by a group of supernovae. In such galactic models one expects a mixed composition near the knee and an increasingly heavy composition above the knee as the acceleration limits of the

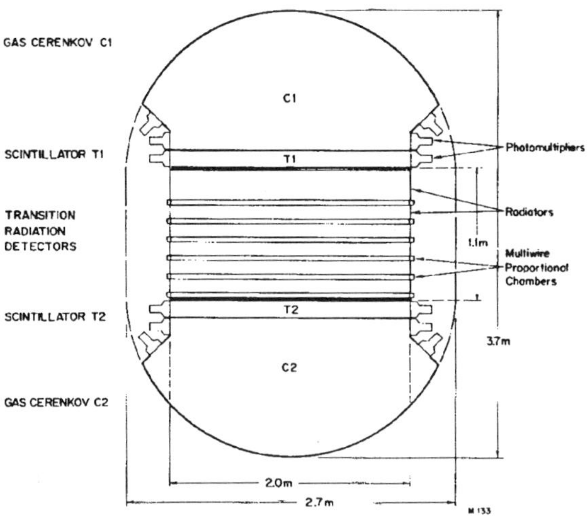

Figure 4: Schematic drawing of the Cosmic Ray Nuclei (CRN) experiment, flown on the Space Shuttle (Spacelab-2). The Lorentz factor of high energy cosmic rays is measured by transition radiation detectors.

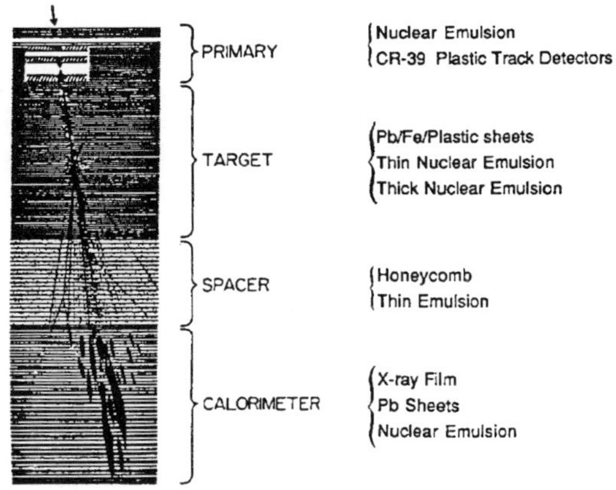

Figure 5: Schematic drawing of a possible passive calorimeter detector to be flown on a high altitude balloon [6].

lighter species are reached. New component models include those in which high energy cosmic rays are extra-galactic in origin and are produced in powerful shocks in the jets of active galactic nuclei (AGN). These models are attractive because AGN are observed to be powerful emitters of high energy gamma-rays and are therefore likely to have sufficient power to accelerate particles to ultra high energies. An extra-galactic component is expected to have a light composition because of strong photoabsorption in the source. Better composition data near the knee can clearly go a long way in settling the question of the cosmic ray origin. Figure 3 shows a summary of the current ideas relating to the question of origin.

2.2 Satellite and Balloon Experiments

Experimental Technique: A variety of experimental methods have been used by satellite and balloon-borne experiments to measure the properties of cosmic ray particles. The experiments fall into two broad classifications. One type of experiment measures the energy and multiplicity distribution of interacting particles in a thin electromagnetic calorimeter (such as stacks of nuclear emulsions). The JACEE balloon payloads are examples of this technique [4]. The other type of experiment uses non-destructive techniques, such as transition radiation, to measure the Lorentz factor of cosmic ray particles. An example of this type is the CRN experiment, shown in Figure 4. [13].

The two types of experiments have different strengths and weaknesses that are somewhat complementary. For example, thin electromagnetic calorimeters require extensive simulations to estimate the energy response of the instrument. On the other hand, the signal in a transition radiation detector saturates at Lorentz factors slightly below 10^5 which corresponds to an energy $\sim 10^{14}$ eV for protons. Both types of experiments share a common problem of flux sensitivity. It is expensive to put very large or massive detectors above the atmosphere for any significant length of time. The exposure for a satellite or balloon-borne experiment is typically on the order of a few m^2-sr-days.

Future Developments: The main direction for the future is increased sensitivity through larger payloads and greater exposure time. Statistically significant measurements of the flux of heavy nuclei above 10^{15} eV will require exposure factors of ~ 1000 m^2-sr-days. Large gains in exposure may be realized from the use of long duration balloon flights in Antarctica. Flights that circumnavigate the South Pole have a number of advantages over those in the northern hemisphere. For one thing, flight durations as long as fourteen days have already been achieved in Antarctica. Another thing is that an Antarctic balloon remains at a nearly constant temperature because of the continuous daylight. Therefore, the balloon stays at a roughly constant altitude and very little ballasting is required.

Balloon detectors can be as heavy as a few tons. An example of a possible passive calorimeter detector is shown in Figure 5. Ideas for low weight transition radiation detectors are also being considered. For example, the use of small (1 cm) proportional counters made of aluminized mylar may permit the elimination of the heavy pressurized gondola [14].

For experiments heavier than a few tons, a rocket launch is needed. In the wake of the Challenger explosion, U.S. agencies have redeveloped various rocket options (e.g. Pegasus). In addition, the Russian government has a surplus of large Proton Launchers. These rockets can place payloads as heavy as 20 tons into orbit and would permit the use of large thick calorimeters (using both passive and active components) for cosmic ray studies.

2.3 Air Shower Experiments

Experimental Technique: High energy cosmic rays impinge on the upper atmosphere and initiate electromagnetic and hadronic cascades. For primary particles with energy above $\sim 10^{14}$ eV the charged particles in these cascades reach ground level where they are observed as *extensive air showers*. Air shower experiments use the earth's atmosphere as a thick calorimeter and sample the particles reaching ground-level. After traversing the 30 radiation lengths of the atmosphere, the showers consist largely of electrons, muons, soft gamma-rays, and Cherenkov photons created by the charged particles. There are also hadrons in the shower near its central core. Above $\sim 10^{17}$ eV the shower particles have enough energy to excite nitrogen fluorescence in the atmosphere that can be detected (the Fly's Eye method).

Ground-based experiments using the air shower technique have the great advantage over space-borne experiments of enormous collection area. The main disadvantage, however, is that air shower experiments do not see the first interaction of the primary cosmic ray. Therefore, detailed Monte Carlo simulations are needed to disentangle information about the primary particle. Energy and composition measurements can only be made on an ensemble of showers rather than on an individual shower basis. An additional problem arises because of uncertainty in the details of the interaction of nuclei and air molecules at high energies. In spite of these difficulties, the air shower technique remains the only feasible way to study cosmic rays above 10^{15} eV.

The shower properties that can be measured by ground-based experiments include:

- Shower size: the number of charged particles at the ground.
- Muon size: the number of muons in the shower. Most muons have energies of a few GeV and can be detected by shielded or buried detectors. TeV muons produced near the first interaction point can be detected by deep underground detectors.
- Longitudinal profile: the depth at shower maximum, X_{max}, and the size at shower maximum, N_{max}, can be inferred from measurements of the Cherenkov or fluorescence radiation created in the shower.

The total energy of the primary cosmic ray can be estimated from any of the above three parameters. The data points in the all-particle spectrum above 10^{16} eV (Figure 1) come from ground-based experiments that use shower size.

Techniques for estimating the primary composition rely on the fact that heavy nuclei interact higher in the atmosphere than lighter species. This effect means that showers created by a heavy nucleus will have a shallower average X_{max}, a smaller shower size at ground-level, and a wider lateral distribution than showers initiated by light nuclei of the same energy. In addition, the number of muons at ground level increases as the primary mass goes up because of 1) the increased fraction of energy going into the hadronic interaction and 2) the greater probability that mesons produced will decay to muons. Measurements of shower parameters, along with help from Monte Carlo simulations, can be used to estimate the primary composition. In principle, an estimate can be made from knowledge of only a single parameter, but recently it has become evident that real progress can only be achieved by experiments (or combinations of experiments) that measure two (or more) parameters simultaneously.

Recent Results: There are numerous earlier studies of cosmic ray composition by air shower experiments. The results at energies above 10^{17} eV are covered at this workshop by Alan Watson. Results at energies near the knee in the spectrum have mostly come from air shower arrays that measure the shower size and lateral distribution. These results are not completely reliable because: 1)the detectors themselves were typically not very sophisticated, 2) the Monte Carlo simulations were somewhat primitive, and 3) the experiments typically lacked the ability to measure several shower parameters relating to composition simultaneously.

Although recent experiments have been more sophisticated, there is still a definite lack on consensus on the cosmic ray composition between 10^{15} and ^{17} eV. Figure 6 shows results on the flux of light primary particles (protons) as a function of energy. High energy data from an air shower experiment on Mt. Norikura (Japan) appear to indicate a decreasing fraction of light particles at high energies. This experiment used emulsion detectors in combination with a particle array.

Figure 7 shows data from the MACRO deep underground muon detector at Gran Sasso (Italy). The multiplicity distribution of TeV muons is compared to Monte Carlo predictions using light and heavy compositions.

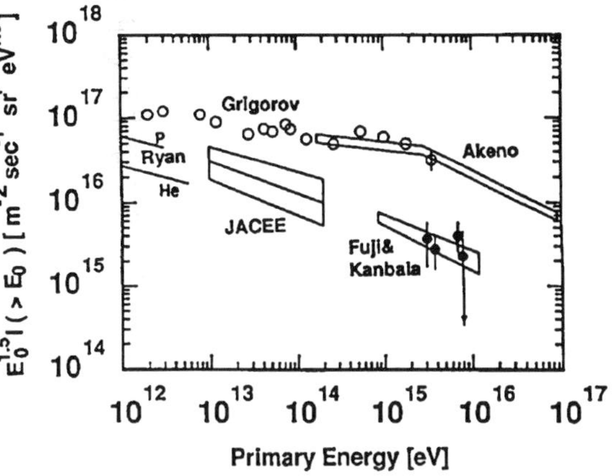

Figure 6: A high energy cosmic ray composition result. The integral cosmic ray flux (multiplied by a flattening factor of $E^{1.5}$) is plotted as a function of energy. The top curve represents various measurements of the all-particle spectrum. The bottom curve represents results on the light (proton) component on cosmic rays from balloon experiments (JACEE) and from an air shower experiment on Mt. Norikura (solid points) [15]). This plot suggests a decrease in the fraction of light primaries at high energies.

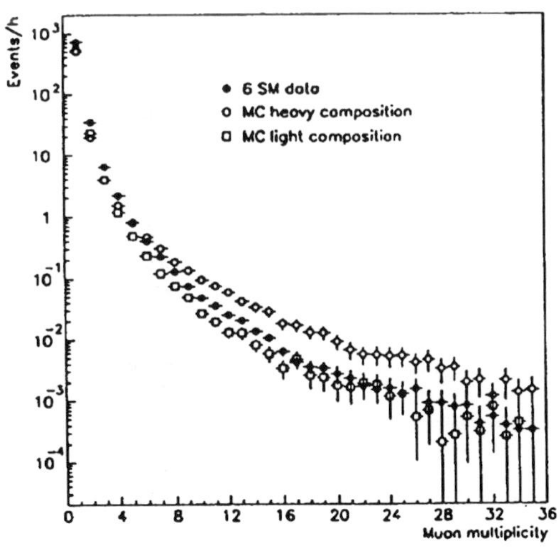

Figure 7: Composition result from a deep underground experiment (MACRO [16]). The TeV muon event rate (solid points) is shown as a function of muon multiplicity. The open points represent the expected multiplicity distributions for light and heavy compositions obtained from simulations. The knee in the spectrum corresponds to a muon multiplicity between 10 and 18. The data appear to indicate a trend towards a lighter composition at high energies.

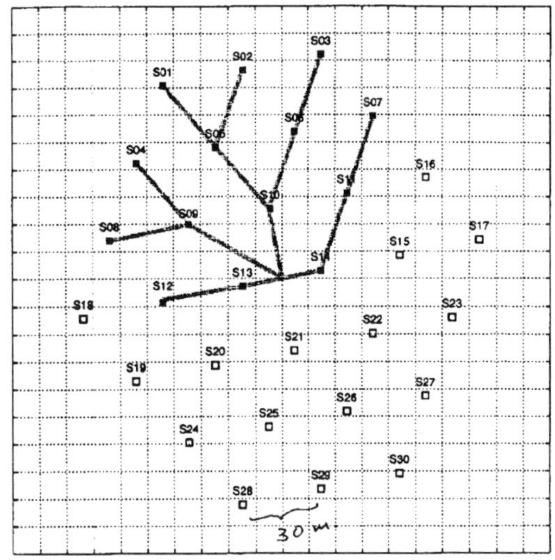

Figure 8: Layout of the new SPASE-2 air shower experiment at the South Pole [17]. Each station contains five scintillation modules; a module contains $0.2\,m^2$ of scintillator and a 2" PMT.

These data appear to support a light composition at high energies. However, there are problems of internal consistency; the event rate at low muon multiplicity is not adequately reproduced by simulation and combined results from the MACRO detector and an extensive air shower array on the surface (EASTOP) indicate a picture of mixed composition (see [1]).

The current data indicate that the question of cosmic ray composition is far from settled at this time.

Existing and Future Experiments: There are a number of existing and planned air shower detectors with sensitivity to cosmic ray composition effects. The experiments are covered in some detail in [1]; here I will only briefly enumerate them.

The MACRO-EASTOP combined experiments have been discussed earlier. MACRO is a large area ($\sim 860\,cm^2$) streamer tube and liquid scintillator detector. The experiment is located in the Gran Sasso laboratory under a minimum overburden of 3.1 km water equivalent. This depth translates into a minimum muon energy of 1.3 TeV. The EASTOP experiment is located on the slope of the mountain at an altitude of 2000 m. It is an array of 35 scintillation detectors spread out over an area of $\sim 10^5\,m^2$. The experiments are recording coincident events and working on a better understanding of systematic effects.

A similar experiment combining a surface array with an underground muon detector is located at the South Pole. The under-ice AMANDA detector currently operating consists of 4 strings of photomultiplier tubes (20/string) deployed on the arc of a circle of radius 30 m. The AMANDA strings are located at a minimum depth of 800 *m* which corresponds to a minimum muon energy of 200 GeV. The SPASE experiment is a small air shower array located on the surface. SPASE consists of 24 scintillation detectors spread out over an area of $6,400\,m^2$. The two experiments have already demonstrated the ability to detect air shower events in coincidence. In the next two years, there are plans to enlarge SPASE to 150 detectors covering an area of $15,000\,m^2$ (Figure 8).

The HEGRA experiment, located on the Canary Island of La Palma, consists of a variety of detector elements. The charged particle array consists of 219 scintillation stations and encloses an area $32,400\,m^2$. Within the same area is an array of wide angle Cherenkov detectors. There are also muon detectors with a combined area of $300\,m^2$. The combination of several different detector elements should allow for a significantly better estimate of the chemical composition than has been done previously.

Located in Dugway, UT, CASA-MIA is the largest air shower array operating in the region near and above the knee. CASA is a surface detector consisting of 1056 scintillation stations and covering an area of $220,000\,m^2$. MIA consists of $2400\,m^2$ of buried scintillator spread out over the same area as CASA and sensitive to muons with energies greater than 0.7 GeV. The two complete detectors have been operating together since early 1992. In addition to measuring the muon and electron components of air showers, images of the longitudinal Cherenkov profile are recorded by the recently installed DICE detector. DICE consists of two detector stations separated by 90 m. Each station contains a bank of 256 photomultiplier tubes viewing a single mirror. Showers above an energy of $\sim 6\times10^{16}$ eV create nitrogen fluorescence in the atmosphere which can be detected by the Fly's Eye HiRes prototype detector. HiRes sits on a hill overlooking CASA-MIA (3.4 km away) and consists of 14 clusters of mirrors and photomultiplier tubes (256/cluster). The combined facilities at Dugway represent the most ambitious attempt to study cosmic ray showers with several detector methods. A example of an air shower seen in coincidence by CASA, MIA, and the Fly's Eye HiRes detector is shown in Figure 9.

The KASCADE experiment is under construction in Karlsruhe, Germany. When completed, the experiment will consist of 256 scintillation detectors covering an area of $40,000\,m^2$. The detectors have absorber to permit a measurement of the total number of muons in the shower (above 0.3 GeV). There will also be a large ($320\,m^2$) hadron calorimeter located at the center of the scintillation array. Combined with the calorimeter will be a system of proportional wire chambers to measure the trajectories of energetic muons (above 2.0 GeV). Data taking with a portion of the apparatus should begin in early 1995.

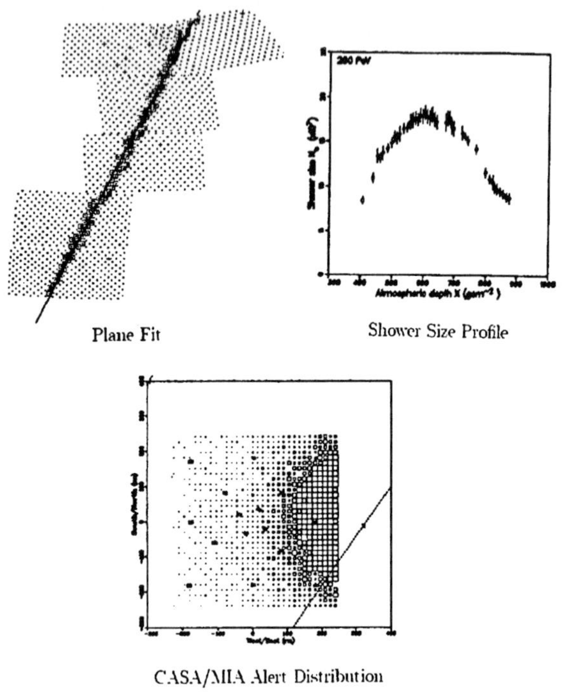

Figure 9: Coincident event recorded by the CASA-MIA experiment and the Fly's Eye HiRes prototype detector. The figure on the top left shows the reconstructed track of the showers from the HiRes detector. On the top right the longitudinal distribution of the nitrogen fluorescence light is shown. The bottom figure shows the distribution of hit detectors in CASA-MIA. The CASA detectors are the uniformly spaced squares; the size of the square is proportional to the number of particles detected. The MIA counters record the number of penetrating muons and are represented by the irregularly spaced rectangles. The cross on the dashed line indicates the best fit location for the shower core from the HiRes fit. The estimate energy of this shower is 2.8×10^{17} eV.

The MILAGRO detector will be built at an altitude of 2,600 m near Los Alamos, NM. MILAGRO, the first water-Cherenkov air shower experiment, will be a large ($4,800\,m^2$) pond of water viewed by 750 photomultiplier tubes. The tubes will be deployed at three depths in the pond making the experiment sensitive to the electrons, hadrons, muons, and photons that hit the surface of the water. MILAGRO is currently under construction and should be operational in a few years.

§3 Gamma-Rays above 20 GeV

3.1 Overview

In discussing the field of high energy gamma-ray astronomy, one must use as a starting point recent exciting discoveries made by the EGRET detector aboard the Compton Gamma Ray Observatory. The results from EGRET [18] are discussed by Brenda Dingus at this Workshop. EGRET, operating in the energy regime between 30 MeV and 20 GeV, has detected gamma-ray emission from ~ 130 point sources. Of these point sources, six are identified as spin-down pulsars, approximately forty are identified as active galactic nuclei (AGN), and the rest are unidentified. The AGN were one of the big surprises of the EGRET observations. Only one source (3C273) had been detected by the earlier generation of gamma-ray instruments (SAS-2, COS-B). Since AGN are distant objects the copious amounts of high energy gamma-ray emission detected as Earth implies tremendous source luminosities, even when possible beaming effects are taken into account.

More than 90% of the gamma-radiation seen by EGRET is in the form of diffuse emission from the galactic plane. This emission comes from the collisions of cosmic rays with gas and dust that produce energetic electrons and neutral pions. The electrons radiate and the pions decay to give gamma-rays. EGRET has also detected diffuse emission from the Large Magellenic Cloud (LMC) (in a $7^\circ \times 7^\circ$ box). Studies of diffuse emission are important because of the way in which the gamma-radiation traces the cosmic ray flux density. For example, recent observations by EGRET of the Small Magellenic Cloud (SMC) [19] provide evidence that cosmic rays are galactic in origin. If cosmic rays were universal (and extended to the SMC) one would expect a diffuse gamma-ray flux of $(2.4 - 4.2) \times 10^{-7}$ photons/cm^2/sec from the SMC. The EGRET flux limit of $< 0.5 \times 10^{-7}$ photons/cm^2/sec argues against a universal picture of cosmic ray origin. This result illustrates the close interplay between cosmic ray and gamma-ray research.

Another exciting result from the Compton Gamma Ray Observatory is the detection of gamma-ray burst sources. Over 1000 gamma-ray bursts (GRB) have been detected by the BATSE instrument [20] and ten have been seen in coincidence by EGRET. The bursts are observed to be isotropically distributed and to have a wide variety of spectral and temporal features that defy classification. Since the bursts have not been demonstrated to repeat and have not been positively identified with lower energy objects, their origin remains a mystery. The general view is that GRB are cosmological or from a galactic halo population with a scale larger than 70 kpc. Very surprising was the detection by EGRET of a burst on February 17, 1994 in which high energy emission was seen roughly eighty minutes after the initial burst.

Above 20 GeV the gamma-ray fluxes become so low as to be undetectable by EGRET. Therefore, only ground-based experiments have sufficient collection area to make gamma-ray detection feasible. The situation here is similar to that for cosmic rays above 10^{14} eV.

3.2 Experimental Technique

Just as in the study of cosmic rays, ground based gamma-ray detectors make use of the method of extensive air showers. Air shower arrays function simultaneously as cosmic ray and gamma-ray instruments; the cosmic rays are the principle background to doing astronomy. There are two major types of ground-based gamma-ray detectors: those that use the atmospheric Cherenkov technique (ACT) and those that use the air shower technique (AST).

The major challenge facing all ground-based experiments lies in their ability to reject the cosmic ray background. This challenge arises because cosmic rays outnumber gamma-rays by three to four orders of magnitude. The preponderance of cosmic rays can be easily understood by considering our conventional picture of cosmic ray origin. We believe that cosmic rays below 10^{14} eV are galactic in origin. Since they are charged, they bend and spiral in the galactic magnetic field. Cosmic are loosely confined by the field and have a lifetime in our Galaxy of approximately ten million years. Gamma-rays are not confined, but are created at a source and propagate outward with a $1/r^2$ fall-off. We have only one chance to intercept a gamma-ray before it continues on its journey out of the Galaxy.

Atmospheric Cherenkov Experiments: When a high energy gamma-ray strikes the upper atmosphere, it creates an air shower of electromagnetic particles that propagate towards the Earth's surface. These relativistic particles are moving faster than the speed of light in the atmospheric medium and therefore emit Cherenkov radiation. Even if the charged particles do not reach the Earth's surface, the Cherenkov radiation is beamed to the ground in a narrow cone of half angle $\sim 1^\circ$ (Figure 10). The radiation is spread out over an area of ~ 100 m in radius, arrives in the space of a few nanoseconds, and is copious: a 1 TeV gamma-ray will produce a shower with roughly 50 photons/m^2 at the ground.

Atmospheric Cherenkov detectors consist of a large mirror, (or several mirrors) that reflects the optical-UV Cherenkov photons into a camera. The Cherenkov image of the shower is converted into an estimate of the incident gamma-ray arrival direction. The angular resolution achievable by Cherenkov telescopes is excellent: 0.2° or even better.

As discussed before, the key feature of ground-based gamma-ray astronomy is the necessity to reduce the copious background of hadronic cosmic rays. Background rejection is difficult because experiments on the ground do not see the first interaction of the primary particle. They only infer something about the nature of the primary from the detection of the secondary shower particles. Atmospheric Cherenkov telescopes achieve background re-

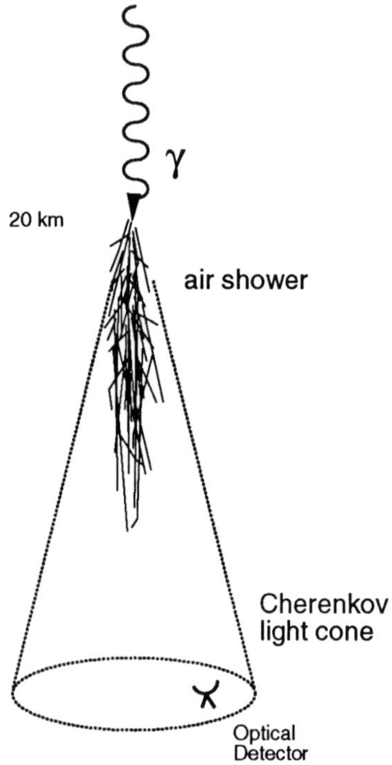

Figure 10: Schematic representation of an air shower initiated by a gamma-ray of energy less than 1 TeV. The gamma-ray interacts in the upper atmosphere and creates an electromagnetic cascade that reaches is maximal extent at an altitude of roughly 10 km. Charged particles in the cascade radiate a cone of Cherenkov light that can be detected by ground-based optical instruments.

jection by either of two methods: imaging or sampling.

The imaging method takes advantage of the fact that showers initiated by hadronic cosmic rays are much more irregular in development than those initiated by gamma-rays. This irregularity comes from the fluctuations in the development of the hadronic core of cosmic ray showers and in the fraction of energy carried away by hadrons and muons. With a multiple-pixel camera (typically an array of photomultiplier tubes), Cherenkov telescopes record an image of the profile of the shower in its lateral and longitudinal extent. Images for gamma-ray showers are elliptical and point towards the center of the field-of-view. Images for showers created by hadronic cosmic rays are typically more irregular in the focal plane (Figure 11). Current state-of-the-art imaging detectors achieve a background rejection of greater than 99%, while keeping the majority of the gamma-ray events.

The sampling method takes advantage of the fact that most of the Cherenkov radiation in a shower reaches the ground in a sharp conical wavefront. By recording the time of arrival of this wavefront in several detectors separated by some distance, the shower direction can be reconstructed with accuracies approaching a few milliradians. In this way, background rejection is achieved through excellent angular resolution. Additional rejection can be obtained from the lateral distribution of the Cherenkov light on the ground.

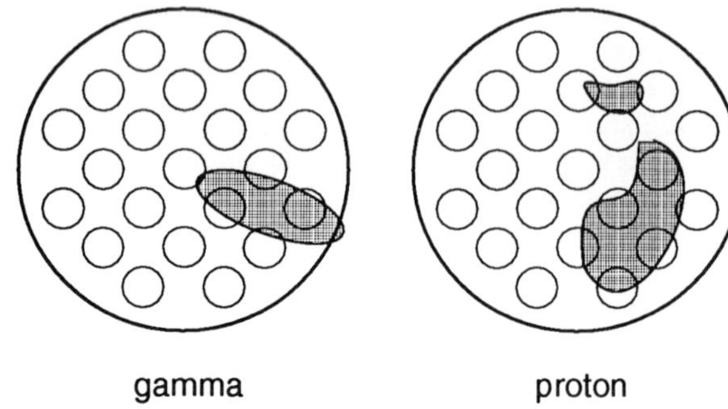

Figure 11: A comparison of images recorded in the focal plane of an atmospheric Cherenkov detector for showers initiated by gamma-rays (left) and by hadronic cosmic rays (right).

The atmospheric Cherenkov technique has been used by ground-based telescopes for more than thirty years. It is only in the last ten years that significant advances have been made in the power of the instruments to reject background. These advances are directly responsible for the compelling detection of three gamma-ray sources by Cherenkov telescopes. State-of-the-art experiments exist on several continents. In the United States, the premier instrument is the Whipple Observatory, located on Mt. Hopkins near Tuscon, AZ. Whipple consists of two mirrors (diameters 10 and 11 meters), separated by ~ 120 m. Each mirror has an imaging camera of roughly 100 PMTs. In Australia, a Japanese-Australian collaboration has built an imaging detector known as CANGAROO. The main element is a 3.5 m diameter mirror with a 95 pixel camera. In France, there are two detectors operating at the site of the Thémis solar facility in the Pyrénées mountains. Each detector has multiple mirrors to make use of the sampling method. The ASGAT experiment consists of an array of seven mirrors (each 4 meters diameter) and THEMISTOCLE is an array of 18 smaller mirrors (each 0.8 meters diameter). The mirrors in ASGAT are equipped with an array of PMTs while those in THEMISTOCLE have a single photomultiplier tube.

Current atmospheric Cherenkov telescopes have typical threshold energies for gamma-ray detection between 0.25 and 2 TeV. The threshold energy is set by the minimal density of Cherenkov radiation needed to trigger the telescope. A relatively straightforward analysis [21] shows that this threshold energy is proportional to the following:

$$E_{th} \propto \sqrt{B\Omega t/A\epsilon} \quad ,$$

where B is the background light level, Ω is the solid angle viewed by each photomultiplier tube, t is the integration time, A is the mirror area, and ϵ is the efficiency of light collection. In the quest for lower energy thresholds, the parameter that is most straightforward to control is the mirror area. Increasing the size of the mirror collection area lowers the energy threshold of the detector.

Air Shower Arrays Charged particle arrays using the air shower technique have been discussed earlier in the context of cosmic ray composition measurements. Up to now, these experiments have been sampling detectors consisting of arrays of elements (e.g. scintillation counters) that are distributed on the ground. The elements typically comprise an active collection area of 1% of the surface area. Relative timing measurements between elements are used to determine the shower direction. The angular resolution is typically 0.5°, depending on the shower size. Pulse-height measurements are used to reconstruct the lateral distribution of the charged particles and hence the primary particle energy.

Air shower arrays have several advantages over atmospheric Cherenkov detectors. Cherenkov telescopes can only operate on clear moonless nights; arrays have close to 100% live-time. The field-of-view of a Cherenkov detector is small, typically a few degrees, while an air shower array views roughly 1 steradian of the overhead sky. On the other hand, air showers arrays require the shower particles to reach the earth's surface. A Cherenkov detector is sensitive to photons produced high in the atmosphere, near shower maximum. This difference leads to significantly higher energy thresholds for air shower arrays. Typical air shower arrays at mountain altitudes have gamma-ray energy thresholds of $\sim$ 100 TeV.

The fact that surface arrays detect the charged particles created in the radiation length right above them, as opposed to the Cherenkov photons created near shower maximum, has another important consequence. The lateral and angular distributions of charged particles contain less information in regards to the nature of the primary particle compared to the same distributions for Cherenkov light. This effect makes it difficult for air shower arrays to achieve the level of background rejection obtainable with Cherenkov detectors. In essence air shower arrays measure the leakage energy out of the back of a calorimeter 30 radiation lengths thick.

We do expect showers initiated by gamma-rays to have significantly fewer muons (by about a factor of 30) than those initiated by hadrons. This lack of muons is a consequence of the fact that gamma-ray showers are almost completely electromagnetic in nature. Recent measurements at the HERA electron-proton collider at DESY in Hamburg, Germany have confirmed this property of high energy photons. The cross section for photo-pion production is found to be much smaller than the cross section for electron-positron pair production (Bethe-Heitler). Several recent experiments (CASA-MIA, HEGRA) have built extensive muon detectors to aid in background rejection. The experiments typically reject 90% of the hadronic cosmic rays.

Motivated in part by reports in the mid 1980's of bright gamma-ray sources at 1000 TeV, a number of second generation air shower arrays were recently constructed for the purposes of gamma-ray astronomy. These arrays are significantly larger, have lower energy thresholds, and have better angular resolution than experiments of the earlier generation. The foremost of the recent arrays are CYGNUS, Tibet, HEGRA, and CASA-MIA. CYGNUS, located at Los Alamos National Laboratory, NM, consists of roughly 200 scintillation detectors covering an area of $70,000\,\mathrm{m}^2$. The experiment has been taking data since 1986 and will be decomissioned and moved to surround the future MILAGRO detector. The Tibet array consists of 49 detector stations on a square 15 m grid. The experiment is located in Yangbajing, Tibet, at an altitude of 4300 m, giving it the lowest energy threshold ($\sim$ 10 TeV for gamma-rays) of any existing air shower array. The HEGRA and CASA-MIA experiments have been described in Section 2.3.

3.3 Recent Results

Very High Energy (0.25-10.0 TeV) : Very High Energy (VHE) gamma-rays are defined as those near 1 TeV in energy. Historically, this energy range has been accessible only to atmospheric Cherenkov telescopes, but air shower arrays (e.g. Tibet, MILAGRO) are beginning to encroach. There are now at least three steady sources of VHE gamma-rays. The Crab and PSR 1706-44 are spindown pulsars that are also detected by EGRET. The third source is Markarian 421– an active galactic nuclei (AGN).

The Crab, first unambiguously detected by Whipple [23], has now been seen by at least five other Cherenkov telescopes. The Crab emission is observed to be unplused, which is somewhat of a surprise because at EGRET energies the dominant emission is pulsed. This fact supports the contention that there must be two components that contribute to the gamma-ray emission at high energies. A model has been proposed in which the pulsed component is associated with synchrotron radiation near the pulsar itself, while the TeV emission comes from inverse-Compton scattering in the nebula [24]. This Self-Sychrotron-Compton (SSC) model is quite successful at describing the high energy emission from the Crab, however it is not overly constrained in the region between 20 GeV and 250 GeV, where no observations have been made. Figure 12 shows some the latest gamma-ray obser-

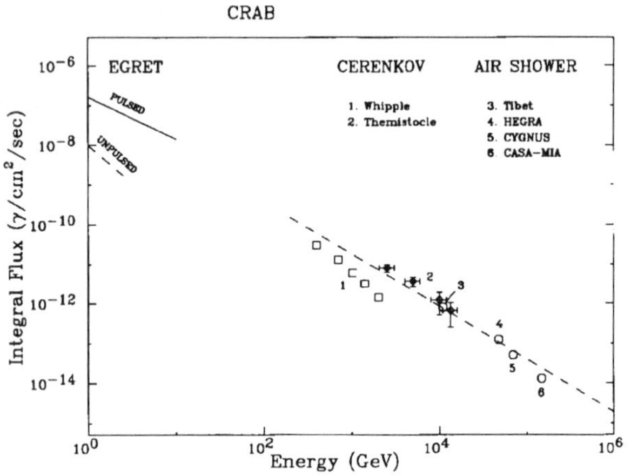

Figure 12: Gamma-ray observations of the Crab at high energies (EGRET), very high energies (Cherenkov telescopes), and at ultra high energies (air shower arrays). The integral gamma-ray flux from the Crab is shown as a function of energy. The Cherenkov telescope results are detections; the air shower array results are upper limits. Shown is the most recent Whipple data [22]. The dashed line is the extrapolated flux from the previous Whipple data [23].

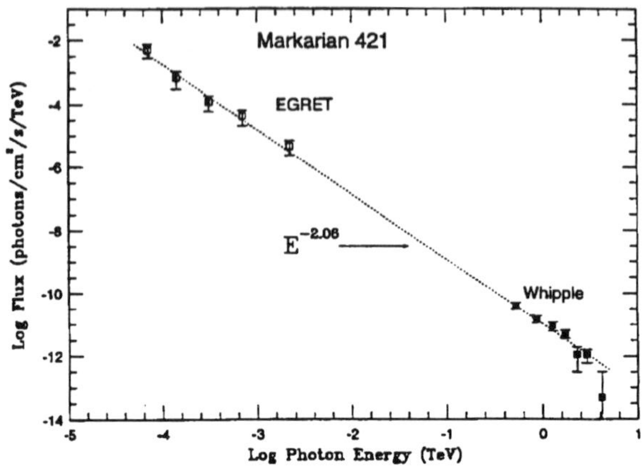

Figure 13: Gamma-ray observations of Markarian 421 at high energies (EGRET) and very high energies (Whipple). The differential gamma-ray flux from Markarian is shown as a function of energy. The dashed line is a single power law fit to the EGRET data.

vations of the Crab spectrum.

There are problems of consistency in the experimental data on the Crab. The Whipple group has recently lowered their flux estimate [22] from their earlier paper. Their new flux disagrees with measurements made by THEMISTOCLE [25]. On the other hand, the Tibet air shower array does not detect the Crab and sets an upper limit [26] which may be in contradiction with the detection level of THEMISTOCLE. One concludes that although that Crab has been convincingly detected at TeV energies, its energy spectrum is not completely understood.

Markarian 421 is an AGN of the Bl-Lac variety. Located at a redshift of 0.031, it is the closest AGN in the EGRET catalog and also the weakest AGN source detected by EGRET. Nevertheless, the discovery by Whipple in 1992 [27] of TeV photons from this object was an important one. Markarian 421 gave proof that there was a source other than the Crab at these energies. The fact that an extra-galactic object some 400 million light-years away could emit TeV gamma-rays detectable on Earth is remarkable. Figure 13 shows the differential energy spectrum of Markarian 421 from MeV to TeV energies. The spectrum is remarkably well fit by a single power law of form $E^{-2.06}$, although more recent data from Whipple [28] does not support this simple fit as well.

AGN are assumed to consist of a massive, compact central source which is presumed to be a super-massive black hole. The central source is surrounded by an accretion disk and jets of material are often seen along the axis of the disk. In most models, gamma-rays are produced by shock acceleration in the relativistic jets. The acceleration is powered by the infall of matter from the accretion disk into the black hole. In the simplest picture, the AGN observed by EGRET are ones having associated jets that are beamed towards the observer.

A very exciting recent result is the detection by Whipple of a flare from Markarian 421 in which the flux increased by about an order of magnitude in a couple of days [29]. The flare was also seen by an x-ray satellite (ASCA) at approximately the same time [30]. The flare detection indicates that the AGN is the source of violent activity on short time scales. Similar short duration outbursts from other AGN were seen by EGRET, which argue for a compact acceleration region.

Ultra High Energy ($>$ 10 TeV) : Gamma-rays above 10 TeV are in the Ultra High Energy (UHE) region. This region is domain of air shower arrays which historically operated at energies near 10^{15} eV (1 PeV). The reports of bright UHE emission from x-ray binary systems like Cygnus X-3 and Hercules X-1 stimulated the construction of a new generation of air shower arrays in the late 1980's. The apparent brightness of Cygnus X-3 stimulated acceleration models in which the origin of cosmic rays could be explained by a few powerful sources. Unfortunately, the earlier results have not been confirmed; the new experiments have been unable to see steady emission from any source, including x-ray binaries.

Figure 14 shows some current limits on UHE gamma-ray emission from Cygnus X-3. Also shown are claimed detections of this object by earlier experiments. The current limits are up to two orders of magnitude lower than

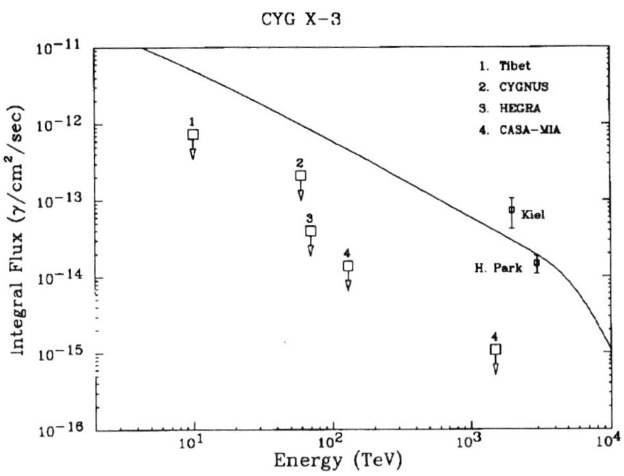

Figure 14: Gamma-ray observations of Cygnus X-3 at ultra high energies. Upper limits from four air shower arrays are shown. Also shown are claimed detections from earlier experiments in the 1980's (Kiel, Haverah Park). The solid curve is a model proposed by Hillas to explain the bright emission from Cygnus X-3 and the origin of cosmic rays below 10^{16} eV.

the earlier detections. In principle, the source (in this case Cygnus X-3) could have shut off its gamma-ray emission in the intervening few years. However, since the earlier detections were marginal in significance (typically four to five standard deviations) and seemed to come and go, it is much more likely that these "detections" were in fact statistical fluctuations at the limit of the experimental sensitivity. Cygnus X-3 has been also ruled out by Whipple as a gamma-ray source at TeV energies.

Air shower experiments have also searched for emission from other sources, including the Crab and Markarian 421 [31][32][33][34]. So far, no steady state emission has been detected and the upper limits to the integral photon flux approach $10^{-14}/\text{cm}^2/\text{sec}$. These limits represent sensitivities comparable to those achieved by Cherenkov telescopes. In Figure 12, for example, upper limits from air shower experiments approach, and in some cases go below, the flux extrapolated from Cherenkov detectors. The UHE limits indicate that either 1) the source spectra roll over at high energies or 2) the photons above 10 TeV are absorbed on their way to Earth. In the case of a source like the Crab, current models indicate that the first possibility is more likely. For extragalactic objects like Markarian 421, there is good reason to expect absorption to be a factor, as we will see in the following section.

There is one very likely astrophysical source of ultra high energy photons: diffuse gamma-ray emission from the galactic plane. These photons come from π^o decay. Detecting diffuse galactic emission is important because it could signal locations of enhanced cosmic ray production. Measurements of diffuse emission can only realistically be made by experiments having some sort of gamma-ray identification. Unfortunately, Cherenkov telescopes, with narrow fields-of-view, are poorly suited for such measurements. Air shower arrays with muon detectors are able to identify gamma-rays by the the paucity of muons in their showers. The current limit on the ratio of the gamma-ray flux to the cosmic ray flux in the direction of the galactic plane is 1.0×10^{-4} [35]. There may also be an isotropic gamma-ray flux that comes from cascading processes resulting from the interaction of extremely high energy cosmic rays with the 3 K microwave background radiation. This flux, being omnidirectional, is more difficult to detect than the galactic radiation, but we should see sensitive limits from a couple of experiments (CASA-MIA, HEGRA) in the near future.

Intergalactic Absorption: The EGRET experiment has detected high energy gamma-ray emission from over 100 point sources. At very high energies, Cherenkov telescopes have detected three steady sources. At ultra high energies, air shower experiments have failed to detect any steady sources. The three types of experiments have approximately the same sensitivity for sources with a differential flux power law spectral index of 2.0. This paucity of sources seen by ground-based experiments could be explained by a number of possibilities:

- The energetics of the source do not permit gamma-ray emission above some energy. Gamma-rays are produced but absorbed in the source or gamma-rays escape but their spectra are soft (spectral index larger than 2.5).
- The sources are episodic. This could explain why Cherenkov telescopes with narrow fields-of-view failed to see the number of sources that EGRET has seen. It would not explain why full-aperture air shower arrays do not see strong sources.
- The gamma-rays interact and are absorbed before they reach the Earth.

We know that the first possibility is very likely to be occurring. For example, there are many sources detected by EGRET that have steep spectra which would make detection at higher energies difficult. The self-synchrotron Compton model for the Crab emission predicts a very rapid fall-off in the spectrum above 10 TeV. The second possibility could explain some, but not all, of the non-observations at very high energies.

The possibility that gamma-rays are absorbed as they cross intergalactic space is motivated by the AGN data. EGRET has detected emission from forty AGN at energies up to 10 GeV. Cherenkov telescopes have searched for emission from more than half of these AGN and, except in the case Markarian 421, have failed to detect any emission. For example, the quasar 3C279 was seen by EGRET to flare to a level thirty times that of Markarian 421, but 3C279 has not been detected at TeV energies. Since 3C279 is much further away than Markarian 421 (redshift of 0.54 as opposed to 0.031), it is natural to suggest that photons from 3C279 are being absorbed on their journey to Earth [36].

The details of intergalactic absorption are reviewed extensively elsewhere in this workshop [8]. The dominant absorption process for TeV photons is pair production ($\gamma\gamma \rightarrow e^+e^-$) off soft photons in the infrared region. The diffuse infrared background is created by normal galaxy formation. Recent estimations of this background place it at approximately the right level to be significant in the absorption of TeV photons from distant sources like 3C279. Unfortunately, at the present time, the diffuse infrared background has not been well measured.

It has been pointed out that high energy gamma-ray astronomy could make use of this absorption process to probe the nature of the intergalactic radiation field. Ideally, one wants to measure the spectral features of several sources at different redshift values. Since the amount of absorption depends on the infrared photon density and the distance to the source, the observation of spectral breaks that are correlated with source distance could be used to estimate the infrared photon density. Conversely, if the infrared density is measured directly, the gamma-ray spectra can be used to infer the distance scale and possibly measure the Hubble constant [37]. Recent calculations have sought to model the formation of galaxies in order to estimate the diffuse infrared photon density from first principles [38]. These calculations assume that the uncertainty in the photon density is dominated by the epoch at which galaxies begin to form. Therefore, high energy gamma-ray observations could be used to probe different models of galaxy formation that depend, for example, on the mixture of cold and hot dark matter.

3.4 Future Directions

With the successes of atmospheric Cherenkov telescopes like Whipple, ground-based gamma-ray astronomy has come into its own. The exciting results from the EGRET instrument motivate better instrumentation on the ground to better understand what happens to the spectra of sources at high energies. It is crucial to realize that detectors on the ground are complementary to those in space. Only in space using direct detection methods will one be able to continuously monitor the entire gamma-ray sky. The great sensitivity of ground-based instruments, resulting from enormous collection areas, will be key in making directed observations of weaker sources. In addition, ground-based experiments will be superior for observations at TeV energies where the source fluxes will invariably be low. Since the gamma-ray sky above 20 GeV is largely uncharted territory there is great motivation for the development of new generations of both space and ground-based instruments.

In the report of the working group on ground-based gamma-ray astronomy [8], three major design goals for new ground-based detectors were identified:

1. High sensitivity and good energy resolution.
2. All-sky coverage with high duty-factor.
3. Lowest possible energy threshold.

Experiments with greater sensitivity will hopefully permit the detection of more sources. For strong sources, increased sensitivity will permit flux monitoring on shorter time scales which is important for studies of variability. An experiment combining the imaging and sampling techniques should reach a sensitivity of 0.01 Crab. There are a number of efforts worldwide to develop finer pixel imaging telescopes and detectors using multiple mirrors with imaging capability. The Whipple group is proposing to reduce the pixel size of their imaging camera by increasing the number of PMTs. The Durham group has a significant upgrade underway in Australia. The most ambitious idea is the Telescope Array, shown in Figure 15. This experiment [39] is proposed to consist of 200 telescopes on a 50 m grid. Each telescope would consist of a 3 m diameter mirror viewed by 16 multi-anode photomultiplier tubes (64 channels per PMT). The experiment would have a flux sensitivity of 5 mCrab and an energy threshold of 100 GeV. Prototype telescopes have already been built for the dual purpose of detecting extremely high energy cosmic rays via the nitrogen fluorescence technique.

EGRET has shown that many gamma-ray sources are highly variable. In addition, a sensitive all-sky survey has not been done at TeV energies. These facts argue for experiments that have wide field-of-view and high-duty cycle. One of the most promising ideas for experiments of this type is that of the extended water Cherenkov detector. Such detectors have lower energy thresholds than conventional sampling air shower arrays. The MILAGRO experiment [40], presently under construction in the mountains of New Mexico, is the first large water Cherenkov detector. The experiment should be the first air shower experiment to operate near energies of 1 TeV. Although the sensitivity of MILAGRO will not approach that of atmospheric Cherenkov detectors, its high duty factor will make it ideal to search for strong transient phenomena, like gamma-ray bursts.

The energy range between 20 and 250 GeV represents one of the last remaining unexplored sections of the elec-

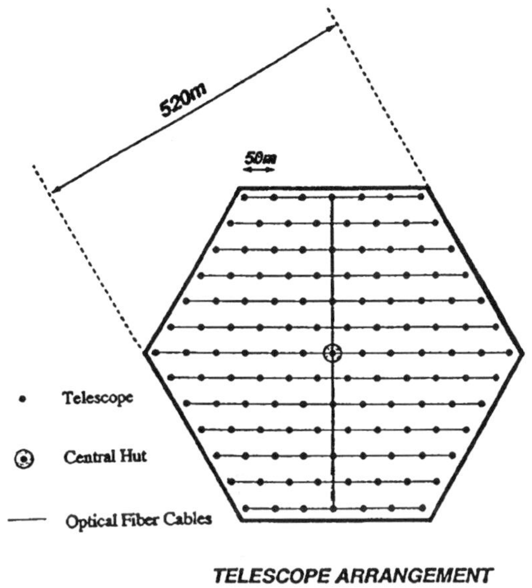

Figure 15: Proposed Telescope Array detector of very high energy gamma-ray astronomy (Japan). The array would consist of 200 telescopes each having 1024 PMT channels.

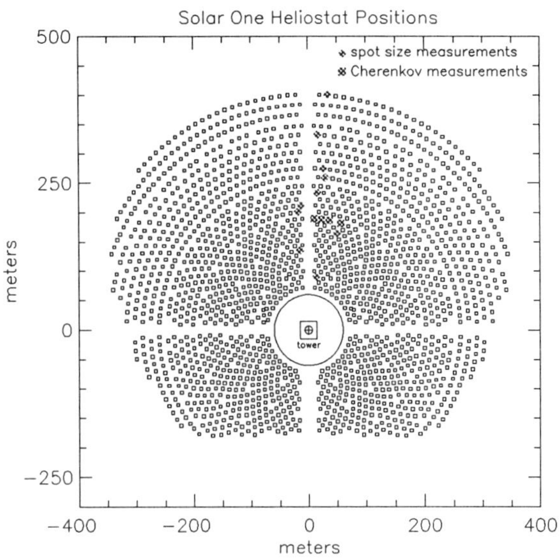

Figure 16: Solar-One heliostat array near Barstow, California. The locations of 1818 heliostats are shown. Each heliostat has an independent altitude-azimuth mount and a mirror surface of $40\,m^2$. Heliostats represented by diamond symbols were used for spot size measurements. The four heliostats indicated by shaded rectangles were used to detect Cherenkov radiation in November 1994.

tromagnetic spectrum. We know that there must be interesting observations to be made there simply because of the disparity in the number of sources seen by EGRET and the number seen by ground-based instruments. In this "unopened window", the Crab emission changes from being dominantly pulsed to dominantly unpulsed in nature (Figure 12). AGN spectra may show features of self-absorption above 10 GeV. We may be able to probe the nature of intergalactic radiation fields from AGN absorption features at different redshift values. Finally, EGRET has detected emission from a gamma-ray burst at 20 GeV. If sensitive observations of bursts like this one could be made up to an energy of 200 GeV we would be in a position to settle the question of the distance scale for gamma-ray bursts. These compelling scientific issues argue for the development of instruments in the energy region between 20 and 250 GeV.

One way to open this window in the electromagnetic spectrum is with a new gamma-ray satellite instrument with a much greater energy reach than EGRET. There are design concepts for several such experiments worldwide. Two such concepts (GLAST, AGATE) are being vigorously pursued in this country. A new gamma-ray satellite experiment would be expensive ($100 M), and would take a decade to fund and construct. It is natural to consider what can be done on the ground. As discussed earlier, the energy threshold of atmospheric Cherenkov telescopes can most effectively be lowered by increasing the mirror area for Cherenkov light collection. A Cherenkov detector using a very large mirror could reach energies as low as 10 GeV. The design for such an experiment (Big Bowl) has been recently formulated [41]. It would consist of an enormous spherical mirror (500 m in diameter) to reflect Cherenkov light into a movable camera box. The main advantage of the design is that almost all the information contained in the Cherenkov radiation is preserved so that low energy response and excellent background rejection can be obtained at the same time. The disadvantage is the long development time and high cost.

Very large mirror arrays have been constructed around the world for solar energy research. In principle, a gamma-ray telescope could be built which would use existing solar heliostats (mirrors) at a fraction of the cost of an experiment which would build mirrors from scratch. Several groups are pursuing the possibility of using solar heliostats to reflect Cherenkov light to detecting elements on a central tower. A French group is conducting tests at the site of the Thémis power plant in the Pyrénées [42]. The Thémis plant, now decomissioned, consists of 200 heliostats each having a mirror surface of $50\,m^2$.

An American-Canadian group has been evaluating the Solar One plant near Barstow, California for the purposes of gamma-ray astronomy. This facility consists of 1818 heliostats (mirror area 40 m^2 each) and 109 newly installed heliostats (mirror area 75 m^2 each) (Figure 16). The group has conducted tests in 1994 which demonstrate that the night sky background conditions are acceptable and that the optical and mechanical characteristics of the heliostats are excellent [43]. In November 1994, the group detected Cherenkov radiation using four heliostats and a large 1 m^2 Fresnel lens as a secondary collector. In 1995, there are plans to image Cherenkov light from as many as ten heliostats using a large 1.7 m parabolic reflector on the central tower. The next step would be to build a prototype detector which would use 45-50 heliostats and which should achieve an energy threshold below 100 GeV.

§4 Summary

The mystery of the origin of high energy cosmic rays continues to confound us. The mystery persists in part because space-based cosmic ray and gamma-ray instruments run out of sensitivity at high energies because of rapidly falling fluxes. Therefore, we must turn to ground-based experiments which historically have been unable to answer our most pressing questions. A new generation of instruments for both cosmic ray and gamma-ray research promises a rich harvest of scientific results in the next millenium. Hopefully, sometime early on in this next millenium, we will be in a position to answer the following questions:

- What is the spectrum and composition of cosmic rays from 10^{14} eV to 10^{17} eV? Can this information tell us anything about the cosmic ray origin and distance scale?
- What are the discrete sources of high energy gamma-rays above 20 GeV? Do we understand the astrophysics of these sources?
- What is the level of galactic and isotropic diffuse gamma-radiation at high energies?
- Can we used tagged gamma-ray beams to probe the nature of interstellar radiation fields?

§Acknowledgments

I would like to thank Corbin Covault for a great deal of help in writing this paper. Thanks also to Roberto Peccei for all his hard work and patience. A number of other people made significant contributions to this paper, including Carl Akerlof, Brenda Dingus, Brian Fick, Tom Gaisser, Alice Harding, Cy Hoffman, Richard Lamb, Nathaniel Longley, Joel Primack, Gus Sinnis, David Smith, Simon Swordy, Shoji Torii and Trevor Weekes. Special thanks is reserved for Alan Watson who generously gave me ten minutes of his allotted time at the plenary session. This work was supported by the National Science Foundation, the Alfred P. Sloan Foundation, and the W.W. Grainger Foundation.

§References

[1] T.K. Gaisser et al. *Report of Group R3B*, this workshop, 1994.

[2] N.L. Grigorov et al. *Proc. 12th Int. Cosmic Ray Conf. (Hobart)*, 2:206, 1971.

[3] Yu.A. Fomin et al. *Proc. 22th Int. Cosmic Ray Conf. (Dublin)*, 2:85, 1991.

[4] K. Asakimori et al. *Proc. 23rd Int. Cosmic Ray Conf. (Calgary)*, 2:21,25, 1993.

[5] S. Yoshida et al. submitted to J. Phys. G, 1994.

[6] J. Waddington et al. NASA Galactic Origin and Acceleration Limit (GOAL) Report, 1992.

[7] J. Wdowczyk and A.W. Wolfendale. *J. Phys. G*, 10:1453, 1984.

[8] Richard C. Lamb and Rene A. Ong. *Report of Group R2*, this workshop, 1994.

[9] E. Fermi. *Phys. Rev.*, 75:1169, 1949.

[10] R.D. Blandford and J.P Ostriker. *ApJ*, 221:L29, 1978.

[11] P.O. Lagage and C.J. Cesarsky. *A&A*, 118:223, 1983.

[12] M. Teshima. *Proc. 23rd Int. Cosmic Ray Conf. (Calgary)*, Rapporteur Vol.:257, 1993.

[13] D. Muller et al. *ApJ*, 374:356, 1991.

[14] D. Muller. private communication, 1995.

[15] To. Saito et al. *Astropart. Phys.*, 1:257, 1993.

[16] O. Palamara et al. *Proc. 23rd Int. Cosmic Ray Conf. (Calgary)*, 2:97, 1992.

[17] T. Gaisser. private communication, 1994.

[18] C.A. Fichtel et al. *ApJ Suppl.*, 94:551, 1994.

[19] P. Sreekumar et al. *Phys. Rev. Lett.*, 70:127, 1993.

[20] G.J. Fishman et al. *ApJ Suppl.*, 92:229, 1994.

[21] M.F. Cawley and T.C. Weekes. submitted to Exp. Astr., 1994.

[22] D. A. Lewis et al. *Proc. 23rd Int. Cosmic Ray Conf. (Calgary)*, 1:279, 1993.

[23] G. Vacanti et al. *ApJ*, 377:467, 1991.

[24] O.C. De Jager and A.K. Harding. *ApJ*, 396:161, 1992.

[25] P. Baillon et al. *Astroparticle Phys.*, 1:341, 1993.

[26] M. Amenomori et al. *Proc. 23rd Int. Cosmic Ray Conf. (Calgary)*, 1:342, 1993.

[27] M. Punch et al. *Nature*, 358:477, 1992.

[28] G. Mohanty et al. *Proc. 23rd Int. Cosmic Ray Conf. (Calgary)*, 1:440, 1993.

[29] A.D. Kerrick et al. *ApJ*, 438:L59, 1995.

[30] T. Takahashi et al. *IAU Circular*, 6063, 1994.

[31] D.E. Alexandreas et al. *ApJ*, 383:L53, 1991.

[32] M. Amenomori et al. *Phys. Rev. Lett.*, 69:2468, 1992.

[33] T.A. McKay et al. *ApJ*, 417:742, 1993.

[34] A. Karle et al. MPI-PhE/95-01, submitted to Astroparticle Phys., 1995.

[35] C.E. Covault et al. *Proc 2nd Compton Symposium (College Park)*, AIP Conf. Proc. 304:499, 1993.

[36] F.W. Stecker et al. *ApJ*, 415:L71, 1992.

[37] M.H. Salamon et al. *ApJ*, 423:L1, 1994.

[38] J. Primack. Plenary talk this workshop, 1994.

[39] M. Teshima et al. *Proc. Tokyo Workshop on Techniques for the Study of EHE Cosmic Rays*, ed. M. Nagano, 109, 1993.

[40] S. Barwick et al. *Proc. 23rd Int. Cosmic Ray Conf. (Calgary)*, 1:436, 1993.

[41] S. Ahlen et al. *Nucl. Inst. and Meth.*, 351A:493, 1994.

[42] E. Pare. *Proc. Towards a Major Atmospheric Cherenkov Detector-II (Calgary)*, ed. R.C. Lamb, 250, 993.

[43] R.A. Ong et al. *Proc. Towards a Major Atmospheric Cherenkov Detector-III (Tokyo)*, ed. T. Kifune, Universal Academy Press, 295, 1994.

The Highest Energy Particles in Nature

A.A. Watson

Department of Physics, The University of Leeds, Leeds LS2 9JT, UK

The methods used to measure cosmic rays, now known to have energies up to 3 x 10^{20} eV, are briefly described. Recent data are reviewed and prospects for new experiments outlined.

§1 Introduction

The highest energy particles found in Nature are cosmic rays which are now known to have energies at least as great as 3×10^{20} eV [1]. The discovery of one particle of this energy, and of two others which are more than 10^{20} eV [2],[3], confirm earlier evidence [4],[5] that cosmic rays are observed at Earth beyond the spectrum cut-off predicted by Greisen [6] and Zatsepin and Kuzmin [7]. This GZK-cut-off is expected to arise if the sources of particles above about 4×10^{19} eV are at cosmological distances so that they suffer significant energy degradation en route to the Earth through the reactions

$$p + \gamma_{2.7K} \rightarrow N + \pi \tag{1}$$

in the case of protons, and

$$A + \gamma_{2.7K} \rightarrow (A - 1) + n \tag{2}$$

in the case of heavier nuclei, where the cosmic background radiation field is denoted by $\gamma_{2.7K}$. The mean free paths for these reactions are, of course, functions of energy. Detailed calculations of the shape of the cosmic ray spectrum resulting from propagation of the particles through the cosmic background radiation-field have been made by many authors under a variety of assumptions about the energy spectrum at the source and have been recently expressed in a particularly clear manner (figure 1 [8]). Here it is seen that protons originating with 10^{20}, 10^{21} or 10^{22} eV end up with an energy of ~ 4×10^{19} eV if their path lengths are ≥1000 Mpc. The rapid fall in energy in the range near 10^{20} eV is associated with the black body spectrum of the 2.7 K radiation field and the resonant nature of the interaction cross-section. Thus the existence of particles with energies ≥10^{20} eV would imply that their place of origin cannot be at a cosmological distance. The actual distance from which the particle can have travelled will depend upon the maximum energy to which sources can accelerate particles. Because of the GZK prediction there has, since 1965, been extensive experimental study of the shape of the cosmic ray spectrum above 10^{19} eV.

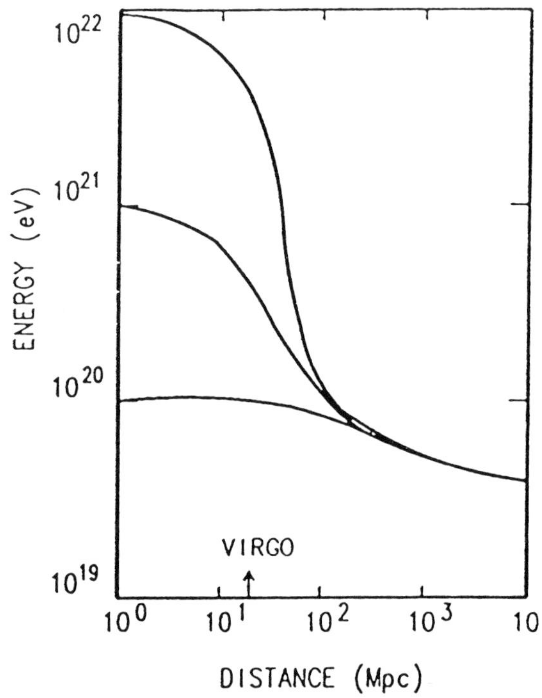

Figure 1: Calculation of the mean energy of protons due to the interaction with the 2.7K radiation as a function of distance for various initial energies of the particles at the source [8].

In addition to this constraint on the location of the sources of the most energetic particles, it is also well-established that cosmic rays of ≥ 10^{20} eV can only be produced in rather exotic objects. A generic analysis of this problem was given by Greisen [9] and subsequently by Cavallo [10]. Greisen showed that the combined constraints (a) that Larmor radius of a proton must be less than the scale length of the magnetic field and (b) the limit on the strength of the field set by the need to keep proton synchrotron losses low, require that $B < 10^{21} / \gamma^2$ G (0.1G for 10^{20} eV) and that the magnetic field energy is $>> 300\gamma^5$ (3.10^{57} ergs at 10^{20} eV), where γ is the Lorentz factor of the particle. A more specific analysis has been made by Hillas [11] in the context of a shock acceleration system in which there are scattering centres with characteristic velocities βc. His analysis shows that, for a particle of charge Z, $B_{\mu G} L_{pc} > 2 E_{15} / Z\beta$ where L_{pc} is the Larmor radius in pc, $B_{\mu G}$ is the field in μG and E_{15} is the energy of the particle in units of 10^{15} eV. Figure 2 is shown a much-referenced diagram from [11] in which many sites where particle acceleration might occur are plotted. Those sites lying below the diagonal line fail to satisfy the inequality for the product of B and L, even for $\beta = 1$ in the case of protons of 10^{20} eV. The dashed line refers to 10^{20} eV iron nuclei and the shaded area is bounded by $\beta = 1/300$. It is clear that the possible acceleration sites which satisfy the inequality are severely limited. Only regions such as the lobes of radio galaxies and the inter-galactic medium have sufficiently weak magnetic fields for synchrotron losses to be negligible; the magnetic energy in such volumes are consistent with the conclusions drawn from Greisen's analysis. Hillas has also pointed out that this restriction applies too to single-shot acceleration where an emf ~ LvB/c may partly be available for acceleration from, for example, a magnetised rotating body.

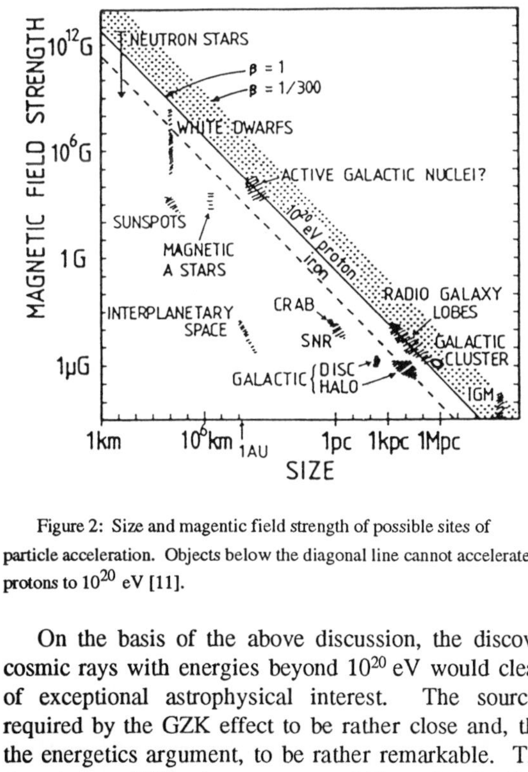

Figure 2: Size and magentic field strength of possible sites of particle acceleration. Objects below the diagonal line cannot accelerate protons to 10^{20} eV [11].

On the basis of the above discussion, the discovery of cosmic rays with energies beyond 10^{20} eV would clearly be of exceptional astrophysical interest. The sources are required by the GZK effect to be rather close and, through the energetics argument, to be rather remarkable. There is thus the possibility that sources could be uniquely identified and the intensity and spectrum of the observed particles could be used to constrain the electrodynamic processes therein. Studies made so far have failed to isolate the acceleration regions although it does seem that they must be within 50 Mpc of our own galaxy. This has led to speculation that exotic processes, such as the collapse of cosmic strings, may be responsible - a deceleration process - or that the properties of neutrino interactions at ultra high energies may be rather different than predicted by conventional implementations of the standard model. Unfortunately, the very low flux of such particles (~ 1 km^{-2} sr^{-1} $century^{-1}$) makes it necessary to plan a giant detector (aperture ~ 10^4 km^2 sr) to extend our very rudimentary knowledge of particles which are the most extreme example of a departure from thermal equilibrium in the Universe. In what follows, the methods so far used to detect these rare events and the information about them will be summarised. Future plans will be outlined.

§2 Detection of Cosmic Rays above 10^{18}eV

Cosmic rays above energies even as low as 10^{15} eV cannot be detected directly by, for example, instruments flown in balloons or satellites. The rate of these cosmic rays is simply too low. However, since the studies of Auger and his group in the late 1930's, it has been known that such particles initiate cascades, or extensive air showers, within the atmosphere from which a significant number of secondaries (mainly muons, electrons and photons) survive to ground level. At 10^{18} eV the secondaries are spread over an area of about 1 km^2 and thus a relatively small number of detectors placed over such an area suffices to detect the shower and can be used to measure the arrival direction of the progenitor and its energy.

Giant shower detectors making use of this principle have been run at Volcano Ranch (US), Haverah Park (UK), Sydney (Australia), Yakutsk (Siberia) and AGASA (Japan). The techniques used at such detectors are well known and have been discussed recently elsewhere [12]. For such particle detector arrays the arrival directions can be determined to within 2.5° and a ground-parameter appropriate to the detection method determined to ~ 15%. The relationship of this ground-parameter, which is typically the density measured in a water-Cerenkov or scintillation counter at about 600 m, to the primary energy has been discussed in [12]. Not surprisingly, in view of the extrapolations involved, there has been considerable scepticism about claims made for the energies of the highest energy events deduced in this way and in particular the evidence for the existence of events beyond the GZK-cut-off has often been disputed. However final results from a quite independent approach have recently been reported [ref 1 and discussion below] and they provide strong support, with superior statistics between 3.10^{17} eV and 10^{19} eV, for the original claims. At higher energies, such differences as do exist, can be understood in terms of statistical fluctuations and uncertainties in energy estimates alone.

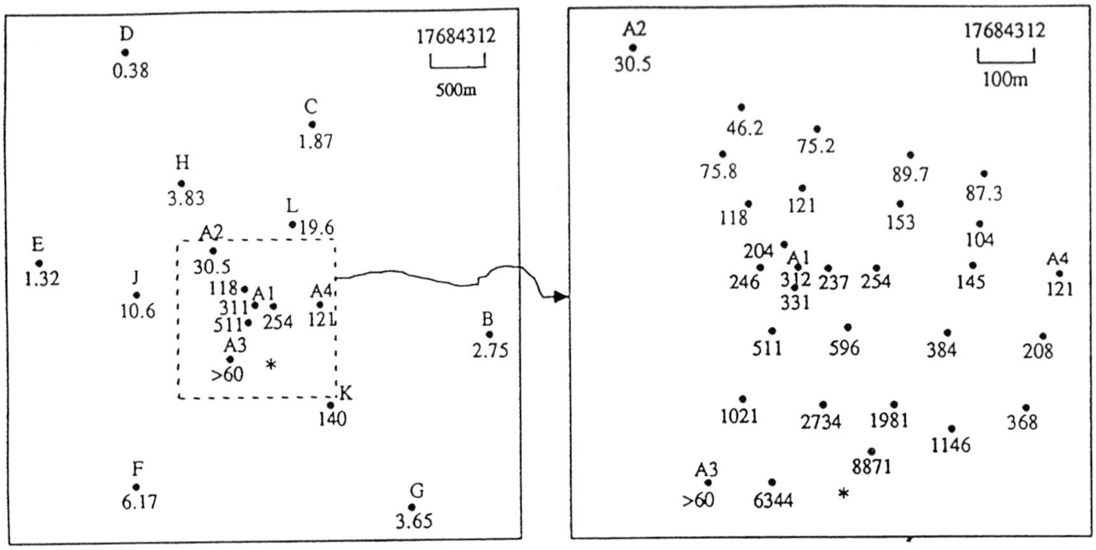

Figure 3: Density maps for event 17684312 (zenith angle = 37°) recorded at Haverah Park [5]. The area enclosed by the dashed square in the left-hand map is enlarged on the right where the unlettered detectors are 1m² water-Cerenkov detectors. The other detectors are of larger area (see [5] for details). The core is marked with a star in both maps.

The array with which I am most familiar is the one which operated at Haverah Park, near Leeds, for twenty years between 1967 and 1987. The best measured event recorded during this period is displayed in figure 3 which serves also to show the configuration of the detector array. The best estimate of the energy of this event is $(1.0 \pm 0.1) \times 10^{20}$ eV and three other events, nominally above 10^{20} eV, were recorded during the twenty-year run time of this experiment. In addition to the data shown in the figure, the temporal spread of the shower disk (10 - 50% risetime) was measured at the detector A2, 878 m from the core, and found to be 185 ns, entirely consistent with what was expected from a detailed study of this characteristic in showers an order of magnitude less in energy [13]. Unfortunately the 10 m² muon detector at the central detector was saturated by the strength of the signal.

A unique method of recording the extensive air showers produced by primary cosmic rays relies upon the scintillation light produced through interactions of the charged particles of the cascade with the nitrogen in the atmosphere. The light is produced dominantly in the 300-400 nm band and although the fluorescent yield is small (typically 4.5 photons per metre of electron track) a detector has been built to observe the emission. The light can be seen above the sky background on clear moonless nights. This feat was achieved by the University of Utah [14] group who have built a device containing two separated "Fly's Eyes" made up of 880 and 460 photomultiplier tubes respectively, 3.3 km apart. With these units it has been possible to map out the longitudinal development of individual shower events for the first time. An example of a shower reconstructed steroscopically by the two eyes is shown in figure 4.

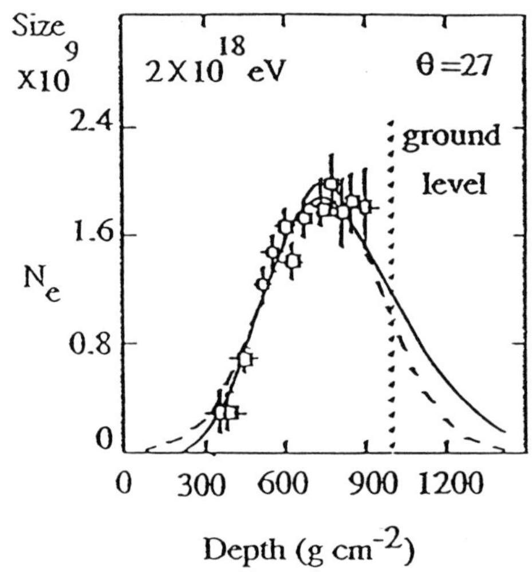

Figure 4: A cascade curve produced by a 2×10^{18} eV primary as observed by Fly's Eye I and II. The depth of maximum is estimated as 740 ± 40 gcm^{-2} [14].

The unique feature of this instrument is that it allows direct measurements of the depth of maximum of the shower and of the shape of the cascade curve. The energy of the particle which initiates each cascade is obtained from the track length integral of the shower-development curve, ie

$$E_{em} = \frac{\varepsilon_0}{X_0}\int N_e(x)dx \qquad (3)$$

where E_{em} is the total energy dissipated in the electromagnetic channel, ε_0 / X_0 is the ratio of the critical energy of electrons to the radiation length and $N_e(x)$ is the observed size of the shower as a function of atmospheric depth, x. A correction of about 10% must be made for energy which goes into muon, neutrino and hadronic channels. The measurement of N_e depends on accurate knowledge of the factors responsible for light production and transmission through the air, calibration of the optical detection system and determination of the shower trajectory. Allowances for these factors have been discussed in detail by the Fly's Eye group.

In the Fly's Eye detector there are clear analogies to the calorimeters used in particle physics experiments with the absorber of the calorimeter being air rather than lead or uranium.

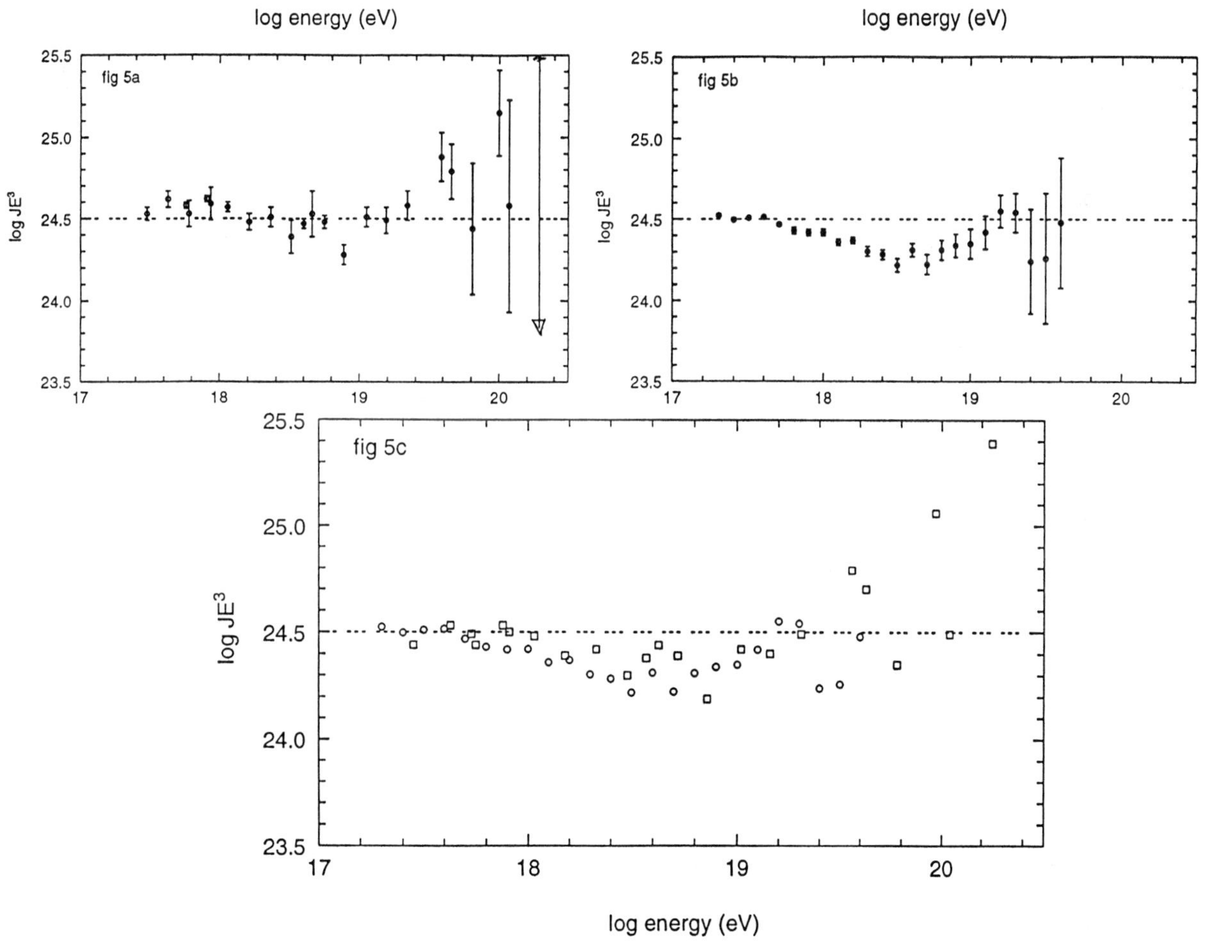

Figures 5(a), (b), (c): The energy spectra reported by Haverah Park [5] and Fly's Eye (stereo data) [1] are shown in figure 5(a) and figure 5(b). The dotted lines are drawn at log JE^3 = 24.5 to guide the eye. The agreement between the two spectra, after a 7% reduction in energy has been made to the Haverah Park events, is clear in figure 5(c) (squares, Haverah Park, circles, Fly's Eye).

§3 Review of Measurements of the Energy Spectrum

The final version of the Haverah Park energy spectrum [5] and the Fly's Eye stereo spectrum [1] are shown in figures 5(a) and 5(b). Both spectra show common features with the spectrum slope steepening above about 10^{18} eV but then recovering beyond about 10^{19} eV. The Fly'e Eye data set contains nearly 4 times as many events as that from Haverah Park. This reflects the fact that below ~ 5.10^{18} eV the Haverah Park group used a highly selected subset of events (~ 1%) so that systematic effects arising in event reconstruction could be tightly controlled. Furthermore the Fly's Eye stero data have allowed the determination of the energy resolution for monocular events. It is 36% below 2×10^{18} eV and 27% above 2×10^{18} eV. Above 10^{19} eV the Haverah Park sample is nearly 60% larger than the Fly's Eye stereo sample. The two spectra do not agree absolutely in intensity at a given energy (as is clear from comparison with the arbitrary reference line shown horizontally in both figures) but if the spectra are normalised at the lowest energies - an adjustment of only 7% is required - the resultant comparison (figure 5(c)) shows very good agreement to the highest energies. This systematic shift is well within the systematic uncertainty of 20% ascribed to the Fly's Eye energies and which arises from that lack of detailed knowledge of the transfer of cascade energy into the fluorescence output. The conversion from the ground parameter measured at Haverah Park to the primary energy undoubtedly contains an unknown systematic error as well.

Given the entirely different techniques which have been adopted and the different systematic errors which are inevitably associated with these approaches, it is reasonable to claim that the cosmic ray spectrum is well measured from 3.10^{17} to (at least) 3.10^{19} eV and that the systematic uncertainties remaining are less than 20%. This result, in my view, justifies the reality of earlier claims for features in the spectrum over this energy range and also lends strong support to the assessments given previously to the energies of showers recorded at Haverah Park.

Beyond 3.10^{19} eV the measurements are dominated by the statistical uncertainties arising from the limited exposure. From these two measurements alone, existence or otherwise of events beyond the GZK-cut-off is not completely clear although the Haverah Park experiment, in which the exposure (AΩt) was nearly twice that of the stereo Fly's Eye detector, has reported four events for which E is claimed to be > 10^{20} eV. The spectrum reported by the Yakutsk group [15] and the preliminary version of the spectrum from the 100 km^2 AGASA array [16] (figures 6(a) and 6(b)) are in broad agreement with features of the Haverah Park and Fly's Eye data, although shifts in energy of rather more than 7% are required to align these two spectra with those of figure 5 at 3.10^{17} eV.

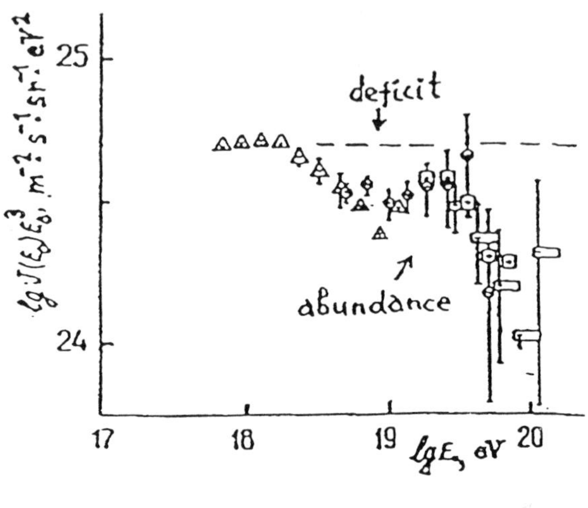

Figure 6(a): The Yakutsk energy spectrum [15]. The axes are the same as adopted in figure 5.

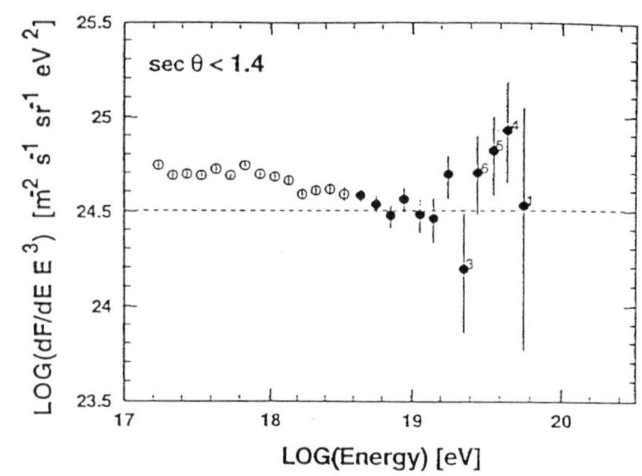

Figure 6(b): Preliminary AGASA energy spectrum [16]. Filled circles : AGASA data. Open circles: Akeno array data.

Considerable efforts have been made to make independent cross-calibrations between different experiments. For example at Haverah Park eight scintillators, similar to those at the Yakutsk array, were operated for a period and energies estimated using the Yakutsk algorithm, which is less dependent on model calculations than the Haverah Park algorithm. Good agreement (better than 15%) was found for a sample of 20 events at 2.10^{18} eV and a further 20 events ranging in energy upto 5.10^{19} eV [17].

§4 The Highest Energy Events

The early claims by Volcano Ranch [4] and Haverah Park [18] that events with energies close to 10^{20} eV do exist have recently been supported by the other groups. In 1990 the Yakutsk group described [3] an event with an energy of $(1.1 \pm 0.4) \times 10^{20}$ eV. While undoubtedly an event which has been produced by a very energetic primary, the almost completely muonic nature of the particles detected and its relatively large zenith angle (58.7°) make it hard to be definite about the energy assignment. The Volcano Ranch and Haverah Park events are at least as convincing.

In 1993 the Fly's Eye group reported [1] a monocular event for which the assigned energy is $\left(3^{+0.4}_{-0.5}\right) \times 10^{20}$ eV. This is the highest energy event ever recorded and the cascade curve reconstructed for this event is shown in figure 7. The cascade is quite 'normal' in terms of what has been observed for the bulk of the energetic particles at lower energies. Marked on the curve are the average depth of maxima expected for iron and proton primaries of this energy.

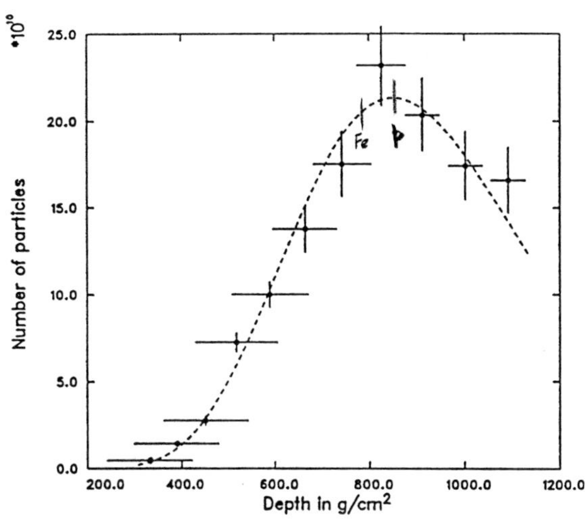

Figure 7: Shower profile for the most energetic event observed by the Fly's Eye detector. The expected positions of shower maximum for showers initiated by iron and proton primaries are shown.

Detector	Date	Reference No.	Energy /10^{20} eV	θ degrees	RA degrees	δ degrees	b degrees	ℓ degrees
Volcano Ranch	22.04.62	4472	1.4	11.7	306.7	46.8	4.8	84.3
Haverah Park	31.12.70	8185175	1.02 ± 0.3	35	353	19	-40	99
	05.12.71	9160073	1.05 ± 0.3	30	199	44	73	107
	18.04.75	12701723	1.2 ± 0.1	29	179	27	78	212
	12.01.80	17684312	1.05 ± 0.08	37	201	71	46	119
Yakutsk	07.05.89	890507133	1.1 ± 0.4	58.9	75.2	45.5	2.6	162.2
Fly's Eye	15.10.91		$3^{+0.36}_{-0.54}$	43.9 (+1.4 -0.6)	85.2	48.0	9.6	163.4
AGASA	03.12.93	25400-0296	(1.7 ~ 2.6)	22.9	18.9	21.1	-41	131

Table 1: Details of events for which the energy of the progenitor has been claimed to be about 10^{20} eV (from Northern Hemisphere Experiments)

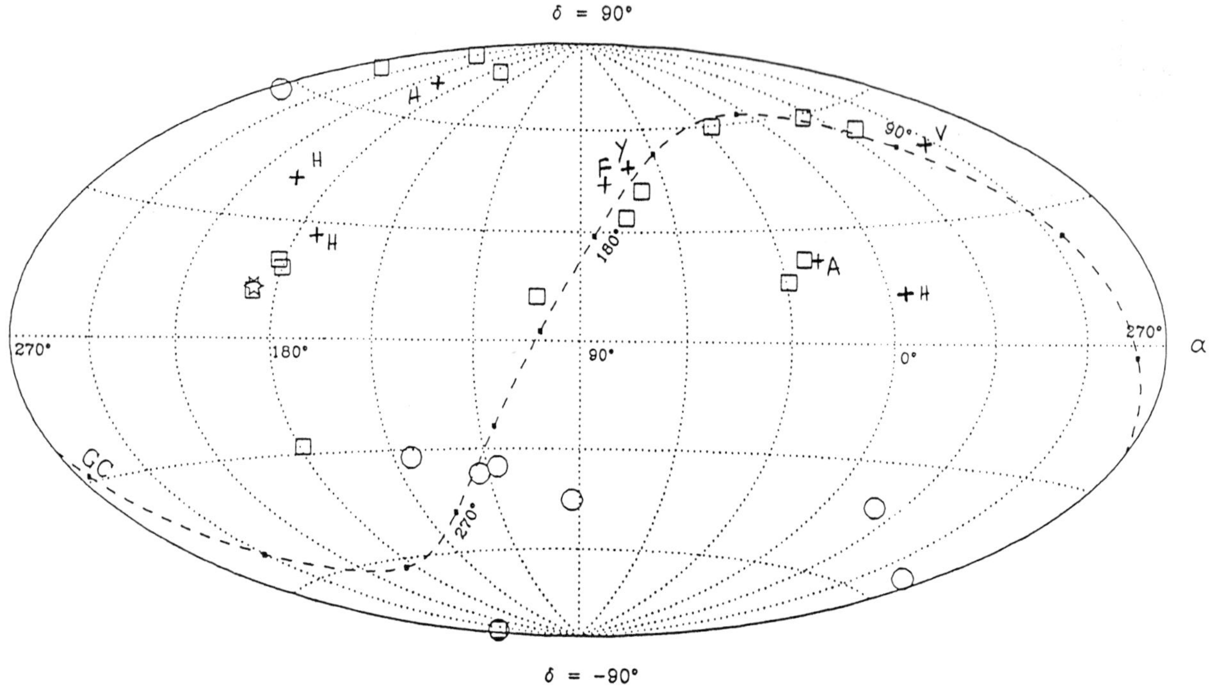

Figure 8: Aitoff plot of 'groups' of more than 4 CR events (E > 10^{19} eV) within 6° of an event (plotted as squares) with E > 3.10^{19} eV and groups listed in the table (plotted as circles). The Galactic Plane is indicated. The star indicates the Virgo cluster (M87) direction. The directions of events from Haverah Park, Volcano Ranch, Agasa, Yakutsk and Fly's Eye (table 1) believed to be above 10^{20} eV are represented by H, V, A, Y and F.

In December 1993 the AGASA group observed [2] a very clean event for which the energy assignment is (1.7 ~2.6) x 10^{20} eV. This event fell in such a position that the central region, or core, of the shower is in a part of the array which is relatively densely populated with detectors. The main signal is detected in unshielded scintillation counters and the fall-off of density with distance from the core position is no way anomalous by comparison with expectation from lower energies. Furthermore muons are detected in 7 detectors at a normal density level (although two stations are saturated) and the temporal distribution of the signal in a 30 m^2 scintillation counter at 1920 m from the shower core is also consistent with expectation.

A summary of the directions and energies of the events which are thought to be close to or greater than 10^{20} eV is given in table 1. The directions of these events are shown in figure 8 superposed on a plot made by Chi et al [19] of clusters of cosmic rays above 10^{19} eV about the directions of events with energies of > 3.10^{19} eV. It is interesting that the largest AGASA event is within 4° of one of the Chi et al clusters. However other of the most energetic events (table 1) are not so closely associated with the cluster directions but it has been suggested [26] that at least some of the clusters are random associations. In figure 9 the directions of the events of table 1 are superposed on a plot made by Covault (Chicago) in which energetic astronomical objects are displayed. There are no obvious associations of the directions of these energetic events with any unusual astronomical object.

§5 Implications of the Highest Energy Cosmic Ray Events

Rather straight forward statements about the distances which a cosmic ray of a given energy can travel from its birthplace before detection can be made with the aid of figure 1. Thus if the event of 3 x 10^{20} eV is a proton[1] and has originated in a source with an initial energy of 10^{21} eV, the average distance which it will travel before its energy is reduced to 3 x 10^{20} eV is ~30 Mpc. Hence the source of this particle must be nearby and is certainly not at a cosmological distance. Fluctuations in the travel distance can occur and

[1] There is evidence from the Fly's Eye measurements [1] that the mass composition is becoming lighter as the energy increases from 10^{17} to 10^{19} eV but no certain statements can be made about the primary mass above 10^{20} eV.

have been investigated (Cronin, private communication) but they do not radically alter this conclusion. Of course this statement does not imply that particles of lower energy, less than ~10^{20} eV, are produced in the same source: it is apparent from Figure 1 that particles of 10^{20} eV could come from sources as distant as ~100 Mpc. However I find the similarity of the flux associated with the Fly's Eye event $\left(\left(4^{+9}_{-3}\right)\times 10^{-17}\,\mathrm{m}^{-2}\mathrm{s}^{-1}\mathrm{sr}^{-1}\right)$ and the flux deduced from the AGASA detection $\left(\left(2^{+4.6}_{-1.6}\right)\times 10^{-16}\,\mathrm{m}^{-2}\mathrm{s}^{-1}\mathrm{sr}^{-1}\right)$, when coupled with the fluxes predicted by extrapolation of the spectra observed above 10^{19} eV by each group ($4.2 \times 10^{-17}\,\mathrm{m}^{-2}\,\mathrm{s}^{-1}\,\mathrm{sr}^{-1}$ and $1.0 \times 10^{-16}\,\mathrm{m}^{-2}\,\mathrm{s}^{-1}\,\mathrm{sr}^{-1}$ respectively) to be so close as to be indicative of a common, and therefore local, origin for all of the events above the GZK-cut-off (see table 2).

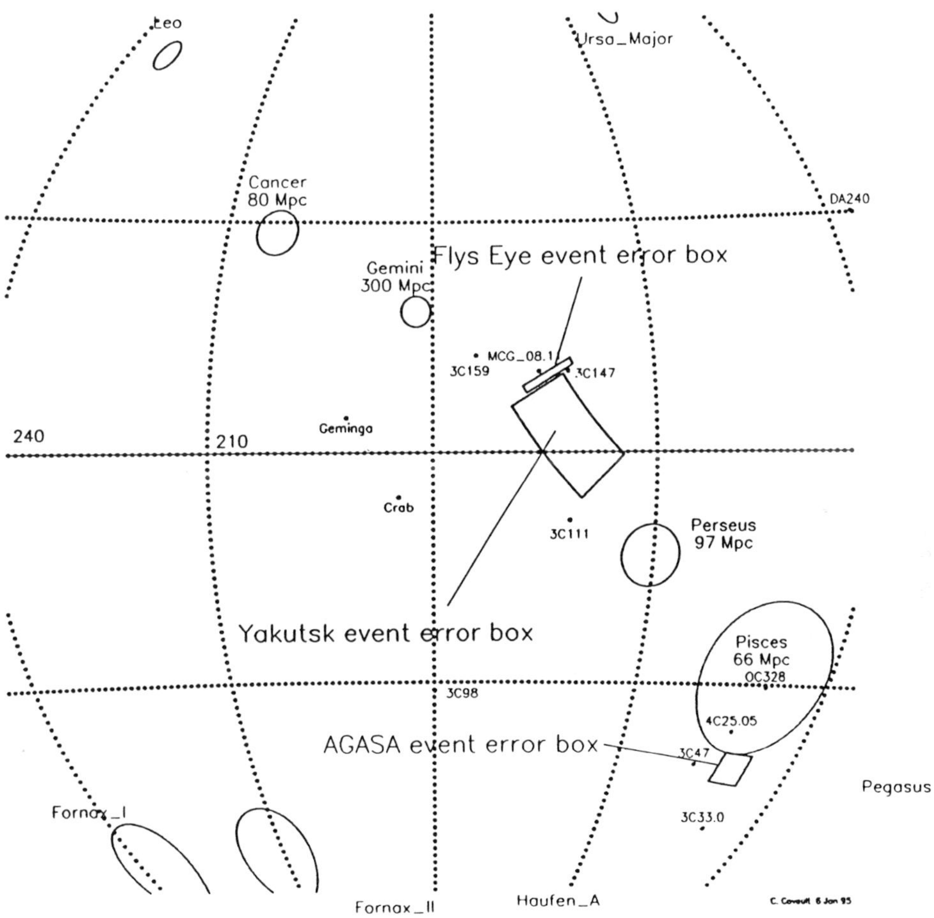

Figure 9: Plot due to C Covault (EFI, University of Chicago) of the region of sky containing the most recently recorded giant showers and some interesting extragalactic objects. The galactic anti-centre is at the centre of the plot.

Detector	Assumed Energy of Super-High Energy Event	Exposure (m^2 s sr)	Integral Flux ($m^{-2} s^{-1} sr^{-1}$)	Integral Flux Predicted from Extrapolation (a) ($m^{-2} s^{-1} sr^{-1}$)
AGASA	2×10^{20} eV	5×10^{15}	$\left(2\begin{matrix}+4.6\\-1.6\end{matrix}\right)\times 10^{-16}$	1.0×10^{-16}
Fly's Eye	3×10^{20} eV	26×10^{15}	$\left(2\begin{matrix}+9\\-3\end{matrix}\right)\times 10^{-17}$	4.2×10^{-17}

(a) In making this prediction I have used $J(E) = 7.36 \times 10^{18} E^{-2.71} m^{-2} s^{-1} sr^{-1} eV^{-1}$ for Fly's Eye stereo spectrum (obtained from [1], tables 2 and 7) and $J(E) = 5.8 \times 10^{18} E^{-2.7} m^{-2} s^{-1} sr^{-1} eV^{-1}$ for AGASA (see [17]) as the differential intensities at 10^{19} eV. The uncertainty in the extrapolated fluxes, arising from statistical uncertainties in the slopes and intensities near 10^{19} eV are small by comparison with the uncertainty range (a 68% confidence band is given) assigned to the observed fluxes at the very highest energies.

Table 2: Comparison of observed fluxes of super-high energy events and the prediction from extrapolation of spectrum measured at lower energies.

A detailed study of the implications of the 3×10^{20} eV event has been made by Elbert and Sommers [21]. In an exhaustive analysis they identify a number of possibilities. They find no likely astrophysical source within 50 Mpc in the direction of the Fly's Eye event. If the intergalactic magnetic field is 10^{-9} Gauss and is in such a direction as to bend the particle maximally from its original direction then a proton would be deflected by only ~3° over this distance. In fact (eg [22]) the intergalactic magnetic field is likely to be in cells with, perhaps, a characteristic dimension of ~ 1 Mpc. This will substantially reduce the deflection. If the cells are of this size and contain magnetic fields of 10^{-9} G, randomly oriented, then the rms deflection over 50 Mpc is about 0.6° [23].

Another possibility, which would allow a distant source, is that the existence of some the new particle which has a low cross-section for interaction in the 2.7 K radiation field but a relatively high cross-section for interaction in the atmosphere. Similarly if the neutrino cross-section was to be very much larger than the extrapolated value expected from accelerator observations [24] there might be a way out of the demands of a local source. However the developments of the AGASA and Fly's Eye event are quite consistent with hadronic primaries (see figure 7) interacting normally in the atmosphere.

The difficulty in accelerating particles to beyond 10^{20} eV electromagnetically, which has already been briefly described (see ref [11] for details), has led to suggestions that other processes might be involved. In particular the idea has been discussed that ultra high energy cosmic rays might be produced by processes other than electromagnetic ones. Bhattacharjee [25] and colleagues have examined the possibility that the collapse of massive cosmic string loops will lead to production of X particles which decay into quarks and leptons. If f is the fraction of the loop energy going into X particles then application of QCD allows an estimate of the cosmic ray flux which could arise from loop collapses. If the strings are massive enough to explain the large scale structure of the Universe, then existing measurements of cosmic rays of 10^{20}eV set a limit of $f \leq 10^{-3}$. An updated discussion of this possibility has been developed by Sigl et al [26]: the possibility of monopolonium decay has been examined by Hill [27] and by Bhattacharjee and Sigl [28].

A more conservative suggestion of Elbert and Sommers [21] is that the source of the highest energy event is M82 which is about 3.5 Mpc distant. If the energy of the Fly's Eye event was actually 2.3×10^{20} eV (at the -1σ limit), then the primary could have been a nucleus of Z ~ 18 surviving from an iron nucleus which had been accelerated to 3.4×10^{20} eV in M82. The 37° deflection between the arrival direction and that of M82 would take place in the magnetic field of our galaxy.

An alternative proposal [21] is that the nearby intergalactic magnetic field is very much stronger than has normally been supposed. They suggest that Cen A and M87 are possible sources and that the lack of correspondence of the arrival directions of the giant event with the direction of these sources (it is 136° from Cen A and 87° from M87) is because there exists a field of (0.05 - 0.5)μG in the nearby intergalactic medium. This field is substantially stronger than expected but it is difficult to infer its value directly (Kronberg, private communication).

If such a field did exist there might by an explanation for the apparent lack of events seen by Fly's Eye (and possibly AGASA) between 7.10^{19} eV and 2 - 3×10^{20} eV. The gap corresponds to the energy at the which the mean free path for interaction is ~ 100 Mpc [29] and the path length of the lower energy particles in the hypothetical field might be sufficiently long to account for their (possible) absence in the spectrum. Uncertainties in the energy estimates of the few Haverah Park events and the single events recorded from Volcano Ranch and Yakutsk (table 1), in either direction, could shift them out of this energy gap. If there are local sources then they would be expected to contribute to the spectrum below 10^{19} eV: it would be interesting to explore what the field requirements would be for the particles to be contained long enough so that the steepening and flattening of the spectrum in the 10^{18} - 10^{19} eV decade was actually a 'bite' taken from a continuous spectrum as a consequence of the pair production process

$$\gamma + p \rightarrow p + e^+ + e^- \quad (3)$$

which has an energy threshold of ~ 10^{18} eV. The mean free path for this process is ~ 1 Mpc but as the fractional loss in each interaction is only ~ 10^{-3} the particles would need to be stored for ~ 10^9 years for significant losses to occur.

The speculation that M82, Cen A and M87 might be sources of the highest energy cosmic rays has been made long ago from a quite independent analysis by Cavallo [10]. His argument was based purely on the energetics of processes of production of cosmic rays of 10^{20} eV. However it is worth recalling that the supernova origin of cosmic rays of less than ~ 10^{15} eV, which was first argued, very largely on energetic grounds by Ginzberg in the 1950's, is currently thought to be correct now that first order Fermi acceleration has been shown to work at relativistic shocks [30]. Perhaps history will be repeated and these three objects will turn out to play a major role in the production of the most energetic cosmic rays.

§6 The Auger Project

Speculations about the origin of the highest energy cosmic rays could be greatly restrained, and there would be a realistic prospect of identifying sources, if there were data on the energy, mass composition and arrival directions of substantially more high energy events. Since 1991 a number of people, led by J.W Cronin, have been promoting the concept of a giant detector of ~ 5000 km^2 which would allow about 50 events above 10^{20} eV to be recorded per year. This project has been named the Auger project, after Pierre Auger, the discoverer of extensive air showers. Here I will describe only the concept behind recent discussions of how to build such a device.

The design concept for the ground array of the Auger project, is to deploy some 3,000-5,000 detectors operated in an autonomous mode, with local intelligence, at a location where solar power use is feasible and line-of-site radio communication possible across large segments of the array. The idea of autonomous detectors was first advanced some 30 years ago and implemented by the University of Sydney group [31] at their site in Narrabri. Some 54 stations were deployed on a one-mile grid covering 60km^2. To reduce the background rate, each station comprised a pair of 6m^2 scintillators which were buried 1.7 metre below ground and so responded preferentially to muons. The relative arrival time at each detector was derived from a timing signal beamed across the array. Each station was powered by gas cylinders and data were logged on audio recorders from which the tapes were collected at approximately weekly intervals.

The Auger project seeks to update the Sydney approach by taking advantage of 30 years of technological development. A number of detailed designs have been considered. A favoured one consists of detectors deployed on three nested grids of spacing 500, 866 and 1500 m [32]. Well over 70% of the area will be covered by the largest spacing, with arrays at the two smaller spacings covering ~10 and ~20% of the area respectively. These sections will provide detailed information on the rare, large, events that fall therein in addition to extending the lower energy threshold in a cost-effective manner. The running time of such a device is envisaged as ten years so that the detectors must be extremely reliable and robust in addition to being virtually maintenance free. Such considerations lead naturally to the choice of plastic scintillation counters as the detectors although the use of other devices such as water-Cerenkov detectors or resistive plate counters will be evaluated.

A promising approach to tying down the mass composition is to measure the muon content of showers. Ideally this would be done at fixed primary energy but in practice it would be advantageous to measure the muon-electron ratio at large (> 700m) distances. A novel method [33] is being explored in which a scintillator-lead-scintillator sandwich is used, each scintillator being equipped with flash ADC circuitry. At the axial distances in question the electron and photon energies are low (average energy ~10 MeV [34]), and it may be possible to separate muons, electrons and photons by comparing signals in the same flash ADC time slots in the lower and upper scintillators of the sandwich. About one radiation length of lead is probably optimum. Some Monte Carlo calculations [33] have been made which suggest that separation of the electromagnetic and muon components may be possible.

The flash ADC's will additionally allow the time profile of the shower front to be measured. The risetime of signals in water-Cerenkov detectors have been used to measure the elongation rate [13], which has been measured directly by

Fly's Eye [1]. The elongation rate is mass-dependent and the same risetime technique can be extended to scintillators. At distances from the axis of ~1km the risetimes are several 100ns and so are easily measurable. Fluctuations in the risetimes will supplement the muon information to improve prospects for mass discrimination.

The problem of datalogging and data transmission is a particularly interesting one and has been explored in some detail [35] by Kelly (Telecom Australia), Nitz (Michigan) and Lloyd-Evans (Leeds). The techniques of cellular radio look applicable and it has been shown that provided the local rate is below 20 Hz, three fold triggers between nearest neighbours can be accommodated with the VHF frequencies likely to be available. So far the design studies have assumed that the detector areas at each station will be ~ 10 m² and to achieve a low density threshold at a rate less than 20 Hz would require that a station area be deployed in two or three blocks separated by a few tens of metres. Alternatively, and perhaps more elegantly, a dual mode of triggering can be envisaged. High densities would be rare and all can be recorded while, as slow pulses are characteristic of distant showers, which are similarly rare, risetime discrimination might additionally be adopted. It has been demonstrated experimentally [36] that pulses with risetimes greater than 100 ns occur at a rate of 1 in 10^3 for densities of 0.4 m^{-2} on 6 m² of scintillator. The direction of each event is determined by measuring the relative arrival times of the shower front at the widely spaced detectors. It has been shown [37] that adequate timing accuracy can be achieved using GPS receivers at such stations.

The solar power required for each station is expected to be in the region of 20-30 watts. The technology is available to supply this but careful electronic design will be necessary to keep down the power and thus the cost. A site at which solar power can be used effectively is also likely to hold out the possibility of being useful for night studies of air-Cerenkov light; the potential of this has not yet been investigated fully. Site surveys are about to take place.

Detailed work suggests that, although there are significant design challenges to meet, the construction and operation of such an array is possible. The Chicago group have successfully operated 1089 scintillation detectors on a 10^5 m² grid [38] for a number of years and shower instruments with 12-15,000 photomultipliers are modest extensions of this. The novelty lies in the use of GPS receivers in an unconventional mode, the design of complex circuitry requiring very low power and the telecommunication aspects. The detectors themselves are somewhat brute-force tools; the instrumental challenge and elegance of the project lies in the novelty of the electronics.

The conceptual design studies have also shown [39] that there would be significant advantages in running a hybrid device with an array of particle detectors and one or more fluorescence detectors in coincidence. The fluorescence detectors envisaged are adaptations of the Fly's Eye Hi-Res instrument [14] and could be located at the centre of perhaps three sites within the boundaries of the particle detector array.

The proposed device would allow maps to be made of the cosmic ray arrival direction distribution as a function of particle mass. Broad anisotropies in the arrival directions, such as might be expected if there is diffusion of particles in a large scale magnetic field would be detectable with ten years of exposure. At a significance level of 10^{-3} a 2% departure from uniformity would be established with 5 x 10^4 events (10^{19} eV) and a 25% anisotropy with ~ 500 events (10^{20} eV). It might be reasonably expected that the highest energy events would cluster on small angular scales (as has been claimed by Chi et al [19]) and if possible sources lie close to such clusters direct correlation will be achievable and inferences about near-extragalactic magnetic fields can be drawn. However it is quite possible that the sources will show a high degree of isotropy, as with gamma-ray bursts, and with the added constraint that an origin at cosmological distances is impossible. Then the topological defect models must be considered seriously. It is highly desirable that two devices be built, one in each hemisphere, so that the sky can be completely surveyed. This is particularly important as there may be very few sources or, in the case of topological defects, uniformity on a 50 Mpc scale is not expected because of gravitational clustering.

What can be promised from the construction of the device envisaged is that we will learn an outstanding new astrophysics, and possibly new particle physics, from the observations which will be acrue.

§Acknowledgements

It is a pleasure to thank the organisers for inviting me to give the talk on which this paper is based. Professor Nagano kindly agreed to let me describe the latest AGASA results prior to publication. Many discussions with Jim Cronin, Jeremy Lloyd-Evans, Dave Nitz, Pierre Sokolsky and Paul Sommers are gratefully acknowledged.

§References

[1] D.J. Bird et al. *Physical Review Letters*, 71:3401, 1993.

D.J. Bird et al. *Ap. J.*, 424:491, 1994.

[2] N. Hayashida et al. *Physical Review Letters*, 73:3491, 1994.

[3] N.N. Efimov et al. Proceedings of International Workshop on Astrophysical Aspects of the Most Energetic Cosmic Rays. M. Nagano and F. Takahara, editors. World Scientific. p20, 1990.

[4] J. Linsley. *Physical Review Letters*, 10:146, 1963.

[5] M.A. Lawrence, R.J.O. Reid and A.A. Watson. *J.Phys.G.*, 17:773,1991.

[6] K. Greisen. *Physical Review Letters*, 16:748, 1966.

[7] G.T. Zatsepin and V.A. Kuzmin. *Sov. Phys - JETP Lett.*, 4:78, 1966.

[8] J.W. Cronin. *Nucl. Phys.B. (Proc Supp)*, 28B:213, 1992.

[9] K. Greisen. Proceedings of 9th International Cosmic Ray Conference, London, 2:609, 1965.

[10] G. Cavallo. *Astron. Astrophys.*, 65:415, 1978.

[11] A.M. Hillas. *Ann. Rev. Astron. Astrophysics*, 22:425, 1984.

[12] A.A. Watson. *Nucl. Phys.B. (Proc Supp)*, 22B:116, 1991.

P. Sokolsky, P. Sommers and B.R. Dawson. *Physics Reports*, 217:227, 1992.

A.A. Watson. *Nucl. Phys.B. (Proc Supp)*, 28B:3, 1992.

[13] R. Walker and A.A. Watson. *J.Phys.G.*, 7:1297, 1981.

[14] R.M. Baltrusaitas et al. *Nucl. Inst. Method*, A240:410, 1985.

[15] B.N. Afanasiev et al. Proceedings of Tokyo Workshop on Techniques for the Study of Extremely High Energy Cosmic Rays (Institute of Cosmic Ray Research, University of Tokyo), p35, 1993.

[16] S. Yoshida et al. *Astroparticle Physics*, 3: (in press).

[17] A.J. Bower et al. *J.Phys.G.*, 9:L53, 1983.

[18] G. Brooke et al. Proceedings of 19th International Cosmic ray Conference, La Jolla, 2:150, 1985.

[19] X. Chi et al. *J.Phys.G.*, 18:539, 1992.

[20] M.A. Lawrence and A.A. Watson. *J.Phys.G.*, 18:1827, 1992.

[21] J.W. Elbert and P. Sommers. *Ap.J.*, (in press, 1 March 1995).

[22] P.P. Kronberg. *Rep. Prog. Phys.*, 325, 1994.

[23] K.O. Thielheim. *J.Phys.A.*, 7:444, 1974.

[24] G. Domokos and S. Kovesi-Domokos. *Phys. Rev.D.*, 38:2833, 1988.

G. Domokos and S. Nussinov. *Phys.Lett.B.*, 187:372, 1987.

[25] P. Bhattacharjee. *Phys.Rev.D.*, 40:3968, 1989.

P. Bhattacharjee, C.T. Hill and D.N. Schramm. *Phys. Rev. Lett.*, 69:567, 1992.

F.A. Aharonian, P. Bhattacharjee and D.N. Schramm. *Phys. Rev.D.*, 46:4188, 1992.

[26] G. Sigl, D.N. Schramm and P. Bhattacharjee. *Astroparticle Physics*, 3 (in press).

[27] C.T. Hill. *Nucl. Phys.*, B224:469, 1983.

[28] P. Bhattacharjee and G. Sigl. *Submitted to Phys.Rev.*

[29] F.W. Stecker. *Phys. Rev. Lett.*, 21:1016, 1968.

[30] W.I. Axford et al. Proceedings of 15th International Cosmic Ray Conference, Plovdiv, 11:132, 1979.

R.D. Blandford and J.P. Ostriker. *Ap.J.Lett.*, 221:L19, 1978.

G.F. Krymsky. *Dok. Acad. Nauk. SSSR*, 234:1306, 1977.

A.R. Bell. *MNRAS*, 182:147 & 443, 1978.

[31] C.B.A. McCusker and M.M. Winn. *Il Nuovo Cimento Ser X*, 28:175, 1963.

[32] K. de Souza et al. *Nucl. Phys.B. (Proc Supp)*, 28B:135, 1992.

J. Lloyd-Evans. Adelaide Design Workshop on Techniques for Study of Cosmic Rays with Energies $>10^{19}$ eV (unpublished).

[33] A. Letessier-Selvon. Adelaide Design Workshop on Techniques for Study of Cosmic Rays with Energies $>10^{19}$ eV, 1993.

J.W. Cronin, these proceedings.

[34] E.W. Kellermann and L. Towers. *J.Phys.A.*, 3:284:1970.

[35] G. Kelley, D. Nitz and J. Lloyd-Evans. Adelaide Design Workshop on Techniques for Study of Cosmic Rays with Energies $>10^{19}$ eV, 1993.

D. Nitz, these proceedings.

[36] S.P. Hart. MSc Thesis, University of Leeds, 1993.

[37] C.L. Pryke and J.Lloyd-Evans. *Nucl. Inst. Method A.* (in press).

J. Lloyd-Evans, these proceedings.

[38] A. Barione et al. *Nucl. Inst. Method A*, 346:329, 1994.

[39] B.R. Dawson. Proceedings of Tokyo Workshop on Techniques for the Study of Extremely High Energy Cosmic Rays (Institute of Cosmic Ray Research, University of Tokyo), p125, 1993.

P. Sommers, these proceedings.

Neutrino Overview

W.C. Haxton

Institute for Nuclear Theory, NK-12, and Dept. of Physics, NK-15, University of Washington, Seattle, WA 98195

I discuss some of the open issues in neutrino physics, emphasizing areas of intersection with astrophysics, that occupied the participants of the Snowmass Workshop on Nuclear and Particle Astrophysics and Cosmology in the Next Millenium.

Although neutrinos have been part of physics for over 70 years, a surprising number of basic questions about their properties and their roles in astrophysics remain open:

*What are the neutrino masses and why are neutrinos so much lighter than the other fermions of the standard model?

*Do the three light flavors account for all of the neutrinos, or do more massive neutrinos exist?

*How do neutrinos behave under particle-antiparticle conjugation?

*Are there nonzero electromagnetic moments, e.g., magnetic moments, anapole moments, electric dipole moments, nonzero charge radii?

*Is there mixing between the neutrino families?

*What accounts for the deficit of neutrinos from the sun and the suppressed $\phi(\mathrm{Be})/\phi(^8\mathrm{B})$ flux ratio deduced from current experiments? Do these results demand new particle physics, such as the MSW mechanism?

*We detected the $\bar{\nu}_e s$ from SN1987A. What are the fluxes and spectra of supernova $\nu_e s$ and heavy-flavor neutrinos? What is the total lepton number radiated by a collapsing star?

*What is the role of neutrinos in the Type II supernova mechanism? Are they essential to mantle ejection?

*Are there cosmic background neutrinos and do they contribute significantly to dark matter?

*What is the role of neutrinos in stellar nucleosynthesis mechanisms such as the neutrino process and r-process? What can nucleosynthesis and stellar cooling arguments tell us about neutrino properties?

I would like to review current work addressing these and other open questions. I will not discuss terrestrial neutrino oscillation experiments or the atmospheric neutrino problem, since these are the subjects of another opening talk.

§1 Neutrino Mass Limits

Existing limits on neutrino masses are

$$\begin{aligned} m_{\bar{\nu}_e} &\lesssim 7\ eV \\ m_{\nu_\mu} &\lesssim 250\ keV \\ m_{\nu_\tau} &\lesssim 31\ \ MeV. \end{aligned} \qquad (1)$$

The $\bar{\nu}_e$ mass limit was established in a series of careful measurements of the shape of the tritium β-decay spectrum near the endpoint [1]. The parameters in that fit include the maximum energy release W_0 and m_ν^2. The surprising result that has emerged from the five experiments is that the best mass value is $m_\nu^2 \sim -100\ eV^2$, with this unphysical result differing from zero by 5σ. The Livermore measurements are shown in Fig. 1. However, the fits give a value for W_0 that agrees with the Seattle Penning trap measurement [2] of the ^{3}H - ^{3}He mass difference to ~1 eV, the approximate accuracy of both determinations. This suggests that the source of the negative m_ν^2 result is an effect that distorts the spectrum shape without perturbing its endpoint, which eliminates many instrumental explanations.

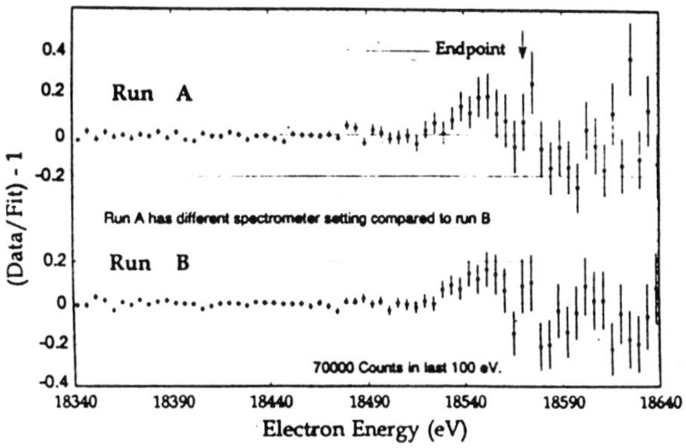

Figure 1: Deviations of the tritium β decay endpoint measurements of the Livermore group [1] from the Kuri plot ($m_\nu = 0$) are shown as a function of the electron energy. The excess counts near the endpoint generate a negative m_ν^2 in the fits.

It may be that calculations of the contributions of excited final states in the T_2 molecule are in error. For instance, if the ground state excitation probability were reduced from ~57% to about ~40%, with the difference appropriately redistributed to excited molecular states, the experimental results can be reproduced [3]. This is not an explanation that the molecular theorists find plausible.

Some of the experimentalists who have done tritium experiments are concerned that the negative m_ν^2 problem calls into question the limits on m_{ν_e} that have been quoted. While there is a standard procedure for deriving such limits, its appropriateness is unclear given that several experiments find negative m_ν^2 at high confidence levels. Bowles and Robertson will discuss this later in this meeting.

From the theoretical side, there is some hope that the smallness of neutrino masses may provide some insight into the general problem of fermion masses in the standard model. In the enlarged multiplets of extended models it is natural to expect that a single scale characterizes the masses of fermions in the same family (e.g. ν_e, e, u, d), an expectation clearly in conflict with observation. However, while the other fermions in the multiplets are Dirac particles, the neutrinos lack obvious additive quantum numbers that reverse sign under particle-antiparticle conjugation. Thus neutrinos can have both Dirac and Majorana mass terms. There are no hints from the known Dirac masses of the standard model about the scale of Majorana masses.

The seesaw mass matrix [4]

$$M \sim \begin{pmatrix} 0 & m_D \\ m_D^T & m_R \end{pmatrix} \tag{2}$$

produces a light neutrino, $m_{light} \sim \frac{m_D^2}{m_R}$, and one unobservably heavy, $m_{heavy} \sim m_R$, where m_R is a new heavy right-handed Majorana mass scale. We see that $\left(\frac{m_D}{m_R}\right)$ is the needed "small parameter" that could account for neutrinos being anomalously light. The masses of the ν_e, ν_μ, and ν_τ, the isospin-up members of the lepton doublets, are then related to the masses of the corresponding members of the quark doublets, m_u, m_c, and m_t. Taking $m_R \sim 10^{16}$ GeV, the grand unification scale, the seesaw mechanism yields the following neutrino mass scales

$$m_{\nu_e} : m_{\nu_\mu} : m_{\nu_\tau} \leftrightarrow 2 \cdot 10^{-12} : 2 \cdot 10^{-7} : 3 \cdot 10^{-3}\ eV.$$

Thus experiments probing small neutrino masses may provide insight into new mass scales far beyond the reach of direct accelerator physics. This is a theme that Gelmini will develop later in this meeting.

§2 The Solar Neutrino Problem, Neutrino Oscillations, and Supernovae

The lack of high energy solar neutrinos, relative to the predictions of the standard solar model, has been apparent since the early days of the ^{37}Cl experiment, two decades ago. The combined results of the ^{37}Cl [5], Kamioka II and III [6], and GALLEX/SAGE [7] experiments now indicate

$$\begin{aligned} \phi(pp) &\sim \phi^{SSM}(pp) \\ \phi(^7Be) &\sim 0 \\ \phi(^8B) &\sim 0.5\, \phi^{SSM}(^8B) \end{aligned} \tag{3}$$

where the ϕ^{SSM} are standard solar model flux predictions [8]. Perhaps the most difficult aspect to understand is the ratio

$$R = \frac{\phi(^7Be)}{\phi(^8B)} \tag{4}$$

which is expected to vary as T_c^{-10}, where T_c is the central core temperature, in solar models. A lowering of T_c to reduce the flux of high-energy neutrinos will raise R, in contradiction to the results in (3). For this reason, it has become more difficult to accommodate experiment by varying the physics or parameters of the standard solar model. [See talk by Bahcall.]

An attractive alternative solution is matter-enhanced neutrino oscillations. The favored MSW explanation [6] of the solar neutrino problem corresponds to $\delta m^2 \sim 10^{-5}$ eV^2 and $\sin^2 2\theta \sim 10^{-2}$. Assuming $m_{\nu_1} \ll m_{\nu_2} \ll m_{\nu_3}$, this could mean $\nu_e \leftrightarrow \nu_\tau$ oscillations with $m_{\nu_\tau} \sim$ few $\cdot 10^{-3}$ eV, a value we have noted is natural in the case $m_R \sim m_{GUT}$. If this is the correct solution, one consequence is that neutrino masses are likely too low to play a role in dark matter, the atmospheric neutrino problem, tritium β decay, and double beta decay. Heroic very long baseline experiments would be needed to reach such δm^2, and the small mixing angle would pose an additional obstacle to any measurement.

Alternatively, perhaps the MSW solution corresponds to $\nu_e \leftrightarrow \nu_\mu$ oscillations, indicating new physics at an intermediate scale $m_R \sim 10^{12}$ GeV. Then a plausible seesaw $m_{\nu_\tau} \sim$ few $\cdot 10$ eV, so the ν_τ could play a crucial role in dark matter, large scale structure formation, and supernova explosions.

Both of these scenarios, plus other possibilities, will be discussed by Kayser. I would like to mention one implication of a cosmologically interesting m_{ν_τ}, oscillations of supernova neutrinos, that Fuller, Qian, and collaborators have explored.

Very general arguments govern the temperatures of supernova neutrinos in the absence of neutrino oscillations. Through most of their migration out of the hot protoneutron star neutral current interactions keep the neutrinos in flavor equilibrium, e.g.

$$\nu_e + \bar{\nu}_e \underset{Z_0}{\leftrightarrow} \nu_\tau + \bar{\nu}_\tau \tag{5}$$

This results in an approximate equipartition of energy between the flavors. After neutral weak decoupling the ν_es and $\bar{\nu}_e s$ remain, for a short time, in equilibrium with the surrounding matter because of the charge current reactions

$$\nu_e + n \leftrightarrow e^- + p \quad (6)$$
$$\bar{\nu}_e + p \leftrightarrow e^+ + n$$

as well as the charge-current enhancement of $\nu_e + e^- \leftrightarrow \nu_e + e^-$. This stronger coupling of the $\nu_e s$ and $\bar{\nu}_e s$ to matter is reflected in the neutrino temperatures found in realistic calculations [9]

$$T_{\nu_e} \sim 4\ MeV \quad (7)$$
$$T_{\bar{\nu}_e} \sim 5\ MeV$$
$$T_{\nu_\mu,\nu_\tau} \sim 8\ MeV$$

The difference between the T_{ν_e} and $T_{\bar{\nu}_e}$ temperatures is due to the surrounding matter being neutron rich, so that the first of the reactions in (6) is more important.

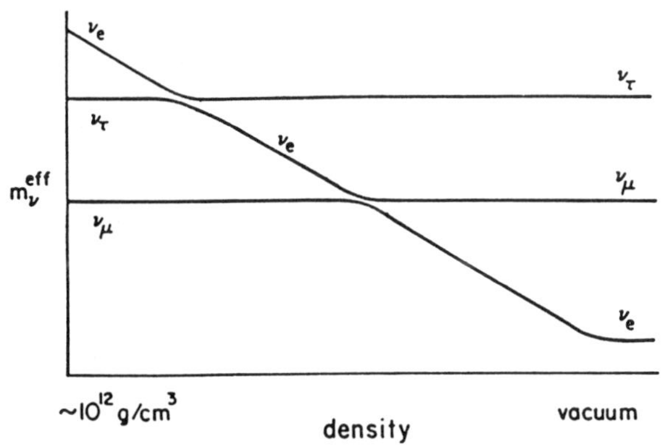

Figure 2: Schematic illustration of the $\nu_e \leftrightarrow \nu_\tau$ MSW level crossing that might occur at densities encountered in a supernova.

A closure-mass ν_τ will experience a level crossing with a ν_e at a density of $\rho \sim 10^9 - 10^{10}$ g/cm^3, as illustrated in Fig. 2. Since this crossing is outside the neutrinosphere, a distinctive hot ν_e flux will result

$$T_{\nu_e}^{MSW} \gg T_{\bar{\nu}_e}$$

provided that the level crossing is adiabatic. Generally this is the case if $\sin^2 2\theta_{e\tau} \gtrsim 10^{-5}$.

One quickly sees that a supernova is a less complicated laboratory for studying MSW oscillations than the sun, since this temperature inversion is so distinctive. As cross sections off nucleons vary as E_ν^2 (while those off nucleon targets with large thresholds rise even more steeply), event rates are proportional to T under the assumption of energy equipartition among the flavors. [For the same reason, oscillations have important consequences for neutrino energy deposition in the star's mantle. See the talks by Fuller and Qian.] Thus an important component of investigations of cosmologically interesting neutrinos must be the maintenance of supernova neutrino observatories that can distinguish $\nu_e s, \bar{\nu}_e s$, and heavy flavor neutrinos. As Cline will stress in a later talk, we must be willing to maintain these observations for times long compared to the galactic supernova rate, $\sim 1/35$ years.

§3 Neutrinoless Double Beta Decay

Neutrinoless $\beta\beta$ decay, our most stringent test of lepton number conservation and of Majorana masses, will be reviewed by Rosen. It can be studied in a number of even-even nuclei (A,Z) where, due to the nuclear pairing force, the first-order β decay to the odd-odd daughter is kinematically forbidden, but the second-order decay to the daughter (A, Z+2) open.

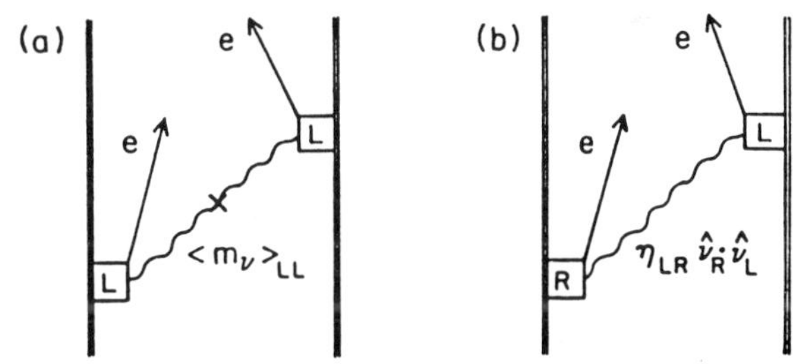

Figure 3: The two-nucleon neutrino mass (a) and right-handed current (b) mechanisms for neutrinoless $\beta\beta$ decay.

Two important mechanisms for neutrinoless $\beta\beta$ decay are shown in Fig. 3. As the neutrino omitted in the first β decay has the wrong handedness to be reabsorbed on the second nucleon, contributions arise only from a nonzero neutrino mass (which breaks the γ_5 invariance by an amount $\sim \left(\frac{m_\nu}{E_\nu}\right)$, where $E_\nu \sim 30$ MeV is the energy of the virtual neutrino) or through an explicit right-handed β decay vertex. The corresponding rates are proportional to [10]

$$\langle m_\nu \rangle_L^{Maj} = \sum_{i=1}^{2n} U_{ei}^{L^2}\ m_i\ \lambda_i^{CP} \quad (8a)$$

$$\langle n_{RL} \rangle = n_{RL} \sum_{i=1}^{2n} U_{ei}^L\ U_{ei}^R\ \lambda_i^{CP} \quad (8b)$$

where m_i indicates the ith neutrino mass eigenstate, n is the number of generations, $U_{ei}^{L^2}$ is the coupling probability of the ith eigenstate to the electron neutrino, and λ_i^{CP} is the relative CP eigenvalue of the ith mass eigenstate. (We assume a CP conserving theory.) The strength of the right-handed current is n_{RL} and U_{ei}^R is the amplitude of the ith mass eigenstate in the right-handed electron neutrino.

Both of the lepton-number-violating parameters in Eqs. (8) vanish in the Dirac limit: when Majorana terms in the neutrino mass matrix are turned off, the neutrino mass eigenstates become pairwise degenerate with opposite λ_i^{CP}.

The most stringent limits on neutrinoless double beta decay have come from ^{76}Ge→^{76}Se: a Ge crystal can serve as both source and detector. Two efforts are now underway to build the first large (~ 10 kg), isotopically enriched detectors (~ 87% ^{76}Ge), the IGEX [11] and Moscow/Heidelberg [12] experiments. [Natural Ge is 7.44% ^{76}Ge.] The Moscow-Heidelberg collaboration has reported first results corresponding to 9.7 kilogram-years,

$$\tau_{1/2}^{0\nu}(0^+ \to 0_1^+) \gtrsim 2.7 \cdot 10^{24} y, \; 90\% \; c.l. \qquad (9a)$$

$$\tau_{1/2}^{0\nu}(0^+ \to 2_1^+) \gtrsim 1.3 \cdot 10^{24} y, \; 90\% \; c.l. \qquad (9b)$$

The first result constrains the Majorana electron neutrino mass to $\langle m_\nu \rangle_L^{Maj} \lesssim (0.5-1.0)$ eV, depending on the choice of nuclear matrix elements. The limit on decays to the 2_1^+ first excited state in ^{76}Se is interesting since this process is nonzero only if right-handed currents contribute. The IGEX $0^+ \to 0_1^+$ limit, corresponding to 10.9 mole-years of ^{76}Ge, is $\tau_{1/2}^{0\nu} \gtrsim 1.7 \cdot 10^{24} y$ at 90% c.l.

While these experiments will establish lifetime limits above 10^{25} years, the challenge facing the field is to develop very large volume clean detectors that can push well beyond this level. Double beta decay provides our best test of Majorana mass terms that are frequently invoked in extended models that attempt to account for small neutrino masses.

§4 Neutrino Electromagnetic Moments and Stellar Cooling

Neutrinos can interact with static magnetic and electric fields if they have magnetic or electric dipole moments. They can also interact with virtual photons if they have a nonzero charge radius or an anapole moment.

As an electric dipole moment requires CP nonconservation, perhaps the most interesting of these is the magnetic moment. Dirac neutrinos can have diagonal magnetic moments: using standard model couplings and assuming an $m_\nu \sim 10$ eV, the one-loop radiative correction generates a magnetic moment $\sim 10^{-19}\mu_B$, where μ_B is an electron Bohr magneton. Majorana neutrinos can have transition but not diagonal magnetic moments.

The most stringent limits on neutrino magnetic moments come from stellar cooling arguments. Raffelt has recently discussed [13] neutrino emission from red giant and horizontal branch stars. A red giant contains a degenerate ^{4}He core, which supports itself by the degeneracy pressure, surrounded by a thin H burning shell that supports the mantle of the star. As the hydrogen shell burning proceeds, more He is synthesized and added to the core mass, which increases the gravitational potential at the burning shell. Thus the burning must intensify, and the star reddens.

Conditions are eventually reached in the degenerate core which allow carbon burning to start, $3\,\alpha \to {}^{12}$C. The rate of energy production by this process goes roughly like $\rho^2 T^{30}$, where ρ and T are the core density and temperature. This sharp dependence on T renders the core mass at ignition extremely sensitive to neutrino cooling mechanisms operating in the core. The more effective the cooling, the longer the delay to He ignition, and thus the larger the degenerate core at the time of the He flash. Not only red giant evolution, but the subsequent evolution of the star along the horizontal branch is affected by the size of the core at ignition.

The dispersion relation for a photon is modified at high density by interactions with the surrounding electron gas. The transverse photon components acquire a mass proportional to the plasma frequency, which varies as $\sqrt{\rho}$. A longitudinal mode (charge density wave) also arises. The most important neutrino cooling mechanism is the decay of these collective plasmon excitations into escaping $\nu\bar{\nu}$ pairs. The coupling of the plasmon to the $\nu\bar{\nu}$ pair occurs through an electron particle-hole intermediate state. Since the core temperature is on the order of 10 keV, neutrino flavors with masses well below 10 keV contribute to the cooling at full strength.

If the neutrino carried a magnetic moment (or electric dipole moment), the plasmon can couple directly to the $\nu\bar{\nu}$ pair, and additional cooling occurs governed by μ_ν^2. The resulting He ignition delay will change the evolutionary paths of red giant and horizontal stars, causing, for instance, an increase in the number of stars on the red giant branch to those on the horizontal branch. This observation can limit μ_ν^2. Raffelt deduced a limit of $\mu_\nu \lesssim 3 \cdot 10^{-12}\mu_B$ from these and similar arguments. This limit applies to diagonal and transition magnetic moments, with the latter the only possibility in the case of a Majorana neutrino.

This limit is important to the conversion $\nu_L \to \nu_R$ of solar neutrinos due to their interactions with the solar magnetic field. Although the conversion $\nu_{e_L} \to \nu_{e_R}$ was discussed some time ago as a possible solution of the solar neutrino problem, this scenario was made less likely by MSW interactions, since matter effects break the degeneracy of the ν_{e_L} and (presumably sterile) ν_{e_R}, rendering the off-diagonal $\mu_\nu B_\odot$ coupling less effective. Lim and Marciano [14] then pointed out that a crossing could occur for a spin-flavor transition, $\nu_{e_L} \to \nu_{\mu_R}$: the vacuum mass difference can be compensated by the matter contributions to the effective mass of the ν_e. Even this scenario requires very large ($B_\odot \gtrsim 10^6 G$) solar fields, given the red giant limit on μ_ν. Furthermore, since the position of the level crossing depends on E_ν, the solar field $B_\odot$ must be extensive to affect an appreciable portion of the solar neutrino spectrum (e.g. ^{7}Be and ^{8}B neutrinos). For these reasons, perhaps this solution to the solar neutrino problem is somewhat contrived.

Similar arguments constraining neutrino magnetic moments and large Dirac neutrino masses have been constructed for supernova neutrinos: transitions to sterile states can shorten the protoneutron star cooling time from ~10 seconds to a value inconsistent with the observed duration of the SN1987A neutrino pulse.

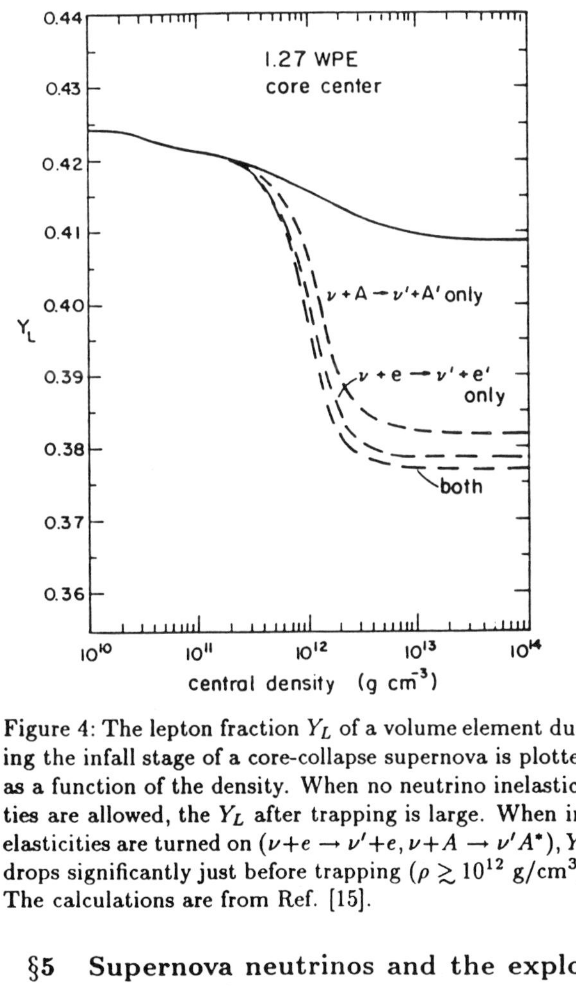

Figure 4: The lepton fraction Y_L of a volume element during the infall stage of a core-collapse supernova is plotted as a function of the density. When no neutrino inelasticities are allowed, the Y_L after trapping is large. When inelasticities are turned on ($\nu + e \rightarrow \nu' + e, \nu + A \rightarrow \nu' A^*$), Y_L drops significantly just before trapping ($\rho \gtrsim 10^{12}$ g/cm^3). The calculations are from Ref. [15].

§5 Supernova neutrinos and the explosion mechanism

Neutrinos are believed to play a crucial role in the core-collapse supernova mechanism. During the infall stage preceding core bounce the initial conditions of the protoneutron star are determined by neutrino emission. As the matter falls into the gravitational potential, the work done on it increases the matter temperature. The competing cooling process is neutrino emission by

$$e^- + p \rightarrow \nu_e + n\,, \qquad (10)$$

the rate of which increases as the electron Fermi level rises. By this reaction the star begins to radiate its lepton number.

Weak interaction rates depend on T^5, when T is the matter temperature. Thus the rate of reaction (10) rapidly increases as the collapse proceeds. Yet neutrino emission only occurs if the density of the matter $\rho \lesssim 10^{12} g/cm^3$: above this density the neutrinos are trapped, i.e., their diffusion time out of the star's core is longer than the collapse time. Thus, once trapping is reached, the initial conditions of the core bounce are fixed.

There is a very narrow window (see Fig. 4) in density where the rate for reaction (10) is high yet the neutrinos are untrapped, $\rho \sim 10^{11} - 10^{12} g/cm^3$. This window determines the trapped lepton fraction Y_L that will characterize the core bounce. This is a crucial parameter in the explosion model: the smaller the value of Y_L, the smaller the inner portion of the iron core that participates in the core bounce. A small iron core produces a weak shock wave and more shock wave losses as it propagates through the outer iron core. For this reason numerical supernova models fail to produce successful explosions for small Y_L.

The prompt hydrodynamic explosion model (one powered only by the shock wave) was found to fail because of lepton number losses associated with reaction (10) and with downscattering, the name given to inelastic reactions such as

$$\nu + e \rightarrow \nu' + e$$

$$\nu + A \rightarrow \nu' + A^*$$

by which energetic neutrinos can lose their energy [16]. (The downscattering is important because, as neutrino coherent scattering cross sections vary as E_ν^2, lower energy neutrinos escape the core more readily than higher energy ones. Thus β decay neutrinos, which tend to be somewhat hotter than the surrounding matter, escape more readily if they are cooled by inelastic reactions.) The inclusion of inelasticities dropped the trapped lepton number fraction from ~ 0.42 to about 0.35, a value incompatible with a prompt explosion.

There are continuing efforts to produce collapse with very low entropies, thereby slowing reaction (10) and hopefully increasing Y_L. One possibility is the neutral current β decay rate

$$Fe^* \rightarrow Fe + \nu + \bar{\nu} \qquad (11)$$

which cools the star without lepton number loss. Yet no effect so far included has significantly raised Y_L.

As Colgate will discuss at this meeting, a more plausible explosion model may be the delayed mechanism: the shock waves fails, stalling at ~200-300 km from the center of the star, but is later regenerated, as Wilson first claimed, [17] by neutrino heating of the nucleon soup that lies in the wake of the shock wave. The most important reactions are

$$\nu_e + n \rightarrow p + e^- \tag{12}$$

$$\bar{\nu}_e + p \rightarrow n + e^+ .$$

Even this mechanism is difficult to make work: net heating occurs only between a "gain radius" [17] and the undissociated He layer lying immediately behind the shock front, and the net energy deposited is small.

There has been a great deal of recent interest in two-dimensional and three-dimensional models in which convection is treated semirealistically. Part of the motivation is that if convection makes neutrino heating more effective, it could be an essential element of the explosion mechanism. A couple of ideas have been put forth to enhance the neutrino luminosity by transporting neutrino-rich matter across the neutrinosphere (prompt convection and neutron finger convection). A recent analysis has called both into question: the transport of lepton number by $\nu_e s$ and $\bar{\nu}_e s$ appears too rapid to allow stable neutron fingers, while the negative entropy gradients that drive deep, prompt convection are quickly destroyed by the convection [18]. Herant, Benz, and Colgate [19] have argued negative entropy gradients drive convection at larger radii, between the neutrinosphere and the shock front. Cold material falls toward the neutrinosphere, where it is reheated and rises. This convection has little effect on the luminosity, but it does harden the neutrino spectrum by allowing hot volume elements to radiate neutrinos at larger radii. This could enhance the rate of neutrino heating both behind and in front of the stalled shock wave.

Another boost to the heating mechanism could result from $\nu_e \leftrightarrow \nu_\tau$ oscillations, as discussed in Sec. 2. The conversion of a hotter ν_τ spectrum into ν_es outside of the neutrinosphere would enhance energy deposition through $\nu_e + n \rightarrow p + e^-$, as Fuller and collaborators have discussed [20]. Fuller and Qian [21] have also pointed out that this would force the matter in the shock wave's wake to become proton rich, which would destroy conditions necessary for the "hot bubble" r-process we will discuss later.

§6 Neutrinos and Nucleosynthesis

One of the triumphs of big bang nucleosynthesis is the constraint it places on the number of light neutrino species, $N_\nu \sim 3$, as Schramm will discuss. Neutrinos also play an important role in the later nucleosynthesis that is reflected in stellar metallicities.

Supernova neutrinos can alter the composition of the matter ejected by the explosion. For example, the product of the neutrino neutral current inelastic cross section off ^{20}Ne and the neutrino fluence through the Ne shell, which resides ~30,000 km from the center of the star, is ~1/300. That is, 0.3% of the matter is affected by such interactions. The cross sections are dominated by the higher temperature ν_μ and ν_τ neutrinos, which deposit ~20 MeV in the scattering. Thus the target is generally excited above the continuum,

$$\begin{aligned} {}^{20}Ne(\nu,\nu')^{20}Ne^* &\rightarrow {}^{19}F + p \\ &\rightarrow {}^{19}Ne + n \\ &\quad \hookrightarrow_{\beta} {}^{19}F , \end{aligned} \tag{13}$$

leading to its breakup and the production of a new nucleus, ^{19}F. When this physics is incorporated into explosive nucleosynthesis codes, which follow the processing of the coproduced neutrons and protons and the heating associated with passage of the shock wave, it is found that this mechanism can account for the galactic abundance of ^{19}F, given a heavy flavor neutrino temperature ~8 MeV. Similarly, as Woosley will show at this meeting, this "neutrino process" can explain the origin of galactic ^{11}B (including its evolution as deduced from measurements of low metallicity stars) and much of the observed ^{7}Li [22]. These nucleosynthesis results can be considered a constraint on the T_{ν_μ/ν_τ} temperature.

Woosley and collaborators [23] have also shown that the expanding "hot bubble" between the neutrinosphere and the shock wave is an attractive site for the r-process, the mechanism by which about half of the heavy elements and all of the transuranics are thought to have been formed. The neutron-rich nucleon soup, as it expands off the star and cools, undergoes an alpha-rich freezeout and then an r-process, using the excess neutrons. The amount of material that is estimated to be ejected could reasonably account for metals found in nature. This r-process appears to require very large entropies/baryon, and thus progress on the supernova mechanism is important to achieve a better understanding of this process. Present calculations do not fully include the effects of the neutrino process on the neutron-proton chemistry in the bubble or on the survival of the synthesized nuclei. As the production occurs at small r ~ 1000 km, the neutrino physics is clearly important.

§7 Neutrino Detection

One of the exciting prospects for the field of neutrino physics is the likelihood that the solar neutrino problem will be solved in the next few years. Two sophisticated new detectors, the Sudbury Neutrino Observatory (SNO) [24] and Superkamiokande [25], should begin to take data near the end of 1995. SNO, a heavy-water Cerenkov detector, will allow experimenters to measure the neutrino flux independent of flavor via the reaction

$$\nu + d \rightarrow \nu' + n + p.$$

The thermalized neutrons will be captured and counted. It will also utilize the hard spectrum of electrons from

$$\nu_e + d \rightarrow e^- + p + p$$

to reconstruct the spectrum of incident ν_es. Superkamiokande, a conventional water Cerenkov detector, will exploit its enormous fiducial volume, ~22 kilotons, to reconstruct the spectrum of incident neutrinos from electrons scattered by

$$\nu + e \rightarrow \nu' + e,$$

a reaction that is sensitive to ν_e and ν_μ (or ν_τ) neutrinos in the ratio ~ 7:1. Both the SNO neutral current measurement and the SNO and Superkamiokande ν spectrum reconstructions could provide a convincing demonstration that the MSW effect is modifying the solar neutrino spectrum. These detectors will inaugurate a new generation of high statistics active experiments.

The SNO and Superkamiokande detectors will measure the high energy portion ($\gtrsim$ 5 MeV) of the ^{8}B spectrum. Ken Lande will be describing efforts on the radiochemical ^{127}I experiment, [26] which is also scheduled to be operating by the end of 1995. This detector will test the ^{8}B and ^{7}Be neutrino fluxes.

Later in this meeting Bob Lanou will describe efforts to develop active detectors sensitive to the lower energy ^{7}Be and pp neutrinos, including liquid and gaseous He detectors and ultrapure scintillation detectors (Borexino). Although these projects are very difficult, it is clear that the detection of low energy neutrinos ($\lesssim$ 1 MeV) by active detectors is the next step in the development of solar neutrino physics.

Additional future challenges include the detection of solar thermal neutrinos of all flavors, which have a flux/flavor $\sim 2 \cdot 10^8/\text{cm}^2/\text{sec}$ and a peak flux density at ~ 1 keV comparable to that of the peak of the pp spectrum ($\sim 10^{11}/\text{cm}^2$ sec/MeV) [27], and of cosmic background neutrinos.

Francis Halzen, Vic Stenger, and Steve Barwick will describe the DUMAND and AMANDA efforts to detect high energy astrophysical neutrinos, as well as the prospects for building very large (~ km^3) water or ice Cerenkov detectors for measuring neutrinos with $E_\nu \gtrsim 10^{15}$ eV. Such detectors could allow us to see into the centers of active galactic nuclei, just as solar neutrinos allow us to probe the solar core. The mechanisms by which nature accelerates particles to energies of 10^{15} to 10^{20} eV are presently unknown.

The work was supported in part of the U.S. Department of Energy and by NASA under grant #NAGW2523.

§References

[1] R.G.H. Robertson et al., *Phys. Rev.* 67 (1991) 957; D. Decman and W. Stoeffl, submitted to *Phys. Rev. Lett.*; Ch. Weinheimer et al., *Phys. Lett.* B300 (1993) 210; E. Holzschun, M. Fritschi, and W. Kündig, *Phys. Lett.* B287 (1992) 381; H. Kawakami et al., *Phys. Lett.* B256 (1991) 105.

[2] R.S. Van Dyck, Jr., D.L. Farnham, and P.B. Schwinberg, *Phys. Rev. Lett.* 70 (193) 2888.

[3] W. Stoeffl, private communication.

[4] M. Gell-Mann, P. Ramond, and R. Slansky, unpublished and in Supergravity, ed. P. van Nieuavenheizen and D. Z. Freedman (North Holland, Amsterdam, 1979) p. 315; T. Yamagida, in Proc. of the Workshop on Unified Theory and Baryon Number in the Universe, ed. O. Sawada and A. Sugamoto (KEK, Tsukuba, 1979) P. 95.

[5] R. Davis, *Prog. Part. Nucl. Phys.* 32 (1994) 13; K. Lande for the Homestake Collaboration, to be publishedin the Proc. of Neutrino '94.

[6] K.S. Hirata et al, *Phys. Rev.* D44 (1991) 2241; Y. Suzuki for the Kamiokande III Collaboration, to be published in the Proc. of Neutrino '94.

[7] P. Anselmann et al, *Phys. Lett.* B327 (1994) 377; J.N. Abdurashitov et al., *Phys. Lett.* B328 (1994) 234.

[8] J.N. Bahcall and M.H. Pinsonneault, *Rev. Mod. Phys.* 64 (1992) 885; S. Turck-Chieze and I. Lopes, *Astrophys. J.* 408 (1993) 347.

[9] See, for example, R. Mayle and J.R. Wilson, *Ap. J.* 334 (1988) 909.

[10] W.C. Haxton and G.J. Stephenson, Jr., *Prog. Part. Nucl. Phys.* 12 (1984) 409; M. Doi, T. Kotani, H. Nishiura, and E. Takasugi, *Prog. Theor. Phys. Suppl.* 83 (1985) 1.

[11] F. Avignone, private communication.

[12] M. Beck, talk presented at the Workshop on Detecting Dark Matter Particles, Berkeley, Feb., 1994.

[13] G.G. Raffelt, *Phys. Rev. Lett.* 64 (1990) 2856.

[14] C.-S. Lim and W.J. Marciano, *Phys. Rev.* D37 (1988) 1368.

[15] S.W. Bruenn and W.C. Haxton, *Ap. J.* 376 (1991) 678.

[16] S.W. Bruenn, *Ap. J. Suppl.* 58 (1985) 771.

[17] H.A. Bethe and J.R. Wilson, *Ap. J.* 295 (1985) 14; R. Mayle and J.R. Wilson, *Ap. J.* 334 (1988) 909.

[18] S.W. Bruenn and A. Mezzacappa, submitted to *Ap. J.* (1994); S.W. Bruenn, A. Mezzacappa, and T. Dineva, to appear in *Phys. Rep.* (1994).

[19] M.E. Herant, W. Benz, and S.A. Colgate, *Ap. J.* 395 (1992) 642; M.E. Herant, W. Benz, W.R. Hix, C.L. Fryer, and S.A. Colgate, UCSC preprint 22 (1994).

[20] G.M. Fuller, R.W. Mayle, B.S. Meyer, and J.R. Wilson, *Ap. J.* 389 (1992) 517; G.M. Fuller, *Phys. Rep.* 227 (1993) 149.

[21] Y.-Z. Qian, G.M. Fuller, R. Mayle, G.J. Mathews, J.R. Wilson, and S.E. Woosley, *Phys. Rev. Lett.* 71 (1993) 1965; Y.D. Qian and G.M. Fuller, submitted to *Phys. Rev. D*

[22] S.E. Woosley and W.C. Haxton, *Nature* 334 (1988) 45; S.E. Woosley, D.H. Hartmann, R.D. Hoffman, and W.C. Haxton, *Ap. J.* 356 (1990) 272.

[23] S.E. Woosley and R. Hoffman, *Ap. J.* 395 (1992) 202; B.S. Meyer, G.J. Mathews, W.M. Howard, S.E. Woosley, and R.D. Hoffman, *Ap. J.* 399 (1992) 656.

[24] G.T. Ewan et al., Sudbury Neutrino Observatory Proposal No. SNO-87-12 (1987).

[25] See, for example, K. Nakamura for the Superkamiokande Collaboration, to be published in the Proc. of the 3rd NESTOR Int. Workshop, Pylos, Greece, Oct., 1993.

[26] B.T. Cleveland et al., to be published in the Proceedings of the 22nd Int. Conf. on Cosmic Ray Physics (1993); W.C. Haxton, *Phys. Rev. Lett.* 60 (1988) 768.

[27] W. Lin and W.C. Haxton, submitted to *Phys. Rev. Lett.*

Neutrino Oscillations

J-P. Revol

PPE Division, CERN, 1211 Geneva 23, Switzerland

Neutrinos represent a new frontier in particle physics, a necessary complement to the $p\bar{p}$ collider at Fermilab, the SLC at SLAC and the LEP and LHC at CERN, which constitute today and in the near future the high energy frontier. The next step in the study of neutrino oscillations will rely both on improved or new detector technologies and on larger sensitive masses. The ICARUS program which led to the development of an "electronic bubble chamber" is presented as an example of detector R&D. It is then natural to review the present status and the future of proton decay searches before describing future neutrino oscillations projects. We can hope that the new generation of neutrino experiments will bring new light into particle physics and Grand Unification schemes, into Astrophysics from the observation of the sun and perhaps supernovae and into Cosmology in clarifying the contribution of neutrinos to the density of the universe.

§1 Introduction

The neutrino sector illustrates well the complementarity between particle physics and nuclear astrophysics and cosmology which are the subject of this workshop.

Recent progress in the field came from the observation of solar neutrinos, supernova neutrinos (SN1987A) and atmospheric neutrinos and from considering the possibility that neutrinos could be a major fraction of dark matter in the universe. The knowledge of neutrino properties, in particular their mass, is necessary to hope to understand the fermion masses. In fact, if the see-saw mechanism is relevant, there is a direct connection between the unification of forces and the nature of neutrino masses. Neutrinos play a fundamental role in cosmology. They are a prime candidate dark matter constituent. Their clear advantage over many other possibilities is that we do know that neutrinos exists, even though we still do not know their mass. In the Standard Model it is arbitrarily assumed that they are massless. A key question is whether their mass is sufficient to contribute significantly to the closure of the universe. In that case, they may well represent the bulk of matter in the whole universe. Neutrinos are also one of the main actors in the supernova mechanism. Detection of neutrinos from the supernova neutronization burst together with visual observation through a telescope would allow one to weigh the resulting neutron star. Yet we still know very little about neutrino properties and progress with direct mass measurements ($m_{\nu_e} < 4.5^{(1)}$ eV [2], $m_{\nu_\mu} < 160$ KeV [3] and $m_{\nu_\tau} < 29$ MeV [4]) is still far from giving us access to the supposed region of interest. The presumed mass sensitivity needed, $m_{\nu_e}, m_{\nu_\mu} \sim 10^{-3}$ to 10^{-1} eV and m_{ν_τ} smaller than a few eV's, can probably only be reached through neutrino oscillation measurements. Solar and atmospheric neutrino experiments have excited our curiosity to the point that there is today an unprecedented effort in preparation [5] to provide definite answers to some of our most burning questions. This entire workshop at Snowmass is devoted to the subject, and we heard from J. Peoples [6] that Fermilab is now seriously considering an "aggressive" long baseline neutrino oscillation program which could start at the turn of the millennium. I would like to first illustrate some of the new developments in detector R&D with the example of ICARUS and then review the main aspects of the future neutrino oscillation program.

§2 Detector R&D towards neutrino oscillation studies.

Bubble chambers have played a fundamental role in particle physics because they provide non-biased images, in three dimensions and with high resolution. Because of the high density of the liquid medium, bubble chambers can combine target and detector functions, an ideal characteristic for detectors to be used in the search for rare events

[1] A recent mass limit for Majorana neutrinos, from neutrinoless double β-decay, is 0.9 eV [1].

contained in the sensitive volume.

However, bubble chambers have a number of drawbacks, they are triggered blindly, they are slow and data analysis is fastidious. Hence, they were progressively replaced by electronic detectors, for which, because of the low density of the gaseous medium used, detector and target functions had to be separated. The ideal detector for the search for rare events, such as proton decays or neutrino interactions is probably a high resolution liquid detector equipped with an electronic readout. This could be referred to as an "electronic bubble chamber". This is precisely what was developed in the ICARUS R&D program (Figure 1).

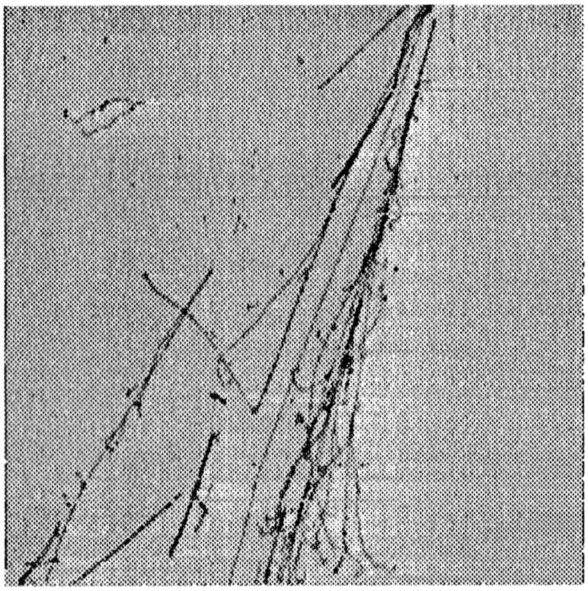

Figure 1: Cosmic ray shower observed in the ICARUS 3-ton prototype at CERN. The drift direction is along the horizontal axis spanning a distance of about 40 cm. The vertical axis corresponds to 192 sense wires with a 2 mm pitch. Each electron in the conversion pair visible at the upper left corner of the picture has about 5 MeV of energy and corresponds to the energy range of ^8B solar neutrinos in ICARUS. Also clearly visible near the center of the figure is a $\pi^+ \to \mu^+(\nu_\mu) \to e^+(\nu_e\bar{\nu}_\mu)$ candidate as well as a clear electromagnetic shower.

The ICARUS R&D program [7] has shown that it is possible today to build an ionization chamber (TPC), using a very large liquid argon volume and with the following properties: (a) it is continuously sensitive; (b) it is self-triggering; (c) it is able to provide three-dimensional images of any ionizing event; (d) it is able to identify particles (dE/dx vs range) and their direction, a feature most precious in the search for rare events. Such a detector is also, naturally, an excellent calorimeter with high granularity and resolution.

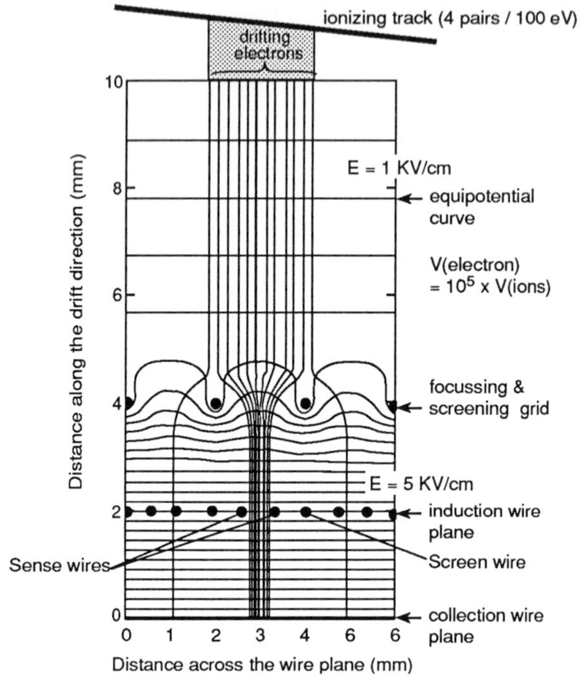

Figure 2: Structure of the ICARUS 3-ton prototype chamber. The induction and collection sense wire pitch is 2 mm. The maximum drift distance is 40 cm. The electric field in the sensitive volume of the chamber is 1 KV/cm.

The choice of liquid was driven by the following considerations: (a) it must be an excellent insulator and available at an extremely high purity level[(2)] better than 0.1 ppb of oxygen equivalent, so that free electrons produced by ionization can drift in the liquid over long distances; (b) it must have a high electron-ion pair yield with respect to the energy deposited in the liquid (typically, 4 electron-ion pairs are produced in liquid argon for every 100 eV of deposited energy); (c) it must be easily available in large quantities, which is the case for argon, a natural component of the earth's atmosphere ($\sim$ 1%) and be relatively cheap ($\sim 1SF$ per liter of liquid argon). The principle of the chamber is rather simple (Figure 2).

An electrically charged particle produces free ions and electrons along its path. As soon as they are produced they start drifting in the electric field. The drift velocity of ions is about 10^5 times smaller than that of electrons, they therefore produce no detectable effect. The focusing grid allows both a "squeezing" of the electric field lines and a shielding of the readout part of the chamber from the sensitive volume part. Information is read both by electric

[2]Note that in this respect, liquid argon has the additional advantage that at 90^0 K most impurities are solidified.

charge induction on the first readout plane encountered by drifting electrons and by electric charge collection on the second readout plane where drifting electrons are collected. The main features of this type of chamber is that there is no charge amplification in the chamber. Therefore high quality electronics is required to obtain a good signal over noise ratio. An important advantage is that even though the collected charge is small it has very small fluctuations [8] (Fano factor of 0.1), it has a small diffusion in the liquid, and can be used several times. A second induction plane could be added for instance to introduce redundancy of information, an important feature for the search for rare events. The ICARUS R&D effort has clearly opened a potentially new and broad physics program [9] with the possibility of a second generation experiment on proton decay, atmospheric neutrinos, long baseline neutrino oscillations, solar neutrinos, supernovae and any other unexpected source of neutrinos. I would like to review briefly the various aspects of this physics program with emphasis on neutrino oscillations. However, I would like first to discuss proton decay which in many ways is a closely related subject and is an important part of the Superkamiokande and ICARUS programs.

§3 Proton Decay

3.1 Status of the field

There is no known reason to believe that baryon number conservation is exact, unlike, for instance, the conservation of electric charge which is directly linked to gauge invariance and to the existence of long range interactions. Eötvös-type experiments found no evidence for the existence of a long-range field coupled to the baryon number down to a level of 10^{-9} times the gravitational force [10]. The Universe is a gigantic example of baryon number violation! To explain the observed predominance of matter over antimatter and the current ratio between photons and baryons ($N_\gamma/N_B \sim 10^{10}$), cosmological theories must assume that, in the very hot initial stages of the Big Bang, there were frequent baryon non-conserving interactions coupled at some level to CP violating processes. Those interactions must lead to proton decay. The main question is: at which level? If the relevant scale is the Planck mass, then the proton lifetime would be of the order of 10^{44} years, and detecting it would be a hopeless experimental task. Statistically it would take at least 10^{10} Superkamiokande's, but even that would not work because of the overwhelming background. However, most Grand Unification Theories (GUTs), in which baryon non-conservation occurs naturally, require some intermediate mass scale (determined by the mass of the Grand Unification bosons). With LEP precision measurements of $sin^2\theta_W$ and $\alpha_S(M_Z)$, the extrapolation of the gauge coupling constants to higher energy scales shows that they meet at a point when evolved within the context of a minimal supersymmetric theory (MSSM) ($M_X = 10^{15.8\pm0.3\pm0.1}$ GeV) [11], while the couplings fail to unify within the Standard Model [12] (SM), which rules out the simplest one-step GUTs, such as "desert-scenario" $SU(5), SO(10)$, etc. Such a colossal energy is still far above that of any known elementary process occurring at present and will probably never be reached by means of accelerator techniques. However, it must have been reached in the very early stages of the life of our Universe (10^{-40} to $10^{-35}s$ after the Big Bang). Therefore, GUTs are naturally relevant to the creation and decay of matter. Proton decay, if detectable, represents a direct way to investigate physics at the Grand Unification scale, and could play a role analogous to β-decay which taught us almost everything about the Standard Model. Early proton decay searches, in the mid 1970's, were stimulated by the first and simplest GUT, based on SU(5). The "desert-scenario" non-supersymmetric $SU(5)$ GUT predicts a value of M_X at $O(10^{14})$ GeV, which, via the dominant $p \to e^+\pi^0$ decay mode, yields a proton lifetime:

$$\tau_p \sim \frac{M_X^4}{\alpha_G^2 m_p^5} \sim 10^{31\pm1} years \qquad (1)$$

a value somewhat below the current experimental limit $\tau_p > 5.5 \times 10^{32}$ years. With the $SU(5)$ GUT in trouble, proton decay experiments were no longer considered fashionable; however, great progress had been accomplished with the pioneering experiments, bringing the proton lifetime limits from 10^{29} to 10^{32} years! Unfortunately, in globally supersymmetric GUTs, the proton lifetime which depends not only on the GUT physics but also on the low energy superparticle spectrum cannot yet be quantified (137 free parameters) and lies generally in a wide range of values: $\tau_p \sim 10^{33}$ to 10^{38} years. Operators exist (so-called $dimension - 5$) leading to proton decays which must change generation. Such channels are characterized by the presence of strange particles in the final state, with the following selection rule:

$$\frac{\Delta S}{\Delta B} = 0, -1 \qquad (2)$$

which tells us that $p \to \bar{\nu}\, \pi^+$ or $p \to \bar{\nu}\, K^+$ are allowed, but that $p \to \bar{\nu}\, K^-\ \pi^+\ \pi^+$ or $n \to e^+\ K^-$ are forbidden.

The situation is remarkably different in the supergravity GUT [13] (SUGRAGUT) which generates its own spontaneous breaking of supersymmetry. The soft supersymmetry breaking terms in the low energy theory are completely determined in terms of only four parameters. Such a theory will definitely be tested by a combination of LEP and future proton decay experiments, since it predicts that either $\tau(p \to \bar{\nu}K^+) < 1.5 \times 10^{33}$ years or one of the gauginos or one of the Higgs particles will be discovered at LEP200 [14]. This is a really exciting challenge both for proton decay and LEP200 experiments.

The discovery of proton decay in the supersymmetric channel would obviously be of great importance. However, from the experimental point of view, it is clear that all the possible decay modes should be searched for, without too much theoretical prejudice. Let us not forget that everything mentioned so far is only pure speculation since there is no evidence whatsoever that Supersymmetry is relevant to Nature and since our dream of the unification of forces is only driven by our hope or belief that Nature meets our criteria for simplicity and beauty. In addition, the present experimental situation has room for substantial improvement, even for decay modes with $\tau/Br < 10^{32}$ years. For instance, for the apparent anomaly first observed by Kamiokande in atmospheric neutrino data [15], the possibility that the source of the effect is in fact proton decay has been considered [16].

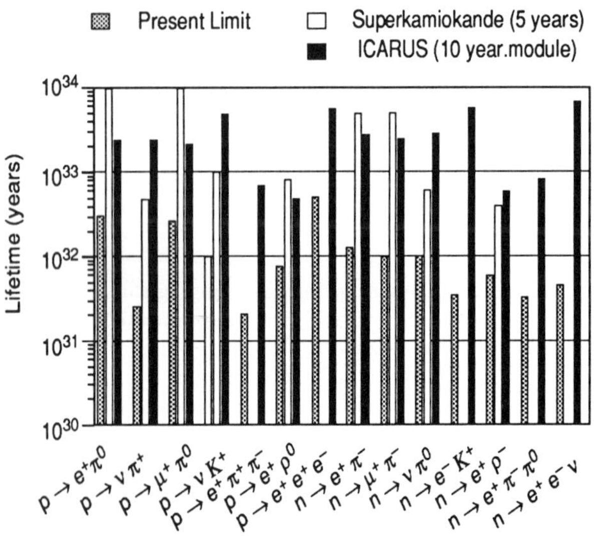

Figure 3: Sensitivity for Superkamiokande and ICARUS and present proton and neutron decay limits in a selected number of channels.

The current generation of detectors has reached the proton lifetime range of 10^{32} years (Figure 3), but only for a limited number of channels. There is the possibility that the current Soudan II detector [17], with a sensitive mass of 1 Kton, may still explore some additional channels, in that range. In order to go further and investigate the region $\tau_p \sim 10^{32}$ to 10^{34} years which seems to be indicated by SUGRAGUT, two approaches are presently considered: (a) increase the mass of the sensitive medium using current detector technologies (the Superkamiokande approach, factor 30 larger than Kamiokande). Due to the relatively large atmospheric neutrino background, the signal-over-noise ratio is small and the sensitivity is proportional to $\sqrt{N_n}$, (N_n is the number of nucleons); (b) use new detector technologies to obtain a negligible background over a more modest detector size (the ICARUS approach). The signal-over-noise ratio is large and the sensitivity is proportional to N_n.

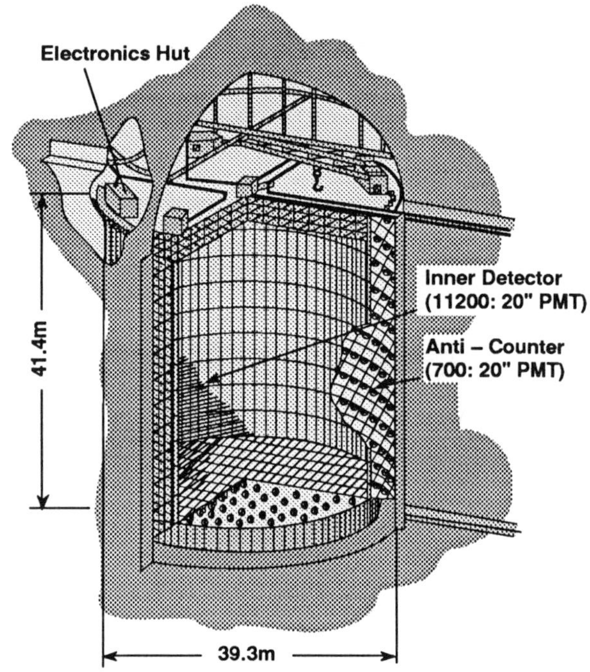

Figure 4: Schematic view of the Super-Kamiokande detector.

3.2 *Superkamiokande*

Superkamiokande [18] (Figure 4) is a spectacular extension of the Kamiokande technique (water Cerenkov) to a much larger fiducial mass: 22,000 tons with a significant improvement in low energy event reconstruction mainly due to a better PMT coverage, 40% versus 20%. The detector is built in the same Mozumi Zinc and Lead Mine as the present Kamiokande experiment in Japan (200 m west-southwest of Kamiokande). Excavations began in November 1991 and has just been completed. Data taking is due to start in 1996. In order to reduce the background for the SUSY preferred decay mode, a delayed (12 ns) coincidence will be used between the 236 MeV (Cerenkov threshold) μ^+ from K^+ decay and the 6.3 MeV γ from the decay of the $^{15}N^*$ formed in the reaction: $^{16}O \to \bar{\nu}K^+ + {}^{15}N^*$. The photomultiplier tube timing resolution is expected to be 2.5 ns. Superkamiokande will, in five years, reach proton or neutron decay lifetimes of up to 10^{34} years for certain channels such as $p \to e^+\pi^0$ and 10^{33} years for $p \to \bar{\nu}K^+$ (Figure 3).

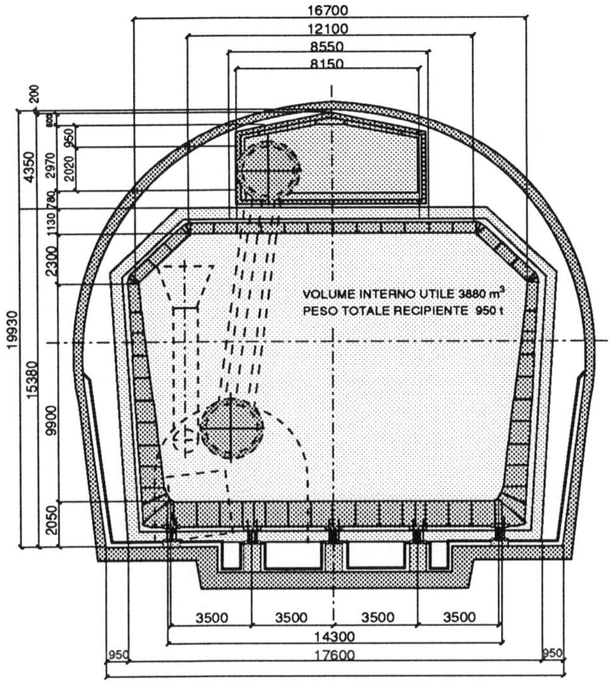

Figure 5: Schematic view of the ICARUS detector to be built in the Gran Sasso Laboratory.

3.3 *Imaging Cosmic And Rare Underground Signals (ICARUS)*

ICARUS [19] (Figure 5) at the Gran Sasso underground Laboratory in Italy, is a large (three 5,000-ton modules) liquid argon TPC. It is an "electronic" bubble chamber with self-triggering capability and with many precise dE/dx samplings along particle paths. A 3-ton prototype is presently operating successfully at CERN [7] and has demonstrated excellent space and energy resolution and particle identification. The redundancy of information allows negligible backgrounds and discovery at the single event level, corresponding to lifetimes of up to 10^{34} years, in channels such as $p \to \bar{\nu}K^+$ (Figure 6) or $n \to \bar{\nu}K^0$. The unique strength of ICARUS is the possibility of studying many exclusive nucleon decay channels. The first module is expected to be completed in 1998.

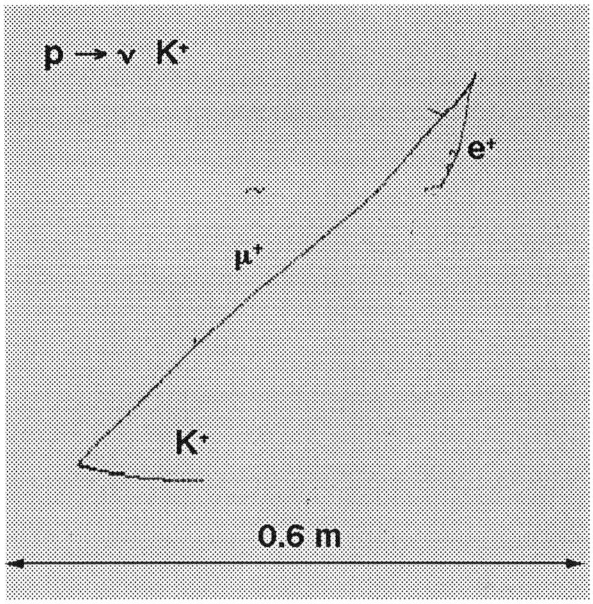

Figure 6: Simulation in the ICARUS detector of a proton decay in the Supersymmetrically preferred mode $p \to \bar{\nu}K^+$, where a μ^+ is produced in the K^+ decay, followed by an e^+ produced in the subsequent μ^+ decay.

§4 Neutrino oscillations

Proton decay and neutrino oscillations have in common that they may both allow physics at the GUT scale and, in addition, they are often studied in the same detector. Twenty years of solar neutrino experiments pioneered by R. Davis have promoted neutrino physics to a new frontier comparable in importance to the high energy frontier (SLC, LEP, TEVATRON and LHC) and the high precision frontier (LEP, Φ, B, K, τ-charm factories). Yet, today, neutrinos are still poorly known and our main question is simple: "Do they have a mass?". In the Standard Model neutrinos are massless but without compelling reason (arbitrariness). Theories beyond the Standard Model which may explain some of its arbitrary features lead us to believe that neutrinos should be massive. For instance, "see-saw" models [20] provide a natural extension of the Standard Model, in which neutrino masses originate from couplings with right-handed Majorana partners, too heavy to be observed.

In these models and for three flavors, one is generally led to three light left-handed Majorana neutrinos (the observed ν_e, ν_μ, ν_τ) with mass matrix $m_\nu = m_D M_N^{-1} m_D^T$ (each entry in the previous expression is a 3×3 matrix) and three superheavy right-handed Majorana neutrinos, corresponding to unobserved massive particles. Both quarks and leptons are expected to have masses of order

m_D. We can distinguish two simplifying cases:

(a) M_N is proportional to the identity matrix resulting in the quadratic see-saw: $m_\nu = m_D^2/M_N$; (b) M_N is proportional to m_D resulting in the linear see-saw: $m_\nu = m_D/M_N$.

To some extent, we may say that we do understand the smallness of neutrino masses from the fact that the M_N terms are very large. However, in general the question remains of knowing whether we should use a linear or a quadratic see-saw, with quarks or with leptons. Considering ratios of neutrino masses eliminates the overall unknown mass scale $M_N(X)$. Assuming, for instance, that we know the muon neutrino mass, we can compute the tau neutrino mass, modulo radiative corrections! Unfortunately, the answer for m_{ν_τ} still varies by three orders of magnitudes! A more definite theory is needed to provide clear predictions. Most predictive are GUTs which naturally give $m_D \sim m_{u_{quarks}}$. Conversely, within these models, the knowledge of m_ν implies the knowledge of the M_N mass scale. For instance, in the $SO(10)$ minimal SUSY GUT [21], the LEP precision data on the interaction couplings, are consistent with a unification scale $M_X = 1.6 \times 10^{16\pm0.4}$ GeV and a SUSY symmetry breaking scale $\leq$ 1 TeV. Perhaps the main feature to retain from these GUTs is a definite mass hierarchy: $m_{\nu_e} : m_{\nu_\mu} : m_{\nu_\tau} = m_u^2 : m_c^2 : m_t^2$.

4.1 Present solar neutrino experiments

The solar neutrino problem stems from four experiments, still actively taking data: Kamiokande using water Cerenkov techniques, and Homestake, SAGE and GALLEX using radio-chemical techniques. The results from GALLEX ($79 \pm 10^{+8}_{-5}$ SNU) confirm a certain deficit compared to the predictions of J. Bahcall ($132\ ^{+20}_{-17}$ SNU) and Turck-Chièze (123 ± 7 SNU). It is not established whether this deficit is due to experimental errors, a poor understanding of the sun or to new neutrino physics. It is therefore crucial to obtain additional information. A first step in that direction is an absolute calibration of the present detectors. In this respect, the GALLEX collaboration is performing such a test with a powerful (1.7 MCi) radioactive ^{51}Cr source. As we write this text we have heard from GALLEX that the ratio of observed events to expected events they detect from their radioactive source is 1.02 ± 0.14 [22]. This important result strengthens GALLEX conclusions on pp neutrinos from the sun.

The most popular explanation for the apparent solar neutrino deficit, if it is to be attributed to new physics, is the Mikheyev-Smirnov-Wolfenstein (MSW) mechanism [23]. However, there may not be a solar neutrino deficit [24], in case flaws exist in present solar models, i.e. several effects could decrease T_c; plasma effects could affect the $p + {}^7Be \rightarrow {}^8B + \gamma$ cross-section, the cross-section for this process may not be known accurately enough in the range 1 to 10 KeV relevant in the sun, diffusion effects, etc. Recently, a solar model consistent with some of the experimental results, was proposed [25] but it is the object of fierce controversy between expert theoreticians [26]. It is clear that the ball is more than ever in the experimentalist's camp!

It is unlikely that existing experiments will settle the issue, since none of them individually can distinguish between effects coming from the nature of neutrinos and those due to the neutrino production mechanism inside the sun. Therefore one will have to wait for the next generation of solar neutrino experiments.

4.2 Future solar neutrino experiments

Superkamiokande

With a solar neutrino rate of 23 elastic events ($\nu_x + e^- \rightarrow \nu_x + e^-$) per day, almost two orders of magnitude larger than Kamiokande, Superkamiokande will carry the first high statistics real time solar neutrino measurement.

The statistical precision is sufficient to detect the solar neutrino flux modulation of 7% due to the eccentricity (0.0167) of the earth's orbit. The experiment should be sensitive to rather small time variations and easily clarify issues such as possible correlation with sun spots etc. However, since Superkamiokande is only sensitive to ^8B neutrinos scattering on electrons and does not measure their energy, it is not expected that they can separate out solar model effects.

ICARUS

Two reactions are used for the direct observation of solar neutrinos from ^8B: (a) elastic ν_e-electron scattering; (b) ν_e absorption ($\nu_e + {}^{40}Ar \rightarrow {}^{40}K^* + e^-$), mainly a superallowed transition followed by ${}^{40}K^* \rightarrow {}^{40}K$ + 1 or 2 γ [2 MeV]. With a threshold at 5 MeV on the recoil electron energy, the Standard Solar Model predicts about 2700 and 3000 (Fermi transition only) events/year/module for reactions a) and b) respectively. The ratio R of the rates for a) and b) provides a direct measurement of the fraction of ν_e's which eventually converted to ν_μ's or ν_τ's through oscillations, independent of the magnitude of the initial solar neutrino flux, hence of the solar model. With two years of data ICARUS could detect a neutrino oscillation probability of 20% or larger.

The Sudbury Neutrino Observatory (SNO)

SNO is a 1000-ton D_2O Cerenkov detector [27], surrounded by a H_2O anti-coincidence blanket, to be installed in the Creighton mine in Canada. Taking data with pure D_2O, SNO will measure the charged current process $\nu_e + d \rightarrow p + p + e^-$ (Q = −1.4 MeV) and the elastic scattering process $\nu_x + e^- \rightarrow \nu_x + e^-$. Under these conditions the sensitivity is low for the neutral current process $\nu_x + d \rightarrow n + p + \nu_x$, (Q = −2.2 MeV). Therefore the experimental scheme alternates data taking periods with the addition of 2.5 tons of NaCl in order to boost the sensitivity to the neutral current process through neutron capture on Chlorine. The 8.5 MeV photon produced gives Cerenkov light. Cycling between the two configurations requires the removal of NaCl (3-4 weeks). Like Superkamiokande, SNO is only sensitive to solar ^8B neutrinos. However, if the NaCl scenario works, they also have a handle on a solar model independent measurement.

HELium at Liquid AZote temperature (HELLAZ)

HELLAZ [28] which is to be proposed at the Gran Sasso Laboratory intends to measure the solar neutrino energy spectrum using the elastic scattering channel ($\nu_x + e^- \rightarrow \nu_x + e^-$) down to $E_\nu \geq 217$ KeV and with high statistics (5×10^3 events/year). It is mainly sensitive to solar pp and ^7Be neutrinos. The technique consists in measuring recoil electron energies from their range and electron count in a high pressure (5 bar) He + 2% CH_4 single electron sensitive TPC at liquid nitrogen temperature (77^0 K). It is a low mass detector (∼ 6 tons, 2000 m^3) and the first step should be a 40 m^3 prototype in the Gran Sasso Laboratory.

The originality of HELLAZ is its precise reconstruction of the neutrino energy from the measurement of the electron energy ($\Delta E_e/E_e = 2-4\%$ at 300 KeV) and direction with respect to the sun: $E_\nu = m_e/[cos(\theta)(1+2m_e/T_e)^{1/2} - 1]$. The 100 KeV threshold on the electron (217 KeV for the neutrino) requires very high purity ($< 10^{-13}$ g/g).

BOREXINO

This 100-ton liquid scintillator detector [29] proposed at Gran Sasso for detection of low energy solar neutrinos to observe the ^7Be line should record 50 events per day in a 250 to 700 KeV electron energy window. It is intended to make use of the ^7Be "Compton edge" at 0.66 MeV in the kinetic energy distribution of ($\nu_e + e^- \rightarrow \nu_e + e^-$) electrons. However, unlike HELLAZ, Borexino makes no neutrino energy measurement, since it only measures electron recoil energies but not directions (at least not with sufficient precision).

Borexino could make a high statistics study of solar neutrinos but the main difficulty is to control the background because this low threshold technique requires an ultra pure scintillator (U and Th $< 10^{-16}$ g/g; $^{14}C/^{12}C < 10^{-18}$ etc.). A crucial background measurement in the Gran Sasso Laboratory will determine the fate of the experiment.

Note that both Superkamiokande and ICARUS can study the variation of the solar neutrino flux due to the variation of the distance earth-sun. That modulation of the rate with a period of one year will be modified if the vacuum oscillation is the answer to the solar neutrino deficit. This would mean that we live in the lucky but unlikely situation where the distance earth-sun is just what is takes to suppress the neutrino rate. This requires a $\Delta m^2 \sim 10^{-10}$ eV^2.

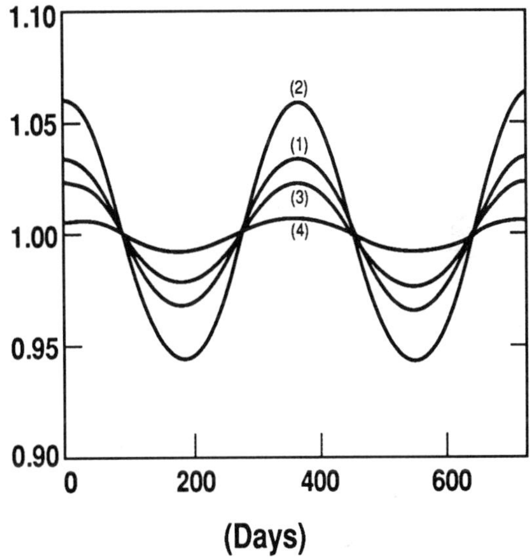

Figure 7: Variation with time of the relative rate of solar neutrino events for: (1) no oscillation; (2) $\Delta m^2 = 5 \times 10^{-11} eV^2$ and $sin^2(2\theta) = 1$; (3) $\Delta m^2 = 8 \times 10^{-11} eV^2$ and $sin^2(2\theta) = 0.8$; (4) $\Delta m^2 = 1.1 \times 10^{-10} eV^2$ and $sin^2(2\theta) = 0.9$.

Without vacuum oscillations the number of events follows the variation of the solid angle:

$$N \propto \frac{1}{L^2} \qquad (3)$$

where L is the distance earth-sun. If we turn on vacuum oscillations we have :

$$N \propto \frac{1}{L^2} \times [1 - P(\nu_e \rightarrow \nu_x)] \qquad (4)$$

$$\frac{N(L_{max})}{N(L_{min})} = \left(\frac{L^2_{min}}{L^2_{max}}\right) \times A \qquad (5)$$

$$A = \frac{1 - sin^2(2\theta) \times \langle sin^2(1.27\Delta m^2 \frac{L_{max}}{E})\rangle}{1 - sin^2(2\theta) \times \langle sin^2(1.27\Delta m^2 \frac{L_{min}}{E})\rangle} \qquad (6)$$

where $L_{min} \sim 1.465 \times 10^{11}$ m and $L_{max} \sim 1.516 \times 10^{11}$ m and $\langle\ \rangle$ indicates the average over the neutrino energy spectrum. The above ratio is 0.934 if there is no oscillation. Figure 7 shows that the variation of Δm^2 and $sin^2(2\theta)$ within the present vacuum oscillation area is important and that detecting the effect may require, in the best case, two years of data for either Superkamiokande or ICARUS.

4.3 Atmospheric neutrino studies

Other very puzzling data come from the study of atmospheric neutrinos. The apparent deficit of ν_μ's produced by cosmic ray interactions in the earth's atmosphere can be interpreted as evidence for oscillations. However this interpretation is not unique [16] as already mentioned in Sect. 3.1; the issue is still open. Results depend on the cosmic ray model and on systematics in electron and muon recognition. A water Cerenkov detector calibration test is presently being carried out in KEK electron and muon beams. Its outcome should be most interesting. The first results [30] indicate that the detector is understood as far as distinguishing between electrons and muons.

The latest data on atmospheric neutrinos, from Kamiokande [31], suggest that the neutrino oscillation length involved may be of a few thousand kilometers since there is a zenith angle variation for upward going multi-GeV neutrinos (Figure 8), which went through the entire earth, and no such effect for downward going neutrinos.

Another very interesting aspect of atmospheric neutrinos is that, if the MSW mechanism is relevant to solar neutrinos, matter effects may very well be present also in the case of atmospheric neutrinos. The MSW resonance condition:

$$\frac{E(MeV)}{\Delta m^2(eV^2)} = \frac{cos2\theta}{2\sqrt{2}G_F N_e} \tag{7}$$

shows that if matter effects exist in the sun, they should also occur in the earth [5]. Therefore, matter effects, so crucial to our understanding of solar neutrinos, can be in principle tested on earth. If resonance occurs in the sun for neutrino energies of 1 $\sim$ 15 MeV, since the density of the earth is 10 $\sim$ 100 times smaller than at the center of the sun, we could expect resonance in the earth for neutrino energies from 10 to 1500 MeV, which is approximately the energy range of atmospheric neutrinos. Neutrino oscillations could also be induced by the gravitational MSW effects [32] in violation of the equivalence principle. These exotic effects will naturally be studied by the new experiments.

Superkamiokande and ICARUS will definitely make an extensive high statistics study of atmospheric neutrinos. ICARUS will detect of the order of 1200 atmospheric neutrino events per year per 5 Kton module in 4π steradians with high resolution and perfect electron muon recognition. The sensitivity of the ICARUS experiment is shown in Figure 10. Superkamiokande will have the advantage of a higher rate with about 3000 events per year. Matter effects studies will benefit from the possibility in ICARUS to distinguish between neutrinos and antineutrinos. In charged-current neutrino interactions, a recoiling proton is associated to the μ^- or e^- lepton in the final state. In antineutrino charged-current interactions, a single lepton is present with some nuclear activity from neutron interactions. In addition, the different capture-to-decay ratio in liquid argon for μ^+ and μ^- allows us to recognize stopping muon charges.

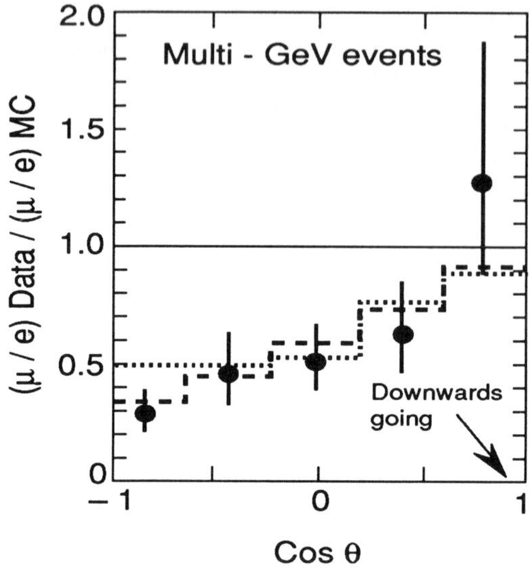

Figure 8: Kamiokande [31], zenith angle distribution of (μ-like/e-like) data events over (μ-like/e-like) Monte Carlo events, for multi-GeV data, where both the fully and partially contained events are included. Also shown are predictions from Monte Carlo simulations with neutrino oscillations with $\Delta m^2 = 1.8 \times 10^{-2}$ eV^2 and $sin^2(2\theta) = 1.0$ for $\nu_\mu \leftrightarrow \nu_e$ oscillations (dashed line) and $\Delta m^2 = 1.6 \times 10^{-2}$ eV^2 and $sin^2(2\theta) = 1.0$ for $\nu_\mu \leftrightarrow \nu_\tau$ oscillations (dotted line).

4.4 Search for $\nu_\mu \leftrightarrow \nu_\tau$ oscillations with ν_μ beams

In the CERN SPS ν_μ beam, the relative fraction of ν_τ's produced directly from hadron decays (for instance $p + N \to D_s +$ anything followed by $D_s \to \tau\ \nu_\tau$) should be $\sim 10^{-7}$, allowing in principle detection of ν_τ's produced through $\nu_\mu \leftrightarrow \nu_\tau$ oscillations down to a sensitivity of the same order. The neutrino beam in the West area of CERN has been entirely rebuilt and two new experiments have started data taking in April this year (1994) and the first significant results should be available by the end of 1995.

WA-95, known as CHORUS [33] from CERN Hybrid Oscillation Research ApparatUS, is using a system of 106 scintillating fibers to point tracks of selected events into an emulsion for computer-assisted search for ν_τ interactions. The signal for a τ decay in the emulsion is an apparent 'kink' in, for instance, a muon track extrapolated from the fiber information. The events to be measured are preselected with kinematics cuts.

WA-96, known as NOMAD [34] for Neutrino Oscillation MAgnetic Detector, uses a complementary technique based on a very fine-grained target made of 150 drift chamber planes inside the magnet from UA1. The extraction of a ν_τ signal is based entirely on kinematics criteria. The main background is hadrons from the neutral-current process: $\nu_\mu + N \to \nu_\mu +$ hadrons, faking a τ decay, and which can be rejected by requiring large momentum components $P_\tau > 1.6$ GeV and $P_{h^-} > 1.6$ GeV together with topological cuts: the direction of the missing transverse energy in the event must be away from the hadron jet; the lepton or hadron produced in τ decay must not be back to back with the hadron jet in the azimuthal angle direction. Charm also constitutes a background; the rejection is based on the fact that it is produced in pairs.

CHORUS and NOMAD have similar sensitivities for the mixing parameter which combine to $sin^2(2\theta) > 2. \times 10^{-4}$ (90% C.L.) for $m_{\nu_\tau} > 7$ eV (Figure 9). This is about 20 times better than the present limit. E803 [35], an approved experiment at Fermilab using the same emulsion technique as CHORUS, should be 10 times more sensitive; however it will not start before the end the decade since it needs the Main Injector.

Another interesting possibility may exist to reach $sin^2(2\theta) \sim 10^{-5}$ by using the ICARUS technique with a 100-ton liquid CH_4 TPC [36, 5] in a 2-tesla magnetic field. The idea is to use the four quasi-free protons to completely determine the kinematics of ν_τ−proton collisions from the additional constraint that the incident neutrino direction is known. One could then reconstruct invariant masses and the tau lepton would appear as a mass peak over a very low background reduced by cuts *à la* CHORUS/NOMAD. This technique would also provide the energy of the incoming neutrino needed to determine the oscillation length, hence the tau neutrino mass. Such a technique, presently under study, would close the window indicated by the MSW interpretation of solar neutrino data (Figure 9).

4.5 Long Baseline Neutrino Oscillations

One interesting possibility to increase the sensitivity of accelerator neutrino oscillation experiments is to extend the baseline outside the limits of the laboratory. Long baseline neutrino beam projects are under discussion at CERN, Fermilab, Brookhaven, Serpukhov and KEK. A feasibility study [37] has shown that it is technically possible to send a CERN ν_μ beam to Gran Sasso or even

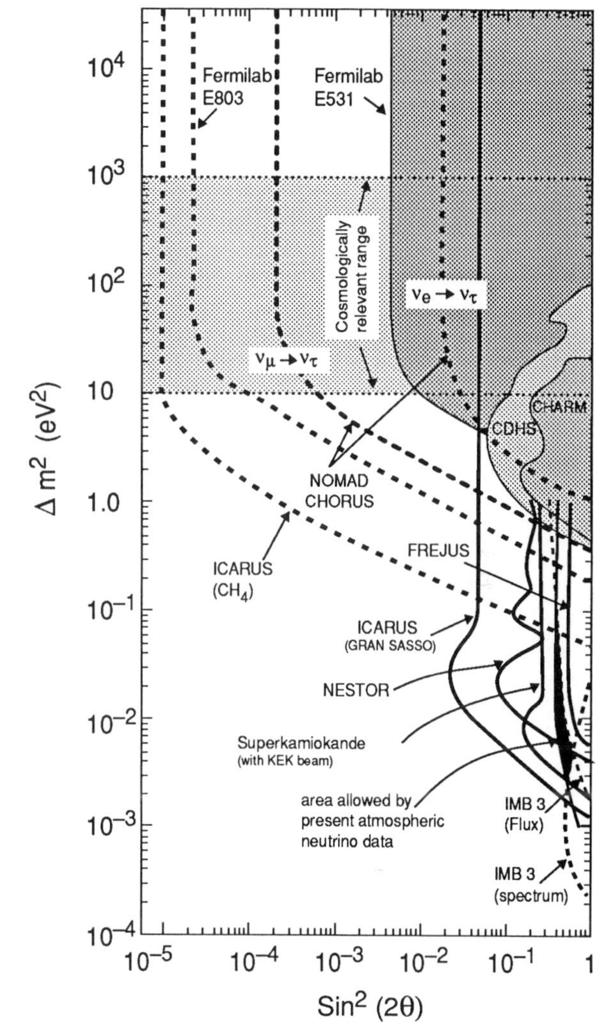

Figure 9: Sensitivity of future $\nu_\mu \leftrightarrow \nu_\tau$ oscillation experiments.

to Superkamiokande which both happen to be in favorable azimuthal directions, considering existing or planned beam lines.

Neutrino beams offer several clear advantages: (a) the energy spectrum is known, peaked and can be tuned to the desired range; (b) known direction and timing allow efficient background rejection; (c) high rate for statistical accuracy and known initial beam composition (almost entirely ν_μ or $\bar{\nu}_\mu$), small ν_e content at large distances; (d) possibility to switch between neutrinos and anti-neutrinos (interesting for matter effects); (e) as most neutrino detectors are also proton decay detectors, an accelerator neutrino beam can in principle be used to calibrate the neutrino background to proton decay, this may require running at lower beam energies than envisaged for the standard long baseline physics.

The optimal conditions for $\nu_\mu \leftrightarrow \nu_e$ oscillations with ICARUS and a CERN SPS beam are obtained with a low proton energy ($\sim$ 120 GeV) allowing 3×10^{13} protons on target every 3.6 s. There are two main types of background to charged-current interactions of a genuine ν_e coming from a ν_μ oscillation: (a) ν_e's in the initial beam: they mostly come from K decays and to a smaller extent from muon decays. Because their transverse momentum is larger than ν_μ's coming from π decays, as the distance increases, their proportion decreases and at Gran Sasso they are practically negligible for 120 GeV protons (3×10^{-4}). The contribution from Charm production is also negligible $\sim 10^{-7}$; (b) Neutral-current interactions of a ν_μ producing a recoiling jet with π^0's faking electrons ($\sim$ 1.5 π^0/jet). In ICARUS π^0's can be identified from their decay properties (separation of the photon showers), from the shape of the shower, and from ionization information. A detailed Monte Carlo simulation has shown that the π^0 background can also be made in practice negligible in the 120-GeV proton case ($\leq$ 0.1%). This type of background increases with energy as it becomes harder to identify π^0's. The ICARUS sensitivity (Figure 10) extends to $sin^2(2\theta) = 1.5 \times 10^{-3}$ and $\Delta m^2 = 1.5 \times 10^{-4}$ eV^2. For the $\nu_\mu \leftrightarrow \nu_\tau$ oscillation study, the highest beam energy is better suited.

The CERN accelerator complex could deliver 3×10^{13} 450 GeV protons on target every 7.2 s. The first method consists of measuring the ratio R_A of neutral current events without a muon in the final state to charged current events with one muon in the final state. One can show that such a ratio can be expressed as:

$$R_A \approx 0.31[1 + 1.95\, sin^2(2\theta) \times (\Delta m^2 1.27 L)^2 \langle E^{-2}\rangle] \quad (8)$$

where, for a 450 GeV proton beam, we used:

$$\frac{\sigma_{NC}}{\sigma^{\mu}_{CC}} \approx 0.31; \qquad f_\tau = \frac{\sigma^{\tau}_{NC}}{\sigma^{\mu}_{CC}} \approx 0.5 \qquad (9)$$

This so-called appearance method has the merit that it does not require a near detector and it is least sensitive to the magnitude of the ν_μ flux, which cancels to first order in the ratio.

A disappearance method can also be used which consists in measuring the ratio of the number of muons produced in the rocks near the detector and coming from the direction of CERN (proportional to the number of ν_μ's in the beam) to the total number of neutrino interactions in the detector (which gets contributions from all neutrino flavors).

The combined sensitivity of ICARUS (Figure 9) for the two methods extends to $sin^2(2\theta) = 5.0 \times 10^{-2}$ and $\Delta m^2 = 2.0 \times 10^{-3}$ eV^2.

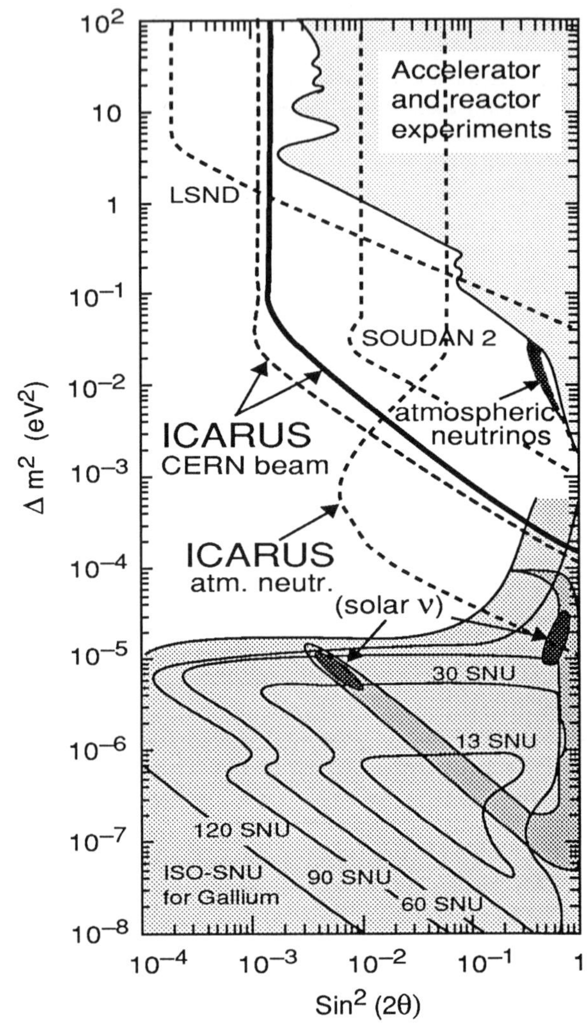

Figure 10: Sensitivity of future $\nu_\mu \leftrightarrow \nu_e$ oscillation experiments.

There are two main long baseline proposals in the USA. E889 [38], at Brookhaven, is a ν_μ disappearance experiment proposing to use a high intensity 1 GeV average energy neutrino beam from the AGS. There are three Cerenkov (probably water) detectors arranged along the neutrino beam line at distances of 1, 3 and 20 km respectively. The assumed intensity is 2×10^{20} protons on target for 120 days of running. At Fermilab, P822 [39] is a proposal for a neutrino beam from the Main Injector, aimed at Soudan II, 730 km away. An upgrade of the Soudan detector is needed.

A comparison [40] between existing simulations of neutrino beams shows that the SPS at CERN and the Main Injector at Fermilab offer equivalent capabilities. These

American projects could be completed by the end of this decade.

A third American proposal is the GENIUS [41] detector for the Gran Sasso Laboratory. However, at this workshop it became clear that there is a strong move to focus the American effort around the Fermilab long baseline program with an upgraded Soudan II as a main target. The good news is that there will almost certainly be an American long baseline project. Indeed, the issue of neutrino oscillations is too important not to have a strong American neutrino program, especially now that the SSC project has disappeared.

The long baseline proposal at KEK [42] would rely on an upgraded 12 GeV proton synchrotron, and make use of both a near detector at KEK and Superkamiokande 250 km away. It is not known at this time whether the project will be approved.

A proposal [43] also exists to send a neutrino beam from UNK at Serpukhov in Russia to the Gran Sasso Laboratory, 2200 km away.

Matter effects can also in principle be studied with a long baseline neutrino beam, since most of the neutrino path is in the earth. In the case of a CERN beam to Gran Sasso, or a Fermilab neutrino beam to Soudan, the neutrino path in matter ($\sim$ 730 km) is too short to reach a spectacular resonance; however, asymmetry measurements [44] between ν_μ and $\bar{\nu}_\mu$ could perhaps allow us to detect the effects. With the Serpukhov neutrino beam to Gran Sasso, if the neutrino flux turns out to be sufficient, the advantage will clearly be an enhanced sensitivity to possible matter effects compared to the CERN or Fermilab neutrino beams [44].

4.6 Supernova neutrinos

With the future generation of neutrino detectors in operation we can hope that the next supernova in our galaxy will be observed in the neutrino channel with great details. Superkamiokande would record a thousand $\bar{\nu}_e$ events for a supernova at the galactic center but with little sensitivity to ν_e's from the neutronization burst. The corresponding burst in ICARUS would only produce 76 events (44 ν_e absorption on liquid argon and 32 electron scattering by all six neutrino species). On the other hand ICARUS would record 11 ν_e events from the neutronization burst [9] and open the possibility of weighing the neutron star produced. In addition, the possibility to observe separately charged and neutral current events as well as identifying neutrino types will result in further sensitivity to neutrino oscillations. Clearly supernovae will offer the longest baseline possible. Also in the energy window above terrestrial anti-neutrinos and below those of atmospheric origin is the exciting possibility of detecting relic supernova neutrinos whose mean energy is predicted to be around 10 MeV.

§5 Conclusion

The magnitude of the experimental effort in preparation for neutrino studies is a clear indication of the importance of the physics at stake. The new solar neutrino experiments should provide a definite answer to the present solar puzzle. They will have an improvement by about two orders of magnitude in the rate of data and can provide results which are independent of the Solar Model. The study of $\nu_\mu \leftrightarrow \nu_\tau$, $\nu_\mu \leftrightarrow \nu_e$ and $\nu_e \leftrightarrow \nu_\tau$ oscillations will also be carried out with short and long baseline accelerator neutrinos and with atmospheric neutrinos.

It is likely that these new experiments will change our perception of the Universe, since they are dealing with some of the most intriguing questions on the nature of neutrinos, the nature of dark matter and the unification of forces. The overall scientific program is even richer because it also increases the live time for observation of possible supernovas or other unforeseen cosmological events.

Proton decay was mentioned here because it is linked to neutrino studies in the sense that both may investigate physics at the GUT scale and both are often studied in the same detector. Proton decay and neutrino studies constitute an important complement to the high energy and high precision frontier programs. The prospect for exciting discoveries is high, and the time window seems to be optimal.

New detector technologies such as the one developed by the ICARUS Collaboration ("electronic bubble chamber") are available and open an important new physics program, especially when combined with larger detector sizes.

This future program should lead to a test of GUTs and perhaps also of Supersymmetry. We can certainly hope that significant progress will be achieved in our knowledge of neutrino properties within the coming decade.

§ Acknowledgments

I would like to thank R. Peccei for inviting me to the plenary session of this most interesting workshop and C. Rubbia for exciting my interest in non-accelerator physics. I am grateful to P. Picchi, F. Pietropaolo and the entire ICARUS Collaboration for their help and to F. Cavanna for in addition editing this text on TEX. I would also like to express my thanks to S. Tracy for her careful reading of this document and to C. Rigoni for drafting most of the figures and transparencies.

§ References

[1] K. Zuber et al., (Heidelberg-Moscow Collaboration), "Enriched ^{76}Ge neutrinoless double-β decay from the Heidelberg-Moscow Experiment", Proc. 27th Int. Conf. on HEP, Glasgow, Scotland, 20-27 July, 1994.

[2] A.I. Belesev et al., "First Results of Troitsk experiment on search for $\bar{\nu}_e$ rest mass in Tritium β-decay", Preprint *INR*-**862/94**, July 1994 and V. Lobashev, Proc. 27th Int. Conf. on HEP, Glasgow, Scotland, 20-27 July, 1994. The recommended and conservative limit is 7.0 eV, Particle Data Group, *Phys. Rev.* **D 50**, (1994) 1173.

[3] K. Assamagan, *PSI PR* **94-19**.

[4] The ALEPH Collaboration, "Determination of an Upper Limit of the Mass of the Tau Neutrino at LEP", *CERN-PPE* **94-107**, submitted to *Phys. Lett.* **B**.

[5] C. Rubbia, "The Renaissance of Experimental Neutrino Physics", Proc. of Coversaciones de Madrid, Sept. 5, 1992, El Escorial, Madrid, Spain; *CERN-PPE* / **93-08**; J.-P. Revol, "Future Neutrino oscillation Experiments", Proc. of "Les Rencontres de Physique de la Vallèe d'Aoste", La Thuile, Aosta Valley, Ed. by M. Greco (1993).

[6] J. Peoples' letter to Snow mass participants, June 27, 1994.

[7] P. Benetti et al., (ICARUS Collaboration), "A three-ton liquid argon time projection chamber", *Nucl. Instrum. Methods Phys. Res.* **A 332**, (1993) 395 and "Performance of a three-ton Liquid Argon Time Projection Chamber", *Nucl. Instrum. Methods Phys. Res.* **A 345**, (1994) 230.

[8] T. Doke, *Portgal. Phys.* **Vol. 12**, (1981) 9.

[9] The ICARUS Collaboration, "ICARUS II, a Second-Generation Proton Decay Experiment and Neutrino Observatory at the Gran Sasso Laboratory, *LNGS*-**94/99-I** (Vol.I), May 1994.

[10] R.V. Eötvös, D. Pekar, F. Fekete, *Ann. Physik* **68**, (1922) 11; J. Renner, *Mat. Term. Ert.* **53**, (1935) 542; P.G. Roll, R. Krotkov, R.H. Dicke, *Ann. Phys.* **26**, (1964) 442; V.B. Braginsky, V.I. Panov, *Sov. Phys.* **JETP 34**, (1972) 463.

[11] See for instance: W. de Boer, "Grand Unified Theories and Supersymmetry in Particle Physics and Cosmology", *IEKP-KA* /**94-01**, *hep-ph/***9402266**, 1994, to be published in *Prog. in Part. and Nucl. Phys.*, and references therein.

[12] U. Amaldi et al., *Phys. Rev.* **D36**, (1987) 1385; U. Amaldi, W. de Boer and H. Fürstenau, *Phys. Lett.* **B 260**, (1991) 447.

[13] P. Nath, R. Arnowitt, and A.H. Chamseddine, *Phys. Rev.* **32 D**, (1985) 2348; J. Hisano, H. Murayama and T. Yanagida, *Nucl. Phys.* **B 402**, (1993) 46.

[14] R. Arnowitt and P. Nath, *Phys. Rev.* **D 49**, (1994) 1479.

[15] K.S. Hirata et al., (The Kamiokande Collaboration), *Phys. Lett.* **B 280**, (1992) 146.

[16] W. A. Mann, T. Kafka and W. Leeson, *Phys. Lett.* **B 291**, (1992) 200.

[17] J.L. Thron et al., (Soudan II Collaboration), *Nucl. Inst. and Meth.* **283**, (1989) 642.

[18] M. Takita, *Proc. Int. Symp. on Neut. Astr.*, Takayama, Japan, (1992); Y. Totsuka, "Superkamiokande", *ICRR-Report*-**227-90-20**, *Proc. Int. Symp. on Underground Phys. Exp, Science Council of Japan*, April, 1990; K. Nakamura, *ICRR-Report*-**309-94-4**, *Proc. 3rd NESTOR Int. Workshop*, Pylos, Greece, Oct. 1993.

[19] The ICARUS Collaboration, "ICARUS II, a Second-Generation Proton Decay Experiment and Neutrino Observatory at the Gran Sasso Laboratory, *LNGS*-**94/99-II** (Vol.II), May 1994.

[20] S. Bludman et al., *Phys. Rev.* **D 45**, (1992) 1810.

[21] P. G. Langacker and M. X. Luo, *Phys. Rev.* **D 44**, (1991) 817.

[22] The GALLEX Collaboration, announcement at INFN (Rome, Italy), Friday 18th of November 1994.

[23] L. Wolfenstein, *Phys. Rev.* **D 17**, (1978) 2369; **D 20**, (1979) 2634; S.P. Mikheyev and A.Yu. Smirnov, *Yad. Fiz.* **42**, (1985) 1441; *Nuovo Cimento* **9 C**, (1986) 17; H. Bethe, *Phys. Rev. Lett.* **56**, (1986) 1305.

[24] D. R. O. Morrison, "The Steady Vanishsing of the Three Solar Neutrino Problems", Proc. 27th Int. Conf. on HEP, Glasgow, Scotland, 20-27 July, 1994, *CERN-PPE/***94-125**.

[25] A. Kovetz and G. Shaviv, to be published in *Ap. J.*, May 1994; A. Dar and G. Shaviv, "A Standard Solution to the Solar Neutrino Problem", *Technion PH* **94-5** and "Has a Standard Model Solution to the Solar Neutrino Problem Been Found? A Response", Asher Space Research Institute, *Israel Institute of Technology*, Haifa, Israel, 1994.

[26] J. Bahcall et al., "Has a Standard Model Solution to the Solar Neutrino Problem Been Found?", *IASSNS-AST* **94/13**, April 1994, unpublished.

[27] G. T. Ewan et al., "The Sudbury Neutrino Observatory Proposal", *SNO*-**87-12**, Oct. 1987; H.C. Evans et al., "Scientific and Technical Description of the Mark II SNO Detector", *SNO*-**89-15**, Oct. 1989.

[28] G. Laurenti et al., "A Study of a High Rate Solar Neutrino Detector with Neutrino Energy Determination", *CERN/LAA/PC/***93-10**, Feb. 1993; T. Ypsilantis, "HELLAZ: a High Rate Solar Neutrino Detector with Neutrino Energy Determination", *Proc. of The Dark Matter School*, Valencia, Spain, 1993.

[29] C. Arpesella et al., "Proposal for a Real Time Detector for Low Energy Solar Neutrinos", **Vol. 1** & **2**, ed. G. Bellini et al., Dept. of Physics, University of Milano

and INFN Milano, Italy and R. Raghavan, *AT&T Bell Lab.*, Murray Hill, NJ, USA.

[30] The Kamiokande Collaboration, "Results from a Test Beam Calibration of a Water Cerenkov Detector at KEK", to be published.

[31] Y. Fukuda et al., "Atmospheric ν_μ/ν_e Ratio in the Multi-GeV Energy Range", presented at the ICHEP 94, June 1994, Glasgow, Scotland, to be published.

[32] M. Casperini, *Phys. Rev.* **D 38**, (1988) 2635; *Phys. Rev.* **D 39**, (1989) 3606; A. Halprin and C.N. Leung, *Phys. Rev. Lett.* **67**, (1991) 1833; *Nucl Phys. B* (Proc. Suppl.) **28 A**, (1992) 139; I. Kazuhito et al., "A Test of the Equivalence Principle by Long Baseline Neutrino-Oscillation Experiments", *TMUP-HEL-* **9207**, (1992); J. Pantaleone et al., "Neutrino Mixing due to a Violation of the Equivalence Principle", *IUHET*-**228**, *UDHEP*-**10-92**; J. N. Bahcall et al., "Solar Neutrinos and the Principle of Equivalence", *IASSNS-AST* **94/54**, Oct. 1994, submitted to *Phys. Rev.* **D**.

[33] N. Armenise et al., (CHORUS Collaboration), *CERN-SPSC/***90-42**, (1990).

[34] P. Astier et al., (NOMAD Collaboration), *CERN-SPSLC/***91-21**, (1991), *CERN-SPSLC/***91-48**, *SPSLC/***P 261**, Add. 1.

[35] K. Kodama et al., "Muon Neutrino to tau Neutrino Oscillations", Fermilab *P803 Proposal*, Oct. 1989.

[36] C. Rubbia, private communication.

[37] A. Ball, B. Bianchi, J.P. Revol, G. R. Stevenson and E. Weisse, "CERN Beams for Long Baseline Neutrino Oscillations Experiments", *CERN-SL/*Note **92-75** (BT).

[38] A.K. Mann et al., "Proposal for a Long Baseline Neutrino Oscillation Experiment at the AGS", Brookhaven National Laboratory, Jan. 1993.

[39] W.W.M. Allison et al., "A Long Baseline Neutrino Oscillation Experiment Using the Soudan II Neutrino Detector", Fermilab *P822 Proposal*, Mar. 1991.

[40] D. Michael, "Comparison of Long Baseline Neutrino Beam Calculations", California Institute of Technology preprint, 1994 and private communication, to be published.

[41] G.J. Alner et al., "Expression of Interest for a Long Baseline Neutrino Oscillation Experiment Using a CERN Beam and a Large Magnetic Sampling Calorimeter at Gran Sasso", *LNGS Proposal*, May 1994.

[42] K. Nishikawa, "A Possible Long Base Line Neutrino Oscillation Experiment Using KEK PS and Super Kamiokande", *INS-Rep.*-**924**, April 1992, submitted to *J. of the Phys. Soc. of Japan.*

[43] Yu. S. Fedotov et al., "On Possibility to Study Neutrino Oscillations by Detectors Located at Gran Sasso (Italy) Using Beams from 600 GeV UNK-1 Machine", Protvino preprint, January 1994.

[44] G. Fiorentini and B. Ricci, "Old and New Interactions in Long Baseline Neutrino Experiments", INFN Ferrara **06-93**, Proc. Workshop on Neutrino Telescopes, Venice, March 1993, ed. M. Baldo-Ceolin.

Gravitational Waves

Kip S. Thorne

Theoretical Astrophysics, California Institute of Technology, Pasadena, CA 91125, USA

This article reviews current efforts and plans for gravitational-wave detection, the gravitational-wave sources that might be detected, and the information that the detectors might extract from the observed waves. Special attention is paid to (i) the LIGO/VIRGO network of earth-based, kilometer-scale laser interferometers, which is now under construction and will operate in the high-frequency band (1 to 10^4 Hz), and (ii) a proposed 5-million-kilometer-long Laser Interferometer Space Antenna (LISA), which would fly in heliocentric orbit and operate in the low-frequency band (10^{-4} to 1 Hz). LISA would extend the LIGO/VIRGO studies of stellar-mass ($M \sim 2$ to $300 M_\odot$) black holes into the domain of the massive black holes ($M \sim 1000$ to $10^8 M_\odot$) that inhabit galactic nuclei and quasars.

§1 Introduction

According to general relativity theory, compact concentrations of energy (e.g., neutron stars and black holes) should warp spacetime strongly, and whenever such an energy concentration changes shape, it should create a dynamically changing spacetime warpage that propagates out through the Universe at the speed of light. This propagating warpage is called a *gravitational wave*—a name that arises from general relativity's description of gravity as a consequence of spacetime warpage.

Although gravitational waves have not yet been detected directly, their indirect influence has been seen and measured with such remarkable accuracy that their reality has been blessed even by the Nobel Prize Committee (that bastion of conservatism which explicitly denied Einstein the Prize for his relativity theories [1]):

The 1993 Prize was awarded to Russell Hulse and Joseph Taylor for their discovery of the binary pulsar PSR 1913+16 [2] and for Taylor's observational demonstration (with colleagues) [3] that the binary's two neutron stars are spiraling together at just the rate predicted by general relativity's theory of gravitational radiation reaction: from the observed orbit, one can compute the rate at which orbital energy should be emitted into gravitational radiation, and from this rate of energy loss one can compute the rate of orbital inspiral. The computed and observed inspiral rates agree to within the experimental accuracy, better than one per cent.

Although this is a great triumph for Einstein, it is not a firm proof that general relativity is correct in all respects. Other relativistic theories of gravity (theories compatible with special relativity) predict the existence of gravitational waves; and some other theories predict the same inspiral rate for PSR 1913+16 as general relativity, to within the experimental accuracy [4, 5]. Nevertheless, the experimental evidence for general relativity is so strong [4], that I shall assume it to be correct throughout this lecture except for occasional side remarks.

There are a number of efforts, worldwide, to detect gravitational radiation. These efforts are driven in part by the desire to "see gravitational waves in the flesh," but more importantly by the goal of using the waves as a probe of the Universe and of the nature of gravity. And a powerful probe they should be, since they carry detailed information about gravity and their sources.

There is an enormous difference between gravitational waves, and the electromagnetic waves on which our present knowledge of the Universe is based:

- Electromagnetic waves are oscillations of the electromagnetic field that propagate through spacetime; gravitational waves are oscillations of the "fabric" of spacetime itself.
- Astronomical electromagnetic waves are almost always incoherent superpositions of emission from individual electrons, atoms, or molecules. Cosmic gravitational waves are produced by coherent, bulk motions of huge amounts of mass-energy—either material mass, or the energy of vibrating, nonlinear spacetime curvature.
- Since the wavelengths of electromagnetic waves are small compared to their sources (gas clouds, stellar atmospheres, accretion disks, ...), from the waves we can make pictures of the sources. The wavelengths of cosmic gravitational waves are comparable to or

larger than their coherent, bulk-moving sources, so we cannot make pictures from them. Instead, the gravitational waves are like sound; they carry, in two independent waveforms, a stereophonic, symphony-like description of their sources.

- Electromagnetic waves are easily absorbed, scattered, and dispersed by matter. Gravitational waves travel nearly unscathed through all forms and amounts of intervening matter [6, 7].
- Astronomical electromagnetic waves have frequencies that begin at $f \sim 10^7$ Hz and extend on *upward* by roughly 20 orders of magnitude. Astronomical gravitational waves should begin at $\sim 10^4$ Hz (1000-fold lower than the lowest-frequency astronomical electromagnetic waves), and should extend on *downward* from there by roughly 20 orders of magnitude.

These enormous differences make it likely that:

- The information brought to us by gravitational waves will be very different from (almost "orthogonal to") that carried by electromagnetic waves; gravitational waves will show us details of the bulk motion of dense concentrations of energy, whereas electromagnetic waves show us the thermodynamic state of optically thin concentrations of matter.
- Most (but not all) gravitational-wave sources that our instruments detect will not be seen electromagnetically, and conversely, most objects observed electromagnetically will never be seen gravitationally. Typical electromagnetic sources are stellar atmospheres, accretion disks, and clouds of interstellar gas—none of which emit significant gravitational waves, while typical gravitational-wave sources are the cores of supernovae (which are hidden from electromagnetic view by dense layers of surrounding stellar gas), and colliding black holes (which emit no electromagnetic waves at all).
- Gravitational waves may bring us great surprises. In the past, when a radically new window has been opened onto the Universe, the resulting surprises have had a profound, indeed revolutionary, impact. For example, the radio universe, as discovered in the 1940s, 50s and 60s, turned out to be far more violent than the optical universe; radio waves brought us quasars, pulsars, and the cosmic microwave radiation, and with them our first direct observational evidence for black holes, neutron stars, and the heat of the big bang [8]. It is reasonable to hope that gravitational waves will bring a similar "revolution".

In this lecture I shall review the present status of attempts to detect gravitational radiation and plans for the future, and I shall describe some examples of information that we expect to garner from the observed waves. I shall begin, in Section 2, with an overview of all the frequency bands in which astrophysical gravitational waves are expected to be strong, the expected sources in each band, and the detection techniques being used in each. Then in subsequent sections I shall focus on (i) the "high frequency band" which is populated by waves from stellar mass black holes and neutron stars and is being probed by ground-based instruments: laser interferometers and resonant-mass antennas (Sections 3, 4, 5, and 6), and (ii) the "low-frequency band" which is populated by waves from supermassive black holes and binary stars and is probed by space-based instruments: radio and optical tracking of spacecraft (Sections 7 and 8). Finally, in Section 9 I shall describe the stochastic background of gravitational waves that is thought to have been produced by various processes in the early universe, and prospects for detecting it in the various frequency bands.

§2 Frequency Bands, Sources, and Detection Methods

Four gravitational-wave frequency bands are being explored experimentally: the high-frequency band (HF; $f \sim 10^4$ to 1 Hz), the low-frequency band (LF; $f \sim 1$ to 10^{-4} Hz), the very-low frequency band (VLF; $f \sim 10^{-7}$ to 10^{-9} Hz), and the extremely-low-frequency band (ELF; $f \sim 10^{-15}$ to 10^{-18} Hz).

2.1 High-Frequency Band, 1 to 10^4 Hz

A gravitational-wave source of mass M cannot be much smaller than its gravitational radius, $2GM/c^2$, and cannot emit strongly at periods much smaller than the light-travel time $4\pi GM/c^3$ around this gravitational radius. Correspondingly, the frequencies at which it emits are

$$f \lesssim \frac{1}{4\pi GM/c^3} \sim 10^4 \mathrm{Hz} \frac{M_\odot}{M} , \quad (1)$$

where $M_\odot$ is the mass of the Sun and G and c are Newton's gravitation constant and the speed of light. To achieve a size of order its gravitational radius and thereby emit near this maximum frequency, an object presumably must be heavier than the Chandrasekhar limit, about the mass of the sun, $M_\odot$. Thus, the highest frequency expected for strong gravitational waves is $f_{\max} \sim 10^4$ Hz. This defines the upper edge of the high-frequency gravitational-wave band.

The high-frequency band is the domain of Earth-based gravitational-wave detectors: laser interferometers and resonant mass antennas. At frequencies below about 1 Hz, Earth-based detectors face nearly insurmountable noise

(i) from fluctuating Newtonian gravity gradients (due, e.g., to the gravitational pulls of inhomogeneities in the Earth's atmosphere which move overhead with the wind), and (ii) from Earth vibrations (which are extremely difficult to filter out mechanically below ~ 1 Hz). This defines the 1 Hz lower edge of the high-frequency band; to detect waves below this frequency, one must fly one's detectors in space.

A number of interesting gravitational-wave sources fall in the high-frequency band: the stellar collapse to a neutron star or black hole in our Galaxy and distant galaxies, which sometimes triggers supernovae; the rotation and vibration of neutron stars (pulsars) in our Galaxy; the coalescence of neutron-star and stellar-mass black-hole binaries ($M \lesssim 1000 M_\odot$) in distant galaxies; and possibly such sources of stochastic background as vibrating loops of cosmic string, phase transitions in the early Universe, and the big bang in which the Universe was born.

I shall discuss the high-frequency band in detail in Sections 3–6.

2.2 Low-Frequency Band, 10^{-4} *to 1 Hz*

The low-frequency band, 10^{-4} to 1 Hz, is the domain of detectors flown in space (in Earth orbit or in interplanetary orbit). The most important of these are the Doppler tracking of spacecraft via microwave signals sent from Earth to the spacecraft and there transponded back to Earth (a technique that NASA has pursued since the early 1970's), and optical tracking of spacecraft by each other (laser interferometry in space, a technique now under development for possible flight in $\sim$ 2014 or sooner).

The 1 Hz upper edge of the low-frequency band is defined by the gravity-gradient and seismic cutoffs on Earth-based instruments; the $\sim 10^{-4}$ Hz lower edge is defined by expected severe difficulties at lower frequencies in isolating spacecraft from the buffeting forces of fluctuating solar radiation pressure, solar wind, and cosmic rays.

The low-frequency band should be populated by waves from short-period binary stars in our own Galaxy (main-sequence binaries, cataclysmic variables, white-dwarf binaries, neutron-star binaries, ...); from white dwarfs, neutron stars, and small black holes spiraling into massive black holes ($M \sim 3 \times 10^5$ to $3 \times 10^7 M_\odot$) in distant galaxies; and from the inspiral and coalescence of supermassive black-hole binaries ($M \sim 100$ to $10^8 M_\odot$). The upper limit, $\sim 10^8 M_\odot$, on the masses of black holes that can emit in the low-frequency band is set by Eq. (1) with $f \gtrsim 10^{-4}$ Hz. There should also be a low-frequency stochastic background from such early-universe processes as vibrating cosmic strings, phase transitions, and the big-bang itself.

I shall discuss the low-frequency band in detail in Sections 7 and 8.

2.3 Very-Low-Frequency Band, 10^{-7} *to* 10^{-9} *Hz*

Joseph Taylor and others have achieved a remarkable gravity-wave sensitivity in the very-low-frequency band (VLF) by the timing of millisecond pulsars: When a gravitational wave passes over the Earth, it perturbs our rate of flow of time and thence the ticking rates of our clocks relative to clocks outside the wave. Such perturbations will show up as apparent fluctuations in the times of arrival of the pulsar's pulses. If no fluctuations are seen at some level, we can be rather sure that neither Earth nor the pulsar is being bathed by gravitational waves of the corresponding strength. If fluctuations with the same time evolution are seen simultaneously in the timing of several different pulsars, then the cause could well be gravitational waves bathing the Earth.

By averaging the pulses' times of arrival over long periods of time (months to tens of years), a very high timing precision can be achieved, and correspondingly tight limits can be placed on the waves bathing the Earth or the pulsar. The upper edge of the VLF band, $\sim 10^{-7}$ Hz, is set by the averaging time, a few months, needed to build up high accuracy; the lower edge, $\sim 10^{-9}$ Hz, is set by the time, $\sim$ 20 years, since very steady millisecond pulsars were first discovered.

As we shall see in Section 3.2, strong gravitational-wave sources are generally compact, not much larger than their own gravitational radii. The only compact bodies that can radiate in the VLF band or below, i.e., at $f \lesssim 10^{-7}$ Hz, are those with $M \gtrsim 10^{11} M_\odot$ [cf. Eq. (1)]. Conventional astronomical wisdom suggests that compact bodies this massive do not exist, and that therefore the only strong waves in the VLF band and below are a stochastic background produced by the same early-universe processes as might radiate at low and high frequencies: cosmic strings, phase transitions, and the big bang.

Of course, conventional wisdom could be wrong. Nevertheless, it is conventional to quote measurement accuracies in the VLF band and below in the language of a stochastic background: the fraction $\Omega_g(f)$ of the energy required to close the universe that lies in a bandwidth $\Delta f = f$ centered on frequency f. The current 95%-confidence limit on Ω_g from pulsar timing in the VLF band is $\Omega_g(4 \times 10^{-9}\text{Hz}) < 6 \times 10^{-8} H^{-2}$ where H is the Hubble constant in units of 100 km sec^{-1} Mpc^{-1} [9]. This is a sufficiently tight limit that it is beginning to cast doubt on the (not terribly popular) suggestion, that the Universe contains enough vibrating loops of cosmic string for their gravitational pulls to have seeded galaxy formation [10, 11].

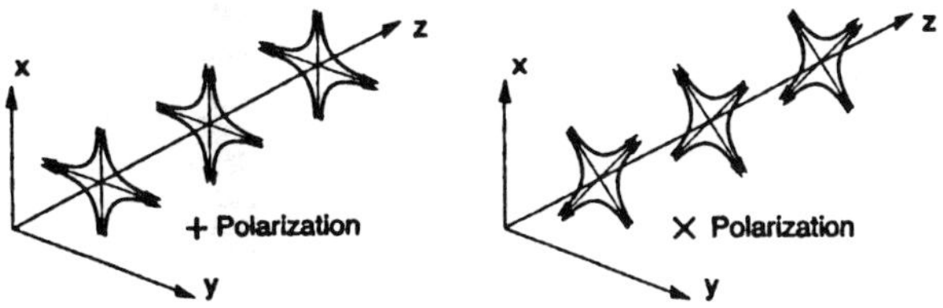

Figure 1: The lines of force associated with the two polarizations of a gravitational wave. (From Ref. [14].)

2.4 *Extremely-Low-Frequency Band, 10^{-15} to 10^{-18} Hz*

Gravitational waves in the extremely-low-frequency band (ELF), 10^{-15} to 10^{-18} Hz, should produce anisotropies in the cosmic microwave background radiation. The tightest limit from microwave observations comes from the lower edge of the ELF band $f \sim 10^{-18}$ Hz, where the gravitational wavelength is about π times the Hubble distance, and the waves, by squeezing all of the space inside our cosmological horizon in one direction, and stretching it all in another, should produce a quadrupolar anisotropy in the microwave background. The quadrupolar anisotropy measured by the COBE satellite, if due primarily to gravitational waves (which it could be [12, 13]), corresponds to an energy density $\Omega_g(10^{-18}\text{Hz}) \sim 10^{-9}$. In Section 9 I shall discuss the implications of this impressive ELF limit for the strength of the early-Universe stochastic background in the HF and LF bands.

2.5 *Other Frequency Bands and Other Detection Methods*

A large number of other methods have been conceived of, for searching for gravitational radiation. Some of them would operate best in the HF, LF, VLF, and ELF bands described above; others would operate best at other frequencies. However, none has shown anywhere near the promise or the achievements of the methods described above (laser interferometry on Earth and in space, resonant mass antennas, Doppler tracking of spacecraft, timing of pulsars, and anisotropy of microwave background). For some references to other methods, see, e.g., [6].

§3 Ground-Based Laser Interferometers

3.1 *Wave Polarizations, Waveforms, and How an Interferometer Works*

According to general relativity theory (which I shall assume to be correct in this paper), a gravitational wave has two linear polarizations, conventionally called + (plus) and × (cross). Associated with each polarization there

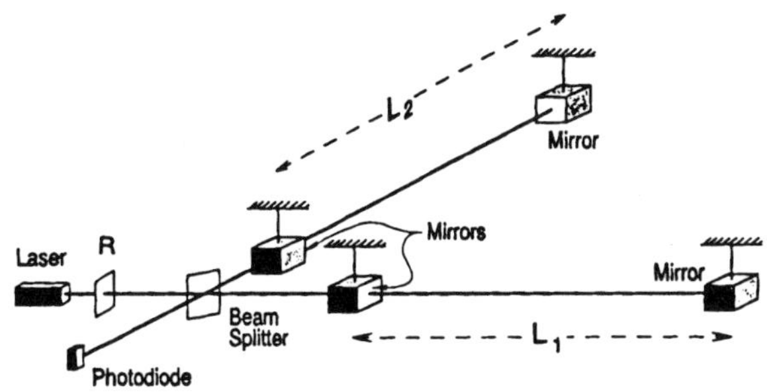

Figure 2: Schematic diagram of a laser interferometer gravitational wave detector. (From Ref. [14].)

is a gravitational-wave field, h_+ or $h_\times$, which oscillates in time and propagates with the speed of light. Each wave field produces tidal forces (stretching and squeezing forces) on any object or detector through which it passes. If the object is small compared to the waves' wavelength (as is the case for ground-based interferometers and resonant mass antennas), then relative to the object's center, the forces have the quadrupolar patterns shown in Figure 1. The names "plus" and "cross" are derived from the orientations of the axes that characterize the force patterns [6].

A laser interferometer gravitational wave detector ("interferometer" for short) consists of four masses that hang from vibration-isolated supports as shown in Figure 2, and the indicated optical system for monitoring the separations between the masses [6, 14]. Two masses are near each other, at the corner of an "L", and one mass is at the end of each of the L's long arms. The arm lengths are nearly equal, $L_1 \simeq L_2 = L$. When a gravitational wave, with frequencies high compared to the masses' ~ 1 Hz pendulum frequency, passes through the detector, it pushes the masses back and forth relative to each other as though they were free from their suspension wires, thereby changing the arm-length difference, $\Delta L \equiv L_1 - L_2$. That change is monitored by laser interferometry in such a way that the variations in the output of the photodiode (the interferometer's output) are directly proportional to $\Delta L(t)$.

If the waves are coming from overhead or underfoot and the axes of the + polarization coincide with the arms' directions, then it is the waves' + polarization that drives the masses, and $\Delta L(t)/L = h_+(t)$. More generally, the interferometer's output is a linear combination of the two wave fields:

$$\frac{\Delta L(t)}{L} = F_+ h_+(t) + F_\times h_\times(t) \equiv h(t)\,. \tag{2}$$

The coefficients F_+ and $F_\times$ are of order unity and depend in a quadrupolar manner on the direction to the source and the orientation of the detector [6]. The combination $h(t)$ of the two h's is called the *gravitational-wave strain* that acts on the detector; and the time evolutions of $h(t)$, $h_+(t)$, and $h_\times(t)$ are sometimes called *waveforms*.

Interferometer test masses at present are made of transparent fused silica, though other materials might be used in the future. The masses' inner faces (shown white in Fig. 2) are covered with high-reflectivity dielectric coatings to form the indicated "mirrors", while the masses' outer faces are covered with anti-reflection coatings. The two mirrors facing each other on each arm form a Fabry-Perot cavity. A beam splitter splits a carefully prepared laser beam in two, and directs the resulting beams down the two arms. Each beam penetrates through the antireflection coating of its arm's corner mass, through the mass, and through the dielectric coating (the mirror); and thereby—with the length of the arm's Fabry-Perot cavity adjusted to be nearly an integral number of half wavelengths of light—the beam gets trapped in the cavity. The cavity's end mirror has much higher reflectivity than its corner mirror, so the trapped light leaks back out through the corner mirror, and then hits the beam splitter where it recombines with light from the other arm. Most of the recombined light goes back toward the laser (where it can be returned to the interferometer by a "light-recycling mirror" labeled R), but a tiny portion goes toward the photodiode.

When a gravitational wave hits the detector and moves the masses, thereby changing the lengths L_1 and L_2 of the two cavities, it shifts each cavity's resonant frequency slightly relative to the laser frequency, and thereby changes the phase of the light in the cavity and the phase of the light that exits from the cavity toward the beam splitter. Correspondingly, the relative phase of the two beams returning to the splitter is altered by an amount $\Delta\Phi \propto \Delta L$, and this relative phase shift causes a change in the intensity of the recombined light at the photodiode, $\Delta I_{\rm pd} \propto \Delta\Phi \propto \Delta L \propto h(t)$. Thus, the change of photodiode output current is directly proportional to the gravitational-wave strain $h(t)$. This method of monitoring $h(t)$, which was invented by Ronald Drever [15] as a modification of Rainer Weiss's [16] seminal concept for such an interferometer, is capable of very high sensitivity, as we shall see below.

3.2 Wave Strengths and Interferometer Arm Lengths

The strengths of the waves from a gravitational-wave source can be estimated using the "Newtonian/quadrupole" approximation to the Einstein field equations. This approximation says that $h \simeq (G/c^4)\ddot{Q}/r$, where $\ddot{Q}$ is the second time derivative of the source's quadrupole moment, r is the distance of the source from Earth (and G and c are Newton's gravitation constant and the speed of light). The strongest sources will be highly nonspherical and thus will have $Q \simeq ML^2$, where M is their mass and L their size, and correspondingly will have $\ddot{Q} \simeq 2Mv^2 \simeq 4E^{\rm ns}_{\rm kin}$, where v is their internal velocity and $E^{\rm ns}_{\rm kin}$ is the nonspherical part of their internal kinetic energy. This provides us with the estimate

$$h \sim \frac{1}{c^2}\frac{4G(E^{\rm ns}_{\rm kin}/c^2)}{r}\;; \qquad (3)$$

i.e., h is about 4 times the gravitational potential produced at Earth by the mass-equivalent of the source's nonspherical, internal kinetic energy—made dimensionless by dividing by c^2. Thus, in order to radiate strongly, the source must have a very large, nonspherical, internal kinetic energy.

The best known way to achieve a huge internal kinetic energy is via gravity; and by energy conservation (or the virial theorem), any gravitationally-induced kinetic energy must be of order the source's gravitational potential energy. A huge potential energy, in turn, requires that the source be very compact, not much larger than its own gravitational radius. Thus, the strongest gravity-wave sources must be highly compact, dynamical concentrations of large amounts of mass (e.g., colliding and coalescing black holes and neutron stars).

Such sources cannot remain highly dynamical for long; their motions will be stopped by energy loss to gravitational waves and/or the formation of an all-encompassing black hole. Thus, the strongest sources should be transient. Moreover, they should be very rare — so rare that to see a reasonable event rate will require reaching out through a substantial fraction of the Universe. Thus, just as the strongest radio waves arriving at Earth tend to be extragalactic, so also the strongest gravitational waves are likely to be extragalactic.

For highly compact, dynamical objects that radiate in the high-frequency band, e.g. colliding and coalescing neutron stars and stellar-mass black holes, the internal, nonspherical kinetic energy $E^{\rm ns}_{\rm kin}/c^2$ is of order the mass of the Sun; and, correspondingly, Eq. (3) gives $h \sim 10^{-22}$ for such sources at the Hubble distance (3000 Mpc, i.e., 10^{10} light years); $h \sim 10^{-21}$ at 200 Mpc (a best-guess distance for several neutron-star coalescences per year; see Section 5.2), $h \sim 10^{-20}$ at the Virgo cluster of galaxies (15 Mpc); and $h \sim 10^{-17}$ in the outer reaches of our own Milky Way galaxy (20 kpc). These numbers set the scale of sensitivities that ground-based interferometers seek to achieve: $h \sim 10^{-21}$ to 10^{-22}.

When one examines the technology of laser interferometry, one sees good prospects to achieve measurement accuracies $\Delta L \sim 10^{-16}$ cm (1/1000 the diameter of the nucleus of an atom). With such an accuracy, an interferometer must have an arm length $L = \Delta L/h \sim 1$ to 10 km, in order to achieve the desired wave sensitivities, 10^{-21} to

Figure 3: Artist's conception of one of the LIGO interferometers. [Courtesy the LIGO Project.]

10^{-22}. This sets the scale of the interferometers that are now under construction.

3.3 LIGO, VIRGO, and the International Interferometric Network

Interferometers are plagued by non-Gaussian noise, e.g. due to sudden strain releases in the wires that suspend the masses. This noise prevents a single interferometer, by itself, from detecting with confidence short-duration gravitational-wave bursts (though it might be possible for a single interferometer to search for the periodic waves from known pulsars). The non-Gaussian noise can be removed by cross correlating two, or preferably three or more, interferometers that are networked together at widely separated sites.

The technology and techniques for such interferometers have been under development for nearly 25 years, and plans for km-scale interferometers have been developed over the past 14 years. An international network consisting of three km-scale interferometers, at three widely separated sites, is now in the early stages of construction. It includes two sites of the American LIGO Project ("Laser Interferometer Gravitational Wave Observatory") [14], and one site of the French/Italian VIRGO Project (named after the Virgo cluster of galaxies) [17].

LIGO will consist of two vacuum facilities with 4-kilometer-long arms, one in Hanford, Washington (in the northwestern United States; Fig. 3) and the other in Livingston, Louisiana (in the southeastern United States). These facilities are designed to house many successive generations of interferometers without the necessity of any major facilities upgrade; and after a planned future expansion, they will be able to house several interferometers at once, each with a different optical configuration optimized for a different type of wave (e.g., broad-band burst, or narrow-band periodic wave, or stochastic wave). The LIGO facilities and their first interferometers are being constructed by a team of about 80 physicists and engineers at Caltech and MIT, led by Barry Barish (the PI) and Gary Sanders (the Project Manager). Robbie Vogt (who directed the project during the pre-construction phase) is in charge of the final design and construction of LIGO's first interferometers, Stan Whitcomb is in charge of interferometer R&D, and Albert Lazzarini is the system engineer and Rai Weiss the cognizant scientist for integration of all parts of LIGO.

A number of other research groups are making important contributions to LIGO: Bob Byers' group at Stanford is developing Nd:YAG lasers, Peter Saulson's group at Syracuse and Vladimir Braginsky's group in Moscow are developing test-mass suspension systems and studying noise in them; Jim Faller's group at JILA is developing active vibration isolation systems; Ron Drever's group at Caltech is developing advanced interferometers; and Sam Finn's group at Northwestern and my group at Caltech are developing data analysis techniques. A number of other groups are likely to join the LIGO effort in the next few years. A formal association of LIGO-related scientists (the *LIGO Research Community*, an analog of a "user's group" in high-energy physics) is being organized, and a LIGO Program Advisory Committee will be formed soon, with voting membership restricted to people outside the Caltech/MIT LIGO team, to advise the LIGO management.

The VIRGO Project is building one vacuum facility in Pisa, Italy, with 3-kilometer-long arms. This facility and its first interferometers are a collaboration of more than a hundred physicists and engineers at the INFN (Frascati, Napoli, Perugia, Pisa), LAL (Orsay), LAPP (Annecy), LOA (Palaiseau), IPN (Lyon), ESPCI (Paris), and the University of Illinois (Urbana), under the leadership of Alain Brillet and Adalberto Giazotto.

Both LIGO and VIRGO are scheduled for completion in the late 1990s, and their first gravitational-wave searches are likely to be performed in 2000 or 2001.

LIGO alone, with its two sites which have parallel arms, will be able to detect an incoming gravitational wave, measure one of its two waveforms, and (from the time delay between the two sites) locate its source to within a $\sim 1^\circ$ wide annulus on the sky. LIGO and VIRGO together, operating as a *coordinated international network*, will be able to locate the source (via time delays plus the interferometers' beam patterns) to within a 2-dimensional error box with size between several tens of arcminutes and several degrees, depending on the source direction and on the amount of high-frequency structure in the waveforms. They will also be able to monitor both waveforms $h_+(t)$ and $h_\times(t)$ (except for frequency components above about 1kHz and below about 10 Hz, where the interferometers' noise becomes severe).

The accuracies of the direction measurements and the

ability to monitor more than one waveform will be severely compromised when the source lies anywhere near the plane formed by the three LIGO/VIRGO interferometer locations. To get good all-sky coverage will require a fourth interferometer at a site far out of that plane; Japan and Australia would be excellent locations, and research groups there are carrying out research and development on interferometric detectors, aimed at such a possibility. A 300-meter prototype interferometer called TAMA is under construction in Tokyo, and a 400-meter prototype called AIGO400 has been proposed for construction north of Perth.

Two other groups are major players in this field, one in Britain led by James Hough, the other in Germany, led by Karsten Danzmann. These groups each have two decades of experience with prototype interferometers (comparable experience to the LIGO team and far more than anyone else) and great expertise. Frustrated by inadequate financing for a kilometer-scale interferometer, they are constructing, instead, a 600 meter system called GEO600 near Hannover, Germany. Their goal is to develop, from the outset, an interferometer with the sort of advanced design that LIGO and VIRGO will attempt only as a "second-generation" instrument, and thereby achieve sufficient sensitivity to be full partners in the international network's first gravitational-wave searches; they then would offer a variant of their interferometer as a candidate for second-generation operation in the much longer arms of LIGO and/or VIRGO. It is a seemingly audacious plan, but with their extensive experience and expertise, the British/German collaboration might pull it off successfully.

3.4 Interferometer Development and Noise Sources

It is not possible, in the short vacuum systems now available (arm lengths $\leq$ 40 meters), to develop and test a multikilometer interferometer as a single unit. This is because the various noise sources that plague an interferometer scale differently from each other with length L and with gravity-wave frequency f. As a result, the various components of the multikilometer interferometers, and the various techniques to be used in them, are being developed and tested separately in a number of different laboratories, and will only be combined together into a single interferometer when the LIGO/VIRGO vacuum systems are completed.

The best known of the LIGO-Project laboratories in which components and techniques are being developed is the 40-meter prototype interferometer at Caltech (Fig. 4). This prototype focuses on the development of methods and components to control "displacement noise," i.e., those noise sources that push the mirrored test masses back and forth as would a gravity wave. The principal sources of displacement noise are *seismic vibrations* of the ground beneath the interferometer (which are filtered out by the masses' suspension wires and by "isolation stacks" made of successive layers of steel and rubber), and *thermally-induced vibrations* of the test masses and of the wires that suspend them (vibrations that are controlled by designing the test masses and suspensions with great care and constructing them from low-loss, i.e. high "Q", materials).

Figure 4: The LIGO Project's 40-meter "Mark II" prototype interferometer at Caltech. This prototype went into operation in 1993. It has much larger vacuum chambers, to accommodate bigger and better seismic isolation stacks, than those of the previous "Mark I" prototype (which operated from the early 1980s to 1992). [Courtesy the LIGO Project.]

Among the LIGO Project's other laboratories, there is a shorter-armed prototype-interferometer facility at MIT, which is devoted to developing methods and components for controlling noise in the phase of the interferometer's light beams. Since the gravity wave makes itself known by the phase shift that it puts on the light of one interferometer arm relative to the other, this phase noise can simulate a gravity wave. Among the various causes of phase noise, the one that is the most fundamental is *photon shot noise* due to the random times at which the light's photons arrive at the photodiode (cf. Fig. 2).

Once the myriad of other noise sources have been brought under control, *shot noise*, *thermal noise* (i.e., thermally induced vibrations), and *seismic noise* (i.e., ground vibrations) are likely to be the ultimate impediments to detecting and studying gravitational waves. Figure 5 shows the spectra expected for each of these three noises in the first interferometers that will operate in LIGO. At frequencies above 200 Hz, shot noise dominates; between 200 Hz and 40 Hz, thermal noise in the suspension wires dominates; and below 40 Hz, seismic noise dominates.

During LIGO's operations, step-by-step improvements

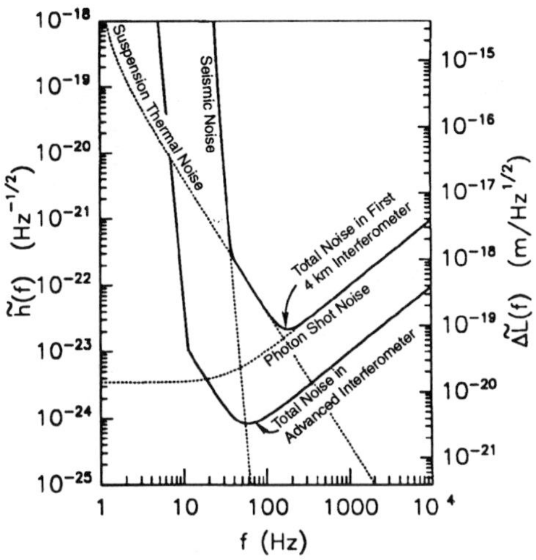

Figure 5: The expected noise spectrum in each of LIGO's first 4-km interferometers (upper solid curve) and in more advanced interferometers (lower solid curve). The dashed curves show various contributions to the first interferometers' noise. Plotted horizontally is gravity wave frequency f; plotted vertically is $\tilde{h}(f)$, the square root of the spectral density of the detector's output $h(t) = \Delta L(t)/L$ in the absence of a gravity wave. The rms h noise in a bandwidth Δf at frequency f is $h_{\rm rms} = \tilde{h}(f)\sqrt{\Delta f}$. (From Ref. [14].)

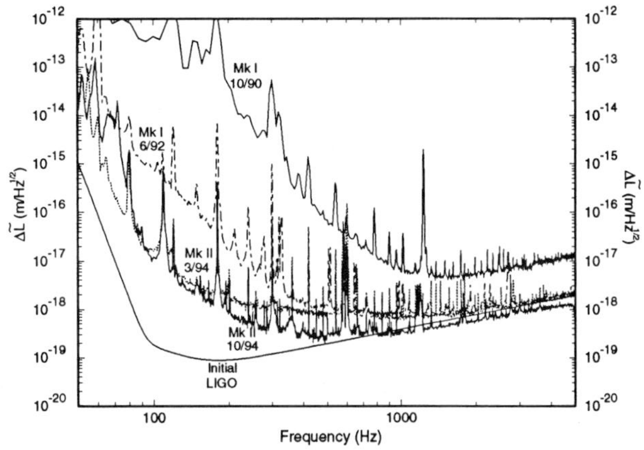

Figure 6: Measured noise spectra in the Caltech 40-meter prototype interferometer (Fig. 4). Since this prototype is devoted to learning to control displacement noise, the spectra shown are $\Delta\tilde{L}(f)$, the square root of the spectral density of the measured arm-length difference. Each of the many spectral lines is well understood, and most could be removed if their removal were of high priority (e.g., they are multiples of the 60 Hz line frequency sneaking into the apparatus by known routes). Those few, very narrow lines that cannot be physically removed by cleaning up the instrument (e.g., thermally-driven violin-mode resonances of the wires that suspend the test masses) will be filtered out in the data analysis. Thus, the interferometer sensitivity is characterized by the continuum noise floor and not the lines. (From Ref. [18].)

will be made in the control of these three noise sources [14], thereby pushing the overall noise spectrum downward from the "first-interferometer" level toward the "advanced-interferometer" level shown in Figure 5. As we shall see below, the sensitivity of the first interferometers might be inadequate to detect gravitational waves. However, we are quite confident that at some point during the improvement from first interferometers to advanced, a plethora of gravitational waves will be found and will start bringing us exciting information about fundamental physics and the Universe.

Notice from Figure 5 that the advanced LIGO interferometers are expected to have their optimal sensitivity at $f \sim 100$ Hz, and rather good sensitivity all the way from $f \sim 10$ Hz at the low-frequency end to $f \sim 500$ Hz at the high-frequency end. Below 10 Hz, seismic noise, creeping through the isolation stacks, will overwhelm all gravitational-wave signals; and above 500 Hz, photon shot noise may overwhelm the signals.

Figure 6 gives an impression of the present state of interferometer technology and the rate of progress. This figure shows a sequence of noise spectra in the 40-meter prototype interferometer at Caltech, during 1990–94 when the 40-meter R&D emphasis was on improving the low-frequency noise performance. The top two (noisiest) spectra are snapshots of the original "Mark I" prototype performance in October 1990 and June 1992; the lower two (quieter) spectra are from the rebuilt, "Mark II" prototype of Figure 4, in March and October 1994. The smooth, solid line, for comparison, is the displacement noise goal for the first 4-km interferometers in LIGO (i.e., it is the upper solid curve of Fig. 5).

Note that the prototypes's arm-length difference was being monitored, in October 1994, to within an rms noise level (in a bandwidth equal to frequency) $\Delta L_{\rm rms} = \sqrt{f}\Delta\tilde{L}(f) < 8\times10^{-16}$ cm over the frequency range 200 to 1000 Hz. This corresponds to an rms gravitational-wave noise level $h_{\rm rms} = \Delta L_{\rm rms}/40{\rm m} < 2\times10^{-19}$, the best that any gravitational-wave detector has yet achieved.

3.5 Semiquantitative Discussion of Interferometer Noise

The LIGO and VIRGO interferometers are expected to have rms noise levels $h_{\rm rms} \lesssim 10^{-22}$ corresponding to test-mass position noises $\Delta L_{\rm rms} \lesssim hL \sim 10^{-16}$ cm. 10^{-16} cm is awfully small: 1/1000 the diameter of the nucleus of an atom, and 10^{-12} the wavelength of the light being used to monitor the masses' motions. How can one possibly monitor such small motions? The following estimate explains how.

One adjusts the reflectivities of the interferometer's corner mirrors so the two arms store the laser light on average for about half a cycle of a ~ 100 Hz gravitational wave, which means for ~ 100 round trips. The light in each arm thereby acquires a phase shift

$$\Delta\Phi \sim 100 \times 4\pi\Delta L/\lambda \sim 10^{-9}\ , \qquad (4)$$

where $\lambda \sim 10^{-4}$ cm is the wavelength of light. If the interference of the light from the two beams is done optimally, then this phase shift (equal and opposite in the two arms) can be measured at the photodiode to an accuracy that is governed by the light's photon shot noise, $\Delta\Phi \sim 1/\sqrt{N}$, where N is the number of photons that enter the interferometer from the laser during the half-cycle of photon storage time. (This $1/\sqrt{N}$ is the usual photon fluctuation in a quantum mechanical "coherent state" of light.) Thus, to achieve the required accuracy, $\Delta\Phi \sim 10^{-9}$, in the face of photon shot noise, requires $N \sim 10^{18}$ photons in 0.01 second, which means a laser power of $\sim$ 100 Watts.

By cleverness [15], one can reduce the required laser power: The light is stored in the interferometer arms for only a half gravity-wave period ($\sim$ 100 round trips) because during the next half period the waves would reverse the sign of ΔL, thereby reversing the sign of the phase shift being put onto the light and removing from the light the signal that had accumulated in the first half period. In just 100 round trips, however, the light is attenuated hardly at all. One therefore reuses the light, over and over again. This is done by (i) operating the interferometer with only a tiny fraction of the recombined light going out toward the photodiode, and almost all of it instead going back toward the laser, and by (ii) placing a mirror (marked R in Fig. 2) between the laser and the interferometer in just such a position that the entire interferometer becomes an optical cavity driven by the laser—with its arms as two subcavities. Then the mirror R recycles the recombined light back into the interferometer in phase with the new laser light, thereby enabling a laser of, say, 5 Watts to behave like one of 100 Watts or more.

Turn from photon shot noise to thermal noise. How, one might ask, can one possibly expect to monitor the mirrors' motions at a level of 10^{-16} cm when the room-temperature atoms of which the fused-silica mirrors are made vibrate thermally with amplitudes $\Delta l_{\rm rms} = \sqrt{kT/m\omega^2} \sim 10^{-10}$ cm? (Here k is Boltzmann's constant, T is room temperature, m is the atomic rest mass, and $\omega \sim 10^{14}$ s^{-1} is the angular frequency of atomic vibration.) The answer is that these individual atomic vibrations are unimportant. The light beam, with its $\sim$ 5cm diameter, averages over the positions of $\sim 10^{17}$ atoms in the mirror, and with its 0.01s storage time it averages over $\sim 10^{11}$ vibrations of each atom. This spatial and temporal averaging makes the vibrations of individual atoms irrelevant. Not so irrelevant, however, are the lowest-frequency normal-mode vibrations of the mirror-endowed masses (since these modes experience much less time averaging than the faster atomic vibrations). Assuming a mass $m \sim$ (a few tens of kg), these normal modes have angular frequencies $\omega \sim 10^5$s^{-1}, so their rms vibration amplitude is $\Delta l_{\rm rms} = \sqrt{kT/m\omega^2} \sim 10^{-14}$ cm. This is 100 times larger than the signals we wish to monitor; but if these modes have high quality factors (high Q's; low losses), then the vibrations will be very steady over the interferometer's averaging time of 0.01 s, and correspondingly, their effects will average down by more than a factor 100. Similar considerations apply to the thermal noise in the masses' suspension wires. For detailed discussions of fascinating and not-fully-understood physics issues that influence the thermal noise, see, e.g., Refs. [19, 20, 21].

Figure 7: The seismic isolation stack that was recently installed in the LIGO Project's Mark II prototype interferometer at Caltech. When the interferometer is in operation, a small tower is mounted on the top steel plate and from the tower hangs one of the interferometer's mirror-endowed masses. [Courtesy the LIGO Project.]

At the LIGO sites, and most any other quiet location on Earth, the ground is continually shaking with an rms displacement $\Delta l_{\rm rms} \sim 10^{-8}$ cm $(100\ {\rm Hz}/f)^{3/2}$. This is 10^7 times larger than the motions one seeks to monitor. At frequencies above 10 Hz or so, one can protect the masses from these seismic vibrations by simple (but carefully designed) passive isolation stacks. Each element in the stack is a mass and a spring (a harmonic oscillator) with normal-mode frequency $f_0 \sim$ (a few Hz). When seismic noise tries to drive this harmonic oscillator far above its resonant frequency [in our case at $f \gtrsim$ (a few tens of Hz)], the amplitude of its response is attenuated relative to the driving motion by a factor $(f_0/f)^2$ [in our case a factor $\gtrsim 10^2$]. Thus, each oscillator in the stack will provide a reduction $\gtrsim 10^2$ in $\Delta l_{\rm rms}$, so a stack of four or five oscillators is enough to provide the required isolation. Figure 7 shows an isolation stack—made of four steel plates and four sets of viton rubber springs (not quite visible between the plates)—that is now operating in the Mark II prototype interferometer of Figure 4. This stack and the pendulum wires that suspend the mirror-endowed test masses provide five layers of isolation. The installation of this new stack was responsible for the sharp drop in low-frequency noise in Figure 6 between June 1992 (Mark I) and March 1994 (Mark II).

The above rough estimates suggest (as Weiss realized as early as 1972 [16]) that it is possible for interferometers to achieve the required sensitivities, $h_{\rm rms} \sim 10^{-22}$ and $\Delta L \sim 10^{-16}$ cm. However, going from these rough estimates to a real working interferometer, and doing so in the face of a plethora of other noise sources, is a tremendous experimental challenge—one that has occupied a number of excellent experimental physicists since 1972.

§4 Resonant-Mass Antennas

A resonant-mass antenna for gravitational radiation consists of a solid body that (heuristically speaking) rings like a bell when a gravitational wave hits it. This body (the resonant mass) is usually a cylinder, but future variants are likely to be spheres or sphere-like, e.g. a truncated icosahedron gravitational-wave antenna or TIGA [22]. The resonant mass is typically made from an alloy of aluminum and weighs several tons, but some have been made of niobium or single-crystal silicon or sapphire (but with masses well below a ton). To control thermal noise, the resonant mass is usually cooled cryogenically to liquid-helium temperatures or below.

The resonant-mass antenna is instrumented with an electromagnetic transducer and electronics, which monitor the complex amplitude of one or more of the mass's normal modes. When a gravitational wave passes through the mass, its frequency components near each normal-mode frequency f_o drive that mode, changing its complex amplitude; and the time evolution of the changes is measured within some bandwidth Δf by the transducer and electronics. Current resonant-mass antennas are narrow-band devices ($\Delta f/f_o \ll 1$) but in the era of LIGO/VIRGO, they might achieve bandwidths as large as $\Delta f/f_o \sim 1/3$.

Resonant-mass antennas for gravitational radiation were pioneered by Joseph Weber about 35 years ago [23], and have been pushed to ever higher sensitivity by Weber and a number of other research groups since then. For references and an overview of the present and future of such antennas see, e.g., Ref. [24]. At present there is a network of such antennas [25], cooled to 3K, and operating with an rms noise level for broad-band gravity-wave bursts of $h_{\rm rms} \simeq 6 \times 10^{-19}$. The network includes an aluminum cylinder called EXPLORER built by a group at the University of Rome, Italy (Edoardo Amaldi, Guido Pizella, et. al.); an aluminum cylinder at Louisiana State University, USA (Bill Hamilton, Warren Johnson, et. al.); and a niobium cylinder at the University of Perth, Australia (David Blair et. al.). This network has been in operation, searching for waves, for several years.

The next generation of resonant-mass antennas is now under construction at the University of Rome (NAUTILUS) and at the University of Legarno, Italy (AURIGA). These are several-ton aluminum bars cooled to 0.05K; their rms design sensitivities for wave bursts are (several)$\times 10^{-20}$ [24].

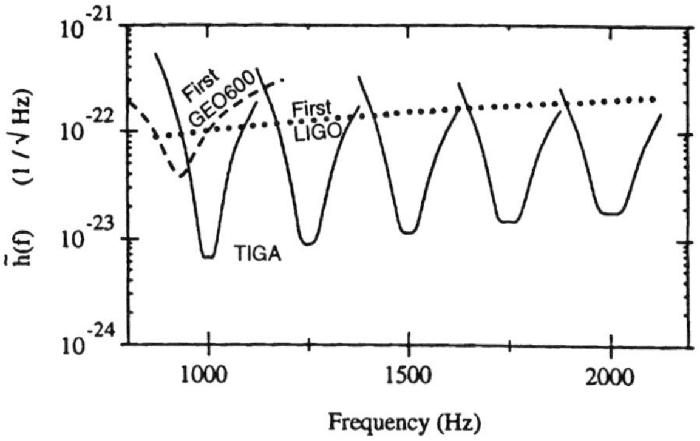

Figure 8: The rms noise curves $\tilde{h}(f)$ (measured in strain per root Hz) for a "xylaphone" of TIGA gravitational-wave detectors (solid curves) for signals of random polarization and direction [22]. The TIGA's are presumed instrumented and cooled sufficiently well that their sensitivity is at the standard quantum limit. Their central frequencies, radii and masses (assuming aluminum material) are {1.0 kHz, 1.30 m, 25.1 ton}, {1.25 kHz, 1.04 m, 12.8 ton}, {1.50 kHz, 0.87 m, 7.4 ton}, {1.75 kHz, 0.74 m, 4.7 ton}, {2.0 kHz, 0.65 m, 3.1 ton}. Shown for comparison are the noise curves for the first LIGO interferometer with random wave polarization and direction multiplied by $\sqrt{5}$ (dotted curve; Fig. 5), and for the first GEO600 detector operated in a narrow-band mode (dashed curve; Ref. [26]).

A subsequent generation, which hopefully would operate in the LIGO/VIRGO era, is being discussed and planned [24]. These are 1 to 100 ton spheres or TIGA's cooled to $\sim$ 0.01—0.05K, with sensitivity goals of $\sim 10^{-21}$. Such antennas might be built by an American collaboration, a Brazilian collaboration, an Italian collaboration called "Omega", and/or a Dutch collaboration called "Grail". Their spherical or near-spherical shapes make them omnidirectional and should give them several-times higher sensitivities than can be achieved by cylinders at the same frequency.

The attractiveness of such antennas in the LIGO/ VIRGO era lies in their ability to operate with impressive sensitivity in the uppermost reaches of the high-frequency band, $\sim 10^3$ to 10^4 Hz, where photon shot noise debilitates the performance of interferometric detectors (cf. Fig. 5). Figure 8 shows the projected rms noise curves of a family of TIGA detectors, each instrumented to operate at the "standard quantum limit" for such a detector (a nontrivial experimental task). Shown for comparison is the rms noise of the first LIGO interferometer—which, of course, is not optimized for the kHz band. The GEO600 interferometer, with its advanced design, can be operated in a narrow-band, high-frequency mode (and probably will be so operated in $\sim$ 1999. Its rms design sensitivity in such a mode is also shown in Figure 8. The TIGA sensitivities are sufficiently good in the kHz band, compared to early LIGO and GEO interferometers, that, although they probably cannot begin to operate until somewhat after the beginning of the LIGO/VIRGO era, they might be fully competitive when they do operate, and might play an important role in the kHz band.

§5 High-Frequency Gravitational-Wave Sources: Coalescing Compact Binaries

The best understood of all gravitational-wave sources are coalescing, compact binaries composed of neutron stars (NS) and black holes (BH). These NS/NS, NS/BH, and BH/BH binaries may well become the "bread and butter" of the LIGO/VIRGO diet.

The Hulse-Taylor [2, 3] binary pulsar, PSR 1913+16, is an example of a NS/NS binary whose waves could be measured by LIGO/VIRGO, if we were to wait long enough. At present PSR 1913+16 has an orbital frequency of about 1/(8 hours) and emits its waves predominantly at twice this frequency, roughly 10^{-4} Hz, which is in the low-frequency band—far too low to be detected by LIGO/VIRGO. However, as a result of their loss of orbital energy to gravitational waves, the PSR 1913+16 NS's are gradually spiraling inward. If we wait roughly 10^8 years, this inspiral will bring the waves into the LIGO/VIRGO high-frequency band. As the NS's continue their inspiral, the waves will then sweep upward in frequency, over a time of about 15 minutes, from 10 Hz to $\sim 10^3$ Hz, at which point the NS's will collide and coalesce. It is this last 15 minutes of inspiral, with $\sim 16,000$ cycles of waveform oscillation, and the final coalescence, that LIGO/VIRGO seeks to monitor.

5.1 *Wave Strengths Compared to LIGO Sensitivities*

Figure 9 compares the projected sensitivities of interferometers in LIGO [14] with the wave strengths from the last few minutes of inspiral of BH/BH, NS/BH, and NS/NS binaries at various distances from Earth. The two solid curves at the bottoms of the stippled regions (labeled $h_{\rm rms}$) are the rms noise levels for broad-band waves that have optimal direction and polarization. The tops of the stippled regions (labeled $h_{\rm SB}$ for "sensitivity to bursts") are the sensitivities for highly confident detection of randomly polarized, broad-band waves from random directions (i.e., the sensitivities for high confidence that any such observed signal is not a false alarm due to Gaussian noise). The upper stippled region and its bounding curves are the expected performances of the first interferometers in LIGO; the lower stippled region and curves are

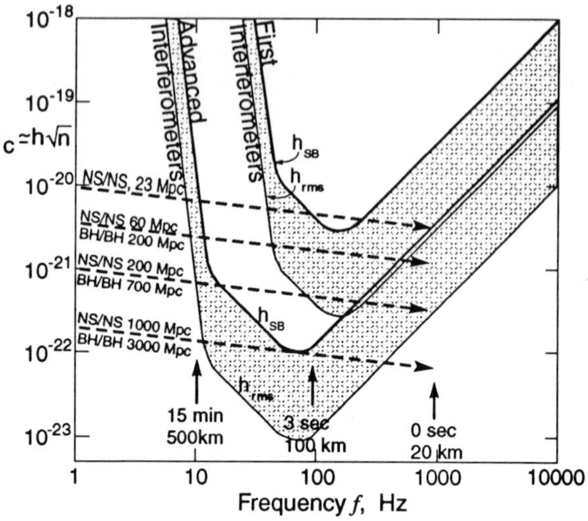

Figure 9: LIGO's projected broad-band noise $h_{\rm rms}$ and sensitivity to bursts $h_{\rm SB}$ (Fig. 5 and Ref. [14]) compared with the strengths of the waves from the last few minutes of inspiral of compact binaries. The signal to noise ratios are $\sqrt{2}$ higher than in Ref. [14] because of a factor 2 error in Eq. (29) of Ref. [6].

performances of more advanced LIGO interferometers; cf. Figure 5.

As the NS's and/or BH's spiral inward, their waves sweep upward in frequency (left to right in the diagram). The dashed lines show their "characteristic" signal strength h_c (approximately the amplitude h of the waves' oscillations multiplied by the square root of the number of cycles spent near a given frequency, $\sqrt{n}$); the signal-to-noise ratio is this h_c divided by the detector's $\sqrt{5}h_{\rm rms}$, $S/N = h_c/(\sqrt{5}h_{\rm rms})$, where the $\sqrt{5}$ converts $h_{\rm rms}$ from "optimal direction and polarization" to "random direction and polarization") [14, 6]. The arrows along the bottom inspiral track indicate the time until final coalescence for an NS/NS binary and the separation between the NS centers of mass. Each NS is assumed to have a mass of 1.4 suns and a radius ~ 10 km; each BH, 10 suns and ~ 20 km.

Notice that the signal strengths in Figure 9 are in good accord with our rough estimates based on Eq. (3); at the endpoint (right end) of each inspiral, the number of cycles n spent near that frequency is of order unity, so the quantity plotted, $h_c \simeq h\sqrt{n}$, is about equal to h—and at distance 200 Mpc is roughly 10^{-21}, as we estimated in Section 3.2.

5.2 *Coalescence Rates*

Such final coalescences are few and far between in our own galaxy: about one every 100,000 years, according to 1991 estimates by Phinney [27] and by Narayan, Piran, and Shemi [28], based on the statistics of binary pulsar searches in our galaxy which found three that will coalesce in less than 10^{10} years. Extrapolating out through the universe on the basis of the density of production of blue light (the color produced predominantly by massive stars), Phinney [27] and Narayan et. al. [28] infer that to see several NS/NS coalescences per year, LIGO/VIRGO will have to look out to a distance of about 200 Mpc (give or take a factor ~ 2); cf. the "NS/NS inspiral, 200 Mpc" line in Figure 9. Since these estimates were made, the binary pulsar searches have been extended through a significantly larger volume of the galaxy than before, and no new ones with coalescence times $\lesssim 10^{10}$ years have been found; as a result, the binary-pulsar-search-based best estimate of the coalescence rate should be revised downward [29], perhaps to as little as one every million years in our galaxy, corresponding to a distance 400 Mpc for several per year [29].

A rate of one every million years in our galaxy is ~ 1000 times smaller than the birth rate of the NS/NS binaries' progenitors: massive, compact, main-sequence binaries [27, 28]. Therefore, either 99.9 per cent of progenitors fail to make it to the NS/NS state (e.g., because of binary disruption during a supernova or forming TŻO's), or else they do make it, but they wind up as a class of NS/NS binaries that has not yet been discovered in any of the pulsar searches. Several experts on binary evolution have argued for the latter [30, 31, 32]: most NS/NS binaries, they suggest, may form with such short orbital periods that their lifetimes to coalescence are significantly shorter than normal pulsar lifetimes ($\sim 10^7$ years); and with such short lifetimes, they have been missed in pulsar searches. By modeling the evolution of the galaxy's binary star population, the binary experts arrive at best estimates as high as 3×10^{-4} coalescences per year in our galaxy, corresponding to several per year out to 60 Mpc distance [30]. Phinney [27] describes other plausible populations of NS/NS binaries that could increase the event rate, and he argues for "ultraconservative" lower and upper limits of 23 Mpc and 1000Mpc for how far one must look to see several coalescence per year.

By comparing these rate estimates with the signal strengths in Figure 9, we see that (i) the first interferometers in LIGO/VIRGO have a possibility but not high probability of seeing NS/NS coalescences; (ii) advanced interferometers are almost certain of seeing them (the requirement that this be so was one factor that forced the LIGO/VIRGO arm lengths to be so long, several kilometers); and (iii) they are most likely to be discovered roughly half-way between the first and advanced interferometers—which means by an improved variant of the first interferometers several years after LIGO operations begin.

We have no good observational handle on the coalescence rate of NS/BH or BH/BH binaries. However, theory suggests that their progenitors might not disrupt during the stellar collapses that produce the NS's and BH's, so their coalescence rate could be about the same as the birth rate for their progenitors: $\sim 1/100,000$ years in our galaxy. This suggests that within 200 Mpc distance there might be several NS/BH or BH/BH coalescences per year. [27, 28, 30, 32]. This estimate should be regarded as a plausible upper limit on the event rate and lower limit on the distance to look [27, 28].

If this estimate is correct, then NS/BH and BH/BH binaries will be seen before NS/NS, and might be seen by the first LIGO/VIRGO interferometers or soon thereafter; cf. Figure 9. However, this estimate is far less certain than the (rather uncertain) NS/NS estimates!

Once coalescence waves have been discovered, each further improvement of sensitivity by a factor 2 will increase the event rate by $2^3 \simeq 10$. Assuming a rate of several NS/NS per year at 200 Mpc, the advanced interferometers of Figure 9 should see ~ 100 per year.

5.3 Inspiral Waveforms and the Information They Can Bring

Neutron stars and black holes have such intense self gravity that it is exceedingly difficult to deform them. Correspondingly, as they spiral inward in a compact binary, they do not gravitationally deform each other significantly until several orbits before their final coalescence [33, 34]. This means that the inspiral waveforms are determined to high accuracy by only a few, clean parameters: the masses and spin angular momenta of the bodies, and the initial orbital elements (i.e. the elements when the waves enter the LIGO/VIRGO band).

Though tidal deformations are negligible during inspiral, relativistic effects can be very important. If, for the moment, we ignore the relativistic effects—i.e., if we approximate gravity as Newtonian and the wave generation as due to the binary's oscillating quadrupole moment [6], then the shapes of the inspiral waveforms $h_+(t)$ and $h_\times(t)$ are as shown in Figure 10.

The left-hand graph in Figure 10 shows the waveform increasing in amplitude and sweeping upward in frequency (i.e., undergoing a "chirp") as the binary's bodies spiral closer and closer together. The ratio of the amplitudes of the two polarizations is determined by the inclination ι of the orbit to our line of sight (lower right in Fig. 10). The shapes of the individual waves, i.e. the waves' harmonic content, are determined by the orbital eccentricity (upper right). (Binaries produced by normal stellar evolution should be highly circular due to past radiation reaction forces, but compact binaries that form by capture events, in dense star clusters that might reside in galactic nuclei [35], could be quite eccentric.) If, for simplicity, the orbit is circular, then the rate at

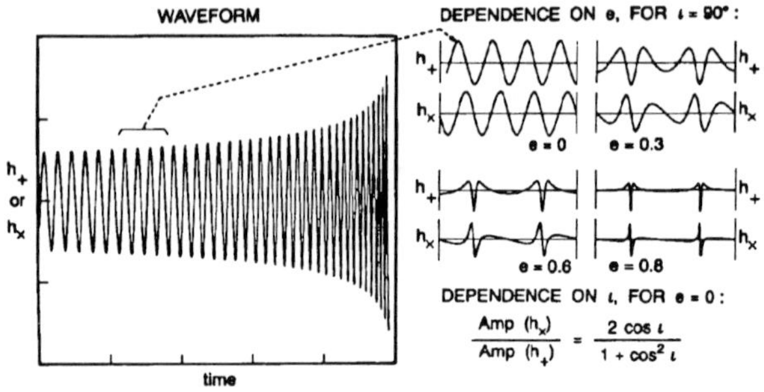

Figure 10: Waveforms from the inspiral of a compact binary, computed using Newtonian gravity for the orbital evolution and the quadrupole-moment approximation for the wave generation. (From Ref. [14].)

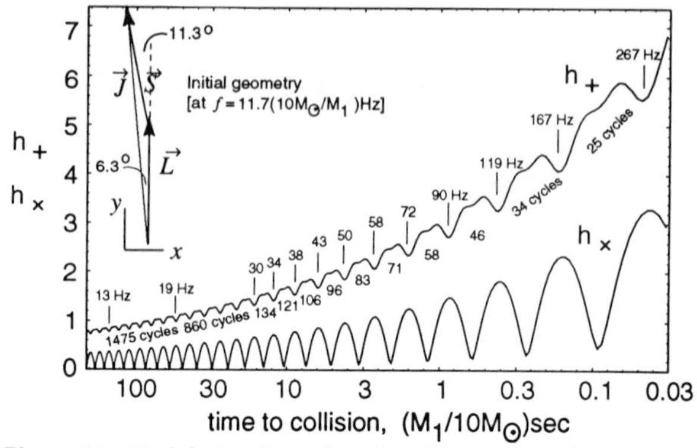

Figure 11: Modulational envelope for the waveform from a $1M_{\odot}$ nonspinning NS spiraling into a $10M_{\odot}$, rapidly spinning Kerr black hole (spin parameter $a = 1$). The orbital angular momentum $\mathbf{L}$ is inclined by $\alpha = 11.3$ degrees to the hole's spin angular momentum $\mathbf{S}$, and the two precess around $\mathbf{J} = \mathbf{L} + \mathbf{S}$, whose direction remains fixed in space as $L = |\mathbf{L}|$ shrinks and $S = |\mathbf{S}| = M_{\mathrm{BH}}a$ remains constant. The precession modulates the waves by an amount that depends on (i) the direction to Earth (here along the initial $\mathbf{L} \times \mathbf{S}$, i.e. out of the paper) and (ii) the orientation of the detector's arms (here parallel to the figure's initial $\mathbf{L}$ and to $\mathbf{L}\times$(direction to Earth) for h_+, and rotated 45 degrees for $h_\times$). The figure shows the waveforms' modulational envelopes (in arbitrary units, the same for h_+ and $h_\times$), parametrized by the wave frequency f and the number of cycles of *oscillation* between the indicated f's. The total number of *precessions* from f to coalescence is $N_{\mathrm{prec}} \simeq (5/64\pi)(Ma/\mu)(\pi M f)^{-2/3} \simeq 20(f/10\mathrm{Hz})^{-2/3}$. (From [38, 39].)

which the frequency sweeps or "chirps", df/dt [or equivalently the number of cycles spent near a given frequency, $n = f^2(df/dt)^{-1}$] is determined solely, in the Newtonian/quadrupole approximation, by the binary's so-called *chirp mass*, $M_c \equiv (M_1M_2)^{3/5}/(M_1+M_2)^{1/5}$ (where M_1 and M_2 are the two bodies' masses). The amplitudes of the two waveforms are determined by the chirp mass, the distance to the source, and the orbital inclination. Thus (in the Newtonian/quadrupole approximation), by measuring the two amplitudes, the frequency sweep, and the harmonic content of the inspiral waves, one can determine as direct, resulting observables, the source's distance, chirp mass, inclination, and eccentricity [36, 37].

As in binary pulsar observations [3], so also here, relativistic effects add further information: they influence the rate of frequency sweep and produce waveform modulations in ways that depend on the binary's dimensionless ratio $\eta = \mu/M$ of reduced mass $\mu = M_1M_2/(M_1+M_2)$ to total mass $M = M_1 + M_2$ [40] and on the spins of the binary's two bodies [41]. These relativistic effects are reviewed and discussed at length in Refs. [38, 42]. Two deserve special mention: (i) As the waves emerge from the binary, some of them get backscattered one or more times off the binary's spacetime curvature, producing wave *tails*. These tails act back on the binary, modifying its inspiral rate in a measurable way. (ii) If the orbital plane is inclined to one or both of the binary's spins, then the spins drag inertial frames in the binary's vicinity (the "Lense-Thirring effect"), this frame dragging causes the orbit to precess, and the precession modulates the waveforms [38, 39, 43]. Figure 11 shows the resulting modulation for a $1M_{\odot}$ NS spiraling into a rapidly spinning, $10M_{\odot}$ BH.

Remarkably, the relativistic corrections to the frequency sweep will be measurable with very high accuracy, even though they are typically $\lesssim 10$ per cent of the Newtonian contribution, and even though the typical signal to noise ratio will be only ~ 9 even after optimal signal processing. The reason is as follows [44, 45, 38]:

The frequency sweep will be monitored by the method of "matched filters"; in other words, the incoming, noisy signal will be cross correlated with theoretical templates. If the signal and the templates gradually get out of phase with each other by more than $\sim 1/10$ cycle as the waves sweep through the LIGO/VIRGO band, their cross correlation will be significantly reduced. Since the total number of cycles spent in the LIGO/VIRGO band will be $\sim 16,000$ for a NS/NS binary, ~ 3500 for NS/BH, and ~ 600 for BH/BH, this means that LIGO/VIRGO should be able to measure the frequency sweep to a fractional precision $\lesssim 10^{-4}$, compared to which the relativistic effects are very large. (This is essentially the same method as Joseph Taylor and colleagues use for high-accuracy radio-wave measurements of relativistic effects in binary pulsars [3].)

Preliminary analyses, using the theory of optimal signal processing, predict the following typical accuracies for LIGO/VIRGO measurements based solely on the frequency sweep (i.e., ignoring modulational information) [46, 44, 45, 47, 38]: (i) The chirp mass M_c will typically be measured, from the Newtonian part of the frequency sweep, to $\sim 0.04\%$ for a NS/NS binary and $\sim 0.3\%$ for a system containing at least one BH. (ii) *If* we are confident (e.g., on a statistical basis from measurements of many previous binaries) that the spins are a few percent or less of the maximum physically allowed, then the reduced mass μ will be measured to $\sim 1\%$ for NS/NS and NS/BH binaries, and $\sim 3\%$ for BH/BH binaries. (Here and below NS means a $\sim 1.4M_\odot$ neutron star and BH means a $\sim 10M_\odot$ black hole.) (iii) Because the frequency dependences of the (relativistic) μ effects and spin effects are not sufficiently different to give a clean separation between μ and the spins, if we have no prior knowledge of the spins, then the spin/μ correlation will worsen the typical accuracy of μ by a large factor, to $\sim 30\%$ for NS/NS, $\sim 50\%$ for NS/BH, and a factor ~ 2 for BH/BH [46, 44]. These worsened accuracies might be improved somewhat by waveform modulations caused by the spin-induced precession of the orbit [39, 43], and even without modulational information, a certain combination of μ and the spins will be determined to a few per cent. Much additional theoretical work is needed to firm up the measurement accuracies.

To take full advantage of all the information in the inspiral waveforms will require theoretical templates that are accurate, for given masses and spins, to a fraction of a cycle during the entire sweep through the LIGO/VIRGO band. Such templates are being computed by an international consortium of relativity theorists (Blanchet and Damour in France, Iyer in India, Will and Wiseman in the U.S., and others) [42, 48], using post-Newtonian expansions of the Einstein field equations. This enterprise is rather like computing the Lamb shift to high order in powers of the fine structure constant, for comparison with experiment. The terms of leading order in the mass ratio $\eta = \mu/M$ are being checked by a Japanese-American consortium (Nakamura, Sasaki, Tagoshi, Tanaka, Poisson) using the Teukolsky formalism for weak perturbations of black holes [49, 50]. These small-η calculations have been carried to very high post-Newtonian order for circular orbits and no spins [51, 52], and from those results Cutler and Flanagan [53] have estimated the order to which the full, finite-η computations must be carried in order that systematic errors in the theoretical templates will not significantly impact the information extracted from the LIGO/VIRGO observational data. The answer appears daunting: radiation-reaction effects must be computed to three full post-Newtonian orders [six orders in v/c =(orbital velocity)/(speed of light)] beyond the leading-order radiation reaction, which itself is 5 orders in v/c beyond the Newtonian theory of gravity.

It is only about ten years since controversies over the leading-order radiation reaction [54] were resolved by a combination of theoretical techniques and binary pulsar observations. Nobody dreamed then that LIGO/VIRGO observations will require pushing post-Newtonian computations onward from $O[(v/c)^5]$ to $O[(v/c)^{11}]$. This requirement epitomizes a major change in the field of relativity research: At last, 80 years after Einstein formulated general relativity, experiment has become a major driver for theoretical analyses.

Remarkably, the goal of $O[(v/c)^{11}]$ is achievable. The most difficult part of the computation, the radiation reaction, has been evaluated to $O[(v/c)^9]$ beyond Newton by the French/Indian/American consortium [48] and as of this writing, rumors have it that $O[(v/c)^{10}]$ is coming under control.

These high-accuracy waveforms are needed only for extracting information from the inspiral waves, after the waves have been discovered; they are not needed for the discovery itself. The discovery is best achieved using a different family of theoretical waveform templates, one that covers the space of potential waveforms in a manner that minimizes computation time instead of a manner that ties quantitatively into general relativity theory [38]. Such templates are in the early stage of development [55, 56, 57].

LIGO/VIRGO observations of compact binary inspiral have the potential to bring us far more information than just binary masses and spins:

- They can be used for high-precision tests of general relativity. In scalar-tensor theories (some of which are attractive alternatives to general relativity [58]), radiation reaction due to emission of scalar waves places a unique signature on the gravitational waves that LIGO/VIRGO would detect—a signature that can be searched for with high precision [59].

- They can be used to measure the Universe's Hubble constant, deceleration parameter, and cosmological constant [36, 37, 60, 61]. The keys to such measurements are that: (i) Advanced interferometers in LIGO/VIRGO will be able to see NS/NS out to cosmological redshifts $z \sim 0.3$, and NS/BH out to $z \sim 2$. (ii) The direct observables that can be extracted from the observed waves include the source's luminosity distance r_L (measured to accuracy $\sim$ 10 per cent in a large fraction of cases), and its direction on the sky (to accuracy $\sim$ 1 square degree)—accuracies good enough that only one or a few electromagnetically-observed clusters of galaxies should fall within the 3-dimensional gravitational error boxes, thereby giving promise to joint gravitational/electromagnetic statistical studies. (iii) Another direct gravitational observable is $(1+z)M$ where z is redshift and M is any mass in the system (measured to the accuracies quoted above). Since the masses of NS's in binaries seem to cluster around $1.4M_\odot$, measurements of $(1+z)M$ can provide a handle on the redshift, even in the absence of electromagnetic aid.

- For a NS or small BH spiraling into a massive ~ 50 to $500M_\odot$ BH, the inspiral waves will carry a "map" of the spacetime geometry around the big hole—a map that can be used, e.g., to test the theorem that "a black hole has no hair" [62]; cf. Section 8.3 below.

5.4 Coalescence Waveforms and their Information

The waves from the binary's final coalescence can bring us new types of information.

BH/BH Coalescence

In the case of a BH/BH binary, the coalescence will excite large-amplitude, highly nonlinear vibrations of spacetime curvature near the coalescing black-hole horizons—a phenomenon of which we have very little theoretical understanding today. Especially fascinating will be the case of two spinning black holes whose spins are not aligned with each other or with the orbital angular momentum. Each of the three angular momentum vectors (two spins, one orbital) will drag space in its vicinity into a tornado-like swirling motion—the general relativistic "dragging of inertial frames," so the binary is rather like two tornados with orientations skewed to each other, embedded inside a third, larger tornado with a third orientation. The dynamical evolution of such a complex configuration of coalescing spacetime warpage (as revealed by its emitted waves) might bring us surprising new insights into relativistic gravity [14]. Moreover, if the sum of the BH masses is fairly large, $\sim$ 40 to $200 M_\odot$, then the waves should come off in a frequency range $f \sim 40$ to 200 Hz where the LIGO/VIRGO broad-band interferometers have their best sensitivity and can best extract the information the waves carry.

To get full value out of such wave observations will require [63] having theoretical computations with which to compare them. There is no hope to perform such computations analytically; they can only be done as supercomputer simulations. The development of such simulations is being pursued by several research groups, including an eight-university American consortium of numerical relativists and computer scientists called the Two-Black-Hole Grand Challenge Alliance [64] (Co-PIs: Richard Matzner and Jim Browne, U. Texas Austin; Larry Smarr, Ed Seidel, Paul Saylor, Faisal Saied, U. Illinois Urbana; Geoffrey Fox, Syracuse U.; Stu Shapiro and Saul Teukolsky, Cornell U.; Jim York and Charles Evans, U. North Carolina; Sam Finn, Northwestern U.; Pablo Laguna, Pennsylvania State U.; and Jeff Winicour, U. Pittsburgh). I have a bet with Matzner, the lead PI of this alliance, that LIGO/VIRGO will discover waves from such coalescences with misaligned spins before the Alliance is able to compute them.

NS/NS Coalescence

The final coalescence of NS/NS binaries should produce waves that are sensitive to the equation of state of nuclear matter, so such coalescences have the potential to teach us about the nuclear equation of state [14, 38]. In essence, LIGO/VIRGO will be studying nuclear physics via the collisions of atomic nuclei that have nucleon numbers $A \sim 10^{57}$—somewhat larger than physicists are normally accustomed to. The accelerator used to drive these nuclei up to the speed of light is the binary's self gravity, and the radiation by which the details of the collisions are probed is gravitational.

Unfortunately, the final NS/NS coalescence will emit its gravitational waves in the kHz frequency band ($800\text{Hz} \lesssim f \lesssim 2500\text{Hz}$) where photon shot noise will prevent them from being studied by the standard, "workhorse," broad-band interferometers of Figure 5. However, a specially configured ("dual-recycled") interferometer invented by Brian Meers [65], which could have enhanced sensitivity in the kHz region at the price of reduced sensitivity elsewhere, may be able to measure the waves and extract their equation of state information, as might massive, spherical, resonant-mass detectors [38, 66]. Such measurements will be very difficult and are likely only when the LIGO/VIRGO network has reached a mature stage.

A number of research groups [67, 68, 69, 70, 71] are engaged in numerical astrophysics simulations of NS/NS coalescence, with the goal not only to predict the emitted gravitational waveforms and their dependence on equation of state, but also (more immediately) to learn whether such coalescences might power the γ-ray bursts that have been a major astronomical puzzle since their discovery in the early 1970s.

NS/NS coalescence is currently a popular explanation for the γ-ray bursts because (i) the bursts are isotropically distributed on the sky, (ii) they have a distribution of number versus intensity that suggests they might lie at near-cosmological distances, and (iii) their event rate is roughly the same as that predicted for NS/NS coalescence ($\sim$ 1000 per year out to cosmological distances, if they are cosmological). If LIGO/VIRGO were now in operation and observing NS/NS inspiral, it could report definitively whether or not the γ-bursts are produced by NS/NS binaries; and if the answer were yes, then the combination of γ-burst data and gravitational-wave data could bring valuable information that neither could bring by itself. For example, it would reveal when, to within a few msec, the γ-burst is emitted relative to the moment the NS's first begin to touch; and by comparing the γ and gravitational times of arrival, we might test whether gravitational waves propagate with the speed of light to a fractional precision of $\sim 0.01\text{sec}/3 \times 10^9\,\text{lyr} = 10^{-19}$.

NS/BH Coalescence

A NS spiraling into a BH of mass $M \gtrsim 10 M_\odot$ should be swallowed more or less whole. However, if the BH is less massive than roughly $10M_\odot$, and especially if it is rapidly rotating, then the NS will tidally disrupt before being swallowed. Little is known about the disruption and accompanying waveforms. To model them with any reliability will likely require full numerical relativity, since the circumferences of the BH and NS will be comparable and their physical separation at the moment of disruption will be of order their separation. As with NS/NS, the coalescence waves should carry equation of state information and will come out in the kHz band, where their detection will require advanced, specialty detectors.

Christodoulou Memory

As the coalescence waves depart from their source, their energy creates (via the nonlinearity of Einstein's field equations) a secondary wave called the "Christodoulou memory" [72, 73, 74]. Whereas the primary waves may have frequencies in the kHz band, the memory builds up on the timescale of the primary energy emission profile, which is likely to be of order 0.01 sec, corresponding to a memory frequency in the optimal band for the LIGO/VIRGO workhorse interferometers, $\sim$ 100Hz. Unfortunately, the memory is so weak that only very advanced interferometers have much chance of detecting and studying it—and then, perhaps only for BH/BH coalescences and not for NS/NS or NS/BH [75].

§6 Other High-Frequency Sources

6.1 Stellar Core Collapse and Supernovae

When the core of a massive star has exhausted its supply of nuclear fuel, it collapses to form a neutron star or black hole. In some cases, the collapse triggers and powers a subsequent explosion of the star's mantle—a supernova explosion. Despite extensive theoretical efforts for more than 30 years, and despite wonderful observational data from Supernova 1987A, theorists are still far from a definitive understanding of the details of the collapse and explosion. The details are highly complex and may differ greatly from one core collapse to another [76].

Several features of the collapse and the core's subsequent evolution can produce significant gravitational radiation in the high-frequency band. We shall consider these features in turn, the most weakly radiating first.

Boiling of the Newborn Neutron Star

Even if the collapse is spherical, so it cannot radiate any gravitational waves at all, it should produce a convectively unstable neutron star that "boils" vigorously (and nonspherically) for the first $\sim$ 0.1 second of its

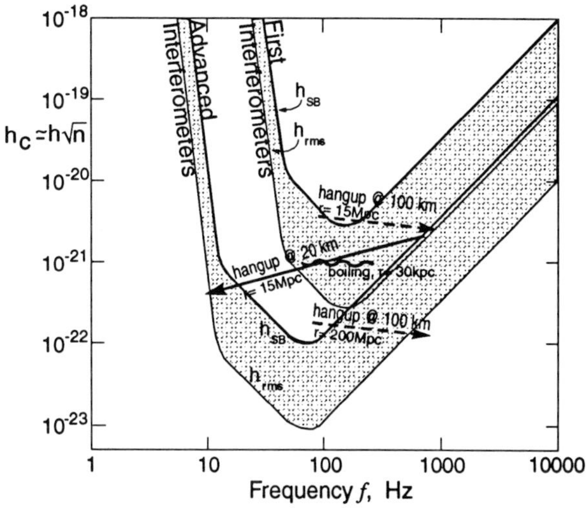

Figure 12: Characteristic amplitudes of the gravitational waves from various processes accompanying stellar core collapse and supernovae, compared with projected sensitivities of LIGO's interferometers.

life [77]. The boiling dredges up high-temperature nuclear matter ($T \sim 10^{12}$K) from the neutron star's central regions, bringing it to the surface (to the "neutrino-sphere"), where it cools by neutrino emission before being swept back downward and reheated. Burrows estimates [78, 79] that the boiling should generate $n \sim 10$ cycles of gravitational waves with frequency $f \sim 100$Hz and amplitude $h \sim 3 \times 10^{-22}(30\text{kpc}/r)$ (where r is the distance to the source), corresponding to a characteristic amplitude $h_c \simeq h\sqrt{n} \sim 10^{-21}(30\text{kpc}/r)$; cf. Figure 12. LIGO/VIRGO will be able to detect such waves only in the local group of galaxies, where the supernova rate is probably no larger than $\sim$ 1 each 10 years. However, neutrino detectors have a similar range, and there could be a high scientific payoff from correlated observations of the gravitational waves emitted by the boiling's mass motions and neutrinos emitted from the boiling neutrino-sphere.

Axisymmetric Collapse, Bounce, and Oscillations

Rotation will centrifugally flatten the collapsing core, enabling it to radiate as it implodes. If the core's angular momentum is small enough that centrifugal forces do not halt or strongly slow the collapse before it reaches nuclear densities, then the core's collapse, bounce, and subsequent oscillations are likely to be axially symmetric. Numerical simulations [80, 81] show that in this case the waves from collapse, bounce, and oscillation will be quite weak: the total energy radiated as gravitational waves is not likely to exceed $\sim 10^{-7}$ solar masses (about 1 part in a million of the collapse energy) and might often be much less than this; and correspondingly, the waves' characteristic amplitude will be $h_c \lesssim 3 \times 10^{-21}(30\text{kpc}/r)$. These collapse-and-bounce waves will come off at frequencies $\sim$ 200 Hz to $\sim$ 1000 Hz, and will precede the boiling waves by a fraction of a second. Like the boiling waves, they probably cannot be seen by LIGO/VIRGO beyond the local group of galaxies and thus will be a very rare occurrence.

Rotation-Induced Bars and Break-Up

If the core's rotation is large enough to strongly flatten the core before or as it reaches nuclear density, then a dynamical and/or secular instability is likely to break the core's axisymmetry. The core will be transformed into a bar-like configuration that spins end-over-end like an American football, and that might even break up into two or more massive pieces. In this case, the radiation from the spinning bar or orbiting pieces *could* be almost as strong as that from a coalescing neutron-star binary, and thus could be seen by the LIGO/VIRGO first interferometers out to the distance of the Virgo cluster (where the supernova rate is several per year) and by advanced interferometers out to several hundred Mpc (supernova rate $\sim 10^4$ per year); cf. Figure 12. It is far from clear what fraction of collapsing cores will have enough angular momentum to break their axisymmetry, and what fraction of those will actually radiate at this high rate; but even if only $\sim 1/1000$ or $1/10^4$ do so, this could ultimately be a very interesting source for LIGO/VIRGO.

Several specific scenarios for such non-axisymmetry have been identified:

Centrifugal hangup at $\sim$ 100km radius: If the pre-collapse core is rapidly spinning (e.g., if it is a white dwarf that has been spun up by accretion from a companion), then the collapse may produce a highly flattened, centrifugally supported disk with most of its mass at radii $R \sim 100$km, which then (via instability) may transform itself into a bar or may bifurcate. The bar or bifurcated lumps will radiate gravitational waves at twice their rotation frequency, $f \sim 100$Hz — the optimal frequency for LIGO/VIRGO interferometers. To shrink on down to $\sim$ 10km size, this configuration must shed most of its angular momentum. *If* a substantial fraction of the angular momentum goes into gravitational waves, then independently of the strength of the bar, the waves will be nearly as strong as those from a coalescing binary. The reason is this: The waves' amplitude h is proportional to the bar's ellipticity e, the number of cycles n of wave emission is proportional to $1/e^2$, and the characteristic amplitude $h_c = h\sqrt{n}$ is thus independent of the ellipticity and is about the same whether the configuration is a bar or is two lumps [37]. The resulting waves will thus have h_c roughly half as large, at $f \sim 100$Hz, as the h_c from a NS/NS binary (half as large because each lump might be half as massive as a NS), and the waves will chirp upward in frequency in a manner similar to those from a binary.

It is rather likely, however, that most of excess angular momentum does *not* go into gravitational waves, but instead goes largely into hydrodynamic waves as the bar or lumps, acting like a propeller, stir up the surrounding stellar mantle. In this case, the radiation will be correspondingly weaker.

Centrifugal hangup at $\sim$ 20km radius: Lai and Shapiro [82] have explored the case of centrifugal hangup at radii not much larger than the final neutron star, say $R \sim 20$km. Using compressible ellipsoidal models, they have deduced that, after a brief period of dynamical bar-mode instability with wave emission at $f \sim 1000$Hz (explored by Houser, Centrella, and Smith [83]), the star switches to a secular instability in which the bar's angular velocity gradually slows while the material of which it is made retains its high rotation speed and circulates through the slowing bar. The slowing bar emits waves that sweep *downward* in frequency through the LIGO/VIRGO optimal band $f \sim 100$Hz, toward $\sim$ 10Hz. The characteristic amplitude (Fig. 12) is only modestly smaller than for the upward-sweeping waves from hangup at $R \sim 100$km, and thus such waves should be detectable near the Virgo Cluster by the first LIGO/VIRGO interferometers, and at distances of a few 100Mpc by advanced interferometers.

Successive fragmentations of an accreting, newborn neutron star: Bonnell and Pringle [84] have focused on the evolution of the rapidly spinning, newborn neutron star as it quickly accretes more and more mass from the pre-supernova star's inner mantle. If the accreting material carries high angular momentum, it may trigger a renewed bar formation, lump formation, wave emission, and coalescence, followed by more accretion, bar and lump formation, wave emission, and coalescence. Bonnell and Pringle speculate that hydrodynamics, not wave emission, will drive this evolution, but that the total energy going into gravitational waves might be as large as $\sim 10^{-3}M_\odot$. This corresponds to $h_c \sim 10^{-21}(10\text{Mpc}/r)$.

6.2 *Spinning Neutron Stars; Pulsars*

As the neutron star settles down into its final state, its crust begins to solidify (crystalize). The solid crust will assume nearly the oblate axisymmetric shape that centrifugal forces are trying to maintain, with poloidal ellipticity $\epsilon_p \propto$(angular velocity of rotation)^2. However, the principal axis of the star's moment of inertia tensor may deviate from its spin axis by some small "wobble angle" θ_w, and the star may deviate slightly from axisymmetry about its principal axis; i.e., it may have a slight ellipticity $\epsilon_e \ll \epsilon_p$ in its equatorial plane.

As this slightly imperfect crust spins, it will radiate gravitational waves [85]: ϵ_e radiates at twice the rotation frequency, $f = 2f_{\rm rot}$ with $h \propto \epsilon_e$, and the wobble angle couples to ϵ_p to produce waves at $f = f_{\rm rot} + f_{\rm prec}$ (the precessional sideband of the rotation frequency) with amplitude $h \propto \theta_w \epsilon_p$. For typical neutron-star masses and moments of inertia, the wave amplitudes are

$$h \sim 6 \times 10^{-25} \left(\frac{f_{\rm rot}}{500{\rm Hz}}\right)^2 \left(\frac{1{\rm kpc}}{r}\right) \left(\frac{\epsilon_e \text{ or } \theta_w\epsilon_p}{10^{-6}}\right) . \quad (5)$$

The neutron star gradually spins down, due in part to gravitational-wave emission but perhaps more strongly due to electromagnetic torques associated with its spinning magnetic field and pulsar emission. This spin-down reduces the strength of centrifugal forces, and thereby causes the star's poloidal ellipticity ϵ_p to decrease, with an accompanying breakage and resolidification of its crust's crystal structure (a "starquake") [86]. In each starquake, θ_w, ϵ_e, and ϵ_p will all change suddenly, thereby changing the amplitudes and frequencies of the star's two gravitational "spectral lines" $f = 2f_{\rm rot}$ and $f = f_{\rm rot}+f_{\rm prec}$. After each quake, there should be a healing period in which the star's fluid core and solid crust, now rotating at different speeds, gradually regain synchronism. By monitoring the amplitudes, frequencies, and phases of the two gravitational-wave spectral lines, and by comparing with timing of the electromagnetic pulsar emission, one might learn much about the physics of the neutron-star interior.

How large will the quantities ϵ_e and $\theta_w\epsilon_p$ be? Rough estimates of the crustal shear moduli and breaking strengths suggest an upper limit in the range $\epsilon_{\rm max} \sim 10^{-4}$ to 10^{-6}, and it might be that typical values are far below this. We are extremely ignorant, and correspondingly there is much to be learned from searches for gravitational waves from spinning neutron stars.

One can estimate the sensitivity of LIGO/VIRGO (or any other broad-band detector) to the periodic waves from such a source by multiplying the waves' amplitude h by the square root of the number of cycles over which one might integrate to find the signal, $n = f\hat\tau$ where $\hat\tau$ is the integration time. The resulting effective signal strength, $h\sqrt{n}$, is larger than h by

$$\sqrt{n} = \sqrt{f\hat\tau} = 10^5 \left(\frac{f}{1000{\rm Hz}}\right)^{1/2} \left(\frac{\hat\tau}{4{\rm months}}\right)^{1/2} . \quad (6)$$

This $h\sqrt{n}$ should be compared (i) to the detector's rms broad-band noise level for sources in a random direction, $\sqrt{5}h_{\rm rms}$, to deduce a signal-to-noise ratio, or (ii) to $h_{\rm SB}$ to deduce a sensitivity for high-confidence detection when one does not know the waves' frequency in advance [6]. Such a comparison suggests that the first interferometers in LIGO/VIRGO might possibly see waves from nearby spinning neutron stars, but the odds of success are very unclear.

The deepest searches for these nearly periodic waves will be performed by narrow-band detectors, whose sensitivities are enhanced near some chosen frequency at the price of sensitivity loss elsewhere—e.g., dual-recycled interferometers [65] or resonant-mass antennas (Section 4). With "advanced-detector technology," dual-recycled interferometers might be able to detect with confidence all spinning neutron stars that have [6]

$$(\epsilon_e \text{ or } \theta_w\epsilon_p) \gtrsim 3 \times 10^{-10} \left(\frac{500{\rm Hz}}{f_{\rm rot}}\right)^2 \left(\frac{r}{1000{\rm pc}}\right)^2 . \quad (7)$$

There may well be a large number of such neutron stars in our galaxy; but it is also conceivable that there are none. We are extremely ignorant.

Some cause for optimism arises from several physical mechanisms that might generate radiating ellipticities large compared to 3×10^{-10}:

- It may be that, inside the superconducting cores of many neutron stars, there are trapped magnetic fields with mean strength $B_{\rm core} \sim 10^{13}$G or even 10^{15}G. Because such a field is actually concentrated in flux tubes with $B = B_{\rm crit} \sim 6 \times 10^{14}$G surrounded by field-free superconductor, its mean pressure is $p_B = B_{\rm core}B_{\rm crit}/8\pi$. This pressure could produce a radiating ellipticity $\epsilon_e \sim \theta_w\epsilon_p \sim p_B/p \sim 10^{-8}B_{\rm core}/10^{13}$G (where p is the core's material pressure).

- Accretion onto a spinning neutron star can drive precession (keeping θ_w substantially nonzero), and thereby might produce measurably strong waves [87].

- If a neutron star is born rotating very rapidly, then it may experience a gravitational-radiation-reaction-driven instability. In this "CFS" (Chandrasekhar, [88] Friedman, Schutz [89]) instability, density waves propagate around the star in the opposite direction to its rotation, but are dragged forward by the rotation. These density waves produce gravitational waves that carry positive energy as seen by observers far from the star, but negative energy from the star's viewpoint; and because the star thinks it is losing negative energy, its density waves get amplified. This intriguing mechanism is similar to that by which spiral density waves are produced in galaxies. Although the CFS instability was once thought ubiquitous for spinning stars [89, 90], we now know that neutron-star viscosity will kill it, stabilizing the star and turning off the waves, when the star's temperature is above some limit $\sim 10^{10}$K [91] and below some limit $\sim 10^9$K [92]; and correspondingly, the instability should operate only during the first few years of a neutron star's life, when $10^9{\rm K} \lesssim T \lesssim 10^{10}$K.

6.3 *Stochastic Background*

There should be a stochastic background of gravitational waves in the high-frequency band produced by processes in the early universe. Because this background will extend over all gravitational-wave frequencies, not just high frequencies, we shall delay discussing it until Section 9.

§7 LISA: The Laser Interferometer Space Antenna

Turn, now, from the high-frequency band, 1—10^4 Hz, to the low-frequency band, 10^{-4}—1 Hz. At present, the most sensitive gravitational-wave searches at low frequencies are those carried out by researchers at NASA's Jet Propulsion Laboratory, using microwave-frequency Doppler tracking of interplanetary spacecraft. These searches are done at rather low cost, piggy-back on missions designed for other purposes. Although they have a possibility of success, the odds are against them. Their best past sensitivities to bursts, for example, have been $h_{SB} \sim 10^{-14}$, and prospects are good for reaching $\sim 10^{-15}$—10^{-16} in the next 5 to 10 years. However, the strongest low-frequency bursts arriving several times per year might be no larger than $\sim 10^{-18}$; and the domain of an assured plethora of signals is $h_{SB} \sim 10^{-19}$—10^{-20}.

To reach into this assured-detection domain will almost certainly require switching from microwave-frequency tracking of spacecraft (with its large noise due to fluctuating dispersion in the troposphere and interplanetary plasma) to optical tracking. Such a switch is planned for the 2014 time frame or sooner, when the European Space Agency (ESA) and/or NASA is likely to fly the *Laser Interferometer Space Antenna*, LISA.

7.1 *Mission Status*

LISA is largely an outgrowth of 15 years of studies by Peter Bender and colleagues at the University of Colorado. In 1990, NASA's Ad Hoc Committee on Gravitation Physics and Astronomy selected a LISA-type gravitational-wave detector as its highest priority in the large space mission category [93]; and since then enthusiasm for LISA has continued to grow within the American gravitation community. Unfortunately, the prospects for NASA to fly such a mission did not look good in the early 1990s. By contrast, prospects in Europe looked much better, so a largely European consortium was put together in 1993, under the leadership of Karsten Danzmann (Hannover) and James Hough (Glasgow), to propose LISA to the European Space Agency. This proposal has met with considerable success; LISA might well achieve approval to fly as an ESA Cornerstone Mission around 2014 [94]. Members of the American gravitation community

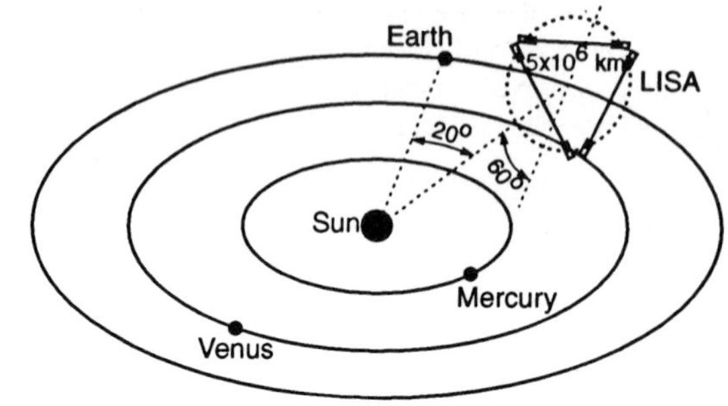

Figure 13: LISA's orbital configuration, with LISA magnified in arm length by a factor ~ 10 relative to the solar system.

and members of the LISA team hope that NASA will join together with ESA in this endeavor, and that working jointly, ESA and NASA will be able to fly LISA considerably sooner than 2014.

7.2 *Mission Configuration*

As presently conceived, LISA will consist of six compact, drag-free spacecraft (i.e. spacecraft that are shielded from buffeting by solar wind and radiation pressure, and that thus move very nearly on geodesics of spacetime). All six spacecraft would be launched simultaneously by a single Ariane rocket. They would be placed into the same heliocentric orbit as the Earth occupies, but would follow 20° behind the Earth; cf. Figure 13. The spacecraft would fly in pairs, with each pair at the vertex of an equilateral triangle that is inclined at an angle of 60° to the Earth's orbital plane. The triangle's arm length would be 5 million km (10^6 times larger than LIGO's arms!). The six spacecraft would track each other optically, using one-Watt YAG laser beams. Because of diffraction losses over the 5×10^6km arm length, it is not feasible to reflect the beams back and forth between mirrors as is done with LIGO. Instead, each spacecraft will have its own laser; and the lasers will be phase locked to each other, thereby achieving the same kind of phase-coherent out-and-back light travel as LIGO achieves with mirrors. The six-laser, six-spacecraft configuration thereby functions as three, partially independent but partially redundant, gravitational-wave interferometers.

7.3 *Noise and Sensitivity*

Figure 14 depicts the expected noise and sensitivity of LISA in the same language as we have used for LIGO (Fig. 5). The curve at the bottom of the stippled region is $h_{\rm rms}$, the rms noise, in a bandwidth equal to frequency, for waves with optimum direction and polarization. The top of the stippled region is $h_{SB} = 5\sqrt{5}h_{\rm rms}$, the sensitivity

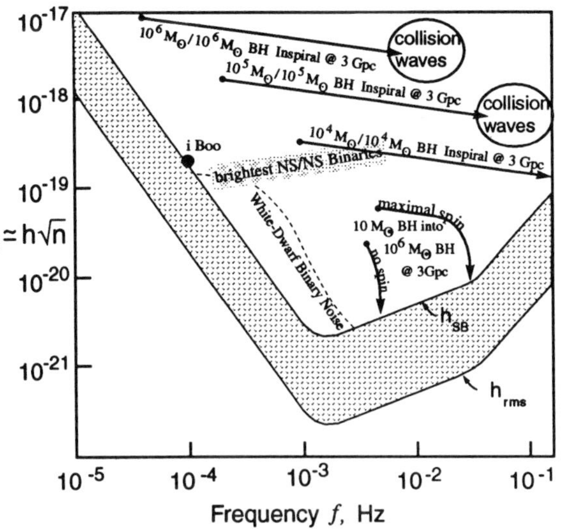

Figure 14: LISA's projected broad-band noise $h_{\rm rms}$ and sensitivity to bursts $h_{\rm SB}$, compared with the strengths of the waves from several low-frequency sources. [*Note:* When members of the LISA team plot curves analogous to this, they show the sensitivity curve (top of stippled region) in units of the amplitude of a periodic signal that can be detected with $S/N = 5$ in one year of integration; that sensitivity to periodic sources is related to the $h_{\rm SB}$ used here by $h_{\rm SP} = h_{\rm SB}/\sqrt{f \cdot 3 \times 10^7 \text{sec.}}$]

for high-confidence detection ($S/N = 5$) of a broad-band burst coming from a random direction, assuming Gaussian noise.

At frequencies $f \gtrsim 10^{-3}$Hz, LISA's noise is due to photon counting statistics (shot noise). The noise curve steepens at $f \sim 3 \times 10^{-2}$Hz because at larger f than that, the waves' period is shorter than the round-trip light travel time in one of LISA's arms. Below 10^{-3}Hz, the noise is due to buffeting-induced random motions of the spacecraft that are not being properly removed by the drag-compensation system. Notice that, in terms of dimensionless amplitude, LISA's sensitivity is roughly the same as that of LIGO's first interferometers (Fig. 9), but at 100,000 times lower frequency. Since the waves' energy flux scales as f^2h^2, this corresponds to 10^{10} better energy sensitivity than LIGO.

7.4 Observational Strategy

LISA can detect and study, simultaneously, a wide variety of different sources scattered over all directions on the sky. The key to distinguishing the different sources is the different time evolution of their waveforms. The key to determining each source's direction, and confirming that it is real and not just noise, is the manner in which its waves' amplitude and frequency are modulated by LISA's complicated orbital motion—a motion in which the interferometer triangle rotates around its center once per year, and the interferometer plane precesses around the normal to the Earth's orbit once per year. Most sources will be observed for a year or longer, thereby making full use of these modulations.

§8 Low-Frequency Gravitational-Wave Sources

8.1 Waves from Binary Stars

LISA has a large class of guaranteed sources: short-period binary stars in our own galaxy. A specific example is the classic binary 44 i Boo (HD133640), a $1.35M_\odot/0.68M_\odot$ system just 12 parsecs from Earth, whose wave frequency f and characteristic amplitude $h_c = h\sqrt{n}$ are depicted in Figure 14. (Here h is the waves' actual amplitude and $n = f\hat{\tau}$ is the number of wave cycles during $\hat{\tau}$ =1 year of signal integration). Since 44 i Boo lies right on the $h_{\rm SB}$ curve, its signal to noise ratio in one year of integration should be $S/N = 5$.

To have an especially short period, a binary must be made of especially compact bodies—white dwarfs (WD), neutron stars (NS), and/or black holes (BH). WD/WD binaries are thought to be so numerous that they may produce a stochastic background of gravitational waves, at the level shown in Figure 14, that will hide some other interesting waves from view [95]. Since WD/WD binaries are very dim optically, their actual numbers are not known for sure; Figure 14 might be an overestimate.

Assuming a NS/NS coalescence rate of 1 each 10^5 years in our galaxy [27, 28], the shortest period NS/NS binary should have a remaining life of about 5×10^4 years, corresponding to a gravitational-wave frequency today of $f \simeq 5 \times 10^{-3}$Hz, an amplitude (at about 10kpc distance) $h \simeq 4 \times 10^{-22}$, and a characteristic amplitude (with one year of integration time) $h_c \simeq 2 \times 10^{-19}$. This is depicted in Figure 14 at the right edge of the region marked "brightest NS/NS binaries". These brightest NS/NS binaries can be studied by LISA with the impressive signal to noise ratios $S/N \sim 50$ to 500.

8.2 Waves from the Coalescence of Massive Black Holes in Distant Galaxies

LISA would be a powerful instrument for studying massive black holes in distant galaxies. Figure 14 shows, as examples, the waves from several massive black hole binaries at 3Gpc distance from Earth (a cosmological redshift of unity). The waves sweep upward in frequency (rightward in the diagram) as the holes spiral together. The black dots show the waves' frequency one year before the holes' final collision and coalescence, and the arrowed lines show the sweep of frequency and characteristic amplitude $h_c = h\sqrt{n}$ during that last year. For simplicity, the figure is restricted to binaries with equal-mass black holes: $10^4 M_\odot/10^4 M_\odot$, $10^5 M_\odot/10^5 M_\odot$, and $10^6 M_\odot/10^6 M_\odot$.

By extrapolation from these three examples, we see that LISA can study much of the last year of inspiral, and the waves from the final collision and coalescence, whenever the holes' masses are in the range $3 \times 10^4 M_\odot \lesssim M \lesssim 10^8 M_\odot$. Moreover, LISA can study the final coalescences with remarkable signal to noise ratios: $S/N \gtrsim 1000$. Since these are much larger S/N's than LIGO/VIRGO is likely to achieve, we can expect LISA to refine the experimental understanding of black-hole physics, and of highly nonlinear vibrations of warped spacetime, which LIGO/VIRGO initiates—*provided* the rate of massive black-hole coalescences is of order one per year in the Universe or higher. The rate might well be that high, but it also might be much lower.

By extrapolating Figure 14 to lower BH/BH masses, we see that LISA can observe the last few years of inspiral, but not the final collisions, of binary black holes in the range $100 M_\odot \lesssim M \lesssim 10^4 M_\odot$, out to cosmological distances.

Extrapolating the BH/BH curves to lower frequencies using the formula (time to final coalescence) $\propto f^{-8/3}$, we see that equal-mass BH/BH binaries enter LISA's frequency band roughly 1000 years before their final coalescences, more or less independently of their masses, for the range $100 M_\odot \lesssim M \lesssim 10^6 M_\odot$. Thus, if the coalescence rate were to turn out to be one per year, LISA would see roughly 1000 additional massive binaries that are slowly spiraling inward, with inspiral rates df/dt readily measurable. From the inspiral rates, the amplitudes of the two polarizations, and the waves' harmonic content, LISA can determine each such binary's luminosity distance, redshifted chirp mass $(1+z)M_c$, orbital inclination, and eccentricity; and from the waves' modulation by LISA's orbital motion, LISA can learn the direction to the binary with an accuracy of order one degree.

8.3 *Waves from Compact Bodies Spiraling into Massive Black Holes in Distant Galaxies*

When a compact body with mass μ spirals into a much more massive black hole with mass M, the body's orbital energy E at fixed frequency f (and correspondingly at fixed orbital radius a) scales as $E \propto \mu$, the gravitational-wave luminosity $\dot{E}$ scales as $\dot{E} \propto \mu^2$, and the time to final coalescence thus scales as $t \sim E/\dot{E} \propto 1/\mu$. This means that the smaller is μ/M, the more orbits are spent in the hole's strong-gravity region, $a \lesssim 10GM/c^2$, and thus the more detailed and accurate will be the map of the hole's spacetime geometry, which is encoded in the emitted waves.

For holes observed by LIGO/VIRGO, the most extreme mass ratio that we can hope for is $\mu/M \sim 1M_\odot/300M_\odot$, since for $M > 300M_\odot$ the inspiral waves are pushed to frequencies below the LIGO/VIRGO band. This limit on μ/M seriously constrains the accuracy with which LIGO/VIRGO can hope to map out the spacetime geometries of black holes and test the black-hole no-hair theorem [62] (end of Section 5.3). By contrast, LISA can observe the final inspiral waves from objects of any mass $M \gtrsim 0.5M_\odot$ spiraling into holes of mass $3 \times 10^5 M_\odot \lesssim M \lesssim 3 \times 10^7 M_\odot$.

Figure 14 shows the example of a $10M_\odot$ black hole spiraling into a $10^6 M_\odot$ hole at 3Gpc distance. The inspiral orbit and waves are strongly influenced by the hole's spin. Two cases are shown [96]: an inspiraling circular orbit around a non-spinning hole, and a prograde, circular, equatorial orbit around a maximally spinning hole. In each case the dot at the upper left end of the arrowed curve is the frequency and characteristic amplitude one year before the final coalescence. In the nonspinning case, the small hole spends its last year spiraling inward from $r \simeq 7.4GM/c^2$ (3.7 Schwarzschild radii) to its last stable circular orbit at $r = 6GM/c^2$ (3 Schwarzschild radii). In the maximal spin case, the last year is spent traveling from $r = 6GM/c^2$ (3 Schwarzschild radii) to the last stable orbit at $r = GM/c^2$ (half a Schwarzschild radius). The $\sim 10^5$ cycles of waves during this last year should carry, encoded in themselves, rather accurate values for the massive hole's lowest few multipole moments [97]. If the measured moments satisfy the "no-hair" theorem (i.e., if they are all determined uniquely by the measured mass and spin in the manner of the Kerr metric), then we can be sure the central body is a black hole. If they violate the no-hair theorem, then (assuming general relativity is correct), either the central body was not a black hole, or an accretion disk or other material was perturbing its orbit [98]. From the evolution of the waves one can hope to determine which is the case, and to explore the properties of the central body and its environment [62].

Models of galactic nuclei, where massive holes reside, suggest that inspiraling stars and small holes typically will be in rather eccentric orbits [99]. This is because they get injected into such orbits via gravitational deflections off other stars, and by the time gravitational radiation reaction becomes the dominant orbital driving force, there is not enough inspiral left to fully circularize their orbits. Such orbital eccentricity will complicate the waveforms and complicate the extraction of information from them. Efforts to understand the emitted waveforms are just now getting underway.

The event rates for inspiral into massive black holes are not at all well understood. However, since a significant fraction of all galactic nuclei are thought to contain massive holes, and since white dwarfs and neutron stars, as well as small black holes, can withstand tidal disruption as they plunge toward the massive hole's horizon, and since LISA can see inspiraling bodies as small as $\sim 0.5 M_\odot$ out to 3Gpc distance, the event rate is likely to be interestingly large.

§9 The Stochastic Gravitational-Wave Background

Processes in the early universe should have produced a stochastic background of gravitational waves that extends through the entire frequency range from extremely low frequencies $f \sim 10^{-18}$ Hz to the high-frequency band $f \sim$ 1—10^4Hz and beyond.

9.1 Primordial Gravitational Waves

The most interesting background would be that produced in the big bang itself. Zel'dovich and Novikov have estimated [100] that the optical thickness of primordial matter to gravitational waves has been small compared to unity at all times since the Planck era, when space and time came into being, and that therefore primordial gravitational waves (by contrast with electromagnetic) should not have been thermalized by interactions with matter. On the other hand (as Grishchuk has shown [101]), whatever might have been the state of the graviton field when it emerged from the big bang's Planck era, it should have interacted with the subsequent, early-time expansion of the universe to produce, via parametric amplification, a rich spectrum of stochastic waves today. The details of that spectrum depend on what emerged from the Planck era and on the evolution $a(t)$ of the universal expansion factor at early times.

The gravitational-wave spectrum is generally described by the quantity $\Omega_g(f)$ = (energy density in a bandwidth equal to frequency f)/(energy density required to close the universe); cf. Section 2.3. The observed quadrupolar anisotropy of the cosmic microwave radiation places a limit $\Omega_g \lesssim 10^{-9}$ at $f \sim 10^{-18}$Hz (Section 2.4). It is fashionable to extrapolate this limit to higher frequencies by ***assuming*** that the graviton field emerged from the Planck era in its vacuum state, and ***assuming*** that the universal expansion $a(t)$ was that of an inflationary era $a \propto e^{Ht}$ for some constant H, followed by a radiation-dominated Friedman era $a \propto t^{1/2}$, followed by the present matter-dominated era $a \propto t^{2/3}$. This standard model produces a flat spectrum Ω_g independent of f for all waves that entered our cosmological horizon during the radiation-dominated era, which means at all frequencies from $\sim 10^{-16}$Hz up through the high-frequency band and somewhat beyond. The observational limit at 10^{-18}Hz implies that this constant value is $\Omega_g \lesssim 3 \times 10^{-14}$ [12, 102]. So weak a background cannot be detected by LIGO in the high-frequency band, nor by LISA at low frequencies, nor by pulsar timing at very low frequencies. LIGO's limiting sensitivities will correspond to $\Omega_g \sim$ (a few) $\times 10^{-7}$ at $f \sim 10^2$ Hz for the first interferometers, and $\Omega_g \sim$ (a few) $\times 10^{-10}$ for advanced interferometers [103]; LISA's sensitivity will correspond to $\Omega_g \sim$ (a few) $\times 10^{-10}$ at $f \sim 10^{-3}$ Hz; and the present pulsar timing measurements correspond to $\Omega_g \sim$ (a few) $\times 10^{-8}$ at $f \sim 4 \times 10^{-9}$Hz (Section 2.3 and Ref. [9]).

On the other hand, if the graviton field did not begin in its vacuum state, or if the equation of state in the very early Friedman era was stiffer than that of radiation, then the primordial backgrounds at high, low, and very low frequencies could be significantly stronger than $\Omega \sim 3 \times 10^{-14}$, and could be strong enough to detect.

9.2 Waves from Phase Transitions in the Early Universe

A stochastic background could also have been produced by phase transitions in the early universe [104, 105]. No known phase transition would put its waves into the high-frequency band, and even hypothetical phase transitions, optimized at high-frequencies, can be only strong enough for marginal detection by advanced LIGO interferometers. The prospects for LISA are a little better: A strongly first-order electroweak phase transition could produce low-frequency waves strong enough for LISA to detect [105].

9.3 Waves from Cosmic Strings

If cosmic strings [106] were produced in the early universe in as large numbers as some theorists have suggested [10, 11], their vibrations would produce a gravitational wave spectrum that is frequency independent, Ω_g = const, from below the very low-frequency band where pulsar timing operates, through LISA's low-frequency band, and on into and through the high-frequency band. Theory suggests [107] that such waves could be as strong as $\Omega_g \sim 10^{-7}$—a level that is already being constrained by pulsar timing observations (Section 2.3). LIGO's first interferometers will operate at this same level, and by the time LIGO's advanced interferometers and LISA reach $\Omega_g \sim$ (a few) $\times 10^{-10}$, pulsar timing might be in that same ballpark.

To summarize: There are known mechanisms that could easily produce a measurable stochastic background in the high, low, and very-low frequency bands. However, the odds of the background being that large, based on currently fashionable ideas, are not great. Despite this, a vigorous effort to detect background waves and to map their spectrum will surely be made, since the cosmological implications of their discovery could be profound.

§10 Conclusion

It is now 35 years since Joseph Weber initiated his pioneering development of gravitational-wave detectors [23] and 25 years since Forward [108] and Weiss [16] initiated work on interferometric detectors. Since then, hundreds of talented experimental physicists have struggled to improve the sensitivities of these instruments. At last, success is in sight. If the source estimates described in this review article are approximately correct, then the planned interferometers should detect the first waves in 2001 or several years thereafter, thereby opening up this rich new window onto the Universe.

§11 Acknowledgments

For insights into the rates of coalescence of compact binaries, I thank Sterl Phinney. My group's research on gravitational waves and their relevance to LIGO/VIRGO and LISA is supported in part by NSF grants AST-9417371 and PHY-9424337 and by NASA grant NAGW-4268. Portions of this review article were adapted from my Ref. [109].

§References

[1] A. Pais. *'Subtle is the Lord...' The Science and the Life of Albert Einstein.* Clarendon Press, Oxford, 1982.

[2] R. A. Hulse and J. H. Taylor. *Astrophys. J.*, 324:355, 1975.

[3] J. H. Taylor. *Rev. Mod. Phys.*, 66:711, 1994.

[4] C. M. Will. *Theory and Experiment in Gravitational Physics.* Cambridge University Press, 1994.

[5] T. Damour and J. Taylor. *Phys. Rev. D*, 45:1840, 1992.

[6] K. S. Thorne. In S. W. Hawking and W. Israel, editors, *Three Hundred Years of Gravitation*, pages 330–458. Cambridge University Press, 1987.

[7] K. S. Thorne. In N. Deruelle and T. Piran, editors, *Gravitational Radiation*, page 1. North Holland, 1983.

[8] W. J. Sullivan. *The Early Years of Radio Astronomy.* Cambridge University Press, 1984.

[9] V. M. Kaspi, J. H. Taylor, and M. F. Ryba. *Astrophys. J.*, 428:713, 1994.

[10] Ya. B. Zel'dovich. *Mon. Not. Roy. Astron. Soc.*, 192:663, 1980.

[11] A. Vilenkin. *Phys. Rev. D*, 24:2082, 1981.

[12] L. M. Krauss and M. White. *Phys. Rev. Lett.*, 69:969, 1992.

[13] R. L. Davis, H. M. Hodges, G. F. Smoot, P. J. Steinhardt, and M. S. Turner. *Phys. Rev. Lett.*, 69:1856, 1992.

[14] A. Abramovici et. al. *Science*, 256:325, 1992.

[15] R. W. P. Drever. In N. Deruelle and T. Piran, editors, *Gravitational Radiation*, page 321 and references therein. North Holland, 1983.

[16] R. Weiss. *Quarterly Progress Report of RLE, MIT*, 105:54, 1972.

[17] C. Bradaschia et. al. *Nucl. Instrum. & Methods*, A289:518, 1990.

[18] S. E. Whitcomb and the LIGO R&D science team. *LIGO Report Number 94-7*, November 1994.

[19] P. Saulson. *Phys. Rev. D*, 42:2437, 1990.

[20] V. B. Braginsky, V. P. Mitrofanov, and O. A. Okhrimenko. *Phys. Lett. A*, 175:82, 1993.

[21] A. Gillespie and F. Raab. *Phys. Lett. A*, 190:213, 1994.

[22] W. W. Johnson and S. M. Merkowitz. *Phys. Rev. Lett.*, 70:2367, 1993.

[23] J. Weber. *Phys. Rev.*, 117:306, 1960.

[24] M. Bassan. *Class. Quant. Grav.*, 11 Supplement:A39, 1994.

[25] G. Pizella. *Nucl. Phys. B*, Supplement 28A:48, 1992.

[26] J. Hough and K. Danzmann et. al. GEO600, Proposal for a 600 m Laser-Interferometric Gravitational Wave Antenna, unpublished, 1994.

[27] E. S. Phinney. *Astrophys. J.*, 380:L17, 1991.

[28] R. Narayan, T. Piran, and A. Shemi. *Astrophys. J.*, 379:L17, 1991.

[29] M. Bailes. In J. van Paradijs, E. van den Heuvel, and E. Kuulkers, editors, *Proceedings of I. A. U. Symposium 165, Compact Stars in Binaries.* Kluwer Academic Publishers, 1995. in press.

[30] A. V. Tutukov and L. R. Yungelson. *Mon. Not. Roy. Astron. Soc.*, 260:675, 1993.

[31] H. Yamaoka, T. Shigeyama, and K. Nomoto. *Astron. Astrophys.*, 267:433, 1993.

[32] V. M. Lipunov, K. A. Postnov, and M. E. Prokhorov. *Astrophys. J.*, 423:L121, 1994. and related, unpublished work.

[33] C. Kochanek. *Astrophys. J.*, 398:234, 1992.

[34] L. Bildsten and C. Cutler. *Astrophys. J.*, 400:175, 1992.

[35] G. Quinlan and S. L. Shapiro. *Astrophys. J.*, 321:199, 1987.

[36] B. F. Schutz. *Nature*, 323:310, 1986.

[37] B. F. Schutz. *Class. Quant. Grav.*, 6:1761, 1989.

[38] C. Cutler, T. A. Apostolatos, L. Bildsten, L. S. Finn, E. E. Flanagan, D. Kennefick, D. M. Markovic, A. Ori, E. Poisson, G. J. Sussman, and K. S. Thorne. *Phys. Rev. Lett.*, 70:1984, 1993.

[39] T. A. Apostolatos, C. Cutler, G. J. Sussman, and K. S. Thorne. *Phys. Rev. D*, 49:6274, 1994.

[40] C. W. Lincoln and C. M. Will. *Phys. Rev. D*, 42:1123, 1990.

[41] L. E. Kidder, C. M. Will, and A. G. Wiseman. *Phys. Rev. D*, 47:3281, 1993.

[42] C. M. Will. In M. Sasaki, editor, *Relativistic Cosmology*, page 83. Universal Academy Press, 1994.

[43] L. E. Kidder. *Phys. Rev. D*, 1995. in press.

[44] C. Cutler and E. E. Flanagan. *Phys. Rev. D*, 49:2658, 1994.

[45] L. S. Finn and D. F. Chernoff. *Phys. Rev. D*, 47:2198, 1993.

[46] E. Poisson and C. M. Will. *Phys. Rev. D*, 1995. submitted.

[47] P. Jaranowski and A. Krolak. *Phys. Rev. D*, 49:1723, 1994.

[48] L. Blanchet, T. Damour, B. R. Iyer, C. M. Will, and A. G. Wiseman. *Phys. Rev. Lett.*, 1995. submitted.

[49] E. Poisson and M. Sasaki. *Phys. Rev. D*, 1995. in press.

[50] M. Shibata, M. Sasaki, H. Tagoshi, and T. Tanaka. *Physical Review D*, 1995. in press.

[51] H. Tagoshi and T. Nakamura. *Phys. Rev. D*, 49:4016, 1994.

[52] H. Tagoshi and M. Sasaki. *Prog. Theor. Phys.*, 92:745, 1994.

[53] C. Cutler and E. E. Flanagan. Phys. Rev. D. paper in preparation.

[54] A. Ashtekar, ed. In N. Deruelle and T. Piran, editors, *Gravitational Radiation*, page 421, Amsterdam, 1983. North Holland.

[55] B. S. Sathyaprakash. *Phys. Rev. D*, 50:R7111–R7115, 1994.

[56] A. Królak, K. D. Kokkotas, and G. Schäfer. *Proceedings of the 17th Texas Symposium on Relativistic Astrophysics, Ann. N. Y. Acad. Sci*, 1995. in press.

[57] T. A. Apostolatos. *Phys. Rev. D*, 1995. in press.

[58] T. Damour and K. Nordtvedt. *Phys. Rev. D*, 48:3436, 1993.

[59] C. M. Will. *Phys. Rev. D*, 50:6058, 1994.

[60] D. Markovic. *Phys. Rev. D*, 48:4738, 1993.

[61] D. F. Chernoff and L. S. Finn. *Astrophys. J. Lett.*, 411:L5, 1993.

[62] F. D. Ryan, L. S. Finn, and K. S. Thorne. *Phys. Rev. Lett.*, 1995. in preparation.

[63] E. E. Flanagan and S. A. Hughes. *Phys. Rev. D*, 1995. in preparation.

[64] Numerical Relativity Grand Challenge Alliance, 1995. References and information on the World Wide Web, http://jean-luc.ncsa.uiuc.edu/GC.

[65] B. J. Meers. *Phys. Rev. D*, 38:2317, 1988.

[66] D. Kennefick, D. Laurence, and K. S. Thorne. Phys. Rev. D. in preparation.

[67] M. Shibata, T. Nakamura, and K. Oohara. *Prog. Theor. Phys.*, 89:809, 1993.

[68] F. A. Rasio and S. L. Shapiro. *Astrophys. J.*, 401:226, 1992.

[69] T. Nakamura. In M. Sasaki, editor, *Relativistic Cosmology*, page 155. Universal Academy Press, 1994.

[70] X. Zhuge, J. M. Centrella, and S. L. W. McMillan. *Phys. Rev. D*, 50:6247, 1994.

[71] M. B. Davies, W. Benz, T. Piran, and F. K. Thielemann. *Astrophys. J.*, 431:742, 1994.

[72] D. Christodoulou. *Phys. Rev. Lett.*, 67:1486, 1991.

[73] K. S. Thorne. *Phys. Rev. D*, 45:520–524, 1992.

[74] A. G. Wiseman and C. M. Will. *Phys. Rev. D*, 44:R2945, 1991.

[75] D. Kennefick. *Phys. Rev. D*, 50:3587, 1994.

[76] A. G. Petschek, editor. *Supernovae.* Springer Verlag, 1990.

[77] H. A. Bethe. *Rev. Mod. Phys*, 62:801, 1990.

[78] A. Burrows, J. Hayes, and B. A. Fryxell. *Astrophys. J.*, 1995. in press.

[79] A. Burrows, 1995. private communication.

[80] L. S. Finn. *Ann. N. Y. Acad. Sci.*, 631:156, 1991.

[81] R. Mönchmeyer, G. Schäfer, E. Müller, and R. E. Kates. *Astron. Astrophys.*, 256:417, 1991.

[82] D. Lai and S. L. Shapiro. *Astrophys. J.*, 442:259, 1995.

[83] J. L. Houser, J. M. Centrella, and S. C. Smith. *Phys. Rev. Lett.*, 72:1314, 1994.

[84] I. A. Bonnell and J. E. Pringle. *Mon. Not. Roy. Astron. Soc.*, 273:L12, 1995.

[85] M. Zimmermann and E. Szedenits. *Phys. Rev. D*, 20:351, 1979.

[86] S. L. Shapiro and S. A. Teukolsky. Wiley: Interscience, 1983. Section 10.10 and references cited therein.

[87] B. F. Schutz, 1995. private communication.

[88] S. Chandrasekhar. *Phys. Rev. Lett.*, 24:611, 1970.

[89] J. L. Friedman and B. F. Schutz. *Astrophys. J.*, 222:281, 1978.

[90] R. V. Wagoner. *Astrophys. J.*, 278:345, 1984.

[91] L. Lindblom. *Astrophys. J.*, 438:265, 1995.

[92] L. Lindblom and G. Mendell. *Astrophys. J.*, 444:804, 1995.

[93] J. Armstrong, P. Bender, D. Eardley, C.W.F. Everitt, R. Hellings, K. Nordtvedt, H.J. Paik, R. Reasenbert, I. Shapiro (chair), J. Taylor, K. Thorne, R. Weiss, and C. Will. *Report of the Ad Hoc Committee on Gravitation Physics and Astronomy.* Astrophysics Division, NASA Headquarters (Washington DC), 1990.

[94] P. Bender, A. Brillet, I. Ciufolini, K. Danzmann, R. Hellings, J. Hough, A. Lobo, M. Sandford, B. Schutz, and P. Touboul. *LISA, Laser interferometer space antenna for gravitational wave measurements: ESA Assessment Study Report.* R. Reinhard, ESTEC, 1994.

[95] D. Hils, P. Bender, and R. F. Webbink. *Astrophys. J.*, 360:75, 1990.

[96] L. S. Finn and K. S. Thorne. *Phys. Rev. D*, 1995. in preparation.

[97] F. D. Ryan. *Phys. Rev. D*, 1995. in preparation.

[98] D. Molteni, G. Gerardi, and S. K. Chakrabarti. *Astrophys. J.*, 436:249, 1994.

[99] D. Hils and P. Bender, 1995. preprint.

[100] Ya. B. Zel'dovich and I. D. Novikov. *Relativistic Astrophysics Volume 2, The Structure and Evolution of the Universe.* U. Chicago Press, 1983. pages 160–163.

[101] L. P. Grishchuk. *Soviet Physics–JETP*, 40:409, 1974.

[102] L. P. Grishchuk. *Class. Quant. Grav.*, 10:2449–2477, 1993.

[103] E. E. Flanagan. *Phys. Rev. D*, 48:2389, 1993.

[104] Ed Witten. *Phys. Rev. D*, 30:272, 1984.

[105] A. Kosowsky and M. S. Turner. *Phys. Rev. D*, 49:2837, 1994.

[106] A. Vilenkin. In S. W. Hawking and W. Israel, editors, *Three Hundred Years of Gravitation*, pages 499–523. Cambridge University Press, 1987.

[107] T. Vachaspati and A. Vilenkin. *Phys. Rev. D*, 31:3052, 1985.

[108] G. E. Moss, L. R. Miller, and R. L. Forward. *Applied Optics*, 10:2495, 1971.

[109] K. S. Thorne. In M. Sasaki, editor, *Relativistic Cosmology*, page 67. Universal Academy Press, 1994.

Black Hole Astrophysics

I. Novikov[1,2*,3,4]

[1] *Astro Space Center of P.N.Lebedev Physical Institute, Profsoyuznaya 84/32, Moscow, 117810, Russia*

[2] *University Observatory, Øste Voldgade 3, DK-1350 Copenhagen K, Denmark*

[3] *NORDITA, Blegdamsvej 17, DK-2100 Copenhagen, Ø Denmark*

[4] *Theoretical Astrophysics Center, Blegdamsvej 17, DK-2100, Copenhagen Ø, Denmark*

* Permanent address

§1 Introduction

The idea that in the universe might be stars, that the world cannot see (now we call them black holes) has been with physicists and astronomers for over 200 years. Nowadays the problem of black holes is at the heart of all astrophysics and physics.

Modern astrophysics considers three types of black holes in the Universe: a) Black holes of stellar masses, which were born when massive stars died, b) Supermassive black holes with masses up to $10^9 M_\odot$ and more ($M_\odot = 2 \times 10^{33}$g is the mass of the Sun) at centers of galaxies, c) Primordial black holes which might appear from the large-scale inhomogeneities at the very beginning of expansion of the Universe. Their masses can be arbitrary, but primordial black holes with $M \leq 10^{15}$g will have radiated away their mass by the Hawking quantum process in a time $t \leq 10^{10}$years (the age of the Universe). Only primordial black holes with mass $M > 10^{15}$g could exist in the contemporary Universe.

In this paper I shall focus on the stellar black holes.

§2 How heavy should a star be to turn into a black hole?

We start from the discussion of the origin of the stellar black holes. "When all the thermo-nuclear sources of energy are exhausted a sufficiently heavy star will collapse" - it is the first phrase of the abstract of a wonderful paper by Oppenheimer and Snyder [1]. Every statement of this paper accords with ideas that remains popular today (including terminology). The authors conclude in the abstract: "... an external observer sees the star shrinking to its gravitational radius."

This was the modern prediction of the formation of the black holes when massive stars die. How heavy should a star be to turn into a black hole? The answer is not simple. A star light enough ends up as a white dwarf or a neutron star. Both these types celestial bodies have the upper limits of their masses. For the white dwarfs it is the Chandrasekhar limit which is about 1.2 - 1.4 $M_\odot$ (see Shapiro and Teukolsky [2], Kippenhahn and Weigert [3]). For the neutron stars it is the Oppenheimer-Volkoff limit (Oppenheimer and Volkoff [4]). The exact value of this limit depends on the equation of state at matter density larger than the density of nuclear matter $\rho_0 = 2.8 \times 10^{14}$g cm^3 . The modern theory gives for this maximum mass $M_{max,OV} \approx 2M_\odot$ (see Baym and Pethick [5], Lamb [6]). Some authors discussed the possibility of existing of so called "quark stars" and "hadronic stars" [see Alcook, Farhi and Olinto[7], Bahcall, Lynn, and Selipsky [8]. At present, there is no evidence for such stars.

Rotation can increase $M_{max,OV}$ only slightly (up to 25%; Friedman and Ipser [9]). Thus one can believe that the upper mass limit for neutron stars should be not greater than $M_0 \approx 3M_\odot$ (Lamb [6], Cowley [10], McClintock [11]). If a star in the very end of its evolution has mass greater than M_0 it must turn into a black hole. It does not mean that black hole progenitors are all normal stars (on "main sequence" of the Herzsprung-Russell diagram, see for example Bisnovaty-Kogan [12], Kippenhahn and Weigert [3] with masses $M > M_0$. The point is that the final stages of evolution of massive stars are poorly understood. Steady mass loss, catastrophic mass ejections and even disruption at supernovae explosions are probably possible (see for example Kippenhahn and Weigert [3]). These processes can essentially reduce the initial stellar mass at the end of its evolution. Thus the initial mass of black hole progenitors could be essentially greater than M_0.

There are different estimates for the minimal masses of progenitor stars M_* that form black holes. For example $M_* \approx 10M_\odot$ (Shapiro, Teukolsky [2]) and $M_* \approx 30M_\odot$ (Lipunov [13]) or even $\geq 40M_\odot$ (van den Heuvel and Habets [14], Schild and Maeder [15]). Note that the evolution of stars in close binary systems differs from the evolution of sole stars because of mass transfer from one star to another (see Novikov [16], Masevich and Tutukov [17]), and conclusions about masses of black holes progenitors in this case could be essentially different (see the end of section 4). In our further discussion in section 3 we shall focus an attention on the fate of single stars.

§3 The total number of stellar black holes

One can estimate how many black holes were created by stellar collapse in our Galaxy during its existence. The spectrum of stellar masses at present for stars on the "main sequence" in the solar neighborhood is roughly known from the observational data together with theory of stellar structure.

The lifetime of massive stars is less than T_0 - the age of the Galaxy ($T_0 \approx 10^{10}$ years). We assume the constant birthrate and the same constant deathrate for massive stars during the lifetime of the Galaxy. Now if we suppose that all stars with $M > M_*$ on "main sequence" (progenitors) must turn into black holes, we can calculate the birthrate of black holes in the solar neighborhood. If this rate is the same everywhere we can estimate total number of black holes in the Galaxy and total mass of all stellar black holes in it.

Many works were devoted to such estimates. But because of great uncertainties the progress here is very slow, starting from pioneering works by Zwicky [18], Schwarzshild [19], Hoyle and Fowler [20], Novikov and Ozernoy [21], Hoyle, Fowler, Burbidge and Burbidge [22]. A review of more recent estimates is given by Shapiro and Teukolsky [2]. Because of the large uncertainty in the lower mass limit M_* for the stars collapsing to black holes we give following expression for the rate of black hole formation in the Galaxy (see Novikov [16])

$$\frac{dN}{dt} \approx 0.1 \left(\frac{M_*}{3M_\odot}\right)^{-1.4} \text{year}^{-1}. \tag{1}$$

At present, we can probably repeat the conclusion of Novikov and Thorne [23]: "For stars with masses greater than ~ 12 to $30 M_\odot$ the [supernovae] explosion may produce a black hole. If this tentative conclusion is correct, then no more than ~ 1 per cent of the [visible] mass of the Galaxy should be in the form of black holes today; and new black holes should be created at a rate not greater than ~ 0.01 per year." Thus, the total number of stellar black holes in the Galaxy can be of the order of $N \approx 10^8$ or less. What is their observational appearance? Investigation of this problem started at the beginning of the 60's.

§4 Physics of disk accretion onto black holes

The most important physical process which leads to observable manifestations of a black hole's presence is gas accretion (Zeldovich [24], Salpeter [25]). For the purpose of finding out and investigating black holes, two specific cases of accretion are of a particular importance: accretion in binary systems and accretion onto supermassive black holes which probably reside at the centers of galaxies.

In both cases the accreting gas has angular momentum $\tilde{L} \gg r_g c$. As a result the gas elements go into Keplerian orbits around the hole, forming a disk or a torus around it. The inner edge of the disk is in the region of the last stable circular orbit. Close to a rotating black hole the Lense-Thiring precession drags the gas around it in the equatorial plane (see Bardeen and Petterson [26]). Viscosity plays a crucial role in the accretion. It removes angular momentum from each gas element, permitting it to gradually spiral inward toward the black hole. As the gas reaches the inner edge of the disk it then spirals down the hole practically in the state of free fall. At the same time the viscosity heats the gas, causing it to radiate. The sources of viscosity are probably turbulence in the gas disk and random magnetic fields. Unfortunately, we are no nearer to having a good physical understanding of the effective viscosity.

Large-scale magnetic fields can play an important role in physics of accretion also.

The properties of the accreting disk are determined by the rate of gas accretion. It is worth noting the following. In the problem of the spherical accretion that was discussed in the section 11.3, the rate of accretion was the result of the solution of the equations governing the gas flow. In the case of disk accretion the rate is an independent (external) parameter and is determined by the evolution of a binary system or the conditions in a galactic nucleus. As a result the rate of accretion can be much higher than in the spherical case, and observational manifestations are much more prominent.

An important measure of any accretion luminosity of a black hole is provided by the Eddington critical luminosity

$$\begin{aligned} L_E &= 4\pi G M_h \mu m_p c/\sigma_T \\ &= (1.3 \times 10^{38} \text{ergs/sec})\, \mu \left(\frac{M_h}{M_\odot}\right). \end{aligned} \tag{2}$$

Here μ is the molecular weight per electron, m_p is the rest mass of the proton, and σ_T is the Thomson cross section. It is the luminosity at which the radiation pressure just balances the gravitational force of mass M_h, for a fully ionized plasma. Eddington [27] derived this value for the discussion of stellar equilibrium, Zeldovich and

Novikov [28] introduced it for the discussion of black holes in quasars and accretion.

If $L \ll L_E$ gravity dominates and photon pressure is not important.

The useful measure of the accretion rate is a so called "critical accretion rate":

$$\dot{M}_E = L_E c^{-2} \tag{3}$$

where L_E is given by (2). We shall use the dimensionless expression $\dot{m} \equiv \dot{M}/\dot{M}_E$. Excluding the innermost parts of the disk relativistic effects are not important in physical processes of accretion.

Lynden-Bell [29] was the first to propose the model of gaseous disk accretion onto black hole. Shakura [30], Pringle and Rees [31], Shakura and Sunyaev [32] built Newtonian models of accretion disk. At last, Page and Thorne [33], Novikov and Thorne [23] gave the theory of disk accretion in the framework of General Relativity.

The first models were rather simple. They focused on the case of moderate rate of accretion $\dot{m} < 1$. Subsequently the theory with $\dot{m} \sim 1$ and $\dot{m} \gg 1$ were developed. They take into account various complex processes in radiative plasma and various types of instability. The review is given by Lamb [6]. However these processes have no direct relation (or at least almost no relation) with specific properties of space-time in the vicinity of a black hole. Thus, we will not describe them here in detail. We only give some theoretical estimates of observational properties of accreting disks around black holes in binary systems and in galactic nuclei.

The source of luminosity at disk accretion is gravitational energy whish releases when gas elements spiral down in the disk. The most of the gravitational energy is released and most of the luminosity is emitted from the inner parts of the disk. The total energy radiated by the gas element must be equal to the gravitational binding energy of the element when it is at the last stable circular orbit. This energy for the mass m_* is

$$\begin{aligned} E_{bind} &\approx 0.057 m_* c^2 && \text{for nonrotating hole,} \\ E_{bind} &\approx 0.42 m_* c^2 && \text{for maximally rotating hole.} \end{aligned} \tag{4}$$

The total luminosity of the disk is

$$\begin{aligned} L &\approx 0.057 \cdot \dot{M} c^2 \\ &\approx \left(3 \times 10^{36} \frac{\text{erg}}{\text{s}}\right) \left(\frac{\dot{M}}{10^{-9} M_\odot/\text{yr}}\right) && \text{for non-rotating hole} \\ L &\approx 0.42 \cdot \dot{M} c^2 \\ &\approx \left(3 \times 10^{37} \frac{\text{erg}}{\text{s}}\right) \left(\frac{\dot{M}}{10^{-9} M_\odot/\text{yr}}\right) && \text{for maxi-mally rotating hole.} \end{aligned} \tag{5}$$

Here $\dot{M}$ is an accretion rate. This parameter is an arbitrary external parameter, which is determined by the source of gas (for example by flux of the gas from the upper atmosphere of the companion star in the binary system). We normalized $\dot{M}$ by the value $\dot{M}_0 = 10^{-9} M_\odot/\text{year}$ because it is probably the typical rate at which the normal star is dumping gas onto its companion black hole (the next section). The ratio $\dot{M}_0/\dot{M}_{crit} \equiv \dot{m}_0 \approx 0.1 \left(\frac{M_h}{M_\odot}\right)^{-1} \leq 1$ for cases of interest (see next section 11.5). As estimates showed (see Lamb [6]), under this condition probably a geometrically thin disk (with heights $h \ll r$) might be formed. This is so called the "standard disk model" (Shakura and Sunyaev [32], Novikov and Thorne [23]). In this model the electron and ion temperatures are equal, the disk is effectively optically thick. Temperatures of the gas in the inner parts of the disk reach $T \approx 10^7 \div 10^8$K. In this region electron scattering opacity modifies the emitted spectrum so it is no longer blackbody. The total spectrum of the disk radiation has the power law $F \sim \omega^{1/3}$ with the exponential cut off at high frequencies (Shakura and Sunyaev, [32]. The innermost regions of such "standard" disks are probably unstable.

In the model proposed by Thorne and Price [34] the ions in the inner region are hot $T_i \approx 10^{11}$K but the electrons are considerably cooler $T_e \approx 10^9$K. This inner disk is thicker than in the "standard" model and produces most of the X-ray emission.

Further development of the theory of disk accretion led to more complex models. Abramovicz *et al.* [35, 36] have demonstrated that when the luminosity reaches the critical one (specifically at $2 < \dot{m} < 80$) radiation pressure in the inner parts of the disk dominates the gas pressure and the disk is thermally and viscously unstable. For $\dot{m} > 80$ the essential part of the energy of plasma is lost by advection into the black hole horizon. This process stabilizes the gas flow against perturbation.

For high mass accretion rate when $\dot{m} \geq 10$ the height of the accretion disk becomes comparable to the radius. In this case the inner edge of the disk can be closer to the black hole than the marginally stable circular orbit because of the essential pressure gradient.

In conclusion we note that in some models of disk accretion the pair production can be important. The review of these aspects is given by Lamb [6].

New development see, for example, Björnsson and Svensson [37, 38] and Artemova et al. [39]. Many aspects of the physics of accretion disks, including the development of instabilities were discussed in the works: Mineshige and Wheeler [40], Wheeler, Soon-Wook Kim and Moscoso [41], Moscoso and Wheeler [42] and Wheeler [43].

§5 Black holes in stellar binary systems

Probably the best evidence that black holes exist comes from studies of X-ray binaries. Galactic X-ray sources were discovered by Giacconi *et al.* [44]. Hayakawa and Matsuoko [45] pointed out that X-rays might be produced by accretion of gas in close binary systems. However, they discussed the accretion into the atmosphere of a normal companion star, rather than onto a compact companion. Novikov and Zeldovich [46] were the first to point out that the accretion onto compact relativistic objects (neutron stars and black holes) in binary systems should produce X-rays. They inferred also that Sco X-1 which had been just discovered, might be a neutron star in a state of accretion. After that the observational data were analyzed by Shklovsky [47]. Models for X-ray sources have been discussed in some detail by Prendergast and Burbidge [48]. They argued that the gas flow forms a disk around the compact object, with approximately a Keplerian velocity distribution.

A new era started in December 1970 when the X-ray satellite "UHURU" was launched. This satellite has provided much new data about the sources (see Giacconi *et al.* [49]). For the history of the early period of the investigations of the observational evidence for black holes in stellar binary systems, see in Novikov [16]. In this section we summarize the observational evidence for black holes in stellar binary systems which, at present, appear to be the best documented cases (for review see McClintock [11] and Cowley [10].

The argument is as follows:

1) The X-rays-emitting object in a binary system is very compact, and therefore cannot be an ordinary star, thus it is either a neutron star or a black hole. This argument comes mainly from the analysis of the features of X-ray emission.

2) The analysis of the observational data to determine the orbital motion in the binary system that makes it possible to obtain the mass of the compact object. Here the most important are data on observed velocity of the optical companion star. Note that the Newtonian gravitational theory is always enough for the analysis. The technique of weighing of stars in binaries is well known in astronomy.

If the mass of the compact component is greater than the maximal possible mass of neutron stars $M_0 \approx 3M_\odot$ (see above), then it is a black hole.

It is worth noting that the evidence is somewhat indirect because it does not confront us with the specific relativistic effects that occur near black holes and which are peculiar to black hole alone. However, it is the best that modern astronomy could proposed so far. In spite of these circumstances we believe that this logic of arguments is reliable enough.

According to the generally accepted interpretation only three presently known systems have the necessary observational confirmation and are strongly believed that the compact X-ray emitting companions of the systems are black holes. Some characteristics of the three leading black hole candidates are summarized in Table 1.

The plausible masses of the compact objects in these systems are essentially larger than $M_0 \approx 3M_\odot$. In Table 2 there are estimates of the minimum masses of these three candidates. These estimates were obtained by various methods and have different reliability. In the most cases even the minimum masses are greater than M_0. Thus, these three objects are strong black hole candidates. Using data from Table 1 and the expressions (2), (3), (5), one can estimate $\dot{m} \equiv \dot{M}/\dot{M}_E$ (see above). In all cases $\dot{m} < 1$ or $\dot{m} \approx 1$.

Several other systems have been suggested as black hole candidates (the review is given by Cowley [10] and Cominsky [50]). Three systems: LMC X-1 (Mass of a compact companion $M_c = 2.91 \pm 0.08$), Nova Muskae 1991 ($M_c = 3.1 \pm 0.4$) and V404 Cyg ($M_c = 6.26 \pm 0.31$), see Cominsky [50], could be considered as good candidates. The total number of systems that are frequently mentioned as possible candidates for black holes is about 15. All seriously discussed candidates are X-ray sources in binary systems. Some of them are persistent, other are transient. For example, Cyg X-1 and LMC X-3 are persistent, and A0620-00 is transient.

During the past more than 20 years since the discovery of the first black hole candidate Cyg X-1 only a few new candidates have been added. This is in contrast to the rapid increase of the number of identified neutron stars. At present many hundreds of neutron stars have been identified in the Galaxy. About 100 of them are in binary systems (Lamb [6]). One might conclude that black holes in binary systems are exceedingly rare objects. Probably that is not true, and the small number of identified black hole candidates is related with the specific conditions which are necessary for their observable manifestation (summary see Cowley [10]. Inoue [51] estimates the total number of only the soft X-ray transient black-hole candidates in the Galaxy to be $(1-3) \times 10^3$. Thus such systems may be as frequent as neutron star binaries. The estimate of the formation rate for black- hole binaries see in van den Heuvel and Habets [14].

At present we know only one persistent X-ray source in the Galaxy and one in another galaxy the LMC which are strong candidates for black holes (Cyg X-1 and LMC X-3 correspondingly). The probable explanation was provided by van den Heuvel [52]. He estimates that the evolutionary stage when a black hole binary continuously X-ray radiates may last only 10^4 years. We can detect it during this short period only.

Thus the population of black-hole binaries may be much

Table 1: Black-Hole Candidates in Binary Systems (McClintock [11])

Property	Cyg X-1	LMC X-3	A0620-00
X-ray luminosity (ergs s^{-1})	$2 \cdot 10^{37}$	3×10^{38}	10^{38}
Spectral type of the optic companion	O9.7Iab	B3V	K5V
Distance (Kpc)	2.5(?)	55	1(?)
Orbital eccentricity	0.00 ± 0.01	0.13 ± 0.05	0.01 ± 0.01
Orbital period (days)	5.6	1.7	0.32
Plausible masses ($M_{\odot}$)			
a) Optical companion	33	6	0.5
b) Compact companion	16	9	9

Table 2: Estimates of the Minimum Mass (in $M_{\odot}$) for the Compact Objects

Cyg X-1	Ref	LMC X-3	Ref.	A0620-00	Ref.
3	(1,2)	6	(5)	3.2	(11)
7	(3)	2.5	(6)	2.90±0.8	(12)
$3.4\left(\frac{d}{2Kpc}\right)^2$	(4)*)	2.3	(7)	4.5	(12)
		6	(7)	3.30 ± 0.95	(13)
		3	(8,9)	6.6	(13)
		4	(10)	3.1 ± 0.2	(14)
				3.82 ± 0.24	(15)
				6	(16)

*) d - distance from the Solar System.

References

1. Webster and Murdin [53]
2. Bolton [55, 56, 57]
3. Gies and Bolton [59]
4. Paczynski [61]
5. Paczynski [63]
6. Mazeh *et al.* [65]
7. Cowley *et al.* [67]
8. Kuiper *et al.* [69]
9. Bochkarev *et al.* [54]
10. Treves *et al.* [58]
11. McClintock [60]
12. McClintock and Remillard [62]
13. Johnston *et al.* [64]
14. Johnston and Kulkarni [66]
15. Haswell and Shafter [68]
16. Cowley *et al.* [70]

larger than we can presently see.

§6 Conclusions

In conclusion I would like to emphasize the comment made by one of the giants of modern astrophysics about the importance of investigating black holes. He said that what is involved is not merely the investigation of yet another, albeit remarkeable celestial body, but a test of the correctness of our understanding of the properties of space and time in the presence of extremely strong gravitational fields.

§Acknowledgements

I acknowledge P. Diener for useful remarks and his help in the preparation of this paper. This investigation was supported in part by the Danish Natural Science Research Council through grant 11-9640-1 and in part by Danmarks Grundforskningsfond through its support for the establishment of the Theoretical Astrophysical Center.

§References

[1] J.R. Oppenheimer and H. Snyder. *Phys. Rev.*, 56:455, 1939.

[2] S.L. Shapiro and S.A. Teukolsky. *Black Holes, White Dwarfs, and Neutron Stars. The Physics of Compact Objects.* John Wiley, New York, 1983.

[3] R. Kippenhahn and A. Weigert. *Stellar Structure and Evolution.* Springer-Verlag., 1990.

[4] J.R. Oppenheimer and G. Volkoff. *Phys. Rev.*, 55:374, 1939.

[5] G. Baym and C.J. Pethick. *Ann. Rev. Astr. Ap.*, 17:415, 1979.

[6] F.K. Lamb. In D.L. Lambert, editor, *Frontiers of Stellar Evolution*, page 299. Astronomical Society of the Pacific, 1991.

[7] C. Alcook, E. Farhi, and A.V. Olinto. *Astrophys. J.*, 310:261, 1986.

[8] S. Bahcall, B.W. Lynn, and S.B. Selipsky. *Astrophys. J.*, 362:251, 1990.

[9] J.L. Friedman and J.R. Isper. *Astrophys. J.*, 314:594, 1987.

[10] A.P. Cowley. *Ann. Rev. Astr. Ap.*, 30:287, 1992.

[11] J.E. McClintock. In E.P.J. van den Heuvel and S.A.Rappoport, editors, *X-ray Binaries and Recycled Pulsars*, page 27, 1992.

[12] G.S. Bisnovaty-Kogan. *Physical Problems of the Theory of Stellar Evolution.* Nauka, Moscow, 1989.

[13] V.M. Lipunov. *Astrophysics of Neutron Stars.* Nauka, Moscow, 1987.

[14] E.P.J. Van den Heuvel and G.M.H.J. Habets. *Nature*, 309:598, 1984.

[15] Schild and Maeder. *Astron. Astrophys.*, 143:L7, 1985.

[16] I.D. Novikov. In L'Universite de Bruxelles, editor, *Astrophysics and Gravitation, Proceedings of the 16th Solvay Conf. on Physics, 1973*, page 317, 1974.

[17] A.G. Masevich and A.V. Tutukov. *Stellar evolution: theory and observations.* Nauka, Moscow, 1988.

[18] F. Zwicky. *Hdb. d. Phys*, 51:766, 1958.

[19] M. Schwarzshild. *Structure and Evolution of the Stars.* Princeton University Press, Princeton, N.J, 1958.

[20] F. Hoyle and W.A. Fowler. *Nature*, 197:533, 1963.

[21] I.D. Novikov and L.M. Ozernoy. 1964. Preprint FIAN, A-17.

[22] F. Hoyle, W.A. Fowler, G. Burbidge, and E.M. Burbidge. *Astrophys. J.*, 139:909, 1964.

[23] I.D. Novikov and K.S. Thorne. In C. Dewitt and B.S. DeWitt, editors, *Black holes, Les Astres Occuls*, page 343. Gordon and Breach Science Publishers, 1973.

[24] Ya.B. Zeldovich. *Dokl. Akad. Nauk USSR*, 155:67, 1964.

[25] E. Salpeter. *Astrophys. J.*, 140:796, 1964.

[26] J.M. Bardeen and J.A. Petterson. *Astrophys. J. Lett.*, 195:L65, 1975.

[27] A.S. Eddington. *The Internal Constitution of the Stars.* Cambridge University Press, Cambridge, 1926.

[28] Ya.B. Zeldovich and I.D. Novikov. *Dokl. Akad. Nauk USSR*, 158:811, 1964.

[29] D. Lynden-Bell. *Nature*, 233:690, 1969.

[30] N.I. Shakura. *Astr. Zh.*, 16:756, 1972.

[31] J.E. Pringle and M.J. Rees. *Astron. Astrophys.*, 21:1, 1972.

[32] N.I. Shakura and R.A. Sunyaev. *Astron. Astrophys.*, 24:337, 1973.

[33] D.N. Page and K.S. Thorne. *Astrophys. J.*, 191:499, 1974.

[34] K.S. Thorne and R.H. Price. *Astrophys. J. Lett.*, 195:L101, 1975.

[35] M. A. Abramowicz, B. Czerny, J.P. Lasota, and E. Szuzkievicz. *Astrophys. J.*, 332:646, 1988.

[36] M.A. Abramowicz, Szuskiewicz, and Wallinder. In F. Meyer, W.J. Duschl, J. Frank, and E. Meyer-Hofmeister, editors, *Theory of Accretion Discs*, volume 290, page 141. NATO ASI Series, Kluwer, 1989.

[37] G. Björnsson and R. Svensson. *Astrophys. J. Lett.*, 371:L69, 1991.

[38] G. Björnsson and R. Svensson. *Astrophys. J.*, 394:500, 1992.

[39] J. Artemova, G. Bisnovatyi-Kogan, G. Björnsson, and I. Novikov. In *Proceding of Vulcano Workshop, in press*, 1994.

[40] S. Mineshige and J.C. Wheeler. *Astrophys. J.*, 343:241, 1989.

[41] J.C. Wheeler, Soon-Wook Kim, and M. Moscoso. *Cataclysmic Variables and Related Physics, 2nd Technion Haifa Conference, Eilat, Israel*, volume 10, page 180. Annals of the Israel Physical Society, 1993.

[42] M.D. Moscoso and J.C. Wheeler. In A.W. Shafter, editor, *Interacting Binary Stars*, volume 56, page 100. Astronomical society of the Pacific, Conference Series, 1994.

[43] J.C. Wheeler. In K.-T. kimi, editor, *Proceeding of Daejon Korea Workshop on Supernovae and the Evolution of the Galaxy*. in press, 1994.

[44] R. Giacconi, H. Gursky, F.R. Paolini, and B.B. Rossi. *Phys. Rev. Lett.*, 9:439, 1962.

[45] S. Hayakawa and M. Matsuoko. *Progr. Theor. Phys. Suppl.*, 30:204, 1964.

[46] I.D. Novikov and Ya.B. Zeldovich. *Nuovo Cim. Suppl.*, 4:810, 1966. addendum 2.

[47] I.S. Shklovsky. *Astrophys. J. Lett.*, 148:L1, 1967.

[48] K.H. Prendergast and G.R. Burbidge. *Astrophys. J. Lett.*, 151:L83, 1968.

[49] R. Giacconi, S. Murray, H. Gursky, E. Kellogg, E. Schreier, and H. Tananbaum. *Astrophys. J.*, 178:281, 1972.

[50] L. Cominsky. In *Proceding of the Snowmass Workshop, in press.*, 1994.

[51] H. Inoue. In K. Koyama, editor, *28th Yamula Conf. The Frontiers of X-ray Astronomy*. In press., 1991.

[52] E.P.J. Van den Heuvel. page 303, Cambridge, 1983. Cambridge University Press.

[53] B.L. Webster and P. Murdin. *Nature*, 235:37, 1972.

[54] N.G. Bochkarev, R.A. Sunyaev, T.S. Kruzina, A.M. Cherepashchuk, and N.I. Shakura. *Sov. Astron. J.*, 32:405, 1988.

[55] C.T. Bolton. *Nature*, 235:271, 1972.

[56] C.T. Bolton. *Nature Phys. Sci.*, 240:124, 1972.

[57] C.T. Bolton. *Astrophys. J.*, 200:269, 1975.

[58] A. Treves, T. Belloni, R.H.D. Corbet, K. Ebisawa, and R Falomo *et al. Astrophys. J.*, 364:266, 1990.

[59] D.R. Gies and C.T. Bolton. *Astrophys. J.*, 304:371, 1986.

[60] J.E. McClintock. In Minas Kafatos, editor, *Supermassive Black Holes*, page 1. Cambridge University Press, 1988.

[61] B. Paczynski. *Astron. Astrophys.*, 34:161, 1974.

[62] McClintock and R.A. Remillard. *Bull. Am. Astron. Soc.*, 21:1206, 1990.

[63] B. Paczynski. *Astrophys. J. Lett.*, 273:L81, 1983.

[64] H.M. Johnston, S.R. Kulkarni, and J.B. Oke. *Astrophys. J.*, 345:492, 1989.

[65] T. Mazeh, J. van Paradijs, E.P.J. van den Heuvel, and G.J. Savonije. *Astron. Astrophys.*, 157:113, 1986.

[66] H.M. Johnston and S.R. Kulkarni. *Accretion-Powered Compact Binaries*. Cambridge University Press, Cambridge, 1990.

[67] A.P. Cowley, D. Crampton, J.B. Hutchings, R. Remillard, and J.E. Penfold. *Astrophys. J.*, 272:118, 1983.

[68] C.A. Haswell and A.W. Shafter. *Astrophys. J. Lett.*, 359:L47, 1990.

[69] L. Kuiper, J. van Paradijs, and M. van der Klis. *Astron. Astrophys.*, 203:79, 1988.

[70] A.P. Cowley, P.C. Schmidtke, D. Crampton, and J.B. Hutchings. *Astrophys. J.*, 350:288, 1990.

Neutrinos

KINEMATICAL PROBES OF NEUTRINO MASS

The N1 Working Group

E. Adelberger, D. Ahluwalia, D. Caldwell, P. Cushman, M. Diwan, L. Durand, J. Engel, G. Gelmini, R. Gibbs, M. Hindi, B. Kayser†, R. Kozub, D. Lalanne, R. Lanou†, J. Laplanche, W. Palmer, R. Peccei, A. Piepke, H. Robertson, S. Rolli, G. Segre, A. Sonnenschein, C. Sutton

§1 INTRODUCTION

The most direct searches for neutrino mass are the experiments which seek a kinematical consequence of this mass in some physical process. These experiments include the study of the β spectrum in tritium β decay as a probe of the mass of ν_e, the measurement of the muon momentum in pion decay as a probe of the mass of ν_μ, the study of the hadronic invariant mass and energy distribution in $\tau \rightarrow \nu_\tau$+hadrons as a probe of the mass of ν_τ, and other studies. The present analysis will be concerned with kinematical probes of this kind, and with neutrinoless double beta decay ($\beta\beta_{o\nu}$), a probe of neutrino mass which, like some of the kinematical probes, involves a nuclear decay. We will survey the present situation, argue that interesting physics might be revealed by experiments with somewhat improved sensitivity, and consider some ideas for achieving the heightened sensitivity.

In Sec. 2, we review the present bounds on neutrino mass from both kinematical probes and $\beta\beta_{o\nu}$. In Sec. 3, we discuss the fact that the tritium β decay probes of the mass of ν_e, M_{ν_e}, reveal an anomalous β spectrum whose character is such that the value of $M_{\nu_e}^2$ that fits it best is negative. We describe a neutrino-mass explanation of this anomaly which does not work, and suggest future tests of molecular-chemistry explanations which might work. In Sec. 4, we turn to future neutrino-mass experiments with somewhat improved sensitivity. We explain how current hints of neutrino mass suggest, albeit speculatively, that the neutrino masses might be big enough to be visible in some of these experiments. Other improved experiments, when combined with astrophysical arguments, could serve to significantly restrict some of the possibilities for neutrino masses. We discuss some ideas for more sensitive kinematical probes and for more sensitive $\beta\beta_{o\nu}$ measurements. In Sec. 5, we describe certain new types of experiments, and in Sec. 6, we state our conclusions.

§2 WHERE WE STAND

2.1 The M_{ν_e} Bound

In Table 1, we show the bounds on the ν_e mass yielded by the recent tritium β decay experiments. These bounds come from study of the shape of the β spectrum near its upper endpoint. It is seen that the β spectrum has an excess of events near its upper endpoint, rather than having the deficit of events that would correspond to a positive $M_{\nu_e}^2$. We note from Table 1 that in all of the experiments, the value of $M_{\nu_e}^2$ that fits this shape best is negative, in some experiments significantly so. This excess can be seen clearly in the data from the Livermore experiment.[5]

Since the β spectrum has an anomalous shape which is not understood — a shape that corresponds neither to a vanishing value of $M_{\nu_e}^2$ nor to a physical (positive) nonvanishing value — it could be that M_{ν_e} is actually somewhat larger than the bounds quoted in Table 1. However, if it were, say an order of magnitude larger, its effects would have been seen. Thus, the bounds in Table 1 do indicate that current experiments are sensitive to values of M_{ν_e} roughly in the 10 eV range.

2.2 The M_{ν_μ} Bound

Upper bounds on M_{ν_μ}, the mass of ν_μ, have come from studies of the decay $\pi \rightarrow \mu\nu_\mu$ using pions at rest or ones in flight, and from studies of the decay $K_L \rightarrow \pi\mu\nu_\mu$. The most stringent existing bound is $M_{\nu_\mu} < 0.16$MeV at 90% confidence level.[7] This bound comes from measurement of the muon momentum in the decay $\pi \rightarrow \mu\nu_\mu$ of pions at rest, combined with information on the pion and muon masses.

2.3 The M_{ν_τ} Bound

Limits on M_{ν_τ}, the mass of ν_τ, have come from studies of the invariant mass and energy of the hadronic system in $\tau \rightarrow \nu_\tau$+hadrons. The bigger the invariant mass is, the less energy is available to make the mass of ν_τ in the final state. The most stringent existing published bound is $M_{\nu_\tau} < 24$ MeV at 95% confidence level.[8]

2.4 The $M_{eff}(\beta\beta_{o\nu})$ Bound

The measurement of the neutrinoless double beta ($\beta\beta_{0\nu}$) decay is at present a more sensitive test of a non–zero neutrino mass than are the kinematic probes. Unlike the latter, $\beta\beta_{o\nu}$ may be one of the best tools to directly explore sub–eV neutrino masses. However one has to pay a price for this extraordinary sensitivity: the mass limits only apply for Majorana neutrinos. Excellent reviews have been given on this subject (see e.g. [9]).In double beta decay two neutrons are simultaneously converted into two protons. The decay becomes observable in those nuclides where single beta decay is either energetically forbidden or at least strongly suppressed. $\beta\beta$–emitters are even-even isotopes. Three decay modes are in general discussed:

$$(\beta\beta 2\nu) : A(Z,N) \rightarrow$$

$$A(Z+2,N-2) + 2e^- + 2\overline{\nu}_e \qquad (2.1)$$

$$(\beta\beta 0\nu) : A(Z,N) \rightarrow A(Z+2,N-2) + 2e^- \qquad (2.2)$$

$$(\beta\beta 0\nu\chi) : A(Z,N) \rightarrow A(Z+2,N-2) + 2e^- + \chi \qquad (2.3)$$

Although extremely rare, decay (2.1) is allowed as a second order effect in the (S)tandard (M)odel of the electro–weak interaction. Decay channel (2.2) and (2.3) are the much more interesting scenarios because they test physics beyond the SM since lepton number conservation is violated. The right–handed anti–neutrino, emitted in the decay of the first neutron, has to be re-absorbed as a left–handed neutrino in an inverse beta decay of the second neutron. This obviously requires the neutrino to be a massive Majorana particle. The question whether $\beta\beta_{o\nu}$ can happen spontaneously (through an incomplete polarization of the massive neutrino) or is due to the coupling of the neutrino to a massless Majoron χ allows us to draw conclusions on the underlying physics.

In the SM, minimally extended to include Majorana neutrino masses, the amplitude for $\beta\beta_{o\nu}$ is proportional to an effective neutrino mass for double beta decay, $M_{eff}(\beta\beta_{o\nu})$, given by

$$M_{eff}(\beta\beta_{o\nu}) = \sum_m CP(\nu_m) \mid L_{em} \mid^2 M_{\nu_m}. \qquad (2.4)$$

Here, the sum runs over all the neutrino mass eigenstates ν_m. The quantity CP(ν_m) is the intrinsic CP parity of ν_m, and M_{ν_m} is its mass. The quantity L_{em} is the element of the leptonic mixing matrix (the leptonic analogue of the quark mixing matrix) coupling ν_m to the electron. In writing Eq. (2.4), we have neglected any CP violation in the leptonic sector. In the SM, the matrix L is unitary, so that $|L_{em}|^2 \leq 1$. Note that owing to the factor $|L_{em}|^2$ and the CP parities in Eq. (2.4), an upper bound on $M_{eff}(\beta\beta_{o\nu})$ does not imply an upper bound on the actual neutrino masses M_{ν_m}. (If CP is violated in leptonic mixing, then the CP(ν_m) in Eq. (2.4) is replaced by a more general phase factor.)

In order to draw conclusions from a measured neutrinoless–decay rate (or a limit of this rate) on the effective Majorana mass of the neutrinos (or a limit on M_{eff}), obviously, one has to unfold the leptonic phase space integral and the nuclear matrix element first. While the calculation of the phase space seems to be well under control, the determination of nuclear matrix elements is a difficult and often controversial task. So far, $\beta\beta_{o\nu}$ has not been seen, and the most stringent existing bound on $M_{eff}(\beta\beta_{o\nu})$ is $|M_{eff}(\beta\beta_{o\nu})| < 0.68$ eV at 90% confidence level.[10] This limit comes from a lower bound of 5.1×10^{24} yr on the half-life for $^{76}Ge \rightarrow ^{76}Se + 2e^-$.

§3 THE PROBLEM OF NEGATIVE $M^2_{\nu_e}$

3.1 Neutrino Physics Possibilities

Could the excess of events near the upper endpoint of the tritium β spectrum, relative to the number predicted by the standard formula either for vanishing or positive $M^2_{\nu_e}$, be a consequence of neutrino physics not included in this formula? One possibility we have explored is that the excess results from the existence of a $\bar{\nu}_e$ magnetic dipole moment. If the outgoing $\bar{\nu}_e$ in a tritium β decay has such a moment, then it will interact electromagnetically with the charged particles it leaves behind. At first sight one would not expect this interaction to be significant, since there is a laboratory upper bound of 4×10^{-10} Bohr magnetons on the $\bar{\nu}_e$ magnetic moment,[11] and an astrophysical upper bound of 3×10^{-12} Bohr magnetons.[12] However, the propagator, $1/q^2$, for the photon whose exchange carries the electromagnetic interaction between the $\bar{\nu}_e$ and the charged particles in the final state is large for small q^2. Since one will encounter small values of q^2 near the upper endpoint of the β spectrum, where the momentum carried by the neutrino is small, the dipole moment possibility was examined. Unfortunately, it was found that the large photon propagator is offset by small phase space factors. It was estimated that if the $\bar{\nu}_e$ magnetic moment is 10^{-10} Bohr magnetons, then the interaction between it and the charged final-state particles in tritium β decay modifies the decay amplitude near the upper endpoint of the β spectrum by only one part in 10^{16}! Clearly, a neutrino dipole moment is not the explanation of the observed excess of events.

A second neutrino-physics possibility, which has been considered previously, is that some of the tritium "β decay" events are actually ν_e capture events induced by capture of a ν_e in the cosmic neutrino background. For this mechanism[13] to produce as many excess events as are seen, the local density of background neutrinos would have to be $10^{16} cm^{-3}$.

3.2 Molecular Chemistry Possibility

As is clear from Sec. 2.1, the issue of negative experimental results for the square of the electron antineutrino mass needs further attention. (This problem has also been summarized nicely by Wilkerson [14].) It should be noted that experiments which have yielded significantly negative $M^2_{\nu_e}$ have all involved a T_2 source. Now, the β decay of this molecule creates a $^3HeT^+$ molecular ion. While there is presently no specific, known problem with the 2-electron molecular spectrum, there is a lack of experimental data on the electronic final states of this molecular ion and it is reasonable to seek a possible solution of the negative $M^2_{\nu_e}$ problem by spectroscopic studies of events involving $^3HeT^+$. The branching ratios and energies for all possible final states must be known precisely in order to extract the neutrino mass from the beta decay spectrum. The theoretical calculations pertaining to these states [15] are believed to be very accurate, but the negative-mass-squared results indicate that something is not understood completely. It is thus important to remove as many model-dependent components as possible from the data analysis procedures.

At least one measurement of the vibrational-rotational (infrared) spectrum of the $^3HeT^+$ molecule has been carried out [16]. Transitions were observed from the de-excitation of vibrational states which were excited during the beta-decay process. However, no transitions originating from excited electronic orbitals in the $^3HeT^+$ molecule have been observed. These transitions are in the vacuum ultra-violet range, with wavelengths less than 100 nm, and the development of efficient, high-resolution detectors of photons in this range is still in its infancy. Most of the recent detector development has centered on semiconductor photodiodes [17], although superconducting tunnel junction devices also sound promising [18]. Some detector development would no doubt be necessary before a successful experiment to obtain electronic transition spectra could be performed.

Ideally, the measurement of electronic photon spectra from $^3HeT^+$ should be done in event mode and in coincidence with the emitted beta particles. It would seem that the sites at which the neutrino mass measurements (using T_2) were performed, i.e., LANL [3] or/and LLNL [5], would be good locations to pursue such measurements. However, the source tubes would probably have to be redesigned to permit efficient observation of VUV photons. Further, even with the development of an efficient VUV detector, the counting rate may be too low to use high precision (and low solid-angle acceptance) beta-ray spectrometers—the use of large solid-angle (and lower energy resolution) beta detectors as coincidence triggers would have to be considered. Photon spectra obtained in this way should, nevertheless, be quite useful.

Another problem which may be encountered in such measurements arises from interactions of photons from $^3HeT^+$ with the T_2 source gas. The desired photon spectrum may thusly be contaminated with transitions corresponding to the de-excitation of T_2. Measurements may have to be done at several gas pressures in order to positively identify the transitions of interest.

Several other experimental problems would, no doubt, be encountered in trying to measure $^3HeT^+$ photon energies and intensities. However, if such measurements are not pursued, the tremendous efforts already undertaken to obtain high quality beta spectra for neutrino mass measurements may be partially wasted, merely for lack of understanding of how to properly analyze the data.

§4 NEAR-TERM FUTURE EXPERIMENTS WITH IMPROVED SENSITIVITY

4.1 Motivation to improve sensitivity (Dirac and/or Majorana)

In considering neutrino-mass experiments with somewhat improved sensitivity, it is very interesting to note that present hints of neutrino mass suggest, albeit speculatively, that there may be neutrino masses within the range to which these experiments would be sensitive. Let us briefly review the present hints, see what neutrino-mass range they point to, and then ask what experiments would be sensitive to mass in this range.

The first existing hint of neutrino mass is the solar neutrino deficit. The solar neutrino data, while difficult to explain without invoking neutrino mass,[19] are elegantly explained by the M(ikheyev), S(mirnov), W(olfenstein) effect,[20] which does invoke neutrino mass. The MSW effect converts a ν_e produced in the solar core into another kind of neutrino, ν_x, which is invisible (or at least largely invisible) to the present terrestrial solar neutrino detectors. Neglecting the possibility of so-far-unknown sterile neutrinos, ν_x is a ν_μ, ν_τ, or some combination of these neutrinos. The $\nu_e \to \nu_x$ conversion occurs while the ν_e produced in the solar core is traveling outward through solar material. The MSW explanation of the solar neutrino data requires that the mass of ν_x, M_{ν_x}, be related to that of ν_e, M_{ν_e}, by

$$M_{\nu_x}^2 - M_{\nu_e}^2 \sim 10^{-5} eV^2. \qquad (4.1)$$

Let us recall enough of the physics of the MSW effect to understand why.

Neglecting mixing, the total energy of a ν_e of momentum p_ν in the sun is

$$E_{\nu_e} = \sqrt{p_\nu^2 + M_{\nu_e}^2} + \sqrt{2} G_F N_e. \qquad (4.2)$$

The second term in this expression, in which G_F is the Fermi coupling constant and N_e is the density of electrons at the location of the ν_e, is an interaction potential energy resulting from the charged-current interaction between the ν_e and the electrons. Now, the ν_x, which does not have a ν_e component, does not have a charged-current interaction with electrons. Consequently, the total energy of a ν_x of momentum p_ν in the sun is simply

$$E_{\nu_x} = \sqrt{p_\nu^2 + M_{\nu_x}^2} \simeq \sqrt{p_\nu^2 + M_{\nu_e}^2} + \left(M_{\nu_x}^2 - M_{\nu_e}^2\right)/2E_\nu. \qquad (4.3)$$

Here, in the second step we have assumed that M_{ν_x} and M_{ν_e} are both tiny compared to p_ν, and denoted $(p_\nu^2 + M_{\nu_e}^2)^{1/2}$, which is an excellent approximation to the neutrino energy, by E_ν.

The MSW $\nu_e \to \nu_x$ conversion requires that, somewhere in the sun, the energy levels E_{ν_e} and E_{ν_x} given by Eqs. (4.2) and (4.3) cross. Thus, this conversion demands that

$$\sqrt{2} G_F N_e = \frac{M_{\nu_x}^2 - M_{\nu_e}^2}{2E_\nu}. \qquad (4.4)$$

Now, we can understand the observation that the solar neutrino flux is below the predicted amount, but not zero, by supposing that the level- crossing condition (4.4) is satisfied for, say, neutrino energies E_ν above ~ 0.7MeV (hence for the ^7Be and ^8B neutrinos), but not for energies below this point (hence, not for the *pp* neutrinos). Noting that the electron density rises from zero at the outer edge of the sun to a maximum of $\sim 10^{26}/cc$ at the center of the sun, we find that Eq. (4.4) will hold somewhere in the sun for $E_\nu \gtrsim 0.7\text{MeV}$, but not for $E_\nu \lesssim 0.7\text{MeV}$, if $M^2_{\nu_x} - M^2_{\nu_e}$ obeys condition (4.1). This is the origin of that condition.

Detailed calculations of the MSW effect show that it can explain existing solar neutrino data if $M^2_{\nu_x} - M^2_{\nu_e}$ satisfies condition (4.1), and $\sin^2 2\theta \simeq 5 \times 10^{-3}$ or 0.6,[21] where θ is the $\nu_e - \nu_x$ mixing angle. Of these two solutions, the one with small θ is favored because it distorts the detectable solar neutrino spectrum in the manner that is suggested by the observations. The *pp* electron neutrinos do not undergo conversion, and are detected on earth. Both the ^7Be and ^8B neutrinos undergo MSW conversion to the undetectable ν_x particles, but in the case of the ^8B neutrinos, this conversion is not fully efficient because the region in the sun where the level-crossing condition (4.4) is satisfied is too thin. Thus, the ^8B flux is not suppressed as much as the ^7Be flux.

The second hint of neutrino mass is the anomalously low value of the ν_μ/ν_e ratio in the flux of neutrinos generated in the earth's atmosphere by cosmic ray interactions. This flux of atmospheric neutrinos is detected by underground detectors, each of which observes not only neutrinos generated in the atmosphere directly above it, but also neutrinos produced in much more distant parts of the atmosphere, such as those on the side of the earth opposite to the detector.

The anomalously low atmospheric ν_μ/ν_e ratio has been interpreted in terms of vacuum neutrino oscillations. The present data are consistent both with $\nu_\mu \to \nu_\tau$ and with $\nu_\mu \to \nu_e$ oscillation.[22,23] To see what neutrino mass difference must be involved, we note that the probability that a neutrino of one flavor, ν_f, will be converted by neutrino oscillation into one of another flavor, $\nu_{f'}$, is proportional to

$$\sin^2 \left\{ 1.27 \left(\frac{\delta M^2}{1eV^2} \right) \frac{(L/1km)}{(E_\nu/1GeV)} \right\}. \tag{4.5}$$

Here, δM^2 is the difference between the squared masses of the two mass eigenstates of which ν_f and $\nu_{f'}$ are composed, and we are assuming that mixing between the $\nu_f - \nu_{f'}$ system and the third, remaining flavor may be neglected. The quantity L is the distance traveled by the neutrino, and E_ν is its energy.

From (4.5), we see that for $\nu_f \to \nu_{f'}$ oscillation to be appreciable, $(\delta M^2/1eV^2) * (L/1km)/(E_\nu/1GeV)$ must not be small compared to unity. Now, for neutrinos created in the atmosphere directly above a detector, $L \sim 20$km, while for those created in the atmosphere on the side of the earth opposite to the detector, $L \sim 10^4$km. Thus, if E_ν is typically 1GeV for the detected neutrinos, then if oscillation is to be appreciable at least for those neutrinos for which $L > 10^3$km, we must have $\delta M^2 \gtrsim 10^{-3}eV^2$.

Recent data[23] show that for neutrinos with a mean energy of 6GeV, the ν_μ/ν_e ratio is anomalously small for upward-going neutrinos (which have travelled a large distance L to reach the detector), but not for downward-going neutrinos (which have only travelled $\sim$ 20km from the atmosphere directly above the detector). From (4.5) these data imply that $\delta M^2 \lesssim 10^{-1}eV^2$, or even the downward-going neutrinos would show an anomaly. Thus, we see that $10^{-3}eV^2 \sim \delta M^2 \lesssim 10^{-1}eV^2$. The best fit to all the KAMIOKANDE data is obtained for $\delta M^2 \sim 10^{-2}eV^2$, and near maximal mixing of the two flavors involved in the oscillation.[23]

The third hint of neutrino mass is the existence of dark matter in the universe, combined with the belief that this dark matter includes a "hot" component, and the very natural view that this component is quite likely comprised of neutrinos. Let us call the neutrino mass eigenstates of which ν_e, ν_μ, and ν_τ are made ν_1, ν_2, and ν_3, and let $\sum \equiv M_{\nu_1} + M_{\nu_2} + M_{\nu_3}$, where M_{ν_m} is the mass of ν_m. Further, let Ω_H be the fraction of critical mass density contributed to the universe by hot dark matter, and h be the Hubble constant in units of 100 km/sec/Mpc. Then, if the hot dark matter consists of neutrinos,

$$\Omega_H = \frac{(\sum/1eV)}{91h^2}. \tag{4.6}$$

If we take $\Omega_H = 0.2$,[24] and $h = 0.5$, this relation implies that

$$\sum \equiv M_{\nu_1} + M_{\nu_2} + M_{\nu_3} \cong 5 \ eV. \tag{4.7}$$

The previous value of h is somewhat uncertain. Interesting new measurements, some of them made with the aid of the Hubble Space Telescope, yield for this parameter 0.80 ± 0.17,[24] 0.73 ± 0.09,[25] 0.52 ± 0.08,[26] and 0.67 ± 0.07.[27] Now, values larger than approximately 0.65, combined with the non-cosmological lower bounds on the age of the universe, would make the current mixed hot- and cold-dark-matter picture[28] of the universe nonviable.[29] Since this picture is an input to our considerations, we assume in this report that

$$0.50 \lesssim h \lesssim 0.65. \tag{4.8}$$

Then, from Eq. (4.6),

$$4.5 \ eV \lesssim \sum \lesssim 7.7 \ eV. \tag{4.9}$$

Suppose, now, that the explanation of all three of the hints we have discussed—the solar neutrino deficit, the atmospheric neutrino anomaly, and the hot dark matter—is indeed neutrino mass as we have described. To what values of the neutrino masses does this point?

In exploring this question, we shall assume that the only existing neutrinos are the ones we already know: ν_e, ν_μ, and ν_τ, so that there are no as-yet-undiscovered sterile neutrinos. In addition since some observations by reactor experiments[30] presently indicate that much of the mass-mixing phase space allowed by the atmospheric experiments is unlikely for $\nu_\mu \to \nu_e$ transitions we shall suppose, as an example, that it is indeed $\nu_\mu \to \nu_\tau$ which is occuring. In that case, the atmospheric neutrino data then suggest the $\nu_\mu - \nu_\tau$ mixing is nearly maximal,[23] which implies that the mixing between ν_e and the $\nu_\mu - \nu_\tau$ system is small. Neglecting the latter mixing, each of ν_μ and ν_τ is a roughly 50–50 mixture of two neutrino mass eigenstates, which we shall call $\widetilde{\nu}_2$ and $\widetilde{\nu}_3$. The data are best fit with $|M^2_{\nu_3} - M^2_{\nu_2}| \sim 10^{-2}eV^2$, as we have said.

In this scenario, the small-mixing MSW mechanism can still operate more or less as we have described, despite the large mixing between ν_μ and ν_τ.[31] The MSW effect may then be pictured as resulting from the circumstance that, in the absence of mixing between ν_e and the $\nu_\mu - \nu_\tau$ system, the total energy of ν_e crosses, somewhere in the sun, that of $\widetilde{\nu}_2$ or $\widetilde{\nu}_3$. Let us say it crosses that of $\widetilde{\nu}_2$. The neutrino ν_x into which the MSW effect converts a ν_e is then $\widetilde{\nu}_2$, a 50–50 $\nu_\mu - \nu_\tau$ mixture. As for the case where all mixing angles are small, so here the level crossing requires that, somewhere in the sun, Eq. (4.4) hold. To explain the apparent fact that some solar electron neutrinos undergo MSW conversion while others do not, we must then have Eq. (4.1), as before. Since the degeneracy between ν_x and ν_e required by Eq. (4.1) is much greater than that between $\widetilde{\nu}_2$ and $\widetilde{\nu}_3$, it is clear that only one of the latter two neutrinos can be involved in the MSW effect. (When the mixing between ν_e and the $\nu_\mu - \nu_\tau$ system is not neglected, the true mass eigenstates of the three-neutrino system are $\nu_1 \sim \nu_e$, and two states, ν_2 and ν_3, which are not too different from $\widetilde{\nu}_2$ and $\widetilde{\nu}_3$.)

Taken together, the neutrino-mass interpretations of the hot dark matter, the atmospheric neutrino anomaly, and the solar neutrino deficit imply, under our assumptions, the constraints

$$M_{\nu_1} + M_{\nu_2} + M_{\nu_3} \simeq 6\ eV, \tag{4.10.a}$$

$$|M^2_{\nu_3} - M^2_{\nu_2}| \sim 10^{-2}\ eV^2, \tag{4.10.b}$$

and

$$M^2_{\nu_2} - M^2_{\nu_1} \sim 10^{-5}\ eV^2 \tag{4.10.c}$$

Obviously, these constraints require the three neutrino mass eigenstates to be approximately degenerate in mass, with each of them having a mass of approximately $6eV/3$. or 2eV.[32] In summary, in this scenario the neutrino mass spectrum is described by

$$M_{\nu_1} \cong M_{\nu_2} \cong M_{\nu_3} \simeq 2\ eV, \tag{4.11.a}$$

$$|M^2_{\nu_3} - M^2_{\nu_2}| \sim 10^{-2}\ eV^2, \tag{4.11.b}$$

and

$$M^2_{\nu_2} - M^2_{\nu_1} \sim 10^{-5}\ eV^2. \tag{4.11.c}$$

What experiments does this scenario suggest? First, since in this scenario the electron neutrino, $\nu_e \sim \nu_1$, has a mass of 2eV, tritium β decay experiments with somewhat better sensitivity than those whose results are reported in Table 1 would be capable of seeing the ν_e mass! Thus, such experiments are worth considering, even though they may be very difficult. Secondly, since the small mixing between ν_e and the $\nu_\mu - \nu_\tau$ system implies that the mixing matrix elements $|L_{em}|^2$ in Eq. (2.4) are all small except for $|L_{e1}|^2 \simeq 1$, we see from Eq. (2.4) that the effective neutrino mass for double beta decay, $M_{eff}(\beta\beta_{0\nu})$, is given by

$$|M_{eff}(\beta\beta_{0\nu})| \cong M_{\nu_1} \cong 2\ eV. \tag{4.12}$$

Of course, this assumes that neutrinos are Majorana particles; if they are Dirac particles there will be no $\beta\beta_{0\nu}$.[33] Now, the current upper bound on $M_{eff}(\beta\beta_{0\nu})$, while nominally somewhat below 2eV, is also somewhat uncertain due to its dependence on uncertain nuclear matrix elements (see Table 3 and discussion in Sec. 4.3). Thus, an $M_{eff}(\beta\beta_{0\nu})$ of 2eV is not in clear conflict with present $(\beta\beta_{0\nu})$ experiments, but suggests that future experiments which are somewhat more sensitive than the present ones may actually see a signal! Finally, although future solar neutrino and neutrino oscillation experiments are outside the scope of this report, it is obviously highly desirable to carry out solar neutrino experiments which can test the MSW explanation of the solar neutrino fluxes and long-baseline $\nu_\mu \to \nu_\tau$ oscillation experiments with accelerator-generated ν_μ beams which can test the $\nu_\mu \to \nu_\tau$ explanation of the atmospheric neutrino anomaly. It is also interesting to carry out searches, such as those being performed by the Los Alamos LSND and the Rutherford KARMEN experiments, for $\nu_\mu \to \nu_e$ oscillation with $\delta M^2 \geq 1$ eV^2 and small mixing angle. Confirmation of such oscillation would falsify the neutrino mass spectrum of Eq. (4.11).

What if only two of the three hints of neutrino mass we have discussed are actually due to neutrino mass? Let us consider the possibilities in turn.

i) Neutrino mass is responsible for the hot dark matter and the solar neutrino deficit:

In this case, we may, to focus our thinking, assume that neutrino mixing is small, and that

$$M_{\nu_1} \leq M_{\nu_2} \leq M_{\nu_3}, \tag{4.13}$$

where ν_1, ν_2 and ν_3 are, respectively, the dominant mass eigenstate components of ν_e, ν_μ, and ν_τ. The masses M_{ν_m} must obey the constraint (4.10.a) and, assuming that the solar neutrino deficit is due to MSW conversion of ν_e into ν_μ, the constraint (4.10.c). A possible scenario which is not possible when neutrino mass is assumed to be responsible for all three of the hints we have discussed is then

$$M_{\nu_1} \ll 10^{-3}eV,\ M_{\nu_2} \simeq 3 \times 10^{-3}eV,\ M_{\nu_3} \simeq 6\ eV. \tag{4.14}$$

Observation of the mass of $\nu_e \sim \nu_1$ in tritium β decay is then out of the question. However, for $M^2_{\nu_3} - M^2_{\nu_1} \simeq (6eV)^2$, as implied by Eq. (4.14), the experimental bound on the neutrino mixing parameter $|L_{e3}|^2$ appearing in Eq. (2.4) for $M_{eff}(\beta\beta_{0\nu})$ is 0.024.[34] If $|L_{e3}|^2$ is not much smaller than this bound, then from Eqs. (4.14) and (2.4)

$$|M_{eff}(\beta\beta_{0\nu})| \simeq |L_{e3}|^2 M_{\nu_3} \simeq 0.1\ eV. \tag{4.15}$$

Assuming that neutrinos are Majorana particles, neutrinoless double beta decay would then be visible in experiments roughly one order of magnitude more sensitive (in terms of $M_{eff}(\beta\beta_{0\nu})$) than the present ones.[35]

With the neutrino masses as in Eqs. (4.14), the CHORUS and NOMAD searches for $\nu_\mu \rightarrow \nu_\tau$ oscillation can see a signal if the $\nu_\mu - \nu_\tau$ mixing angle $\theta_{\mu\tau}$ obeys $\sin^2 2\theta_{\mu\tau} \gtrsim 4 \times 10^{-4}$.[36]

ii) Neutrino mass is responsible for the atmospheric anomaly and the solar neutrino deficit:

The ~eV scale of neutrino mass coming from the hot dark matter constraint (4.10.a) is now gone. Let us assume that the atmospheric neutrino anomaly is due to oscillation of ν_μ into ν_τ, and the solar neutrino deficit to MSW conversion of ν_e into ν_μ. Then, denoting as before the dominant mass eigenstate of ν_e by ν_1, and the mass eigenstates which together are the dominant components of the $\nu_\mu - \nu_\tau$ system by ν_2 and ν_3, we have the constraints (4.10.b) and (4.10.c). It is natural to satisfy these constraints by supposing that

$$M_{\nu_3} \cong \left|M_{\nu_3}^2 - M_{\nu_2}^2\right|^{1/2} \sim 10^{-1}\ eV, \qquad (4.16.a)$$

$$M_{\nu_2} \cong \left[M_{\nu_2}^2 - M_{\nu_1}^2\right]^{1/2} \sim 3 \times 10^{-3}\ eV, \qquad (4.16.b)$$

and

$$M_{\nu_1} \ll 3 \times 10^{-3}\ eV. \qquad (4.16.c)$$

All neutrino masses are now too small to have visible effects in forseeable tritium β decay experiments. Turning to $\beta\beta_{0\nu}$, we see from Eq. (2.4) that M_{ν_3} is large enough to produce an $M_{eff}(\beta\beta_{0\nu})$ of order 0.1eV (which may be observable in future experiments as we shall discuss) if the mixing parameter $|L_{e3}|^2$ is near unity. However, the maximal $\nu_\mu - \nu_\tau$ mixing implied by the $\nu_\mu \rightarrow \nu_\tau$ interpretation of the atmospheric anomaly implies that $|L_{e3}|^2$ is probably not bigger than 0.1. Then the contribution of ν_3 to $M_{eff}(\beta\beta_{0\nu})$ is not bigger than $(0.1) \times 10^{-1}eV = 10^{-2}eV$. We do not know of plans or ideas for any experiments sensitive to such a small $M_{eff}(\beta\beta_{0\nu})$ in the near future. (See, however, Sec. 5.3 with regard to the distant future.)

Of course, if this scenario is true, long-baseline searches for $\nu_\mu \rightarrow \nu_\tau$ oscillation in accelerator ν_μ beams, aimed at confirming the $\nu_\mu \rightarrow \nu_\tau$ interpretation of the atmospheric neutrino anomaly, would prove very interesting.

iii) Neutrino mass is responsible for the hot dark matter and the atmospheric neutrino anomaly:

The hot dark matter constraint (4.10.a) must once again be obeyed. Assuming that the atmospheric anomaly is due to $\nu_\mu \rightarrow \nu_\tau$ oscillation, and denoting as before the dominant mass eigenstate component of ν_e by ν_1 and the dominant components of the $\nu_\mu - \nu_\tau$ system by ν_2 and ν_3, we also have the constraint (4.10.b). Let us assume that

$$M_{\nu_1} \leq Min\left\{M_{\nu_2}, M_{\nu_3}\right\}, \qquad (4.17)$$

so that the neutrino mass hierarchy is not inverted. Then, to satisfy the constraints (4.10.a) and (4.10.b), we must have

$$M_{\nu_2} \cong M_{\nu_3} = (2-3)\ eV. \qquad (4.18)$$

In the absence of the constraint (4.10.c) from solar neutrinos, a possible scenario is

$$M_{\nu_1} \ll M_{\nu_2} \cong M_{\nu_3} \simeq 3\ eV. \qquad (4.19)$$

Observation of the mass of $\nu_e \sim \nu_1$ in tritium β decay is then hopeless. Turning to $\beta\beta_{0\nu}$, we note that the contributions of ν_1 and ν_2 to the $M_{eff}(\beta\beta_{0\nu})$ of Eq. (2.4) will be below 0.1eV. However, if neutrinos are Majorana particles, then the bounds[37] on $|L_{e3}|^2$ for the case where $|M_{\nu_3}^2 - M_{\nu_1}^2| \simeq (3\ eV)^2$ allow ν_3 to contribute as much as $\sim 0.1\ eV$ to $M_{eff}(\beta\beta_{0\nu})$, so that experiments with sensitivity in this range could see a signal.

In this scenario, there may be $\nu_\mu \longleftrightarrow \nu_e$ oscillation with $\delta M^2 = M_{\nu_2}^2 - M_{\nu_1}^2 \simeq (3\ eV)^2$. Despite the large $\nu_\mu - \nu_\tau$ mixing, the oscillation would still depend on the distance L travelled by the neutrino as it does in the case where just two neutrinos mix and all neutrino mixing angles are small. The mixing angle $\theta_{e\mu}$ governing the $\nu_\mu \longleftrightarrow \nu_e$ oscillation probability would be below the present bound[38] of $\sin^2 2\theta_{e\mu} \lesssim 4 \times 10^{-3}$ for $\delta M^2 \sim (3\ eV)^2$.

In summary, the existing hints of neutrino mass suggest that tritium β decay or $\beta\beta_{0\nu}$ experiments with somewhat greater sensitivity to neutrino mass than the present ones could actually see positive signals. These hints also suggest that certain searches for neutrino oscillation could bear fruit.

The scenarios which the existing hints have suggested have involved neutrino masses which are all in the eV range or below. What if these scenarios are all wrong, and the heaviest of the neutrinos, presumably ν_τ, has a mass in the keV or even MeV range? A ν_τ with mass in the MeV range has interesting cosmological and astrophysical consequences (for a review see [39]). Upper bounds on the allowed total energy density of the universe during nucleosynthesis (at $t \simeq 10^{-2} - 10^2$ sec after the Big Bang) exclude the mass range $0.2\ MeV - 33\ MeV$ for a Dirac ν_τ, and $0.4\ MeV - 30\ MeV$ for a Majorana ν_τ, if the ν_τ lifetime is longer than $O(100\ sec)$ (i.e. if ν_τ does not decay before the end of nucleosynthesis) [40,41,42]. Once combined with the new laboratory upper bounds of less than $30\ MeV$ (the bound of $23.8\ MeV$ from ALEPH, or even the preliminary Beijing bound of $29\ MeV$), this nucleosynthesis argument would push the upper mass bound for these long-lived ν_τ ($\tau > 100\ sec$) to $0.4\ MeV$ (and astrophysical arguments would push the limit much further for Dirac neutrinos, as explained below). Let us remark, however, that these nucleosynthesis bounds apply only if the main annihilation modes of the ν_τ are those of the Standard Model. If the ν_τ are assumed to have additional interactions which increase their annihilation cross sections in the early universe, their relic abundance decreases and the nucleosynthesis bounds are loosened. In most models predicting neutrino masses, neutrinos have extra interactions, and in many of them neutrinos have annihilation cross sections larger than weak.

Another interesting bound coming from the supernova $SN1987A$ excludes a Dirac ν_τ of mass from about $10keV$ to $1MeV$. The reason is that the mainly inert right-handed helicity states of massive Dirac neutrinos can be produced in weak interactions due to their admixture in the left-handed chirality neutrino states. Therefore, right-handed ν_τ with masses in the forbidden range produced in the core of the supernova, would have escaped from the star carrying off too much energy and consequently shortening the observed $\bar{\nu}_e$ pulse [43,44]. Thus, a ν_τ lighter than $1MeV$ must be Majorana, unless it is lighter than $O(10keV)$.

The mentioned nucleosynthesis constraints are based on the observational upper limit on the abundance of ^4He, because larger densities during nucleosynthesis lead to overproduction of ^4He. A non-relativistic neutrino species can contribute more to the energy density of the universe than a relativistic one. The number of light neutrino species allowed by the abundance of ^4He is $N_\nu \lesssim 3.3$ [45] and, for the ranges rejected, stable ν_τ with weak annihilation cross sections would contribute more than 1.3 relativistic neutrino species. We can understand, therefore, that if ν_τ decays into relativistic particles before the end of nucleosynthesis, these bounds are loosened and become dependent on the nature of the decay products. "Visible" modes ($\nu_\tau \to \nu'\gamma$, $\nu_\tau \to \nu_e^+ e^-$) are ruled out, as the dominant decay mode for MeV tau-neutrinos, through a combination of laboratory bounds, nucleosynthesis bounds and constraints from supernova observations [46,44]. "Invisible" modes are not only allowed but may have very interesting consequences. These modes are $\nu_\tau \to 3\nu$'s and $\nu_\tau \to \nu'\phi$, where ϕ is a light or massless boson, most naturally a Goldstone boson. A ν_τ with mass $< 1MeV$ and short lifetime $\tau < 30$ sec (producing ν_e or ν_μ [41 and 47], but in particular the effect of ν_e decay products is important [41]) would actually decrease the amount of ^4He produced, with respect to a relativistic stable ν_τ. Therefore ν_e, ν_μ and this particular ν_τ would count in nucleosynthesis as *less* than three effective neutrino species, $N_\nu < 3$. This would rescue the Big Bang from a potentially important inconsistency if the observational data would indicate (as sometimes argued) that the abundance of ^4He is actually lower than usually thought at present (what would lead precisely to a limit $N_\nu < 3$). Alternatively, keeping the present bound, $N_\nu < 3.3$, the constraint on additional "new" contributions from exotic particles would be loosened. These possibilities would be eliminated by a laboratory upper bound of around $1MeV$.

There are still other possibilities to be tested or rejected by lowering the upper bound on m_{ν_τ}. A ν_τ with an annihilation cross section faster than weak by a factor of 10 and with a mass of $20MeV - 30MeV$ and lifetime 200 sec - 1000 sec, decaying into ν_e would allow to increase the upper bound on the baryonic density by a factor of 10 [48]. The upper bound on the density of baryons is due to overproduction of ^4He and ^7Li and underproduction of D. Overproduction of ^4He is avoided because the abundance of these ν_τ is so low that the effective number of neutrinos is $N_\nu \simeq 2$, and the other two abundances come out right because the very effective disappearance of free neutrons during nucleosynthesis with large baryon densities, is avoided by the production of neutrons in the interactions of the $\bar{\nu}_e$ produced in the ν_τ decay with protons ($\bar{\nu}_e\, p \to e^+ n$). Thus, this mechanism produces with a large $\eta \equiv n_B/n_\gamma$ the results otherwise obtained for a lower η, and it allows for $\Omega_B h^2 \simeq 0.2$. Notice that for $h \lesssim 0.5$, within the accepted range $0.4 \leq h \leq 1$, one could have even $\Omega_B = 1$. In this case the nucleosynthesis constraint would <u>not</u> require the existence of non-baryonic dark matter.

Finally, it has been speculated that the residual annihilations of a ν_τ with a mass of 15 MeV to $25MeV$, in the outer core of a supernova could help power the supernova explosion [44] (at present the mechanism of supernova explosion is not well understood), and that a $1MeV - 10MeV$ ν_τ with lifetime 0.1 sec - 100 sec whose decay products include ν_e, could remedy the problems that standard cold dark matter scenarios of structure formation in the universe have to produce acceptably small inhomogeneities on small scales [49].

4.2 Can ∼2 eV be Reached in Beta Decay?

Setting aside for the moment the anomaly seen at the endpoint of the tritium spectrum, one can ask, what is the smallest mass detectable in beta decay?

In order to detect the effect of a neutrino mass of m_ν at the endpoint of a beta spectrum of total kinetic energy E_0, instrumental resolution of order $\frac{m_\nu}{E_0}$ is needed. The spectral fraction per decay that falls in the last m_ν of the beta spectrum is approximately $(\frac{m_\nu}{E_0})^3$.

For spectrometric experiments in which the source and the detector are physically separated, a limit on source thickness is set by the cross section for inelastic interactions of outgoing electrons, such that one must have $\sigma n \leq 1$, where σ is the inelastic cross section (about 5 x 10^{-22} cm^2 for molecular T_2). The dimensions of the source (area A) and the dimensions of the spectrometer (length L) are related through the resolution needed, specifically,

$$\frac{\Delta E}{E} \simeq \frac{\sqrt{A}}{L}.$$

In these conditions, the rate in the last m_ν of the spectrum is of order

$$\frac{dN}{dt} = \left(\frac{\sqrt{A}}{L}\right)^3 \frac{\lambda A}{\sigma},$$

where λ, the mean decay time, is approximately given in seconds (for allowed beta decay) by

$$\lambda = 10^{-33} ft E_0^5,$$

where ft is the beta-decay reduced matrix element (e.g., $\log_{10} ft = 3$ for tritium decay) and E_0 is in eV. Then,

$$\frac{dN}{dt} = 10^{-33} m_\nu^5 ft L^2 \sigma^{-1}$$

when background is absent and detection efficiency is unity. For example, if one sets as a minimum practical limit one event per day and $L = 10$ m, then

$$m_{\nu(ultimate)} = 0.4 \text{ eV}.$$

at 1 standard deviation. A 2-eV limit is then about 3000 times "easier" as a result of the exponent 5. In practice that large factor is eroded quickly by efficiency, background, and the need to explore the rest of the spectrum both for extraction of neutrino mass and for measurements related to systematic uncertainties.

Calorimetric experiments technically evade this limit by virtue of permitting much more intense sources (because energy loss is no longer an issue), but in practice they run into a different limit set by the presence in the data of the entire beta spectrum. The equivalent branching ratio of endpoint data to total is about 3 x $10^{-13} m_\nu{}^3$.

The experiment that most nearly approaches the ideal conditions appears to be the T_2 apparatus of Lobashev et al. [50] A retarding-field spectrometer with magnetic gradients is coupled to a Los Alamos type gaseous molecular tritium source to give very high resolution, large acceptance, and low backgrounds. Initial results are already at the 2.5-eV statistical level [51].

While both the idealized statistical arguments and experimental data indicate that the 2-eV level can be reached, and even slightly surpassed, the anomaly observed at the endpoint must be quantitatively understood before a convincing value for or limit on the rest mass of ν_e can be given.

The structure of the two-electron molecule (T_2) is presumably calculable to arbitrary accuracy given enough effort, but it would clearly be desirable to be able to work with the T atom, or even the T^+ ion, for which the atomic-structure calculations are different and simpler. Space-charge limits preclude a useful T^+ source, but it may not be out of the question to imagine an atomic tritium source of worthwhile density. The Los Alamos source [52] was originally designed to use atomic tritium prepared by dissociation in an RF discharge. The gas was prevented from recombining immediately through the use of a highly polished aluminum source volume (tube), since it had been found that aluminum hindered surface recombination and volume recombination was negligible. The atomic spectrum was never taken, however, because the molecular contamination, even at less than 10%, had to be measured extremely precisely since its endpoint is about 8 eV higher than that of the atomic spectrum. The technology for vacuum ultraviolet absorption molecular spectroscopy using frequency doubled and tripled Nd-YAG pumped dye laser light was still primitive.

Initially a somewhat different approach had been considered, namely the use of the developing technology for preparing dense spin-polarized hydrogen at very low temperatures. The advantage of this method would be the essentially negligible contamination by molecular tritium; the technology was not sufficiently advanced in the 1980s to make this option attractive. Now it is possible to prepare atomic densities in excess of 10^{14} cm^{-3}, within striking distance of the 10^{15} to 10^{16} needed [53]. To establish whether this could form the basis for a successful atomic tritium experiment would require a significant research effort.

4.3 ββ Decay with Improved Sensitivity

Many groups are working in this field using different methods. Because the calculations of $\beta\beta 2\nu$– and $\beta\beta 0\nu$–matrix elements are substantially different, there is no way to test the reliability of neutrino mass limits determined from measured limits of $\beta\beta 0\nu$–decay rates, except to measure this non–standard model decay for several isotopes.

In the absence of any rigorous test, a big experimental effort has been devoted world wide to the measurement of two neutrino double beta decay rates. They can be considered as an indirect probe of our understanding of second order weak decays. It is generally believed that the $\beta\beta 0\nu$–matrix elements are less influenced by the various uncertainties of those calculations than the $\beta\beta 2\nu$ case. If this is in fact true the agreement between calculated and measured $\beta\beta 2\nu$ decay rates can be considered as an upper limit for the uncertainties of the matrix elements of the neutrinoless decay mode.

Almost 40 experiments are studying $\beta\beta$–decay all over the world. The allowed two neutrino $\beta\beta$–decay has now been observed in 9 different isotopes and measured half lives seem to agree reasonably well with the theoretical predictions.

In this Section we would like to outline experiments capable of sensitivities approaching 0.1 eV. Therefore, only the two types of experiments giving the most stringent $M_{eff}(\beta\beta_{o\nu})$ limits today will be discussed. For a more complete description of the field the reader is referred to reference [9]. As will be shown, sub-eV neutrino masses are already within reach of the experiments discussed here.

We shall start with a brief description of the experimental challenges of this field. In fact the problems to be solved are very similar although the ways to solve them might be completely different from experiment to experiment.

Because of the extremely long half lives of $\beta\beta$–decays, often resulting in only a few decays per year, the reduction of background radiation is the major issue in this field of research. Reaching background levels of a few events per year is a non-trivial and time consuming problem if the detector should have kg–size. This goal can be achieved either by reducing the level of background–creating impurities or by detecting their decay with high sensitivity. Two examples for this will be discussed here. In order to eliminate cosmic radiation these experiments are in general performed in underground laboratories.

Within the direct experiments (detecting the electrons emitted in the decay), those using the $\beta\beta$–decaying source simultaneously as detector have at present the highest sensitivity towards a Majorana neutrino mass. In this approach self absorption losses of the electrons in the source can be minimized leading to high detection efficiencies. On the other hand the number of isotopes which can be studied in this way is limited.

The use of large amounts (several kg) of isotopically enriched source material, concentrating the source strength in compact detectors with a low sensitivity towards background radiation, has become customary. ^{76}Ge and ^{136}Xe are the two isotopes studied most successfully so far.

The inclusive geochemical method, detecting the decay

product and not the electrons, will not be discussed here. However it can yield stringent limits if the $\beta\beta 2\nu$–half life is sufficiently long, as in the case of ^{128}Te [54].

Calorimetric ^{76}Ge Experiments

Two experiments, done as international collaborations, are currently under way. Both experiments have around 20 kg of isotopically enriched ^{76}Ge at their disposal. In both projects the extremely valuable enriched material was supplied from Russia. It has an isotopical abundance of 86% of ^{76}Ge compared to only 7.8% in natural Ge.

The idea of using high resolution Ge semiconductor detectors as source and detector has been successfully used before with natural Ge. The 11–fold higher concentration of the source strength allowed however a significant step in sensitivity.

This purely calorimetric approach completely relies on the good energy resolution of Ge detectors. The existence of the non–standard model $\beta\beta 0\nu$–decay would lead to a narrow peak at 2.038 MeV well within the energy range of the natural radioactivity. The extremely good energy resolution allows however a precise determination of the peak energy and concentrates hypothetical $\beta\beta 0\nu$ events into a narrow energy interval.

The detection efficiency ϵ for electrons created inside the Ge crystal is limited only through the leakage into the inactive layer of the Ge detector and small bremsstrahlung losses. It is around $\epsilon = 0.9$.

Attempts to use characteristic differences in the charge collection times of strongly localized $\beta\beta$–events and multi site background events, caused by Compton scattering [55], have so far not entered the stage of data taking.

The Heidelberg–Moscow Experiment This experiment, performed as a collaboration between the Max–Planck–Institut für Kernphysik in Heidelberg and the Kurchatov Institute in Moscow, is located in the Gran Sasso underground laboratory in Italy at a depth of 3500 m w.e.. The experiment is taking data since 1991. Three enriched Ge detectors of 6.0 kg active mass are in regular operation right now. Large Ge crystals, the biggest one having 2.9 kg, are used. They represent a source strength of N=68.7 mol of ^{76}Ge. Two new detectors are in preparation. The total source strength of the experiment will be around 10 kg.

The Ge crystals are mounted in low background cryostats custom made from electrolytical Cu. The detectors have a common passive shield to suppress the γ–radiation coming from ^{232}Th and ^{238}U contained in the laboratory walls, reaching up to 2.6 MeV. It is composed of 10 cm ultra pure so called LC2 grade lead followed by 20 cm of less clean Boliden Pb. The shield has recently been upgraded with 10 cm of boron loaded polyethylene to suppress neutrons coming from (α, n) reactions in the laboratory wall, or residual μ interactions with the walls. The setup has no cosmic veto system. All construction materials of the cryostats and the shield have been carefully selected for their radiopurity using an existing low background Ge detector. Their exposure time to the cosmic radiation above ground has been limited to a minimum, to prevent the production of long lived activation products. The passive shield is mounted in an airtight steel container continuously flushed by high purity nitrogen gas to displace radioactive Rn gas contained in the air.

The large body of data accumulated by this experiment during the last three years of operation allowed for the first time to reach a neutrino mass limit below 1 eV. The group has published [56] a half life limit of $T_{1/2}^{0\nu} > 5.1 \cdot 10^{24}\ y$ (90% c.l.) corresponding to $M_{eff} < 0.7\ eV$ if the nuclear matrix element of [57] is used. This August 1994 result is based on $8.6\ kg \cdot y$ of measuring time corresponding to an exposure of $98.3\ mol \cdot y$ for ^{76}Ge. Available now and being analyzed are $157\ mol \cdot y$ of measuring time. Earlier tentative indications for an excess of events at the $\beta\beta$–decay energy seem to have disappeared with longer measuring time. Figure 1, which was taken from reference [56], shows the measured spectrum around 2.04 MeV.

The large source strength together with the low background of the detectors allowed the first high statistic measurement of the $\beta\beta 2\nu$–decay channel [58]. The half life measured is $T_{1/2}^{2\nu} = (1.43 \pm 0.03^{stat} \pm 0.13^{syst}) \cdot 10^{21}\ y$. A stringent limit for the Majoron–neutrino coupling of $\langle g_{\nu\chi} \rangle < 1.8 \cdot 10^{-4}$ [59] has been determined as well.

Taking into account the background of 0.021 $counts/keV \cdot y \cdot mol$ measured in the interval from 2.00 to 2.08 MeV and assuming that the experiment will continue for two more years after the full source strength has been implemented one can estimate the $T_{1/2}^{0\nu}$ range which can be eventually tested. A square root approach, discussed later, yields a half life limit of around $2 \cdot 10^{25}\ y$.

The International Germanium Experiment The IGEX collaboration is formed by groups from Russia, Spain and the US. The experiment has just recently started taking data. The experimental ansatz is quite similar to the project described before. Five enriched Ge detectors having an active mass of 5.8 kg are in operation in different underground locations. Two of them having 2 kg each, installed in the Homestake mine in SD at $\sim 4000\ m.w.e.$, can be considered to be in their final configuration. The cryostats are made underground of electro-formed Cu. The electro-chemical production of Cu parts called electro-forming is another cleaning step since elements as U and Th tend to be depleted in this way. If done underground it prevents the activation of the Cu. The detectors are in a passive shield made of selected lead. In addition one detector of 0.65 kg fiducial mass is operated in the Canfranc tunnel in Spain and another one of 0.7 kg in the Baksan underground laboratory in Russia.

To date, the experiment has accumulated 15 $mol \cdot y$ of ^{76}Ge exposure time. The background of 0.014 $counts/keV \cdot y \cdot mol$ around 2.04 MeV is comparable to the level of the experiment described before. The group used 11 $mol \cdot y$ to derive a half life limit of $T_{1/2}^{0\nu} > 1.7 \cdot 10^{24}\ y$ (90% c.l.) for the neutrinoless decay mode. The experiment sees evidence for the $\beta\beta 2\nu$–decay of ^{76}Ge but with a somewhat shorter half life of $T_{1/2}^{2\nu} = (1.12 \pm 0.17) \cdot 10^{21}\ y$ [60] than that of the Heidelberg–Moscow experiment. The goal of the collaboration is to produce as many Ge crystals as possible from their enriched Ge. Perhaps they will also end up with around 10 kg in form of Ge detectors. In this respect the reachable sensitivity towards neutrino masses will be similar to that of the previously discussed experiment.

The Caltech–Neuchâtel–PSI ^{136}Xe TPC

The investigation of ^{136}Xe is at present the only other method to match the sensitivity level reached with Ge detectors. The fact that Xe can be used as a counting gas has stimulated many experiments to study this topic. The most successful instrument is the Caltech–Neuchâtel–PSI TPC using isotopically enriched (62.5 %) ^{136}Xe gas at 5 bar pressure. The spatial resolution of this instrument allows to distinguish two–electron events from one–electron events and various other backgrounds. This is a powerful tool to reduce backgrounds even below the astonishing levels reached with Ge detectors. The Q-value of the decay is 2.48 MeV slightly higher than that of ^{76}Ge but still within the energy range of the natural radioactivity.

The TPC is operated in the St. Gotthard underground laboratory in Switzerland at a depth of 3000 m.w.e., by a collaboration between Caltech in Pasadena, the University of Neuchâtel and the Paul Scherrer Institut in Villingen [61]. A source strength of N=24.2 mol of ^{136}Xe is used. The TPC itself and all its metallic components are made from selected OFHC Cu. All technical components such as e.g. resistors have been selected by monitoring with a low background Ge detector. The TPC is shielded against the radioactivity of the laboratory by 5 cm of Cu and 20–30 cm of Pb. The shield is continuously flushed with dry nitrogen gas to remove the Rn. The purity of the counting gas itself is maintained on the 0.1 ppm level in electro-negative contaminants by continuously pumping it through a low Rn getter. The setup is equipped with a cosmic ray veto to tag the muons.

Two–electron events are identified through differences in dE/dx at the beginning and end of a track, respectively. The discrimination efficiency for single electrons is 98 %. Multiple scattering in the high pressure gas prevents the identification of the vertex of the decay. The TPC has an energy resolution of 6.6 % at 1.6 MeV. The powerful background rejection capability of this instrument results in the lowest background level among all existing $\beta\beta$–decay experiments of $B = 0.002\ counts/keV \cdot y \cdot mol$. A drawback is the relatively small detection efficiency of $\epsilon = 0.23$ for electron pairs, which is a folding of the probability that both electron tracks are completely contained in the TPC (30 %) and the finite analysis efficiency (76 %).

The first phase of this experiment was completed in 1992. Figure 2 depicts the measured sum energy spectrum of the two–electron events, together with the excluded $\beta\beta0\nu$–signal.

An exposure time of 18.9 $mol \cdot y$ allowed to exclude a neutrinoless decay half life of $T^{0\nu}_{1/2} \leq 3.4 \cdot 10^{23}\ y$ (90% c.l.) which corresponds to a neutrino mass limit of 2.8 eV. Due to the steep rise of the TPC background at lower energies the $\beta\beta2\nu$–decay has not been measured yet. A half life limit of $T^{2\nu}_{1/2} > 2.1 \cdot 10^{20}\ y$ has been determined which is the best limit obtained so far for this isotope. The group published also a limit for the Majoron–neutrino coupling of $\langle g_{\nu\chi} \rangle < 2.4 \cdot 10^{-4}$ using the matrix element of ref. [62].

The TPC has recently undergone remodelling [63]. A new anode has been constructed. Resistors were replaced by those having lower background. To avoid soldering the TPC wires are now crimped to an acrylic frame. The XY plane was reconstructed as well.

After installation of these new components about 1000 h of data has been taken by mid August 1994. These data show an encouraging factor 5 reduction in background. Meaningful new results should be available within a few months. The group plans to reach a half life limit of $10^{24}\ y$ in one year of operation. Another important goal is to measure the two neutrino $\beta\beta$–decay. To do this the TPC will be operated at lower pressure to improve the track quality at low energies.

Xe gas depleted in the isotope ^{136}Xe is available and could be used to measure the background. This method, however, relies on the assumption that the background of the enriched source and the depleted sample is identical.

The measurement of the $\beta\beta2\nu$–channel is interesting because the theoretical predictions show a considerable discrepancy, while reference [57] gives $1.4 - 21.1 \cdot 10^{21}\ y$, ref. [62] predicts $8.2 \cdot 10^{20}\ y$.

Discussion of the Results

If we combine the technical parameters of both detection concepts into a figure of merit S we can directly compare the TPC to the calorimetric Ge experiments.

$$S = \epsilon \cdot \sqrt{\frac{N}{\Delta E \cdot B}} \tag{4.9}$$

ΔE [keV] is the energy resolution (FWHM) at the $\beta\beta$–decay energy. The sensitivity S is directly connected to the half life limit to be determined from the data if no peak can be found after measuring for the time t [y]. This relation which does not depend on the used isotope, is

$$T^{0\nu}_{1/2} > (3.2 \cdot 10^{23}\ mol^{-1}) \cdot \frac{S}{f} \cdot \sqrt{t} \tag{4.10}$$

f depends on the chosen confidence level of the limit.

The Caltech–Neuchâtel–PSI TPC has a sensitivity of $S = 1.98\ mol \cdot \sqrt{y}$ whereas the Heidelberg–Moscow experiment has $S = 27.9\ mol \cdot \sqrt{y}$ in its present configuration. From this we see that the superior background of the TPC is more than compensated by its worse energy resolution and efficiency.

A comparison in terms of the neutrino mass limits which can be reached is more difficult because the nuclear matrix element calculations of refs. [57,62] are similar to each other for ^{136}Xe but differ considerably for ^{76}Ge. Table 2 gives a compilation of these sensitivities.

As can be seen from Table 2 the technology available today will allow us to explore neutrino masses around 1 eV. Because these experiments are already using tens of kg of highly enriched, and therefore very expensive isotopes, it is difficult to imagine how to gain another order of magnitude by scaling up these ideas. Some of the new ideas are discussed in the next section. To set the scale we estimate the lifetimes corresponding to 0.1 eV neutrino mass.

How good will detectors have to be to see neutrino masses $M_{eff} \approx 0.1$ eV? The Table 3 contains estimates of the lifetimes that correspond to 0.1 eV in several potentially useful nuclei. The table was constructed by culling calculated rates from the available literature. The "best values" reflect a certain degree of personal bias towards one calculation or another. The uncertainties are obviously difficult to estimate; what they really represent is a conservative assessment of the spread in values from respectable calculations. The possibility that none of the calculations is good in some particular nucleus cannot be completely ruled out. The ranges of uncertainty are usually least reliable where they are smallest (e.g. in ^{136}Xe), most often because only one method of the several in circulation (shell model, QRPA, pseudo-SU(3) ...) has been used. When no uncertainty is given it is because only one believable result was found.

§5 PROSPECTS FOR FURTHER IMPROVEMENT

5.1 M_{ν_μ}: *Ideas for going below 160 keV*

Historically, $M^2_{\nu_\mu}$ has also suffered from having negative experimental values yielded from experiments with stopping pions. However, recently this difficulty has been removed by improved extraction of the pion mass from studies of pionic x-rays. Nonetheless efforts should continue to seek other means to lower the limits especially by in-flight methods.

In this workshop, preliminary studies were carried out. The first involves the use of $K^\circ_{\mu 3}$ decay spectra and the second a novel suggestion using pion decays in a $(g-2)$ apparatus.

Limit on the m_{ν_μ} from $K^\circ\mu 3$ Decay Spectra Recent upgrades to the BNL-AGS have made it possible to make many high statistics measurements. Here we will consider the statistics and the resolution needed to improve the limit on the muon neutrino mass using decays. The method relies on the shape of the invariant mass distribution of the pion and the muon, $m_{\pi\mu}$, near the end point where the neutrino energy is low in the center of mass [64]. The use of simultaneously detected $K^\circ_L \to \pi^+\pi^-$ events to calibrate the mass scale and understand the resolution of the spectrometer reduces the sensitivity of this method to previously measured particle masses and to the magnetic field in the spectrometer. An additional check on the experimental method can be obtained by measuring the endpoint of $K^\circ_L \to \pi^\pm\mu^\mp\nu_e$ decays in which one does not expect to observe a non-zero neutrino mass at the sensitivity level of this experiment. Finally, by comparing the end point spectra for neutrinos and antineutrinos one can obtain the most stringent limit on a neutrino-antineutrino mass difference, a non-zero value of which would violate CPT invariance. The only other method that limits a neutrino-antineutrino mass difference is the measurement of π^+ and π^- lifetimes, if one attributes the difference in the lifetimes to the dependence of the decay amplitude on neutrino mass [65,66].

The V-A theory of weak interactions predicts little dependence of the $K_{\ell 3}$ matrix element on the neutrino mass. The boundary of the Dalitz plot does, however, depend on the neutrino mass. In the case of zero neutrino mass the $m_{\pi\mu}$ distribution approaches m_K with the first derivative zero, and in the case of non-zero neutrino mass the same distribution terminates before approaching m_K. The measured quantity, $m_{\pi\mu}$, at the endpoint is simply related to the neutrino energy in the kaon center of mass by the following formula:

$$\begin{aligned} E_\nu &= \tfrac{m_K^2 - m_{\pi\mu}^2 + m_\nu^2}{2m_K} \\ &\approx \tfrac{(m_K + m_{\pi\mu})}{2m_K}(m_K - m_{\pi\mu}) \\ &\simeq m_K - m_{\pi\mu}\,. \end{aligned} \tag{5.1}$$

Since m_K is effectively measured in the same experiment by using $K^\circ_L \to \pi^+\pi^-$ decays, the neutrino energy measurement is independent of the kaon mass. Furthermore, the dependence on pion mass is only in second order, and therefore the pion mass uncertainty (0.4 keV) contributes only about 0.23 keV to the neutrino energy uncertainty.

The previous best experiment of this type achieved mass limits of 725 keV for the muon neutrino and 500 keV for the electron neutrino at 90% confidence level [64]. The experimenters collected 2130 well identified $K^\circ_L \to \pi^\pm\mu^\mp\nu_\mu$ events with $m_{\pi\mu} > 493.5$MeV (the fraction of the total spectrum in the Dalitz plot that is above this mass without dilution due to resolution smearing is 3.1×10^{-5}). They used the invariant mass spectrum including the non-Gaussian tails of $K^\circ_L \to \pi^+\pi^-$ events to calibrate the absolute mass scale and also to simulate the experimental resolution. The lower limit on the mass window is dictated by the need to suppress background arising from $K^\circ_L \to \pi^+\pi^-$ events in which one of the pions either decays in flight or is misidentified as a muon. Kinematically this background should not populate events with $m_{\pi\mu}$ above 489 MeV. This is seen most simply from the use of the following approximation for two body invariant mass in the laboratory frame:

$$m_{12}^2 \approx m_1^2 + m_2^2 + \frac{p_2}{p_1}m_1^2 + \frac{p_1}{p_2}m_2^2 + p_1 p_2 \theta^2 \tag{5.2}$$

where p_1, p_2 and θ are the laboratory momentum of particle 1, particle 2 and the laboratory angle between particles 1 and 2, respectively. In the extreme case that the momentum of the misidentified muon is much larger than the momentum of the pion the invariant mass will be: $m^2_{\pi\mu} \approx m^2_K - (m^2_\pi - m^2_\mu)$. Errors in momentum measurements cause background events to leak out of this kinematic limit to populate the end point region.

One must collect much higher number of events with $m_{\pi\mu} > 493.5$MeV with better mass resolution and muon identification to improve the muon neutrino mass limit. Experiment E871, a search for $K^\circ_L \to \mu^\pm e^\mp$, at the BNL-AGS offers such a possibility [67,68]. E871 is a two arm magnetic spectrometer optimized for two body charged decays of neutral long lived kaons. Two magnets provide nearly independent measurements of the momentum of each particle. A Cerenkov counter, lead glass, and a fine grained muon range system provides redundant particle identification. The trigger and data aquisition system is built to take data at very high rates to achieve sensitivities of 10^{-12} for two body charged decays of the K_L°.

The estimated mass resolution for E871 is 1.13 MeV for $K_L^\circ \to \pi^+\pi^-$ events. The number of $K_L^\circ \to \pi^\pm\mu^\mp\nu_\mu$ events with $m_{\pi\mu} > 493.5$ MeV necessary as a function of the limit on m_{ν_μ} assuming a Gaussian mass resolution with $\sigma_m = 0.5$, 1.0 and 1.5 MeV have been computed; see Figure 3. Barring systematic error the number of events needed to achieve a sensitivity of 100 keV (a significant improvement on the best 90% confidence level limit of 160 keV from pion decay at rest [7]) is about 3.0×10^5. The number of events needed increases very rapidly as the limit is lowered mainly because the V-A shape approaches the kaon mass with zero slope so that in the case of zero neutrino mass there are no events at the end point. Most possible systematic errors, such as variation in the spectrometer acceptance over the fitted mass range or background contamination, will either change the shape smoothly or cause the overall normalization to shift. A non-zero neutrino mass should cause a relatively abrupt change in the shape only at the endpoint, therefore the systematic errors should not be a major difficulty. This, of course, will have to be studied with actual data. The geometric dimensions of the E871 spectrometer, the appropriate kaon spectrum, and the requested intensity have been used to compute the rate of $K_L^\circ \to \pi^\pm\mu^\mp\nu_\mu$ events with $m_{\pi\mu} > 493.5$MeV to be about 300 per hour. E871 should be able to collect more than enough events in a 4000 hour run for this measurement.

Limit on M_{ν_μ} from $(g-2)$ Apparatus

Kinematic tests of the neutrino mass which rely on nuclear beta decay can provide very sensitive limits for the electron neutrino mass, but need well understood resolution functions for the spectrometer and model-dependent atomic corrections as discussed above. A complementary kinematic test of neutrino masses which does not suffer from these limitations is provided by the decay $\pi \to \mu\nu$. As a simple 2-body decay, it is more accessible to experimental measurements of the neutrino mass than the 3-body purely leptonic muon decay $\mu \to e\nu\nu$. The current lowest bounds on the muon neutrino mass were measured in experiments of this kind: from the pion-stopping experiment of Assamagan et al. [7] mentioned above, and $m_\nu < 500$ keV (90% CL) from the pion in flight experiment of Anderhub et al. [69].

Experiments where the pion is brought to rest rely on a very accurate measurement of the momentum of the daughter muon, from which the neutrino mass is inferred using the relation $p_\mu^2+m_\mu^2 = (m_\pi^2+m_\mu^2-m_\nu^2)^2/4m_\pi^2$. The other ingredient in the formula is the mass of the pion, which contributes about equally to Δm_ν^2. Uncertainties in the mass of the muon are a much less significant contribution. Two earlier papers from the same group ([70] and [71]), used essentially the same technique, but with a less efficient production mechanism which necessitated wider collimator openings in their spectrometer and thus a poorer momentum resolution. In the first paper, Abela et al., an upper limit of 250 keV (CL=0.9) is calculated by assigning the probability function to zero for $m_\nu^2 < 0$, a necessary step since plugging directly in the values in the formula above and using the pion mass defined in the Review of Particle Properties (e.g. $m_\pi = 139.5675 \pm 0.0004$ MeV [1990]) yields a negative value for m_ν^2. In the second paper by Daum et al., the uncertainties on the muon momentum were reduced, measuring $p_\mu = 29.79206\pm 0.000684$ MeV, but the problem of a non-physical neutrino mass squared term remained, with $m_\nu^2 = -0.154\pm 0.044$ MeV$^2/c^4$. If the momentum measurement is correct, there are only 2 other alternatives (barring the interesting suggestion that neutrinos are tachyons): either the π^- mass measurements are wrong or $m_{\pi^-} \neq m_{\pi^+}$. Since the second alternative implies a violation of CPT, it is more likely that the pion mass is the problem, since the Particle Properties value is heavily weighted by one experiment [72] which is based on pion capture (4f→3d transitions in ^{24}Mg) and therefore subject to a vacuum polarization correction of 0.2% for a 0.0002% measurement. Thus, the authors choose to quote their result as a determination of the π^+ mass, a result which is 3σ larger than the previous world average. Since that time, the Jeckelmann pionic X-ray spectrum has been reanalyzed [73], resulting in two solutions for the pion mass, depending on whether the strongest component in the 4f-3d transition is assumed to be one or two K-electrons. With the larger pion mass (2 K-electrons), Assamagan et al. find $m_\nu^2 = -0.022 \pm 0.023 MeV^2$, consistent with zero. Since the smaller pion mass still gives a negative mass-squared term by six standard deviations, and since earlier pion mass experiments (e.g. [74]) which observed pionic X-rays in other materials (e.g. phosphorus and titanium) are consistent with the smaller mass, there is still a model-dependent element in this analysis.

If, instead, the pion decay is observed using high energy pions in flight, the neutrino mass determination becomes almost independent of the value of the pion and muon masses. Measuring the difference in momenta $p_\pi - p_\mu$ for forward going muons gives the neutrino momentum, from which the mass can be extracted: $m_\nu^2 = [\sqrt{p_\mu^2+m_\mu^2} - \sqrt{p_\pi^2+m_\pi^2}]^2 - p_\nu^2$. In the Anderhub experiment, 350 Mev pions were momentum analyzed in a 180° spectrometer, passed through a decay region, and the daughter muons were reanalyzed in the same spectrometer to reduce any systematic error in the momentum difference measurement. Time of flight was used to identify the decay muons. It is interesting, but not significant, that the Anderhub experiment also measured a negative $m_\nu^2 = -0.14 \pm 0.20$ MeV$^2/c^4$.

For 350 MeV pions, the effect of a massive 300 keV neutrino is to reduce the forward going daughter muon momentum of 355.7 MeV by ~ 3 keV. Such a precision momentum difference measurement requires superb determination of the decay angle and is ultimately limited by multiple scattering and energy loss along the flight path. Since small angle multiple scattering goes as the square root of the number of radiation lengths of material, and we are measuring the square of m_ν, then to achieve a factor of 10 improvement in m_ν one must be able to reduce the thickness of the chambers by 10^4

while maintaining the same position resolution. This is an extremely difficult, but not necessarily impossible task, and would require new technologies. One possibility is the use of high yield secondary emission foils. The secondary electrons can be electrostatically accelerated across the vacuum chamber, using the magnetic field of the spectrometer to maintain their coherence, then read out by a microchannel plate and image intensifier/CCD chain. These types of detectors have been in use for a long time, usually as beam monitors in accelerators [75]. In order to achieve the required efficiency for very thin foils, it may be necessary to go to NEA (negative electron affinity) foils such as *GaAs* [76]. Although requiring a considerable amount of detector R&D, it now seems possible to repeat the Anderhub experiment with thinner detectors in order to improve the muon neutrino mass limit. Since the uncertainties in the pion mass will still be negligible, even with a factor of ten improvement in precision, this experiment would be worthwhile.

Performing an experiment to measure the pi-mu momentum difference requires an extremely precise and well-understood spectrometer. This has led one of us [77] to consider using the new storage ring being built at Brookhaven for the g-2 experiment for a neutrino mass limit experiment using pions in flight. The ring is 7 meters in radius and will be shimmed to 1 ppm for a median field of 1.47 Tesla. Stationary NMR probes, an NMR trolley which maps the field every day, and absolute NMR calibration probes will ensure that the field is known (and controlled via current feedback coils) to 0.1 ppm. Pions or muons of 3 GeV can be injected into the ring from the AGS. There are several techniques that could work. The most straightforward method would be to measure the pion momentum after the production target, allow the pions to decay in the straight section of the injection line to select for forward-going muons, then measure the muon momentum in the g-2 storage ring using three very thin secondary emission foils. Another interesting possibility is to allow both pions and muons to circulate in the storage ring and measure the time of flight after each turn. Since the period of revolution in a fixed radius, fixed B-field ring is inversely proportional to the energy of the particle, a forward going muon produced in a decay with an associated massive neutrino, having a lower momentum than one produced in conjunction with a massless neutrino, will circle the ring faster. The period in the g-2 storage ring is about 150 ns. Three GeV pions will travel 35 ps/revolution slower than their daughter muons, and muons associated with 300 keV neutrinos will be 1 ps/rev quicker than any of them. This type of measurement is extremely difficult and is only possible because some of the dilated lifetime muons will survive 50 turns or so. In addition, the time of flight measurement must be made in a non-destructive way; the proposed method being the detection of a change in the light output of a laser beam passing through an electrooptic crystal situated a millimeter or so from the particle beam. The polarization of the material is changed by the electric field of the passing particle and is detected by a photodiode through a polarized filter. As with the secondary emission foils, this will require some detector R&D, since the required time resolution (several psec) and sensitivity (single particles at 1mm distance) are close to the current limitations of this method.

5.2 M_{ν_τ}: *What is needed*

No explicit work was done by this group on methods to improve limits on M_{ν_τ}; however, we note in passing that studies made for future "*B*-factories"[78] or "Tau-Charm-factories" with improved detector resolution suggest that levels of $\sim$ 2.5MeV might be reached as a consequence of increased statistics and background subtractions.

5.3 *Double Beta Decay: Ideas for approaching 0.1 eV*

The NEMO project The international NEMO collaboration is exploring a new way to reach sub–eV sensitivities in neutrino mass [79]. The experiment [80], which is fully funded, will be done as a collaboration between French, Russian, Ukrainian and US groups in the Frejus underground laboratory in Modane (France) at a depth of 4800 m.w.e.. From the previous discussion it is obvious that the combination of a detector having tracking capabilities and a good detection efficiency for the electrons would be desirable. Because these requirements are difficult to be met if the source and the detector are identical, the NEMO collaboration is preparing a detector using Geiger tubes filled with a *He*-ethyl alcohol mixture at atmospheric pressure as a tracking device while the electron energy will be measured with plastic scintillators covering the walls of the setup. Because the density of the gas in the sensitive volume is only about 0.2 mg/cm^2, the average energy loss of the electrons in the tracking device will be limited to 14 keV for a 1 MeV electron. The detection efficiency and therefore the calorimetric features of the detector are determined by the fraction of the solid angle covered by the plastic scintillators. The compound efficiency for the detection of two–electron events, estimated through a Monte Carlo simulation, will be $\epsilon = 0.4$. The energy resolution of these scintillators is around 9 % at 3 MeV. Because of the separation of source and detector one is now free to choose a $\beta\beta$–emitter having a Q-value above the energy range of the natural radioactivity. The limited energy resolution results in $\beta\beta 2\nu$–events being scattered into the analysis interval of the $\beta\beta 0\nu$–decay, which is hence an irreducible source of background.

The collaboration prepares 10 kg of isotopically enriched ^{100}Mo (98 % abundance), which has with 3.03 MeV a high Q-value. The source will be put as a thin foil inside the cylindrical tracking detector to limit the energy loss of the electrons in the foil. The requirement of kg quantities for the $\beta\beta$–sources hence results in a relatively large detector (3m in diameter and 3m in height). Figure 4 shows the planned detector and its shielding components.

The design goal of the setup is to be sensitive to $T_{1/2}^{0\nu} = 10^{25}\ y$. Depending on the detector background which can be achieved eventually a sensitivity of a few tenths eV for the Majorana neutrino mass could be reached. The spatial resolution of the instrument as well as the strong segmentation of the calorimetric part of the detector will help to identify backgrounds. A magnetic field (30 Gauss) will allow the measurement of the electron momenta as well as to discriminate against e^+e^- events coming from high energy γ–radiation interacting with the Mo–foil. Delayed particles crossing the tracking device will be tagged to identify e.g. ^{214}Bi whose daughter ^{214}Po is a short-lived α emitter.

Through a measurement of the $\beta\beta2\nu$–decay in the prototype detector NEMO II the group has already shown that the strict background requirements can be met by such a complicated and big device if the enriched Mo is of sufficient radiopurity [80]. Because the strong point of this project is that *any* isotope can be used, the group is investigating whether larger quantities of enriched ^{82}Se, ^{96}Zr, ^{116}Cd or ^{150}Nd can be made with sufficient purity.[81] This project will perhaps allow the next major step in sensitivity.

Other ideas. Thermal detectors offer the possibility of using a wide range of $\beta\beta$–emitters in the detector=source approach. A clever selection of the isotope can yield improved sensitivities.

The Milano group uses a 334 g crystal of natural Tellurium oxide in the Gran Sasso laboratory to study ^{130}Te which has a natural abundance of 34.5 % and a relatively high decay energy of 2.53 MeV. This instrument has an energy resolution of 16 keV at the decay energy coming close to the values of Ge detectors. They have reached a half life limit of $T^{0\nu}_{1/2} > 1.4 \cdot 10^{22}\ y$.

Another project following the conventional calorimetric approach is a collaboration between the $\beta\beta$ groups from Heidelberg and from Kiev preparing $CdWO_4$ scintillators. The idea is to use such scintillators made from isotopically enriched $\beta\beta$–unstable ^{116}Cd. The Q-value of 2.8 MeV (beyond the natural radioactivity) should help to achieve low background. The energy is on the other hand close enough to the prominent γ–line at 2.615 MeV (coming from ^{232}Th) to be contaminated. If the energy resolution is worse than 6.6 % at 1 MeV the $2\cdot\sigma$ intervals around the background peak and the $\beta\beta$–energy will overlap, which sets stringent requirements to the resolution of the detector. Another drawback is the fact that only 32 % of the scintillator mass is contributed by Cd leading to a relatively low abundance of the decaying isotope even if enriched Cd is used. The group is investigating these problems using a 1 kg detector made from natural $CdWO_4$, installed in the Gran Sasso tunnel.

Another potentially promising way to get to the 0.1 eV level as a limit for neutrinoless double beta decay is the use of ^{150}Nd in a cryogenic detector. Until recently this favorable isotope with its 3.4 MeV energy release could be used only as a source, and because multiple scattering of the emerging electrons required that it be used as a thin foil, the amount of ^{150}Nd had to be quite small. As NdF_3 cooled to milliKelvin temperatures, it can be both source and detector, permitting use of as much material as one can afford. Since NdF_3 has a high Debye temperature (θ_d = 390K), its small specific heat provides a large signal and an energy resolution of less than 1 keV at 3.4 MeV in a 10^2 g crystal. This is a factor of at least three improvement in resolution compared to a Ge semiconductor detector.

Another advantage compared to Ge is the energy available. In this energy range backgrounds go down about an order of magnitude per MeV, and Ge has only 2 MeV energy release. Indeed, 3.4 MeV is above any naturally occurring beta or gamma background. It appears possible using ballistic phonons to utilize pulse shape so that alpha backgrounds could also be eliminated. With the good energy resolution and the low radioactivity backgrounds, it is likely that at a deep site during the needed counting time there would be no counts at the endpoint energy unless neutrinoless double beta decay were seen. The lifetime limit would then increase directly with counting time, instead of as the usual square root of that time.

The main issue regarding the feasibility of reaching the 0.1 eV level is the ^{150}Nd matrix element. The QRPA calculation by Staudt, Muto, and Klapdor [57] gives a factor 70 improvement in rate, for all other factors equal, with respect to ^{76}Ge. If that turns out to be true – and the main part of that improvement comes from the larger energy release – then achieving the 0.1 eV level is assured. However, Hirsch, Castaños, and Hess in a recent publication[82] get a very suppressed matrix element because the nucleus is deformed, making it less favorable even than ^{76}Ge.

All of these experiments will probably, on a certain level, be limited by residual background. An elegant way to reach the zero background regime could be to detect the electrons in coincidence with the decay product created at the same location in the detector, to have an additional signature for $\beta\beta$–decays. This requires a detector with spatial resolution. The big problem here is, that the requirements of a large detector containing a reasonable number of source atoms and the identification of single atoms in this "ocean" are hardly compatible. The double beta group in Irvine, which pioneered in the field of the measurement of $\beta\beta2\nu$–decays in several isotopes, is working on this problem.

All discussed projects are dealing with kg quantities of the investigated isotope. If the source should be increased to several tons the only practical $\beta\beta$–emitter left seems to be Xe. The idea of using Xe gas dissolved in a large liquid scintillator, having very low intrinsic radioactive contaminations, has been proposed [83]. Large low background facilities as the test detector of the BOREXINO project (FV: 2 tons) in the Gran Sasso tunnel could be used. The source strength of this configuration would be 27 mol of ^{136}Xe if natural Xe would be used or 733 mol for the final BOREXINO detector (FV: 60 tons).

Because the solubility of Xe gas in liquid scintillator is limited to $\sim$ 2 % the fraction of source to active detector would be far from ideal. The use of enriched Xe would again allow a much more compact detector.

A serious problem would be the poor energy resolution of the device spilling two neutrino double beta events into the zero neutrino analysis interval. Any estimate of the achievable sensitivity depends hence on the unknown half life of the allowed background decay channel.

An even more avantgardistic idea would be the use of a large (several kton) liquid Xe detector similar to the liquid

Ar TPC developed by the ICARUS collaboration. A fiducial volume of 1 kton of natural Xe corresponds to a source strength of $6.7 \cdot 10^5$ mol of ^{136}Xe. In the most optimistic case of no background a neutrinoless half life of 10^{29} y could be tested within one year. This corresponds to a neutrino mass limit of $4 \cdot 10^{-3}$ eV. A similar background and energy resolution as in the previously discussed Xe TPC reduces the sensitivity to 10^{26} y or 0.1 eV. The truth may be somewhere in between these two numbers. The questions whether the high density of the liquid Xe will allow the identification of tracks of MeV electrons and whether long lived radioactive noble gases as e.g. ^{42}Ar, (produced in atmospheric nuclear bomb tests) emitting electrons with energies of up to 3.52 MeV, are a serious problem, remain to be answered.

5.4 Other Topics

100 keV Neutrinos. If neutrinos have mass then it is of course possible for the different types of neutrinos to mix with each other. Such an admixture would exhibit itself in the kinematics of weak two-body and three-body decays in which the neutrino is one of the final products. The spectra of the accompanying particles in these decays would feature an extra peak (in the case of two-body decays) or a kink (in the case of three-body decays) the location of which will depend on the mass of the admixed neutrino and the strength of which will depend on the mixing angle. A wide variety of experiments has been conducted to search for the admixture of a heavy neutrino with the electron neutrino (ν_e) by looking for such extra peaks or kinks in decays involving the electron neutrino (e.g. [84]). These experiments have established varying upper limits on the strength of the admixture of ν_e to heavy neutrinos with masses ranging from a few keV to several GeV. The least stringent limit for the mixing of ν_e with ν_τ is for $100 <$ mass of $\nu_\tau < 1000$ keV, where the current limit is $\sim 0.5\%$. A collaboration between a Nuclear Physics group from Tennessee Technological University (Hindi, Kozub, Robinson) and a Surface Science group from Montana State University (Avci, Zhu, Lapeyre) is currently conducting an experiment which aims to bring down these limits to $< 0.1\%$.

The basic idea is to measure the spectrum of recoil velocities of ^{37}Cl ions following the electron capture (EC) decay of ^{37}Ar. For an isolated ^{37}Ar atom energy and momentum conservation dictate that the recoil energy be 9.54 eV when ν_e is emitted. If a neutrino with mass m_ν is emitted in a certain fraction of the decays then the recoil energy of that fraction will be shifted down in energy by a fractional amount of $\frac{\Delta E}{E} = \left(\frac{m_{\nu_\mu}}{Q}\right)^2$, where Q is the decay energy (814-E_b) keV, where E_b is the excitation energy of the hole left in the daughter ^{37}Cl atom). Thus the highest mass neutrino that could be emitted in the decay is 814 keV; if a 250 keV neutrino were to be emitted, for example, the change in energy of the recoiling ^{37}Cl would be 9.5%. The lowest neutrino mass to which the method could be sensitive depends on our ability to resolve the peak associated with the emission of the massive neutrino from the dominant peak associated with the emission of the electron-type neutrino in the recoil spectrum. The factors which lead to a broadening in the energy spectrum are dependent on the experimental technique, a brief account of which will be given next. Full details of a preliminary measurement have been published recently [85].

The spectrum of recoil velocities from a monolayer consisting of a mixture of one part ^{36}Ar and $\sim 5 \times 10^{-5}$ parts ^{37}Ar physisorbed on a gold-plated Si wafer cooled to 16 K was obtained under ultra-high vacuum conditions by measuring the time-of-flight of the recoiling ^{37}Cl ions. The time-of-flight spectrum was obtained by starting a time-to-amplitude converter by detecting one of the Auger electrons emitted after the EC decay, and stopping it by detecting the ^{37}Cl ion after it had travelled a distance of ~7 cm. The observed energy distribution was found to have a maximum close to the expected energy of 9.5 eV; the FWHM of ~3 eV is about 50% larger than that expected for isolated ^{37}Ar atoms. The factors contributing to energy broadening for an isolated ^{37}Ar atom are the thermal motion of the ^{37}Ar atom prior to decay (~3% of the recoil energy at 16 K), recoil from Auger electrons emitted by the ^{37}Cl atom after the decay (12.2% for KLL Augers, 3.5% for LMM Augers) and various instrumental effects such as variation in the flight distance over the finite size of the detector and source. A Monte Carlo simulation which takes into account all of these effects gives an energy broadening of about 14% if the coincident Auger electron was required to be a KLL Auger and 20% if any Auger electron was accepted. Several additional experimental observations [85] have led to the belief that the extra broadening observed is due to inelastic charge exchange reactions between the recoiling Cl ions and the neighboring ^{36}Ar atoms and possibly with the Au substrate. Until a full quantitative understanding of the width of the observed peak is obtained, mixing limits from this preliminary measurement cannot be quoted.

The experimental group is currently working on several improvements and modifications which it is hoped will reduce the broadening of the main peak and increase the sensitivity to admixtures with heavy neutrinos. These improvements are 1) The production of ^{37}Ar from the $^{40}Ca(n,\alpha)^{37}Ar$ reaction instead of from the $^{36}Ar(n,\gamma)$ reaction. This will eliminate the ^{36}Ar from the surface and allow a much higher activity at a much smaller coverage and source area. 2) The use of position sensitive detectors to better define the flight path. 3) The use of a graphite substrate for adsorbing the ^{37}Ar. This should hopefully lower the charge exchange probability with the substrate and reduce the fraction of backscattered Auger electrons. 4) Possibly detect the Cl ions in coincidence with a K x-ray instead of with Augers. Detection of a K x-ray insures that the Cl ends up with a charge state of +1 (in contrast with the Auger filling of the K hole, which leaves the Cl, on average, with a charge state of +3); thus if the Cl were to pick up an electron it would become neutral and hence would not be detected by the system. This would eliminate the problem of inelastic charge exchange.

The above improvements should enable the experiment to set limits of $\leq 0.1\%$ on the admixture of heavy neutrinos with the electron neutrino. The measurement will be sensitive to neutrinos with masses in the range 200–800 keV. It would be difficult to push the mass below 200 keV with adsorbed ^{37}Ar atoms because of (the insurmountable) broadening due to the initial zero-point motion of the trapped Ar.

§6 CONCLUSION

We have considered the potential impact and the feasibility of improved kinematical probes of neutrino mass. Existing hints of neutrino mass lead one to speculate that moderately improved probes of M_{ν_e} via tritium β decay, and of the effective Majorana neutrino mass M_{eff} via neutrinoless double β decay, may actually observe neutrino mass signals. The requisite improvement in the tritium experiments appears possible in principle, but very difficult. In addition, the tendency of current tritium experiments to yield $M^2_{\nu_e} < 0$ must be understood; a possible role for molecular spectroscopic studies which might help is also presented. The required improvement in the $\beta\beta_{0\nu}$ experiments appears to be attainable. Indeed, several different approaches to significantly improving the sensitivity of $\beta\beta_{0\nu}$ searches are being considered or implemented. Turning to M_{ν_μ}, we examined the possibility of improving sensitivity to this quantity through experiments on pion or kaon decay. We did not examine in any detail the prospects for improving the laboratory bounds on M_{ν_τ}, but argued from astrophysics that improved bounds would be quite valuable. Finally, we considered improving the sensitivity to neutrinos which have masses in the 200 keV - 800 keV range, and are admixed in the electron neutrino, via experiments on electron capture.

The present hints of neutrino mass, and the general theoretical prejudice that neutrinos have nonvanishing masses, give to experimental searches for neutrino mass significant and important discovery potential. We conclude that improvements to the kinematical searches are well-motivated and promising.

†: Conveners

§References

[1] H. Kawakami et al., *Phys. Lett.* **B256**, 105 (1991).

[2] E. Holzschuh, M. Fritschi and W. Kundig, *Phys. Lett.* B287, 381 (1992).

[3] R.G.H. Robertson et al., *Phys. Rev. Lett.* **67**, 957 (1991).

[4] C. Weinheimer et al., *Phys. Lett.* **B300**, 210 (1993).

[5] W. Stoeffl and D. Decman, Lawrence Livermore National Lab. preprint. Submitted to *Phys. Rev. Lett.*.

[6] V. Lobashev, talk presented at the 27th Int. Conf. on High Energy Physics, Glasgow, July 20–27, 1994; A.I. Belesev et al., *Inst. Nucl. Phys.* preprint INR 862–94.

[7] K. Assamagan et al. *Phys. Lett.* **B 335**, 231(1994). We note that the *Review of Particle Properties*, *Phy. Rev.* **D50** 1173(1994) lists the present upper limit as 270 keV.

[8] D. Buscolic et al. (ALEPH Collab.), CERN preprint PPE/95–03.

[9] M. Moe and P. Vogel, *Ann. Rev. Nucl. Part. Sci.* 44, (1994) 247.

[10] K. Zuber, talk presented at the 27th Int. Conf. on High Energy Physics , Glasgow, July 20–27, 1994.

[11] F. Reines, H. Gurr, and H. Sobel, *Phys. Rev. Lett.* **37**, 315 (1976), as interpreted by W. Marciano and A. Sirlin, Brookhaven National Laboratory preprint (September 1988). We are assuming that the $\bar{\nu}_e$ does not have both magnetic and electric dipole moments whose contributions to the experiment of Reines, Gurr and Sobel happen to cancel.

[12] G. Raffelt, *Phys. Rev. Lett.* **64**, 2856 (1990).

[13] J. M. Irvine and R. J. Humphreys, *J. Phys. G* **9**, 847 (1983); R.G.H. Robertson et al., *Phys. Rev. Lett.* **67**, 957 (1991).

[14] J. F. Wilkerson, *Nucl. Phys. B* (Proc. Suppl.) **31**, 32–41(1993).

[15] See, e.g., Hans Agren and Vincenzo Carravetta,*Phys. Rev. A* **38**, 2707 (1988) and references therein.

[16] R. Raitz Von Frentz, K. Luchner, H. Micklitz, and V. Wittwer, *Phys. Lett.* **47A**, 301 (1974).

[17] See, e.g., L. R. Canfield, J. Kerner, and R. Korde, *Appl. Opt.* **28**, 3940 (1989).

[18] M. A. C. Perryman, C. L. Foden, and A. Peacock, *Nucl. Instr. Meth. Phys. Res.* **A325**, 319 (1993).

[19] See, for example, S. Bludman, N. Hato, and P. Langacker, *Phys. Rev. D* **49**, 3622 (1994).

[20] S. Mikheyev and A. Smirnov, Nuovo Cimento **C9**, 17(1986). See also L. Wolfenstein, *Phys. Rev. D* **17**, 2369 (1978).

[21] N. Hata and P. Langacker, *Phys. Rev. D* **50**, 632(1994).

[22] E. Beier, private communication

[23] For a report on interesting new data, see Y. Fukuda et al. (The KAMIOKANDE collaboration), *Phys. Lett.* **B335**, 237(1994).

[24] W. Freedman et al., *Nature* **371**, 757 (1994).

[25] B. Schmidt et al., *Astrophys. J.* **432**, 42(1994).

[26] A. Saha et al., *Astrophys. J.*, to be published.

[27] A Riess, W. Press, and R. Kirshner, *Astrophys. J. Lett.*, to be published.

[28] J. R. Primack, J. Holtzman, A. Klypin and D. O. Caldwell, *Phys. Rev. Lett.*, **74** (1995) 2160.

[29] A Gould, private communication.

[30] G. S. Vidyakin et al (Krasnoyarsk Collab.), *JETP Lett.* **55**, 206 (1992) and *JETP Lett.* **59**, 390(1994); B. Achar et al (Bugey Collab.), *Nucl. Phys.* **B434**, 503(1995); V. Zacek et al(Gosgen Collab.), *Phys. Rev.* **D34**, 2621(1986); P. Vogel, *Nucl. Phys. B* (Proc. Suppl.) **38**, 204(1995).

[31] Our discussion of the MSW effect in the presence of large mixing between two neutrino flavors is preliminary.

[32] D. Caldwell and R. Mohapatra, *Phys. Rev. D* **48** 3259(1993) and *Phys. Rev. D* **50**, B477(1994).

[33] With three degenerate neutrinos of mass $\sim$2 eV, it is possible *in principle* to have $|M_{eff}(\beta\beta_{0\nu})| \ll 2$ eV if the mixing angles are different than in our scenario and CP-violating phases are present in Eq. (2.4). In R. Mohapatra and S. Nussinov, *Phys. Lett.* **B346**, 75 (1995), there is described a model of this kind which, however, cannot accommodate the small-angle MSW explanation of the solar neutrino data and must employ a vacuum oscillation explanation.

[34] C. Angelini et al., *Phys. Lett.* **B179**, 307(1986).

[35] Generation of $\beta\beta_{0\nu}$ by a relatively heavy ν_3 when ν_1 and ν_2 have masses suggested by the solar neutrino problem, and hence are too light to contribute a visible $\beta\beta_{0\nu}$ signal, has been discussed by S. Petcov and A. Smirnov, *Phys. Lett.* **B322**, 109(1994).

[36] The capabilities of these two experiments are indicated in, for example, K. Winter, CERN preprint CERN-PPE/94-147.

[37] F. Boehm, *Proceedings of PASCOS-91*, p. 96, Editors: P. Nath and S. Reucroft, World Scientific (1992); E. Nagy, talk presented at the 27th Int. Conf. on High Energy Physics, Glasgow, July 20–27, 1994.

[38] L. Borodovsky et al., *Phys. Rev. Lett.* **68**, 274(1992). See also B. Armbruster et al., (KARMEN collaboration), *Nucl. Phys. B* (Proc. Suppl.) **38**, 394 (1995).

[39] G. Gyuk and M. Turner, contribution to Neutrino 94, Eilat, Israel, May 1994.

[40] E. Kolb, M. Turner, A. Chakravorty and D. Schramm, *Phys. Rev. Lett* 67, 533 (1991).

[41] S. Dodelson, G. Gyuk and M. Turner, *Phys. Rev.* D49, 5068 (1994)

[42] A. Dolgov, K. Kainulainen and I. Rothstein, Fermilab preprint Pub-94/1994.

[43] A. Burrows, R. Gandhi and M.Turner, *Phys. Rev. Lett.* 68,3834 (1992); R. Mayle, D. N. Schramm, M. Turner and J. Wilson, *Phys. Lett.* B317, 119 (1993); G. Raffelt and D. Seckel, Phys. Rev. Lett. 60, 1793 (1988), Phys. Rev. Lett. 67, 2605 (1991) and *Phys. Rev. Lett.* 68, 3116 (1992).

[44] G. Sigl and M. Turner, Fermilab preprint Pub-94/001-A (submitted to *Phys. Rev. Lett.*).

[45] T.P. Walker, G. Steigman, D.N. Schramm, K. Olive and H.S. Hang, *Ap. J.* 376, 51 (1991).

[46] K. Babu, T. Gould and I. Rothstein, *Phys. Lett.* B321, 140 (1994)

[47] M. Kawasaki, P. Kerman, H. Kang, R. Scherrer, G. Steigman and T. Walker, *Nucl. Phys.* B419, 105 (1994).

[48] G. Gyuk and M. Turner, Fermilab preprint Pub-94/059-A (submitted to *Nature*).

[49] S. Dodelson, G. Gyuk and M. Turner, *Phys. Rev. Lett.* 72, 3754 (1994).

[50] V.M. Lobashev and P.E. Spivak, Nucl. Instr. Methods in Phys. Research A240, 305 (1985).

[51] A.I. Belesev et al., INR Preprint 862/94 (June 1994: A contribution to the Eilat Neutrino '94 International Conference).

[52] R.G.H. Robertson et al., Phys. Rev. Lett. 67, 957 (1991).

[53] I.F. Silvera, B.A.P.S. 40, 1003 (1995).

[54] T. Bernatowicz, J. Brannon, R. Brazzle, R. Cowsik, C. Hohenberg, and F. Podosek, *Phys. Rev.* C47 (1993) 806.

[55] F. Petry, A. Piepke, H. Strecker, H.V. Klapdor-Kleingrothaus, A. Balysh, S.T. Belyaev, A. Demehin, A. Gurov, I. Kondratenko, D. Kotel'nikov, D. Landis, N. Madden and R.H. Pehl, *Nucl. Inst. Meth.* A332 (1993) 107.

[56] A. Balysh, M. Beck, S.T. Belyaev, J. Bockholt, A. Demehin, M. Eskef, D. Glatting, A. Gurov, G. Heusser, J. Hellmig, M. Hirsch, Ch. Hoffman, H.V. Klapdor-Kleingrothaus, I. Kondratenko, D. Kotel'nikov, V.I. Lebedev, B. Maier, A. Müller, F. Petry, E. Scheer, H. Strecker, M. Völlinger and K. Zuber, to be published in the proceedings of the 'XXVII Int. Conf. on High Energy Physics', Glasgow (UK) August 1994.

[57] A. Staudt, K. Muto, H.V. Klapdor-Kleingrothaus, *Europhys. Lett.* 13 (1990) 31.

[58] A. Balysh, M. Beck, S.T. Belyaev, F. Bensch, J. Bockholt, A. Demehin, A. Gurov, G. Heusser, H.V. Klapdor-Kleingrothaus, I. Kondratenko, D. Kotel'nikov, V.I. Lebedev, B. Maier, A. Müller, F. Petry, A. Piepke, A. Pronsky, H. Strecker, M. Völlinger and K. Zuber, *Phys. Lett.* B322 (1994) 176.

[59] M. Beck, F. Bensch, J. Bockholt, G. Heusser, M. Hirsch, H.V. Klapdor–Kleingrothaus, B. Maier, F. Petry, A. Piepke, H. Strecker, M. Völlinger, K. Zuber, A. Balysh, S.T. Belyaev, A. Demehin, A. Gurov, I. Kondratenko, V.I Lebedev and A. Müller, *Phys. Rev. Lett.* 70 (1993) 2853.

[60] F.T. Avignone III, private communication.

[61] J.C. Vuilleumier, J. Busto, J. Farine, V. Jörgens, L.W. Mitchell, M. Treichel, J.L. Vuilleumier, H.T. Wong, F. Boehm, P. Fisher, H.E. Henrikson, D.A. Imel, M.Z. Iqbal, B.M. O'Callaghan–Hay, J. Thomas and K. Gabathuler, *Phys. Rev.* D48 (1993) 1009.

[62] J. Engel, P. Vogel and M.R. Zirnbauer, *Phys. Rev.* C37 (1988) 731.

[63] F. Boehm, private communication and
F. Boehm, J. Busto, J. Farine, K. Gabathuler, G. Gervasio, H. Henrikson, V. Jörgens, K. Lou, R. Lüscher, D. Schenker, M. Treichel, J.-C. Vuilleumier, J.-L. Vuilleumier and J.-M. Vuilleumier, Caltech internal report CALT–63–687 (1994).

[64] A.R. Clark, et al., *Physical Review* **D9** (1974) 533. The authors have used $\triangle\chi^2 = 1.6$ for evaluation of a 90% confidence level. We have recalculated their limits using the correct value, $\triangle\chi^2 = 2.7$.

[65] D.S. Ayres, et al., *Physical Review* **D3**(1971) 1051.

[66] M. Goldhaber points out that one could get a limit of few 10's of eV on $\nu_e - \bar{\nu}_e$ mass difference using the neutrino events from supernova SN1987A. For this analysis one would classify the events that point to the supernova as ν_e induced and the rest as $\bar{\nu}_e$ induced. Observation of a neutrino pulse with thousands of events from a nearby supernova would clearly tell us a lot about masses of all species of neutrinos.

[67] A. Schwartz, BNL-E791 internal note KL-275, unpublished, (1990).

[68] A. Heinson, et al., "A New Search for Very Rare K_L° Decays", Proposal for E871, September 25, 1990.

[69] H.B. Anderhub et al., *Physics Letters* **B114** (1982) 76.

[70] R. Abela et al., *Physics Letters* **B146** (1984) 431.

[71] M. Daum et al., *Physics Letters* **B265** (1991) 425.

[72] B. Jeckelmann et al., *Nucl. Phys.* **A457** (1986) 709.

[73] B. Jeckelmann et al., preprint ETHZ-IPP PR-94-9 (1994).

[74] D.C. Lu et al., *Physical Review Letters* **45** (1980) 1066.

[75] Original suggestion by Alvarez and Cone found in D.R. Cone, University of California Radiation Lab. Report no. 1030 (1950). Application to weakly ionizing particles: C. Chianelli et al. NIM A273 (1988) 245. See also Hardek, Wells, and Lai, IEEE Proceedings (1981) 2219.

[76] R. Rusack, private communication.

[77] P. Cushman, Letter of Intent in progress.

[78] J. J. Gomez-Cadenas, CERN-PPE/94–12 (Jan. 1994).

[79] D. Lalanne, *Nucl. Phys. B* (Proc. Suppl.) 35 (1994) 369.

[80] NEMO III Proposal, by the NEMO collaboration, Laboratoire de l'Accélérateur Linéaire, LAL 94–29 (1994).

[81] D. Dassié et al., *Phys. Rev.*, **D51**, 2090(1995).

[82] J. G. Hirsch, O. Castanos and P. O. Hess, *Nucl. Phys. A* **582**, 124(1995).

[83] R.S. Raghavan, *Phys. Rev. Lett.* 72 (1994) 1411.

[84] See, for example, F. Boehm and P. Vogel, "Physics of Massive Neutrinos", (Cambrdige University Press, London, 1992), and references therein.

[85] L. Zhu, R. Avci, G.J. Lapeyre, M.M. Hindi, R.L. Kozub and S.J. Robinson, *J. Vac. Sci. Technol.* **A12**, 2037 (1994).

Experiment	Sources	$M^2_{\nu_e}$ (eV^2) for Best Fit	M_{ν_e} Bound (eV) at 95% CL
INS, Tokyo	Cd Arach.-T	$-65 \pm 85 \pm 65$	13[1]
Zurich	OTS-T	$-24 \pm 48 \pm 61$	11[2]
Los Alamos	T_2 gas	$-147 \pm 68 \pm 41$	9.3[3]
Mainz	T_2 solid	$-39 \pm 34 \pm 15$	7.2[4]
Livermore	T_2 gas	$-130 \pm 20 \pm 15$	7[5]
INR, Moscow	T_2 gas	-18.5 ± 6	4.5[6]

Table 1: Bounds on M_{ν_e} from tritium β decay experiments.

Experiment	$T^{0\nu}_{1/2}$ *reached* [y] 90% c.l.	M_{eff} [eV] *using* Ref. [29]	Ref. [34]	$T^{0\nu}_{1/2}$ *final* [y] 90% c.l.	M_{eff} [eV] *using* Ref. [29]	Ref. [34]
H-M ^{76}Ge	$5.1 \cdot 10^{24}$	0.7	2.3	$2 \cdot 10^{25}$	0.4	1.2
IGEX ^{76}Ge	$1.7 \cdot 10^{24}$	1.2	4.0	-	-	-
C–N–P ^{136}Xe	$3.4 \cdot 10^{23}$	2.6	4.3	10^{24}	1.5	2.5

Table 2: Half life and neutrino mass sensitivities of the discussed experiments using different nuclear matrix element calculations. H–M and C–N–P denotes the Heidelberg–Moscow and Caltech–Neuchâtel–PSI experiments, respectively. The mass limits already reached are listed together with the estimates for the final sensitivity. Because the IGEX project is still in a relatively early stage no final sensitivity is given for it.

Isotope	Range			Preferred value
^{48}Ca	?	—	?	3.2×10^{26}
^{76}Ge	1.7×10^{26}	—	2.7×10^{27}	3.0×10^{26}
^{82}Se	5.8×10^{25}	—	1.1×10^{27}	1.0×10^{26}
^{96}Zr	5.3×10^{25}	—	7.8×10^{26}	3.8×10^{26}
^{100}Mo	2.6×10^{25}	—	1.9×10^{26}	1.0×10^{26}
^{116}Cd	?	—	?	5.1×10^{27}
^{150}Nd	5.0×10^{24}	—	1.0×10^{26}	9.0×10^{25}
^{130}Te	1.6×10^{25}	—	1.3×10^{26}	3.3×10^{26}
^{136}Xe	1.0×10^{26}	—	6.3×10^{26}	2.5×10^{26}

Table 3: Lifetimes for $M_{eff} = 0.1$ eV.

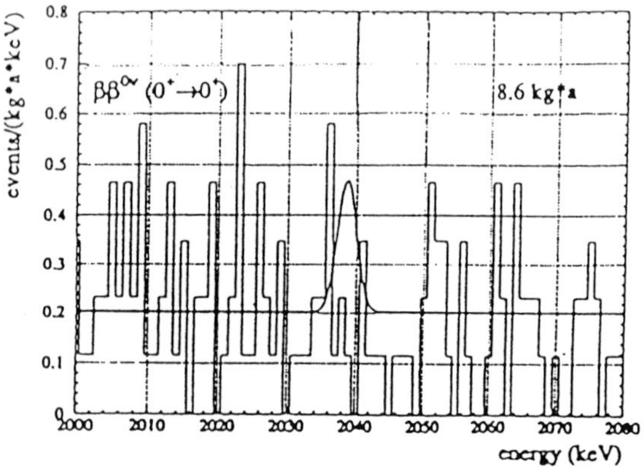

Figure 1: **Measured spectrum of the Heidelberg–Moscow experiment** around 2.04 MeV after 98.7 $mol \cdot y$ of ^{76}Ge exposure time. The $\beta\beta$–signal excluded with 90 % c.l. is also indicated.

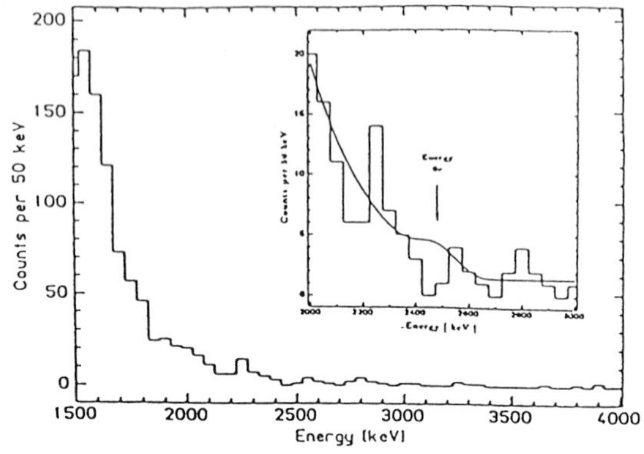

Figure 2: **Measured spectrum of the TPC** Sum energy spectrum of the two-electron events measured with the Xe TPC, taken from ref. [61]. The exposure time is 18.9 $mol \cdot y$. In the insert the energy range around the $\beta\beta$–decay energy is shown together with the excluded signal.

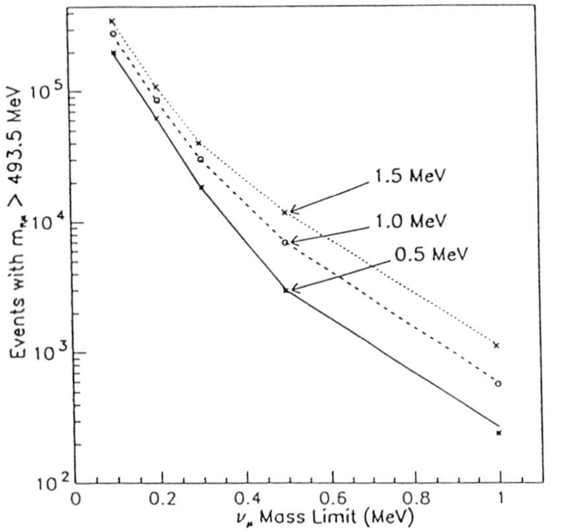

Figure 3: Number of $K_L^0 \rightarrow \pi^{\pm}\mu^{\mp}\nu_{\mu}$ events needed with $m_{\pi\mu} > 493.5$ MeV versus the muon neutrino mass sensitivity for various mass resolutions. This was calculated assuming that the data will be binned with a binsize equal to or smaller than the mass sensitivity. A maximum likelihood method using the small event statistics at the endpoint should yield somewhat better sensitivity.

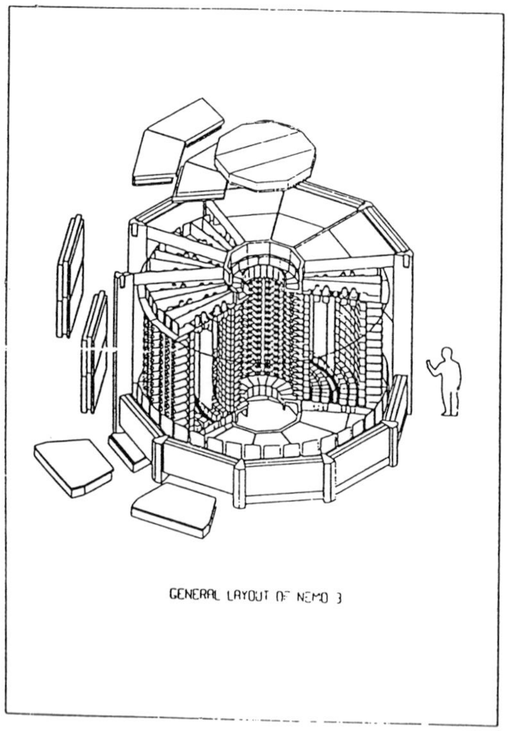

Figure 4: **Picture of the planned NEMO III detector** depicts the NEMO III detector and its shielding. The $\beta\beta$–source is surrounded by plastic scintillators covering the walls of the gas–filled detector. Electron and positron tracks have to be reconstructed over more than 50 cm.

Double Beta Decay

F. Avignone

University of South Carolina, Department of Physics, Columbia, SC 29208

1 Introduction

It is quite evident that the subject of double-beta decay is gaining momentum both in theoretical and experimental activities. In this workshop, recent results and progress and projections on experiments investigating ^{100}Mo ^{76}Ge, ^{136}Xe, and ^{150}Nd were presented. The direct observation of $2\nu\beta\beta$-decay and the measurement of half-lives have by this time been confirmed in the isotopes ^{82}Se, ^{76}Ge, ^{100}Mo, and ^{150}Nd. The present bound on the Majorana neutrino mass parameter is $\langle m_\nu \rangle \leq 1eV$ from ^{76}Ge experiments. The Heidelberg-Moscow-INFN experiment reports the best bound ($\langle 0.9eV \rangle$); while the IGEX experiment, in an earlier stage, reports a bound ($\leq 1.2eV$). Four groups reported on projected experiments to achieve sensitivities of $\sim 0.1eV$ using techniques employing ionization detectors, giant time projection chambers, and a cryogenic bolometric approach.

While the calculations of the nuclear matrix elements were not extensively discussed in this workshop, it is appropriate to present a brief status report here. By the time of the Erice School of September 1993, it became evident that the major differences between the calculated results of the various quasi-particle random phase approximations (QRPA), were in agreement, and in fair agreement with shell model calculations in the f-p-g shell. The Tübingen and Heidelberg groups have agreed that the addition p-n pairing forces to p-p and n-n forces, ameliorates the strong dependence of $2\nu\beta\beta$-decay matrix elements on parameters, and renders $o\nu\beta\beta$-decay matrix elements almost completely insensitive to them [1]. The extraction of bounds on $\langle m_\nu \rangle$ from experimental $o\nu\beta\beta$-decay data are now widely claimed to be fairly reliable and not strongly dependent on one's selection of models. This greatly enhances the utility of $o\nu\beta\beta$-decay experiments as probes of the fundamental properties of neutrinos. The widely held confidence in the results of the QRPA calculations has stimulated strong interest in extending $o\nu\beta\beta$-decay searches to heavier nuclei, for example, ^{136}Xe and ^{150}Nd. While results concerning these isotopes have already been reported elsewhere, projections for next generation experiments were discussed at SNOWMASS-94.

2 ^{76}Ge Experiments

Most of the searches for the neutrinoless $\beta\beta$-decay of ^{76}Ge have been discontinued because they were rendered obsolete by the Heidelberg-Moscow-INFN and International Germanium Experiment (IGEX) Projects employing kilogram quantities of germanium, isotopically enriched to > 86% in ^{76}Ge. This gives an enhancement of a factor of more than eleven in sensitivity over natural abundance detectors.

Germanium experiments enjoy two strong advantages over other $\beta\beta$-decay techniques: they employ tens of moles of the candidate isotope, have counting efficiencies within a few percent of 100%, and have the best energy resolutions of any $\beta\beta$-decay experiments thus far ($\delta E/E \cong 0.15\%$ at the decay energy).

2.1 The MPIH-Kurchatov-INFN ^{76}Ge Experiment

A report on the Heidelberg-Moscow-INFN experiment was presented by Andreas Piepke. The status was essentially that reported at Neutrino-94 [2]. For several years 3 detectors comprising ~ 6 kg of isotopically enriched germanium have been counting in the MPIH-Kurchatov-INFN installation in the Gran Sasso Laboratory at 3500 mwe. The data presented corresponded to 3658.86 kg$\cdot$ d or 113 mole $\cdot$ y.

In the region of the $\beta\beta$-decay full energy peak, 2038 keV, the background is 0.25 ± 0.02 counts/keV/kg/y. From these data they deduced

$$T^{o\nu}_{1/2}(^{76}Ge) > 2.94 \times 10^{24}y \quad (90\%CL)$$

which corresponds to

$$\langle m_\nu \rangle < 0.9eV.^*$$

From 13 December 1991 to 26 March 1992 the bound on the half-life grew from $\sim 5 \times 10^{23}y$ to $\geq 10^{24}y$ and then leveled out and did not change significantly until late 1993, due to the growth of a peak containing excess counts at 2037 keV and 2038 keV. This peak has since ceased to grow as fast and the bound on the half-life has continued to grow slowly again. The suggestion that these counts could be from $^{56}Co(T_{1/2} = 73d)$ were rejected because this contamination would leave 2035 keV if coming from an external source, and 2042 keV if coming from decay internal to the detector. The source of this "peak" has not been identified. It could be an improbable statistical conspiracy. No claims have been made concerning the nature of the peak.

*Since this report this group has pushed this bound to 0.68 eV.

The data from the second detector (M=2.76 kg) was analyzed to extract the half-life and energy spectrum of the $2\nu\beta\beta$-decay of ^{76}Ge. A data set comprising 6125 kg.d ($\sim 19.1 mole \cdot y$) was used. The background peaks were used with a complex Monte-Carlo model employing GEANT-3 to remove the corresponding continua.

The corresponding theoretical $2\nu\beta\beta$-decay spectrum was fit to the corrected spectrum to yield:

$$T_{1/2}^{2\nu}(^{76}Ge) = [1.42 \pm 0.03(stat.) \pm 0.13(syst.)] \times 10^{21}y.$$

Discussion during the presentation conveyed that new data imply that the half-life may be closer to $1.5 \times 10^{21}y$. No official presentation of this particular result was made; however, it was verified by a recent reprint.

2.2 *The IGEX ^{76}Ge Experiment*

Recent results of the International Germanium Experiment, IGEX, were reported and updated at the time of this writing. The collaboration involves three small isotopically enriched detectors (~ 0.7 kg) and two large ones (~ 2.1 kg), operating underground. A third 2.1 kg detector is in preparation while funding for a fourth and fifth detector are being proposed at present; the isotopically enriched material ($\sim 86.5\%^{76}Ge$) is available for these detectors also [3].

The background of all these detectors was rather high during the first few months underground and reduced to ~ 0.2 counts /kg/keV/y. The best background to date is ~ 0.16 in the above units. There is still some residual thorium background (0.05 counts/kg/d) in the peak at 2614 keV.

A total of 12.92 mole · y of data were used from two 2.1 kg detectors to extract the half-life for $2\nu\beta\beta$-decay. Data were analyzed for the two detectors separately, together and also for 6.37 mole years of data collected with two of the smaller detectors in the Canfrac tunnel. The results are in Table 1.

Rico Grande -I	$1.86kg$	9.29mole $\cdot y$	$1.06 \times 10^{21}y$
Rico Grande-II	$2.10kg$	2.93mole $\cdot y$	$1.13 \times 10^{21}y$
Rico Grande-I&II	$4.2kg$	12.92mole $\cdot y$	$1.11 \times 10^{21}y$
Rico I and Rico II	$1.3kg$	6.37mole $\cdot y$	$(0.9 - 1.1) \times 10^{21}y$

Table 1: Rico Grande I & II Data

The Rico Grande I & II data contain 4817 ± 189 $2\nu\beta\beta$-decay events, so the statistical error is only 1.4%. The uncertainties in the Monte Carlo techniques used are typically $\sim$ 15% in all cases. Therefore we would conclude from the IGEX data:

$$T_{1/2}^{2\nu}(^{76}Ge) = (1.10 \pm 0.18) \times 10^{21}y$$

The central value continues to be 23% or 30% shorter than that of the MPIH-Kurchatov-INFN experiment. While this disagreement is not of great interest in comparison to nuclear structure calculations, it could have an interesting impact on the interpretation of $o\nu\beta\beta$ data. If there is as much as 40% difference in the measured $2\nu\beta\beta$-decay half-life, there could be errors in one or both of the methods of correcting for internal and external background. If no significant errors exist, most likely there are systematic errors in the measurements of the detector fiducial volumes. In this case, probably the error is in the experiment associated with the longer half-file, because if the dead layers are underestimated there are in fact fewer atoms than estimated and the extracted half-life is actually shorter than originally claimed ($N\lambda$ = rate $\propto 1/T_{1/2}$). The upper bound on the number of atoms is fixed by the total detector masses which are all known within a fraction of 1%.

In the IGEX experiment there was only one count in the 10 keV window centered at 2038 keV in 10.86 mole·y of data. To a 90% confidence level, there are fewer than 2.7 counts, hence

$$T_{\frac{1}{2}}^{o\nu} > \frac{(\ell n)Nt}{2.7} = \frac{4.53 \times 10^{24}y}{2.7} = 1.7 \times 10^{24}y \ \ (90\%CL)$$

and

$$\langle m_\nu \rangle < 1.2eV \ \ (90\%CL).$$

The same matrix elements were applied as were used in the analysis of the MPIH-Kurchatov-INFN experiment which has already achieved an exclusion $\langle m_\nu \rangle < 0.9eV$.

It is interesting to note that when both large Ge experiments run their course, their data could in principle be combined to yield hundred of mole-years of data and improve the bound on $\langle m_\nu \rangle$ by as much as 40%. While there is no agreement to do this, it is a future option.

3 The Nemo ^{100}Mo Double Beta Decay Experiment

During the past 4 years two gas tracking detectors called NEMO-1 and NEMO-2 have been built in the Fréjus underground laboratory at 4800 m.w.e. The tracking is accomplished in 1 m^3 wire chamber filled with He gas with 4% ethyl alcohol at atmospheric pressure. The volume is separated into 8 octagonal cells with one central wire surrounded by 8 ground wires which are 100μ diameter nickel wires one meter long [4].

The volume is divided by the source plane splitting the tracking detector into two equal parts, each containing 5 frames of 2 × 32 vertical Geiger cells. Each frame is two crossed planes of cells, one vertical and one horizontal. The tracking volume is terminated by aluminized mylar (36μ) just in front of a calorimeter of scintillators on each end. The gas detector has capability to measure delayed signals due to the decay of ^{214}Bi to ^{214}Po which then emits an α-particle with a half-life of $164\mu s$. The tracking detector cells have a transverse spatial resolution of 0.5 mm and longitudinal resolutions of 4.7 mm.

The calorimeters are essentially a wall of $64(12\times12\times 2.25cm)$ scintillators, 2 cm of plastic covered with 2.5 mm of CsI(Na). The photon detection efficiency is 15% at 500 keV, while electrons up to 4 MeV are stopped in the scintillator. The scintillators are covered with polycarbonate and Teflon to protect them from ambient light and to reflect back scintillation light, respectively. A 500 keV electron looses 45 keV in this covering. The scintillators then have a 50 keV threshold.

The source foil consisted of two parts, one enriched to 98.4% ^{100}Mo with a fiducial mass of 172g and one of natural isotopic abundance ($6.9\%^{100}Mo$) having a fiducial mass of 163g. Each had an average thickness of 40μ. The energy resolution is 17.4% at 1 MeV.

There were 1947 events observed in 6140 hrs. of operation, identified as coming from the enriched foil. The background was evaluated as 514 events from the natural foil, yielding 1433 events attributable to $2\nu\beta\beta$-decay. The quoted result is:

$$T^{2\nu}_{1/2}(^{100}Mo) = \{0.95 \pm 0.04(stat.) \pm 0.09(syst.)\} \times 10^{19}y,$$

in good agreement with the previous results of the Irvine group, $(1.16^{+0.34}_{-0.08}) \times 10^{19}y$, and the Osaka group $(1.15^{+0.3}_{-0.2}) \times 10^{19}y$. The half-life for $o\nu\beta\beta$-decay was $> 0.64 \times 10^{22}y$ (90%CL), somewhat shorter than bounds reported earlier; however, this detector is being continuously upgraded and shows every promise of being competitive with other techniques in the long run.

The NEMO-3 detector is a large drift chamber with a planned source of 10 kg of ^{100}Mo intended to operate for five or more years. The projected sensitivity is $T^{2\nu}_{1/2} > 10^{25}$y corresponding to $\langle m_\nu \rangle \sim 0.1$eV.

3.1 Cyrogenic Detectors for Double Beta Decay UCSB, UCB, LBL, Stanford, INR Planned ^{150}Nd Experiment.

A brief report was given by David Caldwell on the progress in the development of cryogenic detectors and on a planned experiment with ^{150}Nd. This detector would be similar to those resulting from the pioneering work by the Milano group.

The detectors developed at Milano by E. Fiorini's group are low temperature bolometers of TeO_2 with a germanium thermister of $\sim 1mm^3$. A crystal of 334 grams has already been used to search for $o\nu\beta\beta$-decay of ^{130}Te. With a smaller crystal ($\sim$ 73g) they were able to obtain an energy resolution of about 5 keV at 1333 keV with ^{60}Co calibration source. The resolution in the larger crystal for the entire experiment was between 15 and 18 keV at the 2614 keV peak from the decay of ^{232}Th and $\sim$16 keV for a running time of 4104.6h.

The background at the total decay energy of 2528.8$\pm$ 1.3 keV was $\sim$ 4.4 counts/keV/kg/yr. This is still a factor $\sim$20 higher than those of the two large Ge experiments, nevertheless they were able to significantly improve the bound on $o\nu\beta\beta$-decay:

$$T^{o\nu}_{1/2}(^{130}Te) > 1.4 \times 10^{22}y \ \ (90\%CL).$$

The previous best bound was $\sim 10^{21}$y from geochemical measurements. This yields a bound of $\langle m_\nu \rangle$ <5.9 eV. A small bump at the decay energy was reported last year and new experiment with four crystals of about 350g each is underway.

The planned ^{150}Nd experiment involves a 100g crystal of NdF_3 isotopically enriched to 90% in ^{100}Nd. The total $\beta\beta$-decay energy of ^{150}Nd is $\sim$ 3367 keV and has a very large theoretical nuclear matrix element, such that the theoretical decay rate is 70 times larger than that for ^{76}Ge. From the above discussion, it appears at first glance that this type of experiment will not compete with the many kg of 86% ^{76}Ge already in operation. Between the larger phase space and nuclear matrix element of ^{150}Nd, a 100 gm detector could compete with 10 kg of enriched Ge. If superior energy resolution and background could be achieved, this experiment might well surpass the large Ge experiments; however, this would require a significant improvement over the performance of the Milano ^{130}Te detectors. Nevertheless, the technique is promising and could result in a very sensitive search, mainly because in principle the energy resolution could significantly surpass that of Ge detectors that are near their fundamental limit. In addition, this group has been successful in observing signals by observing phonons and ionization charge in Ge crystals. If this technique could be applied to other crystals, it might result in significant background reduction. This technique is in its infancy though very interesting.

3.2 Proposed Cryogenic Detection of Double-k Capture and Double Position Emission in ^{58}Ni and ^{96}RU.

Thus far, $\beta\beta$-decay experiments have focused on $\beta^-\beta^-$-experiments rather than on $\beta^+\beta^+$, KK, or Kβ^+ experiments because of the large loss in phase space whenever β^+ particles are produced and because pure x-ray experiments are intrinsically more difficult because the energies detected are smaller and in higher background regions of the spectrum. Nevertheless, $\beta^+\beta^+$ experiments can have six detectable signals in principle, Kβ^+ can have four and KK again only two. The energy release in KK capture is, however, much larger. Boris Pritychenko proposed the consideration of several candidates.

The integrated kinematical factors for Kβ and KK decays are given in Table 2 below. The process of KK-

Table 2: Gamow-Teller Kinematic Factors for $K\beta^+$ and KK Decays.

Isotope	$G(K\beta)y^{-1}$	$G(KK)y^{-1}$
^{78}Kr	1.17×10^{-21}	1.96×10^{-21}
^{124}Xe	4.35×10^{-21}	5.10×10^{-20}
^{106}Cd	1.97×10^{-21}	1.60×10^{-20}
^{96}Ru	1.15×10^{-21}	7.0×10^{-21}
^{130}Ba	1.4×10^{-21}	4.1×10^{-20}
$136Ce$	6.4×10^{-22}	4.3×10^{-20}
^{58}Ni	2.9×10^{-24}	4.3×10^{-23}
^{162}Er	1.7×10^{-23}	4.6×10^{-20}

capture is more probable than $K\beta^+$ or $\beta^+\beta^+$ decays, because the value of the release energy for such processes is much greater. The response of a detector to the two-neutrino mode will be a single peak with twice the K_α binding energy. This energy is ideal for the small cryogenic detectors if background can be controlled.

There is a broad choice of materials for cryogenic detectors, for example, compounds of ^{58}Ni and ^{96}Ru which are of current interest. Calculations of the sensitivities of cryogenic detectors of KK(2ν)-decay are presented in Table-2 assuming the backgrounds already achieved in intrinsic Ge detectors.

Ru is a good material for cryogenic detectors because it is a superconductor with $T_c = 0.49 \pm 0.15^oK$. The signal would appear at $\sim$ 35 keV, while ^{106}Ru has a β-end-point energy of $\sim$0.04 keV and a half-life of $\sim$367d. ^{106}Rh, which follows in the chain, is dangerous because its decay energy is $\sim$3.54 MeV. This contamination can be made small.

The technique looks particularly promising for further studying nuclear matrix elements in 2ν decay modes.

4 Possible Use of the Large Time Projection Chamber Helaz for Double Beta Decay.

A description of the High Rate Solar Neutrino Detector, HELAZ, was presented in the solar neutrino section and also as a possible giant ^{136}Xe $\beta\beta$-decay Time Projection Chamber [5]. While no specific predictions were made one could imagine that such a device could possibly reach a sensitivity of $\langle m_\nu \rangle \sim$0.01 eV. While such an application may be some years away, it could well represent the most sensitive direct search.

The detector would consist of 2000 m^3 of gas which can be a mixture of 20-30% Xe with the basic $HetCH_4$ gas, operated at 77 ^oK which would allow the condensation of radon and the complete elimination of the associated background. While this experiment is only in its design stages mainly for another purpose, it is worth mentioning as a future new technology to move the sensitivity further from the level, $\langle m_\nu \rangle \sim$0.1 eV, the target for Ge experiments, to a new level possibility better by a factor of 10. This was presented by G. Bonvicini.

5 Concluding Remarks

Not all of the results of operating and completed experiments have been presented. To complete this report Professor Angel Morales of the University of Zaragoza has allowed me to update his current table. Table 3 clearly shows that Ge experiments are the most sensitive thus far in placing bounds on the Majorana mass parameter.

In conclusion, there is a great deal of evidence from this workshop, other international conferences, and the literature that double beta decay is an endeavor producing interesting results in both nuclear and particle physics. Since the publication of the first direct observation of $2\nu\beta\beta$-decay by UCI group (^{86}Se), there have been measurements reported of the half-lives of ^{76}Ge, ^{100}Mo, ^{116}Cd, ^{150}Nd, and ^{238}U. This has put great pressure on theoretical models and has lead to a number of theoretical developments which are still evolving.

The most interesting potential value of $\beta\beta$-decay experiments is their ability to detect evidence of Majorana mass at levels of $\sim$0.1 eV with the perfection of present technology. The possibility of using giant Xe detectors such as HELAZ to improve this sensitivity one order of magnitude is intriguing, but its feasibility is yet to be established. Thus far Ge experiments employing kilogram quantities of ^{76}Ge are the most effective with respect to the sensitivity to $\langle m_\nu \rangle$; however, competing techniques are less developed and should be pursued. In the case that neutrinoless $\beta\beta$-decay would be observed in any of these experiments, confirmation would require the observation of a positive result in other isotopes. Accordingly, the continuation of all experiments is desirable, as long as they continue to improve in sensitivity.

Finally, some version of a giant experiment, involving ^{136}Xe for example, should be studied to determine if a future sensitivity is feasible that is significantly better than the projections using current technology.

6 References

1. A. Faessler, Prog. Part. *Nucl. Phys.*, **32**, 289 (1994).
2. H. V. Klapdor-Kleingrothaus, Prog. Part *Nucl. Phys.*, **32**, 261 (1994).
3. F. T. Avignone, III, Prog. Part. *Nucl. Phys.*, **32** (1994).
4. D. Dassie et al., *Phys. Rev.* **D51**, 2090 (1995).
5. F. Arzarello et al., Collége de France Preprint LPC 9428.

Table 3: Sensitivities to KK(2ν) Decay.

Isotope	Compound	Mass kg	Counting Time,y	$T_{1/2}y$ 68% CL	Nuclear Matrix Element
^{58}Ni	NiO	1.0	1.0	1.4×10^{23}	≤ 0.45
^{96}Ru	RuO_2	1.0	1.0	9.0×10^{21}	≤ 0.13

Table 4: Limits on Neutrinoless Double Beta Decay.

Isotope	Experiment	$T^{o\nu}_{1/2}$ Bound Years	C.L(%)	$\langle m_\nu \rangle$ Bound (eV)
^{48}Ca	Beijing	1.1×10^{22}	68	8
^{78}Ge	MPIH/KIAE	5.1×10^{24}	90	0.68
	IGEX	1.7×10^{24}	90	1.2
^{82}Se	UCI	2.7×10^{22}	68	5
^{100}Mo	LBL/MHC/UNM	3.5×10^{22}	68	6
	UCI	2.6×10^{21}	90	24
	Osaka	7×10^{21}	68	13
	NEMO	6.4×10^{21}	90	13
^{116}Cd	Kiev	1.5×10^{22}	90	5.7
	Osaka	1.7×10^{21}	68	17
^{128}Te	Washington U*	7.7×10^{24}		1.0
^{130}Te	Milano	1.4×10^{22}	90	5.9
^{136}Xe	Caltech/Un/Psi	4.2×10^{23}	90	2.3
	Milano	1.2×10^{22}	95	13.6
^{150}Nd	UCI	2.1×10^{21}	90	4
^{78}Kr	Zaragoza/INR	5.8×10^{21}	68	850

* Not a direct measurement. Geocronological measurements do not discriminate between $\beta\beta$-decay modes.

NEUTRINO OSCILLATION EXPERIMENTS WITH ATMOSPHERIC NEUTRINOS*

Tom Gaisser,
Bartol Research Institute, University of Delaware, Newark, Delaware 19716
Maury Goodman
Argonne National Laboratory, Argonne, IL 60439

Abstract

The possible atmospheric neutrino deficit is reviewed with an emphasis on recent data. The regions of Δm^2 implied by a neutrino oscillation hypothesis are considered.

1 Introduction

The atmospheric neutrino beam is a mixture of approximately $\frac{1}{3}$ $(\nu_e + \bar{\nu}_e)$ and $\frac{2}{3}$ $(\nu_\mu + \bar{\nu}_\mu)$. It is therefore possible in principle to do both ν_μ and ν_e disappearance and appearance experiments with this beam. Various classes of atmospheric neutrino interactions are defined:

- Fully contained events are those with all tracks inside the fiducial volume of the detector;
- Partially contained events have the vertex inside the fiducial volume but at least one exiting track;
- Upward, stopping neutrino-induced muons are produced outside the detector, but stop inside a fiducial volume;
- Upward, throughgoing neutrino-induced muons pass through the detector having more than some minimum track-length inside a fiducial volume.

When considering the atmospheric neutrinos as a beam for oscillation searches, it is useful to know the distribution of neutrino energies responsible for these classes of interactions. These are shown in Fig. 1. The neutrino energies for most contained events are below threshold for τ-production, making a ν_τ appearance experiment very difficult.

Low statistics limit the reach in $\sin^2\theta$, while E_ν and R_{Earth} limit the reach in δm^2 of neutrino oscillation experiments with the atmospheric neutrino beam. Uncertainties in normalization of the primary cosmic ray beam, together with uncertainties in the production of pions and kaons in proton collisions with light nuclei, translate into significant uncertainties in the normalization and shape of the atmospheric neutrino spectrum at production. For this reason, conclusions based on ratios are the most secure.

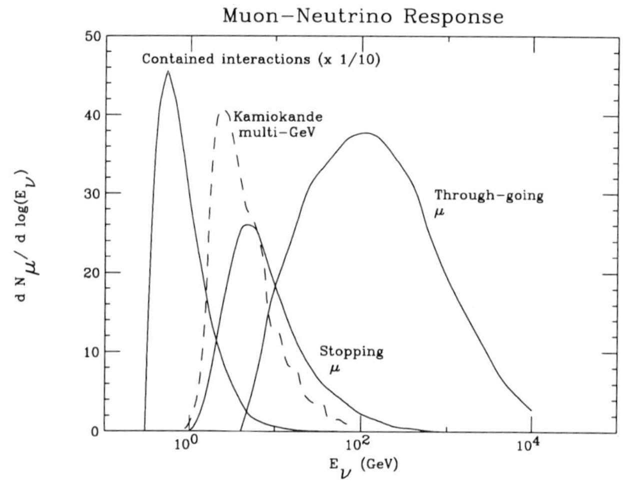

Figure 1: Kamiokande response for ν_μ as a function of energy.

In the next four sections we consider the data for the 4 classes of events. Implications for neutrino oscillations are then discussed.

2 Contained Events

The neutrino flavor ratio from experiments which measure contained events is shown in Table 1 while the data and calculations themselves are shown in

---

*Work supported in part by U.S. Department of Energy, Division of High Energy Physics, Contract W-31-109-ENG-38.

Table 2. The flavor ratio is

$$R \equiv \frac{\nu_\mu/\nu_e)^{data}}{\nu_\mu/\nu_e)^{MC}} \tag{1}$$

while the experiments actually measure

$$R' \equiv \frac{track/shower)^{data}}{track/shower)^{MC}} \tag{2}$$

which only differ if there are neutrino oscillations in the presence of neutral current background to the track and shower samples.

The high statistics water Cerenkov detectors in Table 1 have two ways to distinguish quasi-elastic ν_μ and ν_e events. In one method, the shape of the Cerenkov ring on the wall of the phototubes is used. In the other method, they look for excess phototube hits in a window several microseconds after the event, which would indicate μ or π decay. The two methods have different systematic errors, but are statistically correlated. Experimental tests of e/μ detection efficiency are underway at KEK. Results from the Kamioka group presented at Eilat [1] show no sign of misidentification. Results from the IMB exposure at KEK can be expected in a few months. The Frejus experiment prefers to analyze their contained and partially contained events together and this is discussed in the next section.

Experiment	Exposure kT-year	R'
IMB1	3.8	0.68 ± 0.08
Kamiokande ring	7.70	0.60 ± 0.06
Kamiokande decay		0.69 ± 0.06
IMB-3 ring	7.70	0.54 ± 0.05
IMB-3 decay		0.64 ± 0.07
Frejus contained	2.0	0.87 ± 0.13
Soudan	1.01	0.64 ± 0.19
NUSEX	0.5	0.99 ± 0.29

Table 1: The Atmospheric Neutrino Flavor Ratio. Quoted errors are statistical and systematic added in quadrature.

The results on contained atmospheric neutrino events are consistent with each other if there is a 30-40% deficit of ν_μ events. Such a deficit could be the result of either $\nu_\mu \leftrightarrow \nu_\tau$ or $\nu_\mu \leftrightarrow \nu_e$ oscillations. Since the interpretation of the atmospheric neutrino anomaly in terms of neutrino oscillations depends on the normalization of the neutrino flux, it is desirable to check the normalization as well as possible. In a conventional two-detector neutrino oscillation experiment, this is done by measuring the interaction rate in the neutrino beam close to the source. The nearest analog for the atmospheric neutrino beam of a close-in detector is a measurement of the intensity of muons at the same altitudes where both the neutrinos and muons are produced (10 to 20 km). New, higher precision measurements of this flux are being made [2, 3], and preliminary results [2] are consistent with a relatively high normalization of the neutrino flux [4]. This favors a $\nu_\mu \leftrightarrow \nu_\tau$ interpretation of the atmospheric neutrino puzzle.

Experiment	TR^{see} (ν_μ^{see})	TR^{MC} (ν_μ^{MC})	SH^{see} (ν_e^{see})	SH^{MC} (ν_e^{MC})
IMB1	104	136	297	265
Kam. ring	234	356.8	248	227.6
Kam. decay	182	277.5	300	313.9
IMB-3 ring	182	268	325	257.3
IMB-3 decay	208	261.5	402	348.5
Frejus contained	94	100	89	82
Soudan	33.5	42.1	35.3	28.7
NUSEX	32	36.8	18	20.5

Table 2: The data on the Atmospheric Neutrino Deficit. TR refers to single tracks, SH to single showers

3 Partially contained events

The most recent new result on atmospheric neutrinos was presented during the Snowmass meeting [5]. This was the Kamiokande analysis of their "multi-GeV" events, which include fully contained events with $E_{\rm vis} > 1.33$ GeV as well as partially contained events. The mean neutrino energy responsible for this class of events is $\sim$ 6 GeV, as shown in Fig. 1. This is nearly an order of magnitude higher than for the independent sample [6] of contained events with $E_{\rm vis} < 1.33$ GeV. In this new data sample, there is again evidence for a ν_μ deficit, with a measurement of $R' = 0.59 \pm 0.08$.

The most remarkable feature of the data is the appearance of an angular dependence of the muons of the kind that would be expected if downward muons have path-lengths too short to oscillate; that is, the upward muons are suppressed relative to the downward muons. Given the range of E_ν and R involved for the multi-GeV sample, Eq. 3 below implies an upper limit on δm^2. In contrast to the muons, the angular dependence of the electrons is closer to what is expected if the ν_e are not affected by oscillations. The angular-dependence alone therefore suggests an interpretation in terms of ν_μ disappearance.

On the other hand, a preliminary re-analysis of the Frejus data presented at Snowmass [7] seems to be even more divergent from the results of Refs. [6, 8] than the original Frejus publication [9]. A simultaneous analysis of their contained, partially contained, and throughgoing muons leads to a neutrino oscillation exclusion plot which contradicts a neutrino oscillation interpretation of the data in Table's 1 and 2. It is seen that by itself, the Frejus fully contained data is not in strong disagreement with the other experiments. However, their partially contained events and neutrino induced throughgoing muons seem to be in conflict with the new Kamiokande result. An analysis of all Frejus neutrino data leads to the result $R = 0.96 \pm 0.18$.

One puzzling feature of the new Kamiokande result is that there appears to be a large excess of electron neutrinos and a relatively small deficit of muon neutrinos as compared to the calculation. Using the neutrino flux of Ref. [10], they find

$$e - \text{like} \left(\frac{\text{measured}}{\text{calculated}}\right) = 1.47$$

and

$$\mu - \text{like} \left(\frac{\text{measured}}{\text{calculated}}\right) = 0.83.$$

The corresponding numbers for the low-energy, fully contained event sample are 1.09 (electrons) and 0.66 (muons). Thus, while the low energy sample suggests $\nu_\mu \leftrightarrow \nu_\tau$, the high energy sample looks more like $\nu_\mu \leftrightarrow \nu_e$. However, the ν_e excess may be just a statistical fluctuation.

We also note a statistical point regarding Kamiokande's angular dependence. They plot R = $(\nu_\mu/\nu_e)^{data}/(\nu_\mu/\nu_e)^{MC}$ versus zenith angle. Part of that ratio, ν_e^{data}/ν_e^{MC} is for flux normalizaiton only, and should be flat within statistics in both the no-oscillation case and the $\nu_\mu \rightarrow \nu_\tau$ hypothesis. Binning the ν_e data adds statistical fluctuations to the angular dependence plot which contribute no information. A better representation is $(\nu_\mu/<\nu_e>)^{data}/(\nu_\mu/\nu_e)^{MC}$ where the ν_e average is over all angular bins. If this is done, there is still evidence for an angular distribution due to oscillations, but from a mixing angle which is not maximally mixed.

4 Upward stopping Muons

The measured ratio of stopping to throughgoing upward neutrino-induced muons is consistent with expectation [11]. The data is shown in Table 3. This means *either* that there are no oscillations of ν_μ with mixing angle and δm^2 large enough to be seen with upward neutrinos *or* that there is large mixing of ν_μ, but with δm^2 large enough so that the lower energy stopping muons as well as the higher energy throughgoing muons are affected. The measured stopping/throughgoing ratio is used to rule out a range of $10^{-3} < \delta m^2 < 10^{-2}$ eV^2 at large mixing angle for ν_μ disappearance [11].

The ratio of upward stopping muons to throughgoing muons is sensitive to the absence of background from events which are not initiated by neutrinos. In particular, downgoing muons which initiate hadronic cascades in the rock could mimic upward going stopping muons in some cases. There is evidence for such events from Soudan, Kamioka, and MACRO[12, 13, 14], but it is not straightforward to calculate the possible contamination of the IMB data sample. We do note, however that a background of 12 events out of the 85 would lead to the absence of any neutrino oscillation limit from the IMB analysis, and also that the comparison of upward showers to expectation in IMB may be evidence that such a background in fact exists.

	Observed	Calculated
Stopping tracks	85	84
Exiting tracks	532	516
Showers	49	17

Table 3: Neutrino induced upward event rates in IMB.

5 Upward Muons

Because the implication of upward muon measurements have been somewhat controversial, it is worth going into a little more detail. The IMB group [11] calculated the rate of upward, throughgoing muons starting from a particular neutrino flux [15] and a particular representation of the neutrino cross section [16]. Their calculated rate agreed with their measured rate, apparently ruling out most of the available parameter space needed for a $\nu_\mu \leftrightarrow \nu_\tau$ interpretation of the low energy, contained event anomaly. Frati et al. [17] showed that the Kamiokande data on upward, throughgoing muons [18] is similarly consistent with a calculation assuming the same input. On the other hand, they found that, starting from different, more recent calculations of the neutrino flux [19, 20] and a different representation of the cross section [21], the predicted rate is some 15% higher than the observed rate. They further showed that a discrepancy of this size is consistent with a range of parameters needed to explain the contained event anomaly as $\nu_\mu \leftrightarrow \nu_\tau$. Recently reported results from MACRO [22] similarly show a significant deficit of the measured upward rate as compared to a calculation with the "high" input assumptions. This matrix of measurements and calculations is indirect evidence that these experiments and calculations are all consistent with each other. The same cannot be said for Baksan [23] where there is agreement between measurement and expectation for "high" input assumptions, but an *excess* of measurements over calculation when the low input is used. In all cases the differences are only of marginal statistical significance, as indicated by the summary of measurements/calculations in Table 4.

Experiment	Observed	Calculated "high"	"low"
IMB [11]	0.47 ± 0.02	—	0.455
KAM [18]	2.04 ± 0.13	2.36	2.18
Baksan [23]	161	162	142
MACRO [22]	$74 \pm 9 \pm 8$	101 ± 15	—

Table 4: Upward muons. For Kamioka the units are upward muons in units of $10^{-13}\mathrm{cm}^{-2}\mathrm{s}^{-1}\mathrm{sr}^{-1}$. For IMB, upward events per day and for Baksan and MACRO total number of observed upward events.

6 Discussion

Experiments with atmospheric neutrinos can be ranked in the following order as regards their potential for discovery of new physics:

1. The ratio ν_e/ν_μ. The principal uncertainties in the input to the calculation of this ratio all cancel. Three completely independent calculations [24, 10, 25] agree within 5% of each other for the expected value of this ratio.

2. Angular distributions of neutrino fluxes in a particular energy region. Again, the principal uncertainties cancel in the calculation of the beam, but the statistical significance is poorer because the data must be subdivided into angular bins.

3. The ratio of stopping to throughgoing neutrino-induced upward muons. The cancellation of uncertainties in the calculations here is less complete because the two classes of events in this ratio correspond to different ranges of energy.

4. The absolute intensity of ν_μ or ν_e.

It is noteworthy that the experiments with the greatest statistical significance [6, 8] show a significant deviation from the most secure feature of the atmospheric neutrino beam—the ν_e/ν_μ ratio. The measured value is apparently greater than the ratio at production. The possibility that the discrepancy might result from an oversimplified use of the Fermi Gas Model (FGM) in calculating the neutrino interaction rate is not borne out by careful investigations [26], which conclude that the energy is high enough so that any deviation from the FGM affects both flavors of neutrinos similarly.

If the observed effect is a signal of neutrino oscillations, the lack of observed angular dependence for low energy ($\lesssim$ 1 GeV) events [6, 8] implies $\delta m^2 > 10^{-3}$ eV^2. The large size of the discrepancy requires $\sin^2 2\theta > 0.5$. Much (not all) of available phase space for a $\nu_e \leftrightarrow \nu_\mu$ explanation is ruled out by reactor limits.

In a simple two-flavor oscillation scheme, the oscillation probability for a muon-neutrino is

$$P_{\nu_\mu \to \nu_2} = \sin^2 2\theta \sin^2 \left[1.27\, \delta m^2(\mathrm{eV}^2) \frac{L(\mathrm{km})}{E_\nu(\mathrm{GeV})} \right]. \tag{3}$$

A ν_μ disappearance interpretation (e.g. $\nu_\mu \leftrightarrow \nu_\tau$) of the ratio ν_e/ν_μ in low-energy contained events with $\delta m^2 > 10^{-2}$ eV^2 and large mixing angle would imply disappearance of $\sim$ 100 GeV ν_μ with pathlengths $L > R_{Earth}$. This in turn predicts a deficit of ν_μ-induced, upward, throughgoing muons. Measurements of this rate are inconclusive because of uncertainty in the normalization of the calculated neutrino flux. There is also some uncertainty in the neutrino cross section.

Combining all their data (sub-GeV and multi-GeV), the Kamiokande group find a rather limited allowed region, either for $\nu_\mu \leftrightarrow \nu_\tau$ or for $\nu_\mu \leftrightarrow \nu_e$ with nearly full mixing and with $\delta m^2 \sim$ 1 to 2×10^{-2} eV^2. Simultaneous analysis of all atmospheric and solar neutrino data sets (as of 1993) in the space of 3-flavor neutrino oscillations [27] favors oscillation predominantly in the $\nu_\mu \leftrightarrow \nu_\tau$ sector to explain the atmospherics and in the $\nu_\mu \leftrightarrow \nu_e$ sector for the solar neutrinos. This in turn implies a "natural" order for the neutrino masses, but with $m(\nu_\tau) \sim 0.1$ eV, rather than the value of ~ 10 eV expected from a quadratic see-saw starting from $m(\nu_\mu) \sim 0.003$ eV to solve the solar neutrino puzzle.

In the next year, we expect higher statistics from the Soudan experiment and further results from the KEK beam test. After that, SuperKamiokande will contribute to our understanding of atmospheric neutrinos. Confirmation of the atmospheric neutrino deficit as a manifestation of neutrino oscillations may need to await one of the accelerator based long baseline neutrino experiments.

References

[1] Kamiokande presentation at Neutrino '94 (Eilat).

[2] M. Circella et al. *Proc. 23rd Int. Cosmic Ray Conf.* (Calgary) 4 (1993) 503.

[3] D. Müller et al. (HEAT Collaboration) Proc. 22nd Int. Cosmic Ray Conf. (Dublin, 1991) vol. 2, 177.

[4] D.H. Perkins, Astroparticle Physics 2 (1994) 249.

[5] Kamiokande Collaboration, ISSN 1340-3745 (submitted to Physics Letters B).

[6] K.S. Hirata et al. (Kamiokande Collaboration) Phys. Letters B 280 (1992) 146. This paper reported on a 4.92 kT-year exposure. The exposure for sub-GeV, fully contained events is now 7.7 kT-year [5].

[7] H. Meyer, for the Frejus collaboration, talk presented at Snowmass, July 7, 1994.

[8] R. Becker-Szendy et al. (IMB Collaboration) Phys. Rev. D 46 (1992) 3720.

[9] Ch. Berger et al. (Frejus) Physics Letters B 245 (1990) 305 and 227 (1989) 489.

[10] M. Honda, K. Kasahara, K. Hidaka & S. Midorikawa, Phys. Lett. B 248 (1990) 193.

[11] R. Becker-Szendy et al., Phys. Rev. Letters 69 (1992) 1010.

[12] M. Goodman, Proceedings of the Atmospheric Neutrino Workshop, 6-8 May 1993, LSU-HEPA-93-6.

[13] T. Kajita, Proceedings of the Atmospheric Neutrino Workshop, 6-8 May 1993, LSU-HEPA-93-6.

[14] D. Michael, private communication.

[15] L.V. Volkova, Sov. J. Nucl. Phys. 31 (1980) 1510.

[16] E. Eichten, I. Hinchcliffe, K. Lane & C. Quigg, Revs. Mod. Phys. 56 (1984) 579 [56 (1986) 1065(E)].

[17] W. Frati, T.K. Gaisser, A.K. Mann & Todor Stanev, Phys. Rev. D 48 (1993). 1140.

[18] M. Mori et al., Phys. Letters B270 (1991) 89.

[19] A.V. Butkevich, L.G. Dedenko & I.M. Zheleznykh, Sov. J. Nucl. Phys. 50 (1989) 90.

[20] V. Agrawal, T.K. Gaisser, Paolo Lipari & Todor Stanev (unpublished).

[21] J.F. Owens, Phys. Lett. B266 (1991) 126.

[22] D. Michael for the MACRO Collaboration, talk presented at Snowmass, July 7, 1994.

[23] M.M. Boliev et al. (Baksan Collaboration) in *Proc. 3rd Int. Workshop on Neutrino Telescopes*, Venice, 1991 (ed. M. Baldo-Ceolin) p. 235.

[24] G. Barr, T.K. Gaisser & T. Stanev, Phys. Rev. D 39 (1989) 3532.

[25] E.V. Bugaev & V.A. Naumov, Phys. Lett. B 232 (1989) 391.

[26] J. Engel, E. Kolbe, K. Langanke & P. Vogel, Phys. Rev. D48 (1993) 3048.

[27] G.L. Fogli, E. Lisi & D. Montanino, Phys. Rev. D (1994) (to be published).

LONG-BASELINE NEUTRINO OSCILLATION EXPERIMENTS

Dan Crane and Maury Goodman
Argonne National Laboratory, Argonne, IL 60439

Abstract

The status and capabilities of long-baseline neutrino oscillation experiments are reviewed.

1 Introduction

There is no unambiguous definition for long baseline neutrino oscillation experiments. The term is generally used for accelerator neutrino oscillation experiments which are sensitive to $\Delta m^2 < 1.0\ eV^2$, and for which the detector is not on the accelerator site. One earlier review of possible long-baseline (LBL) experiments is out of date.[1] The Snowmass N2L working group met to discuss the issues facing such experiments. The Fermilab Program Advisory Committee adopted several recommendations concerning the Fermilab neutrino program at their Aspen meeting immediately prior to the Snowmass Workshop.[2] This heightened the attention for the proposals to use Fermilab for a long-baseline neutrino experiment at the workshop.[3, 4, 5, 6] The plan for a neutrino oscillation program at Brookhaven was also thoroughly discussed.[7, 8] Opportunities at CERN were considered, particularly the use of detectors at the Gran Sasso laboratory.[9, 10] The idea to build a neutrino beam from KEK towards Superkamiokande was not discussed at the Snowmass meeting, but there has been considerable development of this idea since then.[11] Brookhaven and KEK would use low energy ($< E_\nu > \sim\ 1\ GeV$) neutrino beams, while FNAL and CERN would plan have medium energy beams ($< E_\nu > \sim\ 15\ GeV$).

Long-baseline Neutrino Oscillation Experiments are motivated in part by the atmospheric neutrino deficit.[12] This is well matched to the capabilities of LBL experiments, $2 \times 10^{-3} eV^2 < \Delta m^2 < 1 eV^2$; $\sin^2(2\theta) > 0.1$. However, there is also theoretical interest in lower mixing angles,[13] ($\sin^2(2\theta) \sim 0.03$) which provides motivation to design experiments with higher statistics in order to achieve lower statistical and systematic errors.

This report will summarize a few topics common to LBL proposals and attempt to give a snapshot of where things stand in this fast developing field.

2 Generic Issues for Long-baseline Experiments

The probability that a neutrino will oscillate is given as

$$P_{\nu_1 \to \nu_2} = \sin^2(2\theta) \sin^2 \frac{1.27 \Delta m^2 L}{E} \qquad (1)$$

with L given in kilometers, E_ν in GeV and $\Delta m^2 = m^2_{\nu_1} - m^2_{\nu_2}$ in eV^2. The figures of merit for comparing one neutrino oscillation experiment to another are the sensitivity curves in the $\Delta m^2 - \sin^2(2\theta)$ plane. Usually these are constructed as 90%CL contour curves for limits or favored regions, and unless there is only sensitivity to high mixing angle, a log-log plot is used. For most limits, the curve corresponds to a minimum probability of oscillation, defined as P_{min}. It depends on the details of the test, neutrino source, energy distribution, possible backgrounds and systematic errors, and statistics. Adequate statistics is a key challenge of long-baseline experiments. A complete comparison of one experiment vis-a-vis another, however, goes somewhat beyond the limit curve, as discussed in the summary.

For the purpose of designing an "optimum" experiment it is useful to consider an approximation to the 90% CL limit curves. At high Δm^2 where the oscillations from different parts of the energy spectrum average out, $P_{min} = 0.5 \sin^2(2\theta)$ for all Δm^2, so that $\sin^2(2\theta_0) = 2P_{min}$. Likewise, at maximal mixing, $\Delta m_0^2 = (< E_\nu > \times \sqrt{P_{min}})/1.27L$. Also at maximal mixing the slope on the log-log plot is

-1/2 because

$$\left.\frac{\partial \Delta m^2}{\partial \sin^2(2\theta)}\right)_{\sin^2(2\theta)=1} = -\frac{1}{2} \qquad (2)$$

The two lines intersect at $\Delta m_1^2 = \Delta m_0^2/\sqrt{2P_{min}}$. Such a two line curve is shown in Figure 1. (The approximation to the actual limit curve is quite good.)

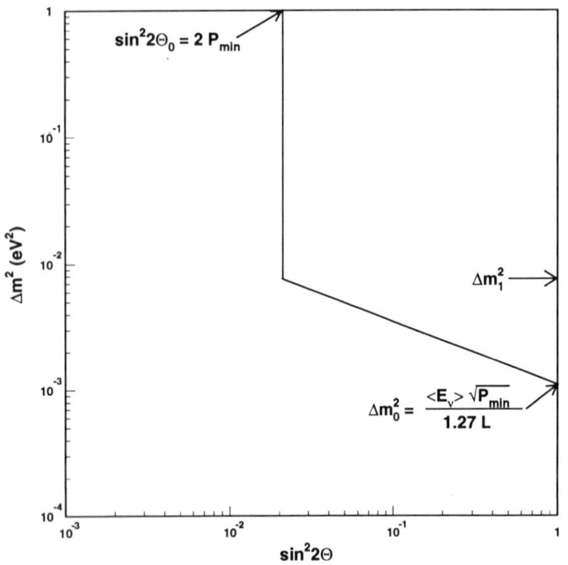

Figure 1: Approximation to a neutrino oscillation limit curve showing the relationship between P_{min} and the intercepts for $\sin^2(2\theta) = 1$ and high Δm^2.

P_{min} can be derived exactly for a variety of tests. We list here the results of these calculations for long-baseline neutrino oscillation tests which have been discussed. We consider the case with no cosmic ray backgrounds or systematic errors:

- Background-free test $P_{min} = 2.3/N_{cc} \times \eta \times \varepsilon$
- Disappearance $P_{min} = 1.29/\sqrt{N_{cc}}(1 - \eta B)$
- NC/CC ratio

$$P_{min} = \frac{1.29R^{true}(1 + R^{true})}{\sqrt{NR^{true}} \times [\eta(1-B) + R^{true}(1 - B\eta)]}$$

- Kinematic cuts $P_{min} = 1.29\sqrt{\kappa N_{cc}}/\varepsilon\eta N_\mu$

where N_{cc} is the total number of charged current events, N is the total number of neutrino events (CC+NC), R^{true} is the expected NC/CC ratio, η is the ν_τ to ν_μ CC cross section ratio, B is the branching ratio for τ decay to muons, ε is the τ event analysis efficiency including branching ratios, and κ is the ν_μ CC background rejection from the kinematic cuts being used, expressed as a fraction from 0 to 1. Numerical constants such as 1.29 depend on the choice of 90% CL limits. $\eta(E)$ is energy dependent, but we estimate $<\eta> \sim \eta(<E_\nu>)$. For $\nu_\mu \to \nu_\tau$ neutrino experiments at Brookhaven or Fermilab energies, typical values for P_{min} are:

FNAL disappearance	$1.35/\sqrt{N_{cc}}$
FNAL $R_{``nc"/``cc"}$	$1.87/\sqrt{N}$
FNAL kinematic cuts	$1.43/\sqrt{N_{cc}}$
BNL disappearance	$1.29/\sqrt{N_{cc}}$
BNL $R_{``nc"/``cc"}$	$3.03/\sqrt{N}$

No neutrino oscillation test is the clear preference based on statistical power alone. The extension to include systematic errors and backgrounds in the formulae is relatively straightforward.[3]

An accurate curve can be calculated in lieu of the approximate lines in Figure 1 by integrating over the neutrino energy spectrum. The approximation is useful, however, for seeing how the limit scales as parameters are changed. For example, to see the effect of placing a fixed size detector at different distances (L) from an accelerator, the statistics $N \propto 1/L^2$. Then $P_{min} \propto 1/\sqrt{N} \propto_L$, so $\Delta m_0^2 \propto \sqrt{P_{min}}/L \propto L^{-1/4}$ for a fixed size detector. Also, $\sin^2(2\theta_0) \propto L$.

In contrast, the variation as a function of beam energy is quite complicated. As the proton beam energy E_p is raised, there are more secondaries per proton, the secondary beam is more forward, and the average neutrino energy rises logarithmically so that there are more events due to the higher ν cross section, as well as relatively more ν_τ events as $\eta(E)$ increases. However, there would be less proton spills in a fixed calendar time, less secondaries would decay in a fixed length beam pipe, and Δm^2 sensitivity decreases as $<E_\nu>$. The result of detailed calculations for the $R_{``nc"/``cc"}$ test show a broad optimum from 100 GeV to 400 GeV for a detector at 700 km.

One important facet of the neutrino energy is the relationship to ν_τ charged current threshold. Below or just above $< E_\nu > \sim 5GeV$, an experiment can measure the change in ν_μ CC event rate, or the NC event rate which does not change. At higher energies, ν_τ CC events contribute and a larger number of neutrino oscillation signatures is possible.

3 Review of Specific Proposals

Plans for long-baseline neutrino experiments are being made for nuclear reactors at San Onofre and Chooz and for accelerators at Brookhaven, KEK, Fermilab, and CERN. The reactor experiments are sensitive to $\nu_e \rightarrow \nu_\mu$ (or $\nu_e \rightarrow \nu_\tau$) oscillations. Accelerator experiments are sensitive to $\nu_\mu \rightarrow \nu_\tau$ or $\nu_\mu \rightarrow \nu_e$ oscillations.

3.1 Reactor Experiments

Nuclear reactors are abundant sources of $\bar{\nu}_e$'s with energies up to about 8 MeV. Several neutrino oscillation experiments have been conducted at nuclear reactors.[14] The event rate of $\bar{\nu}_e p \rightarrow e^+ n$ is measured at two or more locations. Systematic errors probably restrict this kind of experiment to $\sin^2(2\theta) > 0.05$. At present, there is no evidence for oscillations from these experiments. Recent limits from Krasnoyarsk[15] and Bugey[16] have pushed the limits to lower Δm^2 below $10^{-2} eV^2$. New experiments are being planned at San Onofre and Chooz. The San Onofre experiment plans a .13 ton target 24m from the core, along with a 12 ton target at a distance of 650 m. Either experiment should see a large effect if the atmospheric ν_μ deficit is due to $\nu_e \rightarrow \nu_\mu$ oscillations.

3.2 Brookhaven

There are two ideas to use the AGS at Brookhaven to do a long-baseline disappearance experiment. One involves a collaboration led by Al Mann which plans to build four two-kiloton water Cerenkov detectors.[7] Two of them would be 24 km away on the shore of Long Island at Northville. The other two would be on site, one and three km away respectively. In 120 days of running the AGS, they would expect 4400 $\nu_\mu n \rightarrow \mu^- p$ events and 1100 $\nu N \rightarrow \nu N \pi^o$ neutral current events at the far detectors. The systematic errors would be small and verifiable through the use of four identical detectors. Statistical errors would be on the order of 1%. The sensitivity to neutrino oscillations has been increased by aiming the beam off axis in order to favor low energy neutrinos in the beam. The experiment has been approved, but not yet funded. The cost estimate when the proposal was made was $33M.

An alternative idea for a long-baseline experiment at Brookhaven has been developed by Ted Kycia.[8] He would use steel plates and optical spark chambers with video camera readout, and has operated a prototype at Brookhaven. Beam which is extracted from the AGS could be made into a neutrino beam aimed at Connecticut without a large bend. A detector in Collinsville Connecticut would be about 100 km away.

3.3 CERN

An attractive possibility for CERN is to aim a neutrino beam from the SPS to the Gran Sasso Underground Laboratory, 732 km away. The targeting for a neutrino beam to Gran Sasso could be done in the extraction tunnel from the SPS to the LHC.

One proposal has been made by the ICARUS collaboration[9]. They plan to build a 5 kiloton time projection chamber in Hall C of the Gran Sasso. They have built a successful prototype ionization chamber using liquid Argon that is continuously sensitive, self-triggering, able to provide three-dimensional images of ionizing events, and able to identify particles (by dE/dx) and their direction. Such a detector is also an excellent calorimeter with high granularity and resolution. The cost estimate for the ICARUS detector is 62M Swiss Francs.

Another idea suggested by the Genius collaboration led by Barry Barish and Doug Michael is for a 17 kT magnetized iron calorimeter in Hall B of the Gran Sasso.[10] The detector would have 4 cm iron plates and use resistive plate chambers. Studies have shown that a 2 cm pitch for the readout strips give reasonable separation and resolution for tracking. The estimated cost is $28M.

3.4 Fermilab P-822

A collaboration including the Soudan 2 group proposed to use the existing Soudan 2 Detector with a neutrino beam from the Fermilab Main Injector 733 km away (P-822).[3] The Soudan 2 detector is a fine-grained 1 kiloton iron calorimeter designed to measure 1 GeV proton decay events. It is thus well suited for the 2-10 GeV visible E_{had} from neutral current events in the Fermilab beam. The best test for $\nu_\mu \rightarrow \nu_\tau$ oscillations is the neutral current to charged current test. Some Soudan modules would be placed behind the short baseline experiment (E-803) at Fermilab, and $R_{``nc"/``cc"} = \frac{NC}{CC}$ would be measured in two locations by detecting the presence or absence of a muon in the neutrino event. In the far detector, the number of events in the denominator would go down for $\nu_\mu \rightarrow \nu_\tau$ oscillations, while the number of events in the numerator would rise due to ν_τ charged current events.

P-822 could measure 1200 events per Main Injector year with a 320 meter decay pipe. Enhancements such as a toroid magnet for muon momentum measurement and the near detector would cost about $4M.

3.5 Fermilab Expressions of Interest

The Fermilab PAC was intrigued with P-822, but preferred a more ambitious program. They established two broad characteristics for a Long-baseline experiment: "1. It must extend the limit in Δm^2 or $\sin^2(2\theta)$ by a significant factor. This suggests that an acceptable sensitivity in the exclusion plot for such an experiment is $\Delta m^2 \sim 10^{-2} eV^2$ and $\sin^2(2\theta) \sim 10^{-2}$. *and* 2. It will be necessary that a positive signal capability be convincing, which implies the capability of measuring backgrounds and quantitatively determining systematics."[17]

In March of 1994, Fermilab called for Long-baseline Expressions Of Interest (EOI) assuming a Main Injector intensity of 5×10^{13} protons in a pulse every 1.9 s. Three were received. Stan Wojcicki (EOI-7) proposed a narrow band beam and several large water Cerenkov detectors located at SLAC.[4] Barry Barish and Doug Michael (EOI-8) assembled a large group with the idea of constructing a 17 kiloton magnetised iron calorimeter at either IMB or Soudan.[5] Maury Goodman and the P-822 group also suggested a large iron calorimeter with magnetic measurement at a new hall at Soudan (EOI-9) to supplement the existing detector.[6] The large detector cost estimates were $30M - $40M.

At their 1994 Aspen meeting, the Fermilab PAC endorsed a joint short-baseline(E-803)/long-baseline program at the Fermilab Main Injector, and encouraged the three long-baseline groups to merge.[2] A schedule for experiment approval and beam design was spelled out, in order to obtain DOE approval and a congressional line item in the FY1998 budget.

3.6 KEK

A neutrino beam from the KEK 12 GeV PS could be used in conjunction with the SuperKamiokande experiment. Superkamiokande is a 30 kT water Cerenkov detector under construction 245 km from KEK. They will detect 500 ν events for 10^{20} protons on target (*pot*). Another 2kT detector at 25 km located on the surface would be 1.34^o off axis from the neutrino beam, and could measure 1700 ν events for 10^{20}*pot*.

4 Summary and Prospects

Options for Long-baseline experiments for the United States are at Brookhaven and Fermilab, or to participate in programs abroad. The high energy (FNAL, CERN) and low energy (BNL, KEK) opportunities are complementary, but fiscal constraints may well prevent major participation in all options. It is likely that a HEPAP subpanel will consider the alternatives in the next year. As mentioned above, the major figure of merit is the limit plot on the $\sin^2(2\theta) - \Delta m^2$ plane, and the related sensitivity for discovery of oscillations should they exist. This is affected not only by the statistics, but also detailed consideration of possible systematic effects and backgrounds, and also the number of independent tests which could measure neutrino oscillations. There are also several other important factors to be considered, which are more difficult to quantify:

1. Other physics capabilities of the experiment, relating to low energy neutrinos or cosmic rays.
2. Civil construction feasibility and costs.
3. Detector feasibility and costs.
4. Complementarity or incompatibility of an experiment within an accelerator program.
5. Possible public concerns about off-site accelerator facilities.
6. The schedule for getting a compelling positive or negative result.

It seems likely that one or more of the proposed long-baseline experiments will proceed. If ν oscillations are lurking in the area of parameter space $10^{-3}\ eV^2 < \Delta m^2 < 1\ eV^2; \sin^2(2\theta) > 0.02$, as suggested both by the atmospheric neutrino anomaly and independently by several theoretical ideas, long-baseline neutrino oscillation experiments will become a long and active part of our experimental high energy physics program well into the next millenium.

Acknowledgments

Over 100 individuals participated in the N2L session of the Snowmass Workshop, including 34 presentations. Special thanks to Al Mann and Douglas Michael for their work in organizing the sessions. This work was supported in part by U.S. Department of Energy, Division of High Energy Physics, Contract W-31-109-ENG-38.

References

[1] M. Goodman, in the Proceedings of the Workshop on Long-baseline Neutrino Oscillations, Fermilab, 1991.

[2] Letter from Fermilab Director John Peoples to Participants in Particle and Nuclear Astrophysics and Cosmology in the Next Millennium, dated June 17, 1994.

[3] W.W.M. Allison *et al.*, "Proposal for a Long-baseline Neutrino Oscillation Experiment Using the Soudan 2 Neutrino Detector", March 19, 1991; Revised October 12, 1993 and updated March 8, 1994.

[4] S. Wojcicki, "Long-baseline Neutrino Oscillation Experiment Based on Fermilab Main Injector", May 13, 1994.

[5] B. Barish *et al.*, "Expression of Interest for a Long-baseline Neutrino Oscillation Experiment Using a Main Injector Beam and a Large Magnetic Sampling Calorimeter", May 1994.

[6] W.W.M. Allison *et al.*, "Expression of Interest for a long-baseline neutrino oscillation experiment from Fermilab to Soudan using a 16 kT iron calorimeter", May 16, 1994.

[7] I. Chiang *et al.*, "Long-baseline Neutrino Oscillation Experiment at the AGS", E889, August 30, 1993.

[8] T. Kycia *et al.*,, "Proposal to Search for Neutrino Oscillations", P911, November 3, 1994.

[9] P. Cennini *et al.*,, "ICARUS II, A second generation proton decay experiment and neutrino observatory at the Gran Sasso Laboratory", September 1993.

[10] B. Barish *et al.*, "Expression of Interest for a Long-baseline Neutrino Oscillation Experiment Using a Main Injector Beam and a Large Magnetic Sampling Calorimeter", May 1994.

[11] K. Nishikawa, KEK, private communication October 1994.

[12] Tom Gaisser and Maury Goodman; these proceedings.

[13] Dimopoulous, Hall and Raby, *Phys. Rev* **D47** R3697 (1993).

[14] F. Boehm & P. Vogel, Physics of Massive Neutrinos, Cambridge University Press (1987).

[15] Vidyoka *et al.*, JETP Lett. 59, 6, (1994).

[16] P. Vogel, Proceedings of Neutrino '94, Nuclear Physics B Proceedings Supplements, June 1994.

[17] Report of the Fermilab PAC from the June 1993 meeting.

Report from the Short Baseline Neutrino Oscillations Group

NEVILLE W. REAY

Department of Physics, Kansas State University, Manhattan, KS 66506-2601

The following is a short report on work accomplished during and after Snowmass by a short-baseline neutrino oscillations group whose nucleus was constituted from the Fermilab E803/COSMOS collaboration. Emphasis was placed not on setting limits, but on seeing positive evidence for $\nu_e \to \nu_\tau$ and $\nu_\mu \to \nu_\tau$ oscillations, should either exist. Key in this endeavor is reconstructing the momentum components of tau decay products which are perpendicular to the direction of the parent tau. At Snowmass, the particular two-body decay $\tau \to \rho\nu_\tau \to \pi^-\pi^0\nu_\tau \to \pi^-\gamma\gamma\nu_\tau$ was chosen because of its large branching ratio and because of the stress it places both on tracking and electromagnetic calorimetry. Based on monte carlo simulations and information gained from reports made by running CERN experiments CHORUS and NOMAD, strategies were developed which successfully tracked neutral decay components through three radiation lengths of target.

1 Rational for Studying Short-Baseline Neutrino Oscillations:

Interest in oscillations has been stimulated by an increasing body of circumstantial evidence: there is an apparent deficit both of ν_e [1], [2], [3] coming from the sun and of ν_μ coming from atmospheric cosmic-ray interactions [4], [5], [6]. Theory also plays a role: the "Big Bang" model predicts that approximately 300 neutrinos inhabit every cubic centimeter of the universe; were they to have even a small mass, the universe could be closed by gravitational attraction. A see-saw mechanism relating masses of neutrinos to those of associated quarks and/or leptons in the same generation suggests that ν_τ would be the heaviest of known neutrinos, and hence is the leading candidate to provide this gravitational attraction [7]. LEP results limiting to three the number of standard neutrinos strengthens this hypothesis [8].

The probability for two-component neutrino oscillations goes as

$$P(\nu_a \to \nu_b) = \sin^2(2\alpha) \int \sin^2\left(1.27\,\delta m^2 \frac{L}{E_\nu}\right) d\left(\frac{L}{E_\nu}\right)$$

with $\delta m^2 = m_a^2 - m_b^2$ in eV^2, the source-to-detector distance L in meters and the neutrino energy E_ν in MeV. E803/COSMOS[a] short-baseline measurements made close to the source have ultra-high statistics for investigating smaller mixing angles α for values of δm^2 of cosmological interest. Most post-COBE cosmological models require a mix of cold and hot dark matter to explain the evolution of structure on observed distance scales; too much cold dark matter results in too much structure at small scales. Masses of order 1 to 7 eV/c^2 are favored for a hot component of massive neutrinos, and the see-saw mechanism favors $\Delta m^2 \approx m_\tau^2$. If mixing is strongest between adjacent neutrino generations, then $\nu_\mu \to \nu_\tau$ is the most likely source of oscillation into ν_τ's.

The Fermilab Main Injector Upgrade, now scheduled for completion in 1999, will provide 120 GeV proton beams (hence neutrinos) with unprecedented intensity, motivating a comprehensive neutrino-oscillation program featuring the short-baseline experiment COSMOS and the long-baseline experiment MINOS. COSMOS is the focus of this article.

2 COSMOS Goals:

COSMOS is a ν_τ appearance experiment which will demonstrate an unambiguous τ signal if oscillations exist within a factor of five of its 90% confidence level sensitivity. The apparatus is a hybrid-emulsion electronic spectrometer: candidates for τ decays are directly observed in the emulsion target, while information from the spectrometer is used to select and locate events to be scanned, to provide momentum, energy and particle identification

[a]Aichi University of Education, Kariya, Japan; University of Athens, Athens, Greece; University of California (Davis), Davis, CA; University of California (Los Angeles), Los Angeles, CA; Chonnam National University, Korea; Columbia University, New York, NY; Fermilab, Batavia, IL; University of Gifu, Gifu, Japan; Gyeongsang National University, Korea; Hirosaki University, Hirosaki-city, Japan; Illinois Institute of Technology, Chicago, Illinois; Indiana University, Bloomington, Indiana; Kansas State University, Manhattan, KS; Kinki University, Higashi-Osaka, Japan; Kobe University, Kobe, Japan; Korea Advanced Institute of Science and Technology, Korea; Korea University, Seoul, Korea; University of Michigan, Ann Arbor, Michigan; University of Minnesota, Minneapolis, Minnesota; Nagoya Institute of Technology, Nagoya, Japan; Nagoya University, Nagoya, Japan; Okayama University, Okayama, Japan; Osaka City University, Osaka, Japan; Osaka University of Commerce, Japan; Science Education Center of Osaka Prefecture, Osaka, Japan; Soai University, Osaka, Japan; Seoul National University, Seoul, Korea; University of South Carolina, Columbia, SC; Technion -Israel Inst. of Tech.; Toho University, Funabashi, Japan; Tufts University, Medford, MA; Yokohama University, Yokohama, Japan; Utsunomiya University, Utsonomiya, Japan.

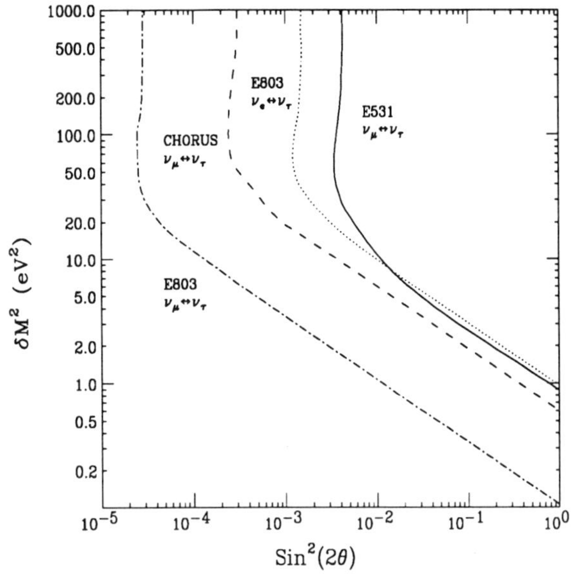

Figure 1: Comparison of 90% CL limit sensitivities of E803/COSMOS, CHORUS, and E531. The estimated E803/COSMOS sensitivity is based on four years of running. The CHORUS curve is their proposal estimate for two years.

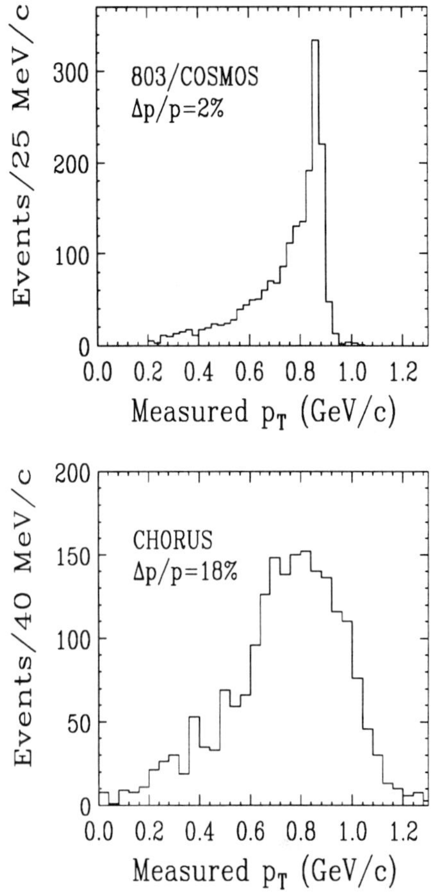

Figure 2: Monte Carlo simulations of the Jacobian peak in transverse momentum for $\tau^- \to \pi^- \nu$ as a function of momentum resolution.

for kinematic analysis of τ candidates, and to reject background. When running is complete, COSMOS will have a proposed 90% confidence level sensitivy 10× better at large and 30× better at small δm^2 than proposed by the present CERN experiments CHORUS and NOMAD.

Present and proposed limits for short-baseline oscillations are given in Fig. 1. These curves are based on 4 eight-month runs at 3×10^{13}, or 2.5 runs at 5×10^{13} protons per pulse. Present limits include those determined by Fermilab experiment E531 [9]. The sensitivity increase of COSMOS is due to several factors: higher neutrino flux and target mass, better background rejection due to higher μ detection efficiency, scanning only for single-prong τ decays, and minimizing emulsion analysis time by incorporating fully automatic techniques, thus raising scanning capability from 25,000 to 300,000 events per year. The increase in scanning capacity will result in a more timely extraction of results, and would permit COSMOS to take more beam if it should become available.

If a positive signal is observed for $\nu_\mu \to \nu_\tau$ or $\nu_e \to \nu_\tau$ oscillations, COSMOS will be able to determine whether the resulting τ candidates have the correct mass and lifetime. Key to this endeavor is determining momentum components of τ decay products transverse to the original τ direction, which requires precise knowledge both of the τ direction and of daughter vector momenta. Emulsion measurements determine the former typically to within a milliradian, while determination of the latter requires both good electromagnetic calorimetry and precision tracking of charged particles through a magnetic field. The Monte Carlo distributions of Fig. 2 show the Jacobian peak in transverse momentum for a typical τ decay as a function of resolution. Note that the sharp edge deteriorates as resolution degrades. Of competing experiments worldwide, only COSMOS will have the apparatus to deliver a sharp edge.

3 Apparatus and method of approach:

An elevation view of COSMOS is shown in Fig. 3. Scintillating fibers near the emulsion provide an intermediate coupling between the precision of emulsion and the coarser resolution of a conventional air-gap magnetic spectrometer. The final sections of the apparatus consist of a precision electromagnetic calorimeter followed by a simple range identifier for muons.

The experiment will run in an intense horn beam

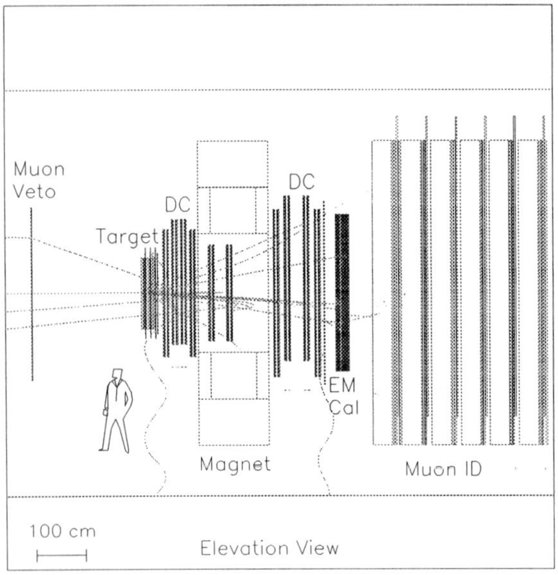

Figure 3: The COSMOS hybrid emulsion-electronic spectrometer.

with an approximately 1 millisecond spill and a cycle time of 1.9 seconds. Because of the high intensity, many spills will contain more than one neutrino interaction in the emulsion. Accordingly, the experiment will run in a highly buffered mode with triggered events stored locally during each spill, and full readout occurring only between spills.

Analysis will proceed in several stages. First, recorded events will be reconstructed and a 10% subset will be selected for emulsion scanning. Using the fully automatic scanning systems based at Nagoya University, candidate events will be found in the emulsion and searched for secondary one-prong decay vertices. Finally, a third pass will be made which combines electronic and emulsion information for interesting events. Selection criteria for τ's include rejecting any event containing a muon, and demanding that each event have at least one track with momentum greater than 1 GeV/c. Further cuts are then made based on discriminant analysis, which maximally correlates kinematic selection criteria such as visible event energy and missing P_T.

The primary interaction vertex for selected events is found by linking one track from the spectrometer into changeable sheets located near the downstream face of each emulsion block. The excellent spatial resolution and two-track separation of scintillating fibers is essential in this linking process. Subsequently, the track is followed upstream from the changeable sheet into the main emulsion stack and back to the primary. One-prong τ-decay candidates are found by following down along all primary charged tracks and searching for a kink of more than 10 milliradians.

When a kink is found, the secondary track from the kink must be followed all the way to the exit of the emulsion and linked to the spectrometer. Kinks are rejected if they are positively charged, have momentum less than 1 GeV/c, or have a P_T of less than 0.25 GeV/c relative to the candidate τ direction. For events meeting these criteria, it must be established that none of the tracks from the primary vertex are identified as electrons (ν_e CC event). Finally, the τ decay modes consistent with the event must be identified, and the event must pass particle identification and kinematic cuts for at least one of these modes.

The ability to measure the transverse momentum components of charged tracks and γ's, when combined with the measured direction of the τ candidate track in the emulsion, provides several powerful kinematic handles for rejecting tau-decay backgrounds. These are illustrated in Fig. 4 for $\tau^- \rightarrow \pi^-\nu$. In the plane perpendicular to the beam, the direction of the τ should be nearly back-to-back with the momentum of the rest of the event ($\Delta\phi_{\tau R} \approx 180°$). The direction of the hadron from the τ decay should be nearly back-to-back with the neutrino ($\Delta\phi_{DK} \approx 180°$), when both are referred to the measured τ direction; the direction of the unseen neutrino is obtained from the missing transverse momentum in the event. Note that use of $\Delta\phi_{DK}$ requires knowledge, available in COSMOS, of both the τ direction and of the missing p_T vector. An independent signature is the transverse mass $M_{min} = \sqrt{p_T^2 + m_{vis}^2} + p_T$, which has a sharp cusp at the τ mass. As shown by the plots in Fig. 4, these distributions from simulated $\tau \rightarrow \pi\nu$ are very different from those for white-star kinks (WSK's). WSK's are interactions of charged tracks coming from the primary vertex in which a single track leaves the point of interaction and there is no evidence of nuclear breakup. Such a topology mimics tau decay. Finally, the decay length of τ candidates falls off exponentially, while that for WSK's is flat.

The COSMOS tracking emphasis is motivated by efficient linkup to the emulsion and maximum use of kinematic constraints to reduce background and display a τ signal. The muon identifier is used to reject events with muons from the primary vertex; it is especially important in reducing the background from anticharm produced in $\overline{\nu_\mu}$ CC interactions. Electromagnetic information in τ^- decay modes ($\pi^- n\pi^0\nu$, $e^-\nu\nu$) and in background events must be reconstructed for kinematic use. As indicated by the simulated $\tau^- \rightarrow \pi^-\pi^0\nu$ event in Fig. 5, conversions in the thick (3 RL) emulsion target usually require the original γ's to be assembled from 'pieces': tracks of conversion electrons and secondary as well as primary γ's in the electromagnetic calorimeter. The design of the electromagnetic calorimeter therefore emphasizes good low-energy performance (down to $\approx$ 100 MeV) in a high-multiplicity

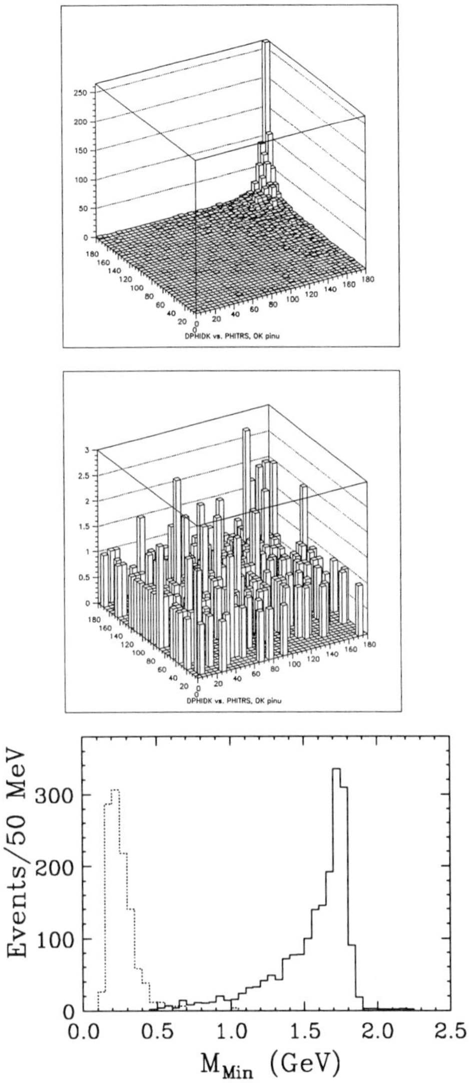

Figure 4: Some kinematic constraints for $\tau^- \to \pi^- \nu$, which separate the signal from whitestar kinks (WSK's). The top is a monte carlo simulation lego plot for $\tau \to \pi\nu$. The left-hand axis is $\phi_{\tau R}$, while the right-hand axis is ϕ_{DK}. The middle plot is a simulation of WSK interactions, which can fake true tau events. The bottom plot shows a simulated minimum mass, which includes experimental resolutions, for both signal (solid) and background (dotted).

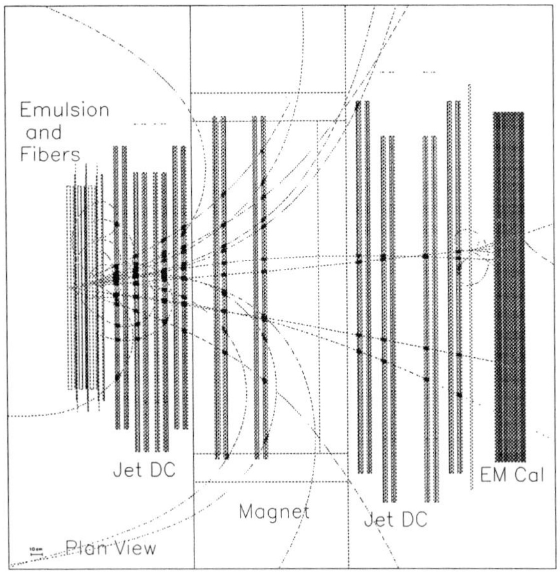

Figure 5: Target, tracking and electromagnetic calorimeter. The excellent pattern-recognition capability of multisampling (jet) drift chambers is evident. The simulated event is a typical $\tau^- \to \pi^- \pi^0 \nu$ in which both γ's have converted in the emulsion. They must be reassembled from e^+, e^- in the drift chambers and secondary γ's in the EMCal.

environment ($\approx 10 - 20$ showers).

Applying these constraints to the more copious $\tau \to \pi^- \pi^0 \nu$ mode is a severe test of the ability of the spectrometer (combined EMCal and tracking) to reconstruct electromagnetic showers. The γ's must be assembled from pieces, as described above, and an algorithm must select the γ's most likely to come from the τ. Preliminary versions of both of these algorithms now exist. As shown by the simulation results in Fig. 6, the challenge can be met: the $\Delta\phi_{DK}$ vs $\Delta\phi_{\tau R}$ distribution is as sharp as that for $\pi\nu$, and the M_{min} cusp is preserved.

Since the signal is approximately exponential in each of the variables decay length, $\Delta\phi_{DK}$, and $\Delta\phi_{\tau R}$, a simple discriminant variable (product of exponentials) does very well in separating signal from WSK's. After this is done, the kink p_T and M_{min} are then still available for further discrimination; backgrounds of much less than one event can be achieved at modest cost in signal.

In summary, algorithms have been developed which preferentially select and reconstruct gammas coming from $\tau^- \to \pi^- \pi^o$ decays even after passing through 3 radiation lengths of target. The next step will be an attempt to use similar techniques to fit the $\tau^- \to e^- \overline{\nu}_e \nu_\tau$ decay mode.

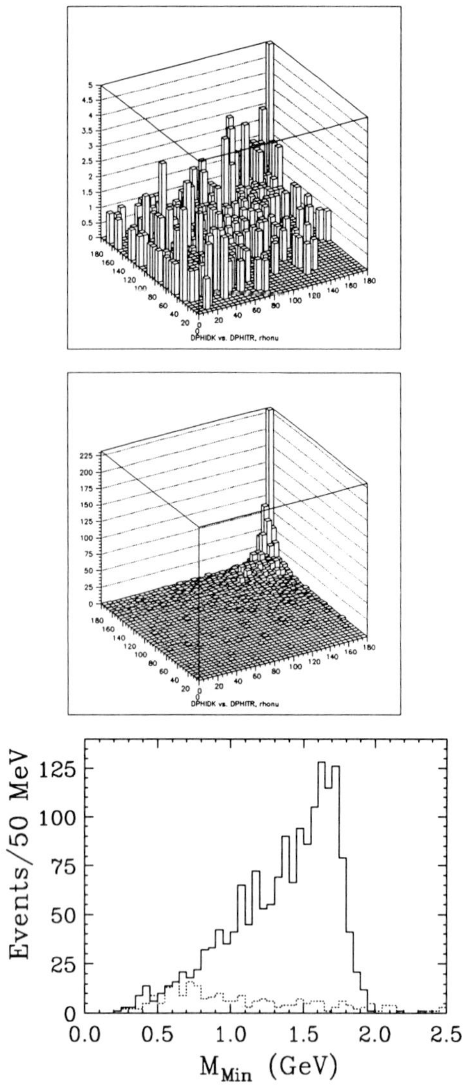

Figure 6: Some kinematic constraints for $\tau^- \rightarrow \pi^-\pi^0\nu$, which separate the signal from whitestar kinks (WSK's). An algorithm has selected the showers most likely to come from the τ. The format for all plots is identical to the corresponding ones in figure 4.

Acknowledgments

I wish to acknowledge the efforts of the COSMOS collaboration and NOMAD representative Sanjib Mishra. This work was supported in part by the U.S. Department of Energy; the U.S. National Science Foundation; the Japan Society for the Promotion of Science; the Japan-U.S. Cooperative Research Program for High Energy Physics; and the Ministry of Education, Science and Culture of Japan.

References

1. R. Davis, Jr., Proc. of the 19th Int. Conf. on Neutrino Phys. and Astrophys. (1989) 518.
2. K.S. Hirata, *et al.*, Phys. Rev. Lett. **65** (1990) 2233.
3. K. Lande, private conversation at Singapore '90.
4. K.S. Hirata, *et al.*, Phys. Lett. **B 205** (1988) 416; T. Kajita, talk at Singapore '90.
5. D. Casper *et al.*, Boston University Preprint **90-23** (1990).
6. Ch. Berger, *et al.*, Phys. Lett. **B 227** (1989) 489.
7. See, e.g., H. Harari, Phys. Lett. B216 (1989) 413.
8. G. S. Abrams, *et al.*, Phys. Rev. Lett. **63** (1989) 2173; LEP results in press.
9. N. Ushida, *et al.*, Phys. Rev. Lett. **57** (1986) 2897.

Solar Neutrinos

W.C. Haxton

Institute for Nuclear Theory, Box 351550, and Dept. of Physics, Box 351560, University of Washington, Seattle, WA 98195-1550

K. Lande

Dept. of Astronomy/Astrophysics, University of Pennsylvania, Philadelphia, PA 19104-6396

S.P. Rosen

Dept. of Physics, University of Texas, Arlington, P.O. Box 19050, Arlington, TX 76019

We summarize the current status of the solar neutrino problem, emphasizing the difficulty of reconciling the results of the chlorine, Kamiokande, and SAGE/GALLEX experiments with plausible variations in the standard solar model. Several new experiments under construction or development will provide important new information on the flux, energy distribution, and flavor of solar neutrinos. We discuss the crucial role these experiments will play in testing the properties of the neutrino, including masses in the range predicted by many extensions of the standard electroweak model.

§1 Introduction

The prospect of quantitatively testing the theory of main-sequence stellar evolution provided much of the original motivation for measuring solar neutrinos: solar neutrinos carry, in their energy distribution and flux, a precise record of the thermonuclear reactions occurring in the sun's core. The predictions of the standard solar model (SSM) are more tightly constrained today than when the the Homestake ^{37}Cl experiment was first mounted (almost three decades ago). Careful laboratory measurements and improved theory have helped to reduce uncertainties in the atomic and nuclear microphysics of the SSM, e.g., in nuclear reaction rates, radiative opacities, and the equation of state. The development of helioseismology has provided a new tool for probing the solar interior. Finally, we better understand our sun in the context of other stars, the observations of which have helped to define the envelope of possibilities for diffusion, mass loss, magnetic fields, etc. This progress has tended to increase our confidence in the SSM and its neutrino flux predictions.

It has also become clear that solar neutrinos play a unique role in particle physics. They may provide our only window on neutrino masses below 10^{-2} eV. In many extensions of the standard model this is the range of predicted masses. The unique sensitivity of solar neutrino experiments to new physics is due in great measure to the Mikheyev-Smirnov-Wolfenstein effect, the amplification of neutrino oscillations in matter [1,2].

The principal neutrino-producing reactions of the pp chain and CNO cycle are summarized in Table 1. The first six reactions produce β decay neutrino spectra having allowed shapes with endpoints given by E_ν^{max}. Deviations from an allowed spectrum occur for ^{8}B neutrinos because the ^{8}Be final state is a broad resonance [3]; much smaller deviations occur because of second-forbidden contributions to the decay. The last two reactions produce line sources of electron capture neutrinos, with widths $\sim$ 2 keV characteristic of the solar core temperature. The resulting solar neutrino spectrum is shown in Figure 1.

In the absence of new neutrino physics, measurements of the pp, ^{7}Be, and ^{8}B neutrino fluxes will determine the relative contributions of the ppI, ppII, and ppIII cycles that comprise the pp chain. As the discusion below will illustrate, the competition between the three cycles is governed in large classes of solar models by a single parameter, the central temperature T_c. The flux predictions of two standard models, those of Bahcall and Pinsonneault (BP) [4] and of Turck-Chièze and Lopez (TCL) [5], are included in Table 1.

Source	E_ν^{max} (MeV)	flux ($cm^{-2}s^{-1}$)		
		BP(with diffusion)	BP(without)	TCL
$p+p \to {}^2H + e^+ + \nu$	0.42	6.00E10	6.04E10	6.03E10
${}^{13}N \to {}^{13}C + e^+ + \nu$	1.20	4.92E8	4.35E8	3.83E8
${}^{15}O \to {}^{15}N + e^+ + \nu$	1.73	4.26E8	3.72E8	3.18E8
${}^{17}F \to {}^{17}O + e^+ + \nu$	1.74	5.39E6	4.67E6	
${}^{8}B \to {}^{8}Be + e^+ + \nu$	~15	5.69E6	5.06E6	4.43E6
${}^{3}He + p \to {}^{4}He + e^+ + \nu$	18.77	1.23E3	1.25E3	
${}^{7}Be + e^- \to {}^{7}Li + \nu$	0.86(90%) 0.38(10%)	4.89E9	4.61E9	4.34E9
$p + e^- + p \to {}^2H + \nu$	1.44	1.43E8	1.43E8	1.39E8

Table 1: Neutrino fluxes predict by the Bahcall/Pinsonneault (with and without He diffusion) and Turck-Chiéze/Lopez standard solar models.

§2 Experiment Results and SSM Uncertainties

Careful analyses of the experiments that will be described in Section 3 indicate that the observed solar neutrino fluxes differ substantially from SSM expectations [6,7,8]:

$$\begin{aligned} \phi(pp) &\sim 0.9\,\phi^{SSM}(pp) \\ \phi({}^7Be) &\sim 0 \\ \phi({}^8B) &\sim 0.43\,\phi^{SSM}({}^8B). \end{aligned} \tag{1}$$

Reduced ^{7}Be and ^{8}B neutrino fluxes can be produced by lowering the central temperature of the sun somewhat. However, such adjustments, either by varying the parameters of the SSM or by adopting some nonstandard physics, tend to push the $\phi({}^7Be)/\phi({}^8B)$ ratio to higher values rather than the low one of Eq. (1),

$$\frac{\phi({}^7Be)}{\phi({}^8B)} \sim T_c^{-10}. \tag{2}$$

Thus the observations seem difficult to reconcile with plausible solar model variations.

In order to assess how serious this problem is, several groups have considered the consequences of SSM uncertainties. These include the reaction cross sections for the pp chain and CNO cycle, the opacities, the deduction of heavy element abundances from solar surface abundances, the solar age and present day luminosity, and the equation of state. The astrophysical S factor for ${}^7Be(p,\gamma){}^8B$ is among the most serious of the uncertainties, with different data sets allowing S(0) to range from approximately 20 to 25 eV barns; a recent measurement using a different technique, the breakup of ^{8}B in the Coulomb field of ^{208}Pb, has produced a preliminary central value of 16.7 ev barns [10]. There exist small differences in the BP and TCL SSMs in the treatments of $S_{33}(0)$, $S_{34}(0)$, the solar lifetime, plasma effects on Thompson scattering, and the composition, as well as in the error assigned to $S_{17}(0)$. Yet the sum of these differences affects the temperature-dependent $\phi({}^8B)$ by less than 10%. In addition, the BP calculation differs from TCL by the inclusion of helium diffusion, which now is the largest contributor to the differences in the resulting flux predictions (12% in $\phi({}^8B)$). (A recently updated BP calculation [4] also included the effects of metal diffusion.)

More important than the "best values" of the fluxes are the ranges that can be achieved by varying the parameters of the SSM within plausible bounds. In order to take into account the correlations among the fluxes when input parameters are varied, Bahcall and Ulrich [11] (BU) constructed 1000 SSMs by randomly varying five input parameters, the primordial heavy-element-to-hydrogen ratio Z/X and S(0) for the p-p, ^{3}He-^{3}He, ^{3}He-^{4}He, and p-^{7}Be reactions, assuming for each parameter a normal distribution with the mean and standard deviation used in their 1988 study. (These were the parameters assigned the largest uncertainties.) Smaller uncertainties from radiative opacities, the solar luminosity, and the solar age were folded into the results of the model calculations perturbatively, using [9] the partial derivatives of the BU SSM.

The resulting pattern of ^{7}Be and ^{8}B flux predictions is shown in Figure 2. The elongated error ellipses indicate that the fluxes are strongly correlated. Those variations producing $\phi({}^8B)$ below $0.8\phi^{SSM}({}^8B)$ tend to produce a reduced $\phi({}^7Be)$, but the reduction is always less than 0.8.

Thus a greatly reduced $\phi(^7\text{Be})$ cannot be achieved within the uncertainties assigned to parameters in the SSM.

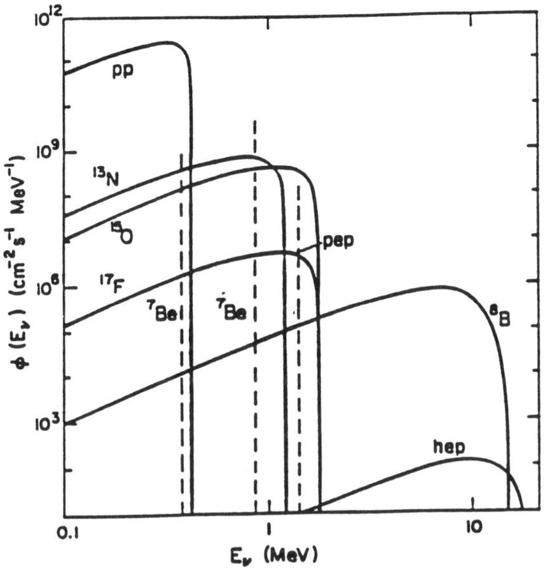

Figure 1: The flux densities (solid lines) of the principal β decay sources of solar neutrinos of the standard solar model. The total fluxes are those of the BP SSM. The ^{7}Be and pep electron capture neutrino fluxes (dashed lines) are discrete and given in units of $\text{cm}^{-2}\text{s}^{-1}$.

A similar exploration, but including parameter variations very far from their preferred values, was carried out by Castellani, Degl'Innocenti, Fiorentini, Ricci, and collaborators [12,13], who displayed their results as a function of the resulting core temperature T_c. The pattern that emerges is striking (see Figure 3): parameter variations producing the same value of T_c produce remarkably similar fluxes. Thus T_c provides an excellent one-parameter description of standard model perturbations. Figure 3 also illustrates the difficulty of producing a low ratio of $\phi(^7\text{Be})/\phi(^8\text{B})$ when T_c is reduced.

The BU 1000-solar-model variations were made under the constraint of reproducing the solar luminosity. Those variations show a similar strong correlation with T_c

$$\phi(\text{pp}) \propto T_c^{-1.2} \qquad \phi(^7\text{Be}) \propto T_c^8 \qquad \phi(^8\text{B}) \propto T_c^{18}. \quad (3)$$

Figures 2 and 3 are a compelling argument that reasonable variations in the parameters of the SSM, or nonstandard changes in quantities like the metallicity, opacities, or solar age, cannot produce the pattern of fluxes deduced from experiment (Eq. (1)).

§3 The Detection Of Solar Neutrinos

Four solar neutrino experiments have now provided data, the Homestake ^{37}Cl experiment, the gallium experiments SAGE and GALLEX, and Kamiokande. The first three detectors are radiochemical, while Kamiokande records neutrino-electron elastic scattering event-by-event.

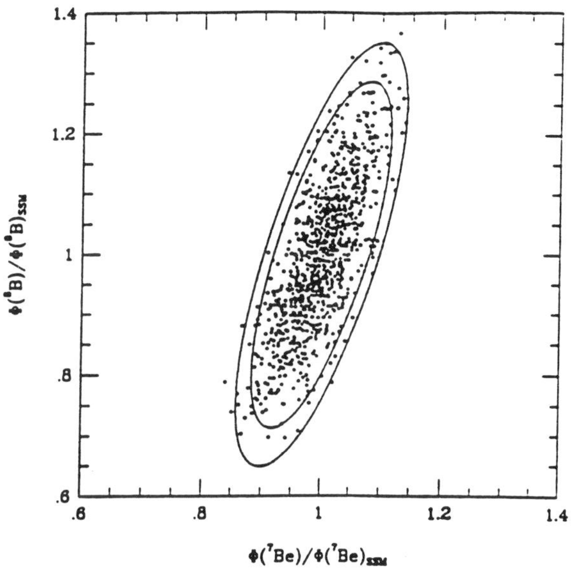

Figure 2: The dots represent the ^{7}Be and ^{8}B fluxes resulting from the 1000 SSMs of BU, with smaller SSM uncertainties added as in Ref. [9]. The 90 and 99% c.l. error ellipses are shown.

3.1 The Homestake Experiment

An experiment to detect neutrinos by the reaction $^{37}\text{Cl}(\nu_e,\text{e})^{37}\text{Ar}$ has operated continuously in the Homestake Gold Mine since 1967 (apart from a 17 month hiatus in 1985/86 caused by the failure of the circulation pumps). The result of 25 years of measurement is [14]

$$\langle\sigma\phi\rangle_{^{37}\text{Cl}} = 2.55 \pm 0.17 \pm 0.18 \ \ SNU \quad (1\sigma) \qquad (4)$$

which can be compared to the BP (with He diffusion) and TCL SSM predictions of 8.0 ± 1.0 SNU and 6.4 ± 1.4 SNU, respectively, all with 1σ errors. (One SNU = 10^{-36} captures/target atom/sec.) As we will discuss below, the ^{8}B and ^{7}Be contributions account for 77% (73%) and 15% (17%), respectively, of the BP (TCL) total. (Unless otherwise specified, BP refers to their 1992 results with He diffusion.)

The experiment [15,16] depends on the special properties of ^{37}Ar: as a noble gas, it can be removed readily from perchloroethylene, while its half life ($\tau_{1/2} = 35$ days) allows both a reasonable exposure time and counting of the gas as it decays back to ^{37}Cl. Argon is removed from the tank by a helium purge, and the gas then circulated through a condensor, a molecular sieve, and a charcoal trap cooled to the temperature of liquid nitrogen. Typically $\sim$ 95% of the argon in the tank is captured in the trap. (The efficiency is determined each run from the recovery results for a known amount of carrier gas, ^{36}Ar or ^{38}Ar, introduced into the tank at the start of the run.) When the extraction is completed, the trap is heated and swept by He. The extracted gas is passed through a hot titanium filter to remove reactive gases, and then other noble gases are separated by gas chromatography. The purified argon is loaded into a small proportional counter along with tritium-free methane, which serves as a counting gas. Since the electron capture decay of ^{37}Ar leads to the ground state of ^{37}Cl, the only signal for the decay is the 2.82 keV Auger electron produced as the atomic electrons in ^{37}Cl adjust to fill the K-shell vacancy. The counting of the gas typically continues for about one year ($\sim$ 10 half lives).

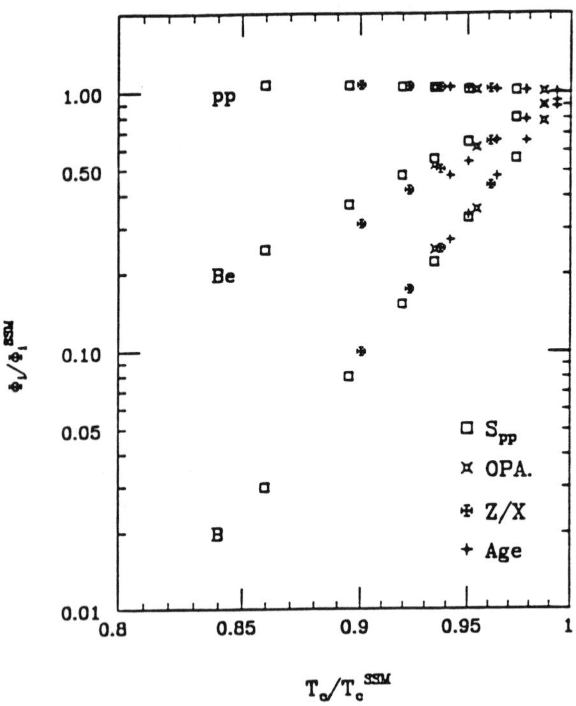

Figure 3: The response of the pp, Be, and ^{8}B fluxes to the indicated variations in solar model input parameters, displayed as a function of the resulting central temperature T_c. From Castellani, Degl'Innocenti, Fiorentini, Lissia, and Ricci [12].

The measured cosmic ray-induced background in the Homestake detector is 0.06 ^{37}Ar atoms/day while neutron-induced backgrounds are estimated to be below 0.03 atoms/day. A signal of 0.48 $\pm$ 0.04 atoms/day is attributed to solar neutrinos. When detector efficiencies, ^{37}Ar decays occurring in the tank, etc., are taken into account, the number of ^{37}Ar atoms counted is about 25/year.

A variety of careful tests of the argon recovery and counting efficiency have been made over the past 25 years. For example, it has been verified that ^{37}Ar produced in the tank by a fast neutron source, which induces (n,p) reactions followed by (p,n) on ^{37}Cl, is quantitatively recovered. Tests have also been made of the recovery of ^{36}Ar produced by the decay of long-lived ^{36}Cl in perchloroethylene. The experiment, performed in a tightly sealed tank to prevent contamination from air inleakage, resulted in a yield of 100 $\pm$ 3%. However, the detector has never been calibrated directly with a neutrino source, despite studies of the feasibility of a ^{65}Zn source [17].

The significance of the Homestake results is due in part to an accurately determined ^{37}Cl cross section. As the 814 keV threshold for exciting the ^{37}Ar ground state is above the pp endpoint, the detector is sensitive primarily to ^{7}Be and ^{8}B neutrinos (see Table 2). The cross section for ^{7}Be neutrinos (and the weaker fluxes of pep and CNO cycle neutrinos) is determined by the known half life of ^{37}Ar. However, ^{8}B neutrinos can generate transitions to many excited states below the particle breakup threshold in ^{37}Ar. The superallowed transition to the 4.99 MeV state, dominated by the Fermi matrix element of known strength, accounts for about 60% of the SSM cross section. The allowed transition strengths can be measured by observing the delayed protons following the β decay of ^{37}Ca, the isospin analog of the reaction ^{37}Cl(ν_e,e) ^{37}Ar [18]. While it was believed that this measurement had been properly done many years ago, the issue was not resolved until kinematically complete measurements were done recently [19,20]. The net result is a ^{37}Cl cross section believed to be accurate to about 3%.

3.2 The Kamiokande Experiment

The Kamiokande experiment [21,22] is a 4.5 kiloton cylindrical imaging water Cerenkov detector originally designed for proton decay searches, but later reinstrumented to detect low energy neutrinos. It detects neutrinos by the Cerenkov light produced by recoiling electrons in the reaction

$$\nu_x + e \rightarrow \nu_x' + e'. \tag{5}$$

Both ν_e and heavy flavor neutrinos contribute, with $\sigma(\nu_e)/\sigma(\nu_\mu) \sim 7$. The inner volume of 2.14 kilotons is viewed by 948 Hamamatsu 20" photomultiplier tubes

(PMTs) providing 20% photocathode coverage, and the surrounding 1.5m of water, serving as an anticounter, is viewed by 123 PMTs. The fiducial volume for solar neutrino measurements is the central 0.68 kilotons of water, the detector region most isolated from the high energy gamma rays generated in the surrounding rock walls of the Kamioka mine.

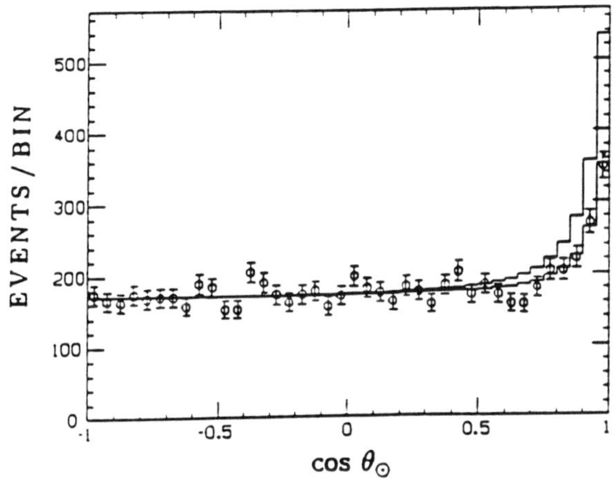

Figure 4: Angular distribution of recoil electrons from Kamiokande II and III showing the excess at forward angles that is attributed to solar neutrinos. Electrons with apparent energies between 7 and 20 MeV are included. The upper histogram is the SSM prediction of BU superimposed on an isotropic background, while the lower histogram is the best fit. From K. Nakamura [24].

In the conversion of the original proton decay detector to Kamiokande II, great effort was invested in reducing low energy backgrounds associated with radon and uranium. This included sealing the detector against radon inleakage and recirculating the water through ion exchange columns. The relatively shallow depth of the Kamioka mine (2700 m.w.e.) leads to an appreciable flux of cosmic ray muons which, on interacting with ^{16}O, produce various short-lived spallation products. These β decay activities are vetoed by their correlation in time with the muons. The experimenters succeeded in lowering the detector threshold to 9 MeV and later to 7.5 MeV. Kamiokande III included improvements in the electronics and the installation of wavelength shifters around the PMTs to increase light collection and currently operates with a threshold of 7.0 MeV.

Kamiokande II/III detects the high energy portion of the ^8B neutrino spectrum. Between December, 1985, and July, 1993, 1667 live detector days of data were accumulated. Under the assumption that the incident neutrinos are ν_es with an undistorted ^8B β decay spectrum, the combined Kamiokande II/III data set gives [23,24]

$$\phi_{\nu_e}(^8B) = (2.89 \pm 0.22 \pm 0.35) \cdot 10^6/cm^2 s \quad (1\sigma) \quad (6)$$

corresponding to 51% of the BP and 65% of the TCL SSM predictions. The total number of detected solar neutrino events is 476^{+36}_{-34}.

This experiment is remarkable in several respects. It is the first detector to measure solar neutrinos in real time. Essential to the experiment is the sharp peaking of the electron angular distribution in the direction of the incident neutrino: this forward peaking, illustrated in Figure 4, allows the experimenters to separate solar neutrino events from an isotropic background. The unambiguous observation of a peak in the cross section correlated with the position of the sun is the first direct demonstration that the sun produces neutrinos as a byproduct of fusion. Finally, although reaction (5) is a soft process where the recoil electron and scattered neutrino share the initial energy, the recoil electron energy distribution provides some information on the incident neutrino spectrum. The recoil spectrum measured by Kamiokande II/III is consistent with an allowed ^8B incident neutrino spectrum, with the overall flux reduced as in Eq. (6). However the statistical accuracy is not high.

3.3 The SAGE and GALLEX Experiments

Two radiochemical gallium experiments exploiting the reaction $^{71}Ga(\nu_e,e)^{71}Ge$, SAGE and GALLEX, began solar neutrino measurements in January, 1990, and May, 1991, respectively. SAGE operates in the Baksan Neutrino Observatory, under 4700 m.w.e. of shielding from Mount Andyrchi in the Caucasus, while GALLEX is housed in the Gran Sasso Laboratory at a depth of 3300 m.w.e. These experiments are sensitive primarily to the low-energy pp neutrinos, the flux of which is sharply constrained by the solar luminosity in any steady-state model of the sun (see Table 2). The gallium experiment was first suggested by Kuzmin [25]. In 1974 Ray Davis and collaborators began work to develop a practical experimental scheme. Their efforts, in which both $GaCl_3$ solutions and Ga metal targets were explored, culminated with the 1.3-ton Brookhaven/Heidelberg/Rehovot/Princeton pilot experiment in 1980-82 that demonstrated the procedures later used by GALLEX [26].

The primary obstacles to mounting the gallium experiments were the cost of the target and the greater complexity of the ^{71}Ge chemical extraction. The GALLEX experiment [27,28] employs 30 tons of Ga as a solution of $GaCl_3$ in hydrochloric acid. After an exposure of about three weeks, the Ge is recovered as $GeCl_4$ by bubbling nitrogen through the solution and then scrubbing the gas through a water absorber. The Ge is further concentrated and purified, and finally converted into GeH_4 which, when mixed with Xe, makes a good counting gas. The overall extraction efficiency is typically 99%. The GeH_4 is inserted into miniaturized gas proportional counters, carefully designed for their radiopurity, and the Ge counted as it decays back to Ga ($\tau_{1/2}$ = 11.43 d). As in the case of ^{37}Ar, the only signal for the Ge decay is the energy deposited by Auger electrons and x-rays that accompany the atomic rearrangement in Ga. An important achievement of GALLEX has been the detection of both the K peak (10.4 keV) and L peak (1.2 keV). While 88% of the electron captures occur from the K shell, many of the subsequent K $\rightarrow$ L x-rays escape the detector and some of the Auger electrons hit the detector walls. This produces a shift of the detected energy of events from the K to the L and M peaks. Thus the GALLEX L-peak counting capability almost doubles the ^{71}Ge detection efficiency.

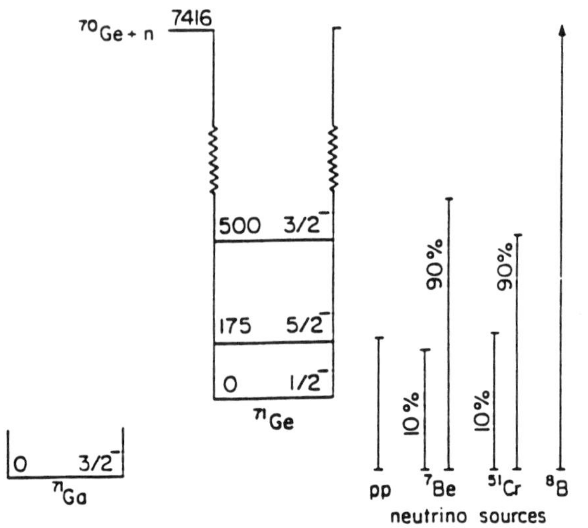

Figure 5: Level scheme for ^{71}Ge showing the excited states that contribute to absorption of pp, ^{7}Be, ^{51}Cr, and ^{8}B neutrinos. The ^{70}Ge + n break-up threshold is 7.4 MeV.

Gallium, like mercury, is a liquid metal at room temperature. SAGE [29,30] uses metallic gallium as a target, separating the ^{71}Ge by vigorously mixing into the gallium a mixture of hydrogen peroxide and dilute hydrochloric acid. This produces an emulsion, with the Ge migrating to the surface of the emulsion droplets where it is oxidized and dissolved by hydrochloric acid. The Ge is extracted as $GeCl_4$, purified and concentrated, synthesized into GeH_4, and further purified by gas chromatography. The overall efficiency, determined by introducing a Ge carrier, is typically 80%. The Ge counting proceeds as in GALLEX.

SAGE began operations with 30 tons of gallium, and now operates with 55 tons. The combined result for stage I (prior to September, 1992) and the first nine runs of stage II (9/92 - 6/93) is [31]

$$\langle\sigma\phi\rangle_{^{71}\mathrm{Ge}} = 69^{+11}_{-11}\,(stat)^{+5}_{-7}\,(sys)\ SNU \quad (1\sigma). \qquad (7)$$

This result includes only events counted in the K peak. The counter and electronics improvements made at the start of stage II should permit L-peak events to be included, but no results have been announced as of December, 1994. The corresponding results for GALLEX I (first 15 runs) and II (8/92-10/93) is [27]

$$\langle\sigma\phi\rangle_{^{71}\mathrm{Ge}} = 79 \pm 10 \pm 6\ SNU \quad (1\sigma). \qquad (8)$$

GALLEX II solar neutrino runs continued until June, 1994, but results for the last portion of this period have not been reported.

GALLEX II solar neutrino runs were interrupted in June, 1994, to permit an overall test of the detector with a ^{51}Cr source, which produces line sources of 746 keV (90%) and 426 keV (10%) neutrinos. The 1.67 MCi source was produced by irradiating $\sim$ 36 kg of chromium, enriched in ^{50}Cr, in the Siloé reactor in Grenoble. Following exposure of the detector and recovery and counting of the produced ^{71}Ge, the ratio of measured ^{71}Ge to expected was calculated [32],

$$R = 1.04 \pm 0.12 \quad (1\sigma). \qquad (9)$$

This is the first test of a solar neutrino detector with an terrestrial, low energy neutrino source. A similar source ($\sim$ 0.5 MCi) has been produced by the SAGE collaboration and was installed in their detector in December, 1994 (J.F. Wilkerson, private communication). The higher Ga density of the SAGE detector increases the effectiveness of the source by about a factor of 2.5, helping to compensate for the weaker neutrino flux.

The nuclear physics of the reaction ^{71}Ga(ν_e,e)^{71}Ge is illustrated in Figure 5. As the threshold is 233 keV, the ground state and first excited state can be excited by pp neutrinos. However, as only those pp neutrinos within 12 keV of the endpoint can reach the excited state, the phase space for reaching this state is smaller by a factor of $\sim$ 100. Thus the cross section is determined precisely by the measured electron capture lifetime of ^{71}Ge. In the BP and TCL calculations these neutrinos account for 54% and 57% of the capture rate, respectively, each predicting 71 SNU. Because of this strong pp neutrino contribution, there exists a minimal astronomical counting rate of 79 SNU [33] for the Ga detector that assumes only a steady-state sun and standard model weak interaction physics. This minimum value corresponds to a sun that produces the observed luminosity entirely through the ppI cycle. The rates found by SAGE and GALLEX are quite close to this bound.

The ^{7}Be neutrinos can excite the ground state and two excited states at 175 keV ($5/2^-$) and 500 keV ($3/2^-$). The source experiment was important not only in checking the overall efficiency of the Ga chemistry, but also in restricting the contributions of the excited states to ^{7}Be capture. As the SSM ^{7}Be capture rates of BP and TCL are both above 30 SNU, the SAGE/GALLEX results alone suggest some reduction in the low-energy pp and ^{7}Be fluxes.

The ^{8}B neutrino capture rates (14 and 11 SNU in the BP and TCL SSMs, respectively) have been calculated from the GT profile deduced by Krofcheck et al. [35]. The corresponding total SSM rates for this detector are 132 SNU and 123 SNU (see Table 2).

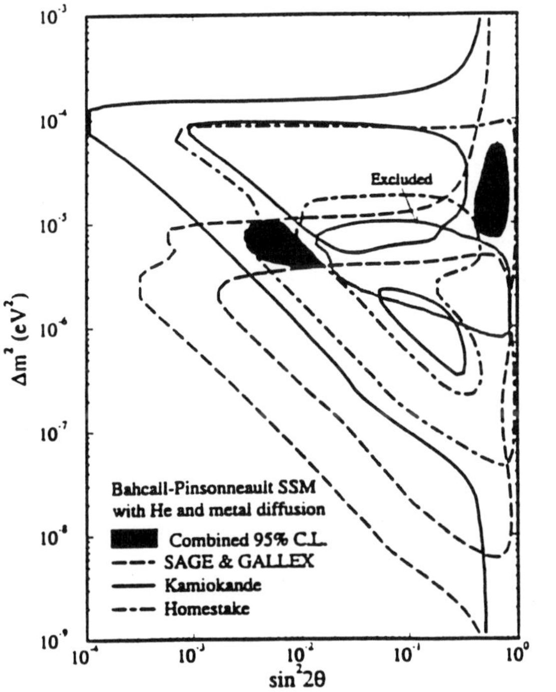

Figure 6: The MSW solutions allowed at 95% c.l. by the combined results of the Homestake, SAGE/GALLEX, and Kamiokande experiments, including Kamiokande II day-night constraints and the GALLEX source experiment results, for the BP flux predictions with He and metal diffusion. From Hata and Haxton [34].

§4 Particle Physics Solutions

In Section 2 it was shown that solar models which reduce the high energy neutrino flux tend to enhance the ^{7}Be/^{8}B flux ratio, contradicting the results of the Homestake, SAGE/GALLEX, and Kamiokande experiments.

An alternative solution, physics beyond the standard model of electroweak interactions, would have far-reaching consequences. Particle physics solutions of the solar neutrino problem include neutrino oscillations, neutrino decay, neutrino magnetic moments, and weakly interacting massive particles. Among these, the Mikheyev-Smirnov-Wolfenstein effect – neutrino oscillations enhanced by matter interactions – is widely regarded as the most plausible.

Neither Dirac nor Majorana neutrino masses can be generated in the standard electroweak model, which contains no right-handed neutrino fields and has only a doublet scalar field. However both neutrino masses and flavor mixing are expected in most extensions of this model. In the seesaw mechanism light neutrinos with $m_{light} \sim \frac{m_D^2}{m_R}$ arise, where m_D is a familiar Dirac mass (often a quark mass m_u, m_c, or m_t) and m_R a heavy right-handed Majorana mass. Taking $m_R \sim 10^{16}$ GeV, a typical grand unification scale for models built on groups like SO(10), the seesaw mechanism gives the crude relation

$$m_{\nu_e} : m_{\nu_\mu} : m_{\nu_\tau} \leftrightarrow 2 \cdot 10^{-12} : 2 \cdot 10^{-7} : 3 \cdot 10^{-3}\text{eV}. \quad (10)$$

The fact that solar neutrino experiments can probe small neutrino masses, and thus provide insight into possible new mass scales m_R that are far beyond the reach of direct accelerator measurements, has been an important theme of the field.

Flavor mixing of massive neutrinos leads to neutrino oscillations: the probability that a ν_e will remain a ν_e after propagating a distance x is

$$P_{\nu_e}(x) = 1 - \sin^2 2\theta_v \sin^2\left(\frac{\delta m^2 x}{4E}\right) \underset{x\to\infty}{\longrightarrow} 1 - \frac{1}{2}\sin^2 2\theta_v \quad (11)$$

where E is the neutrino energy, $\delta m^2 = m_2^2 - m_1^2$, and θ_v the vacuum mixing angle. If the oscillation length

$$L_o = \frac{4\pi E}{\delta m^2} \quad (12)$$

is comparable to or shorter than one astronomical unit, a reduction in the ν_e flux would be expected in solar neutrino experiments. The suggestion that the solar neutrino problem could be explained by neutrino oscillations was first made by Pontecorvo [36], who pointed out the analogy with $K_0 \leftrightarrow \bar{K}_0$ oscillations. From the point of view of particle physics, the sun is a marvelous neutrino source. The neutrinos travel a long distance and have low energies ($\sim$ 1 MeV), implying a sensitivity to

$$\delta m^2 \gtrsim 10^{-12}\text{eV}^2. \quad (13)$$

In the seesaw mechanism, $\delta m^2 \sim m_2^2$, so neutrino masses as low as $m_2 \sim 10^{-6}$eV could be probed. In contrast, terrestrial oscillation experiments with accelerator or reactor neutrinos have been typically limited to $\delta m^2 \gtrsim 0.1\text{eV}^2$.

neutrino source	capture rates (SNU)			
	^{37}Cl(BP)	^{37}Cl(TCL)	^{71}Ga(BP)	^{71}Ga(TCL)
pp	0.0	0.0	70.8	71.1
pep	0.2	0.22	3.1	2.99
^{7}Be	1.2	1.10	35.8	30.9
^{8}B	6.2	4.63	13.8	10.77
^{13}N	0.1	0.063	3.0	2.36
^{15}O	0.3	0.21	4.9	3.66
Total	8.0	6.36	131.5	122.5

Table 2: Predicted capture rates in SNU of the BP and TCL SSMs for the ^{37}Cl and SAGE/GALLEX experiments.

From Eq. (11) one expects vacuum oscillations to affect all neutrino species equally, if the oscillation length is small compared to an astronomical unit. This appears to contradict observation, as the pp flux may not be significantly reduced. Furthermore, the theoretical prejudice that θ_v should be small makes this an unlikely explanation of the significant discrepancies with SSM ^{7}Be and ^{8}B flux predictions.

The first objection, however, can be circumvented in the case of "just so" oscillations where the oscillation length is comparable to one astronomical unit [37]. In this case the oscillation probability becomes sharply energy dependent, and one can choose δm^2 to preferentially suppress one component (e.g., the monochromatic ^{7}Be neutrinos). This scenario has been explored by several groups and remains an interesting possibility. However, the requirement of large mixing angles remains.

The community's view of neutrino oscillations as a solution of the solar neutrino problem changed dramatically when Mikheyev and Smirnov [1,2] showed that the density dependence of the neutrino effective mass, a phenomenon first discussed by Wolfenstein [38], could greatly enhance oscillation probabilities. If the ν_e is primarily comprised of the light eigenstate in vacuum with the ν_μ being heavier, a critical density exists where the matter effects compensate for the vacuum mass difference. This comes about because of the stronger matter interactions of the ν_e, which scatters off solar electrons by both charged and neutral current interactions. Consequently a ν_e produced at sufficiently high densities in the core will be the heavy eigenstate in medium. If it then adiabatically propagates to the solar surface, it will emerge as the heavy eigenstate in vacuum, i.e., primarily a ν_μ. This phenomenon has been investigated numerically by a number of groups. (See, e.g., Rosen and Gelb [39].)

The resulting MSW fit to the ^{37}Cl, Kamiokande, and SAGE/GALLEX experiments, including the constraints from the recent GALLEX neutrino source experiment [32], is shown in Figure 6 [34]. The calculation includes the (correlated) theoretical uncertainties in the SSM flux predictions, terrestrial regeneration, the Kamiokande day-night data, and an improved definition of confidence level contours (see Hata and Langacker [8]). The preferred (in the sense of minimizing the χ^2) solution corresponds to a region surrounding $\delta m^2 \sim 6 \cdot 10^{-6}\mathrm{eV}^2$ and $\sin^2 2\theta_v \sim 6 \cdot 10^{-3}$. It is commonly called the small-angle solution. A second, large-angle solution exists, corresponding to $\delta m^2 \sim 10^{-5}\mathrm{eV}^2$ and $\sin^2 2\theta_v \sim 0.6$, but this region of Figure 6 has shrunk as the precision of the gallium experiments improve.

These solutions can be distinguished by their characteristic distortions of the solar neutrino spectrum. The survival probabilities $P_{\nu_e}^{MSW}(\mathrm{E})$ for the small- and large-angle parameters given above are shown as a function of E in Figure 7.

The calculations of Figure 6 assume flavor oscillations into a ν_μ or ν_τ. This influences the interpretation of the Kamiokande experiment, as heavy flavor neutrinos contribute to elastic scattering. Another possibility is an oscillation into a sterile neutrino. The large-angle solution is then ruled out by the Kamiokande requirement of a large ν_e survival probability.

The MSW mechanism provides a natural explanation for the pattern of observed solar neutrino fluxes. While it requires profound new physics, both massive neutrinos and neutrino mixing are expected in extended models. The preferred solutions correspond to $\delta m^2 \sim 10^{-5}$ eV^2, and thus are consistent with $m_2 \sim$ few $\cdot 10^{-3}$ eV. This is a typical ν_τ mass in models where $m_R \sim m_{GUT}$. On the other hand, if it is the ν_μ participating in the oscillation, this gives $m_R \sim 10^{12}$ GeV and predicts a heavy seesaw $\nu_\tau \sim 10$ eV [40]. Such a mass is of great interest cosmologically as it would have consequences for supernova neutrinos [41,42], the dark matter problem, and the formation of large-scale structure.

If the MSW mechanism proves not to be the solution of the solar neutrino problem, it still will have greatly enhanced the importance of solar neutrino physics: the existing experiments have ruled out large regions in the $\delta m^2 - \sin^2 2\theta_v$ plane (corresponding to nearly complete $\nu_e \rightarrow \nu_\mu$ conversion) that remain hopelessly beyond the reach of accelerator neutrino oscillation experiments.

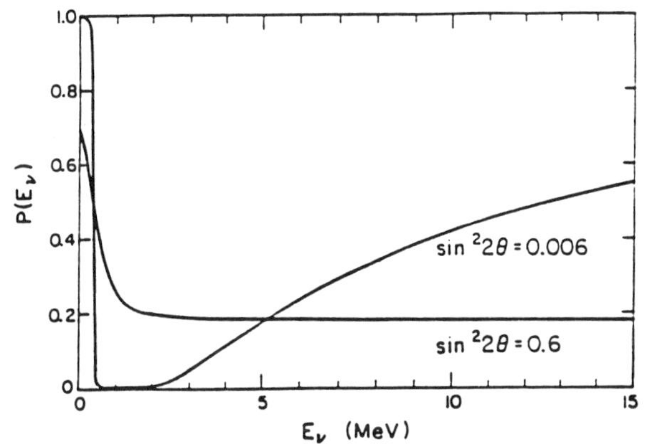

Figure 7: MSW survival probabilities $P_{\nu_e}^{MSW}(\mathrm{E})$ for typical small-angle ($\delta m^2 \sim 6 \cdot 10^{-6}\mathrm{eV}^2, \sin^2 2\theta_v \sim 6 \cdot 10^{-3}$) and large-angle ($\delta m^2 \sim 10^{-5}\mathrm{eV}^2, \sin^2 2\theta_v \sim 0.6$) solutions.

§5 New Experiments

The MSW mechanism has had a particularly strong impact because it was discovered at a time when new data (SAGE/GALLEX, Kamiokande, helioseismology) were eliminating many competing solutions to the solar neutrino problem. The physics of the MSW mechanism is both simple and elegant, which accounts for much of its appeal. But the most important attribute of this solution is that it can be definitively tested. The favored small-angle solution produces a distinctive distortion in the solar neutrino spectrum. Furthermore, if the oscillation is into another flavor (rather than a sterile state), the missing neutrinos can be found through their neutral current interactions. These tests will be made by two high-statistics, direct-counting directors now under construction.

5.1 The Sudbury Neutrino Observatory

A water Cerenkov detector of a different type is under construction deep (5900 m.w.e.) within the Creighton #9 nickel mine at Sudbury, Ontario, Canada [43,44,45]. The central portion of the detector is an acrylic vessel containing 1 kiloton of heavy water, D_2O. This is surrounded by five meters of light water to protect the inner detector from neutrons and gammas. The detector is viewed by 9500 20-cm PMTs, providing 56% photocathode coverage (Figure 8).

The D_2O introduces two new channels. The charged current breakup reaction

$$\nu_e + \mathrm{D} \rightarrow \mathrm{p} + \mathrm{p} + e^-$$

produces a recoil electron which carries off almost all of the final-state kinetic energy. As the Gamow-Teller strength is concentrated very close to the p+p threshold, 1.44 MeV, the electron and neutrino energies are related by $\mathrm{E}_\nu \sim \mathrm{E}_e + 1.44$ MeV. Thus, neutrino spectrum distortions should show up clearly in the measured electron energy distribution. As the GT strength in the deuteron is equivalent to about one-third of a free neutron, the anticipated counting rates are high. For an electron detection threshold of 5 MeV and a ^{8}B neutrino flux equal to about 50% of the BP SSM value, 3300 events will be recorded each year.

A second channel is sensitive equally to neutrinos of any flavor,

$$\nu_x + \mathrm{D} \rightarrow \nu_x' + \mathrm{n} + \mathrm{p}$$

and thus will be crucial in testing whether flavor oscillations have occurred. The anticipated event rate is approximately 2000/year in the BP SSM. The addition of $MgCl_2$ to the D_2O at a concentration of 0.2-0.3% allows the neutrons to be observed by ^{35}Cl(n,γ). The Cerenkov light produced by the showering of the 8.6 MeV capture γ ray will add to the signal from the charged current reaction. By operating the detector with and without salt, the experimenters will separate the charged and neutral current signals. The SNO collaboration also plans to deploy proportional counters filled with ^{3}He to exploit the neutron-specific charge-exchange reaction ^{3}He(n,p)^{3}H. With such detectors, SNO will be sensitive to neutral current events at all times.

The detection of ~ 8 neutrons/day in a kiloton detector places extraordinary constraints on radiopurity. For example, a potentially serious background source is the photodisintegration of deuterium by energetic photons from U and Th chains. The experimental goal is concentrations of $\lesssim 10^{-14}$ grams of U and Th per gram of D_2O.

SNO is scheduled to begin operations late in 1996.

5.2 Superkamiokande

Superkamiokande will be a greatly enlarged version of Kamiokande II/III with improved threshold (5 MeV) and energy and position resolution [46,47,48]. It is currently under construction in the Kamioka mine at a depth of 2700 m.w.e.

The fiducial volume for detecting solar neutrinos will be 22 kilotons, compared to 0.68 kilotons in the existing detector. This plus the improved threshold will increase the detection rate for neutrino-electron scattering by a factor of ~ 90, to 8400/year. Despite the soft kinematics of the $\nu_e - e$ reaction, the experimenters believe the high statistics will allow them to distinguish the spectral distortions produced by competing MSW solutions.

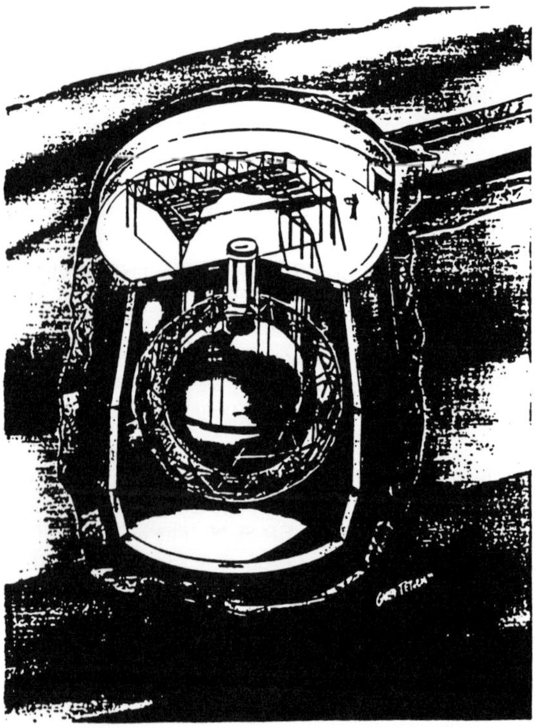

Figure 8: Schematic of the SNO detector now under construction in the Creighton #9 nickel mine, Sudbury. Provided by R.G.H. Robertson and J.F. Wilkerson (private communication).

Because elastic scattering is sensitive to both ν_e and heavy-flavor neutrinos (in the ratio of 7:1), an accurate SNO determination of the ν_e spectrum will allow Superkamiokande to extract the spectrum of ν_μs or ν_τs.

Superkamiokande construction is scheduled to be completed in mid 1996.

5.3 Other Future Detectors

The Borexino collaboration [49,50] has proposed a 0.3 kiloton liquid scintillator for installation in the Gran Sasso Laboratory. The experimenters hope to detect ^{7}Be neutrinos by $\nu - e$ scattering. The detection of very low energy recoil electrons places stringent constraints on U, Th, K, and other activities in the detector, e.g., $\lesssim 10^{-16}$ g U/g, including a requirement for continuous purification. The experimenters will evaluate background problems in a test facility now under construction, and scheduled to be completed by the end of 1995. The anticipated counting rate for the full-scale detector is $\sim$ 18,000 ^{7}Be neutrino events/year for the BP SSM.

A high-counting-rate twin of the ^{37}Cl detector utilizing the reaction $^{127}\mathrm{I}(\nu_e,e)^{127}\mathrm{Xe}$ has been funded recently and is under construction in the Homestake mine [51,52]. With a threshold of 664 keV, the detector is primarily sensitive to ^{7}Be and ^{8}B neutrinos. The initial Homestake detector will contain 100 tons of iodine as a solution of NaI. A smaller version of this detector was recently used at the LAMPF beamstop to measure the ^{127}I cross section for stopped muon decay ν_es. Calibration of the ^{7}Be neutrino cross section by an ^{37}Ar neutrino source is also planned. The 100-ton detector is scheduled to be completed by the end of 1995, with plans for an expansion to 1 kiloton afterwards.

A 5-kiloton liquid argon time projection chamber, ICARUS II, has been proposed for Gran Sasso [53]. In addition to $\nu - e$ scattering, the charged current channel

$$\nu_e +^{40}\mathrm{Ar} \rightarrow e^- +^{40}\mathrm{K}^*$$
$$\hookrightarrow^{40}\mathrm{K} + \gamma$$

will allow the experimenters to measure the shape of the high-energy portion of the ^{8}B ν_e spectrum.

There are a number of important development efforts underway that focus on new technologies for the next generation of solar neutrino detectors. Some of the significant challenges motivating these efforts include neutrino detection by coherent scattering off nuclei and real-time detectors for pp neutrinos, such as the superfluid ^{4}He detector HERON [54] and the high-pressure helium time projection chamber HELLAZ [55].

§6 OUTLOOK

The successes of the Kamiokande and SAGE/GALLEX experiments have led to a more complicated solar neutrino problem. The apparent strong suppression of the ^{7}Be flux (negative in most unconstrained fits!) has led many to conclude that a particle physics solution such as the MSW mechanism is likely.

An MSW solution to the solar neutrino problem would have deep implications for particle physics and cosmology. The most important aspect of this candidate solution is that it can be tested experimentally. The presence of a heavy-flavor solar neutrino component and spectral distortions will be probed by SNO and SuperKamiokande.

If these experiments indicate neutrino oscillations, enormous attention will then be focused on this field. It will be imperative to measure the flux, flavor, and spectral distribution of the ^{7}Be and pp neutrinos in order to exploit this unique window on physics beyond the standard electroweak model.

Under any scenario, one must appreciate the difficulty of the current generation of experiments. For example, the radiopurity goals of SNO and SuperKamiokande are unprecedented. It is therefore important to support a variety of experiments, so that some cross checks on the results are available. In view of the implications these experiments may have, the physics community will demand redundancy.

There are exciting new technologies under development that should be encouraged. These developmental efforts will lead to the detectors of the next decade. There are excellent ideas for active detectors sensitive to pp neutrinos. Cryogenic technologies could lead to low-threshold detectors exploiting coherent scattering.

The progress that has been made in recent years is remarkable: three new experiments have yielded data, and the extraordinary potential of the sun to enhance flavor oscillations is understood. The SNO and SuperKamiokande detectors are nearing completion, and new efforts on Borexino and the iodine detector are underway. Clearly this vital field is on the threshold of new discoveries.

§References

[1] S.P. Mikheyev, A. Yu. Smirnov, *Sov. J. Nucl. Phys.* 42 (1985) 913-7.

[2] S.P. Mikheyev, A. Yu. Smirnov, *Nuovo Cimento*, 9C (1986) 17-26.

[3] J.N. Bahcall, B.R. Holstein, *Phys. Rev. C.* 33 (1986) 2121-7.

[4] J.N. Bahcall, M.H. Pinsonneault, *Rev. Mod. Phys.* 64 (1992) 885-926; An updated calculation that includes the effects of metal diffusion is: J.N. Bahcall, M.H. Pinsonneault, 1995. To appear in *Rev. Mod. Phys.*

[5] S. Turck-Chièze, I. Lopez, *Ap. J.* 408 (1993) 347-67.

[6] M. White, L. Krauss, E. Gates, *Phys. Rev. Lett.* 70 (1993) 375-8.

[7] S.J. Parke, *Phys. Rev. Lett.* 74 (1995) 839-41; W. Kwong and S.P. Rosen, *Phys. Rev. Lett.* 73 (1994) 369.

[8] N. Hata, P. Langacker, *Phys. Rev. D.* 48 (1994) 2937-40.

[9] J.N. Bahcall, W.C. Haxton, *Phys. Rev. D.* 40 (1989) 931-41.

[10] T. Motobayashi, N. Iwasa, Y. Ando, M. Kurokawa, H. Murakami, et al., *Phys. Rev. Lett.* 73 (1994) 2680-3.

[11] J.N. Bahcall, R.K. Ulrich, 1988. *Rev. Mod. Phys.* 60 (1988) 297-372.

[12] V. Castellani, S. Degl'Innocenti, G. Fiorentini, M. Lissia, B. Ricci, *Phys. Rev. D.* 50 (1994) 4749-61.

[13] V. Castellani, S. Degl'Innocenti, G. Fiorentini, B. Ricci, In *Proc. Solar Modeling Workshop*, ed. A.B. Balantekin, J.N. Bahcall, (1995) Singapore: World Scientific, in press

[14] K. Lande, In *Neutrino '94*, ed. A. Dar, G. Eilam, M. Groneau, (1995) Amsterdam: North Holland.

[15] R. Davis Jr., In *Solar Neutrinos and Neutrino Astronomy*, ed. M.L. Cherry, W.A. Fowler, K. Lande. New York: American Institute of Physics, (1985) pp 1-21.

[16] R. Davis Jr., In *Frontiers of Neutrino Astrophysics*, ed. Y Suzuki and K Nakamura 47 (1993) University Academy Press.

[17] L.W. Alvarez, *Lawrence Radiation Lab. Physics Notes*, (1973) memo #767.

[18] J.N. Bahcall, *Phys. Rev. Lett.* 17 (1966) 398-401.

[19] A. Garcia, E.G. Adelberger, P.V. Mangus, H.E. Swanson, H.E. Tengblad, et al., *Phys. Rev. Lett.* 67 (1991) 3654-7.

[20] A. Garcia, E.G. Adelberger, P.V. Mangus, H.E. Swanson, D.P. Wells, et al., *Phys. Rev. C* 51 (1995) R439-42.

[21] K.S. Hirata, T. Kajita, M. Koshiba, M. Nakahata, Y. Oyama, et al., *Phys. Rev. D.* 38 (1988) 448-58.

[22] K.S. Hirata, K. Inoue, T. Ishida, T. Kajita, K. Kihara, et al., *Phys. Rev. D.* 44 (1991) 2241-60.

[23] Y. Suzuki, in *Proc. 16th Int. Conf. on Neutrino Physics and Astrophysics*, to be published (1995).

[24] K. Nakamura, To be published in *Proc. Int. Conf. on Non-Accelerator Particle* Physics (1994) (India).

[25] V.A. Kuzmin, *Sov. Phys. JETP*, 22 (1966) 1051; V.A. Kuzmin, G.T. Zatsepin, *Proc. Int. Conf. on Cosmic Rays* 2 (1966) 1023.

[26] W. Hampel, In *Solar Neutrinos and Neutrino Astronomy*, ed. M.L. Cherry, W.A. Fowler, K. Lande. New York: American Institute of Physics, (1985) pp. 162-74.

[27] P. Anselmann, W. Hampel, G. Heusser, J. Kiko, T. Kirsten, et al., *Phys. Lett. B.* 285 (1992) 376-89.

[28] P. Anselmann, W. Hampel, G. Heusser, J. Kiko, T. Kirsten, et al., *Phys. Lett. B.* 327 (1994) 377-85.

[29] J.N. Abdurashitov, E.L. Faizov, V.N. Gavrin, A.O. Gusev, A.V. Kalikov, et al., *Phys. Lett. B.* 328 (1994) 234-48.

[30] T.J. Bowles, V.N. Gavrin, *Annu. Rev. Nucl. Part. Sci.* 43 (1993) 117-64.

[31] J.S. Nico, In *Neutrino '94*, ed. A. Dar, G. Eilam, M. Groneau, Amsterdam: North Holland and talk presented at the *Int. Conf. High* Energy Physics, (1995) Glasgow.

[32] P. Anselmann, R. Frockenbrock, W. Hampel, G. Heusser, J. Kiko, et al., *Phys. Lett. B.* 342 (1995) 440-50.

[33] J.N. Bahcall, *Neutrino Astrophysics*, Cambridge: Cambridge Univ. Press. (1989) 567 pp.

[34] N. Hata, W.C. Haxton, To appear in *Phys. Lett. B.* (1995).

[35] D. Krofcheck, E. Sugarbaker, J. Rapaport, D. Wang, J.N. Bahcall, et. al., *Phys. Rev. Lett.* 55 (1985) 1051-4; D. Krofcheck, Ph.D. thesis, (1987) Ohio State University.

[36] B. Pontecorvo, *Sov. Phys. JETP.* 7 (1958) 172.

[37] S.L. Glashow, L.M. Krauss, *Phys. Lett. B.* 190 (1987) 199-207.

[38] L. Wolfenstein, 1978. *Phys. Rev. D.* 17 (1978) 2369-74; *Phys. Rev. D.* 20 (1979) 2634-35.

[39] S.P. Rosen and J.M. Gelb, *Phys. Rev. D.* 34 (1986) 969-79.

[40] S.A. Bludman, D.C. Kennedy, P.G. Langacker, *Phys. Rev. D.* 45 (1992) 1810-13.

[41] G.M. Fuller, R.W. Mayle, B.S. Meyer, J.R. Wilson, *Ap. J.* 389 (1992) 517-526.

[42] Y.Z. Qian, F.M. Fuller, R. Mayle, G.J. Mathews, J.R. Wilson, S.E. Woosley, *Phys. Rev. Lett.* 71 (1993) 1965-8.

[43] G.T. Ewan, H.C. Evans, H.W. Lee, J.R. Leslie, H.B. Mak, et al., (1987) Queen's University report SNO-87-12.

[44] G. Aardsma, R.C. Allan, J.D. Anglin, M. Bercovitch, A.L. Carter, et al., *Phys. Lett. B.* 194 (1987) 321-25.

[45] H.H. Chen, *Phys. Rev. Lett.* 55 (1985) 534-6.

[46] Y. Totsuka, In *Proc. 7th Workshop on Grand Unification*, ed. J. Arafune, 118 (1987), Singapore, World Scientific.

[47] Y. Totsuka, In *Proc. Int. Symposium on Underground Physics Experiments*, ed. K. Nakamura, 129 (1990), Univ. of Tokyo.

[48] M. Takita, In *Frontiers of Neutrino Astrophysics*, ed. Y. Suzuki and K. Nakamura, 135 (1993). Tokyo:Universal Academic Press.

[49] R.S. Raghavan, 1991. In *Proc. 25th Int. Conf. High Energy Physics*, ed. K.K. Phua and Y. Yamaguchi, 482 (1991). Japan : South Asia Theor. Phys. Assoc. and Phys. Soc.

[50] M. Campanella, In *Proc. 3rd Int. Workshop on Neutrino Telescopes*, ed. M Baldo-Ceolin, (1992) p. 73. Venice: Univ. di Padova

[51] K. Lande, To appear in the *Proc. 22nd Int. Conf. on Cosmic Ray Physics* (1993).

[52] W.C. Haxton, *Phys. Rev. Lett.* 60 (1988) 768-71.

[53] C. Rubia, *INFN Publication* (1985) INFN/AE-85-7.

[54] S.R. Bandler, *Phys. Rev. Lett.* 68 (1992) 2429-32; J.S. Adams et. al., *Phys. Lett. B.* 341 (1995) 431.

[55] G. Laurenti, S. Tzamarias, G. Bonvicini, P. Krastev, A. Zichichi, et. al., (1994). INFN preprint.

NEUTRINO ASTROPHYSICS

DAVID B. CLINE

University of California, Los Angeles, Department of Physics and Astronomy, 405 Hilgard Ave., Los Angeles, CA, USA

GEORGE M. FULLER

Department of Physics, University of California, San Diego, La Jolla, CA 92093-0319, USA

Here we report on how recent progress in understanding the physics of supernovae and several issues in the early universe and cosmology promise to lead to deeper insights into fundamental neutrino physics (e.g., neutrino masses and mixings) and *vice versa*. Recent and proposed experiments and theoretical studies which bear on our understanding of neutrino oscillations, neutrino masses, and other neutrino properties are discussed in this context. We also report on the progress of supernova modeling, and on the prospects for real time supernova neutrino detection. These topics represent a subset of the discussion issues which arose during the meetings of the Neutrino Astrophysics working group at the 1994 Snowmass Workshop.

1 General Remarks on Neutrinos, Supernovae, and Cosmology

Particle physicists and astrophysicists would very much like to know whether or not any of the neutrino species have rest masses. Indeed, positive evidence for the existence of non-vanishing neutrino masses may well be our first and (absent future definitive results from the proposed LHC and B-factories) perhaps our only indication of the requirement for physics beyond the Standard Model.

Furthermore, the question of neutrino masses ties in with another of the great mysteries of modern physics: the nature of the dark matter. From observations of the relic microwave background radiation and observational inference of the light element primordial abundances, we know that the photon-to-baryon number in our universe is of order $\sim 10^{10}$. On broad theoretical grounds (for example, assuming universality and that the Fermi constant is really constant !) we expect that the neutrino-to-baryon number should be quite similar. From this we can conclude immediately that even very small neutrino masses can be dynamically important for the evolution of the universe [1].

In fact, it would be highly significant for cosmology if the mass of any one neutrino flavor, or the sum of the masses for two or all three flavors, was found to lie in the range of about 1 eV to 100 eV [1]. If a neutrino contributes a fraction Ω_ν of the closure density of the universe it must have a mass $m_\nu \approx 92\Omega_\nu h^2$ eV, where h is the Hubble parameter in units of 100 km^{-1} sec^{-1} Mp^{-1}. Reasonable ranges for Ω_ν and h then give 1 eV to 92 eV as a cosmologically significant range. A neutrino with a mass is the higher end of this range, i.e., $10 \lesssim m_\nu \lesssim 92$ eV, could contribute significantly to the closure density of the Universe. The Cosmic Background Explorer (COBE [2]) observation of anisotropy in the microwave background, combined with observations at smaller scales, the distribution of galaxy streaming velocities, and the results of numerical large scale structure simulations *have been interpreted* as implying that there are two components of dark matter: i.e., hot (HDM) and cold dark matter (CDM) components, with $\Omega_{CDM} \sim 0.6$ and $\Omega_{HDM} \sim 0.3$. The hot dark matter (HDM) component in these models could be provided by a scheme in which the sum of the active light neutrino masses is about 5 - 7 eV [3]. Of course, there may be several other viable models for structure growth which meet the microwave background anisotropy constraints, yet do not invoke neutrino dark matter. Primarily, these would be models for a spatially "flat" universe with $\Omega_{CDM} \sim 0.3$ and vacuum energy density (in units of the closure density) $\lambda \sim 0.6$. Present data on the distribution and sizes of structure in the universe is inadequate to discriminate between these models. But it is with great hope that we note that future observations from the Keck and other large telescopes may give us a definitive answer on the evolution of small scale nonlinear structures with redshift around the epoch corresponding to redshift $z \approx 3$. These observations may or may not point to the need for neutrino dark matter.

However, despite the obvious cosmological and particle physics interest in massive neutrinos, there are very few terrestrial experimental means for measuring neutrino masses in the range of significance for neutrino hot dark matter. The electron neutrino mass is constrained by the tritium end point experiments to be less than about 7.2 eV. It is conceivable that ν_μ to ν_τ accelerator neutrino oscillation experiments, such as the NOMAD and CHORUS experiments at CERN and the E803 experiment at Fermilab, could be used to infer a mass in the cosmologically interesting range for ν_μ or ν_τ. Likewise, the LSND neutrino oscillation experiment at Los Alamos conceivably could infer a cosmologically significant mass (or sum of masses) for the mu and/or electron neutrinos. However, to be successful, all of these laboratory experiments would depend on there being a fairly large (by the standards of, admittedly suspect, theoretical arguments) vacuum mixing between the neutrino flavors probed by each particular experiment. However, there are some astrophysical venues in which small mixing between the neutrino flavors could be amplified by coherent effects.

Reconciliation of the observations of neutrinos from the sun in the Homestake, Gallex, SAGE, Kamiokande II, and (some day) the SNO detectors may well require matter-enhanced neutrino oscillation of some sort in the sun. This would *prove* that at least some of the neutrinos have mass. However, the neutrino mass scale that could be directly inferred from solar considerations would be only on the order of a millivolt, well below the $\sim 1\,\mathrm{eV}$ neutrino mass range of interest for the dark matter question.

But there is hope. We now know that the dynamics and heavy element nucleosynthesis of Type II Supernovae, and in fact nearly every aspect of the behavior of these events, is determined by the interplay of gravitation and neutrino emission and absorption. Very tiny amounts of lepton number violation from, for example, neutrino flavor oscillations, can cause dramatic alteration of supernova explosion models and/or the predicted heavy element nucleosynthesis yield in neutrino-heated ejecta from these objects. If we can understand the supernova explosion mechanism (arguably a very hard problem !) or the nucleosynthesis from neutrino-heated supernova ejecta (arguably an considerably easier problem), then we may gain insights into fundamental neutrino physics that we have otherwise only a poor chance of ever obtaining from laboratory experiments. Much of our discussion in the Neutrino Astrophysics working group centered broadly around these issues: supernova dynamics; nucleosynthesis from neutrino-heated supernova ejecta; and how various aspects of neutrino physics (masses and mixings, lepton number violating processes, majorons, etc.) bear on the first and second issues. In this vein, we now report on progress in understanding these issues and on prospects for direct supernova neutrino detection.

2 Neutrinos and Supernovae

The weak interaction (specifically, neutrino interactions) and gravitation are the twin engines of Type II Supernovae. The progenitors of these events are massive stars ($> 10\,\mathrm{M}_\odot$), which in less than 10 milion years evolve to form Chandrasekhar mass iron cores. The structure of such a core resembles that of a high mean molecular weight white dwarf. The iron cores of the supernova progenitor objects are destabilized by any of a combination of electron capture and associated neutrino energy and entropy losses, photo-disintegration rearrangement and general relativity. These cores subsequently collapse to nuclear density (or beyond) on a timescale of about $\sim 1\,\mathrm{s}$. The core collapse process is halted in a hydrodynamic "bounce," forming a quasi-static hot proto neutron star. The hydrodynamic bounce generates a shock wave, which begins to move out through the infalling material.

The entropy per baryon in the collapsing core is typically (in units of Boltzmann's constant k), $s/k \sim 1$, while the characteristic post-bounce (post shocked) entropy per baryon is $s/k \sim 10$. This large entropy jump across the shock front is quite significant, because it implies that low entropy material (mostly heavy nuclei) flowing through the shock will be photo-disintegrated into free nucleons and alpha particles. In turn, this is an endothermic process, which saps energy from the shock front, typically causing it to stall and evolve into a stationary accretion shock within about $t_{pb} \approx 0.1\,\mathrm{s}$ (here t_{pb} denotes the lapsed time *post core bounce*). We now believe that this stalled shock is revived or "re-heated" by neutrinos emitted from a neutrino sphere (near the edge of the hot proto neutron star). A small fraction of the neutrinos emitted from the neutrino sphere are captured or otherwise interact in the material behind the shock front to deposit energy and so raise the pressure behind the shock and thus do work on it [4]. Producing a shock with $\approx 10^{51}$ ergs of energy is sufficient to produce a Type II Supernova explosion. The shock re-heating process takes on the order of a few tenths of a second to half a second.

It is instructive to examine where the energy resides in this picture of how Type II Supernovae explode. In fact, essentially all the energy resides in the neutrinos. At the outset of core collapse, the strong and electromagnetic interactions are each in equilibrium and the baryons are thus in nuclear statistical equilibrium (NSE). However, initially the weak interaction is not in either thermal or chemical equilibrium. As the core collapses, electrons are captured on protons (most of which reside inside heavy nuclei), driving the material more neutron-rich. Early in the collapse process the neutrinos produced by electron capture escape. But as the density of the core rises, the neutrino mean free path becomes smaller than the size of the core, and the neutrinos become "trapped" and eventually thermalized. The total gravitational binding energy released by the core in collapsing from an iron "white dwarf" to a hot proto neutron star (at the bounce) is about 10^{52} ergs. All but about 10 percent of this energy resides in the thermalized neutrinos which are trapped inside the hot neutron star. The rest resides in the kinetic energy of the infalling material. At bounce this infall kinetic energy is converted with fair efficiency into the initial energy of the outgoing shock ($\approx 10^{51}$ ergs).

At about $t_{pb} \approx 0.1$ s after core bounce we have a hot proto neutron star with a radius of about ~ 45 km, and a dead shock at a radius of at least several hundred kilometers. The neutrinos trapped inside the neutron star have mean free paths which are very small (centimeters near the center of the star). These neutrinos diffuse out of the core with a typical diffusion timescale of 10 to 15 seconds. We can approximate each neutrino species as decoupling from a characteristic neutrino sphere located near the edge of the neutron star, where the density gradient is very steep. Above the neutrino sphere(s) , the neutrinos stream nearly freely, with the material in this region being essentially transparent to neutrinos. The energy spectra for each of the neutrino species at their respective neutrino spheres are roughly Fermi-Dirac black body with near zero chemical potential.

In fact, the mu and tau neutrinos and their antiparticles decouple deepest inside the neutron star (their neutrino sphere has a smaller radius than the neutrino sphere for either the ν_e or $\bar{\nu}_e$), since these species have no charged current contributions to their opacity, while the ν_e or $\bar{\nu}_e$ do. The charged current contributions to the opacities for the ν_e or $\bar{\nu}_e$ come from electron scattering and the charged current absorption processes on free nucleons,

$$\nu_e + \mathrm{n} \rightleftharpoons \mathrm{p} + e^- \tag{1a}$$

$$\bar{\nu}_e + \mathrm{p} \rightleftharpoons \mathrm{n} + e^+ . \tag{1b}$$

In fact, since the material of the neutron star has a neutron excess, the ν_e have a larger opacity contribution from the process in Eq. (1a) than do the $\bar{\nu}_e$ from the process in (1b). As a result, the mu and tau neutrino sphere lies deeper than the neutrino sphere for $\bar{\nu}_e$, which in turn lies deeper than the neutrino sphere for the ν_e. Since the temperature of the material and neutrinos tends to decrease with increasing radius inside the neutron star, generically the average energies of the neutrino species at their respective neutrino spheres and in the region above the neutron star will satisfy $\bar{E}_{\nu_{\tau(\mu)}} = \bar{E}_{\bar{\nu}_{\tau(\mu)}} > \bar{E}_{\bar{\nu}_e} > \bar{E}_{\nu_e}$, so long as there are no significant neutrino flavor transformations.

After bounce, the hot proto neutron star will radiate neutrinos (and hence net lepton number) from its surface and so will quasi-statically contract, releasing $\sim 10^{53}$ ergs of gravitational binding energy as the star shrinks in radius to ~ 10 km over a timescale set, of course, by the neutrino diffusion timescale of about 10 s to 20 s. The neutrino luminosities are about the same for each neutrino species and range from about 10^{52} ergs s^{-1} immediately after bounce to roughly 10^{51} ergs s^{-1} after about 10 or 20 seconds. During this time the average energies of each neutrino species increases, while the disparity between the ν_e and $\bar{\nu}_e$ average energies grows.

From this general overview of the Type II supernova explosion process we can identify three relatively distinct epochs:

1. the collapse or infall epoch with $s/k \sim 1$;

2. the shock re-heating epoch at $t_{pb} \approx 0.1$ s to ~ 1 s where the characteristic entropy per baryon is $s/k \sim 50$;

3. the "hot bubble" r-Process epoch at $t_{pb} \sim 3$ s to ~ 15 s, where $s/k \sim 200$ or more.

The high entropy in the hot bubble/r-Process epoch is a result of the cumulative neutrino heating of the (by this time low density) plasma above the neutron star by the processes in Eqs (1a) and (1b) and by other neutrino interactions.

Let us at this point make some simple generalizations about our overview of the supernova explosion process. First note that there are two numbers of overwhelming signifigance in this picture: the mass of the neutron star; and the neutrino diffusion timescale. These quantities determine the essential features of the supernova and they are, in turn, expected to be known reasonably accurately. The mass of the neutron star must be very close to the mass of the initial iron core ($\sim 1.4\ \mathrm{M}_\odot$), the Chandrasekhar mass. The neutrino diffusion timescale depends on the assumption of universality of the weak interaction, the standard model of the weak interaction, and the saturation density of nuclear matter (i.e., the characteristic density of a neutron star). So really it is the Chandrasekhar mass and the saturation density of nuclear matter which are the only *non weak interaction* quantities which are important in the supernova! This is the principal reason why supernovae are potentially such wonderful probes of the weak interactions and neutrino physics. Of course, do we really know what the density of matter is in the neutron star, or equivalently do we really know the equation of state of matter at high density? And how accurately do we know the initial iron core mass? Furthermore, how accurately do we need to know the answers to either of these questions before we can begin to use supernova nucleosynthesis and dynamics to constrain weak interaction physics? The neutrino astrophysics working group discussed these issues in some detail.

At Snowmass, T. A. Weaver and S. E. Woosley discusssed uncertainties in the initial models for core collapse and the outstanding issues that need to be resolved for the future. We concluded that our understanding of the infall epoch at this point is reasonably good, though uncertainties in electron capture, beta decay, and other weak processes (as discussed by M. Aufderheide, G. Fuller, and T. Ressel) remain. D. Swesty and J. Lattimer presented their views on the effects of kaon condensation and/or a transition from nucleons and mesons to quark soup in the de-leptonizing proto neutron star. This latter issue is quite important, because it bears on the question of whether or not black hole remnants might be left by Type II supernovae. In turn, it may tie in with the expected gravitational radiation signature for supernovae. All of these issues will clearly be subjects of intense research in this and the next millenium!

Stirling Colgate presented a beautiful picture of the "convective engine " which obtains in the shock reheating epoch [5]. In fact, as the material above the neutron star is heated by neutrino interactions a negative entropy gradient develops which is necessarily convectively unstable. Thus, convection *must* accompany neutrino heating during this epoch. Colgate described his recent calculations with Herant and Benz (see also the work of Miller and Wilson), and showed how convection at this epoch neatly transports energy from the region immediately above the neutrino sphere to the region of the shock - effectively pumping neutrino energy into shock energy! With this insight into the role of convection, we know that the re-heated shock will have the requisite energy to produce the Type II Supernova display. We no longer have a nagging uncertainty regarding how the shock gets re-heated. But there are still plenty of other uncertainties surrounding this epoch.

For example, we would very much like to have a better understanding of how neutrinos are transported through the neutrino sphere. Additionally, we need to know what the neutrino energy distribution functions above the neutrino should be. These issues bear on the neutrino luminosity, the neutrino heating rate , and the nucleosynthesis from the neutrino-heated ejecta from the shock re-heating epoch. There are many technical issues in the numerical treatment of neutrino transport in the vicinity of the neutrino sphere. In this region, neutrinos go from optically thick to optically thin in a quite short length scale. At the Snowmass meeting, D. Swesty gave a very good overview of these issues. There was not a great deal of discussion about convection through the neutrino sphere (the so called "salt finger instability") in our working group, though the resolution of this problem may be important for the issues outlined above. Furthermore, the "bubbling" and "boiling" of the material near the neutrino sphere engendered by convection may be a very important source of gravitational radiation in a frequency range in which LIGO may be sensitive (see the references to A. Burrows talk in the report of the Gravitational Radiation working group in these proceedings).

In the course of our working group sessions, R. Sawyer brought up a potentially important issue regarding the possible renormalization of the effective weak coupling constant as a result of matter effects in the neutron star. Sawyer argued that weak intreaction rates of neutrinos on nucleons (as in Eqs (1a) and (1b)) would be reduced by these matter effects. This clearly could be important for our picture of both the shock re-heating epoch and the hot bubble epoch, if these renormalization effects are still significant even at densities characteristic of those of the neutrino spheres at this epoch (less than one percent of nuclear matter density).

In our meeting, S. E. Woosley, Y. Z. Qian, and G. M. Fuller talked about the nucleosynthesis from the neutrino-heated ejecta originating in the hot bubble epoch. In a stunning development in the last two years Woosley and his collaborators have shown that neutrino-heated ejecta from this epoch is an excellent site for the rapid neutron capture process (r-Process) of nucleosynthesis [6]. Most of the elements with masses greater than 100 were created in this process. Subsequently, Y. Z. Qian and G. M. Fuller and their collaborators [7] showed that this r-process nucleosynthesis may be an excellent probe of the mixing of ν_e with either ν_μ or ν_τ (or the mixing of $\bar{\nu}_e$ with either $\bar{\nu}_\mu$ or $\bar{\nu}_\tau$), when one or more of these species has a cosmologically significant mass. The sensitivity of r-Process nucleosynthesis to matter-enhanced neutrino oscillations follows from three points:

1. r-Process nucleosynthesis is sensitive to the ratio of neutrons to protons in the hot bubble and there must be a fair excess of neutrons to get an acceptable abundance pattern;

2. the neutron to proton ratio in the hot bubble is determined by the competition between the reactions in Eqs (1a) and (1b);

3. transformation of, for example, ν_τ into ν_e will result in driving the material less neutron-rich as a result of the average energy hierarchy of the neutrino species outlined above.

This is a subject of intense current research, because the stakes are high. Essentially, the gist of this work is that the known abundance distribution of the heavy elements may provide clues to (or constraints on) the existence of neutrino dark matter and the mixing parameters of the neutrino species!

Finally, it is clear from the above overview of the Type II Supernova explosion process that *any* aspect of the weak interaction that results in lepton number violation can potentially alter our picture of the explosion process. It would be unwarranted bravado to assume that we know everything there is to know about the weak interactions. For example, if neutrinos have masses, how do they acquire them. If physics beyond the standard model is invoked to build these masses then it is legitimate to ask what effect this altered weak physics will have on our picture of the supernova process. In the course of the meetings of our working group, G. Gelmini gave a review of extensions of the stardard model in the weak sector. G.

M. Fuller reviewed how aspects of the supernova problem might depend on such extensions. It became clear from our discussions that any extension of the Higgs sector to include majoron-like Goldstone bosons which couple only to neutrinos has the potential to drastically alter the infall epoch of the supernova (usually in ways which mitigate against ever getting an explosion!).

In any case, the supernova environment is a potentially wonderful neutrino physics laboratory. If we can decipher heavy element nucleosynthesis, we may learn important clues to the neutrino mass and mixing puzzle. Alternatively, terrestrial neutrino detectors proposed for the future may give us a complementary set of insights into such physics, if we would be so lucky a to have a Type II Supernova go off in our galaxy.

3 Supernova Neutrino Observation: Now and in the Future

The issue of whether or not neutrinos have masses is clearly important for astrophysics and cosmology. Unfortunately, as described above, terrestrial experimental probes of neutrino mass, especially the ν_μ and ν_τ masses, remain problematic. Astrophysical considerations may represent the best hope for inferring neutrino masses and mixings. In this section, we examine how proposed neutral-current-based supernova neutrino burst detectors, in conjunction with the next generation water Cerenkov detectors, could use a galactic supernova event to measure (or place costraints on) the ν_μ and/or ν_τ masses if they are in excess of 5 eV. Such measurements would have important implications for our understanding of particle physics, cosmology, and the solar neutrino problem, and would be complementary to constraints derived from proposed laboratory vacuum oscillation experiments.

Conceivably, the longest baseline neutrino beam will be from a supernova. In order to exploit this source, we will need a complement of detectors on Earth to cover the possible time of flight difference for a massive neutrino (ie., 10 eV) or to study the neutrino flavor transformatiom resulting from either the MSW effect occuring in the supernova or vacuum oscillations occuring in flight. Three detectors which can contribute to the study of neutrino interactions from supernovas are Super Kamiokande, ICARUS and SNO [8].

The Super Kamiokonde detector is a scaled up version (40 Ktons) of the Kamioka detector, which with the IMB detector gave the first neutrino detection from a supernova, SN1987A [9]. We note that this detector will be very sensitive to the $\bar{\nu}_e + p \rightarrow e^+ + n$ reaction and therefore to the $\bar{\nu}_e$ flux. This detector is under construction and should be operational in 1996. The ICARUS detector at the Gran Sasso is still in the development stage. The detector is a large liquid Argon calorimeter and it should provide some important information on the ν_e signal from a galactic supernova. The Sudbury Neutrino Observatory (SNO) detector will consist of 1.6 Ktons of $D_2 0$ and should also be ready in 1996. This detector is sensitive to the neutral current breakup of deuterium $\nu_x + d \rightarrow n + p + \nu_x$

At Snowmass there was a lively discussion of the future of supernova neutrino detection. There were presentations on ICARUS, SNO, and the Supernova Burst Observatory (SNBO), as well as discussions of the prospects for obtaining real time coincidences between neutrino detectors and other detectors such as gravity wave detectors [8].

The major physics issues that were discussed revolved around the determination of neutrino masses and/or mixing from the expected neutrino signal from a core collapse supernova event in our galaxy. However, a major problem in this field remains the unknown SNII rate in our galaxy. Furthermore, the feasibility of constructing a supernova detector large enough to be sensitive to extragalactic sources is an issue which, although not settled, must engender pessimism at this point. Let us now turn to some of the specific issues in the field of supernova neutrino detection. Some of this part of the report is adapted from some of our previous work [11].

3.1 Measurement of Neutrino Mass

Perhaps the most straight forward and obvious signature of a massive neutrino would come from the lenghtening in flight time from a distant supernova. For example, the flight time difference between ν_τ and $\nu_e(\bar{\nu}_e)$ in seconds is

$$\Delta t = 0.514 R_{10\ kpc}[(E_{\nu_\tau}/m_{\nu_\tau})^2 - (E_{\nu_e}/m_{\nu_e})^2] \text{ sec}$$

where m_{ν_τ} is in eV, E_{ν_τ} is the neutrino energy in MeV and $R_{10\ kpc}$ is the distance to the supernova in units of 10 kpc. A finite neutrino mass would alter the neutrino spectra in characteristic ways which could result in broadening and flattening of the observed signal.

Thus, neutrino masses might be obtained by comparing the observed neutrino signal with the signal expected from supernova models. Tables 1 and 2 give, respectively, information on supernova neutrino detection methods and estimates of the number of events expected from a galactic supernova at 10 kpc. Since water Cerenkov detectors such as Super Kamiokande are relatively insensitive to ν_μ and ν_τ, they are unlikely to measure cosmologically significant neutrino masses for these flavors. This is perhaps more likely for the neutral-current-based detectors being build at present, like the Sudbury Neutrino Observatory (SNO). The event rates for different neutrino reactions in SNO from a supernova explosion at 10 kpc are shown in Table 3.

Recently there has been real progress in supernova simulations giving an explosion as outlined in section

Table 1: Supernova neutrino detection methods in the 1990s. Adapted from Ref. [11].

Reactions	$\bar{\nu}_e p \to e^+ n$	$\nu_x e \to \nu_x e$	$\nu_x N \to \nu_x N$	$\nu_x N \to \nu_x N^* \to n$
Cross section	Large (KII,SK,IMB,LVD)	Small $\sim E_\nu^2$ (ICARUS)	Large for coherent process	large at high E_{ν_x} SNO/SNBO
Neutrino energy estimate	Yes $E_{\nu_e} \sim E_e$	Partial $E_{\nu_x} \sim f(E_e)$	No	No Threshold may set E_{ν_x}
ν direction	No	Yes	No	No
Time information	Yes	Yes	Yes	Yes
Down time speculations	$\gtrsim 10\%$	$\sim 30\%$	?	Could be small
Maximun detector size	2×10^5 ton (H_2O) LENA $\gtrsim 10^4$ ton liq. scint (LVD)	$\sim 2 \times 10^5$ ton (H_2O) or $\sim 10^3$ ton Cryogenic (ICARUS)	? Kilograms No detector proposed	$\sim 10^5 - 10^7$ ton of NaCl (SNBO) or $\sim 10^3$ ton D_2O (SNO)
Backgrounds	Small if e^+ and n capture detected; OK for H_2O galactic signal	Small if directionally used to reject background	?	Depends on radioactivity of material

Table 2: Events expected from a galactic supernova in future supernova ν detectors at 10 kpc. Adapted from Ref. [11].

	Comparison of future supernova ν detectors					
Process	$\bar{\nu}_e p \to e^+ n$ $\bar{\nu}_e d \to ppe$	$\bar{\nu}_e e \to \bar{\nu}_e e$	$\bar{\nu}_x e \to \bar{\nu}_x e$ $x = \mu, \tau$	$\nu_e N \to N^* \nu_e$ $\to n$	$\nu_x N \to N^* \nu_x$ $\to n$	ν_e prompt
ICARUS		~ 140	25			4
SNO (1.6 kT)	~ 435	3	21	~ 117	~ 200	5–20
LVD (1.8 kT):scint	~ 342					
MACRO (1.kT):scint	~ 220					
Kamiokande II (3 kT)	~ 355	~ 1				
Super Kam (40 kT)	~ 5310	~ 17				~ 5
SNBO (100 kT)	100			~ 100	10000	
Comments	Measure t_ν $E_\nu \sim E_e$ No direction	t_ν E_ν estimated from E_e Θ_ν measured		t_ν only No E_ν! No Θ_ν		$\Delta r \simeq 10$ ms

Table 3: SNO events Rates (10 kpc supernova). Adapted from Ref. [11].

Reaction	Target medium	Events in ν_e burst per kton	Events in cooling phase per kton
$\nu_e + d \to p + p + e^-$	D_2O	10	33
$\nu_x + e^- \to \nu_x + e^-$	D_2O/H_2O	1	16
$\nu_x + d \to \nu_x + p + n$ ($n^{35}Cl \to {}^{36}Cl + \gamma$)	D_2O	6	760
$\bar{\nu}_e + d \to n + n + e^+$	D_2O	0	20
$\bar{\nu}_e + p \to n + e^-$	H_2O	0	120

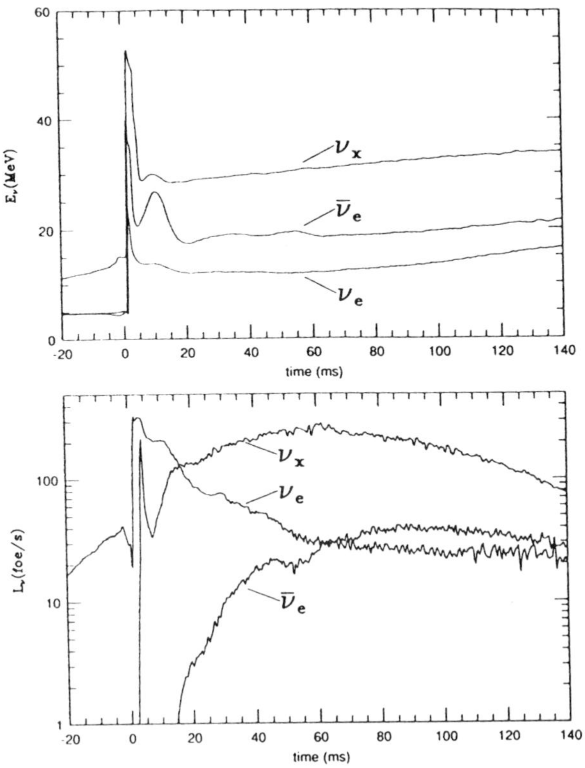

Figure 1: Neutrino luminosity and mean energy as a function of time for the 25 solar mass progenitor. Time 0 corresponds to the bounce of the core. Adapted from Ref. [13].

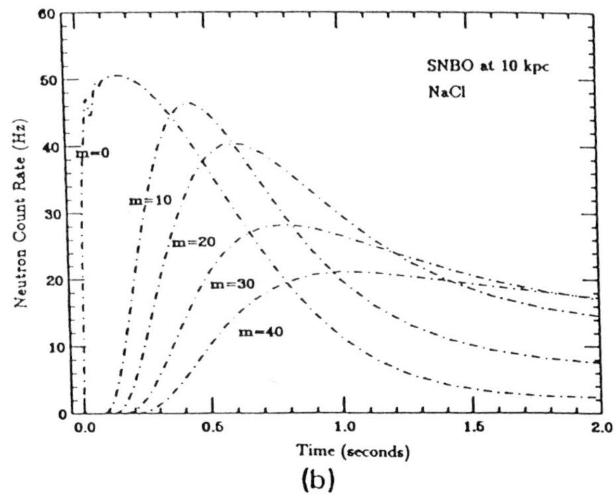

Figure 2: Effects of a massive ν_μ or ν_τ on the signal rate for a NaCl medium for different neutrino masses.

The Super Neutrino Burst Observatory Concept

---

1) GOAL: To measure τ neutrino mass

- direct unambiguous measurement -

2) METHOD: Use time of flight from Galactic SN

3) DETECTOR: Rock ($\nu + A \rightarrow \nu' + A' + n$)

(threshold exists since n is initially bound)

4) ENERGY: Average ν_τ energy is about 3/2 average ($\nu_e, \bar{\nu}_e$) energy

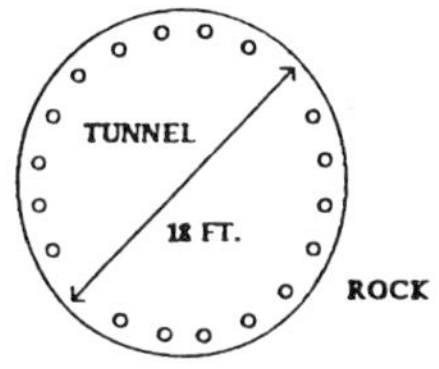

5) COUNTS: More than 10^4 counts expected from Galactic SN

Use "inexpensive" neutron detectors - and

detectors requiring little direct maintenance

6) FROM CALCULATIONS:

$\langle E_{\nu_e} \rangle = 12\text{MeV}$ $\langle E_{\bar{\nu}_e} \rangle = 15\text{MeV}$ $\langle E_{\nu_{\mu,\tau}} \rangle = 25\text{MeV}$

7) RISE TIME: $L_{\bar{\nu}_e}$ and $L_{\nu_{\mu,\tau}}$ rise rapidly at

almost the same time (about 0.1sec)

8) FIRST PREDICTION: Take model output for spectra and $L_{\bar{\nu}_e}$

and $L_{\nu_{\mu,\tau}}$ and feed into detector response

to predict first 0.35 seconds of detection

9) MEASUREMENT DATA: Use Kamiokande II data for supernova

in Milky Way to measure $L_{\bar{\nu}_e}$ rise time

10) LIMITS: Compare ν_μ and/or ν_τ signal to find limits on m_{ν_μ} and/or m_{ν_τ}

LLNL Cal. for SNBO Detector:

No. of Counts $= 2 \times 10^4 \frac{V}{R^2}$ with V in meters^3, R in Mpc;

For Milky Way R = 0.01 Mpc; Let V = 10^4 meters^3

Then No. of Counts = 2×10^4

Figure 3: Sketch of some of the principal concepts for SNBO.

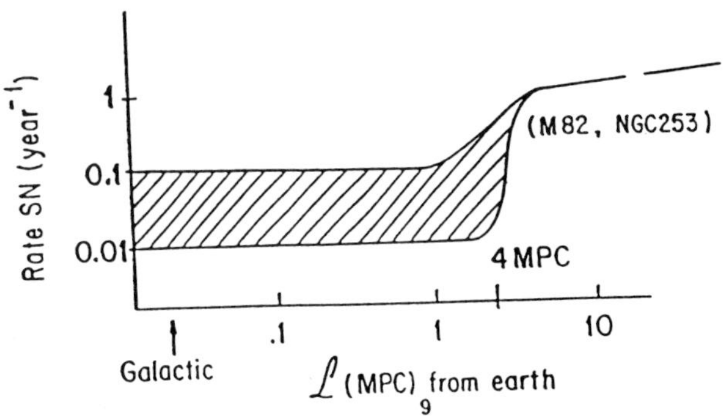

Figure 4: Estimated rate for Supernova explosions as a function of the distance from the Earth.

II. These calculations give interesting predictions for the neutrino spectra as a function of time, as shown in Fig. 1. Detectors like Super Kamiokande and the proposed Supernova Neutrino Burst Observatory(SNBO) may be able to detect such effects. Particularly, the SNBO detector may be of crucial importance for this study [10]. Using these various detectors it should be possible to detect a finite neutrino mass as shown in Fig. 2 [11].

3.2 Impact of Low Energy Neutrino Experiments

Recent measurements of the neutral current processes on nuclei at low energy (20- 50 MeV) by the KARMAN [8] group give additional confidence for the rate estimates for supernova observatories. This group has measured the process $\nu +^{12} C \rightarrow \nu +^{12} C^*$, and found very good agreement with theoretical predictions. They have also measured the charged current processes $\nu_e +^{12} C \rightarrow e^- +^{12} N(gs)$. Reactions like $\nu_e +^{12} C \rightarrow e^- +^{12} N^*$ most likely have been observed as well.

3.3 Ideal Supernova Observatories

The major problem of supernova neutrino detection is the uncertain period of time between supernova explosions in this galaxy. In addition, complementary detectors should be active when the supernova goes off. The comparison of signals from different detectors for the same supernova event should enable us to learn the maximum information about the explosion process and neutrino properties. Some of the requirements of such an ideal supernova observatory include:

- Long operating time, greater than the interval between expected observable supernova explosions (20 years for extra galactic and 60 years for galactic supernovas).
- Effective masses of the detector ranging from Ktons to a Megaton, depending on whether one is aiming at detecting signals from a galactic or an extragalactic supernova.
- capabilities to separate $\bar{\nu_e}, \nu_e$ and $\nu_\mu + \nu_\tau$ interactions
- Some directional ability to be able to point back to the supernova
- The ability to be in coincidence with Gravity Wave detectors

Lacking an ideal observatory, the SNBO group has been studying ideas for a very large detector [10]. Ideally, the SNBO detector should be a neutral current detector able to tag neutrino flavor, with an adequate event rate to be able to meausure m_{ν_x}. Fig. 3 gives a schematic sketch of some of the more detailed steps needed in the development of SNBO. We have located a possible site for the observatory near Carlsbad N.M., the WIPP site. We have studied the radioactive background at this site (measured by the OSU group) and find it acceptable for a galactic supernova detector.

With a detector like SNBO, where up to 10^4 counts would be recorded, it should be possible to detect the supernova within 100 ms of its start, and to alert the WWW to "Supernova in Progress". The real issue in supernova neutrino detection is the long time needed to operate the detector between supernova explosions in the galaxy, longer than most underground detectors have operated for. One possibility is to build a large enough detector to "see" extra galactic supernovas. Fig. 4 shows the possible rate with distance from our galaxy, as estimated by E. Becklin [12]. A 10^7 ton detector may observe a supernova each year, though the feasibility of such a detector is questionable at this point. Nevertheless, aiming for such a rate must be the long term goal of the SNBO prototype study!

Acknowledgements

We wish to thank F. Halzen for helping with the organization of the Neutrino Astrophysics Section. We also would like to thank members of ICARUS, SNBO, and SNO for many discussions.

References

1. R. Cowsik and J. McClelland, Phys. Rev. Lett. **29**, 669 (1972); J. E. Gunn, B. W. Lee, I. Lerche, D. N. Schramm, and G. Steigman, Astrophys. J. **223**, 1015 (1978); J. R. Bond, G. E. Efstathiou, and J. Silk, Phys. Rev. Lett., **45**, 61 (1980); G. R. Blumenthal, S. M. Faber, J. R. Primack, and M. Rees, Nature **311**, 517 (1984).
2. G.F. Smoot, et al., *Astrophys. J.*, **396** L1 (1992).
3. S. A. Bonometto and R. Valdarnini, Phys. Lett. **103A**, 369 (1984); L. Z. Fang, S. X. Li, S. P. Xiang, Astron. Astrophys. **140**, 77 (1984); A. Dekel and S. J. Aarseth, Astroph. J. **283**, 1 (1984); Q. Shafi and F. W. Stecker, Phys. Rev. Lett. **53**, 1292 (1984); P. K. Schaefer and Q. Shafi, Nature **359**, 199 (1992); A. Klypin, S. Borgani, J. Holtzman, J. R. Primack, Astrophys. J., in press (1995); J. R. Primack, J. Holtzman, A. Klypin, D. O. Caldwell, Phys. Rev. Lett., in press (1995); D. Pogosyan and A. Starobinsky, Astrophys. J., in press (1995).
4. H. A. Bethe & J. R. Wilson *Ap J*, **295**, 14, (1985); W. C. Haxton, *Phys. Rev. Lett.*, **60**, 1999, (1988).
5. M. Herant, W. Benz, & S. A. Colgate *ApJ*, **395**, 642, (1992); M. Herant, et al., *APJ* **58**, 1490 (1994); Burrows, A., & Fryxell, B. A. 1993, *ApJ*,**418**, L33 (1993); Miller, D. S., Wilson, J. R., & Mayle, R. W. *ApJ*, **415**, 278 (1993).

6. S. E. Woosley, G. J. Mathews, J. R. Wilson, R. D. Hoffman, & B. S. Meyer, *ApJ*, **433**, 229, (1994).
7. Y.-Z. Qian, G. M. Fuller, G. J. Mathews, R. W. Mayle, J. R. Wilson, & S. E. Woosley, *Phys. Rev. Lett.*, **71**, 1965, (1993); Y.-Z. Qian & G. M. Fuller *Phys. Rev.* **D51**, 1479 (1995).
8. Presentations to the Snowmass Workshop were made by members of the SNO, Super Kamionkande, AMANDA, SNBO and KARMAN groups and some of their results are reported here.
9. K. Hirata, el al., *Phys Rev. Lett* **58**, 1490 (1987); R.M. Bionta, et al., *Phys. Rev. Lett* **58**, 1494 (1987).
10. The SNBO Group - Internal Reports (UCLA)
11. D.B. Cline, G.M. Fuller, W.P. Hong B. Meyer, and J. Wilson *Phys. Rev.*, **D50** 720 (1994).
12. E. Becklin, private communication
13. M. Herant, et al., *ApJ*, **435**, 339, (1994).

High Energy Neutrino Astronomy and its Telescopes: the Case for a Kilometer-Scale Detector

F. Halzen

Department of Physics, University of Wisconsin, Madison, WI 53706, USA

Doing astronomy with photons of energies in excess of a GeV has turned out to be extremely challenging. Efforts are underway to develop instruments that may push astronomy to wavelengths smaller than 10^{-14} cm by mapping the sky using high energy neutrinos instead. Neutrino astronomy, born with the identification of thermonuclear fusion in the sun and the particle processes controlling the fate of a nearby supernova, will reach outside the galaxy and make measurements relevant to cosmology. The field is immersed in technology in the domains of particle and cosmic ray physics to which many of its research goals are intellectually connected. To mind come the search for neutrino mass, cold dark matter (supersymmetric particles?) and the monopoles of the Standard Model. While a variety of collaborations are pioneering complementary methods by building telescopes with effective area in excess of 0.01-0.1 km^2, we show here that the natural scale of a high energy neutrino telescope is 1 km^2. With several thousand optical modules, the scope and price tag of a kilometer-scale instrument is similar to that of experiments presently being commissioned such as the SNO neutrino observatory in Canada and the Superkamiokande experiment in Japan.

1 Introduction

Several world-wide efforts are underway to design and construct a new type of telescope which scans the sky for high-energy neutrinos rather than the photons that are detected by conventional astronomical instruments. As exemplified time and again, the development of a novel way of looking into space invariably results in the discovery of unanticipated phenomena. Large volumes of deep ocean and lake water as well as the Antarctic ice cap are being instrumented to trap cosmic neutrinos. With the present flurry of activity, scientists are hopefully taking the first steps in the future construction of kilometer-scale neutrino detectors, consisting of a million billion liters (a teraliter) of instrumented natural water or ice. These instruments should be able to systematically study the Universe beyond our galaxy and watch cosmic cataclysms without having to wait for a one-in-a-century miracle like a nearby supernova.

The penetrating power of the neutrino has been successfully exploited by particle physicists, who realized some time ago that by illuminating matter with an intense neutrino beam one can glean information on structure deep inside protons. In a very poor choice of words for a physicist, one could say that neutrinos are a diagnostic tool to "X-ray" protons or nuclei. The technique contributed in important ways to the discovery of quarks and is conceptually simple, though technologically challenging. In a high energy physics laboratory a very intense beam of accelerated protons is shot into a "beam-dump" which typically consists of a kilometer-long mound of earth or a 100 meter-long block of stainless steel. Particle physics does the rest. Protons interact with nuclei in the dump and produce a large multiplicity of pions in each collision. Neutral pions decay into two photons, while charged pions decay into a muon and a neutrino. The material in the dump will eventually absorb the photons and muons so that only neutrinos exit at the opposite end, forming an intense and controlled beam ready to perform neutrino microscopy of matter.

Can we use neutrinos to X-ray the Universe? Are there cosmic beam-dumps producing neutrino beams in space? Recall that all that is needed is a proton beam, energetic enough to produce pions, and target material to act as the dump. The answer is definitively positive: cosmic ray experiments have revealed the existence of protons accelerated to energies in excess of 10^{20} eV. The cosmic ray beam will produce neutrinos in any target material. The target may be as ordinary as the Earth's atmosphere and as exotic as the cosmic photons which fill outer space.

Incredibly, we have no clue where the highest energy cosmic rays come from and how they can be accelerated to such energies. Cosmic rays with energies up to some 10^{14} eV are thought to be accelerated by shocks driven into the interstellar medium by supernova explosions. The Lorentz force on a particle near the speed of light in the galactic magnetic field ($\sim 3\mu$G) multiplied by the extent of a typical supernova shock (~ 50 pc) is $\sim 10^{17}$ eV. However, our own Galaxy is too small, and its magnetic fields too weak, to accelerate particles to 10^{20} eV. This energy should require, for instance, a 100 μG field extending over thousands of light years. Such fields exist near the supermassive black holes which power active galactic nuclei (AGNs). The idea is rather compelling as bright quasars are also the dominant source of high energy photons. This is very exciting, implying that high energy cosmic rays are produced in faraway galaxies and carry cosmological information — on galaxy formation, for example. Quasars are the brightest sources in the Universe; some are so far away that they are messengers from the earliest of times.

It is anticipated that beams accelerated near the black hole in active galaxies produce neutrinos on the ambient matter in the active galaxy, e.g. on the photons with an average density of $10^{14}/\mathrm{cm}^3$. Neutrino astronomers believe that the high energy neutrino sky will glow uniformly with bright active galaxies far outshining our Milky Way. But the results may be even more spectacular. As is the case in man-made beam-dumps, photons from celestial accelerators may be absorbed in the dump. It is important to realize that high energy photons, unlike weakly interacting neutrinos, do not carry information on cosmic sites shielded from our view by more than a few hundred grams of intervening matter. The neutrino sky may reveal sources with no counterpart in any wavelength of light. In this way neutrino telescopes may reveal the sources of the enigmatic high energy cosmic rays, thus resolving one of the most lingering puzzles in astronomy.

In this paper we will first review the theorized sources of high energy cosmic neutrinos. We will start with those sources whose detection can be guaranteed by the very existence of high energy cosmic rays. We will then turn to a discussion of active galaxies. Some mixture of elementary particle physics and astrophysics allows us to compute that quasars will emit roughly one neutrino for every proton. As we know the flux of protons reaching Earth, we can estimate the neutrino luminosity associated with quasars. We will conclude that kilometer-size instruments are required to *guarantee* their detection and to do real astronomy. Although we subsequently list the many interdisciplinary science goals of high energy neutrino telescopes, we will concentrate on two particularly exciting challenges: the search for neutrino oscillations and the search for WIMPS, the particles which are pinpointed as the dark matter by an ever-increasing body of evidence. Interestingly, every gedanken-experiment seems to require a kilometer-scale instrument. It is important to keep in mind that, obviously, the most spectacular discoveries may be within reach of the smaller detectors now under construction.

The accelerator physicist's method for building a neutrino detector will typically use lead absorber to filter out all particles but the neutrinos, wire chambers to detect electrons and muons produced in neutrino charge exchange interactions, and electronics with a combined price tag of roughly 10^4 US dollars per m^2. Such a 1 km^2 detector would cost 10 billion dollars. Realistically, we are compelled to develop methods which are more cost-effective by a factor of more than one hundred in order to be able to commission neutrino telescopes with effective area of order 1 km^2. The proven technique developed by existing underground detectors such as Kamiokande and IMB cannot be extrapolated to kilometer scale. In order to achieve large effective telescope area, it is unfortunately necessary to abandon the low-energy thresholds of existing supernova or solar neutrino detectors. There is no free lunch. One simply optimizes the instrument to efficiently catch those neutrinos that are easy to catch, i.e. those with very high energies, typically 1–100 GeV and above, which have a relatively large probability of being absorbed on protons in, or close to the detector and producing telltale electrons or muons. High energy neutrino telescopes, although much larger in effective volume than Superkamiokande and SNO, cannot study the sun because of their GeV-thresholds. A method has however been developed to operate them as a supernova watch, a task which one of them is already performing.

We will finally describe the techniques for detecting neutrinos as well as the four major experiments which are proceeding with construction, each of which has different strengths and faces different challenges. We recall that the interaction of neutrinos with matter is so weak that they will have to penetrate the full thickness of the Earth in order to have some chance of being captured. Detectors of astrophysical neutrinos must therefore be orders of magnitude larger than familiar astronomical instruments. Large volumes of water or ice are needed to trap a few neutrinos. When these collide with nuclei of matter, they will produce muons which will act as observable tracers which travel through kilometers of water and ice thus allowing one to extend the reach of the detector well beyond the actual instrumented volume; see Fig. 1a. Also, high energy muons are nicely aligned with the direction of their parent neutrino. The muons point back to the neutrino sources to better than 1 degree, a deviation which decreases with increasing neutrino energy.

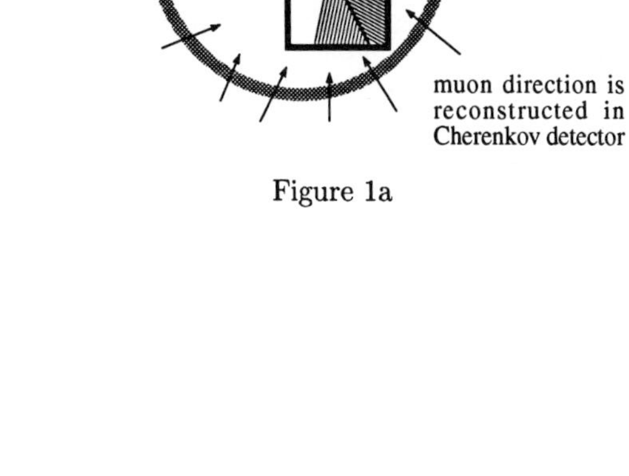

Figure 1a

In all four instruments under construction the proven Cherenkov technique, pioneered by IMB, Kamiokande and others, is extrapolated to kilometer scale. The direction of the neutrino is inferred from the muon track. It is measured by mapping the cone of Cherenkov light radiated by the muon as it travels through the detector. A high energy muon, travelling faster than the speed of light in water or ice, will radiate photons in the form of a Cherenkov cone of light. The arrival times of the Cherenkov photons, recorded by a grid of optical detectors, are used to map the shape of the cone and thus reconstruct the direction of the radiating muon; see Fig. 1b. The photons can be detected with commercial photomultiplier tubes which represent a robust and cheap technology. All this sounds relatively straightforward until one faces the fact that the detector is constantly bombarded with muons produced by cosmic rays in the Earth's atmosphere. These are the decay products of cosmic ray pions and totally unrelated to neutrinos. Near the Earth's surface these muons are over ten billion times more numerous than those signaling the arrival of cosmic neutrinos. In a detector shielded by a kilometer-thick layers of water or ice, the number is reduced and hopefully manageable. Manageable means that the direction of every down-going muon is flawlessly measured so that it can be unambiguously logged as background or as a (much smaller) signal. The trick is to use the Earth as a filter: muons travelling *upwards* through the detector must be spawned by neutrinos because the Earth (about 12500 km of it) will shield out atmospheric cosmic ray muons. Neutrino telescopes point into the ground, with the South Pole telescope studying the northern skies.

Arrival times of Cherenkov wavefront give direction of muon

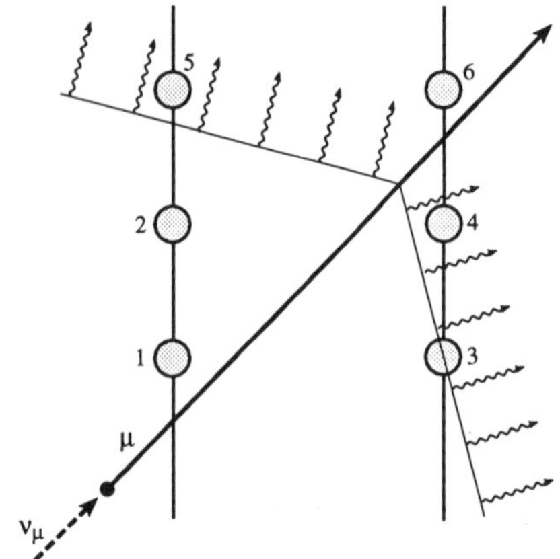

Figure 1b

For the construction of a kilometer-scale detector one can imagine any of the above detectors as the basic building block for the ultimate telescope. The present efforts, described in this article, will hopefully be more than engineering projects. The most exciting and unexpected discoveries may be those made by the small, prototype detectors. Although nothing can be guaranteed, history is on our side. The photon sky has been probed with a variety of instruments sensitive to wavelengths of light as large as 10^4 cm for radio-waves to 10^{-14} cm for the GeV-photons detected with space-based instruments. Astronomical instruments have now collected data spanning 60 octaves in photon frequency, an amazing expansion of the power of our eyes which scan the sky over less than a single octave just above 10^{-7} cm. The lensmakers in Flanders developed the telescope to make an early inventory of the goods on ships crossing the English Channel. Little did they know that Galileo would use the same instrument to discover the moons of Jupiter. The first X-ray telescope was built to study the sun and moon; its successors discovered all kind of wonderful objects like neutron stars and accreting binaries. The gamma ray instruments developed by the US to search for thermonuclear explosions in the Soviet Union discovered the still enigmatic gamma ray bursts. Each wavelength of light has a story like this to tell. Most dramatic was the accidental discovery of the cosmic photon background with apparatus studying sky-interference with telephone communications. Hopefully, in a not-too-distant future, short wavelength neutrinos will contribute their own bizarre tale.

2 Guaranteed Cosmic Neutrino Beams

In heaven, as on Earth, high-energy neutrinos are produced in beam dumps which consist of a high-energy proton (or heavy nucleus) accelerator and a target. Gamma rays and neutrinos are generated in roughly equal numbers by the decay of pions produced in nuclear cascades in the beam dump. For every π^0 producing two gamma-rays, there is a π^+ and π^- decaying into a μ and a ν_μ. If the kinematics is such that muons decay in the dump, more neutrinos will be produced. We want to stress that in efficient cosmic beam dumps with an abundant amount of target material, high energy photons may be absorbed before escaping the source. Therefore, the most spectacular neutrino sources may have no counterpart in high energy gamma-rays.

By their very existence, high-energy cosmic rays guarantee the existence of definite sources of high-energy cosmic neutrinos.[1] Cosmic rays represent a hadron beam of known luminosity, with particles accelerated to energies in excess of 10^{20} eV. They produce pions in interactions with the interstellar gas in our galaxy, the cosmic photon background in our Universe, the sun and moon, and the Earth's atmosphere which represents a well-understood beam dump. These interactions are the source of fluxes of diffuse photons and neutrinos. The atmospheric neutrino beam can be used to study neutrino oscillations over distances of 10 to 10^4 km.[1]

The study of extremely energetic diffuse neutrinos produced in the interaction of ultra high energy, extragalactic cosmic rays on the microwave background is of special interest. The magnitude and intensity of this cosmological neutrino flux are determined by the maximum injection energy of the ultra-high-energy cosmic rays and by the distribution of their sources. If the sources are relatively near, at distances of order tens of Mpc, and the maximum injection energy is not much greater than the highest observed cosmic ray energy (few × 10^{20} eV), the generated neutrino fluxes are small. If, however, the highest energy cosmic rays are generated by many sources at large redshift, then a large fraction of their injection energy would be presently contained in gamma-ray and neutrino fluxes. The effect would be even stronger if the source luminosity were increasing with redshift z, i.e. if cosmic ray sources were more active at large redshifts — "bright phase models".[2]

3 Active Galactic Nuclei: Almost Guaranteed?

Although observations of PeV (10^{15} eV) and EeV (10^{18} eV) gamma-rays are controversial, cosmic rays of such energies do exist and their origin is at present a complete mystery. The cosmic-ray spectrum can be understood, up to perhaps 1000 TeV, in terms of shockwave acceleration in galactic supernova remnants. Although the spectrum suddenly steepens at 1000 TeV, a break usually referred to as the "knee," cosmic rays with much higher energies are observed and cannot be accounted for by this mechanism. This failure can be understood by simple dimensional analysis. The EMF in the supernova shock is of the form

$$E = ZeBRc\,, \tag{1}$$

where B and R are the magnetic field in the galaxy and the radius of the shock. For a proton Eq. (1) yields a maximum energy

$$E_{\rm max} = [10^5\,{\rm TeV}]\left[\frac{B}{3\times10^{-6}\,{\rm G}}\right]\left[\frac{R}{50\,{\rm pc}}\right] \tag{2}$$

and therefore E is less than 10^5 TeV for the typical values of B, R shown. The actual upper limit is much smaller than the value obtained by dimensional analysis because of inefficiencies in the acceleration process.

Cosmic rays with energy in excess of 10^{20} eV have been observed. Assuming that they are a galactic phenomenon, the measured spectrum implies that 10^{34} particles are accelerated to 1000 TeV energy every second. We do not know where or how. We do not know whether the particles are protons or iron or something else. If the cosmic accelerators indeed exploit the 3μGauss field of our galaxy, they must be much larger than supernova remnants in order to reach 10^{21} eV energies. Equation (1) requires that their size be of order 30 kpc. Such an accelerator exceeds the dimensions of our galaxy. Although imaginative arguments exist to avoid this impasse, an attractive alternative is to look for large size accelerators outside the galaxy. Nearby active galactic nuclei (quasars, blazars...) distant by order 100 Mpc are the obvious candidates. With magnetic fields of tens of μGauss over distances of kpc near the central black hole or in the jets, acceleration to 10^{21} eV is possible; see Eq. (1).

One can visualize the accelerator in a very economical way in the Blanford-Zralek mechanism. Imagine that the horizon of the central black hole acts as a rotating conductor immersed in an external magnetic field. By simple dimensional analysis this creates a voltage drop

$$\frac{\Delta V}{10^{20}{\rm volts}} = \frac{a}{M_{\rm BH}}\,\frac{B}{10^4{\rm G}}\,\frac{M_{\rm BH}}{10^9 M_\odot}\,, \tag{3}$$

corresponding to a luminosity

$$\frac{\mathcal{L}}{10^{45}{\rm erg\,s^{-1}}} = \left(\frac{a}{M_{\rm BH}}\right)^2\left(\frac{B}{10^4{\rm G}}\right)^2\left(\frac{M_{\rm BH}}{10^9 M_\odot}\right)^2. \tag{4}$$

Here a is the angular momentum per unit mass taken to be the black hole mass $M_{\rm BH}$.

All this was pretty much a theorist's dream until recently the Whipple collaboration reported the observation of TeV (10^{12} eV) photons from the giant elliptical galaxy Markarian 421.[3] With a signal in excess of 6 standard deviations, this was the first convincing observation of TeV gamma rays from outside our Galaxy. That a distant source such as Markarian 421 can be observed at all implies that its luminosity exceeds that of galactic cosmic accelerators such as the Crab, the only source observed by the same instrument with comparable statistical significance, by close to 10 orders of magnitude. More distant by a factor 10^5, the instrument's solid angle for Markarian 421 is reduced by 10^{-10} compared to the Crab. Nevertheless the photon count at TeV energy is roughly the same for the two sources. The Whipple observation implies a Markarian 421 photon luminosity in excess of 10^{43} ergs per second. It is interesting that these sources have their highest luminosity above TeV energy, beyond the wavelengths of conventional astronomy. During May 1994 Markarian 421 was observed to increase its flux by a factor 10 in one day, strongly suggesting the catastrophic operation of a high energy hadronic accelerator.

Why Markarian 421? Whipple obviously zoomed in on the Compton Observatory catalogue of active galaxies (AGNs) known to emit GeV photons. Markarian, at a distance of barely over 100 Mpc, is the closest blazar on the list. As yet TeV gamma rays have not been detected from any other AGNs. Although Markarian 421 is the closest of these AGNs, it is one of the weakest; the reason that it is detected whereas other, more distant, but more powerful, AGNs are not, must be that the TeV gamma rays suffer absorption in intergalactic space through the interaction with background infra-red photons. TeV gamma rays are indeed efficiently absorbed on infra-red starlight and this most likely provides the explanation why astronomers have a hard time observing much more powerful quasars such as 3C279 at a redshift of 0.54. Production of e^+e^- pairs by TeV gamma rays interacting with IR background photons is the origin of the absorption. The absorption is, however, minimal for Mrk 421 with $z = 0.03$, a distance close enough to see through the IR fog. This implies that all of the AGNs may have significant very high energy components but that only Markarian 421 is close enough to be detectable with currently available gamma-ray telescopes. The opportunities for neutrino astronomy are wonderfully obvious.

This observation was not totally unanticipated. Many theorists[1] have identified blazars such as Markarian 421 as powerful cosmic accelerators producing beams of very high energy photons and neutrinos. Acceleration of particles is by shocks in the jets (or, possibly, also by shocks in the accretion flow onto the supermassive black hole which powers the galaxy) which are a characteristic feature of these radio-loud active galaxies. Many arguments have been given for the acceleration of protons as well as electrons.[1] Inevitably beams of gamma rays and neutrinos from the decay of pions appear along the jets. The pions are photoproduced by accelerated protons on the target of optical and UV photons in the galaxy which reaches densities of 10^{14} per cm^3. The latter are the product of synchrotron radiation by electrons accelerated along with the protons.

Powerful AGNs at distances of order 100 Mpc and with proton luminosities of 10^{45} erg/s or higher are obvious candidates for the cosmic accelerators of the highest energy cosmic rays. Their luminosity often peaks at the highest energies and their proton flux, propagated to Earth, can quantitatively reproduce the cosmic ray spectrum above spectrum 10^{18} eV.[4] A simple estimate of the AGN neutrino flux can be made by assuming that a neutrino is produced for every accelerated proton, i.e. the total luminosities in neutrinos and cosmic rays are equal. This is typical of the astrophysical beam dumps considered[1] and leads to the relation:

$$4\pi \int dE(E\,dN_\nu/dE) \sim L_{CR} \sim 7.2\times 10^{-9}\,\mathrm{erg\,cm^{-2}\,s^{-1}}\,, \qquad (5)$$

which simply states that AGNs generate 1 neutrino for each high-energy cosmic ray. Here the luminosity L_{CR} has been obtained by integrating the highest energy $E^{-2.71}$ component of the cosmic ray flux above 10^{17} eV. Assuming an E^{-2} neutrino spectrum we obtain:

$$E\frac{dN_\nu}{dE} = \frac{1}{4\pi}\frac{7.5\times 10^{-10}}{E\,(\mathrm{TeV})}\,\mathrm{cm^{-2}s^{-1}sr^{-1}}\,. \qquad (6)$$

The flux of Eq. (6) is at the low end of the range of fluxes predicted in models where acceleration is in shocks in the jet[4] and accretion disc;[4,5,6] see Fig. 1. It is also clear that our estimate represents a lower limit, as the proton flux reaching Earth should be corrected for absorption in ambient matter in the source and in the interstellar medium. The neutrino flux corresponds to 300 upcoming muons per year in a neutrino detector with 10^6 m^2 effective area. Detailed model predictions often exceed this estimate by several orders of magnitude and are definitely within reach of the instrument proposed here.

4 Intermezzo: The Case for a Kilometer-Scale Detector

Observing extra-galactic sources of neutrinos has become a pivotal goal in the development of high energy neutrino telescopes. The architecture of detectors is often optimised for the detection of AGN neutrinos. Neutrinos are observed via the muons they produce in the detector volume. At high energy it is possible to enhance the effective volume of detectors by looking for neutrino-induced muons generated in charged-current interactions of ν_μ in the water or ice outside the instrumented detector volume; see Fig. 1a. The effective detector volume is then the product of the detector area and the muon range in rock R_μ. TeV muons have a typical range of one kilometer, which leads to a significant increase in effective detector volume. The average muon energy loss rate is

$$\left\langle \frac{dE}{dX} \right\rangle = -\alpha(E) - \beta(E)\times E\,, \qquad (7)$$

where X is the thickness of material in g/cm^2. The first term represents ionization losses, which are approximately independent of energy, with $\alpha \sim 2$ MeV g^{-1}cm^2. The second term includes the catastrophic processes of bremsstrahlung, pair production and nuclear interactions, for which fluctuations play an essential role. Here $\beta \sim 4\times 10^{-6}$ g^{-1}cm^2. The critical energy above which the radiative processes dominate is

$$E_{\mathrm{cr}} = \alpha/\beta \approx 500\,\mathrm{GeV}. \qquad (8)$$

To treat muon propagation properly when $E_\mu > E_{\mathrm{cr}}$ requires a Monte Carlo calculation of the probability P_{surv} that a muon of energy E_μ survives with energy $> E_\mu^{\mathrm{min}}$ after propagating a distance X.

The probability that a neutrino of energy E_ν on a trajectory through a detector produces a muon above threshold at the detector is

$$P_\nu(E_\nu, E_\mu^{\min}) = N_A \int_0^{E_\nu} \mathrm{d}E_\mu \frac{\mathrm{d}\sigma_\nu}{\mathrm{d}E_\mu}(E_\mu, E_\nu) \times R_{\mathrm{eff}}(E_\mu, E_\mu^{\min}) \,, \tag{9}$$

where

$$R_{\mathrm{eff}} = \int_0^\infty \mathrm{d}X \, P_{\mathrm{surv}}(E_\mu, E_\mu^{\min}, X) \,, \tag{10}$$

and σ_ν is the cross section for a neutrino of energy E_ν to produce a muon of energy E_μ. The flux of ν_μ-induced muons at the detector is given by the convolution of the neutrino spectrum ϕ_ν with the muon production probability (9) as

$$\phi_\mu(E_\mu^{\min}, \theta) = \int_{E_\mu^{\min}} \Big\{ \mathrm{d}E_\nu \, P_\nu(E_\nu, E_\mu^{\min}) \times \exp[-\sigma_t(E_\nu)\, X(\theta)\, N_A]\, \phi_\nu(E_\nu, \theta) \Big\} \,. \tag{11}$$

The exponential factor here accounts for absorption of neutrinos along the chord of the Earth, $X(\theta)$, before they interact to produce muons. Absorption becomes important for $\sigma(E_\nu) \gtrsim 10^{-33}$ cm^2 or $E_\nu \gtrsim 10^7$ GeV.

The event rate in a detector is obtained by multiplying Eq. (11) by its effective area. From Eqs. (6),(11) we obtain order 300 upcoming muon events per year in a $10^6\,\mathrm{m}^2$ detector. It is not a comfortably large rate however as the flux is indeed distributed over a number of sources. There is, however, no competing background. Hopefully one will be able to scrutinize a few nearby sources with good statistics. We should recall at this point that our back-of-the-envelope estimate yields a lower limit well below the range of fluxes predicted by detailed modeling. Optimistic predictions exceed our estimate by over an order of magnitude and are, possibly, within reach of the first-generation telescopes now being commissioned.

5 The Neutrino Sky: Summary

The neutrino sky at GeV-energy and above is summarized in Fig. 2. Shown is the flux from the galactic plane as well as a range of estimates (from generous to conservative) for the diffuse fluxes of neutrinos from active galaxies and from the interaction of extra-galactic cosmic rays with cosmic photons. At PeV energies and above all sources dominate the background of atmospheric neutrinos. In order to deduce the effective area of an instrument required to study the fluxes in the figure, the detection efficiency must be included using Eq. (11). At the highest energies this efficiency approaches unity and 1 event per km^2 per year corresponds to the naive estimate of 10^{-18} neutrinos per cm^2 second. At TeV–PeV energy the one event level per year corresponds to a flux

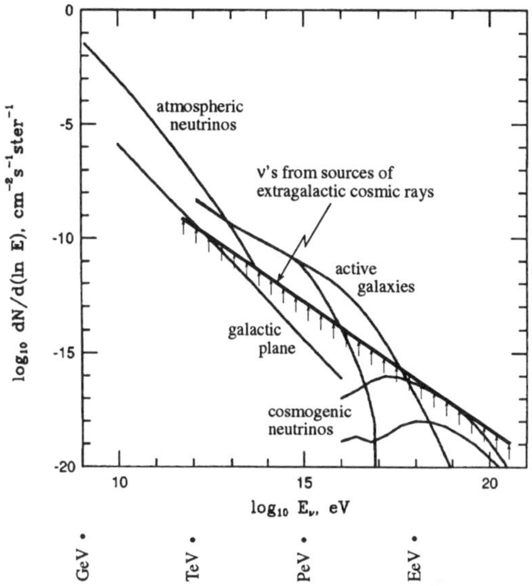

Figure 2

of 10^{-14}–10^{-15} per cm^2 second. As before, we conclude that the diffuse flux from AGN yields order 10^3 events in a kilometer-size detector per year in the TeV-energy range.

It should be emphasized that high energy neutrino detectors are multi-purpose instruments. Their science-reach touches astronomy, astrophysics and particle and cosmic ray physics. Further motivations for the construction of a km^3 deep underground detector include:[1]

1. The search for the t'Hooft-Polyakov monopoles predicted by the Standard Model.
2. Recent underground experiments have given tantalizing hints for neutrino oscillations in the mass range of $\Delta m^2 \gtrsim 3 \times 10^{-2}$ eV^2.[1] Neutrino telescopes may be able to study and extend this mass range by measuring the zenith angle distribution of neutrino-induced muons. As the direction of the atmospheric neutrino beam varies from vertically downward to upward, the neutrino path length (distance from its production to its interaction in the ice) varies from the diameter of the Earth ($\sim 10^5$ km) to the height of the atmosphere ($\sim$ 10 km). With sufficient energy resolution it may even be possible to observe the oscillatory behavior of the flux over several hundred meters suggested by the "atmospheric neutrino anomaly". Only a mature and well-calibrated instrument can do this precision measurement.
3. The search for neutrinos from the annihilation of dark matter particles in our galaxy.
4. The capability to observe the thermal neutrino emission from supernovae[7] (even though the nominal

threshold of the detectors exceeds the neutrino energy by several orders of magnitude!). The method will be discussed in the context of the AMANDA detector which is already monitoring our galaxy and can do so over decades in a most economical fashion.

5. Further study of the science pioneered by space-based gamma ray detectors such as the study of gamma ray bursts and the high energy emission from quasars.

It is intriguing that each of these goals individually point at the necessity to commission a kilometer-size detector. In order to illustrate this I will discuss the search for the particles which constitute the cold dark matter.

6 Indirect Search for Cold Dark Matter

It is believed that most of our Universe is made of cold dark matter particles. Big bang cosmology implies that these particles have interactions of order the weak scale, i.e. they are WIMPs.[8] We know everything about these particles (except whether they really exist!). We know that their mass is of order of the weak boson mass, we know that they interact weakly. We also know their density and average velocity in our galaxy as they must constitute the dominant component of the density of our galactic halo as measured by rotation curves. WIMPs will annihilate into neutrinos with rates whose estimate is straightforward. Massive WIMPs will annihilate into high energy neutrinos. Their detection by high energy neutrino telescopes is greatly facilitated by the fact that the sun, conveniently, represents a dense and nearby source of cold dark matter particles.

Galactic WIMPs, scattering off protons in the sun, lose energy. They may fall below escape velocity and be gravitationally trapped. Trapped dark matter particles eventually come to equilibrium temperature, and therefore to rest at the center of the sun. While the WIMP density builds up, their annihilation rate into lighter particles increases until equilibrium is achieved where the annihilation rate equals half of the capture rate. The sun has thus become a reservoir of WIMPs which annihilate into any open fermion, gauge boson or Higgs channels. The leptonic decays of heavy quarks and weak bosons produced in the annihilation channels $b\bar{b}$ and W^+W^-, turn the sun into a source of high energy neutrinos. Their energies are in the GeV to TeV range, rather than in the familiar KeV to MeV range from its thermonuclear burning. These neutrinos can be detected in deep underground experiments.

We illustrate the power of neutrino telescopes as dark matter detectors using as an example the search for a 500 GeV WIMP with a mass outside the reach of present accelerator and future LHC experiments. A quantitative estimate of the rate of high energy muons of WIMP origin triggering a detector can be made in 5 easy steps.

Step 1: The halo WIMP flux ϕ_χ.
It is given by their number density and average velocity. The cold dark matter density implied by the observed galactic rotation curves is $\rho_\chi = 0.4\,\mathrm{GeV/cm^3}$. The galactic halo is believed to be an isothermal sphere of WIMPs with average velocity $v_\chi = 300$ km/sec. The number density is then

$$n_\chi = 8 \times 10^{-4} \left[\frac{500\ \mathrm{GeV}}{m_\chi}\right]\ \mathrm{cm}^{-3} \tag{12}$$

and therefore

$$\phi_\chi = n_\chi v_\chi = 2 \times 10^4 \left[\frac{500\ \mathrm{GeV}}{m_\chi}\right]\ \mathrm{cm}^{-2}\,\mathrm{s}^{-1}\,. \tag{13}$$

Step 2: Cross section $\sigma_{\rm sun}$ for the capture of WIMPs by the sun.
The probability that a WIMP is captured is proportional to the number of target hydrogen nuclei in the sun (i.e. the solar mass divided by the nucleon mass) and the WIMP-nucleon scattering cross section. From dimensional analysis $\sigma(\chi N) \sim \left(G_F m_N^2\right)^2/m_Z^2$ which we can envisage as the exchange of a neutral weak boson between the WIMP and a quark in the nucleon. The main point is that the WIMP is known to be weakly interacting. Details are not relevant for our rate estimate. We obtain for the solar capture cross section

$$\begin{aligned}\Sigma_{\rm sun} = n\sigma &= \tfrac{M_{\rm sun}}{m_N}\sigma(\chi N)\\ &= \left[1.2\times10^{57}\right]\left[10^{-41}\,\mathrm{cm}^2\right]\,.\end{aligned} \tag{14}$$

Step 3: Capture rate $N_{\rm cap}$ of WIMPs by the sun.
$N_{\rm cap}$ is determined by the WIMP flux (13) and the sun's capture cross section (14) obtained in the first 2 steps:

$$N_{\rm cap} = \phi_\chi \Sigma_{\rm sun} = 3 \times 10^{20}\,\mathrm{s}^{-1}\,. \tag{15}$$

Step 4: Number of solar neutrinos of dark matter origin.
The sun comes to a steady state where capture and annihilation of WIMPs are in equilibrium. For a 500 GeV WIMP the dominant annihilation rate is into weak bosons

$$\chi\bar{\chi} \to WW \to \mu\nu_\mu\,. \tag{16}$$

Each W produces muon-neutrinos with a branching ratio which is roughly 10%. As we get 2 W's for each capture, the number of neutrinos generated in the sun is

$$N_\nu = \frac{1}{5} N_{\rm cap} \tag{17}$$

and the corresponding neutrino flux at Earth is given by

$$\phi_\nu = \frac{N_\nu}{4\pi d^2} = 2 \times 10^{-8}\,\mathrm{cm}^{-2}\mathrm{s}^{-1}\,, \tag{18}$$

where the distance d is 1 astronomical unit.

Step 5: Event rate in a high energy neutrino telescope. For (16) the W-energy is approximately m_χ and the neutrino energy half that by two-body kinematics. The energy of the detected muon is given by

$$E_\mu \simeq \frac{1}{2} E_\nu \simeq \frac{1}{4} m_\chi \,. \tag{19}$$

In the second step we used the fact that, in this energy range, roughly half of the neutrino energy is transferred to the muon. Simple estimates of the neutrino interaction cross section and the muon range can be obtained as follows

$$\sigma_{\nu \to \mu} = 10^{-38}\,\text{cm}^2 \,\frac{E_\nu}{\text{GeV}} = 2.5 \times 10^{-36}\,\text{cm}^2 \tag{20}$$

and

$$R_\mu = 5\ \text{m}\,\frac{E_\mu}{\text{GeV}} = 625\ \text{m}\,, \tag{21}$$

which is the distance covered by a muon given that it loses 2 MeV for each gram of matter traversed. We have now collected all the information required to compute the number of events in a detector of area $10^6\,\text{m}^2$. For the neutrino flux given by (18) we obtain

$$\begin{aligned} \#\,\text{events/year} &= 10^6 \times \phi_\nu \times \rho_{\text{H}_2\text{O}} \times \sigma_{\nu\to\mu} \times R_\mu \\ &\simeq 1000 \end{aligned} \tag{22}$$

for a 1 km^2 water Cherenkov detector, where R_μ is the muon range and $\phi_\nu \times \rho_{\text{H}_2\text{O}} \times \sigma_{\nu\to\mu}$ is the analog of Eq. (11). Notice that this corresponds to 10 events per year for a $10^4\,\text{m}^2$ telescope, an area typical for the instruments presently under construction.

The above exercise is just meant to illustrate that high energy neutrino telescopes compete with present and future accelerator experiments in the search for dark matter and supersymmetry; see below. The above exercise can be repeated as a function of WIMP mass. The result is shown in Fig. 3 (the two branches as well as the structure in the curves are related to details of supersymmetry. These are, for all practical purposes, irrelevant). Especially for heavier WIMPs the technique is very powerful because underground high energy neutrino detectors have been optimized to be sensitive in the energy region where the neutrino interaction cross section and the range of the muon are large. Also, for high energy neutrinos the muon and neutrino are aligned along a direction pointing back to the sun with good angular resolution. A kilometer-size detector probes WIMP masses up to the TeV-range beyond which they are excluded by cosmological considerations. The technique fails for low masses only for those mass values already excluded by unsuccessful accelerator searches. Competitive direct searches for dark matter will have to deliver detectors reaching better than 0.05 events/kg/day sensitivity.

Particle physics provides us with rather compelling candidates for WIMPs. The Standard Model is not a

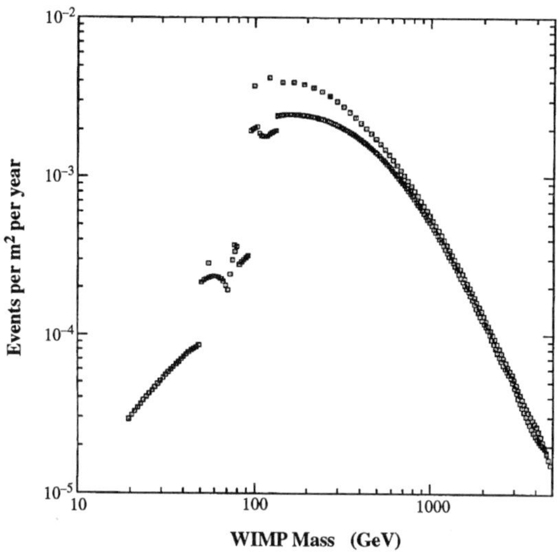

Figure 3

model: its radiative corrections are not under control. A most elegant and economical way to revamp it into a consistent and calculable framework is to make the model supersymmetric. If supersymmetry is indeed Nature's extension of the Standard Model it must produce new phenomena at or below the TeV scale. A very attractive feature of supersymmetry is that it provides cosmology with a natural dark matter candidate in form of a stable lightest supersymmetric particle.[8] This is, excluding the axion, the only candidate because supersymmetry completes the Standard Model all the way to the GUT scale where its forces apparently unify. Because supersymmetry logically completes the Standard Model with no other new physics threshold up to the GUT-scale, it must supply the dark matter. So, if supersymmetry, dark matter and accelerator detectors are on a level playing field. The interpretation of above arguments in the framework of supersymmetry are explicitly stated in Ref. 9.

7 DUMAND et al.: Complementary Technologies

We have presented arguments for doing neutrino astronomy on the scale of 1 kilometer. In order to achieve large effective area it is, unfortunately, necessary to abandon the low MeV thresholds of detectors such as IMB and Kamiokande. One focuses on high energies where: i) neutrino cross sections are large and the muon range is increased; see Equation (11), ii) the angle between the muon and parent neutrino is less than 1 degree and, iii) the atmospheric neutrino background is small.

In a Cherenkov detector the direction of the neutrino is inferred from the muon track which is measured by mapping the associated Cherenkov cone travelling through the detector. The arrival times and amplitudes of the Cherenkov photons, recorded by a grid of optical detectors, are used to reconstruct the direction of the radiating muon; see Fig. 1b. The challenge is well-defined: record the muon direction with sufficient precision (i.e., sufficient to unambiguously separate the much more numerous down-going cosmic ray muons from the up-coming muons of neutrino origin) with a minimum number of optical modules (OM). Critical parameters are the transparancy of the detector medium and the instrument's shielding by water or ice which determines the level of the cosmic ray muon background and the noise rates in the optical modules which will sprinkle a muon trigger with false signals. Sources of noise include radioactive decays such as potassium decay in water, bioluminescence and, inevitably, the dark current of the photomultiplier tube. The experimental advantages and challenges are different for each experiment and, in this sense, they nicely complement one another as engineering projects for a large detector. Each has its own "gimmick" to achieve neutrino detection with a minimum number of OMs:

1. AMANDA uses sterile ice, free of radioactivity;
2. Baikal triggers on pairs of OMs;
3. DUMAND and NESTOR shield their arrays by over 4 km of ocean water.

Detectors under construction will have a nominal effective area of 10^4 m^2. The OMs are deployed like beads on strings separated by 20–60 meters. There are typically 20 OMs per string separated by 10 meters or more. Baikal is presently operating 36 optical modules, 18 pointing up and 18 down, and the South Pole AMANDA experiment started operating 4 strings with a total of 80 optical modules in January 94. All first generation telescopes will eventually consist of roughly 200 OMs.

7.1 Ice Telescopes

AMANDA is operating in deep clear ice. It was first thought that the absorption length of only $\sim$25 m, established in laboratory measurements, would require OMs to be spaced by only 10 to 20 m in polar ice. Operation of the first 4 AMANDA strings and their laser calibration system revealed the surprising result that in-situ natural ice has an extraordinarily long absorption length of more than 250 meters for the blue light to which the PMTs are sensitive. Also, the ice provides a convenient mechanical support for the detector. The immediate advantage is that all electronics can be positioned at the surface. Only optical modules are deployed into the deep ice. Polar ice is a sterile medium with a concentration of radioactive elements reduced by more than 10^{-4} compared to sea or lake water. The low background results in an improved sensitivity which allows for the detection of high energy muons with very simple trigger schemes which are implemented by off-the-shelf electronics. Being positioned under only 1-2 km of ice it is operating in a cosmic ray muon background which is two orders of magnitude larger than deep-ocean detectors such as DUMAND. The challenge is to reject the down-going muon background relative to the up-coming neutrino-induced muons by a factor 10^6. To overcome this problem, a detector has been designed with reduced sensitivity to down-going muons.

Although residual bubbles are found at depths as large as 1 km, their density decreases rapidly with depth. Ice at the South Pole should definitely be bubble-free below 1500 m as it is in all other polar regions. The effect of bubbles on timing of photons has been measured by the laser calibration system deployed along with the OMs. After taking the scattering of the light on bubbles into account reconstruction of muons has been demonstrated by a successful measurement of the characteristic fluxes of cosmic ray muons.

With the discovery of the large absorption length of Cherenkov light in ice, AMANDA has become a superb supernova detector. The passage of a large flux of MeV neutrinos within a few seconds will be detected as an excess of single counting rates in all individual modules. The effective radius of a module for detecting the electrons made by supernova neutrinos is $\sim$8 m, yielding a counting rate of 500 events per module for the duration of a supernova at the center of our Galaxy. This increase over the background counting rate of 1800 Hz in 73 optical modules combines to a 25σ observation for a galactic supernova of the 1987A type. Such a signal is not mimicked by our data acquisition system over the relevant time scale of 100 years. The operating detector also represents an effective volume of over 1 kiloton for the detection of PeV electromagnetic showers initiated by radiating high energy muons or by electron neutrinos, e.g. showers from the production of real intermediate bosons by cosmic electron neutrinos. The showers are triggered by requiring a large amplitude in a cluster of optical modules and may give us a first glimpse at an excess high energy neutrino flux of AGN-origin.

The polar environment turned out to be surprisingly friendly but only allows for restricted access and one-shot deployment of photomultiplier strings. The technology has, however, been satisfactorily demonstrated with the deployment of the first 4 strings. It is clear that the hot water drilling technique can be used to deploy OM's larger than the 8 inch photomultiplier tubes, now used, to any depth in the 3 km deep ice cover. AMANDA will deploy 6 more strings in 1995 at a depth of 1500-1880 meters.

7.2 Water Telescopes

1. The Baikal group has been operating an array of 18(36) Quasar photomultiplier (a Russian-made 15 inch tube) units deployed in April 1993(94). They have reached a record up/down rejection ratio of 2×10^{-4} and, according to Monte Carlo, will reach the 10^{-6} goal to detect neutrinos as soon as the full complement of 200 OMs is deployed. They expect to deploy 97 additional OMs in 1995.

   BAIKAL shares the shallow depth of AMANDA and large background counting rate of tens of kHz from bioluminescence and radioactive decays with DUMAND. The Baikal experiment overcomes these problems with a detector architecture that is conservative in design and highly symmetric. Background noise is suppressed by pairing OMs in the trigger. Half its optical modules are pointing up in order to achieve a uniform acceptance over upper and lower hemispheres. The experiment achieves in this way very large photocathode area per unit volume and is able to scrutinize the downward moving muons with the upward facing OMs in order to assess the mechanisms that lead to misreconstructions.

   A very large number of muon triggers have been collected during the 2 years of operation which were used to measure the the zenith angle distribution of cosmic ray muons in good agreement with Monte Carlo simulations.

2. DUMAND will be positioned under 4.8 km of ocean water, below most biological activity and well shielded from cosmic ray muon backgrounds. Strings of OMs will be moored 25 km from the Island of Hawaii in a octogonal pattern with a string at the center. Deep ocean water is very clear with an absorption length of order 40 m in blue-green wavelengths and an effective scattering length which is expected to be much larger. The background light resulting from radioactive decays, mostly K^{40}, plus some bioluminescence, yields a noise rate of 60 kHz in a single OM.

   The DUMAND group has successfully analysed data on cosmic ray muons from the deployment of a test string. They also briefly operated a string when in December of 95 the power and signal cables from detector to shore were installed along with the junction box for deploying the strings. The group will proceed in two phases with the initial deployment of 3 strings of 24 OMs.

   The DUMAND experiment was conceived 30 years ago and the collaboration has pioneered much of the techniques described in this paper. Deploying the experiment has been challenging because the deep ocean is a difficult location for access and service. Detection equipment must be built to high reliability standards, and the data must be transmitted to the shore station for processing. It has required years to develop the necessary technology and learn to work in an environment foreign to high-energy physics experimentation. The first three strings will serve as a demonstration of the techniques and will allow the colaboration to calibrate natural ocean water as a particle detector.

3. NESTOR is similar to DUMAND, being placed in the deep ocean (the Mediterranean), except for two critical differences. Like Baikal, half of its optical modules point up, half down. The angular response of the detector is tuned to be much more isotropic than either AMANDA or DUMAND, which will give it advantages in, for instance, the study of neutrino oscillations. Secondly, NESTOR will have a higher density of photocathode (in some substantial volume) than the other detectors, and will be able to make local coincidences on lower energy events.

   The NESTOR architecture is imaginative. OMs are deployed on 16 m titanium "umbrellas" which open up when lowered in the water. Each umbrella forms a hexagon with pairs of up-and down-looking OMs fixed at the corners. Hexagons form the "floors" of a tower consisting of 12 hexagonal floors separated by 20–30 meters.

   Measurements at the NESTOR site established an absorption length of about 50 m and revealed tolerable water currents. Muon data has been taken with a scale model of a NESTOR tower. As in all other experiments, cosmic ray muons were reconstructed with a precision in agreement with Monte Carlo expectations.

4. Other detectors have been proposed for near surface lakes or ponds (e.g. GRANDE, LENA, NET, PAN and the Blue Lake Project), but at this time none are in construction.[10] These detectors all would have the great advantage of accessibility and ability for dual use as extensive air shower detectors, but suffer from the 10^{10}–10^{11} down-to-up ratio of muons, and face great civil engineering costs (for water systems and light-tight containers).

8 Sketch of a Kilometer-Size Detector

In summary, there are four major experiments proceeding with construction, each of which has different strengths and faces different challenges. For the construction of a 1 km scale detector one can imagine any of the above detectors being the basic building block for the ultimate 1 km^3 telescope. A kilometer scale detector can e.g. be constructed by deployment of multiple NESTOR towers. Importantly, the characteristics of the detector, including low threshold, are the same as those of the individual towers. The detector does however represent more than the sum of its parts. High-energy muons will be superbly reconstructed as they can produce triggers in 2 or more of the towers spaced by a distance of order 100 m.

The AMANDA group hopes to exploit its large absorption length of blue-green light to built a kilometer-scale detector by using OM-spacings that are a factor 2 larger than other detectors. Such a solution is held hostage to the confirmation that the in-situ effective scattering length is indeed in excess of 100 m as derived from the study of deep ice cores. One of the advantages of building a detector in polar ice is that deployment of OMs is not restricted to a rigid, predesigned frame. Future OMs will be deployed below 1.5 km in order to avoid residual bubbles and improve reconstruction of muon trajectories. Larger spacings between strings, as well as between OMs on a string, will be implemented in order to exploit the newly discovered ~250-m absorption length. AMANDA is scheduled to be completed in January 1996 by the addition of six strings to form a pyramid in which the existing 4 strings form an apex at 0.8 to 1 km. The base, at 1.5 to 1.88 km, will consist of a large pentagon of 5 strings surrounding a central string on a circle of 60-m radius; see Fig. 4. Each string contains 20 OMs with a vertical spacing of 20 m. Although the 6 deep strings contain only 78 OMs, their effective volume of 2.1×10^7 m^3 is, by itself, an order of magnitude larger than the previous 200-OM AMANDA design.[11] The volumes quoted are for 7/3 triggers (at least 7 hits on 3 strings).

The proposed architecture for AMANDA–Mark II is shown in Fig. 4. Ten long strings (40 OMs spased by 20 m for a total length of 780 m) will be added on a radius of 110 m around the deep AMANDA pentagon. Monte Carlo simulation predicts that it will deliver a detector of $\sim$ 1.3 km^3 and effective area of 7×10^5 m^2 for TeV neutrinos. For atmospheric neutrinos the estimated area of 2×10^5 m^2 yields 50,000 events per year. For neutrinos from active galaxies, the effective area increases with neutrino energy. When cuts to improve the determination of muon direction (amounting to a factor 1/3) are included, the effective volume becomes 0.4 km^3 for 1 TeV neutrinos and grows to 1.3 km^3 for 10 TeV neutrinos. For all neutrinos with energy $>$ 1 TeV, the rate will be $\sim$ 500 events per year, using a conservative prediction of the AGN flux. Such a Mark II AMANDA may represent a relatively inexpensive kilometer-scale detector which could grow by either backfilling it with OMs in order to lower its threshold and improve muon reconstruction or, could grow to achieve volumes in excess of a kilometer cube. Hopefully exciting discoveries will dictate the direction of its growth.

We should also point out that several groups are pursuing the construction of large volume (but very high threshold) acoustic and radio detectors in polar ice.

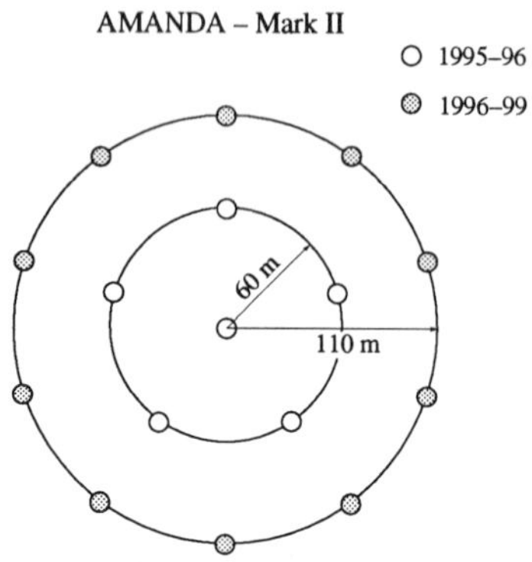

Figure 4

Acknowledgements

I would like to thank my AMANDA collaborators and Tom Gaisser and Todor Stanev, Lori Gray, John Jacobsen, Karl Mannheim, Ricardo Vazquez and Enrique Zas for discussions. This work was supported in part by the University of Wisconsin Research Committee with funds granted by the Wisconsin Alumni Research Foundation, and in part by the U.S. Department of Energy under Grant No. DE-FG02-95ER40896.

References

1. T. K. Gaisser, F. Halzen and T. Stanev, *Physics Reports*, in press.
2. V.S. Berezinsky *et al.*, *Proc. of the Astrophysics of Cosmic Rays*; Elsevier, New York (1991).
3. M. Punch *et al.*, *Nature* **358**, 477–478 (1992).
4. K. Mannheim and P.L. Biermann, *Astron. Astrophys.* **22**, 211 (1989); K. Mannheim and P.L. Biermann, *Astron. Astrophys.* **253**, L21 (1992); K. Mannheim, *Astron. Astrophys.* **269**, 67 (1993).
5. A.P. Szabo and R.J. Protheroe, in *Proc. High Energy Neutrino Astrophysics Workshop* (Univ. of Hawaii, March 1992, eds. V.J. Stenger, J.G. Learned, S. Pakvasa and X. Tata, World Scientific, Singapore).
6. F.W. Stecker, C. Done, M.H. Salamon and P. Sommers, *Phys. Rev. Lett.* **66**, 2697 (1991) and **69**, 2738(E) (1992).
7. F. Halzen, J. E. Jacobsen and E. Zas, *Phys. Rev.* **D49**, 1758 (1994).
8. J. R. Primack, B. Sadoulet, and D. Seckel, *Ann. Rev. Nucl. Part. Sci.* **B38**, 751 (1988).
9. F. Halzen, M. Kamionkowski, and T. Stelzer, *Phys. Rev.* **D45**, 4439 (1992).
10. J.G. Learned, *Proc. of the 13th European Cosmic Ray Symposium*, Geneva 1992, CERN (1992).
11. S. Tilav et al., *Proc. 23rd Inter. Cosmic Ray Conf.*, Calgary, Canada (1993), pp. 4, 461 and 465.

Atmospheric Neutrino oscillations Through the Horizon

Victor J. Stenger

Department of Physics and Astronomy, University of Honolulu, Hawaii 96822, USA

Underground experiments indicate an atmospheric muon neutrino deficit that could be evidence for neutrino oscillations with $\Delta m^2 \simeq 10^2 eV$ and $\sin^2 2\theta \approx 1$. If these parameters are correct, the zenith angular distribution should show a dip just above the horizon in experiments sensitive to neutrino energies of 1 *GeV* and higher. Deep underwater experiments that are sensitive to energies above 10-100 *GeV* will see the main effect below the horizon and by normalizing to their data above the horizon provide an important independent check.

1 Introduction

The deficit in cosmic ray ν_μ interactions compared to ν_e interactions observed in Kamiokande and IMB has a possible interpretation in terms of neutrino oscillations with $\Delta m^2 \approx 10^{-2} eV^2$ and large mixing, $\sin^2 2\theta \approx 1$.[1] Recent results from a Kamiokande analysis of higher energy events in their sample strongly support this interpretation, showing for the first time a zenith angle effect.[2]

In the case where the mixing of only two flavors needs to be considered, the probability for no oscillation is given by

$$P = 1 - \sin^2 2\theta \sin^2(2\pi L/\lambda) \qquad (1)$$

where θ is the mixing angle, L is the path length in km, and λ is the oscillation length. The latter is given by $\lambda = 5E/\Delta m^2$ km where Δm^2 is the neutrino mass square difference in eV^2 and E is the neutrino energy in *GeV*.

Cosmic ray neutrinos provide greater reach than accelerators in the search for neutrino oscillations, as underground experiments dedicated to other purposes have already demonstrated. The "beam" of neutrinos produced by cosmic rays hitting the atmosphere offers both the possibility of lower E, down to a few *GeV*, and greater L, up to 12,000 km.

The current interpretation of observations of a ν_μ deficit rests on calculations that have uncertainties in their estimates of the absolute flux almost as large as the effect, although the ν_μ/ν_e ratio is more reliable. An alternative method that does not depend on absolute fluxes looks at the variation of the cosmic ray neutrino signal with zenith angle.[3]

2 Implications of the Kamiokande Results

The Kamiokande $\cos\theta_z$ distribution for their multi-*GeV* sample, where θ_z is the zenith angle, can be fit to equation (1) with $\Delta m^2 = 10^{-2} eV^2$ and 100 percent mixing by integrating over the atmospheric neutrino spectrum above 1 *GeV*.

In Figs. 1 and 2, the $\cos\theta_z$ distributions for six fixed energies are shown, for the same oscillation parameters.

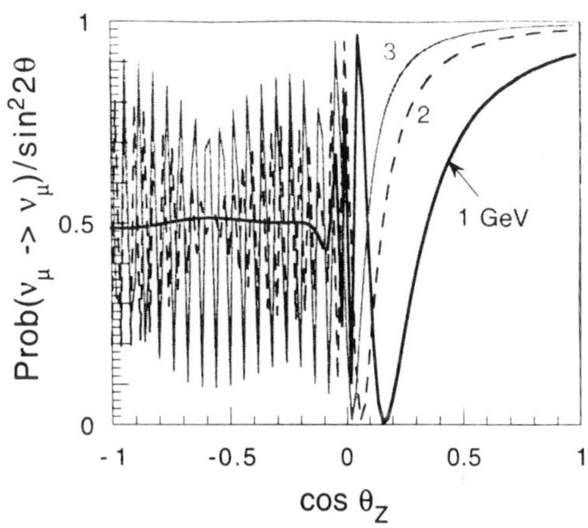

Figure 1: Non-oscillation probability as a function of zenith angle for fixed energies of 1, 2 and 3 *GeV*, for $\Delta m^2 = 0.01\ eV^2$.

These illustrate a very interesting transition effect that could in principle be observed in cosmic ray experiments if these oscillation parameters are true. Below about 3 *GeV*, the oscillation lengths are comparable to the path lengths above the horizon, so this is the important region in this energy range. Below the horizon, the oscillations are very rapid and average out to $P = 0.5$.

Above about 10 *GeV*, the action switches to below the horizon, where very dramatic zenith angle distributions result. To give an ideal of the sensitivity of cosmic ray experiments to neutrino oscillations, Fig. 3 shows the energy at which the path length equals the oscillation length for various Δm^2 as a function of zenith angle.

This suggests that higher energy cosmic rays, in the 10-100 *GeV* range, might provide strong confirming evidence of neutrino oscillations. In Fig. 4, the $\cos\theta_z$ distribution of oscillation probability integrated over the cosmic ray neutrino spectrum with minimum energies of 1 and 100 *GeV* is shown. The lower band corresponds to

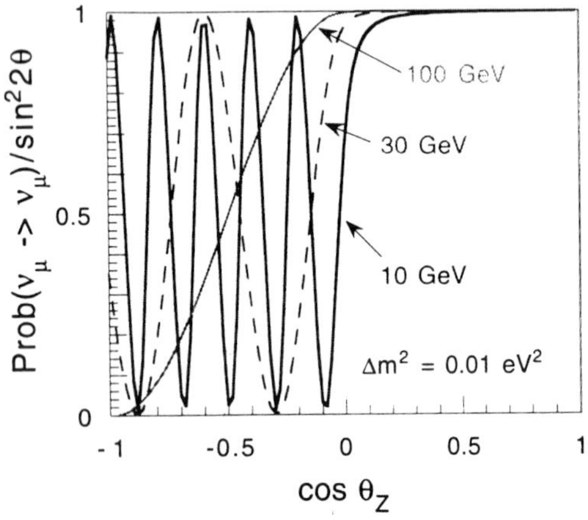

Figure 2: Non-oscillation probability as a function of zenith angle for fixed energies of 10, 30 and 100 GeV, for $\Delta m^2 = 0.01 eV^2$.

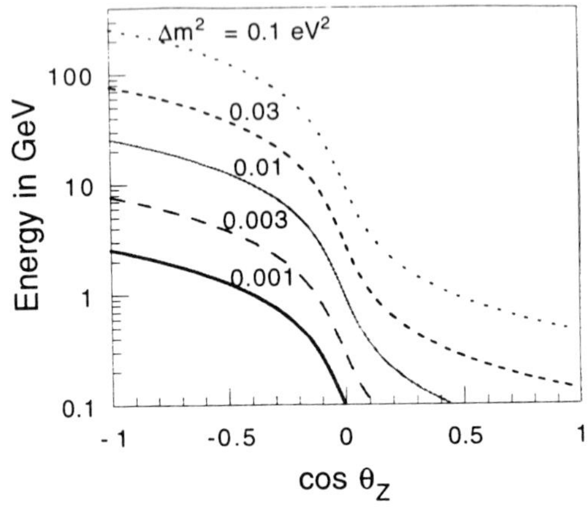

Figure 3: Neutrino energy at which the oscillation length equals the path length as a function of zenith angle, for several values of Δm^2.

the Kamiokande range, while the upper would apply to deep underwater detectors such as DUMAND. Note the interesting dip above the horizon which might be seen with Super Kamiokande.

The higher energy experiments would provide and independent check by looking for a strong zenith dependence in the region below the horizon. Detectors such as DUMAND and NESTOR which are deep enough to look above the horizon for neutrinos can search for a sharp variation in the event rate as the event direction sweeps through the horizon and the neutrino path length in the earth goes through a huge variation. Shallow detectors like Baikal and AMANDA are swamped by cosmic ray muon background at and above the horizon and so are unable to exploit this sensitive technique.

3 Conclusions

New data on the atmospheric neutrino anomaly strongly suggest the neutrino oscillations may be occurring with $\Delta m^2 \approx 10^{-2} eV^2$ and $\sin^2 2\theta \approx 1$. The higher energy data from Kamiokande show a zenith angle effect. If oscillations exist in the parameter range, oscillation effects with atmospheric neutrinos will be most pronounced above the horizon in the energy range above 1 GeV, and below the horizon in the energy range above 100 GeV. Super Kamiokande should see a dip in their $\cos\theta_z$ distribution at 0.2. Underwater experiments should see a fall-off below the horizon. Those that are sufficiently deep to look above the horizon will be able to normalize their data without relying on flux calculations.

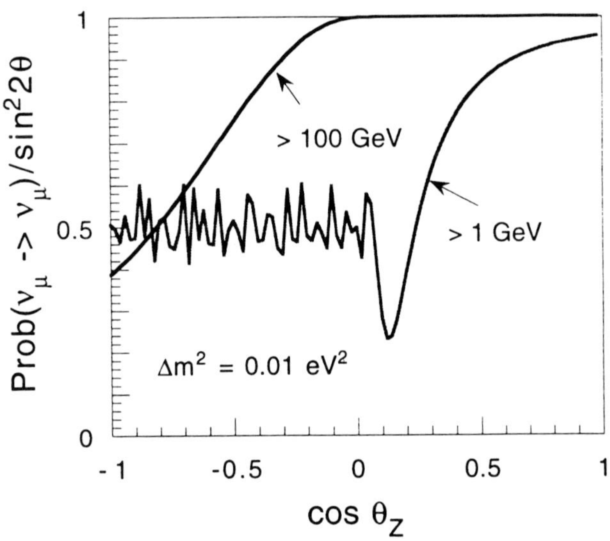

Figure 4: Non-oscillation probability as a function of zenith angle integrated over the cosmic ray neutrino spectrum above minimum energies that correspond to underground and underwater experiments, for $\Delta m^2 = 0.01\ eV^2$.

4 References

1. K. S. Hirata et al., *Phys. Lett.* **B** 205, 416 (1988); D. Casper et al., *Phys. Rev. Lett.***66**, 2561 (1991).
2. Y. Fukuda et al., to be published in *Phys. Lett B.*
3. V. J. Stenger, *Proceedings of the Neutrino Mass Mini-Conference*, Telemark, Wisconsin, 1980, V. Barger and D. Cline ed., p. 174.

Cosmic Rays

THE CHEMICAL COMPOSITION OF COSMIC RAYS IN THE REGION OF THE "KNEE"

T.K. GAISSER[1], K.D. GREEN[2], J. KNAPP[3], N. MAGNUSSEN[4], D. MÜLLER[3], D. NITZ[2], R. PLAGA[5], T. STANEV[1], S. SWORDY[3] and G.B. YODH[6]

[1] *Bartol Research Institute, University of Delaware Newark, DE 19716, U.S.A.*

[2] *Department of Physics, University of Michigan Ann Arbor, MI48109-1120, U.S.A.*

[3] *Enrico Fermi Institute, University of Chicago Chicago, IL 60637, U.S.A.*

[4] *Department of Physics, BUGH Wuppertal 42119 Wuppertal, GERMANY*

[5] *Max-Planck Institute for Physics Foehringer Ring 6, 80805 München, GERMANY*

[6] *Department of Physics, UC-Irvine Irvine, CA 92717, U.S.A.*

ABSTRACT

The flux of cosmic ray particles at earth is measured to decline quickly with increasing particle energy. The spectrum is a power law of nearly constant slope over a wide range from 10^8eV to beyond 10^{20}eV. Measurements above $\sim 10^{10}$eV reflect the population of particles in the interstellar medium, unaffected by solar magnetic fields. The spectral slope steepens significantly in the region around $\sim 3 \times 10^{15}$eV producing a 'knee' in the cosmic ray energy spectrum. The presence of this 'knee' in an otherwise featureless spectrum is made more remarkable by its coincidence with the predicted maximum energies that can be produced by shock acceleration of charged particles in supernova blast waves. The study of the chemical composition in the 'knee' region will provide a firm observational base for these ideas and point the way to the mechanisms of production of cosmic rays at even higher energies – which are at present unknown.

This report concentrates on the motivations for measurements of composition at the 'knee' and discusses the objectives and prospects of direct observations above the atmosphere and indirect techniques using air showers. A description of present activities and possible future directions for these measurements attempts to outline the status of this field. Progress is likely to be rapid since new generations of detectors are expected to link for first time direct and indirect(air shower) composition measurements in the range 10^{14}eV to 10^{15}eV. This provides confidence that the indirect techniques, with their much larger aperture, will extend the composition measurement to energies across and above the 'knee'.

1. Introduction

One hundred TeV is a landmark energy for cosmic-ray physics for at least two reasons, one theoretical and one practical. On the one hand, the theory of diffusive shock acceleration at supernova blast waves has difficulty accounting for the existence of particles above this energy [1]. On the other hand, just where one would like to study the composition in detail to find out what happens above this critical energy, the flux becomes so low that direct experiments with spacecraft or balloon-borne detectors above the atmosphere have so far been virtually impossible. The main thrust of this report is to discuss new opportunities for exploring the energy range above 100 TeV, both with direct experiments and with ground based air shower detectors in ways that achieve greater sensitivity than has been possible so far. This introduction is a brief survey of the astrophysical issues that make the effort of great interest.

It is now well known [2,3,4,5] that shocks driven by supernova remnants have about the right power and produce about the right spectrum to account for the bulk of the galactic cosmic radiation. The mechanism of diffusive shock acceleration involves the collisionless "scattering" of particles back and forth between regions of converging flow at the shock. The scattering centers are magnetic irregularities carried along with the bulk flow of the turbulent, magnetized plasma. Because of the converging flow, each cycle from upstream to downstream and back results in an energy gain. Thus diffusive shock acceleration is a first order Fermi [6] process. This mechanism is observed to energize particles at several sites within the heliosphere where the size and energy scales are relatively low. It is expected to work on larger scales at energetic shocks in the galaxy such as those driven by supernovae.

Diffusive shock acceleration takes time. The acceleration rate depends on the rate at which particles random walk back and forth across the shock, so the finite lifetime of the shock leads to an upper energy cutoff [1]. Understanding acceleration at shocks inside the solar system requires measurement of the spectra up to and beyond the cutoffs characteristic of these processes (e.g. several hundred keV for the Jovian bow shock, 40 MeV for interplanetary waves and several GeV for shocks in solar flares). Similarly, a full understanding of shock acceleration in the galaxy will require measurement through the region of the characteristic cutoff. The goal is to understand the characteristic cutoffs and acceleration sites for galactic cosmic rays.

Figs. 1 shows a summary of data on the high energy cosmic–ray spectrum [39,8,9,10] (the "all-particle" spectrum). Fig. 2 is a compilation of data [11] on the major components of the spectrum as measured with direct observations from balloons and spacecraft. The only measurement that bridges the gap between direct measurements below $\sim 10^{14}$ eV and indirect air shower measurements above that energy is the series of measurements on the Proton Satellites by Grigorov et al. [8].

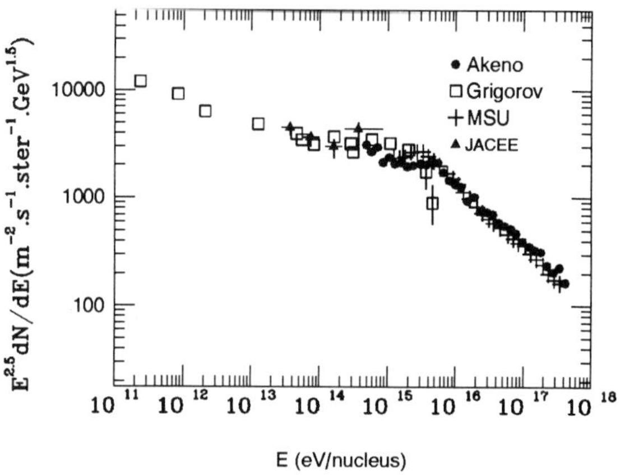

Fig. 1. Summary of data on the high energy cosmic–ray all particle spectrum. The flux is plotted as a function of the total energy per nucleus.

The "knee" of the cosmic-ray spectrum is the feature shown in Fig. 1, where the spectrum steepens at an energy of several thousand TeV. This steepening could be related to the upper cutoff for cosmic-ray acceleration, even though it is somewhat higher than the nominal value of $E_{\max} \sim Z \times 100$ TeV, where Z is the charge of the nucleus being accelerated. This value is obtained for expansion of the blast wave into a generic interstellar medium with a field of 3μGauss. One possibility is that some supernovae, because of their particular environments, could accelerate particles to higher energy. Examples are expansion into the wind of a massive progenitor star [13] and magnetic field configurations predominantly perpendicular to the direction of the supernova blast wave [14].

These examples also suggest that the observed spectrum may not be the result of a single, uniform process, but rather a composite of many individual accelerators. There is some hint of experimental evi-

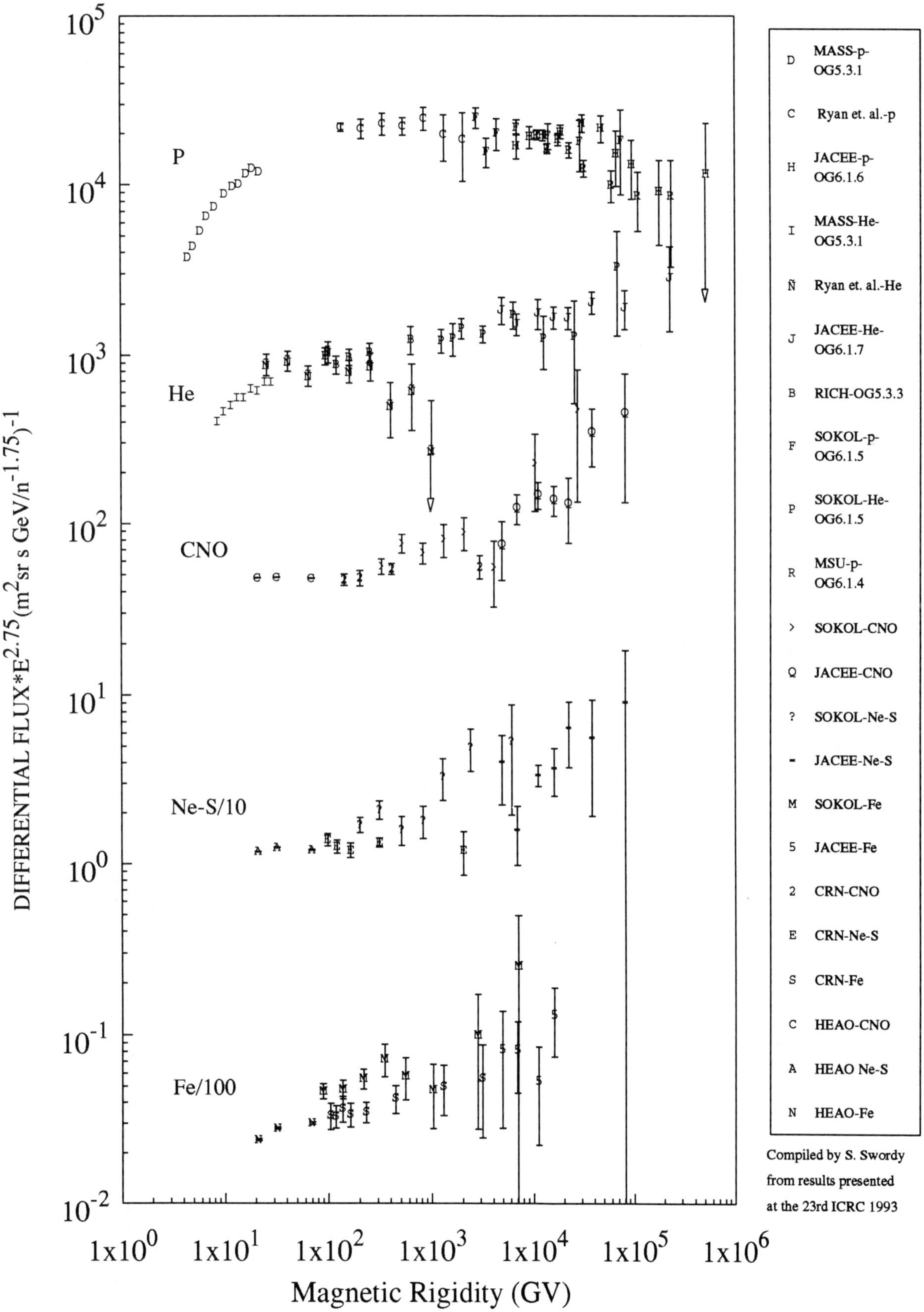

Compiled by S. Swordy from results presented at the 23rd ICRC 1993

dence for this general conclusion from the spectra of the individual components of the primary spectrum, as shown in Fig. 2. In particular, there is a suggestion that hydrogen and helium do not have identical spectra as a function of magnetic rigidity, and it has been pointed out that this could be a reflection of more than one class of supernova at work [15].

There are several difficulties with the picture outlined so far. The natural spectrum associated with diffusive acceleration at strong shocks is a power law with differential energy spectral index $\alpha \lesssim -2$. The difference between the observed value of ≈ -2.7 and the theoretical value is generally explained as the effect of energy-dependent propagation.

If the residence time of cosmic rays in the galaxy τ depends on the particle energy (rigidity) as $\tau \propto E^{-\delta}$, then the observed spectral index is $-(\alpha+\delta)$ with $\delta \lesssim 0.7$. A rough but robust estimate of the expected cosmic-ray anisotropy due to the energy dependence of τ is $A \sim t/(2\tau)$, where ct is the half–thickness of the containment volume. This assumes the cosmic–ray sources are situated in the disk. As the energy increases τ gets smaller and eventually the cosmic rays escape quickly enough so that there should be a significant departure from isotropy. In contrast to a predicted increase in anisotropy expected from this argument, the data [16] seem to show a constant anisotropy between 100 GeV and several hundred TeV, which, at high energies is much smaller than expected if plausible parameter values are used in the above argument. In addition, Colgate (private communication) argues that the high degree of anisotropy may be a problem for the supernova model because of the limited number of supernova remnants that can contribute at any time to the local cosmic–ray flux observed at Earth.

Another problem is the apparent sharpness of the knee; the knee seems to consist of two power laws with different exponents with a very narrow transitional region. There is no theory known to us that would explain this sharpness in a 'natural' way. Possible alternatives to explain these problems include

1. Some experimental results thought to be well-established are actually wrong.
2. The cosmic-ray spectrum looks simpler than it actually is. Indeed, Fig. 2 suggest that there are small difference in the spectra of the different components.
3. The major sources and/or acceleration mechanisms are different from supernova remnants.
4. Some hitherto unknown regulating principle works on galactic cosmic rays to produce a simple power law with a sharp transition.

Clearly, what is needed to resolve these issues is a set of good measurements of the spectra of the major groups of nuclei up to and well beyond the knee of the spectrum. The remainder of this report outlines how this can be achieved.

2. Direct Measurements

As the Earth's atmosphere, roughly 10 proton interaction lengths or 30 radiation lengths deep, does not permit the penetration of primary particles to the ground, it appears obvious that the most detailed information on the cosmic ray composition, on the energy spectra of the individual components, and on the arrival directions, is obtained from *direct measurements* above the atmosphere. Such measurements have shown that the number of elemental species in the cosmic rays is as large as that found in the periodic table and that the relative elemental abundances are similar although not the same as the "universal" abundances in the solar system. Also, electrons, positrons, and antiprotons were discovered in the cosmic rays, and at low energies, even some information on the isotopic composition on cosmic ray nuclei could be derived. Over the entire energy range accessible to direct measurements, the arrival directions of galactic cosmic rays have been found to be isotropic to a very high degree.

Ideally, an instrument performing direct measurements would have to be carried into space, on board a satellite, or a deep space probe. At low energies measurements in space are a necessity, since particles are either deflected by the Earth's magnetic field or, reaching the Earth near the poles, lose their energy rapidly in the stratosphere. However, even space measurements are subject to serious distortions of the energy spectra due to solar modulation - unless a space probe became available that traveled beyond the limits of the solar cavity. Fortunately, solar modulation ceases above rigidities around 10 GV/c and thus is not a problem for the following discussion.

An economic vehicle to perform measurements above *most* of the atmosphere, and particularly suitable for cosmic ray studies at high energies, is the stratospheric balloon. Such balloons float for durations up to several days (or several weeks in circumpolar flights), at altitudes around 40 km, corresponding to a residual atmospheric overburden of less than 5 g/cm^2. For many observations the distortions due to atmospheric interactions are small and can be readily accounted for. The great advantage of balloons is their capability of lifting payload weights of up to $\sim$ 4 tons, i.e., very large detector systems that could be put onto a satellite only at a very great expense. Nevertheless, this weight limit determines the maximum energy up to which direct measurements of cosmic rays can be conducted with balloons. Innovative detector systems with a large area to weight ratio can be expected to perform such measurements up to particle energies around 10^{15}eV. Due to the steeply falling energy spectrum, observations up to much higher energies must remain the domain of *indirect measurements* on the surface of the Earth.

Balloon and space borne instruments often use techniques that determine the Lorentz factor $\gamma = E/mc^2$ of each particle i.e., the energy *per nucleon.* Thus, results from such instruments are usually given in the units GeV/nucleon. Air shower detectors use the atmosphere as a calorimeter and therefore measure the total energy *per particle.* This often leads to confusion, because the results from direct experiments must appropriately shifted and weighted before a comparison with air shower data is made.

2.1. Specific Science Questions

Interactions of Cosmic Rays in the Interstellar Medium. It is commonly assumed that cosmic rays are generated at specific sources throughout the galaxy and that they are subsequently contained for millions of years in the volume of the galaxy (including its halo) before they may escape into extragalactic space. The material available at the acceleration sites consists of protons and of nuclei produced by nucleosynthesis in stars. Thus, the most abundant source nuclei are protons, helium, carbon, oxygen, neon, magnesium, silicon and iron. These are usually referred to as "primary" cosmic rays. When propagating through interstellar gas a variety of interactions can affect these particles. For the energy range of concern here ($\sim 10^{10}$eV to 10^{16}eV) by far the dominant processes are inelastic nuclear collisions with the interstellar gas. For primary protons this leads to antiprotons and to pions which eventually decay into electrons and positrons. Helium produces He^3 and deuterium, and the heavier nuclei fragment into a variety of daughter elements, of which the light elements lithium, beryllium, and boron are studied the most. One usually refers to these as "secondary cosmic rays". From the intensity of these secondary cosmic ray particles one determines some information about the amount of matter encountered by their primary parent particles between injection and observation. Obviously, the cosmic ray flux *arriving* near Earth consists of a mixture of primary and secondary particles. It may be noted that electrons have both a primary and a secondary component while the situation for positrons and antiprotons is not clear at present. Studies of these interstellar reactions require knowledge of the interaction cross–sections. For the spallation cross–sections of heavy nuclei at high energy one usually assumes validity of the geometric approximation which yields cross–sections independent of energy.

Abundances and Spectra of Secondary Cosmic Rays. Observations of the abundances and energy spectra of spallation produced cosmic rays are key to the understanding of the cosmic ray phenomenon. The resolution and precision of direct measurements are essential here. The argument goes as follows:

As has been discussed elsewhere, acceleration of particles in supernova-driven shock fronts appears to be the most likely mechanism for the generation of the bulk of the cosmic-ray flux. However, this mechanism is expected to produce particles with a power law energy spectrum $\sim E^{-\alpha}$ (more exactly, a power law spectrum in rigidity), with a spectral index $\alpha \approx 2.0$. This is *not* what is observed for the cosmic ray flux *arriving* near Earth. While also approximated by a power law, the arriving spectrum is much steeper, with α (observed) ≈ 2.7. This apparent discrepancy finds a natural explanation in measurements of the secondary cosmic ray flux: It has been observed that the secondary to primary intensity ratio decreases with energy, approximately as $E^{-0.6}$. This can be understood in the context of a simple model of propagation as a consequence of the fact that higher energy particles escape from the Galaxy faster than lower energy particles. If one now realizes that the cosmic ray flux in the galaxy must be maintained by equilibrium between the flux produced (presumably by shock acceleration) and the rate of escape from the galaxy ($\propto E^{-0.6}$), it is easy to understand that a source spectrum $E^{-\alpha}$ with $\alpha \approx 2.1$ is required to maintain an ambient flux with a spectrum $\propto E^{-2.7}$. Thus, the energy dependence of particle escape from the galaxy (which was observed before the mechanism of shock acceleration was known in detail), fits in very nicely with the expectation of the shock acceleration model.

Most of the present data on the secondary to primary ratio come from observations of the light cosmic ray nuclei Li, Be, and B, relative to the intensities of their medium-heavy parents C and O (the "L/M ratio"). There is clearly a need to complete the picture through further observations of other secondary particles, for instance the spallation products of iron, and secondaries from the abundant protons and He-nuclei (such as p^-, e^+, He^3). Even more important are measurements of the secondaries at higher energies. At present, the L/M ratio is known up to at most 10^{12}eV per particle, from measurements on the HEAO-3 satellite[17] and on the Space Shuttle (Spacelab-2)[18]. These energies are well below the region of the cosmic ray "knee". There are several models that would predict deviations from the $E^{-0.6}$ energy dependence of the escape path length as one goes to higher energy. To detect such deviations would obviously be essential for any study aimed at understanding the limits of shock acceleration at high energy, i.e., aimed at the most basic understanding of high-energy cosmic rays.

Characteristics of the Cosmic Ray Source. To understand the *origin* of cosmic rays, two observables are essential: The energy spectra of the individual components *at the source* which give evidence for the mechanism of acceleration, and the details of their relative abundances, which characterize the material from which cosmic rays are formed. Thus it is important to derive these quantities from measurements of the arriving cosmic ray flux near Earth. At high energies, this derivation is relatively straight forward. Assuming equilibrium between production and loss of cosmic rays in the galaxy, the simplest propagation model (the "leaky box model" of the galaxy) predicts for the major primary cosmic ray components

$$Q_i(E) \propto N_i(E)\left(\frac{1}{\wedge(E)} + \frac{1}{\wedge_s}\right),$$

where, $Q_i(E)$ refers to the *source spectrum* of component i, $N_i(E)$ is the *observed spectrum*, $\wedge(E)$ the escape path length, and $\wedge_s$ is the mean free path against spallation. (The production rate of component i through spallation of heavier nuclei is small for the most abundant primaries and is ignored here for simplicity.) Thus, the observed spectrum $N_i(E)$ follows the source spectrum $Q_i(E)$ but is modified by the competing action of escape which is energy dependent ($\wedge(E) \propto E^{-0.6}$) and spallation which is taken to be energy independent but depends on the atomic weight of the species. To give numerical examples, $\wedge(E)$ is about $7g/cm^2$ at energies of a few GeV per nucleon, while $\wedge_s$ is 6.4 g/cm^2 for carbon but 2.6 g/cm^2 for iron.

Thus, one expects for all components at sufficiently high energies (where $\wedge(E) < \wedge_s$) the observed spectrum to be steeper than the source spectrum, $N(E) \propto Q(E) \cdot E^{-0.6}$. However, towards lower energies, spallation losses become quite significant for the heavier components like iron, and their spectra should then be flatter than those of the lighter species.

A number of measurements on balloons, HEAO-3, and the Space Shuttle have verified this behavior over the energy range 10 GeV/nucleon to 1000 GeV/nucleon. Quite remarkably, the data show that for all species the cosmic ray source exhibits almost the *same* power law spectrum $Q(E) \propto E^{-\alpha}$ with $\alpha \approx 2.15$. As mentioned above, this spectral shape is a strong circumstantial evidence for the shock acceleration model around supernova remnants. The relative elemental composition at the cosmic ray source appears to be energy–independent. When compared with the generally accepted "local galactic" abundance distribution, one finds, however, that elements with high first-ionization potential are systematically suppressed in the cosmic rays. This feature, while not yet understood in detail, must be seen as a clear signature of the process by which nature selects particles to become cosmic rays. It is astonishing that the same first-ionization potential correlation is also observed in solar flare particles accelerated from the solar surface. Thus, one might speculate that cosmic ray particles originate in stellar winds and flares from pre-supernova stars and are subsequently swept up and accelerated by the supernova shock.

While this discussion might suggest that a great deal is known about the cosmic ray phenomenon, many questions remain. For instance, there is the suggestion that the energy spectrum of silicon does not follow the general trend but is steeper than that of all other components. No plausible explanation of this effect, if it is true, exists. Also, detailed calculations of the shock-acceleration mechanism predict that there should be subtle differences in the energy spectra of nuclei with different charge-to-mass ratio, for instance hydrogen and helium. To study this prediction, a balloon-borne observation (using imaging gas Cherenkov detectors for a very precise energy measurement) is presently being prepared.

Most interesting, however, remains the range beyond a few TeV/nucleon. Is the simple picture described thus far, still correct as we extend the energy range towards the knee ? Indications exist already that this might not be the case: The spectrum of protons appears to become noticeably steeper[10] beyond 10^{14}eV while no such effect is observed for helium at comparable rigidity. Further, there are measurements suggesting that the spectra of the heavier elements, such as iron, above TeV/nucleon energies do not reach the asymptotic shape $\propto E^{-2.7}$ predicted by the simple model. Are we then already witnessing effects due to high-energy limitations of the shock acceleration process, or due to qualitative changes in the processes that govern the leakage of high rigidity particles from the galaxy ? To find an answer to these questions, much more detailed data of the cosmic ray composition, extending to an energy as high as possible, will be required.

2.2. Summary of the Present Observational Status.

At present, measurements of protons and helium nuclei have been reported up to energies around 10^{14}eV. For the heavier nuclei, the spectra of the individual species have been measured to about 1 TeV/nucleon, and to somewhat higher energies for groups of elements (without individual elemental resolution). A compilation of these results is shown in Fig. 2.

Fig. 3 shows the aperture needed for detection of 100 particles above certain energy – the minimum statistics for a rough measurement of the cosmic ray composition. As can be seen, the measurements become more and more statistics limited as the energy increases, precluding an unambiguous interpretation of the trend of the spectra. The data have been obtained with a variety of experimental techniques. They include total absorption calorimeters on balloons and in space, magnetic spectrometers on balloons, calorimetric emulsion chambers on balloons, integrating Gas Cherenkov counters and transition radiation detectors in space, and ring-imaging Cherenkov counters on balloons. We shall discuss, in the next section, the promise of some of these devices for future measurements beyond the TeV/nucleon region.

2.3. Prospects for the Future

It is clear from the preceeding discussion that detailed and statistically significant measurements of the cosmic ray composition and energy spectra beyond the limitations of present data remain a key goal for the future program of direct cosmic ray observations. Such measurements are expected to provide a much clearer picture of acceleration and galactic confinement of cosmic rays than we presently have, and perhaps to identify the onset of new and not hitherto detected sources of high energy particles. Presently available techniques, shown in Fig. 4, permit direct observations to energies of about 10^{15}eV per particle, i.e., one to two orders of magnitude beyond the range of present data.

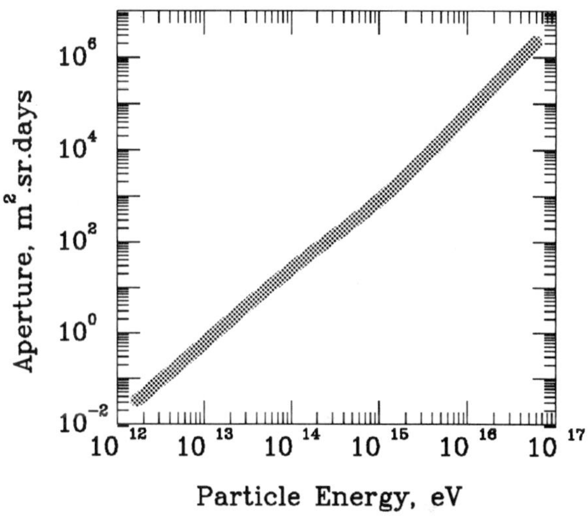

Fig. 3. The aperture required for detection of 100 particles with above energy E.

Extrapolating from existing data, we require exposure factors of about 300 m^2sr days for statistically significant measurements of the proton and helium spectra to 10^{15}eV per particle, and exposure factors of at least 500–1000 m^2sr days for the heavier nuclei. Thus, if detectors of very large area, of the order of 30–100 m^2sr are available, the measurement can be accomplished with a few standard balloon flights yielding a total exposure of at least 10 days. Alternatively, smaller detectors, with geometric factor of a few m^2sr, would require repeated long-duration balloon flights (for instance, circum-polar flights) with a total duration of at least 100 days, or flights in space which may last several years.

These requirements put severe constraints upon the design of the detector. Considering balloon flights first, with a weight constraint of a few tons, it is obvious that total absorption calorimeters are unacceptable (since, for instance, one nuclear interaction length in iron corresponds to a mass of 1.3 tons per m^2 area). Two alternatives have been proposed and successfully tested in previous investigations: "Thin" calorimeters, and devices based on a non-destructive particle detection (such as Gas-Cherenkov counters and transition radiation detectors). Both types of devices have specific strengths and weaknesses but complement each other in several ways. This has been recognized by a NASA study group in 1992 and led to the recommendation that a specific research program be implemented to advance cosmic ray astrophysics through a combined program using both techniques ("GOAL" - Galactic Origin and the Acceleration Limit, a Research Base Enhancement in Cosmic Ray Astrophysics). In the following, we very briefly describe these techniques.

Thin Calorimeters: The incoming particles interact once, and the interaction products are analyzed to determine the particle energy: One either measures, with an *electromagnetic* calorimeter, the total energy of *neutral* pions released in the interaction, or one determines the forward momentum of the *charged* reaction products by measuring their angular distribution. Instruments of this type (see Fig. 4b for a schematic representation) often use passive track detectors such as nuclear emulsions and require extensive scanning efforts after the flight. Geometric factors of a few m^2sr can be realized, and circumpolar long duration balloon flights are required to accumulate long exposure times. Typically, these detectors exhibit moderate charge resolution and are therefore best suited for observations of protons and helium nuclei. Assessment of the energy calibration and of the fluctuations in response depends largely on computer simulations.

Non-destructive Measurements: A non-destructive technique, first used in an instrument on the space shuttle[18], involves the measurement of the particle Lorentz factor γ with transition radiation detectors (Fig. 4a). Here, one observes the X-ray intensity generated when particles traverse radiators of varying dielectric constant. In contrast to thin calorimeters, active electronic devices such as proportional tubes are used for X-ray detection. Such detectors are intrinsically light in weight, requiring only a few grams per cm^2 area. If designed in such a way that no massive pressurized gondola is required, they may have exposure factors of the order of 20 to 100 m^2sr, at a weight of about 2 tons, and therefore can accomplish significant measurements even with a few conventional balloon flights. A limitation of transition radiation detectors is that their response saturates for high Lorentz factors, above about $\gamma \approx 4 \times 10^4$. For the heavier nuclei, this corresponds to energies around 10^{15}eV, i.e., to energies at which simultaneously the particle flux will become so low that statistically significant data can no longer be expected. For light cosmic rays, such as protons and helium, the saturation is achieved much earlier. These devices are also not practical for light cosmic rays , because of large signal fluctuations in these cases. A unique advantage of transition radiation detectors is the fact that their energy response *and* the level of fluctuations can be exactly calibrated on accelerators.

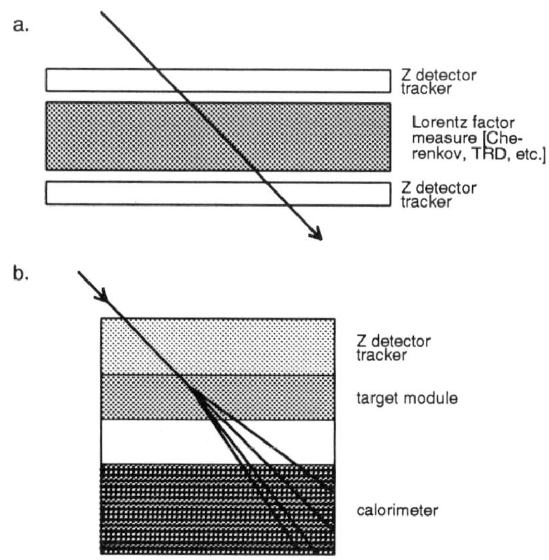

Fig. 4. Schematic representation of non–destructive detectors (a) and thin calorimeters (b).

Other potentially powerful non-destructive techniques have been proposed at this workshop. For instance, one may try to measure the Cherenkov light that is emitted when a cosmic ray particle traverses the upper atmosphere (at a very low index of refraction but a very long path length). Arrays of photomultiplier tubes, suspended by a balloon over an area of $\sim 30m \times 30m$ could have sufficient sensitivity for such a measurement. Another novel approach, would be the use of passive track detector stacks to identify directly produced electron pairs when a particle transverses the detector, and to utilize the energy dependence of this effect for a measurement of the primary particle energy.

For measurements in space, on board a satellite,

total absorption calorimeters still remain a powerful option if a very large mass is permissible. For instance, the Russian Proton Launcher can put payloads of 20 tons or more into orbit. Therefore, the construction of an active thick calorimeter has been proposed as a Russian–American collaboration (Project "Omega"). Such an instrument could also be combined with a transition radiation detector for inter–calibration. After several years of observation, this detector would provide comprehensive measurements of the cosmic ray composition well beyond 10^{15}eV per particle. If one wanted to extend the measurement to even higher energies, one might consider the construction of a very large calorimeter on the surface of the Moon, using natural lunar regolith as material for the calorimeter. While such a device has indeed been proposed, we cannot expect it to become reality within the near future.

3. Air Shower Composition Studies

At energies above 10^{15} eV the cosmic ray flux becomes so low that air showers are the only realistic method for their observation. This method is based on the detection of the particles generated in the cascade initiated by the interaction of the primary cosmic ray in the atmosphere. These secondaries include electrons, photons, nucleons, mesons and muons and can be detected either directly, in particle detectors that work in coincidence, or through the Cherenkov of flourescence light of the charged particles in the atmosphere. The existing shower arrays have a detection threshold of several tens of TeV (total energy per particle), can be built with areas of many square kilometers and have reported events up to energies beyond 10^{20}eV. In summary, in this type of experiment the atmosphere serves as a thick calorimeter sampled by few planes of particle detectors.

Thus while the area of shower arrays is larger by many orders of magnitude than that of detectors above the atmosphere, the evaluation of their response has to rely on Monte Carlo simulations, and the identity of individual primary cosmic rays can in general not be determined. The analysis of the composition is only possible for a statistically significant set of events. In addition, there is a significant degree of uncertainty that comes from the extrapolation of the particle interaction features to an energy range, where they have not been studied in accelerators.

Past studies of the cosmic ray composition with air showers have often reached a point where they could not distinguish between the combination of a cosmic ray composition dominated by heavy nuclei with interaction models obeying Feynman scaling and that of a light composition with strong scaling violation.

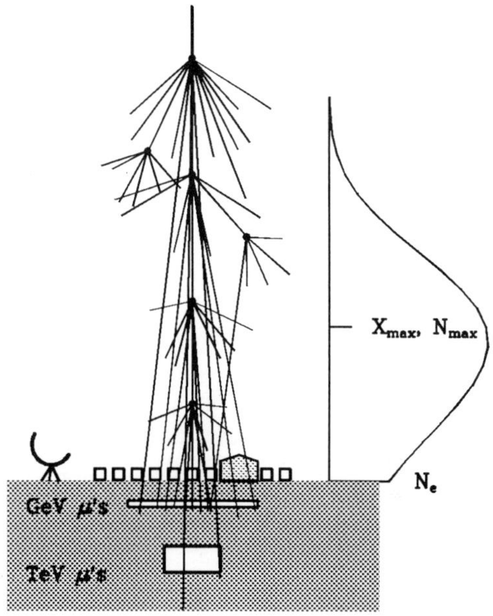

Fig. 5. Air shower development and possible measurements. The skeleton of the air shower – meson production and decay are shown on the lefthand side. The electron size of shower is shown on the righthand side. Shower size is measured by the surface detector, size at maximum – by the Cherenkov dishes. Gev and TeV muons are measured at different depths underground.

Recent advances in understanding the particle interactions at high energy and implementing realistic particle interaction models in the Monte Carlo cascade programs have already led to significant progress in the understanding and analysis of the air shower data. Still further improvements and the performance of massive and detailed cascade and detector simulations are needed for a good measurement of the cosmic ray composition above 10^{15} eV. A schematic picture of the air shower development and of different possible measurements of air shower parameters is shown in Fig. 5.

3.1. Air Shower Sensitivity to the Composition

In general the sensitivity of air showers to the mass of the primary cosmic ray is due to the faster dissipation of the energy in showers initiated by heavy primaries. As in purely electromagnetic showers, the depth of the shower maximum, X_{max}, is proportional to the logarithm of the primary energy per nucleon. A shower initiated by a nucleus of mass A is in the first approximation identical to the superposition of A showers of primary energy E_0/A. This simple fact leads to the following qualitative differences between showers initiated by primary protons (H) and by heavier nuclei:

Shorter longitudinal development:

Shallower X_{max} – showers initiated by primary Fe reach maximum $\sim$ 100 g/cm^2 higher in the atmosphere than H initiated showers.

Smaller shower size (N_e) at ground level. The absorption of the shower after X_{max} leads to smaller surface N_e for the same energy per particle E_0 in showers of heavy primaries.

Softer energy spectrum of the shower hadrons and muons. The multiplicity of the secondary particles depends only logarithmically on the energy. There are more secondary particles with a softer energy spectrum in A–air collisions than in p–air collisions of the same E_0.

Higher number of muons N_μ. Since the showers from heavy nuclei dissipate their energy higher in the atmosphere, i.e. in more tenuous atmosphere, the secondary mesons are more likely to decay and produce muons, rather than interact. The number of muons above a given energy E_μ in Fe showers is higher by a factor of 2–4 than in H showers, the exact value determined by the muon energy and detector altitude.

Wider lateral spread. Showers of heavy nuclei develop earlier in the atmosphere, which has generally an exponential relation between altitude and pressure. Secondary particles, generated with the same transverse momentum, spread wider at the surface observation level. In addition secondaries tend to acquire higher transverse momenta in nucleus–nucleus interactions.

Smaller fluctuations in the shower development. In the simple picture, where a shower of a heavy nucleus is a superposition of A nucleon showers, the fluctuations are smaller than in proton showers by $\sqrt{A}$. When the correct nuclear and particle physics us accounted for[19], the fluctuations increase, but are still significantly smaller than in H induced showers.

3.2. Correlations of the Shower Parameters.

Although the shower parameters depend quite strongly on the primary mass A , it is not easy to distinguish experimentally between showers of different origin. One has to realize that the same measured quantities needed to determine A are used to estimate the unknown primary energy E_0. The steep cosmic ray energy spectrum additionally smears the distinction and the intrinsic fluctuations in the shower development favor the much more numerous lower energy primaries. For these reasons the sensitivity to the composition is greatly increased by experiments that measure several shower parameters simultaneously.

The basic and usually measured shower parameters are:

1. Depth of maximum X_{max} and size N_{max} at maximum. Although for most showers the maximum is achieved well above the observation level, it could be observed by the flourescent or Cherenkov light emitted by the charged shower particles, or by studies of the arrival time of the shower particles at the observation level.

2. Surface size N_e. The surface size is reconstructed from the density profile of the shower particles. It requires a good knowledge (large enough number of detectors) to evaluate the lateral spread of the shower particles. The reconstruction procedure involves determination of the position of the shower core and its zenith and azimuthal angle, which in turn requires a good measurement of the shower front and its curvature.

3. Muon size N_μ, lateral spread, timing and energy spectrum of the shower muons. The muon energy is usually estimated from the amount of overburden that they have to penetrate to reach the detector. GeV muons are detected under a shielding of several meters of rock, TeV muons require detectors covered by kilometers of rock.

4. Number, lateral spread and lateral distribution of the shower hadrons.

Several examples are given below of the improvement of the sensitivity to the composition by simultaneous measurements of several of the shower parameters listed above.

3.3. Strategy of Air Shower Composition Studies

. The best and perhaps unique strategy to make progress in the measurement of the cosmic ray composition with air showers, is to start with shower data in the energy range where composition is known from direct measurements, i.e. around or below 100 TeV. Once the shower behaviour and the detector response are well understood, and all particle physics input to the shower calculation confirmed, the study can be extended to higher energy. It is imperative to recognize the need for a cross–calibration of direct and indirect measurements in the energy region of overlap. This is essential for understanding air shower data which remain the only source of information at the highest energies. Since air shower observations only turn on around 100 TeV, this requires the extension of direct measurements up to 1000 TeV.

4. Operational Air Shower Arrays

In this section we discuss several contemporary air shower arrays that have been constructed for the purpose of the composition measurement and are currently operational or will start operating soon. In the design and construction stage, as well as during the first years of running, all experimental groups want to demonstrate their ability to distinguish between extreme composition models, i.e. H and Fe primaries. After the detectors are well understood and described in Monte Carlo studies, they should be able to correlate the measured experimental parameters with specific, detailed composition models.

4.1. The HEGRA Experiment

The HEGRA experiment is located at the Canary islands at an altitude of 2200 m above sea level (a.s.l.) (790 g/cm^2) and covers an area of $180 \cdot 180\,\mathrm{m}^2$. The array consists of 219 1 m^2 scintillator huts with 15 m grid spacing and a central part with 10 m spacing. Each hut is covered by 5 mm of lead to convert and detect shower photons. A second array of 49 wide angle Cherenkov counters (AIROBICC) covers the same area. Muons are measured by 17 geiger tube 'towers', each 18 m^2 in size. In addition, seven AIROBICC huts were equipped with FADCs to record the pulse spread of the Cherenkov light signal [22].

The separation between primaries of different mass is based on the determination of X_{max}, the light intensity at large radii, and the N_μ/N_e ratio. The composition can also be derived from the fluctuations of various shower parameters.

The lateral distribution of the Cherenkov light density $\rho_\gamma(\mathrm{r})$ can be well described by an exponential $\rho_\gamma(\mathrm{r}) = \mathrm{A} \cdot \exp(-\mathrm{r}/\mathrm{R_l})$ for $30\,\mathrm{m} < \mathrm{r} < 100\,\mathrm{m}$. The so-called light radius $\mathrm{R_l}$ is a parameter sensitive to the depth of the shower maximum X_{max}. According to Monte Carlo studies it can be determined with a precision of around 30 g/cm^2 (0.8 radiation lengths). Due to the large fluctuations, for protons the width of the X_{max} distribution is about three times larger than for iron induced showers. The method can be improved by combining it with the measurement of the lateral density of electrons with the scintillator array and the lateral energy flow as measured by the geiger 'towers'. The lateral distribution of the Cherenkov light for heavy primary particles will be relatively flatter at radii beyond 100 m and will be used in combination with the other parameters.

At distances beyond 150 m from the shower core protons show much broader pulses in time than showers from iron primaries, since the pulse width is a measurement of the height of the shower maximum. Protons again show much larger fluctuations than heavier nuclei. As a test measurement four AIROBICC huts inside the HEGRA array and three additional huts outside up to distances of 250 m from the HEGRA center were equipped with 1 GHz FADCs to measure the shape of the Cherenkov light pulses. The measured arrival times of Cherenkov light for light and heavy nuclei start to separate for distances larger than 80 m from the shower core, the radius of the Cherenkov ring at the HEGRA altitude. Due to their deeper X_{max} the light from primary protons arrives at large radii later than that from iron nuclei. At distances around 100 m the slopes for proton and iron differ by approximately 1.8 ns / 100 m. A measurement resolution of 0.6 ns is achieved from individual AIROBICC huts.

The 17 geiger 'towers' allow a measurement of the number and directions of muons as well as a measurement of the electromagnetic energy density. For a given energy the ratio N_μ/N_e increases with the mass of the primary particle. A study of the angular resolution indicated that that the direction of well contained muons can be determined to better than 1° in one projection (the 'towers' are direction sensitive only in one projection). This good angular resolution allows the calculation of the apparent height of origin of the muons and in this way the determination of X_{max}. The calorimetric measurement of the radial energy flow shows a strong dependence on the primary particle reflecting the different X_{max} distributions and the wider lateral spread in the showers of heavy primaries.

A measurement of the energy dependence of the chemical composition requires the determination of the energy of the primary particle with high precision independent of the nature of the primary. In the past the shower size N_e and/or the number of muons were mostly used. These methods are less reliable because they depend strongly on the primary type and are subject to large fluctuations. The calorimetric aspects of AIROBICC allow one to determine the primary energy nearly independently of the chemical composition after applying corrections for X_{max}, the light yield at larger radii, and for the energy fraction in the shower tail from N_e and N_μ. From Monte Carlo simulations an energy resolution of better than 10% is estimated with less than 10% systematic difference between proton and iron primaries of the same energy per nucleus.

With its unique combination of detectors HEGRA will in the near future use several techniques to determine the chemical composition. The measurement of the Cherenkov light with the AIROBICC array plays an important role for the energy measurement. It has also been proven to be very successful for the γ/hadron separation in the 20 - 100 TeV range[23]. Different measurements of the Cherenkov light with greater than 1 steradian angular acceptance (multiple sampling of the Cherenkov light disk providing a pulse height, arrival time, and pulse shape) combined with scintillator data and information from the geiger 'towers' will provide information to make at least a 4 group (low, medium, high, very high) classification of the chemical composition and its energy dependence.

4.2. The KASKADE Array.

The KASCADE experiment [24] is designed and built with the aim to determine the elemental composition and the flux of the cosmic radiation in the energy range 3×10^{14} eV – 5×10^{16} eV. The simulation code CORSIKA [25] is developed to aid the design and for the analysis of the experimental data. KASCADE stands for KArlsruhe Shower Core and Array DEtector. It is an extensive air shower experiment under construction at Karlsruhe, Germany, at the site of the Nuclear Research Center Karlsruhe. The geographic latitude of KASCADE is 49° N, and it is located 110 m above sea level. The basic configuration of the experiment consists of a large array of detector stations with e/γ and μ detectors and a 320 m^2 central detector for hadron and muon detection.

The array stations enclose 4 conical e/γ detectors, with a layer of 4 *cm* liquid scintillator on the bottom, viewed by a photomultiplier at the top of the cone. Each of these counters has an area of 0.8 m^2. In each second station the four cones carry a converter foil of 3 *mm* of lead to improve the time resolution and increase the signal by conversion of the copious photons in the shower. The photomultiplier and amplifier chain enable a time resolution of better than 1 *ns* and a dynamic range of $\sim$ 10000.

The e/γ detectors are placed on top of a 10 *cm* lead and 4 *cm* iron shield which absorbs the electromagnetic component. Muon detectors are located below the shielding . They consist of 4 plastic scintillators, 3 *cm* thick, of an area of 0.8 m^2 each. The array stations are electronically grouped into 16 clusters of 16 stations each. They cover an area of $200 \times 200 m^2$ with a grid spacing of $\sim$ 13 *m* and provide a coverage of 2% for the e/γ and the array muon detectors. The statistical error of the number of electrons in a shower is dominated by the intrinsic fluctuations of the shower and not by the sampling.

In the center of the array is located the 320 m^2 hadron calorimeter. It is 11 λ_0 thick and consists of iron slabs of varying thickness interleaved with 8 layers of a new type liquid ionization chambers [26] as active elements and a layer of plastic scintillators for fast timing and trigger purposes. The calorimeter is thick enough to absorb even 10 *TeV* hadrons completely and measures the hadron energy with a resolution of better than 30%. The fine spatial segmentation allows the identification of single hadrons and muons. Below the calorimeter 32 multi-wire proportional chambers of a total surface of 300 m^2 measure the location of penetrating muons to better than 1 *cm*. Together with the calorimeter information, the muon tracks can be reconstructed and their arrival direction can be determined with a precision of $\approx 2^o$.

The scintillator trigger of the calorimeter and the muon multiwire proportional chambers is placed below the third absorber layer at a depth of ca. 2.5 λ_0, approximately where the maximum of the hadronic shower development is reached. In total 228 scintillators of 1 m^2 and viewed by photomultipliers are spread over the calorimeter. Their time resolution is better than 2 *ns* allowing a good arrival time determination of muons and hadrons in the calorimeter. On top of the central detector an additional cluster of 16 trigger type scintillators is placed to measure the e/γ density.

By the end of 1994 the physics data taking will begin with 8 array clusters, 100 m^2 calorimeter and all the central muon detectors operational. After the completion of the basic KASCADE detector in 1996, a first upgrade will enlarge the area of muon detectors.

The KASCADE detector will be able to measure simultaneously the electromagnetic component (electron number, lateral distribution and density fluctuations), the muonic component (muon number, lateral distribution, muon tracking) and the hadronic component (hadron number, individual hadron energies and the hadronic energy sum, lateral distribution, density fluctuations, arrival times) with a good precision for each single event.

By means of multivariate and nonparametric analysis methods and the elaborate Monte Carlo program CORSIKA, good and, if possible, model independent estimators for the energy and mass of the primary particle will be formed, using all available observables and their correlations.

At 10^{15} eV we expect a precision of about 30% for the determination of the primary energy, irrespective of the primary mass. The classification in the light (H, He), the medium (CNO), the heavy (Ne–S) and the very heavy (Cr–Fe) group will be possible, the precision depends on the relative abundances and will improve with increasing energy.

4.3. CASA-DICE

The Dual Imaging Cherenkov Experiment (DICE) has been built with the philosophy that any indirect measurement of cosmic ray composition is enormously strengthened by a significant overlap with energies at which the composition is known from direct experiments. This region of overlap can be thought of as a 'calibration' for extending methods of composition determination to higher energies by the indirect technique. The combination of DICE with the existing air shower array is an attempt to measure as many parameters as possible from the same air shower. The importance of multiparameter analysis is stressed elsewhere in this report, but it needs to be emphasized that the nature of air shower fluctuations and the steeply falling cosmic ray spectrum can make attempts to measure composition from only a limited number of air shower features extremely difficult. In particular determined parameters (such as energy and mass estimate) must be overconstrained, to allow an assessment of 'goodness of fit' for each event, this is essential to identify sources of background or unexpected behaviour.

DICE consists of two imaging air Cherenkov detectors located near the center of the CASA/MIA site at Dugway, Utah at an atmospheric depth of 840 g/cm^2. Each DICE detector consists of a 1m diameter mirror and 256 2.5 cm diameter photomultipliers arranged in a hexagonal array in the focal plane. The field of view of the detector is $\sim 15^0$ and each phototube subtends a pixel size of around 1°. The center of the field of view is pointed vertically, so accepted events lie within ~8 degrees of the zenith. The CASA detectors[27] are capable of measuring the shower electron and muon distributions, and core direction and impact position on the ground. For each event the DICE/CASA/MIA combination can therefore measure electron size, muon size, Cherenkov yield, and an estimate of the shower development - including shower max. location - from the image shape. The DICE/CASA/MIA detectors typically trigger down to energies of 10^{14}eV making an overlap with direct measurements[11] possible.

The DICE detectors are separated by 100 m and can make use of showers which have core impact positions up to 200 m from the detectors, resulting in an overall aperture for the experiment of $\sim$2000m^{2}sr. Consequently, a few months of operating DICE during the dark periods should provide $\sim$10,000 m^{2}sr days, enough aperture to study composition up to 10^{16}eV. The important ability to measure the Cherenkov emission image with two independent detectors provides good background rejection. The 1^0 pixels provide a resolution in X_{max} better than 80g/cm^2 for a shower core impact 200 m from DICE. Extensive modelling with the Hillas MOCCA Monte–Carlo program has shown that the shape of the image distribution is also an important tool for mass identification – at the same energy proton showers are much less compact in development than iron nuclei.

This experiment is a collaboration between Kevin Boothby, Dave Kieda, Curtis Larsen, Simon Swordy and the CASA/MIA group with the help of the whole Utah Fly's Eye group who have loaned the detector assemblies and electronics. DICE has been operating in development mode since early 1994. The first well calibrated data set was taken in August 1994 it is expected that some of the first results from this effort will appear in 1995.

4.4. Surface–Underground Coincidence Arrays

Surface–underground coincidence experiments measure the size of the air shower N_e with a surface array and detect the number of high energy muons with a deep underground muon detector. The only difference from a standard shower muon detector is that the energies of the detected muons are high, often in excess of a TeV. Muons of such high energy are generated early in the shower development and are a product of the decay of first or second generation secondary mesons. The idea is not new however both the surface and the underground detectors have to be quite large to combine into a telescope of a sufficient solid angle.

EASTOP–MACRO Coincidence Experiment This coincidence experiment [28] is being performed at the Gran Sasso National Laboratory in Italy. EAS–TOP is at an elevation of 2000 m. a.s.l. Its electromagnetic detector that measures N_e is an array of 35 scintillator modules, each of 10 m^2 area, distributed over an area of $\sim 10^5$ m^2. MACRO is in the underground Gran Sasso laboratory at 963 m a.s.l. with a minimum rock overburden of 3.1 km water equivalent, which corresponds to a minimum muon energy of 1.3 TeV. The lower part of the MACRO detector, used for this work, has a size of 12 $\times$ 72 $\times$ 4.8 m^3. It is equipped with planes of limited streamer tubes with two dimensional digital readout and with liquid scintillator counters. The angular resolution for the muon tracks is 1°, dominated by multiple scattering of the muons in the rock. The aperture of the EASTOP–MACRO coincidence telescope is $\sim$ 30 m^{2}ster. 87.5 live days have been analysed, which contain 1821 coincident events.

The published analysis concentrates on the cosmic ray composition around the knee. Fig. 6 shows the observed correlation between the average number of muons, observed by MACRO, vs. the shower size, measured at the surface. The knee of the cosmic ray spectrum if this figure corresponds to $\log_{10} N_e \simeq 5.75$. The experimental data (stars) are compared to Monte Carlo predictions for three 'pure' compositions – H, He and Fe and to a JACEE–like [10] composition. The dispersion between the Monte Carlo predictions gives an idea of the resolution of the detector. Experimental points tend to agree quite well with the moderately heavy compositions, although there are no predictions for the JACEE–like composition above the knee. The trial compositions are normalized to fit the N_e spectrum observed by the surface array.

There are still, however, problems in the lower energy (N_e) bin, where the experimental points is above all models. Currently the EASTOP–MACRO group is concentrating on a massive computational effort, which should lead to a better understanding of the detector response at relatively low energy, and to a better estimation of the model dependence of the results. A better resolution can be achieved in the future, with higher experimental statistics, when the muon multiplicity distribution is compared with expectations for different N_e bins. EASTOP can also detect coincidences with other large area muon detectors in Gran Sasso, such as LVD.

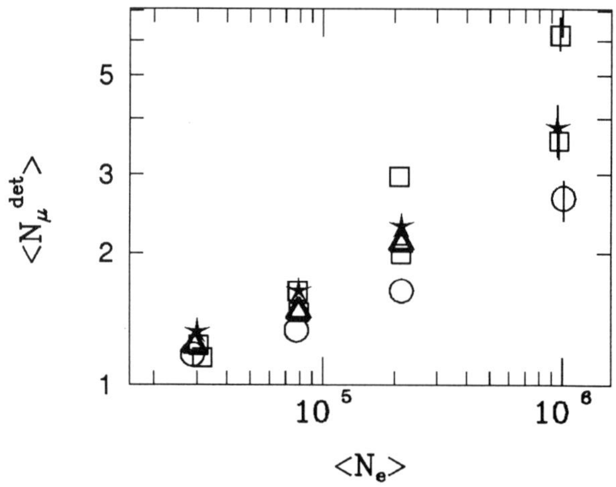

Fig. 6. Average muon multiplicity vs. shower size as measured by EASTOP – MACRO (stars) and from Monte Carlo simulations. Circles indicate pure H flux, diamonds – He and squares – Fe. The heavy triangles correspond to a composition model based on direct measurements below the knee .

SPASE–AMANDA. This is an experiment that is evolving at the South Pole[29]. SPASE is a relatively small air shower array, initially intended for observation of UHE γ–ray sources. It has 24 1 m^2 scintillation counters, enclosing an effective area of 6400 m^2. The atmospheric depth at the South Pole is 700 g/cm^2. The existing AMANDA–4 detector consists of 4 strings, each of 200 m length and containing 20 phototubes, deployed in the ice at a minimum depth of 800 m. They form a sector of a circle of radius 30 m. The linear distance between the centers of the two detectors is 1130 m.

Because AMANDA is bigger than MACRO, the aperture is very similar to that of EASTOP–MACRO. The minimum energy of the muons is lower, $\sim$ 200 GeV, and the corresponding event rate is higher. Fig. 7 shows the angular distribution of coincident events with more than 5 phototubes triggered in AMANDA–4. The actual zenith angle under which SPASE is visible from AMANDA is $38° < \theta < 47°$.

1994 is the first year of operation for the coincidence experiment and the data has not yet been analyzed in terms of cosmic ray composition.

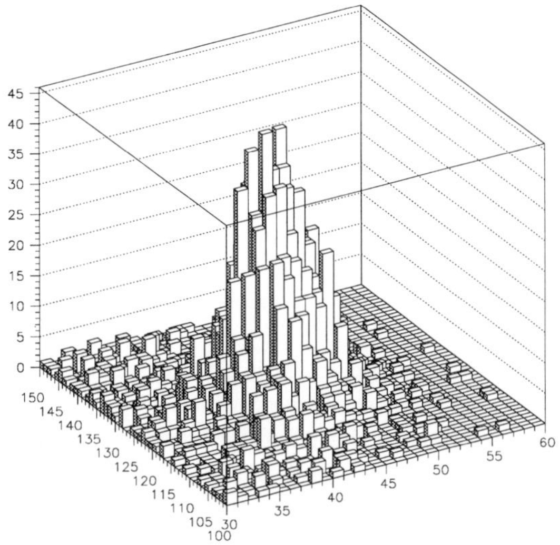

Fig. 7. The arrival directions of the SPASE – AMANDA coincident events confirm that both detectors trigger on the different components of the same air shower .

The capabilities of the coincidence will be significantly improved when the surface array is upgraded by 1996. SPASE–2 will consist of 150 0.2 m^2 scintillation counters, spread over an area of 15 000 m^2. SPASE–2 is designed for surface–underground coincidences and should be able to collect significant statistics above the knee. The shower array will work in coincidence with the existing AMANDA–4 and with the planned AMANDA–6: 6 new strings below 1500 m of ice on a 60 m. radius. The deployment of the new surface array will start in the end of 1994.

5. Future Projects

The following are descriptions of future experiments on indirect measurements of the composition, that are at different stages of planning and design studies.

5.1. *Milagro*

MILAGRO will be the first water-Cherenkov detector to study extensive air showers. It is being built in an existing man-made pond 60 × 80 × 8 m^3, at Fenton Hill at an altitude of 2600 m a.s.l. near Los Alamos, NM. Unlike conventional air shower detectors, which usually sample less than 1% of the particles which reach detector level, MILAGRO will be totally sensitive to the air shower electrons, photons, hadrons, and muons that are contained in the detector.

The threshold of the MILAGRO detector is comparable to atmospheric Cherenkov detectors, however it has several advantages over these optical detectors. MILAGRO is operational 24 hours a day in all weather conditions and it has an open aperture which allows it to view the entire northern sky every day. It will be instrumented with three layers of fast 8 inch phototubes. The Cherenkov light emitted by electrons and positrons, by pairs produced by photons in water, by muons traversing the detector and by hadrons making cascades will be detected by the layer of phototubes under 1.5 meters of water.

A second layer of phototubes, looking upwards, will be sensitive not only to tails of electromagnetic showers but also to hadrons of high energy near the cores of air showers produced by cosmic ray nuclear primaries. A third layer of PMTs, near the bottom of the reservoir, will count muons in the shower. These tubes are shielded from downward coming Cherenkov light by opaque barrier and each tube will have its own 3×3×1 m^3 diffusing muon cell. The muon layer will provide excellent rejection of hadronic showers for the purposes of γ–ray astronomy and a good muon measurement in composition applications at energies above few TeV.

With a trigger requirement of 15 tubes the trigger rate is expected to be about 3 KHz. The trigger efficiency for low energy protons is at least an order of magnitude smaller than that for gamma rays at energies below 300 GeV. The arrival direction is determined by timing. Because of the large number of particles detected per shower, the angular resolution will be $\sim 0.4°$ at 10 TeV, a considerable improvement over present values. For large showers the angular resolution improves to 0.25°.

5.2. *Extending Air Flourescence to Lower Energy*

The Fly's Eye results [30] showing a possible change in slope in the average depth of shower maximum (X_{max}) at 0.5 EeV (EeV = 10^{18} eV) are especially interesting as there also appears to be a break in the spectral slope at the same energy. This spectral break is present in the Haverah Park dataset as well as the Fly's Eye stereo dataset, and the AGASA data do not contradict the observed "dip" structure in the spectrum. The most straightforward interpretation of the dataset is that cosmic rays in the energy range of $10^{16} - 10^{18}eV$ are predominantly galactic and heavy ("iron") whereas the highest energy cosmic rays $E > 10^{18.5}eV$ are predominantly extragalactic and light ("protons").

The data supporting the proposed model are consistent and the reasoning convincing, but it is important to remember that these measurements are at energies above existing accelerator data, and the conclusions depend strongly on Monte Carlo simulations. In addition, the break at $0.5EeV$ is close enough to the Fly's Eye threshold energy that one must asked if the $\langle X_{max} \rangle$ break is not some hidden threshold effect (however unlikely). Both criticisms can be addressed by designing and operating experiments for the $10^{16} - 10^{18}eV$ energy region.

The High Resolution Fly's Eye (HiRes)[31] and the Giant Array are designed to address the physics issues above $10^{19}eV$. Neither of these experiments will cover the region of a change below $10^{17.5}eV$. The new Michigan proposal for a ground detector to run in coincidence with one of the HiRes sites (GEMA) is one possibility. Here we consider the use of a stereo air fluorescence detector. A question of interest is how low in energy can such a detector operate?

Experiments to extend air fluorescence measurements to lower energies are already underway. Since April, 1993, the previously defunct Fly's Eye II was revived and now runs in coincidence with the HiRes prototype. This effort has been a collaboration between Brian Fick (University of Chicago), Kevin Green, and the HiRes Collaboration. The Fly's Eye II detector acts primarily as a geometric constrainer necessary for robust event reconstruction. The event yield is $\approx 200 - 300$/month with threshold near $10^{16.5}eV$, mean energy near $10^{17}eV$, and the data extend up to $10^{18}eV$. This dataset should confirm the break in slope at $10^{17.5}eV$.

Extending air fluorescence to lower energies will require a more sensitive stereo system than is presently in operation. The "simplest" upgrade in terms of additional development and investment would consist of building a HiRes device at the Fly's Eye II site looking back towards the present HiRes device. The incremental capital cost would be \$ 70 K per fully equipped mirror (including pmts and data acquisition electronics) and 12 mirrors would give the necessary aperture. The stereo trigger threshold for two HiRes systems separated by $3.4km$ would be near $10^{16}eV$ as a vertical shower at $1.8km$ should trigger a single site at this energy.

Further reductions in threshold are clearly possible. Signal to noise arguments suggest that a $10^{15.5}eV$ shower could be detected via air fluorescence by a Fly's Eye stereo device if each site consisted of $10m$ diameter mirrors with 1° tube apertures (about the size of the old Fly's Eye tubes for f/1 mirrors) at a distance of $3km$ to the track. While these mirrors would be large, the Whipple collaboration has adequately demonstrated that maintaining and operating them is well within current experience–especially considering the proposal here would keep the mirrors fixed. To extend to such low energy, a site higher in altitude than Dugway, Utah would be desirable.

The principle uncertainty is the effect of Cherenkov light as these showers must be viewed at narrow emission angles. Thus, for the lowest energies, the system could work similarly to DICE albeit at a higher energy. At any rate, stereo viewing will be essential for the Cherenkov "systematics" to be fully understood.

Finally, there is potential for imaging the showers both along (longitudinally) and across (laterally) the shower axis. Shower imaging is only possible in the high signal to noise regime. An upgraded Fly's Eye II detector would provide a large enough aperture to image a handful of $10^{18}eV$ showers each year both longitudinally and laterally at the $40m$ length scale. The possibility of lateral imaging has never been pursued. It could be considered a study of the shower "age" parameter as measured by the lateral extent of the shower as a function of the shower "age" parameter as measured by the depth relative to the shower maximum. Two sites would be essential for a convincing detection of a "bumpy" shower profile.

5.3. GEMA

The proposed GEMA (Gamma, Electron, and Muon Array) array is designed expressly to measure composition up to 10^{18} eV. It consists of 3 concentric graded sub-arrays centered on CASA/MIA [32], each containing 19 stations arranged on an equilateral triangular grid. The spacings of the 3 sub-arrays are 250 m, 500 m, and 1000 m. Shared stations reduce the total number of stations in the array to 43. Each station contains several scintillator–lead–scintillator sandwiches in order to tag γ's, e's and μ's in a shower. The surface area of each station is $10m^2$. The 2500 m^2 MIA consists of scintillation counters, located in 16 "patches" of 64 one m^2

counters each, buried 3 m deep. Each counter is viewed by a single 5" PMT. The arrival time of each pulse is recorded, but no pulse height information is available. The sensitivity of this combination to composition is discussed in the next section.

6. Establishment of Measurement Quality Requirements

The derivation of the cosmic ray composition from air shower data depends heavily on models of the development of atmospheric cascades and of the experimental ability to reconstruct the shower parameters from the measured quantities. This section gives examples of computer analyses aimed at the the establishment of the sensitivity of air shower arrays to the cosmic ray composition. Since the detector simulation is a very important part of the interpretation of the air shower measurements, these studies are by necessity related to the design of particular air shower arrays.

6.1. Simulation of MILAGRO-type Detectors.

We investigate the capabilities of air shower experiments which can measure parameters related to shower size N_e and muon size N_μ at observation level and shower size at maximum N_{max} on event by event basis. We concentrate on MILAGRO type water Cherenkov detectors, that could have an energy threshold as low as 10 TeV

Air showers were generated using Showersim[34] for four effective primary species: H, He, CNO and Fe nuclei. The primaries were sampled from spectra with differential spectral indices - 2.75, - 2.75, - 2.64 and - 2.55 for H, He, CNO and Fe nuclei respectively with a minimum energy of 10 TeV. They were incident upon the atmosphere isotropically at zenith angles up to 60 degrees. These showers had cores hitting within ± 100 m in x and y around a proto-type Milagro water Cherenkov detector with a shower layer of 5000 m^2 and a muon layer of 1500 m^2. The PMTs were placed on a 3 meter grid. The shower layer was at 1.5 m depth and the muon layer(optically isolated) at a depth of 7 meters. In this simulation the number of electrons hitting top layer and number of muons in bottom layer were recorded. Also recorded were the total number of muons above 1 GeV, N_e and N_{max}. N_{max} would have to be determined by judiciously placed wide field of view PMTs observing the Cherenkov light associated with the shower trigger. The observation altitude was chosen to be 700 gm/cm^2. The sample used in the analysis is for energy per nucleus greater than 20 TeV of : 148 K H, 90 K He, 60 K CNO and 30 K iron nuclei. This corresponds to 45 % H, 27 % He, 18 % CNO and 9 % Fe consistent with the numbers corresponding to direct measurements above a few TeV.

In studying whether a given set of air shower measurements has enough sensitivity to distinguish different scenarios, two weighting schemes were chosen: (1) The composition changes according to the spectral indeces given above and (2) H with a bend around 40 Tev as suggested by JACEE and a subsequent rigidity dependent cut off at 500 GV. This is a modification of the Maryland composition model[35].

The events for which we would have information of size at shower maximum limits us to only dark nights. Without accounting for the trigger efficiency one expects an event rate of approximately 44 Hz over an area of 4×10^4 m^2. Trigger efficiency reduces the rate by a factor of 10. In one year of operation some 10 million events are collected.

The knowledge of size at maximum (N_{max}) will give a measure of primary energy. The conversion from N_{max} to energy is atomic mass (A) dependent. To analyze the Monte Carlo data we cut $\log_{10} N_{max}$ into bins of width 0.2 bins starting at 4.2, which is clearly above any threshold effects. The bin between 4.4 and 4.6 corresponds to 60 TeV H and 83 TeV iron and the one between 5.6 and 7.0 explores energies of 908 TeV for protons and 1300 TeV for iron nuclei. We call these Case I and Case II respectively.

Two cases are considered. The first one assumes that the air shower experiment is capable of determining shower size(N_e) and muon size(N_μ) event by event. The second corresponds to a MILAGRO type instrument which measures a certain number of shower particles(Σ) and a certain number of muons (Σ_μ).

Experiments which measure N_e and N_μ have to have the capability of measuring showers with N_e of about 1000 electrons and N_μ of the order of 100 or more. This requires large area electron and muon detectors to be able to sample enough particles to measure the lateral distribution event by event.

For fixed N_{max}, the number of muons and the number of electrons at observation level have somewhat different dependence on primary energy. On a scatter plot of $\log_{10}(N_e/N_\mu)$ vs. $\log_{10}(N_\mu)$ one can see almost complete separation of protons and iron initiated showers. The individual N_e/N_μ distributions for all species overlap, but the shape of the experimental distributions in this variable could be fitted to determine the relative fractions of each component at this energy. One has to fit simultaneously the shape of the N_e/N_μ distribution and the total trigger rate. At low energies, Case I, where the composition is known from direct measurements one should be able to obtain a good agreement with both the absolute rate and the shape of this curve.

At higher energy (Case II) the predictions for the shape of the N_e/N_μ distribution depend on whether protons have or not a bend at 40 Tev. The data should be able to distinguish between the two hypothesis after the simulation has been normalized at the lower energies.

In most experiments it is difficult to measure the true muon size and the true N_e for primaries from 20 TeV to 2000 TeV. In Milagro type detectors one has good efficiency for recording events over this energy range but one does not have enough information to get the total number of muons and electrons. One must use the measured quantities.

Because of the small size of the detector, fluctuations in the detected N_μ can cause problems in Milagro type experiments. One can choose variables similar to N_e/N_μ and N_μ. We use $y = \log_{10}(\Sigma/\Sigma_\mu)$ and $x = \log_{10}(\Sigma_\mu)$. Here Σ represents the number of PMTs hit in the shower layer and Σ_μ is the number of PMts hit in the muon layer. On a scatter plot of the two variables (with two additional cuts ($\Sigma > 400$ and $\Sigma_\mu > 5$), which reduce fluctuations in detector response) a good separation is seen between H and Fe showers. One can enhance the resolving power of the experiment by projecting the two dimensional distribution on an axis t in the xy plane, which separates the populated from the unpopulated area of the plane. Fig. 8 shows separately all four components and their sum, projected on the t axis for Cases I and II. This information should be enough for a composition fit to the magnitude and shape of the detected distribution. It is interesting that in Case I the bend for protons at 40 Tev changes the detected fluxes only slightly. For Case II the separation between the species is better. A cutoff of the proton spectrum at 40 TeV would shift the peak of the distribution (Fig. 8b) and change its shape.

This study suggests that in order to determine composition from ground based arrays it is essential to measure shower size at maximum and then measure as fully as possible the electromagnetic and muonic sizes. Size at maximum is the handle on primary energy. It is furthermore important to find by simulation an optimum variable(s) for maximum separation between species for each experimental arrangement considered. In order to intercompare different experimental techniques it is important to present a resolution function in log(A) for the variable(or variables) considered and curves which clearly indicate whether a change in composition is measurable.

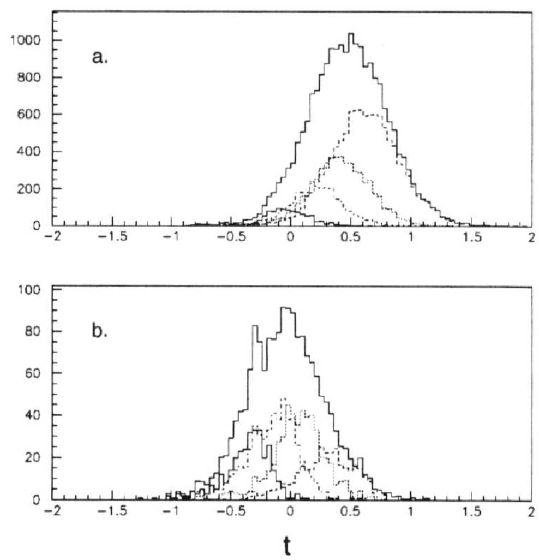

Fig. 8. t–distributions calculated for MILAGRO at lower (case I – a.) and high energy (case II – b.) The dash and dot–dash histograms are for the H, He, CNO and Fe components from left to right. Solid histograms show the sum of all components.

6.2. *Multiparameter analysis of GEMA*

Recent work has emphasized that, because of shower fluctuations, it may not be possible to determine composition adequately when only two ground parameters are used, especially if the primary energy is not accurately measured[36]. A.M. Hillas has suggested that multiparameter classification of showers is a more promising approach[37] The availability of powerful computers now makes possible nearly complete simulation of air showers by Monte Carlo techniques. We describe here Monte Carlo simulations and subsequent multiparameter analysis to examine how well one can extract composition from ground parameters.

Two surface ground array configurations have

been studied. The first is based upon the CASA detector[32] and the second is the proposed surface array denoted GEMA. Both surface arrays in this study make use of the MIA underground muon array[33]. Simulated coincident events between GEMA and the HiRes Fly's Eye[31,30] were also investigated.

Proton and iron showers were simulated using the Monte Carlo program MOCCA[38] This simulation uses a "splitting model" technique to simulate hadronic interactions, somewhat like quark-gluon string models. It accurately reproduces the lateral distributions measured by the Akeno group[39] up to 10^{19}eV, the highest energies that they report. MOCCA generated Feynman x_F distributions agree well with accelerator data at lower energy, although nucleon–nucleon collisions are treated only approximately. All particles are followed down to a fixed energy. Beyond that point, the showers are "thinned" and only a fraction of the particles is followed. Because the thinning increases fluctuations in the showers, we may be making a pessimistic estimate of the arrays' performance. Each shower is randomly placed within the geometrical acceptance of each array 10 times.

The "CASA" array simulation used 1089 $1.44m^2$ stations arranged on a 15m spacing square grid. 154 proton and 154 iron showers were generated above 300 TeV. The showers were thinned at 300 MeV. The stations were considered to be sensitive to electrons and muons. Photons at the detector level were ignored. The detector simulation did not include noise in the counters or limited dynamic range of pulse height measurements. Both will degrade the performance of a real instrument.

For GEMA 600 proton and 600 iron showers were generated above 10^{16}eV. The showers were thinned at 100 GeV. A station is considered to be "alerted" if ≥ 4 particles (μ, e, γ) are detected. Three adjacent alerted stations constitute a trigger. Experimental limitations are included in the detector modeling. Particles (γ's, e's, μ's) are classified according to their measured type, not their actual type. The background of particles uncorrelated with the $> 10^{16}$ eV showers is added to the detectors. The 50% efficiency threshold for the 250 m spacing sub-array is $\approx 3 \times 10^{15} eV$, for 500 m array it is $\approx 3 \times 10^{16} eV$, and for 1000 m array it is $\approx 3 \times 10^{17} eV$. The energies at which the sub-arrays begin to contribute effectively to composition measurement will be higher.

Each patch of MIA muon counters is considered as making a single density measurement, which is statistically corrected for multiple muons hitting a single counter. Random hits are included.

The HiRes Fly's Eye measurements are simulated by fluctuating the measured shower energy according to a gaussian distribution with a 20% standard deviation and fluctuating the measured shower maximum (X_{max}) according to a gaussian distribution with a $25 g/cm^2$ standard deviation. This approximates the resolution expected of a single HiRes Fly's Eye operating in conjunction with a ground array.

The detected γ (GEMA only), e, and μ particle densities in each station and the median muon arrival time T_μ with respect to the first muon detected by each station (GEMA only) are simultaneously fit to determine the shower core position and the density and slope of each measured component. The shower "age" (γ or e density slope + 2), ρ_{mu}, T_{med}, and the slope of T_{med} all show some statistical separation between proton and iron showers. The electron, photon, and muon densities provide a measure of the primary shower energy.

The separate distributions, however, do not address the correlations between the parameters or the complications present with a spectrum of energies. In order to do that we require a prescription for extracting a composition measure from multiple parameters. One relatively simple method is the "K nearest neighbor" (KNN) technique[41]. This method has a solid mathematical basis in that rigorous bounds on the misclassification rate can be derived. In this technique we compare measured shower parameters with simulated showers and compute the "distance" in the multi–dimensional space between the detected shower and each simulated shower. The dimensions are "normalized" relative to one another using the mean covariance matrix. This particular distance measure is called the Mahalanobis distance.

We divide the simulated showers into 2 classes. The first half is used as the simulation "training" set. The second half is used to simulate detected showers. After determining the distance between a detected shower and each shower in the training sample, we examine the K nearest neighbors to note what fraction of these are iron. This fraction (F_I) gives us an empirical measure of the composition. K is chosen to be 5 for the simulation results presented here.

Let us define the "mis-identification fraction" as

$$f = \frac{M_p + M_i}{N_p + N_i}$$

where N_p and N_i are the number of proton and iron showers respectively in the sample, and M_p and M_i are the number of proton and iron showers which are incorrectly identified. Those showers which have $F_I < 0.5$ are flagged as protons. The remainder are flagged as iron. Showers with $X_{max} < 670 gm/cm^2$ are identified as iron. The remainder are identified as protons. Fig. 9 show the misidentification fraction as a function of the shower energy for a GEMA–HiRes combination.

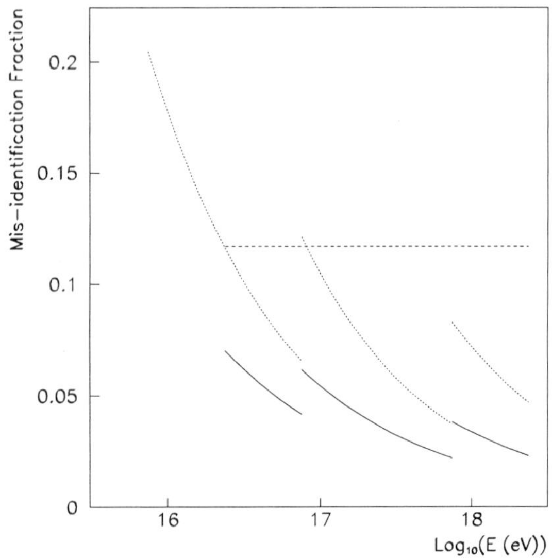

Fig. 9. Discrimination between proton and iron showers in GEMA as a function of primary particle energy. The mis–identification fraction f is plotted versus $log_{10}(E)$). The solid lines display f for HiRes coincident data, and the dotted lines – when only the ground array information is used. The breaks indicate a transition from one sub-array to the next larger. The dashed line indicates the result of using only the HiRes Fly's Eye, determining composition from X_{max} and E.

This is a report on work in progress. The conclusions drawn from that simulation must be tempered by the limited simulation statistics and the overly simplistic detector modeling. We are pursuing a more detailed CASA/MIA study. In addition, model dependencies in the simulations need to be investigated. We plan to generate additional showers using the SIBYLL interaction model[40]. It would be fruitful to also compare with CORSIKA[25] showers. Finally, this study considers only protons and iron. Inclusion of He and CNO showers would be necessary to complete this investigation.

7. Conclusions

We have argued that decisive progress in the understanding of the cosmic ray origin can be expected from direct measurements of cosmic rays beyond the range of present observations, and that well proven experimental techniques exist to accomplish this goal. It is important that such measurements are capable of identifying individual cosmic ray species including both the primary nuclei from protons to iron and the less abundant secondaries such as lithium, beryllium, and boron. Presently available techniques permit measurements up to energies around 10^{15}eV per particle, i.e. one to two orders of magnitude beyond the range of existing data, if the capabilities of modern high–altitude balloons are fully utilized. It should be pointed out that modern balloon-borne instruments are complex and technologically advanced devices.

At higher energies air showers are the only approach to the measurement of the cosmic ray composition. Because of the intrinsic fluctuations in the shower development, the energy and mass of the primary particle cannot be derived on an event by event basis. All results have to be derived from large enough statistical ensembles. The composition measurement will consist of the identification of four or five groups of nuclei and studies of their intensity as a function of the primary energy per nucleus.

Air shower studies can provide a measure of the major groups of nuclei over a wide energy range. This long lever arm will greatly enhance our ability to distinguish between the shape of the source spectra and propagation effects. All possible rigidity dependent effects, related to the particle acceleration at the sources or to propagation, affect both the magnitude of the all particle cosmic ray flux and its mass composition. Thus the knowledge of the energy behaviour of the cosmic ray spectrum and composition above 10^{15} eV, even in several mass groups, will help to establish the acceleration limits and deconvolve the contribution of possible different types of cosmic ray sources.

There are several basic conditions for the success of the composition measurement with air shower data. They have to start with a detailed analysis of showers in the energy range $10^{14} - 10^{15}$ eV overlapping a region where the cosmic ray composition is well known from direct measurements. This overlap will help enormously in the calibration of the air shower technique. The simulation procedures, which are of crucial importance for the shower analysis, will be normalized to fit the measured quantities using the known composition. This will eliminate the duality between composition and particle interactions which has hindered the past composition studies with air showers.

The second requirement is that air shower detectors measure simultaneously as many as possible of the basic shower parameters. Single shower parameters are not very helpful in the derivation of the shower energy and primary mass. As shown above in two examples it is the multi–parameter shower detection and analysis which can improve the resolution of the air shower technique and lead to successful composition measurements.

This is already a trend in all air shower experiments that are under construction or in the proposal stage. The measurement of the shower size at maximum with various techniques is identified as the best model independent way to estimate the shower energy. Surface and deep underground measurements of the shower size and muon number give the basic composition sensitive information about the degree of energy dissipation and the absorption of the shower. The measurement of the lateral and angular spread of the shower particles, the 'image' of the shower, also carry complementary information on the primary mass.

Another area of necessary improvement concerns the methods for shower simulations and data analysis. Contemporary computers allow for simulations that calculate directly the measured parameters. Particle interaction models used should be at the level of the current experimental technique and describe properly all available accelerator data.

There is a noticeable progress in all areas during the past several years. The convergence of the direct and indirect techniques and the many efforts directed to multiparameter measurements and analysis of air shower data are expected to reveal the nature of the knee during the next several years.

8. Acknowledgements

The authors of this report have attempted to summarize the discussions held during the meetings of the subgroup and have greatly benefited from the contributions of all participants. We would like also to thank the organizers of Snowmass'94 and all contributing organizations for the possibility to participate in this important study of the current state and the future of the field.

9. References

1. P.O. Lagage & C.J. Cesarsky, Astron. Astrophys. 118 (1983) 223 and 125 (1983) 249.
2. G.F. Krymsky, Doklady Akad. Nauk. SSSR 234 (1977) 1306.
3. W.I. Axford, E. Leer & G. Skadron, Proc. 15th Int. Cosmic Ray Conf. (Plovdiv) 11 (1977) 132.
4. A.R. Bell, Monthly Notices Roy. Astron. Soc. 182 (1978) 147 and 443.
5. R.D. Blandford & J.P. Ostriker, Ap. J. 221 (1978) L29.
6. E. Fermi, Phys. Rev. 75 (1949) 1169.
7. M. Nagano et al., J. Phys. G 10 (1984) 1295.
8. N.L. Grigorov et al. *Proc. 12th Int. Cosmic Ray Conf.* (Hobart) vol. 2 (1971) 206.
9. Yu. A. Fomin et al., *Proc. 22nd Int. Cosmic Ray Conf.* (Dublin) vol. 2 (1981) 85.
10. K. Asakimori et al. (JACEE Collaboration), *Proc. 23rd Int. Cosmic Ray Conf.* (Calgary) vol. 2 (1993) 21, 25 and *Proc. 22nd Int. Cosmic Ray Conf.* (Dublin) vol. 2, pp. 57 and 97.
11. Simon Swordy, in *Proc. 23rd International Cosmic Ray Conference* (World Scientific, Singapore, ed. D.A. Leahy, R.B. Hicks & D Venkatesan, 1993) p. 243.
12. M. Forman et al., *A.I.P. Conf. Proc. No. 276*, (ed. Lawrence Jones, 1992) p. 620.
13. H.J. Völk & P.L. Biermann, Ap.J. 333 (1988) L65.
14. J.R. Jokipii, Ap.J. 313 (1987) 842.
15. R. Silberberg, C.H. Tsao, M.M. Shapiro & P.L. Biermann, Ap. J. 363 (1990) 265.
16. M. Aglietta et al. (EASTOP Collaboration) Proc. 23rd Int. Cosmic Ray Conf. (Calgary) vol. 2, p. 65 (1993).
17. J.J. Engelmann *et al. Astron. Astrophys.* **233**, 96 (1990).
18. D. Müller *et al. Ap. J.* **374**, 356 (1991); S.P. Swordy, J. L"Heureux, P. Meyer and D. Müller, *Ap. J.* **403**, 658 (1993).
19. J. Engel *et al. Phys. Rev.* **D46**, 5013 (1992).
20. F. Aharonian et al., Proc. of the 23rd ICRC Calgary, Vol.4, 291 (1993); F. Just et al., Proc. of the 23rd ICRC Calgary, Vol.4, 203 (1993).
21. A. M. Hillas, J. Phys .G: Nucl. Phys. **8**, 1461

(1982); A. M. Hillas, J. Phys .G: Nucl. Phys. **8**, 1475 (1982); J. R. Patterson and A. M. Hillas, J. Phys .G: Nucl. Phys. **9**, 1461 (1983).
22. V. Haustein et al., Proc. of the 23rd ICRC Calgary, Vol.4, 375 (1993).
23. A. Karle, Dissertation, Ludwig-Maximilians-Universiät München (1994).
24. P. Doll et al., The Karlsruhe Cosmic Ray Project KASCADE, KfK Report 4686 (1990)
25. J.N. Capdevielle et al., The Karlsruhe Extensive Air Shower Simulation Code CORSIKA, KfK Report 4998 (1992)
26. H.H. Mielke et al., A calorimeter for cosmic ray hadrons up to 10 TeV, submitted to Nucl. Instr. and Meth. (1994)
27. T. A. McKay, PhD Thesis, University of Chicago, 1992.
28. EAS–TOP & MACRO collaborations, Proc. 23rd Int. Cosmic Ray Conf. (Calgary) v. 1, p. 89 (1993).
29. T.K. Gaisser and T. Miller, *Antarctic Journal*, submitted.
30. D.J. Bird, *et al.*, *Phys. Rev. Lett.* **71**, 3401 (1993).
31. E.C. Loh, *Nucl. Phys. B (Proc. Suppl.)* **14A**, 256 (1990).
32. A. Borione *et al.*, *Nucl. Instr. Meth.* **A346**, 329 (1994).
33. D. Sinclair *et al.*, *Nucl. Instr. Meth.* **A278**, 583 (1989).
34. A. Wrotniak, Showersim , U of Maryland preprint PP-85-191 (1985)
35. J.A.Goodman et al , Phys Rev. D26, 1026,1982
36. B. Dawson "Proc. Tokyo Workshop on Techniques for the Study of Extremely High Energy Cosmic Rays" 125 (1994).
37. A.M. Hillas, *Nucl. Phys. B (Proc. Suppl.)* **28B**, 67 (1992).
38. A.M. Hillas, private communication.
39. Yoshida *et al.*, submitted to *J. Phys G: Nucl. Part. Phys.* (1994).
40. R.S. Fletcher *et al.*, *Phys. Rev.* (Nov. 1994).
41. D.J. Hand, *Discrimination and Classification*, 1981.
42. K. Greisen, *Annual Review of Nuclear Science* **10**, 63 (1960).
43. K. Greisen, it op. cit., 78.
44. A.M. Hillas *et al.*, *Proc. 12th ICRC (Hobart)* **3**, 1001 (1971).

GROUND-BASED GAMMA-RAY ASTRONOMY

Richard C. Lamb

Physics and Astronomy Department, Iowa State University, Ames, IA 50011, USA

René A. Ong, Corbin E. Covault

Enrico Fermi Institute, The University of Chicago, Chicago, IL 60637, USA

David A. Smith

LPNHE, Université de Paris VI, 75252 Paris cedex 05, France

Ground-based gamma-ray detectors, operating above 250 GeV, have detected photon emission with good signal-to-noise from two pulsar-driven supernova remnants (Crab Nebula and PSR B1706-44) and from an active galactic nucleus (AGN) of the blazar class (Markarian 421). EGRET, a wide field-of-view satellite detector operating below 20 GeV, has detected an additional ~130 sources, of which six are identified as galactic pulsars and 40 are AGN. The combination of these discoveries plus BATSE observations of gamma-ray bursts raises many detailed scientific questions that may only be adequately addressed by future more sensitive detectors operating both in space and on the ground. Perhaps the major imperative, however, for ground-based gamma-ray astronomy is that the high-energy sky (>20 GeV photon energy) remains a largely unexplored frontier. A full exploration of this region may settle long-standing questions regarding the origins of cosmic rays and could well provide evidence for unexpected phenomena. Some of the many questions raised by the EGRET/BATSE/ground-based observations are: What parameters govern the pulsar process and its apparent cut-off above 1 GeV. What is the role of shell-type supernovae in the production of cosmic rays? What are the production spectra of AGN? Can AGN spectra be used to determine the infrared density of space and thereby calibrate the epoch at which galaxies began to form in the universe? Do gamma-ray burst spectra reach TeV energies? These, as well as other scientific questions, shape the recommendations presented in this report for future, improved ground-based instrumentation. The atmospheric Cherenkov detector has shown itself worthy above 250 GeV. It may be improved in several straightforward ways to produce energy thresholds as low as 20 GeV and thus an overlap with space technology. Water Cherenkov detectors at high altitude offer a practical way to have a full-sky survey instrument for the energy range above 250 GeV. Finally, for surveying the highest possible energies one must utilize extensive air shower arrays.

§1 Overview

High energy gamma-ray astronomy is conventionally divided according to the location of the gamma-ray detector with respect to the earth. Either the gamma-rays are detected in space above the 26 radiation lengths of the earth's atmosphere, or they are detected on the ground using devices which sample portions of the electromagnetic cascades initiated by high energy photons in the atmosphere.

Satellite-based high energy gamma-ray astronomy currently has a practical upper energy limit of $\lesssim$ 20 GeV due to sharply falling photon fluxes of high energy sources and the limited area of satellite-based detectors ($\sim 10^4$ cm^2). In contrast, ground-based detectors have very large effective areas (10^{8-10} cm^2), but are constrained to work above the energy of the minimum detectable shower. Some ground-based instruments have energy thresholds at present as low as $\gtrsim$ 250 GeV. Future developments in both satellite and ground-based gamma-ray astronomy will extend each technique such that they will be able to produce overlapping observations in the 20 to 250 GeV range. Although this section of the Snowmass report is concerned only with ground-based gamma-ray astronomy, much of its scientific motivation is common with that of the space-based gamma-ray community.

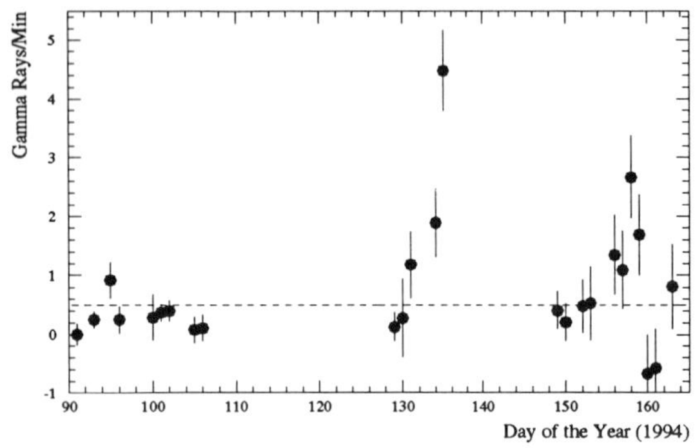

Figure 1: The variation in TeV gamma-rays from Markarian 421 from 1 April, 1994 (day 91) through 12 June, 1994 (day 163) as detected by the Whipple Observatory [1]. The peak intensity of 15.25 May (day 135) occurs 27 hours prior to a reported X-ray high state [2]. Data from 16-28 May (days 136-148) were not obtained due to bright moon conditions. The dashed line shows the 1994 quiescent level.

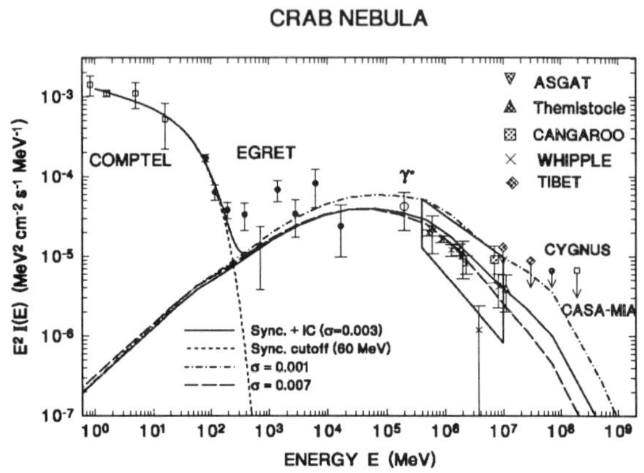

Figure 2: Observations of the photon spectrum from the Crab Nebula. The COMPTEL data are from the PhD thesis of Much [3]; the EGRET data and the model fits are from the work of De Jager et al. [4]. The model assumes a population of relativistic electrons with energies to 3×10^{15} eV. The σ parameter is the ratio of the magnetic field energy density to the electron energy density.

There are two principal driving motivations for ground-based high energy gamma-ray astronomy. The first is that the sky above 20 GeV is still a largely unexplored frontier, *terra incognita*, representing one of the last regions of the electromagnetic spectrum to be explored with sensitive instruments. Although there is yet to be a sensitive survey of the sky at TeV energies, recent high signal-to-background Cherenkov imaging telescopes (pointing at specific objects of promise) have shown that there are at least three established sources of TeV gamma-rays. Two of these, the Crab Nebula and PSR 1706-44, are pulsar driven supernova remnants. The third, Markarian 421, is an active galactic nucleus of the *blazar* class. These sources, taken together with the $\sim$ 130 sources established at lower energies by the EGRET detector of the Compton Gamma Ray Observatory (CGRO) promise a rich harvest of scientific results when this portion of the electromagnetic spectrum is fully explored.

A second major motivation for high energy astronomy is to locate the sources of the cosmic rays. Where and how does Nature accelerate particles to energies which extend to 10^{20} eV and beyond? Particle energies of this magnitude imply the existence of acceleration sites with a range of elementary particle physics phenomena that goes beyond the ability of present-day accelerators to explore. Ground-based gamma-ray astronomy may catch a glimpse of such phenomena.

1.1 Markarian 421

One of the major discoveries of the EGRET instrument on the Compton Gamma Ray Observatory has been its detection of more than 40 active galactic nuclei, all of which are *blazars*. In the simplest, unified picture of the AGN phenomenon, blazars are those AGN which have associated jets that are beamed toward the observer. AGN with jets are among the favored candidates for the origin of high energy cosmic rays.

One of these blazars, Markarian 421, has been observed at energies up to 4 TeV with a ground-based detector. It is clear that this AGN is the source of on-going violent activity on short time scales ($\sim$ days). Observations of this source during the Spring of 1994 are shown in Figure 1. Particularly striking is the flare on day 135 in which the flux increased by an order of magnitude. It is remarkable that the current generation of relatively simple instruments are already capable of probing the high energy emission from Markarian 421 at a distance of 400 million light-years.

1.2 The Crab Nebula: An Example of the Space-Ground-Based Connection

Multi-wavelength observations of the Crab Nebula at photon energies above 1 MeV provide a vivid illustration of how ground-based and space-based astronomy can help each other scientifically. In Figure 2, observations of the Crab Nebula from 1 MeV to 100 TeV are presented. De Jager and Harding [5] have a model for the spectrum which provides a reasonable interpretation to both the data recorded by space instruments (COMPTEL and EGRET) and the data from ground-based instruments.

The model assumes a population of relativistic electrons with energies to 3×10^{15} eV. The synchrotron photons emitted by these electrons spiraling in the magnetic field of the nebula have a cut-off near 60 MeV. Photons of higher energy result when these synchrotron radiation photons inverse-Compton scatter off the electrons. As indicated in Figure 2, measurements made in the unexplored region between 10^4 and 10^5 MeV would help solidify our quantitative understanding of the emission mechanism of the Crab Nebula.

1.3 Scientific Challenges

Discoveries from the Compton Gamma Ray Observatory and the firm TeV gamma-ray detections strongly motivate a combination of space-based and ground-based instruments covering the gamma-ray range from 1 MeV to 100 TeV which will be able to address a number of significant scientific questions, such as:

- What processes determine the cut-off energy for pulsed emission of gamma-rays from pulsars?
- What is the role of shell-type supernovae in the production of cosmic rays?
- What is the geometry of the emission region of AGN?
- At what energies do AGN blazar spectra cut-off?
- Are gamma-ray blazar spectral cut-offs intrinsic to the source or due to intergalactic absorption?
- Is the dominant particle species in AGN jets leptonic or hadronic?
- Can intergalactic absorption of the TeV emission of AGN be a tool to calibrate the epoch of galaxy formation, the Hubble parameter, and the distance to gamma-ray bursts?
- Are there sources of gamma-rays which are "loud" at TeV energies, but "quiet" at lower photon energies?
- Do primordial black holes exist, and, if they do, can the characteristics of their final decay into GeV and TeV photons constrain physics at the Planck mass?

In this report some of the scientific background behind each of these questions will be given. We then present a general discussion of the experimental techniques used in the field of ground-based gamma-ray astronomy and how to extend these techniques in the future to build new, more sensitive detectors. There are three major design objectives of these detectors that we have chosen to concentrate on:

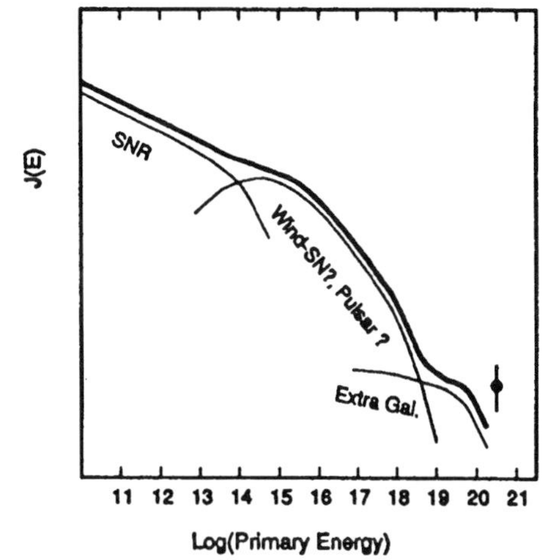

Figure 3: Schematic figure of the cosmic ray total energy spectrum [6]. The low energy cosmic rays are assumed to be produced by supernova remnants (SNR). The highest energy cosmic rays are most likely extra-galactic in nature. In the intermediate energy regime, there are a variety of acceleration mechanisms.

1. Highest possible sensitivity to point-like sources.
2. Lowest possible energy threshold $\sim$ 10 GeV.
3. Widest-aperture, highest duty-cycle.

§2 Scientific Questions

Origin of Cosmic Rays: A long-standing motivation for very high energy gamma-ray astronomy is the desire to establish direct evidence for the sources of cosmic rays. We now know that the cosmic ray spectrum extends to 10^{20} eV and yet the origins of cosmic rays above 1 TeV (10^{12} eV) are not fully understood. Figure 3 summarizes the current ideas of those origins.

There is evidence that cosmic rays below 10^{14} - 10^{15} eV are galactic in origin [7], perhaps produced via shock acceleration at supernova sites. Beyond 10^{15} eV, it is simplest to postulate a continuation of a galactic contribution and, in the process, explain the steepening of the cosmic ray spectrum (the "knee") by the escape of cosmic rays from galactic magnetic confinement or by the termination of the acceleration mechanism itself.

Taken at face value, this model would predict a gradual change in the cosmic ray composition (towards heavier elements) above the knee because of the rigidity dependent breakdown of the confinement or acceleration mechanism. On the other hand, it is generally agreed that basic supernova shock acceleration cannot reach energies above about $10^{14} \times Z$ eV, where Z is the charge of the particle.

Possible extensions to the simplest theory of shock acceleration to reach higher energies have been considered. For example, models in which the shock front expands into the wind of the progenitor star have been proposed. An alternative speculation is that these high energy cosmic rays are extra-galactic in origin, perhaps being produced in the jets of Active Galactic Nuclei (AGN). One expects an extra-galactic component of the cosmic rays, if present, to have a light composition because of photo-absorption off radiation fields at the source and because of spallation of nuclei on their journey through intergalactic space. At the present time, the experimental data cannot distinguish between the galactic and extra-galactic pictures.

Regardless of the particular source model, better data are needed to distinguish between different hypotheses for the location of the sites of cosmic acceleration. These sites produce charged particles with energies well beyond the reach of present terrestrial accelerators. In the vicinity of cosmic ray sources, physical processes occur that are unattainable by other means.

To identify the sources of cosmic rays, we take a two-fold approach: 1) we can observe neutral radiation that comes in a straight line from possible cosmic sources and 2) we can observe charged cosmic rays that reach the earth after deflection in the galactic magnetic field. The first approach will be discussed in this report; the latter topic is covered in the reports from Sections R3 and R4 of this Workshop.

Gamma-Ray Astronomy: The field of high energy gamma-ray astronomy has never been more exciting. There are now several steady sources of TeV gamma-rays. In addition, observations at lower energies made by the instruments on the Compton Gamma Ray Observatory (CGRO) reveal that there is a diverse landscape of astrophysical sources of gamma-rays. These sources, both galactic and extra-galactic, exhibit a rich spectral and temporal structure, most of which is not yet fully understood. Most relevant for the subject of this report are the results from the highest energy instrument on the CGRO, the EGRET detector [8], which is sensitive to gamma-rays with energies greater than 20 MeV.

Existing ground-based experiments have sensitivity to gamma-rays at energies above about 250 GeV. Gamma-rays at these high energies interact in the earth's atmosphere to produce extensive air showers. By detecting the charged particles or the optical Cherenkov light produced by charged particles in the air showers, ground-based experiments determine the arrival direction and energies of incoming gamma-rays. Details of the experimental methods and program are discussed in the Section 3; here we concentrate on our current understanding of the scientific picture.

2.1 Galactic Sources

Spin-Powered Pulsars: High energy gamma-ray emission from the Crab and Vela pulsars has been studied for many years. Four additional pulsars have been seen in gamma-rays by the EGRET experiment (Geminga, PSR 1055-52, PSR 1706-44, PSR 1952+32) and one by BATSE (PSR 1509-58). The total gamma-ray emission from these objects is dominated by the pulsed component, which typically extends to several GeV.

A schematic representation of a gamma-ray pulsar is shown in Figure 4. The energy source is the spinning neutron star, which might have a surface magnetic field as high as 10^{13} Gauss. (All gamma-ray pulsars yet detected are neutron stars without companion stars. Neutron stars with companions, i.e. binary systems, have an additional source of energy, gravitational potential energy, which acts to heat material falling into the neutron star. Binary neutron stars are very luminous in the X-ray band, but have yet to be detected convincingly at gamma-ray energies.)

If the magnetic and spin axes are not co-aligned, a dynamo power source can result. Models for the emission of gamma-rays fall into two broad categories: polar cap models and outer gap models. In the first category, particles are accelerated near the surface of the neutron star at the magnetic poles. In the second, the acceleration takes place in the outer magnetosphere, in the so-called outer gaps where the corotation charge changes sign. In either case, gamma-rays are radiated from the cascades of relativistic electron-positron pairs. One expects the cascade spectrum to fall off at energies around several GeV in the polar cap model, due to magnetic pair production attenuation, and around 10 GeV in the outer gap model, due to the limit on particle acceleration. EGRET has now established high energy cut-offs in the pulsed spectrum at GeV energies for all pulsars except PSR 1706-44, where statistics are still limited.

Remarkably, two of the seven pulsars are now well established sources of 1 TeV gamma-rays (Crab, PSR 1706-44). This emission is observed to be unpulsed and at significantly higher energies than could reasonably be expected from the synchrotron process. One expects the synchrotron component of the Crab Nebula to exhibit a steep falloff at energies above 100 MeV. The data from the Crab confirm this expectation, as shown in Figure 2. The unpulsed emission above 100 MeV, extending beyond TeV energies, can be explained as inverse Compton scattering of synchrotron photons by the relativistic electrons in the nebula. This model describes the data well but is not overly constrained. In particular, there have been no observations in the range between 20 and 100 GeV. In addition, present uncertainties in the gamma-ray energy scale using the atmospheric Cherenkov technique

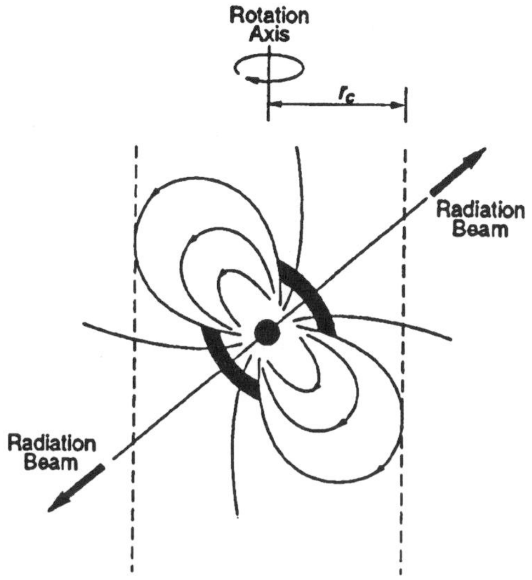

Figure 4: Schematic of a pulsar gamma-ray source. The pulsar consists of a rapidly spinning neutron star with a high magnetic field (from [9]).

make accurate spectra difficult to determine. There is evidence from X-ray [10] and radio [11] data for a plerion surrounding PSR 1706-44. Given its size and age, and thus lower nebular field strength, the TeV gamma-rays from PSR 1706-44 are unlikely to be synchrotron self-Compton emission as in the Crab, but could possible be inverse Compton scattering of the relativistic electrons on other soft photon sources including the microwave background.

What about the other five gamma-ray pulsars? Do they emit TeV radiation? Are there other pulsars not (yet) detected by EGRET but visible at higher energies? Such questions are fertile territory for a new generation of high sensitivity ground-based detectors.

Other Galactic Sources, Supernova Remnants: In addition to pulsars, EGRET has detected more than forty unidentified sources within ten degrees of the galactic plane. Taking into account the numerous extra-galactic sources seen at high latitudes, a significant number of the low latitude sources are likely to be outside of our galaxy. Also, some apparent sources may actually be galactic diffuse emission that is not correctly removed by the model of this background. Nevertheless, a substantial number of the sources are likely to be truly galactic but of unknown nature. Some may be pulsars with no radio counterparts (like Geminga). Others may be sites of cosmic ray production, but many may simply be something new.

Six of the unidentified EGRET sources are coincident in position with shell-like supernova remnants (SNR). These objects have long been considered probable sites of cosmic ray acceleration, and therefore are likely to be detected at very high energies by ground-based experiments. Such detections would provide complementary information to that obtained by space instruments. For one thing, the angular resolution of some ground-based detectors using the atmospheric Cherenkov technique is better than that of EGRET and will improve in the future. Therefore, the potential exists for experiments on the ground to assist in the identification of the unkown EGRET sources. In addition, detection of SNR at very high energies might be easier than at EGRET energies. Charged cosmic rays produced near the SNR should have a significantly harder energy spectrum than the overall spectrum (a power-law exponent of ~-2.0 versus -2.75) because these particles have not diffused through the galaxy. The resulting gamma-ray spectrum near SNR should also be hard, and therefore, the level of very high energy gamma-rays from SNR should be higher relative to the diffuse gamma-ray background than at EGRET energies. These expectations are borne out in recent calculations [12]. For example, Tycho's SNR (distance $\sim$ 5 kpc) has an expected flux that is 10% of the Crab, a level that should be detectable by present ground-based experiments.

2.2 Extra-Galactic Sources

Active Galactic Nuclei (AGN): Active Galactic Nuclei (AGN) are a class of galaxies (Quasars, Blazars, Radio Galaxies, Seyferts, Optically Violent Variables, BL Lac's) which are characterized by large radio output and by large redshifts. They appear as star-like objects, indicating extremely high power output: 10^{42} to 10^{48} erg/s, many orders of magnitude greater than our own galaxy and thus the most luminous objects known in the Universe. They can vary markedly in luminosity over short periods of time, indicating that the emission comes from an object not more than a few light-months across, as confirmed by radio interferometry. They are assumed to consist of a compact, massive central source - presumed to be a super-massive black hole, having as much as 10^9 solar masses - surrounded by an accretion disk and a torus of hot gas. Jets are often seen along the axis of the disk, as sketched in Figure 5. Extended features which appear to be galaxies can sometimes be seen.

In the most common models describing gamma-ray production in AGN the gamma-rays are produced at shocks in the relativistic jets. Power for the emission is supplied by the gravitational potential energy released as matter from the accretion disk falls into the black hole. The broad

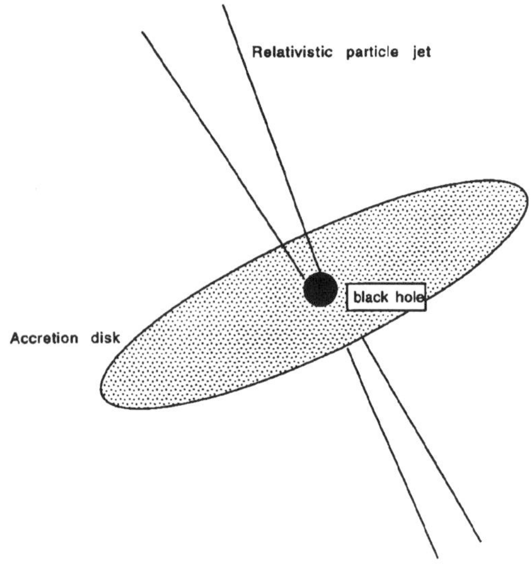

Figure 5: Sketch of the model for an Active Galactic Nuclei. Matter falling from the accretion disk into the black hole powers the collimated jets of relativistic particles. The appearance of an AGN depends on the orientation of the jet relative to the earth.

variety of AGN can be explained simply by the orientation of the jets relative to an earth-bound observer: the optically thick accretion disk viewed edge-on has spectrum characteristic of a radio galaxy. In a blazar the jet axis is presumed to be pointing directly at us. Many of the blazars give off more energy per decade in the gamma-ray band than at any other frequency, implying that significant particle acceleration must be accommodated in any source model.

In models where high energy hadrons are accelerated in the jets, neutrinos are produced with intensities that might be detectable with the next generation of neutrino telescopes. Fewer neutrinos are generated in models where the gamma-rays come mainly from accelerated electrons through inverse Compton scattering. Comparison of results from gamma-ray and neutrino telescopes is therefore an important means of probing acceleration mechanisms and determining the dominant particle constituent of the jets.

The current interest in these sources results from the detection of gamma-rays from about forty AGN by EGRET [13]. The AGN spectra from EGRET run out of statistics around 10 GeV, for the 14-day viewing periods typically used. These spectra are consistent with falling power laws with differential spectral indices between -1.7 and -2.6, although a break from the power law at the highest energies would not be detectable. Any number of models predict a break in the spectra between a few GeV and 1 TeV. Either the acceleration mechanism reaches the end of its range, or gamma-gamma absorption in the source begins to dominate. For example, the synchrotron self-Compton model, successfully applied to the Crab pulsar, can be tested in the case of the AGN by comparing the UV/x-ray intensity to the GeV-TeV flux. Greater flux sensitivity at 10 GeV and beyond would allow spectral measurements in the critical roll-over region.

Another striking feature of many AGN is their violent variability. Figure 1 shows measurements of Markarian 421 during Spring 1994, when the intensity changed by more than a factor of five within a few days, as observed by the Whipple group. EGRET saw a similar flare for the brightest AGN in its catalog, 3C279 (a factor of four in two days). Such large changes in power output on small time scales severely constrain theoretical models of these objects and these flares merit intense study. Conversely, attempts to interpret spectra integrated over several days, when the source conditions are changing wildly, can lead to misleading theoretical conclusions. The EGRET source catalog is probably biased towards the detection of those AGN that flare during the viewing period, since the detection in many cases is at the limit of the instrument sensitivity. Comparison of flare profiles at different wavelengths permits detailed tests of models: in principle, the emission structure of the jets could be mapped out. Again, greater flux sensitivity is needed to enhance the data sample, but a large increase in detector area is unlikely to become available on a space-borne device in the near future.

The energy threshold of ground-based gamma-ray telescopes can in principle be extended down to the range covered by EGRET. At the lower energies, hadron backgrounds are more easily rejected and the flux sensitivities necessary for detailed AGN studies are achievable. Electron backgrounds become relatively more important so that angular resolution becomes the limiting parameter for flux sensitivity.

The recent ground-based detection of gamma-rays up to 4 TeV from the blazar Markarian 421 is remarkable in several respects: first, it confirms that Cherenkov telescopes can overcome the charged cosmic ray background to exploit the very large effective detector areas available using atmospheric showers. Second, the observation of Markarian 421 provides direct evidence of particle acceleration in an AGN to 1 TeV and greater, intriguing in light of the models described above.

The third remarkable point is that Markarian 421 is both the dimmest and the closest extra-galactic source in the EGRET catalog ($z = 0.031$). The blazar 3C279 was seen by EGRET to flare to a level thirty times that of Markarian 421. The spectra for these two sources are shown in Figure 6. Why isn't 3C279 seen at TeV energies, while Markarian 421 is? Perhaps the most reasonable explanation is that TeV gamma-rays above some energy are not able to escape the source region or are absorbed while crossing intergalactic space.

Intergalactic Absorption: The dominant absorption process for high energy photons in intergalactic space is pair creation through collisions with low energy photons: $\gamma\gamma \rightarrow e^+e^-$. A gamma-ray of energy E traveling to earth from an AGN at distance r traverses a column of "soft" photons of energy ϵ and density $n(\epsilon)$. The probability for the gamma-ray to survive the trip is $e^{-\tau}$, where the optical depth τ is, $\tau \propto \sigma r n(\epsilon)$. The threshold energy for pair production occurs when the product of the energies of the two interacting photons equals $(m_e c^2)^2$, and the peak in the cross section occurs when this product is roughly twice the threshold value. Thus for 1 TeV photons, target photons of $\sim$ 0.5 eV (in the infrared region) have the maximal absorbing power. For lower energies ($\simeq$ 50 GeV) and very large redshifts ($z > 3$) there is also a contribution from visible and near UV starlight.

A number of authors [14][15][16] have examined these absorption effects. Some of the incentive to refine such calculations has come from the recent AGN detections by EGRET and Whipple. The calculations have no difficulty in explaining why the bright distant source 3C279 was not seen by Whipple; one does not expect TeV photons to survive the flight distance. Given the current data, however, the calculations are not severely constrained. In addition, as discussed earlier, many models of gamma-ray production in AGN predict a significant softening or cut-offs in the spectrum above 10 GeV due to photoabsorption in the source itself.

Since the absorption depends on the infrared photon density and path length to the source, observation of a break in the gamma-ray spectra correlated with distance for several sources may give a determination of $n(\epsilon)$. Similarly, if $n(\epsilon)$ can be estimated directly by other experiments, breaks in the gamma-ray spectra that vary appropriately with redshift can in principle be used to estimate the distance to the source. The expression for the optical depth depends inversely on the Hubble constant, H_o. A recent paper by Salamon, Stecker, and De Jager [21] has pointed out this dependence and has suggested that from observations of numerous AGN at different values of redshifts the Hubble constant may be determined.

In recent detailed calculations, MacMinn and Primack [16] have sought to model the formation of the galaxies and their evolution. Their calculations basically consider two alternate models for galaxy formation: a cold dark matter (CDM) model representing a moderate era of galaxy formation, and a cold plus hot dark matter (CDHM) model representing a late era of galaxy formation. They conclude that the amount of gamma-ray absorption depends on the model and that, within a given model, the variation of absorption with energy is dominated by the epoch at which galaxies begin to form. In Figure 7, the absorption they calculate for these different scenarios, CDM and CDHM, are shown for a hypothetical source at a redshift of $z = 0.2$.

The possibility of determining the Hubble constant or the epoch of galaxy formation via gamma-ray absorption depends crucially on our understanding of the initial

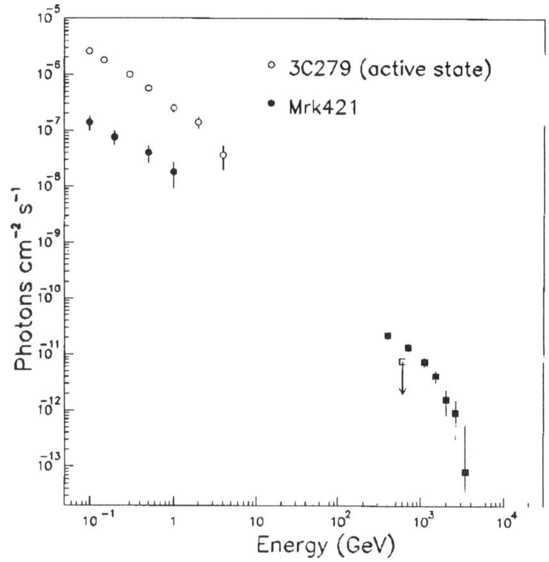

Figure 6: The integral gamma-ray flux for two blazars: Mrk 421 and 3C279 at high energies (circles, EGRET [17][18]) and at very high energies (squares, Whipple [19][20]). The solid symbols represent Mrk 421 and the open symbols represent 3C279. The open square represents an upper limit on the flux from 3C279.

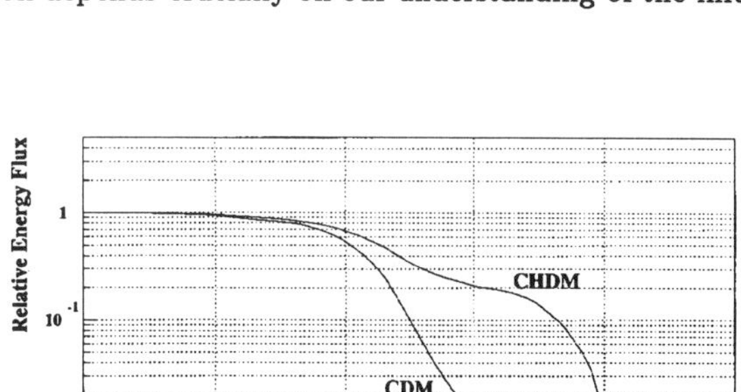

Figure 7: Intergalactic absorption for a source at a redshift of $z = 0.2$ for alternate scenarios of galaxy formation based on two models for the Dark Matter: CDM=Cold Dark Matter, CHDM=Cold and Hot Dark Matter.

gamma-ray spectra at the AGN sources. Clearly one must have a reasonable sample of detected AGN whose intrinsic spectra can be confidently extended to TeV energies. Measuring such spectra would involve both space-based instruments at the lower energies and ground-based detectors at higher energies.

2.3 Other Sources

Gamma-Ray Bursts: Since the discovery of the gamma-ray bursters by the Vela satellite in the early 1970's, every attempt to understand the origin of the bursts has only deepened the mystery. The BATSE detector on the CGRO has seen over 1000 bursts (for the first BATSE catalog see [22]). BATSE has established that these bursts are distributed isotropically over the sky to a very high degree, that they have durations ranging from fractions of a second to minutes, and that the variety of spectral and time structures defy attempts at categorization. The present ensemble of BATSE data gives an indication of a spatial cut-off in the line-of-sight direction. Therefore, only two possible spatial distributions of sources have the characteristics that satisfy the constraints of the observations. Either the bursts are from a galactic population corresponding to a very extended, roughly symmetric halo with a scale larger than 70 kpc, or they are located at cosmological distances.

Thus far, six BATSE bursts have been detected by EGRET, in the GeV domain. If burst spectra extend to GeV energies then why not to TeV? Extrapolating from the brighter EGRET bursts leads to an expectation of more than a thousand photons detected in an atmospheric Cherenkov telescope with a threshold energy of 100 GeV. Even significantly fainter GRB's might also be detected on the ground at high energies.

If GRB's are from a galactic population one would expect no absorption due to pair production in the energy regime covered by ground-based experiments. On the other hand, if they are cosmological, only the nearest ones will be detectable above 100 GeV. Observations in this energy regime have the promise of resolving the distance question decisively. Schemes to slew Cherenkov telescopes to respond to BATSE triggers broadcast by the BACODINE system [23] are being implemented. Another attractive detection possibility will be offered by MILAGRO-like detectors that have broad angular coverage and high duty-cycle of operation.

Primordial Black Holes: Density fluctuations in the early Universe can naturally lead to the formation of black holes with a mass spectrum determined by the equation of state in the early Universe and by the size of the density fluctuations. The primordial black holes may be detectable today as bursts of gamma-rays. Hawking [24] showed that black holes emit black-body radiation characterized by a temperature $T \approx 10^{14}/M$ GeV, with M in grams. The luminosity of the hole is inversely proportional to the mass squared, and the proportionality constant is determined by the number of elementary particle states available at a given temperature. Therefore, as the black hole evaporates, both its temperature and evaporation rate increase. This run-away process leads to a final burst of high energy radiation [25]. Roughly 10% of the rest mass remaining at a given temperature T, is emitted in the form of gamma rays with energies above kT. Thus, observation of the final stage emission of a primordial black hole, should give us information about both the state of the early Universe and the elementary particle spectrum up to the Planck scale.

The apparent duration of the burst is determined by the energy threshold of the detector observing the burst. The higher the energy threshold, the shorter the apparent duration of the burst of radiation. For an extensive air shower array with an energy threshold of 100 TeV the burst would appear to last for $\sim$ 1 second. For a Cherenkov telescope or a MILAGRO type detector at an energy threshold of 500 GeV the burst would appear to last for $\sim$10 seconds. For an instrument such as EGRET the "burst" would last for 10,000 years and would not be visible above the background. Upper limits on the rate-density of primordial black holes have been set by both atmospheric Cherenkov telescopes and extensive air shower arrays.

2.4 Unexpected Phenomena

Major discoveries in astrophysics have been made with instruments intended for completely different purposes. The 3K microwave background, pulsars, and gamma-ray bursts are excellent examples. A ground-based gamma-ray detector with good angular and energy resolution in the energy regime between 20-250 GeV may see something completely new. An example of phenomena not directly motivated by lower energy astronomy would be the detection of resonances in the gamma-ray spectra due to annihilation of primordial supersymmetric particles that have accumulated in galactic halos [26].

Although there have been systematic studies of the TeV gamma-ray sky [27], to date no sensitive all-sky survey has been done. We emphasize the importance of all-sky monitoring instruments that are sensitive to transient and unexpected phenomena. Construction of the first such instrument (MILAGRO) is underway. In addition, future space-based instruments will serve as a crucial complement to ground-based detectors, operating as watchful eyes at lower energies.

§3 Experimental Considerations

Current space-borne experiments are effective at direct detection of astrophysical gamma-rays up to an energy of approximately 10 GeV. Above this energy, the source fluxes of gamma-rays are so low (about 1 photon per square meter per week) that direct detection is not yet achievable.

Fortunately, the techniques of ground-based detectors provide much larger collection area. These techniques take advantage of the fact that a high energy gamma-ray entering the earth's atmosphere will interact and generate showers of relativistic secondary particles. The secondary particles, or the optical-ultraviolet Cherenkov photons they produce, can be detected by instruments on the ground, typically at high mountain altitudes.

The atmospheric Cherenkov technique (ACT) employs optical reflectors to detect the Cherenkov photons produced in the shower. At present, this technique is the only means of detecting astrophysical gamma-rays in the energy range between 0.25-10 TeV. At higher energies, air showers penetrate to mountain altitudes where charged particle detectors (scintillation counters, proportional chambers, etc.) can be used to sample the shower front. Such detectors, using the Air Shower Technique (AST), cover the energy range above 10 TeV, with the upper energy limit dictated by the source flux level.

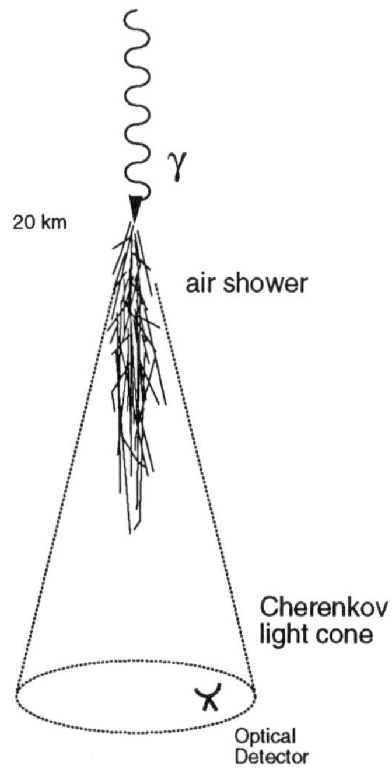

Figure 8: Schematic of Cherenkov light production in the atmosphere. When a energetic gamma-ray interacts at the top of the atmosphere, shower particles emit optical and UV Cherenkov photons into a cone. The Cherenkov light pool can be detected on the ground using optical collectors.

3.1 Atmospheric Cherenkov Technique

The Cherenkov light in an air shower is emitted into a cone with its axis along the direction of the original primary gamma-ray. This is shown schematically in Figure 8. The Cherenkov light on the ground from a typical gamma-ray shower is distributed over an area of about 200 meters diameter. The arrival time of photons at any one location has a width ~ 3 nsec and the arrival direction has an angular spread of ~ 1 degree. Although only a tiny fraction of the energy of the shower results in the Cherenkov light, the number of Cherenkov photons scales with the energy of the primary. Therefore, the Cherenkov light can convey both the direction and the energy information from the original gamma-ray. Figure 9 shows the average density of Cherenkov photons on the ground as a function of primary gamma-ray energy.

The detection of Cherenkov photons requires clear moonless nights. The requirement to detect fast pulses at blue-UV frequencies above the red-dominated night sky background motivates the use of blue-sensitive photomultipliers (PMTs). The early versions of detectors developed in the 1960's were composed of a concentrator, such as a large parabolic mirror, and a single photodetector. The difficulty faced by these early detectors remains a fundamental constraint on the technique: any gamma-ray air

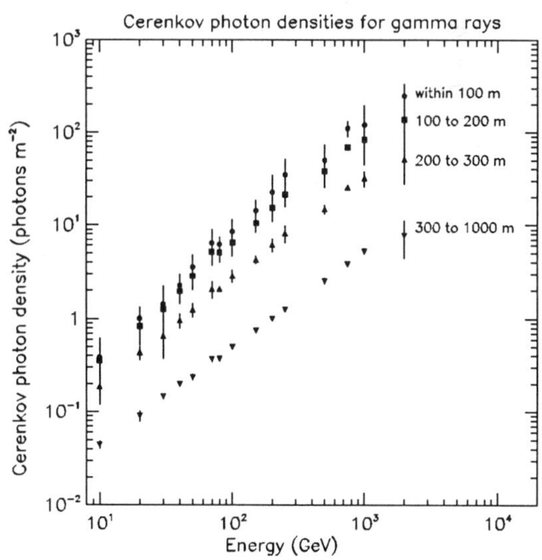

Figure 9: The density of Cherenkov photons on the ground as a function of gamma-ray energy for different radius bins from the shower core. These results come from the simulation MOCCA [28].

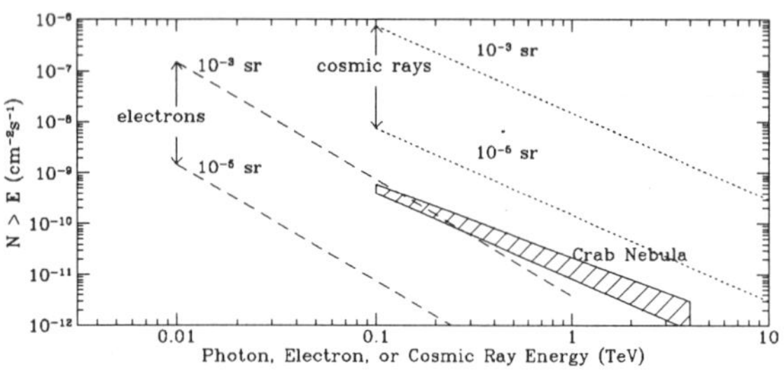

Figure 10: Backgrounds from cosmic ray nuclei and electrons compared with an approximate representation of the spectrum of the Crab Nebula. The backgrounds are calculated for two values of the solid angle: 10^{-5} and 10^{-3} steradians (0.1° and 1° half-angle fields of view). The electron spectrum is taken from Longair[29].

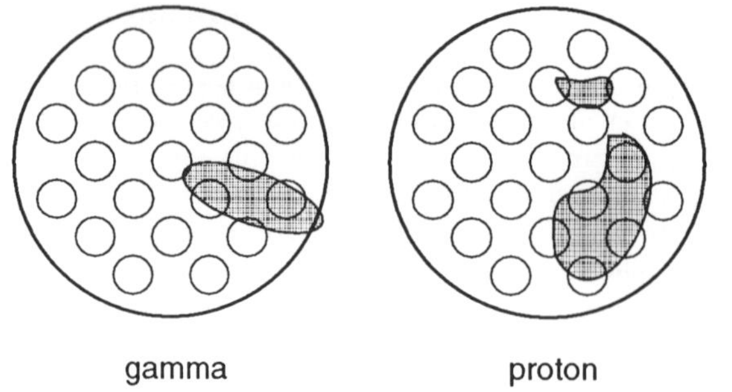

Figure 11: Comparison of Cherenkov image appearances at the focal plane of an imaging collector. Gamma-ray images are elliptical and oriented towards the source in the center of the field of view, while proton images are much less regular and are not oriented.

shower must be detected in the presence of a very large background of air showers events due to the isotropically distributed background of hadronic cosmic rays. This background can be more than 1000 times stronger than the level of signal collected from a typical source.

To place the experimental difficulties in perspective, we compare the Crab Nebula spectrum with the background cosmic ray spectrum in Figure 10 for two different background solid angles, 10^{-3} and 10^{-5} steradians, corresponding to half-angle fields-of-view of 1.0° and 0.1° respectively. The 1.0° field-of-view is appropriate for the simple, early Cherenkov detectors. As shown in Figure 10, at 1 TeV the cosmic ray background is $\sim$ 1000 times that of the Crab signal. (For a completely accurate comparison one must compare a 400 GeV gamma-ray shower with a 1 TeV proton shower, since the efficiency of proton initiated showers for producing Cherenkov light is less than that of gamma-ray showers. Furthermore this difference increases as one moves to lower energies.)

During the 1980's, new efforts were launched to improve the sensitivity of the ACT technique by rejecting as much of the background of cosmic rays as possible. The most recent generation of ACT detectors have developed these techniques to a mature level. Successful ACT experiments rely on one or both of the following two techniques for improved sensitivity: *imaging* and *sampling.*

Imaging Technique: The imaging technique has been most successfully employed for ACT detectors at energies of around 1 TeV. Rather than using a single PMT to detect the Cherenkov light flash, an array of PMT's operating in coincidence is placed at the focus of a single large collector. The distribution of the light on the pixels at the focus of the collector represents an image of the Cherenkov light as seen from the position of the telescope. The PMT array records an image of the angular profile of the Cherenkov light flash. The image that results from a typical gamma-ray shower is elliptical, with an orientation that points toward the axis of the shower relative to the position of the telescope. Monte Carlo simulations have shown that gamma-rays will regularly generate this characteristic oriented elliptical shape. In contrast, the Cherenkov images from hadronic cosmic ray showers are much less regular and homogeneous, since these showers develop much less uniformly. Since hadron arrival directions are isotropic, the orientation of the image in the focal plane will be random. Figure 11 shows a typical image for a gamma-ray shower compared to that from a hadronic cosmic ray shower.

The very different behavior in shower propagation and Cherenkov light arrival patterns is the basis for background discrimination using the imaging technique. By selecting only those events which have images consistent with gamma-rays from the candidate source direction only, imaging telescopes have been able to reject over 99 percent of the hadronic background. The Whipple Observatory was the first to use this technique to successfully detect the Crab [30] but other groups have recently constructed imaging experiments. Recently, the CANGAROO experiment in Australia detected the source PSR 1706-44 using the imaging technique [31]. The imaging experiments have achieved energy thresholds as low as 250 GeV.

Sampling Technique: The second major class of ACT experiments are *sampling* or *timing* detectors. In contrast to the typical single-collector imaging system, sampling detectors depend upon an array of light collectors spread around the ground to sample photons from many locations in the Cherenkov light pool. Each collector focuses Cherenkov light from the shower onto a single photomultiplier. The arrival direction of the primary is inferred from fast timing of the arrival of Cherenkov pulses at each collector. This reconstruction takes advantage of the fact that the Cherenkov light from gamma-ray showers at TeV energies form a shallow conical wavefront. With careful timing, the shower direction can be determined to within a fraction of a degree. With such accurate angular resolution, much of the hadronic background can be removed based upon arrival direction alone. Further rejection can be achieved based upon the uniformity of the light distribution across the Cherenkov pool. The sampling/timing technique has been successfully applied by the ASGAT [32] and THEMISTOCLE [33] collaborations to detect the Crab at energies above 400 GeV and 3 TeV, respectively.

Both the imaging and the sampling/timing techniques have proven to be effective at reducing hadronic background and reconstructing gamma-ray air shower directions to achieve convincing detections of gamma-ray sources, particularly the Crab Nebula. We note that these two effective techniques are largely orthogonal in approach, and therefore we might expect a substantial benefit in rejection by applying both techniques in tandem. Several detectors are indeed planned in such a way as to combine both imaging and sampling/timing techniques in one instrument. For example, the Durham group has a significant upgrade well underway in Australia [34]. To date, the primary difficulties in implementing such hybrid systems have been complexity and cost.

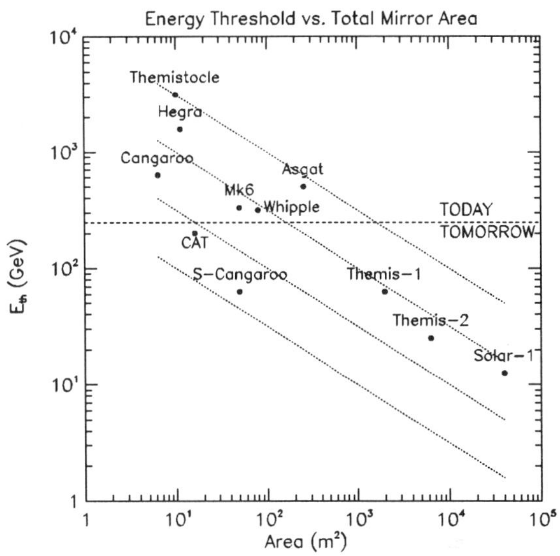

Figure 12: Energy threshold versus total mirror collector area for present and future atmospheric Cherenkov experiments. For a detector of a state-of-the-art, points will lie on the $E \propto A^{-1/2}$ lines as shown.

Sensitivity and Energy Threshold: There are two primary parameters that can be used to specify the performance of atmospheric Cherenkov detectors in terms of capability to detect gamma-rays from astrophysical sources. These two quantities are *energy threshold* and *flux sensitivity*. They are closely related and the most accurate representation of the performance of an instrument is a plot of sensitivity as a function of gamma-ray energy over the full range of energies. However, it is customary to approximate the properties of such curves by stating the energy threshold as a single value below which most photons are not detected. The instrument sensitivity can then be quoted at the median energy of all events in the sample. With such approximate definitions, it is possible to derive an expression for the energy threshold of any arbitrary ACT detector in terms of controllable parameters. Assuming an uniform background of night sky photons, this expression is:

$$E_{th}^{-1} \propto y_\gamma(h)\sqrt{A\epsilon/B\Omega t} \quad ,$$

where E_{th} is the energy threshold, $y_\gamma(h)$ is the gamma-ray Cherenkov photon yield as a function of altitude, A is the mirror area, ϵ the efficiency of light collection and conversion to photoelectron signal, B is the background light level, Ω is the solid angle viewed by each PMT, and t is the integration time. If we wish to consider lowering the energy threshold, we can 1) move the experiment to a high mountain to get a higher Cherenkov photon yield (and possibly lower night sky background), 2) improve the optical efficiency of the system, or 3) increase the collecting area. Of these options, the increase in area holds the most promise for achieving a significant improvement in energy threshold.

Figure 12 shows a plot of energy threshold vs. mirror collecting area for several current and proposed experiments. While improvements in angular resolution, altitude response, and detector efficiency can yield significant lowering in energy threshold, the most substantial improvement is achieved by moving increasing the collecting area (moving diagonally towards the lower right).

3.2 Air Shower Technique

A 100 TeV gamma-ray produces a shower at mountain altitudes of roughly 35,000 charged particles (mostly electrons and positrons). The shower consists of a pancake of relativistic particles, whose width is small and whose orientation preserves the direction of the incident gamma-ray. Relative timing measurements made between detectors separated by distances of ten to twenty meters allow the reconstruction of the gamma-ray direction with an accuracy of roughly one-half degree. Pulse height measurements in these same detectors can be used to estimate the shower size (number of particles) and hence the primary energy.

Since the showers develop high up in the atmosphere, the detectors are typically well past the point of maximum shower development. Therefore, the density of charged particles increases with increasing altitude. As a consequence, for experiments with a fixed detector trigger threshold, the energy threshold decreases with increasing altitude. On the other hand, the energy threshold can be lowered by increasing the density of sampling.

Air shower detectors have several advantages over atmospheric Cherenkov instruments. They can operate close to 100% of the time and monitor roughly 2 steradians of the overhead sky. In contrast, Cherenkov telescopes have a small field of view, typically a few degrees, and operate effectively only on clear moonless nights (10% duty-cycle). There are disadvantages to the Air Shower Technique as well. Unlike the ACT, where most of the Cherenkov light is produced from charged particles near shower maximum, air shower detectors sample the particle content of the shower at a longitudinal depth well past shower maximum. As a result, there are significantly more fluctuations inherent in the charged particle distributions sampled by air shower experiments than those found in the Cherenkov photon distributions. These fluctuations, and the fact that air shower arrays have traditionally been sampling detectors, make background rejection using shower image or shape difficult. However, we expect gamma-ray showers to have significantly fewer muons than showers initiated by cosmic ray primaries of the same energy. Several air shower experiments have been constructed with muon detectors to assist in reducing the cosmic ray background.

Air shower detectors have been in operation around the world for several decades. Originally, these experiments were built to study the cosmic rays themselves. Motivated by reports in the early 1980's of the exciting possibility of bright gamma-ray sources at 1000 TeV, several second generation air shower arrays were constructed for the primary purpose of gamma-ray astronomy: most notably, the CASA-MIA, CYGNUS, HEGRA, and Tibet experiments. The new arrays are significantly larger, have much improved angular resolution, and lower energy thresholds than earlier experiments, leading to an improvement in gamma-ray flux sensitivity of one to two orders of magnitude.

The increased collection area and improved angular resolution have established the viability of air shower experiments to see point sources. For the first time, experiments have detected the shadows cast in the cosmic rays from the finite extent of the Sun and Moon. Observation of these shadows (a typical one shown in Figure 13) validates the

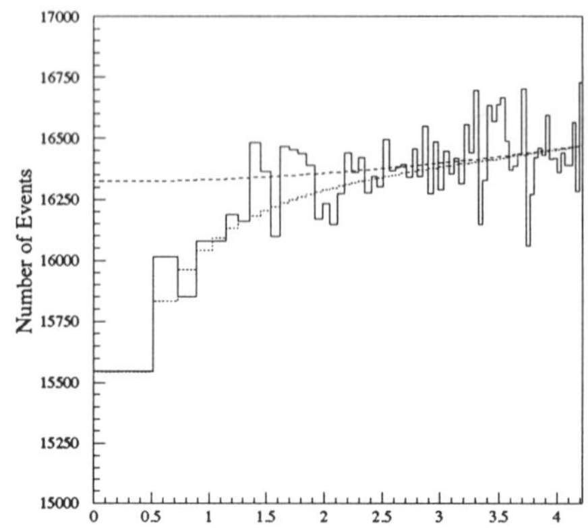

Figure 13: The observed cosmic ray flux seen by the CASA-MIA experiment for equal solid angle annuli centered on the Moon. The dashed curve is the expected number of events, assuming no Moon shadowing. The dotted line is the predicted number of events incorporating the finite size of the moon and an estimate for the angular resolution of the experiment. The statistical significance of the moon deficit is roughly 7.5 standard deviations.

entire chain of shower reconstruction and analysis. The new generation of air shower experiments have established that the northern hemisphere is not anomalously bright in sources of gamma-rays above 10 TeV; the earlier results from the 1980's have not been confirmed. Sensitive limits exist on the emission of gamma-rays from a variety of sources (x-ray binaries, pulsars, AGN), both galactic and extra-galactic [35][36][37][38].

In spite of the absence of bright sources above 10 TeV, the Air Shower Technique remains a powerful and complementary method to the Atmospheric Cherenkov Technique in the search for gamma-ray emission from astrophysical sources. With nearly a 100% duty-cycle of operation and a wide field of view, air shower arrays are ideally suited to observe transient phenomena or emission from unexpected directions. Considerations for third generation air shower experiments have focused on detector improvements that would permit significantly increased gamma-ray sensitivity, especially through a reduction in the detector energy threshold, as well as improved capability for measurements of the cosmic ray composition. The Tibet experiment, by virtue of its high altitude (4300 m) is already the world's lowest energy air shower experiment. An increase in size by a factor of four (to an area of 34,000 m) and additional electronics modifications will lower the gamma-ray energy threshold of this experiment to $\sim$ 6 TeV [39]. The CASA-MIA and HEGRA experiments have recently installed Cherenkov detector elements that will aid in making better composition measurements at energies above the knee [40].

A significant new initiative is the MILAGRO detector, a water Cherenkov detector under construction in the

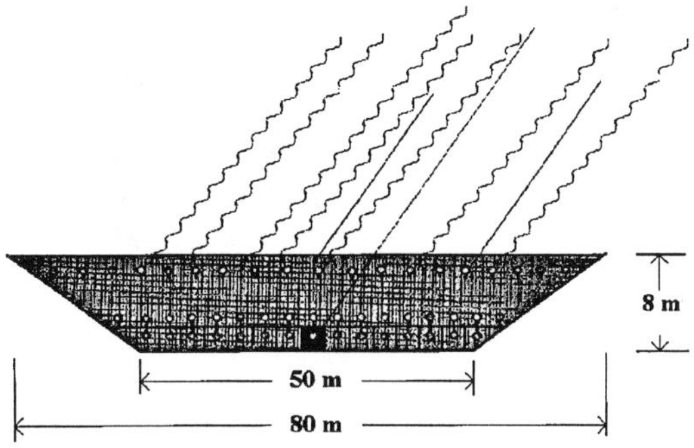

Figure 14: Conceptual drawing of the MILAGRO detector, presently under construction in the mountains of northern New Mexico. This experiment will be the first air shower detector sensitive to gamma-rays with energies near 1 TeV.

mountains of northern New Mexico [41]. This experiment will be the first air shower detector in the sub TeV to few TeV energy regime. The low threshold is achieved by greatly increasing the fraction of the air shower particles sampled by the detector. Unlike existing air shower arrays that sample $\sim 1\%$ of the shower particles, a water Cherenkov experiment is sensitive to essentially all charged particles and photons within its geometric area. An artist's conception of the MILAGRO detector is shown in Figure 14; some of the physics capabilities of this experiment are discussed in the following section.

§4 Summary: The Future

The major purpose of the Snowmass '94 workshops is to summarize the imperatives of the various subfields represented here, to set forth a vision for the future. What are our goals, where are we headed, and what might we learn when we get there?

The centerpiece of the vision for ground-based gamma-ray astronomy must be improved instrumentation. The exploration of this new window of the electromagnetic spectrum has just begun. Discoveries of radiation from two spin-powered pulsars and from a blazar class AGN are a sampling of what lies ahead, as more sensitive instruments are brought on-line. It is also clear that, at least for the atmospheric Cherenkov detectors, significant further improvements can be achieved in a relatively straightforward way without postulating major technological advances.

This vision is already at work driving increased world-wide activity in the area. Initiatives for significant improvements of existing equipment are underway in Europe, Asia, Australia, and North America. A major new facility, MILAGRO gives a new dimension to ground-based instrumentation. Finally, there are some bold proposals, e.g. the Bowl Detector [42] and the Telescope Array [43], which hold the promise of truly large scale advances in sensitivity.

The requirements placed on this instrumentation are driven by the scientific issues. In some cases, the requirements may well be mutually exclusive, e.g. all-sky survey instruments versus detectors with the highest possible sensitivity. Our starting point, therefore, in describing a vision for the field, is the reminder that no one type of instrument will satisfy all requirements. We also recognize that any one instrument can, at best, observe only one-half the sky, i.e. that which passes overhead. Thus, multiple instruments are mandatory for complete coverage.

4.1 High Sensitivity to Point-like Sources

Cherenkov air shower detectors in the region below 1 TeV have already demonstrated good ability to reduce cosmic ray backgrounds so that the signal from the Crab Nebula is larger than the background (cf. Figure 10). It is probable that the most sensitive detectors in the energy range between 100 GeV and 1 TeV will continue to be those using the imaging technique. The logical next step is to combine the imaging and sampling/timing techniques into a single instrument. With two or more reflectors of aperture seven to ten meters, each equipped with imaging cameras of 100 or more pixels, flux sensitivity at the 0.01 Crab level is quite feasible. Detection systems of this kind are already under construction in Arizona, India, the United Kingdom, Russia, Germany, Japan, and Australia. Cameras with as many as 500 pixels are being built by two groups (France, Japan); such cameras will have angular resolutions that will surpass any existing gamma-ray telescope in space or on the ground. Although telescopes of this type can be expanded to arrays of detectors, with energy thresholds as low as 30 GeV, it is unlikely that this approach is economically feasible if it is desired to go as low as 10 GeV.

A detector with sensitivity to sources 1% as bright as the Crab would be a significant milestone in the exploration of the high energy gamma-ray sky. Using the EGRET catalog as a guide, one extrapolates 100 or more sources brighter than this limit, if the spectra from these sources continued to TeV energies. Such a detector would make observations of stronger sources possible under essentially background-free conditions. For these sources the number of detected photons will exceed several/minute. Thus studies of variations on time scales significantly shorter than previously achievable by any gamma-ray instrument will be possible. Intensity variations are one of the fundamental observables associated with a source. A successful theoretical understanding of how the source generates very high energy gamma-rays must necessarily confront and be constrained by such observations.

4.2 Low Energy Threshold

As emphasized throughout this document, the regime between 20 GeV and 250 GeV is unexplored with any type of instrumentation and there may well be surprises. This is also the regime where, for example, the emission from the Crab Nebula, changes from being dominantly pulsed to unpulsed. Theoretical models of the pulsed emission process will be underconstrained until this pulsed cut-off region is fully explored. This is also a region where absorption effects, both intrinsic and extrinsic, are expected to play a role for the AGN and gamma-ray burst sources, if the latter are at cosmological distances. Thus we conclude that there are compelling scientific reasons for developing instruments with sensitivity in this regime. Furthermore, we believe that there is no fundamental reason why ground-based atmospheric Cherenkov detectors cannot achieve thresholds as low as ~10 GeV, primarily by increased photon collection area (Figure 12).

Recognition of these facts has already motivated several new experimental initiatives for lower energy thresholds. At least two groups are exploring the application of large mirror collection areas available from heliostat arrays used for solar power generation. New experiments are being considered for installation at the Solar One site, near Barstow CA, and the Themis site in France. Each site consists of an array of large heliostat mirrors used to focus sunlight onto a central tower receiver. Heliostat arrays would be used at night to collect photons from the full extent of the Cherenkov light pools resulting from gamma-ray air showers. These solar facilities were not originally designed for the purposes of gamma-ray astronomy. Nevertheless, the American and French groups are testing the feasibility of their respective sites for gamma-ray astronomy; the initial results are encouraging. The group working at the Solar One site has reported that the heliostat mechanical and optical performance looks good, as does the night sky background [44]. The French group [45] is planning to conduct site tests in 1995 at Themis. The ultimate energy threshold achievable at these installations has not yet been determined.

A second, more ambitious concept for a low-threshold gamma-ray detector has also been presented at this meeting. Salamon proposed the development of a very large single-reflector fixed-dish telescope. The primary light collector would be a spherical mirror surface, with a diameter as large as 500 meters built into a concave depression in the ground. This "Big Bowl" detector would be similar in design to the Arecibo radio facility, with a movable secondary and detector array at the prime focus. The main advantage of the Big Bowl is that it would collect the maximum number of Cherenkov photons falling anywhere within the enormous mirror area of the primary. This would provide the optimum energy resolution and lowest possible threshold obtainable for a ground-based Cherenkov experiment. The challenge to this approach is the high cost and long development time associated with such a major undertaking. Exploration of the capabilities of such a design is continuing and should prove useful, particularly as a complement to the efforts using solar heliostat arrays. The Big Bowl idea has also motivated new explorations of the possible application of an array of smaller fixed-dish telescopes that could be operated in coincidence to provide lower energy threshold [46]. Such an array of "Little Bowls" might provide a less expensive means to make progress towards lower energy thresholds.

4.3 Wide-Aperture, High-Duty Factor

For many reasons there is a compelling need for a large aperture instrument. The high energy sky revealed by EGRET shows that many of the gamma-ray sources are highly variable and therefore they need to be monitored to determine fully the characteristics of that variability. Such characteristics may provide crucial clues to their understanding. Wide-aperture, sensitive detectors must therefore be developed which not only survey but are capable of monitoring a significant fraction of the sky at any one time. Such instruments will be best utilized if they are in combination with a wide-aperture space detector observing at somewhat lower energies.

At present one of the most promising types of instruments which may accomplish these functions is the extended water Cherenkov detector, exemplified by MILAGRO (Figure 14). Water detectors have lower energy thresholds than conventional sampling air shower detectors. For example, MILAGRO should be able to detect gamma-ray showers with energies as low as 250 GeV. As such it will be an ideal detector for studying episodic and variable phenomena, such as gamma-ray bursts. For an E^{-2} source spectrum MILAGRO will be as sensitive as BATSE, and nearly an order of magnitude more sensitive than EGRET, to the emission from a short duration (one second) burst. This sensitivity to short bursts of emission will also enable MILAGRO to probe a volume of space over two orders of magnitude larger than current instruments for the final stage emission of primordial black holes.

To study a large sample of AGN over all timescales requires a sensitive all sky monitor. MILAGRO will be capable of detecting Mrk 421, the closest AGN in the EGRET catalog. The detection of more distant AGN will be possible if absorption from the intergalactic infrared radiation field is small. If several AGN at various redshifts can be detected, a measurement of the high-energy rolloff in the emission could be a sensitive probe of this radiation field, as discussed in Section 2.

The large area muon detector (1500 m^2) at the bottom of the MILAGRO pond, can also be used to detect solar energetic particles with energies above $\sim$ 10 GeV. Energetic protons and neutrons interact high in the atmosphere to produce muons. MILAGRO will be more sensitive than any other experiment in the detection these solar particles. The level of sensitivity is such that the time structure of solar bursts can be resolved down to a second, as opposed to the present resolution of minutes. This fine time resolution will help determine the acceleration and trapping mechanism responsible for such outbursts.

The technology of the water Cherenkov detector appears to be extendable so that an instrument of this type may well fulfill the need for a sensitive, wide-field, high-duty factor instrument.

4.4 Energy Resolution, Wide-Dynamic Range

One of the dominant themes of the science that we have described is that of the spectral signature of a source. The energy resolution required of future instruments is that there be no significant loss of important spectral information.

Many of the models of AGN spectra predict a cut-off which is relatively rapid, i.e. falling below a simple power-law extrapolation of lower energy behavior by a factor of 10 as the energy is increased by a factor of 10. Pulsar cut-offs, which signal the maximum end-point energies of the parent charged particle spectrum, may be even more rapid. Absorption effects predicted for cosmologically distant AGN have spectral features (cf. Figure 7) whose delineation gives important clues to the distance and time scale of the universe. In any event, energy resolution adequate for these tasks must be built into the specifications of future instruments. With the above in mind, it is reasonable to shoot for an energy resolution, $\sigma_E/E = 20\%$, not yet attained with current single reflector Cherenkov instruments, but achievable with an array of reflectors.

Good energy resolution must be coupled with instrumentation capable of observations over a wide range of energies. A unbroken power-law spectrum and a power-law spectrum with a high energy cut-off are indistinguishable unless there is a sufficient lever arm at lower energies to clearly establish the power-law behavior. Thus all the rich spectral signatures of the AGN and pulsars which have been described in this review will not be detectable unless there is a sufficient dynamic range over which observations can be made. Present ground-based detectors have a dynamic range of less than two decades in energy. This range must be extended. Ideally, one would have a single instrument make such measurements. However, from a practical stand-point this means that observations from ground-based instrumentation, working above 10 GeV, must be coupled with observations from space. Optimally such multi-wavelength measurements should be contemporaneous because of known source variability.

The need for a wide dynamic range also means that a mixture of ground-based instruments will be necessary in order to cover the energy range adequately to 100 TeV. One of the major strengths of the Compton Gamma Ray Observatory is its suite of instruments capable of measuring spectra of strong sources over five decades in energy from 100 keV to 10 GeV. Future ground-based instrumentation will extend this reach to nine decades.

4.5 Next Generation Space Instruments

The discoveries by the Compton Gamma Ray Observatory provide a strong incentive for more sensitive space-borne gamma-ray detectors. However, many of the scientific issues raised by it, particularly by EGRET, can only be fully addressed by a combination of both space and ground-based detectors. Thus the new instrumentation advocated by this document would be greatly diminished in ultimate scientific impact without worthy successors to EGRET and BATSE. For reasons already discussed ideal space instruments should have broad energy coverage with the widest possible field-of-view. Several prototype concepts are being considered, e.g. AGATE, GLAST, and OSCAR.

4.6 The Highest Energy Cosmic Rays

If the sources of ultra-high energy cosmic rays are extragalactic at distances more than twice the distance to the Virgo cluster (20 Mpc) then the photon widow for their exploration effectively closes above 100 TeV due to absorption by the microwave background radiation. Thus the challenge of determining what are the sources of the highest energy cosmic rays could well be beyond the reach of gamma-ray astronomy.

However, at the highest energies ($>10^{20}$ eV), the Larmor radii of charged particles may be significantly larger than the distance to the source. Thus giant air shower arrays (e.g. Auger Project) would give direct information regarding this aspect of the cosmic ray mystery. This information would be unobtainable by photon detectors and therefore a necessary complement to the information obtained by the detectors we have described above.

§Acknowledgments

We would like to express our thanks to Roberto Peccei, Rocky Kolb, Cynthia Sazama, and the entire organizing staff of Snowmass '94. This was truly an outstanding workshop that went a long way in attempting to summarize the current status of the field and our vision for the future. In this report, we have tried to summarize the discussions held during our workshop sessions. The report benefited from many presentations and from the input from all participants, which we gratefully acknowledge.

§References

[1] A.D. Kerrick et al. *ApJ*, 438:L59, 1995.

[2] T. Takahashi et al. *IAU Circular*, 6063, 1994.

[3] R. Much. Ph.D. Thesis, Max-Plank Institute Fur Extraterestriche Physik, 1994.

[4] O.C. De Jager et al. In preparation, 1995.

[5] O.C. De Jager and A.K. Harding. *ApJ*, 396:161, 1992.

[6] M. Teshima. *Proc. 23rd Int. Cosmic Ray Conf. (Calgary)*, Rapporteur Vol.:257, 1993.

[7] P. Sreekumar et al. *Phys. Rev. Lett.*, 70:127, 1993.

[8] C.A. Fichtel et al. *ApJ Suppl.*, 94:551, 1994.

[9] A.G. Lyne and F. Graham-Smith. *Pulsar Astronomy*, Cambridge University Press, 21, 1990.

[10] W. Becker et al. Submitted to A&A, 1994.

[11] D.A. Frail et al. Submitted to ApJ, 1994.

[12] L. Drury et al. *A&A*, 287:959, 1994.

[13] C. von Montigny et al. *ApJ*, 440:525, 1995.

[14] F.W. Stecker et al. *ApJ*, 415:L71, 1992.

[15] E. Dwek and J. Slavin. *ApJ*, 426:696, 1994.

[16] D. MacMinn and J. Primack. *TeV Gamma Ray Astrophysics*, ed. H. Volk, Space Science Reviews, in press, 1995.

[17] Y.C. Lin et al. *ApJ*, 401:L61, 1992.

[18] R.C. Hartman et al. *ApJ*, 385:L1, 1992.

[19] G. Mohanty et al. *Proc. 23rd Int. Cosmic Ray Conf. (Calgary)*, 1:440, 1993.

[20] A.D. Kerrick et al. *Proc. 23. Int. Cosmic Ray Conf. (Calgary)*, 1:405, 1993.

[21] M.H. Salamon et al. *ApJ*, 423:L1, 1994.

[22] G.J. Fishman et al. *ApJ Suppl.*, 92:229, 1994.

[23] S.D. Barthelmy et al. *Proc. 2nd Gamma Ray Burst Workshop (Huntsville)*, AIP Conf. Proc. 307:643, 1993.

[24] S.W. Hawking. *Nature*, 248:30, 1974.

[25] F. Halzen et al. *Nature*, 353:807, 1991.

[26] K. Griest. Plenary talk this workshop, 1994.

[27] H.F. Helmken et al. *Proc. 16th Int. Cosmic Ray Conf. (Kyoto)*, 1:120, 1979.

[28] A.M Hillas. *Proc. 19th Int. Cosmic Ray Conf. (La Jolla)*, 3:445, 1985.

[29] M.S. Longair. *High Energy Astrophysics, Vol 1*, Cambridge University Press, 275, 1992.

[30] G. Vacanti et al. *ApJ*, 377:467, 1991.

[31] T. Kifune et al. *ApJ*, 438:L91, 1995.

[32] P. Goret et al. *A&A*, 270:401, 1993.

[33] P. Baillon et al. *Astroparticle Phys.*, 1:341, 1993.

[34] C.C.G. Bowden et al. *Towards a Major Atmospheric Cherenkov Detector-II (Calgary)*, ed. R.C. Lamb, 230, 1993.

[35] D.E. Alexandreas et al. *ApJ*, 383:L53, 1991.

[36] M. Amenomori et al. *Phys. Rev. Lett.*, 69:2468, 1992.

[37] T.A. McKay et al. *ApJ*, 417:742, 1993.

[38] A. Karle et al. MPI-PhE/95-01, submitted to Astroparticle Phys., 1995.

[39] S. Torii. Private communication, 1994.

[40] T.K. Gaisser et al. *Report of Group R3B*, this workshop, 1994.

[41] S. Barwick et al. *Proc. 23rd Int. Cosmic Ray Conf. (Calgary)*, 1:436, 1993.

[42] S. Ahlen et al. *Nucl. Inst. and Meth.*, 351A:493, 1994.

[43] M. Teshima et al. *Proc. Tokyo Workshop on Techniques for the Study of EHE Cosmic Rays*, ed. M. Nagano, 109, 1993.

[44] R.A. Ong et al. *Proc. Towards a Major Atmospheric Cherenkov Detector-III (Tokyo)*, ed. T. Kifune, Universal Academy Press, 295, 1994.

[45] E. Pare. *Proc. Towards a Major Atmospheric Cherenkov Detector-II (Calgary)*, ed. R.C. Lamb, 250, 993.

[46] C. Akerlof. Talk presented this workshop, 1994.

Galactic Cosmic Rays up to 10 TeV

J.F. Ormes, J. Beatty, W. R. Binns, M. Wiedenbeck, A. Stephens, D. Müller, L. Barbier, A. Stephens, M. Salamon, F. C. Jones, E-S. Seo, R. Streitmatter

§1 Introduction

J.F. Ormes

Measurements of galactic cosmic rays in the energy range from 0.1 GeV to 10 TeV will make important contributions to particle and nuclear astrophysics both in the next few years and following the turn of the millennium. While our understanding of the astrophysical origins of cosmic rays are understood in broad outline, much remains to be done. The details of the nucleosynthetic processes that contribute to this unique sample of matter, the acceleration mechanism, and the effects of the transport of galactic cosmic rays in interstellar plasma are not yet understood. As a complement to studies of these phenomena, precision measurements of the cosmic rays include searches for primordial antimatter and signatures of dark matter that are of fundamental importance to cosmologists and high energy physicists who attended this conference.

New information is coming from gamma ray studies of the galaxy. Observations by experiments on board the Compton Gamma Ray Observatory (CGRO) have shown that nuclear cosmic rays are a natural component of galaxies like ours and the energy density in particles has been shown to be distributed in the same way as matter and magnetic fields on a scale length of 1 or 2 kpc. Observations by GCRO have also found evidence for accelerated carbon and oxygen nuclei in the Orion cloud complex, consistent with their acceleration in regions where star formation is active and stellar winds blow strongly. Of course we have known for some time from the radio maps that energetic electrons are a feature of not only the Milky Way but of spiral galaxies in general.

This section discusses those questions that can be answered by making precision measurements of individual components of the galactic cosmic rays. Such measurements are impossible to do from the ground because the particles interact in the upper atmosphere, and so observations from high altitude balloons and satellites are required to make progress. Balloon borne observations, where viable, are the method of choice because of their lower cost and the speed with which they can be carried out. An important development that greatly extends the capability of balloon borne experiments has recently been demonstrated. Balloon flights have traditionally been one day long but now flights of 10-20 days are possible. For example, several flights launched in Antarctica have lasted 10 days in perpetual sunlight. The balloons circulate around the south pole and the payloads are recovered near the launch site. The major limitation is now the commitment to build the experiments to carry out measurements exploiting this capability. Every effort should be made to convince NASA that balloon borne experiments should be considered as space missions and be provided with funding commensurate with their scientific capabilities[1]. In the coming years when severe budget pressures will exist there are a number of important balloon borne measurements that can be made. In the longer term, however, satellite borne experiments will be needed to carry out those measurements that must be free of atmospheric background or those requiring the statistical precision that can only be obtained by a long observing time.

The topics associated with cosmic rays below 10 TeV are naturally organized into three areas: composition at cosmic ray sources, acceleration processes, and transport processes. The latter is not only of intrinsic interest, but must also be understood so that we know the source spectra and can determine the abundances of nuclei prior to their acceleration and transport.

This chapter covers the potential for measurements from balloon and satellite borne instrumentation that will answer questions about the sources of cosmic rays and their transport. Questions regarding the acceleration mechanisms as probed by observations from balloons, in space and on the ground will be covered in the Chapter entitled "Over the Knee".

§2 Sources

2.1 Ultra-heavy nuclei

J. Beatty and W. R. Binns

Nuclei heavier than iron provide an important window on sources where the nucleosynthesis and acceleration of cosmic rays take place. Cosmic-ray nuclei are produced in the late stages of stellar evolution, and many nuclei are radioactive with mean lifetimes at timescales of interest for probing cosmic ray acceleration and propagation.

[1]It is worth noting that many NASA experiment developers for important space missions such as COBE and GRO cut their teeth scientifically in the balloon and rocket programs. Balloon borne experiments are an excellent training ground for young experimental scientists. They provide an opportunity for students to be involved in a project from start to finish, from conception phase, through implementation to the publication of results.

Present measurements of the elemental abundances of ultraheavy (UH) cosmic rays (Z $\geq$ 30) were obtained by the Heavy Nuclei Experiment (HNE) flown aboard the NASA HEAO-3 spacecraft[1] and by the Bristol University experiment flown aboard Ariel-6[2]. In addition, preliminary data for Z > 75 from the Dublin-ESTEC experiment flown aboard the NASA LDEF satellite has been reported recently. HEAO-3 and Ariel-6 resolved the even charge nuclei for 30 $\leq$ Z $\leq$ 60, and charge groups for $Z > 60$. The major conclusions derived from the HEAO and Ariel data are:

1. The abundance of those even charge elements measured for $30 \leq Z \leq 60$ are in general agreement with solar system abundances after a first-ionization potential (FIP) bias correction is applied, and they do not appear to be strongly enhanced in r-process nuclei.
2. Nuclei in the platinum-lead charge range ($74 \leq Z \leq 80$) and those in the range $62 \leq Z \leq 73$, which are largely secondaries from the platinum-lead region, appear to be enhanced over propagated solar system abundances by about a factor of two. Since these are predominantly r-process nuclei, this suggests an enhancement in the r-process contribution to the source nuclei in this charge range.
3. The abundance of the actinides was found to be higher than the solar system abundance. This result must be interpreted with caution since it is based on the detection of only two actinides in the combined HEAO/Ariel dataset.

To significantly advance the study of ultraheavy nuclei, two developments are required. Resolution must be improved so that the less abundant odd-Z nuclei can be clearly resolved, and large collecting areas must be deployed for long times to improve statistical accuracy of the measurements and provide adequate observations at the higher charges.

One of the major goals of the study of galactic ultraheavy cosmic rays is understanding of their nucleosynthetic origin. Nearly all nuclei with Z > 34 are synthesized by either rapid (r-process) or slow (s-process) neutron capture. Figure 1 shows the result of a propagation calculation which estimates the fraction of each charge which is primary after propagation; elements with Z > 34 are marked to indicate the dominant nucleosynthetic process for that element. Figure 2 shows the predominantly primary r- and s-process nuclei and their first ionization potential. Note that the measured nuclei of each nucleosynthetic class lie in a relatively narrow band of FIP (9-10 eV for r-process, 5-7 eV for s-process). Thus the ubiquitous phenomenon of FIP bias has not been experimentally verified for ultraheavy nuclei. In particular, the r-process nuclei are synthesized under violent astrophysical conditions in supernovae; if accelerated in these events, they might exhibit very different fractionation effects than found in normal stellar atmospheres, where the FIP selection is best understood. Measurements of the low-FIP element ^{37}Rb could address this issue, and can be made using long-duration balloon payloads with the dE/dx vs. time-of-flight technique. A space flight of a similar instrument will be required to provide good statistical coverage of the elements lighter than Pb. Measurement of the U/Th ratio, important for determining the time elapsed from cosmic ray nucleosynthesis to observation[3], remains accessible only to very large experiments with long exposures flown in space, perhaps using passive detector techniques.

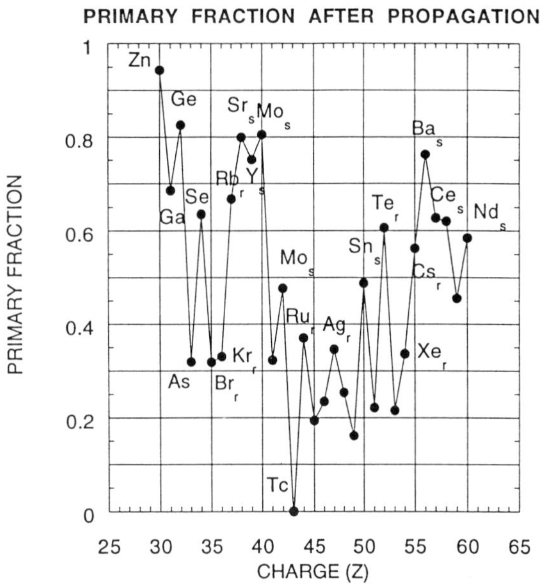

Figure 1: The primary fraction of each element is given after propagation through 7 g/cm^2 of hydrogen. A solar system composition with a "step FIP" was assumed for the source abundances.

§References

[1] W. R. Binns, T. L. Garrerd, P. S. Gibner, M. H. Israel, M. P. Kertzman, J. Klarmann, B. J. Newport, E. C. Stone, and C. J. Waddington. *Ap.J.*, 346, 997, 1989.

[2] P. H. Fowler, R. N. F. Walker, M. R. W. Masheder, R. T. Moses, A. Worley and A. M. Gay. *Ap.J.*, 314, 739, 1987.

[3] J. B. Blake and D. N. Schramm. *Astrophys. and Space Science*, 30, 275, 1974.

2.2 Cosmic-Ray Isotopes

M. Wiedenbeck

The most detailed and precise information obtained from studies of the composition of the nuclear component of galactic cosmic rays has come from measurements of isotopic composition at low energies (near 1 GeV/nucleon). Isotope experiments now in space or under development are expected to measure the abundances of essentially all of the stable and long-lived nuclides up through the iron group ($Z \sim 30$).

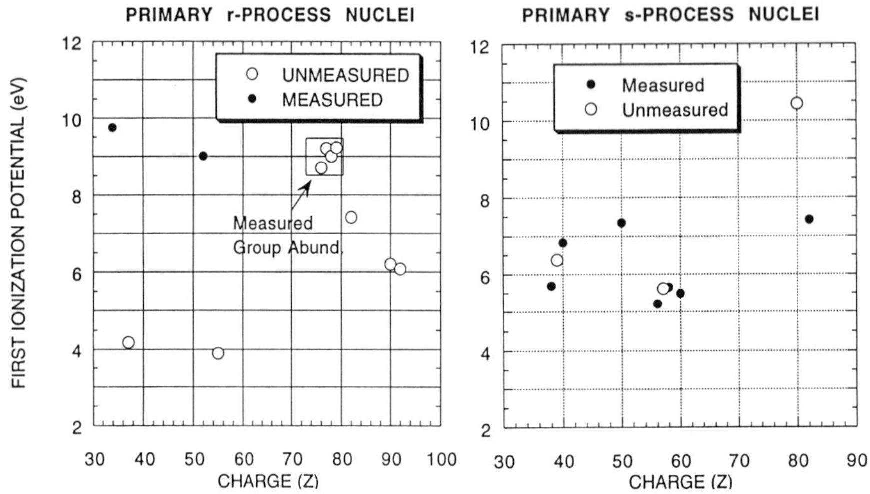

Figure 2: The first ionization potential of the "primary" r- and s-process elements are plotted vs. charge in the left and right panels respectivley. Those elements for which abundances have been individually measured are shown as filled points while those which have not yet been measured are shown as open circles.

Further progress with cosmic ray isotope studies will be particularly challenging because it must combine the high resolution needed to separate isotopes with the very large collecting areas needed to obtain adequate statistics for the rare ultraheavy species.

There is, however, great scientific benefit which can be realized from successfully carrying out such a program. The endothermic nuclear reactions that are responsible for producing the ultraheavies are fundamentally different from the exothermic fusion reactions which make the lighter nuclei[1]. By establishing the isotopic abundances patterns among the ultraheavies one should be able to explore contributions from a wide range of possible nucleosynthesis environments (temperature, density, neutron flux) by establishing the patterns formed by s- and r-process nuclei. To extract such information about the nucleosynthesis of the primary cosmic rays one must rely on species which are not greatly contaminated by secondaries produced by spallation during passage through the interstellar medium. Ultraheavy isotopes are particularly favorable from this point of view since the rapid fall off of abundances with increasing atomic number also means that there are few parent nuclei from which spallation products can be made. Typically elements in the range $30 \leq Z \leq 50$ have three or more isotopes which have significant primary contributions and are suitable for use in nucleosynthesis studies.

This same range of elements is also a rich source for radioactive clock nuclei which can be used to probe the propagation history of galactic cosmic rays[2]. From the lighter species which have already been investigated (particularly ^{10}Be) a mean confinement time in the galaxy has been obtained. However, in order to understand how cosmic rays range through the disk and halo of the galaxy one must take the next step and determine the distribution of interstellar densities sampled by cosmic rays in order to produce this mean. This requires comparing the abundances of beta-decay nuclides with a wide range of half-lives. In just the narrow range of elements from $Z = 34$ to $Z = 43$ one has a suitable set of seven clock isotopes with half-lives ranging from 2×10^4 to 5×10^8 years[2]. expected surviving fraction of those isotopes for various confinement times in the galaxy.

The survival of secondary nuclides which can decay only by electron capture is sensitive to the density of the medium through which the cosmic rays propagate because when a bare nucleus attaches an electron from the interstellar gas its survival or decay depends on the relative times for electron capture and for undergoing a collision which will strip off the electron[3]. The electron attachment cross section is a strong function both of the particle's atomic number and its velocity. At the interstellar energies typical of the cosmic rays for which isotope identification is possible with current techniques, electron attachment becomes significant only for ultraheavy elements. Over the charge range $30 \leq Z \leq 50$, pure electron capture nuclides are available for probing interstellar densities ranging from less than 10^{-2} to greater than 10^6 atoms/cm^3. Using measurements of the abundances of these isotopes it should be possible, for example, to determine whether a significant fraction of the material traversed by cosmic rays during their propagation is contained in high density regions such as one might find surrounding some types of sources.

The scientific questions cited above as amenable to investigation using measurements of the composition of ultraheavy cosmic rays are straightforward extrapolations from questions which have been considered for many years in connection with lighter cosmic ray nuclei. It should be expected that future

[2]See discussion under Transport, especially in the sub-section on "Halo, diffusion, galactic wind".

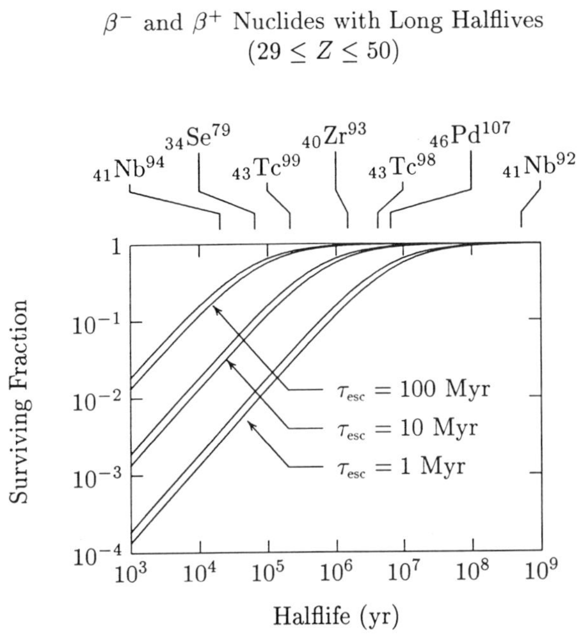

Figure 3: Curves show the expected dependence of the survuving fraction of β-decay secondaries on the half-life and the confinement time (τ^{esc}) in the galaxy for the simplest leaky-box modelof cosmic ray propagation. Specific β-active nuclides in the range $29 \leq Z \leq 50$ are shown along the top of the graph. Comparisons of surviving fractions for nuclides of significantly different half-lives can serve as a discriminater between more complex alternative models of cosmic-ray propagation.

theoretical studies of the origin of these particles will uncover additional, novel problems which are uniquely suitable for investigation using such measurements of the abundances of ultraheavy isotopes.

Meeting the experimental challenge of measuring *isotopic* abundances of ultraheavy elements will require a concerted effort to develop suitable sensor systems. An evolutionary approach is possible, starting with small prototypes which can be tested with accelerator beams of heavy ions to meet the stringent resolution requirements for separating ultraheavy isotopes. Exposure of larger detector systems (~1 m^2 sr), possibly involving mosaics of small, high-resolution devices, can be carried out with long-duration balloon payloads and MidEx-class satellites to study isotopes in the charge range up to $Z \simeq 40$ as well as to further investigate the iron-group with unprecedented precision. Comprehensive studies of ultraheavy isotopes over a broad range of elements could become a reality in the early part of the next century with the exposure of very large detector arrays (10 to 100 m^2 sr) on suitable platforms in space.

§References

[1] J. Adouze and S. Vauclair. *An Introduction to Nuclear Astrophysics.* D. Reidel Publ. (Dordrecht, Holland) 1980.

[2] C. M. Lederer, V. Shirly, E. Browne, J. M. Dairiki, R. E. Doebler, A. A. Shihab-Eldin, L. J. Jardine, J. K. Tali, and A. B. Buyrn. *Table of Isotopes, Seventh Edition.* John Wiley & Sons (New York) 1978.

[3] L. Wilson. *The Nuclear and Atomic Physics Governing Changes in the Composition of Relativistic Cosmic Rays.* Lawrence Berkeley Laboratory Report LBL-7723, PhD thesis, University of California Berkeley, 1978.

2.3 Electrons and Positrons in the Cosmic Radiation

A. Stephens, D. Müller and L. Barbier

Electrons and positrons are a much less abundant component of the cosmic rays than protons and nuclei. However, their origin and propagation through the galaxy are subject to a variety of electromagnetic interactions which do not affect nuclei, and they therefore are unique messengers of some fundamental physical processes in the galaxy.

It is reasonable to assume that the material from which the nuclear cosmic rays originate also contains negative electrons which are accelerated together with the nuclei, presumably by first-order Fermi processes in supernova shocks. However, the subsequent propagation of electrons and nuclei through the galaxy is very different: bremsstrahlung, synchrotron radiation and inverse Compton scattering off interstellar photons lead to rapid energy losses of electrons such that their lifetime in the galaxy decreases with energy, essentially proportional to E^{-1}. Consequently, the energy spectrum of electrons decreases with energy much more rapidly than that of nuclei. Observations have confirmed this behavior for the total electron flux ($e^+ + e^-$) up to energies beyond 1 TeV. At these highest energies, electrons cannot travel galactic distances of more than a few hundred parsec, *i.e.* they must come from rather "nearby" sources. Future measurements of electrons in the 1-10 TeV region therefore offer the exciting prospect of identifying individual cosmic ray sources in the galactic neighborhood of the Solar System.

Much more detail can be derived from separate measurements of electrons and positrons but presently available data do not extend beyond 20 GeV or so. At a few GeV, the measured positron fraction $e^+/(e^+ + e^-)$ seems to be the order of 10% as shown in Figure 4. Evidently, most electrons in this energy region are negative and come from primary acceleration sites. The observed positron fraction can be well explained as due to collisions of nuclear cosmic rays with the interstellar gas, leading to secondary pions which decay into electrons and positrons via the $\pi^\pm \to \mu^\pm \to e^\pm$ chain. However, at energies above ~ 10 GeV, an apparent increase in the positron fraction is observed which cannot easily ascribed to secondary production. Thus, the search for the origin of this apparent positron excess is a major motivation for the current interest in more accurate measurements that also extend to much higher energies.

To explain the excess a number of possible mechanisms have been suggested. For instance, $e^\pm$ pairs could be generated through direct pair production by photons in the strong

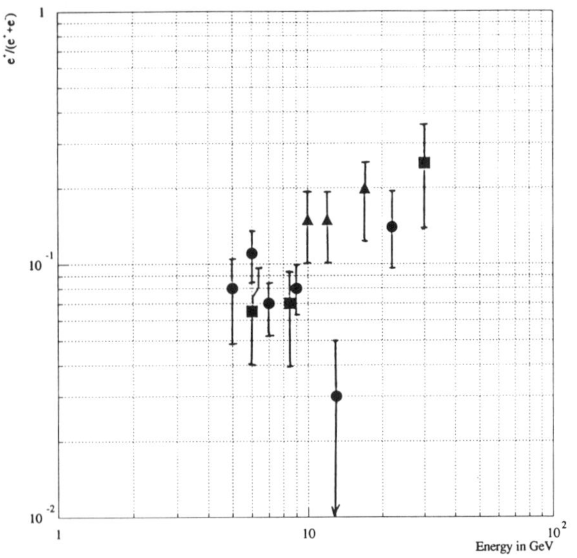

Figure 4: The measured positron - to -(electron + positron) ratio as a function of energy. The data are from: solid circles, reference [1]; solid triangles, reference [2]; solid squares, reference [3]

magnetic field of pulsars. Indeed, electron-positron pair plasmas seems to form a significant part of pulsar winds, and their annihilation could lead to the diffuse galactic gamma ray background of 511 keV. Positrons may also come from the decay of radioactive nuclei produced by explosive nucleosynthesis in supernovae. Most fascinating is a pair production process that may occur in the annihilation of dark matter particles if they exist in the galactic halo. This process would lead to $e^{\pm}$ pairs whose energy is related directly to the mass of the progenitor dark matter particles, and would produce a peak in the positron fraction at energies somewhere between ~50 and several hundred GeV.

Present data on galactic electrons and positrons above 200 MeV come exclusively from balloon borne experiments. Up to ~ 2 GeV however, secondary production in the atmosphere severely restricts the accuracy of positron measurements. Currently, several balloon borne magnet spectrometers are used to improve the existing data set and to extend the energy range of the measurements. However, above 50-100 GeV, the positron flux is likely to become so low that major progress in the important high energy region must be accomplished with measurements in space. Similarly, at low energies, below a few GeV, data that are unbiased by atmospheric background can only be obtained from space. A specific objective in the low energy region are long term observations to study charge-sign dependent effects of solar modulation of cosmic rays.

Summarizing, we expect from ongoing and proposed measurements of cosmic ray positrons and electrons over a wide energy region key information on a great variety of topics in particle astrophysics: the nature and distribution of galactic cosmic ray sources, high energy processes near compact objects, electromagnetic interactions in interstellar space, modulation of cosmic rays in the heliosphere, and perhaps even information on cosmological dark matter particles.

§References

[1] R. L. Golden, S. A. Stephens, B. G. Mauger, G. D. Badhwar, R. R. Daniel, S. Horan, J. L. Lacy and J. E. Zipse. *Astron. and Astrophys.*, 188:145, 1987.

[2] D. Müller and K. Tang. *Ap. J.*, 312:183, 1987.

[3] A. Buffington, C. D. Orth and G. F. Smoot. *Ap. J.*, 199:669, 1975.

2.4 Ultra high energy electrons

A. Stephens

Unlike the nucleonic component, high energy cosmic ray electrons suffer severe energy loss in the galaxy while traversing ambient magnetic and radiation fields, through synchrotron and inverse Compton processes respectively. The life time of electrons against these losses decreases linearly with increasing energy. As a result, high energy electrons are depleted and cannot arrive from sources at large distances.

The lifetime of a 1 TeV electron is about 10^5 yr and that of a 10 TeV electron is about 10^4 yr. At these energies it is expected that the energy spectrum is influenced by nearby sources and provides valuable information on the transport of cosmic rays in the local environment. In order to get 1000 electrons of energy > 1 TeV, one needs an exposure factor of about $1.7\text{x}10^7$ (m^2 sr s) and the required exposure factor to get 100 electrons of energy > 10 TeV is $2.8\text{x}10^8$ (m^2 sr s). It is important to measure the energy spectrum of electrons accurately above 1 TeV using a conventional detector even in a space platform, where large exposure time of the order of a few years can be achieved.

High energy cosmic ray electrons are the only probe of nearby sources where the properties of the interstellar medium through which they diffuse are known in detail. Studies of the balance between cosmic ray pressure, magnetic field energy density and other constituents in this region of interstellar space will be important to understanding the details of energetic particle plasma interactions throughout the interstellar medium.

2.5 Antimatter in the Cosmic Radiation

M. Salamon

One of the most profound questions in cosmology and particle physics arises from the manifest baryon asymmetry of our Universe: why is our observed Universe apparently composed exclusively of matter, with only trace amounts of (secondary) antimatter? At very early epochs after the Big Bang, when temperatures were much higher than the masses of the hypothesized gauge bosons thought to violate B conservation, a state of thermal equilibrium existed among these particle species. This fact, coupled with overall CPT invariance, ensures that equal amounts of matter and antimatter were present in the early Universe. Today, however, we know that the antimatter/matter ratio in our galaxy is less than 10^{-15}, and is less than 10^{-5} within clusters of galaxies, based on the absence of gamma rays that would be otherwise produced in matter-antimatter annihilations[1].

At some point in the history of our Universe, then, baryogenesis occurred. Sakharov originally specified the conditions required for baryogenesis to occur:

1. baryon number (B) violation;
2. both C and CP violation;
3. departure from thermal equilibrium.

Expansion of the Universe allows for (hypothesized) B-violating decays to occur out of thermal equilibrium, and both C and CP violation are known to occur in nature. CP violation has been observed only in the K_0-$\bar{K}_0$ system, however, with a strength orders of magnitude too small to explain the present matter-antimatter asymmetry. Thus an as-yet unidentified CP violating process must be responsible for the observed asymmetry, which may either be intrinsic ("hard") or spontaneous ("soft"). An explicit ("hard") CP-violating term in particles' Lagrangian would result in a Universe in which the baryon asymmetry is the same everywhere. Spontaneous CP-symmetry breaking, however, might allow causally disconnected regions of the early Universe to develop baryon asymmetries of differing signs, so that our Universe might now contain neighboring "domains" of matter and antimatter that are presently causally linked [2].

The observed extragalactic, diffuse gamma ray flux limits the scale size of these domains to be on the order of tens of Mpc or larger. If these domains are indeed of supercluster size, then it is possible that the existence of neighboring antimatter domains can be unequivocally established by the direct detection at Earth of (antimatter) cosmic rays originating in these domains. Such an observation would have the profoundest consequences for cosmology, particle physics, and astrophysics. To conclusively demonstrate the existence of antimatter domains, consisting of antigalaxies and antistars, all that is needed is the detection of a few antinuclei of atomic mass $A > 3$ in the cosmic radiation. Such nuclei could not be produced by any known process within our matter-dominated sector of the Universe, and would therefore signal the presence of antimatter domains.

The numerous searches for cosmic ray antimatter that have been made fall into two classes: those looking for $A > 3$ antinuclei, and those looking for antiprotons and/or positrons. The latter class of experiments do not address the issue of antimatter domains, because both antiprotons and positrons are produced in collisions of cosmic rays with ambient matter, leading to a cosmic ray antiproton/proton ratio of the order 10^{-4} at energies above a GeV. (Positrons are also produced as decay products of radioactive isotopes synthesized by supernovae. In addition, antiproton production by WIMP annihilations in the galactic halo has been hypothesized.) Nevertheless, these antiparticles contain important information relating to the transport and confinement of cosmic rays within the galaxy, and a major goal of current cosmic ray research is to determine the energy dependences of the antiproton and positron fluxes, as described more fully in following sections.

The present experimental limits on the cosmic-ray antimatter/matter ratio (nuclei with $A > 3$) are on the order of 10^{-4} to 10^{-5}, and have all been achieved with balloon payloads, whose short flight times and limited areas have so far restricted the limits to values which are of marginal cosmological significance. One must achieve ratio sensitivities much better than 10^{-4} to probe the extragalactic component, because, from energetics arguments, this component constitutes at most $\sim 10^{-4}$ of the total cosmic ray flux at Earth. Unfortunately, however, we cannot calculate at this time just what antimatter cosmic ray flux we would expect from an antimatter domain at a specified distance. The major uncertainties are in the transport of cosmic rays through extragalactic space, and their penetration into our galaxy. At $\sim$ GeV energies (where the cosmic ray flux is maximal), transport through extragalactic space would probably be diffusive; with the extragalactic magnetic field power spectrum being unknown, estimates of the diffusion distance range from 10^0 to 10^3 Mpc for a Hubble time. In addition, only a fraction of extragalactic cosmic rays reaching our galaxy penetrate it, due to the possible (but as yet unconfirmed) presence of a galactic wind, powered by supernovae, that would convect away most low-energy cosmic rays. Previous estimates of the penetrating fraction give a local antimatter/matter cosmic ray ratio of $\sim 10^{-6}$ at 1 GeV, but in fact this number is effectively unknown. With the advent of long-duration balloon flights and future experiments in space, these limits will hopefully be pushed to levels that do begin to probe neighboring domains, should they exist. There is little, if any, theoretical guidance to dictate what antimatter sensitivities are required, so that even a few orders of magnitude of improvement may yield profound discoveries. Long duration balloon payloads are expected to bring $^4\bar{He}/^4He$ down to $\sim 10^{-7}$ (possibly 10^{-8}). Exposures from space based platforms will be required to achieve ratio limits for $^4\bar{He}/^4He$ of $\sim 10^{-9}$ to 10^{-10}, and for $\bar{Z}/Z$ of $\sim 10^{-8}$, $Z>2$.

An experiment has been proposed to be installed on board the International Space Station Alpha (ISSA) for a duration of 3 years, beginning in late 2000. The spectrometer would consist of a Nd-B-Fe permanent magnet with a bore aperture of 0.8 m^2-sr filled with a 1.7 kG field, yielding an analyzing power of 0.15 T-m^2. The total cost of the magnet is quite small, being < $1M. Inside the magnet bore will either be a TPC (Ne:CH_4), providing $\sim$ 100 3-D measurements of the trajectory with sigma-x = 125 μm, sigma-y = 50 μm, and sigma-z = 400 μm for 4 GeV/nucleon C, or a silicon detector, consisting of six layers of double-sided silicon detectors with 10 micron resolution in the bending plane. Above and below the magnet assembly are time-of-flight (TOF) scintillator pairs and strip detectors for rejection of albedo (upward-moving) cosmic rays that mimic downward-moving anti-nuclei, and for on-line rejection of positive-curvature (normal nuclei) trajectories. At the bottom of the instrument a space is provided for a RICH or other detector to provide independent albedo rejection as well as additional particle identification power and Z and β resolution. Along the magnet sides are anticoincidence (veto) scintillators to reject sprays from particles interacting in the magnet walls that would confuse the TOF measurements. Monte Carlo calculations show that background rejection factors on the order of 10^{-10} are achievable with these designs. Total power consumption will be under 1 kW, and the data output rate will be $\sim$ 200 kbit/s, both compatible with the allocated resources of the ISSA.

Assuming funding for these experiments is forthcoming, in the next several years it is very likely that we will probe the extragalactic low-energy ($\sim$ GeV) cosmic ray component, and with that achievement we can begin to address one of the most fundamental questions in Nature, what is the origin of the excess matter in our Universe?

§References

[1] G. Steigman. *Ann Rev. Astron. and Astrophs.*, 14:339, 1976.

[2] F. W. Stecker. *Nuclear Physics B*, 252:25, 1985.

§3 Acceleration and Transport of Galactic Cosmic Rays

J. F. Ormes

A major theme of this conference has been the importance of the details of the cosmic ray composition and energy spectra in the range of 10^{14} to 10^{16} eV[3]. It is in this energy range that the endpoint of acceleration by supernova shocks is expected to occur. The signature is predicted to be cutoffs in the energy spectra of individual elemental species at a total energy per particle$E_{\rm max} \cdot A = 10^{14} - 10^{16} \cdot Z$ eV. The upper limit to the energy to which particles can be accelerated depends on the geometry of the magnetic field keeping particles in the region of the shock where they gain energy at each crossing of the shock boundary. Precision measurements of the energy spectrum of the individual primary components (H, He, C- O, Si, and Fe) are required to test this theoretically attractive scenario. This will not be easy. The spectra are very steeply falling ($dN/dE \propto E^{-2.7}$) and particle fluxes are very low, so obtaining statistically significant samples of events at energies above 10^{15} eV using detectors in space will be difficult. On the ground, large area detectors can be built to solve the observing power problem. Much can be done by studying the properties of the air showers so that on a statistical basis, some idea of the nature of the composition and shape of the energy spectra can be obtained, but the identity of the incident primary particle is lost in the details of the fluctuations of the properties of individual air shower cascades. The emerging strategy is to make direct measurements in space (*i.e.* on balloons or on satellites) to as high an energy as possible while at the same time trying to measure as many properties as possible of air showers that are sensitive to the particle identities at the top of the atmosphere. If a sufficient overlap can be made (*e.g.* an order of magnitude in energy) demonstrating that air shower techniques give composition consistent with that directly observed above the atmosphere, then the composition at higher energies where space based measurements are out of the question can be inferred from data taken by the ground based arrays.

In order to understand the acceleration mechanism and determine the nuclear composition of cosmic rays at their source, we need to know how to correct for the passage of nuclei through the interstellar medium. The spectra of nuclei absent at the cosmic ray sources are used as tracers of this transport. The spectra of these so called secondary nuclei are steeper than those of their progenitors from the sources, and this is believed to be the result of the energy dependent scattering of these particles in the interstellar magnetic fields that keeps them there approximately 1000 times longer than if there weren't any scattering. It is observed that higher energy particles leave the galaxy more quickly than lower energy ones. There are theoretical reasons to believe that the scattering of particles on magnetic irregularities should be less effective as energy increases, thus making the secondary spectra steeper by a power 1/3, but the observed spectra of secondary nuclei are steeper by 0.5 to 0.6, implying that the source spectra are harder than those observed by a similar amount. It has been proposed that this steepening is due to some continuous acceleration of the particles during their transport. The shape of the spectra could be affected by convective transport as well as diffusive processes, but in any case the processes that change the spectra during the transport must be understood before we can determine the spectrum produced at the cosmic ray sources and hence learn about the acceleration mechanism.

3.1 Transport of Energetic Particles; Theory

F. C. Jones

Most calculations for the propagation of cosmic rays through the galaxy employ the "Leaky Box" model under the assumption that it can simulate most of the important characteristics of the real galaxy insofar as the production of secondary nuclei is concerned. However, when one considers the survival of unstable nuclei results can depend significantly on the particular model of galactic propagation employed.

The basic equation for the leaky box is

$$\frac{\partial N}{\partial t} + v\rho\frac{\partial N}{\partial x} + \frac{N}{\tau} = S\delta(x) \qquad (1)$$

where N is the cosmic ray density, v the particle's speed, ρ is the matter density in g cm^{-3} and x is the path length in g cm^{-2}. The injection into the leaky box is given by the

[3] See Chapter entitled "Over the Knee" for more information on experiments to measure composition in this energy range.

quantity S and occurs at zero path length. The overall escape time of the leaky box is given by the "lifetime" parameter, τ.

If we consider the steady state case equation(1) becomes

$$\frac{\partial N}{\partial x} = -\frac{N}{v\rho\tau} \tag{2}$$

with solution

$$N = \exp\left(\frac{x}{v\rho\tau}\right) = \exp\left(\frac{x}{\langle x\rangle}\right) \tag{3}$$

where the mean path length $\langle x\rangle = v\rho\tau$.

If we now inject a burst of particles at $t = 0$ and ignore (*i. e.* integrate over) the pathlength variable, x we obtain

$$\frac{\partial N}{\partial t} = -\frac{N}{\tau} \tag{4}$$

which immediately yields

$$N = \exp\left(\frac{t}{\tau}\right). \tag{5}$$

Here the mean age and mean path length of the particles are simply proportional to each other and if one can obtain a measure of each quantity (via primary to secondary ratios and surviving fraction of unstable nuclei) one has a simple measure of the mean matter density, ρ in the "leaky box".

Consider a somewhat more complex model, a box in which particles must diffuse to the boundaries before leaking out. The apropriate equation is

$$\frac{\partial N}{\partial t} + v\rho\frac{\partial N}{\partial x} - \kappa\frac{\partial^2 N}{\partial r^2} = S\delta(x) \tag{6}$$

where the simple lifetime or decay term has been replaced with a diffusion operator with a diffusion coefficient, κ and a spatial variable, r. If we impose the usual free escape boundary conditions, *i. e.* $N = 0$ on the boundary, this equation may be solved by expanding N in a complete set of orthonormal eigenfunctions of the diffusion operator with the appropriate behavior on the boundary. If we set

$$N = \sum_i a_i(t, x)N_i(r) \tag{7}$$

with

$$\kappa\frac{\partial^2 N}{\partial r^2} = -k_i^2 N_i \tag{8}$$

and insert this into equation (6) we obtain the set of equations

$$\frac{\partial a_i}{\partial t} + v\rho\frac{\partial a_i}{\partial x} + k_i^2 a_i = S_i\delta(x) \tag{9}$$

where

$$S_i = \int dr N_i(r)S(r). \tag{10}$$

We see at once that for each eigenmode we have an equation that is identical to the basic equation of the leaky box, equation (1). We may therefore immediately write the solutions

$$\begin{aligned}\langle\tau_i\rangle &= k_i^{-2}\\ \langle x_i\rangle &= v\rho\langle\tau_i\rangle\end{aligned} \tag{11}$$

Replacing a simple lifetime parameter with a diffusion operator has not changed the basic relationship between the mean path length and the mean age of the cosmic rays. The path length and age distributions are not necessarily exponentials any longer, however, the smallest value of k_i and hence the longest value of τ_i are associated with the fundamental eigenmode, k_0. This mode is characteristic of the diffusing volume as a whole and hence for times, $t \gg D^2/\kappa$ where D is the scale of the volume, the age distribution is approximately exponential with mean time τ_0 and for path length larger than $v\rho\tau_0$ the path length distribution is also approximately exponential with a mean equal to $v\rho\tau_0$. At short path lengths the exponential shape is broken and it is possible for the age and path length distributions to be richer or depleted in these smaller values depending on whether the source distribution is clumped around or excluded from the point of observation respectively. The important thing is that the mean age and path length are related by $\langle x\rangle = v\rho\langle\tau\rangle$ no matter what combination of eigenfunctions are excited. Therefore, the leaky box model requires only a little modification to fit the data for both primary to secondary ratios and the survival of unstable nuclei.

This is true because in both models the matter density, ρ is considered to be a constant over all space. If we substitute $\rho \to \rho(r)$ in equation (6) we obtain, instead of equation (11) the following relation

$$\begin{aligned}\langle\tau_i\rangle &= k_i^{-2}\\ \langle x_i\rangle &= v\rho_i\langle\tau_i\rangle\end{aligned} \tag{12}$$

where

$$\rho_i = \int dr N_i(r)\rho(r). \tag{13}$$

Now $\langle x\rangle \neq v\rho\langle\tau\rangle$ and the relationship between the mean path length and the mean age is broken.

An example is the diffusive model of the galaxy in which the cosmic rays extend beyond the galactic disk (thickness $\sim$ 200 pc) in a larger spherical volume (radius $\sim$ 10-15 kpc) or "halo" (*eg.* see Berezinskii *et al.* 1990). For simplicity consider a one dimensional model with all of the matter concentrated in a thin disk of thickness $2a$ in the center of a massless halo of thickness $2D$ where $a \ll D$. This model may be described by the equation

$$\frac{\partial N}{\partial t} + 2av\rho\delta(r)\frac{\partial N}{\partial x} - \kappa\frac{\partial^2 N}{\partial r^2} = S(t)\delta(r)\delta(x) \tag{14}$$

where $2a\rho$ is the column density of the thin matter disk. Once again we consider the steady state case and integrate the equation from $r = 0 - \epsilon$ to $r = 0 + \epsilon$ to obtain

$$2av\rho\frac{\partial N}{\partial x} - \kappa\left[\frac{\partial N}{\partial r}\right]_{0-\epsilon}^{0+\epsilon} = S\delta(x) \tag{15}$$

Consider the source to be concentrated at $r = 0$ and require $N(\pm D) = 0$ the jump in the slope at $r = 0$ is just $-2N/D$ and we have

$$2av\rho\frac{\partial N}{\partial x} + \frac{2\kappa N}{D} = 0 \tag{16}$$

This equation also gives an exponential path length distribution as in equation (3) with the mean path length given by

$$\langle x\rangle = \frac{av\rho D}{\kappa} = \left(\frac{av\rho}{D}\right)\frac{D^2}{\kappa} = v\langle\rho\rangle\frac{D^2}{\kappa}. \tag{17}$$

Since D^2/κ is just the mean time for diffusing particles to escape the galaxy it would appear that we have, once again, arrived at the relationship $\langle x\rangle = v\langle\rho\rangle\langle\tau\rangle$. Actually the quantity $\frac{D^2}{\kappa}$ is the time scale for the fundamental mode only and the higher order modes contribute significantly on time scales shorter than this. In fact, for times short compared to the fundamental time scale $\tau_0 = D^2/\kappa$ the age distribution is proportional to $t^{-1/2}$ and decays exponentially with decay time $\approx \tau_0$ for longer times. The surviving fraction, f is given by

$$f = \frac{\int_0^\infty \exp(-\frac{t}{\tau_d})N(t)dt}{\int_0^\infty N(t)dt} \tag{18}$$

where $N(t)$ is the age distribution. If $N(t)$ were a simple exponential as in the previous examples the surviving fraction would be simply

$$f = \frac{1}{1+\tau_0/\tau_d} \tag{19}$$

with a mean age given by

$$\langle t\rangle = \tau_0. \tag{20}$$

For long lived nuclei with $\tau_d \gg \tau_0$ there is little difference since from equation (18) we see that f depends on the overall time scale of $N(t)$ and in this case the surviving fraction is approximately

$$f \approx \frac{1}{3}\exp(\tau_0/\tau_d)\left[2 + \frac{1}{1+\frac{\tau_0}{\tau_d}}\right] \approx 1 - \frac{4\tau_0}{3\tau_d} = 1 - \frac{3\langle t\rangle}{2\tau_d} \tag{21}$$

whereas the leaky box gives

$$f = \frac{1}{1+\tau_0/\tau_d} \approx 1 - \tau_0/\tau_d = 1 - \frac{\langle t\rangle}{\tau_d}. \tag{22}$$

This minor difference becomes major, however, if the nuclei decay time is short compared to the long time scale; for here the surviving fraction is

$$f \approx \sqrt{\frac{\pi\tau_d}{9\tau_0}} = \sqrt{\frac{8\pi\tau_d}{9\cdot 9\langle t\rangle}} \tag{23}$$

which can differ from the leaky box expression,

$$f \approx \frac{\tau_d}{\tau_0} = \frac{\tau_d}{\langle t\rangle}, \tag{24}$$

by a significant amount.

One needs a range of decay times for the unstable nuclei in order to correctly determine the true residence time for cosmic rays in the galaxy. Simply measuring one decay time with great accuracy and assuming a model such as the leaky box can lead to seriously inaccurate results.

Assuming the galaxy has a convective halo does not change things significantly; the outer boundary of the halo is changed to an "effective" boundary D^* whenever $D^* < D$. This effective boundary applies to those particles whose diffusion coefficient is small enough that beyond this boundary diffusion back to the disk cannot compete with the outflowing wind and is given by

$$D^* = \frac{\kappa}{V} \tag{25}$$

where V is the halo wind speed.

3.2 *Stochastic Reacceleration of Cosmic Rays in the Interstellar Medium*

E-S. Seo

Second order Fermi process acceleration of cosmic rays over the whole galaxy cannot serve as the main mechanism for accelerating particles with energies $E > 1$ GeV/nucleon. In this strong continuous acceleration case, particles of higher energy should spend a longer time in the system, which would result in an increase in the abundance of secondary nuclei as energy increases. In fact, observations show that both the abundance of secondary nuclei and the corresponding escape length x_e decrease with energy for $E > 1$ GeV/nucleon. Hence significant continuous interstellar acceleration is excluded for these energies.

Numerical calculations have shown that a modest amount of diffusive reacceleration of cosmic rays can reproduce the observed energy dependence of secondary abundances if it is accompanied by a weaker decrease of the escape length with energy as compared to the case with an absence of any reacceleration. The advantage of all models with reacceleration is a weak energy dependence of cosmic ray anisotropy, evidently compatible with the measurements around 10^{12} - 10^{14} eV. The models also reproduce the observed ratio of secondary to primary nuclei with a single power law dependence of escape length on particle rigidity. This is in contrast to the empirical models without reacceleration, where the escape length must flatten at energies below ~1 GeV/nucleon such as might be expected by the action of a galactic wind and/or turbulent diffusion.

However, it is still not clear whether reacceleration is compatible with data at low energies where the effects of diffusive reacceleration should be most prominent. An energy boost for low energy cosmic rays could help explain the observed elemental and isotopic abundance data. On the other hand, there is controversy about whether or not the H and He spectra below 5 GeV/nucleon are reproduced in the models with reacceleration. Recent studies of the effect of reacceleration on cosmic rays over a wide charge and energy range use models that take into account only inevitable stochastic reacceleration of cosmic rays by the turbulence that supposedly exists in the interstellar medium and is invoked to explain cosmic ray scattering and confinement in the galaxy [1, 2].

Reacceleration effects are stronger for the lighter elements. Both the H spectrum and H/He ratio at ~1 GeV/n are sensitive to the amount of energy gain by reacceleration, so it is important to have accurate measurements of H and He spectra simultaneously over a wide energy range. To determine the difference in source spectra for H and He, there is also a special need for accurate measurements of the H/He ratio in the energy range from 100 GeV/n to 1 TeV/n, where few measurements have yet been made. Very precise measurements of the H and ^{2}H spectra are required to resolve whether or not diffusive reacceleration is an important effect. The ^{2}H production cross section has a feature that should be reflected in the ^{2}H spectrum. Comparision of the ^{2}H spectrum to the H spectrum where the feature is absent provides the experimental test. If distributed acceleration takes place, the feature will be smoothed out. Since solar modulation also might smooth out this feature, measurements from a few hundred MeV/nucleon to a few GeV/nucleon at solar minimum are required to provide a test for the diffusive acceleration hypothesis. Secondary to primary ratios such as B/C up to the highest possible energy can provide a good test for the weak energy dependence of the escape length implied by reacceleration models.

§References

[1] E. S. Seo and V. S. Ptuskin. *Ap. J.* 431, 705, 1994.

[2] U. Heinbach and M. Simon. *Ap. J.* , *In Press* 1995.

3.3 Halo and Galactic Wind

R. Streitmatter

What is the size of the galactic halo occupied by the cosmic radiation? Is there a galactic wind which sweeps cosmic rays, gas and magnetic fields outward from the disk?

As discussed above, the propagation model most often used in the interpretation of data on the cosmic radiation has been the "leaky box" model, an equilibrium model in which sources, gas density, composition, and observed energy spectra are homogeneous in space and constant in time. As a tool for achieving an understanding of the astrophysical setting of the cosmic radiation, the leaky box model is peculiarly unsatisfactory. Its value is in placing constraints on source spectra and providing information on the mean amount of material traversed by the cosmic radiation. However, the model is largely a mathematical scheme for organizing the data by transforming source spectra to observed spectra at Earth. It contains no physical detail and says nothing about the physical character of the propagation, the size of the containment volume, or the manner in which particles escape from it.

There exist models for the propagation of the cosmic radiation which are more physically plausible in their use of astrophysical knowledge of the galaxy. These models incorporate such physical detail as a disk and halo with differing gas densities, explicit propagation modes such as diffusion and convection ("dynamic halo" or "wind"), and clumping of matter into molecular clouds[1, 2].

To constrain and choose among these models, measurements of the radioactive isotopes of the cosmic radiation are valuable, and measurements over a range of energies particularly valuable. These isotopes, created as spallation secondaries of the cosmic radiation, are in effect clocks, requiring the time parameter to be explicitly accounted for in cosmic ray propagation. There are several isotopes whose decay lifetimes are of a magnitude which allows them to be sensitive to cosmic ray propagation lifetimes and matter densities encountered. These include ^{10}Be ($t_{1/2}$ = 1.6 x 10^6 years), ^{26}Al ($t_{1/2}$ = 0.72 x 10^6 years) and ^{36}Cl ($t_{1/2}$ = 3.0 x10^5 years).

The fate of a stable nuclear component of the cosmic radiation is to either escape from the galaxy or undergo fragmentation in collision. For radioactive species, decay competes in parallel with these processes. Measurements of radioactive isotopes can then give a measure of the age of the cosmic radiation by observing what fraction of a radioactive species has not yet decayed, the "surviving fraction".

Production ratios of stable and radioactive components are known from accelerator work. However, the existing data on ^{10}Be is of low statistics and at low energy, a few hundred MeV/nucleon[3, 4, 5].

Interpretation of the data depends upon use of a model. When interpreted in the context of the leaky box model, these data imply lifetimes of the cosmic radiation in the range 15 - 30 million years. They also imply a mean gas density of 0.25 - 0.30 atom/cm^3, a fraction of the known average density in the disk, about one atom/cm^3.

In disk-halo models, the Earth is in the disk of the galaxy, where the majority of matter and, presumably, sources are located. As a result, cosmic rays observed at the Earth contain a larger surviving fraction of radioactive isotopes than is characteristic of the galaxy as a whole. Thus, when the flux of radioactive nuclei is (mis)interpreted in the context of a leaky box model, an erroneously short lifetime is deduced – presuming the disk-halo model is correct. Interpretations of existing ^{10}Be data in the context of disk-halo models yields lifetimes several times larger than leaky box interpretations.

In addition, there are disk-halo models of galactic cosmic ray propagation which include the presence of a galactic "wind", a convective flow of cosmic radiation, ionized gas and magnetic fields being expelled from the disk of the galaxy. The existence of such a wind will cause energy-dependent effects upon the age distribution and composition of cosmic rays, as diffusion and convection compete[6].

Extension of measurements of radioactive isotopes over an energy range extending into the relativistic region will test the viability of the several cosmic ray propagation models and constrain the parameters of those that are viable. At relativistic energies, the decay time of these isotopes is dilated by the Lorentz factor, leading to an energy dependent shift in the playoff of decay against escape and interaction. Additionally, the observed relativistic radioactive isotopes will, being older than the low energy isotopes, have sampled regions of space more distant than the low energy isotopes. In a homogeneous leaky box, this has no consequence, or even meaning. However, in a diffusion model with differing gas densities in the disk and halo, the energy dependence of the surviving fraction of a radioactive isotope will differ from that of a homogeneous model. From this energy dependence, one may constrain such parameters as the diffusion coefficient and the relative sizes of the disk and halo.

Similar remarks apply in the case of galactic wind. Such a

wind will alter the energy dependence of the surviving fraction of radioactive isotopes which reaches the Earth. At low energies, the dependence is different from either the leaky box or simple halo-diffusion models. If there is a galactic wind, one expects its consequences to be manifest at low energies, where convection might dominate over diffusion as a mode of cosmic ray transport. In this instance, the effect is to limit the region of space from which the cosmic rays can reach the Earth, with distant cosmic rays being convected away before having a chance to diffuse to us. At higher energies, where higher diffusion coefficients are expected, diffusion will dominate.

Measurement of the magnitude and energy dependence of the radioactive isotopes which reach the Earth is, then, a significant experimental goal. Valuable high statistics measurements of ^{10}Be, ^{26}Al, and ^{36}Cl in the few hundred MeV/nucleon range should be supplied by the Advanced Composition Explorer (ACE). However, above a few hundred MeV/nucleon, multiple dE/dX rangedown techniques fail and the only viable experimental technique for isotope measurement is use of a magnetic spectrometer.

The most experimentally accessible radioactive isotope for this technique, is the lightest, ^{10}Be. For balloon borne magnetic spectrometers, the atmosphere is a serious source of background for such measurements. For example, in a planned long duration balloon flight of the Isotope Magnet Experiment (ISOMAX), measurements of ^{10}Be up to 3 GeV/nucleon should be able to distinguish predictions of a halo-diffusion model from those of a simple leaky box model at just under three sigma, with half of the experimental uncertainty arising from corrections necessitated by the atmospheric overburden. Atmospheric contamination is such that a balloon-borne spectrometer is not capable of making the precision measurement of surviving fraction necessary to establish the existence of a galactic wind. However, in one year under conditions of moderate solar modulation, an orbiting magnetic spectrometer with only 20 cm^2 sr acceptance could collect several thousand ^{10}Be at energies below a few GeV/nucleon, enough to establish or deny the presence of a galactic wind.

§References

[1] V. L. Ginzburg, Ya. M. Khazan, and V. S. Ptuskin. *Astrophys. Space Sci.*, 68:295, 1980.

[2] V. S. Ptuskin and A. Soutoul. *Astron. Astrophys.*, 237:445, 1990.

[3] M. E. Wiedenbeck, in *Composition and Origin of Cosmic Rays*, M. M. Shapiro (ed.), pps 65-82, D. Reidel Publ. (Dordrecht, Holland) 1983.

[4] J. A. Simpson and M Garcia-Muñoz. *Space Science Reviews*, 46:205, 1988.

[5] A. Lukasiak, P. Ferrando, F. B. McDonald, and W. R. Webber. *Ap. J.*, 423:426, 1994.

[6] F. C. Jones. *Ap. J.*, 229:747, 1979.

3.4 Antiprotons

A. Stephens

Antiprotons have recently been established to be one of the very important, but rare components of cosmic rays.The origin of this component could be attributed to many different physical processes. Some portion of the observed cosmic ray antiprotons should be the secondary particles resulting from high energy interactions of cosmic ray nuclei with ambient matter during their propagation in interstellar space. Part of the observed antiprotons could be the result of the annihilation of super symmetric particles and the evaporation of mini black holes, both of which are remnants of the big bang origin of the Universe. Should the Universe be baryon symmetric, then some antiprotons from antimatter galaxies would be present in cosmic rays. The existing data, which consist of a limited number of measurements on flux values with very large errors and a few upper limits, are not enough at this stage to construct an energy spectrum from which to derive information on the origin of antiprotons. It is of utmost importance to measure the energy spectrum of these particles from about 100 MeV to at least about 50 GeV to establish their origin and their relationship with the observed cosmic rays.

If antiprotons are established as secondary particles produced from interactions of high energy cosmic ray nuclei during their propagation in the galaxy, then an accurate determination of their spectral shape would reveal valuable information on the properties of cosmic ray transport. The expected equilibrium spectrum of antiprotons in the galaxy, on the basis of "Leaky Box Model", which is shown in figure 5, has a characteristic spectral peak around 2 GeV.

Solar modulation resulting from the transport of cosmic rays against the solar wind changes the spectral shape of antiprotons below 10 GeV; this effect increases as the energy decreases. If one can measure the antiproton energy spectrum very accurately to about 50 GeV, a comparison with the expected spectrum in the interstellar space will determine reliably for the first time the absolute magnitude of solar modulation at 1 A.U. as a function of energy. Time variation of the spectral shape over one solar cycle would provide valuable information on the effect of charge sign on solar modulation, when compared with that of the proton component. The magnitude of deceleration during a particle's traversal against the solar wind, which varies with the solar cycle, would be reflected clearly in the spectral peak. Cosmic rays may encounter stochastic acceleration during their propagation in the interstellar space, resulting in a shift in position and spread of this antiproton spectral peak. Thus, an accurate determination of the spectral shape of this peak can provide information on the occurrence of the above processes. The process by which antiprotons are produced as secondaries also creates positrons in cosmic rays. Therefore, one can establish a relationship between these two components to derive very useful information on the physical nature of the confinement region of cosmic rays in the galaxy.

A typical balloon borne magnet rigidity spectrometer for antiproton studies has an area solid angle about 200 cm^2 sr

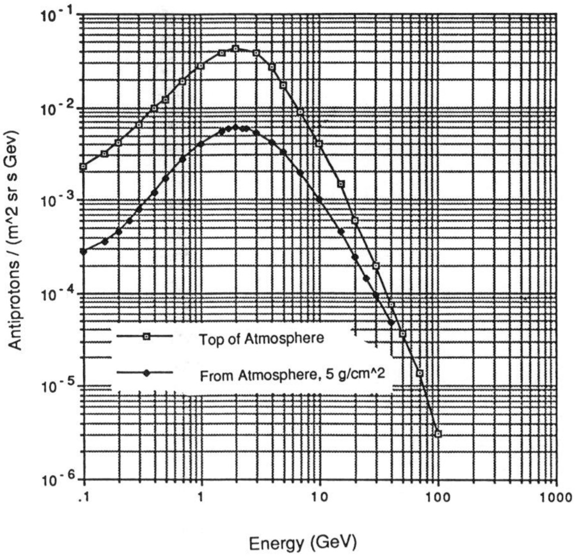

Figure 5: Expected antiproton fluxes. The upper curve is that expected from galactic cosmic rays reaching the top of the atmosphere. The lower curve is that expected from secondaries generated in 5 gm/cm^2 of atmosphere by interactions.

with about 50% efficiency to detect antiprotons. Using such a detector it is possible to collect only about 1500 antiprotons in the energy region between 0.1 and 30 GeV, if it is flown in a long duration balloon flight for a period of 10 days. It can be seen from Figure 5 that while the atmospheric secondaries are not of much concern below about 20 GeV, they are of great concern above 30 GeV. Thus, in the energy region between 30 and 50 GeV, one will be able to detect about 25 primary antiprotons from this flight, out of which about 9 would be atmospheric background. Such a measurement is sufficient to provide conclusive evidence for the origin of the antiprotons in cosmic rays. However, one needs very high statistics to derive information on the other important aspects mentioned above. One requires at least 5000 events in the energy region 0.5 to 5 GeV to determine the shape of the spectral peak, needing an exposure of 50 days. On the other hand, it is possible to measure the absolute flux of antiprotons between 30 and 50 GeV to an accuracy of only 10% with one year of exposure, as the atmospheric background is known to an accuracy of only 15%. Therefore, satellite borne experiments are very essential to construct an accurate energy spectrum of cosmic ray antiprotons and to extract useful information from it.

Gamma-rays From Space: Introductory Overview and Hard Hard X-ray Surveys

J. E. Grindlay

Harvard-Smithsonian Center for Astrophysics, Cambridge, MA 02138

Gamma-ray astronomy below energies of about 10 GeV must be done from space (or high altitude balloons). The recent successes of the Compton Gamma-ray Observatory, following on earlier exploratory missions at high energies (SAS-2 and COS-B), have demonstrated the rich variety of sources and astrophysical phenomena to be studied over the ~6 decades of energy above the x-ray band. The Snowmass Workshop concentrated on the science and missions that should be carried out in the next millenium to followup on CGRO and to complement and extend the only currently planned gamma-ray mission , the ESA/Russian mission INTEGRAL. This introduction reviews briefly the presentations given and the several written contributions which follow. We conclude with the need for the first all-sky hard x-ray imaging survey for the lowest energy band. The recently proposed EXIST mission concept could fulfill this need.

§1 Introduction

Gamma ray astronomy is moving forward on many fronts: the Compton Gamma-ray Observatory (CGRO) is carrying out a highly successful program of observation and discovery from earth orbit (see review by Gehrels, Chipman and Kniffen 1994); high altitude balloon-borne telescopes are pioneering new detectors and technique as long-duration balloons become operational, and ground-based observations at the highest energies have come into their own with the discovery of several sources. At energies below about 10 GeV all gamma-ray studies must be done from at least the highest altitudes attainable by balloons or from space. Space observatories have the advantage of providing the longest exposure times which enable the highest sensitivities to be reached for the increasingly faint higher energy sources. Balloon observations offer a low cost alternative that is becoming again attractive as long-duration flights of 30 days or more become feasible.

At the Snowmass Workshop R1: Space Based Gamma-Ray Astronomy, a broad overview of the exciting science was presented and emerging mission concepts were presented for the possible gamma-ray missions of the coming decade and beyond. In this Introductory Overview, we briefly describe the major scientific drivers for gamma-ray astronomy today and as projected by the several presentations. We then briefly summarize the detector concepts given for the major energy ranges – low (0.01 - 1 MeV), medium (0.5 - 30 MeV) and high (30 MeV- 100 GeV) – where the detection techniques differ significantly.

Several short contributions and one more substantial one were received for these Proceedings. Low energy gamma-ray astronomy and the need for future all-sky monitors is presented briefly by Harmon, and a short summary of the only new gamma-ray mission from space that is currently planned: the International Gamma-ray Astrophysics Laboratory (INTEGRAL) is given by Haymes (and in more detail by Winkler 1994). Following these is a more complete discussion by Hunter and Michelson of current (from CGRO) astrophysics in the high energy band now covered by EGRET and a brief mention of two possible new follow-on mission concepts for this energy range.

Since the Snowmass Workshop was held, NASA has carried out a proposal solicitation and selection for New Mission Concepts in astrophysics, including gamma-ray astronomy. Three gamma-ray mission concepts were selected for further study: a low energy all sky survey mission (EXIST), a gamma-ray burst experiment (BASIS), and a high energy telescope system (GLAST). These and other gamma-ray instruments will likely be proposed for the newly defined Medium Explorer (MIDEX) program at NASA, which aims to launch a new mission each year beginning in 1999. As one such example future mission, the EXIST (Energetic X-ray Imaging Survey Telescope) concept is briefly described here.

We close this Overview with a few comments on implementation of future programs. In particular, we endorse the MIDEX program as the welcome response to calls for more frequent and predictable access to space. We also highlight the newly implemented Long Duration Balloon program as a particularly promising and realistic tool for space gamma-ray missions and prototypes in the immediate future.

§2 Scientific Drivers and Mission Needs

Gamma-ray astronomy has progressed from the realm of initial discovery (as started with the pioneering high energy missions (~50-300 MeV) SAS-2 and COS-B) to an era of intensive study of ever-more faint sources. This is not to say the age of discovery is over; certainly CGRO has shown quite the contrary and has borne out the usual dictum in high energy astrophysics that increased sensitivity and resolution spawn ever more discovery. Rather than enumerate the broad range of scientific problems now being addressed in present and immediately planned (or proposed) space missions, some of which are partially described below, we list some of the obvious scientific needs and questions which should be addressed in future missions:

Need for greatly enhanced sensitivity. Gamma-ray astronomy to the non-practitioners has often seemed like a science of 3σ results. Many, if not most, of these have flowered into far more significant effects and genuine astrophysical measurements with the enhanced sensitivity achieved with CGRO. However, the dynamic range between the brightest cosmic sources (e.g. the Crab) and faintest detected is still only a factor of 10-100 over most of the medium-high energy band and clearly much greater sensitivity is needed. Not only for the detection of sources roughly equivalent to stars fainter than 5th magnitude (i.e. some hundred times fainter than the brightest 0th magnitude stars visible !), but also for the more detailed study of the spectra and time variability of even the brighter sources. Gamma-ray astronomy suffers from photon starvation, and increasingly so with increasing energy. Sometimes referred to as Cocconi's law, that photon spectra are invariably decreasing with increasing energy, ever larger area and lower background detectors must be developed. Except at the lowest end of the hard x-ray band (i.e. $\lesssim$ 100 keV), where multi-layer focusing optics can probably succeed as concentrators or imagers by using multi-layer grazing incidence mirrors, gamma-ray imagers and spectrometers must rely on brute detection area. Thus except for gamma-ray missions designed primarily for bright sources (e.g. gamma-ray bursts or the Sun), the detectors and thus missions to support and fly them will necessarily be large. Thus innovative and low-cost ways must be developed for building and flying the ever larger missions of the future. As mentioned in section 5 below, a particularly low cost and short term way to achieve this is with the Long Duration Balloon (LDB) program; however innovative and low cost solutions for longer term space missions are also needed.

Need for enhanced spectral resolution. Gamma-ray astronomy (as with other wavelength domains) begins with detection, progresses to imaging and culminates with spectroscopy. The best combination is of course imaging spectroscopy, where both spatial and spectral resolution are optimized. Here it is instructive to compare progress with x-ray astronomy, which has progressed from low energy and spatial resolution images (the IPC detector on the Einstein Observatory) to moderately high energy resolution but still low spatial resolution (with the Japanese/US ASCA satellite) to moderately high energy resolution and very high spatial resolution (with the CCD detectors to be flown on AXAF). The low-medium energy and high energy telescopes COMPTEL and EGRET on CGRO have both achieved very roughly the first (IPC/Einstein) stage of spectral and spatial resolution. Clearly higher resolution is needed. The low energy gamma-ray mission INTEGRAL will provide high spectral resolution but low spatial resolution up to a few MeV. Particularly in the hard x-ray and low energy gamma-ray range, where both increasing source fluxes (and thus source counts) and nuclear lines are to be expected, the need for higher spatial and spectral resolution is acute.

Need for enhanced temporal coverage and resolution. As gamma-ray sources are detected over the entire 6 decades of energy generally available for study, they are often found to be variable. In fact, with the exception of diffuse emission from the interaction of cosmic rays with gas clouds in the Galaxy, they are essentially always found to be variable. It is only the degree, and timescale, which is at issue. Thus detectors and missions of the next generation must seek to achieve high enough sensitivity that variability can be measured in enough detail to probe the source physics and basic properties. They should also have broad enough temporal and angular coverage that a broad range of timescales can be studied. Gamma-ray detectors at high energy have the advantage of individual photon tracking so that both wide fields of view and low background (per resolution element) can be achieved. Thus sensitive all-sky monitors can be flown. This can (and should) be done at low energies as well in order to continue and greatly extend the source variability and monitoring studies so effectively begun by the BATSE detector on CGRO (cf. Harmon, these proceedings). As discussed by A. Harding in her discussion of gamma-ray emission from pulsars and the range of outstanding questions future observations may answer, high time resolution (and thus sensitivity) will be increasingly important.

§3 Mission Concepts Presented

We briefly summarize the mission concepts presented during the R1 sessions at Snowmass. These represent a cross section of ideas in the community for what should be done, or might be done, for the future development of gamma-ray missions from space. As mentioned above (and below), additional mission concepts have been proposed in the recent NASA solicitation, and undoubtedly more still will be.

In order of increasing gamma-ray energy, the missions discussed were:

1. *USA* is an x-ray astronomy mission (2-20 keV) which is in the final stages of development by the group at NRL and was described by K. Wood. It will concentrate on long (~1 month) studies of moderately bright x-ray sources (e.g. LMXBs) for high time resolution and long term variability studies. It would allow higher sensitivity searches for underlying pulse periods in prospective gamma-ray pulsars that could be detected at higher energies.

2. *INTEGRAL* is the planned ESA/Russian mission , as mentioned above, and described here (and in the accompanying short paper) by R. Haymes. It will achieve both higher spectral resolution with an array of Ge detectors

as well as spatial resolution using a coded aperture as well as separate coded aperture imager. Since the Snowmass Workshop the instrument selection has been made with the resulting planned detectors even more sensitive than those originally proposed.

3. *TIGRE* is a proposed concept for a low-medium energy tracking detector, as described by T. Turner (University of California, Riverside). It would employ Si-strip detectors for a Compton telescope arrangement that would cover the Compton-pair regime from 1-100 MeV.

4. *AGATE* is the medium-high energy detector concept which would employ drift chambers and is being developed by the GSFC group as a possible successor to EGRET. It was presented at the Workshop by S. Hunter (GSFC) and is motivated by the high energy gamma-ray astrophysics science discussed in the accompanying paper by Hunter and Michelson.

5. *GLAST* is a competing medium-high energy detector concept employing Si-strip detector readouts and described at the Workshop by W. Atwood (SLAC). As a moderately compact detector which would employ techniques developed at SLAC, it would scaleable to various detector sizes and sensitivities and could thus greatly increase the sensitivity and resolution of a follow-on mission to EGRET. As mentioned above, GLAST was selected for further study as a Mission Concept.

6. *OSCAR* is a proposed high energy (1-100 GeV) gamma ray detector system that might bridge the gap between space and ground-based gamma-ray detector systems. As described at the Workshop by D. Mueller (University of Chicago), it would be constructed of a large array of drift tubes and wire chamber readouts to effectively track high energy gamma-ray showers over the detector effective area of several square meters.

Although no specific concepts for gamma-ray burst missions were discussed at Snowmass, these are clearly of great interest and will surely play a prominent role in the future missions for gamma-ray astronomy from space. As one of the most notable problems in astrophysics, the origin and nature of gamma-ray bursts was the primary driver for two of the four Mission Concepts selected for gamma-ray astronomy. GRBs can also be studied in great detail with the proposed all-sky hard x-ray imaging survey mission, EXIST, as described below.

§4 Need for a Hard X-ray Imaging Survey Mission

A critically important region of the spectrum is the hard x-ray band, from ~10-600 keV. In this band, an unusually rich range of astrophysical processes occur in both compact and diffuse sources. The band includes the transition from primarily thermal objects – either optically thin, like supernova remnants or galaxy clusters, or optically thick, like the blackbody emission components of high luminosity x-ray binaries (all of which are generally still detectable at 10 keV) – to objects which are primarily non-thermal, or at least display significantly Comptonized spectra. Examples of the very hottest thermal plasmas directly measurable in astronomical objects, the $\sim 10^{8-9}$K coronae around or above accretion disks in compact binaries and active galactic nuclei, are best studied in this hard x-ray band. In particular, the rapid pace of discovery of hard x-ray emission from black hole (BH) binaries (the so-called x-ray novae) as well as relatively low luminosity neutron star (NS) binaries (the bursters) have shown that hard x-ray spectra offer particularly direct clues to the study of black holes and accretion flows onto BHs vs. NSs.

Synchrotron, or inverse Compton, non-thermal spectra from sources of high energy particles (e.g. pulsars, supernova remnants, and radio galaxies to name a few) are most readily studied in this band since the competing thermal spectra in these objects (from differing source components) are much weaker and do not mask the overall non-thermal spectrum. Since power law non-thermal spectra, or Comptonized thermal spectra, are invariably smoothly decreasing with increasing energy, studies of these objects and processes are best carried out where source photons are most numerous and backgrounds are lowest.

Surprisingly, for such a pivotal region of the astronomical spectrum, the hard x-ray universe is still relatively unexplored. Only one survey (HEAO-A4, in 1978-80) has been carried out in this energy range but at relatively low sensitivity and without the benefit of imaging. Yet the scientific impact of survey missions on astronomy is amply demonstrated at X-ray wavelengths by ROSAT and at IR wavelengths by IRAS. In the hard X-ray and low-energy γ-ray bands (5 - 600 keV), a sensitive *imaging* sky survey is yet to be carried out. The HEAO-A2 and A4 instruments carried out the first all-sky observations above 10 keV, achieving a sensitivity of several mCrab at 2-10 keV energies (A2) and 20-100 mCrab at energies 13-180 keV (cf. Figure 1a). The French coded aperture telescope SIGMA on the Russian GRANAT satellite demonstrated the power of wide-field imaging in the ~35 - 300 keV band but it too was not extremely sensitive and observed only a small fraction of the sky (primarily the Galactic Center). The OSSE instrument on GRO has significantly better sensitivity than either HEAO-A4 or SIGMA above 100 keV, but does not cover the range below 100 keV well, and its relatively narrow field of view and GRO pointing constraints has precluded large-area surveys, except for specific regions such as the Virgo cluster and central Galactic Plane regions. As a consequence, ***no mission has carried out a survey of the sky over the*** $\gtrsim$ ***5 keV to 600 keV energy band with sufficient sensitivity to detect any but the brightest galactic and extra-galactic sources.*** Either the missions have lacked the sensitivity to detect the interesting populations of AGN and galactic compact sources, or the missions sky coverage has been minimal.

It is clear that a 5-600 keV survey will yield significant new scientific results. Observations of selected Seyfert I&II galaxies below 8 keV by ASCA, combined with the results of

OSSE have indicated that the idea of a "canonical" spectra for AGN above 100 keV is too simplistic, with a variety of spectral types including power-laws with exponential cutoffs and Comptonized reflection spectra. To adequately determine the likely contribution of different classes of AGN (i.e. Seyfert I&II and quasars) to the diffuse background will require an ***unbiased*** survey with sufficient sensitivity to detect a large number of AGN and determine their ***time-dependent*** luminosity distribution. A survey is thus needed which covers the entire sky at several different epochs to probe time variability of the luminosity distribution of AGN and to produce a catalogue of hard X-ray selected objects.

A sensitive survey is also critically important for further progress on studies of galactic sources above 10 keV. Even with its very modest sensitivity, effective area (~100 cm^2 at 35-300 keV) and sky coverage, the SIGMA instrument detected tens of hard X-ray and gamma-ray sources above 35 keV including several previously unknown objects. Likewise, Ginga in deep galactic plane scans detected numerous transient and persistent sources at the few mCrab level. This is not unexpected: a source accreting with an Eddington luminosity at 10 kpc will have a flux intensity less than or equal to about 100 mCrab. Most galactic neutron star and black hole systems accrete at substantially less than the Eddington rate, thus requiring sensitivities below 10 mCrab for detection and study.

Finally, a survey mission which incorporates very wide field of view detectors and telescopes can also address the general objectives mentioned above for broad temporal coverage: from short bursts to transient sources. The BATSE detectors on CGRO have shown the power of monitoring for both transients and of course the detailed study of gamma ray bursts. A dedicated survey mission could contribute greatly to both of these major topics in gamma-ray astronomy.

4.1 EXIST Mission Concept Overview

We have begun to study a mission which would conduct the first high sensitivity all-sky imaging survey (5-600 keV) with a wide-field coded aperture telescope. Grazing incidence telescopes, with response limited to $\lesssim$ 80 keV and very small fields of view, cannot do this. The Energetic X-ray Imaging Survey Telescope (EXIST) mission characteristics and baseline instrument, as proposed and accepted for study in the recent NASA solicitation for Mission Concepts, are summarized in Table 1. Full details of the EXIST concept are given by Grindlay et al (1995).

EXIST would be an ***all sky*** survey mission lasting $\gtrsim$ 1 year (for $\gtrsim$ 2 passes through the entire sky). The sensitivities for continuum and line detection are given in Figures 1a and 1b and are quite spectacular: EXIST would be $\gtrsim$ 100$\times$ as sensitive as HEAO-A4 over the whole sky and $\sim$ 10$\times$ as sensitive as typical XTE/HEXTE pointings (in 1° fields) which will cover only ~2% of the sky. Below 200 keV, EXIST could achieve line sensitivities comparable to deep pointings with INTEGRAL in its limited sky coverage.

After an extensive Survey, the EXIST mission could continue as a pointed Observatory of unprecedented sensitivity and resolution or it could continue as a powerful survey instrument, with occasional deep pointings. The Observatory sensitivities are typically a factor of 2-3 more sensitive than the Survey sensitivities (Figure 1), depending on exposure time, and are even more sensitive at lower energies ($\lesssim$ 30-50 keV) because of the reduced background in the smaller co-aligned telescope field of view.

The EXIST mission would include a large array of new-technology Cadmium-Zinc-Telluride (CZT) solid state detectors. These are pixellated to give the high spatial resolution (as well as good energy resolution) needed for a compact coded aperture imaging telescope desisgn. The mission could be launched on the newly defined NASA "Med-Lite" launch vehicle and catapult hard x-ray astronomy into the mainstream of astrophysics.

The basic mission consists of a survey mode with a ROSAT-like scanning strategy, i.e. a rotation scan motion perpendicular to the ecliptic with one complete 360° scan every orbit. The baseline instrument originally proposed consists of 4 coded aperture telescope modules, each of approximately 2500 cm^2 detector area. However, other telescope and detector combinations are possible and are being considered as part of the Concept Study. A key feature of the instrument is an energy dependent field-of-view (FOV): 6° $\times$20° at energies below approximately 40 keV, and 20° $\times$20° at energies above approximately 80 keV. The energy-dependent FOV allows optimization of the survey observations in the two background regimes: diffuse-dominated and internal-dominated. In survey mode, it is desirable that the scan swath width be maximally wide to optimize chances for detection of transient behavior. Our design goal is half-sky coverage over one orbit. Use of offset telescopes allows a wide scan-path while maintaining acceptable signal-to-background ratios. The nominal offset (scan mode) total field of view would be 80° x 20° (FWHM). In the pointed mode, the telescopes could be co-aligned to achieve the optimal on-axis point source sensitivity.

§5 General Needs for Gamma-ray Astronomy from Space

We close with a few remarks to summarize the general needs for the broad field of gamma-ray astronomy from space, as outlined at the Snowmass Workshop.

First, as the broad energy band appropriate to this field (from 10 keV to 10 GeV) suggests, both the science opportunities and experimental challenges are also very broad. The scientific questions range from the physics of hot gas in galaxy clusters and the possible role of magnetic fields which can be measured uniquely by inverse Compton emission spectra at hard x-ray energies; to the physics of the hottest thermal plasmas, at pair temperatures, in compact accretion systems; to the physics of the acceleration of the high energy cosmic rays in the jets of active galactic nuclei which give rise to the prodigious gamma-ray emission detected now from many Blazars at 100 MeV. Thus a wide range of missions are needed to tackle the vast range of scientific questions at hand. At the

Table 1.
EXIST Mission Characteristics

Energy Range	5-600 keV
Angular Resolution	10 arc minutes
Field of View	Survey: 80°×6° (≤40 keV); 80°×20° (≥80 keV) Pointed: 20°×6° (≤40 keV); 20°×20° (≥80 keV)
Orbit	LEO (nominal 30° inclination)
Survey	~50% sky coverage each orbit Full-sky twice per year 0.5-1×10^6 sec per year effective source exposure Survey Sensitivity: < 1 mCrab (5-200 keV)
Scan Motion	ROSAT-like scan once per orbit
Baseline Instrument Parameters	
Number of Telescopes	4
Detectors Detector Area Energy Resolution Spatial Resolution	CdZnTe (0.5 cm thick) 4×2500 cm^2 1.2 keV FWHM @ 10 keV, 3.6 keV FWHM @ 100 keV 1.25 mm
Shield	BGO (2 cm thick)
Collimator	Copper (0.1 mm) and Tungsten (1 mm) Slats
Mask Mask-Detector Spacing	Coded-Aperture (5mm Tungsten) 1.2 m

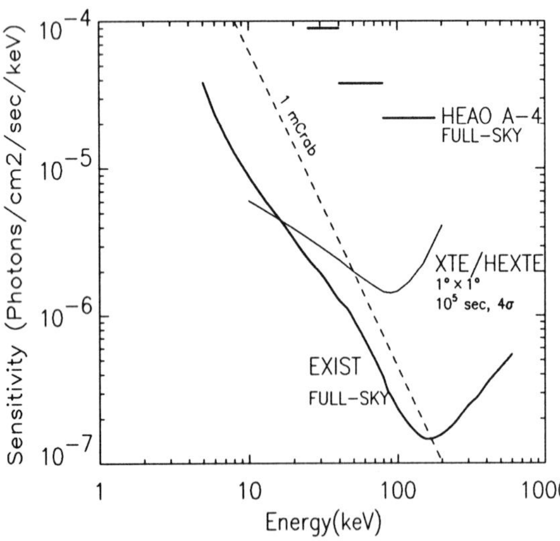

Fig 1a: EXIST Survey continuum sensitivity

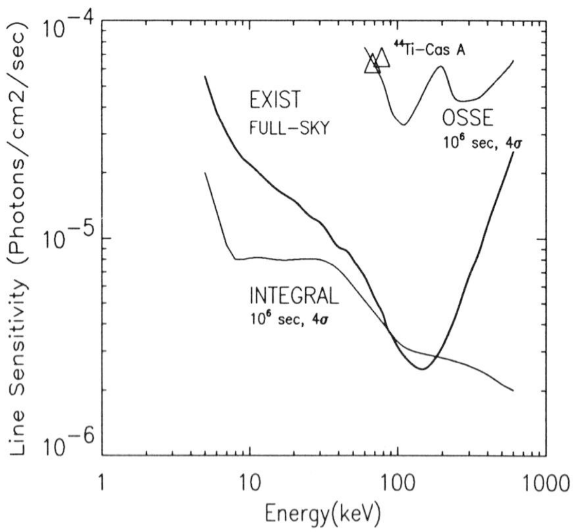

Fig 1b: EXIST Survey line sensitivity

very least, three broad energy ranges of science and technique continue to dictate the needs and probable future missions: low energy (or hard x-ray), which may be defined (here, at least) as from 10 keV to 1 MeV; medium energy (~1 MeV - 30 MeV; and high energy (30 MeV - 10 GeV). Thus we conclude that new missions are needed in each of these energy ranges to greatly extend what has been possible with CGRO.

Second, the missions themselves may span a range of scales of size and complexity (and thus cost). While it is increasingly true that with photon spectra declining with increasing energy that ever larger detectors and telescopes are needed in the gamma-ray band for increasingly sensitive investigations, some experiments can still be modest in size (e.g. the proposed mission concept ETA for gamma-ray burst detectors launched on microsatellites (perhaps 10 kg each) into orbits around the Sun from which they could very accurately time and thus position gamma-ray burst sources). Thus gamma-ray astronomy from space should not be totally dependent on Great Observatory class missions (like CGRO) which in the present climate are less likely to be initiated.

Third, gamma-ray missions must be developed with new technology detectors and mission concepts in order to make them even more affordable for what will still (in many cases) be necessarily detectors or telescopes with very large collecting areas or throughput. Thus low-cost mission opportunities must be sought. The balloon program, long the proving ground for gamma-ray missions and detectors, is even more suitable now with the advent of the Long Duration Balloon (LDB) program. The LDB flights, from 10-30 days, can provide mission opportunities for gamma-ray astronomy (and other disciplines) that are fully competitive with Shuttle flight missions at a very small fraction (perhaps 1%) of the cost.

I thank all the contributors to these sessions on Gamma-ray Astronomy From Space at Snowmass, which proved to be a productive forum to discuss many of the options for this fast-evolving field. I thank the Editors of these Proceedings for being so patient. This work was supported in part by NASA grant NAGW-624.

§References

§References

[1] Gehrels, N., Chipman, E. and Kniffen, D. 194, *Ap. J. Suppl.*, 92, 351.

[2] Grindlay, J.E. et al 1995, *Proc. SPIE*, in preparation.

[3] Winkler, C. 1994, *Ap. J. Suppl.*, 92, 327.

BATSE and Beyond: The Future of All-Sky Monitors

A. Harmon

NASA/Marshall Space Flight Center, Huntsville, AL

June 15, 1995

The value of all-sky monitoring in the x-ray and gamma-ray energy range is unquestioned. We have known since the early years of x-ray astronomy of the transient and rapidly-changing character of high energy emission in binary star systems, quasars and active galaxies, and gamma ray bursters. Many questions about the nature of these objects remain unanswered. Indeed, in the case of gamma ray bursters, we are limited only to a few "clues" with which the theorists must work. In recent years, we have also discovered that transient binary systems contain a compact object with a mass which is greater than that of the canonical neutron star ($1.4M_{\odot}$). We believe that these represent the best evidence for galactic black holes. Nevertheless, our understanding of the compact object and its accretion disk, interactions with the companion star, and the physical and geometric properties of these binary systems is not sufficient to explain the extreme changes in spectrum, luminosity and other phenomena. We learn much from all-sky monitoring instruments which cannot be achieved (or at least is extraordinarily difficult to achieve) exclusively with pointable, limited field-of-view instruments.

The Burst and Transient Source Experiment (BATSE) on the Compton Gamma Ray Observatory (CGRO) is an all-sky monitor which has been operating in low Earth orbit since April 1991. Its projected lifetime was originally 2-5 years; however, with the successful reboost of the CGRO in 1993, and with good management of the onboard power systems, the mission could last to near the turn of the century. A number of scientific advances have been made possible with BATSE's all-sky capability, among those are: the establishment of the concurrent angular isotropy and spatial inhomogeneity of gamma ray bursts, the discovery of new pulsars, x-ray novae, and x-ray transients with relativistic radio jets, the discovery of atmospheric gamma-ray flashes from thunderstorms, and the identification of a soft gamma ray repeater with a supernova remnant. These advances were either made independently by BATSE, or lead other researchers to make important discoveries. The scientific results generated by other experiments and investigations are actually enhanced by the presence of an all-sky monitor.

Early in the next century, following the termination of missions with all-sky monitors such as BATSE, the X-Ray Timing Explorer, and Spectrum-X-Gamma, a need exists to continue and enhance our all-sky monitoring capability. With new technologies that have been developed over the past decade, it appears possible, at reasonable cost, to increase the dynamic energy range, sensitivity, and multi-channel/data flow/telemetry capacity of instruments beyond those now in operation or planned for launch after 1994. New concepts for pointable high energy astrophysical instruments such as those being discussed at this meeting will most certainly benefit from a carefully planned approach to all-sky monitoring.

INTEGRAL and Missions for MeV Astronomy in the Next Millennium

R. Haymes

Department of Space Physics, Rice University, Houston, TX 77251

June 15, 1995

The International Gamma Ray Astrophysical Laboratory, INTEGRAL (Winkler 1994), is the next gamma-ray astronomical space mission that has been approved for development. The scientific objectives of INTEGRAL include simultaneous imagery and spectroscopy of selected astronomical objects, in the 15 keV to 10 MeV gamma-ray energy interval. INTEGRAL will measure selected sources both within and without the Galaxy. These sources will include the central region of the Galaxy, compact objects such as neutron stars and black hole candidates, supernova remnants, novae, radioisotopes in the interstellar medium, high energy transients, Active Galactic Nuclei, and clusters of galaxies.

Table 1: INTEGRAL Characteristics.

Parameter	SPECTROMETER	IMAGER
Energy range	15 keV - 10 MeV	70 keV - 10 MeV
Detector are	327 cm^2	2500 cm^2
E/dE 1 MeV	500	25
Field of view	5.6° fully coded	3.2° fully coded
	13° partially coded	22° partially coded
Angular resolution	1.4° FWHM	17° FWHM
Point source location	10′	1′
(20 sigma source)		
Continuum sensitivity	6×10^{-8} / cm^2-sec-keV	3×10^{-8} / cm^2-sec-keV
(3 sigma in 10^6 s 1 MeV, dE = 1 MeV)		
Line sensitivity	Narrow (2 keV) line	Broad (20 keV) line
(3σ in 10^6s @ 1 MeV)	1.5×10^{-6} /cm^2-sec	1.2×10^{-5} /cm^2-sec
Polarimetry sensitivity (3σ)		10mCrab
Timing accuracy (3σ)	0.1 ms	0.1 ms

INTEGRAL is being developed by the European Space Agency, ESA, in collaboration with the Russian Space Agency which will provide the launch on a Proton rocket in about 2002. The principal scientific instruments are to be aimed at the chosen targets by pointing the spacecraft, which will have a three - axis pointing capability. Its absolute pointing error is to be 15 arc minutes.

Investigators from western countries in Europe will supply INTGRAL's two principal scientific instruments. They are a gamma ray imager and a gamma ray directional spectrometer that looks in the same direction as does the imager. The spectrometer is to be an actively collimated array of germanium crystals, and the imager is to be an array of CsI scintillators and Cd-Te solid state detectors. Both the imager and spectrometer view the sky through coded aperture masks and thus use coded aperture imaging. The Imager pixel and mask element sizes are smaller allowing it to achieve 17 arcmin angular resolution vs. 1.4 degree resolution for the spectrometer. Table 1 shows the performance expected for the imager and the spectrometer. These parameters are already somewhat out of date since the final instrument selection process will result in some changes for the design of both the spectrometer and imager.

There are gamma-ray astronomy problems that will not be attacked by INTEGRAL. One such is the absolute spectrum of the extragalactic background at MeV energies. Another is an all-sky map of MeV gamma ray sources, to be completed before INTEGRAL flies. A small mission called GRAB (Gamma Ray Astronomy Background) is being designed with both of these problems in mind. It is presently being proposed for study, by the author and a group of scientific collaborators. Please contact the author for further information on GRAB or the study.

§References

[1] Winkler, C. Et al 1994, *Ap. J. Suppl.*, 92, 327.

High-Energy Gamma Ray Astrophysics

S.D. Hunter

NASA/Goddard Space Flight Center, Greenbelt, MD 20771

P.F. Michelson

Department of Physics, Stanford University, Stanford, CA 94305

June 15, 1995

The exciting results from the highly-successful Compton Gamma Ray Observatory (CGRO) clearly demonstrate the value of high-energy gamma-ray astronomy and point to the scientific returns that can be expected from follow-on missions to continue gamma-ray observations. The CGRO results have answered a number of important astrophysical questions, yet have also raised several new ones which set the scientific goals of these missions and determine the characteristics of the instruments needed to achieve these goals.

§1 Introduction

In the nearly four years since the launch of the Compton Gamma Ray Observatory (C GRO), high-energy gamma-ray astronomy has evolved from an exploration phase, begun with the OSO-3 (Kraushaar et al. 1972), SAS 2 (Fichtel et al. 1975), and COS B (Swanenburg et al. 1981) instruments, to a rich observational field in its own right, revealing its potential to address a number of the most important problems of astrophysics. The Energetic Gamma Ray Experiment Telescope (EGRET) observations, in the energy range from about 30 MeV to 30 GeV, have produced a broad range of exciting results. These include observations that impact the nature of the mysterious gamma-ray bursts, the collimation of jets of relativistic plasma in active galactic nuclei, the radiation and plasma processes in the super-strong magnetic fields of pulsars as well as in the magnetic loops of the sun, and the physics of relativistic particle acceleration in these and other exotic astrophysical environments such as the remnants of supernova explosions. The EGRET sources from CGRO Phase 1 are listed in the First EGRET Source Catalog (Fichtel et al. 1994).

For many of these phenomena, however, EGRET has provided us with only a glimpse: only a handful of pulsars can be detected; a few tens of active galaxies are observed sporadically during gamma-ray flares; and the identification of many sources of high-energy gamma-rays remains elusive. Major new questions have also been raised that require observations with improved instruments. Future instruments will require greater sensitivity, angular resolution, and energy range than their CGRO counterparts, and will permit a deeper and more profound investigation of the scientific questions that have been raised.

§2 High-Energy Gamma-Ray Astrophysics Scientific Issues

2.1 Active Galaxies

Among the most striking accomplishments of CGRO was the detection of high-energy gamma-rays from active galaxies whose emission at most wavebands is dominated by nonthermal processes. These objects, termed "blazars," are highly variable at most frequencies and are bright radio sources. The radio structure usually consists of knotty jets, with the knots in most cases observed to move outward at speeds that are apparently superluminal. As of this writing, nearly 40 blazars have been strongly detected by EGRET, with another dozen detected at lower significance. A few of these blazars have also been detected with the lower-energy COMPTEL detector (Schönfelder et al. 1994; Bloemen et al. 1995). The remarkable nature of the high-energy gamma-ray emission from blazars is clearly illustrated when the multiwaveband spectrum is displayed on a νF_ν plot. As an example, the observed luminosity of 1633+382 at gamma-ray energies (Figure 1) exceeds that from other wavebands by as much as two orders of magnitude. This is even more astounding when combined with the variability on timescales of less than one week observed in the gamma-ray emission from several blazars: the density of high-energy photons is implied to be so high that they should be unable to escape without being destroyed by photon-photon pair production. The only reasonable solution is for the gamma-rays to be beamed by roughly the same factors as required to explain the superluminal motions. This implies that the gamma rays are produced in the relativistic jet. The timescale of variability of the gamma-ray emission is shorter than at radio frequencies, implying that the gamma-rays originate in the portion of the jet that lies between the central engine and the radio jet imaged with VLBI. We know very little about this portion of the jet. Yet it is this region where the most important physics occurs: the formation of the jet, the acceleration of the energetic particles, the collimation of the flow into a narrow cone, and the acceleration of the flow to Lorentz factors as high as 20.

In many (perhaps most) cases, the gamma rays were detected during periods of enhanced nonthermal activity. In a few well monitored objects there is a close association between flares seen at millimeter to optical wavelengths and high gamma-ray states (e.g., PKS 0528+134 and 3C 279, Reich et al. 1993). Different behavior has been seen in one object, Mkn 421, whose gamma-ray spectrum extends up to TeV energies (as observed by the Whipple Observatory, which detects Cherenkov emission from air-showers). In May 1994 a major flare, with a rise time of only two days, was detected at sub-TeV energies one day before the ASCA satellite observed a very high X-ray state. However, no flare was seen at mm, submm, or IR wavelengths. Any flare at GeV energies is restricted by EGRET observations to have had an amplitude of at least a factor of 2-3 lower. It appears that the flare occurred only for the highest-energy electrons, which emit synchrotron X-rays and inverse Compton TeV gamma rays. The particle injection/acceleration process, which appears to fluctuate rapidly, is readily studied by observing the high-energy gamma rays. Most other blazars do not have synchrotron spectra extending into the X-ray regime and therefore have lower high-energy cutoffs to their electron energy distributions. Fluctuations in the acceleration of these particles should therefore be seen in the GeV range. Some of the variations seen by EGRET may represent this phenomenon, but we do not yet have sufficient simultaneous multiwaveband observations to determine this.

By the end of EGRET's lifetime, some of the questions regarding gamma ray emission from blazars may be mostly answered: the main radiation mechanism, and the site in the jet where the gamma rays originate, relative to that at lower frequencies. However, the sensitivity of EGRET is such that many blazars are detected only sporadically and only a few blazars are detected with sufficiently high signal-to-noise to obtain a light curve. This limited sensitivity restricts considerably our ability to answer many of the profound questions raised. It is clear that the nonthermal luminosity of blazars is dominated by high-energy gamma rays. This strongly suggests that an understanding of the gamma-ray emission from blazars is required to understand the basic physics of blazars.

Simultaneous multiwaveband monitoring of blazars will result in exciting inferences regarding relativistic jets and energetic particle acceleration in blazars. The ability to follow flares, smaller than those which can be detected with EGRET, and to obtain better time and energy resolution (because of the improved sensitivity) will lead to the derivation of the geometry and physical characteristics of the inner jet – the region that lies between the radio jet and the central engine – by observing time delays at different frequencies as the flare propagates along the jet. Tests of shock wave formation can be performed using details in the multiwaveband light curves. In a few years, the VLBA will operate at 90 GHz (where most sources are no longer optically thick as they are at lower frequencies), allowing direct imaging of the innermost region of the radio jet. The time required for a gamma-ray flare to lead to the emergence of a new superluminal component can then be determined accurately, indicating whether the flow accelerates or decelerates as it propagates from the central engine. Such observations will lead to the development of sophisticated models of the jets in the very region where they are focused and accelerated to relativistic flow velocities. This will bridge the large gap that currently exists between jet formation theories and observational data.

In addition to these exciting and expected developments with respect to blazars, we anticipate that other types of active galaxies will be detected, leading to revolutionary studies of high-energy particles in these objects. The nearest Seyfert and radio galaxies (e.g., NGC 4151 and Cen A) should prove to be high-energy gamma-ray emitters at some level. The relation between these galaxy types and blazars is poorly understood; it is of particular importance to determine whether they contain low levels of nonthermal activity or, in the case of radio galaxies, whether the level is the same but simply mis-directed because of the relativistic beaming found to be required in blazars.

With their high supernova rate and high densities of gas with which the resultant cosmic rays can interact, starburst galaxies such as M82 and NGC 253 should be detected as hard gamma-ray sources at levels just below those to which EGRET is sensitive. A comparison can then be made between the levels of cosmic-ray production by supernovae, thought necessary in our Galaxy, and those needed to explain the gamma-ray emission in starburst galaxies, thereby testing the supernova origin of cosmic rays.

With an improvement in flux sensitivity of a factor of 10, the number of blazars which might be visible to a future instrument would increase by a factor of at least 30, from the present 40 sources to about 1000 sources, which could (if there are not basic differences) encompass all the known blazars. Such an instrument could therefore answer the question of whether some blazars are gamma-ray quiet, at least on a statistical basis.

The study of blazar time variability separates roughly into 3 parts: (1) attempting to determine, as a function of energy, the minimum timescale for significant variation; (2) observing variations on timescales of one to a few days, correlated with similar observations at longer wavelengths; and (3) following the longer timescale variations to determine duty cycles and ranges of variation, which are needed in order to be able to calculate accurately the blazar contribution to the unresolved extragalactic gamma radiation. For the shorter timescales, the primary need is for increased sensitivity, again via greater effective area and improved angular resolution. A broader field of view instrument will enhance the ability for following longer timescale variation, because a larger number of interesting objects can be observed simultaneously.

The ability to search for spectral breaks and spectral time variations at energies higher than generally available to EGRET is dependent primarily upon higher effective area. The energy range for which EGRET was instrumented would not need to be extended upward: the sole limitation in this

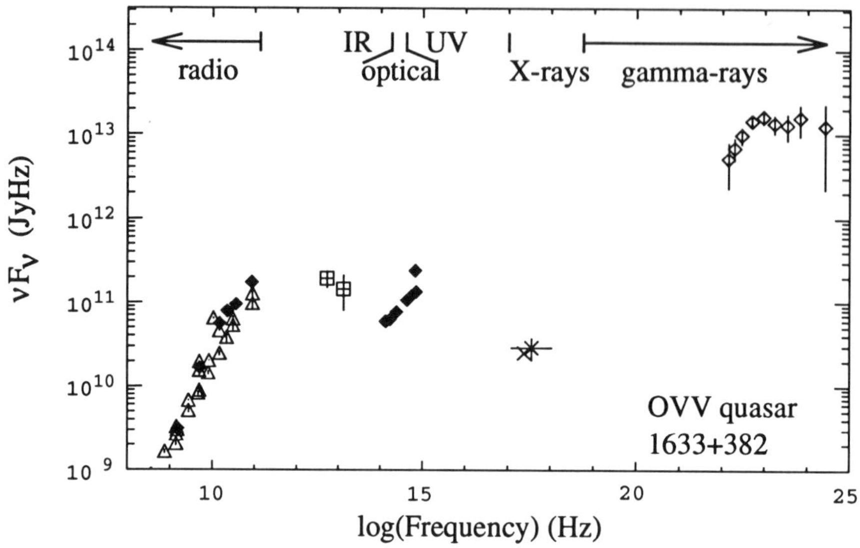

Figure 1: The multiwavelength spectrum of 1633+382. The νF_ν representation shows the observed power per logarithmic frequency interval. The individual observations, which are not contemporaneous, are summarized by von Montigny et al. (1995). The gamma-ray luminosity of 1633+382 dominates over other wavebands by about two orders of magnitude. This gamma-ray dominance appears to be a common feature of the spectra of blazars.

regard for EGRET has been limited photon statistics. Improved angular resolution is not crucial for the spectral studies at high energies (> a few GeV), because in most cases the brighter blazars dominate over the diffuse emission in that energy range. A moderate improvement in energy resolution at high energies would be useful, but is not essential; it is expected to be available from features of the energy-measurement system required for other science goals in any future telescope.

2.2 Bursts

Gamma-ray bursts have been another surprising addition to the detections made by EGRET. EGRET has imaged five bursts above 30 MeV out of the few dozen gamma ray bursts detected by the BATSE instrument within EGRET's field of view. About a dozen high-energy photons were observed in each of these five bursts. Gamma rays have been observed with energies well above a GeV which is more than two orders of magnitude higher than previously observed. Most dramatically, in the burst of 17 February 1994 (Hurley et al. 1994), the high-energy gamma-ray emission extended for an hour or more beyond the sub-MeV emission detected by BATSE, as seen in Figure 2. The highest energy photon from this burst had an energy of 18 GeV and was detected almost an hour after the sub-MeV emission became undetectable.

The existence and long time delay of high-energy gamma rays place many constraints on the nature of the sources of gamma-ray bursts. Spectra extending up to GeV energies rule out thermal spectra of most types. Therefore, particles must be accelerated to energies even higher than the energy of these gamma rays. High-energy gamma rays early in the burst require particles to be accelerated almost simultaneously with the keV gamma-ray emission. Shock acceleration is one possible mechanism.

The lack of attenuation of GeV gamma rays by $\gamma - \gamma$ pair production on the much more plentiful lower energy photons in the source requires beaming and/or bulk Lorentz factors in order to reduce the photon- photon opacity. The Lorentz factors required are much higher than those observed from other astrophysical phenomena, such as the superluminal motion in blazars where the typical bulk Lorentz factors <20. The photon density in burst sources is especially large because of the small size of the source inferred from the rapid time variability and the large distances to bursts. If bursts originate at cosmological distances then bulk Lorentz factors of order 1000 are needed, and even if the sources are located within our Galactic halo, relativistic motion is implied (Harding & Baring, 1994).

If the bright bursts detected by EGRET are representative of all bursts, then a high-energy instrument with greater sensitivity than EGRET will detect more bursts and more photons per burst. It is important to detect more photons in order to see if the high-energy emission is correlated with the MeV photons, or if there is a component with a delayed onset. Extension of burst spectra to higher energies can provide a constraint on the maximum distance to the burst source due to attenuation by photon-photon attenuation.

With detection of more high-energy bursts, it will be possible to learn if the high-energy photons seen in some bursts by EGRET are a common property of most bursts or are an indication of a separate class of bursts. Good sensitivity over the

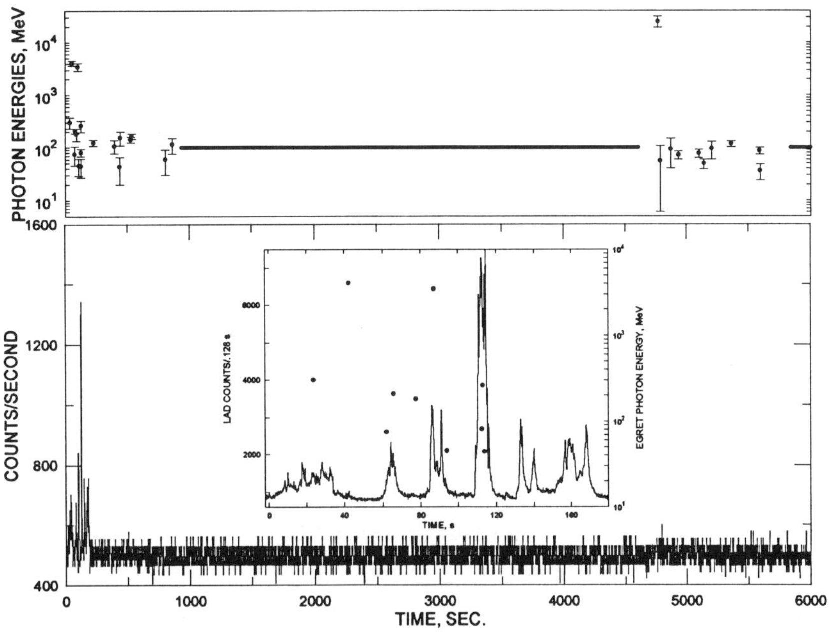

Figure 2: Lightcurve of the gamma-ray burst of 17 February 1994. The top panel shows the EGRET detected gamma-ray energies as a function of time. The bottom panel is the Ulysses 25-150 keV count rates and the insert is the BATSE lightcurve during the first 3 minutes of the burst with high-energy photons also marked (Hurley et al. 1994).

widest practical field-of-view is clearly desirable as a means of detecting more bursts. The best possible angular resolution is also needed to aid in identifying a low energy counterpart. Further, reduction of instrumental deadtime will permit the spectra of strong bursts to be measured during the intense portion of the burst. Finally, the energy measurement system of a next-generation high-energy gamma-ray telescope could be instrumented to detect gamma rays down to a few MeV. Spectra from 1 to a few 100 MeV have been obtained from the EGRET calorimeter by a data collection mode independent of the high-energy gamma-ray energy measurement.

2.3 Isolated Pulsars

The nature of spin-powered pulsars as rapidly rotating, magnetized neutron stars has long been recognized. They are an important astrophysical laboratory for physics in high magnetic and gravitational fields. Inspite of much progress in recent years, fundamental questions about these objects remain unanswered, including the basic radiation mechanism, the structure of the neutron star, the site of particle acceleration, the pulsar birthrate, and the relationship to supernova remnants.

With the results from CGRO, the number of gamma-ray pulsars has risen from 2 to 6. Their light curves at different wavelengths are very diverse, indicating a complex, energy-dependent geometry of emission (Thompson et al. 1994). Possible emission sites range from the surface of the neutron star itself (especially for the X-rays, which in many cases are thermal rather than non-thermal sources) out to nearly the speed of light cylinder. The detection of Geminga as a radio-quiet pulsar (Halpern & Holt, 1992; Bertsch et al. 1992) shows that radio and high-energy pulsars are overlapping subsets of the total population of neutron stars. The beaming information at high energies can, therefore, complement the radio observations to help address the questions of pulsar birthrate (and its relationship to the supernova rate).

Pulsar energy spectra, when viewed over the entire electromagnetic spectrum (e.g., Figure 3 shows the spectrum of PSR B1706−44), illustrate the interest in gamma rays. The observed power per decade from these pulsars is strongly dominated by the gamma-ray emission. The fact that the gamma radiation may represent 10% or more of the spin-down energy of some pulsars (assuming a narrow gamma-ray beam; much more if the beam is broad) indicates an extremely efficient transformation of rotational energy to high-energy electromagnetic radiation. For such pulsars, the fundamental assumption of a neutron star losing energy primarily by dipole radiation may be compromised by the large gamma-ray production efficiency. Phase-resolved energy spectra for the brightest gamma-ray pulsars show considerable variation among the different components of the light curves. All the pulsars show some indication of spectral steepening in the gamma-ray range, with the break energies ranging from about 1 MeV for PSR B1509−58 to around 10 GeV for PSR B1706−44.

The observations have provided considerable detail about

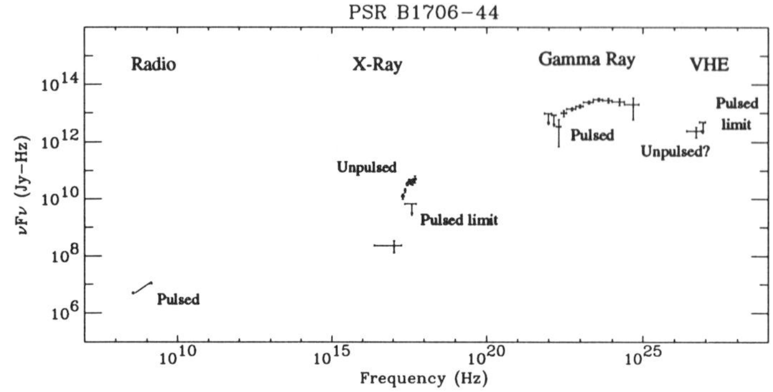

Figure 3: Multiwavelength spectrum of PSR B1706−44. The ν vs νF_ν representation shows the observed power per decade of frequency. The power per decade of this pulsar is dominated by the gamma-ray emission. References: Radio (Johnston et al. 1992); X- Ray (Becker et al. 1995); Gamma Ray (Thompson et al. 1992); VHE (Ogio et al. 1993; Nel et al. 1994).

a few pulsars and minimal information about a few more. This combination has made modeling of high-energy pulsars difficult. There is no theoretical consensus and pulsar modeling continues to be an active field. It is fair to say that no single model yet offers convincing explanations of all the optical through gamma-ray observations of pulsars.

Finding more gamma-ray pulsars, both radio-loud and radio-quiet, will be essential to answering the outstanding questions. The predictions for future pulsar detections made following the SAS 2 and COS B missions were largely unfulfilled: the pulsars which were seen were unexpected, and some which were expected were not seen. A crucial test of pulsar models and their descriptions of how and where in the pulsar's magnetosphere particles are accelerated will be their ability to predict which radio pulsars will be detected as gamma-ray pulsars in the future. Are there other radio-quiet pulsars like Geminga? How many? Answering these questions is important to understanding whether the absence of radio emission is the result of beaming or other physical processes. This information is also important for inferring the fraction of neutron stars which form pulsars. Clearly, a sample of one is insufficient.

The key to finding the gamma-ray pulsation in Geminga was the detection in the X-ray band. A recent reanalysis of EGRET data suggests that the Geminga pulsation might have been detectable in the EGRET data without the guidance of the ROSAT-detected X-ray pulse period. A future instrument with improved sensitivity could have pulse-searching capability. There are 10 or more unidentified Galactic sources in the First EGRET Source Catalog with intensities within a factor of four of Geminga's, thus a count rate four times higher than EGRET's would at least open the possibility of finding pulsations in some of these sources.

A second need is for additional details about the light curves and phase-resolved energy spectra of gamma-ray pulsars, in order to resolve the differences in gamma-ray pulsar models. A count rate improvement of at least a factor of four is needed. Additionally, the high-energy cutoffs seen in gamma-ray pulsars argue for an effective energy range which extends modestly upward from that of EGRET. EGRET detects a few photons above 10 GeV, and a future instrument should be designed to improve on that by an order of magnitude. Low-energy spectral breaks, such as those seen in Vela and PSR B1509−58, also point to the need for good detector response below 100 MeV.

2.4 Solar Flares

Solar flare emission of neutral particles (gamma rays, and neutrons) above 10 MeV provide important new information for understanding the acceleration and transport mechanism(s) of high-energy particles in the solar atmosphere. The gamma-ray spectrometer on the Solar Maximum Mission satellite observed, during intense flares, electron Bremsstrahlung to several hundred MeV, meson decay gamma rays and high-energy neutrons (>100 MeV). Neutron monitors at ground level have responded to solar neutrons with energies above 1 GeV (Chupp, 1990).

Dramatic new high-energy gamma-ray observations were made by EGRET during a solar flare on June 11, 1991 (Kanbach et al. 1993). High-energy gamma radiation was detected for at least 8 hours after the flare onset. The energy spectrum indicates a combination of electron Bremsstrahlung and π^0 decay. The high-energy protons which produced the meson component may have been trapped without re-acceleration in the magnetic field of the flare region for an extended period.

The high-energy neutral emissions from flares give directly the properties of the highest energy electrons and ions produced, information that cannot be obtained in any other manner. In order to further the study of high-energy flare phenomena, high sensitivity, high-energy resolution measurements from MeV energies to several GeV will be important. Such observations will permit, for the first time, a broad band study of the time variation of the acceleration/transport properties of electrons and ions throughout a flare. It is also conceivable that photons with energies much higher than a few GeV could be observed from flares since underground meson telescopes have responded to atmospheric secondaries produced by 30 GeV protons (Swinson & Shea, 1991).

A next generation gamma-ray telescope, with higher sensitivity and broader energy range than EGRET, will open new vistas into solar flare particle acceleration studies. New high-energy observations of flares with simultaneous images of the flaring radio and X-ray emissions (with magnetograph images) will resolve several current outstanding problems in solar flare physics.

2.5 Galactic and Extragalactic Diffuse Radiation, and Molecular Clouds

The Galactic diffuse gamma-ray emission is the dominant feature of the gamma-ray sky. This emission is strongest in a region about 100° wide centered on the Galactic center and is generally associated with spatial structures in the Galaxy, such as the tangent directions of the spiral arms. Based on this association and the intensity and energy spectrum of the emission, the diffuse emission is generally considered to be produced primarily by cosmic ray electron and proton interactions with the matter (via Bremsstrahlung and nucleon-nucleon interactions) and photons (via inverse Compton interactions) in the interstellar medium. Since the Galaxy is essentially transparent to gamma rays up to energies of about 10^{14} eV, and the radiation transfer is relatively simple, this association has prompted many authors to develop models of the Galactic plane gamma-ray emission (see Bloemen, 1989, for a general review). A three dimensional model of the Galactic gamma-ray emission, based on the recent radio surveys of the Galactic matter distribution and assumptions on the cosmic ray coupling (Bertsch et al. 1993, Hunter et al. 1995), has been developed to deduce the Galactic cosmic ray distribution and spectrum. This model, which assumes that the cosmic rays are coupled to the matter and that the cosmic ray electron/proton ratio is uniform in the Galaxy, shows good agreement, both spatially and spectrally, with EGRET observations (Hunter et al. 1995). This agreement provides direct insight into cosmic ray acceleration and propagation. However, the angular resolution of EGRET limits the angular scale of the spectral comparison to 20° × 4° regions in longitude and latitude. A new instrument, with better angular resolution, will permit more detailed searches for cosmic ray gradients including variations in the electron to proton ratio, cosmic ray contrast between the Galactic arm inter-arm regions, and evidence for regions in which the cosmic ray spectrum differs from the spectrum observed locally. Increased sensitivity coupled with improved angular resolution will also allow the flux from fainter gamma-ray point-sources to be more accurately separated from the Galactic plane diffuse emission.

At high latitudes, $|b| > 10°$, the Galactic diffuse emission is due almost entirely to cosmic ray interactions with material in the local solar neighborhood. However, comparison of the observed diffuse emission with a model of this emission provides evidence of a residual emission, presumably of extragalactic origin (Sreekumar et al. 1995). This extragalactic emission, possibly due to unresolved AGN emission, was first inferred from SAS 2 observations and is being studied in detail with the EGRET instrument. The COS B instrument was unable to detect this emission due to a high instrumental background. Improved angular resolution and increased sensitivity will permit more detailed studies of the extragalactic diffuse emission, possibly indicating spatial or spectral variations.

The gamma-ray emission from molecular clouds arises from the same cosmic ray interactions with matter which produce the general Galactic diffuse emission. Molecular clouds provide a means to study these processes and the Galactic cosmic rays in localized regions of the Galaxy. In fact, if the matter density of the molecular cloud is known, then the cosmic ray density in the cloud can be inferred. Nearby molecular clouds which subtend large solid angles, such as the Orion (Digel et al. 1994) and Ophiuchus (Hunter et al. 1994) molecular clouds, have been studied in detail using EGRET data. The gamma-ray emissivity derived for these clouds is consistent with the local emissivity and with the assumptions that the clouds are entirely permeated by cosmic rays and that there are no sources of cosmic rays within the clouds. The gamma-ray emission from these clouds is spatially resolved by EGRET; however, only average spectra could be determined. An instrument with improved angular resolution and sensitivity will permit the gamma-ray emission from more distant molecular clouds to be studied providing a more direct probe of the Galactic cosmic ray density in those regions. The nearby molecular clouds could also be studied on smaller spatial scales to look for variations in the cosmic ray spectrum, possible exclusion of cosmic rays from the dense cores of molecular clouds, and for regions of cosmic ray acceleration.

2.6 Normal Galaxies

The Magellanic clouds are our closest neighboring galaxies and hence are of great importance in studying many astrophysical aspects of external galaxies. As in the case of our Galaxy (§2.6), gamma-ray observations provide one of the most important avenues to study the cosmic ray distribution in external galaxies via their interactions with the interstellar gas.

The EGRET observation of the Large Magellanic Cloud (LMC) is the first detection of a normal galaxy outside the Milky Way in high-energy gamma rays (Sreekumar et al. 1992). This observation has been combined with a knowledge of the matter distribution in the LMC to provide a direct measure of the cosmic-ray density. Although the integrated flux above 100 MeV, $(1.9 \pm 0.4) \times 10^{-7}\,\mathrm{ph\,cm^{-2}\,s^{-1}}$, is barely above the EGRET threshold, it suggests a cosmic ray density that is comparable to the local density and is consistent with that expected from a quasi-static equilibrium model. As discussed by Sreekumar and Fichtel (1991), the derived cosmic ray density is also consistent with that derived from the non-thermal radio emission.

Observations of the Small Magellanic Cloud (SMC) provide only an upper limit (95% confidence) on the gamma ray emission above 100 MeV of $0.5 \times 10^{-7}\,\mathrm{ph\,cm^{-2}\,s^{-1}}$ (Sreekumar et al. 1993). The expected flux, if the cosmic rays are universal (and equal to the local density) rather than galactic in origin, is $(2.4 \pm 0.5) \times 10^{-7}\,\mathrm{ph\,cm^{-2}\,s^{-1}}$ (Sreekumar & Fichtel, 1991). This observation indicates that the cosmic rays are almost certainly not metagalactic, and therefore are galactic in origin. The gamma ray observations also support the idea that the SMC is being disrupted and is unable to confine its cosmic rays efficiently. Neither the energy spectrum nor the spatial distribution of the gamma rays in the LMC and SMC can be studied with EGRET because of the limited number of photons and angular resolution of EGRET.

Gamma-ray observations of the Magellanic Clouds with increased angular resolution and sensitivity would permit the cosmic ray density to be mapped and answer the question of whether the cosmic rays are coupled to the matter density as is suggested by models of the gamma-ray emission from the Milky Way. Improvements in sensitivity should also permit several other nearby galaxies (M31, M82, NGC 253), with predicted fluxes a factor of 3 to 10 below the EGRET threshold, to be detected. Some of these are particularly interesting because there is indirect evidence that they may have higher cosmic-ray densities than our Galaxy.

2.7 Unidentified Sources

Many of the gamma-ray point sources observed with EGRET have been identified with sources known at lower frequencies. However, a comparable number cannot be readily identified. The distribution of these unidentified sources suggests a largely Galactic population; however, there are some at very high Galactic latitudes. Many of the high-latitude unidentified sources show spectra and time variations similar to those of the identified gamma-ray emitting blazars. It is reasonable to anticipate that those sources are similar to, or at least related to, the gamma-emitting blazars identified so far. In order to verify this it will be necessary to reduce the size of the position error contours substantially, because many of the position error contours for the unidentified sources contain no bright (> 1 Jy) flat-spectrum radio sources. If dimmer radio sources are the sites of the gamma-ray production, much smaller gamma-ray source position error contours will be required for firm identification because of the much higher sky density of the dimmer radio sources. This requires better photon statistics, especially at the higher energies, and substantial improvement in the instrument angular resolution.

The unidentified sources in or near the Galactic plane are a more difficult problem. The only identified high-energy gamma-ray sources near the Galactic plane are the five pulsars. The distribution of the unidentified Galactic sources suggests that some may have luminosities greater than those of the known gamma-ray pulsars and a few of the unidentified sources are spatially coincident with radio pulsars. However, since gamma-ray pulsations have not been detected from these sources, identifications cannot be claimed. Improved angular resolution will be needed to resolve these cases. Although some of the unidentified sources may be radio-quiet, Geminga-like pulsars, some also show substantial time variations (unlike the detected gamma-pulsars), and thus probably represent a new class of Galactic gamma-ray objects. The identification of such a class, which will require improved sensitivity and angular resolution, would potentially be one of the most exciting discoveries to come from a future gamma-ray mission.

2.8 Supernovae Remnants

The origin of the Galactic diffuse emission in terms of cosmic rays interacting with the matter in the Galaxy is well understood. Although the mechanism(s) which accelerates cosmic rays to high energies is thought to be understood, it is lacking experimental verification. The favored model, shock acceleration due to supernovae, could be verified by measurement of the energy spectra of gamma-rays from supernova remnants. The supernova remnant cosmic ray spectrum is expected to have a spectral index in the range 2.0 to 2.4, which differs from the 2.7 spectral index of the local cosmic rays because of propagation effects. If supernovae are the injectors and accelerators of high energy cosmic rays then the gamma-ray spectrum would be expected to be harder (flatter) than the Galactic diffuse emission, assuming the local spectral index is representative of the Galaxy as a whole.

There is strong statistical evidence that some of the EGRET unidentified sources in the Galactic plane are supernova remnants. Unfortunately, the identifications are somewhat ambiguous, and improved angular resolution is needed to confirm them. Greater sensitivity at high energies with improvement in energy resolution is needed to more accurately determine the gamma-ray spectrum.

§3 Summary of Instrumental Requirements

The rich variety of astronomical phenomena in the high-energy gamma-ray sky place strong performance demands on the next generation of high-energy gamma-ray telescopes. They must have sensitivity and angular resolution that will meet the requirement to identify point sources with objects at other wavelengths, yet have a sufficiently wide field-of-view that will permit the study of sources that exhibit extreme intensity variations on timescales from seconds to months or longer.

There are a number of technical requirements that any high energy gamma-ray telescope must satisfy in order to meet these objectives. The most basic requirement is that the telescope detect and clearly identify incident gamma rays. In the energy range of interest this is best accomplished by detecting the vertex of the electron-positron pair and observing their subsequent separation. Obtaining good track resolution is also important for determining the gamma-ray arrival direction. The telescope must also discriminate in a very effective way against the much higher fluxes of cosmic rays, earth albedo gamma-rays and trapped radiation that will be incident on the instrument. This requires efficient on-board processing in order to not saturate the available telemetry or introduce an unacceptable dead time. Measurement of the energy of incident gamma rays over a wide range is important for understanding the physics of the emission processes.

It is also important to extend the range upward from the highest energies measured by EGRET. The highest energy gamma rays will also provide the best direction information. Obtaining spectral information at higher energies and improving the sensitivity in the EGRET energy range requires an instrument with a large effective area (detection efficiency times area) and as large an angular acceptance as possible. To allow temporal studies such as searches for fast pulsars, the telescope must be capable of providing precise timing of the detected gamma rays. Finally, the telescope must be reliable, capable of long term operation, and be constructed for the space and launch environments.

At the Snowmass Workshop, several high energy gamma-ray telescope concepts were discussed. All of the new telescope designs are based on the principles outlined above. These designs are aimed at achieving substantially greater sensitivity and improved angular resolution, compared with the capabilities of EGRET. In particular, various research groups are pursuing the application of newer technologies (compared with EGRET) for the track imaging system. These include designs based on large area drift chambers and solid-state trackers based on silicon-strip detectors.

§4 Summary

The success of the Compton Gamma Ray Observatory has demonstrated the enormous value of observational gamma-ray astronomy and points to important astrophysical problems meriting further investigation. NASA's mission to advance scientific knowledge depends on gleaning information from all possible sources. Recognizing the growing importance of multi-wavelength observations as a means to learn about the Universe, the recent NASA document *Space Science for the* 21st *Century* called for having observatories across the electromagnetic spectrum. However, at present, no gamma-ray missions are included in NASA's planning. Astrophysics would be incomplete without a continued study of gamma rays, especially in such topics as AGNs, bursts, the diffuse extragalactic radiation, and pulsars.

§References

§References

[1] Baring, M.G. 1990, MNRAS, 244, 49

[2] Becker, W., Brazier, K.T.S., & Trümper, J. 1995, A&A, submitted

[3] Bertsch, D.L., et al. 1992, Nature, 357, 306

[4] Bertsch, D.L., et al. 1993, ApJ 416, 587

[5] Bloemen, J.B.G.M. 1989, Annu. Rev. Astron. Astrophys., 27, 469

[6] Bloemen, H., et al. 1995, in press

[7] Chupp, E.L. 1990, Science, 250, 229

[8] Derdeyn, S.M., et al. 1972, Nucl. Inst. Meth. 98, 557

[9] Digel, S.W., et al. 1995, ApJ, 441, in press

[10] Fazio, G.G., & Stecker, F.W. 1970, Nature, 226, 135

[11] Fichtel, C.E., et al. 1975, ApJ, 198, 163

[12] Fichtel, C.E., et al. 1994, ApJS, 94, 551

[13] Fishman, G.J., et al. 1994, ApJS, 92, 229

[14] Harding, A.K. 1994, in The Second Compton Symposium, ed. C.E. Fichtel, N. Gehrels, J.P. Norris, New York: AIP Conf. Proc. 304, 30

[15] Harding, A.K. & Baring, M.G. 1994, to appear in Gamma-Ray Bursts, ed. G.J. Fishman, AIP Conf. Proc.

[16] Hunter, S.D., et al. 1995, ApJ, in preparation

[17] Hunter, S.D., et al. 1995, ApJ, 436, 216

[18] Hurley, K., et al. 1994, Nature, in press

[19] Johnston, S., et al. 1992, MNRAS, 255, 401

[20] Kanbach, G., et al. 1993, A&A, 97, 349

[21] Kraushaar, W.L., et al. 1972, ApJ, 177, 341

[22] von Montigny, C., et al. 1995 ApJ, 440, 525

[23] Morrison, P. 1958, Nuovo Cimento, 7, 858

[24] Nel, H.I., et al. 1994, ApJ, 418, 836

[25] Ogio, S., et al. 1993, 23rd ICRC, 1, 392

[26] Paczynski, B. 1988, ApJ, 335, 525

[27] Reich, W., et al. 1993, A&A, 273 65

[28] Schönfelder, V., et al. 1994, ApJS, 92, 593

[29] Schmidt, W.K.H. 1978, Nature, 271, 525

[30] Sreekumar, P., et al. 1992, ApJ, 400, L67

[31] Sreekumar, P., et al. 1993, Phys. Rev. Lett. 70, 127

[32] Sreekumar, P., et al. 1995, in preparation

[33] Sreekumar, P., & Fichtel, C.E. 1991, A&A, 251, 447

[34] Swanenburg, B.N., et al. 1981, ApJ, 243, L69

[35] Swinson, D.B. & Shea, M.A. 1990, Geophys. Rev. Lett. 17, 1073

[36] Thompson, D.J., 1994, in The Second Compton Symposium, ed. C.E. Fichtel, N. Gehrels, J.P. Norris, New York: AIP Conf. Proc. 304, 57

[37] Thompson, D.J., et al. 1992, Nature, 359, 61

HiRes (High Resolution Fly's Eye) Neutrino Aperture*

Yunsil Ho

Dept. of Physics, Columbia University, New York, NY 10027, USA

HiRes experiment is to explore the highest energy cosmic ray interactions in the atmosphere, well above the energy which can be obtained from any existing accelerators. New FADC electronics to be employed to HiRes will greatly enhance our ability to look at the highest energy cosmic rays. By looking at deeply penetrating showers in the atmosphere we can study ultra-high energy (UHE) neutrino initiated interactions in the atmosphere. An improved neutrino aperture calculation with HiRes detector is presented.

1 Introduction

The HiRes (High Resolution Fly's Eye) experiment is to explore the highest energy cosmic ray interactions in the atmosphere. One of the tantalizing possibilities in the study is the detection of ultra-high energy neutrinos.

Several years ago the neutrino aperture for the HiRes was calculated. Since then, we became aware of several factors that improve the calculation. They are the increase in the ultra-high energy total cross section due to small x_{BJ} effects, recent neutrino flux calculation, and a better understanding of the HiRes detector perfomance due to a new type of electronics to be employed in the experiment. Because of these reasons, we have decided to revisit the HiRes neutrino aperture calculation.

2 Neutrino Aperture

The neutrino event rate can be calculated as follows.

$$N \ \ (Detected\ \ Neutrino\ \ Events)$$

$$= \ \{\sigma_\nu(E) \times \begin{pmatrix} fiducial \\ mass \end{pmatrix} \times N_A \times \int \epsilon(\theta) d\Omega \times Acceptance\} \times \frac{d^2 N_\nu}{dA d\Omega} \qquad (1)$$

$$= \ G(E) \times \frac{d^2 N_\nu}{dA d\Omega}.$$

N_A is Avogadro's number, $\epsilon(\theta)$ is the earth shadowing factor as a function of zenith angle. Here G(E) is called "effective neutrino aperture",

$$G(E) : Effective\ \ Neutrino\ \ Aperture$$

$$\equiv \{\sigma_\nu(E) \times \begin{pmatrix} fiducial \\ mass \end{pmatrix} \times N_A \times \int \epsilon(\theta) d\Omega \times Acceptance\}. \qquad (2)$$

If we know the neutrino spectrum, then by multiplying the neutrino differential flux by the effective neutrino aperture of the HiRes, we can estimate the neutrino event rate with the HiRes.

3 Charged Current UHE Neutrino Interaction Cross section

The early calculation on the UHE neutrino cross section was carried out by M. H. Reno and C. Quigg [1]. They noticed that at these high energies the neutrino cross sections are an order of magnitude larger than the ones used before in many astrophysical applications due to the enhancement of structure functions at small Bjorken x region.

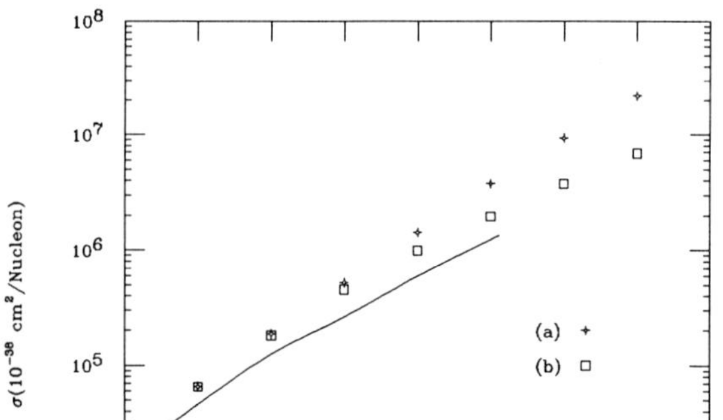

Figure 1: Charged current neutrino cross section with GRV HO parametrization, (a) $Q^2_{min} = .3GeV^2$, (b) $Q^2_{min} = 5GeV^2$. Solid line is for Reno and Quigg's calculation with EHLQ parametrization.

* Based on a talk given at Snowmass Workshop

Table 1: List of recent sets of NUCLEON structure functions as well as old EHLQ structure function.

Name of Set	Λ^4_{QCD} (MeV)	Q^2_{min} (GeV^2)	Reference
EHLQ	290	5.0	Rev. Mod. Phy. 56 (1984) 579 Rev. Mod. Phy. 58 (1986) 1065
MRS D0	215	5.0	Phys. Lett. 306B (1993) 145
MRS D-	215	5.0	Phys. Lett. 306B (1993) 145
GRV HO	200	0.3	Phys. Lett. 306B (1993) 391

Experimental knowledge of the structure functions was then limited to values of $x > 8 \times 10^{-3}$ [2]. Recent experiments H1 [3] and ZEUS [4] from HERA extended their measurements down to $x \sim 10^{-4}$ [5]. A clear rise of the structure function with a smaller value of x has been observed, as predicted by the theory. With these new data, a better parametrization of the structure functions is available.

In our calculation, we need the structure function at x as small as 10^{-9}. In addition we want to get to as small Q^2 as possible, as we are interested in the total event rates. For these reasons, we used GRV HO parametrization and we used Double Logarithmic Approzimation [7] which was used by Reno and Quigg. The charged current neutrino cross sections with GRV HO parametrization are shown in Figure 1 along with the calculation by Reno and Quigg. Comparison between different sets of parametrizations can be found in Table 1. Only GRV HO parametrization is valid for Q^2 value down to 0.3 GeV^2 compared to 5 GeV^2 of others.

To examine the Q^2_{min} effect on the total cross sections, we compared cross sections using GRV HO parametrization using Q^2_{min} of 0.3 GeV^2 and 5 GeV^2. The results, shown in Figure 1, show significant differences in two cases at very high energies. We also plot, in Figure 2, $x \times d\sigma/dx$ for the neutrino energy of 10^{20}eV for the two cases. As one can see, the differences come from very small x regions. Differential cross section, $x \times d\sigma/dx$, for two different neutrino energies of 10^{17}eV and 10^{20}eV (using GRV HO set) is shown in Figure 3. As one can see in Figure 3, the higher the neutrino energy is, the more contribution to the total cross section comes from the smaller x.

Based on these calculations, we have chosen to use GRV HO parametrization which is valid for Q^2 value down to 0.3 GeV^2. By combining GRV HO quark distribution functions and explapolating them for the small $x(< 10^{-4})$ using the DLA (Double Logarithmic Approximation) approximations [7], we calculated the UHE neutrino interaction cross section used for the HiRes neutrino aperture calculation.

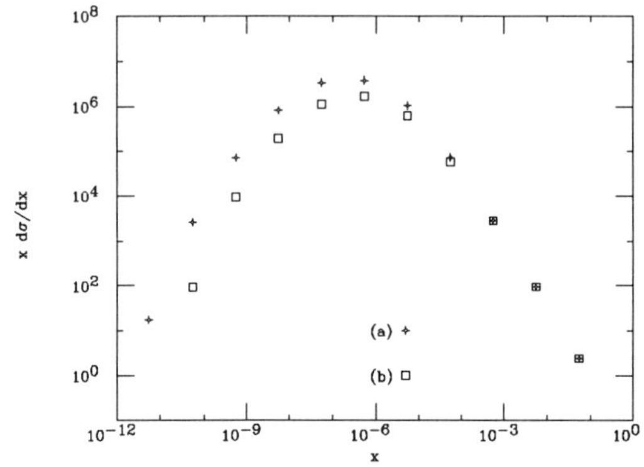

Figure 2: Differential neutrino cross section $x \times d\sigma/dx$ using GRV HO parametrization with (a) $Q^2_{min} = 0.3$GeV^2, and (b) $Q^2_{min} =$ 5GeV^2 for $E_\nu = 10^{20}$ eV.

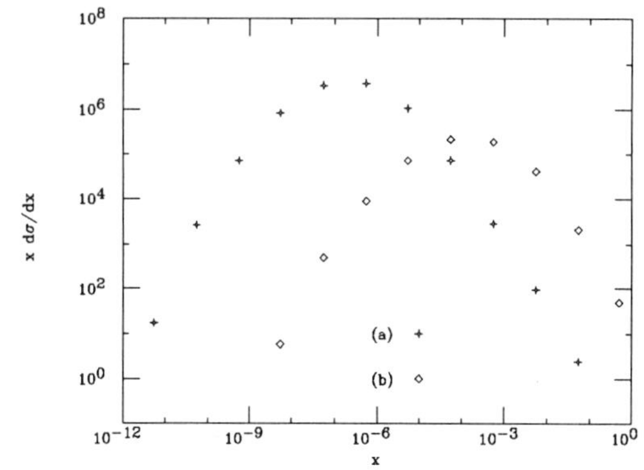

Figure 3: Differential neutrino cross section $x \times d\sigma/dx$ using GRV HO parametrization for (a) $E_\nu = 10^{20}$ eV, and (b) $E_\nu = 10^{17}$ eV.

4 Neutrino Events in the HiRes Detector

In order to calculate neutrino aperture, first we want to isolate neutrino events from the rest of the hadron induced showers. To the best of our knowledge, there is very little difference in the shower shape distributions in both longitudinal and transverse directions in neutrino induced and hadron induced showers. However, we can take advantage of the fact that the weak cross section is significantly smaller than the hadron cross sections, even at these high energies. So we have decided to call a shower which passed through matter more than 30 interaction lengths as the neutrino candidates. Based on known hadron flux at these energies, we estimate that hadron faking neutrino with the conditions above is less than 1.4×10^{-6}/year. We will be basically looking for horizontal showers.

Table 2: Fiducial radius parameter R_0 at the gound ($R_z = R_0 e^{z/H}$) determination, A is an acceptance at $z = 7.5km$ the neutrino event rate is propotional to $R_z^2 A$.

R_0	A	$R_z^2 A$
10 km	88 %	642
20 km	51 %	1487
30 km	28 %	1837
40 km	15 %	1782

We have also decided to have the fiducial volume for the neutrino events to be significantly different than the fiducial for ordinary hadrons. Note that we detect flourescent light whose yield depends on the track length. For example, for a horizontal track at the height of $7.5km$ (air density is down by 1/3), the shower is longer by a factor of 3, consequently generates 3 times more light. Therefore, the radius of fiducial needs to be larger as height increases. We actually chose fiducial radius as $R(z) = R_0 e^{z/H}$ (H is scale height, which is $7.5km$.), since the air density goes as $\rho_0 e^{-z/H}$. In order to determine R_0, we calculated the acceptance of events. As we vary R_0 large enough, the detected number of events should stay constant. Table 2 shows the results of event rate changes at the fixed height as we change the R_0. We chose fiducial radius at ground, R_0, to be 40 km.

When we actually calculate the acceptance for the entire fiducial volume, we also looked at acceptance with two different trigger requirements. So we will call '$A1$', for the acceptance with the trigger requirement of more than 20 PMT's each with more than 5 photo-electrons, and '$A2$', for the acceptance with the less strict trigger requirement of more than 10 PMT's each with more than 3 photo-electrons. Actual fiducial volume used was conical shape with $R = 40km$ at ground level, and $R = 400km$ at $z = 22.5km$ for simplicity. The radius at higher altitude is smaller than the value $R_z = R_0 e^{z/H}$, but this wouldn't affect the neutrino aperture since neutrino event rate will stay the same since anything that far would be out of the detector sensitivity. HiRes effective neutrino aperture for the two different trigger conditions is summarized in Table 3.

Table 3: Effective neutrino aperture for the two different trigger conditions for the various energies: A_1 is the acceptance for ≥ 20 PMT's with ≥ 5 p.e./PMT, and A_2 is the acceptance for ≥ 10 PMT's with ≥ 3 p.e./PMT. G_1 is the corresponding aperture with A_1, and G_2 is the corresponding aperture with A_2.

	10^{17} eV	10^{18} eV	10^{19} eV
$\sigma(10^{-38}cm^2)$	5.2×10^5	1.4×10^6	3.7×10^6
Fiducial Mass	7.8×10^{17} gram		
Solid Angle	$\int \epsilon(\theta) d\Omega(\theta : 70^\circ \sim 90^\circ)$		
A_1	0.3 %	2.0 %	5.0 %
A_2	0.9 %	3.4 %	8.4 %
$G_1(km^2 sr)$	1.5×10^{-3}	2.4×10^{-2}	1.6×10^{-1}
$G_2(km^2 sr)$	4.5×10^{-3}	4.1×10^{-2}	2.7×10^{-1}

5 UHE Neutrino Flux

The neutrino flux can be estimated with three parameters[8]. They are:

1. Energy distribution of the protons at the source is parameterized by 'γ'

   $\frac{dN}{dE_p} \propto E_p^{-\gamma}$.

2. The number of protons produced as a function z (the red shift parameter) is parameterized by 'm';

   $\frac{dN}{dz} \propto (1+z)^m$.

3. The distribution in z is integrated from $z = 0$ (present) to $z = z_{max}$. The 'z_{max}' is the parameter.

Given these three parameters, we calculate both ν and proton event rates. Then the number of proton is normalized to the observed number of cosmic ray above 10^{18} eV.

If we choose $m = -5/2$, $\gamma = 2.5$, and $z_{max} = 5$, then we don't expect to observe any neutrino events. If we choose a optimistic case of $m = 5$, $\gamma = 1.5$, and $z_{max} = 5$, then we expect to detect 7 ν_e CC interactions per year with 'A2' trigger condition, and 4 per year with rather conservative 'A1' trigger condition using 10% duty cycle of HiRes.

For every ν_e CC interaction, there are two associated ν_μ CC interactions, and neutral current interactions. For the additional interactions, we only observe hadron showers. Assuming the acceptance of ν_μ initiated showers is 50% of that of ν_e initiated showers, the total number of ν interactions including neutral current interactions is estimated to be 2.6 times of ν_e charged current interactions.

Using the actual 10% duty cycle of HiRes, the total number of neutrino interactions we expect to observe with HiRes is;

- 18 ν's/year (A2 trigger condition)
- 11 ν's/year (A1 trigger condition).

6 Conclusion

HiRes neutrino aperture has been calculated and neutrino event rate with HiRes has been estimated using the recent UHE Greisen neutrino flux calculation.

The largest uncertainty in event rate estimate comes from the neutrino flux calculation. We make assumptions on the early universe and its spatial propagation. So if we choose a certain set of assumptions, we wouldn't have any neutrino events with HiRes. But if we choose an optimistic set of assumptions of $m = 5$, $\gamma = 1.5$, and $z_{max} = 5$, then we expect to observe $\sim$ 10-20 ν events per year with essentially no backgrounds.

So HiRes can reach the sensitivity to check some of these cosmological assumptions, and we hope to observe ultra-high energy neutrino interactions with HiRes.

Acknowledgments

I would like to thank W. Lee and J. Boyer for useful discussions. This work was supported by the National Science Foundation, NSF PHY 92-15987.

References

1. M.H. Reno and C. Quigg, *physical Review* **37D** (1988) 657.
2. P. Amaudruz et al, *Physics Letters* 295B (1992) 159
3. T. Ahmed et al, *Physics Letters* **299B** (1993) 385.
4. M. Derrick et al, *Physics Letters* **303B** (1993) 385.
5. ZEUS collaboration, DESY-93-110, (1993)
6. L.N. Lipatov et al, *Physics Letters* **60B** (1975) 50.
7. D.W. McKay et al, *Physics Letters* **167B** (1986) 103.
8. J. Boyer et al, To be submitted to *Physical Review* **D** (1995)

Physics of the Highest Energy Cosmic Rays

P. Sokolsky

High Energy Astrophysics Institute, University of Utah, Salt Lake City, Utah, USA

A. A. Watson

Department of Physics, University of Leeds, Leeds LS2 9JT, UK

The observational background which motivates the drive to construct larger and better detectors for the highest energy cosmic rays is described and future plans for new instruments are outlined.

§1 Introduction

The study of cosmic rays with energies well above the knee (10^{15} eV) in the energy spectrum has a long history. Beginning with the pioneering work of Linsley at the Volcano Ranch array, followed by operation of the Haverah Park, SUGAR, Yakutsk, Akeno and AGASA arrays, evidence has been steadily accumulating that the spectrum exhibits a change in slope near 10 EeV (1 EeV = 10^{18} eV) [1]. This flattening may indicate that the spectrum in this region is largely extragalactic in origin. If the sources of this spectrum are sufficiently distant, a cut-off due to the interaction of near 100 Eev protons with black body photons should be observed. This is the Greisen-Zatsepin-Kuzmin (GZK)-cut-off [2]. Recent data have brought to bear a combination of information on spectral shape, composition and anisotropy to this study. In addition to clarifying the nature of the spectral structure, these new results lead to the surprising conclusion that nearby cosmic ray sources must exist that produce particles with energies well in excess of 100 EeV. The recent data have spurred efforts to design and build experiments that have the enormous apertures necessary for detecting these very rare events.

§2 Recent Results from the Fly's Eye and AGASA

Recent results from the Fly's Eye and AGASA detectors have confirmed, with better statistics, some of the claims made initially by the Haverah Park and Yakutsk groups. A general overview of the situation is given by Watson in the report of his invited talk in these proceedings. Here we concentrate on the most recent results.

The Fly's Eye group have reported on their recent results on the spectrum and composition of cosmic rays above 0.1 EeV [3]. The Fly's Eye is an experiment that uses air fluorescence to detect UHE cosmic rays. The extensive air shower (EAS) produced by the interaction of the primary particle with the atmosphere is imaged using the fluorescence light produced by excitation of the nitrogen molecules by the secondary EAS particles. The shape of the EAS longitudinal development is reconstructed from this light curve and the energy of the parent particle and some information on its composition can be determined.

The Fly's Eye group break their data up into better energy resolution stereo data, observed by two independent detectors (FE I and FE II) located 3.4 km apart, and poorer resolution monocular data, observed by FE I alone. The aperture for monocular data is approximately five times larger than for the stereo data and hence this data set has a longer reach in energy. Figures 1 and 2 show the resulting stereo and monocular spectra. The stereo data shows a clear break in the spectrum near 3 EeV with a spectral power index of (-3.18 ± 0.01) below this point and (-2.7 ± 0.10) above. Below 0.3 EeV, the spectrum slope reverts to the universally observed -3.0 power law index above the knee. This stereo data set runs out of statistics above 30 EeV.

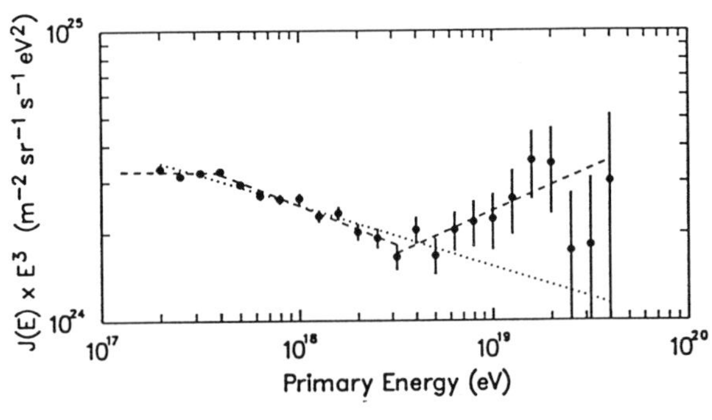

Figure 1: Fly's Eye stereo energy spectrum. Points:data.
Dashed line: best fit in each region.
Dotted line: best fit up to $10^{18.5}$ eV.

The monocular data show a similar structure. The shape of the monocular spectrum is consistent with the stereo spectrum once the poorer energy resolution is folded in. This monocular spectrum would be consistent with a GZK cut-off near 60 to 100 EeV were it not for an outstanding event at 320 EeV.

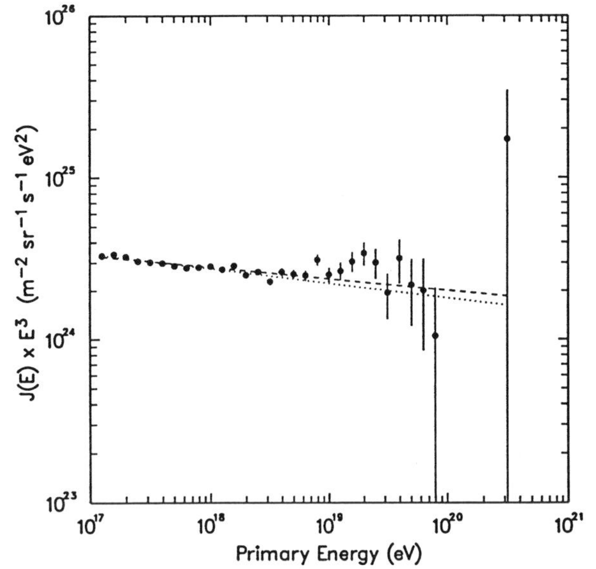

Figure 2: Fly's Eye monocular energy spectrum. Points: data.
Dashed line: best fit of the total spectrum.
Dotted line: best fit up to $10^{18.5}$ eV

Ground arrays determine the energy of an EAS by measuring the particle density at the surface. At approximately 600 m from the core of the shower the effect of shower fluctuations is minimized. This so-called ρ(600) density can be related to the primary energy via Monte Carlo calculations. Recent data from the AGASA ground array (described below in more detail) is shown in Figure 3 [4]. This spectrum shows qualitatively similar features to the Fly's Eye data - a break in the slope and flattening above 10 EeV although the intensity at a particular energy is higher in the AGASA experiment. A reduction in the energy estimates by 0.8 would bring the two spectra into reasonable agreement Again, an outstanding event with energy of 220 EeV is observed.

The features in the spectra observed in the new experiments (Fly's Eye, AGASA) confirm with superior statistics, particularly at the lower energies, the results from the older experiments (Harverah Park, Yakutsk). The four experiments are in agreement on the following points:-

1. There is a steepening of the spectra from -3.0 to about -3.2 just below 10^{18} eV.

2. Above 10^{19} eV a flattening of the spectra is observed with the slope becoming ~ -2.6.

3. All groups have observed one or more events which appear to have energies above the GZK-cut-off but the limited number make it uncertain as to what the rate above 10^{20} eV is.

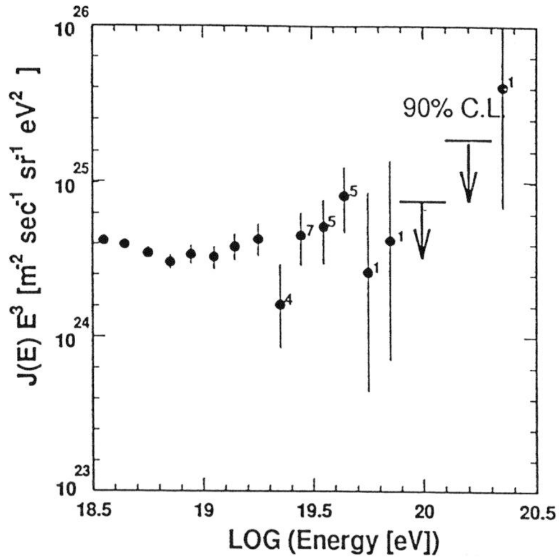

Figure 3: AGASA energy spectrum.

The stereo Fly's Eye data can be used to study the composition of cosmic rays. This is of particular interest in the energy region of the spectral break. Such a break could be a signature for a new, possibly extragalactic, source beginning to dominate the galactic flux observed at lower energies. The distribution of depth of individual shower maxima (X_{max}) has been shown in Monte Carlo calculations to be sensitive to composition. Briefly, one expects EAS produced by heavy nuclei to have their X_{max} at smaller atmospheric depths while protons will have X_{max} which are deeper in the atmosphere. The separation between the peaks of proton and X_{max} distributions for protons and iron is approximately 100 gm/cm^2 independent of the hadronic model chosen. Figure 4 summarizes the experimental situation. Shown are expected mean values of X_{max} as a function of energy for protons and Fe for a specific hadronic model as well as the stereo F.E. data. In this context, the data indicates that the composition is changing, from a predominantly heavy composition near 1 EeV to a predominatly light one by 10 EeV. The change occurs precisely at the point where the spectrum shows a break. While the details of the cosmic ray composition are hadronic model dependent, widely different hadronic models all predict that the slope of the mean X_{max} vs log(E) curve (the elongation rate) must be near 50 gm/cm^2 per decade while the observed slope is 78.9 pm 3.0 gm/cm^2 per decade above 0.3 EeV. At least in the context of the hadronic models investigated so far, the data strongly suggests a changing composition. (See the accompanying contribution by T K Gaisser for a further discussion of this point.)

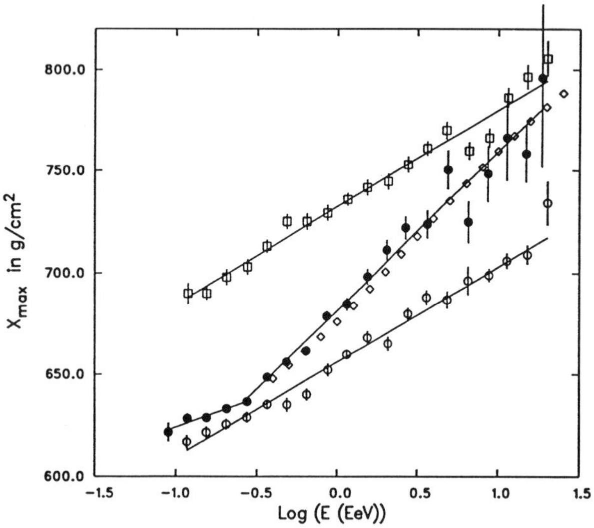

Figure 4: Fly's Eye stereo elongation rate. Black dots: Fly's Eye data. Open squares: proton X_{max} distribution based on QCD Pomeron model. Open circles: iron distribution. Diamonds: expected distribution based on a simple two-component assumption abut the origin of cosmic rays.

One picture which can explain the Fly's Eye and other spectra is the dominance of a new, flatter spectrum over the lower energy one. A two component fit to the Fly's Eye stereo data can yield the slopes of each spectrum (- 3.2 for the lower energy one and -2.6 for the higher energy one). If the lower energy spectrum is primarily heavy in composition and the higher energy one primarily light, one can predict the ratio of heavy to light composition as a function of energy. Using the predicted proton and Fe elongation rates as shown in Figure 4, one can calculate the elongation rate for the two component mixture. The diamonds in Figure 4 indicate the resultant prediction. They are in excellent agreement with the data and support this simple picture.

2.1 Implications of Observing a Continuing Flux Past 100 EeV

One of the signatures of distant extragalactic cosmic ray sources is the GZK-cut-off. This effect, due to the onset of inelastic photoproduction by protons on the 2.7 K black body radiation should produce a cut-off in the spectrum near 60 EeV, if the extragalactic sources are universally distributed. Both the Fly's Eye monocular and the AGASA spectrum would have been consistent with such a cutoff (though not statistically strong enough to prove that it existed) where it not for the observation of two events well beyond 100 EeV. To escape the GZK cutoff, these events must have sources less than 50 Mpc from us [5]. At these energies, there should be very little bending in the extragalactic magnetic fields and the particles should point back to their sources. No obvious candidates, such as AGN's or strong radio galaxies can be identified within the error boxes at distances of less than 50 Mpc. Either new types of sources or much stronger than expected extragalactic fields are required to explain these data. Since the cosmic ray flux at 100 EeV is near one event per square kilometer per century, the origin of these particles and how they relate to the bulk of the cosmic ray flux above 10 EeV and the GZK-cut-off can only be settled by constructing much larger aperture detectors. It is clear that the ability to measure composition above 10 EeV is also vital for any new detector.

§3 Existing and Approved Detectors

There are three detectors presently operating or under construction: AGASA, the Yakutsk array, and the HiRes detector. In what follows we concentrate on AGASA and HiRes.

3.1 The HiRes Detector (Stage I)

The Utah Fly's Eye group has entered into a collaboration with the University of Adelaide, Columbia University and the University of Illinois to construct a next-generation air-fluorescence detector [6]. The first stage of construction of this detector has been funded and is underway. Two sites, twelve kilometers apart, have been chosen. A total of 56 two meter diameter mirrors will be installed at the sites. The mirrors are arranged in rings, each ring subtending 14 degrees of elevation. Two complete rings at one site and one partial ring at the second site will be built. Each mirror will have a 256 phototube array at its focal plane. Each phototube will view a one degree by one degree section of the sky. The signals from the phototubes will be digitized using FADC electronics so that both the amplitude and the detailed pulse shape will be available for later analysis.

All events used will be recorded by both sites in stereo for good control of errors and redundancy of measurement. The aperture for this Stage I detector is optimized for greater than 10 EeV energy and approaches 7500 km^{2}sr at 100 EeV. With a 10% duty factor, this detector should record 10 events per year above 100 EeV and 200 events per year above 10 EeV if the stereo spectrum observed by the Fly's Eye were to continue. This is an order of magnitude increase in statistics over the monocular Fly's Eye. The resolution in energy and X_{max} is also much improved over the Fly's Eye detector, approaching 10% statistical error in energy and 15 gm/cm^2 mean error in X_{max}. This detector can also search for a possible gamma ray and neutrino flux. The method and object of such a search will be described below.

3.2 The HiRes Detector (Stage II)

Since the HiRes detector is modular, it can be easily expanded. If the Stage I detector confirms a continuing UHE flux above the GZK-cut-off, the aperture of the detector can be almost doubled to 15000 km^2str by adding thirty more mirrors. This will complete the two lower rings at the second site. To increase the 100 EeV aperture further would require the construction of additional stations. A third site may in fact be constructed by the proponents of the Telescope Array (see below). If the flux is found to terminate near 100 EeV, the detector can be tuned to optimize the aperture at lower energies (1-100 Eev) by adding the same number of mirrors to higher elevation rings.

3.3 The AGASA experiment

The Akeno Giant Air Shower Array (AGASA) has been operating since the spring of March 1991 [7]. 111 particle density detectors are arranged with inter-detector separation of about 1km covering an area of about 100 km^2. This aperture for collecting the highest energy cosmic rays is the largest in the world at present. About 100 well reconstructed events (whose core is well within the boundary of the array; and whose zenith angle satisfy sec theta leq 1.4) have been observed above 10 EeV. It is expected that more than 300 events with energies greater than 10 EeV will be accumulated by the end of 1995. This should provide further solid evidence for the absence or presence of the GZK-cut-off.

§4 Plans for Future Experiments

Much larger aperture detectors will be required to investigate the cosmic ray flux in detail if it continues to near 1000 EeV. This issue is addressed by two proposals: the Telescope Array, and the Giant Hybrid Array (the Auger project). We discuss them briefly in turn.

4.1 The Telescope Array

The Telescope Array is a proposal by a consortium of Japanese universities and the University of Utah, lead by the University of Tokyo [8]. The goal is to build an air fluorescence detector with a pixel size of 1/4° by 1/4° using multianode photomultiplier tubes. With this small a pixel size, sky noise is essentially negligible. The array is designed to also be sensitive to low energy gamma rays above 100 GeV by detecting Cherenkov light from these showers. It includes 120 fixed and 120 steerable alt-azimuth three meter diameter mirror dishes positioned at two stations 40 km apart. The aperture of such an array at 100 EeV is estimated to be approximately 40,000 km^2sr. The energy resolution is 10% and the resolution in X_{max} is 10 gm/cm^2. A prototype mirror and pmt cluster are being tested at Dugway. There are plans to build ten prototype mirror units at a site near the HiRes detector sites.

4.2 The Giant Hybrid Detector

This proposal, led by groups from the University of Chicago and the University of Leeds, combines the strength of ground arrays (high statistics due to near 100% duty cycle, good sensitivity to possible anisotropies) with the advantages of the air fluorescence technique (model independent energy measurement, measurement of cosmic ray shower development and composition) [9]. The main array, with 1.5 km detector spacing, is expected to have an aperture of 5000 km^2sr near 100 EeV. A densely packed 400 km^2 central array would have 0.75 km detector spacing and have a threshold below 1 EeV. The central portion will collect about 40,000 events per year with energy greater than 1 EeV while the full array would detect 5000 events per year above 10 EeV. Detectors would measure both the electron and muon components at the surface. Two such arrays are proposed, one in the Northern and one in the Southern hemisphere. This will guarantee good coverage for anisotropy measurements. At the center of each array will be an air fluorescence detector. Approximately ten percent of the data will thus have both ground array and longitudinal shower development information. The energy scale, as determined by the fluorescence technique, can be used to calibrate the ground array. Similarly muon information determined by the ground array can be compared to composition measurements from the longitudinal shower development measurements.

§5 Gamma Ray and Neutrino Flux

It is a truism that when opening up a new energy region it is important to be sensitive to the unexpected as well as to resolve specific questions. An unexpectedly high flux of photons or neutrinos is one such possibility. A photon and neutrino flux is predicted by the GZK mechanism though its magnitude is model dependent. Cosmic string models predict a larger neutrino flux and a photon to proton ratio of greater than one at the highest energies [10]. Detectors using atmospheric fluorescence are sensitive to near horizontal and upward going EAS. Calculation indicates that for certain assumptions about the GZK neutrino flux, a HiRes type of detector may be able to identify a significant number of neutrino induced events [11].

Gamma rays are difficult to distinguish from protons by EAS longitudinal development for energies below 100 EeV, but their EAS may have a significantly smaller muon content at the surface. Calculation indicates that the turn-on of the

Landau-Pomeranchuk-Migdal effect near 100 EeV causes striking elongation of gamma ray shower development [11]. Air fluorescence detectors may be able to distinguish such showers from normal hadronic showers.

Observations of (or limits on) such fluxes will clearly be very important in understanding the nature and origin of the UHE cosmic ray flux.

§6 Conclusion

Recent results indicate that there is clear and striking structure in the UHE cosmic ray spectrum. Composition studies support a picture of an emerging extragalactic flux near 10 EeV. The apparent continuation of this flux beyond the expected GZK cutoff energies and the lack of obvious close-in source candidates for such a flux raises many interesting astrophysical questions. This area of astrophysics can only make significant progress by further support for construction of the sophisticated large aperture detectors that have sensitivity to all the relevant quantities of interest.

§References

[1] P. Sokolsky et al. *Physics Reports,* 217:227, 1992.
A.A. Watson. *Nuclear Phys B (Proc. Suppl.)* 28B:3, 1992.

[2] K. Greisen. *Phys. Rev. Lett.,* 16:748, 1966.
A.M. Zatsepin and V.A. Kuzmin. *JETP Lett.,* 4:78, 1966.

[3] D.J. Bird et al. *Phys. Rev. Lett.* 71: 3401, 1993.
D.J. Bird et al. *Astrophys. J.,* 424:491, 1994.
T.K. Gaisser et al. *Physical Review D,* 47:1919, 1993.

[4] S. Yoshida, personal communication.
S. Yoshida et al. *Astroparticle Physics,* in press.

[5] J.W. Elbert and P. Sommers. *Ap. J.,* in press.

[6] Construction Proposal for the High Resolution Fly's Eye, 1993.

[7] N. Chiba et al. *Nucl. Inst. Methods,* A311:338, 1992.

[8] M. Teshima. Proceedings of the Tokyo Workshop on Techniques for the Study of Extremely High Energy Cosmic Rays, 109, 1993.

[9] Papers by J.W. Cronin and P. Sommers, in these proceedings.

[10] F.A. Aharonian et al. *Phys. Rev.* D46:4188, 1992.

[11] Y.Ho, talk presented at this meeting.

Y. Mitzumoto. Proceedings of the Tokyo Workshop on Techniques for the Study of Extremely High Energy Cosmic Rays, 194, 1993.

Superhigh Energy Cosmic Rays

P. Sommers

Department of Physics, University of Utah, Salt Lake City, Utah 84112, USA

The status of measurements on the highest energy cosmic rays as reported at the Snowmass meeting is summarised and possible origins for the Fly's Eye event of 3 x 10^{20} eV are reviewed. The potential of a hybrid ground array - fluorescence detector system for the study of the highest energy cosmic rays is discussed.

§1 Status of Measurements

The universal microwave background radiation constitutes a hostile environment for particles with energies >100 EeV (10^{20} eV). Attenuation lengths are less than about 30 Mpc for superhigh energy protons, nuclei, or photons. This means there can be superhigh energy particles (above the Greisen-Zatsepin-Kuzmin spectral cutoff near 100 EeV) only if there are nearby sources of such particles in the contemporary universe.

Although numerous superhigh energy particles have been reported over the years by ground array experiments, special attention has recently focused on an exceptional air shower detected by the Fly's Eye detector via atmospheric scintillation. The measured energy is 320 EeV, more than twice as great as any previously reported particle, and 4 times the energy of any other particle detected by the Fly's Eye itself.

The high magnetic rigidity of such an energetic particle means that it should suffer little deflection over its longest possible pathlength in intergalactic space, and the Galaxy's magnetic field should not alter its direction much either. It is therefore an intriguing puzzle that its arrival direction ($b \sim 10°$, $\ell \sim 163°$) points toward no likely astrophysical source within the propagation distance limit imposed by the background radiation.

Curiously, the only superhigh energy event reported by the Yakutsk experiment has an arrival direction which differs only by 8° from the direction of the Fly's Eye shower. Allowing for the exposures of the two experiments, the chance probability is 0.008 for a randomly sampled event from one experiment to be within 8° of a randomly sampled event from the other experiment. The arrival direction uncertainties are large enough that the two arrival directions might actually be identical. This underscores the need for improved angular resolution in future detectors of superhigh energy particles. It should be noted that pion photoproduction interactions cause a nucleon to spend a substantial fraction of its time as a neutron. Since an arriving neutron is not deflected at all by the Galaxy's magnetic field, superhigh energy particle astronomy can exploit the best possible detector angular resolution.

S. Yoshida reported preliminary analysis of the only superhigh energy particle so far detected by the AGASA experiment. The energy is approximately 200 EeV, and it is a high quality event because the core lies near the center of the well-instrumented (Akeno) branch of the array. The arrival direction ($b = -41°$, $\ell = 131°$) is not close to the arrival directions of the superhigh energy showers of Yakutsk and the Fly's Eye nor to the directions of the events of about 10^{20} eV reported by the Haverah Park and Volcano Ranch teams.

The Fly's Eye experiment has found evidence for correlated changes in the cosmic ray energy spectrum and composition occurring near 3 x 10^{18} eV. As reported by other experiments, the spectrum is found to flatten (become less steeply falling). Based on the energy dependence of the depth-of-maximum distribution, P. Sokolsky (speaking for the Fly's Eye group) reported that the composition seems to be changing at that energy from a predominantly heavy nucleus population to a light (presumably proton) population. The correlated changes in spectrum and composition support the view that the ``ankle" of the spectrum signifies a transition from dominance by cosmic rays of galactic origin to predominantly extragalactic particles at the highest energies. This conclusion (as well as the fact that the arrival direction of the superhigh energy Fly's Eye shower does not correspond to any remarkable galactic object) suggests that the source of the remarkable particle is extragalactic.

The absence of an obvious acceleration site for the 320 EeV Fly's Eye particle means that there is some new astrophysics (or perhaps some new fundamental physics) to be learned from the study of such superhigh energy particles. The high resolution Fly's Eye (HiRes) is under construction in Utah and will be the first detector of significantly higher sensitivity to study the superhigh energy regime. H.Y. Dai reported on its status and early results. S. Yoshida described prototype work being done for the Japanese Telescope Array Project, whose sensitivity would exceed that of HiRes. A giant detector with a 5000 km^2 ground array operating together with an optical detector of comparable aperture (described elsewhere in these proceedings by Cronin) will be especially valuable for studying the special population of superhigh energy particles.

§2 Origin of 3 x 10^{20} eV Event

A search for possible sources of the 320 EeV Fly's Eye particle was reported by P. Sommers. The starburst galaxy M82 is 3.5 Mpc away and its direction differs by 37° from the particle's arrival direction. That deflection could be due to the Galaxy's magnetic field, but only if the particle had a large charge (Z~20). Propagation simulations show that a nucleus of energy 320 EeV could not be a remnant surviving a 3.5 Mpc journey. However, an iron nucleus spectrum, after propagating 3.5 Mpc, develops a bump at 230 EeV, and the mean Z value for the nuclei in that bump is 18. That energy (230 EeV) is the 1-sigma experimental lower limit for the detected particle's energy, so it is possible that the detected particle was the remnant of a nucleus which started as iron at M82. Various models are conceivable for accelerating nuclei in a starburst galaxy like M82, but none has been developed in detail. It would be surprising if M82, which is not a strong radio galaxy or a source of low energy gamma rays, were the source of superhigh energy particles.

Another conceivable source is the radio galaxy Centaurus A. If the particle's trajectory was bent through 135° in intergalactic space while in transit from Cen A, then it would have originated at an earlier epoch of Cen A when it might have had the powerful radio hot spots which some workers regard as the most likely sites for particle acceleration to superhigh energies. That scenario requires strong extragalactic magnetic fields (~0.1 μG with at least a 5 Mpc cell size, or stronger fields on smaller scale sizes). Such strong extragalactic fields are not expected, but they cannot be ruled out observationally. Superhigh energy astroparticle studies would benefit greatly from specific information about extragalactic magnetic fields. Turning the analysis around, sensitive studies of superhigh energy particles may someday be used as probes of extragalactic fields.

The puzzle of the origin of the 320 EeV particle could have other solutions if there were some particle types which were immune to the cosmic background radiation but which would produce a normal air shower. In that case, the stringent distant limits on the source would not apply. The Seyfert galaxy MGC 8-11-11 is at a redshift 0.0205 (at least 62 Mpc away) and within 2° from the nominal arrival direction, but outside the narrow 2-sigma arrival direction error box. The more distant quasar 3C147 (z = 0.545) is easily within the 1-sigma error box.

The difficulty in identifying a source for this particle has prompted some workers to regard it as evidence that superhigh energy cosmic rays are produced by the annihilation of topological defects, rather than being accelerated. G. Sigl emphasized this possibility in his comparison of theoretical source spectra with the observed cosmic ray spectrum. Since gamma rays from topological defect annihilations are expected to have a higher intensity than protons from the same sources at superhigh energies, this interpretation is favored if the detected particle can be shown to be a photon. Its depth of maximum (815±40gcm^{-2}) does not uniquely identify the primary particle type. T. Stanev showed results of simulating a gamma ray's pair production in the Earth's magnetosphere, magnetic bremsstrahlung of the resulting electrons, and the ensuing superposed electromagnetic cascade in the atmosphere (incorporating the LPM effect). His plot showed the expected depth of maximum to be 1030 gcm^{-2} with little fluctuation. If correct, that work suggests that it may be difficult to reconcile the Fly's Eye measured depth of maximum with the photon hypothesis.

There is certain to be intensified investigation into the possible origins of the superhigh energy particles since there is a guarantee of new astrophysics or new fundamental physics to be learned. There is a clear need for more sensitive detectors to gather high resolution data with good statistics.

§3 Giant hybrid detector

A "hybrid" detector here means a ground component of particle counters operating together with an atmospheric fluorescence component (using the Fly's Eye technique). Plans for a 5000 km^2 detector (to be duplicated in the northern and southern hemispheres) are described elsewhere in these proceedings. This section focuses on the enhanced capability during clear dark nights when the optical component operates. The hybrid detector makes 3-dimensional shower measurements. The ground component measures the two-dimensional lateral distributions of gamma rays, electrons, and muons (separately). The optical component measures the longitudinal development of the charged particle cascade. The hybrid detector is not just two experiments in parallel. There is a strong synergy between the components. The first stage of analysis -- the geometrical reconstruction -- is done by minimizing a chisquare function involving all timing data from both components, and the geometrical resolution is far better than can be achieved by either component operating alone.

At ground level, the two components both measure the charged particle shower size, so a calibration cross check is possible on each shower. For nearby showers, the optical detector can also measure the lateral distribution. For example, suppose the pixel size (for the lowest elevation phototube clusters) is 1/3°. At 5 km, each pixel sees a 30 m segment of the shower, so the lateral distribution can be determined in 30-meter intervals. For the nearby showers, therefore, the ground component and optical component can be cross-checked in detail at ground level, and the optical component measures the lateral distribution at atmospheric depths above the surface also. With reliable cross calibration of the components, the ground component can be used to "normalize" a longitudinal profile seen at very large distance where atmospheric attenuation variation would otherwise cause uncertainty in the optical component's measurements of the shower size.

The hybrid detector will have good sensitivity to the cosmic ray composition because it measures the muon lateral distribution together with the longitudinal development profile. A histogram over the 2-dimensional space of X_{max} and N_{μ} is more sensitive to composition than either its projection onto the X_{max} axis (as determined by an air fluorescence experiment alone) or its projection onto the N_{μ} axis (as determined by a ground array experiment alone). Even more sensitive composition indicators should be constructible from the full longitudinal profile (not simply X_{max}) together with the lateral distributions of electrons, gammas, and muons (not simply the total muon number).

Accurate geometrical reconstruction of air showers is crucial for accurately determining shower parameters (as well as essential for superhigh energy particle astronomy as discussed above). The hybrid array will achieve excellent reconstructions. An angular resolution of 0.1° is obtained using shower front arrival times from the ground array together with the pointing directions and timing information from the pixels of the optical component. The HiRes detector has demonstrated that the track-detector plane can be determined to approximately 0.1°. For a monocular detection (as opposed to the HiRes stereoscopic design), it is difficult to determine the shower axis within that plane just from the eye's phototube times. By incorporating the arrival times at the ground stations, the shower axis can be determined in that plane also to an accuracy of about 0.1°. The shower core is determined to about 25 m.

The challenge in designing the optical component is to match its aperture to that of the ground component (5000 km^2 area) while minimizing construction and operation costs. A single eye at the center of a giant hexagonal array would be simplest to operate. However, it is then necessary for the eye to measure accurately showers which are 40 km away and seen at low elevation angles. This is a daunting challenge in view of the average horizontal extinction length of about 9 km for light (due primarily to scattering of the light by aerosol particles, whose concentration is variable). Such atmospheric transmission problems can be minimized by building multiple eyes, each monitoring the air above its own portion of the 5000 km^2 area. Each eye could be much simpler and cheaper than a single, central eye. Since showers would be detected at closer range, the mirror size could be smaller and the pixel size increased (fewer phototubes and fewer electronics channels per eye). A multi-eye design would presumably require remote operation of the eyes (requiring, in particular, a fail-safe method for closing the eyes before dawn).

Optical component techniques are being refined by the HiRes experiment at Dugway, Utah. A prototype of the Japanese Telescope Array Project has also been approved for construction at Dugway. An essential step toward the giant hybrid detector is to operate prototype ground stations in conjunction with those optical detectors. A small prototype array of detectors could do important physics by making detailed shower measurements near 1 EeV of the type that the giant hybrid array will make at higher energies.

The optical component of the giant hybrid detector will have a duty cycle of about 10%. The ground array will operate independently the rest of the time. Its performance can be optimized by "training" it with the 10% of the events which are precisely (and redundantly) measured by the full hybrid detector.

Design Considerations for a Ground Array

J. W. Cronin

Enrico Fermi Institute, University of Chicago, Chicago, Illinois 60637-1433, USA

A comparison of the calculation of a variety of shower parameters with observations is made for the purpose of demonstrating that a model of shower development exists which is adequate for the design of the detector array of the Auger project.

§1 Ground Arrays

1.1 Introduction

The design of a ground array depends crucially on the properties of the air showers that are to be observed. Fortunately there is a great deal of existing data on air showers with energies $\geq 10^{19}$ eV and simulations largely confirm these results. The lateral distribution and the energy spectra of the shower components are the crucial parameters that determine the design of a surface array which is efficient for energies $\geq 10^{19}$ eV and has the capability to distinguish the muonic component from the electromagnetic component of the air shower at distances far from the core.

1.2 Lateral Distribution

The lateral distribution of charged particles for air showers with energy above 10^{18} eV has recently been measured by the Akeno group at the 100 km^2 AGASA array [1]. Their lateral density distribution for all charged particles is well described by an empirical function:

$$\rho(R) = C\left(\frac{R}{R_M}\right)^{-1.2}\left(1+\frac{R}{R_M}\right)^{-2.6}\left(1+\left(\frac{R}{1\text{km}}\right)^{2}\right)^{-0.6} \quad (1)$$

Here R is the distance from the core and R_M is the Molière radius. This formula applies for showers with zenith angles less than 25° and over the range of energies 10^{18} - 10^{20} eV. The curve is anchored at a distance of 600m by the ground parameter S(600) which is the density (particles/m^2) measured in a 5 cm thick scintillator. The number of particles is defined as the energy deposit in the scintillator divided by 10 MeV: ie. the energy deposited by a minimum particle traversing the scintillator vertically. The energy is proportional to S(600) and is given by:

$$E(\text{eV}) = 2.0 \times 10^{17} \times S(600) \quad (2)$$

This energy conversion constant [2] has been established by simulations and gives good agreement for the cosmic ray flux at 10^{19} with the Yakutsk [3] and Haverah Park [4] ground arrays and the Fly's Eye fluorescence detector [5]. The cosmic ray flux at 10^{19} eV for Akeno and these other detectors is found to be, in units m^{-2} s^{-1} sr^{-1} eV^{-1} x 10^{33}, (3.9 ± 0.3), (3.8 ± 0.3), (2.7 ± 0.3), and (2.4 ± 0.3). Since the spectra fall as E^{-3} a 50% discrepancy in the flux corresponds to a 15% error in the energy determination. That these diverse experiments, with quite different methods of energy determination, agree as well as they do is quite remarkable.

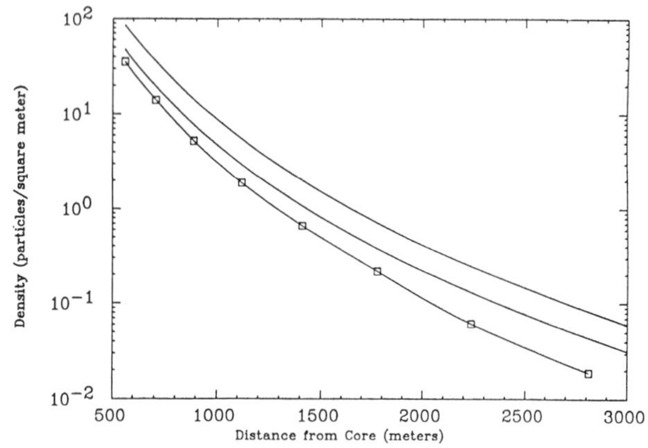

Figure 1: The lateral distribution of charged particles as calculated for 10^{19} eV primary protons. The results of calculations with MOCCA are shown by the squares. The upper curve is the Akeno lateral distribution transformed to 850 gcm^2 (1500 m). The middle curve is the lateral distribution transformed to 1000 gcm^{-2}.

In Figure 1 we plot (open squares) the lateral distribution as calculated for 10^{19} eV primary protons by the MOCCA simulation program of Hillas [6]. The mean energy deposited per meter squared in a 5 cm thick scintillator was calculated from the density and energy spectra of the photons, electrons, and muons provided by the MOCCA simulation. The lower of the two solid lines is the Akeno lateral distribution transformed to a depth of 1000 gm/cm^2. This transformation depends on the ratio of the Molière radii. This length is 9.3 gm/cm^2/(density of air at the surface). For moderate (10°C) temperatures the Molière radius is 91.6 m at Akeno and 84 m at 1000 gm/cm^2. S(600) transforms as:

$$\left(\frac{91.6}{84}\right)^2 \left(\frac{91.6}{84}\right)^{-3.8} \times \exp(-80/500) \qquad (3)$$

The last term in the product is the attenuation with depth of S(600) measured by the Akeno group [2]. The upper solid line (Figure 1) is the Akeno lateral distribution transformed to an atmospheric depth of 850 gm/cm^2 corresponding to an elevation of 1500 m. The simulation predicts a density of 26 charged particles per m^2 at 600 m where S(600) is found to be 36 from the Akeno measurements. The simulation results are low by a factor of 0.7 to 0.45 with respect to the AGASA result as the distance from the core varies from 600 to 2500 meters. The Akeno data points reach to a distance of 3 km and fit their empirical curve very well. There is technical evidence that the MOCCA simulation ,which uses a "thinning algorithm", may underestimate the densities at large distances. On the physical side, low energy neutrons may enhance the apparent density at large distances. In any case we will use the experimental Akeno lateral density distribution to predict the efficiency of the giant array. For the composition and energy distribution of the shower particles at the ground we will rely on the simulation.

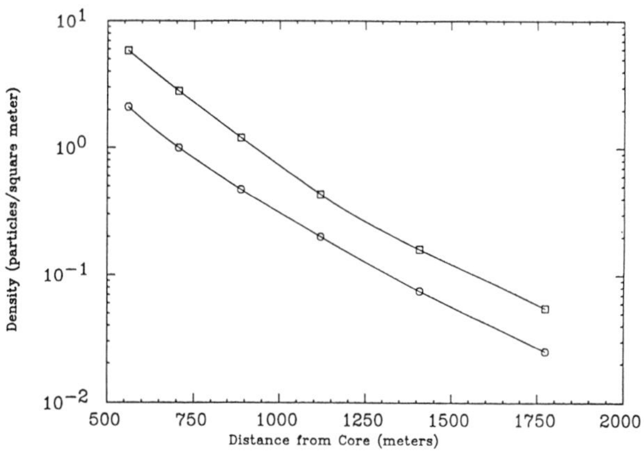

Figure 2: Comparison of MOCCA calculation (circles) of muons (> 1.0 GeV) in showers of 10^{19} eV with AGASA data for showers in the energy interval (0.3 - 1.0) x 10^{19} eV.

The Akeno group also has measured the lateral distribution of muons [7]. Only muons with energy ≥ 1.0 Gev are accepted. In Figure 2 we plot with open circles the Akeno muon density for showers in the energy interval (0.3 - 1.0) x 10^{19} eV. For comparison we plot the muon density found by the MOCCA simulation. Here we have made no attempt to correct for the difference in atmospheric depth of the experiment (920 gm/cm^2) and the simulation (1000 gm/cm^2), nor for the difference in energy. The simulations were for a fixed energy of 10^{19} eV. The experimental data are systematically 2.5 times higher. It seems difficult to account for this factor by the difference of the conditions of the simulation and the experiment.

1.3 Simulation of the Properties of an Array

We assume the Akeno lateral distribution and calculate the efficiency of an array with detectors on the corners of equilateral triangles with 1.5 km sides. We assume detector elements of 10 m^2 area and accept zenith angles out to 45°.

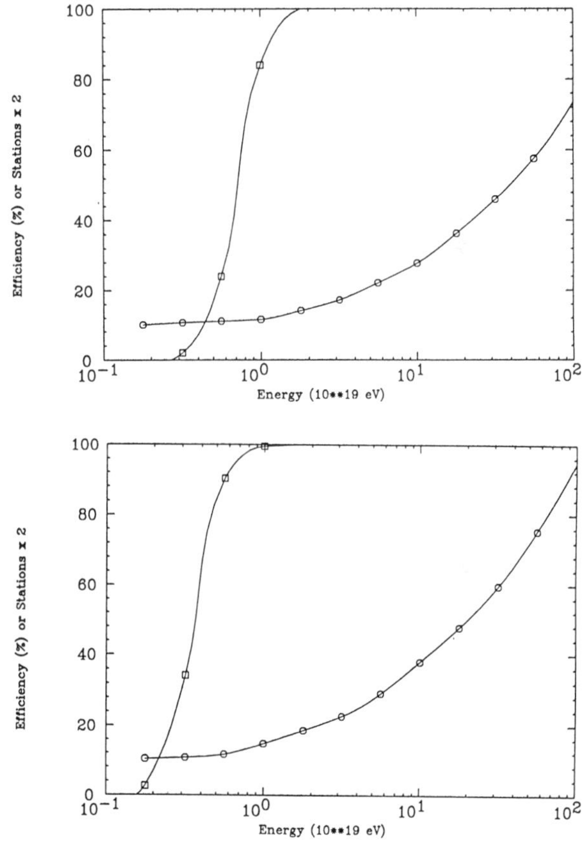

Figure 3a, b: The efficiency of triggering an array in which 10 m^2 detectors are on a triangular grid with 1.5 km spacing. Five stations must be alerted to trigger the array: the Akeno lateral distribution has been used to callculate the efficiency. (a) 1000 gcm^2 (b) 850 gcm^2.

For a station to be alerted we demand that four or more charged particles be detected. At least five stations must be alerted to trigger the array. We assume that the showers strike within the array and neglect edge effects. The distribution of time of arrival of the particles is taken from the measurements of Linsley [8]. In the reconstruction the mean time of arrival is taken as the time at which a station is hit. We reconstruct the density at 1000 m ($\rho(1000)$) assuming a simple power law for the lateral distribution. The charged particle density at 1000 m, like the density at 600 m, is a ground parameter that is rather insensitive to the primary particle. At 10^{19} eV the MOCCA simulation finds $\rho_{Fe}(1000)/\rho_p(1000) = 1.26$. The same ratio for 600 m is 1.03. In Figure 3a we plot the efficiency and mean number of stations for an atmospheric depth of 1000 gm/cm^2. In Figure 3b we plot the same quantities for a depth of 850 gm/cm^2. The advantage of the smaller atmospheric depth is evident. Also plotted on the same curve is the mean number of stations alerted as a function of energy. We have also calculated the precision of some of the reconstructed parameters. These values are indicative of what can be expected, but more careful simulations are really required.

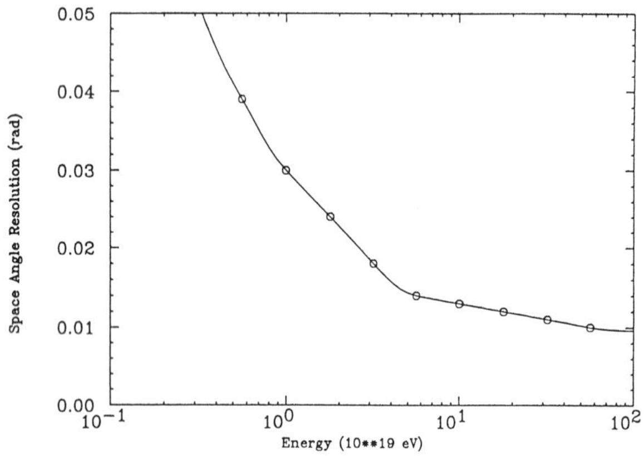

Figure 4: Accuracy of reconstruction of shower direction. The RMS space angle precision is plotted as a function of energy.

In Figure 4 we plot the RMS space angle precision as a function of energy for the array at 1000 gm/cm^2 depth. At 10^{20} eV the precision is of the order of one degree.

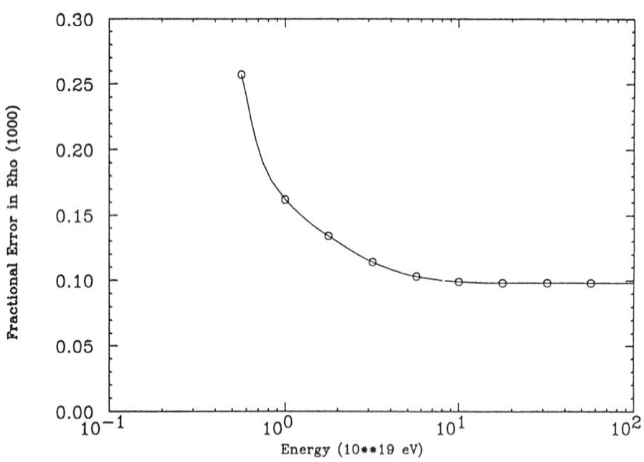

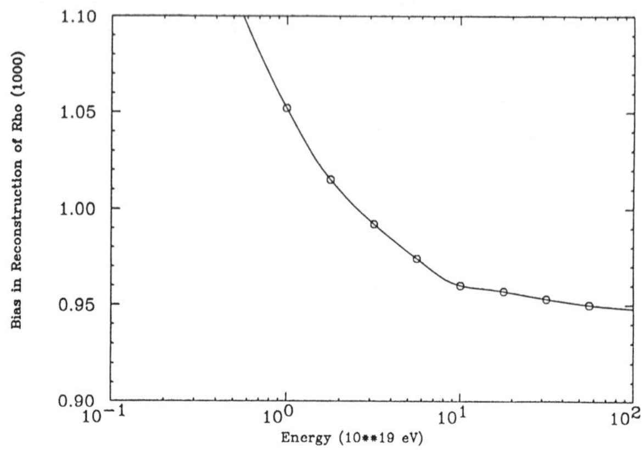

Figure 5a, b: The reconstruction of the ground parameter, $\rho(1000)$, which is proportional to the primary energy of the primary. (a) the fractional RMS uncertainty in $\rho(1000)$, (b) the systematic bias of the increased value from the input value.

In Figure 5a the fractional RMS error in the reconstruction of the ground parameter $\rho(1000)$ is plotted, and in Figure 5b its systematic bias from the input value is shown as a ratio to the input value. The additional errors due to fluctuations in the shower development are not included in these figures.

1.4 Properties of the Showers at Large Distance from the Core

In an array with station spacing of 1.5 km, at most only one detector can be closer than 0.75 km from the core. Thus, it is the properties of the shower particles far from the core which will determine the response of the array. In the following figures we plot some of these properties as determined by the MOCCA simulation. All the simulations are for vertical primaries at 10^{19} eV.

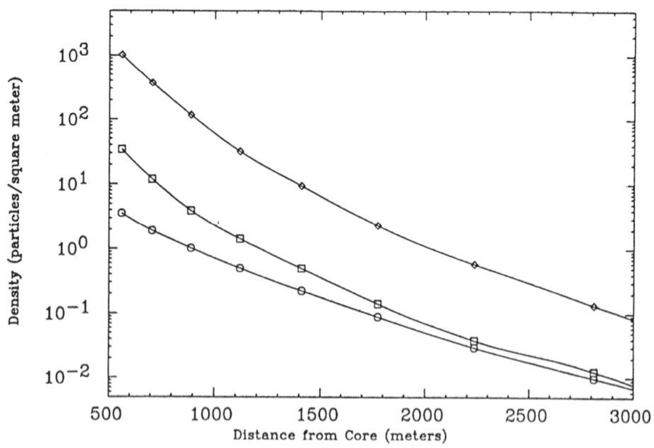

Figure 6: The average lateral distribution of photons (◊), electron/positrons (□) and muons (O) for proton primaries of 10^{19} eV entering vertically.

In Figure 6 we plot the average lateral distribution of the components, photons, electron/positrons, and muons beyond 500 meters. These are shown respectively as diamonds, squares, and circles. Only the muon density is a strong function of the primary for a fixed energy. The lateral distributions of the muons for photon, proton, and iron induced showers are plotted in Figure 7, indicated respectively by the diamonds, squares, and circles.

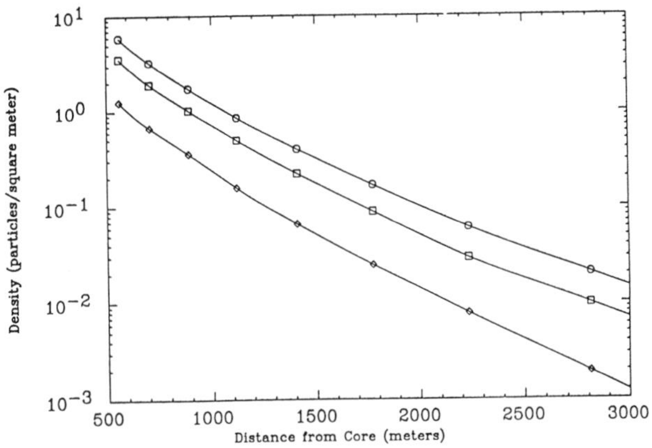

Figure 7: The lateral distribution of muons for photon (◊), proton (□) and iron primaries (○) at 10^{19} eV for vertical showers.

Far from the core the energy spectra of the components are very soft with the exception of the muons. These spectra are shown in Figure 8 for a core distance of 900 m. The diamonds, squares, and circles correspond respectively to photons, electron/positrons, and muons. In Figure 9 we plot the integral distributions of arrival time of the components (with the same symbols). The muons arrive the earliest and at 900 meters the arrival times of all the particles are spread over 2 μsec. This property scales with distance so that the spread in arrival times is ~ 4 μsec at 1800 meters.

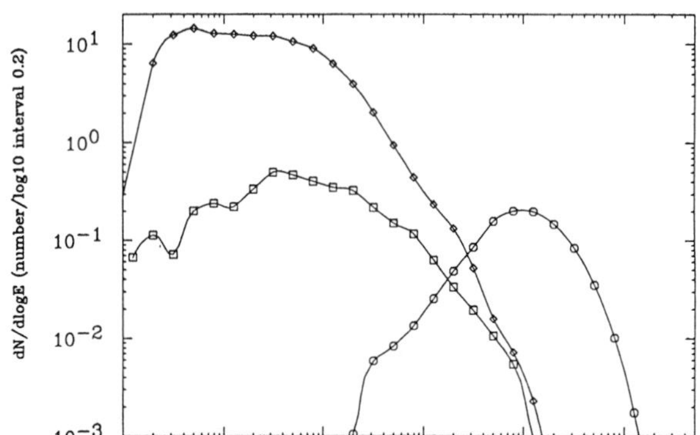

Figure 8: The energy spectra of photons (◊), electrons/positrons (□) and muons (○) at 900 m for proton primaries at 10^{19} eV.

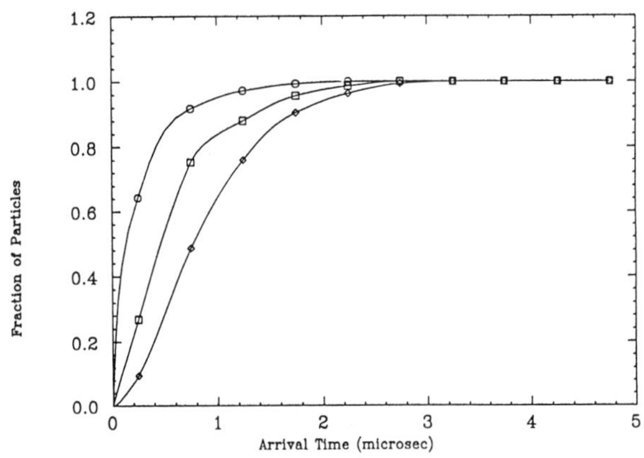

Figure 9: The integral distribution of arrival times of photons (◊), electron/positrons (□) and muons (○) at 900 m for proton primaries at 10^{19} eV.

There is an experimental verification of this time structure. Linsley [8] has measured the RMS spread in arrival times of charged particles for showers. In Figure 10 we compare the MOCCA results (open circles) with the Linsley measurements (solid line). The agreement is reasonable; the descrepancies at short distances are partly due to the binning of the MOCCA results in course time intervals of 0.5 μ sec.

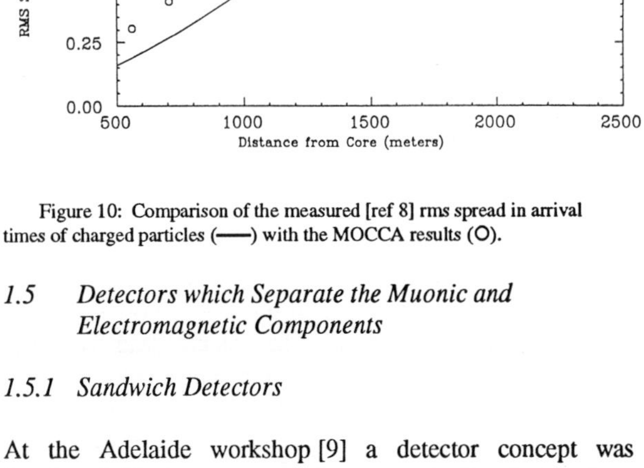

Figure 10: Comparison of the measured [ref 8] rms spread in arrival times of charged particles (——) with the MOCCA results (○).

1.5 Detectors which Separate the Muonic and Electromagnetic Components

1.5.1 Sandwich Detectors

At the Adelaide workshop [9] a detector concept was developed that can separate the muons in the shower from the electromagnetic component. It uses the fact that, far from the core, the shower particles, with the exception of muons, have very low energies. It also uses the fact that the shower particles arrive at the surface spread over a time interval of a few μsec. A detector consisting of one radiation length of lead sandwiched between two thin scintillators or resistive-plate counters has the following properties. Beyond 600 meters about 92% of the electrons are stopped in the lead and register only in the top layer of the sandwich. About 4% of the photons (the higher energy ones) are converted in the lead and register in the bottom layer. The muons register as a coincidence in both layers. If the density is not too high or the detector is sufficiently segmented, the individual particles can be resolved electronically because they are spread out in time. Beyond 1 km about 30 individual particles can be time-resolved by (for example) flash ADC's. In the following two tables, where densities are quoted in particles/m^2, the performance of the sandwich detector is given. The calculations are for vertical proton showers of energy 10^{19} eV.

Distance from core (m)	562	1120	1780
Top only density	26.7	2.0	0.11
Bottom only density	27.1	1.2	0.13
Coincidence density (e/γ)	2.4	0.22	0.019
Coincidence density (muons)	3.6	0.48	0.091
Muon concidence fraction	0.61	0.68	0.82

Table 1: Performance of a 1 rl Lead-Scintillator Sandwich

Distance from core (m)	560	1120	1780
Top only density	28.1	2.1	0.12
Bottom only density	24.9	0.91	0.092
Coincidence density (e/γ)	1.3	0.10	0.006
Coincidence density (muons)	3.6	0.48	0.091
Muon concidence fraction	0.74	0.83	0.93

Table 2: Performance of a 2 rl Lead-Scintillator Sandwich

The non-coincident counts are entirely electromagnetic, while the coincidences have a purity of 70 to 90% depending on the conditions. Clearly the thicker lead is better. Prototypes of the sandwich detector are under construction in Paris, Leeds and Fermilab. More realistic simulations of the performance of this type of detector need to be made, and, more importantly, tests in the field need to be performed. One advantage of the detector is that the bottom-only counts are a sample of part of the photon component of the shower such that the effective density is almost twice as large as the charged particle density. This fact will lower the energy threshold for an array of given spacing.

1.5.2 Water Čerenkov Detectors

Water Čerenkov detectors were successfully used at Haverah Park [10]. These were detectors of large area (in 2.5 m^2 segments) and 1.2 m deep. Most muons deposit at least 240 MeV in such a counter while the electromagnetic particles deposit on average about 10 MeV each. The 1.2 m of water gives a probability of about 90% for the conversion of 10 MeV photons, mostly by Compton effect. If the response time of a large counter can be made to be less than 100 nsec then the rate of energy deposit during the passage of the shower front could be resolved showing sharp peaks corresponding to the passage of a muon. In Figure 11 we plot the lateral distribution of energy flux which shows clearly that the bulk of the energy in the shower at large distances is in the muons. Diamonds, squares and circles correspond to photons, electron/positrons, and muons respectively. While there have been suggestions that water detectors be used for a giant array, there is much work that needs to be done to develop this idea.

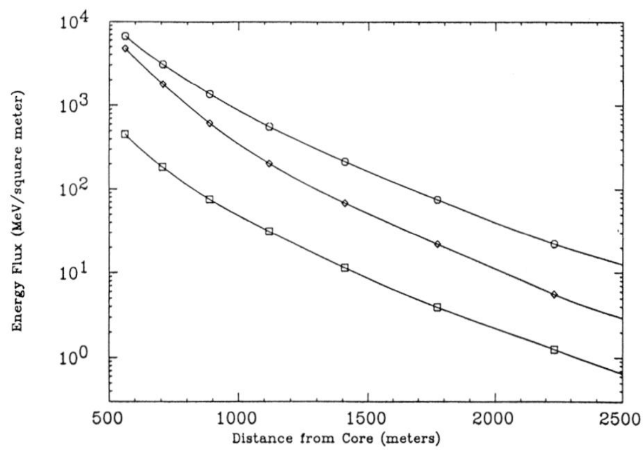

Figure 11: Lateral distribution of the energy flux carried by photons (◊), electron/positrons (□) and muons (○) for proton primaries at 10^{19} eV.

1.6 Shower Fluctuations

In the presence of shower fluctuations one can ask how effective the techniques of muon identification can be in the determination of composition. There are two questions to be considered. The first concerns the mean ratio of muon density to electron density as a function of the primary. This depends on the nature of the first few interactions. It is quite likely that the relative ratios of muons will remain about the same for the different primaries. The second question is the size of the fluctuations. A lower limit for these fluctuations can be obtained from the simulations. The thinning process introduces non-physical fluctuations. The quantity measured is the ratio of muon density to electromagnetic density. In Figure 12 we plot muon density versus photon density at a distance of 900 meters for vertical showers of energy 10^{19} eV. Here the diamonds, squares, and circles correspond respectively to photon, proton, and iron primaries. The distributions are well separated showing that the sensitivity to primary composition is not compromised by shower fluctuations.

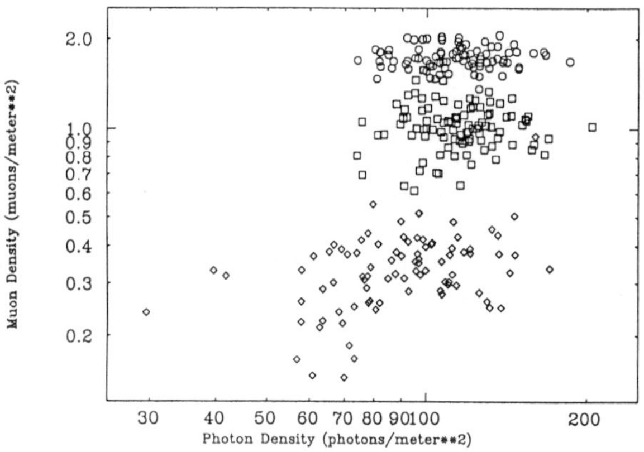

Figure 12: The muon density vs photon density at 900 m for vertical showers produced by photons (◊), protons(□) and iron (O) of 10^{19} eV.

1.7 Geometry of the Array

In Figure 13 a possible plan of a ground array is presented. In this scheme the center of the array is filled with a region where the spacing is 0.75 km instead of 1.5 km. Such an array of area 400 km^2 has a threshold well below 10^{18} eV. A fluorescence detector will be placed in the center and the properties of this hybrid detector are discussed the report by Sommers in these proceedings. The full array will detect ~ 5000 events per year with energy $\geq 10^{19}$ eV. The denser portion will detect ~ 40,000 events per year with energy $\geq 10^{18}$ eV. Such an array would have 3180 detectors. If the spacing must be reduced to 1.25 km then 4590 detectors are required. There are many factors which go into a cost optimization, not the least of which are the basic properties of the showers, the subject most emphasized in this discussion. Fortunately the data from AGASA are available. Also the characteristics of the possible sites play an important role in cost minimization. In an realistic design, ruggedness and reliability are crucial. All of these issues will be explored when the six month long design group begins its work at Fermilab on Jan 30, 1995.

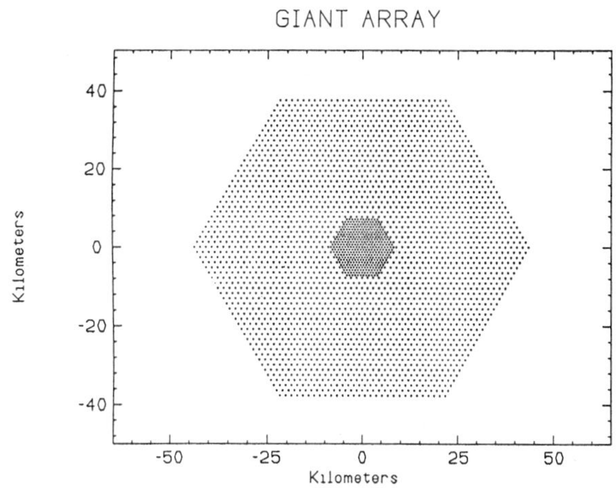

Figure 13: A possible geometry for the ground array.

§References

[1] S. Yoshida et al. *J. Phys. G* 20:651, 1994.

[2] M. Nagano et al. *J. Phys. G* 118:423, 1992

[3] N. N. Efimov et al. *Proc. Int. Symp. on Astrophysical Aspects of the Most Energetic Cosmic Rays*, ed. M. Nagano and F. Takahara, Singapore: World Scientific, p 20 1991.

[4] M.A. Lawrence, R.J.O. Reid, and A.A. Watson. *J. Phys. G.* 117:733, 1991.

[5] D.J. Bird et al. *Phys Rev. Lett.* 71:3401, 1993.

[6] A.M. Hillas. Proceedings of 17th International. Cosmic Ray Conference, Paris, 8:193, 1981

[7] N. Chiba et al. Proceedings of 23rd International Cosmic Ray Conference,Calgary, 4:307, 1993.

[8] J. Linsley and L. Scarsi. *Phys. Rev* 28:2384, 1962.

[9] A. Letessier-Selvon and D. Nitz. Proceedings of Adelaide Workshop, Jan 1993, unpublished.

[10] D.M. Edge et al. *J. Phys.* A6:1612, 1973.

Event Time Tagging for a Giant Air Shower Array

J. Lloyd-Evans

Physics Department, University of Leeds, Leeds LS2 9JT, UK

The possibility of using local GPS reception to time tag the arrival of air shower fronts at remote detector locations is examined. The coherence time required for relative timing stability in air shower reconstruction is < 10 μs - considerably less than the "scrambling" timescale induced by selective availability. Consequently GPS reception, in time solving scheduled common view mode, can achieve relative timing accuracy better than 10 ns over kilometre baselines with high reliability and virtually 100% duty cycle. Experimental verification of this possibility is described.

§1 Introduction

The need for high duty cycle, high quality, exposure at the highest energies (> 5 10^{19} eV) dictates that the next generation air shower array should, in conceptual design, consist of many (upto five thousand) large area detector stations spaced over an area of perhaps 70 km x 70 km on a 1km grid spacing. Furthermore construction logistics dictate that these stations be autonomous, stand-alone, devices with minimal power consumption, sited in a remote (desert) location with radio dataway communication [1],[2].

Conventional air shower arrays record the passage of the air shower disc by relative front edge detector timing transmitted via high bandwidth cable to a central receiving station, with subsequent commercial or in-house TDC electronic registration with respect to a common fiducial time marker (usually the n-fold detector coincidence time).

The autonomous detector scheme of the new giant array requires a radical departure from this conventional approach. Three alternative techniques are being studied: a) distribution (by microwave transmission) of a master time stamp from a central location with local microwave reception and "flywheel" local oscillator interpolation; b) local VHF reception of a suitable satellite transmission (e.g. TV line synchronisation pulses [3]); c) local reception of Global Positioning Satellite (GPS) time (and position) information [4].

This report gives details of conceptual design considerations for the latter possibility, and recent experimental measurements performed at Leeds to verify the expected timing performance.

§2 Design Considerations

At the highest energies an air shower front consists of a swarm of low energy particles (electrons, photons, muons) which is approximately planar (perpendicular to the shower progenitor's direction of motion), with a particle density which falls rapidly (approximately as r^{-3}) and a thickness which increases with increasing distance from the shower axis. The time registration of this front at three or more widely spaced detector stations permits reconstruction of the shower front plane (and second order curvature) and hence the arrival direction of the incoming primary particle. The directional resolution and systematic accuracy of this technique is a complicated function of the detector layout, shower front particle densities, and shower zenith angle. High particle density measurements, close to the core, allow the shower front leading edge to be located with high precision but detectors at large distances, sampling only a few particles, have a greater geometrical moment in shower front reconstruction. Fortunately, these two competing effects result in an angular resolution capability which does not, to first order, have a strong detector spacing dependency. Consequently most air shower arrays have achieved very similar angular resolution - in the PeV range, where the characteristic detector sampling distance is about 50m, the shower front is only tens of nanoseconds thick, and 0.6° is typically achieved; in the EeV range, where the characteristic distance is 500m or more and where the shower front is upto a few microseconds thick, an angular resolution of 2 to 3° has been routinely achieved.

Simulations show that if a time tagging system has an electronic measurement accuracy of ±25 ns for all the relative times of the n detectors struck by a shower, then this makes a negligible contribution to the angular resolution achieved in comparison to that from the inherent shower front sampling uncertainty which depends on the area of the detector and the distance of the detector from the core. A directional resolution of 2°, at least for zenith angles less than 60°, is obtainable, and much greater electronic measurement accuracy is unnecessary. Absolute event timing is not required to anything like this accuracy (milliseconds will suffice, unless other astrophysical goals, such as pulsar timing or short burst phenomena are to be studied). Most importantly, a relative time tagging system must be demonstrably reliable (for many years) in a harsh environment with low failure rates, high duty cycle and minimal power consumption. The GPS system meets these practical criteria. The rapidly increasing commercial exploitation, and long term US commitment, will ensure a reliable system for at least the next decade, and hopefully ever decreasing technological costs.

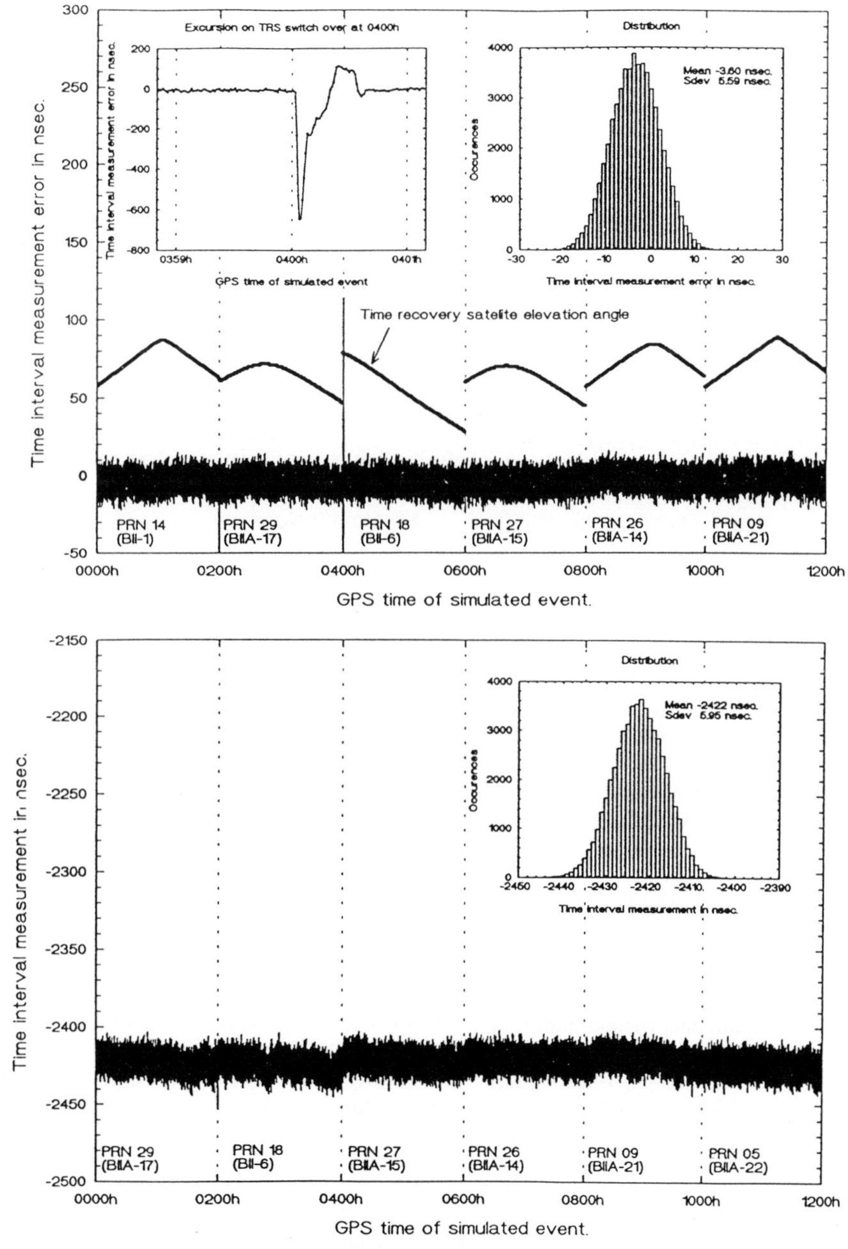

Figure 1. Relative timing stability over 12 hours. A timing scatter plot, the satellite elevation, and a time residual histogram is shown for systems co-located (upper) and 500 m apart (lower). A rare switching excursion is shown in the upper insert in detail.

GPS is a US DOD project allowing the dissemination of precise time and position information anywhere on or above the surface of the Earth at all times [5].

Unrestricted civilian access to the standard service (SPS) is allowed subject to deliberate degradation (selective availability (SA)) for military reasons. Currently, the system consists of twenty four uniquely identified satellites which transmit using spread spectrum techniques and are hence highly immune to interference and jamming. In most commercial applications, a user receives at least four satellite transmissions to enable the reconstruction of user position to an accuracy of better than 100m. Alternatively, if user position is known, a single satellite can be tracked to determine time. Selective availability degrades this time information by many hundreds of nanoseconds in a pseudo-random manner for each detector separately. Fortunately however, the time difference between detectors in an air shower array recording the shower front passage is less than $r \sin\theta/c$ where r is the maximum baseline of detectors struck, and θ is the shower zenith angle. For even the largest air showers, near 10^{20} eV, the footprint of struck detectors has a baseline which is typically only about 5 km and certainly never more than 10 km. Consequently, all of the struck detectors require a time stamp from the GPS satellite within $<< 30$ μs. If all detectors track the same satellite, or satellite group, and the pseudo-random time degradation is slowly varying (timescales $>> 30$ μs) then the selective availability degradation is common to all registered times and cancels out for the relative detector times.

For a practical array design, we then have to demonstrate that a) the selective degradation really is slowly varying and b) that the residual jitter (from ionospheric propagation effects, antenna sensitivity, intrinsic GPS precision, etc.) is less than 25 ns for long baselines, and is constant with high reliability.

§3 Experimental Prototype Performance

Dual systems have been constructed at Leeds, each consisting of a low cost GPS receiver module (Magnavox GPS Engine™ (scientific model), antenna, custom built (pc) time tagging board, and a processor (pc) for control and data acquisition. In a full design, of course, the processor chosen to perform other detector controls/data acquisition would also perform, in interrupt mode, the role of the pc used here. Details of the experimental set up, and features of the GPS engine used, are given in [4]. Briefly, the one pulse per second (1 pps) GPS time marks are used to latch a 100 MHz, 28 bit, counter whilst a second latch records asynchronous "events" (simulated air showers). Intrinsic jitter on the 1 pps of ± 50 ns is improved by subsequent software correction provided by the engine, and by pps interpolation. A direct event link (high bandwidth cable) provides an event trigger to both systems when not co-located.

The two systems have been tested both co-located, and at a separation of 500 m. Tests on kilometre baselines using the UK MERLIN Radio Link are planned for the near future. The systems have been tested in varying atmospheric conditions, over many weeks and show very high reliability. Figure 1 shows detail from just two such runs. No significant biases have been observed for changes in temperature, day-night, which satellite is tracked, or satellite elevation angle. There are occasional excursions (insert of Figure 1(a)) of transient time deviations up to 500 ns, lasting for less than half a minute. These occur very rarely, and coincide with the time at which the satellite being tracked is changed. Figure 1(a) insert also shows the residual time difference histogram when the systems are co-located, and Figure 1(b) when 500 m apart. The standard deviation over a twelve hour run is typically 5.6 ns when co-located, and 5.9 ns when 500 m apart. The mean offset in Figure 1(b) is consistent with the measured event pulse travel-time down the remote cable link. The largest excursions observed during a four day run were +21 ns and -34 ns.

§4 Conclusions

We have tested the feasibility of a GPS relative timing system for a giant air shower array. It has been demonstrated that the system is extremely reliable and provides a relative time jitter with $\sigma = 5.95$ ns over a 500 m baseline - easily adequate for the proposed array. The system has now been in regular use at Haverah Park for more than six months with no problems, and will soon be tested over longer baselines.

Finally it should be noted that a proposed array data communications system [2] will require a reliable relative timing system for time sliced data transmission, which the GPS system advocated here already provides.

§Acknowledgements

We thank Colm O'Sullivan , Paul Ogden and Peter Daly for technical advice. Particular thanks to Clem Pryke, who performed all of the experimental work described here.

§References

[1] J.W. Cronin. Design Concept for a 5000 km^2 Air Shower Array, *University of Chicago Preprint*, EFI 92-08, 1992.

[2] D. Nitz. Contribution To Snowmass 1994 - Highest Energy Cosmic Ray Session, these proceedings.

[3] G. Brook et al. Proceedings of 19th International Cosmic Ray Conference, La Jolla, 3:426, 1985.

[4] C.L. Pryke and J. Lloyd-Evans. A High Performance GPS based autonomous event time tagging system with applications in a next generation Extensive Air Shower Array, submitted to *NIM*, September 1994.

[5] ICD-GPS-200, Navstar GPS Space Segment/Navigation User Interfaces, Rockwell International Space Operations and Satellite Systems Division, 1981.

Communications on a Giant Array

D. Nitz

Department of Physics, University of Michigan, Ann Arbor, Michigan 48109,USA

A conceptual solution to the problem of communication of trigger information and data around the 5000 km^2 envisaged in the Auger project is described.

A giant air shower array poses intriguing problems for triggering the array and reading out the data. Conventional schemes for sending signals between adjacent stations or to a central collection point over cables or optical fibers are cumbersome at best and are likely to be prohibitively expensive. The immense size of the array relative to the extent of interesting events on the ground favors a trigger that incorporates inherently the locality of the showers.

A system based upon cellular radio techniques provides an elegant and cost-effective solution. Each station communicates with its nearest neighbors via packet VHF radio. This simultaneously provides a locality based trigger and provides a robust, fault-tolerant, data-readout mechanism. This system would handle the array triggering in a distributed fashion, be used to transfer the triggered event data and detector diagnostic data to a central collection point, and be used to deliver control commands to the detector stations.

The concept described here was largely developed at the Adelaide Giant Array Workshop in January, 1993. G. Kelley (Telecom Australia) provided the initial design and played a pivotal role in the development of the concept [1]. B. Dawson (University of Adelaide), J. Lloyd Evans (Leeds University), and myself contributed to the effort.

Trigger requirements drive the design of the communications network. For the sake of concreteness, we assume a 3000 station array arranged on a 1.5 km spacing triangular grid. Each station has 6 equidistant nearest neighbors. (See Fig. 1) A station is said to be alerted when a local requirement, such as ≥4 particles detected by a $10m^2$ station within 5μs, is satisfied. Requiring that the detected particles have a modest time spread (say >50 ns) reduces the alert rate by rejecting alerts caused by small showers near the station. In this manner we can ensure that the alert rate will be less than 20 Hz, which we will use as an upper limit in our considerations. Every time a station satisfies its local alert condition, it will queue for broadcast a data packet containing the time of the alert (obtained, for example, from its GPS receiver [2]). The data will be buffered sufficiently so that only the mean rates need be considered. For the trigger we require 3 stations of any 7 station cell to be alerted within a specified coincidence window. This is the minimum number of stations required to reconstruct a shower direction. In reality, more stations will likely be required in the trigger to reduce the fraction of random triggers. A 3 station coincidence, however, provides a convenient conservative point at which to estimate bandwidth requirements. Each station listens to the trigger data from its 6 nearest neighbors, and searches the 7 data streams for a triple coincidence within the coincidence window between itself and 2 neighbors. Note that we are not dependent upon the time of arrival of the data packets, rather we use the alert time which forms the data within the packet. Every time a station finds a triple coincidence, it queues a trigger packet for broadcast which contains the time of the trigger. Each station that has an alert in coincidence with the trigger queues its data for transmission to the collection point. To include nearby less densely connected alerted stations in the event, each alerted station that receives a trigger from one of its neighbors would repeat the trigger broadcast, substituting its own alert time in order to extend naturally the coincidence window. At the collection point the data from the various stations is combined, using the alert times, into a single event stream before output to storage media. All of the trigger logic beyond the station alert occurs at a sufficiently low rate to be performed in software by the station microprocessors.

Let us consider the trigger rate and corresponding bandwidth requirements for triggering and data readout. The rate of accidental triggers is given by the following relationship (to first order)

$$R = NS(S\Delta T)^2 \frac{6!}{(6-2)!2!} \qquad (1)$$

where N is the number of stations in the array, S is the single station rate, and ΔT is the coincidence window. With a maximum station spacing of 1.5 km, the time difference between any 2 stations within the group of 7 is always less than 9 μs (eg. horizontal shower) for a plane shower front.

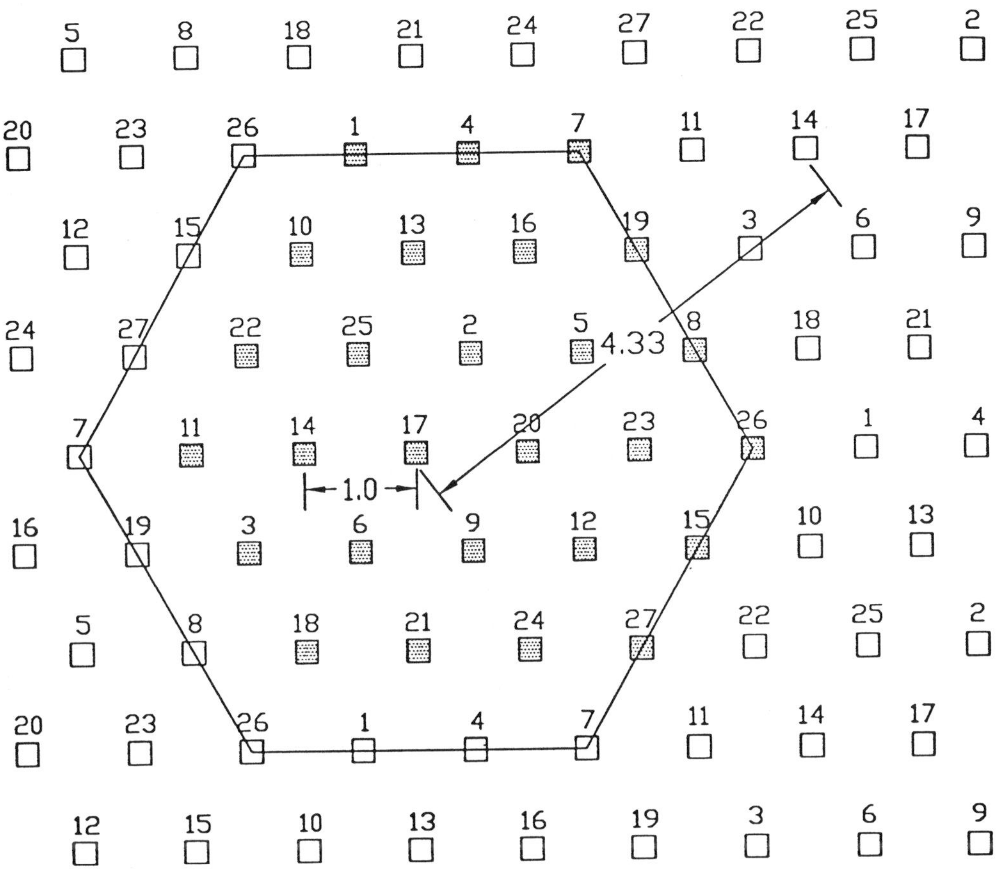

Figure 1: Layout of radio cells. The shaded boxes indicate one 27 element cell. These cells are replicated to tile the whole array. The numbers above the boxes denote the transmit time slot allocated to that station. Each station listens to its 6 nearest neighbors. Note that the closest replication of a nearest neighbor station is 4.33 times farther away, resulting in 25db attenuation of the interfering signal.

Taking a very conservative 20 μs coincidence window and limiting the single station rate to less than 20 Hz, we find that the random trigger rate is less than 0.2 Hz over the entire 3000 station array. By comparison, the trigger rate of real showers above 10^{18} eV is ~ 0.002 Hz, so much smaller that it does not impact bandwidth considerations.

20 bits are sufficient to encode the alert time with 1 μs least count and 1 second overflow. 24 bits would probably be used to keep the information an integral number of bytes. This corresponds to a digital bandwidth requirement of 20 Hz · 24 bits = 480 bits/second to broadcast the station alert information.

The bandwidth required to broadcast the triggers is minuscule, being

$$0.2\ \text{Hz}/3000\ \text{stations} \cdot 24\ \text{bits} = 1.6 \times 10^{-3}\ \text{bits/second.}$$

What bandwidth is required to read out the data? The average number of alerted stations in any trigger will be (3 + epsilon), since the random coincidences totally dominate. Assuming 200 bytes of data per station, we find that the bandwidth required to read the data from any one station is less than

$$0.7 \times 10^{-4}\ \text{Hz} \cdot 1600\ \text{bits} \cdot 3\ \text{stations} = 0.3\ \text{bits/second.}$$

The data packets would be transmitted to the collection point through the network formed by the array of stations. The data rate at the collection point would be less than

$$3000\ \text{stations} \cdot 0.3\ \text{bits/second} = 900\ \text{bits/second.}$$

However, if the collection point is near the center of the array, the rate transmitted through any one of the nearest neighbors to the center would be less than 150 bits/second. This argues that the transceiver for the collection point should be placed at the center of the array in order to minimize the transmit data rate requirement for any single station. This also minimizes the number of hops required to reach the farthest station. The topology of the network provides naturally for fault tolerant data transmission. Each station knows where it is (from, for example, its GPS receiver) and where it wants to send a packet. The regular geometry of the array makes it straightforward to determine an ordered set of nearest neighbors closer to the destination. A station sends a packet enroute to a more distant point via its preferred neighbor. Since the station always listens for packets from its 6 nearest neighbors, it will naturally hear the data packet when the neighbor forwards it towards the destination. If it receives the forwarded packet correctly, the station has acknowledgment that the data have safely reached the target neighbor enroute to its destination. If it does not hear the packet within a time out period, it can then retransmit the packet via the next most preferred neighbor, and so forth. This allows data to be transmitted around inoperative stations.

A certain amount of bandwidth is required for control operations. We will need to be able to download updates to the station control software from a central site as it woulld be impractical to travel to every station to change PROMs. We will also want to be able to run remote diagnostics on the stations. A worst case would be where we would need to download the complete control program. It is reasonable to expect that the control software can be contained within 250K bytes. Due to the nature of the packet network we can download to all stations in parallel. If we can download at 500 bits/second, then it would take 4000 seconds, or 1 hour to reload the entire array, which is acceptable.

The total bandwidth required at each station is then

$$480 + 150 + 500 = 1130\ \text{bits/second.}$$

Adding a generous packet overhead of 20% we find that we require the communications system to provide a raw bit rate of 1500 bits/second. The bandwidth required for triggering varies only as the first power of the station alert rate. However, the bandwidth required for data readout (which is dominated by random trigger coincidences) increases with the 3rd power of the station alert rate and the 2nd power of the coincidence window. The bandwidth for control operations is independent of the station alert rate and will dominate for station alert rates below 10 Hz. We conclude that we will need a communications bandwidth of 750 bits/second per station for alert rates of a few Hz or less and 1500 bits/second for a 20 Hz alert rate.

In the scheme envisioned here every station communicates on the same frequency. The array is divided into hexagonal cells containing 27 stations each. Each station is assigned one of 32 time slots during which it can transmit information. There are 5 left over time slots which might be used for slow control information. Across the array the time slots are reused many times. (See Fig. 1.) The system depends upon the fact that other stations transmitting in the same time slot as a nearest neighbor will be sufficiently far away that their signals will be attenuated to an acceptable level. In this scheme the next closest station using the same time slot will be 4.33 times farther away. Radio waves attenuate more quickly near ground than in free space. For a simple VHF whip antenna, the attenuation near ground is approximately $40 \log_{10}$ (D2/D1). This implies that the unwanted signal will be attenuated by 25 dB with respect to the wanted signal, ample for digital applications. (Typical commercial cellular networks are designed with 18 dB.) A digital transmit bandwidth of $32 \cdot 1500 = 48{,}000$ bits/second is easily sufficient. A factor of 2 or 3 less will be adequate if we accept tighter limits on the station alert rate and lower bandwidth for control functions.

Another technique that allows frequency reuse is the spread spectrum technique, in which the data from each transmitter is encoded with a pseudo random number. A critical feature in commercial systems is a transmit power control feedback loop to maintain constant received signal strength from all stations. Due to the mesh structure of the array, in which each station listens to its 6 nearest neighbors, such power control is not feasible.

Grading of the array introduces complications in the trigger/communications scheme outlined above. At present we don't have a concept that works across the grading boundaries with a single frequency. However, there is a straightforward, brute-force, approach that we could take. Conceptually one could do this as follows. Consider the large spacing grid. Each station has one transceiver. Now make the next gradation by filling in between stations on the inner portion of the array. Each of these additional stations would have one transceiver operating on a 2nd frequency. Those stations which are members of both the large spacing and the in-filled array would have 2 transceivers. For a 3rd gradation we add a 3rd frequency in the same manner. With typical graded array sizes (eg. 1:10:100) the number of stations with double or triple frequency operation is small with insignificant impact on overall cost. This brute force scheme has an advantage in that adding the in-filled stations will not eat up communications bandwidth in the large spacing array.

Power consumption of the transceivers will be minimal. 1 watt of transmit power is more than sufficient to reach nearest neighbors 1.5 km away. Furthermore, since each station transmits only 1/32 of the time, the average transmit power will be less than 30 milliwatts.

This work was supported in part by the U.S. Department of Energy.

§References

[1] R.W. Clay et al. Prospects for 10^{19} eV Cosmic Ray Studies in South Australia, *Nucl Phys (Proc Supp)* 28B:101, 1992.

[2] J. Lloyd-Evans, these proceedings.

Simulation Programs for the Giant Hybrid Array

M. Boratav
Laboratoire de Physique Nucléaire et des Hautes Energies, Université, Paris 6, Paris, France

and H.Y. Dai
High Energy Astrophysics Institute, Department of Physics, University of Utah, Salt Lake City UT 84112, USA

We review the simulation programs used for the design of the Giant Hybrid Array project (the Auger project). These programs should provide a realistic description of the extensive air showers produced by ultra high-energy cosmic rays (UHECR, of energies > 10 EeV) through the secondary particles they produce at ground level, as well as the fluorescence light they create in the atmosphere.

§1 Introduction

The principal aim of the Auger observatory is to explore the domain of the cosmic ray spectrum roughly beyond the Greisen-Zatsepin-Kuzmin cut (a few tens of EeV). In this region the data accumulated in the past are scarce (of the order of ten events beyond 100 EeV). Moreover, one of the main questions about the UHECR, namely their identity (photons or hadrons?, light or heavy nuclei?) remains more or less open.

To design a detector optimized for the study of the UHECR, one needs reliable simulation programs. The crucial detector parameters such as the station spacings, the time resolution of the synchronisation devices, the design of the trigger electronics the energy calibration etc... are strongly dependent on the ability of the simulation programs to provide data as close to physical reality as possible. Thus the Monte Carlo programs must not only be thoroughly checked for their internal consistency but also confronted with available data whenever possible.

The Auger project is presently conceptually defined (the finalisation is the task of the design group) as a giant ground array of about 3000 scintillator detectors covering a surface of 5000 km^2 [1] combined with an optical device (Fly's Eye or Hi-Res type detector) sensitive to the fluorescence light emitted by the atmospheric nitrogen excited through the interaction with air of charged particles produced during the shower development. The two detectors use independent Monte Carlo programs that we shall successively describe.

§2 Ground particles: the MOCCA program

A typical EAS of 10 EeV produces a few times 10^{10} particles (mainly photons and electrons in the MeV region) at sea level. It can be easily understood that a complete simulation and tracking of all those secondaries is impossible even with the most powerful computers. On top of that, a thorough study of the detector behaviour in the detection of such EAS needs the generation of hundreds of showers with several types of incident particles (at least photons, protons and a heavy nucleus such as iron), various incident angles and energies. To accumulate data with such a set of constraints, one needs a simple and fast simulation program which does not, however, sacrifice the physics content to efficiency. MOCCA, written by A.M.Hillas [2], is such a program. It includes a cascade code, an event generator and detector simulation. The physics underlying the event generator part has been described [3], [4].

The main feature of MOCCA is an algorithm which ensures energy conservation in the cascade development in a simple, fast and adaptable way. For example, the algorithm used to describe the interaction of a nucleon with an atmospheric nucleus consists in splitting the available energy in the collision randomly into two parts. One part is assigned to the final state leading nucleon. The remaining energy is again split randomly into four parts and each part in two, one of them assigned to a pion, the other again split in two etc... The process is repeated until some pre-selected value of the energy is reached below which the tracking is stopped. The random splitting of the energy at each step may be done following a flat spectrum or adapted to whatever data are available.

A second time-saving feature of this program is a technique called "thin sampling" or "thinning" process for the tracking of the particles, without which the simulation of showers with initial energies of 10 EeV or more would be impossible. The user pre-defines a "thinning energy" E_{thin} before starting the shower generation. For a shower with an initial energy E_0, the level of thinning is usually defined by the parameter E_{thin}/E_0. All particles are individually tracked down to the thinning energy. When their energy E goes below this value only a small proportion is tracked (roughly E/E_{thin}), each particle being given a weight > 1. The effect of thinning on the quality of the output has been checked. All relevant physical quantities (particle densities, directions, time of arrival...) were shown to be independent of the thinning used, as far as average values are concerned[1]. However, the more or less heavy thinning of the simulation has a strong effect, as can be expected, on the fluctuations around these average values. As an example, let us take the muon densities at a distance of 560 m from the shower axis for 10 EeV showers with Fe nuclei as incident cosmic rays. With a thinning of 10^{-6} the average value is 5.82 m^{-2} with $\sigma = 0.39$; with a heavier thinning of 10^{-5}, the average value is 5.77 m^{-2} but the dispersion becomes much more important: $\sigma = 1.31$. The effect of the thinning should hence be properly evaluated whenever fluctuations of shower parameters are being estimated.

MOCCA uses a fairly realistic model of atmosphere defined by 5 bands of variable thicknesses. All relevant electromagnetic processes are included with elaborate cross sections. The output of the program gives the useful variables (identity, position, direction, energy) for all the surviving particles with energies above a given value (e.g. 200 keV) at a pre-defined altitude. It also provides Cerenkov photons at given mirror positions (but this feature is not used in our simulations). In a recent updated version, MOCCA was adapted to UNIX and VM workstations. It now provides a packed binary output. It was also modified to provide a flexible detector simulation by re-injecting all the ground particles into a medium composed of an arbitrary number of layers of various material. In particular, this version gives the energy deposited in scintillator plates in the so-called "leadburger" geometry (a scintillator-lead-scintillator sandwich). The quality of the output of MOCCA was tested on several points. Comparison of the lateral charged particle distributions to data from Akeno, as well as other results on the simulation of array properties, can be found in these proceedings [1]. Another test was to check the quality of the low energy electromagnetic interactions, which is of great importance for the detector simulation part. For that, we used a small linear electron accelerator (NEPAL) at the Laboratoire de l'Accélérateur Linéaire (Orsay, France). NEPAL is normally used to test accelerating cavities and gives clean electron beams with energies up to 30 MeV. We used the electron beam (between 15 and 25 MeV) on various thicknesses of lead and plastic installed between a set of coincidence counters. A series of electron absorption rates were thus measured and compared to the values provided by MOCCA. There again the results were in good agreement with the simulation output.

The photoproduction cross sections at the highest energies we need are experimentally unknown. The model used in MOCCA is close to that of Donnachie and Landshoff [5], which gives a very good agreement with the HERA data (the highest energy where the model could easily be compared to experiment). We made a test to check if the results of MOCCA were critically dependent on the parametrisation used there. For that a set of showers were simulated with a factor two on the coefficient of the terms in the cross section rising with the energy. We compared the photon, electron and muon densities obtained with the two parametrisations. No significant differences were found (less than 10% at the distances where the data are statistically significant).

§3 A Specific Event-Generator: SYBILL

Since the comparison of the simulation with data at the energies we want to explore is difficult, it is wise to check the systematic effects of the models used by the event generators on the output of the Monte Carlo programs. SYBILL is no doubt the most appropriate alternative we can use to study such effects together with MOCCA. A detailed description of the SYBILL event generator is given by Fletcher et al [6], and the physical basis used therein is described in [4], [7] and [8].

The physical processes which contribute to the development of an EAS resulting from the interaction of UHECR with air molecules cover an energy domain of several orders of magnitude and involve the interaction of a wide range of particles (going from photons to heavy nuclei). The event generators used in the current high-energy accelerator experiments (such as the ones in the LUND or HEMAS programs) are not optimized for such an utilization, however good they may be for describing fixed target or collider data. SYBILL is an event generator specifically aimed to fulfill the needs of cosmic ray shower calculations. It makes use of models reproducing the features of interactions at fixed target energies (KNO scaling of multiplicities, constant height rapidity plateau, Feynman scaling of longitudinal momentum, energy independent average transverse momentum of secondaries) and of collider data at higher energies (wider multiplicity distributions than predicted by KNO scaling, logarithmic increase of the rapidity in the central region, faster than logarithmic increase of multiplicity, increasing average p_T with energy, existence of high-p_T minijets etc...). It then extrapolates these models to energies beyond the collider range where we still observe a quantity of UHECR.

In the low energy range, SYBILL uses the dual parton model [9] for the non-diffractive component of the interactions and a minijet model to describe the diffractive dissociation [8]. The minijet model is also used to bring in the higher energy features described above [10]. Finally the interactions with (or between) nuclei are determined by using the hadron-nucleon scattering parameters.

The predictions of SYBILL were compared to a set of fixed target, collider and UHECR data [6]. The authors make

[1] True at least for the thinning values used, (< 10^{-5} for 10 EeV showers).

a critical analysis of the results. Although the event generator presents some limitations (the program, in particular, should not be used for incident energies below 100 GeV), it gives a satisfactory agreement with the average depths of shower maxima as measured by the Fly's Eye experiment between 10^{17} and 10^{19} eV. SYBILL is currently being implemented inside MOCCA to check the sensitivity of the ground data to the interaction models used.

§4 The Monte Carlo Simulation for the Air Fluorescence Detector

The Fly's Eye technique [11], [12], unlike the ground arrays which sample the shower particles at the ground level to measure the shower lateral distribution, offers a very different approach, measuring the longitudinal profile by collecting the atmospheric nitrogen fluorescence light produced by air shower particles. The capability of detecting the shower longitudinal development allows an estimation of the primary energy and composition less dependent on interaction model.

The Monte Carlo simulation for the air fluorescence detector can be divided into three parts: (a) Geometry and shower generation; (b) Light production; (c) Detector response. Like the Monte Carlo simulation for other types of techniques, the detector response part should be fairly straightforward after one understands how the hardware works. Here we will concentrate on the light production and propagation in the atmosphere and briefly mention the other parts.

4.1 Geometry and Shower Generation

We use four parameters to determine the track geometry: θ, ϕ, R_p and γ. θ is the zenith angle of the track, ϕ is the azimuth angle. θ and ϕ completely determine a plane which is perpendicular to the track and includes the location of the detector (the eye). R_p gives the radial distance from the eye and γ specifies the azimuthal angle within that plane. The number of events in the phase-space $d\theta d\phi dR_p d\lambda$ should be proportional to $d\cos\theta d\phi dR_p^2 d\gamma$. The shower longitudinal profile is generated either by Gaisser and Stanev's full Monte Carlo or analytically calculated for a given energy and primary mass by the Gaisser-Hillas function with proper fluctuations both in the first interaction point (X_0) and depth of shower maximum (X_{max}).

4.2 Light Production

Four basic mechanisms contribute to the generation of the light signal seen by the Fly's Eye detector [11]: nitrogen fluorescence, direct Cerenkov light, air molecule (Rayleigh) scattered Cerenkov light and aerosol (Mie) scattered Cerenkov light. Of these, fluorescence relates most directly to the local number of charged particles in a shower. The intensity of direct and scattered Cerenkov light depends on the history of the charged particles, the incident angle to the atmosphere and the viewing angle from the detector.

4.2.1 Scintillation light

Most of the energy of the primary cosmic ray is dissipated in the atmosphere. Nearly all of the resulting optical fluorescence comes from the 2P band system of molecular nitrogen and the 1N band system of the N_2^+ molecular ion. Most of the light in the fluorescence spectrum is emitted in the spectral region 310 to 440 nm. The air fluorescence efficiency has been measured by a number of groups. It is typically measured either for pure N_2 at low pressures by molecular spectroscopists or in air at pressures that correspond to high altitude air-glow phenomena of interest to atmospheric physicists. Simple kinetic theory arguments relate the fluorescence efficiencies at various pressures. The resultant fluorescent yield integrated over our spectral response is only mildly altitude and temperature dependent. Bunner [13] has combined all known data up to 1963 and extracted best fits for the fluorescence spectrum of air at tropospheric pressures of interest to this experiment. The scintillation light yield is only 4.03 photons per meter at the altitude of Fly's Eye I. However, the large number of particles in the shower makes the detection feasible for extremely high energy cosmic rays. The fluorescence light is emitted isotropically from the shower, allowing for detection of showers at large distances. The estimated uncertainty in the light yield integrated over the wavelength region of 300 to 400 nm is 20 percent. More recent experimental work [14], [15] has tended to confirm Bunner's numbers, but no recent systematic reevaluation has been done.

4.2.2 Direct Cerenkov Light

Electrons in an air shower generate Cerenkov light that is beamed primarily in the forward direction. The amount of Cerenkov light at any point along the shower front depends on the previous history of the shower, and thus is not proportional to local shower size as is the case for fluorescence light. The Cerenkov beam builds up quickly around shower maximum and remains intense, even after the shower dies away. Direct Cerenkov light dominates the light seen by the Fly's Eye detector at emission angles less than 25°. The Cerenkov photons emitted by the charged shower particles in the air essentially maintain the same direction as the charged particles, thus the angular distribution of shower Cerenkov light is determined by the angular distribution of charged particles. Studies of multiple scattering show that the angular distribution of ionizing particles averaged over the length of the EAS can be represented by:

$$\frac{dN}{d\Omega} \propto \frac{e^{-\theta/\theta_0}}{2\pi\theta_0 \sin\theta} \tag{1}$$

where[2] $\theta_0 = aE_T^{-b}$. The Fly's Eye Group have measured θ_0 directly using stereo data [19] and find $\theta_0 = 4.0 \pm 1.2°$, in agreement with the above calculations. The parameters from Stanev et al are the ones currently used by Fly's Eye and HiRes Monte Carlo and reconstruction programs.

4.2.3 *Scattered Cerenkov Light*

As the Cerenkov light builds up with the propagating shower front, the resultant intense beam can generate enough molecular and aerosol scattered light at low altitudes to compete with the locally produced scintillation light.

Rayleigh scattering (scattering of light from air molecules) is proportional to $\rho d\ell/\lambda^4$ where ρ is the local air density and λ is the wavelength of the scattered light, ℓ is the path length. The amount of light Rayleigh scattered from a beam of N_γ photons is

$$\frac{dN_\gamma}{d\ell} = -\rho\frac{N_\gamma}{x_R}\left[\frac{400}{\lambda_{nm}}\right]^4 \tag{2}$$

here $x_R = 2794$ g/cm^2 is the mean free path at sea level for 400 nm photons. For the purpose of estimating the amount of Cerenkov light scattered out of the Cerenkov beam accompanying the shower towards the eye at the angle θ within a solid angle of $d\Omega$ we have

$$\frac{d^2N_\gamma}{d\ell d\Omega} = \frac{dN_\gamma}{d\ell}\frac{3}{16\pi}\left(1+\cos^2\theta\right). \tag{3}$$

Now we discuss the aerosol scattered light. For the Fly's Eye simulation, a simple aerosol model is used. The aerosol profile is described by an exponential function with a scale height of $H_M = 1.2$ km. The light scattered by aerosol from a beam of N_γ photons is

$$\frac{dN_\gamma}{d\ell} = -\frac{N_\gamma}{L_M}e^{-h/H_M} \tag{4}$$

with $L_M = 8.4$ km for 488 nm photons (L_M has a slight wavelength dependence). The normalized angular distribution (so-called phase function) of the aerosol scattered light depends on the assumption of the aerosol size (see for example reference [22]). Since most of the Fly's Eye events have an optical path less than the aerosol extinction length, the simplified aerosol model is a proper approximation. For the HiRes and the Hybrid Array simulation, we are going to apply a widely accepted and well verified detailed atmospheric Monte Carlo program (Modtran VII [20]) for calculating light scattering (both Rayleigh and aerosol) and transmission in the air. In Modtran, six geographical-seasonal model atmospheres are provided with 10 choices of aerosol profiles. In our case, the combination of the "1976 U.S. Standard" atmospheric model with the desert aerosol profile for altitudes less than 4 km (in this case, the aerosol scale height below 4 km is 3 km) is most appropriate.

[2] E_T is the threshold energy (in MeV) for Cerenkov emission for electrons: $E_T = 0.511[2n-1]^{-1/2}$. Here n is the refractive index. There have been three calculations of parameters a and b and they result in (0.83, 0.67) [16], (0.77, 0.65) [17] and 0.85, 0.66) [18] respectively.

4.2.4 *Attenuation*

The light received by the detector must be corrected by the attenuation due to Rayleigh and aerosol scattering (for wavelengths less than 300 nm, one should also take into account the loss due to ozone absorption). The Rayleigh transmission is:

$$T_R = \exp\left\{-\left(\frac{\Delta x}{x_R}\right)\left(\frac{400}{\lambda_{nm}}\right)^4\right\} \tag{5}$$

here Δx is the slant depth in g/cm^2 from the source to the detector. For aerosol scattering, we obtain the corresponding transmission factor for light which travels from height h_1 to $h_2 (h_1 \geq h_2)$:

$$T_A = \exp\left\{\left[\exp(-h_1/H_M) - \exp(-h_2/H_M)\right]\frac{H_M \sec\theta}{L_M}\right\} \tag{6}$$

The overall transmission factor can be expressed as $T = T_R R_A$.

4.2.5 *Sky Noise*

Noise mechanisms limiting detector sensitivity as well as contributing to the tube signal include star light, zodiacal light, diffuse radiation from the galaxy, light produced by photochemical processes in the earth's ionosphere and scattered man-made light. This background light induces a DC signal in any phototube aimed at the night sky and fluctuations in this dc signal constitute the sky noise (signals from phototubes are AC coupled to the electronics in the Fly's Eye and HiRes cases). An analysis of the sky noise in existing Fly's Eye data has been performed. The noise (counted in photo-electrons at the phototube cathode) in a

10 μs window can be well described by a Gaussian function with a standard deviation of about 50 photo-electrons, which corresponds to a dc noise of 0.25 photoelectrons per ns per tube at Fly's Eye I. The night sky brightness from other measurements [21] is about 8×10^5 photons m^{-2} μs^{-1} sr^{-1}. This corresponds to 0.36 photoelectrons per ns per tube after considering the Fly's Eye optical reflectivity, tube quantum efficiency, mirror area, tube opening angle and the optical filter transmittance. The number is in reasonable agreement with our measurement. With reduced tube aperture, the HiRes tubes receive DC noise of 0.02 photoelectrons per ns per tube.

4.3 Detector Response

Detector response is hardware related, so we only mention some techniques which maybe useful also with other hardware. One technique we use in the HiRes simulation is ray tracing to account for the spherical aberration of the mirrors. Another one is the FFT (Fast Fourier Transformation) to convert the light pulse to electronic pulse.

In summary, we have discussed in some detail the light production for the air fluorescence simulation. Like the HiRes, the Hybrid Array needs an accurate aerosol model since most of the events will be distant from the optical detector and the light has to travel through a large amount of aerosol to get to the detector.

§References

[1] J.W. Cronin, these proceedings.

[2] A.M. Hillas. Proceedings of 17th International Cosmic Ray Conference, Paris, 8:193, 1981.

[3] A.M. Hillas. Proceedings of the International Workshop on Techniques to Study Cosmic Rays with Energies Greater than 10^{19} eV, Paris, *Nucl. Phys. B (Proc Supp)* 28B:67, 1992.

[4] T.K. Gaisser. Cosmic Rays and Particle Physics, *Cambridge University Press,* 1992.

[5] A. Donnachie and P.V. Landshoff. *Nucl. Phys.* B231:189, 1984.

[6] R.S. Fletcher, T.K. Gaisser, P. Lipari and T. Stanev, to be published in *Phys. Rev. D.*

[7] R.S. Fletcher, T.K. Gaisser, T. Stanev and P. Lipari. Proceedings of 22nd International Cosmic Ray Conference, Dublin, 4:193, 1991.

T.K. Gaisser. Proceedings of the International Workshop on Techniques to Study Cosmic Rays with Energies Greater than 10^{19} eV, Paris, *Nucl. Phys. B (Proc.Suppl.)* 28B:61, 1992.

J. Engel, T.K. Gaisser, P. Lipari and T. Stanev. *Phys. Rev.*, D46:5013, 1992.

[8] R.S. Fletcher. *Phys. Rev.,* D46:187, 1992.

[9] A. Capella and J. Tran Tanh Van. *Zeit. Phys.,* C10:249, 1981.

[10] T.K. Gaisser and T. Stanev. *Phys. Lett.,* B219:375, 1989.

[11] R.M. Baltrusaitis et al. *Nuclear Instruments and Methods,* A240:410, 1985.

[12] D.J. Bird et al. Proceedings of 23rd International Cosmic Ray Conference, Calgary, OG 458, 1993.

[13] A.N .Bunner. Ph.D. Thesis, Cornell University, Ithaca NY, 1964.

[14] P.L. Hartman. *Planet Space Sci.,* 16:1315, 1968.

[15] R.H. Hughes. *J. Opt. Soc. Am.,* B2:362, 1985.

[16] T. Stanev et al. Proceedings of 17th International Cosmic Ray Conference, 6:256, 1981.

[17] J.W. Elbert, T. Stanev and S. Torii. Proceedings of 18th International Cosmic Ray Conference, 6:227, 1983.

[18] A.M. Hillas. *J. Phys. G: Nucl. Phys.,* 8:1461, 1982.

[19] R.M. Baltrusaitis et al. *J. Phys. G: Nucl. Phys.,* 13:115, 1987.

[20] F.X. Kneizys et al. Users Guide to LOWTRAN 7, AFGL-TR-88-0177, *Air Force Geophysics Laboratory,* 1988.

[21] C.W. Allen. Astrophysical Quantities, *Athlone Press, University of London,* 134, 1973.

[22] M.Z. Luo, Ph.D. Thesis, University of Utah, 1992.

Model-Dependence of Determination of Composition in Giant Air Shower Experiments

T.K. Gaisser

Bartol Research Institute, Universidy of Delaware, Newark, DE 19716, USA

Deduction of the mass composition of cosmic rays requires the use of Monte Carlo calculations which depend on models of nuclear interactions. The sensitivity of the interpretation of the Fly's Eye data on X_{max} as a function of energy is addressed.

One of the main handles on the primary composition of ultra-high energy cosmic rays as observed with giant air shower detectors is the "age" of the shower relative to its energy. By "age" here is meant a generalization of the age of an electromagnetic cascade, which indicates whether a shower is before, at or after its maximum at a particular depth. The way such a quantity is measured by an air shower detector, and with what precision, depends on the nature of the detector. But the goal is to find a measurable quantity which reflects the fact that a shower of a given energy produced by a heavy primary, in which the energy is already subdivided into A nucleons, will "age" quickly as compared to a proton shower of the same energy. The latter will penetrate relatively deeply into the atmosphere and reach maximum development late.

Assuming that the technical problem of defining and measuring an appropriate age parameter has been solved, we must next ask with what certainty and with what precision the primary mass can be ascertained. Given a model of hadronic and nuclear interactions at ultra-high energies, it is a straightforward exercise in Monte Carlo simulations to establish the precision. The answer will likely be that it is not possible to determine the composition on a shower-by-shower basis, but that the relative contributions of two or three major groups of nuclei can be established for a class of events with similar energy (e.g. protons plus helium, CNO and Fe). In addition, it may be possible to assign some events a significant probability of belonging to a particular group of nuclei.

The certainty of the mass determination is model-dependent. This systematic error arises because we wish to explore an energy region that is some two orders of magnitude higher in center of mass energy than the highest hadron-hadron collider experiments ($\sqrt{s} = 200\text{TeV}$ as compared to $\sqrt{s} = 1.8\text{TeV}$). The problem is further compounded by the fact that we need the properties of hadron-nucleus and nucleus-nucleus collisions, especially in the fragmentation region, while the collider experiments measure secondaries of hadron-hadron collisions in the central region.

Figures 1 and 2 [1] illustrate the problems. On the diagram in Fig. 1 the first vertical line is the energy of the CERN-ISR. The second corresponds to the bar $\bar{p}$ p-collider at Fermilab. Since there is no fundamental theory of "minimum bias" strong interactions, it is necessary to rely on phenomenological models

a) to relate the fragmentation region to the central region where both are measured (I on the diagram);

b) to use collider data (II on the diagram) to fix parameters that determine the energy dependence of interactions;

and

c) to extrapolate into region III up to $\sqrt{s} > 100\text{TeV}$ for interpretation of giant air shower experiments.

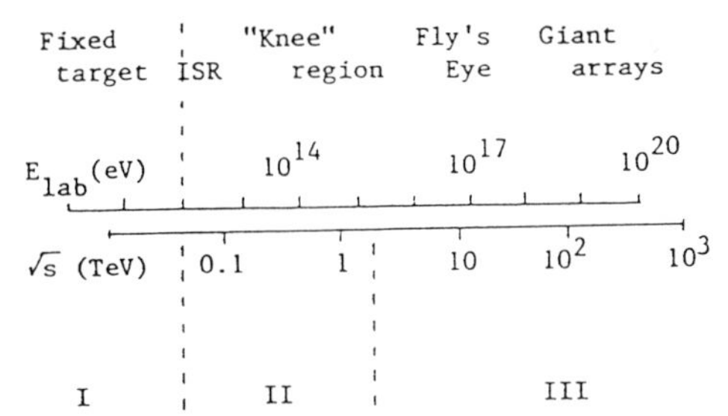

Figure 1:

The three parts of Fig. 2 show

a) a proton-proton collision in the CM system;

b) the CM rapidity distribution of secondary particles;

c) the distribution in CM rapidity of the lab system energy of the outgoing particles.

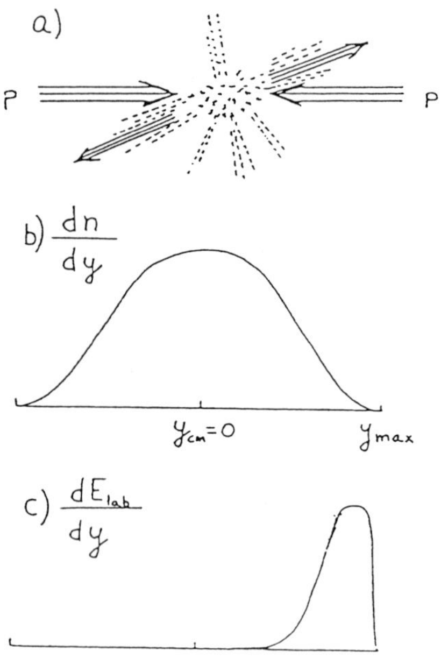

Figure 2:

Only a narrow region of rapidity around $y_{cm} = 0$ is studied in detail in the high energy collider experiments, which is of little importance for cascade development except insofar as it is related (by kinematics and by models) to the fragmentation region where the lab system energy is.

One way to estimate the systematic error associated with the use of models of hadronic interactions is to compare the results of using different models for analysis of the Fly's Eye Experiment. Because that experiment measures the longitudinal development of showers directly, it is easy to define an appropriate measure of shower development--the depth of maximum, X_{max}. This is the depth along the shower axis at which the shower front contains the largest number of ionizing particles. In general,

$$< X_{max} >= R_A \log(E/A) + b_A \qquad (1)$$

for showers generated by nuclei of mass A and total energy E. The coefficient of log(E/A) in Eq. 1 is a partial elongation rate, R_A. If the hadronic and nuclear interactions that determine the cascade development obey Feynman scaling in the fragmentation region, then

$$R_A\,(\text{naive}) = 37.1\,g/cm^2 \times \ell n(10) = 85\,g/cm^2 \qquad (2)$$

as for a purely electromagnetic cascade. (Recall that 37.1 g/cm² is the radiation length in air). The experimentally measured elongation rate [2] is a weighted sum of the R_A for each species. If the composition is independent of energy then $R = R_A$, but $R < R_A$ in a region of energy where the composition becomes heavier and vice versa if the composition becomes lighter. Thus a measured elongation rate that differs from Eq. 2 can indicate either a changing composition or energy-dependence of the hadronic physics that governs shower development (e.g. energy-dependent cross section and/or scaling violation so that $R_A < R_A$(naive)). One known source of energy-dependence is the increase of hadronic cross-sections with energy which reduces the naive elongation rate somewhat (e.g. from 85 to 72 g/cm² for one particular extrapolation of the cross section).

It has been known for a long time that showers of relatively low energy ($<10^{17}$ eV) have depth of maximum much smaller than can be accounted for by scaling models and "normal" composition. In some analyses it appears necessary to have both heavy composition and increase of cross section and inelasticity to account for the data. In one class of models [3],[4],[5],[6], QCD processes become increasingly important at high energy. In these processes, partons (mostly gluons) in the incident hadrons interact to produce small jets which remove energy from the fragments of the projectile and target nucleons. There is a threshold, $x_{min} = 2\,Q_{min}^{\;2}/s$, for the QCD processes. Thus, as energy increases, the allowed x decreases and the QCD processes become increasingly important because the gluon distributions grow rapidly at small x. The rate of this growth, which causes an increase with energy of the inelasticity, is a major source of model-dependence. Another source of model-dependence is the way in which the QCD processes are amplified by effects of nuclei, which further enhances the growth of inelasticity.

Increasing the inelasticity causes a shower to dissipate its energy more rapidly and therefore to have a smaller X_{max}. The effect is similar to that of an increasing cross section. Note that in this class of interaction model, the changes which make X_{max} smaller at a particular energy, will also have the effect of decreasing the elongation rate because the inelasticity increases with energy. The Fly's Eye experiment finds [7]

$$< X_{max} >= 640\,g/cm^2 \;\text{ at } E = 3 \times 10^{17}\,\text{eV},$$

increasing to approximately 760g/cm² at 10^{19} eV, which corresponds to an elongation rate of about 80g/cm² per decade of energy, which is almost the naive scaling value of Equation 2. On the other hand, the elongation rate in the QCD models needed to be consistent with the observed small magnitude of $<X_{max}>$ is R_A = 50 to 55 g/cm². The interpretation made [7] is that the composition undergoes a transition from heavy to light over this energy range so that the measured elongation rate, R, satisfies $R>R_A$. This conclusion is reinforced by a hardening of the energy spectrum in the same energy region, which suggests the dominance of a new population of primary particles above about 3×10^{18} eV.

In reference [8] three models were used to generate distributions of depth of maximum: KNP [3], mini-jet [4] and statistical [9]. The latter is a different type of model in which the inelasticity decreases with energy. This model was rejected because it gives showers that penetrate too deeply even if all primaries are heavy (at least in the representation of the model used by the authors of reference [8]. The KNP model is somewhat more inelastic than the mini-jet model, and it was used to fit the observed distributions of depth of maximum in references [7] and [8]. Indications are that a still more inelastic model would fit the data somewhat better.

Discussion of the Fly's Eye example illustrates the importance of a self-consistent interpretation of several different aspects of the same showers (in this case the X_{max} distributions vs. energy and the energy spectrum). Such constraints are essential when interpreting indirect air shower data with simulations in which there are uncertainties in the interaction models.

§References

[1] T.K. Gaisser. *Nucl. Phys. B (Supp)* 28B:61, 1992.

[2] J. Linsley. Proceedings of 15th International Cosmic Ray Conference, Plovdiv, 12:89, 1977.
See also J. Linsley and A.A. Watson, *Phys. Rev. Letters,* 46:459, 1981.

[3] B.Z. Kopeliovich, N.N. Nikolaev and I.K. Potashnikova. *Phys. Rev.,* D39:769, 1989.

[4] L. Durand and H. Pi. *Phys. Rev. Letters,* 58:303 1987.
G. Pancheri and Y. Srivastava. *Physics Simulation at High Energy*, ed. V. Barger, T. Gottschalk, F. Halzen, World Scientific, p56, 1987.
T.K. Gaisser and T. Stanev. *Phys. Letters,* B219:375, 1989.
R.S. Fletcher et al. (SIBYLL) *Phys. Rev. D* (to be published).

[5] J. Ranft, preprint, 1994.
See also A. Capella and J. Tran Thanh Van. *Z. Phys.,* C10:249, 1981.

[6] H. Bengtsson and T. Sjöstrand. *Comp. Phys. Comm.* 46:43, 1987.
See also T. Sjöstrand and M. Van Zijl. *Phys. Rev.* D36:2019,2019.

[7] D.J. Bird et al. *Phys. Rev. Letters,* 71:3401, 1993.

[8] T.K. Gaisser et al. *Phys. Rev.,* D47:1919, 1993.

[9] G.N. Fowler, A. Vourdas, R.M. Weiner and G. Wilk. *Phys. Rev.,* D35:870, 1987.

Gravitational Phenomena

Black Hole Astrophysics

Peter Bender (JILA, U. Colorado), Elliott Bloom (SLAC), Lynn Cominsky (Sonoma State), Holland Ford (Johns Hopkins), Alan Harmon (NASA/Marshall), Peter Michelson (Stanford), Igor Novikov(Lebedev Physical Institute, Moscow; University Observatory, Nordita, and the Theoretical Astrophysics Center, Denmark), Konstantin Postnov (Moscow University), Jeffrey Scargle (Editor; NASA/Ames), Jean Swank (NASA/Goddard), J. Craig Wheeler (U. Texas), Kent Wood (NRL)

Black hole astrophysics is not just the investigation of yet another, even if extremely remarkable type of celestial body, but a test of the correctness of our understanding of the very properties of space and time in extremely strong gravitational fields. Physicists' excitement at this new prospect for testing theories of fundamental processes is matched by that of the astronomers at the possibility to discover and study a new and dramatically different kind of astronomical object. Here we review the ways that are currently known by which black holes can be identified by their effects on their neighborhood – since, of course, the hole itself does not yield up any direct evidence of its existence or information about its properties. The two most important empirical considerations are determination of masses, or lower limits thereof, of unseen companions in binary star systems, and measurement of luminosity fluctuations on very short time scales.

§1 Black Hole Theory

The reader is no doubt familiar with the fundamental physical considerations by which the attractive nature of the self-gravity of spatially distributed matter inevitably leads to the possibility of collapse to a singularity (see *e.g.* Chapter 33 of [45]). This section outlines the reasoning behind the belief that black holes have indeed been produced in three different locations, and then summarize the ways in which it is thought these objects might reveal themselves.

1.1 Overview

In order for black holes to become an astronomical as well as a physical reality, evidence that Nature can and has produced this state must be obtained. The current situation is much as it was with neutron stars, after physicists had established their theoretical possibility, while Baade and Zwicky ([1]) were arguing that supernovae were the furnaces in which these strange objects could be produced, but before the discovery of pulsars led to the universally accepted conclusion that the Galaxy is littered with neutron stars.

There are three types of black holes:

TYPE	MASS ($M_\odot$)	LOCATION
Dead stars	> 3.2	Throughout galaxies
Supermassive	$> 10^6$	Galaxy centers
Primordial	$> 5 \times 10^{-19}$	Throughout Universe?

The first kind of black hole is the inevitable end-product of the evolution of a sufficiently massive (greater than about 30 $M_\odot$) ordinary star. Section 1.2 contains some considerations that lead to this conclusion (plus similar remarks about the other kinds of black holes). Estimates of the black hole birth rate, based on stellar evolution and related considerations, are on the order of 0.003/year for main sequence stars of mass $M > 30M_\odot$; the rate scales as $M^{-1.4}$. Summing these rates, we find that there should be roughly 10^7 to 10^8 black holes in the Galaxy. This may be an overestimate, as there may be significant mass loss by massive stars in due course of their evolution (up to 50%, [16]).

1.2 Black Hole Formation

Each of the three types of black holes listed above originates in a very different way.

Stellar-Mass Black Holes

All sufficiently massive stars are thought to evolve into black holes. The maximum mass of a neutron star is the well-known Rhoades and Ruffini limit of $3.2M_\odot$ ([55]). Steady or catastrophic mass loss can possibly intervene in some cases, allowing stars that initially exceed this limit to evolve to a final state other than a black hole. Estimates for the minimum mass of stars that cannot escape evolution into black holes are in the tens of solar masses (10 according to [62], 30 according to [35], 50 according to [70] and [59]).

Stars of initial mass in a range $\sim 10-30M_\odot$ often make use of the neutron star final state available to them, as we know from the many pulsars in the Galaxy. The theory of supernovae indicates that there is probably no way that black holes of masses $\leq 1M_\odot$ can be formed. But an alternative scenario invovles ordinary stellar evolution to a neutron star, followed by gradual accretion of matter to the point where the neutron star mass limit is exceeded.

Those stars which are formed in a binary system and evolve into the black hole state (before the companion has time to evolve to its final state) are the best candidates for detection, through their influence on the companion. In particular, the quest for binary stars with unseen companions deduced to be more massive than the Rhoades and Ruffini limit has been a major research thrust (see Section 2.1). It should be kept in mind that this approach automatically excludes detection of black holes that are less massive than this limit, as well as those that are simply not in binaries.

The evolutionary scenario of massive binary systems predicts the existence of a subclass of binary radio pulsars consisting of a neutron star and a black hole, with a density of ~ 1 per 1000 single pulsars ([35]) . In such systems the mass of the unseen companion would be $> 3-4M_\odot$, and the pulsars themselves would be similar to standard isolated ones. The recently discovered binary 1-s pulsar PSR B0042-73=PSR J0045-7319 in the Small Magellanic Cloud, with a massive companion in a highly elongated (eccentricity $e = 0.8$) 51-day orbit ([32]) may be the first such black hole/pulsar. The paradox that the first pulsar discovered in the SMC proved to be in a binary system can be naturally understood if its companion actually is a $10-30M_\odot$ black hole, as illustrated in [36], with calculations of the evolution of radio pulsars after a star formation burst.

Supermassive Black Holes

It is not yet known how supermassive black holes form, although a great deal has been written on this subject. One consideration discussed recently by Haehnelt and Rees ([22]) is that, in most hierarchical models for formation of structure in the Universe, the build-up of structures of galactic size occurs at a time similar to the peak of the quasar activity. They argue that the time-scale for the formation of a black hole in a newly forming dark matter halo is short, and that there need not be a time lag between the formation of a protogalaxy and the "switching on" of a quasar. If a density concentration of the order of $10^8M_\odot$ occurs in a region 1 pc across, they conclude that it will have no non-relativistic equilibrium state which can be supported for long, and will collapse to a supermassive black hole.

Another approach taken by Quinlan and Shapiro ([53]) is to start from an assumed dense cluster of stars in a galactic nucleus and follow the build-up of $100M_\odot$ or larger seed black holes by collisions. The further growth to a supermassive black hole is possible via several mechanisms, including the feeding of the black hole by tidal disruption of stars, pre-existing gas, and gas from solar winds ([18, 14, 47]).

Alternate ways of forming seed black holes of perhaps 10^5 to $10^7M_\odot$ have been investigated recently. Umemura, Loeb, and Turner ([69]) include inverse Compton coupling with the microwave background at redshifts above roughly 160 to dissipate angular momentum from partially ionized density fluctuations near the Jeans mass, allowing them to collapse. Eisenstein and Loeb ([19]) consider the statistical distribution of tidal torques for density fluctuations in different regions of protogalaxies; they find that the number of such regions which will have low angular momentum, and thus be able to collapse rapidly, is large. The estimated density of massive black holes formed by such collapses is comparable with or larger than the density of bright galaxies.

It also is of interest to consider ways in which binaries containing massive black holes with masses in the range of roughly 10^4 to $10^8M_\odot$ may have formed. The possibility of detecting gravitational wave signals from such objects is discussed in the chapter of this volume for **Group G3**. It is generally believed that mergers of pre-galactic structures occurred frequently at earlier times, and binary black holes could result if black holes had already formed in each of two merging structures. Studies to investigate the conditions under which the black holes will spiral together in less than a Hubble time are in progress. Another possibility is for a compact star in the density cusp around a massive black hole to become tightly enough bound so that it will lose a few percent of its rest energy by gravitational radiation before it plunges in. Observations of gravitational wave signals from sources involving massive black holes would give information about the demographics of such objects and about their surroundings.

Primordial Black Holes

Members of this last class of black holes form from primordial density perturbations, either directly, through bubble collisions, or from the collapse of cosmic loops. These black holes will evaporate in the time

$$\tau = 10^{10}(\frac{M}{10^{15}\text{gm}})^3\text{years}. \qquad (1)$$

1.3 Indirect Observations of Black Holes

The current challenge is to find incontrovertible observational evidence of black holes. This is difficult because of the well-known facts concerning the inability of radiation, or information of any kind, to escape from within the Schwarzschild radius of a black hole. Nevertheless a number of viable proposals have been put forth to detect and characterize black holes through the behavior of matter in their vicinity. Two effects that have been examined so far are the orbital motion of a star around a black hole (see Section 2.1), and luminosity variations on the short time scales associated with material orbiting just outside the Schwarzschild radius and/or falling into the black hole (see Section 2.2).

For example, basic physical considerations ([60], [61]) indicate that matter in the vicinity of a black hole will form into a disk, which gradually feeds material into the central maw. In the meantime, however, the hot gas in the disk will radiate energy that readily escapes the system and can reach an observer. This release of energy can be extremely efficient, as the potential energy that can be released, or "binding energy," is $0.057Mc^2$ for a nonrotating black hole and $0.42Mc^2$ for a maximally rotating black hole. The material can come from a companion star in the case of binary systems (stellar wind or Roche lobe overflow), or from material surrounding the galactic nucleus in the case of supermassive black holes. The resulting luminosity is on the order of

$$L_{nr} \approx (3 \times 10^{36}\text{ergs/sec})\frac{\dot{M}}{10^{-9}M_\odot/\text{year}} \qquad (2)$$

for a nonrotating black hole, or about 10 times more for rotating black holes. Section 3.4 gives more detail about the structure of disks around black holes.

A second approach is to try to sense the gravitational potential of unseen black holes in galaxies, by observing brightness and radial velocity profiles in the central regions of galaxies (see Section 2.1).

In the case of black holes that happen to be in binary systems, it should be possible to deduce the presence of a black hole companion by studying the motion of the visible component of the system. Further, one can hope to weigh the black hole, although in practice one determines only the *optical mass function*, which provides a lower limit on the mass (see Section 2.1 below).

§2 Observations

This section reviews the ways in which optical, X-ray, and γ-ray observations are being used to probe the Universe for the presence of black holes.

2.1 Optical Observations

Historically, optical observations of X-ray sources have led to the identification of most of the currently viable black hole candidates.

Dynamical Measurements of Binary Systems

In view of the notion that black holes are the natural end-product of stellar evolution, it is reasonable to expect that many such objects will be members of binary systems, in which the other star has not yet evolved into its final state. This is fortunate, because a binary system is an ideal place to seek the effects of the gravitational field of an otherwise unseeable black hole. The basic idea is to try to find evidence for the presence of a star that should be visible were it an ordinary star, but which is not seen.

To rule out the possibility that the unseen star is a neutron star, it is necessary to show it to have a mass larger than the theoretical upper limit for neutron star masses, namely about 3 $M_\odot$. So the observational problem becomes that of establishing a large lower limit on the mass of unseen binary companions, as was firstly proposed by Zel'dovich in 1965 [85]. (Naturally this method cannot be effective at the detection of black holes that happen to have masses less than this limit.)

As is known from classical binary star studies, if one can measure the period and inclination of a binary orbit, plus the radial-velocity amplitudes due to the orbital motion of each star, then the sum of the two masses is determined. Various other considerations can lead to estimates of the mass ratio, and thence of the individual masses. The motion of a black hole cannot be directly observed. (However, it can be inferred by observations of radial velocity lines in the surrounding accretion disk). It turns out that what is determined from observations of the motion of a visible star of mass M_{opt} in the presence of an unseen companion of mass M_X is the *optical mass function*:

$$f(M) = \frac{(M_X sin\ i)^3}{(M_X + M_{opt})^2} = \frac{PK_{opt}^3}{2\pi G}, \qquad (3)$$

where P is the orbital period, K_{opt} is the semi-amplitude of the optical radial velocity variation, i is the orbital inclination angle, and G is the gravitational constant. This observable provides a lower limit on M_X, because no matter what M_{opt} or the inclination are, it is always true that

$$M_X \geq f(M). \qquad (4)$$

The main observational uncertainty comes from the radial velocity amplitude, K_{opt}, which can be affected by X-ray heating, tidal distortion, non-synchronous rotation of the companion, and spectral contamination due to the surrounding gas and disk. On the other hand, P is most often known more accurately than is G!

If the lower limit (4) is not good enough, then it may be possible to provide an actual estimate of M_X by determining the approximate value of M_{opt} and i, but there are almost always rather large uncertainties in these values.

The best cases for the existence of stellar-mass black holes at present are the binary X-ray sources ([39]). The presence of a large (10^{38} ergs/sec) and rapidly variable X-ray source implies the presence of an object at least as dense and massive as a neutron star. In many cases the optical mass function and additional data imply that the unseen mass $M_X > 3.2\ M_\odot$, the limit necessary to establish that the star should have collapsed to a black hole. It is important that a system not exhibit bursts or pulsations, which are tell-tale signs that the star is a neutron star. Appendix A contains a list of black hole candidates, most of which are the kind of system just described.

Cygnus X-1 is the best-studied case and considered by some to be the best hope for conclusively establishing the existence of a black hole. The optical companion has been identified, and detailed studies of the millisecond variability of the X-ray emission from this system provides some further positive evidence (Section 2.2; [29, 46, 79, 21]).

Two systems in the Large Magellanic Cloud are also good black hole candidates. LMC X-1 may be similar to Cyg X-1, in that its companion is more massive than the unseen object, but uncertainty of the identification of the companion star and the presence of a surrounding nebula of ionized gas provide complications. ([37, 52]). LMC X-3 is interesting in that the X-ray star is more massive than its optical companion! No "normal" star can hide in this system, as it would outshine the companion, which is a B star ([78, 37, 12, 13, 68, 72, 73]).

There is a class of objects called *X-ray transients* – detected only a small fraction of the time that their position has been observed by one or another X-ray telescope. In some cases the presence of short-time-scale bursts or pulsations implies that the compact star is a neutron star, but otherwise it is possible that the X-ray source is a black hole. The systems A0620-00, Nova Muscae 1991 and V404 Cyg are examples of such. There is a considerable body of observational evidence on these cases (for A0620-00: [11, 4, 50, 82, 40, 31, 25]; for Nova Muscae 1991: [15, 54, 2, 9]); for V 404 Cygni: [74, 6]).

In summary, there are six good candidates for binary systems containing stellar mass black holes, as shown in Figure 1, based on [39]. Three systems (Cyg X-1, LMC X-1, and LMC X-3) are persistent X-ray emitters with massive ($> 5M_\odot$) companions. Three systems (A0620-00, Nova Muscae 1991, and V404 Cyg) are transient X-ray sources with low mass ($< 3\ M_\odot$) companions and bright accretion disks. The actual mass determinations are uncertain, and depend sensitively on the assumed relationship between mass and luminosity, the fraction of the Roche lobe which is filled, the assumption of synchronism of the stellar rotation, the contribution of the disk and gas to the optical luminosity of the system, and the details of the stellar evolutionary history of the system.

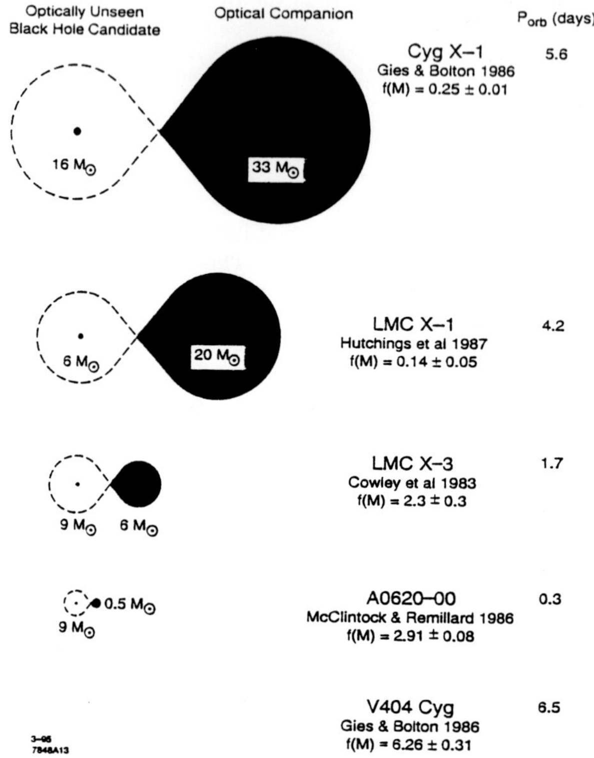

Figure 1: Sketches of the orbital configurations for six of the leading black hole candidates.

Kinematics of Galaxy Cores

As mentioned earlier (Section 1.3) the gravity of unseen black holes in galaxies will produce central cusps in the distribution of mass, characteristic features in radial velocities of stars near the singular nucleus, and perhaps other observable effects. Figure 2 reveals details of both a brightness cusp and large asymmetric Doppler shifts near the center of M 87. These Hubble Space Telescope ([20, 24]) observations of unprecedented spatial resolution provide perhaps the strongest evidence at present in favor of the existence of supermassive black holes in at least

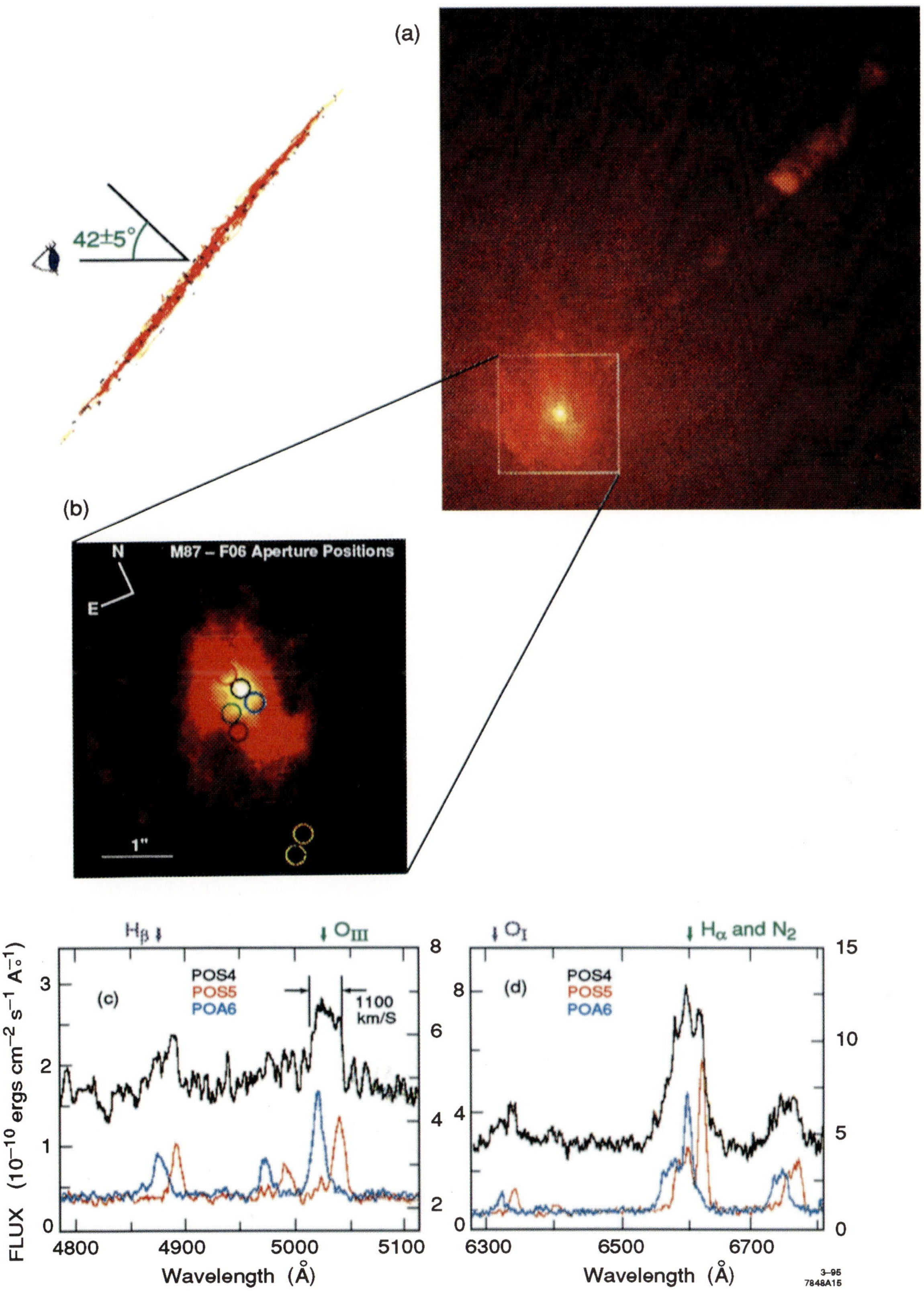

Figure 2: Hubble Space Telescope observations of the gas disk in the nucleus of the galaxy M 87. Wide Field Planetary Camera images (top) and spectroscopy at six locations (bottom).

some nuclei of ordinary galaxies – *i.e.* those that are not currently violently active.

There are also observations of brightness cusps and increased velocity dispersion at the centers of other relatively nearby galaxies. One important case is M 32, where HST observations of the brightness cusp ([33]) can be interpreted most easily as due to a $3 \times 10^6 M_\odot$ black hole. M 31 is more complicated because it shows two cusps ([34]), but is consistent with the existence of either one or two massive black holes. Since M 32 and M 31 are both in the Local Group, and since there are only four spirals or ellipticals other than the Milky Way which are as luminous as M 32, the statistics are of strong interest. If it turns out that both of these galaxies do indeed contain massive black holes, this would suggest that a substantial fraction of all ellipticals and spirals above a certain size may contain such objects.

Another important source of observational information concerning the amount of matter in massive black holes comes from the integrated light of quasars ([10, 63]). From the amount of matter which has to have been fed in to produce the light, at least 3×10^{-5} of all the baryons in the universe are contained in massive black holes.

2.2 X-Ray Observations

This section reviews the X-ray observations that contribute evidence for the existence of stellar-mass black holes. Good references for this material are [65, 66]. It is clear that there is no clean division based on wavelength. X-ray observations are important in interpreting the optical data on binary systems, as discussed in the previous subsection.

We here concentrate on the evidence provided more or less directly by the X-ray variability and spectral data. A large amount of observational and theoretical work has led to the standard model for binary X-ray sources, in which mass flows from one star and accretes onto the other, which is presumed to be a compact star with a deep gravitational potential well. Figure 3, adapted from Figure 25.2 of [45], shows such an effective potential for the motion of a test particle in the Schwarzschild geometry of a concentrated mass.

In most cases the gas forms a repository in the form of a flat disk centered on the compact star, but may also flow more or less directly onto it. This process yields irregular luminosity fluctuations associated with the unsteadiness of the flow of gas onto the disk, the inhomogeneity of the disk itself, and perhaps most important the disordered way in which the gas at the inner edge of the accretion disk falls into the compact central object. While some of the variability and spectral properties of these accretion systems may be independent of the nature of the compact star, there is good reason to hope that distinctive features

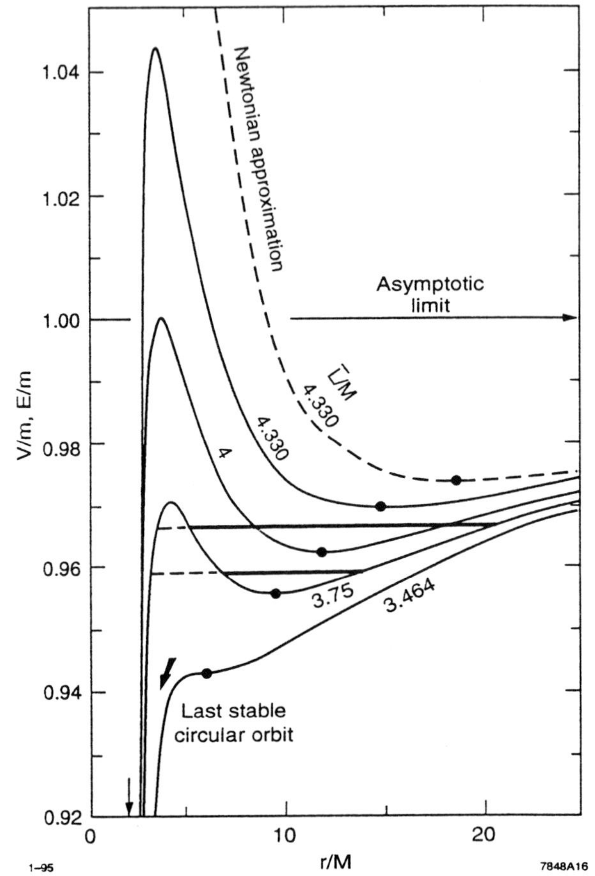

Figure 3: Plot of the effective potential for the motion of a test particle around a non-rotating black hole.

can be found that will serve to identify it – as a black hole, neutron star, or white dwarf.

The primary observational handle on this kind of system is through measurements of the time variability of the emission, presumably connected with the unstable, turbulent flow patterns described just above, and its spectrum – which, by virtue of the high temperatures generated by the dissipative processes in the disk, is concentrated in the X-ray part of the spectrum. Figure 4(a) is a cartoon of this kind of accretion flow for a neutron star system, but could just as well depict that for a black hole. Figure 4(b) is a similar sketch of a possible scenario for accretion and the formation of jets in active galactic nuclei, or AGNs (see Section 2.3).

As described in Section (3.4), complete dynamical models of accretion disks are now being constructed, and comparison of detailed predictions with high-quality observations will be possible in the foreseeable future.

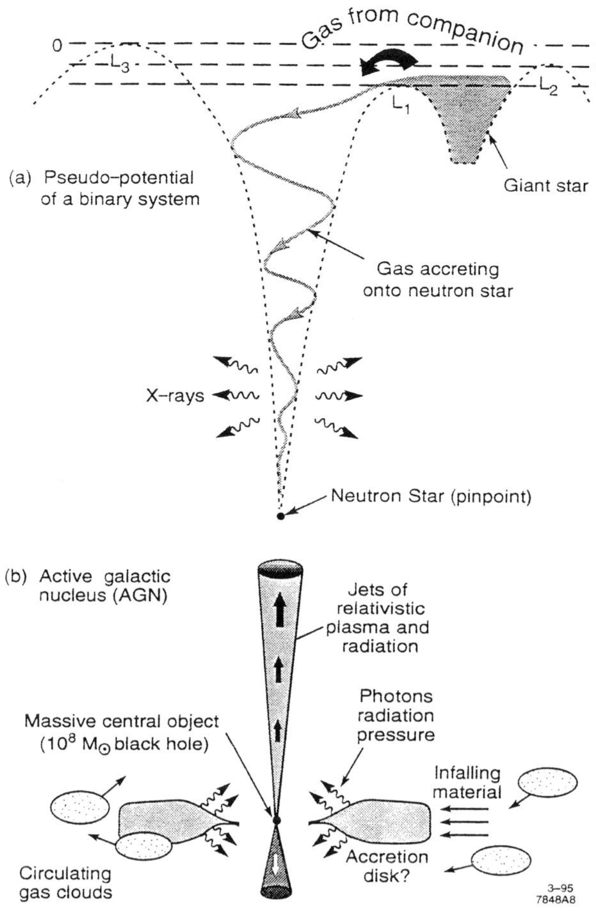

Figure 4: (a) Sketch of the potential due to an ordinary giant star and a companion neutron star, showing how gas from the giant falls into the well around the compact neutron star, emitting X-rays. (b) Sketch of the accretion disk and jet emanating from the nucleus of an active galaxy.

2.3 γ-Ray Observations

The Burst and Transient Source Experiment (BATSE) of GRO was designed to obtain light-curves and spectral information about the so-called γ-ray bursts that had been discovered earlier by the Vela nuclear surveillance satellites, as well as any other transitory sources of γ-ray radiation.

Since BATSE continuously monitors the whole sky, it provides reasonably accurate ($\sim$0.2 deg) positional data as bright sources pass behind the limb of the Earth and allows investigation of spectral and temporal changes in the detected sources. The BATSE Occultation Team ([23, 84]). has used the long-term light curves and spectral information provided by this technique to study black hole candidates over the whole sky. The results are limited to relatively bright objects, and there is potential for source confusion.

These observations are important because they provide information about aspects of black hole candidates that are not well-understood, and in particular, the hard X-ray and gamma ray emission in the $\sim$10 keV - 1 MeV energy range. It is not certain where the high energy emission originates, but generally it is thought that optically-thin regions are responsible, either in the disk, such as in the inner region close to the black hole, or nearby, as in a disk corona or in cooling clouds. The extreme spectral variation within a single source and large changes in luminosity (4 or more orders of magnitude) strongly suggest that we may be dealing with more than one emission region.

The greatest success in describing the high energy spectrum of black hole candidates has been achieved by invoking inverse Compton scattering of soft photons in hot electron plasmas [64]. This model has worked well for Cygnus X-1, and can describe the complex changes in the high energy portion of the spectrum [51]. Broad-band spectra of Cygnus X-1, including soft X-rays, however, show that relativistic corrections should be incorporated into the Comptonization model. A more general analytical form has been developed recently by Titarchuk [67].

It is found that the spectral behavior in "persistent" black hole sources, such as the appearance and evolution of the spectra, and the transitions between the so-called high and low states, are also present in the black hole candidates which are transient X-ray sources. This is a comforting observation, in that there is significant overlap between the black hole candidates deduced from optical studies as described in Section 2.1, and those solely identified on the basis of their broad-band X-ray behavior (see Appendix A) regardless of their persistent or transient nature. So far, however, it has been a difficult task to relate our standard physical and geometric picture of X-ray binary sources to the sudden appearance of the high energy emission and the variable behavior observed in the growing database of transients. It is presumed that the transient outbursts are due to instabilities in the mass transfer or in the accretion disk itself, but a complete understanding of these systems remains as a future goal.

BATSE and the other GRO instruments are collecting observational data that is starting to elucidate these matters. We now summarize some of the empirical results on the light curves and spectra of the black hole candidates that have been observed so far.

The ultrasoft or soft X-ray transients, sometimes called X-ray novae because of the resemblance of their light curves to those of classical novae, usually have a fast rise of about 1-7 days in the hard X-ray band. The spectral evolution and light curves vary from source to source; Nova Persei 1992 had an exponentially decreasing X-ray intensity following the primary maximum, with a time constant of approximately 40 days, and persisted in hard X-rays for nearly 8 months – see Figure 5. In contrast, the

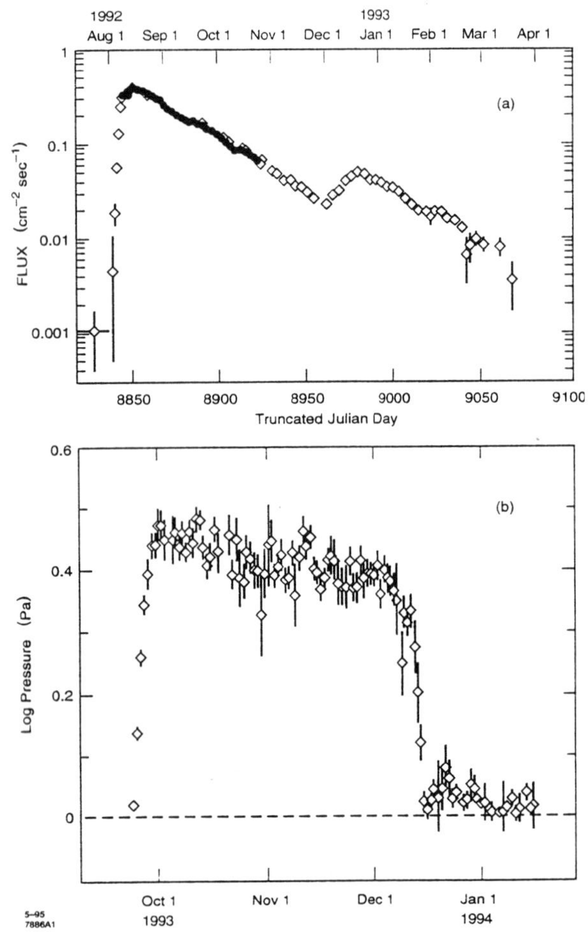

Figure 5: Light curves for (a) Nova Persei (= GRO J0422+32) and (b) Nova Ophiuchi (= GRO J1719-24 = GRS 1716 - 249).

X-ray emission from Nova Ophiuchi 1993 (also shown in Figure 5) had a very slow decay (time constant in excess of 300 days) for about 3 months, then decreased rapidly over a period of 1 week to being nearly undetectable with BATSE. Long-lived secondary maxima in X-ray emission (lasting several weeks to months) have appeared in several of the nova transients, such as Nova Muscae 1991, Nova Persei 1992 (the secondary peak of the latter also decayed exponentially), and most recently in X-Ray Nova Ophiuchi 1993. There may be subtle spectral changes associated with these secondary maxima, but the changes occur over long periods of time and will have to be studied further. Other objects such as Nova Vel 1993 and 4U 1543-47 had rather short-lived primary maxima, with time constants on the order of 2-10 days, and spectra that first soften, then harden much as with Nova Persei 1992.

Other transients with notably hard spectra that resemble those of the stronger black hole candidates, such as GX 339-4 and GRS 1915+105, do not exhibit the fast rise to maximum intensity of the X-ray novae. It is not known what causes this difference in the appearance of the light curves. The hard-state outbursts of GX 339-4 observed by BATSE in 1991, 1992 and 1993-94 are very similar in intensity, duration and time spacing, which suggest an underlying "clock" governing when outbursts may occur. GRS 1915+105, first becoming visible in BATSE data in May 1992, is extremely variable and has been visible in hard X-rays for over two years. It has also been recently identified with the source of superluminal matter ejection (apparent velocity exceeding the speed of light) as observed in high resolution radio interferometry measurements [44]. This is the first galactic source of its kind to exhibit superluminal motion in jets similar to those well-known in active galaxies and quasars. In contrast to the bright X-ray novae, which may have outburst repetition rates of one per decade or even lower, GX 339-4 and GRS 1915+105 seem to have higher "duty" cycles, and may help us to understand better the driving mechanism behind the transient behavior.

EGRET Observations

The Energetic Gamma Ray Experiment Telescope (EGRET) on the Compton Gamma Ray Observatory has the highest energy band of all the GRO instruments ($> 50 MeV$), and can localize a strong point source to 5 to 10 arcminutes. We now summarize what has been learned with this instrument about the spectra and time dependence of active galactic nuclei (AGN).

Curiously, many nearby active galaxies, including Seyferts and some prominent blazars, have not been detected at all in high energy γ-rays. Of the extragalactic sources which are detected, most are quasars and five are BL Lacertae objects. They lie at a wide range of distances, with redshifts from 0.03 to 2.17.

In many of these sources the energy in the γ-ray band, say above 100 MeV, is much larger than the total energy received in all lower bands – and ranges from 10^{44} to 10^{49} ergs/s. This range corresponds to possibilities for the geometry of the emission – principally, whether or not the source is beamed. The spectra in the EGRET energy range are generally well-represented by power laws, with photon spectral indices ranging from -1.6 to -2.6. In some cases there is a suggestion of steepening of the spectrum at the highest energies observed.

Many of the sources exhibit variability, with the γ-ray flux varying by a factor of 3 in a few days in some cases. The most dramatic case of such variability is 3C

279, which displays enormous changes in luminosity, in intervals possibly as short as a few days.

There are a variety of models for the emission of GeV radiation, most of which postulate the presence of a supermassive black hole at the center of the active nucleus. The emission mechanisms invoked include synchrotron self-Compton, inverse Compton scattering off external low-energy photon sources, and cascades initiated by electrons, protons, or heavy nuclei.

§3 Interpretation of Observations

This section describes several of the crucial issues connected with the interpretation of astronomical observations of systems possibly containing black holes.

3.1 Differentiation Between Black Holes and Neutron Stars

One of the key observational difficulties is the possibility that objects suspected as black holes from considerations summarized in Section 2 above, may in fact be neutron stars, and *vice versa*. In binary star accretion systems direct observations of the compact component are hindered by the presence of the other star and the disk. The gravitational fields produced by a black hole and a neutron star of the same masses are largely indistinguishable at the distances at which the companion and the disk lie from the center of the compact object. The optical mass function can be used to identify black holes that greatly exceed the theoretical mass limits (see Section 2.1), but the case of smaller masses is much more difficult.

This section outlines, for the best black hole candidates and firmly identified neutron stars, some observed characteristics which may serve as definitive (or at lease useful) discriminants between these two kinds of compact stars. These considerations are useful to the extent that there is a plausible theoretical explanation of the observed phenomenology, based on known physical properties of these two types of star.

Currently we find clues to the identity of the compact star in gross features of the time-dependence of the radiation from the system: bursts, pulses, quasi-periodic oscillations (QPO's), and transient flares, to name the most important ones.

Type I X-Ray bursts seem to be positive proof that the object is a neutron star, but not all neutron stars exhibit them. Similarly, neutron stars can be pulsars, but black holes cannot. While the electrodynamics of a rotating black hole embedded in a messy magnetized plasma environment is not rigorously understood – nor is that of an ordinary neutron-star pulsar – is seems unlikely that a black hole can function as a pulsar.

The phenomenology of QPO's is rich in features that are probably diagnostic of details in the accretion flows (see e.g. [71]). However, in the absence of more definitive observational studies and detailed dynamical models of the accretion process (see Section 3.4), QPO's are not yet definitive diagnostics.

The phenomenon of a *superhump*, a bright extension of the normal dwarf-nova-like outburst behavior ([28]) is thought to be due to a temporary enhancement of the accretion (and therefore luminosity) due to a tidal instability. Theory indicates that superhumps are easily produced by massive black holes, but are difficult for neutron stars. They have been definitely been observed in Nova Muscae [2] and probably in A 0620-00 [25].

Both neutron stars and black holes exhibit transient outbursts, but it may be possible to discriminate between them based on the recurrence times or other phenomena connected with the details of instabilities occurring within accretion disks. An interesting result from spectral studies such as Tanaka (1992) is the indication that the radius of the inner edge of the accretion disk is a constant, independent of the X-ray luminosity of the system, at about $3R_S$, three Schwarzschild radii. This can be associated with the last stable orbit around a black hole (see Section 2.2 and Figure 3).

Let us turn now to the phenomenology of the radiation spectrum. The indication from various studies is that the spectral signature of a black hole consists of the combination of an ultra soft spectral component plus a power law, while that of a neutron star is a hard thermal spectrum, commonly assumed to be a Comptonized thermal spectrum. This picture is a simplification in a number of ways. Some black hole candidates (*e.g.* V 404 Cyg and Nova Persei 1992 = J 0422+32) have no soft component, and some neutron stars have power laws extending to > 100 keV, but only when the X-ray luminosity is below what may be a critical value, on the order of 10^{36-37} ergs/sec ([5]). Figure 6. presents a summary comparison of the spectral (and timing) characteristics of various systems.

An important question is: what is needed if these important but mostly phenomenological and inconclusive considerations are to become definitive identity tests? The answer, we feel, is observations with broad coverage of wavelength (radio to γ-ray) and time variability (μ-second to decades), coupled with theoretical understanding of the physics of the accretion and radiation processes.

3.2 Wavelet Methods for Analysis of Timing Data

One of the features of the luminosity time series from most X-ray sources, including both the black hole candidates and neutron star accretion sources such as the low-mass

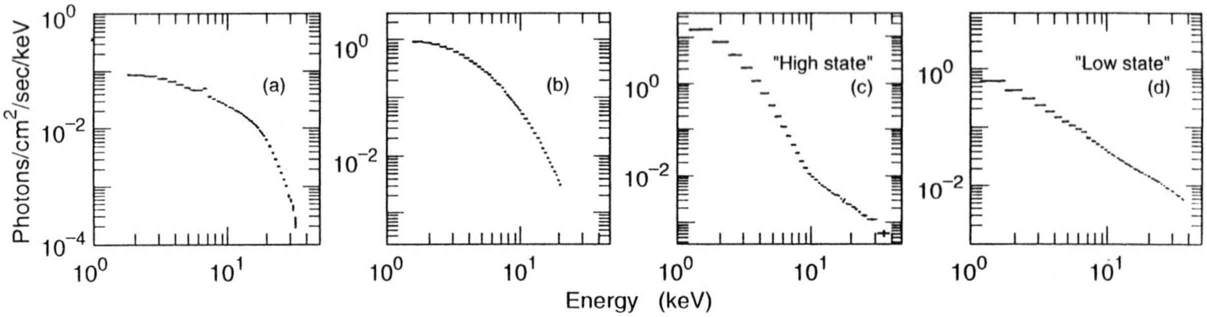

Energy	• Power law • Sharp cutoff Cyclotron lines	• X–ray bursts • Thermal brems. • Spectral shape	• Ultra soft • Hard power law tail	• Single power law
Timing	• Regular pulsations (pulsar)	• Aperiodic variability • QPOs • Small amplitude at rapid time scales	• Large time variability of hard power law tail	• Large variability at all time scales <1sec • Considerable amplitude at ms time scales
Result	• Compact object is a neutron star, $B \approx 10^{12}$ G	• Compact object is a neutron star, $B \approx 10^{8}$ G	• Compact object is a Black hole candidate	• Compact object is a Black hole candidate

2-95 7848A28

Figure 6: Table illustrating the characteristic spectral and power-spectral differences that differentiate black hole candidates, low mass X-ray binaries, and high mass X-ray binaries.

X-ray binaries (LMXB's) is the presence of fluctuations on time scales that span a large range. Indeed, fluctuations are seen from the shortest time scales permitted by the observational time resolution, up to the longest time intervals over which the data are obtained (up to 10's of hours of continuous exposure, or the much longer intervals corresponding to the history of X-ray astronomy as an observational science). Hence we really have only limits on the total extent of the range of time scales.

The techniques of *multiresolution analysis*, and especially the special case of *wavelets*, are perfect tools for studying such data, as well as for representing solutions to the physical equations describing the dynamical evolution of the accretion process. This is because a wavelet basis forms a hierarchy of scales, with each scale level differing from its neighbors by factors of two. In particular, given a specific choice for the *mother wavelet* $\psi(t)$, the basis comprises this set of functions

$$\psi_{s,l}(t) = 2^{-\frac{s}{2}}\psi(\frac{t}{2^s} - l) \tag{5}$$

where s is the *scale index* and l is the *location index*. Note that the width of the function $\psi_{s,l}(t)$ is proportional to 2^s; in data analysis applications the scale index s varies over a range of values so that the shortest scale is on the order of the interval between samples, and the longest scale is on the order of the total interval over which the data is sampled.

The representation of data, or of a function, as a linear superposition of the functions in the wavelet basis takes the form

$$X(t) = \sum_{s,l} c_{s,l}\psi_{s,l}(t), \tag{6}$$

where the

$$c_{s,l} = \int X(t)\psi_{s,l}(t)dt, \tag{7}$$

are the wavelet coefficients of $X(t)$ with respect to the wavelets $\psi_{s,l}(t)$.

There are several key points. First, the wavelets are localized functions. They are non-zero only over a finite subrange of the total time interval over which the observations extend. This allows easy representation of localized features – jumps, discontinuities, or bumps in the data. Compare against the situation with Fourier components, where local features require the superposition and delicate cancellation of the completely global basis functions.

Second, the scales of the wavelets extend over the complete range – from the smallest scale (the sampling interval) to the largest (the sample range), and they do so in a convenient time-scale hierarchy in which the levels differ by a factor of two. Wavelets are thus perfect for representing data (or functions) that are *self-similar* (sometimes called *scaling* or *fractal*). This multi-scale feature can be achieved in Fourier representations by using a logarithmic frequency scale – but the weighting which wavelets naturally apply to the levels in their dyadic hierarchy often turns out to be more suitable to the signal-to-noise present in the time series data.

Finally, there is a simple relationship between the size of the wavelet coefficients and the smoothness of the function they represent through equation (6). Basically, if you diminish the absolute value of some of the wavelet coefficients, you are guaranteed that the resulting function $X(t)$ will be smoother than it was. (The relationship between smoothness and the size of the expansion coefficients does not hold in Fourier analysis.) The practical result of this is a cornucopia of denoising and smoothing algorithms based on truncation or "shrinkage" of the wavelet coefficients ([17]). A *denoising* procedure is a way of correcting data for the presence of noise of knpown or assumed character. Wavelet denoising methods are particularly good at removing noise without smoothing out edges, bumps, or other localized features in the data.

A simple tool that has proven useful in analyzing stochastic data from X-ray sources ([58]) is the wavelet analog of the power spectrum, sometimes called the *scalegram*. If the $C_{s,l}$ are the wavelet coefficients of some discrete data X_n, defined much as in Equation (7) for a continuous function $X(t)$, then the scalegram of these data is defined to be

$$V^X(s) = \frac{S}{N}\sum_l (C_{s,l})^2, \tag{8}$$

where the l-sum is over the allowed values at scale s.

The scalegram can be easily corrected for the Poisson noise present in X-ray data, as in any photon counting data. A simple computation shows that the scalegram of time series data subject to an additive normally distributed observational noise of variance σ_R (usually zero in X-ray astronomy), plus Poisson noise (due to the statistics of counting a finite number of photons), satisfies

$$E[V^{X_{obs}}(s)] = V^{X_{true}}(s) + \sum_n X_n^{true} H_j(n) + \sigma_R^2; \tag{9}$$

$H_j(n)$ depends on the wavelet used, and is just $\frac{1}{N}$ (independent of both n and j) for the Haar wavelet. Thus the scalegram of X-ray data can be corrected for counting statistics simply by subtracting the mean count rate – a very good estimate for the correction term in this equation.

Further, the scalegram can be evaluated for time series data that are arbitrarily spaced in time, including point process data (*e.g.* for such data as the arrival times of individual photons).

Putting all of these features together, the scalegram and related tools seem to be very well-suited for studying fluctuations of X-ray sources on the shortest possible time scales. This is the kind of information that is needed for the identification and characterization of black hole accretion. The next section discusses the attempt to analyze HEAO data on Cygnus X-1 for this very signature.

3.3 Millisecond Time Variability in Cygnus X-1

The accretion of matter onto black holes is almost certain to be accompanied by rapid and violent fluctuations in luminosity. Although there are not yet definitive and reliable theoretical models of the accretion process – see Section (3.4) – the expectation is that there will be fluctuations on a broad range of time scales, all the way down to the approximate millisecond scale associated with the dynamical time expected for accretion onto the black hole itself (millisecond). In addition, if the luminosity comes from blobs of gas that are in orbits around the central black hole, one would expect the power spectrum to contain these orbital frequencies. This would then lead to a truncated "chirp" signal – the frequency rising as the blob spirals into the maw, but abruptly cut off as it passes through the last stable such orbit.

The attempt to study the short time scale behavior of a black hole candidate has already been made ([41] – see also earlier work in [56, 80, 81]), for the bright X-ray source Cygnus X-1. Meekins *et al.* (in [41]) constructed a special algorithm to deal with the problem that variability on longer time scales voids the assumption that the short-time-scale fluctuations are a constant-mean Poisson process. In effect, fluctuations on longer time scales may leak into the measured power at shorter time scales. These authors constructed an "equivalent power" from the comparison of the actual and expected χ-squared values. Figure 7 shows their plot of this power as a function of time scale. Note that there is strong evidence for a cutoff in the power at around 1 ms, in qualitative agreement with the picture described above.

As described above, in Section (3.2), wavelet techniques in general, and the scalegram in particular, are likely to be of great use in this kind of problem. Scargle, in unpublished work carried out at this workshop, demonstrated such an approach using synthetic data generated to be much like the actual Cygnus X-1 data analyzed by Meekins *et al.*, in terms of its signal-to-noise, sampling, *etc.* The data consisted of a chirp signal as described

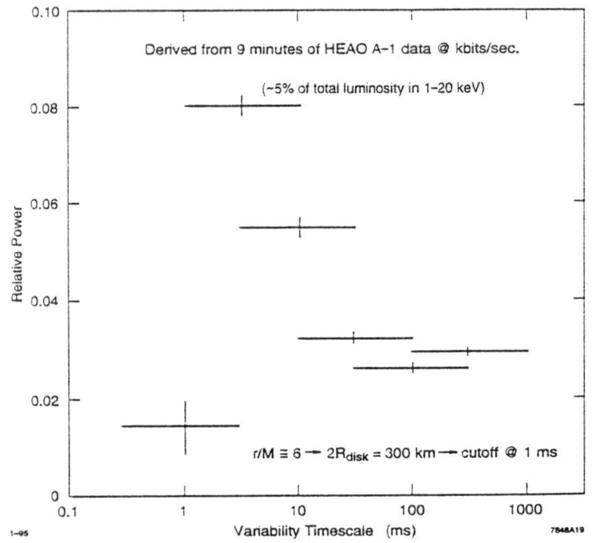

Figure 7: Plot of the power (derived from a χ-squared analysis by Meekins and coworkers, from special high time resolution data on Cyg X-1 by HEAO. This analysis shows that the intrinsic variability of this black hole candidate extends down to a few ms, but then cuts off at around 1 ms – as expected from the accretion of blobs into a black hole.

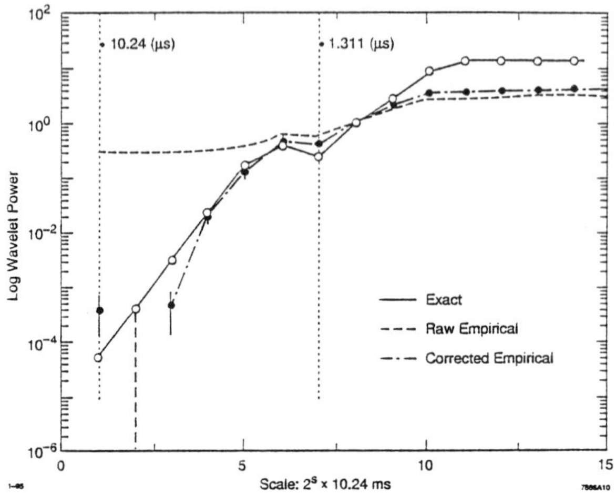

Figure 8: Plot of the wavelet power it vs. time scale index, or *scalegram*, for synthetic data with the same signal-to-noise as the Meekins *et al.* data for Cyg X-1 discussed in the caption to Fig. 7. The synthetic data consisted of randomly occuring chirp signals – cutoff at frequencies above about 1 kHz (time scale = 1.311 ms), to mimic a signal like that possibly present in Cyg X-1. The scalegram shows a steep cutoff starting at about this cutoff (1 ms) and extending down to possibly on the order of 10 μ sec. The solid line with open circles is the scalegram of the known chirp signal, the dashed line is that of the synthetic noisy data, and the solid line with filled circles is this empirical scalegram corrected for the presence of Poisson noise in the data.

above, truncated at the frequency corresponding to the 1 millisecond cutoff found by Meekins *et al.*, and then made noisy by sampling it as a (Poisson-distributed) photon process with the same mean arrival rate as in the actual observations. Figure 8 shows that the scalegram not only senses the cutoff at around 1 millisecond, but also faithfully tracks the known ("true") scalegram of the underlying process for time scales at least an order of magnitude smaller.

In the near future we plan to reanalyze this HEAO data [obtained on May 7, 1978 with the NRL Large Area Sky Survey instrument (A-1)] with the scalegram technique. In the future these and other new techniques will be used to probe the short time scale variability of the black hole candidates that will observed by new X-ray timing experiments – see Section 4.

3.4 Models of Accretion Disks

It is clear from our discussions of the observational issues that future progress in the search for black holes will be based on improved theoretical understanding of both the structure and time-variability of accretion, directly and via disks, onto compact objects. We need to study the differences between accretion onto black holes, neutron stars, and white dwarfs.

As with stellar structure, the starting point is the merging of a body of observational data relevant to the physical conditions (spatial distribution, temperature, large scale motions, and ionization/excitation of the gas) with mathematical representation of the relevant fundamental physical laws.

An understanding of the average, or time-independent, equilibrium configuration of the accretion systems begins with the structure equations:

1. Hydrostatic equilibrium in the vertical direction
2. Conservation of angular momentum
3. Energy balance: Bremsstrahlung and Comptonized Bremsstrahlung cooling, and energy exchange between protons and electrons
4. Pair Balance
5. Pressure = f(temperature)

The theoretical analysis of accretion disks by solving such structure equations is a problem of current interest (see [38, 26] for overviews, and [3, 30, 48, 49, 75, 76, 77, 61, 7, 8, 83, 57, 27] for detailed results).

Thermal instabilities have been found in a number of cases (*e.g.*, because the ionization of hydrogen causes a very sensitive dependence of opacity on temperature at $T \sim 6,000K$). The resulting instability modulates the X-ray flux from the inner disk. Scargle and co-workers ([58]) have developed a simple dynamical model, called the "dripping handrail," for accretion that produces both quasi-periodic oscillations (QPO) and low-frequency noise (LFN) as different aspects of the same physical process. They arbitrarily assumed the existence of a density threshold that may be produced by the kind of instability just discussed, or the basic orbital instability close to the central black hole. Mineshige and co-workers ([42, 43]) have similarly developed dynamical models based on the so-called sandpile models of self-organized criticality, which has many of the same properties as the dripping handrail. In particular, both these models explain the apparently stochastic behavior of the system as intrinsic to its dynamics, without the need to postulate the existence of some kind of random process – such as a random accretion rate.

§4 The Future: New Missions

Since X-ray time series data is so important for probing black hole candidates, there is a need for more data with better signal-to-noise and time resolution. Fortunately there are two space missions that should provide such "timing" data in the next few years.

4.1 X-Ray Timing Explorer

The X-Ray Timing Explorer (XTE) shown in Figure 9 has been designed to study many of the problems decribed in this report. XTE is a NASA mission scheduled to be launched in late 1995. The instruments on XTE are a collaboration of Goddard Space Flight Center, Massachusetts Institute of Technology, and University of California San Diego. It will make pointed observations with both large medium and high energy detectors (2 keV-200 keV), and scanning observations with a smaller All Sky Monitor. The pointed observations will be selected by NASA peer review from proposals submitted in response to a NASA Research Announcement. The large detector area, the data acquisition, and the telemetry have been specifically designed for high resolution timing and high X-ray photon count rates.

XTE will have the following scheduling features for scientific reasons: Sources more than 30° off the ecliptic can be observed at any time. Time-constrained observations will be supported, *e.g.* in order to observe objects at specific epochs, to allow participation in collaborative multi-wavelength observations, and for repeated observations.

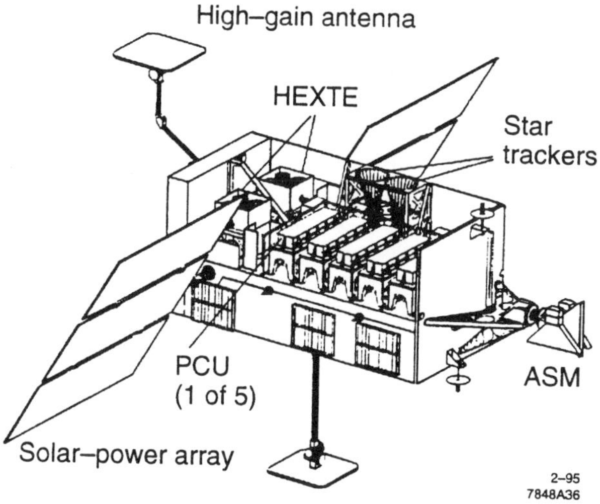

Figure 9: NASA's X-Ray Timing Explorer Satellite (XTE).

There will be provision to monitor given sets of objects, such as active galactic nuclei or Be binaries, with short observations. Transients will be observable with the pointed detectors within 7 hours of detection by the all-sky monitor or notification from another observatory.

The studies to be carried out by XTE include the nature of black holes (including those which may be in active galactic nuclei), neutron stars, and white dwarfs – their interactions with their environs, the systems in which they are formed, and their ultimate fates. The objectives will include, where appropriate, the interior composition and properties, and relationships between magnetic and rotation axes. Of special interest are radiation generation mechanisms, disk and wind instabilities, X-ray transients, the end points of binary system evolution, and the formation of the X-ray background due to cosmologically distributed active galactic nuclei.

For more details about XTE see the World Wide Web at this URL:

http://heasarc.gsfc.nasa.gov/0/docs/xte/xte.html

or contact the Project Scientist Jean Swank at this Internet address:

swank@lheavx.gsfc.nasa.gov.

4.2 USA

The "Unconventional Stellar Aspect" (USA) depicted in Figure 10 has also been designed to study problems described in this report. A collaboration of the Naval Research Laboratory and Stanford University scientists, with representation also from NASA-Ames, Sonoma State University, the University of Oregon, the University of

Washington, the University of Calgary, and Saddleback College, USA is scheduled for launch on the ARGOS satellite in early 1996. It is designed to make long, pointed observations of selected sources, with emphasis on lower X-ray energies. USA's 1 keV - 30 keV energy range extends that of XTE toward low energies, where for many sources the count rates are larger. These instruments, as well as the data acquisition and telemetry systems, have been specifically designed for high resolution timing – with a "workhorse" mode having 4 μsec time resolution and 16 channels of energy resolution – and high X-ray photon counts.

The scientific goals of USA include many of the same as listed above for XTE, but USA will be used in a largely different observation mode – namely dedicating large amounts of observing time to a few important objects. This will allow the elucidation of these black hole and neutron star accretion systems, and is scientifically complementary to the planned functioning of XTE as a general purpose X-ray timing observatory, to be used by guest investigators in a great variety of observing programs.

For more details about USA see the World Wide Web at this URL:

http://xweb.nrl.navy.mil/www_hertz/ usa.html

or contact the Project Scientist Kent Wood at this Internet address:

wood@ssd0.nrl.navy.mil.

4.3 Prospects for the Future

The developments summarized above lead to the conclusion that the identification of black holes with a great degree of certainty is not far off. The observational case for many or all of the binary black hole candidates listed in Appendix A is growing stronger as the data accumulate, in most cases. One must keep in mind, however, the caution provided by the failed black hole candidates in Appendix B.

There has also been much progress in theoretical understanding of the physical conditions and processes in accretion systems connected with black holes and other compact stars.

In addition to improved optical, X-ray and gamma-ray observations, gravitational wave detectors may give new information on black holes (see Chapter G-3). If ground-based detectors being built in the US and Europe reach their full sensitivity, which is well beyond the expected initial sensitivity, coalescence signals for neutron star- black hole binaries would be observable out to well beyond the Virgo cluster. A Laser Interferometer Space Antenna, such as the one being considered by the European Space Agency, would be able to look for and study signals from several types of sources involving massive black holes.

§5 Acknowledgments

We are grateful to Andrew Lee and Han Wen, of the Stanford Linear Accelerator, who served as scientific secretaries of the Snowmass G2 sessions.

7848A34

Figure 10: The Unconventional Stellar Aspect (USA) X-Ray Telescope to be launched on the ARGOS Satellite.

§References

[1] Baade, W., and Zwicky, F. 1934, "Supernovae and Cosmic Rays," *Phys. Rev.*, **45**, 138

[2] Bailyn, Charles D. 1992, "Periodic Variability in the Optical Counterpart of X-ray Nova Muscae 1991," *Ap. J.* , **391**, 298-305

[3] Bao, G., and Ostgaard, E., 1994, "Variability Spectra of Orbiting Blobs around a Black Hole," *Ap. J.*, submitted.

[4] Boley, F., Wolfson, R., Bradt, H., Doxsey, R., Jernigan, G., and Hiltner, W. A. 1976, "Optical Identification of A0620-00," *Ap. J. L.*, **203**, L13

[5] Barret and Vedrenne 1994

[6] Casares, J., Charles, P. A., and Naylor, T. 1993, "A 6.5-day periodicity in the recurrent nova V404 Cygni implying the presence of a black hole," *Nature*, **355**, 614 - 617

[7] Chen, X. and Taam, R. E. 1994a, "Low-Frequency Quasi-Periodic Oscillations in Low-Mass X-Ray Binaries and Galcatic Balck Hole Candidates," *Ap. J.*, **431**, 732 - 741

[8] Chen, X. and Taam, R. E. 1994b, "Variability of Accretion Disks Surrounding Black Holes – The Role of Inertial-Acoustic Mode Instabilities," preprint

[9] Cheng, F. H., Horne, Keith, Panagia, N., Shrader, C. R., Gilmozzi, R., Paresce, F. and Lund, N. 1992, "The Hubble Space Telescope Observations of X-ray Nova Muscae 1991 and its Spectral Evolution," *Ap. J.* , **397**, 664-673

[10] Chokshi, A., and Turner, E. L., 1992, "Remnants of the quasars," *MNRAS*, **259**, 421

[11] Cominsky, L., Jones, C., Forman, W. and Tananbaum, H. , 1978, "Transient X-ray Sources in the Galactic Plane," *Ap. J.*, **224**, 46 - 52

[12] Cowley, A. P., Crampton, D., Hutchings, J. B., Remillard, R., and Penfold, J. E. 1983, "Discovery of a Massive Unseen Star in LMC X-3," *Ap. J.*,**272**, 118 - 122

[13] Cowley, A. P. et al. 1991, "Discovery of a Long-Term Periodic Variation in LMC X-3," *Ap. J.* , **381**, 526 - 533

[14] David, L. P., Durisen, R. H., and Cohn, H. N. 1987, "The evolution of active galactic nuclei. II. Models with stellar evolution," *Ap. J.*, **316**, 505

[15] Della Valle, M. and Pakull, M. 1991, *IAU Circ. No. 5167*

[16] De Jager, C., 1980, "The Brightest Stars," D.Reidel: Dordrecht

[17] David Donoho and Ian Johnstone, of Stanford University, have written a number of papers on this subject that are available in Internet at PLAYFAIR.STANFORD.EDU; theoretical papers are in */pub/reports*, abstracts in *pub/reports/Abstracts*, and a MatLab software package developed with Scargle is in *pub/software/wavelets*. **MatLab** is a trademark of The MathWorks, Inc.

[18] Duncan, M. J., and Shapiro, S. L., 1983, " Monte Carlo simulations of the evolution of galactic nuclei containing massive central black holes," *Ap. J.*, **268**, 565

[19] Eisenstein, D. J., and Loeb, A., 1994, "Origin of quasar progenitors from the collapse of low-spin cosmological perturbations," *Ap. J.*, in press.

[20] Ford, H., Harms, R., Tsvetanov, Z., Hartig, G., Dressel, L., Kriss, G., Bohlin, R., Davidsen, A., Margon, B., and Kochhar, A., 1994, "Narrowband HST Images of M87: Evidence for a Disk of Ionized Gas Around a Massive Black Hole," *Ap. J. Lett.*, **435**, L27-30.

[21] Gies, D. R. and Bolton, C. T. 1986, "The Optical Spectrum of HDE 226868 = Cygnus X-1. II. Spectrophotometry and Mass Estimates," *Ap. J.*, **304**, 371 - 388

[22] Haehnelt, M. G., and Rees, M. J. 1993, "The formation of nuclei in newly formed galaxies and the evolution of the quasar population," *MNRAS*, **263**, 168

[23] Harmon, B. A., et al., 1992, *Compton Observatory Science Workshop*, NASA CP-3137 ed. by C. R. Shrader, N. Gehrels and B. Dennis, "Occultation Analysis of BATSE Data- Operation Aspects," p. 69

[24] Harms, R., Ford, H., Tsvetanov, Z., Hartig, G., Dressel, L., Kriss, G., Bohlin, R., Davidsen, A., Margon, B., and Kochhar, A., 1994, "HST FOS Spectroscopy of M87: Evidence for a Disk of Ionized Gas Around a Massive Black Hole," *Ap. J. Lett.*, **435**, L35-38.

[25] Haswell, Carole A. and Shafter, Allen W. 1990, "A Detection of Orbital Radial Velocity Variations of the Primary Component in the Black Hole Binary a0620-00 (= V616 Monocerotis)," *Ap. J. L.*, 359, L47 - L50

[26] Hawley, J. F., 1986, "General Relativistic Hydordynamcis Near Black Holes," in *Dynamical spacetimes and numerical relativity*, ed. Joan M. Centrella, Cambridge U. Press: Cambridge, 101 - 122

[27] Hirotani, K., Tomimatsu, A., and Takahashi, M., 1993, "Time Variation of Magnetohydrodynamic Accretion onto a Black Hole," *Pub. Astro. Soc. Japan*, **45**, 431 - 441

[28] Hirose, M. and Osaki, Y., 1990, "Hydrodynamic Simulations of Accretion Disks in Cataclysmic Variables: Superhump Phenomena in SU Uma Stars," *PASJ*, **42**, 135-165. See also Osaki, Y., Hirose, M. and Ichikawa, S., "Tidal Effects in Close Binary Systems," 1993, in *Accretion Disks in Compact Stellar Objects*, ed J. C. Wheeler, (Singapore: World Scientific), 272 - 302

[29] Hjellming, R. M. and Wade, C. M. 1971, "Radio Emission from X-ray Sources" *Ap. J. Lett.*, 168, L21 - L24

[30] Ichimaru, Setsuo, 1977, *Ap. J.*, **214**, 840

[31] Johnston, Helen M., Kulkarni, Shrinivas R. and Oke, J. B. 1989, "The Black Hole A0620-00 and its Accretion Disk," *Ap. J.* , 345, 492 - 497

[32] Kaspi, V.M. et al., 1994, "A Massive Radio Pulsar Binary in the Small Magellanic Cloud," *Astrophys. J.*, **423**, L43

[33] Lauer, T. R., *et al.*, 1992, "Planetary camera observations of the central parsec of M32," *Astron. J.*, **104**, 552

[34] Lauer, T. R., *et al.*, 1993, "Planetary camera observations of the double nucleus of M31," *Astron. J.*, **106**, 1436

[35] Lipunov, V.M., Postnov, K.A., Prokhorov, M.E. and Osminkin, E.Yu., 1994, "Binary Radiopulsars with Black Holes," *Astrophys. J.*, **423**, L121

[36] Lipunov, V.M., Postnov, K.A., and Prokhorov, M.E., 1995, "On the Nature of Binary Radiopulsar PSR B0042-73 in SMC," to appear in **Astrophys. J**, March 10 issue

[37] Long, K. S., Helfand, D. J., and Grabelsky, D. A. 1981, "A Soft X-ray Study of the Large Magellanic Cloud," *Ap. J.*, **248**, 925 - 944

[38] Longair, M. *High energy Astrophysics, Vol. 2 Stars, the Galaxy and the interstellar medium*, second edition, Cambridge University Press: Cambridge

[39] McClintock, J. E., 1992, "Black Holes in the Galaxy," in *Frontiers of X-Ray Astronomy*, eds. Y. Tanaka and K. Koyama, (Tokyo: Universal Academic Press), p. 333.

[40] McClintock, Jeffrey E. and Remillard, Ronald A., 1986, *Ap. J.* , 308, 110 - 122, "The Black Hole binary A0620-00"

[41] Meekins, J.F., Wood, K.S., Hedler, R.l., Byram, E.T., Yentis, D.J., Chubb, T.A., and Friedman, H., 1984, "Millisecond Variability of Cygnus X-1," *Ap. J.*, **278**, 288

[42] Mineshige, S., Ouchi, N. Bob, and Nishimori, H. 1994, "On the Generation of $1/f$ Fluctuations in X-Rays from Black-Hole Objects," *PASJ*, **46**, 97 - 105

[43] Mineshige, S. 1994, "Secondary Maxima of Black Hole X-Ray Transients: A Clue to Understanding X-Ray Irradiation," *Ap. J. Lett.*, **431**, L99 - L101

[44] Mirabel, I. F. and Rodriguez, L. F., 1994, *Nature*, **371**, 46

[45] Misner, C. W., Thorne, K. S., and Wheeler, J.A., 1973, *Gravitation*, W. H. Freemen: New York

[46] Murdin, P. and Webster, B. L., 1971, "Optical Identification of Cygnus X-1" *Nature*, **233**, 110

[47] Murphy, B. W., Cohn, H. N., and Durisen, R. H. 1991, "Dynamical and luminosity evolution of active galactic nuclei: Models with a mass spectrum," *Ap. J.*, **370**, 60

[48] Nowak, M., and Wagoner, R., 1992, "Diskoseismology: Probing Accretion Disks. II. G-Modes, Gravitational Radiation Reaction, and Viscosity," *Ap. J.*, **393**, 697 - 707

[49] Nowak, M., and Wagoner, R., 1993, "Turbulent Generation of Trapped Oscillations in Black Hole Accretion Disks," *Ap. J.*, **418**, 187 - 201

[50] Oke, J. B. 1977, "Further Spectrophotometry of the Transient X-ray Source A0620-00," *Ap. J.* , 217, 181 - 185,

[51] Paciesas, W. S. *et al.*, 1994, *Les Houches Conf. on The Gamma-Ray Sky with Compton GRO and Sigma*, "BATSE Observations of Galactic Black Hole Candidates," preprint

[52] Pakull, M. W. and Angebault, L. P. 1986, "Detection of an X-ray-ionized nebula around the black hole candidate binary LMC X-1," *Nature*, **322**, 511 - 517

[53] Quinlan, G. D., and Shapiro, S. L. 1990, "The dynamical evolution of dense star clusters in galactic nuclei," *Ap. J.*, **356**, 483

[54] Remillard, Ronald A., McClintock, Jeffrey E. and Bailyn, Charles D. 1992, "Evidence for a Black Hole in the X-ray Binary Nova Muscae 1991," *Ap. J. L.*, 399, L145 - L149

[55] Roades, C. E., and Ruffini, R., 1974, "Maximum Mass of a Neutron Star," *Phys. Rev. Lett.*, **32**, 324

[56] Rothschild, R. E., Boldt, E. A., Holt, S. S., and Serlemitsos, P. J., 1974, *Ap. J. Lett.*, **189**, L13

[57] Sanbuichi, K., Yamada, T., and Fukue, J., 1993, "Self-Irradiated Accretion Disks around Schwarzschild Black Holes," *Pub. Astro. Soc. Japan*, **45**, 443 - 448

[58] Scargle, J.D., Donoho, D.L., Crutchfield, J.P., Steiman-Cameron, T. Y., Imamura, J. N., and Young, K., 1993, "The Quasi-Periodic Oscillations and Very Low Frequency Noise of Scorpious X-1 as Transient Chaos: A Dripping Handrail?," *Ap. J. Lett.*, **411**, L91

[59] Schield, H., and Maeder, A., 1985, "The initial mass limit for neutron star and black hole formation," *Astron. Astrophys.*, **143**, L7-L10.

[60] Shakura, N.I., 1972, "Disk Model of Gas Accretion onto Relativistic Star in a Close Binary System," *Sov. Astron.*, **49**, 921

[61] Shakura, N.I. and Sunyaev, R.A., 1973, "Black Holes in Binary Systems. Observational Appearance," *Astron. Astrophys.*, **24**, 337 - 355

[62] Shapiro, S.L. and Teukolsky, S.A. 1983, *Black Holes, White Dwarfs, and Neutron Stars*, Wiley and Sons: New York

[63] Soltan, A., 1982, "Masses of quasars," *MNRAS*, **200**, 115

[64] Sunyaev, R. A. and Titarchuk, L. G., 1980, *Astron. Astrophys*, **86**, 121

[65] Tanaka, Y., 1992, Ginga Memorial Symposium, ed. F. Machino and F. Nagase, International Space and Astronautical Science Symposium, Feb. 4-5, 1992, "Black Hole X-Ray Binaries," p. 19.

[66] Tanaka, Y. and Lewin, 1994, "Black Hole Binaries," Chapter 3 of *X-Ray Binaries*, Cambridge U. Press, in preparation

[67] Titarchuk, L., 1994, "Generalized Comptonization Models and Application to the Recent High-Energy Observations," *Ap J*, **434**, 570

[68] Treves, A., Belloni, T., Bouchet, P., Chiappetti, L., Falomo, R., Maraschi, L. and Tanzi, E. G., 1988, "The Ultraviolet-Optical-Infrared Energy Distribution of LMC X-3: Observation of an Accretion Disk around a Black Hole," *Ap. J.* , 335, 142 - 146

[69] Umemura, M., Loeb, A., and Turner, E. L. 1993, " Early cosmic formation of massive black holes," *Ap. J.*, **419**, 459

[70] van den Heuvel, E.P.J. and Habets, G.M.H.J. 1984, "Observational Lower Mass Limit for Black Hole Formation derived from Massive X-Ray Binaries," *Nature*, **309**, 698

[71] M. van der Klis, 1994, "A Comparison of the Power Spectra of Z and Atoll Sources, Pulsars and Black Hole Candidates," *A. and Ap.*, **283**, 469 - 474

[72] van der Klis, M., Tjemkes, S. and van Paradijs, J. 1983, "UBV Photometry of the optical candidate for LMC X-3," *A. Ap.*, **126**, 265 - 268

[73] van der Klis, M., Clausen, J. V., Jensen, K., Tjemkes, S. and van Paradijs, J. 1985, "The optical binary light curve of LMC X-3," *A. Ap.*, **151**, 322 - 328

[74] Wagner, R. Mark, Starrfield, Sumner G., Howell, Steve B., Kreidl, Tobias, J., Bus, S. J., Cassatella, A., Bertram, R. and Fried, R. 1991, "Optical Identification of the X-ray Source GS 2023+338 as V404 Cygni," *Ap. J.*, **378**, 293 - 297,

[75] Wallinder, F. H. 1991, "The stability of slim accretion disks," *A. and Ap.*, **249**, 107

[76] Wallinder, F. H. 1993, "Towards Understanding Active Galactic Nuclei," *Comments Astrophys.*, **16**, 331

[77] Wallinder, F. H. 1994, "The stability of slim accretion disks – the radial dependence," *MNRAS*, in press

[78] Warren, P. R. and Penfold, J. E. 1975, "Photometry of LMC X-ray Candidates," *MNRAS*, **172**, 41P - 47P

[79] Webster, B. Louise and Murdin, Paul 1972, "Cygnus X-1 – a Spectroscopic Binary with a Heavy Companion?" *Nature*, **235**, 37 - 38.

[80] Weisskopf, M. C., Kahn, S. M., and Sutherland, P. G., 1975, *Ap. J.*, **199**, L147.

[81] Weisskopf, M. C., Sutherland, P. G., Katz, J. I., and Canizares, C. R., 1978, *Ap. J. Lett.*, **223**, L17.

[82] Whelan, J. A. J. et al. 1977, "Spectroscopic observations of the X-ray Nova A0620-00," *MNRAS*, **180**, 657 - 673,

[83] Yamada, T., and Fukue, J., 1993, "Relativistic Patchwork Disks around a Schwarzschild Black Hole: Light Curves and Spectrum," *Pub. Astro. Soc. Japan*, **45**, 97 - 102

[84] Zhang, S. N., et al., 1993, "Imaging High-Energy Astrophysical Sources Using Earth Occultation," *Nature*, **366**, p. 245.

[85] Zel'dovich, Ya. B. and Guseinov, O.H. 1965, "Collapsed Stars in Binaries," *Astrophys. J.*, **144**, 84.

§6 Appendix A: Table of BH candidates

SOURCE	POS	TYPE	MASS FUNCTION	OPTICAL	ORBITAL PERIOD	WAVELENGTHS
LMC X-3	0538-641	HMXB	2.3 ± 0.3; $M_X > 7$	B3 V; $V \approx 17$	1.70 d	IR, UV
LMC X-1	0540-697	HMXB	0.14 ± 0.05; $M_X > 2.6$ $M_X = 6$	O7-9 III;$V \approx 14.5$	4.22 d	UV
V616 Mon	0620-003	LMXB, RT	3.18 ± 0.16; $M_X > 7.3$	K5 V; $V \approx 12$-18	7.75 h	radio, IR, UV
Nova Mus	1124-684	LMXB, T	3.1 ± 0.4; $M_X > 3.1$	K0 V-K4 V;$B \approx 13$-21	10.4 h	radio, UV
	1354-645	LMXB, T		$V \approx 17$		
TrA X-1	1524-617	LMXB, RT		$B \approx 17 \rightarrow 19.5$		
	1543-475	LMXB, RT		A1-2 V, $V \approx 15 - 17$		UV
	1630-472	LMXB, RT				
GX 339-4	1659-487	LMXB, RT		$V \approx 15.4 \rightarrow 20$	14.8 h	IR, UV
Nova Oph	1705-250	LMXB, T		$B \approx 16.5 \rightarrow 21$		
	1740.7-2942					radio
	1741-322	LMXB, T				
	1755-338	LMXB		$V \approx 19$	4.4 h?	UV
	1758-258	LMXB?				
	1826-238	LMXB?, T				
	1846-031	LMXB, T				
V1408 Aql	1957+115	LMXB		$V \approx 18.7$	9.3 h	UV
Cyg X-1	1956+350	HMXB	0.24 ± 0.01; $M_X > 7$ $M_X = 16 \pm 5$	O9.7Iab;$V \approx 9$	5.6 d	radio, IR, UV
Nova Vul	2000+251	LMXB, T		K;$V \approx 16 - 21$	8.3 h	radio
V 404 Cyg	2023+338	LMXB, RT	6.26 ± 0.31; $M_X \approx 8$-12	K0 IV;$V \approx 12$-21	6.47 d	radio

Note: LMXB = Low Mass X-ray Binary, HMXB = High Mass X-ray Binary, T = Transient, RT = Recurrent Transient.

§7 Appendix B: Table of FAILED BH candidates

SOURCE	X-RAY CHARACTERISTICS (Reasons for Once Being Considered a Black Hole Candidate)	EVIDENCE FOR NOW BELIEVING SOURCE TO BE A NEUTRON STAR
Cir X-1	RAPID (MILLISECOND) FLICKERING SOFT SPECTRUM SOFT SPECTRUM	TYPE I X-RAY BURSTS (FROM NUCLEAR BURNING ON NEUTRON STAR SURFACE)
X 0331+53 V 0332+53	ERRATIC, FAST (MILLISECOND) VARIABILITY, CHARACTERISTICS SIMILAR TO CYG X-1 LOW STATE	4.4 s ROTATION PERIOD (IN X-RAYS)

Gravitational Wave Astrophysics

A. Abramovici (Caltech), P. Bender (JILA), R. Drever (Caltech), L.S. Finn (Northwestern), R. Flaminio (Pisa), L. Grishchuk (Washington U.), W. Johnson (LSU), K. Kawabe (Tokyo), P. Michelson (Stanford), N. Robertson (Glasgow), A. Rüdiger (MPQ), R.J. Sandeman (ANU), P. Saulson[a](Syracuse), D. Shoemaker (MIT), K. Thorne (Caltech), M. Tinto (JPL), M. Tobar (UWA), R. Weiss (MIT), S. Whitcomb (Caltech)

The search for gravitational waves from astrophysical sources is advancing on a number of fronts. We explain the basic scientific issues involved, and review the present status of research with resonant-mass detectors, interferometers, and detectors in space. We describe in detail both near-term and long-term plans for gravitational wave searches, demonstrating why there is optimism that direct detection of gravitational waves may soon be achieved.

1 The Search for Gravitational Waves

The prediction that gravitational waves should exist comes from the early part of this century, with the formulation of special relativity and the need to recast Newtonian mechanics and gravitation into a relativistically consistent form. All relativistic theories of gravitation, including general relativity, predict the radiation of gravitational waves by accelerating masses. In all theories the waves are so weak that they require astrophysical scale masses moving at relativistic velocities as well as exquisitely sensitive instruments for their detection.

1.1 Goals

The notion that gravitational waves could be detected directly was pioneered by Joseph Weber in the 1960s using resonant-mass detectors. [1] At the time the technology was not sufficiently sensitive to intersect even the most optimistic estimates for the strength of the waves from astrophysical sources. Since these initial attempts, there has been substantial improvement in the sensitivity of resonant-mass detectors, as well as the evolution of alternative concepts: laser interferometry on the ground, and both Doppler tracking and interferometric techniques in space, all of which hold the promise of opening the field of gravitational wave astrophysics in the coming decade.

A critical confirmation of the existence of gravitational waves has come from the observations of gravitational radiation damping in the binary neutron star system PSR 1913+16, discovered in 1974 by Taylor and Hulse. [2] The discovery of this and similar astrophysical systems has had a profound influence on the field, by providing a class of sources with known event rates and wave strengths. These serve as sensitivity targets for the design of gravitational wave detectors.

The direct detection of gravitational waves from astrophysical sources offers a unique opportunity for the study of the gravitational interaction and the exploration of the universe. Successful detection of gravitational waves could offer many insights into the gravitational interaction, including:

- Measurement of the propagation speed and polarization properties of the waves, and from them the rest mass and spin of the graviton. This provides a test of whether they agree with general relativity's predictions, $m = 0$ and $s = 2$
- Verification (by comparing theoretical and observed wave forms) that black holes exist and that their dynamics are as predicted by general relativity. By this, test general relativity for the first time in the domain of highly nonlinear, dynamic gravity.

The astrophysical information potentially derived from the direct detection of gravitational waves includes:

- the behavior of neutron stars in highly dynamical situations. By this, extract information about the uncertain physics that governs neutron stars.
- use of the waves from the coalescence of black hole and neutron star binaries as "standard candles" for an alternative method to determine the Hubble expansion rate and deceleration parameter of the universe . [3]
- the detection of primordial gravitational waves from the big bang and from them information about the initial conditions and earliest stages of evolution of the universe.

Gravitational and electromagnetic waves differ greatly in the nature of their sources and their penetration in the universe. Gravitational waves should be

[a] Editor. E-mail address: saulson@suhep.phy.syr.edu.

emitted by coherent bulk motions of matter (e.g., collapsing stellar cores) and coherent, nonlinear vibrations of spacetime curvature (e.g., collisions of black holes). By contrast, astronomical electromagnetic waves are usually incoherent superpositions of emission from individual atoms, molecules, charged particles and dust. Gravitational waves are emitted most strongly in regions of spacetime where gravity is relativistically strong, whereas electromagnetic waves come almost entirely from weak-gravity regions, since strong-gravity regions tend to be obscured by surrounding matter. Because of these differences, the information carried by gravitational waves is almost "orthogonal" to that carried by electomagnetic waves; our present electromagnetically-based understanding of the universe is inadequate to predict with confidence the strengths of the gravitational waves bathing the Earth. Obversely, if gravitational waves are detected and studied, they may create a revolution in our view of the universe comparable to that wrought by radio astronomy.

1.2 Nature of the waves

In general relativity, the propagation speed of gravitational waves is the same as that for light (the graviton has zero rest mass); the wave's force field is transverse to its propagation direction and has quadrupolar symmetry (i.e., the graviton has spin two). The two polarization states of the waves are called + ("plus") and × ("cross") and are characterized by two dimensionless fields h_+ and $h_\times$. The wave force field causes an antisymmetric change in separation between objects placed along two orthogonal directions transverse to the propagation direction of the wave. The difference in the displacements along the two orthogonal directions $\Delta L/L \equiv h$ is given by

$$h = [1/2(1+\cos^2\theta)\cos 2\phi]h_+ + [\cos\theta\sin 2\phi]h_\times \quad (1)$$

Here the angles θ and ϕ give the wave's propagation direction. Since the relative displacement is proportional to the separation of the objects, h is sometimes called the gravitational-wave strain.

A rough value for the scale of h is worth keeping in mind. For example, a galactic source at 10 kpc distance emitting 1% of a solar mass into gravitational waves in 10 msec would produce a strain $h \sim 10^{-17}$ at the Earth. The same source at a distance of the Virgo Cluster of galaxies, roughly 10 Mpc distance, would give an $h \sim 10^{-20}$. The reduction in h is accompanied by an increase in the number of such sources accessible for detection on the assumption that there are roughly equal numbers of sources per galaxy. For distances comparable to or larger than the Virgo Cluster distance, the rate of detection of a specific type of source then grows roughly as the inverse cube of the minimum detectable h. [4]

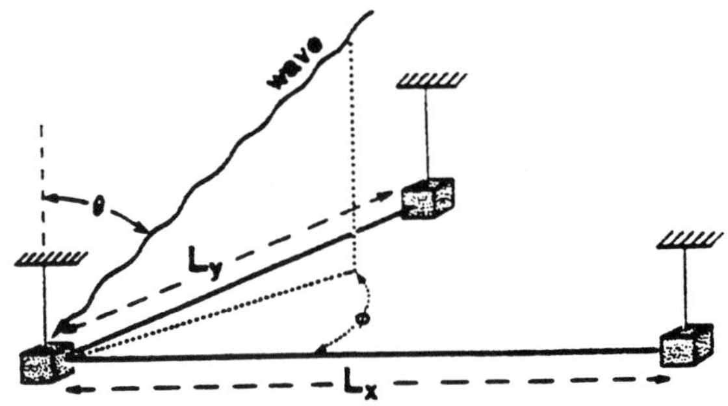

Figure 1: Gravitational wave arriving at a detector.

1.3 Detector concepts

All detection schemes for gravitational waves can be described as measurements of h. The methods differ in their most sensitive frequency bands and in their bandwidths. Ground based detectors are limited to gravitational wave frequencies above approximately 1 Hz due to naturally occurring perturbations in the local gravitational fields from density fluctuations in the ground and the atmosphere. Terrestrial detectors are usually smaller than the gravitational wavelength (except for electromagnetic cavity detectors proposed to search for MHz and higher frequencies, not discussed in these proceedings). Detectors placed in Earth orbit and interplanetary space extend the spectral coverage to 10^{-4} Hz or below, and may be larger than the gravitational wavelength. The ratio of the detector size to the gravitational wavelength determines the signature of the detected waves and the optimal signal processing strategies.

In addition to experiments *per se*, astrophysical observations have been applied to the direct measurement of h. The fluctuation in the arrival rate of radio pulses from pulsars has been used to set limits on gravitational waves in the 10^{-7} to 10^{-9} Hz band. The recent observations of anisotropy in the cosmic background radiation have resulted in upper limits in the 10^{-17} to 10^{-18} Hz band.

Resonant-mass detectors

The most mature detector designs are resonant-mass detectors, based on the original concepts of Weber. These consist of a high Q elastic body of mass M (usually in the form of a cylindrical bar of length L) isolated from external perturbations by a vibration isolation system and by operation in vacuum. A mode of the resonator (with non-vanishing quadrupole moment) is driven into oscillation by a passing gravitational wave, due to an effective

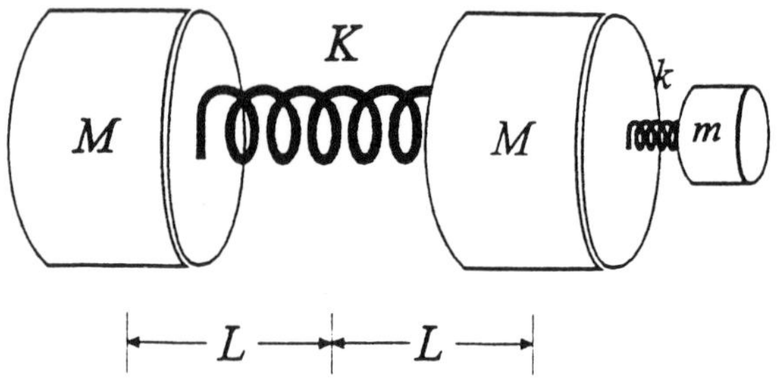

Figure 2: Schematic resonant mass detector. The two large masses and large spring represent the monitored mode of the detector. The small mass and spring represent the resonant transducer.

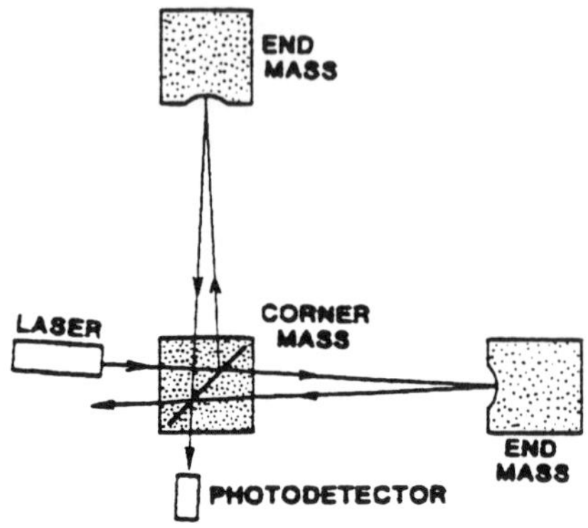

Figure 3: Schematic diagram of a Michelson interferometer.

force given by

$$F_{gw} \sim ML\ddot{h}, \qquad (2)$$

The small motions, of order hL, are sensed by displacement transducers mounted on the resonator. Typically, these transducers are actually mounted so that they reveal the motion of a small mechanical oscillator tuned to the same resonant frequency as the main resonator. The motions modulate an electrical quantity (such as an inductance in a superconducting bridge), allowing a low-noise amplifier to detect a signal.

The cryogenic versions of such detectors have now achieved rms sensitivities better than $h \sim 10^{-18}$ in detection bandwidths of a few Hz near 1 kHz. The resonant-mass detectors have improved by many orders of magnitude in h sensitivity since the original experiments. The best limits on gravitational wave events have come from coincidence searches made with these detectors. By reducing the operating temperature and further improvement in displacement sensors, it is hoped to achieve another factor of 100 improvement in sensitivity and about a factor of 10 increase in detection bandwidth. There is now intensive study of more massive and isotropic detector geometries, which may yield another factor of 10 improvement in overall detection sensitivity.

Laser interferometric detectors

A laser interferometer gravitational wave detector is a Michelson interferometer whose mirrors are hung as pendulums, so that they function as test masses free to move under the influence of a gravitational wave. The simplest configuration, shown in Figure 3, consists of beam splitter at the vertex and two mirrors placed at the ends of the arms of an "L". A gravitational wave pushes the masses back and forth relative to each other, changing the difference $L_x - L_y$ in the length of the detector's two arms by an amount ΔL that is proportional to arm length $L \simeq L_x \simeq L_y$, and to the linear combination of h_+ and $h_\times$ given in Eq. 1. Light from a laser illuminates the beam splitter, which directs half of the light toward each of the two end masses. The mirrors on those masses return the light to the beam splitter, where the two beams recombine. The gravitational-wave-induced change in the arm-length difference ΔL produces a relative phase shift in the recombining beams, thereby changing the proportion of light directed either to the photodetector or else returned to the laser. The photodetector output varies in direct proportion to ΔL and thus proportionally to the combination of $h_+(t)$ and $h_\times(t)$.

In practical variants of such a detector, the phase shift is increased by placing additional suspended mirrors between the beam splitter and the end masses to bounce the light back and forth many times in each arm. The longer optical paths give larger relative phase shifts for the beams that propagate down the arms, and thus a more sensitive interferometric detector. The bandwidth of maximum sensitivity is determined by the inverse of the light storage time in the arms. In most of these instruments, this folding of the optical path is done by installing resonant Fabry-Perot cavities in the interferometer arms, although systems with discrete numbers of reflections are also in use.

Additional features are necessary for proper functioning of an interferometer. To prevent spurious optical phase shifts from fluctuating gas density along the arms of the interferometer, the entire optical path must be evacuated. The test masses must also be very well isolated from external non-gravitational disturbances, such as seismic noise. Finally, numerous feedback control loops must be employed to keep the isolated test masses precisely in place and aligned, and to control the laser wavelength, amplitude, and beam geometry.

More sophisticated interferometric configurations

have been developed. By placing a partially transmitting mirror between the laser and the beam splitter, it is possible to make the entire interferometer into a resonant cavity, in effect matching the laser to the losses in the interferometer so that no light is returned to the laser when the detector is at a dark fringe. The optical power circulating in the interferometer and correspondingly the phase sensitivity is increased. The technique, called power recycling, maintains the detection bandwidth of the interferometer and is equivalent to using a more powerful laser.

By placing still another partially transmitting mirror in the system, this time between the detector and the beam splitter (enclosing the entire interferometer into a cavity that is resonant at the gravitational-wave-induced shift in frequency of the laser light) the spectral response of the interferometer to gravitational waves can be altered. By choice of the transmission and location of this additional mirror it becomes possible to tune the entire interferometer to a narrow gravitational wave frequency in such a manner that the sensitivity is enhanced inversely as the detection bandwith. The technique, called signal recycling, will find application in tailoring the detection sensitivity to source spectra, among other uses.

The interferometric detectors have evolved from table top experiments to medium size prototypes over the past two decades. The most sensitive detector is currently the 40 meter system at Caltech which has achieved an $h_{\rm rms}$ sensitivity of 2.5×10^{-19} near 450 Hz and a bandwidth of several hundred Hz. Major expansions of laser interferometers into long baseline systems, as long as 4 km, are currently under construction by collaborations based in the United States (LIGO), in France and Italy (VIRGO), and in Germany and the U.K. (GEO). Active development programs preparatory to the construction of long baseline interferometers are being pursued in Australia and Japan as well. By extending the baseline and further near-term improvements in the operating parameters of the interferometers, it is expected to achieve sensitivities $h \sim 10^{-21}$ in the initial detectors operated in the large baseline systems by the end of this decade.

To establish confidence in the detection of gravitational waves as well as to extract all the information conveyed by the waves, it is necessary to cross-correlate the outputs of gravitational wave detectors at multiple locations on the Earth. Thus, through scientific necessity, gravitational wave astrophysics is truly an international effort.

Spaced based detectors

Doppler tracking of spacecraft The search for gravitational waves by Doppler tracking of interplanetary spacecraft has been an active part of the NASA Planetary Program for the past two decades. The signature of the waves is a fractional frequency shift in the tracking signals proportional to h. The searches have been primarily executed for gravitational wave bursts with duration shorter than the propagation time between the Earth and the spacecraft, i.e. in the short wavelength limit. Current best upper limits are $h \approx$ few $\times\ 10^{-14}$ near 10^{-3} Hz. The limits on sensitivity are set by oscillator stability and propagation fluctuations in the interplanetary plasma and the Earth's atmosphere. The CASSINI mission to Saturn at the turn of the century offers the next opportunity for improvements in these measurements by about a factor of 10, with the use of more precise oscillators and the application of multiple and higher tracking frequencies to reduce the influence of the interplanetary plasma.

Space based interferometric detectors The operation of a laser-interferometer gravitational wave detector in space, with baselines of order 10^6 km, offers the best opportunity to extend the search for gravitational waves to low frequencies, and to provide sufficient sensitivity for assured detection. The useful observation band lies between 10^{-1} to 10^{-4} Hz. The primary factors limiting the sensitivity are expected to be: 1) at low frequencies, random forces on the test masses, and 2) at high frequencies, the noise in measuring the optical phase due to the limited laser power. Optical and near infrared wavelengths are not strongly affected by the interplanetary plasma. In contrast to the Doppler tracking observations, which have been small additions to existing interplanetary missions motivated by planetary science, a laser interferometer in space is a major new mission dedicated to gravitational wave astrophysics. Such a mission (called LISA, for **L**aser **I**nterferometer **S**pace **A**ntenna) has been considered by NASA and is now under active study by ESA.

2 Present Status of Gravitational Wave Astronomy

2.1 Results of Searches with Resonant-mass Detectors

The experimenters who joined the search for gravitational waves after Weber's announced detection failed to confirm his claim to have seen gravitational wave bursts. Instead, their searches set upper limits on the flux of gravitational wave bursts with substantial power near 1 kHz. The most stringent limit from this first epoch of room temperature bars was set by the group at the Max Planck Institut, then led by H. Billing. [4] This limit appears as the rightmost curve in Figure 4. Because of the long duration of the experiment, it remains to this day the most stringent limit at low event rates.

By 1981, the next generation of resonant mass gravitational wave detectors had begun to operate. The detec-

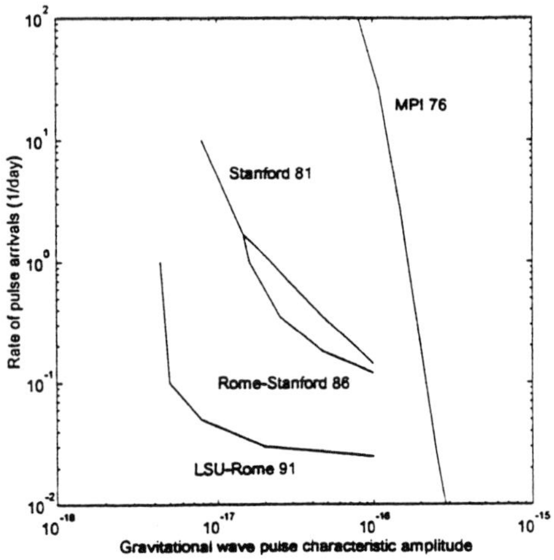

Figure 4: Upper limits on the rate of arrival of gravitational wave bursts.

tors in this class were cooled by liquid helium to a temperature of about 4 K. With the resultant reduction in thermal noise, and other improvements such as low noise superconducting transducers, the rms strain noise of these bars reached (and later went below) $h_{\rm rms} \approx 10^{-18}$. The first example of this class to operate continuously was the bar at Stanford. Based just on the histogram of events in this single bar, Boughn *et al.* [5] were able to set an upper limit on frequent burst events lower by about an order of magnitude in strain than the MPI coincidence experiment. Their limit is also graphed in Figure 4.

In 1986, the groups at Stanford, Louisiana State University, and Rome (whose bar was located at CERN in Geneva) carried out a three-way coincidence experiment among cryogenic bars. [6] This limit is shown as the third curve in Figure 4.

By 1991, the LSU bar had set a new record (which still stands) for rms sensitivity to kHz band bursts, $h_{\rm rms} \approx 6 \times 10^{-19}$. Subsequently, a new coincidence run with the Rome group was carried out. (The Stanford 4 K detector was irretrievably damaged in the Loma Prieta earthquake of 1989.) The upper limit derived from this run represents the most stringent yet set; it is shown as the lowest curve in Figure 4. [7] Observations have continued, and a new result is expected soon.

2.2 Interferometer Status

Present Sensitivity

The state of the art in interferometry is represented by the performance of three large (but still laboratory-scale) prototypes, those at Garching, Glasgow, and Caltech. The MPQ group at Garching has a 30-meter delay-line interferometer that has achieved test mass differential position noise $x(f)$ of approximately 3×10^{-18} m/$\sqrt{\rm Hz}$, or equivalently strain noise $h(f) \approx 1 \times 10^{-20}/\sqrt{\rm Hz}$, in the vicinity of 1 kHz. Glasgow's 10-meter Fabry-Perot interferometer has reached $x(f) \approx 6 \times 10^{-19}$m/$\sqrt{\rm Hz}$, or $h(f) \approx 6 \times 10^{-20}/\sqrt{\rm Hz}$, at frequencies near 700 Hz. The Caltech 40-meter Fabry-Perot prototype interferometer recently reached a sensitivity of 2.5×10^{-19} m/$\sqrt{\rm Hz}$ near 450 Hz. This would translate into an rms sensitivity of $h_{\rm rms} \approx 2 \times 10^{-19}$ to a burst whose power was concentrated between 300 and 1000 Hz. With these latest results, the rms sensitivity of the best prototype interferometers has moved into the same class as that of the best 4 K resonant mass detectors. This is a milestone, but interferometers have yet to be used in searches as sensitive as those of the best resonant-mass detectors.

The major effort in interferometric sensors has been to develop the techniques applicable to multi-km interferometers. By far the best search carried out to date using interferometers was the coincidence run of 100 hours duration by the interferometers at Glasgow and Garching in 1989. It demonstrated that interferometers can function as observing instruments at high duty cycle. [8]

2.3 Status of Space-based Techniques

Spacecraft Doppler tracking and the timing of pulsar signals both search for gravitational waves in the region 10^{-4} to 10^{-1} Hz. This optimum band is determined by the fact that current hydrogen maser reference clocks have their best stability, 1 part in 10^{15}, at integration times around 1000 seconds; as a consequence the Doppler tracking experiments have been performed when the distance to the spacecraft is of the order of a few Astronomical Units. The upper frequency limit comes from Johnson noise in the receivers, which limits sensitivity starting around 1/30 Hz. The other noise terms that set limits to the observations are the fluctuations in the propagation of the microwaves in the Earth's troposphere and the plasma of the ionosphere as well as the plasma of the interplanetary medium.

Doppler Tracking: Results to date

Most of the searches for bursts performed so far have used data from spacecraft tracked at S-Band. A reference sensitivity is that achieved by the Pioneer 11 spacecraft [9] in 1983; the measured noise was about $h \approx 7 \times 10^{-14}$, limited mainly by plasma scintillation. In March 1993, the spacecraft Mars Observer was tracked for about three weeks at X-Band (8.4 GHz). Due to the higher frequency of the radio link, the plasma noise was reduced by almost a factor of 20 with respect to the S-Band experiments. From a preliminary analysis of the data, an upper limit on burst strength of $h \leq 10^{-14}$ has been determined;

Mars Observer is now the most sensitive Doppler tracking experiment to date. [10]

The best upper limits on sinusoids and chirps have been set by the Pioneer 10 experiment in 1993, and by the spacecraft Ulysses during 1992. The Ulysses spacecraft was tracked for three weeks at S-Band, and it also had an X-Band radio signal in the down-link. Because of the two frequencies in the down-link, a subtraction of the dispersive plasma noise in the down-link was possible. The Pioneer 10 spacecraft was tracked for almost 30 days, but only at S-Band. The one-sigma limits set by these experiments were almost equal, $h \approx 3.5 \times 10^{-15}$. [11] An analysis of the data taken in 1993 with Mars Observer will be performed soon, in search for this class of signals. The expected sensitivity at one sigma level is expected to be about 10^{-15}. [12]

The best upper limit on a stochastic background has been deduced recently with the 1993 Pioneer 10 experiment. This spacecraft was tracked at S-band in December 1993 for almost 30 days; the round-trip-light-time was about 16 hours. It was possible to search for a gravitational wave background at low frequencies, slightly below 10^{-5} Hz. Since the local average of the spectrum, around the frequencies where the propagation noise is minimum, has been measured to be 3.5×10^{-12} $\mathrm{Hz}^{-1/2}$, the corresponding one sigma sensitivity has been estimated to be a few parts in 10^{-14}. [13]

Pulsar Timing: Method and results to date

It was pointed out by Detweiler [14] that a pulsar and the Earth could be considered as two freely falling test particles, and that one could monitor their relative motion by recording the arrival times of the radio pulses. This allows a search for gravitational waves in a manner similar to the Doppler tracking technique discussed in the previous section. In this case, however, there is only a "down-link" radio signal, and the pulsar itself is the reference clock of the radio signal. The other major distinction is that by tracking a pulsar one can search for very-low-frequency gravitational waves (between 10^{-9} and 10^{-7} Hz), since the typical distance to a pulsar is four to five orders of magnitude larger than the typical distance to a Doppler-tracked interplanetary spacecraft.

In 1982, a new highly-stable millisecond pulsar was discovered (PSR 1937 + 21); over the past ten years or so radio data from this system were recorded at the Arecibo Observatory. Because this pulsar has much higher intrinsic stability than any one seen previously, interesting cosmological upper bounds to the energy content of a gravitational wave background have been set. The values reported, [15] after seven years of continuous tracking of the pulsar PSR 1937 + 21, set a 2σ upper limit to the energy density per logarithmic frequency of a gravitational wave background at a frequency of 4.4×10^{-9} Hz, or about 4×10^{-7} times the energy density required to close the Universe.

2.4 Indirect Detection of Gravitational Waves

The existence of gravitational waves has been convincingly demonstrated by the reliable, though indirect, observations of emission of gravitational waves by the binary system PSR 1913 + 16, specifically in the predictable change of its orbital parameters. In addition, other astronomical systems have their dynamics determined in crucial ways by gravitational waves. These include cataclysmic variables, low-mass X-ray sources, and progenitors of type I supernovae. [16]

Another area of astrophysics where we may be facing the observational manifestations of gravitational waves is the large-angular-scale anisotropy in the cosmic microwave background radiation (CMBR) recently discovered by the Cosmic Bacground Explorer satellite, COBE. The anisotropy is most likely caused by primordial cosmological perturbations with wavelengths comparable to or longer than the present day Hubble radius. In general, these perturbations are a superposition of density perturbations, rotational perturbations, and gravitational waves. The mechanism that generates the cosmological perturbations is the parametric amplification of their zero-point quantum oscillations by strong variable gravitational field of the very early Universe. [17] The theoretical predictions for the amplitudes and spectra of relic gravitons are uncertain to the same degree to which the knowledge of the rate of expansion of the very early Universe is uncertain. Currently popular inflationary models do not predict gravitational waves in the ground-based interferometer frequency band (10 Hz to 10 kHz) sufficiently strong to be detected by even advanced interferometers. However, these predictions are a matter of fashion, not physics. The observation, or lack of observation, of relic gravitational waves by LIGO or other detectors will provide information, or will put sensitive limits, on the possible rate of expansion of the very early Universe. At the same time, consistent calculations of the quantum mechanical generation of cosmological perturbations and their effects on the CMBR indicate that the contribution of relic gravitational waves to the large angular scale anisotropy may be important. If so, we may be seeing the effects of the relic gravitational waves right now in the form of measured anisotropies of CMBR. It is very important to continue high-precision measurements of the angular anisotropy with an emphasis on the multipole distribution of the angular correlation function and its statistical properties.

3 Gravitational Wave Research Program

3.1 *Plans for Resonant-mass Detectors*

The technology of resonant-mass detectors of gravitational waves has been under development for the longest time of all methods. It also has set the pace in sensitivity, and has been used for deeper, more sustained searches than any other technique. In the next section, we discuss incremental improvement of the present second generation (4 K operating temperature) detectors, which should allow burst sensitivities to improve to about 10^{-19}. Development of a third generation of resonant-mass detectors (the so-called ultracryogenic detectors) is well advanced; these instruments, which should be able to reach $h \sim 10^{-20}$, are the subject of the subsequent section. Finally, we describe the concepts that will lead to a fourth generation of detectors (in the form of large spheres) of even better sensitivity.

Incremental Improvement of Cryogenic Detectors

A massive bar, along with its cryogenic dewar and vibration isolation system, can be thought of as the capital equipment of a resonant-mass detector system. These items, once built well, are not lightly abandoned. On the other hand, mechanical and electrical transducers, and the post-transducer signal processing, are inexpensive enough that an ongoing development program may represent the best way to get the most value out of a detector. Several groups are at work on such transducer development, with the goal of reducing the noise of the 4 K bars from $h \approx 6 \times 10^{-19}$ to about 1×10^{-19}. Two subsystems that will repay improvement efforts are the ultra-low-noise pre-amplifiers for the electrical signals, and the small resonators that serve as mechanical transformers.

All but one of the cryogenic bars in operation now use SQUIDs (Superconducting QUantum Interference Devices) as their electrical pre-amplifiers. The noise level in available devices continues to improve and, as quieter amplifiers are installed on gravitational wave detectors, overall detector sensitivity will improve as well. For example, the LSU bar could improve its rms sensitivity by about a factor of 2.5 simply by replacing its present SQUID with a newer commercially available model (from Quantum Design) that has lower noise, about 3000 $\hbar$; even 1200 $\hbar$ devices are now available. Nearly quantum limited devices have been demonstrated in the laboratory, although these latter devices are not yet commercially available. An effort has been undertaken at the University of Maryland to make such state-of-the-art amplifiers available for use on gravitational wave detectors.

The sole cryogenic bar that does not use a SQUID is the one at the University of Western Australia in Perth; it uses instead a microwave parametric amplifier. Its performance is comparable to the SQUIDs in use today, and also has substantial room for improvement. A factor of 3 reduction in noise in h may be possible with foreseen improvements in the stability and locking of the fundamental oscillator to the resonator cavity that forms the transducer proper.

All resonant detectors in operation use a small mechanical resonator, tuned to the frequency of the bar's fundamental longitudinal mode, as a transformer to better match the mechanical impedance of the system to the impedance at the mechanical input port of the electromechanical transducer. [18] (Another way to see the benefit is to note that the small resonator mass amplifies the motion of the bar by of order the square root of the ratio of the bar mass to the resonator mass.) Further improvement in matching can be obtained by using two cascaded resonators of decreasing mass; this will yield an improvement in the bandwidth, and hence in the sensitivity to gravitational waves.[19] Bandwidths of 70 Hz seem feasible by adding the extra resonator stage. This relationship has been known for some time. What has held up progress until now has been the difficulty of constructing resonators with quality factors high enough so that their thermal noise did not compromise the sensitivity. New work has shown the possibility of constructing resonators with the needed high Q.

Evolution of Third Generation Detectors

Two groups, at the University of Rome and at Stanford, have pioneered the development of the technology that will enable resonant bars to be cooled to temperatures of order 50 mK. This promises a jump in sensitivity soon to $h \sim 10^{-19}$, and eventually to around 10^{-20}.

The strain noise spectrum $h(f)$ of the present resonant detectors has the shape of two deep nearby "valleys" at the frequencies of the two normal modes of the bar-transducer system. If the vibration isolation is sufficient, the minimum value of $h(f)$ is determined only by the detector size and the thermal noise force. For a bar detector made of aluminum, the noise spectrum in the valleys is approximately

$$h(f)|_{\min} = 1.0 \times 10^{-21}/\sqrt{\text{Hz}} \left(\frac{0.6\,\text{m}}{d}\right) \times \sqrt{\left(\frac{T}{4\,\text{K}}\right)\left(\frac{3 \times 10^6}{Q}\right)}, \quad (3)$$

where d is the bar diameter, T is its temperature, and Q is the mechanical quality factor. This expression has been confirmed on second generation detectors, which have T, Q and d similar to the normalization values. It is interesting that the length and frequency dependence end up canceling for a bar, with the size dependence remaining only in the factor d.

The bandwidth, Δf, of the valleys is not always a simple function, but for present detectors it is determined by the inverse ratio of displacement noise to force noise, if we correct for the transformation ratio r. That displacement noise $x(f)$ is typically $1 \times 10^{-17}\ \mathrm{m}/\sqrt{\mathrm{Hz}}$, and r is typically 40, so

$$\Delta f \approx 1\mathrm{Hz}\left(\frac{1\times 10^{-17}\mathrm{m}/\sqrt{\mathrm{Hz}}}{x(f)}\right)\left(\frac{r}{40}\right) \times \sqrt{\left(\frac{T}{4\mathrm{K}}\right)\left(\frac{3\times 10^{6}}{Q}\right)}. \quad (4)$$

The rms burst noise amplitude h_{b} is determined by the ratio of $h(f)|_{\mathrm{min}}$ and the square root of the bandwidth. Because the burst signal is spread over a much broader bandwith than the detector, the result is

$$h_{\mathrm{b}} \approx 7\times 10^{-19}\left(\frac{h(f)|_{\mathrm{min}}}{1\times 10^{-21}/\sqrt{\mathrm{Hz}}}\right)\sqrt{\frac{1\mathrm{Hz}}{\Delta f}}. \quad (5)$$

Making use of a commercially available SQUID with noise of $3000\,\hbar$, a near-term performance of $h_{\mathrm{b}} \approx 1 \times 10^{-19}$ is expected. With continued improvement in transducers, the developers of these devices eventually hope to reach $h_{\mathrm{b}} \sim 10^{-20}$. The University of Rome group has already demonstrated that their ultralow temperature cryostat can cool a large mass antenna to 60 mK. Before reaching a sensitivity of 10^{-20}, a pre-amplifier with noise within a factor of 20 of the quantum limit must be developed and installed.

The other areas of important technology development can be summarized as follows.

- Ultracryogenic reliability: It remains to be seen what engineering adjustments, if any, are required for long-term operation of ultracryogenic systems.
- Mechanical Q: The major question is whether complete detector systems can be made to have quality factors approaching the intrinsic Q of the materials used. Loss in the mechanical joints between pieces is certainly a major contributor. So far, the best results have been the all niobium antenna-transducer combination used in Perth, with a transducer Q of 1×10^7 and an antenna Q that is much higher. Test systems at LSU have recently shown quality factors as high as 3.5×10^7, after careful attention was paid to annealing and joining techniques.
- Electrical dissipation contribution to force noise: Despite the use of superconductors, residual electrical losses remain important in all the types of transducer. The unavoidable fundamental causes of electrical loss are known to be much smaller, so the hunt is on for the "extraneous" causes.
- Losses in resonant transformers: By calculation, there is a significant advantage in adding a second stage of resonant transformation. The full advantage will not be seen if extra losses cannot be avoided.
- Silicon transducers: There is the tantalizing possibility that an all silicon transducer-amplifier combination could be made, largely with techniques borrowed from the semiconductor industry, which might improve performance and reliability tremendously.
- Optical transducers: There is an effort to combine a very high finesse Fabry-Perot optical transducer with the low-temperature, very low power, very high mechanical Q requirements of a cryogenic resonant antenna. [20]

The above estimates indicate why resonant detectors still have tremendous potential to improve. There is proven technology to reduce T by a factor of 80, to 50 mK; there are several demonstrations of quality factors improved by a factor of 12, to 3.5×10^7; the result is that $h(f)_{\mathrm{min}}$ could drop to $3 \times 10^{-23}/\sqrt{\mathrm{Hz}}$. Further, the bandwidth can probably be improved to 50 Hz, if we add one more stage of mechanical matching, to increase r to 300, and successfully transfer to working antennas the reduced noise amplifiers demonstrated in the laboratory. The result is that we can eventually hope for burst noise h_{b} to drop to 3×10^{-21}, down to the region where quantum noise must be considered. This sensitivity would be sufficient to permit a detection of a $10^{-2}M_{\odot}$ burst from a distance of 2 Mpc.

Large Spherical Detectors

There are two more routes to increased sensitivity in resonant detectors: increase the mass of the antenna, and increase the number of antennas. An economical way to do both is to construct a spherical antenna.

A spherical antenna has about 15 times more mass than the typical bar antenna at the same frequency, hence an energy cross-section that is around 15 times larger (about 4 times more sensitive in h), yet the surface area and hence the cost of the cryostat is only 2 or 3 times larger. In addition, a bar has only one quadrupole mode, while a sphere has five, each one aligned with one of the five orthogonal components of the metric tensor; together the five modes detect with equal sensitivity waves of all directions and polarizations. Thus there are five antennas all coexisting inside the same object. The ratio of mode amplitudes can be used to determine the direction and polarization. [21]

The natural advantages of a spherical detector were recognized long ago [22] but never acted upon, probably because several practical problems seemed difficult. What has changed is the conviction that these practical problems have solutions. For example, most of the complications of coupling many resonant transformers to a sphere are eliminated by using a "truncated icosahedral" gravitational antenna configuration, named a TIGA. Another problem, that of fabricating such a large metal structure, may be solved by using explosive bonding to form a large mass from pieces of metal plate.

The extra mass of a sphere reduces the effect of the thermal noise. For aluminum the result is

$$h(f)|_{\min} = 7.0 \times 10^{-24}/\sqrt{\mathrm{Hz}} \left(\frac{f}{1\,\mathrm{kHz}}\right) \times \sqrt{\left(\frac{T}{0.05\,\mathrm{K}}\right)\left(\frac{5.0 \times 10^{7}}{Q}\right)}, \quad (6)$$

where now we have normalized the T and Q to values that should be achieved in the next generation. For a 1 kHz antenna, the back action of optimally coupled sensors will be actually be slightly larger than this until quantum noise becomes important. At the naive quantum limit, using both polarizations (possible with spheres although not with bars) we get

$$h_b \approx \max \begin{cases} 7 \times 10^{-22} \left(\frac{h(f)|_{\min}}{7\times 10^{-24}/\sqrt{\mathrm{Hz}}}\right) \sqrt{\frac{50\,\mathrm{Hz}}{\Delta f}} \\ 7 \times 10^{-22} \end{cases} . \quad (7)$$

This is enough sensitivity to detect a $10^{-2} M_{\odot}$ burst from a distance of 12 Megaparsecs. This is roughly the distance to the Virgo cluster where there is a large population of supernovae.

By constructing detectors of several different sizes, one could have a "xylophone" that would cover a range of frequencies, and could be used to provide some reconstruction of the waveform as well as to increase the detectable distance even further.

The result of this activity is that new collaborations are forming to pursue spherical detectors. In the US, all of the groups currently working on bar detectors (LSU, Stanford, Maryland, Rochester) signed a letter of intent to submit a proposal to the National Science Foundation. Very preliminary cost estimates are that the capital costs for a single 2.6 m 25 ton aluminum TIGA would be in the vicinity of $5 million. In Leiden, the birthplace of low temperature physics, the GRAIL project has announced its interest in collaborating with the Rome group to build two large spherical detectors; they are studying a design with a 3 m copper alloy sphere, weighing 100 tons and cooled to 10 mK. An international study group was formed after the Marcel Grossmann meeting in July 1994, comprising all of the above groups, as well as ones from Brazil and Australia. Many detailed supporting calculations and measurements will be appearing shortly.

3.2 Plans for Interferometric Detectors

The recognition that interferometric detectors could, in principle, be built with sensitivities that matched the astrophysical requirements has led to the development of plans to actually build such detectors. In this section we describe the various plans currently in progress.

Long Interferometers

There is one generic feature that all of these planned high sensitivity interferometric detectors will share. They will need to have test mass separations of the order of one to several kilometers. The most fundamental reason for this requirement is to make the interferometer as immune as possible to the various sources of displacement noise, non-gravitational-wave disturbances to the positions of the nominally free test masses.

Displacement noise is a strong reason for building interferometers with the largest possible distance between the test masses. If shot noise were the only important noise, the situation would be different: the shot noise limit to gravitational wave sensitivity is the same no matter if the light makes one round trip between test masses separated by 1000 km, or if alternatively the light is arranged to bounce back and forth 10^5 times between masses separated by 10 m. But when the masses themselves are moving at a given displacement amplitude $x(f)$, then the equivalent strain noise is

$$h(f) = \frac{2}{L} x(f) \quad (8)$$

where L is the actual distance between the masses. In effect, the displacement noise contributes to a phase shift of the light (thus mimicking a strain signal) by an amount proportional to the number of encounters between the light beam and a test mass.

LIGO The Laser Interferometer Gravitational-Wave Observatory Project, or LIGO, is the effort funded by the U.S. National Science Foundation. It is a joint project of Caltech and MIT. [23] Other U.S. scientists, at Stanford, JILA, Northwestern, and Syracuse, are also participating in aspects of the development of LIGO.

The LIGO Project was approved in 1992. It has begun the construction of two facilities to hold interferometers of 4 km arm length. One site is at Hanford, Washington, while the other is in Livingston, Louisiana. The site-to-site separation of 3000 km ensures that local disturbances of the interferometers will be uncorrelated, and gives a substantial baseline for location of sources on the sky by measurement of signal arrival time differences. Interferometers at the two sites will be aligned with each other, to give the maximum probability of coincident signals.

Several distinctive features have been incorporated in the plans for LIGO's interferometer facilities. One is that the Hanford site will be equipped for the placement of test masses at 2 km separation as well as the full 4 km. This will allow a test of the tidal nature of gravitational waves by the comparison of coincident signals seen in two colocated interferometers of different lengths; this can also supply a veto against non-gravitational-wave disturbances. Another key feature of the design is that the facilities will allow the eventual installation of several interferometers side by side in the same vacuum system, in a way that would allow completely independent operation. This will allow LIGO to serve as a user facility with several independent experiments operating simultaneously.

Construction began at the Washington site in 1994, with work expected to begin in Louisiana in 1995. Interferometer installation will commence in Washington in 1998, in Louisiana the following year. Plans call for the first coincident operation of interferometers at the two sites in 2000.

VIRGO VIRGO is a collaborative effort between the CNRS in France and the INFN in Italy. It involves scientists from Napoli, Frascati, Perugia, Pisa, Annecy, Lyon, Orsay, and Paris. The goal of the project is the construction of an interferometer of 3 km arm length, aimed at detecting gravitational waves in the frequency range from 10 Hz to a few kHz. [24]

The VIRGO project was approved in 1993; a collaborative agreement was signed by the presidents of the INFN and the CNRS in 1994. Land acquisition has begun at the site in Cascina, Italy, 10 km from Pisa. The central building will be completed by the end of 1997; it will initially house an interferometer of 6 meter arm length, on which tests will be performed while the remainder of the vacuum system is completed. The schedule calls for completion of construction by the end of 1999, with operation of the full scale interferometer the following year.

GEO 600 GEO 600 is a joint project of groups of scientists from Germany and the U.K. The main groups involved are the Max-Planck-Institut für Quantenoptik, both at Hannover and at Garching, and the Universities of Glasgow and of Cardiff, with important contributions coming also from Rutherford Appleton Laboratories, Laser-Zentrum Hannover, the MPG Bauabteilung, and several other institutions. [25]

GEO 600 is conceived as a low-cost, intermediate length detector, to be built on a correspondingly short time scale. Several advanced technologies will be implemented early on to achieve, despite the moderate size, a sensitivity approaching that of early phases of the LIGO and VIRGO interferometers. The arm length of the interferometer will be 600 meters. It will be built on a site owned by the University of Hannover in the State of Niedersachsen, Germany.

Pending final approval by the funding agencies, construction could begin in early 1995. Plans call for the first optics test early in 1998, with full-time operation by the end of that year.

Japanese effort The various gravitational wave research groups in Japan have banded together in the TAMA project to jointly build a 300 meter interferometric detector. The plan has a dual purpose: firstly, to establish the techniques necessary for future kilometer-class detectors, and secondly to operate an instrument with good sensitivity, in the hopes that strong enough signals exist for it to register.

A site for the 300 meter interferometer has been selected, on the campus of the National Astronomy Observatory at Mitaka, roughly 20 km from downtown Tokyo. Operations are scheduled to begin in 1998, with the target sensitivity to be reached in 1999.

AIGO A consortium of Australian institutions, consisting of the University of Western Australia, the Australian National University, the University of Adelaide, Monash University, and CSIRO Lindfield, has been formed to promote the construction of AIGO, the Australian International Gravitational-Wave Observatory. A 400 meter interferometer has been planned for construction as soon as full funding can be obtained. A site was identified at Wallingup Plain, north of Perth.

Groups in India, China, and Argentina have joined as international partners, and more are invited. While funding is sought for the large scale facility, R&D activities are proceeding on a variety of features for advanced detectors.

Initial Interferometers

In the next two sections, we give a description of the initial interferometer design for LIGO, and an analysis of its noise budget. Succeeding sections will describe the planned interferometers of the other major projects.

LIGO's initial interferometer The LIGO Project has adopted a design philosophy that its initial interferometer should be based on the most reliable existing technology; the emphasis is on promptly integrating a completely characterized and understood instrument.

The essential features of the optical design of the initial LIGO interferometer are that it is a Michelson interferometer using Fabry-Perot cavities in each arm, equipped with power recycling. We now describe each of these features in turn, followed by a description of some of the major subsystems.

Michelson interferometer The use of the Michelson interferometer is at the heart of all interferometric detectors of gravitational waves. Its key idea is the interference between two light beams that have taken paths that are perpendicular to one another, thus coupling in an optimal way to the quadrupolar symmetry of a gravitational wave. The use of an equal-arm Michelson interferometer reduces the sensitivity to light wavelength fluctuations.

Fabry-Perot cavities Fabry-Perot cavities are chosen as the most practical means of folding an optical path of the optimum length into a vacuum system of only a few kilometers in length. Use of this resonant technique adds an extra control requirement to the interferometer, that of maintaining the resonance condition in the arms. But it yields a crucial advantage: the need for only one mirror, of minimum size, at each end of the arm. In the cavity design of the initial LIGO interferometer, light will in effect be stored for 11 msec. This creates arms with an optimum phase response to all gravitational waves whose frequencies are large compared with 90 Hz.

Power Recycling Very little of the light injected in the interferometer is actually lost in one complete trip through the Fabry-Perot Michelson interferometer. Rather than waste it, the idea of power recycling is to reflect this light back into the interferometer in superposition with fresh light from the laser. [26] A practical implementation is to use a partly transmitting mirror to return the "used" light while allowing the injection of "fresh" light. In the initial LIGO interferometer, the power enhancement effect will be about a factor of 30.

Laser An Argon ion laser will provide 1 to 2 W of effective input power to the interferometer, at a wavelength of about $0.5\,\mu$m. Several prestabilization subsystems are required to condition the input light. The laser's wavelength will be actively stabilized to the length of a rigid Fabry-Perot cavity that serves as a length reference. Transverse and angular wandering of the beam are filtered by passing the light through another long resonant cavity, called a mode cleaner.

Test masses and suspensions LIGO's test masses will be cylinders of fused silica 25 cm in diameter and 10 cm thick. The mirror surfaces will be polished flat (for cavity input mirrors) or with very large radius of curvature (6 km for far mirrors), to a precision of $\lambda/600$ in their central 8 cm diameter. Multi-layer dielectric coatings will give reflection losses of less than 5 ppm.

Each test mass will be held in a sling consisting of a single loop of fine wire to make a pendulum. Positioning and alignment will be performed by small magnetic actuators; these consist of small permanent magnets fixed to the mirrors, each near a small coil through which a control current can be sent.

Vibration isolation The initial LIGO interferometer will use a vibration isolation "stack" consisting of four layers of stainless steel masses separated by fluorocarbon rubber springs. Such a design offers substantial isolation at high frequencies without troublesome high Q resonances. Furthermore, it provides isolation of comparable amounts in all 6 rigid body degrees of freedom, valuable since any real interferometer has non-negligible sensitivity to motions of the mirrors in directions other than along the optic axis. With such a design, sufficient isolation is achieved for frequencies above 100 Hz.

LIGO Noise Budget It is convenient to characterize the strength of various noise terms using the language of power spectra. All of the most important components of the noise budget have non-trivial frequency dependences.

Shot noise When a gravitational wave is present, the light beams travelling through the two interferometer arms acquire different phases on their paths. Once the beams are recombined, the phase difference determines the optical power at the interferometer's output port. A photodetector at the output port thus registers the arrival of a gravitational wave by measuring the power in a beam of light. That is, an interferometric gravitational wave detector is a transducer that converts perturbations to the space-time metric into optical power variations. The fact that light occurs in the quanta called photons leads to a fundamental limit to the precision of gravitational wave measurements, called shot noise.

Shot noise has an amplitude spectral density that is frequency independent, or white, with a magnitude of

$$h_{\mathrm{shot}}(f) = \frac{\pi}{2FL}\sqrt{\frac{\hbar c\lambda}{2\pi P_{\mathrm{out,max}}}}. \tag{9}$$

This formula holds for signal frequencies f that are small compared with the inverse of the storage time $\tau_{\mathrm{s}} \equiv (F/\pi)(L/c)$. But the storage time in LIGO's arm cavities will be made long, about 11 msec, in order to maximize signals to as low a frequency as possible. Then, the effective change δL in interferometer arm length is no longer Lh, but is instead

$$\delta L = Lh\frac{1}{\sqrt{1+(4\pi f\tau_{\mathrm{s}})^2}}. \tag{10}$$

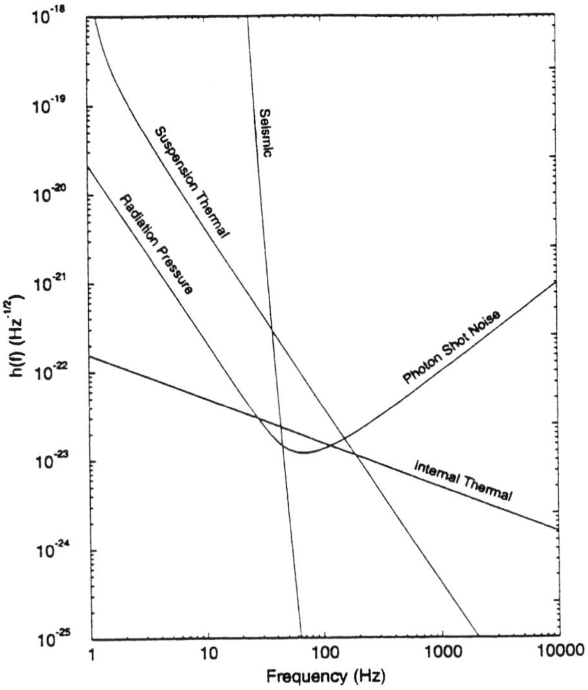

Figure 5: Noise budget of the initial LIGO interferometer.

The level of shot noise in the initial LIGO interferometer is shown in Figure 5.

Thermal Noise One of the most important displacement noise sources is also the one most strongly linked to fundamental physics. This is thermal noise, a generalization of the phenomenon of Brownian motion. The thermal noise displacement power spectrum for a mode of resonant frequency f_0, mass m, and quality factor Q is given by [27]

$$x^2(f) = \frac{k_B T}{2\pi^3 m Q f_0^2 f} \frac{1}{[(1-f^2/f_0^2)^2 + 1/Q^2]}. \quad (11)$$

In the initial LIGO interferometer, one of the most important degrees of freedom will be the simple pendulum mode of the test mass on its wire suspension. With a mass of 10 kg, a resonant frequency of 1 Hz, and a Q of order 10^6, it is expected that the resulting strain noise above the resonance will be about

$$h_{\rm th,pend}(f) = 1.3 \times 10^{-18}/\sqrt{\rm Hz} \left(\frac{1\ \rm Hz}{f}\right)^{5/2}. \quad (12)$$

The level of this noise expected in the initial LIGO interferometer is shown in Figure 5, in the curve labelled "Suspension Thermal".

A second class of modes that will also make an important contribution to the noise budget is the set of internal modes of vibration of the test masses. Even though the lowest of these modes will have a resonant frequency of about 15 kHz, the gently-sloping $1/f$ spectrum below resonance will nevertheless be one of the strongest noise sources near 100 Hz. Quite a few modes actually contribute to the sum. [28] Assuming the modes are characterized by quality factors of order 5×10^6, then the strain noise will be given by

$$h_{\rm th,int}(f) = 1.5 \times 10^{-22}/\sqrt{\rm Hz} \left(\frac{1\ \rm Hz}{f}\right)^{1/2}. \quad (13)$$

This is graphed in Figure 5, in the curve labelled "Internal Thermal".

Seismic noise Seismic noise is the strongest displacement noise with which gravitational wave interferometers have to contend. At the sites chosen for the LIGO interferometers, the ambient spectrum is approximately

$$x(f) = \begin{cases} 10^{-9}\rm m/\sqrt{Hz}, & \text{from 1 to 10 Hz} \\ 10^{-9}\rm m/\sqrt{Hz}(10\ Hz/f)^2, & \text{for } f > 10 \text{ Hz}. \end{cases} \quad (14)$$

The LIGO vibration stack will have a transmissibility that can be approximated, in the region above the resonances, as

$$T(f) = 6 \times 10^{-6} \times (1\,{\rm Hz}/f)^{6.7} \quad (15)$$

This isolation will be supplemented by the 1 Hz low-pass response of the test mass pendulum suspensions. The expected seismic contribution to the initial LIGO interferometer's strain noise spectrum is given by the curve labelled "Seismic" in Figure 5.

VIRGO Interferometer VIRGO will construct a Michelson interferometer with Fabry-Perot cavities for arms, equipped with power recycling. The 3 km arms will have a finesse of 50, so it too is optimized for signals in the vicinity of 100 Hz.

Although these features are similar to those of the initial LIGO interferometer, many important aspects of the design are different. VIRGO will use an infrared Nd:YAG laser, with a wavelength of 1.06 μm. Such lasers can give larger output power than Ar lasers, and with much better "wall-plug" efficiency. The VIRGO design calls for the coherent addition of several lasers to give a total power of 20 W. Power recycling by a factor of 50 should give an effective power of about 1 kW incident on the interferometer beam splitter.

The most dramatically unique feature of the VIRGO interferometer is its vibration isolation system, the seven stage Super Attenuator. [29] The Super Attenuator design is based on a cascade of seven pendulum stages, each approximately 1 m in length. Vertical compliance is added to each stage by means of pre-stressed metal cantilever blades, whose spring constant is softened with magnetic anti-springs. Thus each stage has resonant frequencies of 1 Hz or below in all six rigid body degrees of freedom. This yields an isolation system with transmissibility below 10^{-9} at 10 Hz. Seismic noise has thus been essentially eliminated from the VIRGO noise budget for all frequencies above 4 Hz.

Thermal noise will contribute at roughly the same levels as in LIGO. Here, though, it is the dominant contribution to the instrument noise all the way from 200 Hz down to only a few Hz. Thus, VIRGO will benefit much more from research, discussed below, that might show the way to reduce the contribution of thermal noise.

The GEO 600 Initial Interferometer The GEO 600 facility is being designed with flexibility in mind, to allow a wide variety of interferometer schemes and auxiliary components to be developed and tested. We describe here the key features of the early plans.

The laser to be used will be, like VIRGO's, the all-solid-state Nd:YAG operated at $\lambda = 1.06\,\mu$m. Early plans call for use of a single laser with an output power of 5 W. Power recycling will be implemented more aggressively; GEO will attempt to achieve an effective power enhancement factor of approximately 2000.

One distinctive feature of the optical design of GEO 600 is that the optical paths in the interferometer arms will have an especially simple form – a classic Michelson delay line involving only four one-way passes through the arm, called the "DL4" configuration. This is implemented by a small mirror at the far end of the arm, and a second small mirror adjacent to the beam splitter to return the beam directly back after its first round trip. Combined with the shorter overall length of only 600 meters, this gives the arms less than the optimal storage time for all frequencies below 50 kHz. This choice has been made because of its simplicity, as the GEO team hope to get "on the air" quickly, as early as the latter part of 1998.

Several strategies have been adopted to make up for the short optical path length. First is the plan to achieve the highest possible optical power using power recycling. Secondly, the GEO team plan to implement another advanced optical technique, signal recycling, at the earliest possible moment.

Note that both power recycling and signal recycling are techniques that reduce the level of shot noise in the gravitational wave measurement. Displacement noise sources are not affected. For this latter class of noise sources, GEO 600 will have a disadvantage due to its smaller overall length, a factor of 5 compared with VIRGO and 6.7 with respect to LIGO.

Japanese plans The Japanese 300 meter interferometer will employ Fabry-Perot cavities in a Michelson interferometer, equipped with power recycling. Cavity finesse of 520 will optimize the shot noise for signals near 100 Hz. The interferometer will use a laser-diode-pumped Nd:YAG laser operating at $\lambda = 1.06\mu$m, with an effective power of 30 Watts.

Use of higher laser power will make the sensitivity above 1 kHz nearly comparable to that of the initial LIGO interferometer. Thermal noise dominates the noise budget between 30 and 300 Hz; there, the instrument pays the full penalty of the shorter arm length. An ambitious vibration isolation system is expected to help compensate for the short arms in the band below 30 Hz.

AIGO The AIGO 400 meter interferometer will use the DL4 design developed for GEO 600. Illumination will come from a 1.06 μm Nd:YAG laser. Test masses of single crystal sapphire, diffusion-bonded to Nb membrane flexures, are designed to give low thermal noise. An aggressive vibration isolation system, employing metal cantilever springs and inverted-pendulum pre-isolators, will give especially good low frequency performance.

Upgrades to Initial Interferometers

The period of initial operation of these instruments will be followed, whether or not gravitational waves are discovered right away, by a process of refinement and improvement. Each project will have its own strategy for implementing changes. The LIGO project has made provision for multiple interferometers to be installed and operated at both sites with minimal mutual disturbance. Initially, this capability will consist of the ports for the half-length interferometer at the Washington site. Eventually, this could be expanded to a capability to operate three independent interferometers at the Louisiana site and six (three of each length) at the Washington site. At that stage, there would be little conflict between taking data with the best well-characterized instrument and efforts to develop better instruments, or ones with different optimization strategies.

Phase sensitivity improvements Since the amplitude of the shot noise is proportional to $1/\sqrt{P}$, one straightforward strategy for improving the basic interferometric phase sensitivity is to increase the laser power P. Replacing the prototypes' Argon ion lasers with Nd:YAG lasers would deliver power measured in the tens of Watts, and excellent "wall-plug" efficiency as well. The present state of the art is 20 W of single frequency TEM_{00} light at 1.06 μm, with 30-35 W believed to be within easy reach. There do not appear to be any insuperable obstacles in the way of developing YAG lasers delivering several hundred Watts. [30]

When such powerful lasers are installed in interferometers in which power recycling by factors of 30 to 50 is employed, the amount of power incident on the beam splitter and arm input mirrors will be several to tens of kW. At these power levels, it will be important that these crucial components absorb very little energy. Otherwise, optical distortions will destroy the single TEM_{00} modal character of the beam, destroying the matching into the Fabry-Perot cavities of the interferometer arms, and also ruining the quality of the interference at the beam splitter. [31] To avoid these problems, it will be necessary to construct the crucial components out of fused silica substrates that absorb less than 1 part per million per cm of thickness, and to apply coatings which themselves absorb less than 1 ppm per reflection. It might also be possible to construct compensators for small amounts of thermal distortions, by specially figuring optical surfaces or by servo control of the temperature distribution in crucial components.

In addition to increasing the raw laser power incident on the beam splitter, there are other ways to increase the phase sensitivity of an interferometer, at least over some band of signal frequencies. One of these techniques, signal recycling, was discussed above. Signal recycling might be used to provide a factor of three improvement in SNR for all frequencies below 100 Hz, for example, equivalent to nearly an order of magnitude in laser power. [32] Other combinations of parameters yield narrow-band enhanced sensitivity centered on a non-zero signal frequency.

An alternative arrangement of some of the same components used for signal recycling yields another way to increase phase sensitivity, called resonant sideband extraction. [33] This can be thought of as a scheme for broadband sensitivity enhancement, although it will have of necessity some frequency dependence. The idea is to use Fabry-Perot cavities in the interferometer arms that have storage times well in excess of the periods of the gravitational waves of interest. Ordinarily, this would give no better sensitivity than an interferometer whose arms have storage times comparable to the signal periods. But in resonant signal extraction, the mirror between the beam splitter and the photodetector combines with the input mirror of the arm cavity to form a coupled mirror whose reflectivity at the laser wavelength is high (to give the long storage time), but has a low reflectivity at the gravitational-wave-induced sideband wavelengths. Heuristically, one can think of the effect as follows: high reflectivity at the carrier causes large power buildup in the arms, with proportionally large sideband power from a gravitational wave. The lower reflectivity at the sideband wavelength allows the signal to "escape" and reach the photodetector.

Improvements to seismic noise If it functions during interferometer operations as well as it has during diagnostic measurements, VIRGO's Super Attenuator may be as good a vibration isolator as is needed for measurement of any gravitational wave above 4 Hz. If so, one possible improvement strategy of other projects may simply be to adopt or adapt the Super Attenuator for their own interferometers.

One awkward feature of the Super Attenuator is its substantial height (10 meters), although a version not much more than half that height might work almost as well. Another isolation system embodying similar principles and goals, but without such daunting size, is the four-stage vibration isolator under development by the University of Western Australia group.

Several clever new ideas are under development to produce so-called pre-isolators that have very low resonant frequencies for horizontal motion. The X-pendulum, developed by members of the Japanese group, consists of a rigid pendulum attached to a platform that is suspended by two pairs of crossed wires from another platform above. [34] The wires constrain the motion of the rigid parts below in such a way that the mass moves in a very shallow potential well. The UWA group has constructed an isolator consisting of a platform suspended at one end by a conventional pendulum and at the opposite end by an inverted pendulum. [35] This arrangement also creates a shallow potential well for the platform's motion. Periods in excess of 20 s have been achieved in both devices.

The isolation schemes described above all rely on the natural inertia of large masses supported with low compliance. In distinction to this passive form of isolation, an alternative approach, called active isolation, uses control systems to cause an object one wants to isolate to follow an inertial reference mass suspended from it. Several active isolation systems are under development, seeking one or more of the following attributes: isotropic isolation at very low frequencies, ease of control of resonance quality factors without short-circuiting the isolated parts to the noisy ground, or supplemental isolation for a previously engineered system that can give a reduction in noise input in the low frequency region where mechanical modes give resonant amplification.

The basic structure of an active isolation system consists of an inertial proof mass which is resonantly suspended from a movable platform. A position sensor measures the displacement between the platform and the proof mass. The difference between the instantaneous separation and the nominal DC spacing gives an error signal that is amplified and filtered, then applied to a mechanical actuator that can move the platform. The feedback nulls the error signal, causing the platform to be "locked" to inertial space.

One early use of active isolation may come in the

LIGO Project, which is investigating the use of commercial isolation units from Barry Controls. Sensor noise is low enough to make this a useful pre-isolator outside the vacuum system, for frequencies between a few Hz and 100 Hz. Application of active techniques for the purpose of isolation in a band centered on 1 Hz is being investigated by the VIRGO team. Perhaps the most intensive study of a system for this purpose is being performed at JILA; their work is discussed below in the Advanced Detector section.

Improvements to thermal noise In a well-designed pendulum, the factor determining the Q of the pendulum mode is internal friction in the flexing members (e.g. wires) that support the test mass. First generation interferometers will use wires made from steel or other metals, materials with quality factors of a few thousands; these allow one to achieve $Q = 10^6$ or greater in the pendulum because the flexing wire contributes only a small restoring force compared with the gravitational force on the pendulum mass. At least two strategies suggest themselves for reducing this contribution to thermal noise: reduce the elastic restoring force in the pendulum, and make the flexing element from a material with higher Q. The elastic force can be reduced by replacing round wires with thin ribbons (or short hinges) of equivalent cross-sectional area. Higher Q materials under investigation include fused silica, niobium, and sapphire.

The Q of the test mass modes depends on the material of which it is made, but unlike the pendulum mode there are no advantageous mechanical factors to improve it. Fortunately, fused silica can have Q of order 10^6 or even substantially higher, while simultaneously exhibiting excellent optical properties; this makes it the universal choice for test masses in the first generation interferometers. Hopes for improving the test mass thermal noise follow two paths: finding the best possible kind of fused silica, or finding a completely different material with lower losses. The prospects of the second path would be improved greatly if it would prove practical to build interferometers that don't require transmission through any optical elements, including beam splitters. (For further discussion of this idea, see below.)

Science with first generation interferometers

The significance of the noise spectrum we have been discussing above can only be appreciated when we compare it with the strengths of likely signals. In the following section we will examine the appropriate way to compare signal strengths and noise amplitudes, then discuss what are the likely scientific outcomes of observations with the first generation of interferometers. We will consider both the operation of the interferometer network itself and its use in conjunction with other tools of astronomical observation.

Data Analysis Strategies When we talk about the detection of a signal in the presence of noise, we mean that the detector's output is more like what we would expect if a signal is present added to the noise than what we'd expect if only detector noise is present. In other words, the judgement that a signal has been detected is based on an estimate of the ratio of two probabilities: that the detector output could have been produced by the combination of noise plus a signal, compared with the probability that the output could be the result of detector noise alone. This means that to have a quantitative understanding of signal extraction we need to have a good model of the detector noise, as well as of the detector's response to a signal.

Bursts A burst can be defined as a signal that has a definite beginning and end. Limitation in temporal duration carries with it an inevitable breadth of frequency content, by virtue of the classical uncertainty relation $\Delta f \Delta t > 1$.

A number of astrophysical processes are expected to result in gravitational waves that would qualify as bursts under the above definition. They include: stellar core collapse (the process that leads to a Type II supernova), formation of a black hole, response of a black hole to a perturbation, the coalescence of two neutron stars or black holes, and the inspiral of a binary system consisting of neutron stars or black holes. The last of these, an intermediate form between a burst and a periodic signal, is called a chirp, since it consists of a nearly sinusoidal oscillation that sweeps up in frequency with time.

These sources vary in the degree to which the signal is calculable in advance from what we know. The classic technique for optimum detection of a signal is to construct a matched filter. The output of the matched filter contains only that component of the noise that truly "looks like" the signal; by rejecting other noise components, it allows the most sensitive discrimination between signal and noise. Even when the exact signal waveform is not known, analogous techniques can emphasize "signal rich" frequencies at the expense of those that are "signal poor" to create an approximate optimum filter.

Periodic signals Steady periodic waves represent a rather different class of signals than bursts. Perhaps as simple as a pure sinusoid, more commonly they will also have harmonics and/or sidebands in a pattern characteristic of the source. The archetype for this class of sources is the precessing neutron star. A rotating non-axisymmetric neutron star would also emit periodic gravitational waves. A variant of this class of signals could be termed quasi-periodic – signals that are approximately periodic, but with a phase or frequency that wanders stochastically.

The Fourier transform acts as a set of matched filters for pure sinusoids. This makes clear why the amplitude signal-to-noise ratio of a measurement of a periodic signal grows as the square root of the observation time (as long as the observation time is long compared with the period of the signal.) The frequency resolution of a Fourier transform is inversely proportional to the observation time; and the amplitude of the noise in a single resolution bin is proportional to the square root of the bin width.

One feature is likely to complicate the search for periodic signals. Most predictions based on actual or suspected sources give signal strengths that are quite small compared with the noise obtainable from observations over short intervals. This leads to consideration of long integration times, perhaps of order a year. But the Earth's motion is non-negligible on these time scales – it causes a Doppler shift that could be resolved in one day's integration.

For observations of a source in a known location, the Doppler effect adds technical complexity but no real hurdle in principle. It is quite straightforward to correct for this effect for a given position on the sky. But the correction is different for different sky positions, and naive algorithms quickly become impractical for any envisioned all-sky search of substantial duration. Discovery of an efficient algorithm would be a boon to this field of research.

Stochastic backgrounds Yet a third class of signal is a stochastic background. A steady hiss of gravitational noise akin to electrical "static", it is broadband like a burst, yet always present like a periodic wave. Some mechanisms are known that would produce such a background directly; processes in the early Universe could do so, including the production of an exact gravitational analog to the cosmic microwave background radiation. A stochastic background can also be produced as the confusion limit of a class of burst or periodic sources – so many bursts that there is always one going off, or so many periodic sources that the spectrum is filled with them.

The feature that would let us detect a stochastic background of gravitational waves is the fact that such a background will give identical "noise" in any detector. Thus, the detection strategy is to cross-correlate the outputs of different detectors. A value of the cross-correlation large enough to be distinct from that expected based on noise alone indicates some external influence to which the interferometers respond in common. The burden is then to show that there is no other plausible non-gravitational cause for the correlation. The most obvious strategy here is to cross-correlate signals from instruments separated by hundreds to thousands of kilometers.

Networks and coincidences Simultaneous observation with a network of widely-separated gravitational wave detectors is a crucial feature of any attempt to detect astrophysical signals. We explained in the previous section why this is so for stochastic signals. But it is also essential for burst signals.

One can make simple estimates of the probability that Gaussian noise can give large outputs; such estimates are encouragingly small. But most real instruments have a component of non-Gaussian noise as well that will dominate the probability distribution of rare events. Such "glitches" might not be distinguishable from gravitational wave signals based on evidence from a single detector alone. But the use of multiple detectors gives a crucial extra discriminant – a true gravitational wave will interact with all gravitational wave detectors, while most glitches are internal events limited to just one detector.

The detectors in a network need to be far apart to give assurance that some local effect is not responsible for simultaneous excitation. Global separation does add an extra complication, since then the interferometers are not pointed in the same direction. Although the beam patterns of gravitational wave interferometers are remarkably close to isotropic, detectors in different places will respond differently at the factor of two level; the beam patterns also have a few narrow nulls. So it will be necessary to have a redundant network to make sure that coincidences can be required for all real events, regardless of their location on the sky.

Local coincidences, between two interferometers at the same location, also have a valuable role to play. In that case, one requires that for an event to be real, it must give exactly the same signal at precisely the same time. [36] This gives the strongest possible discriminant against internal glitches, although it might give a spurious coincidence due to a common external but local non-gravitational influence. Local coincidences are thus most valuable as a rigorous veto – unless the local coincidence occurs, the event cannot be caused by a gravitational wave.

A widely separated network also gives a second crucial advantage to the search for gravitational waves. Signals will propagate across the network at the speed of light. This means that careful measurement of the difference in times that signals appear in different detectors will indicate the direction from which the waves came. This is essential if gravitational wave detection is ever to become a branch of astronomy.

What can be seen with first generation interferometers? There are, unfortunately, no gravitational wave sources that are guaranteed to give frequent signals strong enough to be seen in the first generation of large interferometers. The sources we can confidently predict will give easily detectable signals in the second generation of interferometers sometimes referred to as Advanced Detectors, in which sensitivity near 100 Hz has improved by about a factor of 10. Our best-understood source is the inspiral and coalescence of a neutron star binary. Present estimates for their abundance imply that the several strongest events per year will register at only unit signal-to-noise ratio in the initial LIGO interferometer, for example.

These estimates might be overly pessimistic; there are indications that we may have underestimated the abundance of neutron stars by our reliance on finding them as pulsars. Also, even stronger signals might come from binary black hole inspirals (or from mixed binaries consisting of one black hole and one neutron star); bounds on the abundance of such systems are very weak.

Another class of sources whose rates we know well may yet surprise us with unexpected gravitational wave strength. Stellar core collapse in a Type II supernova has been a candidate source for years, but the amplitude of the gravitational wave will depend strongly on aspects of the collapse that are poorly understood even today. Specifically, the departure of the collapsing core from spherical symmetry is crucial in the prediction of the gravitational wave strength – the wave amplitude h is proportional to the second time derivative of the mass quadrupole moment of the source. But the initial conditions of the stellar core, and the physical mechanisms that may come into play during collapse, are so complicated that most attempts at calculating them have assumed by fiat that spherical symmetry is maintained. The possible influences of core angular momentum, magnetic fields, or convection have only been surveyed superficially. Recent observation of remarkably large peculiar velocities of pulsars ($v_{\rm rms} \approx 450\,{\rm km/s}$) [37] may be a clue that core collapse is often a dramatically messy and unsymmetric event. If so, core collapses become much stronger sources.

Hopes for success of the first generation of instruments come from the possibility of registering rare events of the type already understood, but also from the good chance that unpredicted classes of objects may be the strongest sources of gravitational waves. Our ignorance of the Universe is still enormous. Since gravitational waves are a completely different kind of information carrier from those that have been used to study the sky so far, it seems quite plausible that the strongest sources may be objects not yet seen (or at least noticed) by other means.

Correlations with other astronomical observations With LIGO and with the next generation of resonant-mass gravitational wave detectors, there will be abundant opportunities to search for sources of gravitational waves that are correlated with either electromagnetic or neutrino signals expected from the same sources. These potential sources include supernovae, gamma-ray bursts, and low-mass X-ray binaries.

Supernovae One goal of the LIGO Project is to detect gravitational waves from a supernova outburst in the nearest large cluster of galaxies, the Virgo Cluster, and compare the waves' arrival time with the beginning of the optical outburst. Since the distance to Virgo is about 30 million light years, even with an optical time resolution of order one day, one could infer that the light and gravitational waves propagated with the same speed to within a fractional error of (one day)/(30 million years), or of order 10^{-10}, and one could thereby place on the graviton rest mass a limit of order 10^{-5} times the energy of a 1000 Hz graviton, or about 10^{-16} eV. If neutrinos could also be detected, the time resolution for both neutrinos and gravitational waves could be about 1 ms; and if their onsets were that close together, one would infer equal speeds to within about 10^{-18} and corresponding rest mass limits of $\sim 10^{-20}$ eV for the graviton, and $\sim 10^{-2}$ eV for the neutrino.

This limit for the graviton is about the same as one obtains from the validity of the inverse square law for the sun's gravity by study of planetary orbits. Therefore, if gravitational waves were found to propagate at a different speed from light and neutrinos, we would infer that the gravitons which carry the waves are different from the gravitons which mediate the solar system's gravity—an unlikely but not impossible outcome. An alternative explanation could be that gravitons, like photons, have zero rest mass, and they propagate on the geodesics of a different metric from that which governs photons.

Gamma-ray bursts Observation of gravitational waves from a gamma-ray burst (GRB) could provide a clear signature of the nature of these mysterious cosmic explosions. All that is securely known is that they arise from a population that is isotropic to a high degree, is radially bounded, and so far has escaped detection in all wavelength bands except hard X-rays and gamma-rays.

There are at least three different source distribution models to consider. In the local models GRBs originate from comets in the Oort cloud. While these models are not conclusively ruled out by the observations of the BATSE experiment on the Compton Observatory, they must invoke a highly non-standard radial distribution for comets. Extended Galactic halo models can fit the BATSE count data but to avoid anisotropy (since we are not at the center of the galaxy), the core radius of the distribution must be greater than about 100 kpc and the sources must be sampled out to at least 300 kpc. This is larger than the dark matter halo and does not correspond to any known distribution of objects, with the possible exception of the newly discovered high-velocity pulsars. In addition the sampling distance must be less than about 400 kpc to avoid detection of a peak in the spatial distribution due to bursts from M31. Thus the phase space for viable halo models is getting uncomfortably small. Cosmological models offer the simplest angular and spatial distribution consistent with the BATSE data. Beyond $z \approx 1$, redshift effects will exhibit a natural decrease in the source density. The simplest cosmological models imply average source luminosities of 5×10^{46} W. Because of the scale of the energies involved, coalescing binaries consisting of compact neutron stars and/or black holes are an attractive possibility. Since these objects represent strong and understandable sources of gravitational waves, correlated observations by LIGO and orbiting gamma-ray burst detectors might lead to the resolution of the gamma-ray burst mystery.

Low-mass X-ray binaries X-ray sources in low mass binary systems are of interest because they likely contain neutron stars with weak magnetic fields that are spun up to short spin periods by Roche lobe accretion from their low-mass companion stars. These LMXBs are also believed to be the progenitors of millisecond radio pulsars. The magnetic field in a typical LMXB neutron star may be weak enough ($< 10^4$ T) that the accretion disk extends to the surface of the neutron star. In this case, accretion could spin up the neutron star to the Kepler frequency at its equatorial surface. However, other effects may limit the spin frequency well below the breakup limit. Wagoner [38] described a scenario in which the accretion process continually provides angular momentum to the neutron star that is then radiated as gravitational waves due to a nonaxisymmetric instability that may occur at high spin frequency. The accreting neutron star would then be a continuous source of gravitational waves. The onset of this Chandrasekhar-Friedman-Schutz (CFS) instability [39] occurs when the gravitational radiation instability growth time equals the viscous damping time. In the Wagoner scenario most of the gravitational energy of the accreting matter is released in X-rays, while most of the angular momentum is removed by the gravitational waves. Because of the nonaxisymmetric distortion of the star, the X-ray flux is expected to be weakly modulated at the frequency of the gravitational wave. Thus if the CFS instability is excited it can manifest itself with a unique signature: X-ray flux modulation at both the stellar spin frequency and at the lower CFS pattern frequency.

If viscosity damps the CFS modes then, in principle, the neutron star can be spun up to near break-up at which point the Kepler orbital frequency at the star's equatorial radius equals the spin frequency. Under these conditions the neutron star may develop other non-axisymmetric distortions (e.g., the formation of sizable "mountains" on the surface may be enhanced as the neutron star approaches the break-up frequency) that will result in gravitational wave emission; alternatively, significant mass outflows may develop involving essentially no gravitational wave emission.

Observational opportunities to search for rapidly-spinning neutron stars in low-mass X-ray binary systems will come about with the expected 1995 launch of the NASA X-ray Timing Explorer (XTE) and with the launch of the USA experiment on the Air Force Advanced Research and Global Observation Satellite (ARGOS) in late 1995 or early 1996. Discovery of candidate systems by either of these X-ray detectors would be an important input to subsequent gravitational wave searches.

The Black Hole Grand Challenge Alliance A major research effort in computational relativity has been undertaken, to complement the experimental work on gravitational wave interferometers. The Black Hole Grand Challenge Alliance is a collaboration of researchers from eight institutions in the United States, whose goal is to numerically integrate the equations of general relativity in order to predict the gravitational waveform from the coalescence of two black holes. This work recently obtained funding from the National Science Foundation's High Performance Computing and Communications initiative.

This project is interesting from several points of view. Successful completion would mean a new level of competence had been reached in our treatment of Einstein's field equations, which are a notoriously difficult set of coupled partial differential equations. The black hole coalescence problem is one with not much symmetry to simplify the problem. Nevertheless it is a problem of pure vacuum gravitation, without the complications of understanding the equation of state of matter under extreme conditions.

Detailed knowledge of the coalescence waveform would be of substantial benefit to a search for black hole events from gravitational wave detectors, and would let any possible detections make direct contact with unambiguous theoretical predictions. It is hard to imagine a richer scientific result to come out of gravitational wave observations than a test of general relativity in the extreme conditions of a black hole.

Second generation interferometers

Since there is no guarantee that the initial generation of interferometers will detect gravitational waves, it is pru-

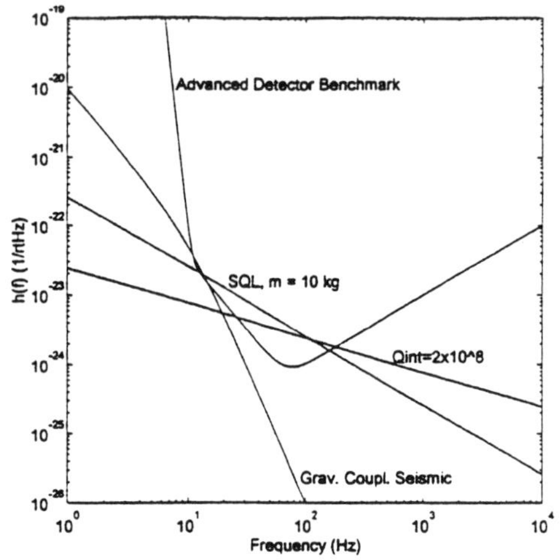

Figure 6: The LIGO Advanced Detector Benchmark, and some important noise spectra.

dent to look ahead to what kinds of major improvements could be made to their sensitivity. After a discussion of fundamental limits to performance, we will describe what technical features would be required to construct an interferometer whose performance could meet or exceed the "Advanced Detector Benchmark" curve shown in Figure 6.

Fundamental limits We start this examination by discussing two fundamental limits to the sensitivity of interferometers of this style, the quantum limit and the gravitational coupling of seismic noise. In both cases, we will see that they aren't truly "hard" limits; still, they set levels that will be difficult to surpass.

The standard quantum limit We alluded earlier to a fundamental limit to interferometric measurement precision that is a sort of dual to shot noise. This is the radiation pressure noise that comes from the fluctuating number of photons reflecting from the test masses' mirrored surfaces. The spectral density of this noise can be expressed as

$$h_{\rm rp}(f) = \frac{2F}{\pi L m f^2}\sqrt{\frac{2\hbar P_{\rm out,max}}{\pi^3 c\lambda}}. \qquad (16)$$

The crucial fact is that this noise grows as $\sqrt{P}$, the opposite of the scaling of the shot noise.

The total optical noise is the combination of shot noise and radiation pressure noise. For fixed interferometer parameters, the spectrum will be characterized by a low frequency region $h(f) \propto f^{-2}$ where radiation pressure noise is dominant, and a high frequency region $h(f) \propto (1 + (4\pi f \tau_s)^2)^{1/2}$ in which shot noise is dominant. The different frequency dependences mean there is a frequency at which this noise is a minimum. As long as this frequency is below the cavity corner frequency $f_c = (4\pi\tau_s)^{-1}$, then the total noise is just what you would expect if you assumed there was a fundamental position noise given by the quantum mechanical uncertainty principle $\Delta x \Delta p > \hbar$.

As the optical power is varied, the two components of the optical noise shift; the set of the minima of these curves traces out a locus called the "pseudo-spectral density" of the quantum limit,

$$h_{\rm QL}(f) = \frac{1}{\pi f L}\sqrt{\frac{2\hbar}{m}}. \qquad (17)$$

The level of this locus for test masses of the size used in the inital LIGO interferometer is shown in Figure 6, in the curve labelled "SQL, $m = 10$ kg". There is, of course, no prohibition on using larger masses, thus lowering the level of the standard quantum limit. For further discussion of the possibility of using larger masses, see the discussion on thermal noise, below.

At high frequencies, reaching the quantum limit will require very high optical power in the interferometer. But even the modest power proposed for some of the first generation interferometers will cause the quantum limit to be reached at interesting frequencies. For example, the 1 kW of power incident on the VIRGO beam splitter means the quantum limited sensitivity is reached at around 20 Hz; below that frequency, shot noise would be hidden by radiation pressure noise. Conceivable improvements in the thermal noise of the VIRGO pendulums would reveal this noise.

No naive extension of the sort of interferometer we've been describing will have optical noise below this curve. Sophisticated use of quantum mechanical principles may allow some progress beyond it, however. The key is the use of specially prepared squeezed light, injected into the output port of the interferometer. Ordinary squeezed light will only change the relative proportion of shot noise and radiation pressure noise at a given power, without enabling the standard quantum limit to be surpassed. But by taking advantage of the correlation between position and momentum errors one can, in principle, make measurements beyond the standard quantum limit. [41]

Gravitationally-coupled seismic noise There is a fundamental limit to the useful amount of seismic isolation that can be applied to a test mass in a terrestrial laboratory. This comes from the direct gravitational coupling of seismic noise to test mass motion, in effect making a "short circuit" around the vibration isolation system.

Proper instrument design and operating procedures can minimize the number of moving masses that are close enough to the test masses to matter. But there will always be a component of gravitational noise due to the seismic noise motions of the ground around the interferometer, and to density fluctuations in the atmosphere. [40] The coupling to the seismic component (which is the dominant one) is, up to a factor of order unity, the same as if the test masses were attached to the ground with a spring that gave a resonant frequency of $f_{\rm grav} \approx (1/2\pi)\sqrt{G\rho} \approx 70\,\mu$Hz, where G is Newton's constant,and where ρ is the density of the surface layer of the Earth near the test masses.

A graph of the resulting interferometer noise spectrum, based on the generic seismic noise spectrum of Eq. 14, is shown in Figure 6. It is small, but not so low as to be below the amount of seismic noise that will be transmitted through as good an isolator as the Super Attenuator. At all frequencies above about 4 Hz, the net amount of isolation is dominated by this gravitational effect, not the performance of the isolator itself.

Note that this noise source will be stronger than the standard quantum limit for 10 kg masses at all frequencies below about 10 Hz. No clever schemes have yet been proposed to reduce this noise.

How can we reach the Advanced Detector benchmark? We consider in turn the three strongest contributors to the noise spectra of the initial interferometers: noise in the measurement of optical phase shift (shot noise), seismic noise, and thermal noise.

Phase sensitivity Reaching the LIGO Advanced Detector benchmark will require phase sensitivity substantially better than that in the initial interferometer. One possible combination of parameters that could reach the target is:

- a doubled (i.e. $\lambda = 0.503\mu m$) Nd:YAG laser powerful enough to deliver 20 W of stable optical power to the input side of the power recycling mirror,
- power recycling by a factor of 30, yielding 600 W incident on the main beam splitter, and
- either signal recycling or resonant signal extraction sufficient to provide another factor of two to three in sensitivity improvement at the frequencies of interest.

Note that, at the power levels in an interferometer such as the one described above, it would be necessary for the crucial fused silica components (the beam splitter and Fabry-Perot arm input mirrors) to have absorption levels below 1 ppm per cm of thickness, and to have coatings with absorption below 1 ppm per reflection. Otherwise, thermal lensing effects would destroy the quality of interference necessary for proper functioning of the interferometer.

Seismic noise Nowhere has progress been more encouraging than in the development of seismic isolation. The carefully tested initial version of the Super Attenuator (incorporating air springs for vertical compliance) has already demonstrated performance that would meet the Advanced Detector benchmark. The newer version, using metal cantilever springs and magnetic anti-springs, is both more stable and a better isolator, judging from all tests that have been conducted so far. It appears that seismic noise need not be a significant part of the noise budget for all frequencies above 4 Hz.

Seismic noise can almost certainly be tamed to lower frequencies still. The JILA active isolation project has as its goal the reduction of seismic noise at 1 Hz to below the level of the gravitational coupling to seismic noise. [42] Active isolation is well-suited to isolation at such a low frequency, beyond the reach of even the heroic engineering incorporated in the Super Attenuator. The JILA group's design philosophy is to constuct several cascaded active stages, each capable of providing 40 dB of isolation from 1 Hz to 10 Hz in all six rigid body degrees of freedom. With the successful demonstration of JILA's first active stage, prospects seem excellent for extending the gravitational wave window down to 1 Hz, at least at strain levels of order 10^{-20}.

Thermal noise Substantial advances are required before thermal noise in the pendulum mode can be reduced to the level of the Advanced Detector benchmark. A quality factor of 10^9 would required to reach the goal at 100 Hz, and $Q = 10^{10}$ is necessary to reach it at 10 Hz, assuming that the test masses are 10 kg. Fused silica and niobium flexures, among others, offer the possibility of achieving such quality factors, especially when one imagines using fine ribbons instead of round wires for flexures. Experiments to be carried out in the near future should be able to verify whether such specifications can be attained.

Reducing the thermal noise from the test mass internal modes may prove to be the most challenging aspect of reaching the Advanced Detector benchmark. While quality factors of only 10^7 are required to meet this goal at 10 Hz, a Q of 10^{10} is necessary to make thermal noise negligible at 100 Hz.

Scaling current designs up in mass is a discouraging prospect; although there is an explicit dependence of $h_{\rm th}(f) \propto m^{-1/2}$, the reduction in frequency of modes as the mass is scaled up transforms the scaling of the noise from a particular mode to $h_{\rm th}(f) \propto m^{-1/6}$. An important additional effect is that modes of higher order will contribute to the total noise in a larger mass; all modes with wavelength longer than the beam diameter need to be considered. This would nullify any improvement in noise by increasing the mass, [28] as long as the beam diameter is kept fixed.

Note, though, that the quantum limit is in excess of the Advanced Detector benchmark at 100 Hz, unless the test masses have masses of 200 kg or greater. If we are limited to 10 kg masses, then there is no benefit to be gained from Q values in excess of about 10^9. There needs to be further exploration of optical arrangements matched to large test masses.

It is uncertain whether grades of fused silica can be found that would exhibit 10^9 at 100 Hz. The purest grades of fused silica can probably achieve the 10 Hz requirement with $Q \approx 3 \times 10^7$.[43] Better understanding of what level of internal friction is achievable in fused silica is urgently needed.

Substantial variations on present designs may be required. One possible solution would be to independently measure the internal thermal vibrations of the test masses, for subtraction from the interferometer output. A small multi-pass interferometer that measured the motion of the mirror's front surface by shining light in through the back of the test mass is one possibility. There is at least one major hurdle, however. Unless the measurement of internal vibration is made to accurately reveal the motion of the mirror surface where the main interferometer beam reflects, then accurate subtraction cannot be made. In the interferometer just described, internal variation of the index of refraction will also cause phase shifts, by an amount that varies from mode to mode. Even this would not be an especially difficult problem if only one or a few modes dominated the noise. But careful consideration of the modal sum appropriate to cylindrical masses has shown that only a factor of two reduction would be possible without accounting for dozens of high order modes.

Another possibility might be to lower the temperature of the test masses, since $h_{\rm th}(f) \propto T^{-1/2}$. Going from room temperature to that of liquid helium offers a possible improvement of more than a factor of 8. One difficulty with this idea that has yet to be overcome is to design a way to conduct the substantial heat load away from the test masses – dissipated laser power may be measured in hundreds of mW, and the heat load from the warm walls of the vacuum system will also be significant. (The latter problem can be overcome by building a cold shield around each test mass.) What makes this difficult is that thermal conductivity must be supplied while meeting the need to keep the mechanical quality factors of the various modes high – this need is what dictates the use of very fine wires or other flexures as the only attachments to the masses.

Even if this tricky bit of cryogenic engineering could be accomplished, there is another difficulty with the application of cooling to present designs. The losses in fused silica have a temperature dependence that would tend to negate the benefits of cooling. Fused silica losses rise steeply as the temperature is lowered below room temperature, reaching a peak (with $Q < 10^3$) at about 50 K. Below that temperature losses decline somewhat, but have not been seen to return anywhere near the room temperature values; at 4 K, fused silica has $Q \approx 10^4$. So to achieve the benefits of cooling, one would almost certainly have to abandon the use of fused silica as the test mass material.

The prospects for finding a simple substitute for fused silica are uncertain at best. Not only does fused silica have a high mechanical quality factor, but its very low optical absorption and birefringence and excellent homogeneity in its index of refraction all play important roles in ensuring that present optical designs can work well. Sapphire has excellent transparency and high mechanical Q. However, it has substantial birefringence, and without better control of the orientation of crystal axes would cause large wavefront distortions that would degrade the quality of the dark fringe and thus the performance of power recycling.

Optical configurations for low thermal noise Transmission through optical components has been built into the initial interferometer designs in several ways. One is the transmission through the input coupling mirror of the Fabry-Perot cavities adopted by most of the large projects. (The delay line design of GEO 600, as well as more ambitious delay lines that can do substantial amounts of beam folding, make no use of transmission through any of their cavity mirrors.) In addition, all interferometer designs adopted so far make crucial use of transmission through the beam splitter, as a way of generating coherent beams for the two arms and of then recombining the beams to generate the interference fringe. Without the need for light to pass through the bulk of an optical component, one would have a much wider choice of materials. This might include crystalline silicon, whose mechanical Q has been shown to reach 3×10^7 at 300 K. Its Q improves as temperature is reduced, so that if it could be cooled successfully one would actually reap the benefits that one would naively expect. Sapphire has exhibited even higher Q, 2×10^8 at room temperature. For non-transmissive applications, its birefringence is not a problem. Sapphire also exhibits reduced losses at low temperature; $Q \approx 4 \times 10^9$ has been reported at 4 K. The thermal noise from 10 kg sapphire test masses at room temperature is drawn in Figure 6, in the curve labelled "$Q_{int} = 2 \times 10^8$".

One way to obtain substantial benefits from the use of these alternative mirror materials might be to abandon the use of Fabry-Perot cavities as beam folding elements. A Michelson delay line could be made of multiple individual mirrors, each of the size of one of the mirrors in the first generation interferometers, but made of a high Q material. (This optical geometry has been named a *schellenbaum* by the MPQ group.) In this de-

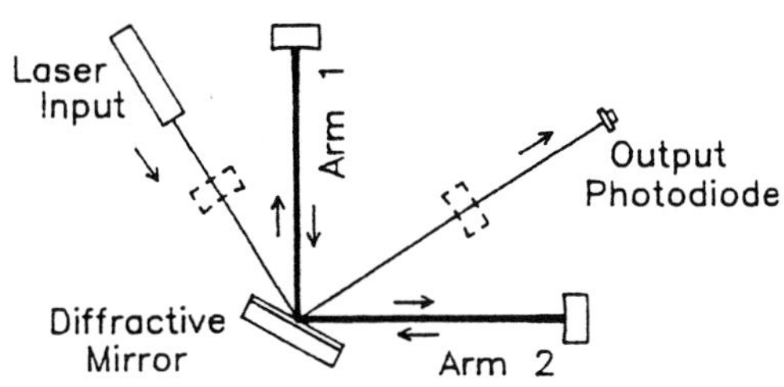

Figure 7: Schematic diagram of a Michelson interferometer based on the use of a diffraction grating as a beam splitter.

sign, light would still need to be transmitted through the beam splitter. The latter component does contribute some thermal noise, but at a level down by a factor of the number of bounces in the arms. So if the arms had of order ten round trips, then arm mirrors with Q up to 10^9 would dominate the thermal noise in an interferometer whose beam splitter had.$Q \approx 10^7$. Such a design could reach thermal noise levels approaching the Advanced Detector noise budget.

The possibility of actually constructing a non-transmitting beam splitter have begun to receive some attention. One intriguing idea is to form a diffraction grating on the surface of a mirror. [44] With proper choice of grating parameters, diffraction orders higher than the first could be eliminated. One possible interferometer configuration using such components is shown in Figure 7. Whether such beam splitters can be made that will meet the necessary specifications concerning, for example, optical losses, remains to be seen.

Open Research Questions

For ease of reference, we present here a list of the most pressing open questions for future research on gravitational wave interferometers. They deal mainly with topics relevant to reaching the Advanced Detector benchmark, or in the case of data analysis with getting the most possible information out of data of any given level of sensitivity.

Data analysis Many aspects of the data analysis procedure for gravitational waves may have been solved already in other contexts by signal processing experts in, say, the contexts of radar, sonar, or communications engineering. It would be extremely valuable if experts in these fields were to contribute to this new branch of astrophysics. Among the outstanding needs here are:

- efficient algorithms for handling an all-sky search for periodic sources, or other multi-parameter searches, such as the chirped waveform of an inspiralling binary as a function of masses and spins of stars,
- examination of the optimum way to detect signals whose waveforms can not be predicted in detail, such as "non-deterministic" bursts, or quasi-periodic sources,
- construction of careful models of the multi-interferometer system with realistic description of the noise, for the purpose of calculating statistical confidence of detection at intermediate signal strengths, and
- study of how changes in various aspects of detector noise spectrum will affect ability to detect different classes of sources.

Thermal noise The materials science community may have a great deal of information to offer to our search for the way to achieve the lowest possible levels of mechanical dissipation in our test masses. Possible directions for progress include:

- better varieties of fused silica (perhaps ultra-high purity) than are commercially available at present,
- reliable techniques for bonding pendulum wires or hinges to test masses in a way that preserved high Q factors in both wires and test masses,
- techniques for monitoring of internal modes of vibration at high sensitivity and fidelity to the motion of the mirror surface,
- non-transparent materials with very high Q, for use in alternative optics schemes that don't make essential use of transmission through components, and
- practical designs for cooling test masses to liquid helium temperature without compromising the mechanical features required for high Q suspensions.

Optical materials and components Optics experts may be able to help in several crucial aspects of interferometer design, including:

- materials with very low transmission loss (less than 1 ppm per cm),
- mirror coatings with absorption less than 1 ppm per reflection, and
- development of all-reflective beam splitters that can meet the demanding optical and mechanical specifications of advanced interferometers.

3.3 Plans for Space Experiments

Plans for Doppler Missions

Near-term Plans In past missions scientists did not have the opportunity to optimize the gravitational wave experiments on space probes. For the forthcoming CASSINI mission, however, the components of the radio system on board the spacecraft as well as on the transmitting antenna on Earth are being specifically designed according to requirements imposed by the gravitational wave experiment.

The CASSINI spacecraft is a three-axis-stabilized spacecraft, which will be launched in October 1997 for gathering scientific data from Saturn, Titan and other satellites of the Saturnian system. During part of the spacecraft trajectory to Saturn (cruise and opposition), from the year 2001 until 2003, each year a forty day search for gravitational waves will be performed.

The spacecraft will have X- and Ka-band (about 32 GHz) uplink radio signals, and will be capable of transponding back simultaneously to Earth three frequencies (one X-band signal coherent with the X-band uplink; a Ka-band signal coherent with the X-band uplink; and another Ka-band signal coherent with the Ka-band uplink). At opposition a coherent Ka-band link will minimize the plasma noise to a level below the noise level due to electronics, while off opposition a combination of the three Doppler links would allow one to totally calibrate for the plasma. In order to further minimize the tropospheric noise, a water vapor radiometer will be installed at the Ka-band antenna. This should allow, in principle, calibration of the water vapor fluctuations to $80 - 95\,\%$.

By the year 2000 it is also expected that better clocks will be available. A combined Trapped Ion / Superconducting Cavity Maser Oscillator (TI/SCMO) should be in operation, reaching an Allan deviation at 1000 seconds of about 4×10^{-16}.

As a consequence of such improvements, it is expected that the CASSINI sensitivity to bursts will be about $h \approx 7\times10^{-16}$. Due to the forty days of continuous tracking, the use of high-frequency links and a tropospheric calibration system, the expected sensitivity to sinusoids and chirps will be about $h \approx 2\times10^{-17}$. Finally the sensitivity to a stochastic background should improve, to a level of $h(f) \approx 6\times10^{-15}/\sqrt{\mathrm{Hz}}$.

What else could be done? It seems very likely that by the end of this century space-qualified, high-performance trapped-ion clocks will be available. [45] The expected sensitivities of these clocks will be comparable to the sensitivities of hydrogen masers, but their mass and power consumption will be about ten times smaller. For a small cost increase one could add to the on-board instrumentation a trapped-ion clock, a closed-loop Doppler heterodyne receiver, and include a transponder at the transmitting station. In this, together with the usual Doppler link (Earth-spacecraft-Earth), one would have the complementary Doppler link spacecraft-Earth-spacecraft, with Doppler data recorded on board. As shown by Vessot, [46] these two streams of Doppler data can be linearly combined in the time domain to remove tropospheric and ionospheric frequency fluctuations. Use of high-frequency links would minimize interplanetary scintillation noise, making such experiments the most sensitive to date.

Scientists at the Jet Propulsion Laboratory are studying various technical options for implementing this alternative Doppler tracking technique on the proposed Pluto Fast Fly-By mission. Two small spacecraft would be launched a few weeks apart from each other, in order to get scientific information about the planet Pluto and its moon Charon. The radio hardware on board these two spacecraft will already have a Doppler readout and recorder, because of the planned uplink atmospheric occultation at Pluto. The availability of a new generation of space-qualified clocks in time for this mission would make the Pluto Fast Fly-By the ideal spacecraft for implementing this "combined Doppler" method.

Laser Interferometer Space Antennas

In May 1993, two proposals for laser gravitational wave detectors in space were submitted to the European Space Agency (ESA). One, called LISA, [47] was for an antenna consisting of a cluster of four spacecraft, with two central spacecraft located close together at one corner of an equilateral triangle and one end spacecraft at each of the other two corners. Laser beams are sent out from the two central spacecraft to the end ones through 0.3 m diameter transmit/receive telescopes. The received signals are used to phase-lock lasers in the end spacecraft, and beams from these lasers are sent back to the central spacecraft. The returned signals are beat against the local lasers, and the phases of the two beat notes are measured. The antenna functions like a large Michelson interferometer. An arm length of 5×10^6 km for the interferometer was proposed, with the cluster to be located roughly 45° behind the Earth, in heliocentric orbit.

The second proposal was for a similar detector called SAGITTARIUS. It would have two spacecraft with 0.15 m diameter telescopes at each corner of the antenna, with all six in retrograde circular geocentric orbits in the ecliptic, and with 1×10^6 km interferometer arm length. Both proposals were submitted for the third ESA medium sized mission (M3) under the ESA Horizon 2000 program, with the intention of their being considered as possible joint ESA-NASA missions. However, ESA decided to carry out an Assessment Study of both proposals as possible ESA-only missions. The Study resulted in LISA

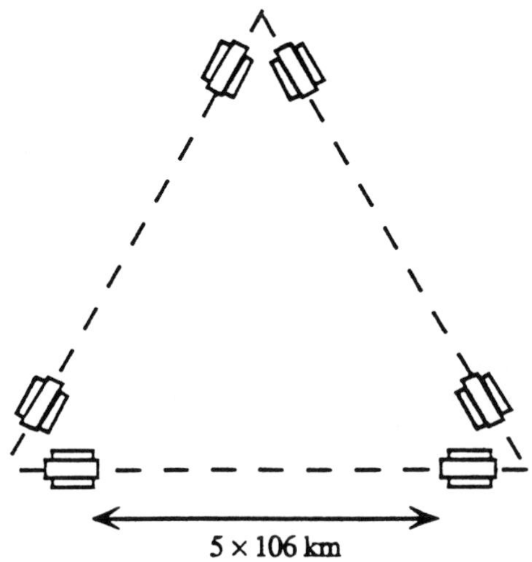

Figure 8: Schematic diagram of LISA.

being recommended to ESA for consideration, but the ESA cost estimate was too high for the M3 mission.

LISA, with six spacecraft (see Figure 8) instead of four to ensure reliability, was also suggested to ESA in October 1993 as a new cornerstone mission for a proposed extension of the ESA program, Horizon 2000+. (The advantage of the six spacecraft configuration is that the antenna can continue to operate as a single two-arm interferometer even if one or two spacecraft fail, as long as they are not at the same corner.) An additional brief technical assessment of LISA as a cornerstone mission was made and presented to the ESA Survey Committee (SC) in November 1994. [48] As a result, LISA was one of three cornerstone missions recommended by the SC. In addition, the SC recommended "*that ESA engage in the necessary technological and system studies with a view to implementing a cornerstone-level programme on the observation of gravitational waves, in particular at low frequencies (< 1 Hz). A proper technological and financial framework should be provided so as to make such a mission possible towards the end of Horizon 2000+. System studies should start when feasible.*"

The characteristics of a LISA mission will be discussed below on the assumption of an ESA cornerstone mission, with some participation by the U.S. and other non-ESA countries. Such participation could aid in the technology development for the mission, and lead to a modest but valuable further increase in the antenna sensitivity. A similar laser gravitational wave mission was the subject of a NASA workshop held in 1990 under the Astrotech 21 program, and some additional research and development efforts have gone into it since then.

Mission concept for LISA The cluster of LISA spacecraft rotates as an equilateral triangle around a center point which is roughly 20° (or, if funding allows, 30° to 45°) behind the Earth. The plane containing the six spacecraft has its normal pointed 30° above the sun, and the antenna rotates once per year in that plane.

The heart of each spacecraft is a roughly 1 kg cubical test mass made of a Pt-Au alloy with low magnetic susceptibility. The laser beam reflects off the face of the test mass to define the end of the interferometer arm. A capacitive position sensor surrounds the test mass and measures its position with a sensitivity of better than 1 nm/$\sqrt{\text{Hz}}$. A 1 mm capacitor plate gap is chosen for the sides of the test mass and a 2–4 mm gap is used on the ends, to reduce perturbing forces due to patch fields. Close attention is paid to keeping other sources of perturbing accelerations of the test mass small as well, particularly at frequencies of 0.1 to 10 mHz. The position sensor signals are used to control microNewton thrusters, thus counteracting solar radiation pressure and other forces on the spacecraft and also controlling the spacecraft attitude. These actuators are Field Emission Electric Propulsion (FEEP) thrusters, devices which use Cs ions to develop high specific impulse.

The lasers to be used to measure the changes in interferometer arm lengths are laser-diode pumped Nd-YAG lasers with 2 W nominal output power at 1.06 μm wavelength. However, they are operated at half this level to give long operating lifetimes, and redundant lasers also are included. Since the spurious accelerations of the test masses are kept small, the apparent changes in the lengths of the interferometer arms give excellent information on the phase noise in the lasers. This information is needed to correct for the laser phase noise when changes in the difference in the arm lengths are determined, and will also be used to detect gravitational waves. LISA would be designed for an operating lifetime of 3 years or more. The data, after preprocessing and compression on the spacecraft, would be sent back by X-band transponders to ground antennas.

Antenna sensitivity There are two main types of noise sources which must be considered. One is noise in measuring the difference in the arm lengths, as defined by the separations between the test masses. The second is the noise due to spurious accelerations of the test masses. The first is expected to be dominant at frequencies above roughly 1 mHz, and the second at lower frequencies. The total frequency range of interest is from somewhat below 0.1 mHz to 1 Hz. However, the highest sensitivity is expected from 1 to 30 mHz.

One obvious noise source in measuring the difference in interferometer arm lengths is shot noise in the received signals. However, another important noise source is due to fluctuations in the direction of the laser beams. Be-

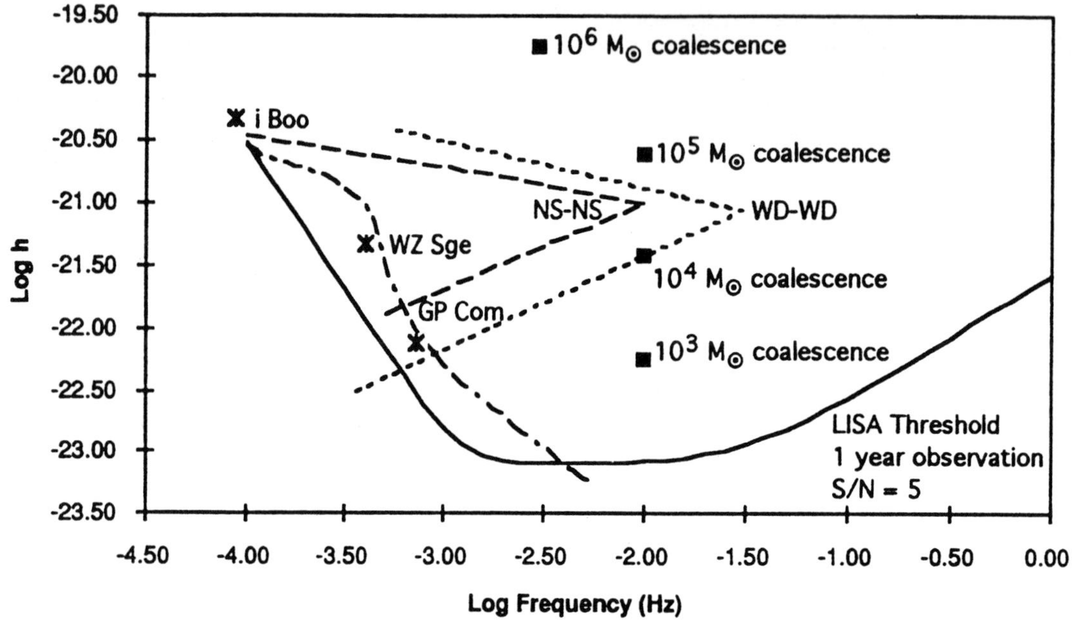

Figure 9: Gravitational wave signal strengths and LISA sensitivity.

cause of effects such as small errors in the surfaces of various optical elements and the finite diameter of the transmitted beam, the wavefronts arriving at a distant spacecraft are not spheres. Thus a change in the attitude of one spacecraft with no change in the test mass positions would lead to a change in the apparent arm length. To reduce this error source, the pointing tolerance is about 5 $\text{nrad}/\sqrt{\text{Hz}}$ for frequencies above 0.1 mHz. When all of the error sources affecting the arm length difference measurements are combined, the total error budget level proposed for LISA is 20 $\text{nm}/\sqrt{\text{Hz}}$ from below 1 mHz to 1 Hz.

There are a large number of potential sources of spurious accelerations of the test masses which have been considered. These include the following:

- fluctuating thermal distortion of the spacecraft, leading to time-varying gravitational force on the test mass,
- fluctuating thermal radiation pressure force due to temperature difference variations across the test mass cavity in the position sensor,
- residual gas impacts on the test mass,
- spacecraft displacement plus gravity gradient from payload and spacecraft,
- spacecraft displacement plus electrical force from patch fields,
- Lorentz force from charge on test mass plus interplanetary magnetic field,
- magnetic force from test mass susceptibility, spacecraft magnetic field gradient, plus fluctuating interplanetary field, and
- cosmic ray impacts on the test masses.

Of the above sources of spurious accelerations, the first may require the most care in the spacecraft design. It is assumed that the power dissipation in the spacecraft will be kept as constant as possible, but fluctuations in the solar optical intensity also will cause temperature changes in the main spacecraft structure. Thus materials with low thermal expansion coefficients will be used. From 0.1 mHz to above 1 mHz, the total error allowed for the second derivative of the arm length difference due to spurious accelerations of the test masses is

$$\Delta \ddot{L}(f) \approx 3\times 10^{-15}\left(\frac{f}{0.1\,\text{mHz}}\right)^{-1/3}\ \text{m}\,\text{s}^{-2}/\sqrt{\text{Hz}}. \quad (18)$$

Scientific objectives The primary scientific objective for LISA is to detect and study in detail signals from sources involving massive black holes (MBHs). The following ideas are gaining widespread acceptance: that many galaxies contain MBHs, and that mergers were common in the past for pregalactic structures or galaxies which may already have contained MBHs. Provided that the MBHs spiral together and coalesce in less than a Hubble time, the resulting signals could be observed by LISA with high S/N ratio out to large cosmological distances for most combinations of MBH masses between roughly 10^3 and $10^7 M_\odot$. The number of such binaries would be enhanced if MBH creation can occur at the earliest times of structure formation. [49]

Another potentially observable type of source is compact star-MBH binaries. If a neutron star or white dwarf in the cusp around a MBH in a galactic nucleus is deflected in close enough to the MBH, substantial gravitational radiation will occur. Some fraction of such highly elliptical binaries will survive and give observable signals for of order 100 yrs before coalescence. Calculations indicate that some signals of this kind may be observable if the background due to many close white dwarf binaries in other galaxies is not too high, but the full sensitivity of LISA would be needed to detect and study them. The case where the compact star is a 5-20 $M_\odot$ black hole also appears to be promising.

The observation of signals from binaries containing MBHs would provide a completely new type of astrophysical information. Particularly important would be data on the abundance of MBHs of different masses and their redshift distribution. Strong signals from MBH-MBH binaries would make possible unique and valuable new tests of the predictions of general relativity for mass motions under extremely strong field conditions. The number of compact star-MBH binaries could give information on conditions near the MBHs.

LISA would certainly observe signals from hundreds to thousands of neutron star binaries in our Galaxy. In addition, comparable numbers of close white dwarf binaries appear likely to be observed. Cataclysmic variables and neutron star-BH binaries probably will be seen unless the confusion limit due to close white dwarf binaries at frequencies near 1 mHz is too high. A few known binaries are likely to be observed as well.

The expected frequencies and signal strengths for a number of the above types of sources are given in Figure 9. Also shown is the expected LISA sensitivity, as discussed in the previous section. The signal-to-noise ratio of 5 for one year of integration time used in calculating the LISA sensitivity is appropriate for many of the galactic sources shown, since locations and expected frequencies are available for only a few of them. However, a signal-to-noise ratio of about 10 would be needed for the compact star-MBH binaries because of the additional complexity in their signals. For coalescing equal mass MBH-MBH binaries, the signal strength shown is adjusted to allow for the observable duration of the signal being less than a year.

4 Conclusion

There has been substantial progress recently in improving the sensitivity of searches for gravitational waves. Even more encouraging is the fact that plans for new detectors with astrophysically interesting sensitivy are well advanced. A milestone was passed recently with the approval of several construction projects for large interferometers. Interesting technical challenges still remain, but it seems likely that gravitational waves from astrophysical sources will be detected within the next decade, thus opening a new window on the Universe.

Acknowledgments

All of the participants in the Gravitational Wave section of the Snowmass 94 Summer Study would like to thank the organizers and the Divisions of Astrophysics, Nuclear Physics, and Particles and Fields of the American Physical Society for putting together such a stimulating program in such pleasant surroundings. We also thank our respective funding agencies for support of our research.

References

1. J. Weber, *Physical Review* **117** (1960) 306; J. Weber, *Physical Review Letters* **22** (1969) 1320.
2. R.A. Hulse and J.H. Taylor, *Astrophysical Journal Letters* **195** (1975) L51; J.H. Taylor and J.M. Weisberg, *Astrophysical Journal* **345** (1989) 434.
3. B.F. Schutz, *Nature* **323** (1986) 310.
4. P. Kafka and L. Schnupp, *Astronomy and Astrophysics* **70** (1978) 97.
5. S.P. Boughn, W.M. Fairbank, R.P. Giffard, J.N. Hollenhorst, E.R. Mapoles, M.S. McAshan, P.F. Michelson, H.J. Paik, R.C. Taber, *Astrophysical Journal Letters* **261** (1982) L19.
6. E. Amaldi, O. Aguiar, M. Bassan, P. Bonifazi, P. Carelli, M.G. Castellano, G. Cavallari, E. Coccia, C. Cosmelli, W.M. Fairbank, S. Frasca, V. Foglietti, R. Habel, W.O. Hamilton, J. Henderson, W. Johnson, K.R. Lane, A.G. Mann, M.S. McAshan, P.F. Michelson, I. Modena, G.V. Pallottino, G. Pizzella, J.C. Price, R. Rapagnani, F. Ricci, N. Solomonson, T.R. Stevenson, R.C. Taber, and B.-X. Xu, *Astronomy and Astrophysics* **216** (1989) 325.
7. P. Astone, *et al.*, Proceedings of the Xth Italian General Relativity Conference, Bardonecchia, September 1-4, 1992, unpublished.
8. D. Nicholson, C.A. Dickson, W.J. Watkins, B.F. Schutz, J. Shuttleworth, G.S. Jones, D.I. Robertson, N.L. Mackenzie, K.A. Strain, B.J. Meers, G.P. Newton, H. Ward, C.A. Cantley, N.A. Robertson, J. Hough, K. Danzmann, T.M. Niebauer, A. Rüdiger,

R. Schilling, L. Schnupp, and W. Winkler, paper in preparation, 1995.

9. J.W. Armstrong in *Gravitational Wave Data Analysis*, ed. B.F. Schutz (Kluwer, Dordrecht, 1989).
10. J.W. Armstrong (1994), private communication.
11. J.D. Anderson, J.W. Armstrong, and E. Lau, *Astrophysical Journal* **408** (1993) 287; B. Bertotti *et al.*, *Astronomy and Astrophysics* **296** (1995) 13.
12. J.W. Armstrong (1994), private communication.
13. J.D. Anderson and B. Mashoon, *Astrophysical Journal* **290** (1985) 445.
14. S. Detweiler, *Astrophysical Journal* **234** (1979) 110.
15. D.R. Stinebring, M.F. Ryba, J.H. Taylor and R.W. Romani, *Physical Review Letters* **65** (1990) 285.
16. S. Rappaport, P.C. Joss, and R.F. Webbink, *Astrophysical Journal* **254** (1982) 616.
17. L.P. Grishchuk, *Annals of the New York Academy of Science* **302** (1977) 439.
18. H.J. Paik, *Journal of Applied Physics* **47** (1976) 1168.
19. J.-P. Richard, *Physical Review Letters* **52** (1984) 165.
20. J.-P. Richard, *Physical Review* **D46** (1992) 2309.
21. W.W. Johnson and S.M. Merkowitz, *Physical Review Letter* **70** (1993) 2367.
22. R. Forward, *General Relativity and Gravitation* **2** (1971) 149; R.V. Wagoner and H.J. Paik in *Proceedings of the International Symposium on Experimental Gravitation*, Pavia 17-20 September 1976 (Roma Accademia Nazionale dei Lincei, Roma, 1977).
23. A. Abramovici, W.E. Althouse, R.W.P. Drever, Y. Gürsel, S. Kawamura, F.J. Raab, D. Shoemaker, L. Sievers, R.E. Spero, K.S. Thorne, R.E. Vogt, R. Weiss, S.E. Whitcomb, and M.E. Zucker, *Science* **256** (1992) 325.
24. C. Bradaschia, R. Del Fabbro, A. Di Virgilio, A. Giazotto, H. Kautzky, V. Montelatici, D. Passuello, A. Brillet, O. Cregut, P. Hello, C.N. Man, P.T. Mahn, A. Marraud, D. Shoemaker, J.-Y. Vinet, F. Barone, L. Di Fiore, L. Milano, G. Russo, S. Solimeno, J.M. Aguirregabiria, H. Bel, J.-P. Duruisseau, G. Le Denmat, P. Tourrenc, M. Capozzi, M. Longo, M. Lops, I. Pinto, G. Rotoli, T. Damour, S. Bonazzola, J.A. Marck, Y. Gourghoulon, L.E. Holloway, F. Fuligni, V. Iafolla, and G. Natale, *Nuclear Instruments and Methods in Physics Research* **A289** (1990) 518.
25. K. Danzmann, J. Chen, P.G. Nelson, T.M. Niebauer, A. Rüdiger, R. Schilling, L. Schnupp, K.A. Strain, H. Walther, W. Winkler, J. Hough, A.M. Campbell, C.A. Cantley, J.E. Logan, B.J. Meers, E. Morrison, G.P. Newton, D.I. Robertson, N.A. Robertson, S. Rowan, K.D. Skeldon, P.J. Veitch, H. Ward, H. Welling. P. Aufmuth, I. Kröpke, D. Ristau, J.E. Hall, J.R.J. Bennett, I.F. Corbett, B.W.H. Edwards, R.J. Elsey, J.R.S. Greenhalgh, B.F. Schutz, D. Nicholson, J. Shuttleworth, J. Ehlers, P. Kafka, G. Schäfer, H. Braun, and V. Kose, *Lecture Notes in Physics* **410** (1992) 184.
26. R.W.P. Drever in *Gravitational Radiation*, eds. N.Deruelle and T. Piran (North Holland, Amsterdam, 1983).
27. P.R. Saulson, *Physical Review* **D42** (1990) 2437.
28. A. Gillespie and F. Raab, *Physics Letters A* (1995) in press.
29. R. Del Fabbro, A. diVirgilio, A. Giazotto, H. Kautzky, V. Montelatici, and D. Passuello, *Physics Letters* **A132** (1988) 237.
30. R. Shine in *Aspen Winter Conference on Gravitational Waves and their Detection* January 22-28, 1995, ed. S. Meshkov, unpublished.
31. W. Winkler, K. Danzmann, A. Rüdiger, and R. Schilling, *Physical Review* **A44** (1991) 7022.
32. B.J. Meers, *Physical Review* **D38** (1988) 2317; K.A. Strain and B.J. Meers, *Physical Review Letters* **66** (1991) 1391.
33. J. Mizuno, K.A. Strain, P.G. Nelson, J.M. Chen, R. Schilling, A. Rüdiger, W. Winkler, and K. Danzmann, *Physics Letters* **A175** (1993) 273.
34. M.A. Barton and K. Kuroda, *Physical Review D* (1995) in press; N. Kanda, M.A. Barton, and K. Kuroda, *Review of Scientific Instruments* (1995) in press.
35. D.G. Blair, J.F. Liu, E.F. Moghaddam, and L. Ju (1994) preprint.
36. R.W.P. Drever, J. Hough, A.J. Munley, S.-A. Lee, R. Spero, S.E. Whitcomb, J. Pugh, G. Newton, B.J. Meers, E. Brooks III, and Y. Gürsel in *Quantum Optics, Experimental Gravity, and Measurement Theory*, eds. S. Meystere and M.O. Scully (Plenum, New York, 1983).
37. A.G. Lyne and D.R. Lorimer, *Nature* **369** (1994) 127.
38. R.V. Wagoner, *Astrophysical Journal* **278** (1984) 345.
39. S. Chandrasekhar, *Physical Review Letters* **24** (1970) 611; J.L. Friedman and B.F. Schutz, *Astrophysical Journal* **222** (1978) 281.
40. P.R. Saulson, *Physical Review* **D30** (1984) 732.
41. M.T. Jaekel and S. Reynaud, *Europhysics Letters* **13** (1990) 301.
42. R. Stebbins, D. Newell, S. Richman, P. Bender, J. Faller, and J. Mason, *Proceedings of the SPIE Conference on Vibration Monitoring and Control* **2264** (1994).
43. V.B. Braginsky (1994) private communication.
44. R.W.P. Drever in *Proceedings of the Seventh Marcel Grossmann Meeting on General Relativity*, Stanford, California, July 1994, eds. M. Keiser and R.T. Jantzen (World Scientific, Singapore, 1995).

45. L. Maleki and J.D. Prestage (1994) private communication.
46. L.L.Smarr, R.F.C. Vessot, C.A. Lundquist, R. Decher, and T. Piran, *General Relativity and Gravitation* **15** (1983) 129.
47. K. Danzmann, A. Rüdiger, R. Schilling, W. Winkler, J. Hough, G.P. Newton, D. Robertson, N.A. Robertson, H. Ward, P. Bender, D. Hils, R. Stebbins, C.D. Edwards, W. Folkner, M. Vncent, A. Bernard, B. Bertotti, A. Brillet, C.N. Man, M. Cruise, P. Gray, M. Sandford, R.W.P. Drever, V. Kose, M. Kühne, B.F. Schutz, R. Weiss, and H. Welling, *LISA: Proposal for a Laser-Interferometric Gravitational Wave Detector in Space*, (Max-Planck-Institut für Quantenoptik, Garching, 1993) Report 177, unpublished.
48. *Horizon 2000 Plus: European Space Science in the 21st Centure*, ESA SP-1180 (ESTEC, Noordwijk, The Netherlands, Nov. 1994).
49. D. Hils and P.L. Bender, *Astrophysical Journal* (1995), in press.

Quantum Aspects of Gravity

Valeri P. Frolov

CIAR Cosmology Program, Theoretical Physics Institute, University of Alberta, Edmonton, Canada T6G 2J1

and Lárus Thorlacius

Institute for Theoretical Physics, University of California, Santa Barbara, CA 93106-4030, USA

The reconciliation of gravity theory and quantum physics is an important goal of theoretical physics and many fundamental issues can only be settled when a theory of quantum gravity is in hand. Superstring theory offers the promise of a unified description of all interactions including gravity, but many aspects of string physics remain mysterious. Theoretical study of the very early universe and black holes, where Planck scale physics comes into play, may reveal important clues about fundamental physical law.

The G1 Working Group focused on quantum aspects of gravitational phenomena. By the nature of the subject, all the talks and discussions were on theoretical problems and most were on formal black hole theory, reflecting the current research interests of the active participants in the Working Group. In these introductory words we will discuss some general features of the endeavor which collectively goes under the heading 'quantum gravity' and attempt to motivate some of the key physical questions that workers in this area are addressing. Our remarks will be brief and as there exists at present no accepted theory of quantum gravity they can only reflect our personal assessment of the subject. The general discussion will be followed by contributions from speakers in the organized lecture sessions of the Working Group. These contain accounts of current research topics but an effort has been made to keep them accessible to non-experts.

§ 1 Gravity and Quantum Physics

Exploring the interface between general relativity and quantum theory has been one of the great intellectual challenges of twentieth century physics and will undoubtedly remain one well into the next millennium. The measured values of the constants of nature c, G, and $\hbar$, which govern the strength of relativistic, gravitational and quantum effects, indicate that the realm of quantum gravity is remote indeed. The characteristic energy and length scales are the Planck energy, $E_{\rm pl} = \sqrt{\hbar c^5/G} \sim 10^{19}$ GeV, and the Planck length, $l_{\rm pl} = \sqrt{G\hbar/c^3} \sim 10^{-33}$ cm, which are very far out of reach in present day experiments.

It is nevertheless important to push forward our understanding at this frontier. For one thing, Planck scale dynamics may have subtle effects on physics at lower energies, which cannot be anticipated without some knowledge of that dynamics. If our attention is restricted to low-energy physics alone we will be left with an incomplete theory, within which many fascinating fundamental questions could never be resolved. Furthermore, the very search for underlying principles can benefit physics in general. Theoretical efforts to understand basic physical laws have in the past led to important insights and new ideas that have had impact on different branches of physics.

It is difficult to achieve theoretical understanding of fundamental physical laws without the guiding light of experimental discovery. The lack of laboratory data has led theorists to rely increasingly on internal consistency and even mathematical aesthetics as tools for shaping their theories. This approach has been remarkably fruitful, culminating in the development of superstring theory, which successfully addresses the short distance problems that undermine a more conventional quantum field theory of gravity.

Superstring models have a number of appealing features. Gravity is united with the other known interactions into a single geometric framework, which can also accommodate multiple generations of chiral fermions in a natural way. Unification is widely believed to be a necessary ingredient in a successful theory of quantum gravity. On the other hand, our understanding of string dynamics is limited. The theory is defined in terms of a perturbation expansion around a classical ground state and there is a multitude of equally valid ground states to choose from, without any *a priori* preference given to one over another. The formalism is well suited to the calculation of scattering amplitudes but is not equipped to address many important issues in quantum gravity, such as initial conditions, quantum mechanical interpretation, or the cosmological constant problem.

Despite its shortcomings, string theory remains the only candidate for a consistent unification of gravity with quantum physics and a high priority should be placed on obtaining a deeper understanding of string physics.

While further mathematical developments are clearly called for it is also important to maintain focus on physical phenomena in quantum gravity. In this respect, black holes and the very early universe provide a useful testing ground for theoretical ideas. Conditions extreme enough to bring Planck scale physics into play are realized in these systems and the various theoretical puzzles of black hole physics and quantum cosmology provide hurdles for candidate theories to pass over. There is also the chance that some observable features of the Universe can be traced to quantum gravity effects, either at the earliest epoch or through primordial black holes.

§ 2 Black Holes

In general relativity a black hole is by definition a region of spacetime which is not in the causal past of future null infinity, or in less precise words, it is a region from which signals cannot escape to the outside world. Since nothing that occurs inside a classical black hole can influence events outside, is it even necessary to develop the physics of black hole interiors? If the black hole mass is large compared to the Planck mass then all invariant features of the geometry outside the event horizon, including the curvature, are similar to those of flat Minkowski space. We could hope to get by without invoking quantum gravity at all, provided we ignore everything that goes on inside the black hole.

There are at least two compelling reasons why this is not so. An obvious one is that if we want to use black holes as tools to learn about strong gravity effects then we have to go where the action is, *i.e.* deep inside a black hole. The classical singularity occurs in a region where the gravitational coupling is strong and quantum effects are expected to be significant. A singularity signals a breakdown of the equations that predict it and it is an interesting question what replaces the black hole singularity in the quantum theory.

A second reason is that the classical picture of a black hole as a stable end-result of gravitational collapse is incorrect. Black holes emit Hawking radiation and gradually evaporate. A large black hole with a smooth external geometry will eventually reduce to a small one with strong curvature at the event horizon. The final stage of the evaporation can only be described by a quantum theory and it is unknown at present whether the black hole completely disappears or whether some strong coupling effects stabilize a Planck scale remnant.

Black hole evaporation precipitates a serious conflict between general relativity and quantum physics. Semi-classical calculations indicate that a black hole emits black body radiation at the Hawking temperature $T_h = \hbar c^3/8\pi k G M$, where M is the black hole mass. Now imagine matter in a pure quantum mechanical state undergoing gravitational collapse. In quantum mechanics a pure initial state always evolves to a pure final state but if a black hole is formed then, according to the semi-classical computations, the final state will contain outgoing thermal radiation and thus be a mixed state. The information about the initial pure quantum state appears to have been lost inside the black hole. This paradox has received a lot of attention in recent years and will be discussed in more detail in some of the lectures in this volume. Its resolution will presumably require us to give up some accepted dogma and this is often the route to progress.

A related issue concerns thermodynamic properties of black holes. The laws that govern classical black hole dynamics can be cast in a form that closely parallels the laws of thermodynamics and the analogy is strengthened when quantum effects are included. A black hole emits black-body radiation at a temperature which is proportional to the black hole surface gravity, and appears to carry an entropy $S = A/4$, where A is the area of the event horizon in Planck units. The black hole entropy must be considered physical if the second law of thermodynamics is to extend to systems which include black holes, for otherwise one could reduce the total entropy by sending thermal matter into a black hole. The ordinary laws of thermodynamics can be understood in terms of statistical physics and we would like to have a corresponding explanation of black hole thermodynamics. An actively pursued question is how black hole entropy can be given a microphysical interpretation in quantum gravity.

At the technical level, much of the work on quantum aspects of black hole physics has been in a semiclassical context, involving quantized matter in a classical background geometry. Some progress has been made in taking into account the back-reaction on the geometry due to Hawking emission, although so far a controlled systematic framework is only available in simplified toy models. It is important to develop this area further but many important questions cannot be answered within a semiclassical approximation and will require a more fundamental approach.

We have by no means exhausted the list of interesting questions that arise in black hole physics. Let us mention a few more: 1) In a quantum theory one expects pair creation of black holes and there is some controversy as to what the rate of pair creation would be. 2) Is the spectrum of black hole states in quantum gravity discrete or continuous? The answer to this question is intimately related to the issues of entropy and information loss mentioned above. 3) To what extent do the uniqueness or no-hair theorems on classical black holes carry over to quantum gravity? 4) A black hole carrying electromagnetic charge has a timelike classical singularity, which becomes visible to observers that enter the black hole and pass through the so called Cauchy horizon. The classical initial value problem breaks down in this case and one may ask how Cauchy horizons affect quantum mechanical evolution.

§ 3 Quantum Cosmology

The other main area where quantum gravity finds application is in cosmology. In the hot big bang scenario the Universe expands from a configuration governed by physics at very high energy, where quantum effects are expected to be important. A quantum theory of cosmology must deal with the earliest epoch in the evolution of the Universe, where the classical cosmological solution breaks down at a singularity. The relevant laws of physics are unknown at present and the nature of the questions to be answered will depend on the form these laws take.

If we assume that the Universe evolves from some initial quantum state we can ask to what extent the theory specifies this initial state and what observations on the present Universe could in principle reveal information about it. Possibly the dynamics of the subsequent evolution is such as to seek out a particular configuration for a large class of different initial states, as for example in inflationary models. In that case the precise nature of the initial state is somewhat less important and will at any rate not be accessible from later observations.

In order to properly address such issues we have to confront some basic conceptual problems that arise when we attempt to give a quantum mechanical description of cosmology. First of all, our Universe is a unique object whereas the statistical predictions of quantum mechanics normally apply to ensembles of identically prepared systems. A closely related concern is the meaning of quantum measurements when all observers are necessarily part of the quantum system itself. Another important question is how the classical Universe we inhabit now, with its irregular matter distribution, evolves from an initial pure quantum mechanical state. Recent work on the decoherence of histories in quantum theory is aimed at addressing some of these problems.

In many models of quantum cosmology our observable Universe is only a small piece of a much larger structure, which contains regions of spacetime that are removed from causal contact with our region, or even topologically separate 'universes'. We may have to settle for a theory that does not uniquely determine the initial configuration of our region of spacetime but only assigns probabilities to different possibilities. Similarly, the very laws of low-energy physics in our observable Universe and the number of observed spacetime dimensions could also be subject to stochastic rules. If the probability distribution is strongly peaked around some given set of low-energy laws and matter configuration we could still claim that our theory predicts these. Another possibility is that the measured values of the constants of nature and the matter content we observe are only favored by the fact that our form of intelligent life can only evolve in regions of spacetime with these properties. This weak form of the so called anthropic principle does not give very satisfying answers but is hard to rule out as a logical possibility.

There is no lack of interesting questions in quantum cosmology. The challenge for workers in this field is to organize the answers into a coherent framework and extract physical predictions that are relevant to the observable world around us. At present, quantum cosmology is a speculative enterprise which borders too closely on the metaphysical for the taste of many physicists. On the other hand, the inquiry into the origin of the Universe has always had a strong appeal and physics brings a unique perspective, rooted in physical observations made on the Universe at large, to the debate.

§ 4 The G1 Working Group

We have touched upon a number of topics that are of current interest in quantum gravity. Some of these are discussed further in the various contributions that follow, and are based on lectures delivered during the Working Group sessions. The black hole information paradox has been actively studied in recent years and a majority of the lectures were concerned with different aspects of this problem. The following is a list of the contributions and the topics covered in each:

S.B. Giddings and L. Thorlacius give an introductory review of the information puzzle.

This is followed by a contribution from A. Strominger, who first discusses the information problem in the simplified context of two-dimensional models and then describes a possible resolution of the paradox based on third quantization.

V. Frolov considers black hole thermodynamics and discusses some recent efforts to give a dynamical interpretation to the entropy carried by a black hole.

L. Thorlacius discusses kinematic requirements for returning quantum information in Hawking radiation and their possible implementation in string theory.

Unusual causal properties of string field theory, which may enable information return, are described in the contribution by D.A. Lowe.

The issue of back-reaction on the geometry of spacetime due to quantum effects is addressed by A.O. Barvinsky, who discusses an approach based on a non-local effective action.

The final contribution from the G1 Working Group to these Proceedings is by P.C. Argyres and deals with recent numerical work, which has exhibited universality and scaling in black hole formation, reminiscent of scaling in statistical systems. As it is based on classical Einstein equations this work does not directly involve quantum aspects of gravity but it hints at an interesting mathematical structure and may have implications beyond the classical theory.

§ Acknowledgements

We thank the organizers and the staff of the Summer Study for their efforts and the participants in the G1 Working Group for lively discussions.

The work of V.F. is supported in part by NSERC and that of L.T. is supported in part by NSF grant No. PHY89-04035.

Introduction to the Information Problem

Steven B. Giddings

Department of Physics, University of California, Santa Barbara, CA 93106-9530, USA

Lárus Thorlacius

Institute for Theoretical Physics, University of California, Santa Barbara, CA 93106-4030, USA

Hawking's black hole information paradox highlights the incompatibility between our present understanding of gravity and quantum physics. The paradox arises in the context of black hole evolution where infalling matter in a pure initial quantum state appears to evolve into outgoing thermal radiation. Its resolution may involve fundamental information loss, subtle athermal correlations in the outgoing radiation, or long-lived black hole remnants.

§ 1 The Puzzle

The black hole information problem has received considerable attention as it identifies a serious conflict between quantum mechanics and general relativity. To illustrate the problem, consider a gedanken experiment in which a black hole of mass M is formed in collapse of matter initially described by a pure quantum state $|\psi\rangle$, with corresponding density matrix $\rho = |\psi\rangle\langle\psi|$. An important quantity is the entropy, which for a general density matrix is given by $S = -\mathrm{Tr}\rho \ln \rho$, and is zero for this pure state. After formation, the black hole decays by emitting Hawking radiation [1]. The original semi-classical calculation involved quantized matter in a classical background black hole geometry. It showed the radiation to be approximately thermal and thus described by a mixed-state density matrix of the form

$$\rho = \sum_n p_n |\psi_n\rangle\langle\psi_n| \,. \tag{1}$$

The entropy in the Hawking radiation is nonzero, and can be shown to be of order $M^2/m_{\rm pl}^2$ where $m_{\rm pl}$ is the Planck mass. This indicates that the final state is missing information as compared to the initial state, in accord with the general formula $\Delta I = -\Delta S$ relating information and entropy. Hawking argued that the source of the missing information is the correlation between the particles that come out in Hawking radiation and those that enter the black hole; without observing the internal state of the black hole, this information cannot be recovered.

An obvious question is what happens to the information. There are several possibilities. The first is that the black hole completely disappears at the end of Hawking evaporation. If the above calculation is to be trusted, this then means that a pure quantum state has evolved into a mixed quantum state. That implies a fundamental loss of information, and conflicts with the basic principles of quantum mechanics which always evolves pure states to pure states. Such a radical conclusion suggests one consider other options. One is that some assumptions that go into the semi-classical calculations are flawed and the information is in fact gradually released from the black hole over its lifetime and carried in the Hawking radiation. Alternately, one might imagine that the information only escapes after the black hole reaches the Planck mass, where quantum gravity becomes important and Hawking's calculation fails in any case. The resulting outgoing state must, however, contain a huge amount of information, $I \sim M^2/m_{\rm pl}^2$, and this information is to be transmitted using only the remaining energy, $E \sim m_{\rm pl}$. The only way to send a large amount of information with a small amount of energy is to use a large number of very low energy particles, and this takes a long time. A simple estimate [2] gives a time

$$\tau \sim \left(\frac{M}{m_{\rm pl}}\right)^4 t_{\rm pl} \,, \tag{2}$$

which substantially exceeds the age of the Universe for an initial black hole formed from a mass as small as that of Pyramid peak. This means that in this case one is left with a slowly-decaying black hole remnant.

Each of these three alternatives – *information loss, information return,* and *remnants* – has been advocated by serious physicists but in each case serious physical objections have also been raised. By considering a sufficiently massive black hole the spacetime curvature and other local invariant features of the geometry can be made arbitrarily close to those of flat Minkowski space everywhere outside the black hole horizon. It is often claimed that the physics of black hole evaporation must therefore be governed by low-energy physics and in particular be independent of Planck scale dynamics. On the other hand, each of the three alternatives appears to violate some principle of low-energy physics and therefore, if one of these scenarios is to be correct, we expect there must be a loophole in some of the low-energy arguments, through which Planck scale physics enters in some way. In what follows we consider each proposal in turn and identify key theoretical challenges that must be met in order to make it viable. We will stick to the three basic scenarios outlined above as other proposed resolutions of the information puzzle usually involve variations on these themes [2,3].

§ 2 Information Loss

Hawking's proposal that pure states evolve into mixed states due to gravitational effects manifestly violates the basic quantum mechanical principle of unitarity. Furthermore, a general connection between information and energy indicates that loss of information should imply violation of energy conservation. In particular, it is natural to expect that virtual processes involving Planck scale black holes cause information and energy loss in low-energy physics. The question is then whether this will be a small effect or one which severely disturbs the low-energy vacuum.

Hawking attempted to generalize quantum mechanics to allow for information loss [4]. He proposed to replace the unitary S matrix, which maps an initial pure quantum state to a pure final state, by a superscattering matrix $\$$ which acts on density matrices rather than state vectors,

$$\rho_{AB} \rightarrow \$_{AB}^{\;CD} \rho_{CD} , \tag{3}$$

and can map a pure quantum state to a mixed one. It was subsequently argued [5] that this generalization, at least when implemented in terms of an evolution equation for density matrices, violates either locality or energy-momentum conservation. Hawking's proposal is undermined by virtual black hole creation leading to unsuppressed Planck scale energy fluctuations. The challenge to advocates of the information-loss scenario is, therefore, to *find an explicit description of information loss consistent with energy conservation.*

§ 3 Information Return

Unitarity can be maintained if all information about the quantum state of the collapsing matter is encoded into the Hawking radiation, but in this case we have to give up some notions of locality and causality [6,7].

Imagine a hapless traveler falling into a large black hole. From the traveler's perspective nothing bad happens until near the singularity long after passing the event horizon. If the traveler's information is to be encoded into the Hawking radiation there must be some physical mechanism that transfers all the information from the infalling traveler to the outgoing radiation leaving nothing behind. Since the Hawking radiation originates outside the black hole, and the local spacetime curvature is weak everywhere in its causal past, this mechanism cannot operate in a local fashion. In the context of a large black hole the non-local effects must operate over long distances in order to reconcile the viewpoints of the traveler, who feels fine until deep inside the black hole, and of outside observers, who claim the traveler is completely disrupted before entering the black hole. The principle of ***black hole complementarity*** [7] states that such a reconciliation need not contradict the known laws of low-energy physics, because in order to compare measurements made by distant observers to measurements made inside a black hole one must either be a superobserver outside spacetime or have knowledge of extreme short-distance physics.

Susskind has suggested that string theory is sufficiently non-local to enable information return [8]. Strings are indeed not point-like objects but naïvely one would only expect non-local effects on the fundamental string scale. According to Susskind, the enormous relative boosts encountered in describing different observers near the event horizon magnify this short distance non-locality so that it becomes relevant on macroscopic scales. However, even if the kinematics of string theory differ sufficiently radically from point-like theories that information from the contents of a black hole is in some sense accessible at the horizon, it remains an open question whether string interactions alter the Hawking radiation significantly enough to imprint this information on the outgoing state. A concrete challenge in this scenario is thus to *find nonlocal dynamics (string or otherwise) sufficient to transfer the information to the Hawking radiation.*

§ 4 Remnants

If remnants solve the information problem, there must be an infinite number of remnant species so that they can encompass the information associated with an arbitrarily large black hole. These remnants should have masses $\sim m_{\rm pl}$. The obstacle here is that an infinite variety of remnants with approximately equal masses leads to the problem of infinite production rates for remnants in ordinary processes. Examples would be in Hawking radiation from large black holes, in thermal ensembles, or if the remnants are charged, in Schwinger pair production in electric fields. The production rate for a given remnant species may be incredibly tiny, but the infinite number of species still gives an infinite total rate.

This logic has been questioned [9] on the grounds that

it considers remnants to be essentially similar to elementary particles and in particular to be described by effective field theory. A concrete model for remnants are extremal Reissner-Nordstrom black holes, which are also expected to have an infinite number of internal states [10]. Studies of their pair production [11]. suggest the possible relevance of Planck-scale physics in computing production rates [12]. This opens the possibility of finite production. A challenge to remnant advocates is then to exhibit the mechanism by which remnants evade the naïve estimates, and in particular to *provide a concrete effective description of remnants that avoids the connection between infinite species and infinite production rates.*

§ 5 Discussion

Theorists have already learned a great deal by thinking about Hawking's paradox. In particular, our picture of semiclassical black hole physics, including the backreaction on the geometry due to Hawking emission, has been considerably clarified in recent years. A conclusive resolution of the paradox nevertheless remains elusive.

It is of course quite possible that none of the above proposals correctly describes the physics of black hole evaporation. Perhaps some combination of these ideas is closer to the truth or perhaps some key observation is missing and a completely new approach required. The correct scenario may or may not be determined in the next six years, but understanding the underlying Planck-scale physics is certain to be an important element of the physics of the next millennium.

§ Acknowledgements

We wish to thank the organizers for organizing an interesting and enjoyable workshop. The work of S.B.G. is supported in part by DOE grant No. DOE-91ER40618 and NSF grant No. PHY91-57463 and that of L.T. by NSF grant No. PHY89-04035.

§ References

[1] S.W. Hawking, *Comm. Math. Phys.* **43** (1975) 199, *Phys. Rev.* **D14** (1976) 2460.

[2] R.D. Carlitz and R.S. Willey, *Phys. Rev.* **D36** (1987) 2327,2336; J. Preskill, *"Do Black Holes Destroy Information?"* Caltech preprint, CALT-68-1819, January 1992, hep-th/9209058.

[3] S.B. Giddings, *Phys. Rev.* **D46** (1992) 1347; D.N. Page, *"Black Hole Information,"* Alberta preprint, ALBERTA-THY-23-93, May 1993, hep-th/9305040.

[4] S.W. Hawking, *Comm. Math. Phys.* **87** (1982) 395.

[5] J. Ellis, J. Hagelin, D.V. Nanopoulos, and M. Srednicki, *Nucl. Phys.* **B241** (1984) 381; T. Banks, M.E. Peskin, and L. Susskind, *Nucl. Phys.* **B244** (1984) 125.

[6] G. 't Hooft, *Nucl. Phys.* **B335** (1990) 138, *Phys. Scr.* **T36** (1991) 247; E. Verlinde and H. Verlinde, *Nucl. Phys.* **B406** (1993) 43.

[7] L. Susskind, L. Thorlacius, and J. Uglum, *Phys. Rev.* **D48** (1993) 3743; L. Susskind and L. Thorlacius, *Phys. Rev.* **D49** (1994) 966.

[8] L. Susskind, *Phys. Rev. Lett.* **71** (1993) 2367; *Phys. Rev.* **D49** (1994) 6606.

[9] T.Banks and M. O'Loughlin, *Phys. Rev.* **D47** (1993) 540; T. Banks, M. O'Loughlin, and A. Strominger, *Phys. Rev.* **D47** (1993) 4476.

[10] A. Strominger and S. Trivedi, *Phys. Rev.* **D48** (1993) 5778

[11] G.W. Gibbons, in *Fields and geometry*, ed. A. Jadczyk (World Scentific, 1986); D. Garfinkle and A. Strominger, *Phys. Lett.* **B256** (1991) 146.

[12] S.B. Giddings, *Phys. Rev.* **D49** (1994) 4078.

Two-Dimensional Black Holes and the Information Puzzle

Andrew Strominger

Department of Physics, University of California, Santa Barbara, CA 93106-9530, USA

Several years ago it was realized that two-dimensional methods provide effective tools for addressing the black hole information puzzle [1]. Vigorous application and improvements of these tools over the last two years have crystallized the puzzle, eliminating some proposed resolutions and greatly constraining the nature of others. In the first half of this report I will briefly review some of the work on these two-dimensional models. In the second half I will briefly describe a possible resolution of the information puzzle that we were led to by these investigations. Discussion of other approaches to the puzzle can be found for example in [2].

§ 1 Two-dimensional Models

Two-dimensional dilaton gravity models may be viewed as the S-wave sector of a four-dimensional theory of gravity. The general spherically symmetric, four-dimensional line element may be written

$$ds^2 = g_{ab}(\sigma)d\sigma^a d\sigma^b + e^{-2\phi(\sigma)}d\Omega^2$$

where $(\sigma^1, \sigma^0) \sim (r, t)$, g_{ab} is a two-dimensional metric and $4\pi e^{-2\phi}$ is the area of a two sphere at fixed σ expressed in terms of the dilaton field ϕ. Thus the dynamics of four-dimensional S-wave gravity are described by a two-dimensional metric *plus* a scalar dilaton.

Models of this general type (when coupled to matter) contain black holes, Hawking evaporation and, consequently, an information puzzle. However, they are far simpler than their four-dimensional cousins. This is in part because two-dimensional quantum gravity is renormalizable: the short distance problems of quantum gravity are successfully untangled from the long-distance information problem. Even better, it is known that two-dimensional quantum gravity is equivalent to conformal field theory and the technology developed in this context over the last ten years can be fruitfully applied to the black hole problem.

Of course one should always bear in mind the possibility that we are being led down the garden path. There is no guarantee that the resolution of the information puzzle for real four-dimensional black holes is the same as that for toy two-dimensional black holes.

Two-dimensional models have been improved in steps over the last four years. The starting point is the classical action [1]

$$S_{c\ell} = \frac{1}{4\pi}\int d^2\sigma\sqrt{-g}\,\left[e^{-2\phi}(R + 4(\nabla\phi)^2 + 4\lambda^2) + \sum\nolimits_{i=1}^{N}(\nabla f_i)^2\right]$$

where λ is a dimensionful constant and the f_i are N matter fields. Classical solutions of this theory describe formation of black holes by collapsing matter.

The quantum theory is described to leading order in a $1/N$ expansion by the effective action [1]

$$S_{\text{eff}} = S_{c\ell} + \frac{N}{24\pi}\int d^2\sigma\partial_+\rho\partial_-\rho$$

in a conformal gauge in which the two-dimensional metric is a $ds^2 = -e^{2\rho}d\sigma^+ d\sigma^-$. The last term is deduced from the trace anomaly, and incorporates the back reaction of the Hawking radiation on the geometry. Solutions of S_{eff} can be found numerically [3]. One finds that black holes shrink due to Hawking emission as expected. Eventually they reach zero size, a naked singularity appears, and the computer crashes. While more input is required to evolve beyond this point, one important lesson can already be gleaned: in large-N, two-dimensional models the information does not come out of the black hole prior to the evaporation endpoint. This was quantified in [4] by direct calculation of the entropy.

A great improvement of this model was found in references [5,6,7], where it was noted that equally sensible toy models for black hole physics can be obtained by modifying $S_{c\ell}$ with counterterms which vanish at large radius (large negative ϕ). These counterterms are constrained by conformal invariance. A judicious choice of counterterms leads to a theory which can be transformed (using field redefinitions) to a free field theory! Furthermore, in terms of the redefined fields, the spacetime can be analytically continued both through the origin and through the singularity.

These improved theories are simple enough that a fully quantum treatment is feasible. Notable progress in this direction has been made in [8], but some issues remain unresolved. In the remainder of our discussion we shall concentrate on the large-N, semiclassical behavior.

A semiclassical analysis reveals a disaster lurking in the improved theories. Collapsing matter forms a black hole which initially evaporates as expected. Unfortunately, black holes never stop evaporating, even when the horizon reaches $r = 0$! The mass of the spacetime asymptotically tends to minus infinity. Of course no one believes that anything like this could happen in the real world. Evidently these two-dimensional models fail to provide a faithful model of four-dimensional black hole physics at (and after) the evaporation endpoint. Thus while two-dimensional models have been of use for studying the flow of information in and out of black holes prior to the endpoint, models developed so far have not been as useful for studying possible types of endpoint behavior.

Attempts to alleviate this problem have been made by imposing reflecting boundary conditions at the origin $r = 0$ (which corresponds to a timelike line of constant ϕ in the two-dimensional field theory) [7,9,10].[1] This is a standard problem in conformal field theory. Interestingly, it turns out that the simplest stable solution in the present context is highly non-linear. With these boundary conditions, all incoming matter is reflected through the origin. However, there is a threshold for black hole formation, above which unphysical behavior reappears. The black hole never stops evaporating and the mass of the spacetime in the region exterior to the black hole goes to minus infinity. The reflected pulse in a sense comes back out, but it does so after the end of time as measured by an observer exterior to the black hole. This is a consequence of distortion of the geometry by the infalling matter.

Indeed both the evaporation endpoint (*i.e.* the point at which the black hole horizon has zero size) and the end of time are (for large infalling pulses) prior to the future lightcone of the region where the pulse reaches the origin. Thus no boundary conditions at the origin can avert this disaster.

Modifications of the two-dimensional model are needed to obtain dynamics which are sensible in all regimes. No fully satisfactory model exists as of this writing. Attempts are being made to modify the theory as follows [11].

At the evaporation endpoint, the black hole horizon has reached zero (or Planckian) size. However, it may have a large interior region. The geometry (at least for some spacelike slices) contains a large black hole interior region (storing lots of information) connected by a Planckian umbilical cord to the exterior spacetime. In our discussion so far, this umbilical cord cannot be broken. However, if topology change is allowed in quantum gravity (as we believe to be the case) eventually the umbilical cord will break and the black hole interior becomes a baby universe. This possibility should be incorporated into the two-dimensional models.

To a string theorist, this process is nothing more than open string (baby universe) emission from the end of a semi-infinite string (the original spacetime) and the technology for describing such a process is at least partially in hand.

[1] The original attempt in [7] had some technical difficulties which were corrected in [9,10].

§ 2 Information Retrieval

Let us assume that black hole evaporation indeed terminates in the decay of the black hole interior into its own baby universe. What does this mean for the issue of information loss?

The answer depends on how one treats the portion of the quantum state which is carried away by the baby universe. Hawking has proposed that one simply throw away (*i.e.* trace over) this portion of the state. This leads to a theory in which information is irretrievably lost. Indeed in this context there is no observable content to the statement that it is "carried away by a baby universe". The information might as well have been destroyed at a singularity.

A second, inequivalent, proposal [11] is to treat the baby universes as indistinguishable quantum particles in their own right, which inhabit a larger "third-quantized" Hilbert space. The indistinguishability means that one does not just trace separately over the state of each baby universe, but one symmetrizes by tying together in all possible ways the traces over all baby universes created at any time in the entire history of the universe.

This proposal was objected to [12] on the grounds that it violated cluster decomposition: the symmetrizing over all baby universe correlates widely separated events. However, it turns out [13] that this violation of clustering is physically unobservable, because the theory decomposes into superselection sectors, in each of which clustering is valid. Even better, the scattering is unitary within each superselection sector! The argument [13] is in essence very similar to those used earlier in wormhole physics [14], but with additional subtleties.

The superselection sectors are labeled by the eigenvalues α_i of the (third-quantized) operators ϕ_i which create and destroy a baby universe in the ith quantum state. The unitary outcome of gravitational collapse depends on the α_i's, but they cannot be measured *except* by forming black holes and watching them evaporate. This requires an enormous number of experiments. Before the α_i's are known, the outcome of gravitational collapse is unpredictable, even if the exact solution of quantum string theory were at hand. Indeed, an averaging over the unknown α_i's for the case of a single black hole formation/evaporation exactly reproduces Hawking's prescription. Thus, in this proposal, information is preserved in principle but lost in practice.

§ References

[1] C. G. Callan, S. B. Giddings, J. A. Harvey and A. Strominger, *Phys. Rev.* **D45** (1992) R1005.

[2] L. Susskind, to appear proceedings of the conference *Particles, Strings, & Cosmology*, Syracuse University, May '94.

[3] D. Lowe, *Phys. Rev.* **D47** (1993) 2446; T. Piran and A. Strominger, *Phys. Rev.* **D48** (1993) 4729.

[4] T. Fiola, J. Preskill, A. Strominger, and S. Trivedi, *Phys. Rev.* **D50** (1994) 3994.

[5] A. Bilal and C.G. Callan, *Nucl. Phys.* **B394** (1993) 73.

[6] S.P. de Alwis, *Phys. Lett.* **B289** (1992) 278; *Phys. Lett.* **B300** (1993) 330; *Phys. Rev.* **D46** (1992) 5429.

[7] J.G. Russo, L. Susskind, and L. Thorlacius, *Phys. Rev.* **D46** (1992) 3444; *Phys. Rev.* **D47** (1993) 533.

[8] E. Verlinde and H. Verlinde, *Nucl. Phys.* **B406** (1993) 43.

[9] T.D. Chung and H. Verlinde, *Nucl. Phys.* **B418** (1994) 305.

[10] A. Strominger and L. Thorlacius, *Phys. Rev.* **D50** (1994) 5177.

[11] A. Strominger, unpublished.

[12] L. Susskind, *Comment on a Proposal by Strominger.* Preprint SU-ITP-94-14; hep-th/9405103.

[13] J. Polchinski and A. Strominger, *A Possible Resolution of the Black Hole Information Puzzle*, Preprint UCSBTH-94-20, hep-th/9407008.

[14] S. Coleman, *Nucl. Phys.* **B307** (1988) 864; S.B. Giddings and A. Strominger, *Nucl. Phys.* **B307** (1988) 854.

Entropy of Black Holes

Valeri Frolov

CIAR Cosmology Program; Theoretical Physics Institute, University of Alberta, Edmonton, Canada T6G 2J1

§ 1 Black-Hole Entropy Problem

According to the thermodynamical analogy in black hole physics, the entropy of a black hole in the Einstein theory of gravity equals $S^{BH} = A_H/(4l_{\rm P}^2)$, where A_H is the area of a black hole surface and $l_{\rm P} = (\hbar G/c^3)^{1/2}$ is the Planck length [1, 2].

The entropy in black hole physics plays essentially the same role an in the usual thermodynamics. In particular it allows one to estimate what part of the internal energy of a black hole can be transformed into work. The generalized second law implies that when a black hole is a part of a thermodynamical system the total entropy (i.e. the sum of the entropy of a black hole and the entropy of the surrounding matter) does not decrease. The success of the thermodynamical analogy in black hole physics allows one to hope that this analogy may be even deeper and it is possible to develop a statistical-mechanical foundation of black hole thermodynamics.

Thermodynamical and statistical-mechanical definitions of the entropy are logically different. *Thermodynamical entropy* S^{TD} is defined by the response of the free energy F of the system to a change of its temperature:

$$dF = -S^{TD} dT. \qquad (1)$$

(This definition applied to a black hole determines its Bekenstein-Hawking entropy.)

Statistical-mechanical entropy S^{SM} is defined as

$$S^{SM} = -\mathrm{Tr}(\rho \ln \rho), \qquad (2)$$

where ρ is the density matrix describing the internal state of the system under consideration. It is also possible to introduce the *informational entropy* S^I by counting different possibilities to prepare a system in a final state with given macroscopical parameters from different initial states

$$S^I = -\sum_n p_n \ln p_n, \qquad (3)$$

with p_n being the probabilities of different initial states.

In standard case all three definitions give the same answer.

Is the analogy between black holes thermodynamics and the 'standard' thermodynamics complete? Do there exist internal degrees of freedom of a black hole which are responsible for its entropy? Is it possible to apply the statistical-mechanical and informational definitions of the entropy to black holes and how are they related with the Bekenstein-Hawking entropy? These are the questions which are to be answered.

Historically first attempts of the statistical-mechanical foundation of the entropy of a black hole were connected with the informational approach [2, 3]. According to this approach the black hole entropy is interpreted as "the logarithm of the number of quantum mechanically distinct ways that the hole could have been made"[3]. The so defined informational entropy of a black hole is simply related with the amount of information lost by stretching the horizon, and as was shown by Thorne and Zurek it is equal to the Bekenstein-Hawking entropy [3].

The dynamical origin of the entropy of a black hole and the relation between the statistical-mechanical and Bekenstein-Hawking entropy have remained unclear. In the present talk I describe some new results obtained in this direction.

§ 2 Dynamical Degrees of Freedom

Calculations in the framework of the Euclidean approach initiated by Gibbons and Hawking[4, 5] relate the entropy of a black hole to the tree-level contribution of the gravitational action, namely the action of the Euclidean black hole instanton. In this approach the entropy of a black hole has pure topological origin, and it remains unclear whether there exist any real dynamical degrees of freedom which are responsible for it. The problem of the dynamical origin of the black hole entropy was intensively discussed recently. The basic idea which was proposed is to relate the dynamical degrees of freedom of a black hole with its quantum excitations. This idea has different realizations[6, 7, 8, 9, 10, 11].

Here I discuss the recent proposal [7, 11] to identify the dynamical degrees of freedom of a black hole with the states of all fields (including the gravitational one) which are located inside the black hole. Such modes (for a non-rotating black hole) have negative energy. In the Hawking process, the creation of a particle outside a black hole (such a particle necessarily has positive energy) is accompanied by a creation of a corresponding particle in a mode with negative energy inside a black hole. As a result these modes with negative energies are permanently excited and their state is described by thermal density matrix. Only very small number of those particles which are created outside a black hole (external particles) can penetrate the potential barrier and reach infinity. Namely these particles form the quantum radiation of a black hole. All other external particles are reflected by the potential barrier and fall down into the black hole. During the time when they are still outside the horizon, the corresponding internal modes (which are described by a thermal density matrix) give the contribution to the black hole entropy.

§ 3 Statistical-Mechanical Entropy

By averaging over states located outside the black hole one generates the density matrix of a black hole and can calculate the corresponding statistical-mechanical entropy S^{SM}. The main contribution to the entropy of a black hole is given by inside modes of fields located in the very close vicinity of the horizon. Contributions of different fields enter S^{SM} additively. The calculations give the following result for the contribution of a chosen field to S^{SM}

$$S^{SM} = \int d\boldsymbol{x} \sum_{\lambda} \mu_{\lambda}(\boldsymbol{x}) s(\beta\omega_{\lambda}), \tag{4}$$

$$s(\beta\omega) = \frac{\beta\omega}{e^{\beta\omega} - 1} - \ln(1 - e^{-\beta\omega}), \tag{5}$$

where $s(\beta\omega)$ is the entropy of a single oscillator with the frequency ω at temperature $T = 1/\beta$. Here

$$\mu_{\lambda}(\boldsymbol{x}) = g^{\tau\tau} g^{1/2} [R_{\lambda}(\boldsymbol{x})]^2 \tag{6}$$

is a phase space density of quantum modes and $R_{\lambda}(\boldsymbol{x})$ are spatial harmonics corresponding to the mode with a collective quantum number $\lambda = (\omega, l, m)$.

The so defined S^{SM} contains a volume divergence, connected with the integration over the space regions near the horizon and is of the form[7]

$$S^{SM} \approx \frac{\alpha}{\pi^2 \varepsilon} \tag{7}$$

where $\alpha \equiv \frac{\pi^2}{90}\left[h(0) + \frac{7}{8}h(1/2) + h(1)\right]$, $\varepsilon = (l/r_+)^2$, $h(s)$ is the number of helicities of field of spin s, and l is the proper distance cut-off parameter. One may expect that quantum fluctuations of the horizon may provide natural cut-off and make S^{SM} finite. Simple estimations[7] of the cut-off parameter show that $S^{SM} \approx S^{BH}$.

§ 4 No-Boundary Wave Function

Another approach to the problem of dynamical degrees of freedom of a black hole was proposed in Ref.[11]. Its basic idea is the following. The study of propagation of perturbations in the spacetime of a real black hole can be reduced to the analogous problem for its 'eternal version', (i.e. in a spacetime of an eternal black hole with the same parameters). The space of physical configurations of a system including a black hole can be related to the space of 'deformations' of the Einstein-Rosen bridge of the eternal black hole and possible configurations of other (besides the gravitational) fields on it, which obey the constrains and preserve asymptotic flatness. In a spacetime of an 'eternal version' of a black hole, perturbations with initial data located on the inner part of the Einstein-Rosen bridge are propagating to the future remaining entirely inside the horizon, and hence the corresponding perturbations in a 'physical' black hole also always remain under the horizon. That is why these data should be identified with internal degrees of freedom of a black hole. A quantum state of a black hole can be described by a wavefunction defined as a functional on the configuration space of deformations of the Einstein-Rosen bridge. In this representation deformations of the external and internal parts of the Einstein-Rosen bridge naturally represent degrees of freedom of matter outside the black hole and black hole's internal degrees of freedom. The no-boundary ansatz (analogous to Hartle-Hawking ansatz in quantum cosmology) singles out a state which plays the role of ground state of the system[11]. By its construction the no-boundary wavefunction of a black hole is symmetric with respect to the transposition of the interior and exterior parts of the Einstein-Rosen bridge. We call this property *duality*. For a 'real' black hole formed in the gravitational collapse, this exact symmetry is broken. Nevertheless, since there is a close relation between physics of a 'real' black hole and its 'eternal version', the duality of the above type plays an important role and allows one, for example, to explain why the approach based on identifying the dynamical degrees of freedom of a black hole with its external modes gives formally the same answer for the dynamical entropy of a black hole as our approach.

For study the fields contribution to the statistical-mechanical entropy in the one-loop approximation it is sufficient to fix mass M of a black hole as a parameter in the wave function, and consider only the part describing fields perturbations. By tracing over the external variables one obtains the density matrix of a black hole and can calculate its statistical-mechanical entropy[11]. The result coincides with (7).

§ 5 Why the Entropy is $A/4$?

In the above approach (as well as in other dynamical approaches) different fields give independent contributions to S^{SM}. Even if the cut-off parameter depends on the number of fields in Nature it is virtually impossible to exclude the dependence of the expression for S^{SM} on the fields properties. This behavior of S^{SM} differs this quantity from the universally defined Bekenstein-Hawking entropy. This was considered by many as the puzzle. In Ref.[12] a simple solution to this puzzle was proposed. The Bekenstein-Hawking entropy is analogous to the thermodynamical entropy, defined by the response of the free energy on the change of the temperature. The standard prove of the equality of the statistical-mechanical and thermodynamic entropy requires commutativity of the operations of tracing over the internal states of the system and differentiating with respect to the temperature. In the case of a black hole, where the number and properties of the internal states depend on the mass of a black hole, which in its turn depends on the equilibrium temperature, this commutativity property is not valid. In order to demonstrate this one can write the contribution of a chosen field to the free energy of a black hole in the form, which is similar to (4)

$$F = \int d\boldsymbol{x} \sum_{\lambda} \mu_{\lambda}(\boldsymbol{x}) f(\beta\omega_{\lambda}) + \ldots, \tag{8}$$

$$f(\beta\omega) = \ln(1 - e^{-\beta\omega}), \tag{9}$$

where $(\ldots)$ denotes the terms independent of β which do not contribute to S^{SM}. After summation over l, m and integration over the spatial volume one gets

$$F = \int d\omega N(\omega|M,\varepsilon) f(\beta\omega_{\lambda}) + \ldots, \tag{10}$$

where $N(\omega|M,\varepsilon)$ is the density of number of states.

For the thermal equilibrium the mass M of a black hole is related with temperature ($\beta \equiv T^{-1} = 4\pi M$). That is why for the calculation of the response dF on the change of the temperature dT one needs to take into account additional dependence of N on T. This additional dependence of N on T reflects the fact that operations d/dT and Tr do not commute for our system. As the result S^{TD} differs from S^{SM} and one has[1] $S^{TD} = S^{SM} + \Delta S$.

It is possible to show that the additional term ΔS exactly cancels the leading (divergent near the horizon) contribution to the black hole statistical-mechanical entropy[12]. As the result of this cancellation one-loop corrections to the thermodynamical entropy of a black hole (describing the contribution to this quantity of the internal dynamical degrees of freedom of a black hole) are small. This explains the universality (independence on number and properties of fields) of the expression for the thermodynamical (Bekenstein-Hawking) entropy of a black hole.

[1] It is interesting to note that this relation can be used to give a simple explanation of the entropy 'renormalization' procedure, which was postulated by Thorne and Zurek[3]

§ Acknowledgements

This work was supported by the Natural Sciences and Engineering Research Council of Canada.

§ References

[1] J. D.Bekenstein, Nuov.Cim.Lett. **4** (1972) 737.

[2] J. D. Bekenstein, Phys.Rev. **D7** (1973) 2333.

[3] W. H. Zurek and K. S. Thorne, Phys.Rev.Lett. **54** (1985) 2171.

[4] G. W. Gibbons and S. W. Hawking, Phys.Rev. **D15** (1976) 2752.

[5] S. W. Hawking, In: *General Relativity: An Einstein Centenary Survey.* (eds. S. W. Hawking and W. Israel), Cambridge Univ.Press, Cambridge, 1979.

[6] G. 't Hooft, Nucl.Phys. **B256** (1985) 727.

[7] V. Frolov and I. Novikov, Phys.Rev. **D48** (1993) 4545.

[8] S. Carlip and C. Teitelboim, preprint gr-qc/9312002 (1993).

[9] L. Susskind and J. Uglum, Phys.Rev. **D50** (1994) 2700.

[10] D. Garfinkle, S. B. Giddings, and A. Strominger, Phys.Rev. **D49** (1994) 958.

[11] A. I. Barvinsky, V. P. Frolov, and A. I. Zelnikov, preprint Thy 13-94, gr-qc/9404036 (1994).

[12] V. Frolov, preprint Thy 22-94, gr-qc/9406037 (1994).

Black Hole Complementarity

Lárus Thorlacius

Institute for Theoretical Physics, University of California, Santa Barbara, CA 93106-4030, USA

§ 1 Introduction

In this talk I shall discuss Hawking's information paradox [1] solely from the viewpoint that the information about the initial quantum state of infalling matter forming a black hole is returned to outside observers and is encoded in the outgoing Hawking radiation as the black hole evaporates. I'll assume that there is no fundamental information loss and that any stable or long-lived black hole remnants are finitely degenerate at the Planck scale. This is a conservative viewpoint in that it assumes unitarity in all quantum processes, even when gravitational effects are taken into account, but it presents a novel view of spacetime physics near an event horizon.

Imagine a team of technologically advanced observers studying the formation of a black hole and its subsequent evaporation from a safe distance. The observers prepare a pure quantum state of infalling matter and then make careful measurements on the Hawking radiation emitted by the black hole over its entire lifetime. To determine the final state, our observers will have to patiently perform an enormous number of such experiments, using an identically prepared initial state, because only mutually commuting observables can be measured in any single run. They also have to be able to make sophisticated observations of correlations between quanta emitted at different times in the life of the black hole, for even if the formation and evaporation process as a whole is governed by a unitary S-matrix the radiation emitted at any given moment will appear thermal.

Only the region exterior to the black hole event horizon is accessible in the reference frame of the asymptotic observers. In this frame the infalling matter must give up all information about its quantum state to the outgoing Hawking radiation. Note that it is not enough for the information to be imprinted onto the Hawking radiation. It must also be removed from the infalling matter as it approaches the event horizon, for otherwise we would have a duplication of information in the quantum state in violation of linear quantum mechanics [2]. A useful analogy is a book set on fire. All the information initially contained on the pages can in principle be gleaned from measurements on the outgoing smoke and radiation but at the end of the day this information is no longer available in book form. There is an important difference between the burning book and matter falling into a black hole. In the former case it is a well understood microphysical process which transfers the information from book to radiation whereas matter in free fall entering a black hole encounters nothing out of the ordinary upon crossing the event horizon. The curvature and other coordinate invariant features of the geometry are weak there if the black hole is large and the region where strong gravitational effects are expected is a proper distance of order the Schwarzschild radius further inside the black hole.

We are thus led to conclude that the physical description of matter approaching the event horizon differs between the asymptotic and free fall reference frames by more than is warranted by the usual behavior of local fields under coordinate transformations. How serious is this apparent contradiction? The gravitational redshift between the two frames is enormous; the relative boost factor grows exponentially with the time measured in the asymptotic frame and, as 't Hooft has emphasized [3], it becomes much larger than anything that has been achieved in experiments. It is therefore legitimate to question whether the usual Lorentz transformation properties of localized objects correctly relate observations made in the two frames [4].

The principle of black hole complementarity [2] states that there is no contradiction between having all the information return to outside observers encoded in the Hawking radiation and having observers in free fall carry information into a black hole. The validity of this principle rests on matter having unusual kinematic properties at high energy but I will argue that it does not conflict with low-energy physics. The basic point is that the apparent contradiction only comes about when we attempt to compare the physical description in different reference frames. The laws of nature are the same in each frame and low-energy observers in any single frame cannot establish duplication of information.

§ 2 The Stretched Horizon

The evaporation of a large black hole is a slow process and on a short timescale compared to the black hole lifetime we can approximate the evolving geometry by a static Schwarzschild geometry. An outside observer, who is at rest in Schwarzschild coordinates sees thermal radiation at a temperature which depends on the spatial position. Near the black hole this temperature goes like $T \sim 1/\delta$, where δ is the proper distance between the observer and the event horizon. The high temperature radiation can be attributed to the acceleration required to prevent the observer from falling into the black hole, which diverges in the $\delta \to 0$ limit.

Our knowledge of the laws governing very high energy physics is limited and for now I'll only deal phenomenologically with the region nearest the event horizon where the local temperature is diverging. Later in the talk I'll motivate the phenomenological description by appealing to string theory. It is well established that, from the point of view of outside observers, the classical physics of a quasistationary black hole can be described in terms of a 'stretched horizon' which is a membrane placed near the event horizon and endowed with certain mechanical, electrical and thermal properties [5]. The nature of this description is coarse grained in that it is dissipative and irreversible in time. One doesn't have to be very specific about how near the event horizon the stretched horizon is placed as long as it is close compared to the typical length scale of the classical problem, which could for example be to describe a black hole interacting with a companion in a binary.

To go beyond this classical picture we postulate that the coarse grained thermodynamic description of the classical theory has an underlying microphysical basis. The quantum mechanical stretched horizon is a membrane, carrying microphysical degrees of freedom, with an area larger than that of the event horizon by one Planck unit [2]. The term Planck unit is being used in a loose sense and simply refers to whatever high-energy scale at which the radical kinematic behavior, required for returning the information, enters. In string theory this would be the fundamental string scale which can be considerably lower than the usual Planck energy. In order to implement black hole complementarity we also have to postulate that the membrane has no substance in the frame of an observer entering the black hole in free fall.

§ 3 Gedanken Experiments

It is important to determine whether black hole complementarity leads to observable duplication of information. The results of measurements performed inside a black hole are not available outside the event horizon so outside low-energy observers cannot establish duplication. Consider, however, a gedanken experiment [6] where an observer first learns the results of measurements made on the outgoing Hawking radiation, which reveal the quantum state of some system that was previously sent into the black hole, and then enters the black hole in order to receive a direct signal from the same system.

It turns out that it is impossible to carry out this experiment employing only low energy physics [6,7]. Outside observers have to carry out correlation measurements on the outgoing Hawking radiation for a very long time before they can hope to recover the information about the system that is sent into the black hole. An observer who waits outside for that information and then enters the black hole should receive the message from the original system before crashing into the singularity. The black hole geometry is such that the measurement on the system must then be made, and the result transmitted, in an extremely short time after the system passes through the event horizon. The timescale is in fact so short that quanta of frequency $\omega \sim \exp(M^2)$, where M is the black hole mass in Planck units, would have to be employed to achieve this task [6] and the back-reaction on the geometry due to such a high-energy pulse would be very violent. Conversely, if the experimenters only have low energy radiation, *i.e.* less than Planck energy, at their disposal then it will be impossible to transmit the signal in time for the later observer to receive it before encountering strong gravitational effects near the singularity.

It is a generic feature of gedanken experiments of this type that short distance physics enters into their analysis in an essential way and apparent contradictions with black hole complementarity can be traced to unwarranted assumptions about physics beyond the Planck scale. Another class of experiments involves attempts by external observers to detect whether quantum information is stored at the stretched horizon by probing the neighborhood of the event horizon [6]. Their analysis also requires short-distance physics even for a large black hole.

§ 4 The Stretched Horizon in String Theory

If the ideas presented above are correct then it is essential to gain understanding of physics at very short distances in order to resolve the issue of information loss. String theory is widely believed to provide a consistent short-distance description of matter and gravity and Susskind has argued that the kinematic behavior of fundamental strings is consistent with the requirements of black hole complementarity [4]. The basis for this claim is that zero-point fluctuations of string modes make the size of a string depend on the time resolution employed [8]. The shorter the time over which the oscillations of a string are averaged the larger its spatial extent will appear.

Consider a string configuration in free fall approaching a black hole event horizon. An observer at rest far away from the black hole measures asymptotic time t but because of the increasing redshift, a unit of asymptotic time corresponds to an ever shorter time interval, $\delta\tau \sim \delta t \exp(-t/4M)$, in the free-fall frame near the event horizon. The distant observer is therefore using a shorter and shorter resolution time to describe the string configuration and, once it passes within a proper distance of order the string scale from the event horizon, the string begins to spread both in the longitudinal and transverse directions [4]. The longitudinal spread cancels out the longitudinal Lorentz contraction caused by the black hole geometry. Meanwhile, the spread in the transverse directions causes the configuration to cover the entire horizon area in a short time compared to the black hole lifetime. In this view, the stretched horizon is made out of the strings in the infalling matter which forms the black hole. On the other hand, the spreading effect is not present in the free-fall frame, where there is no redshift to enhance the time resolution, and from the point of view of an infalling observer there is no stretched horizon, in line with the principle of black hole complementarity.

Since the stringy stretched horizon is formed from the infalling matter itself, it efficiently absorbs the quantum information contained in that matter. The string spreading process also thermalizes the stretched horizon [9]. This comes about because a fixed time resolution in the asymptotic frame translates in the free-fall frame into a time-dependent mode cutoff on the scalar fields, which give the transverse location of the string. As time goes on, new modes emerge below the cutoff, and the random phases of the different modes lead to a classical stochastic evolution of the scalar fields, which can be given an interpretation in terms of a branching diffusion of discrete string bits. This discussion has entirely been in terms of free string theory and there are indications that string interactions significantly enhance the spreading effect [10].

A crucial remaining question in this approach is how the thermalized information stored at the stretched horizon gets imprinted on the outgoing Hawking radiation. Our present understanding of interacting string theory is insufficient to properly address this issue.

To summarize, it appears that black hole complementarity cannot be ruled out on the basis of known principles of low-energy physics. It requires radical kinematic behavior at high energies, which is not a feature of conventional local quantum field theories, but seems to be realized in string theory.

§ Acknowledgements

This work was supported in part by NSF Grant No. PHY89-04305.

§ References

[1] S.W. Hawking, *Comm. Math. Phys.* **43** (1975) 199, *Phys. Rev.* **D14** (1976) 2460.

[2] L. Susskind, L. Thorlacius, and J. Uglum, *Phys. Rev.* **D48** (1993) 3743.

[3] G. 't Hooft, *Nucl. Phys.* **B335** (1990) 138, *Phys. Scr.* **T36** (1991) 247.

[4] L. Susskind, *Phys. Rev. Lett.* **71** (1993) 2367; *Phys. Rev.* **D49** (1994) 6606.

[5] K.S. Thorne, R.H. Price, and D.A. MacDonald, *Black Holes: The Membrane Paradigm,* (Yale University Press, New Haven, CT, 1986), and references therein.

[6] L. Susskind and L. Thorlacius, *Phys. Rev.* **D49** (1994), 966.

[7] J. Preskill (private communication, 1993); H. Verlinde (private communication, 1993).

[8] M. Karliner, I. Klebanov, and L. Susskind, *Int. J. Mod. Phys.* **A3** (1988) 1981.

[9] A. Mezluhmian, A. Peet, and L. Thorlacius, *Phys. Rev.* **D50** (1994) 2725.

[10] L. Susskind, *The World as a Hologram,* Stanford University preprint, SU-ITP-94-33, September 1994, hep-th/9409089.

Causality in String Theory

David A. Lowe

Department of Physics, University of California, Santa Barbara, CA 93106-9530, USA

The extended structure of fundamental strings gives rise to a string field theory with unusual causal properties. This is examined by considering the commutator of string fields in interacting light-cone string field theory. In general, this commutator is non-vanishing outside the string analog of the light-cone. This fact could have important implications for our understanding of localization of information in quantum gravity. This talk was based on work done in collaboration with L. Susskind and J. Uglum [1].

§ 1 Introduction

Conventional string theory provides us with a perturbative S−matrix. However, it is of interest to compute more general observable quantities, such as probability amplitudes at finite times. Here, we will be primarily interested in the causal properties of such amplitudes, and we will use string field theory to try and answer these questions.

In the known covariant formulations of quantized string field theory, the interactions are nonlocal in the center of mass coordinate x^μ, and, as emphasized in [2], the initial value problem breaks down. Therefore, the theory cannot be canonically quantized in the conventional way. Fortunately, one is able to fix light cone gauge, where the interactions are local in the center of mass coordinate $x^+ = \tau$. This allows a conventional canonical quantization, and a second–quantized operator formulation exists [3-6], which allows one to perform the kind of calculations referred to in the preceding paragraph.

Our main concern will be the calculation of the commutator of two string fields on a flat spacetime background. Further details may be found in [1]. In the free theory, this commutator vanishes outside the "string light cone" [7,8], but it will be shown here that this is no longer the case when interactions are included. The conclusion is that measurements could detect the information carried by a string state outside the light cone of the center of mass, but only if these measurements could be performed with resolution times smaller than the string scale. This result supports recent arguments by Susskind concerning the nature of information in string theory [9], and could have important implications for our understanding of localization of information in a theory of quantum gravity.

§ 2 Commutator of String Fields

Let us introduce the light-cone coordinates

$$X^+ = (X^0 + X^{D-1})/\sqrt{2}, \quad X^- = (X^0 - X^{D-1})/\sqrt{2} \quad (1)$$

and parametrize the worldsheet of the string by the variables σ and τ. Light-cone gauge corresponds to fixing $X^+(\sigma) = x^+ = \tau$. In the following we will consider open bosonic strings; the generalization to closed strings and to superstrings should be similar. The transverse coordinates are expanded as

$$\vec{X}(\sigma) = \vec{x} + 2\sum_{l=1}^{\infty} \vec{x}_l \cos(l\sigma) \ . \quad (2)$$

In light-cone gauge, the string field is a physical observable and can be decomposed in terms of an infinite number of component fields. In the absence of interactions, the string field takes the form [4,5]

$$\Phi(\tau, x^-, \vec{X}(\sigma)) = \int \frac{d^{D-2}p}{(2\pi)^{D-1}} \int_0^\infty \frac{dp^+}{2p^+} \Bigg[\sum_{\{\vec{n}_l\}} A(p^+, \vec{p}, \{\vec{n}_l\}) e^{i(\vec{p}\cdot\vec{x} - p^+ x^- - \tau p^-)} f_{\{\vec{n}_l\}}(\vec{x}_l) + h.c. \Bigg] \ . \quad (3)$$

Here the light-cone energy of a string state is given by

$$p^-(p^+, \vec{p}, \{\vec{n}_l\}) = \frac{\vec{p}^{\,2} + 2\sum_{l,i} l n_l^i + m_0^2}{2p^+} \ , \quad (4)$$

where m_0^2 is the mass squared of the ground state of the string. For bosonic strings, the ground state is a tachyon and m_0^2 is negative. In order to effect the light cone quantization and calculate the commutator, we will regard m_0^2 as a positive adjustable parameter [7,8]. Of course, this is inconsistent with Lorentz invariance for the bosonic string, but our results will be essentially unchanged in the superstring case where $m_0^2 = 0$.

The $f_{\{\vec{n}_l\}}(\vec{x}_l)$ are harmonic oscillator wave functions given by

$$f_{\{\vec{n}_l\}}(\vec{x}_l) = \prod_{l=1}^{\infty} \prod_{i=1}^{D-2} H_{n_l^i}(x_l^i) e^{-l(x_l^i)^2/(4\pi)} \ , \quad (5)$$

with $H_{n_l^i}(x_l^i)$ a Hermite polynomial. The A operators obey the canonical commutation relations

$$[A(p^+,\vec{p},\{\vec{n}_l\}),A^\dagger(p^{+\prime},\vec{p}^{\,\prime},\{\vec{n}_l{}'\})]=2p^+ \times(2\pi)^{D-1}\delta(p^+-p^{+\prime})\delta^{D-2}(\vec{p}-\vec{p}^{\,\prime})\delta_{\{\vec{n}_l\},\{\vec{n}_l{}'\}}\ , \quad (6)$$

A component field is obtained from Φ by multiplying by the appropriate wave function (5) and integrating over the normal mode coordinates $\vec{x}_l$. For example, the tachyon field is given by

$$T(\tau,x^-,\vec{x})=\int\frac{d^{D-2}p}{(2\pi)^{D-1}}\int\frac{dp^+}{2p^+}[a_T(\vec{p},p^+)e^{i[\vec{p}\cdot\vec{x}-p^+x^--p^-\tau]} +a_T^\dagger(\vec{p},p^+)e^{-i[\vec{p}\cdot\vec{x}-p^+x^--p^-\tau]}]\ , \quad (7)$$

where $a_T(\vec{p},p^+)=A(\vec{p},p^+,\{\vec{0}\})$ and p^- is given by (4).

Now we want to include a cubic interaction. The light cone Hamiltonian becomes $H=H_0+H_3$, where H_0 is the Hamiltonian for free string field theory and the cubic interaction term H_3 is given by

$$\begin{aligned} H_3 &= g\int\Phi_{\alpha_1}(\vec{X}_1(\sigma))\Phi_{\alpha_2}(\vec{X}_2(\sigma))\Phi_{\alpha_3}(\vec{X}_3(\sigma)) \\ &\quad \delta(\sum_{r=1}^3\alpha_r)\Delta(\vec{X}_1(\sigma)-\vec{X}_2(\sigma)-\vec{X}_3(\sigma)) \\ &\quad \times\mu(\alpha_1,\alpha_2,\alpha_3)\prod_{r=1}^3 d\alpha_r\prod_{r=1}^3\mathcal{D}\vec{X}_r(\sigma)\ , \end{aligned} \quad (8)$$

where g is the open string coupling and $\alpha_r=2p_r^+$. The measure factor μ and the matrix Γ are defined in [1].

Now that interactions have been included, we wish to determine whether the commutator of two string fields vanishes when the arguments of the string fields lie outside the string light cone [7,8]. Suppose that

$$[\Phi(x_1^+,x_1^-,\vec{X}_1(\sigma)),\Phi(x_2^+,x_2^-,\vec{X}_2(\sigma))]=0 \quad (9)$$

when

$$\frac{1}{\pi}\int d\sigma(X_1(\sigma)-X_2(\sigma))^2<0\ , \quad (10)$$

where we are using the mostly minus convention for the spacetime metric. For fixed $X_2(\sigma)$, equation (9) can be regarded as a function of $X_1(\sigma)$, which vanishes in the entire region in which equation (10) is satisfied. Differentiating equation (9) with respect to x_1^+ and setting $x_1^+=x_2^+=\tau$, one obtains

$$[\dot{\Phi}(\tau,x_1^-,\vec{X}_1(\sigma)),\Phi(\tau,x_2^-,\vec{X}_2(\sigma))]=0 \quad (11)$$

in the region in which equation (10) holds. Here $\dot{\Phi}$ denotes $\partial\Phi/\partial x^+$. At equal light cone times, equation (10) reduces to

$$\frac{1}{\pi}\int d\sigma(\vec{X}_1(\sigma)-\vec{X}_2(\sigma))^2>0\ . \quad (12)$$

Therefore, to prove that the string field commutator does not vanish identically outside the string light cone, it is sufficient to prove that equation (11) fails to hold when equation (12) is satisfied.

To proceed, note that we can use the Heisenberg equation of motion to express the field $\dot{\Phi}$ as

$$\dot{\Phi}=i[H,\Phi]\ . \quad (13)$$

Consider now the matrix element

$$\begin{aligned} &\langle 0|\Phi(p_3^+,\vec{X}_3(\sigma_3))\,[\dot{\Phi}(x_1^-,\vec{X}_1(\sigma_1)),\Phi(x_2^-,\vec{X}_2(\sigma_2))]|0\rangle \\ &= i\langle 0|\Phi(p_3^+,\vec{X}_3(\sigma_3))\,[[H,\Phi(x_1^-,\vec{X}_1(\sigma_1))], \\ &\quad \Phi(x_2^-,\vec{X}_2(\sigma_2))]\,|0\rangle\ , \end{aligned} \quad (14)$$

where all fields are evaluated at $\tau=0$. Expanding $H=H_0+H_3$, the terms involving H_0 all vanish by orthogonality. This is a reflection of the fact that the commutator does in fact vanish outside the string light cone in free string field theory [7]. The remaining terms can be expressed as

$$\begin{aligned} &\langle 0|\Phi(p_3^+,\vec{X}_3(\sigma_3))\,[\dot{\Phi}(x_1^-,\vec{X}_1(\sigma_1)),\Phi(x_2^-,\vec{X}_2(\sigma_2))]|0\rangle \\ &= \frac{2ig(2\pi)^{5(D-1)/2}}{(2\pi)^3p_3^+}\int_0^\infty\frac{dp_1^+}{2p_1^+}\int_0^\infty\frac{dp_2^+}{2p_2^+} \\ &\Big(V(2p_1^+,\vec{X}_1(\sigma_1);2p_2^+,\vec{X}_2(\sigma_2);-2p_3^+,\vec{X}_3(\sigma_3)) \\ &-V(-2p_1^+,\vec{X}_1(\sigma_1);2p_2^+,\vec{X}_2(\sigma_2);-2p_3^+,\vec{X}_3(\sigma_3)) \\ &-V(2p_1^+,\vec{X}_1(\sigma_1);-2p_2^+,\vec{X}_2(\sigma_2);-2p_3^+,\vec{X}_3(\sigma_3))\Big) \end{aligned} \quad (15)$$

where V is the vertex factor obtained from (8). The three terms represent the three possible kinematical situations, in which the center of mass of string 3, 2, or 1 lies between the centers of mass of the other two, respectively. The p_1^+ integral may be performed by using the $\delta(\sum_{r=1}^3\alpha_r)$ factor. Then one notes that the functional δ-function in (8) contains a zero mode piece $\delta^{D-2}(\sum_{r=1}^3\alpha_r\vec{x}_r)$. Using one of these delta functions, say for the x^1 component, allows the integral over p_2^+ to be performed, and sets

$$\begin{aligned} \alpha_2 &= -\alpha_3 s\ , \\ \alpha_1 &= (s-1)\alpha_3, \end{aligned} \quad (16)$$

where

$$s=\frac{(x_3^1-x_1^1)}{(x_2^1-x_1^1)}\ . \quad (17)$$

The crucial point to notice is that one is left with a $(D-3)$-dimensional δ-function requiring the $\vec{x}_r$ to be collinear, and that each term in (15) has support on a distinct ordering of the $\vec{x}_r$ on the line connecting them.

We therefore find that the commutator of two string fields is in general non–vanishing outside the string light-cone, $\int d\sigma(\delta X^\mu(\sigma))^2 = 0$, when interactions are included. This also implies the commutator is non–vanishing when the centers of mass of the strings are spacelike separated. It should be stressed here that the non–vanishing of the commutator at spacelike separations has nothing to do with the fact the bosonic string has a tachyon. The same will be true in the tachyon–free superstring case.

Of more direct physical interest is the analogous calculation for the component fields. For simplicity, we will only consider the tachyon field, though the generalization to an arbitrary mass eigenstate is straightforward. Following the previous line of reasoning, we compute the matrix element

$$\langle 0|T(p_3^+, \vec{x}_3)\,[\dot{T}(x_1^-, \vec{x}_1), T(x_2^-, \vec{x}_2)]|0\rangle \tag{18}$$

where, as before, all operators are at time $\tau = 0$. The vertex appearing in equation (18) is the Mandelstam vertex [3], which has the momentum space representation

$$\begin{aligned} V(\alpha_r, \vec{p}_r) &= \delta^{D-2}(\sum_{r=1}^{3} \vec{p}_r)\delta(\sum_{r=1}^{3} \alpha_r) \\ &\times \exp\left(\frac{\tau_0}{2}\sum_{r=1}^{3}\frac{\vec{p}_r^{\,2} + m_0^2}{\alpha_r}\right). \end{aligned} \tag{19}$$

Fourier transforming to coordinate representation, one obtains

$$\begin{aligned} V(\alpha_r, \vec{x}_r) &= \delta^{D-2}(\sum_{r=1}^{3} \alpha_r \vec{x}_r)\delta(\sum_{r=1}^{3} \alpha_r) \\ &\times \left(\frac{\alpha_1\alpha_2\alpha_3}{8\pi^3\tau_0}\right)^{(D-2)/2} \exp\left(\frac{\tau_0 m_0^2}{2}\sum \frac{1}{\alpha_r}\right. \\ &\left. + \frac{\alpha_1\alpha_2\alpha_3}{8\tau_0}\left(\frac{\vec{x}_1 - \vec{x}_2}{\alpha_3} - \frac{\vec{x}_1 - \vec{x}_3}{\alpha_2}\right)^2\right). \end{aligned} \tag{20}$$

As was the case for the general string field vertex, equation (20) contains the factor $\delta^{D-2}(\sum_{r=1}^{3}\alpha_r\vec{x}_r)$, so the result is non–vanishing only when the points $\vec{x}_r$ are collinear. The off-shell vertex corresponds to, say, one tachyon splitting into two others such that all transverse centers of mass lie along the same line at equal times. In addition there is a Gaussian factor depending on the separation of the particles.

Consider a configuration in which $\vec{x}_3$ lies between $\vec{x}_1$ and $\vec{x}_2$, so that only the first term in equation (18) is non–zero. A simple calculation then gives

$$\begin{aligned} &\langle 0|T(p_3^+, \vec{x}_3)\,[\dot{T}(x_1^-, \vec{x}_1), T(x_2^-, \vec{x}_2)]|0\rangle = \\ &\quad -ie^{ip_3^+((1-s)x_1^- + sx_2^-)}\frac{\delta^{D-3}(\vec{x}_3 - \vec{x}_1 - s(\vec{x}_2 - \vec{x}_1))}{8\sqrt{2\pi}(p_3^+)^2 s(1-s)|x_2^1 - x_1^1|} \\ &\quad \exp\left(-\frac{s(1-s)}{2\gamma(s)}(\vec{x}_1 - \vec{x}_2)^2 - m_0^2\frac{\gamma(s)(s^2 - s + 1)}{2s(1-s)}\right) \\ &\quad \times\left(\frac{(2\pi)^2 s(1-s)}{\gamma(s)}\right)^{(D-2)/2}, \end{aligned} \tag{21}$$

where s is given in equation (17) and

$$\gamma(s) = -[s\log(s) + (1-s)\log(1-s)]\,. \tag{22}$$

Note that because of our choice of configuration, $s \in [0, 1]$, and that γ is non–negative. The matrix element (21) depends on the transverse displacement $|\vec{x}_1 - \vec{x}_2|$ through a Gaussian factor with variance

$$\sigma^2 = \frac{\gamma(s)}{s(1-s)}\,. \tag{23}$$

One therefore finds that the matrix element has support over a distance of order σ^2 outside the light–cone of the center of mass. This spread can be made quite large. Indeed, for small s, we have

$$\lim_{s\to 0}\frac{\gamma(s)}{s(1-s)} \sim -\log(s)\,, \tag{24}$$

so for $s \sim \exp(-(\vec{x}_1 - \vec{x}_2)^2)$, the matrix element is appreciable. This can always be achieved by choosing x_3^1 sufficiently close to x_1^1.

The question is whether one is able to resolve this information in practice. To get an estimate of how quickly the matrix element is oscillating in light cone time, we can calculate the matrix element

$$\langle 0|T(p_3^+, \vec{x}_3)\,[\ddot{T}(x_1^-, \vec{x}_1), T(x_2^-, \vec{x}_2)]|0\rangle\,, \tag{25}$$

and divide by the matrix element (18). This is proportional to the frequency of oscillation. To do this carefully, we must multiply both (18) and (25) by a slowly varying function f and then integrate over $\vec{x}_2, \ldots, \vec{x}_{D-2}$ to eliminate the $\delta^{D-3}(\sum \alpha_r\vec{x}_r)$ factors. Performing this calculation leads to the following oscillation time scale

$$\delta t \sim \frac{p_3^+ s}{(\vec{x}_2 - \vec{x}_1)^2\log^{-2}(s) + (D - 2 + m_0^2)} \tag{26}$$

valid for $s \to 0$. For the case of interest, $s = \exp(-(\vec{x}_1 - \vec{x}_2)^2)$, this becomes

$$\delta t \sim \frac{p_3^+\exp(-(\vec{x}_1 - \vec{x}_2)^2)}{(\vec{x}_1 - \vec{x}_2)^{-2} + (D - 2 + m_0^2)}\,. \tag{27}$$

The conclusion is that in order to observe the spread of information over more than a string length, one must perform measurements involving time scales much smaller than the string time.

The information content of these matrix elements exhibits precisely the same type of diffusive behavior as was described in [9], and which was argued to provide a possible resolution of the black hole information paradox. Under conditions relevant to strings propagating near the horizon of a very massive black hole, the spread of the matrix elements (18) can become arbitrarily large. The calculation presented above is relevant because the near horizon geometry can be approximated by Rindler space, which is simply a section of flat Minkowski space. Suppose $T(\tau, p_1^+, \vec{x}_1)$ and $T(\tau, p_3^+, \vec{x}_3)$ represent two strings, closely separated in the transverse coordinates, falling together toward the horizon $\tau = 0$ of a black hole of mass M. As a function of Schwarzschild time t, the radial momentum P of a string, measured by a static Schwarzschild observer, is given at late times by

$$P(t) = e^{t/4M} p^+ , \tag{28}$$

where p^+ is the (conserved) longitudinal momentum of the string.

We would like to sample string 1 with a string near the horizon which is very far from string 1 in the transverse coordinates, and which has radial momentum which is small compared to that of string 1 or 3. To that end, we will choose $P_2(t) = P_3(0)$. This means we must choose p_2^+ as a function of time, given by $p_2^+(t) = e^{-t/4M} P_3(0)$. For the arguments of these fields, the parameter s appearing in the matrix element (21) is given by

$$s = \frac{p_2^+(t)}{p_3^+} = e^{-t/4M} . \tag{29}$$

One thus finds that the spread of the Gaussian is given by $\sigma^2 \sim \frac{t}{4M}$. This diffusive behavior is the same as that found in [9,10] for the mean square transverse radius of a string.

§ Acknowledgements

This work was supported in part by NSF grant No. PHY91-16964.

§ References

[1] D.A. Lowe, L. Susskind and J. Uglum, *Phys. Lett.* **327B** (1994) 226.

[2] D. A. Eliezer and R. P. Woodard, *Nucl. Phys.* **B325** (1989) 389.

[3] S. Mandelstam, *Nucl. Phys.* **B64** (1973) 205; *Nucl. Phys.* **B83** (1974) 413; *Phys. Rep.* **13C** (1974) 259.

[4] E. Cremmer and J.-L. Gervais, *Nucl. Phys.* **B76** (1974) 209; *Nucl. Phys.* **B90** (1975) 410.

[5] M. Kaku and K. Kikkawa, *Phys. Rev.* **D10** (1974) 1110; *Phys. Rev.* **D10** (1974) 1823.

[6] J. F. L. Hopkinson, R. W. Tucker and P. A. Collins, *Phys. Rev.* **D12** (1975) 1653.

[7] D. A. Lowe, *Phys. Lett.* **326B** (1994) 223.

[8] E. Martinec, *Class. Quant. Grav.* **10** (1993) L187; "Strings and Causality", preprint EFI-93-65, November 1993, hepth/9311129.

[9] L. Susskind, *Phys. Rev. Lett.* **71** (1993) 2367; *Phys. Rev.* **D49** (1994) 6606.

[10] A. Mezhlumian, A. Peet, and L. Thorlacius, *Phys. Rev.* **D50** (1994) 2725.

Nonlocal Effective Action and Black Holes

A. O. Barvinski

Nuclear Safety Institute, Bolshaya Tulskaya 52, Moscow 113191, Russia
Lebedev Research Center in Physics, Leninsky Prospect 53, Moscow 117924, Russia

Progress in understanding the quantum physics of black holes and gravitational collapse to an essential extent depends on the issue of singularities in the quantum gravitational domain. While classical Einstein equations typically lead to the development of spacetime singularities and, thus, give rise to the problem of information loss, the inclusion of quantum effects can drastically change the classical picture [1] due to the back reaction of the vacuum polarization and particle creation. In certain situations the back reaction damps the spacetime singularities and even makes the curvature invariants uniformly bounded throughout the whole spacetime [1]. This, in particular, qualitatively changes the setting of the information problem in quantum gravity.

A direct way to the quantum back-reaction problem is the formalism of the effective action and effective equations [2]. Their calculation cannot be done explicitly, but one can develop approximation schemes appropriate to the setting of the quantum gravitational problem. One such scheme, motivated by the potential boundedness of the spacetime curvature, is the covariant expansion in powers of curvatures. A local version of this expansion, known as the Schwinger-DeWitt technique [2,3], relies on the smallness of both the curvatures and their spacetime derivatives and breaks down in massless theories. In the series of papers [4-6] this expansion was generalized to the case of arbitrarily high spacetime gradients of curvature field strengths. This results in the nonlocal expansion of the effective action in powers of curvatures with coefficients – nonlocal covariant form factors. The systematic study of these form factors within the semiclassical expansion theory allows one to reproduce the known conformal (anomalous) part of the effective action and also obtain its previously unknown conformally-invariant part unrelated to local conformal anomalies [4,5]. This, in particular, allows one to obtain the one-loop vertex form factors for the theory of the most general type including arbitrary matter and gravitational fields [5].

Beyond the semiclassical loop expansion, the analysis of the covariant perturbation theory shows that important observable characteristics of collapsing gravitational systems require only the knowledge of certain asymptotic limits of the spectral weights of the above form factors. Numerical values of these limits may have a nonperturbative nature and can be kept as fundamental parameters of the theory determining the qualitative behavior of the system. Within such a model-independent approach to quantum gravity theory [7,8] it was, in particular, shown that in the self-consistent problem the asymptotic flatness of spacetime requires at most logarithmic low-energy behaviour of the nonlocal form factors, the coefficients of this logarithmic behaviour determining the energy flux through future null infinity [8]. It has been also demonstrated that the stable component of the Hawking radiation, which plays an important role in gravitational collapse and the problem of information loss, begins at cubic order in the effective action expanded in powers of the curvatures [8]. This important contribution, which is currently under study, promises to maintain the future asymptotic flatness in gravitational systems due to nontrivial cancellations in nonlocal form factors and also generates a qualitatively new phenomenon – coherent gravitational waves radiated by these systems entirely due to quantum gravitational effects [9].

§ References

[1] V.P.Frolov and G.A.Vilkovisky, *Phys. Lett.* **106** (1981) 307.

[2] B.S.DeWitt, *Dynamical theory of groups and fields* (Gordon and Breach, New York, 1965).

[3] A.O.Barvinsky and G.A.Vilkovisky, *Phys. Rep.* **119** (1985) 1.

[4] A.O.Barvinsky and G.A.Vilkovisky, *Nucl. Phys.* **282** (1987) 163; **B333** (1990) 471; **333** (1990) 512.

[5] A.O.Barvinsky, Yu.V.Gusev, G.A.Vilkovisky and V.V.Zhytnikov, *Covariant Perturbation Theory (IV). Third Order in the Curvature,* Report of the University of Manitoba (University of Manitoba, Winnipeg, 1993).

[6] A.O.Barvinsky, Yu.V.Gusev, G.A.Vilkovisky and V.V.Zhytnikov, *J.Math.Phys.* **35** (1994) 3525; **35** (1994) 3543.

[7] G.A.Vilkovisky, *Class.Quantum Grav.* **9** (1992) 895.

[8] A.G.Mirzabekian and G.A.Vilkovisky, *Phys. Lett.* **B317** (1993) 517.

[9] A.G.Mirzabekian and G.A.Vilkovisky, work in progress.

Universality and Scaling in Black Hole Formation

Philip C. Argyres

School of Natural Sciences, Institute for Advanced Study, Princeton, NJ 08540, USA

This is a brief review of numerical work of Refs. [1–3] on the formation of small-mass black holes in classical general relativity. Their results are described from a point of view which treats Einstein's equations as generating a renormalization group (RG) flow on the space of initial conditions.

The recent intriguing numerical results of Choptuik [1], Abrahams & Evans [2], and Evans & Coleman [3], indicate a connection between classical general relativity and scaling at critical points in statistical systems. The purpose of this talk is to briefly review the results of the above-mentioned authors, and, along the way, to make a few obvious remarks on the relation of scaling solutions in general relativity to scaling in statistical systems.

Choptuik simulated the collapse of spherical wave-packets of a massless scalar field ϕ minimally coupled to gravity. Abrahams & Evans simulated the axial collapse of gravity waves, while Evans & Coleman examined the spherical collapse of a relativistic perfect fluid (satisfying $p = \frac{1}{3}\rho$). In all three cases it was found that for initial conditions depending on a generic parameter s, there exists a critical value s^* of this parameter above which no black hole forms (*i.e.*, all the incoming mass is eventually scattered back to spatial infinity, and the space-time settles down to flat Minkowski space), while for $s < s^*$ a black hole forms with a mass $M_{\rm bh}$ satisfying the universal relation $M_{\rm bh} \sim (s^* - s)^\gamma$, with $\gamma = 0.37p^{-}0.01$. This relation holds only for s sufficiently close to s^*, and is universal in the sense that the exponent γ does not depend on the shape of the initial wave packet (the initial conditions) or on the type of matter or symmetry of the problem (since all three groups find the same exponent). Also, Choptuik has found the same behavior for non-minimally coupled and massive scalar fields in spherical geometries [4].

Before discussing this universality more critically, let me describe in more detail Choptuik's simulation [1]. Consider the initial condition consisting of a spherical shell of scalar field with compact support inside a radius r_0, with total mass M_0. We will let the profile of the scalar field and its first time derivative be arbitrary but fixed, and will make a one-parameter family of initial conditions by varying r_0, *i.e.*, by translating the scalar wave-packet radially. A massless scalar field minimally coupled to gravity is scale-invariant: an overall rescaling of the coordinates will take one solution into another. So, the relevant scale-invariant parameters to measure are $s = r_0/(2M_0)$ and $M_{\rm bh}/M_0$. Then, qualitatively, Choptuik's results are sketched in Fig. 1, where

$$\frac{M_{\rm bh}}{M_0} \sim (s^* - s)^{(0.37)} \qquad (1)$$

for s less than but close to s^*. Note that for $s >> s^*$ the initial data is in the weak-coupling regime, since however wild the shape of the scalar wave-packet, by taking s large enough and rescaling the radial and time coordinates we see that the initial data can be made as close to flat space as we like. In this limit it has been shown analytically that no black hole forms [5]. In the opposite extreme, when $s \leq 1$ then all the mass is inside its Schwarzschild radius and a black hole with mass $M_{\rm bh} = M_0$ is inevitable.

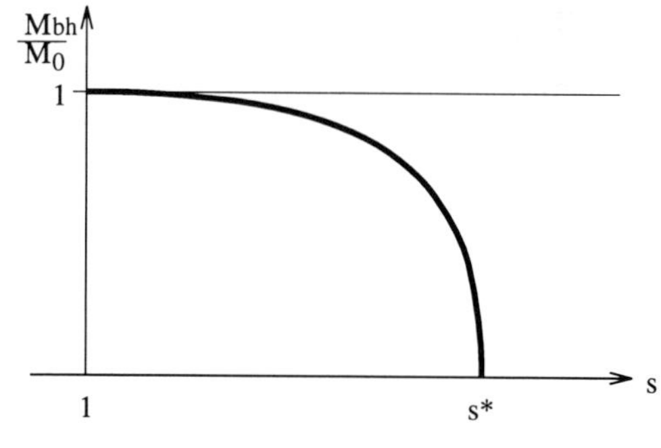

Figure 1:

The formation of a zero-mass black hole in the strongly-coupled regime at $s = s^*$ is strongly reminiscent of the ferromagnetic second-order phase transition where a spontaneous magnetization arises below a critical temperature. Taking this analogy seriously suggests that the value of the critical exponent $\gamma = 0.37p^{-}0.01$, however, must be taken with a grain of salt, since it has been determined numerically by fitting to the curve in Fig. 1 only over two or three e-foldings in $\ln(s^* - s)$. Experience with fitting exponents in numerical simulations in other critical systems indicates that this may only give a determination accurate at the 20% level: there are often universal subleading corrections to the scaling relations which tend to increase the apparent exponent unless one is much closer to the critical point $s = s^*$.

As further evidence supporting this analogy to second-order phase transitions, the numerical simulations also show that solutions approach universal scaling solutions at criticality. In order to describe these solutions I first need to describe the coordinate systems for the resulting space-times. Let us approach the critical point from large s, so that globally space-time is like Minkowski space. Then, for the spherically symmetric situations we can use dynamical Schwarzschild coordinates, for which the line element is

$$ds^2 = -\alpha^2(r,t)dt^2 + \beta^2(r,t)dr^2 + r^2 d\Omega^2. \qquad (2)$$

This fixes the gauge up to an arbitrary reparametrization of the time coordinate (*i.e.* an arbitrary positive function of t). A natural gauge choice is to fix $\alpha(0,t) = 1$, which implies that t measures central proper time. For an arbitrary gauge, the central proper time T is defined by

$$T(t) = \int^t \alpha(0,\tilde{t})d\tilde{t}. \qquad (3)$$

(Note that the origin of central proper time is still not fixed by this definition.) The spherical symmetry keeps the metric functions α and β from being dynamical degrees of freedom. For a scalar field coupled to gravity with spherical symmetry (Choptuik's case) only the scalar field $\phi(r,t)$ itself is dynamical. Einstein's equations can be shown to reduce to a single nonlinear second order partial integro-differential equation for ϕ. Similarly for a perfect fluid with spherical symmetry (Evans & Coleman) the single dynamical degree of freedom can be taken to be the fluid energy density $\rho(r,t)$. In the axially-symmetric pure gravity case (Abrahams & Evans) the line element is more complicated, and includes a function $\eta(r,t,z)$ which is the single dynamical degree of freedom.

We can classify the self-similar (scaling) solutions to Einstein's equations in these situations by viewing the evolution equations as generating a RG flow on the space of initial conditions. This point of view has proven useful in discussions of "intermediate asymptotics" in hydrodynamics and scaling in nonlinear diffusion equations [6]. We define the RG transformation with parameter Λ to be simply

$$\mathrm{RG}_\Lambda : \phi(r,t) \to \phi_\Lambda(r,t) \equiv \Lambda^{\beta_\phi}\phi\left(\frac{r}{\Lambda^{\alpha_r}}, \frac{t^* - t}{\Lambda^{\alpha_t}}\right), \qquad (4)$$

where I have used the scalar field for illustration. A self-similar (scaling) solution of the equations would then correspond to a fixed point of the RG transformations. For example, a continuously self-similar solution is a solution $\phi^*(r,t)$ satisfying $\phi^*(r,t) = \phi^*_\Lambda(r,t)$ for all Λ. Using the definition (4) this has the solution

$$\phi^*(r,t) = (t^* - t)^{\beta_\phi/\alpha_t}\widetilde{\phi^*}\left(\frac{r^{\alpha_t}}{(t^* - t)^{\alpha_r}}\right), \qquad (5)$$

where $\widetilde{\phi^*}$ is a fixed function of one variable. More generally, we can consider the situation where the solution is discretely self-similar. This solution corresponds to a *limit cycle* of the RG flow instead of a fixed point: $\phi^*(r,t) = \phi^*_{\Lambda_0^n}(r,t)$ for all integers n. Such a limit cycle can always be written as

$$\phi^*(r,t) = \sum_n (t^* - t)^{(\beta_\phi/\alpha_t)+(2\pi i n/\ln\Lambda_0)} \times \widetilde{\phi^*_n}\left(\frac{r^{\alpha_t}}{(t^*-t)^{\alpha_r}}\right), \qquad (6)$$

which is now described by the infinite series of functions $\widetilde{\phi^*_n}$.

The results of the numerical simulations in this language are:

scalar field S-wave [1]:	$\alpha_r = \alpha_T = 1$
	$\beta_\phi = 0$
	$\Lambda_0 = 3.4$
axial gravity-wave [2]:	$\alpha_r = 1, \alpha_{t,z} = ?$
	$\beta_\eta = 0$
	$\Lambda_0 = 0.6$
perfect fluid S-wave [3]:	$\alpha_r/\alpha_T = 1$
	$\beta_\rho/\alpha_T = -2$
	$\Lambda_0 = 0$

Thus, only the perfect fluid collapse gives a simple fixed point; the others have limit cycles. Note that the appropriate scaling variables in the spherically symmetric cases are the geometric (areal) radial coordinate r and the central proper time T; and that the exponents all have their naive scaling values. Fig. 2 presents a heuristic sketch of the limit cycle RG flows.

The immediate question presented by these numerical simulations is how one could analytically compute the critical exponent $\gamma = 0.37$. The most direct approach to this problem is to try to get an analytic understanding of the critical (scaling) solution, and then to do linear perturbation theory around this solution to obtain the exponent(s). However, since the critical point is in the strongly-coupled regime, this seems a daunting problem. In this regard our experience with critical phenomena in statistical systems suggests that we look for a simpler (weakly-coupled) system in the same universality class. A broader question is whether this critical behavior persists when we move away from spherical symmetry or add

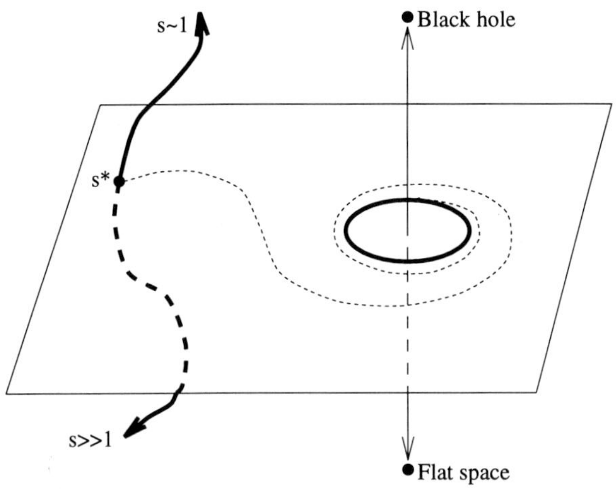

Figure 2: A sketch of the (infinite-dimensional) space of initial conditions. The plane represents the (co-dimension 1) critical surface in which the RG flows towards the critical limit cycle (the circle); off the plane the RG flows either to the black hole or flat space limits. The thick line represents a generic one-parameter family of initial conditions, while the dotted line in the plane is the RG trajectory of the $s = s^*$ initial condition.

more matter fields. When there is more than one dynamical degree of freedom, direct computation becomes prohibitively difficult, so this question would be best approached by the type of linear stability analysis mentioned above.

My thanks to M. Bucher, J. Distler, and A. Shapere for helpful discussions.

§ References

[1] M.W. Choptuik, *Phys. Rev. Lett.* **70** (1993) 9.

[2] A.M. Abrahams and C.R. Evans, *Phys. Rev. Lett.* **70** (1993) 2980.

[3] C.R. Evans and J.S. Coleman, *Observation of Critical Phenomena and Self-Similarity in the Gravitational Collapse of Radiation Fluid,* gr-qc/9402041.

[4] M.W. Choptuik, in *Approaches to Numerical Relativity,* edited by R. d'Inverno (Cambridge Univ. Press, London, 1989).

[5] D. Christodoulou, *Comm. Math. Phys.* **105** (1986) 337.

[6] For an explanation of this point of view and further references, see N.D. Goldenfeld, *Lectures on Phase Transitions and the Renormalization Group,* Addison-Wesley (1992).

Cosmology

Cosmological Models at the Millennium

Joshua A. Frieman

NASA/Fermilab Astrophysics Center, Fermi National Accelerator Laboratory, Batavia, IL 60510

Department of Astronomy and Astrophysics, The University of Chicago, Chicago, IL 60637

This chapter provides an overview of cosmological models, starting with a review of the standard big bang cosmology and ending with scenarios for the early universe. We introduce the expanding, homogeneous and isotropic Friedmann-Robertson-Walker universe and briefly summarize the observational evidence which has led to its adoption as the standard cosmological model. We then discuss scenarios for the very early universe, focusing on the inflationary universe and topological defect models, and their possible role in seeding the observed large-scale structure of the Universe. We emphasize the exciting prospects for testing early universe models, and thus for probing particle physics at an energy scale of order 10^{16} GeV, with astronomical and terrestrial observations in the coming decade.

§1 Introduction

Over the last fifteen years, theoretical cosmology has undergone a renaissance: extrapolating concepts from elementary particle physics, in particular the standard electroweak gauge theory, to very high energies, a framework has emerged in which one can meaningfully speculate about the evolution of the very early universe. This marriage of particle physics and cosmology has led to a number of remarkable developments, including models for the generation of the baryon asymmetry, the inflationary scenario, the notion that topological defects could be created in cosmological phase transitions, and predictions for non-baryonic particle dark matter, to name just a few. In addition, both inflation and topological defects have provided causal frameworks for generating primordial density perturbations in the early universe, thought to be the seeds for galaxy formation.

In recent years, observational cosmology has been undergoing its own rebirth. There has been an explosion of information on the large-scale clustering of galaxies from redshift, peculiar velocity, and photometric surveys gathered by ground-based telescopes. Studies of rich clusters of galaxies via their gravitational lensing effects as well as X-ray emission from hot intracluster gas have started to provide new clues to the distribution of dark matter. The detection of large-angle anisotropies in the cosmic microwave background radiation by the COBE satellite has provided the first probe of structure on very large scales, while a series of ground-based and balloon-borne anisotropy experiments on smaller scales now have positive results as well. On the scale of the universe itself, there has been steady progress in attempts to measure cosmological parameters (in particular, the age, expansion rate, and mean density) as well as the light element abundances more precisely, using the Hubble Space Telescope and the new generation of large ground-based telescopes.

As a consequence of these observational advances, cosmology is becoming data-driven in an unprecedented way: we can now confront cosmological models of the early universe with an impressive array of observations. There is still debate about the reliability and interpretation of much of the data, but we are definitely entering what one might call the 'scientific' age of cosmology: those theories which are sufficiently predictive (and therefore interesting) are becoming increasingly falsifiable and will stand or fall in the coming years. This is a very healthy development for the field. It also implies that a larger portion of cosmologists' time in the future will be spent in the realm of 'phenomenology', developing more precise methods for testing theoretical predictions against observational results and discriminating among competing models.

It is safe to say that at present the big bang framework for the large-scale evolution of the universe rests upon solid foundations. We also have in hand rough architectural sketches of a model which may account for the origin and evolution of structure within this framework: primordial, quasi-scale-invariant perturbations from inflation or defects, with some variant of cold dark matter, which subsequently grow by gravitational instability. However, while the picture is becoming clearer, we still lack firm answers to basic questions about the large-scale universe, among them: what is the origin of the perturbations responsible for large-scale structure? once perturbations are generated, how do luminous galaxies form in detail? what is the universe made of, that is, what is the dark matter(s), how much is there, and how is it distributed? when and how was the baryon (matter/anti-matter) asymmetry

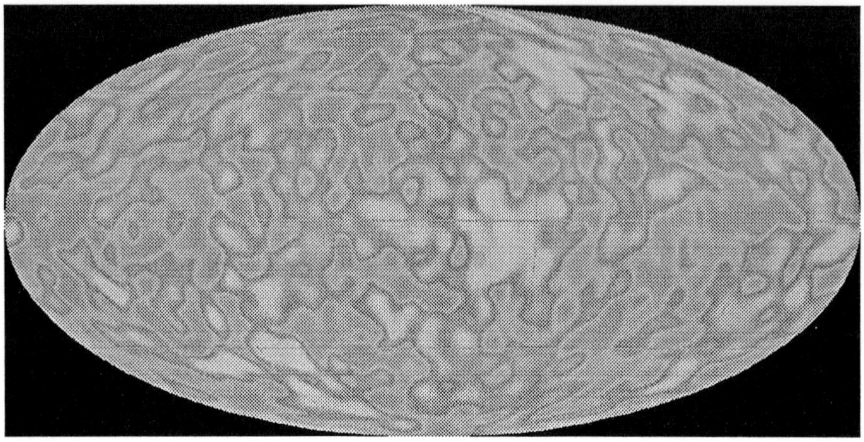

Figure 1: Temperature fluctuations in the CMBR observed by COBE DMR. Lighter areas correspond to low-temperature regions. The typical fluctuation amplitude is of order 10^{-5} [2].

generated? Beyond these basic questions there are others for which more precise observational answers are eagerly anticipated at the millennium, for they involve parameters for the models, in particular: how fast the universe is expanding, the rate at which the expansion is being slowed by gravity, and the age of the universe.

The first section provides an overview of the standard cosmology and the observational evidence in its favor. Subsequent sections focus on scenarios for the early universe, their predictions for the generation of structure, and how they may be tested by observations in the near future. Most of the topics introduced here will be treated in greater depth in subsequent chapters. The Appendix introduces the standard notation and units used.

§2 The Standard Cosmology

2.1 Homogeneity and Isotropy

The standard hot Big Bang model, based on the homogeneous and isotropic Friedmann-Robertson-Walker (FRW) spacetimes, is a remarkably successful operating hypothesis describing the evolution of the Universe on the largest scales. It provides a framework for such observations as the Hubble law of recession of galaxies, interpreted in terms of the expansion of the universe; the abundances of the light elements, in excellent agreement with the predictions of primordial nucleosynthesis; and the thermal spectrum and angular isotropy of the cosmic microwave background radiation (CMBR), as expected from a hot, dense early phase of thermal equilibrium betwen matter and radiation. These three pieces of evidence are the fundamental pillars upon which our confidence in the standard cosmology rests.

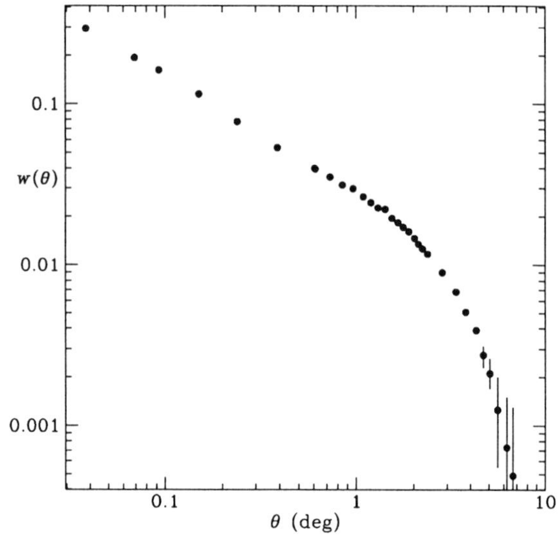

Figure 2: Angular correlation function $w_{gg}(\theta)$ measured in the APM survey, for galaxies in the apparent magnitude range $17 < b_J < 20$ [4].

While homogeneity and isotropy are, strictly speaking, assumptions of the model, they rest on a strong and growing foundation of observational support. The primary evidence for angular isotropy on large scales comes from the smallness of the CMBR large-angle anisotropy detected by the COBE satellite [1, 2] and the FIRS balloon experiment [3]. Fig. 1 shows a map of the CMBR produced from the first two years of COBE DMR data; the rms temperature fluctuation on an angular scale of 10° is $(\Delta T/T) \simeq 10^{-5}$. Additional support comes from the isotropy of radiation backgrounds at other wavelengths, as well as from the isotropy of deep galaxy and radio source counts.

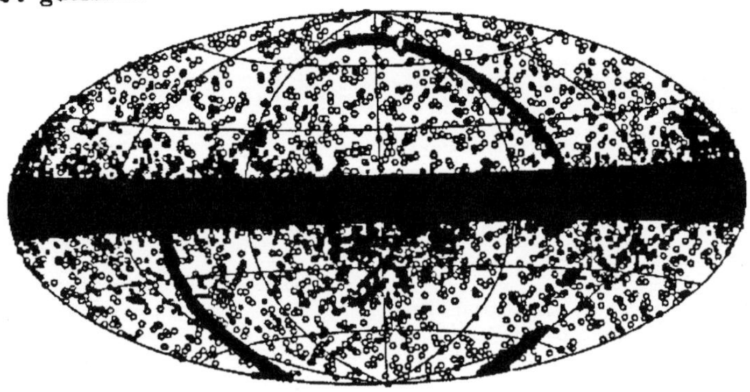

Figure 3: An equal-area projection of the IRAS-QDOT galaxies (open circles) at distances in the range $20 - 500h^{-1}$ Mpc. Solid squares are regions not surveyed, and the black band is an excluded region encompassing the Galactic disk [9] .

These different sources carry information about quite disparate scales and types of matter in the Universe. For example, the APM[4] and EDSGC[5] surveys measured the angular positions of more than one million galaxies over an area covering about 10 % of the southern sky out to an effective depth of roughly 600 h^{-1} Mpc. In these surveys, the rms fluctuation in galaxy number density on an angular scale of 5° is of order 10^{-3} (see Fig. 2). In other words, averaged over sufficently large angular scales, we see roughly the same number of galaxies (brighter than a fixed limit) per steradian in different parts of the sky at this depth. These galaxy surveys give us information about the relatively local distribution of *luminous* matter in the universe. On the other hand, through the Sachs-Wolfe effect, the large-angle CMBR measurements directly probe the gravitational potential over length scales of order 6000 h^{-1} Mpc, and are thus sensitive to the mass distribution itself – the distinction is important to keep in mind, since the evidence for *dark matter* leads us to be wary about identifying the distribution of light with that of mass. Of course, the non-zero fluctuations in galaxy density and CMBR temperature contain interesting information about the clustering of galaxies and mass, to which we return below.

Evidence for large-scale homogeneity comes in part from galaxy redshift surveys. However, compared to the angular photometric surveys, which currently give two-dimensional information for several million galaxies, currently complete redshift surveys yield three-dimensional information for typically ten thousand galaxies in a more local neighborhood. In the redshift surveys, one can verify directly that the rms fluctuations in the spatial number density of galaxies become small when averaged over large enough scales. For example, in the full-sky surveys selected from infrared sources in the IRAS catalog (the 1.2 Jy survey [6] and the 1-in-6 QDOT survey [7, 8] shown in Fig. 3), the rms fluctuation in the number of galaxies in cubical volumes of side $L = 60h^{-1}$ Mpc is of order $\delta N_{gal}/N_{gal} \simeq 0.2$ and decreases with increasing cell volume. This approach to homogeneity is roughly consistent with that seen in the deeper angular surveys. Larger structures such as superclusters, great attractors, great voids, and long filaments do exist, and have received considerable attention. Of particular note in this regard are the Great Wall, extending roughly $170 \times 60 \times 5h^{-3}$ Mpc^3 in the Center for Astrophysics (CfA) survey extension [9, 10] (see Fig. 4), and the peaks separated by $\sim$ 100 Mpc in deep pencil-beam surveys [11]. But in a statistical sense, the net fluctuations in galaxy number become small on the largest scales where they have been reliably counted in large-area redshift surveys. This is consistent with the visual impression from Fig. 3.

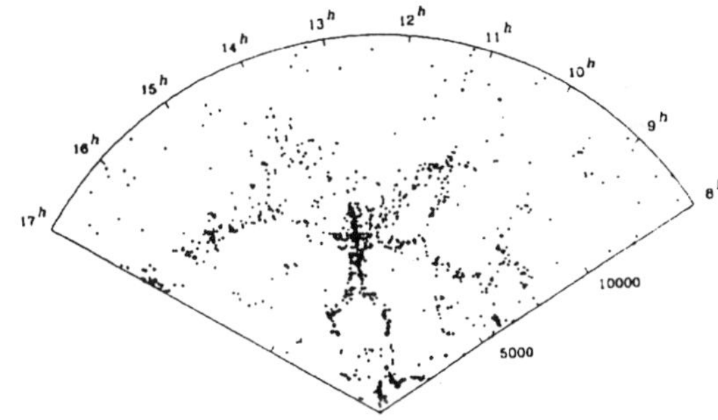

Figure 4: A 6°-thick slice from the CfA redshift survey: the angular coordinate is right ascension (in hours), and the radial coordinate is recession velocity in km/sec [10]. The concentration in the center is the Coma cluster, while the band of galaxies stretching across the figure at $v \sim 7000$ km/sec is part of the Great Wall.

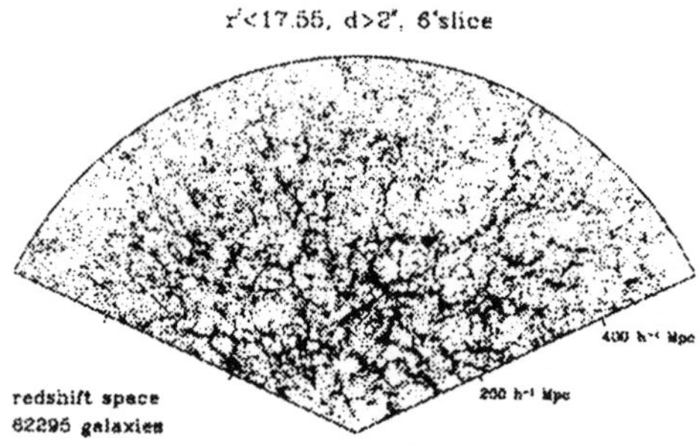

Figure 5: A simulated 6°-thick slice from the Sloan Digital Sky Survey: the radial coordinate is recession velocity in units of the Hubble parameter (redshift) [12]. This slice is roughly four times deeper than that in Fig. 4.

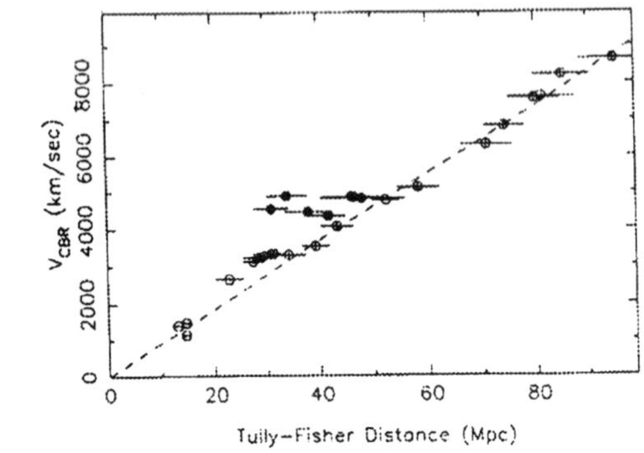

Figure 6: Hubble's law: galaxy recession speed is shown against distance inferred from the Tully-Fisher relation [13].

In the near future, the galaxy maps shown here will be substantially extended by new surveys using multi-fiber spectrographs to simultaneously measure many redshifts in the same field, an ingenious technological development which makes large-area redshift surveys possible to greater depths in a finite survey time. Advances in multi-fiber spectroscopy are being exploited by the ongoing Las Campanas survey, the planned AAT 2-degree-field survey, and by the Sloan Digital Sky Survey, which aims to measure one million galaxy redshifts over a contiguous area of π sr in the northern sky (see Fig. 5 [12]). This survey will use a 640-fiber spectrograph to accumulate redshifts at an unprecedented rate. A concurrent photometric survey will measure angular positions for roughly 50 million galaxies.

Below, we will return to discuss the statistical information about large-scale structure–that is, the departures from homogeneity and isotropy–that can be extracted from such surveys, and its use as a probe of models for the early universe.

2.2 Hubble's Law

The second piece of evidence for homogeneous and isotropic expansion is Hubble's law: if the universe remains homogeneous and isotropic over time, the distance $\ell(t)$ between any two fundamental comoving observers (each of whom defines the local cosmic rest frame) satisfies $\ell(t) = \ell_0 a(t)$, where ℓ_0 is the initial separation and $a(t)$ is the universal scale factor. Then the relative speed of one observer with respect to the other is given by

$$v(t) = d\ell/dt = \dot{a}\ell_0 = (\dot{a}/a)\ell(t) \equiv H(t)\ell(t) \quad , \qquad (2.1)$$

where a dot denotes derivative with respect to time t. For observers sufficiently nearby that the light travel time ℓ/c is small compared to the time over which $H(t)$ changes appreciably, we can replace H by its present value, $H(t_0) \equiv H_0$, where subscript 0 denotes the present epoch. Relation (2.1) then reproduces Hubble's observation in 1929 that the recession speed of a galaxy is proportional to its distance from us. The Hubble parameter, $H_0 = (\dot{a}/a)_0$, measures the expansion rate; it has been estimated by a variety of methods to be in the range 40 – 100 km/sec/Mpc (it is conventional to write this as $H_0 = 100h$ km/sec/Mpc, with $0.4 < h < 1$).

A recent example of this is shown in Fig. 6, from [13]; each circle indicates a cluster of galaxies, with distance estimated using the Tully-Fisher relation for spiral galaxies and recession velocity inferred from the redshift; the horizontal error bars indicate the $1-\sigma$ spread in inferred distances for a number of galaxies per cluster. (The closed circles around 40 Mpc indicate clusters in the vicinity of the Great Attractor region.)

Photon wavelengths are stretched and their frequencies redshifted by the expansion, $\lambda(t) \sim \nu^{-1} \sim a(t)$. This leads to the definition of redshift z:

$$1 + z(t_e) = \frac{\lambda_{obs}}{\lambda_{em}} = \frac{a(t_0)}{a(t_e)} \qquad (2.2)$$

is the ratio of the photon wavelength observed at the present (t_0) to its wavelength at emission (t_e). The redshift z thus plays several roles: through the Doppler shift, it is a measure of recession velocity, $v \simeq cz$ for $z \ll 1$; through the Hubble law, it is a measure of distance, $v = cz = H_0\ell$; and through eqn.(2.2), the redshift $z(t)$ can be used to characterize a cosmological epoch t. The most distant objects directly observed are luminous quasars, which have been seen out to redshifts approaching $z = 5$. By comparison, the photons in the cosmic microwave background radiation were probably last scattered at a redshift $z \simeq 1000$.

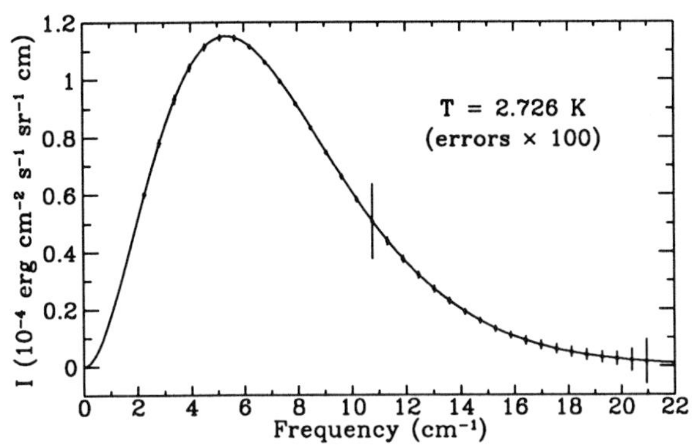

Figure 7: CMBR brightness vs. frequency measured by COBE FIRAS [14]. The error bars on the points have been multiplied by a factor of 100 to make them visible.

2.3 The Hot Big Bang: the CMBR and Primordial Nucleosynthesis

The expanding universe picture naturally traces back to a denser, hotter past–the Big Bang. Two pieces of evidence strongly support this model: the thermal spectrum of the CMBR and the light element abundances in concordance with the predictions from big bang nucleosynthesis. Spectral measurements of the CMBR cover the range of photon wavelengths from roughly 0.05 - 50 cm. The most impressive verification of the thermal Planck spectrum comes from the COBE FIRAS observations in the range 0.05 - 1 cm, shown in Fig. 7, in excellent agreement with a blackbody temperature of $T = 2.726 \pm 0.010$ K [14]. Since the recent universe is apparently transparent to CMBR photons (radio galaxies are seen as point sources out to redshifts of order $z \sim 4$), the CMBR could not have been thermalized at recent epochs and is plausibly the remnant of a much earlier ionized plasma phase when the matter and radiation were in thermal equilibrium.

When the temperature of the universe was comparable to nuclear binding energies, $T \sim 1$ MeV, protons and neutrons fused into nuclei. The predicted abundances of the light elements ^{4}He, ^{3}He, D, and ^{7}Li depend predominantly on the baryon-to-photon ratio at the time of nucleosynthesis. The predictions are in good agreement with the primordial abundances inferred from observations if the fraction of the critical density in baryons, $\Omega_b = \rho_b/\rho_c$, lies in the range $0.01 \lesssim \Omega_b h^2 \lesssim 0.02$ [15].

Taken together, the Hubble expansion, CMBR spectrum and isotropy, and light element abundances indicate that the homogeneous and isotropic Big Bang provides a reliable framework for the evolution of the universe over the approximate temperature range 10 MeV to 2.726 K, corresponding to times from roughly 10^{-2} sec to $13 - 15$ Gyr. In constructing early universe models, one extrapolates from this observational foundation up to temperatures close to the Planck scale, $T \sim 10^{19}$ GeV, corresponding to times $t \sim 10^{-43}$ sec. This would clearly be foolhardy were it not for the fact that these models leave signatures which can be probed by observations of the present Universe. However, these signatures from the early universe are filtered through subsequent evolution that depends upon the global cosmological parameters and especially upon the (dark) matter content of the universe. The current uncertainty in these parameters allows for considerable latitude in the range of allowed models. This provides strong motivation for attempts to substantially improve the observational determination of the cosmological parameters, and there is reason to hope for such improvements as the millennium draws near. The next section provides a brief overview of the current status of the cosmological parameters and dark matter, and some anticipated developments in the near future; again, these topics will be addressed in greater depth in subsequent chapters.

§3 Cosmological Parameters: Expansion Rate and Age of the Universe

The principal observable cosmological parameters of the FRW models are the Hubble parameter, H_0, the age of the Universe, t_0, the ratio of the present (non-relativistic) mass density to the critical density (see the Appendix), Ω_0, the deceleration parameter, $q_0 = -(1 + \dot{H}_0/H_0^2)$, which measures the rate at which the gravitational attraction of the matter is slowing down the expansion, the cosmological constant Λ, and the present CMBR temperature T_0. Of these, only the CMBR temperature is well determined. The age of the universe is related to the other parameters through an expression of the form $H_0 t_0 = 1.02h(t_0/10^{10}\text{yr}) = f(\Omega_0, \Omega_\Lambda)$, where f is a function of order unity and h parametrizes the uncertainty in the value of the Hubble constant. Thus the *Hubble time* $H_0^{-1} = 9.8 \times 10^9 h^{-1}$ yr sets the timescale for the age of the universe, while the *Hubble length* $cH_0^{-1} = 3000h^{-1}$ Mpc sets the lengthscale for the present observable universe. For a matter-dominated universe with $\Lambda = 0$, f falls monotonically with increasing Ω_0, and two useful limits are $f(0,0) = 1$ ($a \sim t$ for $\Omega_0 = 0$) and $f(1,0) = 2/3$ (since $a \sim t^{2/3}$ for $\Omega_0 = 1$). More generally, over the range $0 < \Omega_0 \leq 1$, $k \leq 0$, $\Omega_0 - 3\Omega_\Lambda/7 \leq 1$, an excellent approximation is [16]

$$H_0 t_0 \simeq \frac{2}{3} \frac{\sinh^{-1}(\sqrt{(1-\Omega_a)/\Omega_a})}{\sqrt{1-\Omega_a}} \qquad (3.1)$$

where $\Omega_a = \Omega_0 - 0.3(\Omega_0 + \Omega_\Lambda) + 0.3$.

Some examples of the evolution of the scale factor $a(t)$ are shown in Fig.8, including three cases with $\Lambda = 0$:

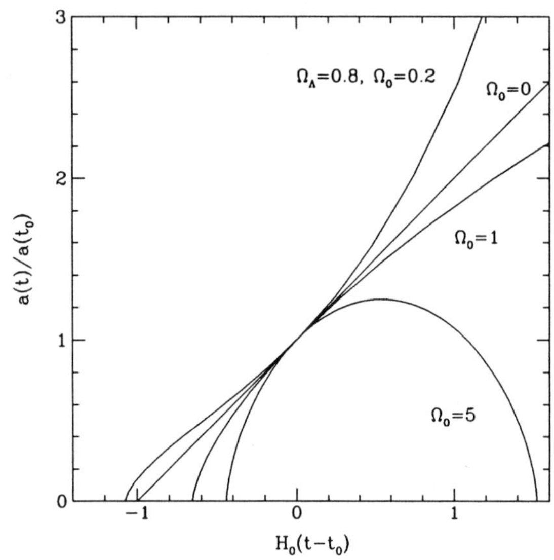

Figure 8: Evolution of the cosmic scale factor in four FRW models: 3 models with $\Lambda = 0$ ($\Omega_0 = 0, 1, 5$) and one with $\Omega_\Lambda = 0.81 - \Omega_0$.

empty ($\Omega_0 = 0$), flat ($\Omega_0 = 1$), and closed ($\Omega_0 = 5$), and one example of a flat ($k = 0$) model dominated by a cosmological constant ($\Omega_\Lambda = 0.8 = 1 - \Omega_0$). This figure displays the decrease in the expansion age $H_0 t_0$ from unity as Ω_0 is increased from zero. It also demonstrates that a significant cosmological constant term can make $H_0 t_0$ large. The solution for the spatially flat ($k = 0$) model with non-zero Λ has the form

$$\frac{a(t)}{a(t_0)} = \left(\frac{\Omega_0}{\Omega_\Lambda}\right)^{1/3} \sinh^{2/3}\left[\frac{3\Omega_\Lambda^{1/2} H_0 t}{2}\right] , \qquad (3.2)$$

where $\Omega_\Lambda + \Omega_0 = 1$. At late times, it approaches the exponential de Sitter solution, $a(t) \sim e^{\sqrt{\frac{\Lambda}{3}}t}$.

Historically, significant effort was spent trying to measure or constrain the parameters q_0 and Ω_Λ through the classical 'cosmological tests', such as the Hubble diagram, angular size as a function of redshift, and galaxy counts as a function of redshift and apparent brightness. For example, to construct the Hubble diagram, one measures the apparent brightness of a well-defined sample of objects (say, the brightest galaxies in clusters) as a function of the object's redshift; for galaxies of fixed intrinsic luminosity, the scaling of apparent magnitude with redshift is a function of the cosmological parameters. Unfortunately, galaxies at large distances ($z \sim 1$), where the distinction between model parameters becomes observationally significant, are seen when they were much younger than their nearby counterparts, so a model for galaxy luminosity evolution must be used to interpret the results. So far, these tests have not placed stringent universally accepted constraints on the cosmological parameters. However, significant progress has recently been made in understanding galaxy evolution, and there is hope that the effects of evolution and cosmology might be disentangled in coming years, particularly with the advent of deep spectroscopic surveys planned for large telescopes (such as the DEEP project on the Keck 10m telescope). Recently, it has been pointed out [17] that the probability that a quasar at a given redshift is gravitationally lensed by a foreground galaxy is a sensitive test for the cosmological constant: in models with $\Omega_\Lambda > 0$, one generally expects a higher lensing probability. Based on current surveys for lensed quasars, such as the HST Snapshot Survey, the bound $\Omega_\Lambda \lesssim 0.6 - 0.8$ has been inferred in the case of a spatially flat ($\Omega_\Lambda + \Omega_0 = 1$) universe[18, 19].

3.1 H_0 and the Distance Scale

The Hubble parameter relates the observed recession velocity v_r or redshift z of a galaxy to its distance d: for $v_r \ll c$, the recession velocity is $v_r = cz = H_0 d + v_p$, where v_p is the peculiar radial velocity of the galaxy with respect to the Hubble flow, usually assumed to arise from gravitational clustering. Galaxy redshifts can be measured quite accurately, so all the difficulty in determining H_0 resides in finding reliable distance indicators for extragalactic objects at distances large enough that the Hubble term dominates over the peculiar motion. Observed peculiar velocities are typically of order several hundred km/sec, so that distance measurements well beyond 50 Mpc or more are required for reasonable accuracy.

A wide variety of techniques has been used to establish an extragalactic distance scale[20], and this is reflected in the spread of results for H_0, roughly 40 – 100 km/sec/Mpc. Distance estimates made using methods such as the Tully-Fisher relation between 21-cm rotation speed and infrared luminosity for spiral galaxies, calibrated by observations of Cepheid variable stars in several nearby galaxies, have yielded high values for the expansion rate, roughly $H_0 = 80 \pm 10$ km/sec/Mpc. Two newer methods, planetary nebula luminosity functions[21] and galaxy surface brightness fluctuations[22] yield values for H_0 in this range as well, and are being further developed. Methods using Type Ia supernovae as standard candles have yielded low values, $H_0 \simeq 50 \pm 10$ km/sec/Mpc, although a recent variation of this method using light curve shapes finds[23, 24] $h \simeq 0.65$. SNe Ia are thought to be the explosions of white dwarfs which accrete matter from binary companions until they reach the Chandrasekhar mass, and there is some evidence that they form a homogeneous class; they also have the advantage that they can be observed to great distances.

There are also a variety of methods being employed

to measure the distances of extragalactic objects directly, bypassing the extragalactic distance ladder built up from Cepheids. The expanding photosphere method[25] has been used to determine the distances to a number of type II supernovae at large distances, generally yielding good agreement with the Tully-Fisher distances for these galaxies. Other 'direct' methods which hold future promise include measurement of the Sunyaev-Zel'dovich effect, due to the Compton upscattering of CMBR photons by hot gas in rich clusters[26, 27], and the differential time delay between images in gravitationally lensed quasars[28].

Recently, observations at the CFHT telescope[29] and with the Hubble Space Telescope (HST)[30] have found respectively 3 and 20 Cepheids in two galaxies in the Virgo cluster. The inferred distance to Virgo (assuming these galaxies are close to the center of the cluster) are in accord with the high values of the Hubble constant, $h \simeq 0.8 \pm 0.17$. In the future, HST observations of Cepheids in other nearby galaxies (as well as other Virgo cluster galaxies) which can be used as secondary distance calibrators should reduce this uncertainty by a factor of order two.

3.2 The Age of the Universe, t_0

Three methods have traditionally been used to infer the age of the Universe, t_0. Nuclear cosmochronology is based on radioactive dating of $r-$process elements, that is, heavy elements formed by rapid neutron capture, most probably in supernovae. The element ratios Re/Os and Ur/Th generally indicate $t_0 = 10 - 20$ Gyr[31], with a large uncertainty due to the unknown element formation history (e.g., the star formation rate over time). A second method involves the cooling of white dwarfs: when low-mass stars exhaust their nuclear fuel, they become degenerate white dwarfs, gradually cooling and becoming fainter. The number of white dwarfs as a function of luminosity drops dramatically for $L_{wd} < 3\times10^{-5}L_{\odot}$, suggesting that there has not been sufficient time for them to fade below this value (assuming the drop is not a selection effect). Coupled with models of white dwarf cooling, this implies that the age of the galactic disk is about $t_0 \simeq 9 - 10 \pm 2$ Gyr[32, 33].

The most extensively studied technique for constraining t_0 is the determination of the ages of the oldest globular clusters in the galaxy. When stars finish burning hydrogen, they turn off the main sequence, characteristically reddening and brightening. By observing the color-magnitude (color vs. apparent brightness) diagram for a cluster, one can determine the apparent brightness of stars in the cluster that are now leaving the main sequence. Knowing the distance to the cluster then gives the absolute luminosity of stars at the turn-off. On the other hand, stellar evolution theory relates stellar luminosity to the time a star spends on the Hydrogen-burning main sequence. Thus, the measurement of the main sequence turn-off luminosity gives an estimate of the age of the cluster. The turn-off luminosity in the oldest globular clusters in the galaxy is slightly below that of the sun, yielding the age estimate $t_{gc} = (13 - 15) \pm 3$ Gyr[34]. In practice, cluster ages are determined by comparing the luminosity ratio of horizontal branch stars to those at the main-sequence turn-off. Theoretical uncertainties include the degree of mass loss, the role of convection, and the initial Helium abundance, but it appears very difficult to bring the oldest globular ages down below 11-12 Gyr [35]. A significant observational source of error for the method is the uncertainty in the distances to the globular clusters, determined by RR Lyrae stars. Observations with the corrected Hubble Space Telescope mirror may reduce the observational uncertainty in t_{gc} to as little 10%.

3.3 The 'Age Crisis'

It is useful to place these numbers for the cosmological parameters in theoretical context. As Fig. 8 shows, cosmological evolution depends on the product

$$H_0 t_0 = 1.06 \left(\frac{H_0}{80\text{km/sec/Mpc}}\right)\left(\frac{t_0}{13\text{Gyr}}\right) \quad . \qquad (3.3)$$

If an extended period of inflation took place in the early universe, then the spatial geometry should now be observationally indistinguishable from the spatially flat Einstein-de Sitter model. If the cosmological constant vanishes, spatial flatness then implies the critical matter density, $\Omega_0 = 1$ (with the concomitant requirement of non-baryonic dark matter–see below), and $H_0 t_0 = 2/3$. Comparing with (3.3), this yields an uncomfortably low age for globular clusters unless $H_0 \lesssim 50$ ($t_0 \gtrsim 13$ Gyr) and is definitely problematic unless $H_0 < 65$ ($t_0 > 10$ Gyr), still on the low side of the majority (but not all) of the H_0 observations.

As shown in Fig. 8, the problem is alleviated if the cosmological constant is non-zero, which opens up the possibility that $H_0 t_0 > 1$. In the spatially flat model, $\Omega_0 + \Omega_\Lambda = 1$, the most stringent constraint on the cosmological constant appears to be that from gravitational lens statistics, $\Omega_\Lambda \lesssim 0.7$. For $\Omega_\Lambda = 1 - \Omega_0 < 0.7$, eqn.(3.2) implies $H_0 t_0 < 0.96$. Thus, from (3.3), with a cosmological constant saturating this bound, the globular cluster age $t_0 \geq 13$ Gyr ($t_0 > 10$ Gyr) implies $H_0 < 72$ km/sec/Mpc ($H_0 < 94$ km/sec/Mpc). The implied value of $\Omega_0 \gtrsim 0.3$ is consistent with dynamical mass estimates from galaxy clusters (within the errors) and with most of the dynamical estimates from large scale flows. However, the fact that we would be observing the universe just at the epoch when Ω_0 is comparable to Ω_Λ might seem to beg for explanation.

A third possibility is that inflation did not occur (or last sufficiently long) and that we live in an open, low-density (perhaps purely baryonic) universe with negligible Λ. As Fig. 8 shows, this also helps alleviate the 'age problem' somewhat compared to the Einstein-de Sitter model. Dynamical mass estimates suggest that $\Omega_0 \gtrsim 0.1$, which implies $H_0 t_0 \lesssim 0.9$ for the open model. Taking $t_0 > 13$ Gyr as before, this implies $H_0 < 68$ km/sec/Mpc. The challenge in this case is to form large-scale structure without violating CMBR anisotropy constraints and to account for the large-scale peculiar velocities. In this case, it also would be a 'coincidence' that we live just at the epoch when the curvature term in (A.5) is becoming appreciable compared to the matter term. Clearly the existence or severity of the age problem must be weighted by the present *uncertainty* in the cosmological parameters. As we approach the millennium, we face the exciting prospect that astronomical observerations may be able to pin down these parameters and thus determine which of the above (or none of the above!) scenarios corresponds to our Universe.

§4 Dark Matter and Ω

The evolution of large-scale structure, and therefore the connecting thread between the early universe and the observed clustering of galaxies, depends critically on the matter content of the universe. The density of luminous matter, that is, of matter associated with typical stellar populations in galaxies, is inferred to be $\Omega_{lum} \sim 0.007$. However, it is well known that the luminous parts of galaxies are not the whole story: there is strong evidence from flat spiral galaxy rotation curves, from a variety of observations of galaxy clusters, and from large-scale peculiar motions that there is substantially more mass in these systems than can be attributed to the luminous matter: the bulk of the matter in the universe is *dark*.

In a typical spiral galaxy, the luminous components (the bulge and the disk) generally cannot account for the observed extended flat rotation curve traced by stars and gas beyond several kpc from the galactic center. To reproduce the velocity observations, one posits a quasi-spherical dark halo truncated at some large radius. If these systems are typical of the universe, this provides a lower bound of roughly $\Omega_0 \gtrsim 0.02h^{-1}$ for the matter associated with galaxy halos.

In clusters of galaxies, there are now a variety of methods for estimating cluster masses. The traditional dynamical method, first used by Zwicky in the 1930's, applies the virial theorem to the measured cluster galaxy velocity dispersion, and typically yields an equivalent $\Omega_0 \sim 0.1 - 0.2$. However, this method assumes that galaxies trace the underlying cluster mass and that the galaxy velocity distribution is isotropic. Independent information on the dark matter distribution in clusters comes from the giant luminous arcs and arclets, high-redshift galaxies gravitationally lensed by rich foreground clusters into extended banana-shaped images [36, 37]. These estimates are generally in reasonable agreement with the cluster masses inferred from the virial theorem in the cores of rich clusters. More recently, this method has been extended into the 'weak lensing' regime, in which one studies the shear distortion pattern induced by a cluster statistically by measuring the image shapes and orientations of a large number of faint background galaxies [38]. This allows one to probe the cluster mass distribution over larger scales and in clusters too weak to produce arcs. In some cases, the weak lensing masses are consistent with the virial masses, while in others the lensing-inferred masses appear to be several times larger. This relatively new method holds considerable promise for the future, as large-area imaging cameras are installed on 4-m class telescopes at sites with excellent seeing. Finally, X-ray satellites have begun to map the density and temperature profiles of the hot gas which permeates many clusters. If the gas is in hydrostatic equilibrium, it can be used to directly trace the cluster mass distribution. Cluster masses inferred by this method are generally comparable to the virial estimates. The X-ray observations also indicate that clusters are surprisingly baryon-rich: the gas constitutes typically $(5 - 10)h^{-3/2}\%$ of the inferred binding mass within approximately $1h^{-1}$ Mpc of the center of a rich cluster like Coma. If this ratio is representative of the baryon mass fraction of the universe, then the nucleosynthesis bound on Ω_B would imply [39] the *upper* limit $\Omega_0 \lesssim (0.2 - 0.4)h^{-1/2}$, well below closure density. As the study of clusters matures, it will be interesting to see whether this implies a sub-critical universe or that clusters are more complex than previously thought.

On still larger scales, peculiar velocities (i.e., deviations from the Hubble flow) have been used to infer the cosmic density [40]. This idea has been applied in a number of ways. One method is to compare samples of the density perturbation field and the divergence of the peculiar velocity field covering the same volume; assuming they arise gravitationally, the proportionality between them depends on the rate of growth of the density fluctuations, which in turn depends on Ω_0. On very large scales, linear perturbation theory is a reasonable first approximation, and one finds the relation

$$\nabla \cdot \vec{v}_p = -H_0 \Omega_0^{0.6} \delta \quad , \tag{4.1}$$

where the density field $\delta(\vec{x}) = (\rho(\vec{x}) - \bar{\rho})/\bar{\rho}$, and the perturbation growth rate enters through $d\ln\delta/d\ln a \simeq \Omega^{0.6}$. If one expresses distances d in terms of their equivalent Hubble velocities, $v = H_0 d$, then H_0 drops out of Eqn.(4.1), so the uncertainty in the Hubble parameter does not undermine this method.

The topographic correlation between the observed galaxy field (from redshift surveys) and the density field inferred from galaxy peculiar velocity surveys suggests that galaxies do broadly trace the mass distribution on large scales, in the sense that the local galaxy density increases approximately linearly with the local velocity divergence. Nevertheless, the galaxy distribution may be 'biased' with respect to the mass. In the simplest linear bias model, the smoothed galaxy and mass density fields are assumed to be proportional, $\delta_{gal}(\vec{x}) = b\delta(\vec{x})$, where b is the bias factor, taken to be constant for a given class of galaxies. Since δ_{gal} is what is observed, the comparison between the galaxy and velocity fields actually constrains the combination $\Omega_0^{0.6}/b$, where the bias factor refers, e.g., to galaxies selected from the IRAS catalog. Recent determinations have found $\Omega_0^{0.6}/b \sim 0.5 - 1$; for a bias factor of order unity, this is consistent with $\Omega_0 \sim 0.3 - 1$. It is well to keep in mind, however, that biasing is presumably a complex process associated with the non-linear stages of galaxy formation, so that the proportionality of the galaxy and density fields may be non-linear and/or scale-dependent. In addition, even with substantial smoothing of the density field, the perturbation amplitude in many regions is not small, and corrections to the linear theory must be taken into account.

An alternative incarnation of this method does not use the peculiar velocity field directly (which requires independent distance measurements for the galaxies) but makes use of the fact that peculiar velocities distort the shape of the galaxy power spectrum in redshift surveys. Measuring the anisotropy of galaxy clustering in redshift surveys then yields an estimate of $\Omega_0^{0.6}/b$; results from, e.g., the IRAS 1.2 Jy survey have recently come in at the low end of the above range,[41] but the sampling errors due to the relatively small surveyed volume are still large. The accuracy of this method to determine Ω_0 should improve substantially with the next round of large-scale redshift surveys. In addition, measurement of the Sunyaev-Zel'dovich effect in a number of clusters at moderate redshift will probe the peculiar velocity field over very large scales.

The observational picture on Ω_0 is currently unclear: the determinations on small scales (clusters and below) consistently yield $\Omega_0 \sim 0.1 - 0.2$ (with the possible exception of weak lensing), while dynamical determinations on large scales show large scatter, $\Omega_0 \sim 0.3 - 1$. Either way, there is a strong dynamical case that over 90 % of the mass in the universe is dark. As the dynamical evidence for the existence of dark matter continues to mount, and as surveys begin to map the distribution of dark matter in different systems and over various scales, it becomes increasingly urgent to determine its composition. In short, we know dark matter exists, we are making progress in understanding *where* it is and how much of it there is, but we now need to know *what* it is. In the complex lexicon of particle astrophysics, the possibilities are divided into two categories: baryons and non-baryonic matter.

4.1 Baryonic Dark Matter

The density of baryons in the universe is an important ingredient in structure formation models, particularly in regard to theoretical predictions for small-angle CMBR anisotropies. Concordance of the observed light element abundances with the predictions of primordial nucleosynthesis restricts the baryon density to the range $0.01 < \Omega_B h^2 < 0.02$ [15]. This range may shift or be narrowed as deuterium measurements from Lyman-α clouds accumulate. The corresponding baryon to photon ratio, $n_B/n_\gamma = 2.7 \times 10^{-8}\Omega_B h^2 \simeq 3 \times 10^{-10}$, should ultimately be a predictable number from models for the baryon asymmetry. Recent theoretical work has concentrated on the possibility of generating this asymmetry in a (weakly) first-order electroweak phase transition. Comparison of the nucleosynthesis range for Ω_B with the density of luminous matter Ω_{lum} suggests that some or all of the baryons are dark or in underluminous populations. On the other hand, comparison with the density inferred from galaxy rotation curves, $\Omega_0 \sim 0.02h^{-1}$, shows that baryons could constitute some or all of the dark matter in galaxy halos.

One possibility for dark baryons in halos are degenerate brown dwarfs, substellar ($M < 0.08M_\odot$) objects which did not reach sufficiently high temperature to burn hydrogen. Currently three independent groups are searching for galactic brown dwarfs (which have been dubbed MACHOS, for massive compact halo objects); the signature is a gravitational microlensing event, in which a background star, say in the LMC or the bulge of the Milky Way, symmetrically brightens and fades as a MACHO passes near its line of sight.[42] Although they are distinguishable from ordinary variable and flare stars by the time symmetry and achromaticity of the light curve, such microlensing events are intrinsically rare–the probability that a given background star in the LMC is lensed by a halo-filling MACHO population is $\tau \sim 5 \times 10^{-7}$, so a large number of stars must be accurately monitored. The American-Australian MACHO project, the French EROS collaboration, and the Polish-American OGLE collaboration first announced the discovery of microlensing events in the fall of 1993 [43, 44, 45]. As of early 1995, they have a combined total of more than 50 microlensing events; the vast majority are in the direction of the galactic bulge, and only several are toward the LMC. [46, 47] The preliminary indication from the low rate toward the LMC is that the Milky Way halo is not predominantly composed of MACHOs in the mass range to which the experiments are sensitive.

4.2 Non-baryonic Dark Matter

If $\Omega_0 > \Omega_B$, as the dynamical observations on scales larger than clusters suggest, some form of non-baryonic matter must be invoked to make up the balance. Particle physics theories provide a host of weakly interacting particles as dark matter candidates. It is convenient to distinguish two broad classes of non-baryonic dark matter, hot and cold, on the basis of their clustering properties. The prototypical hot dark matter candidate is a light neutrino with mass $m_\nu \simeq 20$ eV (more generally, the mass density of light neutrinos is $\Omega_\nu h^2 = (m_\nu/92\text{eV})$). Since they are relativistic until relatively recent epochs, light neutrinos would free-stream out of and damp out density perturbations up to the scale of galaxy clusters, unless there is some mechanism (such as topological defects) which continually seeds the formation of perturbations on small scales. Galaxies would form after clusters via fragmentation ('top down'). Due to phase-space constraints[48], light neutrinos would not cluster significantly on the scale of galaxies: baryons would constitute the predominant dark matter in galaxy halos, while neutrinos would dominate in clusters. Cold dark matter, on the other hand, is non-relativistic at all epochs of interest for structure formation, so it has negligible free-streaming length–it clusters on all scales. In cold dark matter models, structure generally forms hierarchically, with smaller clumps merging to form larger ones ('bottom up'). It is also worth noting the recently surging popularity of a mix 'n match scenario: a combination of 70-80% cold and 20-30% hot dark matter (say, with $m_\nu \simeq 5$ eV) produces a favorable spectrum of large-scale density perturbations in the context of inflation, with some advantages over pure cold dark matter.

The theoretically favorite candidates for cold dark matter are weakly interacting massive particles (WIMPs), with masses generally in the range $20 - 150$ GeV, and the axion, an ultra-light scalar with a mass of order 10^{-5} eV. The most attractive WIMP candidate is the neutralino, the lightest supersymmetric fermionic partner of the standard model bosons; its weak annihilation rate in the early universe naturally leaves it with an abundance comparable to that corresponding to the present critical density. The axion is the pseudo-Nambu-Goldstone boson associated with spontaneous breakdown of a global $U(1)$ symmetry (the Peccei-Quinn symmetry) introduced to explain why the strong interactions conserve CP. The global symmetry is spontaneously broken at some large mass scale f_{PQ}, through the vacuum expectation value of a complex scalar field, $\langle\Phi\rangle = f_{PQ}\exp(ia/f_{PQ})/\sqrt{2}$. At energies below the scale f_{PQ}, the only relevant degree of freedom is the massless axion field a, the angular mode around the bottom of the Φ potential. At a much lower energy scale, $\Lambda_{QCD} \sim 100$ MeV, the symmetry is *explicitly* broken when QCD becomes strong, and the axion obtains a periodic potential of height $\sim \Lambda_{QCD}^4$. In 'invisible' axion models with Peccei-Quinn scale $f_{PQ} \sim 10^{12}$ GeV, the resulting axion mass is $m_a \sim \Lambda_{QCD}^2/f_{PQ} \sim 10^{-5}$ eV and $\Omega_a \sim 1$. Although light, invisible axions interact so weakly, with cross-section $\sigma \sim 1/f_{PQ}^2$, that they were never in thermal equilibrium: they form as a cold Bose condensate.

Accelerator searches for neutrino mixing and beta-decay experiments on neutrino mass should provide useful constraints on the possibility of neutrino dark matter. Active experimental efforts are also underway to detect both WIMPs and axions. Direct WIMP detection looks for the signals produced when a halo WIMP collides with a nucleus in a kg-size cryogenic crystal, depositing of order 10 keV in ionization and phonons (detector schemes based on scintillation and excitations of superfluids and superconductors are also being developed). Indirect WIMP detectors search for high energy neutrinos produced when WIMPs annihilate in the Sun and the Earth; large underground or underwater detectors currently in place or under development with sensitivity to WIMP annihilations include SuperKamiokande, MACRO, AMANDA, and DUMAND. Accelerator searches for supersymmetry also will constrain the neutralino parameter space. A large-scale axion search based at Livermore will search for resonant conversion of halo axions to microwave photons in a cavity with a strong magnetic field. A scaled-up version of an idea originally proposed by Sikivie, this detector should approach the cosmologically interesting region of axion couplings for the first time[49].

Finally, given the paucity of direct evidence for dark matter, it is probably healthy to keep an open mind to alternatives. While it is natural to ascribe flat galaxy rotation curves and large cluster velocity dispersions to unseen matter, Milgrom[50] and others have argued that they may instead signal a breakdown of Newton's law of inertia at very low acceleration. The extent to which Milgrom's modified Newtonian dynamics (MOND) accounts for all the phenomena normally imputed to dark matter is controversial, and a full theory with which one could explore cosmology has been lacking.

§5 The Early Universe

The thermal spectrum of the CMBR and the concordance of primordial nucleosynthesis predictions with light element abundances strongly suggest that the early universe can be accurately described as a dilute, adiabatically expanding gas of relativistic particles in local thermal equilibrium, with instantaneous temperature $T(t) \simeq 1\,\text{MeV}(t/\text{sec})^{-1/2}$. While this may reflect conditions over

T(GeV)	t(sec)	Event
10^{19}	10^{-43}	Planck era: quantum gravity
10^{15}	10^{-35}	Grand unification: inflation, topological defects
10^{3}	10^{-12}	Supersymmetry, technicolor
10^{2}	10^{-10}	Electroweak transition: $SU(2) \times U(1) \rightarrow U(1)_{em}$
0.2	$10^{-4.5}$	Quark-hadron transition; chiral symmetry breaking
10^{-3}	1	Neutrino decoupling
5×10^{-4}	4	e^+e^- annihilation
7×10^{-5}	200	Big Bang nucleosynthesis
10^{-9}	5×10^{11}	Matter-radiation equality
3×10^{-10}	10^{13}	H recombination; photon decoupling
6×10^{-13}	10^{17}	Non-linear structures form

Table 1: Thermal history of the universe.

much of the early history of the universe, we know it is not the full story, because a gas in thermal equilibrium is featureless and structureless. Although thermal equilibrium is a good starting point, the interesting epochs in cosmic history–the ones which leave potentially observable signatures or relics, such as the light elements, the baryon asymmetry, and particle dark matter–are those in which a particle species i goes out of thermal equilibrium, because the rate of interactions keeping it in equilibrium, Γ_i, falls below the expansion rate $H(t)$. This happens at the freeze-out temperature T_F, defined implicitly by $(\Gamma/H)_{T_F} \simeq 1$. It is worth noting that there may also be particle relics, such as magnetic monopoles and axions, which were never in thermal equilibrium.

A timeline for the early universe with some important epochs delineated is shown in Table 1. The CMBR probes conditions back to the time of photon decoupling, while the light element abundances probe conditions at temperatures comparable to nuclear binding energies, $T \sim$ MeV. Going backward in time, QCD predicts that chiral symmetry should be broken at a temperature of order 100 MeV. At about that time, quarks should also become confined inside hadrons, but it is not clear if this is a smooth (second-order) transition or a first-order transition involving the release of latent heat and the nucleation of hadron bubbles. If the quark-hadron transition is first-order, the resulting inhomogeneous baryon distribution might leave observable signatures in the light element abundances.

Going to earlier times, the electroweak symmetry should be restored at a temperature of order 100 GeV. The dynamics of this transition is currently an area of active investigation: if it is a first-order transition, it is possible that the resulting non-equilibrium conditions were ripe for the generation of the baryon asymmetry. If baryons and anti-baryons had been present in exactly equal numbers in the early universe, their annihilation would eventually have driven the baryon to photon ratio down to a value many orders of magnitude smaller than the observed value of $n_B/n_\gamma \simeq 3 \times 10^{-10}$. Thus, when the baryons were relativistic, there must have been a small asymmetry between the density of baryons and anti-baryons, $\eta = n_B/n_b = (n_b - n_{\bar{b}})/n_b \sim \eta$. As first pointed out by Sakharov, the generation of such a baryon asymmetry requires baryon-number and CP-violating interactions as well as a departure from thermal equilibrium (since $n_b = n_{\bar{b}}$ in equilibrium). It is currently thought that baryogenesis takes place at the GUT or electroweak eras.

Going back earlier than 10^{-10} sec, we must invoke physics beyond the standard electroweak model, and the events become increasingly speculative. Particle physics models suggest there may be new physics lurking at the TeV scale–perhaps supersymmetry, technicolor, or various extensions of the standard model. In the simplest grand unified theories, the strong and electroweak interactions are unified (the symmetry between them restored) at an energy scale of order 10^{15} GeV. A cosmological phase transition at that epoch might lead to inflation or to the generation of topological defects such as monopoles, cosmic strings, or textures; the resulting density and gravitational wave perturbations produced could provide the seeds for large-scale structure and leave a signature in the CMBR anisotropy. Classical cosmology runs into a wall at the Planck era, $t \sim 10^{-43}$ sec: at that epoch, quantum fluctuations in the spacetime metric are expected to be large, and a quantum theory of gravity is required. If superstrings provide the fundamental description of nature, inherently stringy effects would become important around that scale.

One of the major developments in theoretical cosmology over the last fifteen years is the discovery of scenarios for the generation of primordial density fluctuations. Currently there are two early universe paradigms for the origin of perturbations: inflation and topological (or non-topological) defects. Both scenarios predict large-angle anisotropies in the CMBR and produce large-scale structure, but their predictions differ in details that are under active study.

§6 The Inflationary Universe

The inflationary scenario originally arose out of the attempt to solve several puzzles encountered when one extrapolates the standard cosmology back to the very early universe, in particular the horizon and flatness problems and the overabundance of magnetic monopoles in grand unified theories [51]. The horizon problem arises from the observation that the CMBR is isotropic over large angular scales (see Fig. 1): in the standard cosmology, regions corresponding to separations larger than about one degree were not yet in causal contact when the photons last scattered (at $t \sim 10^{13}$ sec – see Table 1). Yet the microwave sky is remarkably isotropic over all scales larger than a few arcminutes, to of order one part in 10^5. The flatness problem is the puzzle of why the spatial curvature term in (A.5) does not presently dominate over the matter-density term by a large margin: if $\Lambda = 0$, we can rewrite (A.5) suggestively as $|\Omega(t) - 1|^{-1} = a^2(t)H^2(t) = \dot{a}^2$, which is a decreasing function of time if the universe is dominated by a fluid with non-negative pressure (or, more precisely, from (A.3) by a fluid with $p > -\rho/3$). The natural timescale for the universe to become curvature-dominated is the Planck time, $t_{Pl} \sim 10^{-43}$ sec, yet our universe is still not strongly curvature-dominated at $t_0 \sim 10^{17}$ sec $\sim 10^{60} t_{Pl}$.

It must be stressed that these puzzles are not inconsistencies of the standard cosmological model. Rather they point to features of the observed universe which the standard model does not explain and which moreover appear highly unlikely when the standard cosmology is embedded in a somewhat larger class of cosmological models: the present nearly homogeneous and spatially flat state of the Universe appears to be very sensitive to the initial conditions. Inflation was designed to reduce this sensitivity by widening the class of initial conditions which evolve to a nearly homogeneous and spatially flat state within the observable universe. This is the sense in which inflation is said to 'solve' the horizon and flatness problems.

Inflation is a hypothetical early epoch in cosmic history during which the scale factor of the universe accelerates, $\ddot{a} > 0$. This rapid expansion drives the spatial curvature toward zero, $\Omega(t) \rightarrow 1$. To solve the horizon and flatness problems, this epoch must last long enough for the scale factor to grow by a factor of e^{N_e}, where the number of inflationary e-folds of growth satisfies $N_e > N_{min} \simeq 60 \pm 10$ (N_{min} depends slightly upon initial conditions and logarithmically upon the reheat temperature after inflation). In most inflation models, the growth of the scale factor during inflation is quasi-exponential, $a \sim e^{tH(t)}$, where $H(t)$ decreases on a timescale long compared to the expansion time, $|\dot{H}| \ll H^2$. Because of the quasi-exponential behavior, unless N_e is almost exactly equal to N_{min}, a sufficiently long period of inflation implies that the present universe should be observationally indistinguishable from the spatially flat model, i.e., $\Omega_0 + \Omega_\Lambda = 1 + \epsilon$, where $\epsilon = \mathcal{O}(e^{-2(N_e - N_{min})})$ is exponentially small. (Recently, there has been some interest in 'open' inflation models, where one arranges $\Omega < 1$ by having $N_e = N_{min} \pm$ a few, but this requires fine tuning either of initial conditions or of scalar field parameters [52, 53, 54]. This fine tuning has nothing to do with the small coupling problem (see below), although the latter, which is a feature of *all* inflation models, is often inaccurately stated to be a fine tuning problem.)

Acceleration of the scale factor requires that the energy density be dominated by a fluid with equation of state $p < -\rho/3$; this is most simply achieved with a scalar field ϕ, the *inflaton*, undergoing a classical 'slow-rollover' phase [55, 56], where its stress-energy-momentum is dominated by its potential, $V(\phi)$. For a spatially homogeneous scalar field $\phi(t)$, the energy density and pressure are given by

$$\begin{aligned} \rho &= \frac{\dot{\phi}^2}{2} + V(\phi) \\ p &= \frac{\dot{\phi}^2}{2} - V(\phi) \end{aligned} \tag{6.1}$$

Thus, accelerated expansion takes place if ϕ dominates the stress-energy of the universe and if $V(\phi) > \dot{\phi}^2$; if this inequality is strongly satisfied, $V(\phi) \gg \dot{\phi}^2$, then $p_\phi \simeq -\rho_\phi = -V(\phi)$, and the expansion is approximately exponential.

6.1 Fluctuations from Inflation

Soon after its invention, it was realized that inflation could in principle provide one of the holy grails of cosmology: a causal mechanism for the origin of density fluctuations that later grow to form large-scale structure. Curvature perturbations cannot be causally generated on scales larger than the instantaneous Hubble radius H^{-1}, because of local energy conservation. Perturbations on galaxy scales today were larger than the Hubble radius at redshifts larger than $z \sim 10^6$ ($T \gtrsim 100$ eV). If there is no new physics at atomic scales to produce fluctuations at late times after they enter the Hubble radius, then in the standard cosmology (without inflation) they must be postulated as initial conditions. (Note that this argument applies only to curvature perturbations: an important loophole is the early generation of isocurvature perturbations, the mechanism of defect-mediated structure formation.)

An early epoch of inflation alters this situation: during a phase of accelerated expansion, the Hubble radius H^{-1} grows more slowly than the scale factor. Thus a perturbation of comoving wavelength λ can be created causally when its physical scale $\lambda_{phys} = a(t)\lambda$ is less than H^{-1} during inflation, expand outside the Hubble radius at some time $t_A(\lambda)$ during inflation, and then eventually 're-enter' the Hubble radius in the recent radiation- or matter-dominated era at a time $t_B(\lambda)$. Scales λ corresponding to the range from the galaxy scale ($\sim$ Mpc) to the current Hubble radius (3000 h^{-1} Mpc) crossed outside the Hubble radius in the interval $50-60$ e-folds before the end of inflation, i.e., when $a(t) = a(t_{end})e^{-(50-60)}$.

Quantum fluctuations of the slowly rolling inflaton field provide the dynamical mechanism for generating density perturbations in inflation [57, 58, 59, 60]. Mathematically, one quantizes the spatial fluctuations of the inflaton field (and the corresponding metric perturbation) about its homogeneous classical expectation value. In its vacuum state, such a quantum field has a spectrum of zero-point fluctuations by the Heisenberg uncertainty principle, with an rms amplitude on scale λ at Hubble-radius crossing (t_A) given by $(\Delta\phi)_{\lambda^{-1}=Ha} = H/2\pi$. These spatial variations in ϕ correspond to fluctuations in the inflaton energy density, which are stretched by the accelerated expansion and converted into adiabatic density fluctuations in all species during the reheating phase, when ϕ decays at the end of inflation. The resulting perturbation amplitude when a comoving wavelength λ re-enters the Hubble radius is given approximately by

$$\left(\frac{\delta\rho}{\rho}\right)_{t_B(\lambda)} \simeq \frac{(\Delta\phi)_{\lambda^{-1}=aH}V'(\phi)}{\dot{\phi}^2} = \left(\frac{3H^2}{2\pi\dot{\phi}}\right)_{t_A(\lambda)} \qquad (6.2)$$

Letting ϕ denote the value of the inflaton when a given scale λ crosses outside the Hubble radius and defining $(\delta\rho/\rho)_{t_B(\lambda)} \equiv A_S(\phi)$, this can be written as

$$A_S(\phi) = \sqrt{384\pi}\left(\frac{V^{3/2}(\phi)}{m_{Pl}^3 V'(\phi)}\right)_{t_A(\lambda)} \simeq \left(\frac{H^2(\phi)}{m_{Pl}^2|H'(\phi)|}\right)_{t_A(\lambda)} \qquad (6.3)$$

where a prime denotes differentiation with respect to ϕ.

Since the period of ten e-folds of the scale factor over which structure on astronomical scales is generated is brief, and the Hubble expansion rate, which sets the scale of the fluctuation amplitude, is slowly varying during inflation ($\dot{H} \ll H^2$), the density perturbation amplitude at Hubble-radius crossing $A_S(\phi)$ is nearly independent of scale λ. The density perturbation amplitude at Hubble crossing ($\lambda_{phys} = H^{-1}$) is essentially the gravitational potential perturbation on that scale, $\delta\Phi_\lambda \sim G\delta M/\lambda \sim (\delta\rho/\rho)_\lambda(\lambda_{phys}/H^{-1})^2$, and it sets the scale for the large-angle CMBR anisotropy through the Sachs-Wolfe effect, $\delta T/T \sim \delta\Phi$. Thus, inflation predicts that the gravitational potential fluctuations are nearly independent of scale. Such a scale-invariant spectrum was first proposed on different grounds by Harrison, Zel'dovich, and Peebles and Yu ten years before inflation. Considerable excitement was generated when analysis of maps of the CMBR sky from the first year of COBE DMR data yielded a large-angle spectrum consistent with the scale-invariant spectrum [1].

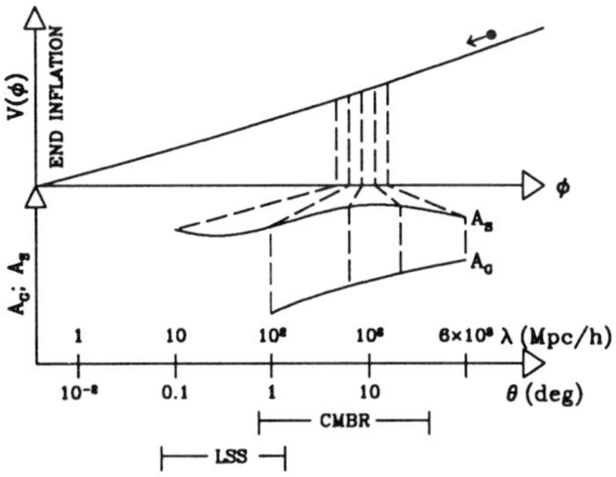

Figure 9: Sketch of an inflaton potential $V(\phi)$ and the scalar $A_S(\phi)$ and tensor $A_G(\phi)$ fluctuation spectra produced by quantum fluctuations. ϕ is the value of the classical field when the comoving scale λ crossed outside the Hubble radius during inflation. Also shown is the corresponding angular scale on the surface of last scattering and the rough angular and lengthscale intervals probed by CMBR anisotropy experiments and galaxy surveys (LSS).

In addition to the scalar metric fluctuations arising from quantum fluctuations in the inflaton field, there are also gravitational wave perturbations associated with quantum fluctuations in the transverse traceless (tensor) part of the metric field [61]. The perturbation equations for the graviton amplitudes $h_{+,x}$ are identical to those for a massless, minimally coupled scalar field with amplitude $\delta\phi = h/\sqrt{16\pi G}$. The resulting amplitude at Hubble-radius crossing is

$$h_{\lambda^{-1}=aH} \equiv A_G(\phi) \simeq \frac{2H(\phi)}{\sqrt{\pi}m_{Pl}} \simeq \frac{V^{1/2}(\phi)}{m_{Pl}} \qquad (6.4)$$

and the associated large-angle CMBR anisotropy is of order $\delta T/T \sim A_G(\phi)$. Since $H(\phi)$ does not vary significantly over the observationally accessible range of lengthscales, the gravitational wave perturbations are also approximately scale-invariant. A sketch of the correspondence between scalar field evolution and the imprint of large-scale density and gravity wave fluctuations is shown in Fig. 9 (from [62]).

While the nearly scale-invariant spectra of scalar and tensor perturbations are generic features of inflation, the amplitude of the perturbations, from Eq.(6.3) and (6.4), are clearly model-dependent, varying with the scale and shape of the inflaton potential. Since these amplitudes are now tightly constrained by CMBR anisotropy measurements on large scales, we already have significant constraints on the form of phenomenologically acceptable inflation models (see below).

Although the perturbations from inflation are *nearly* scale-invariant, the (generally small) deviations from scale-invariance are also model-dependent, and it is useful to characterize them in a phenomenological way. Over the range of scales of interest, the scalar and tensor spectra can be approximated by power laws in wavelength,

$$A_S(\phi) \sim \lambda^{(1-n_S)/2} \quad ; \quad A_G(\phi) \sim \lambda^{-n_T/2} \tag{6.5}$$

where the spectral indices n_S and n_T characterize the spectra; for exact scale-invariance, $n_S - 1 = n_T = 0$. The scalar gravitational potential spectrum corresponds to a density fluctuation spectrum (at fixed time), $P_\rho(k) \sim k^{n_S}$. The spectral indices can be calculated as an expansion in the 'slow-roll' parameters, defined by

$$\begin{aligned} \varepsilon &\equiv \frac{m_{Pl}^2}{16\pi}\left(\frac{V'(\phi)}{V(\phi)}\right)^2 \\ \eta &\equiv \frac{m_{Pl}^2}{8\pi}\frac{V''(\phi)}{V(\phi)} \end{aligned} \tag{6.6}$$

The conditions $\varepsilon, \eta \lesssim 1$ are sufficient for the scalar field energy-momentum to be dominated by the potential, $V(\phi) \gtrsim \dot{\phi}^2$ and thus for inflation to take place. In terms of the slow roll parameters, the spectral indices are

$$\begin{aligned} n_S &\simeq 1 + 2\eta - 6\varepsilon \\ n_T &\simeq -2\varepsilon \end{aligned} \tag{6.7}$$

Since ε is positive definite, $n_T \leq 0$.

The inflation perturbations can thus be characterized by four numbers: an amplitude and spectral index for the scalar and tensor modes. One way to characterize the amplitudes is by their contribution to the (mean square) quadrupole anisotropy through the Sachs-Wolfe effect, denoted by C_2^S and C_2^T for the scalar and tensor modes respectively. The COBE DMR measurements essentially fix the sum $C_2^S + C_2^T$, leaving three undetermined variables: n_S, n_T, and the relative tensor to scalar contribution to the large-angle anisotropy, $r = C_2^T/C_2^S$. From Eqns.(6.3-6), this ratio can be expressed as

$$r = \frac{C_2^T}{C_2^S} = 14\varepsilon = -7n_T \tag{6.8}$$

so there are only two independent quantities to be determined. Note that n_T cannot be arbitrarily large, for then the COBE-normalized density fluctuation amplitude is too small to make non-linear structure on galaxy scales. A likelihood analysis of power law spectra using the CMBR anisotropy detected in two years of COBE DMR data finds $n \simeq 1.17 \pm 0.31 (0.96 \pm 0.36)$ including (excluding) the measured quadrupole in the analysis [63], consistent with a nearly scale-invariant spectrum.

6.2 *Naturalness and the Small Coupling Problem*

The result (6.3) for the density perturbation amplitude provides a severe constraint on the form of the inflaton potential. As an example, consider a chaotic inflation model [64] with a simple quartic potential, $V(\phi) = \lambda\phi^4/4!$. The inflationary slow-roll conditions $\varepsilon, \eta < 1$ are satisfied for $\phi > m_{Pl}\sqrt{3/2\pi}$ and the number of e-folds of accelerated expansion is larger than N_{min} if the initial value of the field satisfies $\phi_i \gtrsim \phi_{min} = 3m_{Pl}$. The density perturbation amplitude on the scale of the present Hubble radius is $(\delta\rho/\rho)_{H_0^{-1}} = (\pi\lambda)^{1/2}(\phi_{min}/m_{Pl})^3 \simeq 50\lambda^{1/2}$. The COBE DMR observation indicates $(\delta\rho/\rho)_{H_0^{-1}} \simeq 10^{-4}$, which implies that the inflaton self-coupling must satisfy $\lambda \lesssim 4 \times 10^{-12}$.

This example illustrates a general conclusion that applies to all inflation models: the inflaton field must be very weakly self-coupled. In different models, this constraint appears in different guises, but it is alway present in one form or another: the potential must contain a very small dimensionless number of order 10^{-12}. Attitudes concerning this problem vary widely among theorists. It is often said that $\lambda_\phi \sim 10^{-12}$ constitutes unacceptable 'fine tuning', but that is a misapplication of the fine tuning concept in this context: we are not interested in restricting small changes to λ but in ensuring that λ itself is small. To others, it is not an issue of great concern, because we know there exist other small numbers in physics, such as lepton and quark Yukawa couplings $g_Y \sim 10^{-5}$ and the weak hierarchy ratio $M_{weak}/M_{Pl} \sim 10^{-17}$.

Nevertheless, in the context of a particular model, one can ask whether such a small value for λ_ϕ is in principle unnatural. A small parameter λ is said to be "technically natural" if it is protected against large radiative corrections by a symmetry, *i.e.*, if setting $\lambda \to 0$ increases the symmetry of the system. For example, in this way, low energy supersymmetry might protect the small ratio M_{weak}/M_{Pl} by cancelling boson and fermion loops. Essentially all inflation models that have been proposed in the last decade satisfy this criterion of technical naturalness. However, in models that are only technically natural, the small coupling λ_ϕ, while stable against radiative corrections, is itself unexplained. Technical naturalness is a useful concept for low energy effective Lagrangians, such as the electroweak theory and its supersymmetric extensions, but it points to a more fundamental level of theory for its origin. For example, the underlying origin of the mass hierarchy M_{weak}/M_{Pl} in supersymmetric theories is thought to be associated with hidden sector physics in supergravity/superstring theories, which start out with no small dimensionless couplings at the Planck scale. Since inflation takes place *relatively* close to the Planck scale, it would be preferable to find the inflaton in such particle physics models which are "strongly natural", that is, which have no small numbers in the fundamental Lagrangian. In this case, one would expect the inflaton coupling to arise from renormalization group factors, e.g., $\lambda_\phi \sim e^{-n/\alpha}$, where α is a gauge coupling constant and n is a number of order a few. A convincing model with these properties that is also desirable from the viewpoint of particle physics phenomenology has not yet been found. The task of constructing such a theory remains a paramount goal for inflation theorists.

§7 Topological Defects

Spontaneous symmetry breaking, the idea that there are underlying symmetries of nature that are not manifest in the the vacuum state, plays a central role in particle physics models that unify the fundamental interactions. In unified gauge theories—including the standard electroweak model—the underlying symmetry of the Lagrangian is larger than $SU(3)_C \otimes U(1)_{EM}$, the symmetry of the vacuum. However, when the temperature of the universe is higher than the characteristic unification scale, spontaneously broken symmetries are restored. Thus, in the course of cosmic evolution, we expect that there was a series of phase transitions associated with the spontaneous breakdown of gauge (and perhaps global) symmetries (see Table 1).

Depending on the topology of the vacuum manifold, spontaneously broken gauge theories can contain stable, quasi-classical configurations of gauge and Higgs fields such as domain walls, cosmic strings, and magnetic monopoles. For example, when a symmetry group G is broken to a subgroup H, strings form if the fundamental group $\pi_i(G/H)$ (loosely, the group of mappings of the manifold of degenerate vacua onto the circle S^1) is nontrivial, i.e., if the coset space G/H is multiply connected. Topological defects are quasi-localized regions where the gauge and scalar fields depart from the vacuum, and the symmetry is effectively restored in the defect core. Kibble first pointed out that, for particle physics models containing defect solutions, defects would inevitably form in symmetry-breaking phase transitions, since the scale over which the fields can be correlated is limited by the particle horizon at the time of the transition [65]. Due to the effective symmetry restoration in their cores, defects carry substantial stress-energy-momentum and can have important gravitational effects. They are thus of significant cosmological interest: the strings formed in the breakdown of a class of grand unified theories may act as seeds for the formation of large-scale structure.

Defects such as strings and monopoles can also arise in the spontaneous breakdown of global symmetries. For global defects, however, the gradient energy density of the Higgs field outside the defect core is not compensated by the gauge field, so that the energy density of global defects is not as strongly localized as that of their local defect counterparts. Moreover, global symmetry breaking can give rise to new structures that are not topologically stable but nevertheless may survive for cosmologically long times. For example, in the spontaneous breaking of the global symmetry $O(4) \to O(3)$, textures form— knots in the Higgs field that arise when the field winds around the three-sphere. These knots are generally formed by misalignment of the field on scales larger than the horizon at the symmetry-breaking phase transition, because of the Kibble mechanism. For the breakdown of larger global symmetries, e.g., $O(N) \to O(N-1)$ with $N > 4$, there is no topological structure; nevertheless, large-scale spatial gradients in the Higgs field will be formed by the Kibble mechanism, and they can also give rise to density perturbations when a given gradient scale enters the horizon.

In potentially viable defect models, such as gauge cosmic strings and global $O(N)$ models with $N \geq 2$, the population of defects or field gradients approaches a scaling solution, in which the energy density of the fields scales as $\rho \propto t^{-2}$. The resulting spectrum of density perturbations seeded by defects is approximately scale-invariant, and the perturbation amplitude when a given lengthscale crosses the Hubble radius is set by the scale of symmetry breaking M, $(\delta\rho/\rho)_{\lambda=H^{-1}} \sim (M/m_{Pl})^2$. To match the amplitude of observed galaxy clustering, the symmetry breaking scale should be comparable to that expected in grand unified theories, $M \sim 10^{16}$ GeV. A pleasing feature of defect models is that the density fluctuations are independent of dimensionless coupling constants. On the other hand, the cosmological evolution of defects and the manner in which they seed density perturbations is much more complex than in inflation models, and the detailed predictions of these models are still being worked out. For example, in addition to scalar density perturbations, defects generate a spectrum of gravitational waves; however, unlike the case of inflation, these tensor perturbations are correlated with the scalar modes. Also, due to the coherence of the defect fields, the density perturbations they generate are generally non-Gaussian. By contrast, in most inflation models (those involving a single, weakly coupled inflaton field) the primordial fluctuations are Gaussian distributed, because they arise from an essentially free quantum field in its vacuum state.

§8 Testing theories of the early Universe

8.1 CMBR anisotropies

The COBE DMR measurements have found anisotropies on large angular scales (the angular resolution of the DMR instruments is several degrees) at approximately the level theoretically expected and with a spectrum roughly consistent with scale-invariant primordial fluctuations. This is an important first test for the primordial gravitational instability picture for structure formation. While it places important constraints on models (e.g., inflation models with excessive tilt, $n_s \lesssim 0.7$, are excluded), it does not yet discriminate strongly between competing models. For example, Figure 10 shows a sample microwave sky generated in the cosmic string scenario; on these large angular scales, the spectrum is broadly consistent with the COBE DMR detection [66]. On smaller scales, the spectrum of inflationary and defect models is expected to differ significantly; in addition, in most defect models, one would expect non-Gaussian features to arise in CMBR maps at arc minute to degree scales.

A number of ground-based and balloon-borne experiments are now probing CMBR anisotropies on smaller angular scales. Although there are still wide variations between different experimental groups, there does seem to be a general trend of a larger anisotropy amplitude on scales of order 1/2 degree than on larger scales. Again, this is expected in most models. In the future, one looks forward to full-sky maps with sub-degree resolution, most probably requiring a satellite mission, in order to probe the anisotropy power spectrum in sufficient detail for model discrimination and to test for non-Gaussian features.

The CMBR temperature autocorrelation function $C(\theta)$ is defined operationally by measuring the temperature fluctuation at some direction (angle) $\boldsymbol{\alpha}$, $\delta T(\boldsymbol{\alpha}) = T(\boldsymbol{\alpha}) - \langle T\rangle$, multiplying by the temperature at angle $\boldsymbol{\alpha}+\boldsymbol{\theta}$, and averaging over positions $\boldsymbol{\alpha}$, $C(\theta) = \langle \delta T(\boldsymbol{\alpha})\, \delta T(\boldsymbol{\alpha}+\boldsymbol{\theta})\rangle_\alpha$. It is useful to expand $C(\theta)$ in Legendre polynomials: $C(\theta) = (4\pi)^{-1}\sum_l (2l+1) C_l P_l(\cos\theta)$. The coefficients C_l are the multipole moments, which specify the angular power spectrum of the temperature fluctuations.

There are some important features to look for in the spectrum of temperature fluctuations. If inflation is correct and the primordial spectrum is approximately scale-invariant, $n_S \simeq 1$, then at large angular scale (multipole $l \ll 100$) the spectrum $l(l+1)C_l$ should be roughly flat. These scales were outside the Hubble radius at last scattering, and the signal is dominated by the gravitational redshift of photons emerging from gravitational potential wells at the time of last scattering (Sachs-Wolfe effect). The multipole $l \sim 200$ corresponds approximately to the angular scale of 1°, roughly the angle subtended by the Hubble radius on the last scattering surface. On scales comparable to and smaller than this ($l \gtrsim 200$), the anisotropy signal is dominated by acoustic oscillations of the photon-baryon fluid. For primordial adiabatic fluctuations, this effect leads to the characteristic 'Doppler peaks' at multipoles $l \gtrsim 200$. At very small angular scales ($l \gtrsim 1000$), these oscillations, and the resulting anisotropy amplitudes, are depressed by Silk damping (photon diffusion) and the finite thickness of the last scattering surface. Since the Hubble radius at the time of last scattering depends on Ω, the position of the first Doppler peak is sensitive to the mean density, $l_{peak} \simeq 200/\sqrt{\Omega}$—it should be shifted to higher l in an open universe [67]. In inflation models, the amplitudes of the peaks depend on the inflation parameters n_S, n_T, and r, as well as on cosmological parameters such as h, Ω_B, and Ω_Λ. (They also depend on whether the universe was reionized at high redshift.) At higher l, the anisotropy spectrum is also sensitive to assumptions about the dark matter. The anisotropy power spectrum from an inflation model with $n_S = 1$ and $r = 0$ is shown in figure 11 for standard cold dark matter and mixed (cold plus hot) dark matter [68]. Note the flat region at small l and the Doppler peaks at $l \gtrsim 200$. In defect models, the Doppler peaks may be less pronounced and shifted in l.

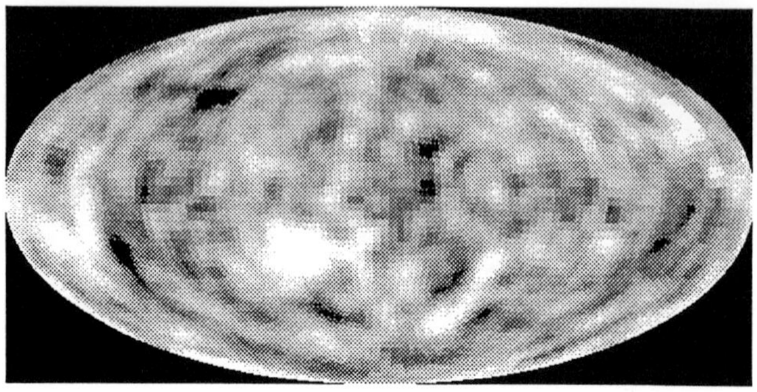

Figure 10: Map of fluctuations in the CMBR predicted in the cosmic string model. The angular resolution is several degrees and the peak amplitude is $\delta T/T = 50 \times 10^{-6}$.

The detection of CMBR fluctuations so far does not strongly discriminate between theoretical models. In the future, however, detailed observational study of the anisotropy power spectrum can provide a significant test for inflation. The first step will be a clearer indication of a Doppler peak in the spectrum and the determination of its amplitude and position. Ultimately, we would like to determine the angular power spectrum to the cosmic variance limit (the accuracy of an all-sky map) for multipoles up to $l = 1000$ or beyond. This would help

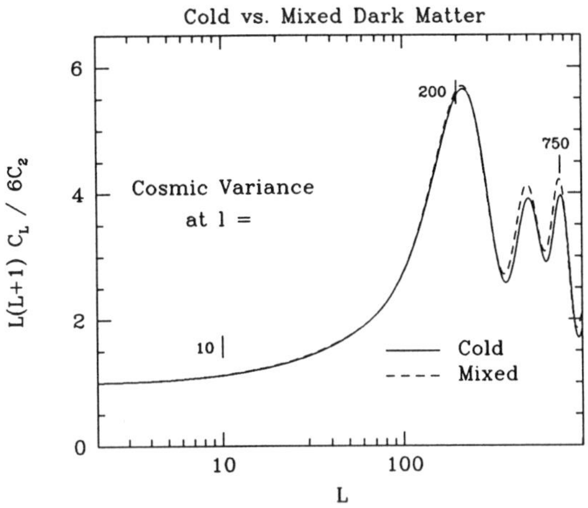

Figure 11: The spectrum of temperature fluctuations predicted in an inflationary model containing no tensor perturbations and a primordial scalar perturbation spectrum of the Harrison-Zel'dovich form, $n_S = 1$, normalized to the quadrupole C_2, for both cold and mixed dark matter. Note the flat region at small l and the Doppler peaks at $l \geq 200$. The vertical marks show the expected cosmic variance for a full-sky map.

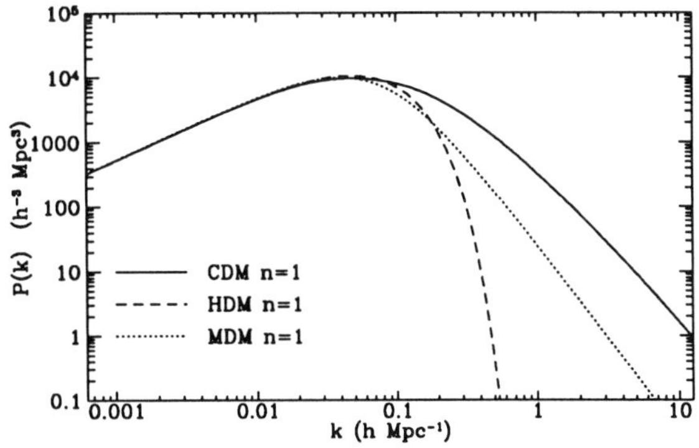

Figure 12: The processed linear power spectrum for a primordial spectrum of the Harrison-Zel'dovich form, $n_S = 1$, for cold, hot, and mixed dark matter.

reduce the degeneracy ('cosmic confusion') between variations in model and cosmological parameters. Full-sky maps with sub-degree scale resolution are also required to test whether the fluctuations in the temperature are Gaussian distributed, as predicted by inflation; non-Gaussian CMBR features in topological defect models would be expected to show up on these scales.

8.2 Large-scale Structure

The clustering of matter in the universe provides another probe of early universe models for structure formation. Defining the density fluctuation field, $\delta(\mathbf{x}, t) = (\rho(\mathbf{x}, t) - \langle\rho\rangle)/\langle\rho\rangle$, it is convenient to consider the Fourier transform,

$$\delta(\mathbf{x}, t) = \int \frac{d^3k}{(2\pi)^3} \delta_{\mathbf{k}}(t) e^{i\mathbf{k}\cdot\mathbf{x}} \ , \tag{8.1}$$

in terms of which the density power spectrum $P_\rho(k)$ is defined as

$$\langle \delta_{\mathbf{k}} \delta^*_{\mathbf{k}'} \rangle = P_\rho(k) \delta_D(\mathbf{k} + \mathbf{k}') \ . \tag{8.2}$$

As noted above, the primordial density spectrum from inflation is generally a power law in wavenumber, $P_\rho(k, t_i) = Ak^{n_s}$, with $n_S \sim 1$. The spectrum subsequently subsequently evolves to produce the density spectrum at late times, $P_\rho(k, t_0) = T^2(k) P_\rho(k, t_i)$, where the transfer function $T(k)$ encodes the scale-dependence of the linear ($\delta \ll 1$) gravitational evolution of the perturbation modes. The transfer function depends on the nature of the dark matter (hot, warm, cold, or a mixture), its density (Ω_{DM}), the baryon density Ω_B, the Hubble parameter h, and the nature of the processes generating the fluctuations (adiabatic, isocurvature, or seeded defects). Power spectra for scale-invariant ($n_S = 1$) models with adiabatic fluctuations and three choices of dark matter are shown in Fig. 12. On small scales, there is an additional correction for non-linear evolution. Finally, the spectrum of luminous objects such as galaxies is related to the density spectrum by a bias prescription: for the simplest linear bias model, $P_{gal}(k, t_0) = b^2_{gal} P_\rho(k, t_0) = b^2_{gal} T^2(k) P_\rho(k, t_i)$ (not including the non-linear power correction). However, as noted above, there is little theoretical reason to expect the relation between mass and light to be that simple. Ultimately, the complexities of non-linear evolution and the inclusion of non-gravitational (hydrodynamical) processes in the formation of galaxies must be addressed by numerical simulations.

There are now a number of catalogs which have been used to determine the galaxy power spectrum $P_{gal}(k)$ down to wavenumbers $k \sim 0.01h$ Mpc^{-1}. These suggest that the standard cold dark matter (CDM) spectrum (with $n_S = 1$, $\Omega_{cdm} h = 0.5$, and linear bias) does not have exactly the right shape to match the observed galaxy spectrum: it has relatively too little (too much) power on large (small) scales, as shown in Fig. 13 (from [69]). Currently popular variants which improve the spectral shape include a mixture of hot and cold dark matter, with $\Omega_{hot} \simeq 1 - \Omega_{cold} \simeq 0.2$, or low-density CDM ($\Omega_{cold} \simeq 0.3$, with or without a cosmological constant to retain spatial flatness). The spectral shape is also improved by increasing Ω_B above the range normally allowed by nucleosynthesis, reducing h to its lower limit of 0.4 or below, or tilting the primordial spectrum to $n_S < 1$. By

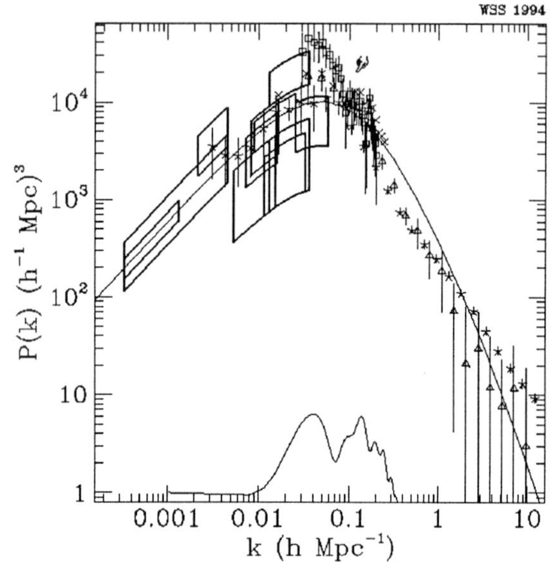

Figure 13: The power spectrum for standard CDM (solid line) is compared with the results from several galaxy surveys (points) and with CMBR anisotropy measurements (the error boxes, assuming $n_S = 1$). Due to the different methods used and corrections applied to the data, this figure should not be overinterpreted.

the turn of the millennium, larger galaxy redshift surveys will extend the measurements of $P_{gal}(k)$ down to smaller wavenumbers, overlapping the scales probed by COBE DMR. This should allow more direct constraints on the bias between galaxies and mass. In addition, we will have other methods that more directly probe the *mass* power spectrum over large scales: large-scale peculiar velocity measurements (e.g., from the Sunyaev-Zel'dovich effect in clusters) and weak gravitational lensing measurements of the shear field over degree scales.

§9 Conclusions

The robustness of the evidence for the standard cosmology has allowed cosmologists to use it as a jumping off point for theoretical exploration of the early universe. Now, cosmology is entering a new era: theoretical models for the evolution of the early universe that have been developed over the last two decades are becoming subject to a variety of observational tests. These observational probes fall roughly into four categories: determination of the cosmological parameters (e.g., H_0, t_0, and Λ); determination of the nature, quantity, and spatial distribution of the dark matter (including dark baryons); the microwave background anisotropy; and the large-scale structure of the (dark and luminous) matter. New measurements planned or underway in each of these areas will allow us to test models for the early universe and discriminate between them.

In each of these areas, we can identify critical advances that are anticipated over the next decade. Improved measurements of the cosmological parameters are necessary to pin down the transfer function for perturbation growth and for CMBR anisotropies. In particular, additional Cepheid distances from HST should significantly advance the calibration of secondary distance indicators, while the alternative direct distance estimators (e.g., the Sunyaev-Zel'dovich effect) should continue to mature. The settling of the H_0 debate will have critical impact on cosmology–e.g., if a consensus emerges on the high value $h = 0.8$, the simple, elegant Einstein-de Sitter model with $\Omega_0 = 1$ will have to be abandoned. If the 'expansion age' $H_0 t_0$ turns out to be 1 or larger and gravitational lens constraints on Λ are not flawed, then the standard cosmology itself would be in crisis. Deep spectroscopic surveys and the discovery of significant numbers of high-redshift supernovae should ultimately provide important information on the deceleration parameter. In the area of dark matter, one looks forward to further developments in WIMP and axion detection, and to neutrino mass experiments. As of this writing, the LSND experiment at Los Alamos has reported preliminary evidence for neutrino oscillations consistent with a light neutrino in the mass range of several eV. In the microwave background, long-duration balloon flights and perhaps a post-COBE satellite will produce CMBR sky maps at sub-degree-scale resolution. These should go a significant way towards testing inflation and defect models as well as provide additional constraints on the cosmological parameters. In the area of large-scale structure, new galaxy surveys will uncover the statistics of galaxy clustering out to scales that overlap those probed by COBE. There will also be much larger samples of quasars and absorption line systems than heretofore, to probe clustering at early times. In addition, the mass power spectrum will be probed by peculiar velocity and weak lensing studies.

Cosmology thus stands hopefully but rather uncertainly on the threshold of the new millennium: we are soon to learn whether the speculative physical ideas developed in the last two decades are consistent with the universe around us. While the initial tests have been passed, the crucial tests of 'cosmology beyond the standard model' are yet to come.

§Acknowledgements

I would like to thank the participants and organizers of Snowmass '94 for an enjoyable, stimulating conference. This research was supported by the DOE and by NASA grant NAG5-2788 at Fermilab.

§Appendix: Friedmann-Robertson-Walker Models

First, a brief word on notation and units. The fundamental unit of distance in extragalactic astronomy is the parsec (pc): 1 pc = 3.26 light-years = 3.09×10^{18} cm. The typical galactic scale is measured in kiloparsec (kpc): the solar system is approximately $R_0 = 8 - 8.5$ kpc from the center of the Milky Way. The nearest large galaxy to our own is Andromeda, at a distance of roughly 700 kpc, and the typical distance between neighboring large galaxies (outside clusters) is a few Megaparsec (Mpc). Absolute extragalactic distances are uncertain by a factor of order 2, and are usually given in terms of h^{-1} Mpc, where the Hubble parameter $H_0 = 100h$ km/sec/Mpc, and observations indicate $0.4 < h < 1$. Cosmological distances are often estimated using redshifts via the Hubble law, $v = cz = H_0 d$, and so are sometimes expressed in terms of recession velocity: 10,000 km/sec = 100 h^{-1} Mpc and corresponds to a redshift $z = 0.033$. Astronomical masses and luminosities are measured in solar units, $M_\odot = 1.989 \times 10^{33}$ gm, $L_\odot = 3.86 \times 10^{33}$ erg/sec. Large galaxies commonly have a luminosity of order $10^{10} - 10^{11} L_\odot$ and mass of order $10^{11} - 10^{12} M_\odot$.

In the context of general relativity, homogeneity and isotropy severely restrict the form of the spatial geometry: a uniform stress-energy momentum tensor implies that the constant-time 3-surfaces have uniform spatial curvature. The spacetime metric thus takes the Friedmann-Robertson-Walker (FRW) form

$$ds^2 = dt^2 - a^2(t)\left[\frac{dr^2}{1-kr^2} + r^2\left(d\theta^2 + \sin^2\theta d\phi^2\right)\right] , \tag{A.1}$$

where $a(t)$ is the global scale factor, r, θ, ϕ are the fixed coordinates carried by fundamental comoving observers, t is proper time measured on the clocks they carry, and $k = 0, 1, -1$ is the sign of the spatial 3-curvature. The case $k = 0$ corresponds to flat, Euclidean 3-space (R^3) written in spherical coordinates, $k = 1$ corresponds to the geometry of the three-sphere (S^3), and $k = -1$ to the 3-hyperboloid (H^3), the three-dimensional analogue of a hyperbolic saddle. Thus, models with $k \leq 0$ are spatially infinite (open), while those with $k = 1$ are spatially finite (closed).

The FRW models are characterized by the global scale factor $a(t)$, whose dynamics is determined by the matter content of the universe through Einstein's equations,

$$H^2 \equiv \left(\frac{\dot{a}}{a}\right)^2 = \frac{8\pi G}{3}\rho - \frac{k}{a^2} + \frac{\Lambda}{3} \tag{A.2}$$

and

$$\frac{\ddot{a}}{a} = -\frac{4\pi G(\rho + 3p)}{3} + \frac{\Lambda}{3} . \tag{A.3}$$

Here ρ is the mean energy density of matter, p is its pressure, and Λ is the cosmological constant, *i.e.*, the effective contribution to the energy-momentum from the vacuum state.

Observations suggest that the fluid energy density of the universe is currently dominated by non-relativistic matter (m), while the early universe was dominated by ultrarelativistic particles, or radiation (r). For non-relativistic matter, the energy density scales as $\rho_m \sim a^{-3}$, while for radiation, $\rho_r \sim a^{-4}$. The transition from radiation to matter domination ($\rho_m = \rho_r$) thus occurred at the epoch

$$1 + z_{eq} = \frac{a_0}{a_{eq}} = \frac{\rho_m^0}{\rho_r^0} = 2.3 \times 10^4 \Omega_0 h^2 , \tag{A.4}$$

where the present matter density $\rho_m^0 = \Omega_0 \rho_{crit} = 1.88 \Omega_0 h^2 \times 10^{-29}$ g cm^{-3}, $\rho_{crit} = 3H_0^2/8\pi G$ is the present density of the spatially flat ($k = 0$) model, and the present radiation energy density in the CMBR and an assumed three species of massless neutrinos is $\rho_r^0 = (\pi^2/30) g_* T_0^4 = 8.1 \times 10^{-34}$ g cm^{-3}.

Two features of the solutions to the Friedmann equation (A.2) with $\Lambda = 0$ are worth noting. In this case, there is a one-to-one correspondence between the spatial geometry and the fate of the universe: open models ($k \leq 0$) expand forever, while closed models ($k > 0$) eventually recollapse, because the energy density for matter and radiation fall off faster than a^{-2}. Conversely, in the early universe, $a \ll a_0$, the matter and radiation terms dominate over the spatial curvature, and the dynamics of the model is well approximated by setting $k = 0$. In this limit, for a matter-dominated universe, $a(t) \sim t^{2/3}$, $\rho_m = 1/6\pi G t^2$, while for radiation-domination, $a(t) \sim t^{1/2}$ and $\rho_r = 3/32\pi G t^2$.

The principal observable cosmological parameters of the FRW models are the Hubble parameter, $H_0 = (\dot{a}/a)_0$, the age of the Universe, t_0, the present (non-relativistic) mass density relative to the 'critical' density of the spatially flat, Einstein-de Sitter ($k = \Lambda = 0$) model, $\Omega_0 = \rho_0/\rho_{crit} = 8\pi G\rho_0/3H_0^2$, the deceleration parameter, $q_0 = -(\ddot{a}a/\dot{a}^2)_0$, which measures the rate at which the gravitational attraction of the matter is slowing down the expansion, and the contribution of the cosmological constant to the present expansion rate, $\Omega_\Lambda = \Lambda/3H_0^2$. From the Friedmann equations, these parameters are related by

$$1 = \Omega_0 + \Omega_\Lambda - \frac{k}{a_0^2 H_0^2} \tag{A.5}$$

$$q_0 = \frac{\Omega_0}{2} - \Omega_\Lambda . \tag{A.6}$$

For vanishing cosmological constant, $\Omega_0 - 1$ determines the sign of the spatial curvature: $\Omega_0 = 1$ for the spatially flat model ($k = 0$), and it is less than one for open models.

§References

[1] G. F. Smoot, C. L. Bennett, A. Kogut, E. L. Wright, J. Aymon, N. W. Boggess, E. S. Cheng, G. De Amici, S. Gulkis, M. G. Hauser, G. Hinshaw, C. Lineweaver, K. Loewenstein, P. D. Jackson, M. Jansen, E. Kaita, T. Kelsall, P. Keegstra, P. Lubin, J. Mather, S. S. Meyer, S. H. Moseley, T. Murdock, L. Tokke, R. F. Silverberg, L. Tenorio, R. Weiss, and D. T. Wilkinson. *Astrophysical Journal Letters*, 396:L1, 1992.

[2] C. L. Bennett, A. Kogut, G. Hinshaw, A. Banday, E. L. Wright, K. Gorski, D. T. Wilkinson, R. Weiss, G. F. Smoot, S. S. Meyer, J. C. Mather, P. Lubin, K. Loewenstein, C. Lineweaver, P. Keegstra, E. Kaita, P. D. Jackson, and E. S. Cheng. *Astrophysical Journal*, 436:423, 1994.

[3] K. Ganga, E. Cheng, S. Meyer, and L. Page. *Astrophysical Journal Letters*, 410:L57, 1993.

[4] S. J. Maddox, G. Efstathiou, W. J. Sutherland, and J. Loveday. *Monthly Notices of the Royal Astronomical Society*, 242:43p, 1990.

[5] C. A. Collins, R. C. Nichol, and S. L. Lumsden. *Monthly Notices of the Royal Astronomical Society*, 254:295, 1992.

[6] K. B. Fisher, M. Davis, M. A. Strauss, A. Yahil, and J. P. Huchra. *Astrophysical Journal*, 402:42, 1993.

[7] G. Efstathiou, N. Kaiser, W. Saunders, A. Lawrence, M. Rowan-Robinson, R. S. Ellis, and C. S. Frenk. *Monthly Notices of the Royal Astronomical Society*, 247:10p, 1990.

[8] H. Feldman, N. Kaiser, and J. Peacock. *Astrophysical Journal*, 426:23, 1994.

[9] V. de Lapparent, M. Geller, and J. P. Huchra. *Astrophysical Journal Letters*, 302:L1, 1986.

[10] M. Geller and J. P. Huchra. *Science*, 246:897, 1989.

[11] T. Broadhurst, R. Ellis, D. Koo, and A. Szalay. *Nature*, 343:726, 1990.

[12] This slice was made by D. Weinberg using N-body simulations of C. Park and J. R. Gott. See J. Gunn and D. Weinberg. in *Wide-Field Spectroscopy and the Distant Universe,* eds. S. Maddox and A. Aragon-Salamanca. World Scientific, Singapore, 1995.

[13] J. R. Mould, L. Stavely-Smith, R. Schommer, G. Bothun, P. Hall, M. Han, J. Huchra, J. Roth, W. Walsh, and A. Wright. *Astrophysical Journal*, 383:467, 1991.

[14] J. C. Mather, E. S. Cheng, D. A. Cottingham, R. E. Eplee, D. J. Fixsen, T. Hewagama, R. B. Isaacman, K. A. Jensen, S. S. Meyer, P. D. Noerdlinger, S. M. Read, L. P. Rosen, R. A. Shafer, E. L. Wright, C. L. Bennett, N. W. Boggess, M. G. Hauser, T. Kelsall, S. H. Moseley, R. F. Silverberg, G. F. Smoot, R. Weiss, and D. T. Wilkinson. *Astrophysical Journal*, 420:439, 1994.

[15] T. Walker, G. Steigman, D. Schramm, K. Olive, and H. Kang. *Astrophysical Journal*, 376:51, 1991.

[16] S. Carroll, W. H. Press, and E. Turner. *Annual Review of Astronomy and Astrophysics*, 30:499, 1992.

[17] M. Fukugita and E. Turner. *Monthly Notices of the Royal Astronomical Society*, 253:99, 1991.

[18] C. Kochanek. *Astrophysical Journal*, 419:12, 1994.

[19] D. Maoz and H. Rix. *Astrophysical Journal*, 416:425, 1994.

[20] G. Jacoby et al. *Publications of the Astronomical Society of the Pacific*, 104:559, 1992.

[21] G. Jacoby et al. *Astrophysical Journal*, 356:322, 1990.

[22] J. Tonry. *Astrophysical Journal Letters*, 373:L1, 1991.

[23] A. G. Riess, W. H. Press, and R. P. Kirshner. *Astrophysical Journal Letters*, 438:L17, 1995.

[24] M. Hamuy, M. Phillips, J. Maza, N. Suntzeff, R. Schommer, and R. Aviles. *Astronomical Journal*, 109:1, 1995.

[25] B. P. Schmidt, R. P. Kirshner, and R. G. Eastman. *Astrophysical Journal*, 395:366, 1992.

[26] M. Birkinshaw and J. Hughes. *Astrophysical Journal*, 420:33, 1991.

[27] T. Wilbanks, P. Ade, M. Fischer, W. Holzapfel, and A. Lange. *Astrophysical Journal Letters*, 427:L75, 1994.

[28] R. Blandford and R. Narayan. *Annual Review of Astronomy and Astrophysics*, 30:311, 1992.

[29] M. J. Pierce, D. Welch, R. McClure, S. van den Bergh, R. Racine, and P. Stetson. *Nature*, 371:385, 1994.

[30] W. Freedman, B. Madore, J. Mould, R. Hill, L. Ferrarese, R. Kennicutt, A. Saha, P. Stetson, J. Graham, H. Ford, J. Hoessel, J. Huchra, S. Hughes, and G. Illingworth. *Nature*, 371:757, 1994.

[31] J. Cowan, F. Thielemann, and J. Truran. *Annual Review of Astronomy and Astrophysics*, 29:447, 1991.

[32] D. Winget et al. *Astrophysical Journal Letters*, 315:L77, 1987.

[33] I. Iben and G. Laughlin. *Astrophysical Journal*, 341:430, 1989.

[34] A. Renzini in. *Proceedings of the 16th Texas Symposium on Relativistic Astrophysics and 3rd Symposium on Particles, Strings, and Cosmology*, eds. C. Akerlof and M. Srednicki. New York Academy of Sciences, New York, 1992.

[35] X. Shi. *Astrophysical Journal*, in press, 1995.

[36] R. Lynds and V. Petrosian. *Bulletin of the American Astronomical Society*, 18:1014, 1986.

[37] G. Soucail et al. *Astronomy and Astrophysics*, 172:L14, 1987.

[38] J. Tyson, F. Valdes, and R. Wenk. *Astrophysical Journal Letters*, 349:L19, 1990.

[39] S. D. M. White, J. Navarro, A. Evrard, and C. Frenk. *Nature*, 366:429, 1993.

[40] A. Dekel. *Annual Review of Astronomy and Astrophysics*, 32, 1994.

[41] S. Cole, K. Fisher, and D. Weinberg. *Monthly Notices of the Royal Astronomical Society*, in press, 1995.

[42] B. Paczynski. *Astrophysical Journal*, 304:1, 1986.

[43] C. Alcock, R. Allsman, T. Axelrod, D. Bennett, K. Cook, H. Park, S. Marshall, M. Pratt, C. Stubbs, K. Griest, S. Perlmutter, W. Sutherland, K. Freeman, B. Peterson, P. Quinn, A. Rodgers, S. Chan, and C. Akerlof. *Nature*, 365:621, 1993.

[44] E. Aubourg et al. *Nature*, 365:623, 1993.

[45] A. Udalski, J. Szymanski, J. Kaluzny, M. Kubiak, W. Krzeminski, M. Mateo, G. Preston, and B. Paczynski. *Acta Astronomica*, 43:289, 1993.

[46] A. Udalski et al. *Astrophysical Journal Letters*, 426:L69, 1994.

[47] C. Alcock et al. *preprint*, 1994.

[48] S. Tremaine and J. Gunn. *Physical Review Letters*, 42:407, 1979.

[49] K. Van Bibber in. *Proceedings of the 7th Meeting of the American Physical Society Division of Particles and Fields.* eds. C. Albright, P. Kasper, R. Raja, and J. Yoh World Scientific, Singapore, 1992.

[50] M. Milgrom. *Astrophysical Journal*, 333:689, 1988.

[51] A. Guth. *Physical Review D*, 23:347, 1981.

[52] D. Lyth and E. Stewart. *Physics Letters B*, 252:336, 1990.

[53] B. Ratra and P. J. E. Peebles. *Astrophysical Journal*, 432:5, 1994.

[54] M. Bucher, A. Goldhaber, and N. Turok. *preprint*, 1994.

[55] A. Linde. *Physics Letters B*, 108:389, 1982.

[56] A. Albrecht and P. Steinhardt. *Physical Review Letters*, 48:1220, 1982.

[57] S. Hawking. *Physics Letters B*, 115:295, 1982.

[58] A. Starobinski. *Physics Letters B*, 117:175, 1982.

[59] A. Guth and S. Y. Pi. *Physical Review Letters*, 49:1110, 1982.

[60] J. Bardeen, P. Steinhardt, and M. Turner. *Physical Review D*, 28:679, 1983.

[61] V. Rubakov, M. Sazhin, and A. Veryaskin. *Physics Letters B*, 115:189, 1982.

[62] E. Copeland, E. Kolb, A. Liddle, and J. Lidsey. *Physical Review D*, 48:2529, 1993.

[63] K. Gorski, G. Hinshaw, A. Banday, C. Bennett, E. Wright, A. Kogut, G. Smoot, and P. Lubin. *Astrophysical Journal Letters*, 430:L89, 1994.

[64] A. Linde. *Physics Letters B*, 129:177, 1983.

[65] T. Kibble. *Physics Reports*, 67:183, 1980.

[66] B. Allen, R. Caldwell, P. Shellard, A. Stebbins, and S. Veeraraghavan. *in preparation*, 1995.

[67] M. Kamionkowski, D. Spergel, and N. Sugiyama. *Astrophysical Journal Letters*, 1994.

[68] S. Dodelson, E. Gates, and A. Stebbins. *in preparation*, 1995.

[69] M. White, D. Scott, and J. Silk. *Annual Reviews of Astronomy and Astrophysics*, 32:319, 1994.

Measuring the Cosmological Parameters

Marc Davis

Depts. of Astronomy and Physics, University of California, Berkeley CA, 94720 USA

A very brief summary of the present status of measurement of the fundamental cosmological parameters H_0, Ω, and Λ is here presented. The current observations would suggest the age of the Universe to be less than that of the oldest stars. This problem cannot be resolved by recourse to a large value of the cosmological constant, as that is precluded by the statistics of gravitational lenses. Estimates of the density parameter consistently suggest a low value for Ω on scales of clusters of galaxies The only evidence for high values of Ω are derived from large scale flow analyses, but not all of these estimates of self-consistent. The present situation is one of total confusion; hopefully the coming decade will witness clarification of these fundamental issues.

§1 Introduction

Astronomers have known since the late 1920's that the Universe is expanding. In a spherically symmetric, homogeneous Robertson-Walker metric, the solution to Einstein's equation of General Relativity is specified by H_0 and q_0. H_0^{-1} is a measure of the current age of the Universe; while q_0 is a measure of the curvature and deceleration rate of the Universe.

Two additional quantities are the cosmological constant, which measures the vacuum energy density of the Universe, and Ω, which is the ratio of the total mass/energy density in the Universe to the critical density required to just bring the expansion to a halt sometime in the future. Consider first $\Lambda = 0$. Then if $\Omega < 1$, the mass/energy density is less than critical and the self-gravity of the Universe is insufficient ever to stop its expansion (an "open" Universe). If $\Omega > 1$, the expansion will eventually stop and the Universe will collapse (a "closed" Universe). The Λ term is a measure of a vacuum energy density, which remains constant as the Universe expands, unlike ordinary matter and radiation whose densities decrease with expansion. A "vacuum energy density" means that energy is present in an empty Universe even though no particles or radiation exist. The key point about is that such energy generates gravity even though there is no normal matter or radiation present, hence gravity from a vacuum. A Λ dominated universe will expand exponentially with time, with positive acceleration, whereas a matter or radiation dominated universe expands as a power law of time, with negative acceleration. If there is currently no vacuum energy density in the Universe, $\Lambda = 0$ and $q = \Omega/2$; most cosmologists assume that the present value of Λ is 0, but understanding why it is so small (more than 100 orders of magnitude smaller than during the inflationary epoch) is a profound question of fundamental physics.

Since inflationary models provide a dynamical mechanism to drive the curvature of the Universe to negligibly small values, current interest in the cosmological constant models is primarily limited to the $k = 0$ case, where $\Omega_B + \Omega_{DM} + \Omega_\gamma + \Omega_\Lambda = 1$, where the subscripts B, DM, γ, and Λ denote baryons, dark matter, radiation (including neutrinos), and cosmological constant, respectively.

§2 The Hubble constant H_0

The value of the Hubble constant is of fundamental importance in many cosmological settings: from extrapolating theories of the earliest epochs of inflationary models of cosmology, to interpreting the evidence from Big Bang nucleosynthesis, to locally setting the energy, luminosity, and size scales for a variety of extragalactic phenomena. The age of the Universe is approximately $(2/3) f h^{-1} \times 10^{10}$ years, where $h = (H_0/100\text{km/s/Mpc})$ and where f ranges from 1 for an $\Omega = 1$, $\Lambda = 0$ model to $f = 3/2$ in an $\Omega = 0$, $\Lambda = 0$ model, to $f = 1.61$ in an $k = 0$, $\Omega_\Lambda = 0.8$ model, a model which has frequently been suggested as a "best" solution to a variety of cosmological tests. A fundamental reality check comes from requiring the oldest stars in our Galaxy to be younger than the age of the Universe. This requirement, an obvious logical necessity, sets an upper limit to H_0. Current estimates of H_0 range between 45 Km/s/Mpc (a Megaparsec (Mpc) is about 3×10^{24} cm) and 90 Km/s/Mpc. A value near 45-55 is considered "low" and 70-90 is considered "high". In the simple Einstein de Sitter Universe with $\Lambda = 0$ and $\Omega = 1$ then $H_0 = 50$ implies an age of 13.3 billion years, but $H_0 = 80$ gives only 8.3 billion years! Given that the best age estimates for globular cluster stars are in the range 13-15 billion years [1], most cosmologists would argue that that high values of H_0 are unreasonable. However, the question is subject to direct experimental test, which must take precedent over preconceptions, no matter how well founded.

The determination of the Hubble constant has occupied observers for several decades. Meanwhile the astrophysical motivation for an accurate value of this critical parameter has become ever stronger. For example, models of large-scale structure evolution need a more accurate value for H_0. The peak in the power spectrum of fluctuations in CDM type models is at a wavelength of approximately 32 $(\Omega h^2)^{-1}$ Mpc, the horizon scale at the epoch when the matter density equaled the radiation density. This corresponding size scale today is an important relic of the Big Bang, but its value is proportional to H_0^{-2} and therefore suffers from the current uncertainty. An accurate measurement of H_0 is crucial for assessing whether the detailed models of the evolution of structure in the Universe can be reconciled with a wide range of observations.

Since astronomers measure distances to objects by their redshifts, one factor of H_0 is absorbed in the distance measure, and the remaining scale factor is now defined by the quantity $\Gamma \equiv \Omega h$. Measurements of large scale power in the galaxy distribution, based on angular correlation analysis of faint galaxies, and direct power spectral analysis of several redshift catalogs of galaxies, suggests a value $\Gamma \approx 0.2 - 0.3$ [2] [3]. The standard CDM model, with $h = 0.5$, and $\Omega = 1$, does not have sufficient large scale power to quantitatively match the large scale correlations in the observed galaxy distribution. ¿From a theoretical perspective, the preferred, simplest cosmological model, would have parameters $\Lambda = 1$, $\Omega = 1$, and $h = 0.25$. Such a Universe would be comfortably older than the oldest stars, and would have adequate large scale power to fit the observations. An accurate measure of H_0 is essential to confirm, or more likely to dispell, this fantasy.

Ground-based facilities and techniques have improved dramatically over the past three decades, yet the value of H_0 remains uncertain within a factor of two. The problem is that different techniques and different research groups get discrepant values for H_0, a sure sign that unknown systematic errors still exist. Which value is correct? The Hubble Space Telescope (HST) is likely to help resolve this long-standing issue.

HST is providing an accurate calibration of the cosmic distance scale relying principally on the well-established astronomical technique using Cepheid variable stars in other galaxies. First results from this work have recently been published for 13 Cepheids in one galaxy in the Virgo cluster, suggesting a best estimate of $H_0 = 80 \pm 17$ km/s/Mpc [4]. The prospects of further observations and of reducing the errors by a factor of two are excellent. This result is fully consistent with a variety of other recent estimates of H_0 [5].

These high values of H_0 are dramatically in conflict with the globular-cluster and other ages in a simple Einstein deSitter universe, and such a universe would have far too little large scale power unless it is open or Λ dominated. This contradiction is likely to be the most perplexing cosmological question of the next decade. Solving this problem could mean major changes to stellar evolution theory, a non-zero value for Λ, or perhaps something completely different.

The successful refurbishment of HST will allow that instrument to make more definitive measurements of the Hubble constant, but because of the conflict it is clearly vital to verify this value by alternative, independent means such as the detailed study of supernova photospheres [6], and supernovae luminosities [7], the measurement of the effects of electron scattering in the hot gas within galaxy clusters (the Sunyaev-Zel'dovich effect) [8] [9], or the time delay of quasar variability as measured in separate components of gravitational lenses [10]. All these approaches offer potential alternative measurements of H_0, and all are under active investigation.

§3 Global tests of Ω and Λ

To measure q_0 directly one must measure the change in expansion rate over a large range of cosmic time. This is done using "global" cosmological tests extending over large distances, where the light travel time is an appreciable fraction of the age of the Universe. The basic idea is that the size and appearance of a distant patch of the Universe, as viewed from our vantage point, depends both on how the Universe expands (its global geometry) and on the bending of light by the gravity from intervening matter.

Extensive efforts in the 1960s and 70s to study the apparent luminosity or size of distant objects hoped to be "standard candles", such as very luminous galaxies, were mostly abandoned after it was realized that the intrinsic luminosity of these standards probably changed strongly with time because of galaxy evolution. Recent studies of the apparent size of features in distant radio galaxies might provide new ways to measure q_0. Direct counts of galaxies as a function of measured red shift (the volume test) can also be a powerful probe of the curvature of space. This test was attempted in the last decade [11], again with ambiguous conclusions. With improved modeling of the evolution of galaxies and a major effort to obtain spectra of a large sample of faint galaxies, this test might prove to be an effective way to measure both q_0 and Λ. Cosmological models with different values of q_0 and Λ predict substantially different volumes of space for a fixed increment of red shift, and the number of galaxies is a measure of the size of that volume. Not only does this volume evolution affect the number of quasars, supernovae, or galaxies at any red shift, but also the number of potential gravitational lenses.

The number of gravitational lenses found in a survey for lensed quasars is a volume test of the cosmological model because the number of lenses depends on the expected number of lens galaxies between the observer and the high redshift quasar. Compared to the theoretically favored $\Omega = 1$ flat cosmological model, both flat comologies with a positive cosmological constant Λ and low density $\Omega < 1$ contain more volume per unit redshift and therefore produce larger numbers of lenses [12]. Flat universes with $\Lambda = 1$ produce 10 times as many lenses, and empty $\Omega = 0$ universes produce twice as many lenses as the Einstein-deSitter model.

Detailed analysis of cosmological constraints derived from the statistics of lenses is given by Kochanek [13]. In an optically selected sample of 680 Quasars, there are 5 lenses. For a range of reasonable models of elliptical and lenticular galaxies (the only types massive enough to form lenses of the observed separations) the statistics suggest $\Lambda < 0.5$ at 90% confidence. The cosmological models with $\Omega = 0.2$ and $\Lambda = 0.8$ that are sometimes suggested to solve the density and Γ problems of large scale structure, are ruled out by the lens statistics with more than 99% confidence. The dominant systematic uncertainties that might bias this strong conclusion, such as a seriously incorrect mass model of the galaxies, are quite unlikely, as two very different mass models lead to similar constraints. This is by far the strongest observational constraint on Λ, much stronger than that derived from the integrated Sachs-Wolfe effect in the analysis of the CMBR fluctuations of COBE.

Poisson statistics limit the constraint on the open cosmological model. The fact that lenses are not more common favors a high density universe, and the only available constraint is $\Omega > 0.2$ at 90% confidence. New searches for lenses are underway using radio and optical selection techniques. As additional lenses are discovered and the statistics improve, we can expect the cosmological constraints to sharpen substantially.

Detection of standard events at cosmological distances is another means to measure q_0. Perhaps the most promising activity in this regard is the effort to detect Type Ia supernovae in high redshift galaxies [14]. If sufficient numbers of these rare events can be detected and followed over the course of their lifetimes, it should be possible to directly measure the q_0. A concerted effort of searching for candidate events using large format, low noise CCD detectors and automated, massive computing is essential to detect these events quickly enough for a successful followup of each candidate. Such a project is underway and results are eagerly anticipated. In one recent observing run, 6 supernovae in the range $0.3 < Z < 0.5$ were detected and followed up. In order to measure q_0 to 5% precision using redshift $Z = .5$ supernovae, it will be necessary to detect and carefully follow the light curves of 100 type 1a events.

§4 Local tests of the density parameter

Without going to cosmological distances, it is possible to measure the density parameter using the so-called "local" tests. Many of the local tests weigh local structures using the virial theorem, which states that the kinetic energy of a self-gravitating system should be approximately equal to its potential energy. Since luminous galaxies must be used to measure the kinetic energy of the system, only the component of the mass density clustered with luminous galaxies can be examined in this fashion. The data has remained consistent for many years: although there is abundant evidence for dark matter in the halos of galaxies, and within groups and clusters of galaxies, none of this evidence suggests $\Omega > 0.2$ [15],[16], [17]. It is important to note that all these estimates based on virial equilibrium apply to scales of order 1 Mpc and smaller.

If the density of the Universe is higher than this value, the galaxy distribution must be somehow biased relative to the mass distribution. Normalization of the mass fluctuations is conventionally expressed as $\sigma_8 = b^{-1}$, where b is the bias in the galaxy distribution. (The rms amplitude of the fluctuations of optically selected galaxies is unity in this volume).

A number of bias mechanisms have been suggested in the past, [18] [19] [20], but the models for *large*, $b \approx 2$ bias are not compelling. Furthermore large bias would have observational consequences that are not detected [21]. Finally, the large amplitude of the fluctuations detected by COBE, when fit to models of large scale structure, do suggest values of $\sigma_8 \approx 1$ [22].

Imaging x-ray satellites permit detailed analysis of the hot gas within clusters of galaxies, which have characteristic temperatures of $3 \times 10^7 - 10^8$ degrees K. Gas at this temperature radiates predominantly as Bremstrallung radiation, and careful analysis of the density and temperature profiles of the x-ray properties of the cluster yields a good measure of the distribution of the hot gas as well as the gravitational potential. With new results coming from the ROSAT and ASCA satellites [23], recent reports are providing good measures of the baryon fraction within clusters of galaxies (dominated by the hot gas but including the luminous galaxies) [24]. The fraction, as measured in several clusters with the ASCA satellite [25], appears to be 20%, and is nearly constant with radius within the clusters. This value, when compared to the nucleosynthesis constraint, $\Omega_B = 0.015h^{-2}$, is consistent with $h = 0.27$ in a flat universe, or h=0.8 in an $\Omega = .12$ open universe.

Neither situation is consistent with all available data.

This is a serious problem because the X-ray data is of very high quality, and for the first time includes good constraints on temperature gradients within the clusters. From a theoretical viewpoint, one does not expect substantial segregation of the baryons from the dark matter on the scale of clusters of galaxies (hundreds of kpc). Resolution of the conflict is eased with a lower value of H_0, but the problem is likely to remain unresolved until we have a much better understanding of cluster formation including all details of the gas dynamical processes. Improvements in computing technology should enable progress on this question in the next several years.

§5 Cosmic flows

On scales larger than a megaparsec, gravitational systems are not in virial equilibrium, so one must turn to other means of measuring the mass density. By analysis of the deviations of galaxies from smooth Hubble flow, combined with analysis of the gravitational field of the local galaxy distribution, one can apply the results of gravitational instability theory to infer the density parameter Ω. In an Einstein deSitter Universe at late times in linear perturbation theory, one expects $\mathbf{v} = \frac{2}{3H_0}\mathbf{g}$, where $\mathbf{v}$ and $\mathbf{g}$ are the velocity and gravity fields [26]. For a more general cosmology, the proportionality constant between $\mathbf{v}$ and $\mathbf{g}$ scales as $\Omega^{0.6}$, and since the gravity field is measured in terms of the galaxy distribution with unknown bias b, the cosmological combination measured by this test is $\beta \equiv \Omega^{0.6}/b$.

Based on the dipole of the observed galaxy distribution compared to the CMBR dipole, one can derive a 90% confidence lower limit, $\beta > 0.4$ [27]. Dekel etal [28] used the POTENT algorithm to compare the deviations from Hubble flow of a set of 493 galaxies to the gravity field predicted by an IRAS selected galaxy sample, concluding $\beta = 1.2 \pm 0.3$. More recent analyses [29], [30] using better, more extensive data both for the velocity and gravity fields, and different algorithms, find a best value of $\beta \approx 0.6 \pm 0.2$.

The results must be considered as preliminary, but the method should be capable of measuring β to an accuracy of 10%. This type of test requires full-sky redshift surveys of the galaxy distribution, and very extensive maps of the deviations of Hubble flow. At present it is possible to compare the two fields out to a redshift cz=6000 km/s. This test is remarkable in that it permits a measure of the fluctuating component of the mass distribution on scales 30 times larger than permitted by the virial theorem. Pushing the comparison to greater distances and more precision at closer distances will require more precise distance indicators for galaxies and more complete sampling of the local galaxy distribution, both of which should occur in this decade.

The only local observational evidence for a high density universe derives from this test. If indeed the improved data does suggest a high value of β, it will be strong evidence for high Ω, as values of $b < 1$ are very unlikely. If that turns out to be the case, cosmologists will still have to explain why the virial tests consistently give much lower values of Ω on smaller scales. The data would seem to be telling us that the dark matter does not fully cluster with the luminous galaxies except on scales in excess of a few Mpc. Yet this is not how CDM, or a mixture of hot and cold dark matter, is expected to behave; the dark matter, unless thermally hot today, should cluster with the galaxy distribution on scales in excess of 100 kpc. Is there something in the physics of the dark matter that we have not included in our models?

§6 Conclusions

In spite of 60 years of effort, we are still uncertain of the basic cosmological parameters that describe our Universe. The present data does not seem to give a consistent picture. The best values of H_0 suggest an age that is too small, or a cosmological constant Λ incompatible with the relative paucity of lensed quasars. Measures of the mass density Ω are consistent with low values on virialized scales, but on the scale of cosmic flows, the inferred values are considerably larger, and some of these analyses suggest $\Omega = 1$. The observed extent of large scale galaxy clustering strongly suggests $\Omega(H_0/100) = 0.2-0.3$, suggesting either a low density Universe or a low Hubble constant, or both. The resolution of these problems is not obvious. Either some of the data is incorrect, or our understanding of the thermal history of the Universe is incomplete, or the dark matter behaves in a more complicated fashion than we assume, or perhaps the Universe is open after all. The coming decade is likely to bring a variety of new observational constraints that will either suggest a solution to the present dilemma, or lead to even greater confusion.

§References

[1] A. Sancisi. Observational tests of inflation. Kluwer, Boston, 1991.

[2] J.A. Peacock and S.J. Dodds. *Monthly Notices of the Royal Astr. Soc.*, 267:1020, 1994.

[3] G. Efstathiou, R. Bond, and S.D.M. White. *Monthly Notices of the Royal Astr. Soc.*, 258:1P, 1992.

[4] W.L. Freedman et al. *Nature*, 371:757, 1994.

[5] G.H. Jacoby et al. *Publ. of the Astron. Soc. of the Pacific*, 678:599, 1992.

[6] B.P. Schmidt et al. *Astrophysical Journal*, 432:42, 1994.

[7] A.G. Riess, W.H. Press, and R.P. Kirshner. Using SN Ia Light Curve Shapes to Measure the Hubble Constant. *Astrophysical Journal*, submitted, 1994.

[8] M. Birkinshaw and J.P. Hughes. *Astrophysical Journal*, 420:33, 1994.

[9] M.L. Fischer and A.E. Lange. *Astrophysical Journal*, 419:433, 1993.

[10] W.H. Press, G.B. Rybicki, and J.N. Hewitt. *Astrophysical Journal*, 385:416, 1992.

[11] E.D. Loh and E.J. Spillar. *Astrophysical Journal*, 307:L1, 1986.

[12] E.L. Turner. *Astrophysical Journal*, 365:L43, 1990.

[13] C.S. Kochanek. *Astrophysical Journal*, 419:12, 1993.

[14] S.H. Perlmutter et al. *Astrophysical Journal*, in press, 1994.

[15] M. Davis and P.J.E. Peebles. *Astrophysical Journal*, 267:465, 1983.

[16] M. Geller. *Large Scale Structure in the Universe*. Saas-Fee, 1987.

[17] G.P.E. Efstathiou. *Observations of Large Scale Structure in the Universe*. Les Houches Lectures 1993, Elsevier Science Publishers, 1995.

[18] N. Kaiser. *Astrophysical Journal*, 284:L9, 1984.

[19] R. Carlberg. *Astrophysical Journal*, 433:468, 1994.

[20] S.D.M. White, M. Davis, G. Efstathiou, and C.S. Frenk. *Nature*, 330:451, 1987.

[21] D. Vallsgabaud, J.M. Alimi, and A. Blanchard. *Nature*, 341:215, 1989.

[22] K.M. Gorski et al. *Astrophysical Journal*, 430:L89, 1994.

[23] Y. Fukazawa et al. *Publ. of the Astron. Soc. of Japan*, 46:L55, 1994.

[24] S.D.M. White, J.F. Navarro, A.E. Evrard, and C.S. Frenk. *Nature*, 366:429, 1989.

[25] T. Ohashi. Dark matter in the universe. 5th U.Maryland Oct. Workshop, 1994.

[26] P.J.E. Peebles. *Principles of Physical Cosmology*. Princeton, 1993.

[27] M.A. Strauss et al. *Astrophysical Journal*, 397:395, 1992.

[28] A. Dekel. *Astrophysical Journal*, 412:1, 1993.

[29] J. Willick et al. *Astrophysical Journal*, in preparation, 1994.

[30] A. Nusser and M. Davis. *Astrophysical Journal*, 421:L1, 1994.

Primordial and Stellar Nucleosynthesis in the Next Millennium

David N. Schramm

University of Chicago, 5640 S. Ellis Avenue, Chicago, IL 60637, and NASA/Fermilab Astrophysics Center, Fermi National Accelerator Laboratory, Batavia, IL 60510-0500

G. J. Mathews

Department of Physics, University of Notre Dame, Notre Dame, IN 46556, and University of California, Lawrence Livermore National Laboratory, Livermore, CA 94550

The nucleosynthesis working group discussed both primordial and stellar nucleosynthesis. Discussions were held on abundance observation and experimental nuclear measurement as well as theory. Although primordial nucleosynthesis is expected to continue to be one of the pillars of big bang cosmology well into the next millennium, the field is currently limited by the state of the observed abundances. The most urgent need is for more quantitative definitive observations of primordial abundances along with supporting theoretical investigations of chemical evolution, stellar processing, stellar model atmospheres, and recombination/collisional excitation rates so as to reduce the systematic uncertainties. For stellar evolution and supernovae the most important areas for development are for multi-dimensional hydrodynamics and radiation transport simulations which are reliable over a large range of length scales and time scales along with the development of supporting experimental studies of effects of turbulence and convection as well as the measurement of key nuclear reaction rates, possibly with radioactive nuclear beams.

§1 Introduction

Nucleosynthesis has been a cornerstone of nuclear and particle astrophysics and cosmology for decades. Primordial nucleosynthesis is the key probe of the early universe at times of order 1 to 10^4 sec. It provides an independent determination of the baryon to photon ratio and constrains numerous physical parameters and processes which might have occurred before during or after the epoch of nucleosynthesis. Stellar and supernova evolution and nucleosynthesis is also key to understanding the properties of stars and the chemical and dynamical evolution of galaxies as well as providing important chronometers and links in the cosmic distance ladder.

Nucleosynthesis continues to be an area of much activity involving astronomical spectroscopists, experimental nuclear physicists and theorists of varied backgrounds. The strength of nucleosynthesis is the ability to make very quantitative predictions and compare them with observations. The Snowmass workshop focused primarily on primordial nucleosynthesis and light element abundances as well as stellar nucleosynthesis, galactic evolution and experimental nuclear astrophysics. Galactic evolution studies are particularly of interest in the next millinneum both as a means to understand galaxies and as a means to deconvolve present observations to determine the primordial abundances.

Although nucleosynthesis studies have been pursued for decades, there are numerous advances yet to be made which will occupy our thoughts and effort well in to the next millennium. This paper is a summary of some of the areas which were identified during the workshop where efforts and resources might be best spent in the advance of nuclear and particle astrophysics and cosmology into the coming decades. We will first discuss primordial nucleosynthesis issues and then briefly summarize the discussions concerning stellar and supernova nucleosynthesis.

§2 Primordial Nucleosynthesis

The power of the standard homogeneous big bang nucleosynthesis (BBN) model comes from the fact that essentially all of the physics input is well determined (with the possible exception of ^{3}H (α,γ) ^{7}Li and ^{3}He (α,γ) ^{7}Be nuclear reaction rates). The appropriate temperature regimes, 0.1 to 1 MeV, are well explored in terrestrial nuclear physics experiments. In fact, the nuclear reaction rates for BBN are known far more accurately than the rates are known for the centers of stars like the Sun where the temperature is only ~ 1 keV.

Overview talks on the subject of standard homogeneous big bang nucleosynthesis were presented by D. Schramm and T. Walker. Discussions of nonstandard models were made by K. Jedamzik and G. Mathews. It is one of the triumphs of modern cosmology that standard primordial nucleosynthesis is able to account so well for the inferred primordial abundances of hydrogen, deuterium, ^{3}He,^{4}He, and ^{7}Li with only a single adjustable parameter, the baryon-to-photon ratio η.

To calculate what happens in the big bang, all one has to do is follow the density of relativistic particles (photons, electrons, neutrinos, etc.) and baryons as the universe expands and cools. For nuclear reactions, the only relevant temperatures are from a little above 1 MeV ($\sim 10^{10}$K) down to $\sim$ 10 keV ($\sim 10^8$ K). At higher temperatures, no complex nuclei other than free neutrons and protons can exist, and the ratio of neutrons to protons, n/p, is just determined by $n/p = e^{-Q/T}$, where $Q = (m_n - m_p)c^2 \sim 1.3$ MeV. Equilibrium applies because the weak interaction rates are much faster than the expansion of the universe at temperatures much above 10^{10}K. At temperatures much below 10^8K, the nuclear reaction rates become slow compared to the cosmological expansion rate.

After the weak interaction drops out of equilibrium, a little above 10^{10}K, the ratio of neutrons to protons changes more slowly due to free neutrons decaying to protons, and similar transformations of neutrons to protons via interactions with the ambient leptons. By the time the universe reaches 10^9K (0.1 MeV), the ratio is slightly below 1/7. For temperatures above 10^9K, no significant abundance of complex nuclei can exist due to the continued existence of gammas with energies greater than MeV. Note that the high photon to baryon ratio in the universe ($\sim 10^{10}$) enables significant population of the high energy ($\sim$MeV) Boltzmann tail until $T \lesssim 0.1$MeV.

Once the temperature drops to about 10^9K, sufficient abundances of nuclei can exist in statistical equilibrium through reactions such as $n + p \leftrightarrow {}^2\mathrm{D} + \gamma$ and $^2\mathrm{D} + p \leftrightarrow {}^3\mathrm{He} + \gamma$ and $^2\mathrm{D} + n \leftrightarrow {}^3\mathrm{H} + \gamma$, which in turn react to yield ^{4}He. Since ^{4}He is the most tightly bound nucleus in the region, the flow of reactions converts almost all the neutrons that exist at 10^9K into ^{4}He. (See Figure 2 for an example of the reaction rate network [1]) The flow essentially stops at ^{4}He because there are no stable nuclei at either mass-5 or mass-8. Since the baryon density during big bang nucleosynthesis is relatively low, only reactions involving two-particle collisions occur. The only nuclei heavier than ^{4}He which is produced in a significant amount are ^{7}Be and ^{7}Li which are produced from reactions of ^{4}He with ^{3}He or ^{3}H. As we will discuss, paths around the mass-8 gap can be found by neutron capture if $n/p > 1$, or in local regions of high baryon density where 3-body reactions and alpha capture can occur.

Eventually, ^{3}H decays to ^{3}He, and any mass-7 made decays to ^{7}Li. Thus, big bang nucleosynthesis makes ^{4}He with traces of ^{2}D, ^{3}He, and ^{7}Li. (Also, all the protons left over that did not capture neutrons remain as hydrogen.) For standard homogeneous BBN, all other chemical elements are made later in stars and in related processes (such as cosmic ray spallation).

The reliability of theoretical calculations seems well established. We heard talks from D. Seckel, T. Walker, L. Krauss [2, 3], and D. Thomas concerning possible higher order corrections, and the numerical and statistical accuracy of standard primordial nucleosynthesis calculations. The codes in use at the present time are certainly accurate to better than a few tenths of a percent, for example, in the abundance of helium.

The major theoretical action in standard BBN has not been in redoing the calculations but has instead focused on the chemical evolution of the light elements and the development of methodology to extrapolate present observations back to early time. For example, the proof in the 1970's that deuterium was of cosmological origin, and only decreased in time since the big bang, enacted to constrain Ω_b. At this meeting much attention focused on lithium evolution and its possible variations.

The major uncertainties at the present time are first and foremost, limits on the observed primordial light-element abundances, and a few nuclear reactions (particularly those critical to the formation of ^{7}Li and ^{7}Be). There is at present no compelling reason to doubt the validity of the standard homogeneous primordial nucleosynthesis model, although interesting alternatives (e.g. baryon inhomogeneities, anisotropies, time varying constants, etc) have been proposed which, although significantly constrained, are not ruled out. These alternatives may even permit a few loopholes (cf. §4) which could slightly alter the basic conclusions of BBN such as the limits on Ω_b.

The degree to which the various conclusions of standard BBN and alternative scenarios are constrained is currently limited by the systematic uncertainties of primordial abundance determinations. This situation will remain until bf more quantitative, definitive, observations of primordial abundances are available. The main theme of the discussions and the driving force of further research in primordial nucleosynthesis into the next millennium is therefore improving the light element abundance determinations and reducing the systematic uncertainties. Much discussion, centered on the present status and future of primordial-element abundance measurements. These we summarize briefly here and in §3.

2.1 Light Element Abundances

The study of the light element abundances has undergone a recent burst of activity on many fronts. The status of current deuterium observations were reviewed in talks be J. Linsky, J. Pasachoff, and D. Lubowich. Helium was reviewed by E. Skillman and G. Shields. Lithium was reviewed by V. Smith and P. Nissen, while Be and B were reviewed by D. Duncan. G. Wallerstein and J. Cowan discussed heavier element observations of extreme metal poor systems. New results on each of the cosmologically significant abundances have sparked renewed interest and new studies. Primordial nucleosynthesis continues along with the Hubble expansion and the microwave background radiation as one of the three pillars of big bang cosmology. Of the three, big bang nucleosynthesis, leads most directly to the interplay of cosmology with nuclear and particle physics. Furthermore, since the Hubble expansion is also part of alternative cosmologies such as the steady state, it is BBN and the microwave background that really drive us to the conclusion that the early universe was at one time hot and dense.

Recent heroic observations of ^{6}Li, Be and B, as well as ^{2}D, ^{3}He and new ^{4}He determinations, have all gone in the direction of strengthening the basic picture of cosmological nucleosynthesis. In particular, theoretical calculations of cosmic ray production of ^{6}Li, Be and B have fit the observations well, thus preventing these measurements from disturbing the standard scenario [4]. In this regard, the recent reports of D/H in quasar absorption systems at redshift $Z \sim 3$ are particularly interesting [5, 6], since BBN requires that fragile deuterium be found in primitive material. However, the possible variation of D/H in different lines of sight at $Z \sim 3$ argues that perhaps hydrogen cloud interlopers may be imitating deuterium, at least in lines of sight having higher apparent D/H. Such an interpretation avoids a potential conflict with ^{3}He arguments. Nevertheless, recent theoretical calculations of inhomogeneous big bang nucleosynthesis may slightly alter some of the basic conclusions of standard BBN. We also briefly discussed the possible impact on BBN of the recent ROSAT cluster results and the recent halo microlensing results. All of this discussion highlights how dramatically robust and active the field of BBN is and will continue to be well into the next millennium.

Let us now briefly review the history, with special emphasis on the remarkable agreement of the observed light element abundances with the calculations. This agreement works only if the baryon density is well below the cosmological critical value even in most nonstandard models.

2.2 The Present Millennium of Big Bang Nucleosynthesis

One way to gain a perspective on where we will be in the next millennium is to look at what has occurred in the last millennium Let us now, therefore, briefly review the history of the present millennium. The consistent theme of the past millennium has been the remarkable agreement of the observed light element abundances with the calculations. This agreement works only if the baryon density is well below the cosmological critical value even in inhomogeneous models.

Aside from some interesting theological discussion, there was no serious discussion of primordial nucleosynthesis until near the latter half of the last century. In the period from 1948 to 1960 there were about a half a dozen [7] pioneering papers on big bang nucleosynthesis. In the initial papers due to Gamow and his associates Alpher and Herman, the important symbiotic connection between BBN and the 3K background was noted for the first time. The initial BBN calculations of Gamow's group[9] assumed pure neutrons as an initial condition, and thus, were not particularly accurate. Nevertheless, their inaccuracies had little effect on the group's correct and insightful prediction of the cosmic background radiation.

Another outstanding advance during this time frame was the work of Hayashi [8] who first recognized the role of neutron-proton equilibration. Since that paper, the essential framework for BBN calculations has not varied significantly.

In the period from 1960 to 1970 the number of papers on the subject roughly doubled [7]. These papers ushered in the era of modern BBN calculations. The work of Alpher, Follin and Herman[9] and Taylor and Hoyle,[10] proceeding the discovery of the 3K background, and Peebles [11] and Wagoner, Fowler and Hoyle [12] immediately following the discovery, all do essentially the same basic calculation as more recent work [4, 13, 14, 15, 16, 17, 18, 19] the results of which are shown in Figure 1. As far as the calculation itself goes, solving the reaction network is relatively simple by the standards of explosive nucleosynthesis calculations in supernovae. The changes over the last 25 years are mainly in terms of more recent nuclear reaction rates as input, not as any great calculational insight, although the current Kawano code [18] is somewhat streamlined relative to the earlier Wagoner code [12]. In fact, the early Wagoner code is, in some sense, a special adaptation of the larger nuclear network calculation developed by Truran [20] for work on explosive nucleosynthesis in supernovae. With the possible exception of Li yields, the reaction rate changes over the past 25 years have not had any major affect (see Yang et al. [17] and Krauss and his collaborators [2], Copi, Schramm and Turner [1], or Smith et al [19] for discussion of uncertainties). The one key improved input is a better neutron lifetime determination [21].

The real excitement for BBN over the last 25 years has been understanding the evolution of the light element abundances and using that information to constrain physical parameters and cosmological models. In the 1960's, the main focus was on ^{4}He which is very insensitive to the baryon density. The agreement between BBN predictions

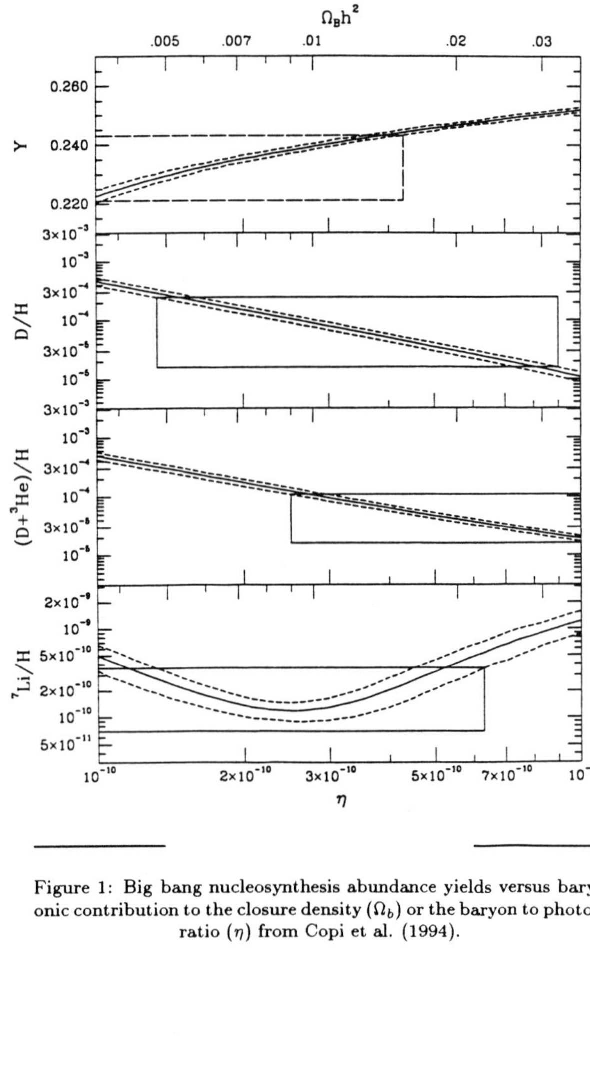

Figure 1: Big bang nucleosynthesis abundance yields versus baryonic contribution to the closure density (Ω_b) or the baryon to photon ratio (η) from Copi et al. (1994).

and observations helped support the basic big bang model but gave no significant information at that time with regard to density. In fact, in the mid-1960's, the other light isotopes (which are, in principle, capable of giving density information) were generally assumed to have been made during the T-Tauri phase of stellar evolution [22] and so, were not then taken to have cosmological significance.

It was during the 1970's that BBN fully developed as a tool for probing the Universe. During the period from 1970 to 1980 some 50 papers on this subject appeared. The possibility that BBN could be used to probe the early universe was in part stimulated by Ryter et al. [23] who showed that the T-Tauri mechanism for light element synthesis failed. Furthermore, ^{2}D abundance determinations improved significantly with solar wind measurements [24, 25] and the interstellar work from the Copernicus satellite [26]. (Recent HST observations reported at this meeting by Linsky have compressed the ^{2}D error bars considerably.) Reeves, Audouze, Fowler and Schramm [27] argued for cosmological ^{2}D and were able to place a constraint on the baryon density excluding a universe closed with baryons. Subsequently, the ^{2}D arguments were cemented when Epstein, Lattimer and Schramm [28] proved that no realistic astrophysical process other than the big bang could produce significant ^{2}D. It was also interesting that the baryon density implied by BBN was in good agreement with the density implied by the dark galactic halos [29].

By the late 1970's, a complimentary argument to ^{2}D had also developed using ^{3}He. In particular, it was argued [30] that, unlike ^{2}D, ^{3}He was made in stars; thus, its abundance would increase with time. Since ^{3}He like ^{2}D monotonically decreased with cosmological baryon density, this argument could be used to place a lower limit on the baryon density [31] using ^{3}He measurements from solar wind [24, 25] or interstellar determinations [32]. Since the bulk of the ^{2}D was converted to ^{3}He in stars.

The other significant development of the 70's for BBN was the explicit calculation of Steigman, Schramm and Gunn [33] showing that the number of neutrino generations, N_ν, had to be small to avoid overproduction of ^{4}He. (Earlier work [10, 34, 35] had commented about a dependence on the energy density of exotic particles but had not done an explicit calculation probing N_ν.) This will subsequently be referred to as the SSG limit. To put this in perspective, one should remember that the mid-1970's also saw the discovery of charm, bottom and tau, so that it almost seemed as if each new detector produced new particle discoveries, and yet, cosmology was arguing against this "conventional" wisdom. Over the years, the SSG limit on N_ν improved with ^{4}He abundance measurements, neutron lifetime measurements, and with limits on the lower bound to the baryon density, hovering at $N_\nu \lesssim 4$ for most of the 1980's and dropping to slightly lower than

4 just before LEP and SLC turned on [4, 13, 36, 37]. This was verified by the LEP results [38] where now the overall average is $N_\nu = 2.99 \pm 0.02$.

If the 70's marked the beginning of the application of BBN to constrain the early universe, the 80's marked its coming into full appreciation. Some 300 papers were written during the decade of the 80's which characterized ever improving refinements on existing constraints as well as application to a number of new ones.

In [17] the basic picture was reviewed and among other things the D + ^{3}He constraint was shown to be quite restrictive. Rood, Bania and Wilson [39] recently have shown that ^{3}He is indeed enriched in planetary nebulae; hence, the argument that ^{3}He increases with time is now strengthened. However, there is nonetheless the worry that interstellar ^{3}He measurements [40] do vary with location more than one might expect for an isotope produced primarily in low mass stars. This latter point has been the subject of much recent discussion.

An important development in the 80's was the observation that the lower boundary from ^{3}He and the upper boundary from ^{2}D yielded the requirement that ^{7}Li be near its minimum of ^{7}Li/H $\sim 10^{-10}$, which was verified by the Pop II Li measurements of Spite and Spite [41, 42, 43]. This, confirmed the situation first hinted by [29] and emphasized by Yang et al. [17] that the light element abundances are consistent over nine orders of magnitude with BBN, but only if the cosmological baryon density, Ω_b, is constrained to be around 5% of the critical value (for $H_0 \simeq$ 50 km sec^{-1} Mpc^{-1}). On Figure 1 the concordant region of $\Omega_b h^2 \sim 0.013$ corresponds to $\Omega_b \sim 0.05$ for $H_0 = 50$ km sec^{-1} Mpc^{-1}. This concordance region is slightly larger than that of Walker et al [13] due primarily to the inclusion of possible systematic errors on Li/H.

Another interesting development in the 80's was the renewed interest in baryon inhomogeneous models [44]-[45], [7] originally motivated [46] by the prospect of a first order QCD phase transition in the early universe, although other processes may lead to such inhomogeneities as well [7]. This lead to many papers and much speculation about altered limits from primordial nucleosynthesis which we return to below.

The decade of the nineties has seen a continuing exponential increase in activity in BBN. Already, as many papers have appeared as in all previous decades combined (cf. [7]). Among the interesting developments, the Li plateau argument has been further strengthened with the observation of ^{6}Li in a Pop II star by Smith, Lambert and Nissen [47]. Since ^{6}Li is much more fragile than ^{7}Li, and yet it survived, the possible nuclear depletion of ^{7} Li is constrained [48, 49]. The ^{6}Li seen is in reasonable agreement with the spallation-produced Be and B for stars of this metallicity. It has been noted, however, that in one particular stellar evolution model with rotational mixing

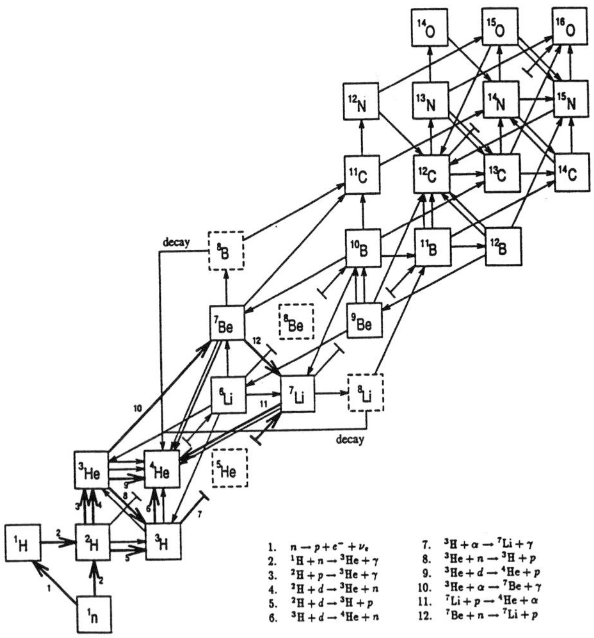

Figure 2: An example of a BBN reaction network (Copi Schramm & Turner 1994). Only lower half of network is important for standard homogeneous BBN.

[50], this abundance of ^{6}Li would not be present unless this star (HD 84937) is a sub-giant. Nevertheless, this observation of ^{6}Li has now been verified by Hobbs and Thorburn [51] and a possible detection in a second Pop II star has been reported. Clearly more work on other stars is needed to firm up the critical issue of Pop II ^{6}Li survival.

An exciting new observation has been reported by Songaila et al. [5] and by Carswell et al. [6] of a possible D/H measurement in a QSO absorption system at $Z = 3.3$. Such a detection would be an extremely important confirmation of the basic BBN argument that deuterium is primordial. However, as pointed out in several papers (e.g. [52]) there is a fair probability that this observation is merely an interloping H cloud with a velocity of 82 km/sec relative to the main cloud which mimics the D absorption. Therefore, the implied D/H of $\sim 2.5 \times 10^{-4}$ should be viewed as an upper limit at the present time. Indeed, there are preliminary unpublished reports from Carswell et al. and from Tytler et al. of measurements in other absorption systems which indicate a value an order

Isotope	Primordial Value	Note	Reference
^{4}He	$Y_0 = 0.232 \pm 0.003^{+0.01}_{-0.005}$	Ex-H-II region	Olive & Steigman [56] Pagel [37] Skillman [57]
D + ^{3}He	$(D + {}^3He)/H\Big\vert_0 \leq 1.1 \times 10^{-4}$	Solar wind with maximum correction for ^{3}He depletion	Geiss [58]
^{2}D	$D/H\Big\vert_0 > 1.4 \times 10^{-5}$	ISM (HST)	Linsky et al. [59]
^{7}Li	$^7Li/H\Big\vert_0 = 1.4 \pm 0.3^{+1.8}_{-0.4} \times 10^{-10}$	Systematic includes possible depletion in stars and spallogenic addition	Spite & Spite [41] Thorburn [60]

Table 1: Current estimates of primordial light element abundances. First $\pm$ is statistical, second is systematic. Limits are extreme of systematic + 2σ statistical.

of magnitude lower.

It should be noted that such a high value would be hard to fit with relatively low ^{3}He values (few $\times 10^{-5}$) observed in the ISM today [53, 54] unless there was significant ^{3}He destruction in the early galaxy [55] (contrary to the current notion that D primarily burns to ^{3}He in stars and that most ^{3}He survives) or the ^{3}He observations are in error. The issue of ^{3}He evolution is now critical and will obviously be an area of continued interest in the near future. If an extragalactic D/H determination can eventually be confirmed in several directions, it would provide the firmest determination of the baryon density and may enable a collapse of the present range for Ω_b to an even narrower band. However, unpublished reports of other observations by Carswell et al. and Tytler et al. of a different QSO system at $Z \sim 3$ appear to set an upper limit on D/H of 6×10^{-5}, which would imply that the higher value was merely a hydrogen interloper.

2.3 The Next Millennium of Primordial Nucleosynthesis

An extrapolation of the exponential growth of papers on this subject for the past 25 years would imply that something like 2000 publications on this subject will appear by the end of the present millennium. Continuing into the next millennium one would estimate that something like 2×10^7 papers will appear in the final decade before the present generation of cosmologists retire (say around 2050). If this trend were to continue there would be over 5×10^{80} papers published in the final decade of the next millennium. Although this absurd number highlights the unreliability of such extrapolation, it nevertheless is an indication that a great deal of activity will continue in this field. The question is then, what will all of these papers be about? The remainder of this primordial nucleosynthesis discussion suggests some possibilities.

§3 Observations and Systematic Errors

Perhaps the most pervasive theme of the primordial nucleosynthesis section of the nucleosynthesis workshop was the focus on potential systematic errors in light element abundance determinations. Comparisons between stellar nucleosynthesis theory and observed abundances usually need only relative abundances of heavy elements so that many systematic errors cancel. Furthermore, most comparisons with stellar nucleosynthesis are content with agreement at the level of a factor of 2. This, in part, has resulted in astronomical spectroscopists not explicitly showing their full range of systematic error with their results. Comments, such as "model atmosphere used," are in the text, but explicit separate error ranges for different model atmosphere assumptions are rarely provided. Unfortunately for BBN, the absolute abundances relative to H are critical, and model atmosphere effects do not cancel!

3.1 Lithium Observations

Lithium observations were discussed by P. Nissen and V. Smith. In observations of lithium one finds the line ratio

Li I/Li II<< 1 which means that (Li I/H) depends upon temperature at the top of the stellar photosphere. One must therefore carefully model the thermal stability, possible non-LTE effects, magnetic heating, cooling by H^- and metals, etc.

The need for an understanding of the systematic uncertainties in model atmospheres was recently brought into focus by the factor of almost two differences in the value of Li/H on the Spite plateau when different atmosphere assumptions were made. While the existence of the plateau itself is not disputed, its absolute value is more uncertain than one might deduce from reading any single lithium observation paper.

Observation of ^{6}Li is made by determining the asymmetry in the λ6707Å line. The problem with such measurements are that they require considerable observing times on large telescopes with high resolution. Also, convection can mimic the line asymmetries and must somehow be distinguished.

Even accepting the apparent ^{6}Li detection, as P. Nissen discussed at Snowmass, the possibility of stellar lithium depletion at a level of a factor of two cannot be ruled out by the ^{6}Li observations. Some Pop I and Pop II depletion models [61, 62, 63] can be consistent with significant lithium destruction for Pop II halo stars observed in the plateau. Indeed, they may even require some depletion to account for the observed dispersion in the data. However, the Pop I depletion model of Faulkner suggests the lack of any need for forcing large depletions on Pop II stars. Note that most systematics do cancel for ^{6}Li/^{7}Li. Clearly, this is an area where improved stellar destruction models and model atmospheres would be of benefit to this important cosmological problem.

There is also the fact that strong lithium lines are observed in AGB stars. However, to understand the contribution of lithium from this source one needs good models for stellar evolution including thermal pulses and envelope burning. This along with a better understanding of possible supernova lithium production mechanisms will be a productive area for future research.

Some proposals for future observations which were discussed were the use of the λ8126Å Li I line, which although weaker involves less blanketing and could provide more reliable abundances. Also, Observations of C,N,O, isotopes and fluorine are excellent probes of the thermal pulses and envelope burning. The main need is for more careful observations and analysis on 4 meter and 8 meter telescopes with fast spectrometers.

3.2 D/H and [D+^{3}He]/H

One of the main constraints on Ω_b is from the abundances of deuterium and ^{3}He. The lower limit to the deuterium abundance sets an upper limit on Ω_b and the upper limit to the sum of deuterium and ^{3}He sets a lower limit to Ω_b. The primordial abundances of these two nuclides is particularly uncertain due to the unknown degree to which they have been destroyed in stars and (in the case of ^{3}He) the possible production in stars.

The status of current deuterium observations were reviewed in talks be J. Linsky, J. Pasachoff, and D. Lubowich. Deuterium is the only light element which one believes is only created in the big bang and destroyed in stars. Thus, there are essentially no contaminating sources for deuterium, only sinks. It is also an isotope of the most abundant element and has the largest isotopic splitting, all of which makes it ideal as a focus for extragalactic study.

Perhaps the light element abundance which is least susceptible to systematics is the D/H value from HST in the ISM. But this only gives a lower limit to the primordial D/H, since D is destroyed in stars. The possibility of new radio work on D/H discussed by Lubowich and by Pasachoff was intriguing. Furthermore, one also has to worry about H interlopers. Naturally, there was considerable discussion about the first reported detections (Songaila et al. [5]; Carswell et al. [6]) of the primordial deuterium Lyman-α absorption line along the line of sight to a high redshift QSO as well as discussion of the recent Lyman-α absorption toward nearby bright stars like Capella and Procyon by Linsky et al [59].

We are in a unique situation at the present time in that the possibility now exists for accurate abundance determinations in material outside the solar system. Previous estimates of the deuterium abundance have relied upon Solar-System determinations which are subject to large corrections for galactic chemical evolution. Indeed, many more such measurements must be made in the coming decades. The interstellar absorption analysis, however, is subject to systematic uncertainties from stellar emission. It also requires an understanding of uncertain chemical evolution effects to determine the primordial value.

The QSO absorption detection towards Q0014+813 [5] may avoid some of the chemical evolution uncertainty, however this detection is subject to the possibility of confusion with a weak interloping H cloud at lower redshift (-82 km s^-1) relative to the strong hydrogen Lyman-α absorption system. Only after a large number of such measurements are made can a reliable lower limit to the primordial deuterium abundance be placed.

There is also considerable future in the determination of abundances for galactic deuterium. The mapping of a galactic deuterium gradient would be an exceedingly useful probe of galactic star formation and evolution.

The helium-3 variations in galactic H-II regions show that ^{3}He destruction in massive stars is probably non-negligible. Thus, it is probably inappropriate to use the H-II region data. However, the observation of ^{3}He/H $\sim$

10^{-3} in planetary nebulae is in good agreement with low mass stellar evolution theory. Therefore, using the Solar System [$^3He + D]/H$ value as determined in the solar wind and in meteorites may be the best value to use.

Many HII regions should be observed to test the depletion of ^{3}He in massive stars. However, such studies must confront the need for long observation time. The development of an EUV high resolution spectrograph for such studies would be exceedingly helpful (cf. Table 3).

3.3 ^4He

The current status of ^{4}He observations and potential systematic errors were discussed in detail by E. Skillman [57]. The random errors in the correlation of helium with metallicity are very small due to multiple exposures, several standard stars, and good linear detectors. In principle, it is possible to obtain line ratios which are accurate to within 2% [57]. The problems are that we need more high quality observations at low [O/H]. The deviations (if any) from the linear regression must be quantified and understood from a galactic chemical evolution stand point [64, 65]. For example it should be understood why the C/N/O ratios in irregulars are not like that of our Galaxy and other spirals. Also, possible contamination from W-R stars in H II regions must be understood. Most importantly, uncertainties in theoretical recombination/cascade calculations must be quantified and reduced.

Based upon the recent analysis [56] of preliminary data from Skillman, the inferred primordial value is $Y_p = 0.232$, but it seems that, in addition to a statistical uncertainty of only ± 0.003, a shift to high values as much as ~ 0.01 and to low by ~ 0.005 by systematics cannot be excluded. This implies an upper limit of 0.245 to the primordial helium abundance. It will be exceedingly important to better quantify these limits in the next millennium.

Given that systematics are so important for light element abundances, we would like to encourage observers to strive toward more accurate abundance determinations and to quote explicitly the possible range on systematics as an additional error in addition to (and separate from) the traditional statistical error. This is now the current practice in particle physics, and since BBN is perhaps approaching a similar level of fundamental impact, we should be similarly cautious in how we quote results. It should also be noted that systematic errors are frequently not gaussian distributed (often they are close to rectangular). Thus, it is inappropriate to add them in quadrature with the statistical errors. Also, as was discussed during the workshop by Walker, Krauss, and Thomas, although complex maximum likelihood calculations of BBN are useful for quantifying abundance constraints, they can also be quite misleading when only gaussian distributions are presumed.

Calculation	D	^{3}He	^{4}He	Li	Be,B
Plasma Spectr.	-	-	X	-	-
Stell. Model Atm.	-	-	-	X	X
Gal. Chem. Evol.	X	X	X	X	X
Stell. Int. Models	-	X	-	X	X
Chem. Models for ISM	X	-	-	-	-

Table 2: Some theoretical developments which would be most helpful for standard big bang nucleosynthesis.

Table 1 summarizes an interpretation of the current light element abundances with our estimate of their uncertainties. The numbers are drawn from the discussions and a recent summary by Copi, Schramm and Turner [1]. The only significant difference with Walker et al [13] is in the opening of the lithium range due to systematic errors that were not appreciated in the observational papers at that time. Also, the new higher helium abundances of Skillman [57] are represented.

The limits shown in Table 1 can be applied to the following fitting formula to obtain a limit on the number of equivalent neutrino species allowed by BBN:

$$N_\nu \lesssim 3.4 + 0.08 \left[\frac{Y_0 - 0.245}{0.001}\right] + \left[\frac{\left.\frac{D+{}^3He}{H}\right|_0 - 1.1\times 10^{-4}}{10^{-4}}\right] \qquad (1)$$

where the subscripts denote the maximum primordial value consistent with observations.

In Tables 2 and 3 we summarize the present and future spectroscopic capabilities along with useful areas for theoretical development from the summary talk by J. Linsky.

§4 Nonstandard Big Bang Nucleosynthesis

As noted above, BBN yields all agree with observations using only one freely adjustable parameter, $\eta = n_b/n_\gamma$, or equivalently, the baryon density, ρ_b (since n_γ is directly related to the temperature of the microwave background.) Thus, BBN can make strong statements regarding ρ_b if the observed light element abundances cannot be fit with any alternative theory. However, a number of alternative big bang nucleosynthesis models have been proposed over the

Wavelength	Telescopes	Date	Δv (km s^{-1})	D	^{3}He	^{4}He	Li	Be,B
Optical	KECK	now	$\lesssim 3$	X	-	X	X	-
	GEMINI	1997?	?	X	-	X	X	-
	VLT (ESO)	1997?	$\lesssim 3$	X	-	X	X	-
	8m, 4m Telescopes	soon	?	X	-	X	X	-
UV + FUV	HST-GHRS(1120-3200 Å)	now	3.5	X	-	-	-	X
	HST-FOS (1150Å- 1 μm)	now	150	-	-	-	-	-
	HST-STIS(1170Å- 1 μm)	1997?	2.5	X	-	-	-	X
	IUE (1170 - 3200 Å)	now?	30	-	-	-	-	-
	IMAPS (900 - 1200 Å)	now	1.5	X	-	-	-	-
	FUSE (900 - 1240 Å)	2000?	10	X	-	-	-	-
EUV	EUVE (80 - 800 Å)	now	500	-	-	-	-	-
	FUSE (450 - 900 Å)	???	10	-	X	-	-	-
Radio	VLA, ARECIBO	now	-	X	X	-	-	-
	GBT	1996?	-	X	X	-	-	-
	New Large Tel. (India)	1995?	-	X	-	-	-	-
	Special Purpose		-	X	X	-	-	-

Table 3: Spectroscopic capabilities and Opportunities (Linsky 1994). An "X" denotes the abundance most affected.

years. Such models can alter the light element abundances as is illustrated in Table 4. At least some of these models may alter the constraints on η or N_ν. And such studies will no doubt continue in the coming decades.

A review of the various nonstandard models is given in [7]. Here we briefly review the alternative that has received the most attention in recent years. It is the study of baryon inhomogeneities which are of sub-horizon scale during the epoch of primordial nucleosynthesis. Such models were initially inspired by the possibility of a first order QCD phase transition [44, 66] although a number of other scenarios for the formation of such inhomogeneities have been proposed [7]. While super horizon-scale baryon inhomogeneity had been considered previously (cf. [67, 17]) and was found to make little difference in the upper limit to the average value of η, in the work of Applegate and Hogan [68] it was appreciated for the first time that sub-horizon baryon inhomogeneous models added the ingredient of variations in the n/p ratio.

The order of this transition as determined from lattice gauge calculations depends sensitively on the number of light quark flavors. The transition is first order for 3 or more light flavors and second order for 2. Because the s-quark mass is so close to the transition temperature it has been difficult to determine the order. At least one recent calculation [69] indicates a clear signature of a first order transition when realistic u, d, s quark masses are included, but others indicate second order or not a phase transition at all. In view of this uncertainty, it has been important to explore the maximal cosmological impact that can occur. This maximal impact requires a relatively strong first-order phase transition if it is to be produced during the cosmological QCD transition. There are, however, some alternative models to produce such inhomogeneities such as electroweak baryogenesis, superconducting cosmic strings, a kaon condensate, or inflation generated isocurvature fluctuations (see [7]).

The preferential diffusion of neutrons versus protons out of the high density regions could lead to big bang nucleosynthesis occurring under conditions with both density inhomogeneities and variable neutron/proton ratios. Initial studies [44, 70] of the impact of baryon inhomogeneities seemed to suggest that $\Omega_b \sim 1$ might be possible, while fitting the observed primordial abundances of ^{4}He, ^{2}D, ^{3}He, but with an overproduction of ^{7}Li. Since ^{7}Li is the most uncertain of the cosmological abundance constraints and has a different observed abundance in Pop I stars versus the, traditionally more primitive Pop II stars, [41, 42, 43] depletion process might be going on to reduce the excess ^{7}Li. However, recent ^{6}Li observation [47] may now constrain the degree to which such depletion should occur to not more than a factor of two.

At first it appeared that if the lithium constraint could be surmounted [66] by the subsequent effect of excess neutrons diffusing back into the high density regions as the nucleosynthesis depletes neutrons and protons in the high density regions. Malaney and Fowler [66] argued that for certain phase transition parameter values (e.g. nucleation site separations ~ 10 m at the time of the transition), this back diffusion could destroy much of the excess lithium.

Also, the work of Alcock et al. [71] showed that the hydrodynamic expansion of the inhomogeneities late in the nucleosynthesis epoch could further deplete the lithium over abundance.

However, Kurki-Suonio et al. [72] Thomas et al. [73], Terasawa and Sato [74, 75] and Mathews et al. [76] have eventually argued that in their detailed diffusion models, that the back diffusion not only affects ^{7}Li, but also the other light nuclei as well. In particular, for $\Omega_b \sim 1$, ^{4}He is overproduced (although it does go to a minimum for similar parameter values as does the lithium).

Added to this has been the seminal work of Jedamzik et al. [77] in which not only the diffusion but also the hydrodynamics is correctly evolved in a one-dimensional model. This work showed that the lithium is not efficiently destroyed during the hydrodynamic expansion of the inhomogeneities.

It is now generally concluded that the upper limit to Ω_b in the baryon inhomogeneous models can not allow for values as large as unity and still satisfy the observed abundance constraints. Nevertheless, inhomogeneous models which are constrained by the abundance limits given in Table 1 do allow slightly higher values of Ω_b, perhaps as large as 0.1 [55] (see Fig. 3). Although several recent papers [72, 73] have concluded that the inhomogeneous models are constrained to agree with the standard homogeneous model limits on Ω_b, these were based upon more restrictive ^{7}Li and ^{4}He abundance limits than is adopted in Table 1.

Figure 3 shows the updated results of Mathews, Kajino and Orito [55] for spherical shell fluctuatiuons of varying diameters l. Contour lines are drawn showing the constraints from different light element abundances. Similar figures can be found in [72, 73, 77, 1], where similar conclusions are reached if allowance is made for the new abundance limits. Notice that the Li, ^{2}D, and even the ^{4}He constraint do not allow $\Omega_b \sim 1$. Note also that with the Pop II ^{7}Li constraint, the results for Ω_b are quite similar to the standard model. However, with the new ^{7}Li ^{4}He and/or ^{2}D to bound the baryon density a slight (factors of < 2) variation in Ω_b is allowed. Allowing for condensed sphere geometry and/or expanding the ^{7}Li abundance upper limit also slightly increases the allowed baryon density [55] to an absolute upper limit of $\Omega_b \sim 0.15$.

Thus, an optimally tuned first order quark-hadron transition is able to slightly alter the basic conclusions of homogeneous BBN regarding Ω_b. (It may also slightly change the N_ν argument.) Indeed, it has recently been demonstrated [78] that it is possible to lower the limit on Ω_b in the inhomogeneous model to a value as low as that of luminous matter and still maintain perfect agreement with observed abundances. This poses the interesting possibility of removing the need to invoke significant contributions from dark baryons on the basis of nucleosynthesis

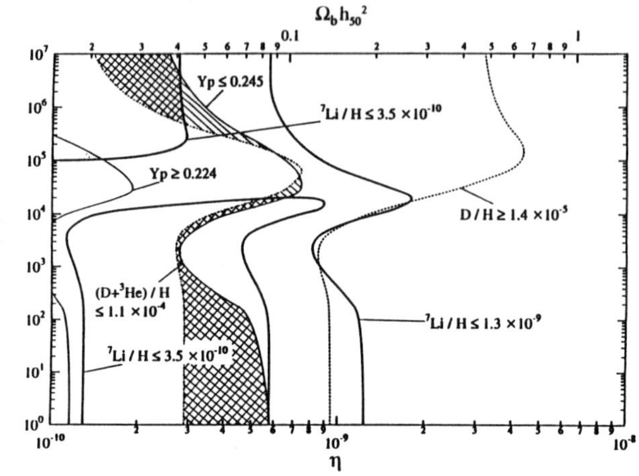

Figure 3: Constraints on fluctuation separation distance and Ω_b (or η) based upon the indicated observed abundance limits for an inhomogeneous nucleosynthesis model with spherical shell fluctuations (Mathews et a. 1994). Note, that even with the presently adopted abundance limit for ^{7}Li it is possible for Ω_b to slightly exceed the limit from standard homogeneous BBN to values as large as 0.1. $\Omega = 1$ is still excluded, however.

(although there are other reasons to believe that they are important).

Clearly, quantifying the degree to which such nonstandard models can affect the conclusions of the standard homogeneous model will be important in future decades. Areas of possible future studies should, among other things, include multi-dimensional simulations of the formation and evolution of inhomogeneities.

§5 Abundance Signatures of Nonstandard Nucleosynthesis

In Table 4 a number of possible alterations of the primordial abundances are noted. One area of future study should be the search for such signatures. These studies could serve to establish or eliminate the viability of alternative big bang scenarios.

The abundance signatures which have received the most attention in recent times are those due to possible baryon inhomogeneities. Inhomogeneous scenarios introduce the possibility of high ^{7}Li or lower ^{4}He [79] (for a given η) as discussed in the previous section. It has also been argued [45, 70] that leakage beyond the mass 5 and 8 instability gaps can occur in neutron rich regions leading to traces of ^{9}Be, ^{10}B, ^{11}B. It has also been shown [80] that for reason-

Effect	Abundances Relative to SBB	Constraints
Baryon Inhomogeneities	$\downarrow\uparrow{}^4$He, $\uparrow{}^7$Li, $\uparrow$ CNO	QCD Physics
Particle Decay	$\uparrow{}^3$He, $\downarrow{}^4$He, $\uparrow{}^6$Li	Properties of new particles
Neutrino Degeneracy	$\downarrow\uparrow{}^4$He	Lepton Charge of Universe
Neutrino Oscillations	$\downarrow\uparrow{}^4$He	Create/Alter Degeneracy
Neutrino Electromagnetics	$\uparrow{}^4$He	Dipole Moment, Charge Radius
Massive Neutrinos	$\uparrow{}^4$He	Neutrino Mass
RH Weak Interactions	$\uparrow{}^4$He	Mass of M_W^R, Coupling Constants
Cosmic Strings	$\uparrow{}^4$He (normal); $\downarrow{}^4$He, $\uparrow{}^7$Li (SCS)	Mass/Length, Currents
Anisotropy	$\uparrow{}^4$He	Shear Forces, Magnetic Fields
Time Varying Constants	$\downarrow\uparrow{}^4$He	Superstrings, Relativity
Primordial Black Holes	$\uparrow{}^3$He, $\downarrow{}^4$He, $\uparrow{}^6$Li	Number Density, Hawking Radiation
Mirror Matter	$\uparrow{}^4$He	Viability, Coupling Constants

Table 4: Effect on abundances and main constraints imposed by various nonstandard models (Malaney & Mathews 1993).

able regions of the parameter space that even r-process elements can be produced. In another recent paper [77] it has also been shown that proton-rich regions could synthesize significant abundances of heavy elements even up to iron without disturbing the agreement with light element abundances.

Thus, detection of nuclei beyond ^{7}Li in primitive objects may be a signature of primordial inhomogeneity. Most of the recent interest in observing primordial abundance anomalies has centered on Be and B, although a number of arguments now exist [73, 74, 75, 81] that the production of these nuclei even in inhomogeneous scenarios may be small.

There has been much recent attention given to observations [82, 83, 84] of beryllium lines extreme Pop II stars. For one very metal poor Pop II star, HD 140283, boron was also observed [85]. These observations showed Be/H and B/H ratios that scaled with metallicity. Thus, contrary to some initial claims, the Be/H and B/H observations do not require cosmological origin, only a scaling with oxygen of the same process that produced Be and B in the Pop I stars. Non-LTE effects in the Be and B determinations [86] do not alter these conclusions, although they may imply a higher B/Be (See [87]).

The process that produced Be and B in Pop I stars (as well as ^{6}Li), is thought to be cosmic ray interactions [88]. For Be and B, such spallation comes from the breakup of heavy nuclei (primarily CNO but also Ne, Mg, Si, S, Ca and Fe) by protons and alphas. As noted by Epstein et al. [89] for lithium one must also include alpha plus alpha fusion reactions. This latter point has been noted in [90, 91] who also emphasized that Be and B spallation production on Pop II abundances would imply a significant enhancement of lithium from alpha-alpha relative to the reduced production of Be and B from depleted heavy nuclei. While the ^{6}Li so produced would be destroyed at the base of the convective zones in most stars observed [92, 61], it did manage to survive in at least one extremely high temperature (thin convection zone) star [47, 51] as anticipated in [92].

Perhaps most critical to any spallation origin is the resultant B/Be ratio. The observed cosmic ray ratio [93] is B/Be $\sim$ 14. The spallation of carbon favors B relative to Be (mass 11 requires only a single nucleon ejected from mass 12), whereas oxygen spallation produces a ratio closer to unity. Thus, cosmic ray observations actually place only an upper limit on the B/Be ratio which one might expect in Pop II cosmic rays (See also [87]. However, of more concern here is the lower limit on B/Be achievable by a spallation process [91]. The resultant limit is B/Be $\simeq$ 7 which means that cosmic ray spallation can yield $7 \lesssim B/Be \lesssim 14$. Furthermore, non-spallation mechanisms such as ν-process [94] in supernovae (see also Dearborn et al. [95]) or a low energy cosmic ray spike [96, 87] can enhance B/Be, but no known process can produce Be without making B. Thus, the lower bound on B/Be is firm, whereas values above 14 would only be indicative of a B enhancing addition.

In standard BBN, the ^{9}Be/H yields are always [97] less than 10^{-14} regardless of η. Thus, homogeneous BBN cannot yield Be/H consistent with the Pop II stellar observations. Even inhomogeneities [73] are not likely to alter these conclusions. Thus, at first glance an observed plateau in Be/H and B/H versus low metalicity might be a problem for BBN. However, recently Yoshii et al. [98] have argued that stellar accretion during galactic evolution can produce such a partial plateau for extremely low metalicities with no cosmological origin for Be and B. This latter point is extremely interesting and deserves further study. Indeed, from the B/Be ratio one should be able to directly distinguish between a primordial or galactic origin for such a plateau. Hence, searches for Be, B, and heavy-element abundance signatures in particular should continue as a valuable area of pursuit in the coming decades.

§6 Limits on Ω_b and Dark Matter Requirements

For the standard homogeneous big bang nucleosynthesis constraints of Table 2, the inferred range of the fraction of the critical density that is in baryons is $0.02 \leq \Omega_b \leq 0.14$ for $0.04 \lesssim h_0 \lesssim 0.7$ [99], where h_0 is the Hubble constant in units of 100 km/sec/Mpc. Even if this constraint can be slightly expanded in nonstandard nucleosynthesis models, it appears at the present time that the universe *cannot be closed with baryonic matter.* (This point was made over twenty years ago [29] and has survived the test of time.) If the universe is truly at its critical density (as preferred by inflation [100, 101]), then nonbaryonic matter is required. This argument has led to what will continue to be one of the major areas of research at the particle-cosmology interface, namely, the search for non-baryonic dark matter.

Another important conclusion regarding the allowed range in baryon density is that it is in very good agreement with the density implied from the dynamics of galaxies, *including their dark halos* [29]. Galaxy dynamics and nucleosynthesis agree at a value for Ω_b which is about 5% of the critical density. The recent MACHO [102] and EROS [102, 103] reports of halo microlensing may well indicate that at least some of the dark baryons are in the form of brown dwarfs in the halo. Note also that if the universe is indeed at its critical density, it requires most nonbaryonic matter not to be associated with galaxies and their halos.

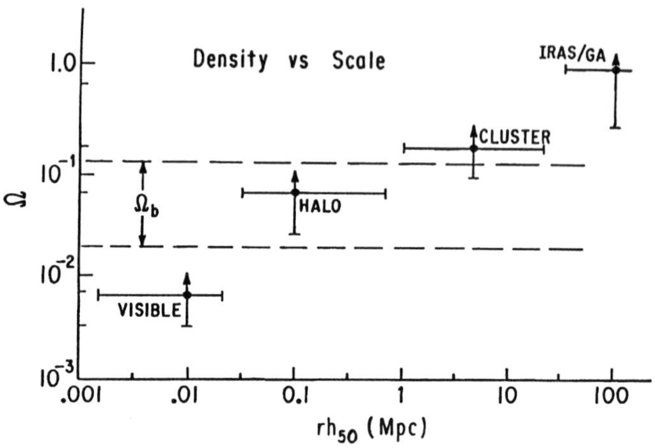

Figure 4: Implied densities versus the scale of the measurements.

The arguments requiring some sort of dark matter fall into separate and quite distinct areas. These arguments are summarized in Figure 4. First, Newtonian mechanics applied to various astronomical systems indicates that galactic halos seem have a mass to light ratio ~ 10.

Note, however, that unless the universe contained large inhomogeneities [78], big bang nucleosynthesis requires [99] that the bulk of the baryons in the universe be dark since $\Omega_{vis} \ll \Omega_b$. Thus, the dark halos could in principle be baryonic [29, 102, 103]. Determinations [104] on scales larger than clusters of galaxies) hint that Ω on those scales is indeed greater than Ω_b (cf. Fig. 4), thus requiring the existence of non-baryonic dark matter. Note that the need for nonbaryonic dark matter is not dependent on the existence of dark galactic halos. This point is occasionally forgotten.

Non-baryonic matter can be divided [71] into two major categories: hot dark matter (HDM) and cold dark matter (CDM). Hot dark matter is relativistic until just before the epoch of galaxy formation. The best example is a low mass neutrinos with $m_\nu \sim 20$ eV ($\Omega_\nu \sim m_\nu(\text{eV})/100h_0^2$). Cold dark matter is moving slowly at the epoch of galaxy formation and can clump on very small scales, whereas

HDM has more difficulty being confined on small scales. Examples of CDM could be massive neutrino-like particles (WIMPS) with masses, $M_x \gtrsim 3GeV$, or the lightest super-symmetric particle which is presumed to be stable and might also have masses of several GeV. Axions, while very light, would also be moving very slowly [105] and thus would clump on small scales. Or, one could also go to non-elementary particle candidates, such as planetary mass black holes or quark nuggets of strange quark matter, possibly produced at the cosmological quark-hadron transition [106, 107]. Another possibility would be any sort of massive topological remnant left over from some early phase transition. Note that CDM would clump in halos, thus requiring the dark baryonic matter to reside between galaxies, whereas HDM would allow baryonic halos. The MACHO [102] and EROS [103] events may favor HDM depending on what fraction of the halos can eventually be explained by the MACHO/EROS events.

While the exact nature of the seeds to stimulate the clumping has yet to be fully determined, COBE's discovery of large scale fluctuations in the microwave background probably shows that some sort of seeds did indeed exist and that the basic scenario is probably on the right track.

Some baryonic dark matter must exist since we know that stellar remnants, dust, rocks, planets, etc. exist. Also, unless a very special class [78] of baryon inhomogeneities existed in the early universe, the lower bound from big bang nucleosynthesis is greater than the upper limits on the amount of visible matter in the universe. However, unless the microlensing objects make up the bulk of the halo, we do not know what form this baryonic dark matter is in. It could be either in condensed objects in the halo, as MACHO and EROS may imply, such as brown dwarfs and jupiters (i.e. low luminosity objects with $\lesssim 0.08 M_\odot$) luminous stars), or in black holes (which at the time of nucleosynthesis would have been baryons). Or, if the baryonic dark matter is not in the halo, it could be in hot intergalactic gas, hot enough not to show absorption lines in the Gunn-Peterson test, but not so hot as to be seen in the x-rays. The exciting report by Jakobsen et al. [108] of a Gunn-Peterson effect observed with HST for He-II at high Z shows that at least some hot IGM exists (and verifies that He seems primordial).

Another possible hiding place for the dark baryons would be failed galaxies, large clumps of baryons that condense gravitationally but did not produce stars. Such clumps are predicted in galaxy formation scenarios that include large amounts of biasing where only some fraction of the clumps shine. Evidence for the presence of a large amount of hot gas is found in clusters of galaxies from ROSAT. In particular, Mushotzky [109] and White et al. [110] have discussed how certain observed rich clusters have $M_{HotGas}/M_{Total} \sim 1/4$, which, with $\Omega_{Cluster} \sim 0.2$ and $\Omega_b \sim 0.05$, would imply no conflict with BBN. Rather, they may imply that clusters are not fair samples of the baryon to non-baryon ratio in the universe. This is a point which deserves clarification in future investigations. Apparent variation [109] in M_{Hot}/M_{Total} for small groups relative to clusters seems to support the point of view that megaparsec scales are not always fair samples. If true, this is difficult to reconcile with cold dark matter models but can be fit with topological defects and HDM or with mixed models.

Hegyi and Olive [111] have argued that dark baryonic halos are unlikely unless they are in the form of low mass objects or of massive black holes. It is worth noting that these loopholes are not that unlikely. The initial mass function for stars with Pop I composition falls off roughly like a power law and there is some sort of lower cut-off near $0.1 M_\odot$. However, we do not know the origin of this mass function and its shape. No current star formation model based on fundamental physics quantitatively predicts it. Predicting and understanding the stellar initial mass function will be one of the important challenges of the next millennium.

Whatever is the origin of this mass function, it is probably related to the metalicity of the materials, since metalicity affects cooling rates, etc. It is not unreasonable to expect the initial mass function that was present in the primordial material which had no heavy elements (only the products of big bang nucleosynthesis) would be peaked either much higher than the present mass function or much lower—higher if the lower cooling from metals resulted in larger clumps, or lower if some sort of rapid cooling processes ("cooling flows") were set up during the initial star formation epoch, as seems to be the case in some primitive galaxies.

In either case, moving the IMF either higher or lower produces the bulk of the stellar population in either brown dwarfs and jupiters or in massive black holes. Thus, the most likely scenarios indeed involve a first generation of condensed objects in a form of dark baryonic matter that could make up the halos, and could explain why there is an interesting coincidence between the implied mass in halos and the implied amount of baryonic material. However, it should also be remembered that a consequence of this scenario is to have the condensation of the objects occur prior to the formation of the disk. If the first large objects to form are less than galactic mass, as many scenarios imply, then mergers are necessary for eventual galaxy size objects. Mergers stimulate star formation while putting early objects into halos rather than disks. Mathews and Schramm [112] have recently developed a galactic evolution model which does just that and gives a reasonable scenario for chemical evolution. Thus, it is possible for the halos to be in the form of dark baryons. The microlensing reported by MACHO and EROS is not a problem but may tell us something about early star formation. All of this remains to be explored in the next millennium.

§7 Stellar Nucleosynthesis

Stellar and supernova nucleosynthesis is the key to understanding the properties of stellar interiors and the chemical and dynamical evolution of galaxies. In contrast to big-bang nucleosynthesis, stellar and supernova nucleosynthesis is an area where there as an abundance of good observational data, but a need for numerous theoretical refinements in the next millennium.

Summary talks on stellar and supernova nucleosynthesis were presented by T. Weaver, F. Timmes, M. Aufderheide, W. Hillebrandt, J. Wilson, and D. Swesty. One essential theme from all of these talks was the need for theoretical refinements of the simulation models as well as the input physics. In a sentence it could be summarized as a need for three-dimensional, high-resolution, hydrodynamics and radiation transport calculations which are accurate over a large dynamic range of time scales and length scales. There is also a need for observational data and ground based experiments as discussed below.

§8 The Stellar Nucleosynthesis in the Present Millennium

Unlike primordial nucleosynthesis for which essentially no papers were published prior to the end of the present millennium, there is a history of excitement about stellar evolution and supernovae which has extended throughout the present millennium and before (in a sense) which gives us perhaps some glimpse as to what to expect in the next millennium.

The first excitement about supernovae in the present millennium was probably the observation of the Crab supernova which was recorded by Chinese astronomers in 1054 AD. Unfortunately this is probably the worst recorded incidence in which stellar evolution theory lagged behind observation (and probably the worst incidence of publish or perish as well) because the astronomers lost their head for failing to predict the supernova. Nevertheless, we count this as the first paper on supernovae. The next excitement probably is attached the Tycho and Kepler supernova remnants which were observed around the middle of the millennium. We count this as two papers.

Serious papers on stellar evolution and supernova theory began later in the millennium around the 19th century and exponentiated in the latter half of the 20th century along with the rest of science. There is a distinguishing characteristic of stellar evolution and supernova papers. In addition to the usual background of papers, an explosion of papers on the subject occurs after each observed nearby supernova. The modern example of that is SN1987A which produced a burst of some $\sim 10^3$ papers within a year or two after the event and produced an overall increase in the activity of the field.

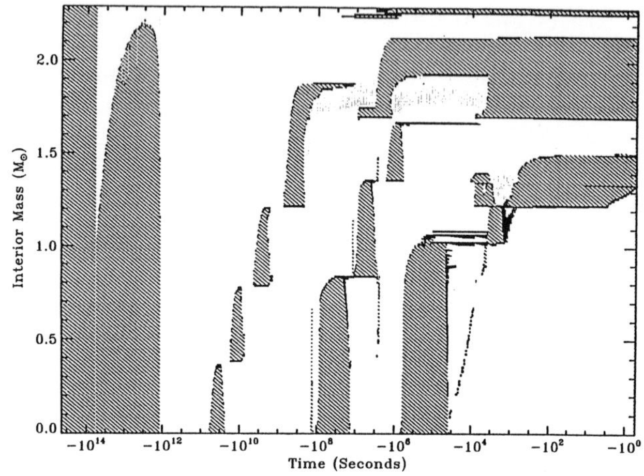

Figure 5: History of thermonuclear burning of a massive star. Shaded regions highlight the occurrence of convective stages.

Using the past as a guide one would expect some 10^6 papers to appear after the next local supernova which will occur sometime in the middle of the next millennium. Subsequently something like 10^9 papers to be written on the supernova which will occur near the end of the next millennium.

The question is then, what will all of these papers be about? The answer was, perhaps, hinted at in the consistent theme of the summary papers presented at the conference.

§9 Stellar Yields, Precollapse, and Galactic Evolution

Discussion on the status of stellar evolution calculations was led by T. Weaver [113]. Figure 5 shows a time history of the core of a massive star. The shaded regions denote the onset of convection. Indeed, convection remains as one of the main uncertainties in such calculations, along with effects of rotation and uncertainties in some key reactions (particularly the $^{12}C(\alpha,\gamma)^{16}O$ rate) which are discussed below. Clearly the future calculations must be ultimately based upon 3-dimensional coupled nucleosynthesis, hydrodynamics, and radiation transport with sufficient resolution to resolve the convective flows and their interaction

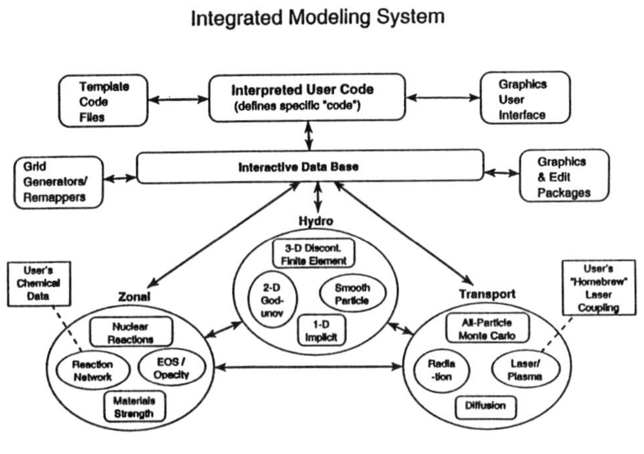

Figure 6: Schematic block diagram of a proposed integrated modeling system by T. Weaver.

with the radiative regimes. Indeed, convective overshoot is probably the largest remaining systematic error in such calculations. Weaver introduced a concept of integrated astrophysical modeling which he and his group are developing. This involves a modular construction of the various physics packages which can be adapted to each physical situation. Figure 6 displays a schematic view of this concept which includes a user options for various hydro and transport packages which can be integrated for any user-specified application.

F. Timmes briefly discused galactic chemical evolution [114] and showed the the importance of good stellar yields as input to such calculations. This discussion also highlights the importance of future development of multidimensional hydrodynamic/stellar orbit calculations for the study of galactic chemical dynamics such as is currently being proneered by . number of groups, e.g. [115]. This is another area where high resolution multi dimensional hydrodynamics and radiation transport will play a key role in the scientific advances of the next millennium. In this case the problem is made even more difficult be the necessity of evolving the stellar component as well.

M. Aufderheide discussed [116, 117] the importance of accurate stellar input physics, particularly the stellar weak reaction rates. Such rates play a key role in the final stages of precollapse as the core begins to neutronize. The quantitative understanding of such rates, however will require a combination of experiments using intermediate-energy radioactive ion beams to measure ground-state transitions, and large-basis shell model calculations to determine the excited state contribution.

§10 Supernovae

In supernovae, a number of needs for improved computational abilities were pointed out. W. Hillebrandt discussed the needs for better modeling of turbulence down to the smallest scales [118]. Indeed, the Reynolds number in the heated supernova material is high implying turbulent behavior. However in calculations the numerical viscosity is usually too high to see the "weather" on small scales. He discussed the example of neutrino driven convection in supernovae which shows no turbulence in 3D but shows turbulence cascading to smaller scales in 3D simulations. Similarly, nuclear flames in type Ia supernovae, and rotation induced shear and circulation can require a range of distance scales from the size of the core down to the 0.01 mm scale. Clearly, achieving such resolution is a major challenge for the next millennium.

J. Wilson discussed the supernova mechanism and the delicate role played by neutrino convection and transport both below the neutrinosphere and above it [119, 120, 121]. It is necessary for about 1% of the 10^{53} erg emitted in neutrinos to deposit their energy above the core in order for the star to explode. Most of this energy is deposited by ν or $\bar{\nu}$ captures on electrons with some contribution from $\nu - e$ scattering as well. Also there is a negative entropy gradient above the core which is unstable to convection. This convection probably has an important influence on the nucleosynthesis just above the core [122] as well as the efficiency of the explosion. The study of late time nucleosynthesis in the hot bubble above the nascent neutron star is currently a topic of considerable interest [122, 123]. Among other things, this may provide the most stringent constraint on the mixing between electron and tau neutrinos [124].

In the interior, convection can boost the neutrino luminosity and bring energy to the surface more efficiently. There is also a question of the development of neutron fingers which can enhance the neutrino energy deposition.

It is clear that both above the neutrino sphere and below it, high resolution 3-dimensional calculations are necessary before a definitive understanding of the supernova mechanism is possible.

A final discussion of the range of time scales involved in the supernova problem was presented by D. Swesty. There are several important time scales which enter the problem. These span 8 orders of magnitude. These range from the Courant stability time (10^{-7} to 10^{-6} sec), the dynamical accretion shock time (10^{-5} to 10^{-4} sec), the timescale for collapse and bounce of the core (0.1 to 0.2 sec), and the neutrino diffusion time (0.1 to 20 sec). Clearly, this large dynamic range is a challenging hydrodynamics problem, particularly given that such calculations are also to be performed in a relativistic, radiating, reactive, and self gravitating imperfect fluid.

This range of time scales can imply an accrued numerical error and because of this there is a lack of understanding of the accuracy of the codes and calculations that have been done. There are also needs for improved microphysics (EOS; neutrino scattering). The field would be much benefited if standard comparison test problems could be run for the various existing and future codes. Swesty is currently developing an implicit code which may be best suited to describing the various time scales in the problem.

Other problems which Swesty pointed out are the need for careful energy and lepton conservation, the elimination of explicit and fixed mesh codes, improved spatial and energy resolution, and the elimination of operator splitting, i.e. the hydrodynamics, radiation transport, and energy equations should be directly coupled. Also, the use of approximate Fokker-Planck solutions to the neutrino diffusion should be eliminated and replaced with direct solutions of the Boltzmann equation. One should of course also consider MHD effects which may be important for a number of aspects [125].

Clearly, there is much theoretical work which will absorb our attention well into the next millennium. There is also a possibility for significant observational and experimental advances.

§11 Nuclear Physics Input

The current prospects for improving the experimental nuclear input physics were reviewed by M. Gai and P. Parker and G. Mathews. A number of improvements of existing nuclear data demand further investigation. Particularly $^{12}C(\alpha,\gamma)^{16}O$ reaction must be clarified including the E2 contribution. This may be studied by coulomb excitation and dissociation. Coulomb dissociation may also provide further insight into the mass 7 producing reactions. It is also important to note that ^{7}Li (and ^{7}Be) producing reactions continue to be the most uncertain in standard homogeneous BBN, and it is hoped the new studies will be carried out. It is also important to note that ^{7}Li (and ^{7}Be) reactions continue to be the most uncertain in BBN and it is hoped that new studies will be carried out. Recent results along this line [126] offer promise.

Two potential new growth areas are identified for the coming millennium. One is the development and use of radioactive nuclear beams which will provide information on nuclear reaction of material away from stability. Such information is important for clarifying nucleosynthesis in inhomogeneous big bang scenarios and supernovae.

Another potential growth area could be the development of confined plasma environments such as Tokomak or inertial confinement fusion as in the proposed National Ignition Facility. In such environments one can hope to explore nuclear reaction rates in the same plasma environments in which they occur in stars, possibly eliminating the uncertainties introduced by plasma screening corrections and extrapolation of the astrophysical S factor. Such environments may more easily also provide measurements of opacity, equation-of-state, and turbulence instabilities which can serve to test and normalize the existing stellar evolution codes [cf. Fig. 6]. It is thus expected that experimental nuclear investigations may continue to be a thriving and productive enterprise well into the next millennium.

§12 Conclusion

Nucleosynthesis continues to be an active topic with major new developments now being reported from astronomical spectroscopists and experimental nuclear physicists as well as theorists. The results currently being obtained (and about to be obtained in the coming decades) have critical bearing on major issues in cosmology and astrophysics now and in the coming millennium.

§13 Acknowledgments

Work performed in part under the auspices of the U.S. Department of Energy at Lawrence Livermore National Laboratory under contract No. W-7405-ENG-48 and DOE Nuclear Theory Grant SF-ENG-48. Work supported in part at the University of Chicago by NSF grant AST 90-22629, DOE grant DEF02-91-ER40606, and NASA grant 1231, and by NASA through grant NAGW 2381 at Fermilab.

§References

[1] C. Copi D. Schramm and M. Turner. *Science*, submitted, 1994.

[2] L.M. Krauss and P. Romanelli. *Astrophysical Journal*, 358:47, 1990.

[3] P. Kernan and L. M. Krauss. *Physical Review Letters*, 72:3309, 1994.

[4] K. Olive D. N. Schramm G. Steigman and T. Walker. *Physics Letters B*, 238:454, 1990.

[5] A. Songailia L.L. Cowie C.J. Hogan and M. Rugers. *Nature*, 368:599, 1994.

[6] R.F. Carswell et al. *MNRAS*, in press, 1994.

[7] R. A. Malaney and G. J. Mathews. *Phys. Rep.*, 229:145, 1993.

[8] C. Hayashi. *Prog. Theor. Phys.*, 5:224, 1950.

[9] R. A. Alpher J. W. Follin and R. C. Herman. *Physical Review*, 92:1347, 1953.

[10] R. Taylor and F. Hoyle. *Nature*, 203:1108, 1964.

[11] P.J.E. Peebles. *Physical Review Letters*, 16:410, 1966.

[12] R. Wagoner W.A. Fowler and F. Hoyle. *Astrophysical Journal*, 148:3, 1967.

[13] T. Walker G. Steigman D. N. Schramm K. Olive and H.S. Kang. *Astrophysical Journal*, 376:51, 1992.

[14] D. N. Schramm and R.V. Wagoner. *Ann. Rev. of Nuc. Sci.*, 27:37, 1977.

[15] K. Olive D. N. Schramm G. Steigman M. Turner and J. Yang. *Astrophysical Journal*, 246:557, 1981.

[16] A. Boesgaard and G. Steigman. *Ann. Rev. of Astron. and Astrophys.*, 23:319, 1985.

[17] J. Yang M. Turner G. Steigman D. N. Schramm K. Olive. *Astrophysical Journal*, 281:493, 1984.

[18] L. Kawano D. N. Schramm G. and Steigman. *Astrophysical Journal*, 327:750, 1988.

[19] M. S. Smith L. H. Kawano and R. A. Malaney. *Ap. J. Suppl.*, 85:219, 1993.

[20] J. Truran. *Doctoral Thesis, Yale University (1965).*

[21] W. Mampe P. Ageron C. Bates J.M. Pendlebury and A. Steyerl. *Physical Review Letters*, 63A:593, 1989.

[22] W.A. Fowler J. Greenstein and F. Hoyle. *Geophys. J.R.A.S.*, 6:6, 1962.

[23] C. Ryter H. Reeves E. Gradstajn and J. Audouze. *Astronomy and Astrophysics*, 8:389, 1970.

[24] J. Geiss and H. Reeves. *Astronomy and Astrophysics*, 18:126, 1971.

[25] D. Black. *Nature*, 234:148, 1971.

[26] J. Rogerson and D. York. *Astrophys. J. Lett.*, 186:L95, 1973.

[27] H. Reeves J. Audouze W. A. Fowler and D. N. Schramm. *Astrophysical Journal*, 179:909, 1973.

[28] R. Epstein J. Lattimer and D. N. Schramm. *Nature*, 263:198, 1976.

[29] J.R. Gott III J. Gunn D. N. Schramm and B.M. Tinsley. *Astrophysical Journal*, 194:543, 1974.

[30] R.T. Rood G. Steigman and B. M. Tinsley. *Astrophys. J. Lett*, 207:L57, 1976.

[31] J. Yang D. N. Schramm G. Steigman and R. T. Rood. *Astrophysical Journal*, 227:697, 1979.

[32] R. Wilson R.T. Rood and T. Bania. *in Proc. of the ESO Workshop on Primordial Healing, eds. P. Shaver and D. Knuth, (European Southern Observatory, Garching, 1983).*

[33] G. Steigman D. N. Schramm and J. Gunn. *Physics Letters B*, 66:202, 1977.

[34] V.F. Schvartzman. *JETP Letters*, 9:184, 1969.

[35] P. J. E. Peebles. *Physical Cosmology.* Princeton University Press, 1971.

[36] D. N. Schramm and L. Kawano. *Nuc. Inst. and Methods A*, 284:84, 1989.

[37] B. Pagel. in *proc. of 1989 rencontres de moriond* (1990).

[38] ALEPH L3 OPAL DELPHI results. *1993 Lepton-Photon meeting at Ithaca, NY (1993).*

[39] R.T. Rood T. Bania and J. Wilson. *Nature*, 355:618, 1992.

[40] D.S. Balser et al. submitted, 1994.

[41] J. Spite and M. Spite. *Astronomy and Astrophysics*, 115:357, 1982.

[42] R. Rebolo P. Molaro and J.Beckman. *Astronomy and Astrophysics*, 192:192, 1988.

[43] L. Hobbs and C. Pilachowski. *Astrophys. J. Lett.*, 326:L23, 1988.

[44] R. Scherrer J. Applegate and C. Hogan. *Physical Review D*, 35:1151, 1987.

[45] R. Boyd and T. Kajino. *Astrophysical Journal*, 359:267, 1990.

[46] E. Witten. *Physical Review Letters*, 30:272, 1984.

[47] V.V . Smith D. L. Lambert and P. E. Nissen. *Astrophysical Journal*, 408:262, 1992.

[48] K. Olive and D. N. Schramm. *Nature*, 360:434, 1992.

[49] G. Steigman B. Fields K. Olive D. N. Schramm and T. Walker. *Astrophys. J. Lett.*, 415:L35, 1993.

[50] B. Chaboyer. *Astrophys. J. Lett.*, 432:L47, 1994.

[51] L. Hobbs and J. Thorburn. *Astrophys. J. Lett.*, 428:L25, 1994.

[52] G. Steigman. *MNRAS*, 269:L53, 1994.

[53] E. Vangioni-Flam K. Olive and N. Prantzos. *Astrophysical Journal*, 427:618, 1994.

[54] M. Edmunds. *MNRAS*, submitted, 1994.

[55] G. J. Mathews T. Kajino and M. Orito. *Astrophysical Journal*, submitted, 1994.

[56] K. Olive and G. Steigman. *U. Minnesota preprint*, UMN-TH=1230/94, 1994.

[57] E. Skillman et al. *Astrophysical Journal*, 431:172, 1994.

[58] J. Geiss. in *origin and evolution of the elements*, eds. n. prantzos et al. (cambridge univ. press, cambridge, 1993), p. 89.

[59] J. Linsky et al. *Astrophysical Journal*, 402:694, 1993.

[60] J. Thorburn. *Astrophysical Journal*, 421:318, 1994.

[61] M. Pinsonneault C. P. Deliyannis and P. Demarque. *Astrophys. J. Suppl.*, 78:179, 1992.

[62] B. Chaboyer and P. Demarque. *Astrophysical Journal*, 433:510, 1994.

[63] C. P. Deliyannis et al. *Astrophys. J. Lett.*, 434:L81, 1994.

[64] M. Balbes et al. *Astrophysical Journal*, 418:229, 1993.

[65] B. Pagel. *Phys. Rep.*, 227:51, 1993.

[66] W.A. Fowler and R. Malaney. *Astrophysical Journal*, 333:14, 1988.

[67] R. V. Wagoner. *Astrophysical Journal*, 179:343, 1973.

[68] J. H. Applegate and C. J. Hogan. *Physical Review D*, 30:3037, 1985.

[69] Y. Iwasaki et al. *Lattice 94*, in press, 1994.

[70] J. Applegate C. Hogan and R. Scherrer. *Astrophysical Journal*, 329:572, 1988.

[71] R. Bond and A. Szalay. *in Proc. Texas Relativistic Astrophysical Symposium, Austin, Texas*, 1992.

[72] H. Kurki-Suonio R. Matzner K. Olive and D. N. Schramm. *Astrophysical Journal*, 353:406, 1990.

[73] D. Thomas D. N. Schramm K. Olive G. J. Mathews B. Meyer and B. Fields. *Astrophysical Journal*, 430:291, 1994.

[74] K. Sato and N. Tarasawa. *Astrophysical Journal*, 361:L47, 1990.

[75] K. Sato and N. Tarasawa. *in The Birth and Early Evolution of Our Universe: Proceedings of Nobel Symposium 79 ed. J. S. Nilsson B. Gustafson and B.S. Skagerstam.* World Scientific: Singapore, 1991.

[76] B. S. Meyer C. R. Alcock G. J. Mathews and G. M. Fuller. *Physical Review D*, 43:1079, 1991.

[77] K. Jedamzik G.M. Fuller G.J. Mathews and T. Kajino. *Astrophysical Journal*, 422:423, 1994.

[78] K. Jedamzik G. J. Mathews and G. M. Fuller. *Astrophysical Journal*, in press, 1994.

[79] G. J. Mathews D. N. Schramm and B. S. Meyer. *Astrophysical Journal*, 404:467, 1993.

[80] T. Rauscher et al. *Astrophysical Journal*, 429:499, 1994.

[81] K. Jedamzik G. M. Fuller and G. J. Mathews. *Astrophysical Journal*, 423:50, 1994.

[82] G. Gilmore B. Edvardsson and P. Nissen. *Astrophysical Journal*, 378:17, 1991.

[83] S. Ryan M. Bessel R. Sutherland and J. Norris. *Astrophys. J. Lett.*, 348:L57, 1990.

[84] S. Ryan J.E. Norris M.S. Bessell and C.P. Deliyannis. *Astrophysical Journal*, 388:184, 1992.

[85] D. Duncan D. Lambert and D. Lemke. *Astrophysical Journal*, 401:584, 1992.

[86] D. Kiselman. *Astronomy and Astrophysics*, in press, 1994.

[87] B. Fields K. Olive and D. N. Schramm. *Astrophysical Journal*, in press, 1994.

[88] H. Reeves W. A. Fowler F. and Hoyle. *Nature*, 226:727, 1970.

[89] R. Epstein W. D. Arnett D. N. and Schramm. *Astrophys. J. Lett.*, 190:L13, 1974.

[90] G. Steigman and T. Walker. *Astrophys. J. Lett.*, 385:L13, 1992.

[91] T. Walker G. Steigman D. N. Schramm K. Olive and B. Fields. *Astrophysical Journal*, 413:562, 1993.

[92] L. Brown and D. N. Schramm. *Astrophys. J. Lett.*, 329:L103, 1988.

[93] J. Simpson. *Ann. Rev. Nuc. and Part. Sci.*, 33:323, 1983.

[94] S. E. Woosley D. H. Hartmann R. D. Hoffman and W. C. Haxton. *Astrophysical Journal*, 356:272, 1990.

[95] D. Dearborn D. N. Schramm G. Steigman and J. Truran. *Astrophysical Journal*, 347:455, 1989.

[96] M. Meneguzzi J. Audouze and H. Reeves. *Astronomy and Astrophysics*, 15:337, 1971.

[97] D. Thomas D. N. Schramm K. Olive and B. Fields. *Astrophysical Journal*, 406:569, 1993.

[98] Y. Yoshii G. J. Mathews and T. Kajino. *Astrophysical Journal*, Submitted, 1994.

[99] K. Freese and D. N. Schramm. *Nuclear Physics B*, 233:167, 1984.

[100] A. Guth. *Physical Review D*, 23:347, 1981.

[101] A. Linde. *Particle Physics and Inflationary Cosmology*. Harwood Academic Publishers, N.Y., 1990.

[102] C. Alcock et al. *Nature*, 365:621, 1993.

[103] E. Aubourg et al. *Nature*, 365:623, 1993.

[104] C. Fisher. *Ph.D. Thesis, University of California at Berkeley, and references therein*, 1992.

[105] M. Turner F. Wilczek and A. A. Zee. *Physics Letters B*, 125:35, 519, 1993.

[106] M. Crawford and D. N. Schramm. *Nature*, 298:538, 1982.

[107] C. Alcock and A. Olinto. *Ann. Rev. Nuc. Part. Phys.*, 38:161, 1988.

[108] P. Jakobsen A. Boksenberg J.M. Deharveng P. Greenfield R. Jedrzewski and F. Paresce. *Nature*, in press, 1994.

[109] R. Mushotsky. *Annals of the N.Y. Academy of Sciences*, 688:184, 1993.

[110] S.D.M. White J.F. Navarro A.E. Evrard and C.S. Frenck. *Nature*, 366:261, 1993.

[111] D. Hegyi and K. Olive. *Astrophysical Journal*, 303:56, 1986.

[112] G. Mathews and D. N. Schramm. *Astrophysical Journal*, 404:468, 1993.

[113] S. E. Woosley and T. A. Weaver. *Phys. Rep.*, 227:65, 1993.

[114] F. Timmes et al. *Astrophysical Journal*, submitted, 1994.

[115] A. Burkert J. W. Truran and G. Hensler. *Astrophysical Journal*, 391:651, 1992.

[116] M. B. Aufderheide et al. *Phys. Rev. C*, 48:1677, 1993.

[117] M. B. Aufderheide et al. *Astrophysical Journal*, 424:257, 1994.

[118] J. C. Niemeyer and W. Hillebrandt. *Astrophysical Journal*, submitted, 1994.

[119] J. R. Wilson and R. W. Mayle. *Phys. Rep.*, 227:97, 1993.

[120] D. S. Miller J. R. Wilson and R. W. Mayle. *Astrophysical Journal*, 415:278, 1993.

[121] M. Herant et al. *Astrophysical Journal*, 435:339, 1994.

[122] S. E. Woosley et al. *Astrophysical Journal*, 433:229, 1994.

[123] B. S. Meyer et al. *Astrophysical Journal*, 399:656, 1992.

[124] Y. Z. Qian et al. *Physical Review Letters*, 71:1965, 1993.

[125] D. Kossl E. Muller and W. Hillebrandt. *Astronomy and Astrophysics*, 229:378, 397, 401, 1990.

[126] C. R. Brune R. W. Kavanagh and C. Rolfs. *Physical Review*, C:Submitted, 1994.

Baryogenesis

Andrew G. Cohen

Department of Physics, Boston University, Boston MA 02215, USA

Robert H. Brandenberger

Physics Department, Brown University, Providence RI 02912, USA

David B. Kaplan

Institute for Nuclear Theory, NK-12, University of Washington, Seattle WA 98195, USA

Rabindra N. Mohapatra

University of Maryland, College Park MD 20742, USA

Scott Thomas

Stanford Linear Accelerator Center, Stanford University, Stanford CA 94309

We review current ideas for the dynamical origins of the cosmic baryon excess. Emphasis is placed on models which involve electroweak physics.

§1 Introduction

1.1 Preliminaries

One of the most striking cosmological observations we can make today is the apparent preponderance of matter (in the form of nucleons carrying positive baryon number) over anti-matter (in the form of anti-nucleons carrying negative baryon number). If the Universe at high temperatures were to have equal numbers of baryons and anti-baryons, then the large cross section for nucleon-antinucleon annihilation would lead to a relic baryon abundance today approximately ten orders of magnitude smaller than what is observed. The quantitative expression of this matter anti-matter asymmetry is the *baryon number density divided by the photon entropy* today, η, whose value is known by direct observation to be approximately 10^{-10}, to within an order of magnitude or so. In fact if nucleosynthesis occurs according to standard ideas, this number is bounded by a range substantially less than an order of magnitude, and in addition the baryon asymmetry must have been in place at all temperatures below a few MeV.[1]

In order to explain this baryon asymmetry, we might postulate a particular initial condition at high temperatures, carefully tuned so that after the era of nucleon-antinucleon annihilation we are left with the observed relic abundance. This however has two deficiencies: firstly, we would like to avoid the unpleasant introduction of *ad hoc*, finely tuned initial conditions imposed at the origin of our Universe (avoidance of such fine tuning has become a common theme in both cosmology and particle physics); secondly, our current ideas of inflationary cosmology do not allow any such initial condition to survive an inflationary epoch (indeed one of the attractive features of inflation is its ability to remove dependence on initial conditions). For these reasons most current approaches focus on a *dynamical* origin for the baryon asymmetry—the evolution of the Universe from early times where the number of baryons is equal to the number of anti-baryons, through an epoch where baryon number is created, to late times where η has its relatively large value and nucleosynthesis can proceed. This dynamical evolution of the baryon asymmetry is referred to as baryogenesis.

1.2 Constraints and Models

In the late 1960s Sakharov realized that the existence of baryogenesis would have striking implications for theories

[1] For an up to date discussion of the constraints nucleosynthesis places on η consult the nucleosynthesis section in this volume.

of fundamental particles and their interactions. These observations have become know as the laws of baryogenesis, and can be easily summarized:

- Non-equilibrium: Strict thermal equilibrium and the invariance of a Hamiltonian under the discrete symmetry CPT guarantees zero baryon number for the Universe. Since our current understanding of relativistic field theory suggests that CPT is always a good symmetry, a departure from strict thermal equilibrium is necessary to accommodate a baryon asymmetry.
- Baryon Violation: Clearly if there is an epoch where baryon number is created than baryon number cannot be an exactly conserved quantum number. To date no violation of this quantum number has been observed.
- CP Violation: The discrete symmetry CP relates the properties of baryons to those of anti-baryons. If the Universe is to favor the production of nucleons with positive baryon number, the interactions which cause this production must violate this symmetry. CP violation has been seen in nature; the standard model of weak interactions incorporates this CP violation in the form of a phase in the KM matrix of quark mixing angles.

The models of baryogenesis that have developed since Sakharov's initial work can be loosely divided according to the source of the non-equilibrium physics (although the boundary is not completely sharp):

Phase Transitions: The thermodynamics of the early Universe allows for the possibility of phase transitions. Since such transitions may involve substantial departures from equilibrium they can also satisfy the first of our conditions. Most models of "electroweak baryogenesis" are of this type.

Relic Decays: As the temperature drops below a particle mass, annihilation or direct decay of this particle may not be rapid enough to keep the particle abundance at its equilibrium value; consequently we may have substantially greater energy in the form of such particles than thermal equilibrium would allow. Subsequent evolution and decay of such a particle can be far from equilibrium, and satisfy the first of our laws. Grand Unified theories of baryogenesis are of this type.

Good reviews of both categories exist [1][2]; due to space limitations we describe only one representative from each class: Electroweak Baryogenesis for the first case, and GUT Baryogenesis for the second.

§2 Electroweak Baryogenesis

2.1 General Features

In the standard model of weak interactions we currently live in a phase where the weak gauge symmetry is spontaneously broken, much like the interior of a superconductor (where the *electromagnetic* gauge symmetry is spontaneously broken). Although the source of this weak symmetry breaking is currently not understood, it is likely that at the high temperatures which existed in the early history of the Universe this symmetry was unbroken.[2] If this is the case, then the weak interactions underwent a phase transition at a temperature of approximately 100 GeV during the early evolution of the Universe. If this phase transition is not second order, it may provide the source of non-equilibrium physics necessary for baryogenesis to occur.

In addition to the possibility of a phase transition, the standard model of weak interactions also contains a source of baryon violation. Although the standard model does not contain baryon violation at the classical level, this symmetry is violated at the quantum level through an "anomaly". The baryon violating effects of this anomaly would appear through quantum barrier penetration phenomena; the barrier is expected to be such that the probability of a baryon violating event at temperatures much below the phase transition temperature is negligibly small. However Kuzmin, Rubakov and Shaposhnikov [3] have suggested that at high temperatures barrier penetration is not necessary for baryon violation to occur, and thermal activation across the top of the barrier is sufficient. This allows us to trade a WKB suppression factor for a Boltzmann factor. In fact the barrier itself disappears in the unbroken phase of the weak interactions. Thus at the weak phase transition baryon violation may be occurring rapidly and in a non-equilibrium way, allowing us to satisfy the first and second of Sakharov's requirements.

The third requirement, CP violation, is somewhat problematic. The standard model of weak interactions is known to violate CP. This would suggest a very economical picture of baryogenesis, in which all of Sakharov's ingredients are found in the weak interactions, and which come together at the weak phase transition to produce the baryon asymmetry we observe. Unfortunately naïve estimates suggest that the CP violation in the standard model (through the Kobayashi-Maskawa matrix) is too small to account for the observed value of η, and consequently most models of electroweak baryogenesis incorporate new sources of CP violation. After discussing models

[2] This is not inevitable. There exist models of symmetry breaking which do not have symmetry restoration at high temperatures. No such model has been proposed for electroweak symmetry breaking, however.

with new CP violation we will return to the subject of standard model KM CP violation to see where we stand.

Since the details of electroweak symmetry breaking are not understood, neither are the details of the electroweak phase transition. If this transition is of the classic first order type, it will proceed by the nucleation of bubbles of broken phase which expand into the sea of unbroken phase, eventually converting the Universe to the form in which we live. Of course the phase transition may be much more complicated than this, and without a detailed understanding of electroweak symmetry breaking we must consider alternatives. We can divide electroweak baryogenesis models loosely into two categories: those that involve a standard first order transition which proceeds through bubble nucleation, and in which baryogenesis arises through the interaction of quarks and leptons with these expanding bubbles; and "topological defect" mediated baryogenesis in which topologically non-trivial field configurations play a significant role in baryon production.

2.2 Bubbles and Electroweak Baryogenesis

These mechanisms are based on the observation that if interaction with the Higgs field during the weak phase transition produces an asymmetry in the density of $SU(2)$ doublet fermions (that is, an asymmetry in quantum numbers that are violated by *non-anomalous* interactions) near the surface of an expanding bubble wall, that asymmetry will in turn bias anomalous weak processes to produce a net baryon number [4][5]. For particle distributions near thermal equilibrium the strength of the bias is computable in terms of the change of the free energy during an anomalous weak interaction, ΔF, which is simply related to the local density of $SU(2)$ fermion doublets. Baryon production is then given by the master equation

$$\frac{dn_B}{dt} = -3\Gamma_a \frac{\Delta F}{T} = -9\frac{\Gamma_a n_D}{T^4}, \qquad (1)$$

where $\Gamma_a \equiv \kappa_{\text{weak}}(\alpha_w T)^4$ is the rate/volume for anomalous weak interactions, α_w is the weak fine structure constant, and n_D is the difference between the total number densities of $SU(2)$ doublet fermions and anti-fermions.[3]

Before examining the quantitative details of how a bias in fermion number may be realized in various models, it is worth recognizing several generic features that all such models must possess:

1. They must involve fermions carrying $SU(2)$ quantum numbers (although not necessarily quarks), since only such particles participate in the $SU(2)$ anomaly responsible for B violation.

2. They must involve particles with a significant coupling to the Higgs field (or other order parameter which develops an expectation value at the transition) since that interaction is the source of non-equilibrium physics during a first order phase transition;

3. The phase transition must be strong enough that baryon violation is negligible in the broken phase at the transition temperature, so that the baryon asymmetry produced is not washed out during the subsequent evolution of the Universe.

Two other features which are plausible *a priori* but have only been found to be (probably) necessary after detailed analyses are:

4. Additional CP violation must be introduced beyond the Kobayashi-Maskawa (KM) model;

5. The physics involved must be *nonlocal*, involving the transport of particles in front of the phase transition boundary. Transport is necessary because the interaction between the Higgs and the fermions occurs inside the bubble wall, while baryon violation is rapid only in front of it [6][7][8].

Mechanisms which incorporate the above five features have been studied in detail in two opposite limits: "non-adiabatic models" in which the domain wall is assumed to be thinner than the mean free path of particles in the plasma; and "adiabatic models" where the opposite is true. In the non-adiabatic case, the required non-locality (attribute (5) in the above list) is due to particle reflection and transport, while in the adiabatic case the non-locality occurs due to diffusion.[4]

2.3 The non-adiabatic (thin wall) case

The first example of an EWB mechanism was presented in [4][5] and provides a simple illustration of how these various features can come together. That mechanism was based on the singlet Majoron model [11][12] with three families of massive neutrinos. In this model the advancing bubble of broken phase reflects neutrinos back into the symmetric phase; because of CP violating interactions between the neutrinos and the Higgs field, a region of net $SU(2)$ doublet number is created in front of the bubble wall which biases anomalous weak interaction events to

[3] For more details, see [2]. The goal is to explain how and why a nonzero bias ΔF arises.

[4] There have been a variety of names used for these two cases. The non-adiabatic mechanism has been called the "charge transport" and "non-local" mechanisms, while the adiabatic case has been called "spontaneous baryogenesis" or "local". Since both mechanisms end up being nonlocal, and because "spontaneous baryogenesis" is historical and obscure we advocate the use of adiabatic/non-adiabatic as the most descriptive monnikers for the thick wall/thin wall cases [9][10].

produce baryon number. The baryons are subsequently engulfed by the broken phase, where they are stable and survive to the present epoch. The model meets the above criteria in the following manner: (i) The fermions biasing baryon production are left handed neutrino doublets; (ii) the right-handed neutrinos have large Yukawa interactions with the Higgs field and drive the non-equilibrium processes (reflection from the bubble wall); (iii) the singlet Majoron model has been shown to accommodate a sufficiently strong phase transition [13]; (iv) the neutrino mass matrix allows for non-KM CP violation; (v) the long mean free path of low energy neutrinos gives rise to significant transport of net neutrino number—reflected from the bubble wall—far into the symmetric phase. Similar models can be constructed for top quark reflection [14], or τ reflection, which can be significant in a two doublet model with a finely-tuned ratio of the two doublet vacuum expectation values [15].

To calculate the baryon number produced in a non-adiabatic model, one first computes CP violating asymmetries for the reflection of particles from the advancing phase boundary. When convoluted with the momentum distribution of the incident flux, and the transport coefficients for the reflected particle distributions, one finds the net density profiles n_{X_i} for approximately conserved quantum numbers X_i which pile up in front of the bubble wall. For example, in Ref. [14] the reflection coefficients for top quarks were calculated by solving the Dirac equation, and the net density profile in front of the wall for axial top quark number was subsequently computed by means of a (crude) Monte Carlo simulation. Provided that the change in X_i during an anomalous event is nonzero, an X_i asymmetry will bias anomalous weak processes to create net baryon number. To compute the bias one must solve for the C_i in the equation

$$\Delta F = \sum_i \Delta X_i \mu_{X_i} = -\frac{\sum_i C_i n_{X_i}}{T^2} , \qquad (2)$$

where n_{X_i} is the X_i-charge density. An example of such a computation for the C_i is found in [14]. The final baryon density at position z may then be computed by integrating Eq. 1 over time, bearing in mind that the rate for anomalous weak interactions drops to zero inside the bubble (for simplicity we replace the actual drop off with a step function near the surface of the bubble; the final result is insensitive to the precise location of this sharp cut-off. We choose the origin of coordinates so that the bubble wall is located at position $v_w t$ at time t):

$$\begin{aligned} n_B(z) &= -\frac{3\Gamma_a}{T}\int_{-\infty}^{\infty} dt\ \Delta F \theta(z - v_w t) \\ &= \frac{3\Gamma_a}{T}\int_{-\infty}^{\infty} dt \sum_i C_i n_{X_i}(z - v_w t)\theta(z - v_w t) \\ &= \frac{3\Gamma_a}{T}\frac{1}{v_w}\int_0^{\infty} dx \sum_i C_i n_{X_i}(x) . \end{aligned} \qquad (3)$$

This expression can account for the observed baryon asymmetry, but typically the factor of α_w^4 in Γ_a requires that CP violation in fermion/Higgs interactions cannot be much below the 1% level.

2.4 The adiabatic (thick wall) case

Particle reflection from the phase boundary is only sensibly described if the domain wall is thin compared to the particle mean free path. This is plausible for models which have $SU(2)$ singlet scalars participating in the weak phase transition (*e.g.* the singlet Majoron model, the minimal SUSY model with an extra singlet, *etc.*), but models with only Higgs doublets tend to have broad domain walls of width $10/T$–$40/T$, where T is the transition temperature. For thick walls it is appropriate to treat the Higgs field as a classical background, where the time variation of a CP-odd field provides the necessary CPT violation necessary to drive baryogenesis. This process was first analyzed for second order phase transitions [9][10] ("spontaneous baryogenesis"); for a first order transition with finite wall width one must also add diffusive effects, which account for the necessary non-locality.

Spontaneous baryogenesis begins with the idea that the presence of a (CP-odd) space-time dependent background field θ may give rise to the interaction $\partial_\mu \theta J^\mu_Y$, where J^μ_Y is fermion current. The presence of this interaction means that the free energy of the system is minimized for nonzero Y charge, and provided that Y is explicitly violated in the Lagrangian, one will tend toward this minimum and create an asymmetry in Y. As in the non-adiabatic case, if there exists some approximately conserved global charge X which is not orthogonal to Y and which changes during an anomalous event, an asymmetry in Y implies a nonzero value for ΔF in Eq. 1, and anomalous baryon production ensues. The task of the model builder is to identify X and to perform the straightforward thermodynamic calculation of ΔF in the presence of the θ field. Examples of this approach are presented in [16][17] [18][19] [20]. In these papers θ arises as a non-Goldstone component of a Higgs field involved in the phase transition. An alternative has recently been suggested in which θ is the Goldstone boson eaten by the Z [21].

As described, all of these approaches fail to produce enough baryon number, as was pointed out in [6][7][8], because they fail criterion (5) above—they attempt to produce the baryon asymmetry locally within the domain wall, and in that region the baryon violation rate is suppressed. Ref. [22] pointed out that diffusive effects could be important, and initially concluded that diffusion completely eliminated baryon production. Since then other analyses [23] [24] [25] [26][27] have suggested the contrary: that in fact diffusion provides an effective way to transport X charge to the region in front of the domain wall. In this manner the adiabatic models are nonlocal, and in fact resemble the non-adiabatic models in their final results.

Without diffusion or other transport, the effect of the background field θ is to cause the density n_i of each particle to locally relax toward the value $\bar{n}_i$ which minimizes the θ dependent free energy. The equation describing the relation is $\dot{n}_i = -\Gamma_{ij}(n_j - \bar{n}_j)$, where Γ_{ij} is the local rate of particle i-type number violation in the theory in the presence of n_j. For example, a left handed top quark density can arise from top-Higgs interactions that flip chirality. The early papers describing the adiabatic mechanism [16][17] [18][19] [20] set n to $\bar{n}$, neglected diffusion and assumed Γ^{-1} to be fast compared to the relevant time scale of the phase transition, $\tau \equiv (wall\ width)/(wall\ velocity)$. To include transport processes, one simply includes currents:

$$\dot{n}_i = -\vec{\nabla} \cdot \vec{J_i} - \Gamma_{ij}(n_j - \bar{n}_j) \ . \tag{4}$$

To make this a useful equation, one must find a constitutive relationship for $\vec{J}$ as a function of n and θ. Reference [23] performed a strict expansion simultaneously in powers of derivatives of θ and powers of the fundamental violation in the Lagrangian of the ith charge, and showed that a standard diffusion equation resulted:

$$\dot{n}_i = D_{ij}\nabla^2 n_j - \Gamma_{ij}(n_j - \bar{n}_j) \ , \tag{5}$$

with a straightforward prescription for computing the $\bar{n}_j$. The parameters D_{ij} are diffusion coefficients and can be estimated easily, although a quantitative computation is difficult, plagued by infrared divergences.

Equation 5 (including the effects of the QCD anomaly [28]) has been numerically integrated for the two Higgs model with thick domain walls discussed in [20][17] [18][19] and shown to provide the non-locality necessary to account for the observed baryon abundance. Two interesting features that were discovered are (i) For a broad range of anomalous interaction rates, the final baryon number produced depends only on the *ratio* of anomalous weak interactions to anomalous *strong* interactions, due to competition between $SU(3)$ and $SU(2)$ anomalous processes. Assuming that $\kappa_{\rm strong} \approx \kappa_{\rm weak}$, this constant factor (which is not well-determined theoretically) drops out of the expression for the final baryon number; (ii) In the two doublet model, the diffusion length is dominated by the mean free path of Higgs fields rather than that of quarks (which is shorter). Thus Higgs particles diffuse far into the symmetric phase, subsequently converting back into top quark doublets which bias baryon production.

There are several effects neglected by the expansion used to derive Eq. 5. For example, in the derivative expansion $\dot{\theta}$ and $\vec{\nabla}\theta$ were treated as being of the same order, while in fact they differ by a power of the wall velocity — a significant factor for very non-relativistic walls. Also absent are classical forces in the diffusion term; for example, if there were an electric field $E = -\nabla\phi$, the diffusion term would have been modified to $D\nabla^2(n-\phi)$ (giving the Einstein relation between mobility and diffusion). To the order in the expansion described above, such forces do not appear. At sub-leading order in this expansion however, they do appear. For example the spatial gradient of the top quark mass (which is responsible for axial top number violation) gives rise to a CP conserving force, and the second derivative may give a CP violating force [29]. The importance of this and other effects are analyzed in Ref. [26][27] for different values of the phase transition parameters, anomalous weak interaction rates, and fermion-Higgs couplings.

In Refs. [23][25][24][26][27][29] it is shown that in reasonable models the bias in the baryon violation rate from transport of either lepton or quark doublet CP violating asymmetries can be sufficiently large to explain the observed baryon asymmetry.

It is remarkable that the physics of the adiabatic and non-adiabatic cases end up looking so similar, even though the former involves a coherent bias and diffusion, while the latter involves single particle reflection and charge transport far from equilibrium. Both regimes can account for the baryon asymmetry with new sources of CP violation that are moderately large, at the $\gtrsim 1\%$ level in the relevant interactions.

2.5 Defect-mediated Baryogenesis

Most realizations of electroweak baryogenesis discussed in the literature start with the assumption that the electroweak phase transition is strongly first order and that the transition proceeds by the nucleation and subsequent expansion of critical bubbles. Unfortunately, however, the order of the electroweak phase transition is not known. From analyses of condensed matter systems it is known that even if a phase transition is first order, the transition may be driven by classical fluctuations [30], rather than by bubble nucleation. Hence, it is of interest to look for realizations of electroweak baryogenesis which are independent of the details of the electroweak transition.

In Ref. [31] it was pointed out that topological defects may play an analogous role to bubble walls in catalyzing electroweak baryogenesis (see also Ref. [32][33] for a discussion of the role of topological defects in GUT scale baryogenesis). The basic point is very simple. In the walls of a topological defect, Sakharov's conditions for baryogenesis are satisfied in the same way as they are in the walls of a bubble of true vacuum expanding into a sea of false vacuum in the context of a strongly first order transition driven by the nucleation of critical bubbles. Baryon number violating processes are unsuppressed in the core of the topological defect, CP violation is enhanced in the defect wall (where the CP violating phase of an extra Higgs doublet has a non-vanishing gradient), and the defects themselves are out of equilibrium.

Since topological defects are produced independent of the order and dynamics of the phase transition [34], a topological defect-mediated baryogenesis scenario is appealing given our present ignorance about the details of the electroweak phase transition. However, in order for the mechanism to work, topological defects within which the electroweak symmetry is restored must be present. In the minimal electroweak theory there are no topological defects. In Ref. [31] it was suggested that semi-topological field configurations such as electroweak strings [35] may act as the seeds for baryogenesis. However, for the standard electroweak theory and for the measured value of the weak mixing angle, electroweak strings are unstable [36].

An improved implementation of the defect-mediated electroweak baryogenesis scenario was put forward in Ref. [37]. It was pointed out that topological defects produced in a phase transition above the electroweak scale and within which the electroweak symmetry is unbroken will satisfy the required conditions. In Ref. [38], this mechanism was studied in more detail. In particular, it was pointed out that dissipative and nonlocal effects can enhance the efficiency by many orders of magnitude.

The net baryon to entropy ratio which can be generated by topological defect-mediated baryogenesis depends on the topology of the defects and on the scale at which they are produced. It also depends on the specific extension of the electroweak theory considered in order to enhance CP violation, but since this dependence is the same as for all other models of electroweak baryogenesis, it will not be discussed any further.

In general, topological defect-mediated baryogenesis is *less* efficient than the usual mechanisms based on bubble walls. The reason is obvious: whereas bubble walls sweep across essentially the entire spatial volume before percolating, topological defects will sweep out only a fraction of space during the course of their evolution. This leads to a volume suppression factor SF. Volume suppression is least if the topological defects are produced at a scale σ close to the electroweak symmetry breaking scale σ_{EW}. If the defects are produced at early times, their number density red-shifts and they only occupy a small fraction of the volume at the electroweak symmetry breaking scale. The factor SF also depends on which type of topological defect is being considered. In a model with domain walls moving through space with a velocity v_w, the factor SF can be as large as γv_w, where γ is a factor of the order 1. Baryogenesis is also more efficient (i.e. SF is larger) if the defects are super-conducting since in this case the electroweak symmetry is restored to a larger radius about the defect [39][40].

Nonlocal defect-mediated electroweak baryogenesis can be understood by studying the reflection of fermions by the defect walls. If there is CP violation in the wall, then right handed and left handed fermions are scattered asymmetrically, producing a chiral current of the same sign from both defect walls into the core. Via sphaleron transitions, this current leads to a non-vanishing baryon number density. As in the case of bubble wall-mediated baryogenesis, the nonlocal mechanism may be more efficient.

As an example, consider cosmic string-mediated electroweak baryogenesis. To obtain an appreciable baryon to entropy ratio, the scale σ cannot be much higher than σ_{EW}. Hence, the strings are still in the friction dominated era. Their initial number density is given by the Kibble mechanism. The net baryon to entropy ratio which can be obtained in this case turns out [38] to be suppressed compared to what can be obtained in bubble-mediated baryogenesis by a factor of

$$\lambda \left(\frac{\sigma_{EW}}{\sigma}\right)^p v_w$$

where the power p is $3/2$ for non-superconducting cosmic strings and $3/4$ for superconducting ones, and λ is the Higgs self coupling constant. In this calculation it has been assumed that sphalerons fit into the string core.

In conclusion, it has been shown that electroweak baryogenesis can be implemented independent of the order and detailed dynamics of the electroweak phase transition. Topological defects which are remnants of a symmetry breaking at a scale above σ_{EW} can play the same role in mediating baryon number violation as the bubble walls in the usual scenarios. The new scenario is most efficient if the defects are formed just above σ_{EW}. Technicolor models of electroweak symmetry breaking may provide an implementation of the scenario.

2.6 *Electroweak Baryogenesis and the Standard Model*

As described earlier the minimal standard model does satisfy Sakharov's three conditions for baryogenesis: i) anomalous processes in the electroweak plasma violate baryon number; ii) the electroweak phase transition is expected to be (at least weakly) first order, thereby giving departure from thermal equilibrium; iii) CP violation is present in flavor changing weak interactions. It is therefore an interesting question as to whether the standard model can produce the observed baryon asymmetry. A number of problems for the minimal model are immediately apparent however. Foremost among these is the small magnitude of CP violation arising from the

Kobayashi-Maskawa quark mixing matrix. Any physical manifestation of this CP violation must involve all three generations (since the two generation model is CP conserving). It is therefore suppressed by multiple mixing angles and mass splittings. In fact, the invariant dimensionless measure of CP violation is given by $\mathcal{J} \equiv \mathrm{Det}[\lambda_U \lambda_U^\dagger, \lambda_D \lambda_D^\dagger] \sim 10^{-21}$, where $\lambda_{U,D}$ are the up and down quark Yukawa matrices. This is to be contrasted with extensions of the standard model which can have order unity CP violation in equivalent dimensionless units. A naïve estimate for η which includes $\mathcal{J}$ leads to an asymmetry many orders of magnitude less than the observed value of $\sim 10^{-10}$. However there may well exist mechanisms, such as that of Farrar and Shaposhnikov discussed below, which might amplify this small CP violation to an observable level.

Another problem for baryogenesis in the minimal model is the requirement that any baryon number produced at the electroweak phase transition not be washed out by anomalous processes after the transition is complete. This puts an upper limit on the Higgs mass. The one loop perturbative limit in the minimal model is $m_h < 35$ GeV [41]. This is in conflict with the experimental bound of $m_h > 60$ GeV. Recently it has been claimed that nonperturbative effects can influence the strength of the phase transition, thereby raising the limit on the Higgs mass [42]. This is necessary if baryogenesis is to take place in the minimal model.

Even if one leaves aside the problems of wash out of baryon number and the strength of the phase transition in the minimal model, the question of how the small CP violation could lead to an acceptable baryon asymmetry must be addressed. Nasser and Turok have recently suggested a mechanism by which this may take place [21]. They claim that a moving bubble wall is unstable to the accumulation of a large net axial charge, which may be described as a Z boson condensate. Baryogenesis results from the accumulated charge as in the usual spontaneous or charge transport mechanisms. At present it is not possible to make quantitative predictions using this mechanism.

The most well developed mechanism for amplifying the CP violation has been proposed by Farrar and Shaposhnikov [43]. They note that a nonzero baryon number can be reflected from the bubble wall in the presence of flavor mixing and CP violation. This occurs because scattering trajectories which acquire different CP phases in the course of scattering can interfere quantum mechanically. Since the mechanism depends so crucially on this quantum mechanical interference it is worth describing how it leads to a baryon asymmetry [43, 44, 45].

A fermion scattering from the bubble wall conserves angular momentum, so a left handed particle, L, is scattered to a right handed particle, R, and similarly for antiparticles, $\bar{L}$ and $\bar{R}$. The important interference appears between trajectories with a CP even amplitude, $\mathcal{A}$, and those with a CP odd amplitude, $\mathcal{B}$. Schematically, the probability for a net particle current to arise in $L \to R$ and $\bar{R} \to \bar{L}$ scattering (the CP conjugate process) from the unbroken phase back into the unbroken phase is then $\mathcal{P}_{LR} - \mathcal{P}_{\bar{R}\bar{L}} = -4\,\mathrm{Im}\mathcal{A}\,\mathrm{Im}\mathcal{B}$, where the CP violation appears in the phase of $\mathcal{B}$. To obtain the total baryon current, the probability for a net particle current from $R \to L$ and $\bar{L} \to \bar{R}$ must be added to this. However, CPT ensures $\mathcal{P}_{RL} = \mathcal{P}_{\bar{R}\bar{L}}$ and $\mathcal{P}_{\bar{L}\bar{R}} = \mathcal{P}_{LR}$. The baryon current reflected is therefore proportional to the net difference in L and R flux incident on the wall $\sim ((\vec{p}_L - \vec{p}_R)/T)\cdot\vec{v}_w$. The only interactions in the plasma which distinguish L and R (and therefore split the respective thermal distributions) are the weak interactions. So the net flux is suppressed by a factor of the weak coupling, $|\vec{p}_L - \vec{p}_R|/T \sim \alpha_w$. Finally, unitarity guarantees that the net baryon current scattered into the unbroken phase is balanced by an equal and opposite baryon current scattered into the broken phase. Now the net baryon number deposited in the unbroken phase can be eliminated by anomalous processes in the electroweak plasma. The rate for such processes is however exponentially suppressed in the broken phase. So a net baryon number can be left behind the moving wall. Putting all this together, the baryon to entropy ratio can be estimated for the simplified problem of scattering from a bubble wall in one dimension to be very roughly

$$\eta \sim 10^{-2}\alpha_w\, v_{\mathrm{W}} \int d\tilde{\omega}\, \frac{\Delta(\omega)}{T}$$

where 10^{-2} arises from the number of degrees of freedom in the plasma, and $\Delta(\omega)$ is the frequency dependent probability for net baryon reflection arising from the interference as described above.

The full calculation of $\Delta(\omega)$ requires solving the Dirac equation for all three flavors in the presence of the wall although the important numerical contribution can be seen from a simple heuristic argument [43]. The dispersion relations for quark excitations have a large gap (or thermal mass) of order $\Omega \sim 50$ GeV due to QCD interactions with the plasma. They also turn out to have (for a limited range of momenta) a small gap splitting L and R states in the broken phase. This gap is of order of the bare quark mass. Quarks incident from the unbroken phase with energy falling in this small gap are therefore totally reflected as no propagating modes with equal energy exist in the broken phase. The CP conserving amplitude for reflection in this energy band is therefore just some phase, $\mathcal{A} = e^{i\varphi}$. The CP violating amplitude arises from mixing with other flavors. The off diagonal pieces of the Hamiltonian coupling different flavors arise from charged current interactions with the plasma. For down type quarks these are in magnitude roughly $\mathcal{H} \sim (\alpha_w T^2/\Omega)(V^\dagger M_u^2 V)/M_{\mathrm{W}}^2$, where V is the Kobayshi-Maskawa matrix. The factor of M_u^2/M_W^2 arises because of the usual GIM suppression for flavor changing neutral currents. Treating the off diagonal interactions as perturbations, the lowest order CP odd amplitude arises in third order (since all three generations must be involved). For (the numerically important) s quark this gives an amplitude for mixing to b, then scattering off the wall, mixing to d, and back to s, of

$$\mathcal{B} \sim \frac{\langle s|\mathcal{H}_{sb}|b\rangle}{m_b - m_s} \langle b|e^{i\varphi}|b\rangle \frac{\langle b|\mathcal{H}_{bd}|d\rangle}{m_b - m_d} \frac{\langle d|\mathcal{H}_{ds}|s\rangle}{m_s - m_d}$$

where $1/(m_{q_i} - m_{q_j})$ are the energy denominators for the splittings between the flavors. Interfering this amplitude with the CP conserving one for total reflection from the wall gives

$$\int d\tilde{\omega} \frac{\Delta(\omega)}{T} \sim \frac{m_s}{T} \left(\frac{\alpha_w T^2}{\Omega M_W^2} \right)^3 \frac{m_t^4 m_c^2 s_{12} s_{23} s_{13} \sin\delta \sin\varphi}{m_b^2 m_s}$$

where the factor of m_s/T arises (in one dimension) since total reflection occurs only for energies which fall within the gap between L and R states. This rough estimate gives $\eta \sim 10^{-13}$. Farrar and Shaposhnikov however claim that a more careful treatment gives an acceptable result.

The enormous enhancement of the intrinsic CP violation in this mechanism arises for two reasons. First, the CP conserving amplitude for total reflection has unit modulus for energies falling within the $L-R$ gap of the broken phase. The gap arises however because of quantum mechanical mixing between L and R states in the broken phase due to the bare quark mass. This mixing proceeds over time scales of order $1/m_q$. In order for the mixing to take place (and therefore for the gap to appear) the quark excitations must remain coherent over this time scale. Second, the CP violating amplitude is enhanced by the small energy denominators, $\sim 1/m_q$. The small energy denominators correspond to quantum mechanical mixing between the unperturbed states on time scales on the order $1/m_q$. Again coherence must be maintained over this time scale in order that this trajectory can interfere with the CP conserving one. The enhancement therefore relies crucially on the coherence of the quark excitations in the plasma. However, the quasi-particles have a finite coherence length and damping time due to interactions with the plasma. Estimates of this coherence length from QCD interactions typically give $\ell \sim \text{few}/T$, which is much shorter than the required $1/m_s$ for coherence to survive.

A careful treatment of the role of decoherence in this mechanism has been given by Gavela, Hernandez, Orloff, and Pene [44] and Huet and Sather [45]. Both groups agree on the parametric dependence of the scattering probabilities. Gavela *et al.* consider the scattering of wave packets with finite width and imaginary contribution to the self energy to mimic the effect of the finite coherence length. Huet and Sather use a real space Green's function with damping along the particle trajectory to solve the scattering problem. To lowest order the momentum space Green's function contains an imaginary part describing the damping and a real part describing propagation and mixing between flavors. The lowest order interference between CP even and odd amplitudes turns out to appear in 3rd order in flavor mixings among left handed particles and 9th order in ℓ (in order to form the invariant $\mathcal{J}$). The resulting scattering probability for baryon number is

$$\int d\tilde{\omega} \frac{\Delta(\omega)}{T} \sim \left(\frac{\alpha_w T^2}{\Omega M_W^2} \right)^3 \mathcal{J} v^{12} \ell^9$$

where v is the Higgs expectation value in the broken phase. For the estimate of ℓ given above this gives $\eta \sim 10^{-25}$. Mixing among right handed quarks turns out to suffer only a logarithmic GIM suppression and is suppressed by only 7 powers of ℓ, but still gives an insignificant contribution. Unless the coherence length for quasi-particles is much larger than the naïve estimate, the loss of coherence seems to destroy the large enhancement.

In addition to the loss of coherence other factors probably also reduce the effectiveness of this mechanism. First, unless the anomalous weak interaction rate, $\Gamma_a = \kappa(\alpha_w T)^4$, is quite large ($\kappa \sim 10^{(3-4)}$) the erasure of the baryon current scattered into the unbroken phase is probably somewhat suppressed. Second, scattering from the wall of particles with momenta larger than the inverse wall thickness is WKB suppressed. Avoiding this requires a strongly first order phase transition; this requires large corrections to the perturbative finite temperature Higgs potential [42]. Third, the suppression in 3 dimensions is almost certainly more severe than that in the 1 dimensional example given above. At present it seems rather difficult to imagine how standard model CP violation can be amplified to a large enough level to account for the baryon asymmetry with the mechanism of Farrar and Shaposhnikov. However the mechanism is quite elegant, and the question of coherence of the quasi-particles deserves further attention.

§3 GUT Baryogenesis

A major boost for the idea of grand unification came in the late seventies from the realization that such models generically have all three ingredients necessary for baryogenesis to proceed [1]. The fact that these models require strong and weak interactions to be part of a single unified symmetry implies that some representation of the GUT gauge group will necessarily contain both quarks and leptons. This in turn implies the existence of super-heavy bosons, X and Y (with masses within a few orders of magnitude of the grand unification scale, M_U) whose interactions with quarks and leptons violate baryon and lepton number. In general these interactions are also CP violating. A generic old GUT scenario for baryogenesis then proceeds as follows: around the epoch when the temperature of the Universe was below the masses of the X and/or Y bosons, their interactions dropped out of equilibrium (*i.e.* the decay rate Γ_X became less than the expansion rate $g_*^{1/2}T^2/M_{Pl}$) and at that point the B-violating decays of these particles coupled with CP violation generated an excess of baryons over anti-baryons. Since the inverse decays at these temperatures are not effective due to Boltzmann suppression of the high energy particles, this net baryon number of the universe survives to lower temperatures. The out of equilibrium condition for this so-called "drift and decay" scenario can be guaranteed for phenomenologically reasonable choices of the coupling parameters of the theory.

The value of η predicted by this mechanism is of course quite model dependent. The reason is that in order to obtain a non-vanishing baryon asymmetry, one needs interference between tree and one-loop graphs; furthermore, the loop graphs must have a non-zero absorptive part, and thus the particles inside the loop must be light enough to appear as final state decay products of the X bosons. This is actually a highly nontrivial restriction which for instance implies that in the minimal non-supersymmetric $SU(5)$ model, the value of η is several orders of magnitude smaller than the observed value. It is however not an insurmountable obstacle and simple extensions of the model cure this difficulty (*e.g.* an extra Higgs multiplet or a fourth generation or extending the gauge group to $SO(10)$).

An important change in our thinking about GUT baryogenesis occurred following the work of Kuzmin, Rubakov and Shaposnikov [3], who showed that there are baryon violating *weak* interactions in equilibrium at all temperatures greater than approximately 100 GeV. Sakharov's third constraint would then prevent the production of a baryon asymmetry through these GUT interactions. However these baryon violating weak interactions do *not* violate $B-L$ (baryon minus lepton number)—if the GUT produces a net $B-L$ than these weak interactions will still allow a baryon asymmetry. In the minimal $SU(5)$ model, $B-L$ is an exact symmetry of the theory (even after symmetry breaking), and as a result no baryon asymmetry is produced. The same result holds for minimal supersymmetric SU(5) as well. On the other hand, the situation is much more promising in $SO(10)$ GUTs. It has been noted [46] that in SO(10) models with the see-saw mechanism for neutrino masses, the right-handed Majorana neutrino (N_R) masses are very near the GUT scale. If all appropriate out-of-equilibrium conditions are satisfied, then decay of the N_R can produce a lepton asymmetry, which is converted to a baryon asymmetry by these baryon violating weak interactions. In view of some recent indications for neutrino masses in solar and atmospheric neutrino data as well as recent enthusiasm for the mixed dark matter picture of the Universe, $SO(10)$ models are currently a focus of much attention. Let us therefore discuss prospects for a healthy baryogenesis in these models.

Case	Constraint
$N_R \to \nu_{Lj}\phi^0$	$M_{N_i} \geq \left(\frac{h_{ii}^2}{160\pi}\right) M_{Pl}$
$N_R\ell_R \to u_R d_R$	$M_{N_i} \leq 800 v_R \left(\frac{v_R}{M_{Pl}}\right)^{1/3}$
$N_R\ell_L \to Q_L T_R$	$M_{N_i} \geq \left(\frac{h_{ii}^2}{10\pi}\right) M_{Pl}$
$N_R N_R \to N_R N_R$	$M_{N_i} \leq v_R \left(\frac{30 v_R}{f M_{Pl}}\right)^{1/6}$
$\nu_L\nu_L \to \phi\phi$	$M_{N_i} \geq g_*^{-1/2} h_{ii}^4 M_{Pl}$

Table 1: Constraints on M_N from baryogenesis

In this picture of baryogenesis via high scale leptogenesis, the following constraints must be satisfied: (i) first, all lepton-number violating interactions involving the relevant right-handed neutrino at the near-GUT temperature (*i.e.* $T \simeq (10^{-1}$ *to* $10^{-5}\ M_U)$) must be out of equilibrium; (ii) second, the couplings of the right-handed Majorana neutrinos must be such that the lepton to photon ratio produced is of order $\simeq 10^{-9}$; (iii) finally, all lepton number violating interactions involving lighter particles (such as neutrinos) must be out of equilibrium from the temperature for leptogenesis down to the weak scale, 100 GeV. Let us briefly recapitulate these constraints and summarize their impact on model parameters [47]. The relevant generic processes are : (A) $N_R \to \nu_{Lj}\phi^0$; (B) $N_R\ell_R \to u_R d_R$; (C) $N_R\ell_L \to Q_L T_R$; (D) $N_R N_R \to N_R N_R$; (E) $\nu_L\nu_L \to \phi\phi$. The out-of-equilibrium conditions for the various processes are are summarized in Table 3.

In the above equations, the h_{ii} denote the diagonal Yukawa couplings of the right-handed neutrinos to the standard model Higgs boson and are therefore related to quark masses. We assume the off diagonal Yukawa couplings to be negligible in this discussion; they will of course be very important in generating the lepton asymmetry.

The above conditions impose constraints on the particle spectrum of a given $SO(10)$ model. To see this, let us first consider the different SO(10) models.

3.1 Non-supersymmetric $SO(10)$ models

In this case, depending on the chain of symmetry breaking, the scale of $SU(2)_R$ is anywhere between 10^9 GeV to 10^{13} GeV [48]. Assuming the N_R masses to be around the $B - L$ symmetry breaking scale, it is easy to see that only the electron or muon type right-handed neutrino can satisfy the out-of-equilibrium conditions (A) thru (D) and the muon type N_R satisfies the condition (D) only if $v_R \geq 10^{12}$ *to* 10^{13} GeV.For a generic SO(10) model, it is safe to assume that the N_{eR} is the lightest of the right-handed neutrinos and it is the decay of the N_{eR} that leads to leptogenesis. The magnitude of the lepton asymmetry ϵ has been calculated by Luty [47] and is given by:

$$\epsilon = \left(\frac{1}{\pi(h^*h)_{11}}\right) \Sigma_j \; Im[(h^*h)_{1j}(h^Th^*)_{j1}]f(M_{Nj}^2/M_{N1}^2) \tag{6}$$

where the function $f(z) = \sqrt{z}[1-(1+z)ln1 + z/z]$ and we have used the fact that the $SO(10)$ Yukawa couplings are symmetric in the generation indices. In the case where $M_{N_{\mu,\tau}} \gg M_{N_e}$, one gets $f \simeq M_{N_e}/2M_{N_{\mu,\tau}}$. Taking only the second generation contribution to ϵ, one can get (using $M_{N_e}/M_{N_\mu} \simeq 10^{-2}$), $\epsilon \approx 10^{-8}$ which is in the desired range to get the correct value for η. Jungman and Gherghetta [47] have evaluated this asymmetry for realistic SO(10) models and obtain similar orders of magnitude for ϵ. In particular, these estimates apply to non-susy SO(10) models with PQ symmetry which are interesting for the reason that not only do they solve the strong CP problem, but also,the axion in this model can provide the cold dark matter of the Universe, and the tau neutrino having a mass in the eV range also provides a good candidate for the hot dark matter of the Universe. Theoretical models of this type have been constructed [49] and are testable in the next generation of proton decay experiments such as super-Kamiokande and Icarus.

3.2 SUSY SO(10) models

Susy SO(10) models have the new feature that no intermediate scales are allowed by the LEP data on low energy gauge couplings; as a result, the value of the $B - L$ breaking scale is also of order 10^{16} GeV. The rest of the discussion in this case parallels the non-susy case and the scenario favored by baryogenesis constraints is that $M_{N_{\mu,\tau}} \gg M_{N_e} \simeq 10^{11} - 10^{12}$ GeV and it is the decay of N_e that is responsible for leptogenesis. If the right-handed neutrino masses in this model arise via the coupling of $\mathbf{1\bar{2}6}$ to the fermions, one needs to fine tune its couplings to be of order 10^{-5}. This is rather unpleasant. On the other hand, this baryogenesis scenario could be taken over to the case of superstring inspired SO(10) models. Due to the highly restricted nature of the allowed multiplets in this case, the right-handed Majorana neutrino masses can arise in this model only via the higher dimensional couplings of the form $\mathbf{16_f 16_f 16_H \bar{16}_H}/M_{Pl}$. Such couplings are expected to occur naturally in these models, since Planck scale physics plays such an important role in determining the low energy physics. Using the fact that the $\mathbf{16}_H$ multiplets have vev's of order 10^{16} GeV, the N_R masses are naturally of order 10^{13} GeV. The downside to this story of the higher dimensional operators is that they can induce large R-parity violating terms which could erase [50] the preexisting lepton asymmetry since they can in principle remain in equilibrium to temperatures far below the leptogenesis temperature $T \simeq M_{N_e}$. One must therefore have symmetries built into the theory such that all R-parity violating higher dimensional operators are sufficiently suppressed. Such discrete symmetries are however expected to arise below the Planck scale in superstring theories.

3.3 Flipped $SU(5) \times U(1)$

Flipped $SU(5) \times U(1)$ is another popular string motivated model, which has been extensively discussed in the recent literature [51]. In this model, the fermions are assigned to $\mathbf{10}_1$, $\mathbf{\bar{5}}_{-3}$ and $\mathbf{1}_5$ representations of the gauge group (denoted by F, f and e^c respectively) and the Higgs superfields transform as $\mathbf{10}_1$, $\mathbf{\bar{10}}_{-1}$, $\mathbf{5}_{-2}$ and $\mathbf{\bar{5}}_2$ representations of the gauge group and denoted by H,$\bar{H}$, h and $\bar{h}$ respectively. In Ref. [51], it was noted that neutrino masses arise via an extended see-saw mechanism [52] where the N_R of each generation pairs up with a gauge singlet fermion, so that $M_{N_R} \simeq 10^{16}$ GeV for each generation. Furthermore, lepton number violation by this matrix is of order $O(m_W/M_U)$. Therefore, the mechanism of Ref. [46] cannot generate enough leptons to explain the observed baryon asymmetry of the universe. Thus the baryon asymmetry would appear to be a problem with the minimal version of the flipped SU(5) model discussed in the literature. One way to cure this problem would be to use higher dimensional operators such as $F.\bar{H}F.\bar{H}/M_{Pl}$ so that one has the usual see-saw mechanism for the neutrino masses and one can then use the mechanism of Ref. [46] for baryogenesis. In order for the lepton asymmetry to survive, the theory must have a matter parity under which F,f and e^c are odd and all Higgs superfields are even so that no higher dimensional lepton number violating terms are allowed in the theory.

§4 Conclusions and Outlook

Although our general ideas of baryogenesis are in quite good shape, many details remain unspecified. The most severe of these uncertainties, which applies to both types of models described here, is the origin of *CP* violation. It now seems likely that the *CP* violation associated with a phase in the KM matrix is insufficient to account for the present baryon asymetry, and therefore new *CP* violation beyond that found in the standard model is required. On the positive side, there are many proposed sources of such new *CP* violation which can produce a large asymmetry and nevertheless be consistent with current experimental data; on the negative side, none of these proposals is overwhelmingly compelling. The best source for new information is likely to come from experiment—almost all of these proposals give electric dipole moments for the electron or neutron which are within an order of magnitude or so of the current experimental limits. Over the next few years we can expect improvment in these experiments; if no dipole moments are seen this would be strong evidence (although not obviously conclusive) of something lacking in our current understanding of baryogenesis.

In addition to uncertainties in *CP* violation, models of electroweak baryogenesis are plagued by our inadequate knowledge of the weak phase transition. Although our current understanding of the weak interactions strongly suggests the existence of such a transition, our ignorance of the details of weak symmetry breaking prevents us from making any conclusive statements; we do not know if the transition is first or second order, if it is driven by bubble nucleation or fluctuations, or even exactly when it occurs. Most research to date has focused on the nature of the phase transition in the minimal standard model, in which electroweak symmetry breaking arises through the generation of a vacuum expectation value for a single fundamental electroweak doublet scalar field. These investigations indicate that for the experimentally allowed values of the Higgs mass the electroweak phase transition in this model would be second order. Unfortunately even with this detailed model these calculations are not wholly reliable. In addition this minimal model cannot accomodate any *CP* violation beyond that in the KM matrix, which we have already remarked appears to be inadequate for baryogenesis. Consequently future investigations are likely to concentrate on other models of weak symmetry breaking, and especially on the development of more reliable techniques for investigating phase transitions in the early Universe.

In addition to understanding the nature of the phase transition, with models of electroweak baryogenesis we also need to understand the interaction of fermions (usually quarks and leptons but also exotic particles contained in extensions of the standard model) with the fluctuations that arise at this phase transition. For example, if the phase transition proceeds through nucleation of bubbles, we have already seen that the interaction of quarks and leptons with these bubbles is the crucial ingredient for baryogenesis. If the phase transition occurs with large thermal fluctuations then quarks and leptons will interact with these fluctuations in a way which will affect baryogenesis. In any case, this interaction is quite complicated and better techniques are necessary for a reliable, quantitative calculation of the resulting baryon asymmetry; interactions of relativistic particles at high temperature is an active subject, and many of the results found useful for QCD at high temperatures may also prove valuable for understanding interactions at the weak phase transition.

We have seen that models of electroweak baryogenesis involve electroweak symmetry breaking and quark and lepton Yukawa couplings in an intimate fashion. These are the two areas of particle physics in which our ignorance is maximal and therefore not surprisingly we must look to experiments for further information. The best hope for investigating these issues will be the LHC, scheduled for completion sometime in the next milleneum. Whatever the results from these experiments, they are likely to have significant implications for our understanding of the origin of the baryon asymmetry.

§Acknowledgments

We would like to acknowledge the help of many people in preparing this document; sadly most of them are unknown or wish to remain anonymous.

§References

[1] E. Kolb and M. Turner. Grand Unified Theories and the Origin of the Baryon Asymmetry. *Ann. Rev. Nucl. Sci.*, 33:645, 1983.

[2] A.G. Cohen, D.B. Kaplan, and A.E. Nelson. Progress in Electroweak Baryogenesis. *Ann. Rev. Nucl. Part. Sci.*, 43:27, 1993.

[3] V.A. Kuzmin, V.A. Rubakov, and M.E. Shaposhnikov. On the Anomalous Electroweak Baryon Number Nonconservation in the Early Universe. *Physics Letters B*, 155:36, 1985.

[4] A.G. Cohen, D.B. Kaplan, and A.E. Nelson. Weak Scale Baryogenesis. *Physics Letters B*, 245:561, 1990.

[5] A.G. Cohen, D.B. Kaplan, and A.E. Nelson. Baryogenesis at the Weak Phase Transition. *Nuclear Physics B*, 349:727, 1991.

[6] M. Dine. Electroweak Baryogenesis: An Overview (Where are we now?). *Yale-Texas Workshop*, page 19, 1992.

[7] M. Dine and S. Thomas. Electroweak Baryogenesis in the Adiabatic Limit. *Physics Letters B*, 328, 1994.

[8] M. Dine, P. Huet, and R. Singleton, Jr. Baryogenesis at the Electroweak Scale. *Nuclear Physics B*, 375:625, 1992.

[9] A.G. Cohen and D.B. Kaplan. Thermodynamic Generation of the Baryon Asymmetry. *Physics Letters B*, 199:251, 1987.

[10] A.G. Cohen and D.B. Kaplan. Spontaneous Baryogenesis. *Nuclear Physics B*, 308:913, 1988.

[11] Y. Chikashige, R.N. Mohapatra, and R.D. Peccei. Spontaneously Broken Lepton Number and Cosmological Constraints on the Neutrino Mass Spectrum. *Physical Review Letters*, 45:1926, 1980.

[12] Y. Chikashige, R.N. Mohapatra, and R.D. Peccei. Are There Real Goldstone Bosons Associated with Broken Lepton Number? *Physics Letters B*, 98:265, 1981.

[13] Y. Kondo, I. Umemura, and K. Yamamoto. First Order Phase Transition in the Singlet Majoron Model. *Physics Letters B*, 263:93, 1991.

[14] A.E. Nelson, D.B. Kaplan, and A.G. Cohen. Why There is Something Rather than Nothing: Matter from Weak Interactions. *Nuclear Physics B*, 373:453, 1992.

[15] M. Joyce, T. Prokopec, and N. Turok. Efficient Electroweak Baryogenesis from Lepton Transport. 1994. hep-ph - 9401352.

[16] N. Turok and J. Zadrozny. Dynamical Generation of Baryons at the Electroweak Transition. *Physical Review Letters*, 65:2331, 1990.

[17] N. Turok and J. Zadrozny. Electroweak Baryogenesis in the Two Doublet Model. *Nuclear Physics B*, 358:471, 1991.

[18] L. McLerran, M. Shaposhnikov, N. Turok, and M. Voloshin. Why the Baryon Asymmetry of the Universe is Approximately 10**-10. *Physics Letters B*, 256:451, 1991.

[19] M. Dine, P. Huet, R. Singleton, Jr., and L. Susskind. Creating the Baryon Asymmetry at the Electroweak Phase Transition. *Physics Letters B*, 257:351, 1991.

[20] A.G. Cohen, D.B. Kaplan, and A.E. Nelson. Baryogenesis at the Weak Phase Transition. *Physics Letters B*, 263:86, 1991.

[21] S. Nasser and N. Turok. Z Condensation and Standard Model Baryogenesis. 1994. hep-ph - 9406270.

[22] M. Joyce, T. Prokopec, and N. Turok. Why Hypercharge Doesn't Make Baryons. 1994. hep-ph - 9401351.

[23] A.G. Cohen, D.B. Kaplan, and A.E. Nelson. Diffusion Enhances Spontaneous Electroweak Baryogenesis. *Physics Letters B*, 336:41, 1994.

[24] M. Joyce. A Note on Spontaneous Baryogenesis. 1994. hep-ph - 9406356.

[25] D. Comelli, M. Pietroni, and A. Riotto. On Spontaneous Baryogenesis and Transport Phenomena. 1994. hep-ph - 9406369.

[26] M. Joyce, T. Prokopec, and N. Turok. Nonlocal Electroweak Baryogenesis. part 1: Thin wall regime. 1994. hep-ph - 9410281.

[27] M. Joyce, T. Prokopec, and N. Turok. Nonlocal Electroweak Baryogenesis. part 2: The classical regime. 1994. hep-ph - 9410282.

[28] G. Giudice and M. Shaposhnikov. Strong Sphalerons and Electroweak Baryogenesis. *Physics Letters B*, 326:118, 1994.

[29] M. Joyce, T. Prokopec, and N. Turok. Electroweak Baryogenesis from a Classical Force. 1994. hep-ph - 9408339.

[30] N. Goldenfeld. Dynamics of Cosmological Phase Transitions: What can we learn from condensed matter physics. In A.-C. Davis and R. Brandenberger, editors, *Formation and Interactions of Topological Defects*, New York, 1995. Plenum Press.

[31] R. Brandenberger and A. Davis. Electroweak Baryogenesis with Electroweak Strings. *Physics Letters B*, 308:79, 1993.

[32] R. Brandenberger, A. Davis, and M. Hindmarsh. Baryogenesis from Collapsing Topological Defects. *Physics Letters B*, 263:239, 1991.

[33] A.-C. Davis and M. Earnshaw. Baryogenesis from Collapsing Cosmic String Loops. *Nuclear Physics B*, 394:21, 1993.

[34] T.W.B. Kibble. Topology of Cosmic Domains and Strings. *J. Phys.*, A9:1387, 1976.

[35] T. Vachaspati. Vortex Solutions in the Weinberg-Salam Model. *Physical Review Letters*, 68:1977, 1992.

[36] M. James, L. Perivolaropoulos, and T. Vachaspati. Detailed Stability Analysis of Electroweak Strings. *Nuclear Physics B*, 395:534, 1993.

[37] R. Brandenberger, A.-C. Davis, and M. Trodden. Cosmic Strings and Electroweak Baryogenesis. *Physics Letters B*, 335:123, 1994.

[38] R. Brandenberger, A.-C. Davis, T. Prokopec, and M. Trodden. Local and Nonlocal Defect-mediated Electroweak Baryogenesis. 1994. BROWN-HET-962.

[39] J. Ambjorn, N. Nielsen, and P. Olesen. Superconducting cosmic strings and w condensation. *Nuclear Physics B*, 310:625, 1988.

[40] W. Perkins and A.-C. Davis. Cosmic Strings and Electroweak Symmetry Restoration. *Nuclear Physics B*, 406:377, 1993.

[41] M. Dine, R. Leigh, P. Huet, A. Linde, and D. Linde. *Physical Review D*, 46:550, 1992.

[42] M. Shaposhnikov. On Nonperturbative Effects at the High Temperature Electroweak Phase Transition. *Physics Letters B*, 316:112, 1993.

[43] G. Farrar and E. Shaposhnikov. Baryon Asymmetry of the Universe in the Minimal Standard Model. *Physical Review Letters*, 70:2833, 1993.

[44] M. Gavela, P. Hernandez, J. Orloff, and O. Pene. Standard Model CP Violation and Baryon Asymmetry. *Modern Physics Letters A*, 9:795, 1994.

[45] P. Huet and E. Sather. Electroweak Baryogenesis and Standard Model CP Violation. 1994. hep-ph 9404302.

[46] M. Fukugita and T. Yanagida. Baryogenesis Without Grand Unification. *Physics Letters B*, 174:45, 1986.

[47] M. Luty. Baryogenesis via Leptogenesis. *Physical Review D*, 45:455, 1992.

[48] D. G. Lee, R. N. Mohapatra, M. Parida, and M. Rani. Predictions for Proton Lifetime in Minimal Non-supersymmetric SO(10) Models: An Update. *Physical Review D*, 50:??, 1994.

[49] R. N. Mohapatra and G. Senjanović. Superlight Axion and Neutrino Masses. *Zeit. fur Phys.*, C17:53, 1983.

[50] B. A. Campbell, S. Davidson, J. Ellis, and K. Olive. Cosmological Baryon Asymmetry Constraints on Extensions of the Standard Model. *Physics Letters B*, 256:457, 1991.

[51] I. Antoniadis, J. Ellis, J. S. Hagelin, and D. V. Nanopolous. Supersymmetric Flipped SU(5) Revitalized. *Physics Letters B*, 194:231, 1987.

[52] R. N. Mohapatra. Mechanism for Understanding Small Neutrino Mass in Superstring Theories. *Physical Review Letters*, 56:561, 1986.

Searching for Machos in the Cosmic Dark Matter

C. Alcock

Lawrence Livermore National Laboratory, Livermore, CA 94550, USA

The search for the elusive dark matter is arguably the most compelling scientific quest of this decade.One of the proposed constituents of the dark matter is the Macho, an object of astrophysical mass that emits no light. The status of the searches for Machos in the galactic dark matter, and the significant progress reported by the EROS, MACHO, and OGLE collaborations, is reviewed. The prospects for future progress is discussed. It is probable that significant progress can be made in determining the quantity and distribution of Machos in our galaxy using techniques that are well understood.

§1 Introduction

This section was written in close collaboration with the MACHO Science Team, and benefitted also from discussions with the EROS and OGLE teams.

Evidence for large quantities of dark matter in the Universe has been accumulating for many years, and it is now well established that most of the mass of the Milky Way and similar galaxies is predominantly in some presently invisible form (see e.g., the review of Fich and Tremaine [9]). This dark matter cannot be in the form of normal stars or gas, which can readily be detected.

A wide range of candidates have been proposed to account for this dark matter. These fall into two main classes: the particle-physics candidates such as massive neutrinos, axions or other weakly interacting massive particles (WIMPs), and the astrophysical candidates, including substellar objects below the hydrogen burning threshold $\approx 0.08 M_\odot$ ('brown dwarfs'), or stellar remnants such as white dwarfs, neutron stars or black holes; these are generically known as massive compact halo objects (Machos). Such objects would be much too faint to have been detected in current sky surveys. The search for Machos is our topic here.

It is worth stressing that the searches for Machos described here, and all of the other dark matter searches described elsewhere in this volume, are looking exclusively for objects located in the dark halo of the Milky Way. These experiments do *not* provide an estimate of Ω, since the amount of dark matter within 50 kpc of spiral galaxies is approximately known from rotation curve data. This represents $\Omega \sim 0.05$: if $\Omega = 1$ either the halos must extend far beyond 50 kpc or there must be intergalactic dark matter.

Paczynski [17] suggested that Machos could be detected by their gravitational 'microlensing' of background stars. This indirect technique does not depend upon light emitted by the Machos. It should be remembered, though, that the objects in question are not required to be dark, merely significantly fainter than the source stars. However, since the Large Magellanic Cloud is located 30° from the galactic plane, the optical depth from known (i.e., visible) stars is no more than $\sim 10^{-8}$ (Bahcall [5]).

§2 The Gravitational Microlens

The gravitational lens is characterized by an angular size θ_E, called the Einstein radius. If the star lay directly behind the Macho, its image would be a ring of light of this size. If the Macho is separated from the source by some finite angle, $\theta = b/D_\mathrm{d}$, where b is the impact parameter of the Macho, it will magnify the star by an amount $A(u) = (u^2+2)/u\sqrt{u^2+4}$ that depends only on the ratio of the separation to the Einstein radius, $u = \theta/\theta_E$. Note that $A > 1.34$ when $u < 1$, and $A \sim u^{-1}$ when $u \ll 1$. The Einstein

radius is related to the underlying physical parameters by

$$\theta_E = \sqrt{\frac{4\pi G M D_\mathrm{ds}}{c^2 D_\mathrm{d} D_\mathrm{s}}}$$

where M is the Macho mass, and D_d, D_s, and D_ds are the (observer-lens), (observer-source), and (lens-source) distances, respectively. The term 'microlensing' is used when θ_E is so small that the two images cannot be separated with current observing equipment, and the image doubling cannot be seen.

Frequently the related quantity $R_E = \theta_E D_\mathrm{d}$ is referred to as the radius of the Einstein ring, where R_E is the physical size of the ring described above, measured at the location of the lens. For the cases that are discussed here, with a source distance of 50 kpc $\approx 10^{10}$ AU and a lens distance of 10 kpc, the Einstein radius is $R_E \approx 8\sqrt{M/\mathrm{M}_\odot}$ AU. The coincidence between this scale and the orbit of the earth around the sun can be exploited in the study of microlenses, as will be described below.

The observable phenomenon, as the lens passes the star at constant relative projected velocity **v**, is the varying magnification of the star.

The magnification is given as a function of time by $A[u(t)]$, where $u(t) = \sqrt{\omega^2(t-t_0)^2+\beta^2}$. Here β is the impact parameter in units of the Einstein radius, t_0 is the time of maximum magnification, and ω^{-1} is the characteristic time of the event.

Fortunately, microlensing has many strong signatures which can discriminate it from stellar variability:

- Since the optical depth is so low, only one event should be seen in any given star.
- The deflection of light is wavelength-independent; hence, the star should not change color during the magnification.
- The events should have lightcurves as predicted by the theory (above).

All these characteristics are distinct from known types of intrinsic variable stars; most variable stars are periodic or semi-regular, and do not remain constant for long durations. They usually change temperature and hence color as they vary, and they usually have asymmetrical lightcurves with a rapid rise and slower fall.

In addition to these individual criteria, if many candidate microlensing events are detected, there are further statistical tests that can be applied:

- The events should occur equally in stars of different colors and luminosities.
- The distribution of impact parameter u_{min} should be uniform from 0 to the experimental cutoff $u(A_{min})$.
- The event timescales and peak magnifications should be uncorrelated.

When a Macho is detected and its light curve is measured, one determines three parameters, ω, t_0, and β. Of these, only ω is related to the Macho's physical parameters:

$$\omega = \frac{v}{D_\mathrm{d}\theta_E}.$$

For microlensing of sources in the Large Magellanic Cloud, $\omega^{-1} \approx 70\sqrt{M/\mathrm{M}_\odot}$ days. One of the principal limitations of present experiments in this area, one which cannot be resolved with purely ground-based work, is that our uncertain knowledge of the quantities D_d, D_ds, and v for an observed microlensing event means that the uncertainty in the inferred mass of the Macho, M, spans more than an order of magnitude [15]. The mass is our only clue to the true nature of the Macho.

Parallax, obtained by the simultaneous observation of a microlens event from telescopes separated by a distance of order R_E, allows one to measure a second parameter, $\tilde{v} = (D_\mathrm{s}/D_\mathrm{ds})v$, (generally one obtains one vector component of $\tilde{v}$, and the absolute magnitude of another), and sometimes the full velocity [11].

The complete solution of a microlensing event requires, in addition, the determination of θ_E. This is very difficult to achieve, since typically $\theta_E <$ milliarcsec. For a handful of cases in which the Macho passes directly in front of the source star (or nearly in front of the star), it will be possible to measure θ_E. This is possible because the magnification equation given above must be modified to take into account the finite angular size θ_S of the star. The modification depends only upon the ratio θ_S/θ_E. (An important complication is introduced by the center to limb variation of brightness over the face of the star.) If θ_S can be estimated spectroscopically, one can in turn infer θ_E. This has been done for one event seen by the MACHO Project.

In the rare cases when one has ω, $\tilde{v}$, and θ_E, the Macho mass, distance, and transverse velocity can each be determined. For example,

$$M = \frac{c^2\tilde{v}\theta_E}{4G\omega}.$$

§3 The "Macho Fraction" from Gravitational Microlensing

The gravitational microlens 'optical depth' is the quantity that probes directly the Macho fraction of the dark matter. If ρ_{total} is known, for instance along the line of sight to the Large Magellanic Cloud, then the experimental estimate of the optical depth τ along this line of sight yields the fraction of the total dark matter that is in the form of Machos.

The 'optical depth' τ for gravitational microlensing is defined as the probability that a given star is lensed with $u < 1$ or $A > 1.34$ at any given time, and is

$$\tau = \pi \int_0^{D_\mathrm{s}} \frac{\rho(D_\mathrm{d})}{M} R_E^2(D_\mathrm{d})\, dD_\mathrm{d}$$

where ρ is the density in Machos. Since $R_E \propto \sqrt{M}$, while for a given ρ the number density of lenses $\propto M^{-1}$, the optical depth is independent of the individual Macho masses. Using the virial theorem, one finds that $\tau \sim (V/c)^2$, where V is the rotation speed of the Galaxy.

More detailed calculations [15] give an optical depth for lensing by halo dark matter of stars in the Large Magellanic Cloud of

$$\tau_\mathrm{LMC} \approx 5 \times 10^{-7},$$

under the assumptions that (1) all of the dark matter is in the form of Machos; and (2) the most naive model of the halo (spherically symmetric, small core radius) is correct. This very low value means that only one star in two million will be magnified by $A > 1.34$ at any given time. (Note that this estimate assumes that all of the dark matter is in Machos, and hence is a crude upper limit to the optical depth.)

The ongoing ground-based surveys follow millions of stars photometrically in order to obtain event rates of a few per year against the Large Magellanic Cloud. The optical depth in principle can be estimated directly from the experimental data, once a statistically significant number of events has been recorded. This interpretation is complicated, in practice, by inefficiencies introduced by the irregular sampling, and by the very crowded star fields that must be observed in order to make the large number of photometric measurements.

The interpretation of microlens optical depths in terms of the fraction of the dark matter in the form of Machos is limited by our poor knowledge of ρ_{total}. The dark halo may be spherical or flattened [19], and if flattened it may or may not be aligned with the plane of the disk [10]. The core radius of the halo is not well understood. Also important is a complete understanding of the contribution of the disk and bulge to the total mass interior to the solar circle. This has turned out to be more significant than expected.

Improving our knowledge of ρ_{total} is clearly important. Improvement can be obtained by measuring the optical depth along many well separated lines of sight, combined with the measured rotation curve of the disk. The variation of τ with location on the sky would provide us with a form of *tomography* of the dark matter distribution, which gives the missing shape information. The rotation curve data provide the normalization for ρ_{total}. To date measurements have only been attempted along two well separated lines of sight, towards the Large Magellanic Cloud and towards the central bulge of the Milky Way. The latter line of sight does not probe the dark halo unambiguously, but is very useful in probing the structure of the Milky Way, and thus helping to determine ρ_{total}.

§4 The Current Experimental Situation

Three groups have reported detections of microlens events: the MACHO Project [1], [2], [3], the EROS Project [4], and the OGLE Project [20], [21]. The MACHO Project follows $\sim 10^7$ stars in each of the Large Magellanic Cloud and the galactic bulge. The MACHO Project has recorded ~ 60 events, of which four are toward the Large Magellanic Cloud, and the rest toward the galactic bulge. The EROS Project has followed $\sim 4 \times 10^6$ stars in the Large Magellanic Cloud, and recorded two microlensing events. The OGLE Project follows $\sim 10^6$ stars in fields near the galactic center, and has recorded ~ 14 events.

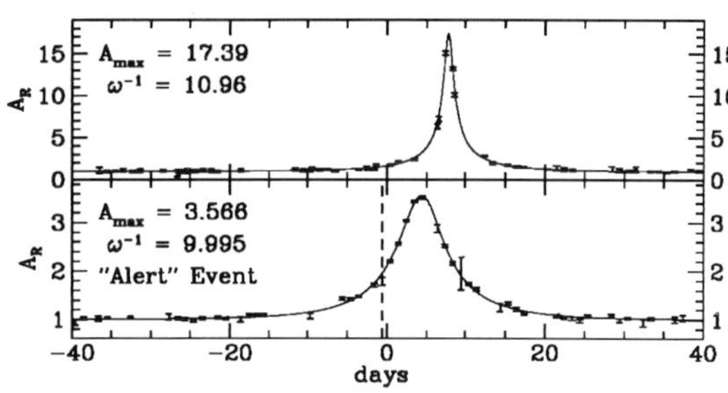

Figure 1: Two microlensing events seen by the MACHO Project, with the discovery date of the latter shown

Two example events are shown in Figure 1. The left panel shows a high magnification event from the MACHO Project (the star is in the galactic bulge). The right panel shows an event that was noticed well before it reached maximum magnification. (The epoch at which the event was recognized is marked on the figure.) This "alert recognition" is a new capability now routinely used by both MACHO and OGLE. The MACHO Project currently uses the alert process to select events for more careful photometric monitoring at Cerro Tololo Interamerican Observatory, and for spectroscopic follow-up. Spectra have been obtained during two alert events, and in each case spectra were taken at epochs spanning the maximum. The spectra showed no evolution during the events, confirming the gravitational lens interpretation.

The early results from these surveys are very surprising. Preliminary estimates of the optical depth toward the Large Magellanic Cloud indicate that (1) the event rate is too high to be accounted for by a previously known population; and (2) the event rate seems to be lower than what was expected for a spherical dark halo in which all of the dark matter is comprised of Machos. The recent work of Bahcall *et al.*[5] has *almost* eliminated low luminosity stars as the objects responsible for the microlensing that is seen.

A quantitative examination of the MACHO observations towards the Large Magellanic Cloud [3] shows that the best-fit Macho fraction, assuming a standard model for the dark matter distribution, is $f \sim 0.2$, but with very large uncertainties. It is not possible to exclude the possibility that all of the dark matter is in the form of Machos (especially considering our poor knowledge of the true distribution of the dark matter), nor is it possible to exclude that *none* of the halo dark matter is in Machos. These results do show that significant answers to the question of the Macho fraction are within reach of the current experiements, perhaps supplemented by new experiements as described below.

The galactic bulge observations have yielded substantial surprises. The estimated optical depths are $\sim 3\times$ higher than the upper limits that were estimated prior to the experimental results. These upper limits [15], [18] were based upon standard models for the galaxy. This higher than anticipated optical depth means that the standard models for the mass distribution in the galaxy are in need of *substantial* revision. Suggestions have included (1) that the galactic bulge is a very massive bar, with the long axis pointing in our direction d spacecraft 1 AU away to resol; (2) that the mass of the disk is much greater than anticipated ([2]); (3) that the dark halo has a low density or even hollow core [8]; and (4) that the dark halo is a very flattened structure, and the high rate toward the galactic center is due to halo Machos (this is tantamount to saying "there is no halo!").

It is important to note that the interpretation of microlensing results toward the Large Magellanic Cloud is contingent upon understanding the galactic bulge results.

§5 What can be achieved from the ground?

The experience of the three groups allows us to project what could, in principle, be achieved from the ground. This assumes that next- generation experiments are exploited fully to realize the limits of ground-based microlens surveys. It will be possible to measure the optical depth toward the Large Magellanic Cloud with reasonable certainty.

It might be possible to do this also for the Small Magellanic Cloud. It will not be possible to obtain useful estimates of the microlens optical depth toward other targets that are high above the galactic plane. All other targets that contain adequate numbers of stars are either so distant that the observations are confusion limited (e.g. M33), or the stars are in globular clusters that are too faint to observe usefully from the ground. This limits the potential usefulness of the gravitational microlens tomography technique that was described above.

Two groups ([7], [6]) will attempt to measure microlens event rates toward M31. Their technique is sensitive only to Machos in the halo of M31, and will in the end have none of the additional information about the nature and prevalence of Machos than can be achieved for our own galaxy.

Ground-based work alone will not be able to determine the Macho fraction of the dark matter. The uncertainties in the shape and radial profile of the dark halo will be the primary limitation. This limitation can be alleviated by using the Macho Parallax Effect, reducing the uncertainty about where along the lines of sight the individual Machos are, and adding some velocity information. This is discussed in the next chapter. The situation with respect to the bulge observations is more promising. It is possible that ground-based observations will be able to distinguish between a large central bar, pointing in our direction, and a "maximal" disk, but Han and Gould [13] have expressed doubts that this can be achieved. It would not be possible, however, to distinguish a "maximal" disk from an extremely flat halo; this would require either the parallax technique or the tomography technique.

Finally, it is possible to search for planetary companions around the Machos by looking for the small distortions in the light curves that are produced [16]; [12]. This search would require a latitudinally spread network of telescopes providing nearly continuous monitoring of events that are found in a survey such as MACHO.

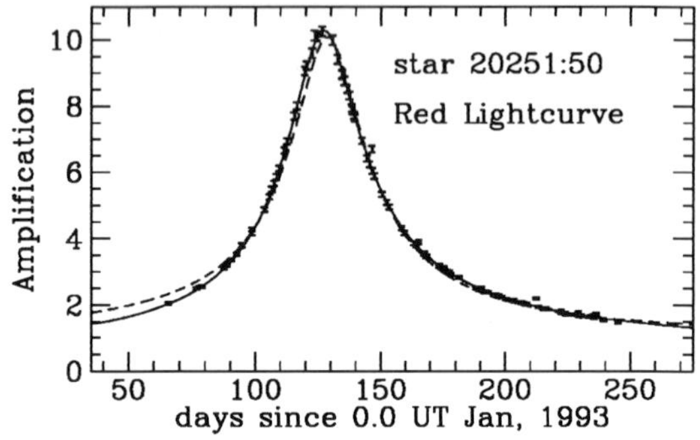

Figure 2: Lightcurve of an event from the MACHO Project showing the parallax effect. The dashed line is the best fit not including the motion of the earth around the sun, the solid line includes the motion of the earth around the sun.

§6 Observing Macho Parallax

In the introduction we showed the great value of measuring the projected velocity $\tilde{v}$, especially in elucidating the masses of Machos. This can be accomplished by the parallax technique: measuring an event from two different lines of sight. From the ground one can measure Macho parallaxes only for events of exceptionally long duration, so that the acceleration of the earth around the sun significantly changes the projected velocity during the event. These long duration events are uncommon.

Figure 2 shows the light curve of one of the more spectacular events discovered by the MACHO project. This is the only event for which the effect of parallax motion has been detected. There is a significant asymmetry in the lightcurve of this star, which is well fit when the motion of the earth is taken into account. The best fit indicates that the velocity of the Macho with respect to the line of

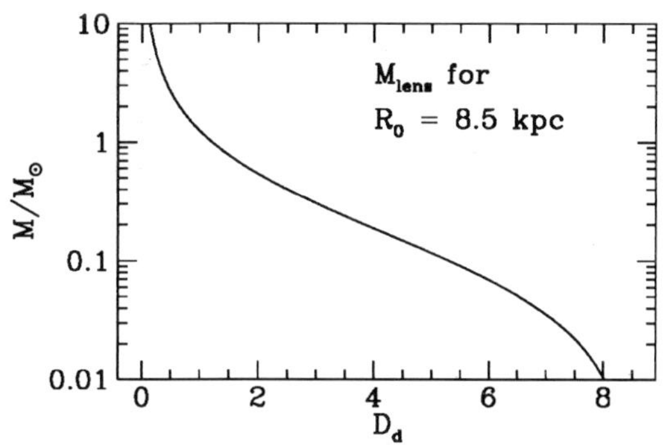

Figure 3: Mass of Macho versus distance to the Macho

sight between source star and the Sun is 54 ± 5 km/sec when projected to the position of the Sun. The knowledge of the projected velocity allows us to derive a relation between the mass of the Macho and its position along the line of sight as seen in Figure 3.

From this figure, we can immediately conclude that the lensing object is either in the Galactic Disk or it is a brown dwarf in the galactic bulge. The latter is unlikely as it requires that the transverse speed of the Macho with respect to the line of sight to the source star be much smaller than average value for bulge lenses. If we had information like this for a large fraction of the microlensing events, then we would be able to tell whether the majority of the lensing objects are in the galactic disk, bulge, or bar. Similarly, toward the Large Magellanic Cloud, routine measurements of microlensing parallaxes would allow us to determine whether the observed events are Machos in the dark halo, the Milky Way disk, or perhaps in the Large Magellanic Cloud.

Ground-based work faces a severe limitation in measuring parallaxes: it is only for the very longest events that the earth will move far enough in its orbit during the event to make the parallax effect visible from the ground. This limits us to measurements made on events for which the Machos are in the galactic disk, and eliminates from useful study bulge and dark halo Machos. For this vast majority of cases, the microlensing events will end before the earth has had a chance to move very far, so the parallax information will only be available if events can be observed simultaneously from the earth and a small satellite in a solar orbit of order ~ 1 AU away from the earth.

Satellite parallaxes would go a long way toward cracking the Macho mystery. For events seen toward the Large Magellanic Cloud, measurement of $\tilde{v}$ would clearly distinguish between Galactic Machos ($\tilde{v} < 300\,\mathrm{km\,s^{-1}}$) and those in the Large Magellanic Cloud ($\tilde{v} \sim 2000\,\mathrm{km\,s^{-1}}$). For Galactic Machos, $\tilde{v} \sim v$, so that one could distinguish between Machos in the disk ($v \sim 50\,\mathrm{km\,s^{-1}}$), the thick disk ($v \sim 100\,\mathrm{km\,s^{-1}}$), and the halo ($v \sim 200\,\mathrm{km\,s^{-1}}$). It would be possible to measure the transverse direction for at least some of these Machos, yielding additional information about their distribution.

The situation is somewhat more complicated for Machos seen towards the bulge. Parallaxes can statistically distinguish between Machos lying in the disk versus those which lie in the bulge itself, and also allow one to estimate the mass of the former [13]. Moreover, for ~ 10% of bulge Machos and ~ 4% of disk Macho events, it should be possible to measure θ_E from the ground and therefore use the parallax to determine M, D_d, and v individually.

§7 Conclusions and Recommendations for the Future

What we know now:

- Gravitational microlensing has been observed.
- The event rate toward the Large Magellanic Cloud is lower (~ 20% but with large errors) than expected for a "standard halo" filled with MACHOs of substellar mass.
- The event rate towards the galactic bulge is 3 times higher than expected.
- The standard models for the inner Milky Way are incorrect.

Priorities for the next few years:

- Get good statistics for the event rate towards the Large Magellanic Cloud (primary handle on halo MACHOs).
- Attempt to get the event rate towards the Small Magellanic Cloud
- Map the event rate as f(l,b) near the galactic center to disentangle galactic models.

Long-term priorities:

- Develop photometric follow-up network to determine lightcurves better (to measure higher order effects such as finite angular size of star, which yields much more information per event, and to search for planetary companions).
- Macho Parallax Satellite (measure lightcurves simultaneously from earth and spacecraft 1 AU away to resolve the lens; this will determine where the Machos are).

§Acknowledgments

This research was supported by the Department of Energy at the Lawrence Livermore National Laboratory and by the National Science Foundation through the Center for Particle Astrophysics at the University of California.

§References

[1] Alcock, C., Akerlof, C.W., Allsman, R.A., Axelrod, T.S., Bennett, D.P., Chan, S., Cook, K.H., Freeman, K.C., Griest, K., Marshall, S.L., Park, H.-S., Perlmutter, S., Peterson, B.A., Pratt, M.R., Quinn, P.J., Rodgers, A.W., Stubbs, C.W., & Sutherland, W., 1993, *Nature*, **365**, 621.

[2] Alcock, C., Allsman, R.A., Axelrod, T.S., Bennett, Chan, S.,

Cook, K.H., Freeman, K.C., Griest, K., Marshall, S.L.,Perlmutter, S., Peterson, B.A., Pratt, M.R., Quinn, P.J., Rodgers, A.W., Stubbs, C.W., & Sutherland, W., 1994, *ApJ*, in press.

[3] Alcock, C., Allsman, R.A., Axelrod, T.S., Bennett, Chan, S., Cook, K.H., Freeman, K.C., Griest, K., Marshall, S.L.,Perlmutter, S., Peterson, B.A., Pratt, M.R., Quinn, P.J., Rodgers, A.W., Stubbs, C.W., & Sutherland, W., 1995, *Phys. Rev. Letters*, **74**, 2967.

[4] Aubourg, E., Bareyre, P., Brehin, S., Gros, M., Lachieze-Rey, M., Laurent, B., Lesquoy, E., Magneville, C., Milsztajn, A., Moscosco, L., Queinnec, F., Rich, J., Spiro, M., Vigroux, L., Zylberajch, S., Ansari, R., Cavalier, F., Moniez, M., Beaulieu, J.-P., Ferlet, R., Grison, Ph., Vidal-Madjar, A., Guibert, J., Moreau, O., Tajahmady, F., Maurice, E., Prevot, L., & Gry, C., 1993, *Nature*, **365**, 623.

[5] Bahcall, J.N., Flynn, C., Gould, A., & Kirhakos, S. 1994 ,

[6] Bouquet, A., Kaplan, J., Melchior, A., Giraud-Héraud, Y., & Baillon, P., 1993, preprint, astro-ph/931209.

ApJ Lett., **435**, L51.

[7] Crotts, A. & Tomaney, A., 1993, Columbia preprint.

[8] Evans, N.W. 1993, *M.N.R.A.S.*, **260**, 191.

[9] Fich, M. & Tremaine, S. 1991 *Ann. Rev. Astron. Astrophys.*, **29**, 409.

[10] Frieman, J. and Scoccimarro, R. 1994, *ApJ*, **431**, L23.

[11] Gould, A., 1994, *ApJ*, **421**, L75.

[12] Gould, A., & Loeb, A., 1992, *ApJ*, **396**, 104.

[13] Han, C., and Gould, A., 1994, preprint, astro-ph/9410052.

[14] Kiraga M., and Paczynski, B. 1994, *ApJ Lett.*, in press.

[15] Griest, K., 1991, *ApJ*, **366**, 412.

[16] Mao, S. & Paczynski, B., 1991, *ApJ*, **374**, L37.

[17] Paczynski, B, 1986, *ApJ*, **304**, 1.

[18] Paczynski, B. 1991, *ApJ Lett.*, **371**, L63.

[19] Sackett, P., and Gould, A., 1993, *ApJ*, **419**, 648.

[20] Udalski, A., Szymanski, M., Kaluzny, J., Kubiak, M., Krzeminski, W., Mateo, M., Preston, G.W., & Paczynski, B., 1993, *Acta Astronomica*, **43**, 289.

[21] Udalski, A., Szymanski, M., Kaluzny, J., Kubiak, M., Mateo, M., & Krzeminski, W., 1994a, *ApJ*, **426**, L69.

Direct Detection of WIMP Dark Matter

Brian L. Dougherty

Department of Physics, Stanford University, Stanford, CA 94305, USA

Direct searches for galactic abundances of weakly-interacting, particulate dark matter began roughly a decade ago. Through the use of essentially conventional detectors, certain candidates have already been ruled out: most notably Dirac neutrinos with masses above about 10 GeV/c^2. Given another order of magnitude in detector sensitivity, supersymmetric (SUSY) partners may be found to constitute dark matter. To improve detectable rates, groups around the world have explored and are now developing newer technologies. Greatly improved limits, or even detections, should be attained within the next decade.

The broad motivational arguments for direct searches of weakly interacting massive particles (WIMPs) as dark matter candidates are condensed into an esoteric nutshell below. It must be understood that the following summarizes many years' work by several people, and glosses over a wealth of detail.

There is strong astrophysical evidence reflecting dynamics at the largest scales that corroborates the theoretically well-motivated notion that our universe is critically closed. The success of Big Bang Nucleosynthesis calculations indicates, however, that roughly 90 % of all matter is non-baryonic. In addition, large-scale structure and fluctuations in the cosmic microwave background suggest that the bulk of this, possibly quite exotic matter was nonrelativistic, or "cold" during freeze-out. Finally, it is argued that generic particles with masses near that of the proton and which freeze-out cold would now be abundant enough to close the universe if their interaction strength is weak, assuming no initial asymmetry of particle-antiparticle densities. And there are a number of particles hypothesized on independent grounds which might provide dark matter candidates.

Observations at smaller astronomical scales reveal that most galaxies are surrounded by a roughly spherical, non-dissipative halo. The mass indicated is about ten times that which shines as stars, but not more than what can be accommodated by Nucleosynthesis limits. Hence, there may exist a large, relatively non-radiative baryonic component to galaxies. It is unlikely however, that slow-moving, weakly-interacting particulate matter does not clump in the same space, especially if it seeds galaxy formation. In our own galaxy, if the local halo were entirely composed of WIMPs, then scattering events due to dark matter particles in devices of moderate size may just be detectable if typical backgrounds could be reduced.

At Snowmass '94, a collection of those involved in the experimental search for dark matter assembled to discuss this field: models, parameters, avenues of research, progress, and the future. In the text that follows, I will try to present a coherent statement which summarizes those discussions, and I will add a few comments of my own. Let me apologize here for any errors made in reporting the efforts of various groups.

Some of the people who took part at these meetings, and who provided notes from which this text was prepared include: D. Akerib, C. Alcock, F. Avignone, P. Barnes, K. Buckland, B. Cabrera, D. Caldwell, B. Chugg, A. DaSilva, J. Engel, G. Gelmini, S. Golwala, M. Gray, K. Griest, F. Halzen, G. Jungman, M. Kamionkowski, R. Lanou, A. Piepke, J. Primack, B. Pritychenko, F. Pröbst, M. Ressell, R. Ross, B. Sadoulet, D. Seckel, T. Shutt, A. Sonnenschein, N. Spooner, W. Stockwell, and C. Tao.

In the following I will survey the current status and prospects for direct detection of particulate dark matter. I will not present a comprehensive review of the theoretical and observational motivations for the subject, nor its experimental history. That was not the purpose of the Snowmass '94 workshop, which was directed much more to the future. Besides, there exist already several excellent reviews of the dark matter problem. (And I rely heavily upon these for the introductory sections, especially that by Primack, Seckel and Sadoulet, [1].) Instead, I will first merely outline the preliminaries: the motivations and evidence for, properties of, and constraints on dark matter, followed by some discussion of candidates. I will then briefly review the important aspects of detectability: event rates, signatures, and backgrounds. My emphasize throughout will be on the possibilities for direct detection of WIMPs as a well-motivated class of dark matter candidate. I will then take a somewhat closer look at the techniques that have been used before, and those which are currently under development. Finally, I will cover some experimental issues of community-wide concern, and finish by discussing possible future advances.

§1 Introduction

To the pure experimentalist, it may be enough that the problem is well motivated, important, and challenging. To mount a direct search, it remains however to understand the particle fluxes that are implied, and the likely cross sections for observable effects. In other words, the probable abundance and distribution of dark matter must first be specified, including its density in both physical space and energy. Then the character of possible particles must be argued: mass ranges, interaction types and strengths, secondary effects, etc.. I now turn to expand briefly upon these details.

1.1 The Abundance of Dark Matter

Observational evidence for the presence of matter in excess of that indicated by radiation is consistently positive at nearly all astronomical scales. And a clear trend has emerged, that the relative amount of dark matter (i.e. the mass to luminosity ratio) increases with the volume of space that is sampled.

The largest-scale evidence (at 100's to 1000's Mpc) includes red-shift surveys of galaxy counts versus distance, and some peculiar velocity maps.[2] These provide sufficient information to derive gravitational potentials, and return the largest estimates of Ω, the overall density of our universe, including values consistent with closure, $\Omega = 1$. However, this data is particularly sensitive to h, which denotes ongoing disagreements over the current Hubble parameter (with $0.4 < h < 1$). There are also geometrical and evolutionary ambiguities which prevent these results from being considered as definitive.[3] On somewhat smaller scales, infall data for galactic collections, especially the Virgo cluster, tracks large-scale flows toward gravitational wells. The lensing of distant objects by foreground clusters can also be used to measure masses.[4] For clusters believed to be in equilibrium, matter densities can be estimated from both velocity dispersion measurements of component galaxies, and from the spectral distributions of x-ray emissions.[5][6] At this point, however, the volume of space sampled is on the order of 1 Mpc^3, and the indicated dark matter densities are lower, around $\Omega \sim 0.2$.

At very small, galactic scales the available information is generally more convincing, but also yields even lower estimates of Ω. Many spirals exhibit flat or rising angular velocities of tracer objects as a function of distance from the galactic center.[7] Such rotation curves extend up to 3 times the optical radius, as indicated by measurements of neutral hydrogen cloud emissions. These results demand that mass increases with distance, in contrast to the light intensity, which tends to fall away exponentially. The total mass implied is often as much as an order of magnitude greater than that suggested by the luminosity. Similar estimates are obtained from velocity dispersion measurements of globular clusters, velocity maps of outlying satellites, and the pattern of disturbances caused by objects which penetrate the galactic plane.

There are strong theoretical arguments that require closure. First of all is the "naturalness" problem, which stems from the way Ω, if it does not start out exactly at unity, rapidly deviates to either infinity or zero as the universe expands. There is an unlikely fine tuning during the Big Bang that would be required to maintain Ω at anything close to unity today. Secondly, the "flatness" problem is solved by inflationary hypotheses, which also fix Ω at unity, provided the cosmological constant vanishes.[8] Finally, the growth of early density fluctuations, which seed structure formation and proceed more slowly for less dense universes, would conflict with currents limits on angular variations in the cosmic background radiation unless Ω is large.[9]

The consensus is that (1) although the relevant data was not mentioned above, luminous objects clearly account for less than about one percent of the mass necessary for closure, (2) galaxies typically contain many times more matter than that which shines, and (3) the overall density of our universe is at or near its critical value. Given the sparsity of galaxies, therefore, any dark matter effects sought by even an indirect-detection experiment are likely to be dominated by dark matter in the immediate neighborhood. Depending on bias in structure formation (or, how well galaxies and clusters trace the underlying dark matter)[10], this may lead to a local, observable density on the order of only $\Omega \sim 0.1$.

1.2 The Nature of Dark Matter

Lighter elements formed as the universe cooled to below the binding energies of nuclei. Since we know the relevant nuclear interaction cross sections, we can predict the average relic amounts of various isotopes. The observed abundances of H, D, ^{3}He, ^{4}He, ^{6}Li, and ^{7}Li all agree remarkably well with these calculations, provided the primordial baryonic mass density is $\Omega_B \sim 0.017/h^2$.[11] A firm upper limit of $\Omega_B < 0.14$ is also indicated. While there have been several suggested modifications of the Big Bang Nucleosynthesis model employed above, the data "prefer" variables in non-standard models to approximate values encountered in the standard analysis. Thus, if $\Omega = 1$, then most the matter in our universe is composed of nonbaryonic particles. But also, much of the "missing" mass in galaxies could be accounted for by non-luminous baryons, presumably in the form of brown dwarfs or black holes. (Dust and gas are disfavored by galactic transparency measurements.) However, to avoid bar-formation instabilities, spiral galaxies are thought to have collapsed within huge, dissipationless halos with much larger mass.[12] It is difficult to understand how a relatively large fraction ($\sim 10\%$) but no more of these baryons fall into the disk.

Large-scale structure arises from primordial density fluctuations, where matter moves only under the calculable influence of gravity. Computer-based studies indicate that a significant density of small-scale fluctuations, or "seeds" are needed to explain the observed angular modulations in cosmic microwave background radiation, as well as most regularities in the distributions and velocities of galaxies and clusters. This in turn implies that the dominant particles must have been "cold", or nonrelativistic during "freeze-out"; i.e. after they fell out of thermal equilibrium with the cooling universe, and when their exponentially dropping number (due to annihilation) was halted by the diluting effects of the Big Bang expansion. Relativistic, or "hot" particles would free stream, smoothing small fluctuations.[13]

It is therefore likely that such slow-moving particles would clump in the gravitational wells of galaxies, especially if they seed galaxy formation. To keep the dark matter from collapsing into the galactic disk as well, it should be dissipationless: i.e. weakly interacting. That idea is bolstered by the following argument.

It is usually, though not compellingly assumed that primordial particle-antiparticle asymmetries are negligible. This case would be guaranteed if the particles in question were Majorana fermions. Then relic dark matter densities are directly related to annihilation cross sections at freeze-out, $< \sigma v >$.[14] The latter are in turn related to the square of the mass: that is, for most weakly interacting particles. A concordance exists such that our universe would be closed by WIMPs of mass around a few GeV/c^2. If an initial asymmetry existed, which could occur for Dirac WIMPs, then this becomes a lower bound.

To summarize, if $\Omega \sim 1$, then the dark matter is almost certainly nonbaryonic, most of it is likely to be cold, and it is arguably as massive as protons but weakly interacting. Now, the annihilation strength is related (by crossing) to the cross section for elastic scattering with ordinary matter. Event rates in search experiments may then be estimated.

The obvious nonbaryonic candidates for dark matter are neutrinos, the only particles that are otherwise known to exist. But there are many laboratory and astrophysical measurements already available which restrict the viable range of neutrino masses. These include neutrinoless double beta decay limits, beta decay end-point measurements, accelerator and reactor-based oscillation data, the solar and atmospheric neutrino deficit problems, and supernova observations. There are also Big Bang arguments which (1) predict the present density and "temperature" of neutrinos, and (2) agree with accelerator measurements (of the Z width) in delimiting the number of light neutrino species to three. In fact, neutrinos are so well characterized that they are now ruled out as a cold dark matter candidate.[15] But they may comprise a hot "additive" which contributes certain necessary features to structure formation simulations.

Supersymmetric partners and cosmions, though speculative, are also "well-motivated" in the sense that they were first introduced on independent grounds. Supersymmetry solves the hierarchy problem in particle physics, and may be required when a quantum theory of gravity is finally achieved.[16] There is a wide, model-dependent range of SUSY scattering processes and strengths which could be obtained. Accelerator experiments limit the lightest superpartner mass to $\sim$ 20 GeV/c^2, but this is valid only within certain parameter regions of the minimal model. Cosmions, on the other hand, were invented to solve the solar neutrino problem, and as such their interactions are more tightly constrained.[17]

Other possibilities include even more hypothetical particles with ill-defined properties. The list includes magninos, light Higgs, and various other exotica. For the experimenter who contemplates a search for particulate dark matter, it may suffice that there exist a plethora of WIMP candidates that are not entirely ad hoc.

1.3 Direct-Detection Rate Estimates

Throughout this and most later sections I will emphasize the possibility of observing WIMP dark matter via direct interaction within terrestrial detectors. Here therefore is a review of WIMP interactions with ordinary matter.

Each type of dark matter candidate comes with a more or less restricted range of intermediary-particle coupling strengths, g_δ. In addition, the coupling to a particular nucleus, g_N, depends on whether WIMPs interact predominantly through axial currents to the spin of nuclei, or includes scalar terms. If the latter obtains (e.g. the WIMP is a Dirac particle), then nuclei recoil coherently when the overall momentum transfer is small. A large increase in scattering cross section is therefore anticipated, proportional to the square of the number of neutrons. If spin-spin interactions dominate (i.e. the WIMP is a Majorana fermion), then coherence is lost since most nucleon spins cancel pairwise. Dimensional analysis and kinematics then fix some of the remaining ambiguities: for low-energy elastic scattering with nuclei of mass m_N, the cross section is

$$\sigma_{\delta N} \sim g_\delta^2 g_N^2 \frac{m_\delta^2 m_N^2}{(m_\delta + m_N)^2} \frac{1}{m_I^4}, \tag{1}$$

where m_δ is the WIMP mass, and m_I is the mass of the intermediary particle.[18] Thus, for $m_N << m_\delta$ the cross section is independent of m_δ, whereas for $m_N >> m_\delta$ it varies as the square of m_δ.

The spin-dependent coupling strengths deserve further comment. In general, the internal nuclear spin structure must be modeled, and form factors assessed. Standard shell-model calculations, however, which are usually reliable for other applications, are inadequate for deriving the nuclear wave functions of many interesting nuclei; e.g. deformed germanium. Another difficulty arises from recent European Muon Collaboration experiments, which suggest that nucleon spins are not carried by the constituent quarks, but by the gluons. Better measurements are needed.[19] Different calculations should be checked against each other and known data: magnetic moments, beta decay strengths, etc.. Overall, form factors can make a difference of up to a factor of five in total cross section, especially for heavy nuclei. It is for these reasons that all experimenters should at least indicate which form factors are assumed, when relevant to the their conclusions.

To estimate fluxes pertinent for any direct search, the Milky Way halo must be understood. Like those for most spiral galaxies, our halo is presumed to be roughly approximated as an isothermal sphere, in which the dark matter density falls off nearly as the square of the distance from the galactic center. This model is supported not only by the rotation curve data, structural arguments (bar-mode stability), and observations of compact, outlying tracers alluded to earlier. Recently, huge clouds of thin gas and other infrared objects have been observed which completely envelope nearby galaxies.[20] The oblate spheroid shape of these otherwise dark objects may directly trace the dark particulate matter, if the latter supports the baryons against dissipational collapse into the galactic disk.

We therefore assume the dark matter density varies as

$$\rho = \rho_o \frac{a^2}{a^2 + r^2}, \tag{2}$$

where $\rho_o \approx 1.0$ GeV/cc is the core density and $a \sim 5.6$ kpc is a scale. Thus, in our local neighborhood, $\rho \approx 0.3$ GeV/cc. Now, the form of dark matter here is unspecified; i.e. it could be composed mostly of baryonic material. Recently initiated microlensing searches for massive, compact halo objects (MACHOs) have indeed begun to accumulate evidence that at least some of our halo is baryonic. In fact, a surprising number of MACHO-like objects are found near the Milky Way bulge.[21] This could indicate that the scale, a, above is too small to characterize the possible WIMP distribution. After all, dynamics at the inner regions of nearly every galaxy are roughly consistent with just the shining matter alone; i.e. no dark matter is needed. However, MACHO searches are as yet uncalibrated with respect to detection efficiencies, and the microlensing sources are still a bit ambiguous. Therefore, our halo may be made mostly of WIMPs with the above density profile.

Our isothermal halo presumably contains particles with the following Maxwell-Boltzman velocity distribution ...

$$\frac{dn}{dv} \propto \frac{v^2}{\sigma_H^3} exp\left(-\frac{v^2}{\sigma_H^2}\right), \; for\; v < v_e, \tag{3}$$

where $\sigma_H \approx 210$ km/sec is the halo velocity dispersion, and $v_e \approx 600$ km/sec is the escape velocity. Average WIMP velocities in a terrestrial reference frame must be computed after accounting for the Sun's movement through our galaxy, and the Earth's orbital motion. The gravitational influence of Earth and Sun must also be acknowledged. This all results in a mean speed of about $< v > \approx 310$ km/sec.

Clearly, more work needs to be done to refine our halo model and parameter estimates. (The same can obviously be said about all aspects of the dark matter problem.) And with continued scientific interest and support, such refinements are very likely to be forthcoming. For now, however, if one accepts what has been argued so far, then elastic WIMP scattering rates in a detector can be estimated ...

$$R \approx 4.3 \frac{X_N}{m_\delta m_N} \sigma_{\delta N} \rho < v > \; kg^{-1} day^{-1}, \tag{4}$$

where X_N is the fraction of detector mass that is active as a target, and when $\sigma_{\delta N}$, ρ and $< v >$ are measured in units of 10^{-38} cm^2, 0.3 GeV/cc, and 300 km/sec respectively. From the above discussion, these units are seen to be typical of WIMPs in our halo, so a small but observable direct-detection rate may be available. Note that since there are mass dependencies in $\sigma_{\delta N}$, event rates do not necessarily fall with target mass as explicitly indicated in equation 4.

§2 Direct Detection

Given the above estimate of very low scattering rates within ordinary matter, it is important to seek as many detection signatures as possible which are attributable to WIMP interactions and not to backgrounds. Then, depending on the available or foreseeable technology, a direct search should emphasize the sensitivity of a detector to one or more signatures. This involves instrumenting large detector masses, improving thresholds and resolutions, reducing or discriminating against backgrounds, etc.. In addition, multiple and redundant detection techniques should be encouraged. Beside providing further information to distinguish signals from background, these can help reduce ambiguities in the properties of the observed WIMPs.

In addition, we should assess technologies against the energy deposited when a dark matter particle interacts within a detector. Upon elastic scattering with a WIMP, a nucleus recoils with energy

$$\Delta E = \frac{m_N m_\delta^2 v^2}{(m_N + m_\delta)^2}(1 - cos\theta), \quad (5)$$

where v is the incoming WIMP velocity and θ is the center-of-mass scattering angle. Since most WIMP cross sections are isotropic, flat recoil spectra from $E = 0$ to $2m_N m_\delta^2 v^2/(m_N + m_\delta)^2$ are returned. Absolute maximum recoil energies occur when WIMP and target masses are equal. For $m_N << m_\delta$ recoil energies are independent of m_δ, whereas for $m_N >> m_\delta$ they vary as the square of m_δ. To set the absolute scale, for $\sim$ 2 GeV/c^2 halo WIMPs traveling at $\sim$ 300 km/sec and hitting $\sim$ 30 GeV/c^2 target nuclei (e.g. silicon), only about 120 eV of energy is deposited on average. This is small when compared to the thresholds of many common particle detectors.

In the following, the general issues of backgrounds and signatures will first be reviewed. Then a more detailed look will be taken at the kinds of detector methods in development today. Finally, specific topics of wide concern will be addressed and some future directions anticipated.

2.1 Backgrounds

Cosmic rays generate backgrounds both directly, via interactions in or near a detector, and indirectly, through the production of long-lived radioactive elements. The most straightforward method of dealing with this is to locate any experiment deep underground, where only neutrinos and the highest energy muons can penetrate. In addition, scintillators or other active shields surrounding a detector can be used to veto events associated with passing muons. To avoid the buildup of cosmogenic radioactivity, which is especially linked with the hadronic component of cosmic rays, materials used in the construction of experiments may be stored in relatively shallow underground facilities. Some materials may also be obtained from ancient sources: e.g. raw ores from selected mines, and shielding lead from long-sunken ships. Finally, muon-generated events are often directly distinguishable and of somewhat less concern since large amounts of energy are typically deposited.

Natural radioactivity is ubiquitous. Crustal abundances of uranium and thorium are measured at the parts per million level. A few substances typically used in equipment contain particular isotopes: ^{40}K in the glass of photomultiplier tubes, ^{115}In in vacuum seals, ^{210}Pb in shielding. As a response to the wide-spread need for low activity materials in other count-rate-limited applications, large radioactive surveys of construction elements and manufactured goods have been undertaken. Certain, "clean" ore bodies and industrial refining techniques have been identified. Special processes have also been developed to address particular needs.

When radioactivity cannot be avoided in materials found outside the detector proper, then active or passive shielding is required. Depending upon the kind of background that dominates, different shielding arrangements are called for. Alpha emissions are easily blocked by thin, clean shields. Gamma and beta fluxes are attenuated by heavy materials like lead, mercury, etc., and by using scintillators or ionization-based vetos. Neutrons can be moderated with lighter, especially hydrogenous shields to energies below the threshold for signals, or occasionally by using certain neutron-capture detectors as vetos. Complete attenuation is almost impossible, however, since (1) higher-energy gammas and neutrons that scatter about in shields down-grade in energy, often with an accompanying increase in flux, (2) cross-conversion of backgrounds can occur, as in bremsstrahlung, alpha-induced fission, and neutron capture excitation followed by gamma emission, and (3) active shields may contain radioactivity which eventually dominates the background.

Even when materials outside the detector are as well shielded as may be practical, residual activity in or effectively at the surface of the detector may be present. Ways to identify these backgrounds must be sought, if only to attempt a statistical subtraction. If the energy resolution is good enough, photopeaks can be identified, and their accompanying Compton continua modeled. Radioactive decay sequences can be identified through spectroscopy or timing, and the associated backgrounds of concern might be measured or simulated. The effectiveness of computer-intensive modeling may depend, however, on a detailed knowledge of the experimental geometry and the location of characteristic backgrounds. This may not be easy to acertain, since, by definition, information is accumulated slowly in a low activity environment. Of course, some checks can be performed by intentionally introducing radioactive sources in select positions. Also worrisome are dead regions or edge effects in detectors which return poor signal-carrier collection efficiency. If events can be localized, fiducial volume cuts should then be used.

Finally, instrumental effects can mimic or obscure signals. Electronics noise, microphonics, and radio-interference are all well-known nuisances which are addressable in a variety of ways.

2.2 Signatures

The first four signatures to be reviewed below are largely statistical in nature. The last involves more direct, event-by-event methods to discriminate against backgrounds.

In calorimetric detectors, the presence of WIMPs is indicated by an exponentially falling spectral shape, as predicted by combining the expected Maxwell-Boltzman velocity distribution of WIMPs with the flat recoil energy response indicated by equation 5. This shape is modified, however, when scattering energies are measured by detecting only particular kinds of signal carriers. It is well established that the ionization yield for nuclear recoils in solids is significantly less than that for electron recoils. This yield is also a function of energy: it decreases for smaller recoils. The source of this variation can be traced to the relatively low velocities and concomitant atomic neutrality that accompanies recoiling nuclei. For some scintillators, total light output also varies with recoil energy in characteristic ways, and this is due to dependencies in the population of excited molecular levels. In general, if electrons or nuclei are scattered, more or less light is produced, respectively. For many materials, these effects have been carefully calibrated, so few ambiguities exist with respect to the expected WIMP spectral shape.

Annual modulations in event rates will also signify the presence of WIMPs. Such modulations are due to the Earth's ≈ 30 km/sec orbital motion through the galactic WIMP "gas" whose velocity distribution is presumably isotropic in the halo reference frame. Since the mean deposited energy also varies, these modulations may be observed by threshold detectors as well as calorimeters. Furthermore, there is a calculable difference in the strength of this effect with latitude, so detectors placed at various locations on the Earth's surface might eventually be used to assess WIMP isotropy.[22]

The Sun's ≈ 230 km/sec motion through the galaxy leads to an apparent WIMP "wind" coming from a particular direction which rotates with the time period of a sidereal day. Detectors are needed in which information on recoil direction is preserved, either by direct tracking or the detection of spatial asymmetries in the signal carriers generated. And of course, isotropic backgrounds may then be discriminated against.

Both event rates and recoil energies should vary with the detector material, and in a correlated fashion, as indicated by equations 1 and 5. Thus, WIMP masses may be determined by obtaining signals from a variety of detectors. Typical background events are expected to have a different material dependence.

Most low-energy backgrounds are due to x-rays, which may (1) not penetrate very far into a detector, and (2) dissipate different amounts of energy into various signal carriers than do nuclear recoils. Thus, fiducial volume cuts can eliminate a large source of interference. And detectors which simultaneously measure, say, ionization and phonon production can discriminate against more penetrating gamma backgrounds on an event-by-event basis. The presence of residual neutrons, however, must be addressed separately.

2.3 Detector Technologies

(Let me take the opportunity here to thank Tom Shutt of the Center for Particle Astrophysics at U. C. Berkeley for contributions to the following text.)

Elastic scattering within a material bulk gives rise to a host of secondary effects. The deposited energy may be dissipated through ionization: including the generation of electron-ion or electron-hole pairs, or via the breaking of Cooper pairs and the creation of quasiparticles. Phonons may be produced which may be sensed promptly or, if one waits long enough, yield a pulse of heat. Scintillation photons may be generated. Finally, microscopic structural damage, or tracks, may be left. In various research groups around the world today, every one of these mechanisms are being studied for possible use in dark matter detectors.

In many cases, the primary energy "carrier" generated at a scattering site is not that which is ultimately detected. Ionization processes in solids yield "hot" electrons, which relax via the emission of phonons. Also, such carriers may be purposely converted in auxiliary sensors, to optimize detection efficiencies. Phonons absorbed by a superconductor at the surface of a solid produce quasiparticles which are more easily counted. The multitude of interdependencies means that every group pursues its own, unique technology, and still more detector varieties are possible.

For the rest of this section, a representative sample of the actual methods in use or being developed will be surveyed in rough, descending order of the energy required to produce the main signal carriers. This, rather arbitrary choice puts to the front techniques which provide some of the best background discrimination methods. Left to the end are methods which can deliver the best resolutions.

Mica Detectors

One U. C. Berkeley group is using ancient or annealed mica as solid state nuclear track detectors.[23] This material is capable of preserving recoils in the microscopic damage left behind when a scattering event occurs. Sensitivities of $\sim$ keV/amu are recorded by carefully measuring the forms of pits produced when clean, cleaved surfaces are washed with anisotropic etchants. The distinct advantage of ancient mica is that $\sim$ 1 billion year exposures are attainable, which makes up for the limited practicable target sizes, measured in the square millimeters. Care must be taken to date the mica sample, and assess whether tracks have faded due to heating in geologic time. Naturally occurring ^{238}U or ^{232}Th activity tends to result in quite large etch pits, since many alphas are emitted through the decay chains. These are distinguishable from the small pits resulting from $\sim$ keV nuclear recoils.

Some muon interactions and fission tracks are harder to recognize. However, beta and gamma backgrounds are completely avoided, since nuclear recoils and the subsequent damage do not result. WIMP limits comparable to the spin-dependent rates in much larger detectors are achievable. Furthermore, real-time track detectors are being developed which utilize much larger samples of freshly annealed, radioactivly screened material of selected composition. In addition, forward/backward recoil direction asymmetries can be resolved, yielding a powerful background discriminant, and confirming the presence of a WIMP wind.

Gaseous Tracking Detectors

Low-pressure, gas-filled time projection chambers were explored a few years ago at Saclay, and one is now under development at U. C. San Diego. Low-energy, secondary charges produced along tracks are drifted into sense planes, and recoils are reconstructed. Large magnetic fields reduce diffusion of these charges transverse to the drift axis and cause high-energy, primary electrons to spiral. The gas mixture and pressure is optimized when short, straight nuclear recoils can be discerned, and yet distinguished from low-energy, tightly-wound electron tracks. Nuclear recoil directions can be determined by the way ionization densities fall toward the end of tracks. (Stopping powers are on the low-velocity side of the Bragg peak.)

The U.C. San Diego group drift the secondary charges through high-voltage grids, causing scintillation that is recorded with a CCD camera. This approach yields position information in the drift direction, but is much simpler and less costly than a full TPC. They have demonstrated good electron/nuclear discrimination (getting rejection factors over 99 %, depending on energy) for recoils above a threshold of about 2 keV in 20 Torr of methane spiked with triethylamine, and still retain direction information in their 2D images. Energy resolution is relatively poor, however.

The main drawback to these detectors is that at low gas pressures a very large volume is needed to observe significant WIMP scattering rates. This raises the shielding and veto requirements, as well as the cost. Also, the radioactive backgrounds, which are typically high in such chambers, must be understood. In some sense, these detectors may be the best suited for an expensive follow-up experiment if a signal is seen using any of the other techniques.

Scintillators

Scintillators are being studied by many groups. The basis of electron/nuclear recoil discrimination has been known for many years. As mentioned above, there are various molecular energy levels that become populated when an event occurs within a scintillant. The associated de-excitation photons can vary both in energy (e.g. visible to ultraviolet) and time structure of emission. And different types of events excite those levels in different ways.

Thus, in room-temperature, thallium-doped NaI, the BPRS group finds that light from nuclear recoils arrives roughly 10 % faster than that from electron recoils.[24] They time the arrival of individual photo-electrons (p.e.) using two, fast photomultiplier tubes attached to each side of a NaI(Tl) crystal, and measure at least 3 p.e. per keV of deposited energy. Discrimination improves with p.e. count (recoil energy): at around 2 keV, rejection factors of about 5 are anticipated, while above 10 keV, rejection factors in excess of 20 are possible. Raw background rates are approaching 1 event/kg/keV/day, quite encouraging for detectors that have yet to be optimized for low activity.

A UK group, on the other hand, works with NaI crystals cooled to ≈ 100 K, and have varied the Tl doping in order to optimize the collection of both visible and UV light.[25] These two emissions have different arrival times, and their intensity ratio can differ by a factor of five, depending on whether the initiating events are electron or nuclear recoils. Thus, color-filtered light retrieval, or photo-electron timing yield rather good discrimination: they aim for a factor of 100 with just 5 or 6 photo-electrons.

Hybrid systems are also being studied. A Milano group has measured scintillation in cryogenic CaF_2 crystals, but do so simultaneously with phonons in order to discriminate against electron recoils. And various groups have looked into ionization and scintillation in other media, like liquid Xe.[26] A group at Brown has simultaneously observed rotons and scintillation created by alphas in liquid He, though the discrimination ability of this method has not been tested.[27]

All techniques involving scintillation, however, suffer from the fact that typically tens of eV are required, on average, to produce each photon, and thus the energy resolution and threshold for such measurements is much worse than for phonon or ionization signals. In addition, there are thorny technical challenges, like the difficulty of calibrating detector efficiencies for events in the bulk, where low-energy x-rays cannot penetrate.

A great advantage of the scintillator-only approaches is that large detector masses (over 100 kg) can be instrumented easily, and at relatively low cost. Moreover, these detectors are simple to operate. In addition, both sodium and iodine nuclei have spin, and are widely separated in mass, providing broad sensitivity to axial couplings.

Quite low radioactive backgrounds appear to be achievable as well. Special manufacturing processes can deliver clean NaI, and silica or quartz light pipes attached with selected optical glues are available. Photomultiplier tubes with negligible amounts of ^{40}K have also been manufactured recently. In fact, even without recoil discrimination, large NaI detectors have been used to set limits on WIMP masses within new event rate regions that had not been excluded. With discrimination capabilities, it is therefore possible that new WIMP limits may be achieved that are roughly comparable to those obtained using detectors with higher resolution.

Solid-State Ionization Detectors

The best current limits on WIMP rates in direct-detection experiments comes from data obtained using Ge and Si detectors which measure ionization only.[28][29] No fiducial volume cuts were employed, and neither were any signal-background discrimination techniques. The systems used were converted double-beta decay set-ups, which had demonstrated some of the lowest, raw backgrounds available: roughly 0.1 event/kg/keV/day around 10 keV. This was achieved through the use of highly selected materials, deep underground sites, and clean active and/or passive shielding. Dirac neutrinos, with their well-characterized and large scattering cross sections, have subsequently been ruled out as WIMP candidates for a wide range of masses above $\sim$ 10 GeV/c^2.

However, most of the energy deposited in semiconductors is dissipated as phonons. For electron (nuclear) recoils, only about $\sim 1/3$ ($\sim 1/10$) of the energy goes into the production of electron-hole pairs. And since state-of-the-art capacitive noise levels in large detectors is at best $\sim$ 0.3 keV rms, ionization thresholds of $\sim$ 2 keV limit sensitivities for nuclear recoils to those above $\sim$ 6 keV. Thus, WIMPs of mass below roughly 10 GeV/c^2 cannot be excluded at any event rate. To lower thresholds substantially might require almost as much effort as the development of new technologies, and without adding any background discrimination method, that effort could not result in sensitivities that would compete with those anticipated with the new detectors.

Superconducting Granule Detectors

At least one group, centered at Bern, is planning a detector based on superheated superconducting granules.[30] The idea is that when a WIMP scatters in a small sphere of Sn or Zn, the deposited energy goes into breaking Cooper pairs, suppressing the Meissner effect. Depending on the temperature, granule diameter and strength of an applied magnetic field, the threshold for a granule "flip" can be adjusted, and sensed using pickup loops and amplifiers or SQUIDs. Recoil energy thresholds of $\sim$ 1 keV have been demonstrated in small systems using $\sim 10\mu$ diameter spheres.

Gamma backgrounds typically do not impart enough nuclear recoil energy to engender phase transitions. Those that do often produce multiple flips, while WIMP scatterings would always be single-sphere events. Alphas and neutrons also produce single flips, however, so such backgrounds might be irreducible.

To achieve significant WIMP event rates, billions of spheres must be monitored at a sensitivity that allows one to measure flip multiplicities for each event. Magnetic interactions among spheres complicate the problem, leading to physically very large arrays or to arrangements which make use of flip avalanches to yield signal amplification. It is not yet clear whether sufficient uniformity in sphere manufacture can be achieved.

Since these are threshold devices, not calorimeters, seasonal event-rate modulations can be detected, but to measure energy spectra, thresholds must be adjusted, either slowly through time or via configuration in different regions of the array.

Prompt Phonon Detectors

In most solids a majority of the energy deposited in any interaction is initially dissipated into phonons. The excitation energies of these "primary" phonons are measured in milli-electron volts, or about 1000 times smaller than those for electron-hole pairs in semiconductors, but also above most Cooper-pair binding energies. Therefore, better energy resolution is possible in principle by measuring phonons, or by converting phonons into broken Cooper pairs (quasiparticles).

A Stanford group is working on superconducting transition-edge sensors as a means to detect the prompt phonons generated by energy depositions in cryogenic, mono-crystalline silicon absorbers.[31] For a significant portion of these signal carriers, mean free paths can be centimeters long, depending on phonon energy and crystal purity. Fast detectors with masses of a few hundred grams can then be instrumented. In addition, since such "ballistic" phonons tend to propagate along crystal axes, patterns of "focused" phonons may be discerned on the surfaces of these detectors. This would help determine event locations, and eliminate backgrounds. There may even be very subtle asymmetries in phonon generation associated with the direction of the scattering event which, if they could be detected, would further enhance WIMP sensitivities.

When phonons are absorbed at the solid surface by a thin superconducting line, the heat imparted tends to drive the metal up its resistive transition edge. In titanium-based sensors, signals are proportional to the length of line driven completely normal. Energy resolutions of roughly 1 keV FWHM for 6 keV x-rays were found. With tungsten, material parameters and low-impedance SQUID-based readout permit feedback, which allows the metal to be held near the steep middle of its resistive transition, while electrical bias power is integrated.

This mode of operation delivers resolutions of $\sim$ 100 eV.

Pulse onset, height and length all vary with event energy, depth, and location with respect to the sensor aspect. Some focusing evidence has also been demonstrated. Thus timing, signal partitioning among closely spaced sensors, and pulse shapes can each be used (redundantly, in some cases) to establish fiducial volume cuts.

Discrimination between nuclear and electron recoils is achievable by simultaneously measuring ionization. This has been accomplished with $\sim$ 1 keV resolution. But small electric fields must be used for charge-collection, else Joule phonons generated during drift obscure those produced by the scattering event itself. This necessitates the use of very pure crystals in order to avoid electron or hole trapping and the subsequent loss of charge resolution. Nevertheless, above a $\sim$ 50 % selection cut, rejection factors have been observed to be in the 1000's. As an alternative, some measurements have indicated that there may be subtle phonon spectral differences associated with electron and nuclear recoils. This suggests the possibility of discrimination based on phonon signals alone, if sensors of proper refinement can be made.

An Oxford group is putting superconducting tunnel junctions on Nb substrates.[32] Scattering events generate phonons and quasiparticles in the Nb. Signal carriers of both types may be converted in the junctions to measurable tunneling currents. This promises much better energy resolution because conversion processes in the junction are very efficient: i.e. little re-radiation back into the substrate occurs. In fact, other groups have reported full widths around 10 eV for 6 keV x-ray events occurring directly in such a sensor.

The main difficulty for instrumenting voluminous detectors is phonon collection efficiency. Large, series arrays of tunnel junctions have been contemplated. Oxford has pursued this and an alternative possibility: of plating on the Nb surface large-area superconducting absorbers with higher transition temperatures, which are then attached to the tunnel junctions. Phonons are first converted into quasiparticles in these absorbers, and in turn diffuse until they are "trapped" in the junction material. Energy resolutions of $\sim$ 3 keV FWHM for 60 keV gamma rays have been measured.

Recoil discrimination may be had by carefully evaluating the pulse shape. Many substrate quasiparticles recombine before reaching the surface absorber, and so produce late, secondary phonons. Recoil-dependent differences in quasiparticle-phonon partitioning, and in the primary phonon energy distributions and decay times may lead to discernible pulse-shape variations.

Bolometers

A Munich group plans to instrument a $\sim$ 1 kg array of sapphire blocks with superconducting phase-transition thermometers.[33] These last are broad-area metal sensors, not thin lines, which are temperature-biased at the steepest portion of their resistive transition. It takes many milliseconds for all of the energy deposited in an event to drain from the substrate through this sensor. The metal heats upon phonon absorption, and the subsequent resistance pulse is read out very accurately using SQUIDs; i.e. the device is a slow, total calorimeter. Energy resolutions of $\sim$ 200 eV FWHM for 6 keV x-rays have been obtained using Ir/Au layers plated on 30 g substrates.

The lightness of sapphire constituents makes this substrate especially responsive to smaller-mass WIMPs. And it is reasonable to assume raw backgrounds can be had that are comparable to those already achieved in solid-state ionization systems. Therefore, it is expected that sensitivities for WIMP masses down to 1 GeV/c^2 will be improved by orders of magnitude. No added background discrimination techniques are required.

A Tokyo group has studied $\sim$ 3 g detectors composed of LiF substrates with small, neutron-transmutation-doped germanium thermistors attached.[34] Such NTD Ge devices have resistances that are rapidly dependant on temperature, throughout a wide dynamic range. Their fundamental resolution limit, dictated by Johnson noise, has nearly been achieved. When measuring energies deposited in a substrate however, phonon collection efficiency and speed are limited by thermistor size and Kapitza boundary resistance with respect to the substrate or interface. The LiF-NTD Ge resolution is only $\sim$ 9 keV FWHM for 60 keV gammas, so more development is planned. Nevertheless, since ^{7}Li and F both have spin, and are light, such a detector would be quite sensitive to low mass WIMPs with axial couplings.

The Center for Particle Astrophysics group centered at Berkeley is planning to instrument a $\sim$ 1 kg array of $\sim$ 5 Ge crystals affixed with NTD Ge thermistors.[35] They have spent considerable time addressing the interface problem and optimizing thermistor parameters. Energy resolutions below $\sim$ 700 eV FWHM for 60 keV gammas have been demonstrated using $\sim$ 100 g substrates.

Recoil discrimination is to be attained by simultaneously measuring ionization. Once again, low drift fields must be maintained. Some difficulties have been found with regions near the crystal edges and surfaces that yield poor charge collection efficiency, but successful methods are employed to reduce these effects. Ionization resolutions comparable to those for phonons are obtained. Recoil discrimination factors of $\sim$ 100 have been reported, above the appropriate thresholds. Since most of the recoil energy goes into phonons, the thermistor signal allows greater sensitivity to lower-mass WIMPs than in comparable solid-state Ge ionization systems. With raw backgrounds at a level of $\sim$ 1 event/kg/keV/day, it is expected that event-rates may just enter the region in which SUSY WIMPs could be found.

2.4 Miscellaneous Considerations

At Snowmass a number of "subgroup" meetings were held to discuss various issues of general, experimental interest. These included: detector material dependencies, WIMP event-rate modulations, background discrimination methods, analyses of data, and WIMP cross sections within SUSY models. Notes from some of these meetings have already been incorporated into the preceding text.

The Use of Multiple Materials

It has long been noted that WIMP masses and velocity distributions may be indicated by the differences in signals recorded by detectors made of dissimilar materials. Little work has been published, however, quantifying the statistical strength of this statement. There appears first of all the question of how, even in the absence of backgrounds, one fully resolves definite signals. If WIMPs come in a multiplicity of masses, then quite convoluted recoil spectra may be recorded. Such a possibility is not only allowed a-priori, but must be entertained to avoid biasing the interpretation of data. Presumably, as many different well-chosen detector materials must be employed as there are WIMP masses, and the latter number is not known.

Assuming dominance by only one WIMP of mass m_δ, then it should suffice to observe clear signals in two target materials of dissimilar atomic masses m_1 and m_2, so long as m_δ is smaller than both m_1 and m_2. The uncertainties in this WIMP's velocity distribution and in m_δ necessarily worsens as the statistical strength of the signal weakens.

A second question concerns how well different detector materials may be used to resolve backgrounds further, and so reach better limits than would be the case for each detector treated individually. Here we suppose two different targets, say silicon and germanium, are run at the same site and within the same local shielding environment. Known backgrounds are then subtracted from each data set, given supplementary measurements and modeling; for example, photopeaks are assayed and the associated Compton continua are computed. There typically remains events extending to higher energies which likely do not contain any WIMP signals. If one irreducible background source, such as neutron scattering, is now assumed to dominate, then the residual spectra from each detector can be compared, and inconsistencies will indicate the presence of further backgrounds or WIMPs.

The above analysis cannot be applied to data taken in different local environments, since backgrounds vary so widely. It is feared, however, that these considerations are moot even for multiple detectors operated within the same cryostat. Due to differences in auxiliary structures like electrical contacts and mechanical supports, identical conditions may never be obtained. The radioactive content in such proximate components may vary in a way that overwhelms any common backgrounds. Indeed, in earlier experiments, different spectra have been observed (often due to trace alpha activity) in otherwise identical detectors fabricated simultaneously by the same processes and from a single source of materials.

Statistics

When backgrounds are present, many questions arise. Recoil spectra from WIMPs are continuous, and presumably fall away gradually at higher energies. Statistical analyses are therefore not as simple as, say, the determination of the number of counts in a photopeak superimposed upon a smooth background. And indeed, as evidenced by the debates over the interpretation of neutrinoless double beta decay data, the treatment of small numbers of counts even for peaks is not yet standardized. Continuous backgrounds and instrumental artifacts like noise cannot simply be extrapolated from signal-less regions of the recorded data. Reasonable ranges of variations for these underlying spectra must become part of the systematic uncertainty in any claim of a WIMP signal or limit. And such ranges must include the errors on any auxiliary measurements of known, and hence subtractable backgrounds or instrumental effects.

Not only are there debates throughout the community on the correct statistical procedures, but the very meaning of conclusions sometimes come under fire. Indeed, it might appear that nearly every group analyzes its experiment with a particular shading of technique, if not a uniquely customized method. Most of the various procedures used today are "justifiable" for some reason(s). And this would not be too large an issue if the numerical results were no different regardless of what method was employed, so long as interpretations were understood. But results do differ, sometimes quite significantly. Nevertheless, even this should not lead to rancor while limits are still being placed on dark matter scattering rates. If the procedures used are legitimate, most physicists might view the differences as a part of the systematic ambiguity in any conclusion to an experiment. When detections are claimed, however, very close examination of methods may be required. Until dark matter signals are demonstrated beyond doubt in a single, very large scale search, statistical questions seem inevitable.

There is little in the way of an authoritative and comprehensive explication of statistical techniques and interpretations which deals especially well with small data sets. The latest Review of Particle Properties, the conventional resource for particle physicists, presents only a limited set of options.[36] Standard texts geared specifically for the physical sciences adopt various particular viewpoints. There remains a lively debate within the literature: recent articles offer opposing arguments for the "best" forms of analysis.[37]

What might be desired is a clear, compellingly argued review which thoroughly treats the advantages and disadvantages of the various, yet reasonable analysis techniques which abound. Certain of these techniques would be shown to be problematic, or compromised by interpretational difficulties, while certain others could be arguably favored. "Accepted" procedures might then become more consensually recognized, and the final results of various experiments may be cast into identifiable forms with commensurate standards of optimality and completeness. Different experiments could then be compared less ambiguously. However, it is not clear how such a "state-of-the-art" review would be received. One could argue that the "science" of statistics is still evolving, and that consensus regarding techniques will solidify or change regardless of such input.

Assumptions and the Treatment of Data

There is considerable latitude available in the arguments and assumptions that characterize the dark matter problem. It is therefore incumbent upon the interpreter of experimental data to specify what range of model and parameter space is being tested. The most important, basic issues include astrophysical assumptions concerning our halo. Then the WIMP properties should be surveyed: interaction types (axial, vectorial, inelastic?), and cross sections, masses, etc.. Form factors may be standardized, but one must be cognizant of possible variations: EMC versus Naive Quark Model spin structures, and nuclear theory controversies.

In the empirical analysis and presentation of data, certain desiderata have been discussed. The aim is to reach some consensus, the better to compare and correlate results. First of all, it has been suggested that experimental groups always display, if not exchange their raw data, thus promoting clarity with respect to the available information. It is to be understood that this obvious(?) step will not encourage abuses, but rather a spirit of community, even cooperation.

That spirit is likely to be essential, at least in the short term. The statistical strength of the first WIMP "detection" will probably be weak, given the size of many projected experiments. Regardless of the "handles" one experiment may have on WIMP properties (e.g. spectral or material dependencies), the first positive results may only indicate a combined range of cross sections and particulate number densities in our halo. The interpretation of such signals will become rather more compelling when different experiments are closely compared.

In addition, the treatment of backgrounds should be clear to all reviewers. Although this subject has largely been covered in the above text, we will re-appraise the procedures here. We may distinguish three modes of reducing background. First, certain backgrounds can be identified and eliminated on a event-by-event basis, either through the use of topological or other discriminators. Such direct methods are based most "closely" on the data, and should be well characterized. The actual techniques in use, and reviewed above include: recoil tracking or distinctions made in track length or direction, and signal partitioning into ionization, scintillation, phonons etc. or via particular spectral differences in a single species of signal carrier.

Secondly, depending on the quality of auxiliary information, a few remaining backgrounds could be estimated. Thus, neutron fluxes may be measured with nearby counters, and their effect simulated via Monte Carlo. The provisos on such statistical procedures have already been mentioned, but especially important is the need to incorporate statistical and systematic uncertainties into one's analysis. There are many such indirect methods in use or planned. These include: evaluation of photopeaks and radioactive decay chains, use of segmented detector arrays, correlated effects with shielding alterations or introduced background sources, detectors made of different materials, etc..

Finally, irreducible backgrounds of unknown origin may persist. These are among the most difficult to treat properly, and a firm starting point for analysis is preferred. Attempts should be made to avoid ad hoc procedures. One could compare "fits" of WIMP signals with data via likelihood ratio calculations or other general methods. Monte Carlo studies may also be tried, provided the techniques and interpretations are well characterized.

It is also desirable to reduce the dependence on models that can complicate the interpretation of results. Thus, in an initial presentation, WIMP events rates (not cross sections) may be plotted versus mass. This ties the conclusions closer to the data, and minimizes the remaining ambiguities to issues like WIMP velocity distributions, form factors, ionization yields, etc..

One should heed the history of the solar neutrino problem. Initial observations of positive effects were weak, and the relevant theory was just entering its childhood. This spurred innovations, both in theory (refinements of the solar model, MSW solutions) as well as in experimental technique (use of proton-decay detectors, development of gallium technology). Later results from different kinds of searches helped bolster confidence that the problem is real, as well as narrow the range of possible solutions. Progress occurred slowly, painstakingly through both individual and community-wide effort.

Monte Carlos

How good are the current Monte Carlo programs used to model backgrounds? Sequential versions of GEANT, a CERN product, differ by as much as a factor of three in output for neutron cross sections.[38] A reliable set of resources is needed that are checked against low energy empirical data.

Present uses for Monte Carlos include the simulation of responses and calibration of detectors, and the evaluation of backgrounds of all sorts. An experimenter may need to understand gammas and betas from radio-isotopes, muon spellation, bremsstrahlung, etc.. Studies of neutron fluxes from fission sources and muon-induced reactions are also of interest. Shielding may then be optimized, and experimental efficiencies predicted before going through the excessively slow and expensive proccess of altering components in, or even whole designs for the physical apparatus.

The best Monte Carlo programs track individual particles as realistically as possible. Multiple scattering, energy loss, straggling, etc. are all simultaneously accounted for as particles are tracked through successive layers of an apparatus. Secondaries are noted and tracked individually. Branching proccesses in radioactive decay are accurately modeled, as are cross sections for inelastic interactions. Finally, the physical layout of an experiment is easily inputted, including shielding geometries which incorporate penetrations for cabling, cryogenic structures, etc.

How does one avoid the use of Monte Carlos? Background sources that are introduced into an apparatus can sometimes be used to measure responses. Altering the shielding, turning off the veto, etc. can also provide information. It is unlikely however, that Monte Carlos can be avoided completely.

SUSY - WIMP Rates

There appears to be no good "toy" model which can accurately approximate most aspects of the SUSY-WIMP theory "space". But there are many parameters that must be chosen before SUSY-WIMP scattering rates can be calculated. And often, rates must be recalculated even as just one parameter is changed. The indicated "overhead" in time and effort is too much for a typical experimenter to invest. Yet that same physicist would like to work out for himself how sensitive various experiments could be.

Luckily, SUSY theory with respect to WIMP rates is not so esoteric that it cannot be automated. Theorists M. Kamionkowski and G. Jungman have written a computer program which will do the necessary calculations.[39] This program is intended as a public resource, and is soon to be made available. It is still being checked: against other programs and specific calculations, as well as for bugs. Their goal is a 10 % level of accuracy for most output, with warnings when the uncertainties expand excessively.

The user may specify halo parameters, form factors, and constraints on WIMP masses obtained from other kinds of experiments. Some detector characteristics, thresholds, and backgrounds may also be incorporated. There are libraries of standard SUSY parameters which can be altered if desired, and scanned according to the user's rules. The code is compartmentalized for easy checking and updating. Different outputs may be chosen: plots of rates versus mass, differential cross sections versus SUSY parameters, etc.. Thus, our experimenter can "explore" SUSY-WIMP implications on his own.

2.5 Future Prospects

Current WIMP limits are determined with few-hundred-gram detectors which measure only one type of signal carrier, immersed in radioactive backgrounds of around 10 events/kg/day. How can backgrounds be lowered in bulk systems, by say, five orders of magnitude, when interactions due to atmospheric neutrinos begin to dominate?

First of all, raw radioactive backgrounds must be reduced. With continuing improvements in underground sites, clean sheilding and construction material, a factor of 10 seems possible. This will require substantial effort in ore identification, refining techniques, trace analysis, stockpiling, etc.. But with foreseeable interest and support, including industrial needs for low-activity systems, this goal is not unrealistic.

The last decade has seen remarkable progress in the development of new, more sensitive detectors. The capabilites of some devices are quite beyond those encountered in traditional particle physics experiments. Resolutions are better, leading to lower thresholds. Multiple signal carriers are measured simultaneously. The locations of low energy events in solids can be determined.

A few of these technologies will mature over the next few years, and relatively large scale experiments will be brought "on line". But exploration of the range of possibilites seems only to have begun. We can be confident that refinements and innovations will lead to even better background rejection methods. Large-scale detectors may are likely to be made which are capable of discriminating nuclear and electron recoils at a level above 99.9 %.

There may remain residual gamma and beta activity. More worrisome however, might be neutron-scattering backgrounds, which pass the discrimination cuts and mimic WIMP signals. Somewhat "heroic" efforts could be entertained, for instance to enclose the experiment with massive, fission-free moderators and absorbers.

Fiducial volume cuts and detector segmentation may, however, provide the means to attain the last factor of reduction that is needed. Low-energy background radiations are typically not very penetrating. And as a consequence, they are also prone to multiple scattering.

Thus, we may indeed reach effective background levels of 10^{-4} events/kg/day. This might enable a search of nearly the entire range of possible SUSY-WIMPs

§References

[1] J. Primack et al. *Ann. Rev. Nucl. Part. Sci.*, 38:751, 1988.

[2] A. Dressler et al. *Ap. J.*, 313:42, 1987.

[3] E. Loh and E. Spillar. *Ap. J. Lett.*, 307:L1, 1986.

[4] J. Tyson et al. *Ap. J. Lett.*, 349:L1, 1990.

[5] J. Hughes. *Ap. J.*, 337:21, 1989.

[6] L. Cowie et al. *Ap. J.*, 317:593, 1987.

[7] V. Rubin et al. *Ap. J.*, 238:471, 1980.

[8] A. Guth. *Phys. Rev.*, D23:347, 1981.

[9] N. Kaiser and J. Lasenby, editors. *The Post-Recombination Universe.* 1988.

[10] A. Deckel and M. Rees. *Nature*, 326:455, 1987.

[11] K. Olive et al. *Phys. Lett.*, B426, 1990.

[12] S. White and M. Rees. *Mon. Not. R. Astron. Soc.*, 183:341, 1978.

[13] M. Davis et al. *Ap. J.*, 292:371, 1985.

[14] B. Lee and S. Weinberg. *Phys. Rev. Lett.*, 39:165, 1977.

[15] L. Krauss. *Phys. Rev. Lett.*, 64:999, 1990.

[16] E. Witten. *Nucl. Phys.*, B166:513, 1981.

[17] J. Faulkner and R. Gilliland. *Ap. J.*, 299:994, 1985.

[18] M. Goodman and E. Witten. *Phys. Rev.*, D31:3059, 1986.

[19] J. Ashman et al. *CERN-EP/87-230*, 1987.

[20] P. Sackett et al. *Nature*, 370:441, 1994.

[21] C. Stubbs. *these procedings.*

[22] F. Avignone et al. *Phys. Rev.*, D47:5238, 1993.

[23] D. Snowdwn-Ifft et al. eds. D. Cline and R. Peccei, (World Scientific). *Trends in AstroParticle Physics, procedings*, page 440, 1990.

[24] BPRS Collaboration. *Astroparticle Physics, to appear*, 1994.

[25] N. Spooner et al. *Nucl. Phys.*, 35:162, 1994.

[26] Belli et al. *Gran Sasso preprint LNGS 93/55*, 1993.

[27] S. Bandler et al. *J. Low Temp. Physics*, 93:785, 1993.

[28] D. Reusser et al. *Phys. Lett.*, B235:143, 1991.

[29] D. Caldwell et al. *Phys. Rev. Lett.*, page 510, 1988.

[30] M. Abplanalp et al. *J. Low Temp. Physics*, 93:809, 1993.

[31] B. Cabrera et al. eds. YSuzuki and K.Nakamura, (Universal Academy Press). *Frontiers of Neutrino Astrophysics, procedings*, 1992.

[32] A. Hahn et al. *J. Low Temp. Physics*, 93:611, 1993.

[33] W. Seidel et al. *J. Low Temp. Physics*, 93:797, 1993.

[34] M. Minowa et al. *J. Low Temp. Physics*, 93:803, 1993.

[35] P. Barnes et al. *J. Low Temp. Physics*, 93:791, 1993.

[36] *Phys. Lett.*, B:1275, 1994.

[37] R. Kraft et al. *Ap. J.*, 374:344, 1991.

[38] R. Brun et al. *CERN Data Handling Div. Rep. No. DD/EE/84-1*, 1984.

[39] jungman@npac.syr.edu.

Axions: Experiment and Phenomenology

Leslie J Rosenberg

Massachusetts Institute of Technology,
Department of Physics and Laboratory for Nuclear Science,
Cambridge, MA 02139, USA

The axion, a hypothetical elementary particle, was proposed originally as a solution to the strong CP problem in QCD. It was later recognized that axions would be good candidates for the dark matter as well. This is an self-contained review of the current status of axion experiment and phenomenology, emphasizing developments reported at recent Snowmass and Berkeley workshops.

§1 Overview

In a by now classic set of observations, rotation speeds of stars in galaxies as a function of distance from the galactic center show a universal feature: the speed is approximately constant for distances outside the core of the galaxy. This "flat rotation curve" and other observations as well imply the existence of large halos of nonluminous matter surrounding galaxies. The density of this dark matter at our position in the galaxy is approximately 0.3 GeV/cm^3, this value uncertain by perhaps a factor of two. The origin of the dark matter remains a mystery, and discovering its identity is one of the most exciting challenges confronting science today. It seems plausible from the success of models of nucleosynthesis in explaining the abundance of light isotopes, and of cosmological inflation in explaining the flatness and homogeneity of the present universe, that the bulk of dark matter is non-baryonic, with the baryonic component accounting for no more than 10-20% of critical density. The leading non-baryonic dark matter candidates are neutrinos, weakly interacting massive particles (such as the lightest supersymmetric particle), primordial black holes, and axions. Axions fall into the category of "cold dark matter," whereas neutrinos are "hot dark matter." Generally speaking, studies of large scale structure and galaxy formation favor substantial amounts of cold dark matter over hot dark matter, although this point is hardly settled. If the dark matter is a mixture of the two, it is the cold component that dominates galactic halos. If our nucleosynthesis arguments are correct and our flat universe prejudice proves well founded, then the dominant component of dark matter is non-baryonic, likely a new and exotic form of matter. The recent microlensing evidence for MACHOs could very well be a substantial baryonic dark matter component. However, it is unlikely MACHOs are the dominant dark matter, even in our own galactic halo.[1] So, it is the non-baryonic component of dark matter—perhaps axions—that is now the subject of intense search.

The existence of the axion is well motivated from a particle physics point of view as it would explain why, contrary to expectations, the strong interactions conserve parity P and CP. This "strong CP problem" is solved by the presence of the axion,[2] a prediction of an elegant solution, Peccei-Quinn symmetry, to the strong CP problem of QCD. The resulting axion mass is a free parameter, with plausible range from 10^{-12}eV to 1MeV, some 18 orders of magnitude. Current limits on the axion mass from astrophysical and cosmological considerations constrain the axion mass to the three decades 1-1000μeV. Laboratory experiments are underway to probe the first decade of allowed mass.

This is an self-contained overview of the current status of axion experiment and phenomenology, emphasizing developments reported at recent Snowmass[3] and Berkeley[4] workshops. The Snowmass workshop was, as far as the axion discussions were concerned, the continuation of the Berkeley workshop, and the combined material presented at both workshops is a coherent snapshot of our current understanding of the axion. The goal of this paper is to report on axion presentations from these workshops and place these presentations in the overall context of axion phenomenology and experiment. Much of the background material is closely adapted from Kolb and Turner's wonderful book "The Early Universe".[5]

§2 Theory I: The Axion and QCD

QCD, the theory of the strong interactions, has an impressive string of successes. The non-Abelian nature of the interaction is experimentally established. Decay rates and quantum statistics support the notion of color. Cross sections and branching ratios are in accord with perturbative predictions. There is, however, one loose end, the strong CP problem. The essential non-Abelian nature of the theory introduces QCD effects that violate CP. This CP violation could surface as, say, an electric dipole moment to the neutron. However, no such moment is seen, with present bounds billions of times below the expected level of CP violation. The source of CP violating effects can be traced to the QCD vacuum. Leaving out considerable detail, the reasoning behind this assertion is as follows: The QCD vacuum is the lowest energy gluon field configuration. In QCD, there are many degenerate vacuua, each labeled by winding number n, and gauge transformations can change one winding number vacuum into another. So, in order to preserve our cherished notion of gauge invariance, we construct a gauge invariant vacuum by a Bloch-wave-like superposition of vacuua, like so:

$$|\Theta\rangle = \sum_n e^{-in\Theta}|n\rangle \quad . \tag{1}$$

Such a vacuum, the Θ vacuum, is gauge invariant. This Θ vacuum resurfaces as a new effective non-perturbative term in the QCD Lagrangian proportional to $\bar{\Theta}G\tilde{G}$, where G is the gluon field strength tensor, $\tilde{G}$ its dual, and with M the quark mass matrix, $\bar{\Theta} = \Theta + \arg\det M$. The $\bar{\Theta}$ factoring the CP violating term contains the QCD vacuum Θ and phases from the quark mass matrix. The $G\tilde{G}$ term in the Lagrangian is a total derivative, and does not contribute to the classical equations of motion or perturbative effects. However, the term is explicitly CP violating, and can contribute non-perturbative CP violating effects. With a $\bar{\Theta}$ of order 1, the neutron in this way would acquire a CP violating electric dipole moment[6] of order 10^{-15}e-cm. The current limits are pushing 10^{-25}e-cm.[7], and this experiment in the context of the Θ vacuum limits $\bar{\Theta}$ to less than 10^{-10}. In a nutshell, this immeasurably small value of $\bar{\Theta}$ is the strong CP problem. Of course, a parameter equal to zero is no cause for alarm. After all, the neutrino mass may very well be zero, with no one losing much sleep over it. However, the immeasurably small $\bar{\Theta}$ is of greater concern. Recall that $\bar{\Theta}$ contains contributions from QCD (through the Θ vacuum) and weak interactions (through the quark mass matrix). It is likely the weak contribution is non-zero, so the Θ vacuum contribution appears with opposite sign and almost precisely cancels. The two contributions, one from QCD and one from weak interactions, are independent. The zero in the strong CP problem is more than a simple zero, it is the near perfect cancellation of two independent but finite effects. There are ways to evade this problem,[8] among them: Unlikely as it seems, perhaps a u- or d-quark is massless, so a chiral rotation can make $\bar{\Theta}$ zero. Lowest order current algebra mass relations forbid a massless light quark, but the issue of whether or not the effective quark mass is zero from higher order corrections is controversial. Perhaps $\bar{\Theta}$ of zero, forcing CP conservation, is a new tree level symmetry of QCD—dubbed soft CP—although no one has yet advanced a workable model. To me and many others, the most compelling solution is the one invoking the axion. However, M.Dine in his presentation said, "I give about the same odds for a massless u-quark and the axion, but less for spontaneous CP option." He also observed that, "A massless u-quark is hard (impossible) to establish experimentally."

The skeptic will wonder if this is a tempest in a teapot. Perhaps there really is no Θ vacuum with its objectionable CP violation, and hence no strong CP problem. The difficulty with this viewpoint is that the concept of the Θ vacuum provides a basis for understanding the observed light meson masses. Ignoring the Θ vacuum and taking for simplicity massless u- and d-quarks, the Lagrangian has symmetry $SU(2)_L \times SU(2)_R \times U(1)_V \times U(1)_A$. The SU(2) symmetries are the usual chiral symmetries, and their spontaneous breakdown gives rise to Goldstone bosons in the form of the pion triplet. The unbroken vector $U(1)_V$ symmetry ensures baryon number is conserved, but the symmetry corresponding to axial $U(1)_A$ is not seen in nature—for instance, there are no opposite parity partners to the $J^P = (1/2^+)$ baryon states such a conserved axial current would generate. This apparent spontaneous breakdown of the axial symmetry gives rise to another Goldstone boson in the form of the η. Now, moving closer to the real world, the u- and d-quarks have mass, and this should be reflected in comparable pion and η masses. However, we find the η is substantially heavier than the pion. This is termed the η mass problem, and its resolution is rooted in the Θ vacuum. The details are technical, but simply, the mass relations between pions and the η come from the usual current algebra. If the axial current couples to an additional light particle, light even when the u- and d-quarks are massive, then naive mass relations no longer hold.[9] The quantized oscillation about the vacuum Θ parameter is a particle that can couple into the axial current, and the oscillation quanta turn out to be axions. This effect is seen elsewhere. For instance, the amplitude for the decay $\eta \to 2\pi$ is zero in the case of even massive u- and d-quarks,[10] unless something like axions and the Θ vacuum is invoked.

It seems inescapable that the QCD vacuum is complicated and gives rise to interesting physical effects. However, in taking the vacuum seriously, we are left with the strong CP problem. Again, the solution I find compelling is that proposed by Peccei and Quinn.[11] They showed that by a slight extension of the standard model Higgs sector, the combined electroweak and QCD theories have a global U(1) symmetry, dubbed the Peccei-Quinn symmetry. At the scale f_{PQ}, the axion appears as the Goldstone boson of this spontaneously broken $U(1)_{PQ}$ symmetry.[12] In considering oscillations in Θ, we have turned it into a dynamical variable. We can then find the vacuum expectation value of this variable, considered in a classical potential. We first note that although the axion, as Goldstone boson of the broken $U(1)_{PQ}$ symmetry, is massless, it nonetheless acquires some effective mass (as does the η) through its axial anomaly. Besides the effective mass term, the axial anomaly introduces a term in the Lagrangian very much like like the Θ vacuum term and proportional to $C(a/f_{PQ})\bar{\Theta}G\tilde{G}$, where C is a model dependent constant related to the anomalies of the PQ symmetry, and a is the axion field. We can treat the Θ and anomaly terms in the Lagrangian as a classical potential, finding it minimized at the axion vacuum expectation value of $-\bar{\Theta}f_{PQ}/C$. At axion vacuum expectation value corresponding to this potential minimum, the CP violating $G\tilde{G}$ terms vanish. Rather than cumbersome, this stikes me as inevitable and elegant: introducing the spontaneously broken $U(1)_{PQ}$ symmetry generates the axion as the Goldstone boson. The axial anomaly endows the axion with mass, as well as introducing an additional CP violating term in the Lagrangian. The new term together with the Θ vacuum term have a classical potential minimum at the axion vacuum field value that removes the CP violating terms. There remains the issue of the value of the axion mass. As a Higgs-like particle, its coupling becomes stronger with increasing mass, and its mass is not very well constrained by the theory. However, we could certainly detect a 1 MeV axion—its presence in accelerator experiments would be hard to miss—but lighter axions are not so easily ruled out, with the corresponding PQ symmetry breaking scale greater than 100 GeV.

§3 Theory II. Properties of the Axion

In the last section, I reviewed how axions emerge as a consequence of the complicated QCD vacuum. This section continues with a discussion of how the axion couples to ordinary matter and radiation. The mass of the axion is substantially unconstrained by theory. The axion mass and the PQ symmetry breaking scale f_{PQ} are related through

$$m_a = \frac{\sqrt{z}}{1+z}\frac{f_\pi m_\pi}{f_{PQ}/N} \quad , \tag{2}$$

where z is the ratio of u- and d-quark masses (presumed near 0.5), and N is the color anomaly of the PQ symmetry, related to the coupling of axions to color. The anomaly N is model dependent, axion couple to color differently in different models in much the same way that the relation between the vacuum expectation value of the axion field and the vacuum angle $\bar{\Theta}$ is model dependent (the theoretical constant is closely related in both cases). The axion can be considered to have effective couplings with normal matter and radiation, with coupling constants $g_{a\gamma\gamma}$, g_{aee}, g_{app}, etc., for the axion coupling to photons, electrons and protons. Since the elementary axion couplings are model dependent, these effective couplings are model dependent as well. For instance, the effective two photon coupling constant is

$$g_{a\gamma\gamma} = \frac{\alpha/2\pi}{f_{PQ}/N}(E/N - 2(4+z)/3(1+z) \quad , \tag{3}$$

with E is the electromagnetic anomaly, related to the coupling of axions to charged leptons, and the factor containing the ratio z of light quark masses $2(4+z)/3(1+z)$ is approximately 2. The tree level coupling of axions to color is fundamental to the role of axions in solving the strong CP problem. A tree level coupling of the axions to charged leptons is optional; here different theories call for different couplings. The extreme cases are axions with no tree level coupling to electrons (termed hadronic axions[13]), and axions where all the tree level quark and electron couplings are of the same strength (these are axions in a simple GUTs scenerio and are termed DFSZ axions[14]). Note that even in hadronic axion models, there is an effective electron coupling through the anomalies. The effective coupling constants of axions with normal matter and radiation all depend in a Higgs-like way on $1/f_{PQ}$, so the larger the axion mass, the stronger the couplings. When the axion is very light, the couplings are very weak, and the axion is hard to detect. Such axions are termed invisible axions. The current round of axion searches looks for these invisible axions through their coupling with two photons. This coupling has strength $g_{a\gamma\gamma}$, resulting in two photon lifetime

$$\tau = 7\times 10^{24}\text{sec}\frac{(m_a/\text{eV})^{-5}}{\{E/N - 2(4+z)/3(1+z)\}/0.7} \quad . \tag{4}$$

There is nothing forbidding the anomaly ratio E/N from having the special value such that the axion has effectively infinite two photon lifetime. However, the E/N ratio in the simple DFSZ GUTs models has value 8/3, and there is no reason to think dynamics would drive E/N to the special infinite lifetime value.

§4 Astrophysical and Laboratory Constraints on the Axion Mass

The original axion, rooted in symmetry breaking in the standard model $SU(2)\times U(1)$, called for the PQ symmetry breaking scale f_{PQ} to be at the 250GeV weak scale, leading to axions with masses near 10keV. Such axions couple relatively strongly and were quickly ruled out by accelerator experiments, e.g., $K \rightarrow \pi + a$ followed possibly by the axion decay into two electrons.[15] Non-observation of such decays constrain the axion mass to less than 10keV, and effectively rules out the original axion. However, at least in retrospect, there is no particularly good reason to think the PQ symmetry breaking scale should be the weak scale, and the search continues for invisible axions, those with f_{PQ} much larger than 250GeV. Through their effects on stellar dynamics and cosmological evolution, mass constraints can be placed on even invisible axions.

4.1 Axions and Stellar Dynamics

The factor regulating the evolution of a star is the rate at which thermal energy is dissipated. Electromagnetic interactions regulate photon energy transport, with the interaction length of nuclear reaction photons at the center of the sun on the order of centimeters; these photons bounce around the sun for millions of years before they escape. The electromagnetic interactions are, in this context, quite strong. On the other hand, invisible axions as well as neutrinos efficiently transport energy out of the star, thereby accelerating the star's evolution.

In our nearest star, the sun, nuclear reactions release a few erg/g/sec. At the 20 or so million degrees at the center, about $10(m_a/\mathrm{eV})^2$erg/g/sec DFSZ axions of mass m_a are created through Compton processes like $e+\gamma \rightarrow e+a$. DFSZ axions more massive than about 1eV would naively transport more energy out of the star than produced in nuclear reactions. In compensation, the star contracts in order to raise its temperature and nuclear reaction rates. It is this higher nuclear reaction rate responsible for accelerating evolution. The epoch in the sun's nuclear evolution can be determined by, e.g., measurements of the sun's ^{4}He content. From the age and nuclear epoch of the sun, and models of solar dynamics, is is unlikely that axions are at this moment could remove more than about half of the liberated nuclear energy, constraining the DFSZ axion mass to less than about 1eV.[16]

There are other ways of constraining the axion mass through solar dynamics. As is well known, solar neutrino production is exquisitely sensitive to the central temperature of the sun. Should there be an axion-induced temperature rise, the observed solar neutrino event rate would increase. If anything, there is evidence for a "solar neutrino deficit", and certainly no evidence for an excess, and these considerations ultimately constrain the DFSZ axion mass to below about 1eV. Photons created via inverse Primakoff production by incident solar axions are detected through their conversions in large magnetic fields[17] or the atomic fields of Ge.[18] The ensemble of solar limits constrain the DFSZ axion mass to less than about 1eV or above the central temperature of a few keV.

Red giants are excursions from main sequence stars where the surface temperature is lower than that of the sun, with luminosity 10,000 times greater. The central temperature of red giants, driven by helium burning, approach 10^8K. The evolution of red giants out of the main sequence has been extensively modeled,[19] and the red giant population relative to main sequence populations is well understood. The evolution of a star into a red giant happens relatively quickly. Also relatively quickly, a red giant exhausts its helium fuel and continues evolving into a compact object. The population of these red giant stars provide stringent constraints on the axion mass. The argument is again based on the effect of axions on stellar evolution. As in the case of the sun, the effect of hadronic axions on red giants is to raise the central temperature, thereby increasing the helium burning rate and reducing the time a star spends as a red giant. The count of red giants in a population of stars thereby goes down, and these considerations result in a red giant constraint on the hadronic axion mass of a few eV or so.[20]

For DFSZ axions, the $\gamma+e \rightarrow a+e$ Compton coupling to electrons dominates. As a star evolves into a red giant, helium waste from hydrogen burning adds to a helium core, the core supported by electron degeneracy pressure. As the helium core builds up, its size shrinks, liberating gravitational binding energy and raising the temperature. When the core becomes sufficiently hot, the helium ignites, and the star continues its evolution as a red giant. Without axions, any core cooling is throttled by neutrino radiation, and the neutrino path length in the degenerate electron core is not long. However, DFSZ axions, through their Compton coupling to electrons, efficiently transport radiation out of the star. The axion production rate is proportional to an inverse power of the axion mass, and much like the neutrino production rate, is proportional to a high power of temperature.[21] If we assume axions dominate energy transport out of the core, then the core temperature varies as the an inverse power of the axion mass. This is somewhat surprising: for DFSZ axions, the higher the axion mass, the lower the core temperature. A sufficiently massive DFSZ axion will prevent the helium from reaching the helium burning temperature, and such considerations constrain DFSZ axions to masses less than about 10^{-2}eV. Also, as for the solar limit, these limits do not apply to axion masses greater than the red giant core temperature near 10keV.

4.2 Axions and Supernovae

In 1987, a rare event occurred, a nearby star in the LMC exploded as a supernova. Dubbed SN1987A, this event had enormous implications for particle and astrophysics. The impulsive release of the approximately 10^{53} ergs of gravitational binding energy eventually appeared as neutrinos with characteristic temperature of around 10MeV. However, the neutrinos released in the core have a relatively short path length; it takes many scatterings before a neutrino can escape. The supernova cooling is throttled by the relatively strong neutrino interactions. The Kamiokande and IMB detectors together recorded 19 neutrinos spread over about 10 seconds, a result consistent with our understanding of Type II supernovae dynamics and the number of light neutrino flavors. An axion would efficiently transport energy out of the explosion, and reduce the spread in neutrino arrival times. The nucleons in the core have an effective coupling to axions, and since for this process, the electromagnetic anomaly does not directly enter, the DFSZ and hadronic axions couple to nucleons with similar strength. A detailed calculation[22] predicts that for axions of mass near 10^{-3}eV, the axion and neutrino energy transport out of the core are about equal. These calculation begins to break down near axion mass so large that they no longer escape without extensive rescattering. Briefly, effects of multiple scattering begin near axion mass 20×10^{-3}eV, and by 2eV, the relatively strongly interacting axions transport less power out of the core than any one neutrino species. These considerations constrain the axion mass to below 10^{-3}eV or above 2eV.

There was much discussion of this supernova bound to the axion mass, driven by G.Raffelt and D.Seckels' suggestion that the axion nucleon bremstrahlung calculation does not adequately consider nucleon density effects. In the nucleon-dense environment of the core, the "formation time" of the axion radiation is much longer than the time between scatterings, and the axion bremstrahlung is suppressed in a manner similar to that of the recently observed Landau-Pomeranchuk-Migdal effect in QED. Raffelt suggested this suppression could be very large—perhaps allowing axions with mass as high as 10^{-1}eV—thus weakening the supernova bound. He noted that, "however, if collective modes dominate (phonons, spin-waves) our argument is not directly applicable", and the issue remained unsettled.

§5 Production of Relic Axions

The laboratory and astrophysics bounds—essentially upper bounds—on the axion mass involve creation and detection of new axions. If axions are the dark matter, they must be a child of the early universe. We know of only a few ways in which substantial numbers of such relic axions are created. These production mechanisms, coupled with the requirement that axions not severely overclose the universe, result in lower bounds on the axion mass. The mass range between the upper and lower bounds is the axion window mentioned in the introduction.

Recall how a relic axion acquires its mass: The invisible axion PQ symmetry breaking occurs at the high temperature f_{PQ}. Above this temperature (that is, at earlier times) the PQ symmetry is unbroken and there is no axion. Just below this temperature, but well above the temperature corresponding to the QCD scale Λ_{QCD}, the PQ symmetry is broken. However, since the axion acquires its mass through the color anomaly, and at high temperatures the quark chiral symmetry is unbroken, the axion is massless. When the temperature drops below Λ_{QCD}, the quark chiral symmetry is broken, and the axion acquires a mass.[23] This is how an axion acquires its mass, but in order to address the question of the density of relic axions, we need to examine specific production mechanisms.

5.1 Relic Axions: Thermal Production

It the near thermal equilibrium of the early universe, axions were created and destroyed in particle interactions. For example, axions participated in Primakoff-like processes involving quarks or nucleons $\gamma + N \leftrightarrow a + N$. The important scales in thermal processes are the reaction rates involving axions and the expansion rate of the universe. For low reaction rates, axions freeze out of equilibrium and their overall density is then fixed in any comoving volume—this scenario is familiar from relic neutrino production. From considerations of this sort, we find that for axions to freeze out at closure density, the axion mass must be greater than about 100eV (with factors of two uncertainty arising from the Hubble constant).[24] Such relatively heavy axions are already excluded by astrophysical bounds, so it is unlikely that thermal axions comprise a substantial component of dark matter.

5.2 Relic Axions: Misalignment Production

In our Θ vacuum picture of QCD, CP conservation is a consequence of the $\bar{\Theta}$ parameter being driven to a classical value corresponding to the potential minimum as a consequence of the axion acquiring mass. Recall, however, that the axion did not start out with mass—it acquired mass near the temperature Λ_{QCD}—so the $U_{PQ}(1)$ angle parameter has arbitrary value in early times. We say the initial value is misaligned with its present value. An analogy is found in the simple pendulum: Without gravity, there is no special pendulum angle. Turn on gravity, though, and the average pendulum angle is driven to zero. Note that although the pendulum average angle is zero, the pendulum oscillates with a non-zero RMS angle. There is energy stored in the oscillations, and, returning to axions from the analogy, quanta of these oscillations are axions. This is the mechanism of misalignment production.[25] You can take this analogy too far—in particular, we have not included the Hubble expansion damping of oscillations—but as for the pendulum case, the quanta form a Bose condensate, and the axions have an extraordinarily tiny velocity dispersion. Misalignment axions are cold dark matter. The present density of misalignment axions is[26]

$$\Omega_a = 0.85 \times 10^{\pm 0.4} (\Lambda_{QCD}/200\text{MeV})^{-0.7} \times (m_a/10^{-5}\text{eV})^{-1.18}/h^2 \quad , \qquad (5)$$

where the Hubble factor h arises from the expansion driven damping, and the QCD scale arises from the temperature where mass appears. This prediction for the present axion density assumes an initial misalignment angle of $\pi\sqrt{3}$—the RMS angle of the interval $-\pi$ to π. This is reasonable as, without inflation, the initial misalignment angle was a composite of independent individual misalignment angles from a great number of causally disconnected volumes. With these assumptions and canonical values for the Hubble and QCD scales, axions with mass near 10^{-5}eV have closure density, and axions with substantially lower mass would severely overclose the universe. This misalignment axion production mechanism therefore provides a lower limit to the axion mass. Should inflation have occurred after axions first appear at the PQ symmetry breaking scale, there is just one initial misalignment angle corresponding to the angle in our pre-inflation volume. Here, the statistics of many causally disconnected volumes cannot be invoked, and one is forced to the weaker argument that a misalignment angle very near zero is unlikely. The axions are produced with a momentum dispersion related to the inverse of the initial Hubble length. Today, the momentum is reduced by the ratio of 2.7K to the temperature of the universe when axions acquired mass. Dividing by mass gives the present axion velocity dispersion of $10^{-22}(m_a/\mathrm{eV})^{-0.82}c$. Indeed, this is very cold dark matter.

5.3 Relic Axions: String Decay

Recall that the axion is the Goldstone boson of the spontaneously broken global U(1) symmetry. Such broken symmetries have strings as solutions to the equation of motion. These strings are nearly one-dimensional objects, either closed loops or infinitely long, arising from the mismatch between the arbitrary choice of broken vacuum in neighboring volumes.[27] Assuming no inflation or that inflation occurred before the breaking of PQ symmetry, a network of strings develop carrying appreciable energy. These strings interact, and in doing so form loops. These loops evaporate via axion radiation, and these radiated axions could form a substantial component of dark matter.[28] Recent calculations differ as to the contribution of string axions, with estimates at the workshop ranging from negligible (P.Sikivie reporting on his work with C.Hagmann) and very significant—perhaps 100 times the contribution from misalignment axions (R.Battye reporting on his work with E.Shellard). If we take Battye and Shellards' calculation seriously, string axions would provide closure density at a mass of perhaps 10^{-4} to 10^{-3}eV, really squeezing the axion window. Shellard, though, cautions that uncertainties remain large. Also unresolved was the question of the phase space structure of string axions; the issue here is how their primordial velocity spectrum compares to a thermal distribution.

5.4 Relic Axions: Phase Space Structure

Our galactic structure is due to gravitational relaxation processes acting on the more uniform distribution of primordial matter. Axions, whether string or misalignment produced, are subject to these same gravitational forces, and relax with normal matter into something resembling an isothermal dark matter halo around our galaxy. There are large deviations from a thermal distribution in that the highest energy particles have discrete values of velocity. The phase space structure of radial infall axions in a halo are shown in upper figure 1, where the radial coordinate is distance from the center of the galaxy.[29] The velocity spectrum of axions at our radius from the center of the galaxy is shown in lower figure 1. The isothermal peak is to the left, and the newer infall peaks are off to the right. Y.Yang, reporting on her work with P.Sikivie and I.Tkachev, suggests that perhaps 10% of the axions are in the first infall peak, a narrow structure with fractional width narrower than 10^{-19}. It was agreed that the radial infall model was too simple, angular momentum needs to be incorporated. Recently, Sikivie, reporting on work by Tkachev, said that hints are emerging that, in a model including angular momentum, a significant fraction of axions still appear in the first infall peak. These results have important positive consequences for present and future rounds of axion searches.

There were several studies suggesting axions condense into axionic balls. I.Tkachev, reporting on his work with E.Kolb, considered the non-linear axion terms in the potential.[30] These terms, negligible at reasonably low axion densities, might become important near early universe large-amplitude axion isothermal fluctuations. Kolb and Tkachev studied this issue in a radial collapse model and found a sufficiently large fluctuation can collapse in the radiation dominated epoch, producing a dense axionic ball. They gave an interesting example of such an object: Take the ball to contain the total axion mass within a Hubble volume at the time when the inverse axion mass equaled the Hubble length, say at a temperature of a GeV. This ball then has mass about 10^{-9} solar masses and radius about 10^{13}cm. Tkachev guessed that 80% of axions are bound in such balls. The Earth encounters one of these balls only very rarely, every 30 million years, but what a signal it would produce in the axion detectors!

Another axion clustering scenario was presented by W.Suen, reporting on his work with E.Seidel[31] They studied the gravitational collapse of a scalar (or pseudoscaler) field by adding gravity as a classical potential to the axion Klein-Gordon equation. As an example, they found that for an axion of mass 10^{-5}eV, the resulting axion ball weighed 10^{29}g with radius 10cm. He guessed that

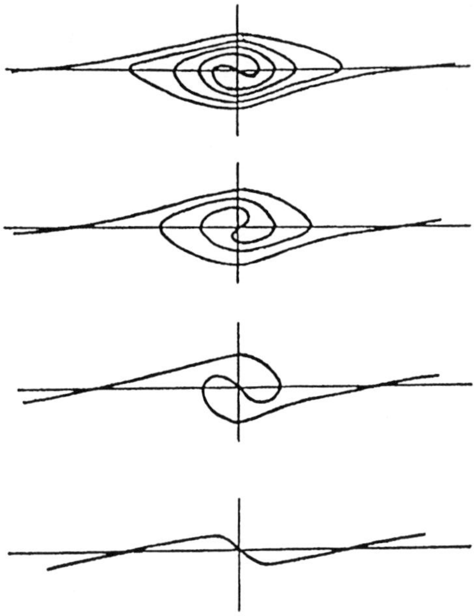

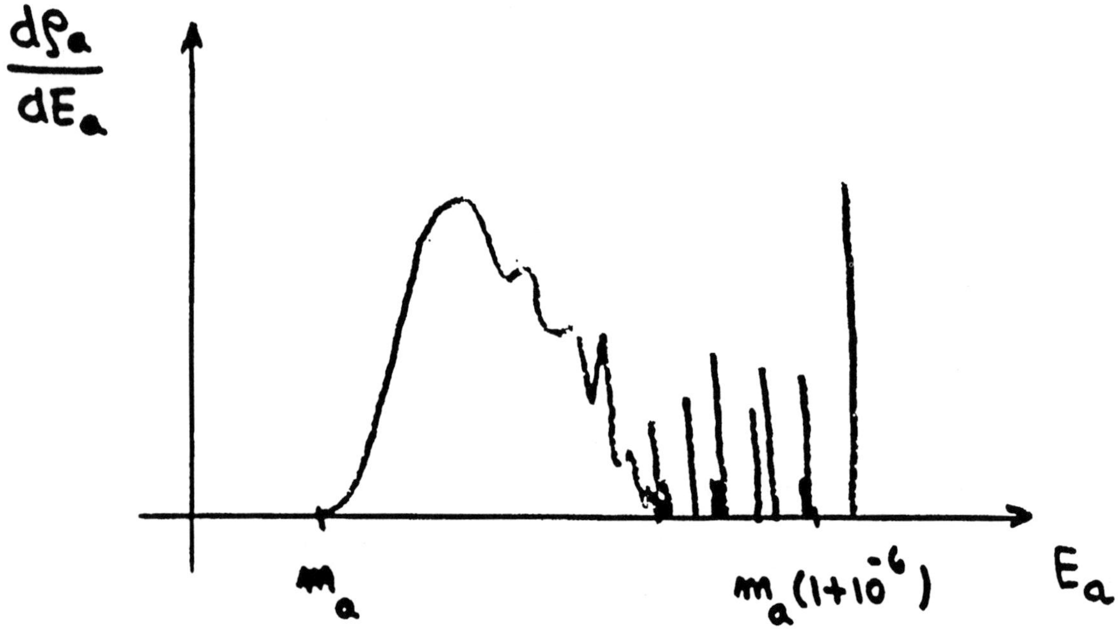

Figure 1: upper: Phase space of axions. The distance coordinate is the distance from the center of the galaxy. lower: Velocity distribution of axions near Earth.

perhaps 95% of axions are bound up in such balls.

Of course, axion balls would be very hard to see in the present round of axion search experiments. However, T.Weiler reported an interesting possibility: at some critical axion ball density, axions lasing into photon pairs could occur. The modeling is complicated, involving spontaneous emission, stimulated emission, surface losses, and back reactions of photons into axions. The crucial issue in discussions was when the lasing occurred. If axion balls dominate dark matter, it seems unlikely that all the axion balls could have already lased away, yet nothing in the formalism prevented the balls from lasing early in their evolution. This issue was not resolved in discussions.

5.5 *Relic Axions: Summary*

There is near unanimous agreement that thermal axions would probably not contribute greatly to closure density. However, axions from misalignment or strings could be significant. Many uncertainties remain, most notably the issue of the velocity spectrum of string axions. There are additional uncertainties coming from Λ_{QCD} and the Hubble constant. The summary of the various laboratory, astrophysical and cosmological constraints on axion mass are shown in figure 2.

§6 Miscellaneous Theoretical Issues

D.Lyth reported on his studies of axions in the broader context of supersymmetry. There is a new parameter in this, the supersymmetric axion mass, layered on the usual models of misalignment and string axion production. He summarized his talk with, "If forced to place a bet, I'd place even money on the four possibilities: 1. Strings no entropy (axions ruled out); 2. Strings no entropy (axion mass between 10^{-4} and 10^{-3}eV, and density in axions $> 10^{-2}$); 3. Strings entropy (axion mass between 10^{-8} and 10^{-3}eV, and density in axions $> 10^{-2}$); 4. No strings (axion mass between 10^{-8} and 10^{-3}eV, and unknown density in axions."

S.Barr reported on his studies on "Saving the axion from the (possible) effects of quantum gravity." The issue is how to ensure the PQ symmetry is respected by quantum gravity. S.Barr showed how this was possible by introducing new symmetries. In discussions, the point was raised as to whether this is a real issue or not. One camp noted that, after all, there is as yet no theory of quantum gravity. The other camp noted that some general statements can be made about any reasonable quantum theory of gravity, and one such statement is that quantum gravity does not respect PQ symmetry. The issue remains very much open.

§7 Detection of Relic Axions

There is a region of allowed axion mass near 2eV, above the SN1987A constraint, but below the hadronic red giant bound. These axions live about the right length of time so that there are still significant numbers of them in halos, yet decay frequently enough to be observed through their nearly monochromatic two photon decay.[32] The signature is a axion line on top of the overall sky glow. Data from our own halo constrains the axion mass to less than 4eV or so. One can look for axion lines in galactic clusters. Here, the axion lines are Doppler shifted, while the airglow lines are not and can therefore be subtracted cleanly away. Also, on- and off-source cluster observations allow subtraction of air glow lines. These considerations constrain the hadronic axion to masses below 2eV, leaving masses between 10^{-6} to 10^{-3}eV as the sole axion window.

7.1 *Detection of Relic Axions: Sikivie-Type Axion Detectors*

The hypothesis that the halo of our galaxy is made of axions can be tested experimentally by looking for the resonant conversion of galactic halo axions to photons in a laboratory magnetic field. To this effect, a tunable high Q cryogenic cavity is placed in the bore of a superconducting solenoid, and the frequency of its lowest TM mode is slowly changed until excess power is detected in that mode from resonant axion—photon conversion.[33] The excess power is

$$P = 4 \cdot 10^{-26}\text{Watt} \left(\frac{V}{0.22\text{m}^3}\right)\left(\frac{B_0}{10\text{Tesla}}\right)^2 \times C_{nl}\left(\frac{g_\gamma}{0.97}\right)^2\left(\frac{\rho_\text{a}}{0.5\cdot 10^{24}\text{g/cm}^3}\right) \times \left(\frac{m_\text{a}}{2\pi\text{GHz}}\right)\min(Q_\text{L}, Q_\text{a}) \quad (6)$$

where V is the volume of the cavity, B_0 is the effective magnetic field strength, C is a mode-dependent form factor of order 0.5, ρ_a the density of galactic halo axions at the earth, m_a is the axion mass, Q_L is the loaded quality factor of the cavity and $Q_a \sim 10^6$ is the "quality factor" of the galactic halo axions, i.e., the ratio of their energy to their energy spread. Finally, g_γ is the coupling strength of the axion to two photons. The value $g_\gamma \sim 0.36$ is predicted for DFSZ axions, and the value is 0.97 for a model of hadronic axions. This is a tiny amount of power; consider that with the nominal value of constants in the above expression, the black body radiation in the appropriate Q_a bandwidth emitted from the 1K cavity is ten times larger than the power from axion conversions.

There is a semi-classical way to derive this expression for the conversion power of axions in the cavity,[34] where

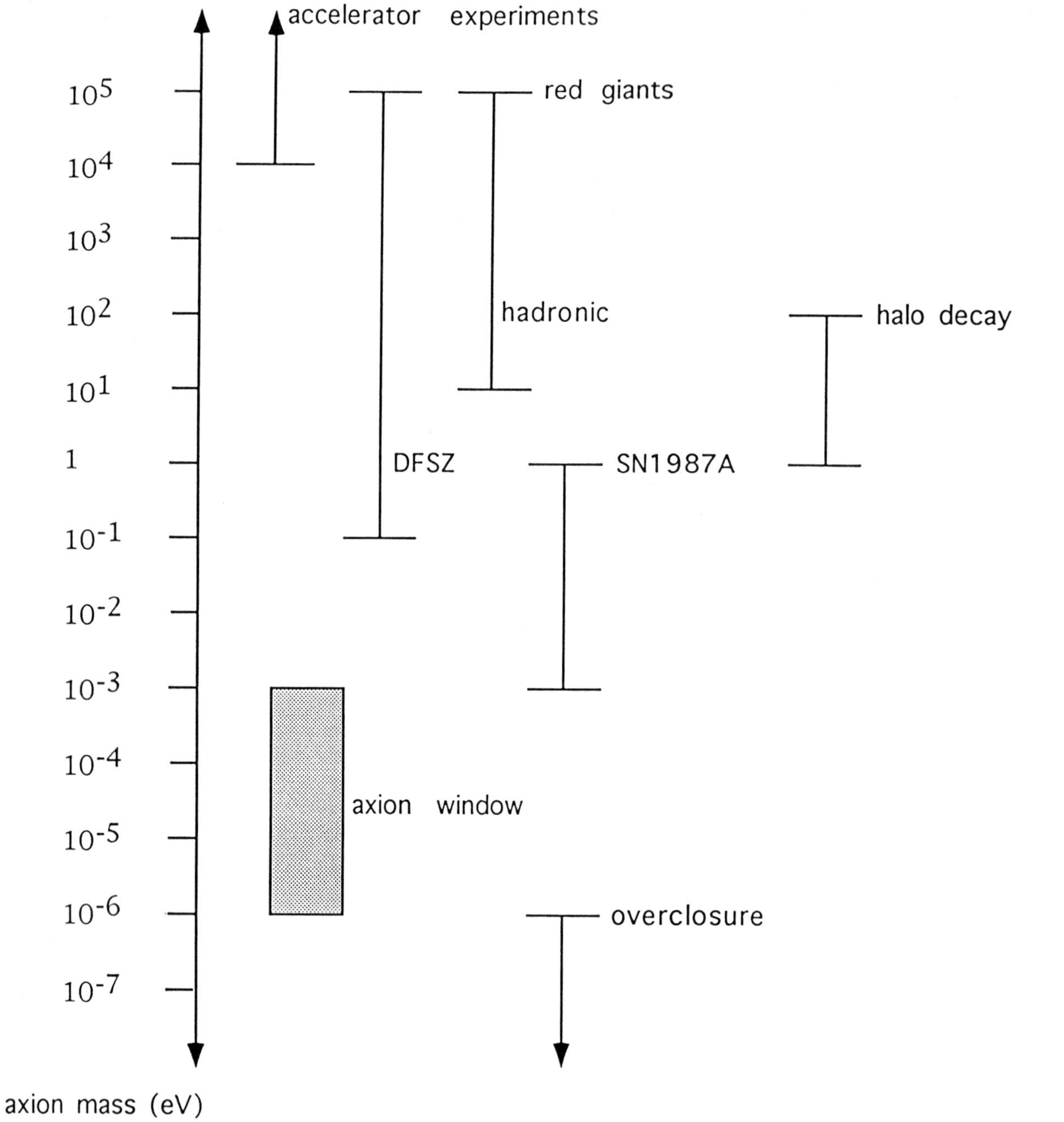

Figure 2: Laboratory, astrophysical and cosmological constraints on the axion mass.

a correspondence is made between effective Lagrangians describing coupling the axion to two photons in the cavity and the classical LRC equivalent circuit. The axion interaction Lagrangian density is $\mathcal{L} = -(g_{a\gamma\gamma}/4\pi)a\vec{E}\cdot\vec{B}$, where $\vec{E}$ and $\vec{B}$ are electromagnetic fields in the cavity. In this density, the axion field (a) has de Broglie wavelength large relative to the cavity size, and can therefore be treated as spatially constant. Assume the cavity is a right circular cylinder of volume V, and take for B an externally applied magnetic field B_0 along the cavity axis. The integral of $\vec{B}_0\cdot\vec{E}$ over the cavity volume depends on the configuration of $\vec{E}$, that is, on the $\vec{E}$ fields of that particular cavity mode. By expanding the E fields into normal cavity modes $\vec{E} = \sum_i E_i\hat{e}$ with normalization $\int \hat{e}_i\cdot\hat{e}_i = V$, we can define a form factor $G_i = 1/V\int \hat{e}_i\cdot\hat{z}dV$, where we have taken the cavity axis and constant B field along the z-direction. The resonant behavior for a particular mode of the cavity may be represented by a series driven LRC equivalent circuit, with interaction Lagrangian $L = \mathcal{E}q$, $\mathcal{E}$ is the driving EMF, and q the charge on the capacitor. We make the correspondence between the cavity and equivalent circuit by equating the cavity energy $(1/8\pi)\int \vec{E}_i\cdot\vec{E}_i dV$ and equivalent circuit energy $(1/2C)q^2$, and equating the cavity and equivalent circuit interaction Lagrangians, resulting in EMF

$$\mathcal{E} = \frac{-g_{a\gamma\gamma}}{\sqrt{4\pi}}B_0G_i(V/C)^{1/2}a \quad . \tag{7}$$

At resonance, the power from axion conversions is

$$P_0 = \mathcal{E}^2/R = \frac{g^2_{a\gamma\gamma}}{4\pi}B_0^2G_i^2\frac{V}{RC}\langle a^2\rangle \quad , \tag{8}$$

where $\langle a^2\rangle$ is the axion density divided by m_a^2. With the cavity resonant frequency ω tuned to the axion energy E_a, and the cavity Q given in the LRC representation as $1/\omega RC$, we are done. The power generated in the cavity from axion to photon conversions is

$$P_0 = \frac{g^2_{a\gamma\gamma}}{4\pi}B_0^2G_i^2VQ\omega\langle\rho_a\rangle/m_a^2 \quad . \tag{9}$$

In the appropriate units, this is exactly the same result arrived at from field-theoretic arguments.

One does not know the value of the axion mass, which determines the resonant frequency $f = m_ac^2/h$ but it is known that misalignment axions with mass near 4μeV provide the critical energy density for closing the universe. The search rate appropriate for a given signal to noise ratio (s/n) is given by the formula:

$$\frac{df}{dt} = \frac{72\text{GHz}}{\text{year}}\left(\frac{4}{\text{s/n}}\right)^2\left(\frac{V}{0.22\text{m}^3}\right)^2\left(\frac{B_0}{10\text{Tesla}}\right)^4 \tag{10}$$

$$\times C^2\left(\frac{g_\gamma}{0.97}\right)^4\frac{\rho_a}{0.5\cdot 10^{-24}\text{g/cm}^3}\left(\frac{5K}{T_n}\right)^2$$

$$\times\left(\frac{f}{1\text{GHz}}\right)^2\left(\frac{Q_W}{Q_a}\right)$$

where T_n is the total noise (cavity plus electronic) of the microwave detector.

Pilot experiments (also called First Generation Experiments) have been carried out using relatively small volume magnets and—by current standards—noisy amplifiers, at Brookhaven National Laboratory (BNL) and at the University of Florida (UF). These experiments had B_0^2V values of 0.36 and 0.45 T^2m^3, respectively. Figure 3 shows the regions in the coupling-squared versus mass plane eliminated by these searches, compared with the mass-dependent predictions of coupling constants. The pilot experiments have demonstrated the principle of cosmic axion detection and have shown that reasonable sensitivity can be achieved over a wide range of frequencies. However, much larger systems are needed to provide meaningful tests of the theoretically favored models.

A second-generation axion search, to be operated initially at the Lawrence Livermore National Laboratory, is now under construction in the U.S. The spokespersons are Leslie J Rosenberg (MIT) and Karl van Bibber (LLNL).[35] The capability of this experiment to either detect axions (with s/n of 4) or exclude them (at the 97.7% C.L.) is shown in figure 3 as the region extending into KSVZ (hadronic) axion couplings. The key goals of the experiment are: (1) To achieve a power sensitivity which is conservatively a factor of 40 improvement (and probably closer to a factor of 100) over the pilot experiments. This is to be achieved by a combination of scaling up in magnet volume, and improving the noise temperature of state-of-the-art microwave amplifiers. (2) To search the mass range $1.5\mu eV < m_a < 12.6\mu eV$. This is to be achieved through filling the magnet volume with multiple cavities.

A sketch of the U.S. experiment is shown in figure 4. The experiment will utilize a commercial magnet with a central field of 8.5T. The volume for the cavities will have an inner diameter of 50 cm, and a length of 100 cm. Hence, $B_0^2V = 12\,T^2m^3$, approximately a factor 25 over the pilot experiments. The experimental volume will be separated from the magnet cryostat by a cold-vacuum wall. The magnet will be operated at 4.2K, and it will be possible to exchange cavity arrays and electronics in the experimental volume whilst the magnet is energized, and more importantly the temperature of the cavity arrays will be independent of that of the magnet. Initially, the cavity will be operated at about 1.5K, lower than the noise temperatures of the best microwave amplifiers available today in the UHF through S-bands (0.5 through 3GHz). The expected total noise temperature of 3K (physical plus electronic) yields another factor 1.6 in improved sensitivity over the pilot experiments.

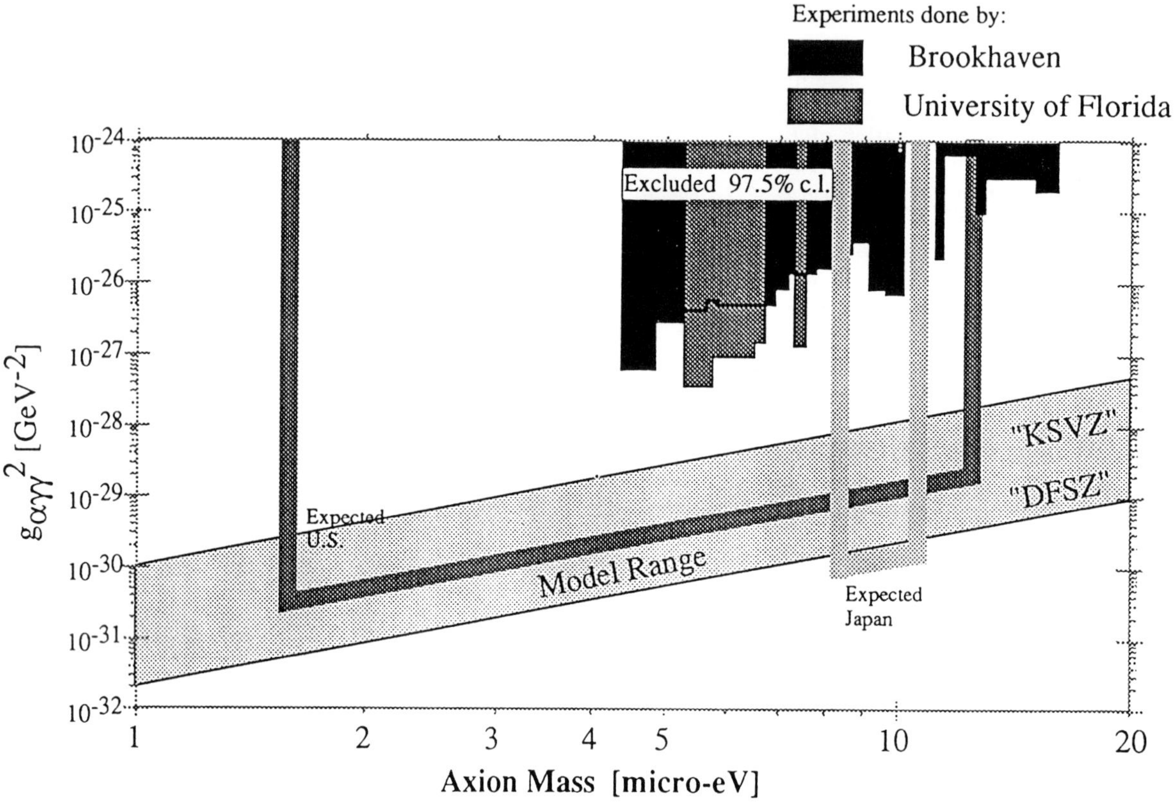

Figure 3: Exclusion regions in the axion mass versus coupling constant plane obtained by the pilot cavity experiments. Also shown are the coupling constants expected in a range (DFSZ and hadronic) of axion models. The area extending into the KSVZ (hadronic axion) region is the expected sensitivity of the U.S. experiment. The narrow region extending into the SFSZ region is the expected sensitivity of the Japanese experiment.

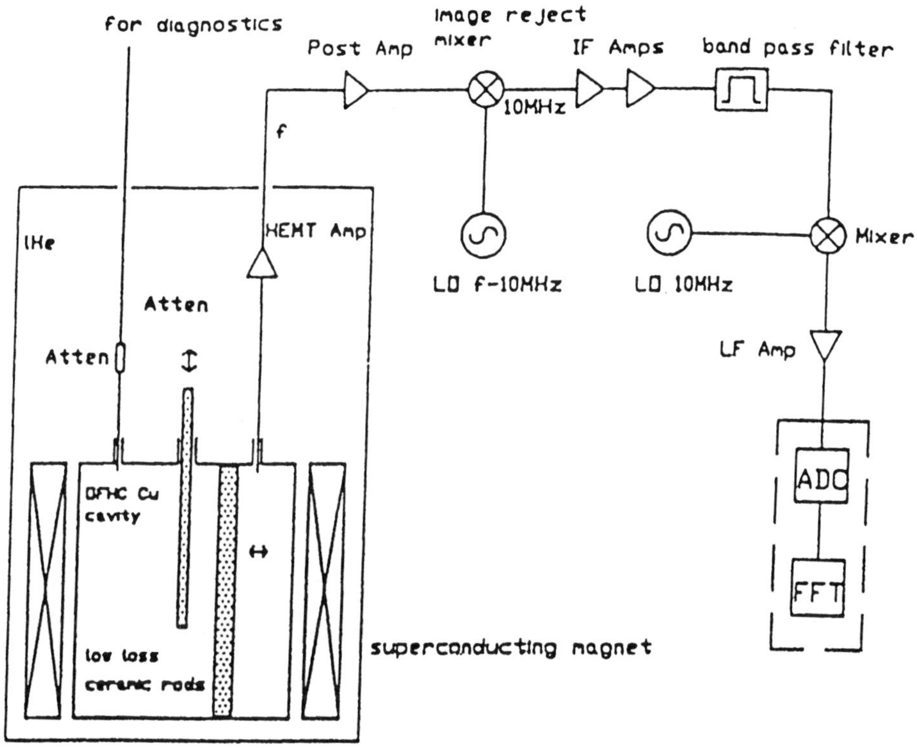

Figure 4: Sketch of the U.S. axion search experiment.

In the U.S. experiment, for the first time, arrays of multiple cavities will be used to extend the mass search range. Each cavity is separately tuned by moving dielectric or metallic rods within the cavity, and in this way the experiment will cover the range $1.5\mu eV < m_a < 12.6\mu eV$. It will start taking data in late 1994, and run for approximately three years. This second-generation experiment will be the first to have sensitivity to a likely dark matter candidate with plausible couplings to matter and radiation.

Also, the possible narrow peaks in the halo axion phase space are exploited to increase the experiment sensitivity. How this works is as follows: The ultimate signal to noise ratio is given by $\mathrm{SNR} = (P_s/T_n)\sqrt{t/B}$, with P_s and T_n the signal power and noise temperature coming out of the receiver electronics, and t and B the signal integration time and bandwidth. The integration time is about 100s, and for isothermal halo axions, the bandwidth is of order 1kHz. Now consider a narrow line of width 10^{-2} (this is actually a rather modest width compared to the intrinsic phase space width of the first infall peak) containing a fraction f_1 of the halo axions. The signal to noise ratio in this narrower bandwidth is better by the factor $f_a\sqrt{10^5} = 300f_1$. However, we now have many more frequency bins to search, and therefore many more candidate peaks. In order to keep the number of candidate peaks constant, we increase the candidate peak threshold from 2σ to 5σ above background. The loss in sensitivity from raising the threshold is then $4/7 = (2+2)/(5+2)$, but the overall sensitivity improves by $(4/7)300f_1$. If we take f_1 to be 10% as indicated by the modeling, then the experiment signal to noise ratio improves by a factor of 18 by looking for the first infall peak in a narrow bin search. In the U.S. experiment, there are separate processing paths for the isothermal and narrow peak searches.

Another second generation axion search, to be operated at Kyoto, is now under construction in Japan. The spokesperson is Seishi Matsuki (Kyoto). It is similar to the U.S. design, except for its dilution refrigerator and Rydberg atom beam system.

D.Kleppner gave an overview talk on Rydberg atoms with application to the axion experiments.[36] Rydberg atoms have the following features: (1) Similar properties to "hydrogen atoms" with massive nuclei. (2) Very large atomic radii, for example, a Rydberg atom excited to the 100th principal quantum number has radius $R_{100} \sim 1\mu$m. (3) Very low binding energy of valence electron, for example, $E_{100} \sim 10^{-3}$eV $\sim$ 300GHz $\sim$ 10cm^{-1}. (4) The interval between energy levels is very small, for example, $\Delta E_{100} \sim 3 \times 10^{-6}$eV $\sim$ 7GHz $\sim 2 \times 10^{-1}$cm^{-1}. (5) Very long lifetimes, for example, $\tau_{100} \sim 10^{-3}$sec. (6) The electric dipole transition moment between $\Delta n = 1$ states is very large, for example, the absorption cross section with $n = 100$ can approach 1cm^2. The above features, especially (4) and (6) suggest Rydberg atoms with principal quantum number $n \sim 100$ can be used as sensitive microwave photon counters.

The Japanese realization of a Rydberg atom microwave single photon counter is shown in figure 5. Here, atoms in the ground state are sent through a detection cavity (not to be confused with the axion conversion cavity). Just before entering the cavity, atoms are excited to a Rydberg state by a laser, where the $n \rightarrow n+1$ transition is tuned to the conversion cavity microwave frequency by a weak external magnetic field. Rydberg atoms that have absorbed a microwave photon are the only ones ionized,[37] and the electrons are detected and counted by the Channeltron.

The conversion cavity, a right circular cylinder 7 cm diameter and 80 cm long, is threaded by a 7T magnetic field, and the tuning range is 2.4GHz $\pm$ 10-20%. Extrapolating from their earlier experience with Rydberg counting, they hope for a sensitivity—even with the small volume of their detector—below the DFSZ model over a narrow mass range for a year of operation, as shown by the lower excluded region in figure 3.

The U.S. second generation has marginal sensitivity for the relatively weak couplings of DFSZ axions from the isothermal peak. Although great sensitivity is possible with a narrow bin search, the modeling of these narrow infall lines is still primitive. There is talk of a third generation experiment, essentially a scaled-up version of the U.S. experiment, to search for the isothermal axion peak. The National High Magnetic Field Laboratory in Tallahassee Florida is planning a new 45T magnet facility, requiring a set of superconducting outer magnets adding to the field of an inner resistive Bitter magnet. These superconducting "outsert" magnets alone could be used in a third generation axion search; as designed, they have inner diameter 70cm, length 1.1m, and 15.7T field, for B^2V increase of a factor of 10 over the U.S. second generation experiment. It is possible a minor redesign of the magnets could increase the B^2V figure of merit by 50% or so. If the total noise temperature (cavity plus electronic) is 2K, then this third generation experiment will have better than DFSZ sensitivity over the axion mass range 1.1 to 13μeV. The detector itself is designed around the mature principles established in the earlier experiments.

There were discussions at both workshops on improving the sensitivity of the Sikivie-type RF cavity axion detectors. Everyone recognized the need for better sensitivity over a wider mass range. D.Tanner discussed ideas for improving the sensitivity through better cavity Q, and ideas for going higher and lower in frequency. He explored possible novel resonant cavity structures. The usual cavity has smooth walls with substantial axion coupling only to the TM_{010} mode. Corrugated reactive walls have higher Q, but the form factor is poor unless the external magnetic field has an unusual geometry. A dielectric resonator with reflecting walls out in the radiation zone has a high Q, but again an unusual external magnetic field geometry is needed in order to get a reasonably large form factor. He explored possible resonant cavity materials. Copper provides the highest cryogenic Q of all normal metals, with Q about 10^5 at 1GHz. If there were no magnetic field, then niobium could replace copper, giving Q of over 10^{11}. However, superconductors in the mixed state are penetrated by fluxoids. When subjected to RF fields, the fluxoids vibrate about their pinning sites and dissipate power, and the Q suffers enormously. While this effect is well known and studied, the issue of whether or not increasing the pinning force yields a commensurate reduction in RF losses is still open.

D.Tanner discussed going to higher frequency by partitioning a large cavity into smaller but higher frequency cavities with the cavity outputs power combined. A room temperature test of this technique with two cavities at the University of Florida confirmed the idea works. The question was raised as to why the cavity tuning can't continue on forever. Tanner noted that any one cavity can be tuned only so far before mode localization unacceptably degrades the form factor. The upper limit to the axion search frequency comes from the "Swiss watch problem," one can partition the volume only so many times before the complexity becomes overwhelming.

M.Pospielzalski gave an overview of the noise limits of conventional junction semiconductor devices. He mentioned the tremendous progress made over the last few years. However, the basic science of sources of noise near 1GHz is not understood. Mainstream interest in ultra low noise amplification seems keenest above 30GHz, driven by microwave background studies, and radioastronomy. At these frequencies, the best amplifiers have noise about five times the quantum limit.

J.Clarke discussed microwave amplification with DC SQUIDSs. He measured noise of DC SQUIDs operating in the low UHF, and compared with predictions. For example, he found a "tuned" SQUID at 4.2K physical tem-

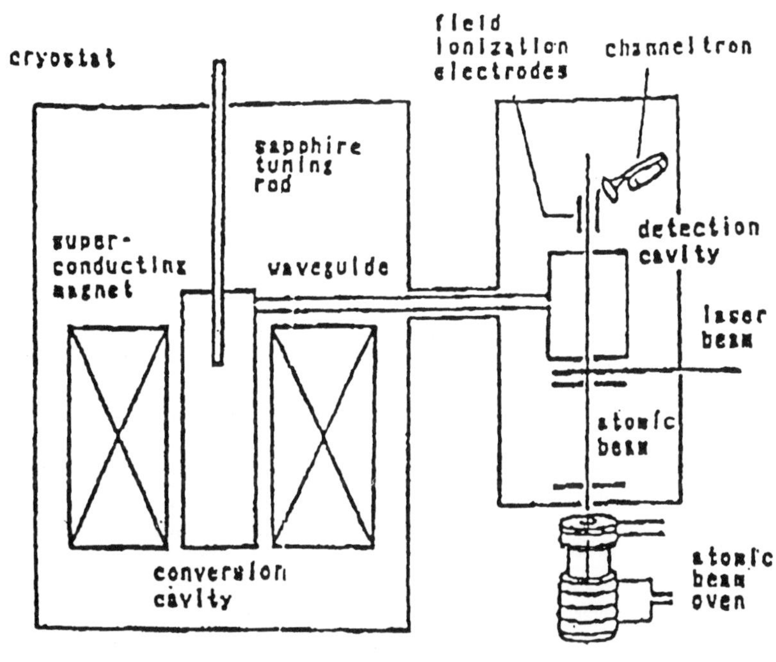

Figure 5: Sketch of the Japanese axion search experiment based on Rydberg atom microwave detection. In the actual experiment, the detection and conversion cavities are directly coupled.

perature had gain 19dB and noise of 1.7 ± 0.5K, compared to predicted noise of 1.1K. In another example, he found an "untuned" SQUID at 4.2K physical temperature had gain 16.5dB and noise of 3.8 ± 0.9K, compared to predicted noise of 3.5K. In a final example, he found an "untuned" SQUID at 1.5K physical temperature had gain 19.5dB and noise of 1.0 ± 0.5K, compared to predicted noise of 0.9K. With professional microwave engineering and a dilution refrigerator, he expects near quantum-limited performance.

S.Matsuki presented his idea for beat frequency detection of axions in a cavity.[38] Galactic halo axions having a long de Broglie wavelength coherently interact with artificially driven microwave photons in a resonant cavity, producing an interference term with low beat-frequency. By calculating the auto-correlation of the low frequency components of the output power picked up from the cavity, this coherent interference term is selectively enhanced while the incoherent background noise arising from the amplifier system is strongly suppressed. That is to say, the long de Broglie wavelength translates into a long coherence time. Noise photons are not coherent, so the auto-correlation function of their time spectra decreases rapidly with increasing time, while the term containing the axion conversion photons beating against externally injected photons is flat.

7.2 Detection of Relic Axions: Other Detection Schemes

K.van Bibber talked about a "solid state plasma detector." In a plasma, the index of refraction n is $n = \sqrt{1 - \omega_p^2/\omega_\gamma^2}$, with ω_p the plasma frequency and ω_γ the incident radiation frequency. By tuning the plasma, the effective volume of the plasma can be adjusted. Perhaps the plasma can be electrons in silicon, excited by laser or injected at a cathode. Although there is a paucity of data on the mobility and recombination time for electrons in silicon at ultra-cold temperatures, the rough numbers are promising.

The most developed alternative idea for searching for 10^{-3}eV axions is the wire grid detector.[39] This axion mass is well above that accessible by cavity techniques. The wire grid detector is a large number of parallel superconducting wires embedded in a material transparent to microwave radiation. The wires carry current producing a static B field within the detector. Axions entering the detector volume convert into a photon, with momentum transfer provided by the inhomogeneity of the magnetic field. Hence, to obtain resonant conversion, the magnetic field is made inhomogeneous on the length scale $1/m_a$. In a more concrete realization of this method, consider an array of parallel superconducting wires fixed in place by a microwave transparent dielectric. The static current through any one plane of wires varies as $\sin(ndq)$ with n the wire number along the plane, d the wire spacing, and $q = 1/m_a$ the momentum transfer. This current generates a spatial cosine variation of magnetic field. Assuming a noise temperature of 10K, a detector of linear dimensions of order a couple of meters is required to search at a reasonable rate near axion mass of 10^{-3}eV. Such a detector might be a cube with sides of length 2m, containing 10^8 wires grouped into 10^4 parallel wire planes. For a search near axion mass of 10^{-5}eV, the detector might be a cube 10m on a side, but requiring only 10^5 wires spaced 2cm apart. The radius of the wires b must be chosen neither too large nor too small. If b is too small, the magnetic field at the wire surface exceeds the critical field. If b is too large, the signal is lost from scattering from wires. However, since the photon wavelength is long relative to the wire diameter, the scattering is less of a problem. In practice, the wire size is probably governed by issues of quench stability. For instance, the 10^{-3}eV axion search calls for 0.05mm diameter wires and a maximum current of 1200A. The 10^{-4}eV axion search calls for 0.4mm diameter wires and a maximum current of 1500A. The wire for the 10^{-4}eV exists, but the 10^{-3}eV search requires improved wire. The plane polarized photons emerging from the ends can be focused with a parabolic mirror or channeled into a waveguide. The electronics and signal analysis techniques are similar to that of the cavity detectors.

Two groups proposed axion searches using the SSC magnet test string. R.Stroynowski presented an overview of the technique. Laser light is sent through a region of high magnetic field, with ellipticity and rotation the physical quantities of interest. Ellipticity depends on phase shift of the light wave and is generated by both pure QED and photon-axion interference. The rotation depends on wave attenuation only and cannot be generated by pure QED processes. Polarized light propagating through a transverse magnetic field will gain ellipticity due to Delbruck scattering, and the ellipticity is multiplied by the number of times the light passes through the field volume. Essentially, the effect of the axion is to contribute another factor to the ellipticity. A pilot experiment by a BNL-Rochester-FNAL-Trieste collaboration based on this technique has already reported on their data.[40] Their magnet was 8.8m long, and B^2 was $4.5\mathrm{T}^2$, the laser frequency was 2.4eV, and there were 250 or so bounces of the light. With a 10^{-4}eV DFSZ axion, the expected ellipticity is 3×10^{-26} with rotation 9×10^{-25}, some 10 or 12 orders of magnitude beyond the sensitivity of their experiment.

Other experiments have reported data based on laser techniques. The same Rochester-BNL-FNAL-Trieste group looked for photon regeneration in a magnetic field.[41] An argon-ion laser was sent through 200 traversals in the bore of a CBA dipole magnet, each 440cm long with 3.7T field. Via the Primakoff effect, the laser photon scatters off the magnetic field producing a real photon. An opaque wall separates the production magnet from another CBA magnet where the axion could scatter back into a photon. The photon was then detected by a photomultiplier tube. Nothing was seen, and the limits are not competitive with astrophysical limits. Members of the same group looked for axions produced in the center of the sun.[42] Here, thermal axions produced in the sun have energies on the 1-10keV range. These axions are detected by their reconversion into x-rays by a magnetic field. The x-rays are detected with a proportional chamber. No signal was seen, and again the technique is not competitive with astrophysical limits.

R.Stroynowski described the proposed new laser experimentsbased on SSC test string magnets.[43] In the experiment's first incarnation, the magnet length is 150m long with 6.76T field, and the laser frequency is 1.2eV with 1000 bounces through the magnetic field. He proposed a clever way of tuning out the QED effects by careful adjustment of gas pressure. The expected ellipticity is greater than that from the pilot experiment, but the issue of noise was not addressed. The pilot experiment was 12 orders or so in magnitude away from sensitivity to DFSZ axions, limited ultimately by noise; the noise issue is therefore paramount. There was much discussion of noise at the workshop, but no conclusions as to the noise limited sensitivity of the new experiment. A draft of another similar proposal circulated at the axion sessions.[44]

§8 Conclusions

The axion, first proposed in the late 1970's as a solution to the strong CP problem, also answers the question of the nature of dark matter. If anything, the case for the existence of axions has grown stronger over time. Interest in axions has also tracked the rising interest in topological defects, and axionic strings could have important cosmological consequences. The primordial nucleosynthesis upper bound to the baryon density of 0.2 of critical density has survived years of scrutiny and the bound is unlikely to topple soon. The amount of visible mass is substantially less than 0.2, allowing room for some baryonic dark matter. The mass density of the universe is within a factor of 10 or so of being the critical density, and since the mass density of the universe in the standard big bang cosmology evolves exponentially away from an exactly closed universe, it is therefore likely that the universe is exactly flat. Experimentally, the total mass density is greater than 0.3. This and the nucleosynthesis bound imply that there is a substantial amount of dark matter. The suggestion has been made that perhaps MACHOS—Jupiter mass objects—form not only the baryonic component of dark matter, but all the dark matter, or maybe all the dark matter in our galactic halo. However, early hints are that MACHO dark matter is not dominant, even in our own halo. This, and the recent negative Hubble telescope search for substantial dark matter in the form of low mass stars, weakens the case for baryons as the primary dark matter. The measurements of the microwave background quadrupole anisotropy tells us that while pure cold dark matter—for example, axions—is not a perfect fit, hot dark matter is a horrific fit. The following point is contentious, but I believe the anisotropy data tells us that dark matter is substantially cold. The present window of allowed axion mass—10^{-6} to 10^{-3}eV—has likewise been under intense scrutiny for years and has survived substantially unscathed. It is intriguing that misalignment axions in this window have the right mass to close the universe. Sikivie-type RF cavity experiments are underway to probe this window with reasonable sensitivity, and other experiments are in the proposal stage. It looks promising for axions: (1) they are on firm theoretical ground; (2) the data calls for substantial amounts of non-baryonic dark matter, even in our own halo; (3) the data hints a substantial component of dark matter is cold, say, in the form of axions; (4) experiments are underway to look for these halo axions.

§Acknowledgments

I thank Pierre Sikivie and Michael Turner for providing the intellectual underpinnings to axion searches. I also thank my colleagues at LBL, Florida, FNAL, Livermore, and MIT for their good ideas and suggestions. Frank Nezrick and Dave Seckel chaired some of the axion sessions and have my thanks as well. Finally, I thank MIT for allowing me the freedom to search for axions; I sense they will not be disappointed.

§References

[1] E.Gates, G.Gyuk and M.Turner, Fermilab preprint FERMILAB PUB 94/381-A.

[2] R.D. Peccei, reviews the axion and the strong CP problem in, C. Jarlskog ,"CP Violation", World Scientific, Singapore (1989) 503.

[3] "Particle and Nuclear Astrophysics and Cosmology in the Next Millennium", Snowmass, Colorado (June 29 to July14, 1994).

[4] "Strategies for the Detection of Dark Matter Particles", Berkeley, California (February 20-24,1994).

[5] E.Kolb and M.Turner, "The Early Universe", Addison-Wesley (1990).

[6] Because of the CPT theorem, I am sometimes careless in noting the distinction between T and CP violation. To get it right, see Robert G. Sachs, "The Physics of Time Reversal", University of Chicago Press (1987).

[7] Altarev et al., Phys.Lett. B276 242.

[8] For example, see T.Cheng and L.Li, "Gauge Theory of Elementary Particle Physics," Oxford University Press, New York (1984).

[9] J.B.Kogut and L.Susskind, Phys.Rev. D11 (1975) 1477.

[10] J.S.Bell and D.G.Sutherland, Nucl.Phys. B4 (1968) 315.

[11] R.D.Peccei and H.R.Quinn, Phys.Rev. D16 (1977) 1791.

[12] S.Weinberg, Phys.Rev.Lett. 40 (1978) 223; F.Wilczek, Phys.Rev.Lett. 39 (1978) 1304.

[13] J.Kim, Phys.Rev.Lett.43(1979)103; M.Shifman, A.Vainshtein and V.Zakharov, Nucl.Phys. B166 (1980) 493.

[14] M.Dine, W.Fischler and M.Srednicki, Phys.Lett. B104 (1981) 199; A.Zhitnitsky, Sov.J.Nucl.Phys. 31 (1980) 260.

[15] Y.Asano et al., Phys.Lett. 107B (1981) 159-162.

[16] J.Frieman, S.Dimopoulos and M.Turner, Phys.Rev. D36 (1987) 2201.

[17] D.M.Lazarus et al., University of Rochester Department of Physics and Astronomy report UR-1253 (1992).

[18] F.Avignone et al., Phys.Rev. D36 (1987) 2752.

[19] I.Iben, Ann.Rev.Astron.Astrophys. 12 (1974) 215.

[20] G.Raffelt and D.Dearborn, Phys.Rev. D36 (1987) 2211.

[21] G.Raffelt, Phys.Rev.D33 (1986) 897, and references therein.

[22] R.Brinkmann and M.Turner, Phys.Rev.D38 (1988) 2338.

[23] This pattern of symmetry breaking can be treated as instanton effects, c.f., D.Gross, R.Pisarski and L.Yaffe, Rev.Mod.Phys. 53 (1981) 43.

[24] M.Turner, Phys.Rev.Lett. 59 (1987) 2489.

[25] J.Preskill, M.Wise and F.Wilczek, Phys.Lett. B120 (1983) 127; L.Abbott and P.Sikivie, Phys.Lett. B120 (1983) 133; M.Dine and W.Fischler, Phys.Lett.B120 (1983) 137.

[26] M.Turner, Phys.Rev. D33 (1986) 889.

[27] H.Nelson and P.Oleson, Nucl.Phys. B61 (1973) 45; A.Vilenkin, Phys.Rep. 121 (1985) 263.

[28] R.Davis, Phys.Lett. B180 (1986) 225.

[29] P.Sikivie and J.R.Isper, Phys.Lett. B291 (1992) 288.

[30] E.Kolb and I.Tkachev, Fermilab preprint FERMI-LAB PUB 94-055-A (1994); E.Kolb and I.Tkachev, Phys.Rev.Lett. 71 (1993) 3051.

[31] E.Seidel and W.Suen, Phys.Rev.Lett. 72 (1994) 2516.

[32] M.Turner, Phys.Rev.Lett. 59 (1987) 2489.

[33] P. Sikivie, *Phys. Rev. Lett.* **51** (1983) 1415.

[34] J.Rogers, University of Rochester, thesis, Department of Physics preprint UR-1021 (1987).

[35] The collaboration includes scientists from MIT, University of Florida, University of California at Berkeley, University of Chicago, Lawrence Livermore National Laboratory, Lawrence Berkeley Laboratory, and Fermi National Accelerator Laboratory.

[36] For a review of Rydberg atoms, see T.Gallagher, "Rydberg Atoms," Cambridge University Press, Cambridge (1994).

[37] The potential of the Rydberg electron in an external electric field "leans over", so that above a critical value of n, the atom spontaneously ionizes when passing the ionization electrodes.

[38] S.Matsuki, I.Ogawa and K.Yamamoto, Preprint ICR-NS, Kyoto University (June, 1994).

[39] D.Tanner and Qang, University of Florida UFIFT preprint HEP-93-2.

[40] Y.Semertzidas et al., Phys.Rev.Lett. 64 (1990) 2988.

[41] G.Ruoso et al., Z.Physik C (1992).

[42] D.M.Lazarus et al., Phys.Rev.Lett. 69 (1992) 2333.

[43] R.Stroynowski and A.Zhitnitsky, Southern Methodist University preprint SMU-HEP-94-12 (August 1994).

[44] S.Lee et al., "Measuring the Speed of Light in a Strong Magnetic Field," a proposal to the U.S. Department of Energy Office of Energy Research, program solicitation number DE-PS05-94ER40863 (1994).

The Cosmic Background Radiation

G.F. Smoot

Lawrence Berkeley Laboratory and Department of Physics, University of California, Berkeley

Observations of the Cosmic Microwave background have provided many of the most powerful constraints we have on cosmology and events in the early universe. The spectrum and isotropy of CBR have long been a pillar of Big Bang models. The discovery of low levels on anisotropy has provided new information and tools for our understanding of the early universe. Further observations promise to enhance greatly our knowledge of processes in the early universe and cosmological parameters. We can anticipate rapid advance in this field up to and through the year 2000 which will dramatically focus our efforts in cosmology during the next millenium. This paper outlines the primary science likely to be discovered and defined by a vigorous airborne and ground-based program which should be strongly supported. If successfully executed, we can anticipate a measurement of the CBR anisotropy spectrum to within a factor of two of the confidence level unavoidably set by cosmic variance. Even so, observations of the CBR are the best and often the only way to obtain information on many critical parameters so that an ambitious satellite experiment that maps the full-sky to the $\delta T/T \sim 10^{-6}$ level is well justified.

§1 Introduction

From its prediction (Gamow 1948 & Alpher & Herman 1948) and discovery (Penzias & Wilson 1965) the cosmic background radiation (CBR) has been a cornerstone of the Big Bang cosmology and led to its widespread acceptance. In the intervening years improved observations have lead to a better understanding of its role in the early universe and the wealth of information that it carries about the early universe. With this work its importance to cosmology has grown rapidly while the technology to exploit it has kept pace. A major advancement came with the COBE satellite in the early 1990's with a very precise measurement of the spectrum (Mather et al. 1993) as being black body and the discovery of anisotropies (Smoot et al. 1992) at the 10^{-5} level. This discovery was followed by an explosion of activity and the reports of anisotropies by several groups and a rapid advance in theoretical work.

1.1 The CBR Spectrum

The spectrum and temperature of the cosmic background radiation are key issues in cosmology. The Big Bang model makes the unequivocal prediction that to high order the spectrum will be thermal and that the temperature will vary with redshift as $T(z) = (1+z)T_0$. The current COBE FIRAS observations now give the best evidence that the spectrum is blackbody and the best estimate of the temperature T_0. The precision of the spectral observations is within a factor of three of predicted distortions based upon known objects in the universe. Future improvements both from better data analysis and improved experiments are possible. There is significant new science possible with improved results.

There is much evidence for the remote presence of the cosmic background radiation from CN observations and the presence of anomalous formaldehyde absorption in distant galaxies and from the Sunyaev-Zel'dovich effect. The only evidence for the redshift dependence of the temperature comes from the Keck telescope observations of Songalia et al. (1994). The remote sensing of the cosmic background radiation is an actively emerging field. We can expect significant progress in the study of the SZ effect and the remote sensing of the CBR in the future.

1.2 Anisotropies in the CBR

Fluctuations in the temperature of the cosmic background radiation have now been detected over a wide range of angular scales and a consistent picture may be emerging. The existence and nature of these anisotropies have a profound impact on cosmology. First they define the kind of universe that we inhabit and how models can be developed as well as constrained. The observations of the CBR angular distribution are the best evidence that we have for the large scale isotropy and homogeneity of the universe. As such they imply that models of the universe can be treated as FRW (Friedmann-Roberstson-Walker) cosmologies - that is an expanding universe slightly perturbed around the Robertson-Walker metric. This simplifies calculations of the development of the universe from very early times to the present. The universe indicates no sign of shear and vorticity or other factors that would force us to more complicated cosmologies.

The natural, and commonly accepted, interpretation of the very small anisotropies discovered is that they result from primordial perturbations in the early universe. All indications are that the large-scale structure in the universe has developed by the process of gravitational instability from these small amplitude primordial perturbations in the energy density. Slightly overdense regions begin collapsing under the influence of gravity, becoming more and more overdense. Under dense regions rarify as they expand and their material flows to more dense regions. Over time the density contrast increases. The result is the formation of large scale structure such as galaxies, galaxy clusters, voids, and features such as the great wall of galaxies.

Observations of the CBR anisotropies provide information on these primordial perturbations and on smaller ($\leq 1^\circ$) scales on the initial development of these perturbations. As a result it is possible to glean much new information about cosmological parameters and the content of the universe through the detailed study of the resulting temperature fluctuations. For example the angular scale ($\leq 1^\circ$) is set by two things, (1) the speed of light - setting the largest region that can be affected by the motion of matter and energy, and (2) the speed of sound - setting the scale for regions in which matter can clump. The speed of sound is roughly $c/\sqrt{3}$ so that these two sizes are closely related. The age of the universe times this speed sets the size of these regions. The angle on the sky is set by the ratio of the age of the universe at the time the CBR last interacted to the time that we observe the CBR and by the geometry of the universe. Since the lengths (or ages) are in ratio, any uncertainty in the distance scale (or Hubble constant) cancel and the angle on the sky is set strictly by the geometry of the universe. A flat universe predicts a peak in CBR fluctuations at an angular scale of about $\theta_{flat} = 0.8^\circ$ (or $\ell \approx 220$). An open universe predicts a smaller angle while a closed universe predicts a larger angle simply as a consequence of the curving of light on its way to the observer. For a last scattering redshift of ≈ 1100 the relation is simply $\theta_{peak} \cong \theta_{flat}\sqrt{\Omega_0}$ where we have assumed the standard relation between Ω and the curvature of the universe. This is as direct a measurement of the geometry of the universe as one can imagine. The corrections to the relationship are all weak.

The geometry of the universe is one of many parameters one could determine with precise measurements of the CBR anisotropies. One can expect to distinguish between models involving inflation and topological defects. Perhaps the easiest way is (see e.g. Albrecht et al. 1995, Crittendon & Turok 1995) by comparing the existence and location of the higher order peaks in the power spectrum. However, there are other possible tests. It is also possible to distinguish among various models of inflation. Precise observations of the CBR anisotropies can in fact provide us with the opportunity to determine the parameters of inflationary models including measuring the ratio of scalar (density) to tensor (gravitational waves) perturbations, the slope of the inflaton potential during the epoch of astronomical importance, as well as the energy scale of inflation during this epoch. It is possible that these observations would provide definite evidence for the existence of gravitons. (See e.g. Knox 1995, Bond et al. 1994 and many others).

Precise observations on angular scales $\leq 1^\circ$ can provide information on the Hubble constant, a limit on Ω_{baryon} that is comparable to and independent of the Big Bang nucleosynthesis results, information on the nature of the dark matter and evidence that the gravitational instability mechanism is working as predicted. There are many theoretical discussions and tutorials of these points (see e.g. Scott, Silk, & White 1995).

1.3 Polarization

The polarization of the CBR can also provide useful information about the early universe. If our current models are correct, then the level of polarization expected is discouragingly low. Existing results are at a level that indicate that there is no great inconsistency in our current models but the limits are still more than one to two orders of magnitude greater than model predictions. At the predicted level experiments are likely to be limited by instrument noise and observation time for decades to come. However, at high frequencies it is possible that the foreground will be sufficiently low as to allow discovery level observations. The polarization is likely to continue to occupy a very distant back seat to anisotropy observations.

§2 Current Status & Anticipated Results by 2000

The standard hot Big Bang model of cosmology is supported by three observational results: the Hubble flow of distant galaxies, the abundances of the light elements, and the existence of an isotropic thermal radiation bath. But the hot Big Bang model is incomplete; the universe today is far more complex than a homogeneous soup of hydrogen, helium, neutrinos, and microwave photons. The observable matter is organized in a hierarchy of dense clumps surrounded by empty voids. Stars group into galaxies, which in turn form groups, clusters, and still larger structures. Explaining our increasingly detailed observations of the large scale structure of the universe is a fundamental problem in cosmology: what is the origin of observed structures and how did they form?

To search for the origins of structure we must turn to relics from that era. The cosmic microwave background (CMB) is one such relic; its photons, through their spatial and frequency distribution, reflect the distribution of mass and energy in the early universe and record subsequent interactions between the evolving matter and radiation fields. The paradigm for structure formation consists of the gravitational infall and collapse of small "seed" perturbations in the density of the early universe. The central result of the *Cosmic Background Explorer (COBE)* was the support of this basic picture through the detection of CMB anisotropy without a corresponding distortion from a blackbody spectrum. Within this broad outline, however, a number of more detailed questions remain unanswered. No cosmological models are able to reconcile in detail the *COBE* measurements of the CMB spectrum and anisotropy with measurements of galaxy counts and peculiar velocities. What is missing from our understanding? Evidence indicates that the dominant component of the universe is "dark" matter, inferred only through its gravitational effects. What is the nature of this dark matter? Interpretation of the CMB anisotropy on co-moving scales directly comparable to the largest structures observed in optical surveys depends on the poorly constrained ionization history of the universe. In standard models the universe became neutral at redshift $z \sim 1100$ when the temperature fell below the ionization potential of hydrogen, yet the local universe is thought to be highly ionized to redshift $z \sim 5$. What is the thermal history of the universe? When did the universe become re-ionized?

2.1 The CBR Spectrum - Science & Status

The observed Hubble recession of galaxies has a natural interpretation as evidence for an expanding universe: in the remote past, the universe was much smaller and denser than today. At the high energies corresponding to very early times, photon-creating processes ($\gamma + \mathrm{e} \rightarrow \gamma + \gamma + \mathrm{e}$, $\mathrm{e} + \mathrm{p} \rightarrow \mathrm{e} + \mathrm{p} + \gamma$, $\mathrm{e}^+\mathrm{e}^- \rightarrow \gamma\gamma$) proceed rapidly with respect to the Hubble expansion, creating a system in local thermodynamic equilibrium. A system in thermal equilibrium is completely characterized by its temperature and any conserved quantum numbers: in the absence of non-equilibrium interactions, the CMB will be an isotropic blackbody radiation field. In an adiabatic expansion, the photon occupation number is constant; hence, the expansion of the universe does not by itself distort the CMB spectrum, but merely scales the thermodynamic temperature $T = T_0(1 + z)$.

Distortions in the CMB spectrum arise from non-equilibrium interactions of the matter and radiation fields in the evolving universe. Many such interactions are known to exist. Figure 1 illustrates an idealized thermal history of the universe. As the CMB temperature falls below the rest mass, particles pair-created at higher temperatures fall out of equilibrium and either decay, annihilate, or "freeze out" at a relic density. The resulting energy releases might be expected to leave distinctive signatures in the CMB spectrum. Unfortunately for high-energy physics, the earliest universe is unobservable to us in photons. Photon-creating processes ($\gamma + \mathrm{e} \rightarrow \gamma' + \gamma' + \mathrm{e}'$) proceed rapidly for times $t < 1$ year, rethermalizing an arbitrary distortion to a new (albeit hotter) Planck spectrum. Processes at epochs with $kT_{CMB} > 2$ keV leave no significant direct signature on the CMB spectrum, and must be studied through any long-lived relics (gravitational potential variations, exotic particle species, topological defects) they might spawn.

The presence today of ordered structures (galaxies, galaxy clusters, and larger-scale structure) in the face of initial isotropy is a powerful argument for such relics. The *COBE* DMR instrument shows anisotropy in the early universe ($z \sim 1100$) to be of order $\Delta T/T \sim 10^{-5}$ (Smoot et al. 1992) compared to the current clustering of order unity. Processes capable of generating structure quickly enough, without violating observational limits on the CMB isotropy, will in general release energy to the matter or radiation fields, which in turn generate distortions in the CMB spectrum. One such mechanism is "dark" non-baryonic matter, whose dynamical properties allow it to clump much more rapidly than the baryons, which only later fall into the dark-matter gravitational potential wells. Such non-baryonic dark matter may be the lightest stable member of a family of particles (e.g. supersymmetric partners to the known particle families), in which case either the decay of the heavier unstable members or rare decay modes of the stable members can distort the CMB spectrum. Broad classes of dark-matter candidates include such decay modes. Other non-equilibrium energy-releasing processes in the early universe include the dissipation of shock and sound waves associated with primordial density or entropy perturbations, the dissipation of primeval turbulence, dissipation of gravitational wave energy, phase transitions in the early universe and associated topological defects, or energy released through isotropization of an anisotropic universe. *Any* transfer of kinetic or thermal energy between the matter and radiation fields at $t > 1$ year must alter the CMB spectrum from a blackbody distribution. The size and shape of the present distortions depend on size, redshift, and mechanism of the energy transfer, allowing observers to distinguish between various physical processes.

The most general form of CMB spectral distortion results from interactions with a hot electron gas at temperature T_e. Three classes of spectral distortions are particularly important, resulting from processes at different epochs. The simplest distortion is photon production from electron-ion interactions (free-free emission or thermal bremsstrahlung): $\mathrm{e} + \mathrm{Z} \rightarrow \mathrm{e} + \mathrm{Z} + \gamma$. The distortion to the present-day CMB spectrum is given by

$$\Delta T_{\mathrm{ff}} = T_\gamma \frac{Y_{\mathrm{ff}}}{x^2} \tag{1}$$

where T_γ is the undistorted photon temperature, x is the dimensionless frequency $h\nu/kT_\gamma$, Y_{ff} is the optical depth parameter for free-free emission

$$Y_{\mathrm{ff}} = \int_0^z \frac{k[T_e(z) - T_\gamma(z)]}{T_e(z)} \frac{8\pi e^6 h^2 n_e^2 g}{3m_e(kT_\gamma)^3\sqrt{6\pi m_e kT_e}} \frac{dt}{dz'} dz' \ , \tag{2}$$

n_e is the electron density, and g is the Gaunt factor (Bartlett & Stebbins 1991). The distorted CMB spectrum is characterized by a quadratic rise in temperature at long wavelengths as the photon distribution thermalizes to the plasma temperature, and is the dominant signature for a warm plasma ($T_e \sim 10^4$ K) at recent epochs ($z < 1100$).

Compton scattering ($\gamma + \mathrm{e} \rightarrow \gamma' + \mathrm{e}'$) of the CMB photons from the hot electron gas also contributes to CMB spectral distortions. Compton scattering transfers energy from the electrons to the photons while keeping the photon number fixed. For recent energy releases ($z < 10^5$), the gas is optically thin, resulting in a CMB distortion

$$\Delta T_{\mathrm{RJ}} = -2yT_\gamma \tag{3}$$

in the Rayleigh-Jeans part of the spectrum where there are now too few photons, and an exponential rise in temperature in the Wien region where there are now too many photons. The magnitude of the distortion is related to the total energy

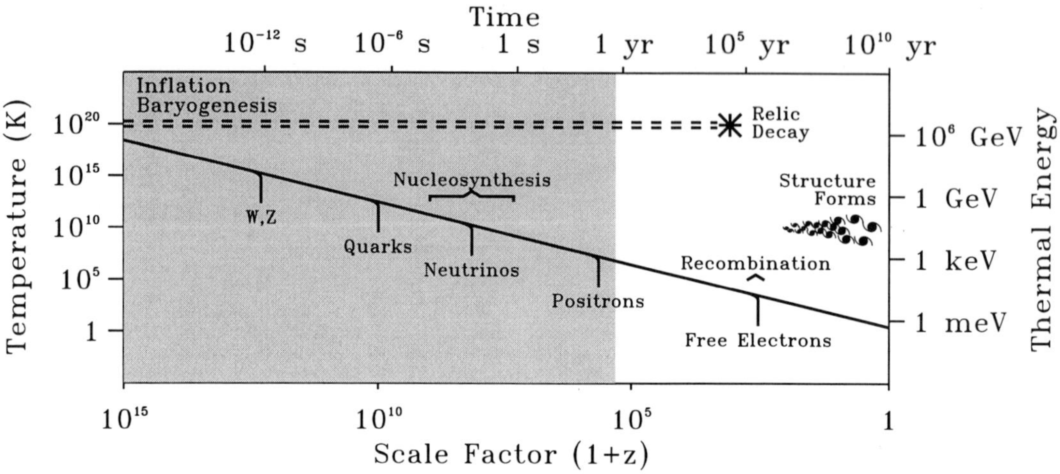

Figure 1: Idealized thermal history of the universe. The epochs at which various particle species annihilate, decay, or decouple are indicated. Events in the shaded region leave no direct signature in the CMB spectrum, but affect it though the decay of long-lived relics.

transfer

$$\frac{\Delta E}{E} = e^{4y} - 1 \approx 4y \tag{4}$$

(Sunyaev & Zel'dovich 1970), where the parameter y is given by the integral

$$y = \int_0^z \frac{k[T_e(z) - T_\gamma(z)]}{m_e c^2} \sigma_T n_e(z) c \frac{dt}{dz'} dz' \ , \tag{5}$$

of the electron pressure $n_e k T_e$ along the line of sight and σ_T denotes the Thomson cross section (Zel'dovich & Sunyaev 1969).

Compton scattering alters the photon energy distribution while conserving photon number. After many scatterings the system will reach statistical (not thermodynamic) equilibrium, described by the Bose-Einstein distribution

$$\eta = \frac{1}{e^{x+\mu_0} - 1} \tag{6}$$

with dimensionless chemical potential

$$\mu_0 = 1.4 \frac{\Delta E}{E}. \tag{7}$$

Free-free emission thermalizes the spectrum to the plasma temperature at long wavelengths. Including this effect, the chemical potential becomes frequency-dependent,

$$\mu(x) = \mu_0 \exp\left(-\frac{2x_b}{x}\right), \tag{8}$$

where x_b is the transition frequency at which Compton scattering of photons to higher frequencies is balanced by free-free creation of new photons. The resulting spectrum has a sharp drop in brightness temperature at centimeter wavelengths (Burigana et al. 1991).

The different distributions probe energy releases at different epochs, and hence different physical processes. Figure 2 shows the distortion to the CMB spectrum corresponding to each distribution. The equilibrium Bose-Einstein distribution results from the oldest non-equilibrium processes ($10^5 < z < 8 \times 10^6$) such as the decay of relic particles or primordial anisotropy. Energy releases at more recent epochs ($z < 10^5$) produce a Comptonized spectrum, a prime candidate being a hot ($T_e > 10^5$ K) intergalactic medium. Free-free emission dominates for recent reionization ($z < 10^3$) from a warm IGM. The different physical processes distort the CMB spectrum at different wavelength regimes. Measurements near the CMB peak at millimeter wavelengths are sensitive primarily to the Compton processes y and μ. Measurements at longer wavelengths are sensitive to plasma bremsstrahlung from a recent reionization or to the earliest energy releases from relic decay. Measurements at both long and short wavelengths are required to constrain fully the thermal history of the universe.

Figure 3 summarizes precise measurements of the CMB spectrum from ground-based, balloon, and space platforms. The *COBE*-FIRAS instrument provides a precise determination of the CMB spectrum at mm and sub-mm wavelengths, while ground-based and balloon-borne radiometers measure the spectrum to much lower precision at centimeter and longer wavelengths. The CMB spectrum is consistent with a blackbody across more than 3 decades of frequency: a least-squares fit to all CMB measurements yields limits

$$T_\gamma = 2.73 \pm 0.01 \text{ K}$$
$$|y| < 2.5 \times 10^{-5}$$

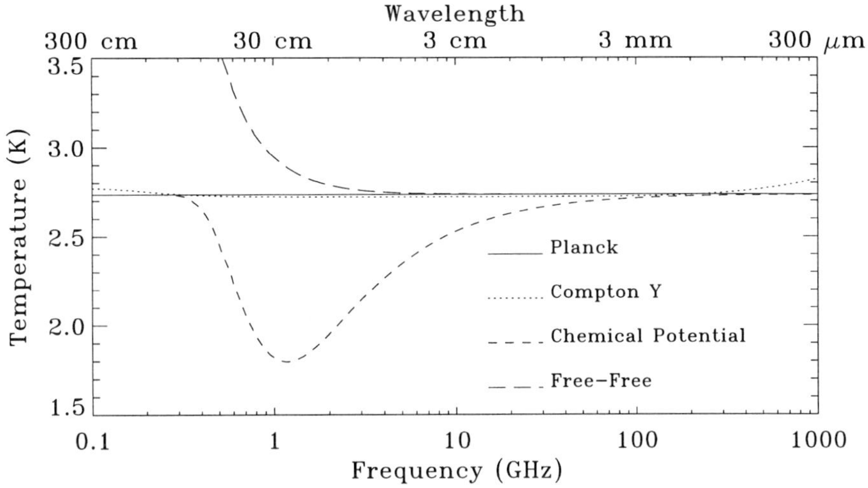

Figure 2: CMB distortions resulting from energy-releasing processes at different epochs.

$$|\mu_0| < 3.3 \times 10^{-4}$$
$$|Y_{ff}| < 1.9 \times 10^{-5}$$

at 95% confidence, corresponding to limits on energetic processes $\Delta E/E < 2 \times 10^{-4}$ occurring between redshifts 10^3 and 8×10^6 (Mather et al. 1994, Wright et al. 1994, Bersanelli et al. 1994). The early universe, apparently, was a quiet place, with most of the energy residing in an isotropic radiation bath and the dark matter. The large scale structure and voids observed in optical galaxy surveys did not result from primeval explosions, since the energy required is more than 2 orders of magnitude greater than allowed. There is no uniform, hot intergalactic medium: the 35 keV electrons that produce the diffuse X-ray background have a volume filling factor below 10^{-4} (Wright et al. 1994).

The *COBE* results limit deviations from a blackbody spectrum to less than 0.03% of the peak CMB intensity (less than 1 mK in temperature units) over the 0.05–5 mm wavelength range, but results at longer wavelengths are much less precise: distortions as large as several percent could exist at centimeter wavelengths or longer without violating existing data. The *COBE* limits on spectral distortions at mm wavelengths do *not* imply correspondingly small deviations at centimeter wavelengths. In fact, results at cm wavelengths lie systematically colder than *COBE* (Kogut et al. 1991). Although suggestive of a possible distortion, the existing long-wavelength data lack sufficient precision to discriminate between a small distortion and an error in the relative calibration between the *COBE* and ground-based results (Kogut 1992).

Nor have the current observational results exhausted the supply of interesting cosmological questions. Among the most urgent questions is the thermal history of the universe: when did the universe recombine? In the simplest model, recombination occurs at redshift $z \sim 1100$ when the CMB temperature falls below the ionization potential of hydrogen, leaving the bulk of the matter in a neutral state. The direct observational evidence for the standard thermal history, though, is weak. The Gunn-Peterson test shows no evidence for significant neutral hydrogen column density in the intergalactic medium, indicating that the local universe is highly ionized to $z \sim 5$. The *COBE* limit requires only that the total energy released during reionization be less than $4y$ times the energy in the CMB. Although the energy per ionization is ~ 10 eV, there are 10^9 CMB photons per electron with characteristic energy $kT = 10^{-3.6}\ (1+z)$ eV. The COBE requirement $\Delta E/E < 10^{-4}$ for a single reionization is easily met. Photoionization from the first generation of collapsed objects could plausibly reionize the universe. Indeed, some reionization at $z \sim 50$ is almost inevitable in cold dark matter models of structure formation and could occur as early as $z \sim 200$ in non-Gaussian isocurvature models, including topological defect models (Tegmark, Silk, & Blanchard 1994).

The ionization history of the universe depends upon the rare high-amplitude peaks in the density field, and hence provides important information for models of structure formation. If evidence for reionization is found, it probes the physics of the heating mechanism, while a negative result confirms the standard thermal history (the universe is neutral for $50 < z < 1100$), ruling out many alternate cosmologies. Fixing the thermal history complements efforts to interpret the CMB anisotropy: at degree angular scales, pre-existing anisotropies will be smoothed by Compton scattering as they pass through the plasma, reducing their amplitude. The reduced amplitude from reionization can be mistaken for evidence of other physical processes, including gravity-waves, a larger Hubble constant, or an intrinsic reduction in primordial power at small angular scales (White, Krauss, & Silk 1994, Bond et al. 1994).

We may observe a recent reionization through its distinctive spectral signatures, requiring measurements of the CMB spectrum at both long and short wavelengths. The *COBE*

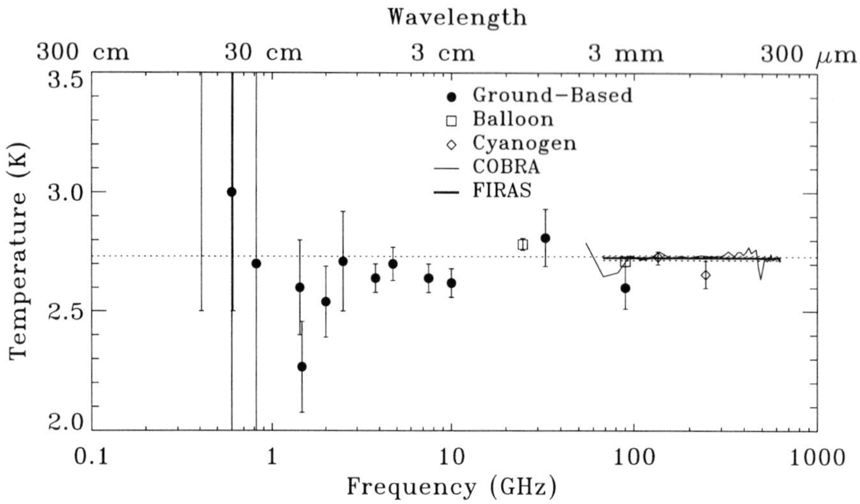

Figure 3: Precise measurements of the CMB thermodynamic temperature. The dotted line represents a 2.73 K blackbody. The CMB spectrum is poorly constrained at centimeter or longer wavelengths.

limits to Compton-distorted spectra at millimeter wavelengths limit the plasma temperature from above, $\Delta T_y \propto n_e k T_e$. Precise new measurements at centimeter wavelengths of the free-free emission from the same reionized gas limit the plasma temperature from below, $\Delta T_{\rm ff} \propto n_e^2/\sqrt{T_e}$. Together, the combined results can either detect the direct signature of reionization or rule out all ionized models except those with temperatures $T_e \sim T_\gamma$ (Bartlett & Stebbins 1991). Given the current *COBE* limits on y, a determination of the thermal history at $z \sim 50$ requires a measurement of the free-free parameter $Y_{\rm ff}<10^{-7}$, corresponding to a 100 μK spectral distortion at 10 cm wavelength. This is obtainable within reasonable experimental sensitivity.

A second question of interest is the power spectrum of primordial density perturbations. The *COBE* detection of CMB anisotropy on angular scales above 7° (Smoot et al. 1992) provides important support for structure formation via gravitational instability. The *COBE* anisotropy data are well-described by a Gaussian primordial density field with power spectrum $P(k) \propto k^n$ per comoving wave number k, with power-law index $n = 1.0 \pm 0.4$ (Górski et al. 1994). Measurements of the CMB spectrum complement direct measurements of the density power spectrum. Short-wavelength fluctuations which enter the horizon while the universe is radiation-dominated oscillate as acoustic waves of constant amplitude and are damped by photon diffusion, transferring energy from the acoustic waves to the CMB spectrum and creating a chemical potential. The energy transferred, and hence the magnitude of the present distortion to the CMB spectrum, depends on the amplitude of the perturbations as they enter the horizon through the power-law index n (Daly 1992; Hu, Scott, & Silk 1994). Although the CMB spectral distortion predicted by the inflationary value $n = 1$ is unobservably small, models with "tilted" spectra $n>1$ can produce observable distortions. Probing the range $n<1.4$ of direct interest to non-standard models requires a determination of the chemical potential $|\mu_0|<10^{-5}$, corresponding to a 0.5 mK spectral distortion at 10 cm wavelength.

Exotic particle decay provides another source for non-zero chemical potential. The dynamics of clusters of galaxies requires a stronger gravitational field than can be inferred by luminous matter, strongly suggesting the existence of a "dark" component of the matter field. Evidence exists that much of this dark matter is non-baryonic. Particle physics provides a number of dark matter candidates, including massive neutrinos, photinos, axions, or other weakly interacting massive particles (WIMPs). In most of these models, the current dark matter consists of the lightest stable member of a family of related particles, produced by pair creation in the early universe. Decay of the heavier, unstable members to a photon or charged particle branch will distort the CMB spectrum provided the particle lifetime is greater than a year. Rare decays of stable particles (e.g., a small branching ratio for massive neutrino decay) provide a continuous energy input and also distort the CMB spectrum. The size and wavelength of the CMB distortion are dependent upon the decay mass difference, branching ratio, and lifetime. Stringent limits on the energy released by exotic particle decay provides an important input to high-energy theories including supersymmetry and neutrino physics.

At present work in the CBR spectrum is relatively limited. The COBE team is still analyzing the FIRAS data but with a reduced staff. We can hope that a successful analysis could lead to as much as a factor of three improvement in the current COBE FIRAS limits or measurements. That is the practical limit and still leaves a significant portion of the spectrum at longer wavelengths to be probed with precision. Currently only a couple of groups are actively making spectrum observations. At the present time, it appears no groups are funded for such measurements for the next few years with the exception of internal GSFC award for the ARCADE balloon-borne instrument as a possible predecessor for a satellite. Suzanne Staggs (U. Chicago) and colleagues are planning a 10 GHz balloon measurement. At this point the CBR spectrum field has been the victim of its own success in terms of precise experiments finding thus far null results. The results have been and continue to be powerful constraints on cosmological models. The need for better measurements in the cm wavelength region and the prospects for a satellite are covered in the longer term future section.

2.2 Anisotropies in the CBR - Science & Status

The origin of large scale structure in the Universe is one of the key fundamental issues in cosmology. Gravitational instability models hold that large-scale structure forms as the result of gravitational amplification of initially small perturbations in the primordial density distribution. These density perturbations leave their imprint as anisotropies in the Cosmic Microwave Background (CMB) radiation. The *COBE* Differential Microwave Radiometers (DMR) maps of the cosmic microwave background anisotropy measure the primordial density distribution on super-horizon scales and offer a unique probe of the"initial conditions" for structure formation. On smaller angular scales there can be significant processing and movement of the primordial perturbations. The resulting CMB anisotropy power spectrum is rich in content and information on these processes and on cosmological parameters.

Anisotropies in the cosmic microwave background (CMB) probe the distribution of mass and energy in the early universe and provide a discriminant of competing models of structure formation. In inflationary models, the large-scale CMB anisotropy results from quantum fluctuations which are stretched in the accelerating expansion of the Universe to astronomical scales. Competing models (topological defects, axions, late-time phase transitions) generally involve relic residual high-energy vacuum state and have higher-order correlations and non-Gaussian distributions.

In general, it is thought that the cosmic background radiation photons travel to us in the present without significant interaction with matter from a redshift of about $z_{ls} = 1100$. This is the redshift for which the high energy tail of the cosmic background photons can ionize hydrogen. Above $z_{ls} = 1100$ the universe is ionized and the photons and baryon-electron plasma interact through Thomson scattering. The transition of the cosmic radiation from a collisional regime to free photons takes place in a time on the order of the Hubble time at that epoch. The last scattering region is a shell of finite thickness in redshift. If the decoupling occurred at a relatively large redshift ($z \geq 1000$), the thickness of the shell is roughly 1/15 of the mean redshift, which is relatively narrow ($\approx 10'$) from the observers point of view. Because of the strong thermal contact between the radiation, there is a compensation that makes the shell essentially equivalent to a surface and it is often treated and named as such. The last-scattering surface is generally taken to be optical depth unity for Thomson scattering. This is about equal to the visibility depth as Thomson scattering is nearly isotropic, so that a single scattering erases most previous anisotropy.

Causally-connected regions at the surface of last scattering, as viewed from the present epoch, subtend an angle $\theta \sim 1^\circ\!.7\, \Omega_0^{1/2}\ (\,1100/(1+z_{ls})\,)^{1/2}$. Anisotropy on larger angular scales thus reflects primordial conditions unaffected by local physics in the photon-baryon fluid and offers a look at the primordial fluctuations. Anisotropy on smaller angular scales provides a glimpse at the processing and other effects allowing a test between competing models.

CMB Temperature Anisotropies from Primordial Perturbations

If the cosmic background has a black-body (Planckian) distribution locally at the last scattering surface, the occupation number for each mode is

$$n(\nu) = \frac{1}{e^{h\nu/kT} - 1}. \tag{9}$$

If the photons undergo a redshift z in traveling from that location of last scattering to the receiver, then the received frequency, ν_R, is related to the emitted frequency, ν_E, according to:

$$\frac{\lambda_R}{\lambda_E} = \frac{\nu_E}{\nu_R} = (1+z). \tag{10}$$

Since the number of photons per mode will be conserved,

$$\begin{aligned} n(\nu_R) &= n(\nu_E) = \frac{1}{e^{h\nu_E/kT_E} - 1} \\ &= \frac{1}{e^{h\nu_R(1+z)/kT_E} - 1} = \frac{1}{e^{h\nu_R/kT_R} - 1}, \end{aligned} \tag{11}$$

which implies that the relation between the temperature on the emitting or last-scattering surface, T_E, and the temperature observed by a receiver, T_R, is

$$\frac{T_R}{T_E} = \frac{1}{1+z} = \frac{(k^\alpha u_\alpha)_R}{(k^\alpha u_\alpha)_E} \tag{12}$$

where u_α is the four-vector velocity of the observer or emitter and k^α is the vector tangent to the null geodesic (photon path) connecting the events of emission and observation. The dominant effect is the cooling by redshift. To first order the variation in observed CMB temperature, T_R, (assuming a Planckian distribution) is set by the variation in emitting temperature and the variation in redshift according to the relation

$$\frac{\delta T_R}{T_R} \cong \frac{\delta T_E}{T_E} - \frac{\delta z}{1+z} \tag{13}$$

where T_E is the temperature on the surface of emission (the last scattering surface in general for the CMB), and $(1+z)$ is the redshift from the observer (receiver) R to the surface of emission E. The primary effect is the expansion of the Universe which causes both the initial E temperatures and temperature variations to decrease by the factor of $(1+z)$ leaving the ratio unchanged. If the redshift to the emitting surface varies with direction, $\delta z(\theta,\phi)$, then there will be a corresponding temperature variation with direction.

The equivalence principle tells us that there will be a gravitational redshift from the matter-energy density perturbations. The Doppler effect produces a frequency shift arising from the difference in velocity between the emitter and observer. Varying velocities on the surface of last scattering will result in changes in different directions. There are thus three effects identified here as causing temperature anisotropy: (1) primordial (at last scattering surface) temperature variation (2) varying redshift with location due to gravitational potential variations, and (3) motion of the emitter and observer. The varying redshift can be caused by scalar, vector, and tensor perturbations (i.e. density variations, vorticity, and gravitational waves respectively) or general anisotropic expansion. One can just write down an equation for the received temperature as the sum of these effects. Often these effects are treated separately, but one must be careful with questions of coordinate system (choice of gauge) and treatment, particularly in the case of unusual cosmologies. A full General Relativistic treatment shows that one can trade the expansion redshift, gravitational redshift, and Doppler shift for each other by choice of coordinates but that the total result should be independent of coordinates with proper treatment. This results from the equivalence principle.

We will focus here on the anisotropies of primordial density perturbations in a near Friedmann-Robertson-Walker (FRW) universe. The high degree of CMB isotropy (e.g. Smoot et al. 1992) indicates that this is an appropriate approach (Stoeger, Maartens, & Ellis 1994). The standard practice is to use conformal time, linearize the problem in an expanding universe, and use the fact that light travels on a null geodesic. The conformal time, η is defined as $d\eta \equiv dt/a(t)$, where $a(t)$ is the scale factor for the Universe. Then $g_{\mu\nu}$, linearly perturbed from the Robertson-Walker metric, $ds^2 = g_{\mu\nu}dx^\mu dx^\nu \cong a^2(\eta)[\eta_{\mu\nu} + h_{\mu\nu}]dx^\mu dx^\nu$, where $\eta_{\mu\nu}$ is the Minkowski metric.

Sachs and Wolfe (1967) gave the first and an excellent treatment of the CMB anisotropies that result from potential variations in a FRW space-time, though their results are some what more general. They derived the relation good to first order:

$$T_R = T_E\frac{\eta_E^2}{\eta_R^2}(1+\frac{\delta T_R}{T_R}), \tag{14}$$

where η is the conformal time and the subscripts R stands for received and E for emitted or last scattered location epochs and

$$\frac{\delta T_R}{T_R} = \frac{1}{2}\int_0^{\eta_R-\eta_E}(\frac{\partial h_{\mu\beta}}{\partial\eta}e^\mu e^\beta - 2\frac{\partial h_{\beta 0}}{\partial\eta}e^\beta)_{(o)}dy, \tag{15}$$

where $\delta T_R/T_R$ includes all the effects by using the null geodesic and e^β is the direction vector of the light the receiver sees. The Sachs-Wolfe calculation is generally separated into parts: gravitational redshift from potential variation on the surface of last scattering, the time variation of potentials along the photon's path, the Doppler effect due to the relative motion of the emitter and observer (to first order in v/c), and variations in the temperature or number density of photons. This can be written as

$$\frac{\delta T_R}{T_R} = \frac{\delta\Phi}{c^2} + \frac{2}{c}\int_E^R d\tau\frac{\partial\Phi}{\partial\tau}(\tau,\mathbf{x}) + \mathbf{n}\cdot\frac{(\mathbf{v}_R-\mathbf{v}_E)}{c} + \frac{\delta T_E}{T_E} \tag{16}$$

where the first term is the cosmological gravitational redshift, the second term is for the time changing gravitational potential along the photons' path, the next is the Doppler effect of the receiver and emitter relative velocities along the line of sight, and the last term is the variation in temperature with emitter location (e.g. see Appendix B of White, Scott, & Silk 1994).

Anisotropy from Adiabatic Density Fluctuations

We now consider the effect of adiabatic density fluctuations. Adiabatic fluctuations are those in which all constituents (photons, baryons, and whatever dark matter) maintain a constant (number) density ratio. Thus the photon density, and therefore temperature, vary in tandem with the potential. Since the photon energy density $\rho_\gamma \propto T^4$, the variation in temperature is

$$\frac{\delta T_E}{T_E} = \frac{1}{4}\frac{\delta\rho_\gamma}{\rho_\gamma}|_E. \tag{17}$$

The potential, Φ, is related to the density field $\rho(x,t)$ via the Poisson equation,

$$\frac{1}{a^2}\nabla^2\Phi = 4\pi G\rho, \tag{18}$$

where a is the cosmological scale factor and ∇^2 is the Laplacian with respect to comoving coordinates. In the Newtonian limit $\Phi = GM/r$ so that fluctuations in density, $\delta\rho$ result in potential fluctuations $\delta\Phi$. The relationship is readily found in the following way:

$$\delta\Phi = \frac{G\delta M}{r} = G\frac{\delta M}{M}\frac{M}{r} = G\frac{\delta\rho}{\rho}\frac{4\pi r^2}{3}\rho. \tag{19}$$

The Hubble expansion gives us the relation between the Hubble expansion rate, H, and the critical density, ρ_c, namely $\rho_c = 3H^2/8\pi G$ which we can substitute into the equation above yielding:

$$\delta\Phi = \frac{1}{2}\frac{\delta\rho}{\rho}(Hr)^2\frac{\rho}{\rho_c} \quad or \quad \frac{\delta\rho}{\rho} = 2\delta\Phi\frac{1}{(Hr)^2}\frac{\rho_c}{\rho} \tag{20}$$

We make the approximation that $\rho \approx \rho_c$ which is more and more accurate as one goes back in time. Even if $\Omega_o \approx 0.01$, which is the lower limit set by visible matter, then ρ will be 90% of ρ_c by a redshift of 1000. At scales $r > c/H$ (larger than the horizon) most of the contribution comes from distances comparable to c/H because causality limits the range over which gravitation can act so that $\delta\rho/\rho \sim 2\delta\Phi/c^2$. (Note than in the weak field approximation for comoving coordinates $h_{00} = 2\delta\Phi$ which indicates the geodesic approach will get the same answer.) Thus for adiabatic perturbations $\delta\rho/\rho \approx 2\delta\Phi/c^2$ and since $\delta\rho_\gamma/\rho_\gamma = 4/3\ \delta\rho/\rho$, then $\delta\rho_\gamma/\rho_\gamma|_E = -8/3\ \delta\Phi/c^2$.

In this case the Sachs-Wolfe calculation gives

$$\frac{\delta T_R}{T_R} = \frac{1}{c^2}\delta\Phi + \frac{2}{c}\int_E^R d\tau \frac{\partial \Phi}{\partial \tau}(\tau, \mathbf{x}) + \mathbf{n}\cdot(\mathbf{v}_R - \mathbf{v}_E)/c + \frac{1}{4}\frac{\delta\rho_\gamma}{\rho_\gamma}|_E \tag{21}$$

$$\frac{\delta T_R}{T_R} = \frac{1}{3c^2}\delta\Phi + \frac{2}{c}\int_E^R d\tau \frac{\partial \Phi}{\partial \tau}(\tau, \mathbf{x}) + \mathbf{n}\cdot(\mathbf{v}_R - \mathbf{v}_E)/c \tag{22}$$

For most circumstances the Sachs-Wolfe effect is dominated by the conditions on the surface of last scattering. The effect of the time varying potentials (called the Rees-Sciama effect) on the photons is generally an order of magnitude less than the last-scattering surface effect of perturbations. When the perturbations are in the linear regime, the time rate of change of the potential is zero and there is no Rees-Sciama effect.

In this section we focus on the CMB anisotropy generated by the gravitational potential differences created by adiabatic fluctuations. We leave the treatment of relative velocities to a later section. This is generally valid for anisotropies produced by adiabatic fluctuations on scales larger than the horizon. Adiabatic fluctuation potential variations, $\delta\Phi$, on the surface of last scattering give anisotropies:

$$\frac{\delta T}{T} = \frac{1}{3}\frac{\delta\Phi}{c^2} \tag{23}$$

which are in turn related to the density field $\rho(\mathbf{x}, t)$ via the Poisson equation above.

If the primordial perturbations are initially small, $\rho(\mathbf{x}, t) = \overline{\rho}\ [1 + \delta(\mathbf{x}, t)]$, we can linearize the treatment and Fourier expand the density perturbations $\delta(\mathbf{x})$,

$$\delta(\mathbf{x}) \ = \ \sum_k \delta_k \ e^{i\mathbf{k}\cdot\mathbf{x}} \ \rightarrow \ \frac{V}{(2\pi)^3}\int_{Vol} \delta_k \ e^{i\mathbf{k}\cdot\mathbf{x}} d^3k \tag{24}$$

where, by the Fourier transform,

$$\delta_k = V^{-1}\int_{Vol} \delta(\mathbf{x}) \ e^{i\mathbf{k}\cdot\mathbf{x}} d^3x \tag{25}$$

The density field is Gaussian if the amplitude probability distribution at any spatial point follows a Gaussian distribution

$$p(\delta) \ = \ \frac{1}{\sqrt{2\pi\sigma^2}} \ \exp(-\frac{\delta^2}{2\sigma^2}) \tag{26}$$

with a uniform distribution in phase, where $\delta_k = A(k)e^{i\phi_k}$,

$$0 < \phi_k \ \leq \ 2\pi. \tag{27}$$

This is expected to be the case for most sources of perturbations and in particular for nearly all inflationary models. Gaussian fields have the desirable property that they are completely specified by the power spectrum $P(k) \ = \ \langle|\delta_k|^2\rangle$ or its Fourier transform, the 2-point correlation function. To this point all tests on the COBE DMR data show them to be consistent with Gaussian statistics (Smoot et al. 1994, Hinshaw et al. 1994, Kogut et al. 1994, Kogut 1995). The significance of these tests for all angular scales is not yet clear; however, we will assume it as a working assumption. If it is not true, then inconsistencies will appear, when new, high-quality CMB anisotropy data become available. Though the power spectrum will show anomalies, it is likely that other tests for non-Gaussian fluctuations and non-random phase will be more powerful.

We can now use the first term in the Sachs-Wolfe effect to calculate the CMB anisotropy from the power spectrum of density perturbations

$$\frac{\delta T}{T} = \frac{1}{3}\frac{\delta\Phi}{c^2} = -\frac{a_0^2 H_0^2}{2(2\pi)^3}\int k^{-2}\delta_k \ e^{i\mathbf{k}\cdot\mathbf{x}} d^3k, \tag{28}$$

where the vector $\mathbf{x}$ points to the last-scattering surface and has length $2cH_0^{-1}$, H_0 is the current Hubble expansion rate, and a_0 is the current scale size of the Universe.

2.3 Spherical Harmonic Decomposition & Power Spectrum

The CMB anisotropy may be decomposed into a spherical harmonic representation

$$T(\theta, \phi) \ = \ \sum_{\ell m} a_{\ell m} Y_{\ell m}(\theta, \phi) \tag{29}$$

where θ and ϕ are the spherical angles on the sky. The spherical harmonic coefficients $a_{\ell m}$ are given by

$$\langle|a_{\ell m}|^2\rangle \approx V^{-1}\frac{H_0^4}{2\pi}\int_0^\infty \frac{dk}{k^2}|\delta_k|^2[j_l(kx)]^2, \tag{30}$$

where j_l is the spherical Bessel function of order l, and the average is over all possible realizations of fluctuations. One expects this is equivalent to an average over all observation positions in the Universe. This is justified by the assumption that we do not occupy a special position - the Copernican Principle. This average over all realizations (observation position) results in random phases for the various components.

If we assume that the initial power spectrum is a power law:

$$|\delta_k|^2 = AVk^n, \tag{31}$$

the integral over $[j_l(kx)]^2$ yields

$$\langle a_{\ell m}^2\rangle \ = \ \frac{AH_0^{n+3}}{16}\ \frac{\Gamma[l + (n-1)/2]\ \Gamma[3-n]}{\Gamma[l + (5-n)/2]\ \Gamma[(4-n)/2]^2} \tag{32}$$

$$= \ (Q_{rms-PS})^2\ \frac{4\pi}{5}\ \frac{\Gamma[l + (n-1)/2]\ \Gamma[(9-n)/2]}{\Gamma[l + (5-n)/2]\ \Gamma[(3+n)/2]} \tag{33}$$

(Kolb & Turner 1991, Bond & Efstathiou 1987). Up to a numerical factor $\langle a_{\ell m}^2\rangle^{1/2}$ is equal to the value of $\delta\rho/\rho$ on the present horizon.

In this case the spherical harmonic coefficients $a_{\ell m}$ are Gaussian random variables with zero mean and ℓ-dependent variance given by $\langle a_{\ell m}^2\rangle$.

The power spectrum from a power law of primordial perturbations is then simply given by the above formula for perturbations larger than the horizon at last scattering. For smaller angular scales it is no longer accurate both because of the approximation used and because of perturbation processing.

Anisotropy from Perturbation Processing

The primary perturbation processing in the early universe is the acoustic oscillation of the photon-baryon fluid. A positive perturbation will provide an extra gravitational attraction on its surrounding material, which will cause material to flow towards the inhomogeneity. The density fluctuation will continue to pull in material until the pressure forces due to the thermal motions act to stop it. The baryons and photons are coupled strongly in the early universe and they both contribute energy density for the gravitational attraction and pressure to counter act it. Any non-baryonic dark matter will contribute to the gravitational attraction also. The coupled system is very much like a mass on a spring. It behaves as an oscillator and there are acoustic oscillations for perturbations on all physical scales.

The oscillations are characterized by the sound speed which is essentially $1/\sqrt{3}$ the speed of light until decoupling of the photons and baryons at the surface of last scattering. This means that perturbations larger than the scale of the horizon at last scattering are still in the first compression stage with a compression roughly proportional to the physical scale over the horizon scale.

Perturbations that are nearly the horizon scale at the surface of last scattering are just maximally compressed. They are slightly smaller than the horizon size because the speed of sound will average to slightly less than the speed of light over $\sqrt{3}$. Smaller perturbations will vary from being maximally compressed to maximally rare and back to maximally compressed depending upon their physical size which determines how many oscillations have taken place. As we know from oscillator theory dissipation will cause the amplitude to die down proportional to the number of oscillations. There is dissipation due to photon viscosity and leakage.

These acoustic oscillations can be viewed as sound waves, not traveling through space but in time. The oscillations all start at the same instant but oscillate with different frequencies depending their physical scale. This results in two effects: bulk motion of the baryons and photons and thus varying velocity on the surface of last scattering, and also a varying density of photons at the surface of last scattering. These two effects are 90° out of phase with each other. Note that in an oscillation the extremes of motion (turning points) are when the velocity is zero and the velocity is highest in the middle of the oscillation.

Receiver Motion Doppler Effect

We consider the effect of emitter and receiver motion. First consider receiver motion as illustrative of the Doppler effect and then we will consider the velocity variations at the surface of last scattering. Receiver motion with velocity $\beta = v/c$ relative to an isotropic Planckian radiation field of temperature T_o produces a Doppler-shifted temperature

$$T(\theta) = T_o \frac{(1-\beta^2)^{1/2}}{(1-\beta cos(\theta))} \quad (34)$$

$$\approx T_o\,(1 + \beta cos(\theta) + (\beta^2/2)cos(2\theta) + ...) \quad (35)$$

The first term is the monopole CBR temperature without a Doppler shift. The term proportional to β is a dipole, varying as the cosine of the angle between the velocity and the direction of observation. The term proportional to β^2 is a quadrupole, varying with cosine of twice the angle with amplitude reduced by $1/2\ \beta$ from the dipole amplitude.

The dipole anisotropy was detected a little more than a decade after the CMB was discovered (Smoot et al. 1977). The COBE DMR maps clearly show a dipole distribution consistent with a Doppler-shifted thermal spectrum (Kogut et al. 1993, Fixsen et al. 1993). The DMR finds the same thermodynamic amplitude for all three frequencies. Using the DMR direction and FIRAS spectrum one finds the difference in spectra taken near the hot and cold poles is described to better than 1% of its peak value by the difference in blackbody spectra. The DMR maps are well-fitted by a dipole cosine dependence. There is annual modulation by the 30 km/s earth orbital velocity. All observations are consistent with the kinematic effect of our motion relative to the CMB rest frame and thus a Doppler shift origin.

The implied velocity for the solar-system barycenter is $\beta = 0.00123 \pm 0.00003$ where we assume a value $T_0 = 2.726$ K, towards $(\alpha, \delta) = (11.17^h \pm 0.03^h, -6.7^\circ \pm 0.3^\circ)$, or $(l, b) = (264.4^\circ \pm 0.3^\circ, 48.4^\circ \pm 0.5^\circ)$. This in turn implies a velocity for the Galaxy and Local Group of galaxies relative to the CMB. The derived velocity is $v_{LG} = 627 \pm 22$ km s^{-1} toward $(l^{II}, b^{II}) = (276^\circ \pm 3^\circ, 30^\circ \pm 3^\circ)$.

For a power-law density power spectrum $P(k) \propto k^n$ (the potential power spectrum goes as $\Phi_k^2 \propto k^{n-1}$ so that in this notation $n = 1$ is scale invariant), the contribution to the rms peculiar velocity $\langle v^2 \rangle^{1/2}$ arising from primordial potential fluctuations of length scale larger than λ is

$$< v^2 >^{1/2} \approx 400 \mathrm{km\ s}^{-1} \frac{Q}{16\mu K} (\frac{50h^{-1}\mathrm{Mpc}}{\lambda})^{(n+1)/2} \quad (36)$$

where h^{-1} is the H_o in units of 100 km s^{-1} Mpc^{-1} and Q is the expected average quadrupole CMB intensity amplitude. Though the primordial perturbation spectrum likely extends down to scales much smaller than 1 Mpc, dissipative effects and virialization dominate at scales of 10 Mpc and may extend out to scales of 30 to 40 Mpc. These effects on average will increase the expected rms velocity. Since the rms velocity is broadly distributed, the observed Galactic velocity of 627 ± 22 km s^{-1} is consistent with being produced via a primordial power spectrum of perturbations at the level detected by the COBE DMR. A specific prediction of this origin is that the rms velocity flow will decrease with volume averaging size λ according to the formula above.

The Doppler effect of this velocity and the velocity of the Earth around the Sun, as well as any velocity of the receiver relative to the Earth, is normally subtracted from the data. It is likely that some intrinsic dipole power is removed as well. Thus power spectrum investigations traditionally skip the dipole ($\ell = 1$) component.

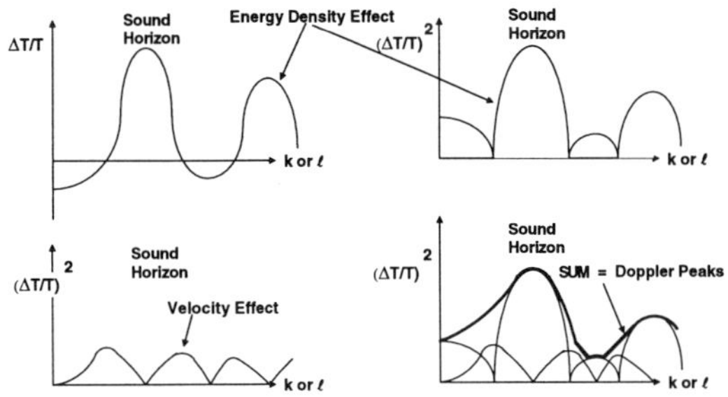

Figure 4: A rough schematic showing the contributions to the "Doppler peaks" from the density enhancement and velocities associated with the acoustic oscillations occurring up until decoupling. The upper left frame shows the temperature variation due to the change in the density of photons as a function of scale. The right frame shows the corresponding power spectrum which is just the rectified mean square power. The lower left frame shows the effect of the motion of the photon fluid during the oscillations. Note the velocity and compression/rarefaction are 90° out of phase with each other. The final panel shows the power spectra for the two components and their sum. The successive peaks and minima in the energy density are at harmonics of the sound horizon.

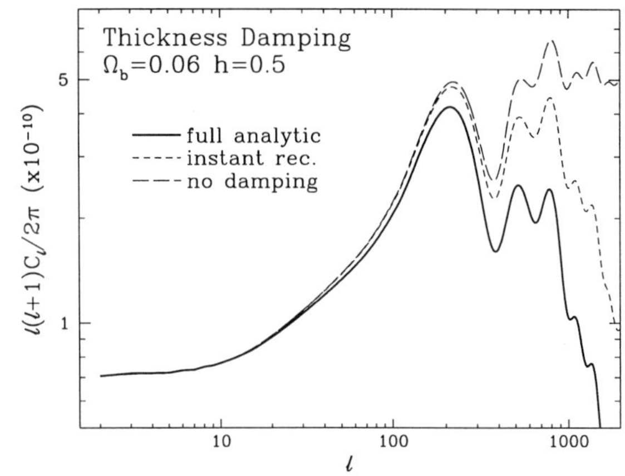

Figure 5: The effect of damping on the "Doppler Peaks". Estimating damping in the instantaneous recombination approximation leads to a significant underestimate of the damping scale (Hu & Sugiyama 1995)

Last Scattering Doppler Effect

Motion of the emitting or last scattering surface will cause CMB anisotropy via the Doppler effect. For primordial density perturbations the major expected motion is due to thermal-acoustic oscillations of the primordial baryon-photon fluid. We expect that all perturbations will undergo oscillation and that as a function of wavenumber k there will be the first full compression at the sound horizon (the distance sound can propagate from the beginning of the universe until the time of last scattering). There will be alternating rarefaction and compression peaks at harmonics of the sound horizon.

We know that the velocity profile will appear as shown in Figure 4. The velocity and density curves are 90 degrees out of phase. The density enhancement wins out over the velocity in amplitude but the velocity effect causes the long slow rise to the first Doppler peak.

Damping of Small Angular Scale Anisotropies

Figure 5 shows the effect of damping of the oscillations including the finite thickness of the last scattering surface for a $\Omega_0 = 1$, $\Omega_b = 0.06$ and $H_0 = 50$ km/s/Mpc universe. This severe damping arises as the photons diffuse through the baryons. The diffusion is most significant at recombination, which is when the photon-baryon fluid decouples. The effect of photon diffusion is to exponentially, $e^{-\tau}$, damp the temperature fluctuations coming from the density fluctuations. There are two parts to the process: (1) diffusion of photons out of the overdense regions which reduces the temperature contrast, and (2) rescattering of the diffusing photons which tends to isotropize them. The diffusion length grows rapidly during recombination but is in effect for a shorter time. This means that the smallest physical scales are most damped. The diffusion of photons out and into the oscillations means that the oscillation is also damped (called Silk damping) which also mostly affects the smaller perturbations up until decoupling. As the photons diffuse they still are affected by the gravitational redshift and they are rescattered and isotropized, further reducing the signal.

The effect of re-ionization

For a reionized universe the last scattering is delayed or stretched out, and the diffusion length grows to be nearly the horizon size at last scattering. If the universe never went through a neutral (decoupled) phase, clearly the oscillations will continue and the Doppler peaks will be shifted to the sound horizon at that last scattering time. Thus observation of the Doppler peak in the expected place can tell us that the universe did not undergo significant reionization until very late ($z < 50$) when the density of available electrons was very low so that the fluctuations survived being erased by Compton scattering. This is complementary to possible spectral measurements for checking the ionization history of the universe.

Anisotropy Results

Since the DMR announcement of the discovery of anisotropy nine groups have reported CMB anisotropies. Figure 6 shows the current observational status of the CMB anisotropy power spectrum.

The 'MIT' FIRS experiment (Meyer et al. 1991, Page et al. 1990) is the only experiment, other than the DMR, to map a significant portion of the sky. The FIRS experiment has a $\sim 3°$ beam width and covered nearly a quarter of the sky with a single balloon flight. The FIRS data correlate well with the DMR data (Ganga et al. 1993) and show a similar power spectrum (Ganga et al. 1994) consistent with scale invariance.

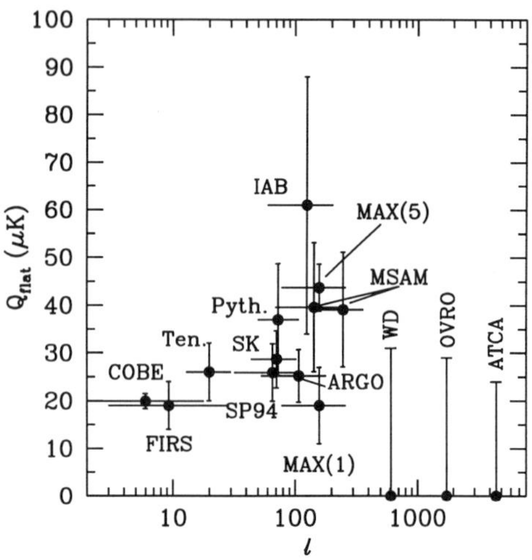

Figure 6: Current status of CMB anisotropy power spectrum observations adapted from Scott, Silk, & White (1995). The amplitudes plotted are the quadrupole amplitudes for a flat (unprocessed scale-invariant spectrum of primordial perturbations, i.e. a horizontal line) anisotropy spectrum that would give the observed results for the experiment. Figure 7 gives an indication of the expected spectrum for a processed spectrum in a CDM model for comparison.

The Tenerife (Watson et al. 1991) is also a large angular scale experiment (beam width 5°) that covers a differenced (8°) strip scanned on the sky by the earth's rotation. The Tenerife experiment has pointed out bumps on the sky as specific locations of anisotropy (Hancock et al. 1994). The ULISSE experiment (de Bernardis et al 1992) reported upper limits on 6° CMB anisotropy using balloon-borne bolometric observations. The Advanced Cosmic Microwave Explorer (ACME South Pole) (Gaier et al. 1992 & Schuster et al. 1993) reported upper limits and detections of fluctuations operating with HEMT amplifiers. The Saskatoon "SK93" experiment (Wollack et al. 1993) used HEMT amplifiers to detect CMB anisotropy from Saskatoon, SK, Canada. Fluctuations were reported from South Pole observations by the Python experiment (Dragovan et al. 1993). The ARGO balloon-borne experiment (de Bernardis et al. 1994) observed a statistically significant signal with a 52′ beam. The Italian Antarctic Base (IAB) experiment (Piccirillo and Calisse 1993) used bolometric techniques with a 50′ Gaussian beam and reports anisotropy. The Millimeter-wave Anisotropy eXperiment (MAX) is a balloon-borne bolometric instrument with high sensitivity in the medium angular scale that has completed five flights detecting significant CMB fluctuations (Alsop et al. 1992, Meinhold et al 1993, Devlin et al 1994, Clapp et al 1994). The Medium Scale Anisotropy Measurement (MSAM) balloon-borne experiment (Cheng et al. 1994) is a very similar balloon-borne medium-scale CMB anisotropy instrument but with a different chopping scheme that allows the results to be reported either as a difference or a triple difference, providing two effective window functions. Also from the South Pole the White Dish experiment (Tucker et al. 1993) reports an upper limit on CMB anisotropy. Arcminute scale anisotropy upper limits were reported using the Owens Valley Radio Observatory (OVRO) (Myers et al. 1993). The Australia Telescope Compact Array (ATCA) was used to place upper limits on CMB anisotropy in a Fourier synthesized image (Subrahmayan et al. 1993).

The field is moving so rapidly that such plots get out of date quickly. At this stage we can begin to see that the reported results, though scattered, are actually in rough agreement with each other and with many models. The goal at the moment is to refine the results and our ability to distinguish between models. It is a hot topic whether the data show evidence for a Doppler peak and then whether that peak is in the flat universe location. The existence of the peak and its location would go far towards telling us not only whether we are on the right track with these models, and if there is a connection with large scale structure formation, but also about various cosmological parameters.

§3 Instruments

We can anticipate significant advances over the next few years as many groups continue ground-based measurements both with the conventional beam switching and interferometer techniques, balloon-borne experiments including long-duration flight instruments BOOMERANG and TOPHAT as well as the idea of ultra-long duration balloon flights, e.g. ACE. There is also significant effort going into the design of potential space missions including COBRAS/SAMBA, PSI, MAP, and FIRE. In looking toward the future, it is instructive to consider these instruments and their likely results. Their likely progress drives both the field and the design of and need for future work. (The space missions are considered in the FUTURE section.)

3.1 MAX/MAXIMA

Consider the MAX/MAXIMA payload as representative of current and currently planned balloon-borne missions.

MAX

The Millimeter-wave Anisotropy eXperiment (MAX) is a balloon-borne bolometric instrument which observes at multiple frequencies with high sensitivity on the 0.5° angular scale. MAX has completed five flights detecting significant CMB fluctuations (Fischer et al. 1992, Alsop et al. 1992, Meinhold et al 1993, Devlin et al 1994, Clapp et al 1994, Tanaka et al. 1995, Lim et al. 1995).

The MAX instrument consists of an off-axis Gregorian telescope and a bolometric photometer mounted on an attitude-controlled balloon-borne platform which makes measurements at an altitude of 36 km. The Gregorian telescope consists of a 1-meter primary and a nutating elliptical secondary. The underfilled optics provides a 0.55° FWHM beam when focused and aligned. The 5.7 Hz nutation of the secondary modulates the beam on the sky sinusoidally though ±0.68° and the attitude control sweeps the beam over a 6° or 8°

path and back in about 108 seconds, producing about 15 to 20 independent temperature differences on the sky. Depending upon the time of observation and location of region under observation sky rotation can cause the observed region to be in the shape of a bow-tie.

On flights 4 & 5 the single-pixel four-band bolometric receiver features negligible sensitivity to radio frequency interference and an adiabatic demagnetization refrigerator to cool the photometer to 85 mK. The dichroic photometer used for MAX has $(\delta\nu/\nu)$ of 0.57, 0.45, 0.35, and 0.25 filter bands at 3.5, 6, 9, and 15 cm^{-1}. MAX covers the high frequency side of the window formed by galactic dust emission rising at higher frequencies and Galactic synchrotron and free-free emission increasing at lower frequencies. The 15 cm^{-1} channel acts as a guard against Galactic dust and atmospheric emission. The multiple frequencies provide sufficient redundancy to provide confidence that the signal is CMB and not a foreground or systematic effect.

MAX is calibrated both by an on-board commandable membrane and by observations of planets, usually Jupiter. The two techniques agree at roughly the 10% level. The calibration is such that the quoted temperature difference is the real temperature difference on the sky.

MAX makes deep CMB observations (typically one hour) on regions generally selected to be low in dust contrast and total emission and free from known radio sources. MAX has made observations on five flights The data from most of the scans are in good agreement but the scan of the mu Pegasi region is significantly lower than the rest. That is why there are points plotted for the average of the agreeing region and a single point for the mu Pegasi data. We are hopeful that the 5th flight will resolve this issue but it seems to be coming out at an intermediate value.

The center of the scan is the same for the three observations of GUM (the star Gamma Ursae Minoris) but the relative geometry is such that the three scans made bow-tie patterns which cross at the star. White and Bunn (1995) have made use of this fact to construct a two dimensional map of the region which is roughly $10^\circ \times 5^\circ$. The title of their paper is "A First Map of the CMB at 0.5° Resolution".

Making maps is clearly the appropriate approach for the current generation of new experiments. MAX is evolving to a new system MAXIMA, which is designed and constructed for the goal of getting the power spectrum around the first "Doppler" peak and making maps covering a significant portion of the sky.

MAXIMA

MAXIMA stands for MAX imaging system. The current one-dimensional scans are very useful data for the discovery phase of CMB anisotropy research. Soon progress will depend upon the availability of two-dimensional maps of low galactic foreground regions (low dust in this case) with several hundred pixels so that sampling variance is less important (see section 4.1). In addition one can look for properties of the sky which are not predicted by theories and could be overlooked in statistical analyses. It also makes it possible to catalog features

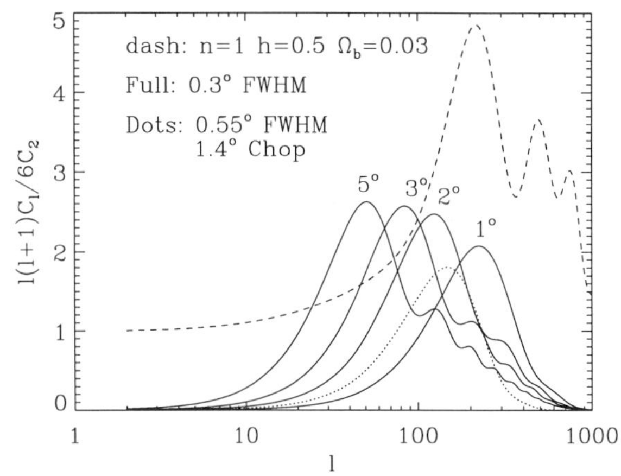

Figure 7: Window functions possible for the new MAXIMA mirror for chop angles of 1, 2, 3, and 5 degrees and the dotted line shows the MAX window function. For reference a theoretical power spectrum for CDM is shown as a dashed line. Figure courtesy Shaul Hanany.

for comparison to or motivation of other experiments.

Under the auspices of the NSF Center for Particle Astrophysics a collaboration consisting of groups from the University of California at Berkeley, Caltech, the University of Rome, and the IROE-CNR Florence have begun work on a new system. To make an imager a new optical system was necessary. The primary feature is a 1.3-meter, off-axis, lightweight primary mirror. The primary will be modulated which allows a much larger beam chop angle on the sky with less spill over and thus more pixels in the focal plane. Cold secondary and tertiary mirrors provide a cold Lyot stop and the field-of-view required for the array of 20 arcminute pixels. The geometrical aberrations in the center of the field-of-view are less than 10 arcminutes.

A larger primary mirror requires a larger gondola which is now constructed. The chop angle can both be increased and varied allowing the instrument to sample the shape of the power spectrum over the range $40 < \ell < 300$. Figure 7 shows some sample window functions possible with the new system.

An additional feature is new detector electronics with AC coupling in order to allow linear scanning in a total power mode, making maps and power spectrum measurements directly. This approach is different than that of making a number of different window functions as shown in figure 7. The idea is to use a scan or raster scan of the CMB anisotropies on the sky directly rather than obtaining a set of differences at different chop angles. One is thus mapping directly and measuring the power spectrum as the fourier transform of the data. At this stage the instrument is designed to operate in this mode either by scanning the primary mirror in a sawtooth pattern or moving the entire gondola in azimuth.

Another major change will be going from a single pixel four-frequency photometer to an eight-pixel receiver. This will allow taking data at eight times the rate and thus make two-dimensional mapping feasible. The receiver design has been completed and the new dewar ordered. The bolometers will be changed to have a spider-web substrate so that cosmic ray transient occurrences will be reduced by more than an order of magnitude. The target date of a first flight of the new gondola is August 1995. We can anticipate that within three years MAXIMA will have made maps and will have measured the anisotropy power spectrum around the location of the first doppler peak.

3.2 MSAM/TOPHAT

MSAM/TOPHAT is similar to MAX/MAXIMA at the present. A notable difference between MSAM and MAX has been that MSAM used a three-position chop analyzed either as a triple beam or double beam (two chop angles on the sky) observation (Page et al. 1994). MSAM angular resolution is 0.5° between 5 and 23 cm^{-1} (150 and 700 GHz or wavelengths 0.4 to 2.0 mm). MSAM has had two flights (June 1992 and May 1994) both from Palestine, Texas.

MSAM is preparing for another flight with expanded frequency coverage in 5 spectral bands between 2.3 and 5 cm^{-1} (70 to 150 GHz or wavelengths 2.0 to 4.3 mm). The instrument is expected to improve its signal-to-noise ratio by about a factor of three over the previous results.

All observations are along a ring surrounding the north celestial pole.

TOPHAT is conceived as a long-duration balloon-borne experiment with the detectors located on the top of the balloon rather than in a gondola hanging below the balloon. It will observe in five spectral bands between 5 and 21 cm^{-1} (150 and 630 GHz or wavelengths between 0.5 to 2.0 mm). The current plans call for the measurement of 40 points on the sky, each with an rms sensitivity of $\delta T_{rms} \approx 1$ μK or $\delta T_{rms}/T_{CMB} \approx 3 \times 10^{-7}$ including removal of the galactic foreground dust emission. It is likely that TOPHAT will evolve towards a less sensitive observation of a larger area of the sky for reasons of sample variance. It will be useful to have some deep (high-sensitivity) scans such as this or what MAXIMA can do to understand the signal well. Eventually sampling variance and other science drives one toward observing a larger fraction of the sky.

3.3 BOOMERANG

BOOMERANG is the long-duration balloon-borne version of MAXIMA and an intermediate step toward a bolometer space mission, e.g. FIRE or COBRAS/SAMBA. It is planned to make a many day flight circumnavigating Antarctica as early as December 1996. BOOMERANG will move more directly towards mapping a significant region of the sky than the experiments thus far discussed.

3.4 ACE

As a follow up to their South Pole HEMT observations the Santa Barbara group has proposed ACE (Advance Cosmic Explorer). It is a large, light-weight (200 kg), system aimed at making flights lasting 90 days or more. It would utilize advanced HEMTs, active refrigerators, and a 2-m diameter composite mirror to cover the frequency range 25 to 90 GHz. In three such flights such a system could map 75% of the sky to an angular resolution of 10 arcminutes at a level of about 20 μK. This project is still in the early phase but is indicative of what with sufficient funding one might achieve by the year 2000.

3.5 Ground-Based Instruments

Ground-based instruments have made a significant contribution to CMB anisotropy observations. They have been more successful than originally envisioned as a result of the observers' clever strategies to minimize and reduce the effect of the atmosphere. These strategies have included going to high, dry sites such as the South Pole and Teide peak on Tenerife and using triple-beam chopping or other similar techniques. These techniques are more difficult to use when going to mapping and making observations over an extended portion of the power spectrum. Here again it is possible that significant progress can be made though it is likely to be eventually limited before the science is exhausted.

An exciting exception is the use of aperture synthesis interferometers. The Ryle Telescope images of the Sunyaev-Zeldovich effect in clusters and the CAT (Cambridge Anisotropy Telescope) results have convinced many that interferometers have a bright future in actually mapping anisotropy on small angular scales over selected regions of the sky. A number of proposals are pending. Noteworthy are the VSA (Very Small Array) in England and the Caltech interferometer. If funded, these interferometers are likely to provide a very good first cut at the CMB anisotropy power spectrum on angular scales less than about 0.5° (ℓ>400).

§4 Limitations: Cosmic Variance, Sensitivity, & Foregrounds

The fundamental limitations to these observations are the cosmic variance, sensitivity of the detectors and the unavoidable foregrounds. Systematic errors are a difficult and thorny issue and must be dealt with carefully. That it is possible to overcome systematics has an existence proof in the form of accepted results to date. The difficulties are not to be minimized and improving by an order of magnitude is likely to uncover new issues. However, we shall assume that observers will be able to overcome these as they are not necessarily fundamental limitations.

4.1 Cosmic & Sample Variance

An important effect that must be considered in understanding the temperature fluctuations expected and how accurately one can determine fundamental parameters is cosmic and sampling variance. In most cosmological models, the observed CMB temperature field is a single realization of a stochastic process. A single realization, e.g. the observable

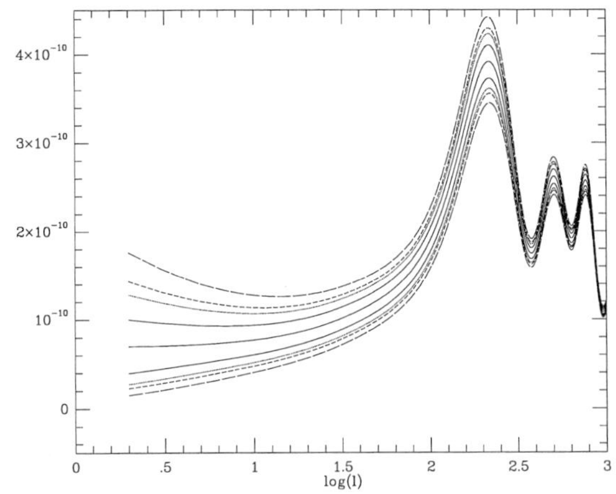

Figure 8: Center line shows the power spectrum for standard CDM ($h = 0.5, \Omega_b = 0.05$, courtesy of Sugiyama) The pairs of lines are the cosmic variance at 68%, 90%, 95%, and 99%, respectively.

universe set by our horizon, will not, in general, exactly follow the ensemble mean of the parent population ("cosmic variance"). If we only cover a portion of the sky, we do less well and we are limited by "sample variance". This problem is most acute at the largest angular scales; the CMB quadrupole, for example, is described by only 5 parameters. At higher multipole moments the larger number of components ensures that cosmic variance becomes less important. Incomplete sky coverage can introduce another uncertainty, that the fraction examined may not be representative of the realization as a whole ("sample variance"). Sample variance is not an issue for the full-sky *COBE* maps, but it is an important limitation for observations at smaller angular scales.

In general the cosmic variance adds an irreducible theoretical uncertainty to each spherical harmonic amplitude with a variance equal to twice the variance of the spherical harmonic amplitude. For the power spectrum this means that the cosmic variance rms error at each ℓ is equal to the power at ℓ times the factor $\sqrt{2/(2\ell+1)}$. This is quite a serious uncertainty for low ℓ. Figure 8 shows the limitation set by cosmic variance (i.e. assuming full sky coverage with uniform accuracy) in determining a sample power spectrum.

If the sky is not covered uniformly or if only a portion of the sky is sampled, then the sample variance must be larger than the cosmic variance. In general the increase in variance is set by the ratio of the whole sky to experiment effective solid angles (Scott, Srednicki, & White 1994). It is appropriate to note that at present sample variance is as large or larger than the instrument noise and calibration uncertainty for many current balloon results.

4.2 Sensitivity

Sensitivity has always been an issue in CBR anisotropy observations. For coherent receivers (those observing a single coherent state) the rms fluctuations in output power is given by the formula

$$\delta T_{rms} = \frac{s(T_{system} + T_{obs})}{\sqrt{B\,\tau}} \tag{37}$$

where T_{system} is the equivalent noise power of the system expressed in units of antenna temperature $P = kT_A B$, T_{obs} is the antenna temperature of the observation target – which when looking at the CBR is often about 5 K which is set by the CBR 3 K and the foreground input noise power. B is the bandwidth, τ is the observation time, and s is the sky observation scheme and is a factor near unity (typically $1/\sqrt{2} \le s \le 2$ to π).

Technology has been the usual limit for coherent receivers but recently the system temperatures have been getting noticeably smaller $10 \le T_{system} \le 100$ K. The future promises to provide receivers that will have $T_{system} \le 10$ K which leaves increasing the bandwidth as the only means to improve sensitivity of a given receiver. That too is limited for practical and observational reasons. The only avenue left is to go to multireceiver systems - e.g. arrays. A good system with 10 K system temperature, a 5 K sky temperature and a 10% bandwidth would have the rms sensitivity and integration time to $\delta T/T = 10^{-6}$ (including the conversion from differential antenna temperature to thermodynamic temperature) shown in the following table:

Sensitivity & Observation Times

Frequency (GHz)	Wavelength (cm)	Sensitivity (μK $s^{1/2}/t^{1/2}$)	Time to 10^{-6} (sec)
30	1	274	10500
45	0.667	237	7500
53	0.567	206	6600
90	0.333	158	5100
125	0.24	134	5300
150	0.20	122	6100
210	0.143	104	11400
300	0.1	87	50000

These times are very long in that they require on the order of an hour or more per pixel and one needs many pixels for reasons of statistics. This means that to archive this sensitivity level over a significant portion of the sky one must utilize both arrays of detectors and one of three approaches: (1) ground-based observations (probably interferometers will prove to be the best approach because of the atmospheric foreground problem), (2) very long duration ballooning, or (3) space-borne (satellite dedicated to these observation) platforms.

Incoherent detectors are the other approach to observing the CBR. They generally work by measuring the total power absorbed (bolometers) and thus can look at multiple modes (quantum states). They gain sensitivity by utilizing a very wide bandwidth, multiple modes, and by cooling the detectors to very low temperatures, typically ≤ 0.3 K. Filter technology and optical efficiency for such systems is not yet a fully mature technology though moderately so. Thus the bandpass and optical efficiency are not likely to be well-described by a square or simple function. The sensitivity of a bolometer is difficult to quantify as precisely as for a coherent receiver since it depends upon the particular configuration, especially the bandpass filter and optical efficiency. The sensitivity of a good bolometer system these days is roughly $200\mu K\ s^{1/2}$ thermodynamic temperature near the peak of the spectrum and falling to either side as the power decreases due to either the high-frequency Wien fall off and the decrease in states and energy at lower frequencies. Ultimately it may be possible to make bolometer detector systems with up to a factor of 10 improvement over this sensitivity.

At the present time, the best detectors (both coherent and bolometers) are roughly comparable in sensitivity but separated in frequency. Generally coherent receivers work better at lower frequencies $\leq$ 100 GHz. and bolometers work better at higher frequencies $\geq$ 80 GHz.

It is interesting to compute the fundamental limit to anisotropy observation sensitivity. It is easiest to understand in terms of photon counting as there are the least number of complicating factors to consider. However, one cannot just take the 415 photons per cubic centimeter arriving at the speed of light multiplied by collecting area ($5.17 \times 10^{15} \times$ Area) to find the number of photons counted. A quantum-limited detector in the sense of one that counts the photons incident and adds no other noise is the best one can do as a detector. However, one must also collect those photons onto the detector. For a coherent detector it is obvious that the number of photons detected per second is set by the integral over the bandwidth, B, of the occupation number and that one collects B such states per second.

$$N_{coherent} = B \int \epsilon(\nu) n(\nu) d\nu = B \int \frac{\epsilon(\nu)}{e^{h\nu/kT_o} - 1} d\nu \qquad (38)$$

where $\epsilon(\nu)$ is the optical efficiency of the system at each frequency and $B = \int \epsilon(\nu) d\nu$. The limit is easily understood in terms of thermodynamic equilibrium. The power available at the terminals of a resistor is simply $P = kT_a B$. If a such a resistor is connected to an antenna system, and if it is looking at a blackbody of temperature T_o, it must come to the same temperature and emit and receive power at the same rate. The number of photons received per unit time is then just $N = kT_a B/h\nu = B/[exp(\frac{h\nu}{kT_o}) - 1]$ which is the same as before. There is a conspiracy set by diffraction that means the solid angle times collecting area is just the wavelength squared $\Omega A = \lambda^2$.

Figure 9 shows the minimum time to a sensitivity of 10^{-6} for an ideal coherent detector that counts photons and for an ideal bolometer.

For bolometer systems the argument is slightly more involved. It is the collection of photons onto the detector that limits the number of photons actually available to be counted. However, now we can not use the diffraction relationship $\Omega A = \lambda^2$. We must actually determine the throughput (étendue) and optical efficiency. The throughput is set by a number of considerations such as the size of the optics but it can be made larger than λ^2 by a substantial amount. The number of photons per unit bandwidth per unit time is

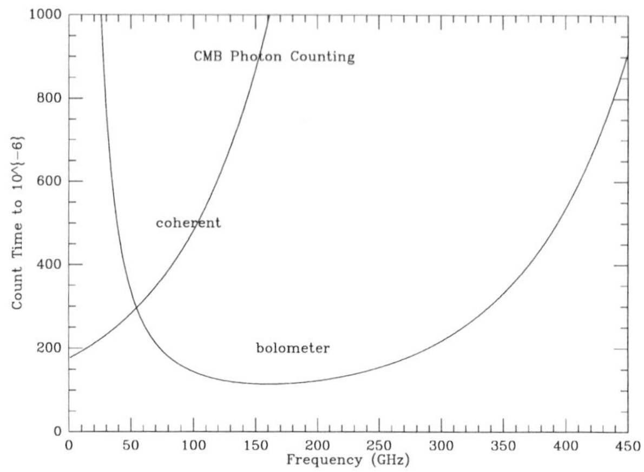

Figure 9: The time to a sensitivity of $\delta T_{rms}/T_{CMB} = 10^{-6}$ set by detectors that count the individual CMB photons. The upper line is for an ideal coherent detector with a 10% bandwidth. The lower line is for an ideal photon counter (e.g. bolometer) with throughput $A\Omega = 0.3$ cm^2 steradian and ideal optical efficiency over a 10% bandwidth. The time for bolometers is likely to rise more rapidly as frequency increases as some systems will obtain higher angular resolution at higher frequency at the expense of $A\Omega$.

then the brightness times $A\Omega$ times the optical efficiency:

$$N = \frac{2}{exp(\frac{h\nu}{kT_o}) - 1} \frac{A\Omega}{\lambda^2} \epsilon(\nu) = 2N_{coherent} \frac{A\Omega}{\lambda^2} \epsilon(\nu) \qquad (39)$$

The factor of two comes from the fact that two polarizations are available. Coherent detectors can gain back the polarization factor of 2 by putting two detectors on the same photon collecting optics. The advantage of bolometers (incoherent detectors) is the polarization factor of two and the possibility of making the throughput $A\Omega$ significantly larger than λ^2.

Bolometers retain the advantage of throughput. Generally for a simple optics system the throughput is constant for frequencies above the lowest modes, since geometrical optics is more relevant than diffraction. The lowest (optics cut off) mode and area is set by the collecting size of the bolometer. For example, on the 5th flight of MAX the throughput was roughly 0.12 cm^2 steradian.

At present the size (collecting area) of the bolometers is limited by the heat capacity and conductivity and by cosmic ray hits. Improvements such as the "spider" absorber used by MAXIMA and BOOMERANG may allow bolometers to overcome the current limits significantly. If so, then bolometers may be able to push their performance margin over coherent receivers such as HEMTs down to frequencies as low as 40 GHz.

One can estimate the combined effect of cosmic (or sample) variance and instrument noise relatively simply given that one expects them to be random and independent of each other. To first order the random fluctuations can be treated

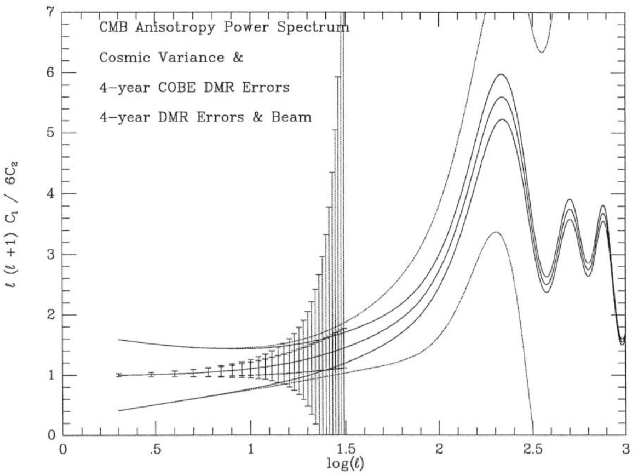

Figure 10: Center line shows the power spectrum for standard CDM ($h = 0.5, \Omega_b = 0.05$, courtesy of Sugiyama) The pair of lines on either side show the effect of cosmic variance alone. The solid error bars show the effect of instrument noise for the COBE DMR 4-year 53 GHz data set alone. The dotted error bars show the effect of DMR instrument noise combined with the beam pattern. The dotted pair of lines shows the anticipated combined effect of cosmic variance and instrument noise without the effect of the beam filter. This shows the importance and the gain from having as high a resolution as possible. It also shows that future experiments will require better net sensitivity than the DMR by more than a factor of 10.

as gaussian and the variance is (see e.g. Knox 1995)

$$\langle (C_\ell^{obs} - C_\ell)(C_{\ell'}^{obs} - C_{\ell'}) \rangle = \frac{2}{2\ell+1} (C_\ell + < \sigma_{lm}\sigma_{l'm} >_m)^2 \tag{40}$$

where σ_{lm} is the error in the spherical harmonic coefficient a_{lm}. Conceptually this is just

$$\frac{\Delta C_\ell}{C_\ell} = \sqrt{\frac{2}{2\ell+1}} \left(1 + \frac{1}{S/N_\ell}\right) \tag{41}$$

where S/N_ℓ is the mean signal-to-noise for an $a_{\ell m}^2$, i.e. $S/N_\ell = \sigma_{\ell m}^2 / a_{\ell m}^2$, since $C_\ell \equiv \langle a_{\ell m}^2 \rangle_m$.

If the process that causes anisotropies is stochastic with random phase, there is a fundamental limitation set by the cosmic or sample variance. One gains little in determining the power spectrum by making observations with S/N_ℓ better than about 10. For a sensitivity limited experiment, generally the signal to noise S/N_ℓ improves with the inverse of the observation time.

If less than the full sky is measured then there will be a larger error because fewer samples are taken, to first order one has

$$\frac{\Delta C_\ell}{C_\ell} = \sqrt{\frac{2}{2\ell+1}} \sqrt{\frac{4\pi}{\Omega_{obs}}} \left(1 + \frac{1}{S/N_\ell}\right) \tag{42}$$

where Ω is the effective solid angle of the sky covered. (Effective means weighted by the total error: sampling and noise.) Or equivalently, if each pixel has the same error σ_{pix}

$$\frac{\Delta C_\ell}{C_\ell} = \sqrt{\frac{2}{2\ell+1}} \sqrt{\frac{4\pi}{\Omega_{obs}}} \left(1 + \frac{4\pi\sigma_{pix}^2}{W_\ell C_\ell \times N_{pix}}\right) \tag{43}$$

where N_{pix} is the total number of pixels observed and W_ℓ is the beam filter function. Clearly one wants W_ℓ as close to unity as possible for the ℓ's of interest. If the sky coverage is not uniform, then an effective weighting by determines the signal-to-noise term which is then approximately $\sigma_{obs}^2/(C_\ell \times t_{total} \times f)$ where σ_{obs} is the detector noise per second, t_{total} the total observation time, and $f \equiv \langle Y_{\ell m} Y_{\ell m}^* \rangle$ which represents the average of the ℓ-spherical harmonics weighted by the sky coverage.

The end conclusion is that for a fixed observation noise and time, the measurement of the power spectrum is improved by covering as uniformly as possible as much sky as possible with as much angular resolution as possible.

There may be credibility and observational reasons for getting a higher signal-to-noise per pixel. However, great concentration on a small portion of the sky only allows one to understand the signal and systematics including possible foregrounds better. If the systematics and foregrounds can be neglected, then the optimum is clearly biased to greater sky coverage.

It turns out also to be true that higher angular resolution also improves the measurement of the power spectrum substantially. This is especially true when one considers fitting the power spectrum. Figure 10 shows this clearly with the anticipated limitation set by the instrument noise and beam size and cosmic variance in determining a sample power spectrum for the full four-year COBE DMR data set. For foreground and practical reasons the bandwidth is usually limited to roughly 10% for coherent receivers and 20 to 30% for bolometer systems. The frequency range that is likely to be most optimal for CBR anisotropy observations is the range $50 \leq \nu \leq 160$ GHz with the best depending upon angular scale of observation. This window is set by unavoidable foreground emissions.

4.3 Foreground Emissions

The microwave sky is dominated by relic emission from the cosmic microwave background and local emission from within our Galaxy. The spatial distribution and frequency spectrum of these components probe physical conditions and processes ranging from the early universe to the local interstellar medium.

Figure 11 shows the spectra of the CMB and Galactic emissions. Galactic emission at millimeter wavelengths is dominated by thermal emission from interstellar dust. Galactic emission at centimeter wavelengths is dominated by two components: synchrotron emission from cosmic-ray electrons accelerated in the Galactic magnetic field, and free-free emission (thermal bremsstrahlung) from the warm ($T_e \sim 8000$ K) ionized interstellar medium. Emission from Galactic sources probes physical conditions in the interstellar medium. An understanding of the combined Galactic foregrounds tests

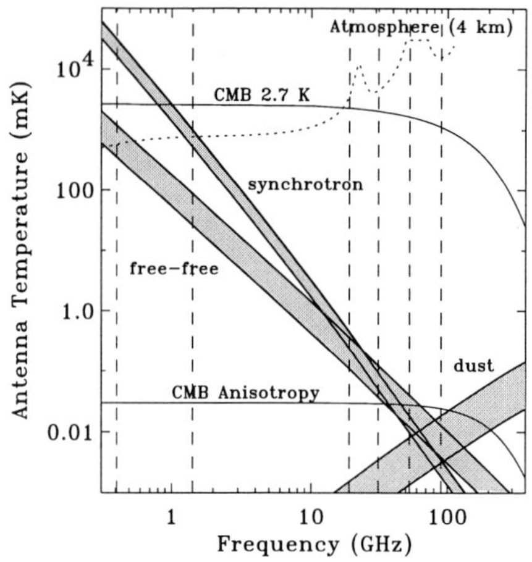

Figure 11: CMB and foreground emission spectra. The shaded regions indicate the range of synchrotron, free-free, and dust emission at Galactic latitude $15° < |b| < 75°$. The dotted line indicates typical atmospheric emission from a mountain or Antarctic site. Solid lines indicate the mean CMB spectrum and rms amplitude of anisotropy. Vertical dashed lines indicate published sky surveys, including *COBE*. There are no sensitive surveys at cm wavelengths.

models of the large-scale structure and energy balance of the Galaxy, and is crucial to mapping the CMB spectrum and anisotropy.

Three phases of the interstellar medium are known to coexist in the Galaxy: a cool phase ($T \lesssim 100$ K) in neutral clouds, a hot ($T \sim 10^6$ K) "coronal" component, and a warm ($T \sim 10^4$ K) ionized component. The cool component, including both dense and diffuse molecular clouds, comprises no more than 2% of the ISM by volume. The volume partition between the hot and warm ionized gas, central to an understanding of the energy balance of the Galaxy, is unknown. Possibilities consistent with existing observational data range from a quasi-homogeneous warm background with isolated cavities or superbubbles of hot gas, to a rough equipartition between the components, to a pervasive and dominant hot coronal medium. The coexistence of the phases is governed by the balance between heating and cooling mechanisms; however, the heating mechanisms of the respective media are poorly understood (Reynolds & Cox 1992). Our knowledge of the ionized ISM derives from dispersion measurements toward pulsars, radio continuum measurements, atomic recombination lines, and methodical observations of Hα and various optical forbidden lines (Reynolds 1990). Much remains to be learned about the ionized interstellar gas – notably, the sources of its ionization far from the disk, its heating, and its distribution throughout the Galaxy. The ionization has been ascribed to ultraviolet flux from the young stellar population of the disk, shocks, Galactic fountain flows, or decaying dark matter. Observed column densities of highly ionized species (Savage & Massa 1987) support a picture involving collisional ionization from a Galactic fountain-type flow together with photoionization either by an extragalactic radiation field or from the disk, or else by ionizing radiation emitted by cooling gas in a Galactic fountain. Ionization from O-star associations within isolated cavities (superbubbles) due to supernovae explosions is also consistent with current observations.

Interstellar Dust Emission

Interstellar dust hides much of the Galactic plane in the optical. Longer wavelength observations are able to penetrate this obscuration, allowing study of individual objects. The obscuration at high Galactic latitudes is two orders of magnitude less but the signals sought by extragalactic astronomers and cosmologists are typically more diffuse and subtle, so high latitude dust remains a concern. In some regions of the sky dust emission can be a significant signal in the mm-wavelength range and affect cosmic microwave background (CMB) observations. As optical infrared measurements have moved deeper into the infrared towards longer wavelengths, the cooler components of interstellar dust have been revealed. In cooler regions dust has been observed at typical temperatures ranging from 200 K down to levels near 20 K. Dense, cold molecular clouds are expected to contain dust at temperatures as low as a few Kelvin.

Cosmologists have long worried that there might be a significant component of cold cosmic or Galactic dust that is affecting and contributing to the observed cosmic signals. We are now beginning to gather sensitive measurements with sufficient wavelength coverage to explore the high latitude dust. So far a consistent picture can be drawn from the observations. There is no evidence for a significant component of interstellar dust colder than 15 K at high Galactic latitudes. However, there is evidence that the emissivity of the dust is not as steep as frequency squared but more like $\nu^{1.5}$ in the mm-wavelength range. There is, in fact, evidence against high-latitude dust in the temperature range $4 < T_{dust} < 15$ K (Smoot 1995).

The good news is that the level of dust emission is sufficiently low as to allow observations of CMB anisotropy to the 10^{-6} level over a reasonable frequency range. Often plots are made showing the confusion versus frequency for various angular scales (e.g. Figure 13). Figure 12 shows the power spectrum of dust emission as observed by the COBE DIRBE instrument. Various regions of the sky were selected and the power spectrum computed and compared. The result is as anticipated. The interstellar dust emission has more power on large angular scales than on small to the extent that even plotted as $\ell(\ell+1) \times T_\ell^2$ (for direct comparison with the CMB power spectrum) the dust power spectrum decreases with increasing ℓ.

Galactic Synchrotron Emission

Synchrotron emission from relativistic electrons accelerated in the Galactic magnetic field dominates the Galactic foreground at long wavelengths. The volume emissivity for a

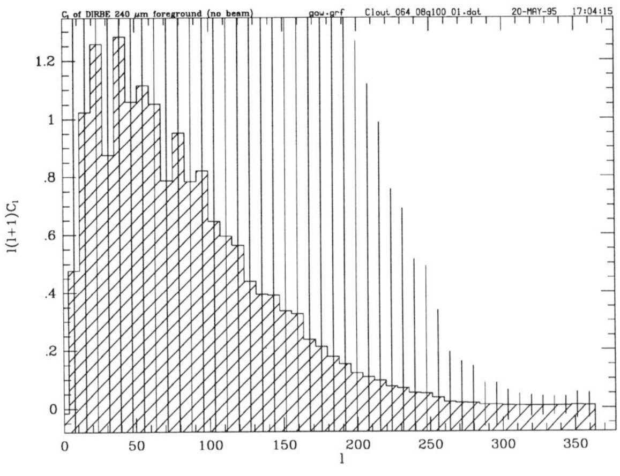

Figure 12: The interstellar dust power spectrum as observed by DIRBE at 240 microns. The errors indicate the scatter in signal from one region to another. The scatter is large because most power is on large angular scales. The signal level is sufficiently low to allow precise CMB anisotropy observations.

power-law distribution of electrons, $N(E) = AE^{-p}dE$, is

$$\epsilon(\omega) = \frac{\sqrt{3}Ae^3B}{2\pi m_e c^2(p+1)}\,\Gamma(\frac{p}{4}+\frac{19}{12})\,\Gamma(\frac{p}{4}-\frac{1}{12})\,[\frac{m_e c\omega}{3eB}]^{-\frac{p-1}{2}} \qquad (44)$$

where B is the magnetic field intensity. At centimeter wavelengths, synchrotron emission can be approximated as a power-law in antenna temperature

$$T_{\rm synch} \propto \nu^{\beta_{\rm synch}} \qquad (45)$$

where the spectral index $\beta_{\rm synch} \approx -2.8 \pm 0.3$ is a function of the electron energy spectrum and magnetic field intensity, and will vary across the sky. The synchrotron spectral index is a local approximation only: over large frequency intervals, the curvature of the synchrotron spectrum can not be ignored, while spatial variations are important even in a restricted frequency range. Bennett et al. (1992) present a model of synchrotron emission based on radio surveys at 408 and 1420 MHz and local measurements of the cosmic ray electron energy spectrum. Existing sky surveys, including *COBE*-DMR, provide only a weak test of this model. Systematic uncertainties and noise combine to limit direct knowledge of the synchrotron spectral index to ± 0.1, insufficient to test the assumption that the local electron spectrum is representative of the Galaxy as a whole. Precise measurements at centimeter wavelengths, tied to a common calibration standard, could detect the predicted steepening of the synchrotron spectrum and map the spatial variations in $\beta_{\rm synch}$ to precision ± 0.02. Such measurements would test models of cosmic-ray acceleration through supernovae shocks and OB associations (Banday & Wolfendale 1990), and verify that the locally-measured electron energy spectrum is representative of the Galaxy as a whole.

Fortunately, the synchrotron emission falls off so steeply with frequency that it allows precise measurements at frequencies of 50 GHz and above. Again one can compute the power spectrum of the synchrotron maps and find that much of the power is on the largest angular scales. The frequency dependence is sufficiently different from the CMB's that separation is relatively easy.

Free-Free Emission

As a Galactic foreground free-free emission arises from the thermal collisions in the interstellar plasma. For temperatures below $T_e < 10^6$ K and frequencies below $\nu < 10^{10} T_e$ Hz, which covers the entire CMB spectrum for most interstellar plasma, the optical depth is

$$\tau \approx 0.08235 T_e^{-1.35} \nu^{-2.1} \int n_e^2 dl \qquad (46)$$

(Lang 1974). The 0.1 in the -2.1 comes from the Gaunt factor and is approximate but fairly accurate. Thus the free-free emission from the warm ISM may be approximated as a power-law in frequency, $T_{\rm ff} \propto \nu^{\beta_{\rm ff}}$ with $\beta_{\rm ff} \approx -2.1$ nearly independent of temperature. Now we can estimate the confusion caused by Galactic free-free emission. Unfortunately, as figure 11 demonstrates, in the frequency range under consideration, at no frequency does the free-free emission dominate so that it can be unambiguously mapped.

Multi-frequency measurements allow separation of the Galactic and CMB components based on their different spectra. Microwave measurements of the warm ISM complement recombination-line mapping of the same gas. The ratio of free-free antenna temperature to Hα line intensity leads to the temperature dependence of the ISM,

$$\frac{T_{\rm ff}\ (\mu{\rm K})}{I_\alpha\ ({\rm R})} \approx 520\ [1 + \sqrt{\frac{T_e}{10^4\ {\rm K}}}\ ]\left(\frac{\nu}{10GHz}\right)^2 \qquad (47)$$

where the Hα intensity is in Rayleighs ($10^6/(4\pi)$ photons cm^{-2} s^{-1} sr^{-1}). The Wisconsin Hydrogen Alpha Mapper (WHAM) has been funded to map the northern sky with a 1° beam to 10% accuracy in Hα (Reynolds, private communication). In combination with a microwave determination of the free-free emission, the temperature of the ISM can be mapped to $\sim$20% precision. The combined data sets will determine whether the intensity variations are attributable solely to density variations, or whether electron temperature (and hence heating mechanism) are significant factors. The combination will allow a good assessment of the free-free component.

However we can estimate how significant the free-free contribution will be as a foreground to CMB measurements and determine that there is still a significant window that allows observations to the 10^{-6} level. Though the free-free emission never dominates the sky, it does have a sufficiently different spectral index (~ -2.1) than the other components. Thus ratios of maps taken in the cm-wavelength range can reveal regions of enhanced free-free emission. Preliminary analysis of the GEM (Galactic Emission Mapper, De Amici et al. 1995), which has been mapping a substantial portion of the sky at four frequencies: 0.408, 1.5, 2.3, and 5 GHz, shows that the regions of high free-free emission tend to be concentrated in knots on the Galactic plane. Surprisingly, there are a number of such HII regions away from the Galactic plane but they cover a relatively small fraction of the sky. Any free-free emission covering large areas is quite reduced. We can limit the free-free confusion at frequencies above 50 GHz to less than about 10^{-6} for all but the largest angular scales.

The low level of free-free power at small angular scales is reminiscent of that for interstellar dust emission probably for the same reason. It is likely that most of the free-free emission is correlated with the density (goes as n_e^2) of hydrogen in the Galaxy as are molecular clouds and interstellar dust. A cross-correlation of the COBE DMR data with the DIRBE three lowest frequency bands (dominated by interstellar dust) shows a significant common spatial structure (Kogut et al 1995). Thus the correlated component will have a similar power spectrum to the dust – decreasing at smaller angular scales. The correlated signal is below the 10 μK level at 50 GHz for Galactic latitudes above 30°. Most of the power is on the largest angular scales, e.g. the quadrupole. At higher ℓ the free-free power is significantly less. This is consistent with the HEMT observations (Tenerife, South Pole, and Saskatoon). Thus we can still be confident that there is a good window through the Galactic foregrounds.

Extragalactic Foregrounds

Figure 13 shows the expected fluctuation level at angular resolutions of 10′ and 30′. At high frequency ($\nu \gtrsim 140$ GHz) the main foreground components will be dust emission from infrared cirrus and from normal spiral galaxies, as well as that from starburst galaxies; towards the low–frequency range ($\nu \lesssim 70$ GHz) galactic synchrotron and free–free emission become increasingly important (Toffolatti et al. 1994). Calculations of the residual temperature fluctuations from unresolved extragalactic sources show that in the range 50–300 GHz they will contribute $\Delta T/T \lesssim 10^{-6}$, i.e. below the expected noise level of the detectors. A check (Gawiser & Smoot 1995) based upon the infrared sources from the IRAS catalog shows a high level of consistency with the theoretical predictions of Toffalati et al. Only nearby sources are strong enough to be important and there is a significant frequency range in which the infrared source confusion is below the 10^{-6} level. These are readily detected by the higher frequency channels, particularly if the angular resolution increases with frequency.

Extragalactic radio sources are more problematic and a significant number must be excised from the data. Some are known to be variable so that subtraction would require monitoring. This would probably require coordination with ground-based instruments, since these sources are stronger at longer wavelengths where the angular resolution is poorer making them more difficult to detect with the same instrument. Still the fraction of the sky covered by significant sources is small so that for frequencies above 30 to 50 GHz it is likely that for most of the sky the confusion from the sources is below the 10^{-6} level.

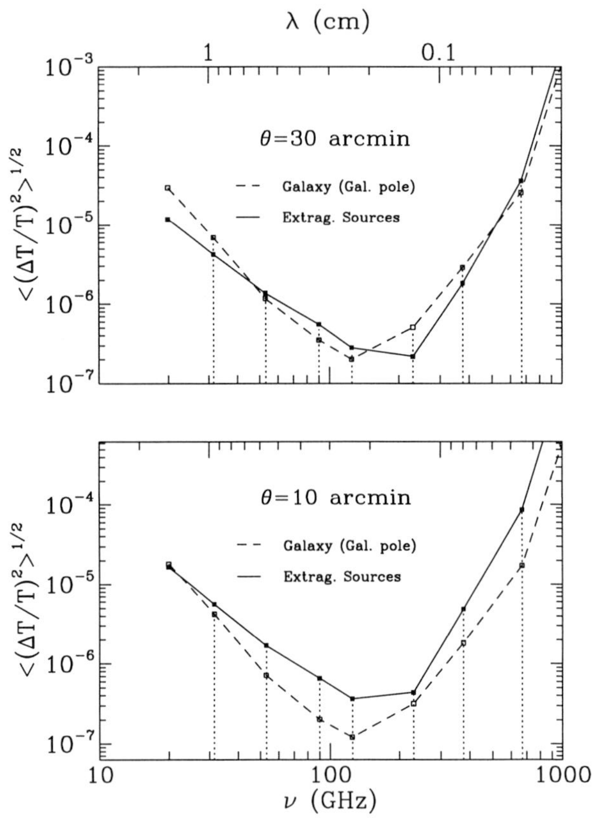

Figure 13: Estimated fluctuation levels due to Galactic polar emission (dashed lines) and to extragalactic sources (solid lines), from Toffolatti et al. (1994). The vertical dotted lines show the COBRAS/SAMBA frequencies.

Foreground Minimization & Removal

The Galactic foregrounds may be separated from the CMB by their frequency dependence, supplemented by knowledge of their spatial morphology. This technique is widely used (c.f. Bennett et al. 1992, Brandt et al. 1994) and depends on the frequency coverage and combined uncertainties of the sky surveys used in the fitting process. A limiting factor has been the large calibration uncertainties in the synchrotron-dominated radio surveys (Haslam et al. 1982, Reich and Reich 1986) and the decade gap in frequency coverage between the radio surveys and CMB measurements above 30 GHz (Figure 11). Several authors have pointed out the need for precisely calibrated microwave sky surveys to map the foreground Galactic emission (Bennett et al. 1992, Kogut et al. 1993, Bersanelli et al. 1994, Brandt et al. 1994, Kogut 1995).

Atmospheric Emission

Though not a fundamental limitation, as satellites can avoid it, atmospheric emission and especially its variability is an issue for suborbital programs. This is a limitation both for spectrum and anisotropy projects. For ground-based (including mountain-top and Antarctic plateau) experiment design and operational frequency range is limited by atmospheric effects. Clearly some sites are better than others and some techniques allow better elimination of atmospheric effects. However, the atmosphere does drive some experiments to balloon-borne platforms. The effects of the atmosphere and its limitations on CMB experiments are being quantified (Smoot et al. 1987, Church 1995, Bersanelli et al 1995). It is still possible to make significant advances in the suborbital environment. Ultimate scientific goals combined with the limitations imposed by the atmosphere will eventually drive experiments to space-borne platforms.

§5 FUTURE: 2000$^+$

5.1 The CBR Spectrum – Future

The analysis and interpretation of present experiments is likely to improve our current knowledge of the spectrum by no more than a factor of three. That still leaves a number of important scientific questions unaddressed.

Answers to this new level of questions require new information on the early universe. It will be necessary to measure the CMB spectrum to 10^{-4} precision at centimeter wavelengths. Precise measurements of the CMB spectrum at centimeter wavelengths probe different physical processes than the *COBE* results at millimeter and sub-mm wavelengths and would provide quantitative information on such processes as:

• The transition from an ionized universe to a neutral state and subsequent heating and reionization by the first generation of collapsed objects

• The abundance, lifetimes, and decay modes of hypothesized non-baryonic particles, including supersymmetric partners of known particles or other particle-physics dark-matter candidates

• The decay of primordial turbulence, including the short spatial wavelength end of the observed power spectrum of primordial anisotropy

CMB photons must traverse the interstellar medium within our Galaxy. Precise CMB measurements require an understanding of the diffuse Galactic foregrounds. Figure 11 illustrates the relative intensity of cosmic and Galactic emission at high Galactic latitude. Over a broad range of wavelengths, the CMB dominates the brightness temperature of the sky, and may be distinguished from Galactic sources by its different spectral signature and spatial dependence.

Several factors point to a multi-frequency survey at centimeter wavelengths to 0.01% or better absolute precision as both scientifically interesting and technologically feasible. Measurements in this band probe a poorly surveyed window in the electromagnetic spectrum lying between radio-dish surveys at decimeter wavelengths and *COBE* at millimeter wavelengths. Previous measurements in this band have been limited by interference from the Earth's atmosphere. A multi-channel space experiment with 0.1 mK precision would improve existing data by a factor of 500, provide a decisive test of reionization for redshift $z>50$, produce important constraints to particle dark-matter candidates, and probe physical processes within the Galaxy including the volume partition and heating mechanism of the interstellar medium and the acceleration mechanism of cosmic-ray electrons.

Galactic radio emission is dominated by synchrotron radiation from cosmic-ray electrons and by electron-ion bremsstrahlung (free-free emission) from the ionized interstellar medium (ISM). Despite surveys carried out over many years, relatively little is known about the physical conditions responsible for these diffuse emissions. Synchrotron emission results from cosmic-ray electrons accelerated in magnetic fields, and thus depends on both the electron energy spectrum and the Galactic magnetic field. Radio surveys, at frequencies below a few GHz where synchrotron emission dominates the diffuse sky brightness, map the synchrotron intensity to limited precision but are unable to separate variations in column density, magnetic field, and energy spectrum. Diffuse free-free emission from the ionized ISM *never* dominates the radio sky and is largely unconstrained by observational results. Radio surveys thus far only detect the quadrupolar component and provide almost no information on the spatial power spectrum. Recombination line (Hα) maps of the same gas suffer from undersampling at high latitudes and self-absorption at low latitudes. A secondary goal of a diffuse emission measurement is to map the spatial dependence of the diffuse Galactic foregrounds, providing answers to outstanding questions on physical conditions in the ISM:

• What is the heating mechanism in the ISM? Is the gas heated by photoionization from the stellar disk, shocks, Galactic fountain flows, or exotic processes such as decaying halo dark matter?

• How is energy injected by supernovae processed? Does the hot gas exist as isolated cavities or chimneys within a pervasive warm background?

• How are cosmic rays accelerated? Is the energy spectrum of local cosmic-ray electrons representative of the Galaxy as a whole?

Achieving 0.01% or better precision across a decade of wavelength requires a design in which instrumental artifacts are minimized or eliminated, imposing several constraints on possible designs. First, it must have an external calibration to compare the power received from the sky with that from an on-board blackbody target, cancelling any instrumental signal to first order. It must have multiple narrow-band channels: separation of a potentially distorted CMB from the Galactic foregrounds based on spectral shape requires at least as many frequency channels as free parameters in any spectral fit. The "interesting" parameters test deviations from a blackbody spectral shape, but are insensitive to the precise value of the undistorted temperature. To this end, the various channels must all observe a common external target: the multiply differential comparison of sky–target between channels is then sensitive only to the target emissivity and temperature gradients but not its absolute temperature. The instrument should be isothermal with the diffuse sky temperature to reduce the effects of small reflection and attenuation within the instrument itself. Finally, it must observe above the bulk of the atmosphere, requiring either a balloon or space platform.

It is clear that for the cm wavelengths that a precision measurement must be made from balloons or space. Attempts from the ground, e.g. GEM, suffer from atmospheric variability in the cm-wavelengths and lack of funding. One proposed instrument to make this measurement is the DIMES (Diffuse Microwave Emission Survey) instrument. DIMES proposes to measure the diffuse cosmic and Galactic emission to 0.1 mK precision in several narrow bands ($\Delta\nu/\nu \sim 10\%$) spanning the poorly surveyed 1–10 cm wavelength range, improving current measurements by a factor of 500. At each frequency, a cryogenic radiometer switched for gain stability between an internal reference load and a beam-defining antenna will measure the signal change as the antenna alternately views the sky and an external blackbody calibration target. By rapidly comparing each channel to the same external target, uncertainties in the absolute target temperature cancel in the derived sky spectra, so that deviations from a blackbody spectral *shape* may be determined much more precisely than the absolute temperature.

The DIMES instrument propose more than two orders of magnitude improvement over the best previous absolute cm-wavelength measurements. Much of this improvement can be attributed to the fully cryogenic design and multiple levels of differences, which cancel instrument gain variations, reflection, and emission to first order. The limiting factor is likely to be the thermal stability of the instrument and the external calibration target: temperature gradients within the target or changes in absolute temperature as the target covers each antenna in turn can mimic spectral and spatial structure in the sky.

These observations will be difficult. Unfortunately, without a careful measurement and und understanding of the

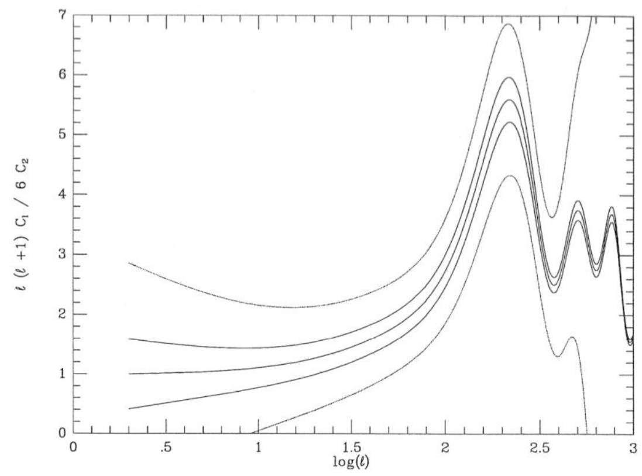

Figure 14: Center line shows the power spectrum for standard CDM ($h = 0.5, \Omega_b = 0.05$, courtesy of Sugiyama, 1995) The pair of lines on either side show the effect of cosmic variance alone. The dotted pair of lines shows the anticipated combined effect of cosmic variance and instrument noise for a single 10-hour flight of the MAX/MAXIMA experiment covering about 10% of the sky. More flights of MAX/MAXIMA, BOOMERANG, and/or ACE should bring down the errors by as much as a factor of 3. The results will come fairly close to cosmic variance limits.

Galactic emission processes, a cosmic free-free distortion cannot be distinguished from the zero level of Galactic free-free emission. The Galaxy has a minimum emission measure of 1 to 2 cm^{-6} pc, which gives on the oder of 1 mK signal at 10 cm wavelength. Thus to achieve a 0.1 mK sensitivity will require a very good modeling of the Galactic emission or the use of data from related processes such as H-α emission observations.

5.2 *The CBR Anisotropy – Future*

Progress in the field of cosmology has been extraordinary in the last decade, both in terms of impressive new observations and of consolidation of the theoretical framework of their interpretation. Accurate observations of the cosmic background radiation have played a central role in this progress. In 1992 the COBE team announced the detection of intrinsic temperature fluctuations in the CBR at angular scales larger than $\sim 7°$, with brightness amplitude $\Delta T/T \sim 10^{-5}$ (Smoot et al. 1992).

As outlined above we can expect very significant and rapid progress in the observation of CMB anisotropies. One can expect that the power spectrum of anisotropies will be measured to a sensitivity of about 10% around the first "Doppler" peak, to within a factor of two of the cosmic variance at small ℓ as a result of various experiments now in the works. Figure 14 shows a sample of the progress one can expect in determining a sample power spectrum in the next few years. To overcome the limit set by sampling variance, it will be necessary to coadd data from multiple balloon flights or observation seasons, since each flight or observing season is likely to cover only 10 to 20% of the sky. Historically it has proven difficult to coadd data in this manner. Seams appear and calibration may vary. The major advantage of a good satellite mission is the ability to cover the full sky in a single flight or season. The most significant problems in reconstructing the power spectrum will be for those portions where the wavelength is greater than or approximately as large as the surveyed patches. If the CMB anisotropies are from a random-phase stochastic process as assumed when utilizing the power spectrum, then coadding the patches can be expected to give nearly as good a result for the higher ℓ's as a single full sky survey. In addition we have a full sky survey from the COBE DMR that gives us a nearly cosmic variance limited power spectrum for $\ell \lesssim 20$. If RELICT 2 is successful, we will have a cross check of the low ℓ portion of the power spectrum. Thus we expect these satellite results and the balloon-borne and ground-based (e.g. interferometers) experiments to provide us a good (nearly cosmic variance limited) power spectrum of anisotropies.

Such results will achieve the first level of scientific goals. The questions are: What science is left? How important is it? And can it be obtained with any approach now foreseen and with existing technologies? The short answer is that there is very significant and vital science left to be explored provided the anisotropies and power spectrum can be measured quite accurately. There is a rich theoretical literature exploring various aspects of the issue. The remaining science puts a very high premium on full sky coverage, high sensitivity, and low systematics and residual foregrounds in the data.

A new mission can be expected to have as a product two things: (1) a power spectrum covering the range from about $10'$ to the full sky ($\ell = 2$) which will check and consolidate the previous results with improved systematics and foreground removal. (2) A map of actual anisotropies with good signal to noise. This requires a sensitivity per pixel at about the $\delta T/T \leq 3 \times 10^{-6}$ level (10 μK) after removal of foregrounds and instrument signature. This is likely to require raw sensitivity about a factor of 3 better than the ultimate quality of the maps and multiple frequency observations spanning the full frequency window and edges for foreground identification.

These are issues that require a space-based platform. There are now four groups actively working on proposed satellite missions to map CMB anisotropies. At the moment three are actively competing directly with each other for a NASA opportunity and much information is kept confidential until the selection. Thus it is instructive to consider the COBRAS/SAMBA mission to understand the scope of a satellite mission, for comparison with the others as they are evaluated, and most importantly to compare with the scientific goals.

COBRAS/SAMBA is currently undergoing Phase A study by ESA. The COBRAS/SAMBA mission is designed for extensive, accurate mapping of the anisotropy of the Cosmic Background Radiation, with angular sensitivity from sub-degree ($\sim 10' - 30'$) scales up to the full sky thus overlapping with the COBE–DMR maps and with signal sensitivity approaching $\Delta T/T \sim 10^{-6}$. This will allow a full identification of the primordial density perturbations which grew to form the large–scale structures observed in the present universe. The COBRAS/SAMBA maps will provide decisive answers to several major open questions relevant to the structure formation epoch and will provide powerful tests for the inflationary model as well as several astrophysical issues. COBRAS/SAMBA will utilize a combination of bolometric and radiometric detection techniques to ensure the sensitivity and wide spectral coverage required for accurate foreground discrimination. An orbit far from Earth has been selected to minimize the unwanted emission from the Earth as a source of contamination.

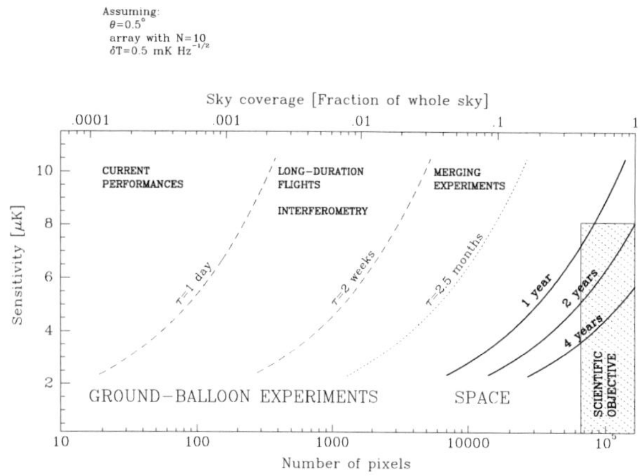

Figure 15: Number of pixels covered versus sensitivity for various potential platforms.

The COBRAS/SAMBA mission is the result of the merging of two proposals presented in 1993 to the European Space Agency *M3 Call for Mission Ideas*: COBRAS (Cosmic Background Radiation Anisotropy Satellite; Mandolesi et al. 1993) and SAMBA (Satellite for Measurements of Background Anisotropies; Puget et al. 1993). The COBRAS/SAMBA team completed the ESA assessment study in May 1994, and the project continued and is currently in the Phase A study within the European Space Agency M3 programme.

COBRAS/SAMBA Scientific Objectives

The COBRAS/SAMBA mission will produce near all-sky maps of the background anisotropies in 8 frequency bands in the range 30–800 GHz, with peak sensitivity $\Delta T/T \sim 10^{-6}$. The maps will provide a detailed description of the background radiation fluctuations. Individual hot and cold regions should be identified above the statistical noise level, at all angular scales from $\lesssim 10'$ up to very large scales, thus providing a high resolution imaging of the last scattering surface.

The COBRAS/SAMBA maps will provide all multipoles of the temperature anisotropies from $\ell = 1$ (dipole term) up to $\ell \simeq 1500$ (corresponding to $\sim 7'$). It is the information contained in this large number of multipoles that can probe the various proposed scenarios of structure formation and the shape of the primordial fluctuation spectrum (for comparison, the COBE–DMR maps are limited to $\ell \lesssim 20$).

The high resolution COBRAS/SAMBA maps will provide a key test for structure formation mechanisms, based on the statistics of the observed $\Delta T/T$ distribution. The inflationary model predicts Gaussian fluctuations for the statistics of the CBR anisotropies, while alternative models based on the presence of topological defects, such as strings, monopoles, and textures, predict non–Gaussian statistics (e.g. Coulson et al. 1994). Due to the different nature of their early history causality constrains primordial perturbations from a source such as inflation and from topological defects to have a different anisotropy power spectra particularly in the region of the "Doppler" peaks (Albrecht et al. 1995). The angular resolution and sensitivity of COBRAS/SAMBA will allow discrimination between these alternatives with tests of both the power spectrum and statistics.

The high-order multipoles will allow an accurate measure of the spectral index n of the primordial fluctuation spectrum:

$$(\delta\phi)^2 \propto \lambda^{(1-n)} \tag{48}$$

where $\delta\phi$ is the potential fluctuation responsible for the CBR anisotropies, and λ is the scale of the density perturbation. This corresponds to CBR temperature fluctuations $(\Delta T/T)^2 \propto \theta^{(1-n)}$ for angles $\theta > 30'\ \Omega_0^{1/2}$. The proposed observations will be able to verify accurately the scale invariant "Harrison–Zel'dovich" spectrum ($n = 1$) predicted by inflation. Any significant deviation from that value would have extremely important consequences for the inflationary paradigm. The COBE–DMR limit on the spectral index after two years of observations ($n = 1.1^{+0.3}_{-0.4}$, 68% CL; Gorski et al. 1994) can be constrained $\sim$ 10 times better by the COBRAS/SAMBA results.

The proposed observations will provide an additional, independent test for the inflationary model. Temperature anisotropies on large angular scales can be generated by gravitational waves (tensor modes, T), in addition to the energy-density perturbation component (scalar modes, S). Most inflationary models predict a well determined, simple relation between the ratio of these two components, T/S, and the spectral index n (Davis et al. 1992, Little & Lyth 1992):

$$n \approx 1 - \frac{1}{7}\frac{T}{S}. \tag{49}$$

The COBRAS/SAMBA maps will be able to verify this relationship, since the temperature anisotropies from scalar and tensor modes vary with multipoles in different ways.

A good satellite mission will be able not only to test the inflationary concept but also to distinguish between various models and determine inflationary parameters. There is an extensive literature on what can be determined about inflation such as the scalar and tensor power spectra, the energy scale of inflation and so on (see e.g. Steinhardt 1995, Knox 1995). Such quality measurements lead also to good observations or constraints for Ω_0, Ω_{baryon}, Λ, H_0, etc. Sub–degree anisotropies are sensitive to the ionization history of the universe. In fact, they can be erased if the intergalactic medium underwent reionization at high redshifts. Moreover, the temperature anisotropies at small angular scales depend on other key cosmological parameters, such as the initial spectrum of irregularities, the baryon density of the universe, the nature of dark matter, and the geometry of the universe (see e.g. Crittenden et al. 1993, Bond et al. 1993, Kamionkowski et al. 1994 Hu & Sugiyama 1994, Scott, Silk, & White 1995)). The COBRAS/SAMBA maps will provide constraints on these parameters within the context of specific theoretical models.

Moreover, COBRAS/SAMBA should measure the Sunyaev–Zel'dovich effect for more than 1000 rich clusters, using the higher resolution bolometric channels. Combined with X–ray observations these measurements can be used to estimate the Hubble constant H_0 as a second independent determination.

Foreground Emissions

In order to obtain these scientific goals, the measured temperature fluctuations need to be well understood in terms of the various components that add to the cosmological signal. In fact, in addition to the CBR temperature fluctuations, foreground structures will be present from weak, unresolved extragalactic sources and from radiation of galactic origin (interstellar dust, free–free and synchrotron radiation).The COBRAS/SAMBA observations will reach the required control on the foreground components in two ways. First, the large sky coverage ($\geq$ 90% of the sky) will allow accurate modeling of these components where they are dominant (e.g. galactic radiation near the galactic plane). Second, the observations will be performed in a spectral range as broad as possible.

In fact, the COBRAS/SAMBA channels will span the spectral region of minimum foreground intensity (in the range 50–300 GHz), but with enough margin at high and low frequency to monitor "in real–time" the effect of the various foreground components (see e.g. Brandt et al. 1994). By using the COBRAS/SAMBA spectral information and modeling the spectral dependence of galactic and extragalactic emissions it will be possible to remove the foreground contributions with high accuracy.

It should be noted that in most channels the final limitation to the cosmological information of the COBRAS/SAMBA maps is expected to be due to the residual uncertainties in the separation of the foreground components rather than statistical noise. This explains why the overall design of the instrument and payload is highly driven by the need of achieving a spectral coverage as large as possible. Performing measurements where the dominant foreground components are different will permit a powerful cross check on residual systematic errors in the CBR temperature fluctuation maps.

The need of accurate characterization of all non-cosmological components, of course, brings the benefit of additional astrophysical information. The very large COBRAS/SAMBA data base, particularly when combined with the IRAS survey, can provide information on several non–cosmological issues, such as the evolution of starburst galaxies, the distribution of a cold–dust component, or the study of low–mass star formation.

Table COBRAS/SAMBA Payload Characteristics								
Telescope	1.5 m Diam. Gregorian; system emissivity $\leq 1\%$ Viewing direction offset $\geq 70°$ from spin axis							
Instrument	LFI				HFI			
Center Frequency (GHz)	31.5	53	90	125	140	222	400	714
Wavelength (mm)	9.5	5.7	3.3	2.4	2.1	1.4	0.75	0.42
Bandwidth ($\frac{\Delta\nu}{\nu}$)	0.15	0.15	0.15	0.15	0.4	0.5	0.7	0.6
Detector Technology	HEMT receiver arrays				Bolometers arrays			
Detector Temperature	$\sim$ 100 K				0.1 - 0.15 K			
Cooling Requirements	Passive				Cryocooler + Dilution system			
Number of Detectors	13	13	13	13	8	11	16	16
Angular Resolution (arcmin)	30	20	15	12	10.5	7.5	4.5	3
Optical Efficiency	1	1	1	1	0.3	0.3	0.3	0.3
$\frac{\Delta T}{T}$ Sensitivity ($1\sigma, 10^{-6}$ units, 90% sky coverage, 2 years)	1.7	2.7	4.1	7.2	0.9	1.0	8.2	10^4
$\frac{\Delta T}{T}$ Sensitivity ($1\sigma, 10^{-6}$ units, 2 % sky coverage, 2 years)	0.6	0.9	1.4	2.4	0.3	0.3	2.7	5000

Table 1: Instrumental Parameters for COBRAS/SAMBA the most important factors are the frequency coverage, the angular resolution, sky coverage, and sensitivity.

The Payload

The COBRAS/SAMBA model payload consists mainly of a shielded, off-axis Gregorian telescope, with a parabolic primary reflector and a secondary mirror, leading to an integrated instrument focal plane assembly. The payload is part of a spinning spacecraft, with a spin rate of 1 rpm. The focal plane assembly is divided into low-frequency (LFI) and high-frequency (HFI) instrumentation according to the technology of the detectors. Both the LFI and the HFI are designed to produce high-sensitivity, multifrequency measurements of the diffuse sky radiation. The LFI will measure in four bands in the frequency range 30–130 GHz (2.3–10 mm wavelength). The HFI will measure in four channels in the range 140–800 GHz (0.4–2.1 mm wavelength). The highest frequency LFI channel and the lowest HFI channel overlap near the minimum foreground region. Table 1 summarizes the main characteristics of the COBRAS/SAMBA payload.

The Main Optical System

A clear field of view is necessary for the optics of a high-sensitivity CBR anisotropy experiment to avoid spurious signals arising from the mirrors or from supports and mechanical mounting. A Gregorian configuration has been chosen, with an primary parabolic mirror of 1.5 meter, and an elliptic secondary mirror (0.57 m diameter). Stray satellite radiation and other off–axis emissions are minimized by underilluminating the low–emissivity optics. The telescope reimages the sky onto the focal plane instrument located near the payload platform. The telescope optical axis is offset by 70° or more from the spin axis. Thus at each spacecraft spin rotation the telescope pointing direction sweeps a large (approaching a great) circle in the sky, according to the sky scan strategy.

Blockage is a particularly important factor since several feeds and detectors are located in the focal plane, and unwanted, local radiation (e.g. from the Earth, the Sun and the Moon) needs to be efficiently rejected. A large, flared shield surrounds the entire telescope and focal plane assembly, to screen the detectors from contaminating sources of radiation. The shield also plays an important role as an element of the passive thermal control of the spacecraft.

The Focal Plane Assembly

The necessary wide spectral range requires the use of two different technologies, bolometers and coherent receivers incorporated in a single instrument. Both technologies have shown impressive progress in the last ten years or so, and more is expected in the near future. The thermal requirements of the two types of detectors are widely different. The coherent radiometers (LFI), operating in the low frequency channels, give good performance at operational temperature of $\sim$ 100 K, which is achievable with passive cooling. The bolometers, on the other hand, require temperatures $\leq$ 0.15 K in order to reach their extraordinary sensitivity performances. The main characteristics of the LFI and HFI are summarized in Table 1.

The LFI consists of an array of 26 corrugated, conical horns, each exploited in the two orthogonal polarization modes, feeding a set of state–of–the–art, high sensitivity receivers. The receivers will be based on MMIC (Monolithic Microwave Integrated Circuits) technology with HEMT (High Electron Mobility Transistor) ultra–low noise amplifiers (see e.g. Pospieszalski et al. 1993). Since the whole LFI system will be passively cooled, it can be operated for a duration limited only by spacecraft consumables (up to 5 years). The three lowest center frequencies of the LFI were chosen to match the COBE-DMR channels, to facilitate the comparison of the product maps.

About 50 bolometers will be used in the HFI instrument, which require cooling at ~ 0.1 K. The cooling system combines active coolers reaching 4 K with a dilution refrigeration system working at zero gravity. The refrigeration system will include two pressurized tanks of ^{3}He and ^{4}He for an operational lifetime of 2 years.

Orbit and Sky Observation Strategy

One of the main requirements for the COBRAS/SAMBA mission is the need of a far–Earth orbit. This choice greatly reduces the problem of unwanted radiation from the Earth which is a serious potential contaminant at the high goal sensitivity and angular resolution. The requirements on residual Earth radiation are basically the same for the LFI and the HFI systems. Adopting a low–earth orbit, such as that used by the COBE satellite, the requirement on straylight and sidelobe rejection would be a factor of 10^{13}, which is beyond the capabilities of present microwave and sub–mm systems and test equipment. Two orbits have been considered for COBRAS/SAMBA: a small orbit around the L5 Lagrangian point of the Earth–Moon system, at a distance of about 400,000 km from both the Earth and the Moon and the L2 Lagrange point of the Earth–Sun system. From the Earth–Moon Lagrange point the required rejection is relaxed by four orders of magnitude, which is achievable with careful, standard optical designs. For the Earth–Sun L2 point the situation for the Earth and Moon is even better and the Sun is basically unchanged but because the Earth, Moon, and Sun are all roughly in the same direction, the spacecraft can be oriented very favorably.

These orbits are also very favorable from the point of view of passive cooling and thermal stability (Farquhar & Dunham 1990). The spacecraft will be normally operated in the anti-solar direction, with part of the sky observations performed within $\pm 40°$ from anti-solar.

Other potential missions considered both a heliocentric orbit and the Earth–Sun L2 point. All concerned seemed to have come to the conclusion that the Earth–Sun L2 point is the best choice. Operationally, it is difficult to find a more optimum location.

The main goal of the mission is to observe nearly the whole sky ($\gtrsim 90\%$) with a sensitivity of 10–15 μK within the two year mission lifetime. Deeper observation of a limited ($\sim 2\%$) sky region with low foregrounds could significantly contribute to the cosmological information. Simulations have shown that these observational objectives can be achieved simultaneously in a natural way, using the spinning and orbit motion of the spacecraft, with relatively simple schemes.

§6 Interpretation, Future

In three short years the field of CMB anisotropy observations and theory has made great strides. Until April 1992 all plots of CMB anisotropy showed only upper limits, except for the $\ell = 1$ dipole. Now we are beginning to trace out the shape of the power spectrum and to make maps of the anisotropies. This promises to deliver a wealth of new information to cosmology and to connect to other fields. The COBE DMR has now released the first two years of its data and the full four-year data set is being processed and prepared for release in fall 1995. We can expect improved results from the DMR on the large angular scales but the scientific interest has moved to covering the full spectrum and learning what the medium and small angular scales will tell us. Already we are seeing plots showing the CMB anisotropy spectrum related to and overlaid on the primordial density perturbation power spectrum and attempts to reconstruct the inflaton potential. These are the first steps in a new period of growth.

Experiments are underway. Nearly every group has data under analysis and is also at work on developing new experiments. The first of these are the natural extensions of the ongoing experiments. Some groups are considering novel approaches. Real long-term progress depends on avoiding the potential foregrounds: fluctuations of the atmosphere, a source of noise that largely overwhelms recent advances in detector technology, and Galactic and extragalactic signals. This requires instruments having sufficient information (usually only through multifrequency observations) and observing frequencies to separate out the various components. It also means going above the varying atmosphere. Collaborations are working on long-duration ballooning instruments. Ultimately, as COBE has shown, going to space really allows one to overcome the atmospheric problem and to get data in a very stable and shielded environment. A number of groups are working on designs for new satellite experiments. The COBRAS/SAMBA mission (Mandolesi et al. 1994) leads the way in the multi-wavelength and benign orbit location. With the new data that are appearing, can be expected, and ultimately will come from the COBRAS/SAMBA mission we can look forward to a very significant improvement in our knowledge of cosmology.

An accurate, extensive imaging of CBR anisotropies with sub–degree angular resolution would provide decisive answers to several major open questions on structure formation and cosmological scenarios. The observational requirements of such an ambitious objective can be met by a space mission with a far–Earth orbit and instruments based on state–of–the–art technologies.

Atmospheric disturbance, emission from the Earth and limited integration time are the main limiting factors which prevent ground–based and balloon–borne experiments from obtaining sufficient sensitivity over very large sky regions, with additional difficulties in reaching accurate foreground removal (see Danese et al. 1995 for a recent discussion). Only a suitably designed space mission can meet the scientific goals outlined in section 2. On the other hand it should be stressed that experiments from the ground or from balloons are not alternative to a space mission like COBRAS/SAMBA, but rather complementary.

§7 Recommendations

A strong vigorous program of CMB observations should be supported. The field, especially CMB anisotropies, is very active and fertile at the present and stands at the threshold of results that will revolutionize cosmology and point the way to future. We can anticipate that critical new observations will result in breakthroughs in our understanding by about the year 2000.

There will remain more high-value science that is best approached by CMB observations. To make a quantum step forward will require space-based missions. Both the CMB spectrum and anisotropy are open for major advances. The polarization of the CMB is likely to move forward as a piggyback effort on high-quality anisotropy experiments. The area theory currently shows to be very rich is the detailed study of CMB anisotropies.

To make the appropriate quantum step over what is achievable by existing and propose CMB anisotropy instruments a satellite mission has to excel in a number of areas:

- full sky coverage $\geq 90\%$
- high sensitivity - $\leq 10\mu$K per pixel
- good angular resolution - $\lesssim 10'$
- low residual foregrounds
- low systematics.

These features will be necessary as one can anticipate that suborbital programs will map out the CMB anisotropy power spectrum to nearly the cosmic variance limit. What the satellite project aims to do is improve the quality of the data and actually map the anisotropies with reasonable signal-to-noise ratios. Thus high quality results in terms of low residuals and low systematics given by uniform sky coverage with a single wide-spectral range, high sensitivity instrument in a single well-calibrated mission is the key issue. This will provide confidence and a check of the results to that point and will provide a map of anisotropies for statistical and morphological study. It should represent a major consolidation of our knowledge of CMB anisotropy and cosmology.

Acknowledgments

Thanks to Douglas Scott and Martin White for providing the plot of the current state of CMB observations and comments, Marco Bersanelli and Al Kogut for wording and plots and Laura Cayon for assistance in providing plots of the results including cosmic variance and instrument noise. This work is supported in part by the Director, Office of Energy Research, Office of High Energy and Nuclear Physics, Division of High Energy Physics of the U.S. Department of Energy under Contract No. DE-AC03-76SF00098.

References

Albrecht, A., Coulsen, D., Ferreira, P., & Magueijo, J., 1995 submitted to ApJ, astro-ph 9505030.

Alsop, D. C., et al. 1992, ApJ, 395, 317

Banday, A.J., & Wolfendale, A.W., 1990, MNRAS, 245, 182

Banday, A.J., et al. 1994 ApJ, 436, L99.

Barnes, W.J., 1994, Ph.D. thesis, Massachusetts Institute of Technology

Bartlett, J.G., & Stebbins, A., 1991, ApJ, 371, 8

Bennett, C.L., et al. 1992, ApJ, 396, 5.

Bennett, C.L., et al. 1992, ApJ, 396, L7

Bennett, C.L., et al. 1994 ApJ, 436, 423.

de Bernardis, P., Masi, S., Melchiorri, F., Melchiorri, B., & Vittorio, N. 1992, ApJ, 396, L57

de Bernardis, P., et al. 1994, ApJ, 422, L33

Bersanelli, M., et al., 1995, ApJ, submitted

Bersanelli, M., et al., 1994, ApJ, 424, 517

Bersanelli, M., et al., 1995, ApJ, June 1.

Bond, J.R., & Efstathiou, G., 1987, MNRAS, 226, 655

Bond, J.R., Crittenden, R., Davis, R.L., Efstathiou, G., & Steinhardt, P.J., 1994, Phys. Rev. Lett., 72, 13

Brandt, W.N., Lawrence, C.R., Readhead, A.C.S, Pakianathan, J.N., & Fiola, T.M., 1994, ApJ, 424, 1

Burigana, C., Danese, L., & De Zotti, G.F., 1991, A&A, 246, 49

Cayon, L. & Smoot G.F., 1995, Vista in Astronomy

Cheng, E. S., Cottingham, D. A., Fixsen, D. J., Inman, C. A., Kowitt, M. S., Meyer, S. S., Page, L. A., Puchalla, J. L., & Silverberg, R. F. 1994, ApJ, 422, L37

Church, S., 1995 MNRAS, 272, 551.

Clapp, A. C., Devlin, M. J., Gundersen, J. O., Hagmann, C. A., Hristov, V. V., Lange, A. E., Lim, M., Lubin, P. M., Mauskopf, P. D., Meinhold, P. R., Richards, P. L., Smoot, G. F., Tanaka, S. T., Timbie, P. T. & Wuensche, C. A. 1994, ApJL, 433, 57-60.

Coulson, D., Ferreira, P., Graham, P., & Turok, N., 1994, Nature, 368, 27

Crittenden, R., Bond, J.R., Davis, R.L., Efstathiou, G., & Steinhardt, P., 1993, Phys. Rev. Lett., 71, 324

Daly, R., 1992, ApJ, 371, 14

Danese, L., & Partridge, R.B., 1989, ApJ, 342, 604

Danese, L., Toffolatti, L., Franceschini, A., Bersanelli, M., Mandolesi, N., 1995, Astroph. Lett. & Comm.,

Davis, R.L., Hodges, H.M., Smoot, G.F., Steinhardt, P.J., Turner, M.S.; 1992, PRL, 69, 1856. erratum 70:1733

DeAmici et al., 1995, Col. Rivista de Fisica

Devlin, M. J., Clapp, A. C., Gundersen, J. O., Hagmann, C. A., Hristov, V. V., Lange, A. E., Lim, M., Lubin, P. M., Mauskopf, P. D., Meinhold, P. R., Richards, P. L., Smoot, G. F., Tanaka, S. T., Timbie, P. T. & Wuensche, C. A. 1994, ApJL, **430**, L1.

Dragovan, M., Ruhl, J. E., Novak, G., Platt, S. R., Crone, B., & Peterson, J. B. 1993, 427, L67-70

Farquhar, R.W & Dunham D.W., 1990; *Observatories in Earth Orbit & Bejond*, Y.Kondo Ed., Kluwer, p.391.

Fischer, M. L., et al. 1992, ApJ, 388, 242

Gaier, T., Schuster, J., Gundersen, J., Koch, T., Seiffert, M., Meinhold, P.,& Lubin, P. 1992, ApJ, 398, L1

Ganga, K., Page, L., Cheng, E., & Meyer, S. 1993, ApJ., 410, L57.

Ganga, K., Page, L., Cheng, E., & Meyer, S. 1994, ApJ,

Gawiser, E., & Smoot G.F. 1995 ApJ in preparation

Górski K.M., Hinshaw, G., Banday, A.J., Bennett, C.L., Wright, E.L., Kogut, A., Smoot, G.F., & Lubin, P., 1994, ApJ, 430, L89

Hancock, S., Davies, R. D., Lasenby, A. N., Gutierrez de la Cruz, C. M., Watson, R. A., Rebolo, R., & Beckman, J. E. 1994, Nature, 367, 333

Haslam, C.G.T., Salter, C.J., Stoffel, H., & Wilson, W.E., 1982, A&A Suppl., 47, 1

Hinshaw, G., et al., 1994, ApJ, 431, 1

Hu, W. & Sugiyama, N. 1995 ApJ May 10.

Hu, W., Scott, D., & Silk, J., 1994, ApJ., 430, L5

Johnson, D.G., & Wilkinson. D.T., 1987, ApJ, 313, L1

Knox, L., 1995, submitted to PR D, astro-ph/9504054, FermiLab Pub 95/008A.

Kogut, A., et al. 1991, in *AIP Conf Proc 222: After The First Three Minutes*, ed. C.L. Bennett & V. Trimble (New York:American Institute of Physics), 62

Kogut, A., 1992, in *2nd Course: Current Topics in Astrofundamental Physics*, ed. N. Sanchez & A. Zichichi (Singapore:World Scientific), 137

Kogut, A., et al. 1992, ApJ, 401, 1

Kogut et al., 1992, ApJ, 401, 1.

Kogut, A., et al., 1993, ApJ, 419, 1

Kogut, A., Banday, A.J., Bennett, C.L., Hinshaw, G., Loewenstein, K., Lubin, P., Smoot, G.F., & Wright, E.L., 1994, ApJ, in press

Kogut, A., 1995, Astroph. Lett. and Comm., in press

Kogut, A., Banday, A.J., Bennett, C.L., Hinshaw, G., Lubin, P.M., & Smoot, G., 1995, ApJ (submitted)

Kogut, A. 1995 Proceedings of Erice School for Fundamental Astrophysics Sanchez & Zuichichi ed.

Lang, K.R., 1974 Astropysical Formulae, Springer-Verlag

Little & Lyth 1992 PLB, 291, 391.

Mandolesi, N., Bersanelli, M., Cesarsky, C., Danese, L., Efstathiou, g, Griffin, M., Lamarre, J.M., Norgaard-Nielson, H.U., Pace, O., Puget, J.L., Raisanen, A., Smoot, G.F., Tauber, J., & Volonte, S. 1995, Planetary & Space Sciences in press.

N. Mandolesi, G.F. Smoot, M. Bersanelli, C. Cesarsky, M. Lachieze-Rey, L. Danese, N. Vittorio, P. De Bernardis, G. Dall'Oglio, G. Sironi, P. Crane, M. Janssen, B. Partridge, J. Beckman, R. Rebolo, J. L. Puget, E. Bussoletti, G. Raffelt, R. Davies, P. Encrenaz, V. Natale, G. Tofani, P. Merluzzi, L. Toffolatti, R. Scaramella, E. Martinez-Gonzales, D. Saez, A. Lasenby, G. Efstathiou; 1993, *COBRAS*, proposal submitted to ESA M3.

Mather, J.C., et al. 1994, ApJ, 420, 439

Meinhold, P., & Lubin, P. 1991, ApJ, 370, L11

Meinhold, P., et al. 1993, ApJ, 409, L1

Meyer S. S., Page, L., & Cheng, E. S. 1991, ApJ, 371, L7

Myers, S. T., Readhead, A. C. S. & Lawrence, C. R. 1993, ApJ, 405, 8

Page, L.A., Cheng, E.S., Golubovic, B., Gundersen, J., & Meyer, S.S., 1994 Appl. Opt. 33, 1

Piccirillo. L., & Calisse, O. ApJ, 411, 529-533.

Pospieszalski, M.W. 1993, 23rd EuMC, Sept 1993, p.73.

Puget, J.L., Ade, P., Benoit, A., De Bernardis, P., Bouchet, F., Cesarsky, C., Desert, F.X., Gispert, R., Griffin, M., Lachieze–Rey, M., Lamarre, J.M., De Marcillac, P., Masi, S., Melchiorri, F., Pajot, F., Rowan–Robinson, M., Serra, G., Torre, J.P., Vigroux, L., White, S.; 1993, *SAMBA*, proposal submitted to ESA M3.

Reich, P., & Reich, W., 1986, A&A Suppl. Ser., 63, 205

Reynolds, R. J., & Cox, D. P., 1992, ApJ, 400, L33

Reynolds, R.J., 1990, *IAU Symp. 139, The Galactic & ExtraGalactic Background Radiation*, ed. S. Bowyer & C. Leinert (Dordrecht:Reidel), 157

Sachs, R. & Wolfe, A., 1967, *Ap. J.* **147**, 73

Savage, B. D., & Massa, D., 1987, ApJ, 314, 380

Scott, D., Srednicki, M., & White, M. ApJ 1995, 421, L5.

Scott, D., Silk, J., & White, M. Science, 1995, 268, 829.

Smoot, G.F., Gorenstein, M., & Muller, R. 1977 PRL 39, 898.

Smoot, G.F., Levin, S., Kogut, A., De Amici, G., & Witebsky, C., 1987, Radio Science 4, 52.

Smoot, G.F. *et al*, 1992, *Ap. J. Lett.***396**, 1

Smoot, G.F., Tenorio, L., Banday, A.J., Kogut, A., Wright, E.L., Hinshaw, G., & Bennett, C.L., 1994, ApJ, 437, 1.

Smoot, G.F. 1995, accepted Planetary & Space Sciences

Songalia et al., 1994 Nature, 371, 43.

Steinhardt, P.J., 1995, Proc. of Snowmass Workshop,

Subrahmayan, R., Ekers, R. D., Sinclair, M. & Silk, J. 1993, MNRAS, 263, 416

Sugiyama, N., 1995, ApJS, in press.

Sunyaev, R.A., & Zel'dovich, Ya.B., 1970, Ap. Space Sci., 7, 20

Tegmark, M., Silk, J., & Blanchard, A., 1994, ApJ, 420, 484

Toffolatti, L., et al., 1994, in Proc. of the Workshop "The Cosmic Microwave Background", Capri (NA), 20-24 September 1993. Astrophys. Lett. & Comm., in press

Tucker, G. S., Griffin, G. S., Nguyen, H. T. & Peterson, J. B. 1993, ApJ, 419, L45

R.A. Watson *et al*, 1992 *Nature* **357**, 660

White, M., Scott, D., & Silk, J., 1994 ARAA 32, 319.

Wollack, E. J., Jarosik, N. C., Netterfield, C. B., Page, L. A., & Wilkinson, D. 1993, ApJ, 419, L49

Weinberg, D.H., & Cole, S., 1992, MNRAS, 259, 652

Wright, E.L., et al., 1992, ApJ, 396, L13

Wright, E.L., et al. 1994, ApJ, 420, 450

Zel'dovich, Ya. B., & Sunyaev, R.A., 1969, Ap. Space Sci., 4, 301

Large-Scale Structure Observations

Richard G. Kron

Fermilab, P.O. Box 500, Batavia, IL 60510, USA

§1 Introduction

In February 1987 the National Optical Astronomy Observatories organized a workshop in Tucson on the topic of "Instrumentation for Cosmology." The goal was to identify the most promising direction for ground-based observational cosmology in terms of scientific priority and practical feasibility. The workshop resulted in a consensus that the top priority program was to explore further large-scale structure as revealed in the distribution of galaxies. To do this, a new redshift survey was proposed that would be at least one order of magnitude larger than any current survey (~100,000 galaxies as opposed to ~10,000). The new survey would require a wide-field telescope dedicated to the task for a number of years, plus instrumentation designed to achieve an effective rate of redshift measurement.

Since 1987 the importance of large-scale structure studies has intensified as results from ongoing redshift surveys have demonstrated yet more intriguing puzzles, other wavebands have been exploited, coherent velocity fields have been identified, and *COBE*'s detection of anisotropy has inspired complementary efforts. The following reviews some of the underlying current programs in the field of mapping large-scale structure, with a view towards suggesting promising avenues for further work.

§2 Redshift Surveys

Galaxy redshift surveys continue to be one of the most fundamental techniques in observational cosmology, as they provide a direct look at the three-dimensional distribution of luminous material over a wide range of spatial scales. The sampling frequency is restricted by the average space density of galaxies, and the maximum scale is limited by the sizes of the surveys themselves, as well as low-level systematics in the selection and measurement of the galaxies across the sampled volume. Priorities for new surveys are accordingly to increase the surveyed volume, and to decrease the amplitude of the systematics.

Redshift surveys are logically constructed in three steps: the production of a list of galaxies with pertinent measurements based on imaging data ("image parameters"); the specification of the rule for the selection of some of these galaxies for spectroscopic observation based on the image parameters; and the actual observations of redshifts. Redshift surveys differ according to the specifics of each of these steps. Choices in observational parameters (such as the size and shape of the area surveyed) are made depending on the sensitivity desired to structure on a particular angular or spatial scale, for example, and depending on an adopted strategy for maximizing the efficiency of use of telescope time and instrument capabilities. As more is learned about large-scale structure, to some extent new surveys can be tuned to investigate particular features.

Galaxies differ not only in their physical scale (mass, luminosity, and size), but also in their distribution of light and mass (centrally concentrated or extended) and their spectral energy distribution (if a young population of stars contributes significantly to the integrated light, the spectrum will be relatively blue, whereas if the bulk of the light originates in generations that are at least 1 Gyr old, the spectrum will be relatively red). Naturally there are many other attributes, such as the characteristics of the spiral arms (the "Hubble morphological type" of a galaxy, which is based on visual inspection of images); and measures that refer to the interstellar medium in the galaxy, such as the far-infrared flux arising from thermal emission from dust and the neutral hydrogen mass deduced from the 21-cm flux. These ways of categorizing the type of a galaxy tend to be correlated, but often only loosely.

There are two reasons why the rich variety of galaxy types is important for redshift surveys and affects our measurement of large-scale structure. First, the characteristic surface brightnesses of galaxies at visible wavelengths are by coincidence comparable to the natural surface brightness of the dark night sky. Whether a particular galaxy is included in a catalogue thus depends on how its flux is distributed in projection on the sky, and catalogue completeness cannot be described by a simple limit in apparent flux. Second, it turns out that galaxy type (defined in one of the ways mentioned above) and environment are related to each other: galaxies in environments with many close neighbors tend to lack light from young generations of stars and are seen to be relatively red, whereas there is a higher fraction of blue galaxies in lower-density environments.

Aside from the redshift, other quantities may be derived from the spectra, such as atomic line strengths and line widths (indicative of the stellar velocity dispersion or systemic rotation). These other data are indirectly important for large-scale structure because they, too, provide an index of the type of galaxy. The precision in the redshift and these other parameters depends on instrumental specifics. Redshifts that are accurate to 60 km/s or so can resolve the internal dynamics of small groups of galaxies. Also, given adequate velocity resolution and signal-to-noise ratio, line widths can be measured for at least the more massive galaxies. The ratios of line strengths measure the degree of ionization and therefore are also indicative of the stellar population.

2.1 Construction of Catalogues

Imaging maps of large areas of sky suitable for establishing statistically complete samples of galaxies can and have been done with a wide range of techniques, including scanning photographic plates, identifying extragalactic sources in the *IRAS* catalogues, and analysis of CCD drift scans. Many of these ongoing and planned surveys have been described in recent conference proceedings, such as Soifer (1993) and MacGillivray *et al.* (1994). There are several desiderata for such surveys from the point of view of large-scale structure studies. Among these are large solid angle; uniform and deep detection limits across the survey area; insensitivity to obscuration by dust in the Milky Way; freedom from other systematics as a function of direction in the sky; ability to discriminate against ordinary stars and other confusing sources; and the availability of a variety of measures of galaxy type, such as spectral indices and image shape parameters.

Surveys in the blue spectral range have the advantage that the sky background is relatively dark in that band. The thresholds for detection can then be set relatively low, enabling classes of "low-surface-brightness galaxies" to be included. However, a trivial amount of mass in a galaxy in the form of hot stars can strongly perturb the integrated light, so the blue light may be a poor indicator of the underlying stellar mass distribution. Moreover, blue light is selectively absorbed by dust, which means that the uncertainties in the distribution of dust in the Milky Way are important.

At the other extreme in wavelength, the *IRAS* far-infrared survey established a photometrically homogeneous, all-sky catalogue of galaxies which is essentially unaffected by dust obscuration. Galaxies that emit in the *IRAS* bands have relatively large quantities of gas and dust, the stuff out of which new generations of stars form. Hence, this sample may not be representative of galaxies in general. The detection threshold depends on the apparent size of the galaxy with respect to the angular resolution of the map; this is true for all surveys, but since the *IRAS* angular resolution is relatively coarse, the limitation occurs at relatively nearby distances.

Surveys at near-infrared wavelengths (2.2 microns, the so-called K band) offer a potentially important compromise. This wavelength is sufficiently long that it is insensitive to dust absorption, but sufficiently short that the radiation measured comes from stellar photospheres, as opposed to the interstellar medium. There are two problems with surveys at near-infrared wavelengths: the Earths atmosphere emits strongly in this waveband, and array detectors have small formats (with respect to CCDs used at visible wavelengths). The first point can be addressed by long integrations and careful subtraction of the sky background, but when coupled with the second point, it is then very time consuming to cover substantial areas of sky. Nevertheless, major efforts are addressing these issues. A survey telescope sited at the South Pole is designed to have a very low thermal background, and two other survey projects (2MASS and DENISE) are planned that will cover the whole sky, but with only a short dwell time at any given place - hence a relatively bright threshold.

2.2 Galaxy Selection Function

Given a catalogue of galaxies detected in some survey as described above, one proceeds to select a subset of these for redshift measurement. The goal is to devise a selection procedure that is amenable to modeling so that the observations can be directly compared to theory. In particular, we want to know the probability that a certain type of galaxy at a certain distance is included in the sample, as a function of distance (this is called the selection function). One does not know the distances prior to obtaining the redshifts, of course, but if enough data are available it may still be possible to pick galaxies for redshift measurement such that the selection function has desirable properties.

If the sample were limited strictly by a threshold in the apparent flux in some fixed waveband, the selection function would be directly related to the frequency distribution function of the intrinsic luminosities of galaxies within a typical volume (the so-called luminosity function) once something is known about the spectra of the galaxies (because the redshift affects the flux that appears in a fixed waveband) and once the cosmological volume element is specified. As already suggested, the variety of galaxy types, and the interaction of their apparent properties with the detection/selection thresholds, make the selection function more complicated.

No selection cut can be sharp because of inevitable random errors and systematic errors. The random errors introduce a bias called Malmquist bias, which can be evaluated if the nature of the random errors is sufficiently well understood. The value of the limiting flux for selection will vary from one direction in the sky to another according to the residual systematic errors in the calibration of the detected count rates. The angular scale of these errors, and the amplitudes at different angular scales, depends of course on the specifics of the galaxy catalogue.

At visible wavelengths our view of the sky is strongly modulated by absorption from dust localized in the plane of the Milky Way. By chance, the Sun is situated almost exactly in the midplane of the disk, but a simple plane-parallel model for the column density of dust as a function of angle from the plane is too crude because the dust is distributed inhomogeneously on smaller scales. Since the absorption is wavelength-dependent, a measurement of the differential amount of absorption at different wavelengths ("reddening") can be used to infer the absolute absorption. This is a relatively mature technique in classical astronomy, but its application does depend on the assumption that the absorption and scattering properties of the dust are similar along all relevant lines-of-sight.

2.3 Spectroscopic Measurement

If there are several suitable galaxy targets within the field-of- view of a telescope, it pays to observe their spectra simultaneously, which can be done by using optical fibers to couple the light at the focal plane of the telescope to the entrance aperture of the spectrograph. Some mechanical system is then required to reconfigure the fibers when the telescope is pointed to a different field. There are two categories of solutions: in one, fibers are plugged, off of the telescope, into a pre-drilled plate (as in the Las Campanas Deep Redshift Survey), and in the other the fibers are reconfigured by a robot that is integral to the focal plane hardware (as in the HYDRA system on the WIYN telescope). The former method is simpler mechanically but is more labor-intensive; other advantages and disadvantages also apply, and the choice between these techniques depends on various practicalities.

As mentioned already, galaxies tend to have low contrast with respect to the sky, which means that the noise is largely due to the sky spectrum obtained along with the galaxy spectrum. The sky spectrum is partly continuum (due to reflected sunlight off of interplanetary dust) and partly emission lines and bands (due to fluorescence in the upper atmosphere). Thus, the sky noise is a strong function of observed wavelength, and its subtraction can be problematic. Faint redshift surveys must naturally deal with this problem at a higher level of accuracy than surveys of brighter galaxies. Optical fibers transform their input beam into their output with a variety of systematics that can affect the accuracy of the sky-subtraction. Hence, faint-galaxy redshift surveys adopt a completely different technique for multi-object spectroscopy. The fibers are dispensed with, and the focal plane is imaged directly through the spectrograph. A mask designed for each field is placed in the focal plane so that only the light from the galaxies passes through small slits that are oriented perpendicular to the direction of the spectral dispersion. This arrangement allows the sky spectrum to be sampled on either side of the galaxy, which yields significantly better sky subtraction. Since there are many more faint galaxies than bright ones, the area of the detector can usually be packed with spectra despite the smaller angular field-of-view accessible in this mode.

Not all spectra yield a redshift, since some spectra may possess lines too weak to be detected. Also, depending on the redshift, the stronger lines may appear outside of the accessible spectral range. There is no simple way to categorize, say, the number of photons required to obtain a redshift, since the spectral information is distributed over wavelength in an intrinsically complicated way, which is moreover different for different sources. In general two lines are required to determine a redshift because the identification of a single line may be ambiguous. The standards for what constitutes a satisfactory determination of a redshift scales with the apparent flux: faint- galaxy redshifts are pushed to lower levels of significance than bright-galaxy redshifts. The selection function, then, depends not only on which galaxies were selected for redshift measurement, but which category of galaxy spectra actually yield a reliable redshift. Typically 90% of targeted galaxies do yield a redshift, but if the remaining galaxies have a different distribution in redshift, then the systematics could be important.

The new, large redshift surveys apply these various approaches. The Sloan Digital Sky Survey (SDSS) will use a 2.5-m telescope with a 3-degree field and 600 fibers configured with drilled plates. The Anglo-Australian survey (2dF) will use a 4-m telescope with a 2-degree field and 400 fibers configured with a robot. The Deep Extragalactic Evolutionary Probe (DEEP) will use the Keck 10-m telescope with a 0.3-degree field and a focal plane mask. DEEP is proceeding with available instrumentation, and will upgrade later to a spectrograph capable of over 100 spectra at a time. The SDSS is designed to cover most of the accessible sky that lies away from strong dust absorption; DEEP is designed to make redshift measurements of small areas to high redshifts in order to study evolutionary effects; and the 2dF survey is intermediate between these cases. These three new major surveys are thus complementary in their capabilities.

§3 Environmental Effects and Galaxy Biasing

It was earlier remarked that the types of galaxy found in a particular place depends on the local space density of galaxies. One interpretation of this phenomenon is that galaxies in dense environments have had their interstellar medium stripped away by ram pressure from hot gas associated with those same environments, hence quenching any new star formation. However, it appears that the systematics of galaxy-type versus environment extends smoothly to densities that are too low for this process to be effective. Thus, there may be a more fundamental connection between large- scale structure and galaxy type (Tucker 1994), a connection that would bear on galaxy-formation mechanisms.

Pictures for the development of structure in the universe are based on the properties of the mass: its density fluctuations, velocity structure, and coupling with radiation as a function of cosmic epoch. The structure we see in the galaxy distribution may or may not be simply related to the underlying distribution of mass. The baryonic material has become luminous in different places according to some more-or-less complicated chain of astrophysical events, which may continue to prove difficult to trace with confidence with hydrodynamical modeling. The alternative approach is empirical, where in different places we probe both the mass (say, by the technique of weak lensing) and the light, and study the connection under different conditions. Thus we need to consider the widest variety of measures of mass on large scales, and the widest variety of ways to categorize the environment on these scales.

§4 Relative Distances and Bulk Motions

The measurement of large-scale coherent streaming motions provides important new constraints on the evolution of structure, the nature of the dark matter, and Ω. The distribution of mass can be deduced from bulk motions due to gravitational instability, and compared to the known distribution of luminous material, providing a direct determination of the quantity $f(\Omega)/b$, where b is the biasing parameter. (The quantity $f(\Omega)/b$ is also measured by the anisotropy of galaxy correlation in redshift - angle space.) To detect individual bulk motions, relative distances that are independent of redshift are required. Because of the presence of random motions, it is also necessary to measure many galaxies. Finally, distance errors that are systematic with direction in the sky or depth must be carefully controlled.

Statistical distance errors generally scale with the distance, which restricts the size of the local volume within which bulk motions can be determined at a particular significance in velocity. For example, a 600 km/s bulk motion is a 10% effect at a distance of only 60 h^{-1} Mpc [on the other hand, by smoothing over larger volumes and taking into account the expected spectrum of peculiar velocities on different scales, Dekel (1994) has argued that bulk motions should be detectable to a few hundred Mpc h^{-1}]. The more massive galaxies tend to be more amenable to techniques for distance determination, and the number of suitable galaxies in a given volume may therefore be limited. The distribution of absorbing dust in the Milky Way affects both the transformation of apparent flux to intrinsic luminosity, and the presumed original sample selection. Finally, the techniques that are easiest to apply and which can efficiently yield the requisite thousands of distance measurements select only a couple of items from the available menu of observables. Redundancy of at least some of the information is necessary for consistency checks, but to obtain this requires much more observational effort.

An important method of galaxy distance determination is the correlation of some measure of the luminosity or physical size with some measure of the kinematical field. There are many variations according to the available data, but the best-known one is the Tully-Fisher technique. [The first application was actually by Öpik (1922)]. In its original Tully-Fisher form, it used the 21-cm line width, the total apparent brightness in the blue band, the projected inclination angle on the sky, and the Hubble morphological type. Ideally, instead of a simple line width, one would like to have the complete rotation curve (to determine the distribution of mass within the galaxy), and instead of the integrated blue light, one would prefer to have the two-dimensional distribution of light in the near-infrared band (since the light at longer wavelengths arises from a stellar population that is expected to trace more faithfully the mass in the form of stars). Although the Hubble classification scheme is venerable, the types do not in fact correlate well with other objective ways to categorize galaxies. Since the critical parameter in the correlation of kinematics to luminosity is the mass-to-light ratio, it is important to pursue the development of a reliable index for this quantity. With this more extensive set of data, one would be in a much better position to investigate the manner in which the mass-to-light ratio varies within galaxies and from one type of galaxy to another.

There are a number of other methods for determining distances to relatively nearby galaxies. These tend to be based on the statistical distribution of brightnesses for luminous things in a given galaxy, such as the brightest stars, H II regions, planetary nebulae, and globular star clusters. A sophisticated variant is the "surface brightness fluctuation" technique (Tonry, Ajhar, and Luppino 1991). When calibrated with respect to nearby galaxies, these techniques can also be used to determine the Hubble constant H_o. Thus there is strong motivation to continue to develop these and other methods that apply to common galaxies at intermediate distances.

Depending on the technique used, relative distances to individual galaxies may still be in error by up to 15% (Courteau *et al.* 1993). If the galaxy is a member of a recognizable group, and if the distance errors to the individual group members are uncorrelated, then one can derive a more accurate distance to the group as a whole. When considering rich clusters of galaxies, the cluster itself may have additional attributes that can be used as distance indicators, such as derived from the X-ray emitting gas and the associated decrement in the microwave background radiation (Fischer and Lange 1993).

§5 Distant Sources, Cosmological Tests, and Galaxy Evolution

Some of the stock-in-trade methods for measuring distances to galaxies can be applied at large redshift, in which case they can be used as a test for the cosmological deceleration parameter q_o. One such method is involves the detection of distant supernovae and the measurement of their light curves, as described in more detail elsewhere. An adequate distinction between different cosmological models requires redshifts of at least 0.3 to be sampled. Depending on the errors in the peak brightness, and how large the redshifts are, several dozen supernovae would need to be discovered, each at an early phase in its development. Such a program is challenging because of the time-critical element, the faintness of the sources, and the rarity of the events (which requires substantial areas of sky to be surveyed, again depending on the depth sampled).

Another method is the high-redshift analogue of the luminosity - line width relation. Emission lines are sufficiently strong in some galaxies at high redshift to enable a determination of the kinematic broadening. In special cases it is even possible to measure the rotation curve, albeit with coarse sampling. If the emission is coming from a rotating disk, the inclination angle of which is known from high-resolution imaging, the mass of the galaxy can be determined modulo its distance. The derived luminosity, on the other hand, depends on the distance squared, enabling the distance itself to be determined for a particular mass-to-light ratio. Thus, the combination of high-angular-resolution imaging and kinematic information allows not only a cosmological test, but a method of tracking the evolution of galaxies in terms of mass and light.

The effectiveness of both the distant-supernova technique and the galaxy dynamics technique depends on the angular resolution. The *Hubble Space Telescope* clearly has great potential for cosmology in this regard, but many problems require more total exposure time than can be reasonably accommodated on this instrument. Hence, there are several initiatives to improve the sharpness of images obtainable with ground-based telescopes. Besides providing for more observing time, these telescopes may also obtain a wider field-of-view and may have the capability to sample the near- infrared spectral range. These initiatives include efforts to minimize temperature inhomogeneities in the air near the telescope (roughly half of the total atmospheric image blur is local to the telescope), the development of fast steering mirrors to remove first-order wavefront tilt, and the pursuit of adaptive optics systems that will remove higher-order wavefront distortions. At visible wavelengths, these last-named techniques have restricted fields-of-view even by comparison with the *Hubble Space Telescope*, but on the other hand the application at near-infrared wavelengths is especially promising.

The properties of distant sources in the radio band are also potentially useful cosmological probes. For example, another method relies on modeling jet dynamics in powerful radio sources, constrained by a time series of multi-frequency maps and polarization measurements. The model yields a physical size for the jet, which can then be compared to the apparent angular dimensions (Kovalev 1994).

§6 Quasar Absorption Lines

The systems of absorption lines seen in the spectra of quasars provide a different look at large-scale structure. An intervening cloud with a column density far below the detection limit for 21-cm emission can still yield detectable absorption in Ly α, and the redshift range extends far beyond that probed by normal visible galaxies. The density of metal lines is of the order of 1 absorber per $\Delta z = 1$, and the density of Ly α absorbers is about 30 times higher. The spatial sampling overlaps with known structures at lower redshift. Thus, measures of correlation in the distribution of quasar absorption lines complements information from galaxy clustering, in the sense that the tracers are different in nature, high redshifts can be sampled, and the selection function is completely different (Crotts *et al.* 1985).

The visibility of an absorption line depends not only on the column density of gas, but also on the ionization state, which in turn depends on the radiation field. It is not even yet known whether the ionizing radiation field is dominated by the mean intergalactic flux, or by sources relatively close to the absorber. The apparent evolution in the comoving number density of absorbers may be driven partly by ionization effects, and, in the case of the metal lines, evolution in the cosmic abundance relative to hydrogen.

Of the various classes of absorption lines seen in the spectra of quasars, namely damped Ly α systems, Lyman limit systems, the Ly α forest, and metal systems, only the metal systems show strong evidence for clustering, on scales of 600 km/s in the rest frame of the absorbers, with an average amplitude of the correlation function of 5-10 between 200 and 600 km/s (Sargent, Boksenberg, & Steidel 1988). Because the scale of the clustering is greater than an individual galaxy halo, the clustering may be probing large-scale structure (although the high amplitude suggests it might be strongly biased).

Only recently has there been good evidence for clustering of the Ly α forest, in Q0055-269 (Cristiani *et al.* 1995). The line distribution deviates from Poisson at the 4-σ level. There is evidence at the 99.5% confidence level for a deviation from Poisson in Q0014+813 as well. The correlation functions of both systems match, with an amplitude of about 0.5 on scales of 100-200 km/s. There is evidence for anticorrelation on scales of 500-1500 km/s in Q0055-269, but this needs confirmation. One can also cross-correlate absorption systems of quasars lying close together in the sky, a technique that has yielded evidence for large-scale voids (Crotts 1989).

Because the Ly α forest clouds are optically thin to ionizing radiation, fluctuations in the radiation field would result in correlations in addition to those from large-scale structure. If quasars dominate the radiation field, large fluctuations would occur for $z > 3.5$, where the attenuation volume due to intervening Ly α clouds and Lyman limit systems is so small that significant local Poisson fluctuations would occur in the radiation field, leading to correlations. The photoionization process of the clouds itself (at $z > 5$) could in principle result in correlations by locally readjusting the baryon to dark matter ratio differently in different H II regions as the clouds form and the H II regions overlap. For these reasons, the Ly α forest may be a problematic tracer of large-scale structure.

Thus, the best systems for measuring large-scale structure may be the Lyman limit and damped Ly α systems, which are most likely to represent galaxies, but these systems are relatively sparse. Even in a wide-field survey designed to find quasars and measure absorption-line systems (such as the Sloan Digital Sky Survey), the number of pairs of quasars, both showing damped Ly α falling within a few Mpc of each other, is expected to be only a few dozen. This is just enough perhaps to measure large-scale structure, especially if the correlation is strong. The situation is improved by about a factor of 5-10 for Lyman limit systems.

A better way to use the absorbers may be as probes of the field galaxy population at high redshift. Once a damped Ly α or Lyman limit system is found, the field surrounding the quasar may be imaged to search for neighboring galaxies. This appears to be a promising approach: although the numbers are still very low, the rate of success so far argues for very strong correlations (Wolfe 1993). Similarly, clustering statistics from imaging fields surrounding known Mg II systems (Steidel *et al.* 1994) appear to provide a powerful diagnostic of both galaxies at large redshift (which are thus selected in a very different way than in flux-limited samples) and the characteristics of the absorbers themselves.

§7 Weak Gravitational Lensing

Gravitational lensing has become an increasingly important tool in observational astrophysics because it offers the opportunity to directly probe mass distributions through the deflection of passing light. Such deflections depend only on the distribution of gravitating mass and not on its composition, making it possible to measure mass directly rather than inferring it from the observed intensity of emitted light.

Observations of strongly lensed systems (where quasars are multiply imaged or background galaxies bent into dramatic arcs) provide information about the mass concentrations in the densest central regions of the lensing object (Kaiser 1992). Recently, however, there has been much theoretical work and some observational study of weakly lensed systems. If the lensing mass is smaller or further from the line of sight, the images of background galaxies are mildly distorted rather than split or bent into arcs. The observed ellipticity and orientation of a galaxy is then a combination of its intrinsic shape and distortions introduced by lensing. Since the intrinsic ellipticity and orientation of a particular galaxy is unknown, a statistical analysis of many galaxy images is necessary to quantify the nature of the lensing mass.

The distortions in background images produced by large galaxy clusters can be used to measure the cluster potential and thus the structure of the dark matter content. In clusters such as A370, the number of strongly lensed arclets is too low to characterize the lensing mass accurately (Kochanek 1990). Theoretically, statistical studies of the ellipticities of background galaxies near a cluster, combined with measurements of arclets, can provide a much better map of the mass distribution in the cluster (Kochanek 1990). Lensing probes the potential in clusters independent of the composition or dynamics of the mass because it is purely a gravitational effect. Recent examples of observational efforts to deduce the mass distribution in distant rich clusters include Smail *et al.* (1994) and Fahlman *et al.* 1994.

On the largest scales, measurements of ellipticities of background galaxies (e.g., Mould *et al.* 1994) may provide a way to measure the density variations in the universe independent of how light traces mass. Since photons rather than galaxies are the tracers of the potential in lensing studies, fluctuations in total mass, not just luminous mass, are measured. Systematic distortions in background images are characterized by an ellipticity correlation function which is similar to the galaxy-galaxy correlation function. Structure on the scale of the observed voids and walls is expected to produce a correlation length of a few degrees with sensitivity to density contrasts of about 0.3% (Blandford *et al.* 1991). For CDM models, the correlations are expected to be detected at the several sigma level for background objects at redshifts greater than 0.3 and to become stronger with increasing redshift (Miralda-Escudé 1991).

The weak lensing technique is one of the few observational tools with the potential to map out structure in the surface mass density in the cosmos. It is however not an easy undertaking, as exemplified by the different conclusions about the degree of concentration of rich-cluster potential wells that have recently been derived. Two major experimental improvements need to be made. First, better understanding of the systematic effects involved with accurate measurements of galaxy shapes is essential. The effects of optical aberrations, telescope guiding errors, effects of finite pixel size, and seeing need to be properly characterized in order to avoid spurious coherent alignment in the images. Second, larger numbers of galaxies at greater depth need to be observed in multiple passbands in order to gain the statistical leverage necessary to disentangle a weak lensing system. These considerations suggest an observing program that is demanding in both telescope time and instrumental characteristics, specifically large collecting aperture, wide field-of-view, and good angular sampling.

§8 Directions for the Future

There are many basic features of large-scale structure that are still poorly determined, among these being the power spectrum at large scales, limits to non-Gaussian fluctuations, the evolution of clustering, the nature of large-scale peculiar velocities, and the connection between the distribution of non-baryonic and luminous matter. Observational progress on these problems will require a coordinated approach because of the common need to survey large volumes systematically.

There are at least two key reasons to be encouraged that the next decade will yield substantive new results concerning large-scale structure. First, the astronomical phenomenology is rich, including many types of observables for galaxies, rich clusters, and larger structures, over a wide range of wavelengths. Moreover, the observables are routinely being exploited in new ways: the study of gravitational lensing of distant galaxy images, for example, is not yet a decade old, but this technique for mapping mass distributions is likely to play a major role in the decade to come. Second, technological developments are proceeding apace, and provide for dramatic gains in our capability to sample distant parts of the universe. NASA's Great Observatories program has been designed to provide windows across the electromagnetic spectrum, and much of the motivation for these satellites (AXAF and SITRF still to come) has been based on providing the capability to address cosmological issues. Technological developments have been aggressively pursued at ground-based observatories as well, among which are adaptive optics, array detectors sensitive at near-infrared wavelengths, and the construction of optical telescopes in the 8-10 meter aperture range with correspondingly powerful instrumentation. This hardware is in general adaptable, so that often new problems can be attacked as quickly as they are identified.

§9 References

§References

[1] Blandford, R.D., Saust, A.B., Brainerd, T., & Villumsen, J.V. 1991, MNRAS, 251, 600

[2] Courteau, S., Faber, S.M., Dressler, A., & Willick, J.A. 1993, ApJ, 412, L51

[3] Cristiani *et al.* 1995, preprint

[4] Crotts, A.P.S. 1989, ApJ, 336, 550

[5] Crotts, A.P.S., Melott, A.L., York, D.G., & Fry, J.N. 1985, Phys. Lett., 155B, 251

[6] Dekel, A. 1994, Annu. Rev. Astr. Ap., 32,

[7] Fahlman, G., Kaiser, N., Squires, G., & Woods, D. 1994, ApJ, 437, 56

[8] Fischer, M.L. & Lange, A.E. 1993, ApJ, 419, 433

[9] Gould, A. & Villumsen, J. 1994, ApJ, 428, L45

[10] Kaiser, N. 1992, ApJ, 388, 272

[11] Kochanek, C.S. 1990, MNRAS, 247, 135

[12] Kovalev, Yu. A. 1994, Astronomical and Astrophysical Transactions, in press

[13] MacGillivray, H.T. *et al.* (eds.), 1994, Astronomy from Wide-Field Imaging, I.A.U. Symp. 161, (Dordrecht: Kluwer)

[14] Mould, J., Blandford, R., Villumsen, J., Brainerd, T., Smail, I., Small, T., & Kells, W. 1994, MNRAS, 271, 31

[15] Miralda-Escudé, J. 1991, ApJ, 380, 1

[16] Öpik, E. 1922, ApJ, 55, 406

[17] Sargent, W.L.W., Boksenberg, A., & Steidel, C.C. 1988, ApJS, 68, 539

[18] Smail, I., Ellis, R.S., Fitchett, M.J., & Edge, A.C. 1994, MNRAS, in press

[19] Soifer, B.T. (ed.), 1993, Sky Surveys: Protostars to Protogalaxies, A.S.P. Conf. Series Vol 43, (San Francisco: A.S.P.)

[20] Steidel, C.C., Dickinson, M., & Persson, S.E. 1994, ApJL, in press

[21] Tonry, J.L., Ajhar, E.A., & Luppino, G.A. 1990, AJ, 100, 1416

[22] Tucker, D. 1994, PhD dissertation, Yale University

[23] Wolfe, A. 1993, Annals of NY Acad. Sci., vol. 688

Large Scale Structure Theory

Joel R. Primack

Santa Cruz Institute for Particle Physics, University of California, Santa Cruz, CA 95064

If Particle Astrophysics can answer the key questions of the nature of the initial conditions for structure formation and the nature of the dark matter, then it can perhaps provide a lasting foundation for physical cosmology. Theoretical models for structure formation can be characterized in several useful ways, and the simplest of the models with Gaussian initial fluctuations have been worked out in considerable detail and compared with observations on various scales. It is on nonlinear scales $\lesssim 10\,h^{-1}$ Mpc that the greatest differences exist between models that have been normalized to agree on the largest scales with the COBE data; here especially there is a need for better statistical tests which are simultaneously *robust*, *discriminatory*, and *interpretable*. The era at which galaxy and cluster formation occurs is also a critical test of some models. Needs for the future include more powerful computers and faster and cleverer codes, better control of cosmic variance in simulations, better understanding of processes leading to galaxy formation. better ways of comparing observational data with models, and better access to observational and simulation data. Because of the tremendous amount of extremely useful data on observational cosmology that is now becoming available, the present period is likely to be regarded as the golden age of cosmology; a modest increase in the quantity and quality of support for theoretical cosmology is likely to pay great dividends now.

§ 1 Goal of Particle Astrophysics

The goal of Particle Astrophysics is to construct a fundamental theory of the material universe — i.e., to explain elegantly and economically observations from the smallest physical scale to the entire cosmological horizon. Cosmology is mainly concerned with understanding the origin and evolution of galaxies and the large scale structures in the universe: clusters, voids, superclusters, filaments and great walls of galaxies, etc. The key questions that must be answered are the nature of the initial conditions for the formation of structure, and the nature of the dark matter (DM) which makes up most of the mass of the universe.

Cosmology today is like physics before Newtonian Mechanics, biology before Darwin, or geology before Plate Tectonics. In cosmology, we do not yet have a fundamental theory, a grand picture of the origin and evolution of the universe. We have only bits and pieces of the story. Almost certainly the account General Relativity provides of the expanding universe is such a piece. Perhaps standard Big Bang Nucleosynthesis (BBN) is, too. BBN covers the period starting about one second after the Big Bang, and it describes how more than 90% of all the fusion in the universe occurred during the first three minutes, producing most of the helium and other light nuclei. But until the discovery by the COsmic Background Explorer (COBE) satellite, launched in November 1989, of the fluctuations in the cosmic background radiation (CBR), we were building cosmological theories on sand. We had little evidence concerning the initial conditions and the events of the crucial first second, and we had several mutually inconsistent speculative theories of the origin of the galaxies and of their large-scale distribution in the universe. The CBR fluctuation data from COBE and many other instruments has at last allowed us to begin to build cosmological theory on a solid foundation.

COBE's measurement of the largest inhomogeneities in the CBR gives us a vantage point from which we can look backward to inquire about the origins of these fluctuations, and forward toward structure formation. The relative magnitudes of the various multipole components of the CBR fluctuations observed by COBE are consistent with a Zel'dovich spectrum (primordial power spectrum $P_p(k) \equiv |\delta_k|^2 \propto k^{n_p}$ with $n_p = 1$) of primordial fluctuations, which is what the simplest models of inflation predict. But higher precision observations on large and smaller scales will be needed before we can tell whether the spectrum is actually Zel'dovich, has a "tilt" ($n_p \neq 1$), or requires two or more parameters to characterize it.

Now that we know the magnitude of the fluctuations that gave rise to structure when the universe was only

~ 0.00002 of its present age, we have a solid starting point for calculations of their evolution into the structures that we see today. Given this evidence, it is possible — but still needs to be verified — that all the structure in the universe from the scale of galaxies to the entire horizon grew by gravitational collapse from early inhomogeneities. If this is true, it requires, among other things, that there be a sufficiently large density of dark matter.

Cold Dark Matter (CDM) was an educated guess regarding such a fundamental theory — it was the simplest theory proposed in the 1980s that had a chance to be true. It assumed that the primordial fluctuations were as predicted by the simplest cosmic inflation models: adiabatic, Gaussian, and with a Zel'dovich spectrum; and it assumed that the dark matter was "cold," i.e. preserved primordial fluctuations on all cosmologically relevant scales. But, like the original SU(5) Grand Unified Theory in particle physics, CDM was apparently too simple to be true (cf. e.g. Davis et al. 1992, Ostriker 1993). In constructing a fundamental theory of cosmology, it now appears that even if the dark matter is mostly cold, the data requires a hybrid theory containing elements of at least two simpler theories, such as Cosmological Constant (Λ) + Low-Ω Cold Dark Matter (ΛCDM) or $\Omega = 1$ Cold + Hot Dark Matter (CHDM). The resulting theory may thus be a little like the $SU(3) \times SU(2) \times SU(1)$ Standard Model of particle physics, which is of course also a hybrid theory. Another possibly useful analogy is elliptical planetary orbits. For millennia, until Kepler and Newton, astronomical prejudice favored circles; now we know that only an unusual accident of planetary formation would give circular orbits (i.e., very small ellipticity).

§ 2 Characterizing Models

There are at least four useful ways to characterize possible cosmological theories:

Negative vs. positive energy structure formation. The usual assumption is that structure grew by gravity — initially denser regions expanded more slowly, became increasingly dense relative to their surroundings, and eventually collapsed to form galaxies, clusters, etc. In this case, it is the minima of the gravitational potential that become visible astronomical objects, and we can characterize this as a negative-energy scenario. The main alternative now being pursued is the possibility that structure formed from Cosmic Strings, Textures, or some other cosmic topological defects. In that case, the defects are the seeds for structure formation, but the seeds themselves form and perhaps move as a result of the dynamics of the underlying fields; for example, long Cosmic Strings have a large string tension which continuously pulls them straight on ever-larger scales. This can be characterized as a positive-energy structure formation scenario. For a discussion of how to tell which is the case in our universe, see e.g. Babul et al. 1994.)

Dark Matter: total density in matter in units of critical density (Ω_0), and fractions in cold (Ω_c) and hot (Ω_ν) dark matter, and in baryons (Ω_b). From the viewpoint of particle physics, the best-motivated candidate dark matter particles are axions or supersymmetric WIMPs for cold dark matter, and light (few-eV) neutrinos for hot dark matter. It is usually assumed that the nature of particles comprising the cold or hot dark matter is irrelevant to their astrophysical consequences, and while this is not necessarily so (cf. Starkman et al. 1994; Bonometto et al. 1994), for simplicity we will follow this usual practice in this section.

Gaussian vs. non-Gaussian fluctuations. Gaussian fluctuations can be completely characterized by giving just the power spectrum of the fluctuations; the phases are random. Non-Gaussian fluctuations require in principle an infinite amount of information to characterize them, although the main non-Gaussian models being considered at present — topological defect models — in fact have one or at most a small number of parameters. Any model, such as cosmic strings, in which the primordial fluctuations arise in connection with extended physical structures necessarily gives rise to non-Gaussian fluctuations, since there must be phase correlations between different wave numbers. These have been called *non-local* non-Gaussian fluctuations (Kofman et al. 1990). Inflationary models with several inflatons can lead to a non-Gaussian random field of perturbations which is a non-linear function of some other Gaussian random field; these are *local* non-Gaussian fluctuations. The problem of telling from observations whether the initial fluctuations were Gaussian remains open. Even if the initial fluctuations were Gaussian, the nonlinearity of gravity makes the distribution non-Gaussian in the nonlinear regime. Statistics such as the local genus (cf. Vogeley et al. 1994) and the skewness of the galaxy probability density field (pdf) (cf. Saunders et al. 1990, Kofman et al. 1994) have been used to quantify non-Gaussianity. The conclusion so far is that if galaxies trace the mass, the observations are consistent with Gaussian initial conditions. However, if $\Omega \lesssim 0.3$, then non-Gaussian initial fluctuations are required to account for the non-Gaussian pdf observed at the present epoch (Nusser and Dekel 1993, Bernardeau et al. 1994).

Adiabatic vs. isocurvature fluctuations. In adiabatic primordial fluctuations, all the components (cold and hot dark matter, baryons, radiation) fluctuate together. This is what is predicted by the simplest inflationary models. But in some inflationary models and in topological defect models, it can also happen that at some epoch the total density is unperturbed but there are isocurvature fluctuations in some components relative to others. In general, there are as many modes of fluctuations as there are different components in the mass-energy density.

There is no problem in physics, no matter how complicated, which cannot be made still more complicated. But of course that should not be our goal! Although many cosmological models have been considered by various authors, we propose to concentrate here on a particular class of such models, namely those inspired by the hypotheses of inflation (hence with $\Omega = 1$, or at least curvature $k = 0$, and a near-Zel'dovich primordial spectrum of adiabatic fluctuations) and (all or mostly) cold dark matter. We do this not only because we believe that such models still have the best prospects of ultimately being found to be consistent with the data. Our main motivation for concentrating on these models is that they are *well specified*, in the sense that they are described by a small number of adjustable parameters (unlike general non-Gaussian models, for example), and very *predictive*, in the sense that many of their consequences can be worked out fairly easily with relatively few uncontrolled approximations. Thus they can be confronted rather directly with the observational data, and eventually most (or all) such theories can actually be ruled out — as standard CDM already has been (or is close to being) ruled out. If a small class of such models survive, they may actually have something to do with the real universe. Even if not, this research should help to develop better statistical methods for comparison of cosmological theories with observational data. The other class of models that we will consider are topological defect models, because they are the best-studied representative of the positive-energy, non-Gaussian, isocurvature class of models. The main other models discussed at the Snowmass workshop were low-Omega CDM without a cosmological constant (see e.g. Ratra and Peebles 1994, Kamionkowski et al. 1994). The best hope for a universe made entirely of ordinary matter was the isocurvature baryon models (see e.g. Cen et al. 1993), but these now appear to be in grave difficulty (e.g. Chiba et al. 1994).

The great advantage of keeping a tentative theory in mind as one thinks about data is that it helps in organizing the facts. If it is a good theory, it will also call attention to particularly important facts — especially those that may contradict it! CDM stimulated the creation of better models of the origin and evolution of galaxies and large scale structure, it helped motivate the acquisition and analysis of crucial data, and it has been a valuable test bed for data analysis tools — allowing, for example, development and testing of the POTENT algorithm (Dekel, Bertschinger, & Faber 1990) for reconstructing the total density field from measured peculiar velocities. Comparison of the original standard $\Omega = 1$ CDM model and its variants (cf. Liddle & Lyth 1993 for a review) with the observational data has certainly been useful during the past decade.

§ 3 Testing Models

It is useful to divide the discussion of how to confront models with the observational data according to the scale of the observations: (a) greater than 100 Mpc, (b) 10-100 Mpc, (c) less than 10 Mpc, and (d) early structure formation. The data on the largest scales is probably useful mainly for establishing the normalization of the fluctuations, measuring the contribution of gravity waves, and testing inflation. Comparing with the data on smaller scales tests the gravitational clustering hypothesis, and — it it is valid — the spectral shape and other features of the cosmological model, including the cosmological parameters H_0, Λ, and Ω, and the nature of the dark matter — e.g., whether it is a mixture of cold and hot dark matter. With a given normalization of the spectrum, the smaller scale data also tests the shape of the primordial spectrum (e.g., whether there is "tilt"). But there are many problems with actually carrying out this program even when large-scale redshift surveys become available, perhaps the worst of which is the uncertainties about galaxy formation which make identification of "galaxies" in the simulations uncertain. Fortunately, there are several ingenious new techniques that promise to improve this situation. For example, weak gravitational lensing (e.g. Kaiser and Squires 1993) or extension of peculiar velocity surveys to larger scales (which requires new ways of measuring distance independent of redshift, or of measuring the peculiar velocity directly e.g. by the Sunyaev-Zel'dovich velocity term — see Wilbanks et al. 1994) may allow direct determination of the distribution of mass on large or even very large scales.

3.1 Very large scales

The shape of the power spectrum deduced from the two-year COBE data and other large-angle CBR measurements is consistent with a power-law power spectrum with $n_p \approx 1.1 \pm 0.2$, and with the rms amplitude quoted as the quadrupole for an $n_p = 1$ spectrum given by $Q_{\text{rms-ps}} \approx 20\,\mu\text{K}$ (Gorski et al. 1994, Bunn et al. 1995).

If a CDM spectrum is normalized to this amplitude for $h = 0.5$, it appears to be consistent with all the available data on large scale structure (LSS) down to scales of approximately 100 h^{-1} Mpc (e.g., Peacock & Dodds 1994, Dekel 1994). But the amplitude of the LSS is presently known at best to about $\pm 25\%$ — that is roughly the uncertainty on the large-scale bulk velocity, for example (e.g., Dekel 1994), which at least measures the mass power spectrum. The amplitude of the galaxy or cluster power spectrum has the further uncertainty that the ratio of the rms fluctuation amplitudes of these objects to that of the mass, i.e. the "bias" $b_{\rm gal} = (\frac{\delta\rho}{\rho})_{\rm gal}/(\frac{\delta\rho}{\rho})_{\rm mass}$, is known rather poorly. Indeed, the extent to which the bias of a given category of astronomical objects can be regarded as a constant even on a given scale is not very well tested.

Within these fairly large uncertainties, the consistency between the CMB data and the LSS data supports the validity of the gravitational clustering hypothesis. In order to really test this both this and the hypothesis of cosmic inflation, it will be necessary to do better. Perhaps the most important issue is the contribution of gravity waves. These tensor modes can contribute to the low spherical harmonics $\ell \lesssim 20$ of the CMB temperature fluctuations, but hardly at all to the higher ℓ ones; and they are of course irrelevant to the formation of structure, to which only the scalar mode density fluctuations contribute. In principle, the tensor contribution can be determined by comparing the large and small angle CMB anisotropies, but in practice this will require more accurate CMB measurements on small scales than are presently available, and also knowledge of Ω_b and the extent of ionization of the universe after the recombination era, both of which have a strong influence on the amplitude of the CMB fluctuations near the first Doppler peak, $\ell \sim 200$. For the time being, it is perhaps best to regard the COBE normalization $Q_{\rm rms-ps} \approx 20$ as an *upper* limit to the density fluctuation amplitude, since the tensor and scalar modes add in quadruature. One of the most urgent issues for CMB studies is to determine a *lower* limit on the density fluctuation amplitude by constraining the tensor mode amplitude. This will allow improved tests of the gravitational clustering hypothesis, and measurement of the "tilt" of the primordial fluctuation amplitude.

All the nearby surveys, such as the CfA2 survey with an effective depth of about 15,000 km/s, have found structures as large as the surveys themselves. This left open the possibility that still larger structures would be found by even larger surveys, which would contradict the gravitational clustering hypothesis (e.g. Piran et al. 1993). Although the very large scale periodicity of peaks in the galaxy distribution with a length scale of $\sim 135\,h^{-1}$ Mpc seen in the BEKS (Broadhurst et al. 1990, Szalay et al. 1993) pencil beam redshift survey was unexpected in any cosmological model, it is significant that no indications of still larger scales were seen in this data (preliminary reports indicate that pencil beams in different directions also have peaks with such separations but not such strong periodicity). Large scale redshift surveys are now in progress, notably the KOSS southern sky redshift survey and the ESO Key Project. Preliminary reports suggest that no larger structures have in fact been found (Kirshner characterizes this as "the end of greatness"), again supporting the gravitational clustering hypothesis. The much larger scale surveys just beginning — the Two Degree Survey at the AAT, and the Sloan Digital Sky Survey at the Apache Point Observatory — will be able to measure the sizes of these large structures and characterize their correlations, shapes, and other statistical properties. These will provide essential constraints on models of cosmic structure formation. These statistical properties appear on the basis of the data available at present to be consistent with the expectations from CDM-type models, but it remains to be seen whether this is true for topological defect models.

3.2 Large scale structure ($\sim 10 - 100$ Mpc)

On these scales a great deal of galaxy redshift data (Giovanelli & Haynes 1991) and peculiar velocity data (Dekel 1994) is already available, although much of it remains unpublished. There are also several redshift surveys for optically selected clusters, and large-scale redshift surveys for X-ray selected clusters (which are likely to be less affected by projection effects and galactic obscuration) are now in progress. Comparison of the spectrum of fluctuations measured with this data and with the small-angle CMB data when it is available will eventually provide a test of the gravitational clustering hypothesis.

Comparison with specific theories must be done on the basis of nonlinear simulations since on these scales linear theory is no longer reliable. All the available tests — power spectra of galaxies and mass (a preliminary POTENT result), galaxy and cluster correlations, skewness and higher moments of the pdf — suggest that CDM, normalized to fit on scales of 100 Mpc and above, fits increasingly poorly on smaller scales: it has too much power. For example, CDM predicts far too many clusters (White et al. 1993), and it predicts that mass autocorrelations become negative for separations beyond about 30 h^{-1} Mpc, while all measurements of cluster correlations show that they remain significantly positive for separations out to $\sim 50\ h^{-1}$ Mpc. These correlations are sensitive to the slope of the power spectrum, and indicate a steeper decline with increasing k than CDM.

Many simulations have been done of inflation-inspired cosmological models of the CDM variety, with adequate resolution to allow at least rough predictions of the locations and velocities of galaxies. It is urgent that similar simulations be done for a wider class of models, including particularly cosmic defect models. These models have several features that make them harder to simulate than CDM-type models; for example, the existence of small-scale large-amplitude structures (e.g. string wakes) even at high redshift, which may require inclusion of hydrodynamics, or the existence of more than one relevant scale. As discussed at Snowmass in connection with CMB data, the versions of cosmic defect models in which it was assumed that the universe would remain at least partly ionized after the recombination era – because of radiation from stars or other collapsed objects formed in early structures – are inconsistent with the indications of a Doppler peak in the CMB data. It will be interesting to see whether more realistic treatments of cosmic defect models will do better in this regard, and what sort of galaxy distribution and velocities they predict at various redshifts.

3.3 Intermediate scales (less than 10 Mpc)

It is on scales less than about 10 Mpc that the greatest differences exist between models that have all been normalized to agree on the largest scales with the COBE data. It is these scales on which galaxy locations and velocities, as revealed by relatively dense redshift surveys (i.e., with fainter galaxies included), have the greatest potential to help discriminate between cosmological models, for example those containing more or less of various mixtures of cold and hot dark matter, with or without a cosmological constant. (Someday there may be enough galaxy peculiar velocities based on accurate distance measurements independent of redshift to allow these to be used to discriminate between theories on small scales, but for the time being it remains necessary to smooth peculiar velocity data over scales of at least 5 Mpc to overcome the large uncertainty in each such measurement.) The statistics that have been used for this purpose include N-point functions, the void probability function VPF and related functions, skewness and kurtosis coefficients S_3 and S_4, multifractal analyses, the genus density, etc. These statistics indicate that galaxies exhibit hierarchical scaling as expected in gravitational clustering models, but most of these statistics (with the possible exception of the VPF, see e.g. Ghigna et al. 1994) do not appear to be able to discriminate very efficiently in redshift space between alternative Gaussian models – although they may discriminate between these and non-Gaussian (e.g., Coles et al. 1993) or scale-dependent-bias models (e.g., Frieman & Gaztenaga 1994, Fry & Gaztenaga 1994).

Simulations are of random patches of the universe, so comparisons with observational data must be statistical. There are broadly two different approaches to making such comparisons; one can work either in the "theoretical plane" – i.e., attempt to "correct" the data for selection effects, redshift space effects, etc. – or in the "observational plane" – i.e., "observe" the simulations. As computational power has grown, it has become increasingly advisable to observe the simulations rather than attempt to correct the data, since simulations have much more information – for example, the velocity as well as location of each object identified as a galaxy. Thus it is possible to construct magnitude-limited redshift surveys from simulations with no ambiguity in the conversion to redshift space, but it is more difficult to recover real space information from only redshift space data.

On the other hand, one should not underestimate the difficulties of simulating observational data. Perhaps the greatest problem is determining which objects in the simulations are to be identified with observed galaxies. In dissipationless simulations, with only dark matter included, perhaps the worst problem is overmerging. Nearby dark matter halos merge into large blobs, even though in the real universe the individual galaxies within a group or cluster can retain their separate identities since the gas can condense a great deal within the larger dark matter halos. Even in hydrodynamical simulations (e.g. Evrard, Summers, and Davis 1994), there are serious limitations; only limited spatial resolution is available with even the largest supercomputers, and many relevant physical processes such as energy input from stars and supernovae are neglected or treated superficially. Although the main strengths and limitations of the several different approaches to hydrodynamical simulations seem to be reasonably well understood (see e.g. Kang et al. 1994), the accuracy and resolution currently currently available is limited.

The necessary art, at the present stage of cosmology, is to invent statistical tests that are both robust and discriminatory. Robust, in the sense that the difficulties of the sort mentioned above – e.g. in galaxy identification or "illumination" (assignment of luminosity to objects identified as galaxies) – do not significantly affect the statistical results. And discriminatory, in the sense that the statistical tests give significantly different results for the various cosmological models that are of interest. That a given statistical test is actually robust and discriminatory can be checked by trying a wide variety of different approaches to galaxy identification and illumination of simulations of a number of different cosmological models (for an example of such statistics, see Nolthenius et al. 1994, 1995).

3.4 Early structure formation

A major difference between cosmological models is in their predictions for the origins of galaxies, clusters, and large scale structure, and the evolution of these with redshift.

Detection of large-mass collapsed objects at high redshift would certainly be contrary to the predictions of models with $\Omega = 1$, especially models such as CHDM in which small-scale fluctuation power is significantly suppressed compared to standard CDM. For example. a possible detection of a large HI cloud was reported but not confirmed; the upgraded Arecibo telescope and the Giant Meter-wave Radio Telescope will soon provide sensitive tools for searching for such clouds at high redshift. Detections of galaxy clusters and quasar superclusters at redshifts $z \sim 2$ have also been reported, and there are some remarkable HST WFPC2 pictures of clusters at redshifts up to nearly unity (e.g. Dressler et al. 1994). Although the striking differences between the galaxies in such clusters and those at lower redshift support earlier indications that cluster galaxies have evolved significantly since redshift 1, galaxies in the field appear to have evolved less dramatically since the universe was half its present age (e.g., Steidel et al. 1994). The most useful data would provide indications of the number densities of the objects (e.g. galaxies or clusters) considered at various redshifts as a function of their mass (indicated for example by internal velocity dispersion or gas temperature), since that is what simulations predict most directly.

One of the most useful data sets for comparison with theories of structure formation is provided by absorption lines in the spectra of high-redshift quasars. The absorption systems with the highest density of neutral gas – known as damped Lyman α absorption systems – are presumably protogalaxies, and the quantity of gas in such systems at redshift $z \sim 3$ is roughly the same as the amount of ordinary matter in all the stars and gas that we can see in the universe at the present epoch (see e.g. Wolfe 1993). An important question that will discriminate strongly between various cosmological models is whether the quantity of gas in such systems peaks at redshifts about 3-4, as expected in models where structure forms relatively late such as CHDM (see e.g. Klypin et al. 1995), or increases to still higher redshift, as expected in models such as ΛCDM where structure forms significantly earlier.

§ 4 Needs for the Future

Perhaps the most important information needed as a basis for constructing the first fundamental theory of cosmology is the values of the fundamental cosmological parameters, especially H_0, Ω, and Λ. (The status of various programs to measure these parameters is summarized in another section.) Below we summarize a number of other areas in which progress is needed.

Bigger computers, faster and cleverer codes, shared software. The greatest dynamical range in force resolution currently available is only a little better than three orders of magnitude in dissipationless simulations, worse in hydrodynamical simulations. This means that in a simulation with a 100 Mpc box, not large enough to simulate large surveys and large structures such as the Great Wall, the resolution is not much better than 100 kpc, an order of magnitude larger than the visible parts of galaxies. Moreover, there is a tradeoff between mass and force resolution: codes that permit better force resolution use fewer particles and thus have poorer mass resolution. A few groups such as the Grand Challenge Cosmology Consortium (GC^3) have recently devoted a great deal of effort to developing new and improved codes that exploit the new generation of massively parallel supercomputers that is now becoming available. It is very desirable that these codes become widely available so that the entire cosmology community can benefit. New technologies for visualization of the results of supercomputer calculations also hold considerable promise.

Cosmic variance. One of the worst problems with all simulations is cosmic variance; since only a random patch of universe is simulated, a number of such simulations must be compared to a number of regions for which galaxy survey data is available, and it is not clear that even the largest redshift surveys yet completed provide fair samples. This problem is exacerbated by the fact that, as many calculations have shown, there is a feed-down of effects from large structures to small; for example, rms pairwise velocities are strongly influenced by the presence of relatively rare rich clusters of galaxies (see e.g. da Costa et al. 1994). Controlling cosmic variance is one of the most important challenges for current theory, until really large simulations can be compared to really large data sets such as the SDSS.

Better understanding of processes leading to galaxy formation. Since galaxy formation includes a number of generations of stars, including supernovae and the resulting chemical evolution, and perhaps often also involves more exotic objects such as the massive black holes thought to power active galactic nuclei, developing a secure understanding of these formation processes is likely to take a long time. It is even possible that a general theory of cosmology, including at least a general outline of the initial conditions and the nature of the dark matter, will precede rather than follow a detailed understanding of galaxy formation. No doubt a great deal of data on galaxies in both early and intermediate stages of formation will be necessary. Fortunately, such data is now coming from the new generation of great telescopes in space and on the ground. But present-epoch galaxies are brightest in the near infrared, and higher-redshift galaxies are expected to be brightest in the several micron band which can only be accessed from space. A large space infrared telescope such as SIRTF has long been seen as a high priority for astronomy, and the need for it was reiterated several times during the Snowmass workshop. Meanwhile, we will need to make use of even indirect data such as the amount of extragalactic background infrared light from early galaxies, which can perhaps be probed by its absorption of TeV photons from AGNs via pair production (MacMinn and Primack, 1995).

Better ways of comparing observational data with models. We need new and better statistical tests, which are both *robust* against the difficulties of galaxy identification in simulations and the biases and selection effects always present in survey data, and *discriminatory* between the classes of cosmological models of interest. On the whole, it is probably better to compare theory with data by "observing" simulations rather than "correcting" data. It is very desirable that standard software become available to theorists and observers, so that standard versions of various statistics can be tried on many datasets from simulations and observations.

Better access to observational and simulation data. It is unfortunate that the only redshift survey covering a reasonably large volume which is publicly available is the 1982 CfA1 survey. Many papers have been published analyzing data from newer and larger redshift surveys in the years since then, but the redshift data remains largely unavailable. It is also desirable that simulation data (e.g. catalogs of objects identified as galaxies) be made available. The journals and funding agencies should ask a committee of observers and theorists to establish reasonable rules regarding access to such data – for example, all data used for a given paper must be made publicly accessible within one year of the publication of the paper – and ask referees to help enforce these rules. The POTENT group has set a good example of the sort of public access advocated here, by making their peculiar velocity dataset available in a timely way and in convenient form.

Need for NSF Astrophysics and ParticleAstro theory program. Recent ten-year National Academy survey reports on Astronomy have called for the creation of programs in theoretical astrophysics in the funding agencies, especially the NSF, in parallel with the NSF theoretical physics program. The most recent NAS report, the Bahcall Report, even included an Appendix on this. Without such a program, theorists, who tend to work in several areas of astrophysics and sometimes also particle physics, must compete for scarce funds with observational astronomers. We would like to repeat this message here. Of course, in addition to increased funding for theoretical astrophysics, alleviating the current severe funding crisis in astronomy will require increased funding for the entire field. For cosmology in particular, the present period is likely to be regarded as the golden age, when for the first time an overall cosmological theory may be within our reach. A modest increase in the quantity and quality of support is likely to pay great dividends in such a period.

§ References

Babul, A. et al. 1994, ApJ, 427, 1.

Bernardeau, F., Juszkiewicz, R., & Dekel, A. preprint 1994

Bonometto, S.A., Gabbiani, F., & Masiero, A. 1994, Phys. Rev. D49, 3918.

Broadhurst, T.J., et al. 1990, Nature 343, 726.

Bunn, E.F., Scott, D., & White, M. 1995, ApJ 441, L9.

Cen, R.Y., Ostriker, J.P., & Peebles, P.J.E. 1993, ApJ 415, 423.

Chiba, T., Sugiyama, N., & Suto, Y. 1994, ApJ 429, 427.

Coles, P. et al. 1993, MNRAS, 264, 749.

da Costa, L.N., Vogeley, M.S., & Geller, M.J. 1994, ApJ 437, L1.

Davis, M., Efstathiou, G., Frenk, C., & White, S. D. M. 1985, Ap. J. 292, 371.

Davis, M., Efstathiou, G., Frenk, C., & White, S. D. M. 1992, Nature 356, 489.

Dekel, A., Bertschinger, E. & Faber, S. 1990, ApJ 364, 349.

Dekel, A. 1994, Ann. Rev. Astron. Astrophy. 32, 371.

Dressler, A. et al. 1994, ApJ 435, L23

Evrard, A.E., Summers, F.J., & Davis, M. 1994, ApJ 422, 11

Frieman, J.A., & Gaztanaga, E. 1994 ApJ 425, 392.

Fry, J.N., & Gaztanaga, E. 1994, ApJ 425, 1.

Ghigna, S. et al. 1994 ApJ 437, L71

Giovanelli, R., & Haynes, M.P. 1991, Ann. Rev. Astron. Astroph. 29, 499

Gorski, K.M. et al. 1994, ApJ 430, L89

Kaiser, N., & Squires, G. 1993, ApJ 404, 441.

Kamionkowski, M., et al. 1994, ApJ 434, L1.

Kang, H. et al. 1994, ApJ 430, 83.

Klypin, A., et al. 1995, ApJ (in press).

Kofman, L., et al. 1991, in *Proc. Workshop on Large Scale Structure and Peculiar Motions in the Universe*, D.W. Latham & L.N. da Costa, eds. (Astron. Soc. of the Pacific), p. 251.

Kofman, L., et al. 1994, ApJ 420, 44.

Liddle, A. R. & Lyth, D. 1993, Phys. Rep. 231, 1.

MacMinn, D. & Primack, J.R. 1995, Space Sci. Rev. in press.

Nolthenius, R., Klypin, A., & Primack, J.R. 1994, ApJ 422, L45.

Nolthenius, R., Klypin, A., & Primack, J.R. 1995, ApJ in press.

Nusser, A., & Dekel, A. 1993, ApJ 405, 437.

Ostriker, J.P. 1993, Ann. Rev. Astron. Astroph. 31, 689.

Peacock, J.A., & Dodds, S.J. 1994, MNRAS 267, 1020.

Piran, T. et al. 1993, MNRAS 265, 681.

Ratra, B., & Peebles, P.J.E. 1994, ApJ 432, L5.

Saunders, W. et al. 1990, Nature 349, 32.

Starkman, G.D., Kaiser, N., & Malaney R.A., 1994, ApJ, 434, 12.

Steidel, C.C., Dickinson, M., & Persson, S.E. 1994, ApJ 437, L75.

Szalay, A.S. 1993, Proc. NAS 90, 4853.

Vogeley, M.S., et al. 1994, ApJ 420, 525.

White, S.D.M., Efstathiou, G., & Frenk, C.S. 1993, MNRAS 262, 1023.

Wilbanks, T.M., et al. 1994, ApJ 427, L75.

Wolfe, A.M. 1993, *Texas/PASCOS Proceedings*, Ann NY Acad Sci 688, 281.

THE SEARCH FOR PROTON DECAY

Todd Haines

Physics Division, Los Alamos National Laboratory

Kenji Kaneyuki

Tokyo Institute of Technology

Clark McGrew

Department of Physics, University of California, Irvine

Rabindra Mohapatra

Department of Physics, University of Maryland

Earl Peterson

Department of Physics, University of Minnesota

David B. Cline

Department of Physics, University of California, Los Angeles

June 20, 1995

The conservation of the quantum number called baryon number, like lepton (or family) number, is an emperical fact even though there are very good reasons to expect otherwise. Experimentalists have been searching for baryon number violating decays of the proton and neutron for decades now without success. Theorists have evolved deep understanding of the relationship between the natural forces in the development of various Grand Unified Theories (GUTs) that nearly universally predict baryon number violating proton decay, or related phenomena like n- $\overline{n}$ oscillations. With this in mind, the Proton Decay Working Group reviewed the current experimental and theoretical status of the search for baryon number violation with an eye to the advancement in the next decade.

§1 Introduction

The (particle physics) standard model is remarkably successful at describing all observed reactions among elementary particles, withstanding the tests of dozens of experiments over the last decade or so. It is commonly held that the standard model is not a complete theory of particle interactions, but instead the low-energy limit of a complete theory (such as a GUT) that has the particles and their interactions unified at a very high-energy scale. The search for new particles or new interactions of known particles is perhaps the only way that a deeper understanding will emerge.

The search for proton decay (baryon number violation) has occupied experimenters for decades, unfortunately without success. General theoretical principles suggest that baryon number may not be a conserved quantum number. If it were, there ought to be a force (and gauge boson) that couples to baryon number just as the electromagnetic force couples to electric charge; no such composition dependant force has ever been observed. We therefore expect that baryon number violating decays or oscillations will occur, although with unknown frequency. It is with this in mind that the subgroup met to discuss the current status of the search for proton decay and look at the progress that might be made in the next decade. This summary of those meetings begins with a brief theoretical overview. There are then synopses of the proton decay searches from the active experimental groups.

§2 A Theoretical Perspective

As the standard model of electroweak interactions based on the gauge symmetry $SU(3)_c \times SU(2)_L \times U(1)_Y$ keeps accumulating its trophies of success from one experiment after another, a key question facing particle physicists today is the nature of new physics beyond the standard model. There are many unanswered questions in the standard model such as the stability of the weak scale, origin of fermion mass hierarchies, solution to the strong CP problem, origin of matter in the universe, massive neutrinos, etc., that make it imperative that there be new physics beyond the weak scale. Specific issues of interest for this working group are twofold: (i) first, are there compelling scenarios of new physics that contain baryon violation as an integral part and (ii) second, do they lead to *predictions* for proton lifetime that are accessible to the next round of experiments? In this section, we will argue that the answers to the first two questions are definitely *yes*. We will then argue that there are baryon non-conserving signatures in near-electroweak TeV scale theories that involve low mass lepto-quarks as well as diquarks; therefore searching for those decay modes will complement the ongoing search for lepto-quarks in collider experiments.

Before discussing the detailed predictions/expectations for proton lifetime in various scenarios of new physics, we wish to point out the importance of the selection rules in $\Delta B \neq 0$ processes as a "barometer" of the scale of new physics [1, 2, 3, 4]. The point is that any baryon violating operator that arises from physics beyond the standard model must respect $SU(3)_c \times SU(2)_L \times U(1)_Y$ invariance and in the effective field theory below the TeV scale, these operators will manifest themselves as non-renormalizable operators. Denoting the quark and lepton doublets by Q and L and the singlets by u^c, d^c *and* e^c, one finds that in non-supersymmetric theories, the lowest dimensional operator (i. e. $QQQL$ etc.) has dimension $d = 6$ and conserves $B - L$ quantum number. It therefore scales like M^{-2} and the present limits on proton lifetime imply that $M \geq 10^{15}$ GeV or so. Thus observation of proton decay modes that respect the $B-L$ quantum number (e. g. $p \to e^+ + \pi^0$) will be a proof of the existence of new physics at scales of order of the typical grand unification scale ($\simeq 10^{15}$ GeV). On the other hand, there are processes such as neutron-anti-neutron oscillation[2], $p \to \ell\ell\ell$ decay[3], double proton decay (i.e. $pp \to e^+e^+$)[4] which scale respectively as M^{-5}, M^{-5} and M^{-8} and therefore probe new physics at much lower scales such as 100 TeV to a few TeV respectively. Therefore discovery of any baryon non-conserving process thru its selection rule will provide evidence not only for new physics but also about the scale of this new physics.

2.1 Grand unification and proton lifetime; Generalities:

Grand unified theories (GUTs) are attractive from many points of view such as unification of forces (and therefore the gauge symmetries and couplings), unification of fermions (which provide the hope to understand the hierarchy and the structure observed in the fermion sector), understanding of electric charge quantization(ECQ) and origin of matter etc. After an initial explosion of interest in GUTs in the late 70's and early 80's, lack of any evidence for proton decay as well as the emergence of new theoretical ideas which could explain both the origin of matter as well as the quantization of electric charge led to a lull in the activity in this field.

A recent revival of interest in GUTs has been caused for two reasons: one is the precision measurement of the three gauge couplings at LEP and second, the indications of possible non-vanishing masses for neutrinos in order to understand the deficit neutrinos from the sun. Two amazing facts emerge from the precision measurements of the three gauge couplings: one is that if we assume the scale of supersymmetry breaking to be at the weak scale (in the TeV range), then the rate of variation of the gauge couplings with energy (or the slopes of the beta function) is completely fixed and implies a unification of couplings at a scale of about 2×10^{16} GeV; alternatively, if we take a nonsupersymmetric theory but supplement it with the hypothesis that the gauge symmetry expands at some scale M_I to $SU(2)_L \times SU(2)_R \times G$ where $G \equiv U(1)_{B-L} \times SU(3)_c$ or $SU(4)_c$, then again the the slopes of the beta function are fixed and unification of gauge couplings occur at about the same scale as above with $M_I \simeq 10^{12}$ GeV or so. The reason the first result is of interest is that the SUSY scale being around a Tev is one attractive way to stabilize the weak scale against radiative corrections. On the other hand, if solar neutrino results are an indication of non-vanishing neutrino masses, the values of neutrino masses that fit the solar neutrino data in the very elegant MSW scheme imply (via the so-called see-saw mechanism) a scale of $B - L$ symmetry breaking which is precisely in the range predicted by LEP data in conjunction with an SO(10) grand unified theory.

Taking these arguments seriously, one is led to contemplate two possible scenarios for grand unification: one is the supersymmetric SU(5)[5] and the second one is non-supersymmetric SO(10). The predictions for proton lifetime in both these theories has been studied extensively in recent days (for SUSY-SU(5) in [6] and for SO(10) in[7]).

2.2 SUSY SU(5) and proton decay:

In general supersymmetric theories, the existence of superpartners of the quarks and leptons allows one to write down a dimension 5 operator that leads to $\Delta B \neq 0$ and $\Delta(B - L) = 0$ processes. In the context of the specific SUSY SU(5) model, these operators arise from the exchange of heavy color triplet Higgsinos and the above mentioned dimensionality means that they scale like $\simeq {M_U}^{-1}$. As a result, they dominate over the conventional gauge boson exchange graphs, which gave the dominant proton decay mode in non-SUSY theories such as $SU(5)$ and moreover, the symmetry properties of the $d = 5$ operator imply that the $B-L$ conserving decay mode so familiar in non-SUSY theories such as $SU(5)$ are not allowed; instead the leading decay mode becomes $p \to K^+\bar{\nu}_\mu$. The box graph that converts the raw dim 5 operator into a four-Fermi operator involves the exchange of gluinos, charginos etc. It has been shown by Arnowitt and Nath[6] that the dominant graph arises from chargino exchange and is directly proportional to the chargino mass. So a heavier chargino would lead to a more unstable proton. This dependence has been shown for proton lifetime in fig.1. One can conclude from this figure that if the decay mode $p \to K^+\bar{\nu}_\mu$ is not seen at the level of $\tau_p = 5 \times 10^{33}$ *years*, then the chargino must be light enough to be seen at LEP II; otherwise, SUSY-SU(5) will be ruled out. One caveat is that the minimal SUSY-SU(5) which leads to all these conclusions does not have a realistic fermion spectrum (e.g. it predicts the experimentally inconsistent relation $m_e/m_\mu = m_d/m_s$) and any extension of the model to correct this feature will introduce uncertainties into these predictions since unlike in the non-SUSY GUT models the proton decay predictions arise from the same coupligs in the theory that predict fermion masses.

2.3 Non-SUSY SO(10) and proton decay:

As mentioned earlier, an alternative scenario that leads to unification of gauge couplings with a well-motivated physics input is the non-SUSY SO(10). In the minimal version of the theory, there are four possible models depending on the four possible intermediate symmetries, of the form $SU(2)_L \times SU(2)_R \times G_a$ (where G_a = *(I)* $SU(4)_c \times D$; *(II)* $SU(4)_c$ *(III)* $SU(3)_c \times U(1)_{B-L} \times D$ and *(IV)* $SU(3)_c \times U(1)_{B-L}$, D denotes the discrete Z_2 gauge symmetry which is contained in SO(10)) that SO(10) can break to, depending on the choice of Higgs multiplets . All these models have enough free parameters in the Yukawa couplings so that all fermion masses can be accomodated. The dominant proton decay mode in this model is the classic decay mode $p \to e^+\pi^0$; the predictions for proton lifetime in this case has been studied in two papers recently[7] and the results are (quoted from the first paper of ref. 7):

$$Model\ I : \tau_p = 1.44 \times 10^{32.1\pm.7\pm1\pm1.8} years$$

$$Model\ II : \tau_p = 1.44 \times 10^{37.4\pm.7\pm1^{+.5}_{-5}} years$$

$$Model\ III : \tau_p = 1.44 \times 10^{34.2\pm.7\pm.8\pm1.7} years$$

$$Model\ IV : \tau_p = 1.44 \times 10^{37.7\pm.7\pm.9^{+.5}_{-2}} years$$

In the above predictions, the first uncertainty is due to hadronic matrix elements, second is due to uncertainty in the gauge couplings and the last one is due to the heavy particle threshold corrections. In case I, we have presented an extremely conservative estimate of the threshold uncertainties but it is more likely to be of order $\simeq .2$ so that, if no evidence for proton decay mode $p \to e^+\pi^0$ is found at SuperKamiokande, then this model will be ruled out. Another signature of the inherent left-right symmetry of the SO(10) models is the equality of the branching ratios $B(p \to e^+\pi^0) = B(p \to \bar{\nu}\pi^+)$.

Finally, there is a different appraoach to minimal SO(10)[8], where a one step breaking of SO(10) down to standard model is made consistent with unification and LEP data by using the heavy particle threshold corrections. A prediction of this model is that proton lifetime must be less than 10^{34} years after all threshold corrections are taken into account. Therefore, this model will also be ruled out if no proton decay is seen at SuperKamiokande at the expected level. We conclude this section by pointing out that even null results at SuperKamiokande and ICARUS are going to provide us with important information about the future directions in grand unified theories by ruling out several very interesting GUT models.

2.4 Baryon non-conservation as a probe of new physics at low scale:

Baryon number violating processes mediated by higher dimensional operators (such as neutron-anti-neutron oscillation, $pp \to e^+e^+$ decay or $p \to \ell\ell\ell$ decay) provide important information about physics beyond the standard model which supplement that gained from the study of the B-L conserving decay models discussed earlier. Since such processes become observable if there is new physics in the mass ranges of one to a hundred TeV, in the minimal versions of the GUT theories discussed, such processes would be unobservable. Thus any evidence for them would be an indication of physics at a nearby scale and would be an indication against simple GUT theories. Furthermore, these processes invariably arise from the exchange of either diquark Higgs bosons or leptoquark Higgs bosons of TeV range mass; therefore, any positive evidence for the any of the above processes is an indication of the existence of low mass leptoquarks /or diquarks. It is possible to construct consistent and interesting low mass scale theories [3, 2, 4] where these higher dimensional operators can arise with observable strength for natural values of the parameters.

§3 Experimental Search with IMB

IMB-3 was a large ring imaging water Čerenkov detector with a fiducial mass of 3.3 kton. During a 7.6 kton year exposure, 935 contained events were collected at a rate of approximately one event per day. This sample is largely consistent with the predicted atmospheric neutrino flux and is been used to set preliminary limits on the nucleon decay rate[10].

Unfortunately, there is significant uncertainty in the estimated background due to atmospheric neutrinos. This uncertainty comes in two forms, experimental systematic uncertainties and fundamental theoretical uncertainties. The experimental systematics have been estimated to be ~20%. The fundamental theoretical uncertainties are highlighted by the current discussion of the "missing" muon type events in

IMB-3 and Kamiokande[10, 9]. In particular, IMB-3 measures the ratio of muon neutrino type events to electron neutrino type events to be 0.54 ± 0.06 when the expected ratio is 1.04 ± 0.2. This may reflect a fundamental uncertainty in the neutrino flux at the detectors, or some new physical phenomenon like neutrino oscillations.

Since there is a large uncertainty in the neutrino flux at the detector, results are presented for two different assumptions. The first assumption is that the measured muon to electron ratio represents the neutrino flux present at the detector, e.g. due to a mistake in the predicted neutrino flux. The second assumption is that the ratio of muon neutrinos to electron neutrinos present at the detector has been correctly predicted, e.g. the measured ratio is due to some experimental error. Preliminary results are presented in table 1[11]. A careful examination of table 1 shows that for many modes the statistical uncertainty is less that the systematic uncertainty in the background estimate. Unless this situation is resolved, background subtracted searches for nucleon decay will be severely hampered in future nucleon decay detectors.

§4 The Status of Soudan-II Search

The Soudan 2 detector is a calorimeter with 1 meter drift tubes occupying the voids in a stack of corrugated 1 meter square 1.6mm thick steel plates. A proportional wire system measures the charge deposited in the tubes. The detector consists of 224 modules each containing 240 layers of steel and tubes, with a total mass of 963 metric tons. Although the final modules were installed in November of 1993, the partially-completed detector has been acquiring data on throughgoing muons, atmospheric neutrino interactions and nucleon decay since 1989, when 1/4 of the final detector was operational. Approximately 1.7 Kty of exposure has been accumulated to date (July 1994), but only the first 0.5 Kty portion of this has been analyzed for possible nucleon decay signatures.

After the data have been scanned for contained event selection (the basic requirements are that the event come no closer than 20cm to the exterior of the detector and that no in-time shield hit be recorded), the events are sorted into various topological categories. Each distinct topology is then tested against a possible nucleon decay mode. Only three possible combinations of mode and topology have been analyzed thus far:

$$p \to \nu K^+ (K^+ \to \nu\mu^+, \mu^+ \to \nu\nu e^+)$$
$$n \to e^- K^+ (K^+ \to as\ above) \text{ and}$$
$$n \to \mu^- K^+ (K^+ \to as\ above).$$
(We assume the μ^- is captured before decay.).

The first of these has a 3-"track" topology, while the latter is 4-"track". The full power of the detector in the form of hit counting, track/shower separation and ionization measurement is then used to discriminate between these modes and possible neutrino-induced backgrounds.

Initially each event is rejected or accepted for further analysis based on a total hit requirement, a requirement that each track/shower assignment be appropriate, and that each real track direction is consistent with the mode. Further analyses then involve the measured kaon range and ionization, the muon range and ionization, the decay-electron's shower size, and the reconstructed momentum flow at both the K-mu and mu-e vertices. Four algorithms that utilize this information have been investigated: simple cuts (any failure rejects the candidate), a 7-dimensional nearest-neighbor analysis, a combined liklihood analysis and a neural-net determination. All of these lead to rejection conditions which can be eased (to improve the acceptance) or tightened (to reject more background). While all methods of analysis are comparable, the neural net technique is slightly superior to its rivals, and is the basis of the current (preliminary) limits (which have been set to reject background at the 1 event/5 Kty level): (1) $p \to \nu K^+$ with lifetime limit $> 4.5 \times 10^{30} yr$ and 0.10 events background, (2) $n \to e^- K^+$ with lifetime limit $> 7.5 \times 10^{30} yr$ and 0.03 events background, and (3)$n \to \mu^- K^+$ with lifetime limit $> 6.5 \times 10^{30} yr$ and 0.04 events background.

§5 Kamiokande Results and the Future: SuperKamiokande

Kamiokande is a ring imaging water Cherenkov detector that is still in operation. Its fiducial volume for nucleon decay analysis is 1,040 ton. A total of 948 20" PMT are attached inside the inner detector. The total detector exposure of $3.62kt-yr$ of KAMIOKANDE-II data are analyzed for $(B-L)$ violating nucleon decay processes. A total of 325 fully contained events were observed.

We analyzed 13 nucleon decay channels using total invariant mass, momentum imbalance and shape of momentum distributions. No statistically significant evidence of nucleon decay were obtained. Therefore, we set 90 % confidence level lower limits as shown in Table 2. We obtained 90 % C.L. life time limits as 3.3×10^{32} years ($p \to e^+\pi^0$) and 1.3×10^{32} years ($p \to \overline{\nu} K^+$) based on KAMIOKANDE-I+II $4.92kt-yr$ data on the $\Delta(B-L) = 0$ decay modes.

Presently the total detector exposure of Kamiokande is $7.7kt - yr$. Unfortunately, we don't have statistically significant evidence of nucleon decay and we will have at most only to 8.0 or $8.5kt-yr$ of data for the nucleon decay analysis.

Therefore, we are constructing SuperKamiokande. SuperKamiokande is a 50,000 ton water Cherenkov detector with 11,200 20" PMT. It is essentially just a scaled up version of Kamiokande. The inner detector volume is 32,000 ton, and 40% of the surface of the inner detector is covered by PMTs. The fiducial volume will be 22,000ton. SuperKamiokande will be accumulate events at 22 times rate of the present Kamiokande (and about 7 times that of IMB). We will start data taking in April 1996, and will have $100kt - yr$ of data at 2002.

The decay modes such as $e^+\gamma\gamma$... or $\mu^+\gamma\gamma$... will be background free in SuperKamiokande owing to its good momentum and position resolution. If we assume that ϵBm is same as KAM-II and neutrino B.G. is 0, the detector sensitivities on each decay mode are 10^{33-34} years. We will have to generate and analyze at least $1Mton \cdot yr$ or $10Mton \cdot yr$ of M.C. for these modes.

The nucleon decay mode, $p \to \overline{\nu} K^+$, is background limited mode in Kamiokande. However, this mode can be background free utilizing the 6.3MeV gamma coming from the decay of the daughter ^{15}N. The characteristics of this mode is 6.3MeV gamma and a delayed signal of 236 MeV μ^+. If we can distinguish γ from μ^+ with 10n sec time difference, the detector sensitivity will be 10^{33} years. There are a lot of possibilities to improve on detecting nucleon decay signal, such as $K_S(\pi^+, \pi^-)$, K_L with regeneration or charge exchange.

In conclusion, we will have $100kt - yr$ of data at the beginning of the next century, and will be accessible to 10^{34} years for $p \to e^+\pi^0$, and 10^{33} years for $p \to \overline{\nu} K^+$ with SuperKamiokande.

§6 The Future at ICARUS

The search for proton decay must continue. This being one of the few ways to probe the 10^{16} GeV energy scale. The ICARUS detector is partially designed as a dedicated proton decay detector.

ICARUS is ideally suited to take up such a challenge by providing both a large sensitive mass (5,000 tons of liquid argon per module) and a new detector technique which can best be thought of as a modern version of the bubble chamber. As will be described in detail below, a proton decay will provide unambiguous, background-free, spectacular, metre-long tracks easy to reconstruct with superb particle identification. The proton decay search will be carried out in many exclusive channels simultaneously. A clear advantage of the ICARUS technique is that discovery will be possible down to a single event, while any other existing or proposed experiment will have to rely on statistical background subtraction to eventually show the existence of a signal.

In view of the importance of the scientific issue at stake and armed with a technological breakthrough in experimental technique now to hand, the ICARUS collaboration considers the search for the proton decay as the priority of the experiment.

There exist classes of supergravity models which will obviously be tested by a combination of LEP and ICARUS, since they predict that either $\tau(\mathrm{p} \to \overline{\nu}\mathrm{K}^+)$ will be smaller than 1.5 x 10^{33} years (which is well within ICARUS range) or that one of the gauginos or one of the Higgs particles will be discovered at LEP200.

Particle identification benefits greatly from the ability to measure the ionization loss (dE/dx). In particular, using dE/dx versus range only, an excellent separation is obtained between pions and kaons. The separation between kaons and pions is obvious even here where we make use of the dE/dx information only. If one takes into account that, in addition, energy and topology information are also available, it is easy to understand that the proton decays can clearly be identified event per event, and, for most channels, it is not necessary to relay on statistical methods to eventually extract a signal.

In the absence of background, the limit on the nucleon lifetime reachable in T years of observation is given by the simple formulae:

$\tau_{\mathrm{P}} > 1.2\mathrm{xMxTx}\eta(10^{32}\mathrm{year})$ (90% C.L.) for the proton

$\tau_{\mathrm{n}} > 1.4\mathrm{xMxTx}\eta(10^{32}\mathrm{year})$ (90% C.L.) for the neutron

where M is the detector mass in kton and η is the overall detection efficiency. Proton decay events are characterized by a definite value of the total energy and by the fact that the total momentum of the decay products must be zero. These features, which are true for a free nucleon, are also approximately verified for a nucleon bound in a nucleus, provided the decay products do not rescatter before escaping the nucleus. As a consequence, we also include in our definition of detection efficiency $\eta = \epsilon_D \hat{\epsilon}$ the probability ϵ that the decay products do not interact with the nucleus in which they were produced, and the reconstruction efficiency ϵ_D. These nuclear effects, the distortions of the energy and momentum distributions due to the nucleon Fermi motion, and the reinteraction of decay particles with the nucleus have been studied by Monte Carlo simulation methods.

For example, for the decay mode $\mathrm{p} \to \overline{\nu}\mathrm{K}^+$ we obtain $\epsilon =$ 0.85 and a corresponding lifetime limit $\tau_{\mathrm{p}} > 4.7\mathrm{x}10^{32}$ (90% C.L.) years for one year of data-taking. This would increase the present limit by about two orders of magnitude. For this decay mode in particular (as for many others) we don't expect to have any significant background. Kaon production by atmospheric neutrinos is very rare. The probability that a charged-current neutrino interaction with emission of a proton from the argon nucleus can simulate a $\mathrm{p} \to \overline{\nu}\mathrm{K}^+$ decay is also expected to be negligible, because we can either distinguish the direction of motion of the charged hadron due to the increase of ionization at the end of the range, or directly recognize the kaon from the dE/dx and range measurements. However, since for this particular channel the discovery requires only one event, because of the negligible background, a lifetime of 10^{34} years is in fact reachable.

We have performed a complete study of only some of the proton decay modes of interest. Similar studies are under way. However, our detailed study of three characteristic decay modes indicates, that, except for channels such as $\mathrm{p} \to \mathrm{e}^+\nu\nu$ for which the topology is identical to that of atmospheric ν_{e} charged-current events, the background corresponding to a data-taking period of one year is negligible. It is interesting to note here that the availability of a CERN neutrino beam aimed at the Gran Sasso Laboratory would offer the possibility to directly simulate, if needed, some of the nucleon decay backgrounds.

In summary, many nucleon decay modes can be searched for simultaneously and, after only one year of data taking, ICARUS will reach or exceed most present limits. The equivalent of five years of data will take us to the unexplored region between 10^{33} and 10^{34} years for some of the most relevant channels.

§7 An Aside: $n - \overline{n}$ Oscillations

Baryon number violating processes mediated by higher dimensional operators (such as neutron-antineutron oscillation) provide important information about physics beyond the standard model that supplement that gained from the study of proton decay. Such processes become observable if there is new physics in the mass range of one to a thousand TeV; at higher mass scales, n-$\overline{n}$ oscillations are unobservable. Thus evidence for neutron-antineutron oscillations would be an indication of physics near the TeV scale. Furthermore, neutron-antineutron oscillations invariably arise from the exchange of either diquark Higgs bosons or leptoquark Higgs bosons of TeV range mass; therefore, any positive evidence for neutron-antineutron oscillations is an indication of the existence of low mass leptoquarks or diquarks.

The search for n-$\overline{n}$oscillations has been carried out to date using both a beam of free neutrons from the ILL reactor and using neutrons bound in either oxygen or iron nuclei. The best limit on the n-$\overline{n}$ oscillation time using free neutrons is $1\text{x}10^7$ sec; it is thought this can be extended to about 10^8 sec in the future. The current limit on the bound neutron's oscillation time is $1.2\text{x}10^8$ sec, from the Japanese Kamiokande water-Cherenkov (bound lifetime limit $> 4.3\text{x}10^{31}$ years) and Frejus iron calorimeter (bound lifetime limit $> 6.5\text{x}10^{31}$ years) proton decay detectors.

The signature for a bound n-$\overline{n}$ conversion event is the subsequent annihilation of the antineutron in the nucleus (on either another neutron or on a proton). Such an event has a distinct signature because it is annihilation essentially at rest. Several pions are created and they carry away the nearly 2 GeV of energy produced in the annihilation. This search can be continued into the future.

The first step could be to take advantage of the 7.7 kt-yr of data that now exists (on tape) from IMB-3. These data can extend the n-$\overline{n}$ oscillation time in oxygen to $6\text{x}10^{32}$ years, corresponding to a free lifetime of about $4\text{x}10^8$ sec. SuperKamiokande will have a total mass of 50 kt and a fiducial mass of 20 kt, about a factor of 7 larger than IMB. Data collected from SuperK can be used to set the world's most stringent limit on the bound n-$\overline{n}$ oscillation time in oxygen to $4\text{x}10^{33}$ years, corresponding to a free lifeimt of about 10^9 sec, from 2 years of SuperKamiokande data.

There is the concern that the effect on the n-$\overline{n}$ oscillation time due to the neutrons being bound in the oxygen nucleus has been poorly estimated. The deuteron is the most weakly bound nucleus so such effects should be more reliable. The SNO detector, the world's only underground heavy water-Cherenkov detector, can search for bound n-$\overline{n}$ oscillations in deuterium to a bound lifetime sensitivity of $4\text{x}10^{31}$ years, corresponding to a free lifetime of about $2\text{x}10^8$ sec, using 2 years of SNO data.

§8 Summary

Observation of nucleon decay (or another baryon number violating process) would revolutionize our view of particle physics by providing evidence for the unification of the forces. It would also establish a deep connection between quarks and leptons, providing insight into a (presently unknown) deep symmetry that connects them. Unfortunately, progress over the last decade has only been in extending the limits on the proton lifetime by one or more orders of magnitude. The next decade will see the sensitivity of the experiments extended by at least another order of magnitude, with progress beyond that difficult to predict.

Progress beyond the current experiments is hard to foresee due to the sheer size of the detectors needed to extend the sensitivity by another order of magnitude. These detectors will need to be 100 to 1,000 kilotons in size. A 1,000 kiloton detector will observe about 250 atmospheric neutrino events per day (100,000 per year). Clearly excellent detector resolution will be required to reject the atmospheric neutrino backgrounds. However, at this rate, understanding the neutrino backgrounds will be of utmost importance, perhaps even ultimately limiting the sensitivity of the search for nucleon decay.

§References

[1] S. Weinberg, Phys. Rev. Lett. *Phys. Rev. Lett.* **43** , 1566 (1979); F. Wilczek and A. Zee, *Phys. Rev. Lett.* **43**, 1571 (1979).

[2] R. N. Mohapatra and R. E. Marshak, *Phys. Rev. Lett.* **44**, 1316 (1980).

[3] J. C. Pati, A. Salam and U. Sarkar, *Phys. Lett.* **B133**, 330 (1983); P. J. O'donnell and U. Sarkar, Toronto Preprint (1994).

[4] R. N. Mohapatra and G. Senjanović, *Phys. Rev. Lett.* **49**, 7 (1982); L. Arnellos and W. Marciano, *Phys. Rev. Lett.* **48**, 1708 (1982).

[5] S. Dimopoulos and H. Georgi, *Nucl. Phys.* **B193**, 150 (1981); N. Sakai, *Z. Phys.* **C11**, 153 (1981).

[6] R. Arnowitt and P. Nath, *Phys. Lett.* **287B**, 89 (1992); CTP-TAMU-52/93; J. Hisano, H. Murayama and T. Yanagida ,Tohuku Preprint TU-400 (1992).

[7] D. G. Lee, R. N. Mohapatra, M. K. Parida and M. Rani, Univ. of Md Preprint UMD-PP-94-106 (1994); F. Acampora, G. Amelino-Camelia, F. Buccella, O. Pisanti , L. Rosa and T. Tuzi, DSF Preprint 93/52.

[8] L. Lavoura and L. Wolfenstein, *Phys. Rev.* **D48**, 264 (1993).

[9] R. Becker-Szendy et al., Phys. Rev. D 46, 3720(1992).

[10] T.J.Haines et al., PRL 57, 1986(1986).

[11] C. McGrew, PhD. Thesis, University of California Irvine (unpublished), 1994.

Decay Mode	Recovery Efficiency	Candidates	Background assuming $\frac{\nu_\mu}{\nu_e} = 0.54$	Lifetime Limit	Background assuming $\frac{\nu_\mu}{\nu_e} = 1.04$	Lifetime Limit
$n \to e^+ e^- \nu$	0.57	5	7.5	518	9.3	518
$n \to e^+ K^-$	0.14	35	29.4	20	26.4	17
$n \to e^+ \pi^-$	0.30	3	5.0	270	4.8	270
$n \to e^+ \pi^- \pi^0$	0.44	38	34.2	67	30.6	53
$n \to e^+ \rho^-$	0.49	4	4.8	293	6.5	446
$n \to \mu^+ e^- \nu$	0.42	25	29.4	219	34.1	378
$n \to \mu^+ K^-$	0.10	20	28.4	90	34.0	90
$n \to \mu^+ \mu^- \nu$	0.81	100	145.0	734	188.9	734
$n \to \mu^+ \pi^-$	0.14	1	1.9	122	3.7	122
$n \to \mu^+ \pi^- \pi^0$	0.29	17	20.8	173	25.2	266
$n \to \mu^+ \rho^-$	0.36	3	9.5	324	11.5	324
$n \to \nu \eta^0$	0.17	0	1.2	158	1.1	158
$n \to \nu \gamma$	0.80	163	144.7	44	123.6	28
$n \to \nu \gamma \gamma$	0.49	5	7.5	441	9.3	441
$n \to \nu K^0$	0.21	34	34.1	48	30.6	35
$n \to \nu K^{*0}$	0.51	40	50.0	455	56.4	455
$n \to \nu \omega^0$	0.28	12	22.5	252	26.7	252
$n \to \nu \pi^0$	0.30	6	6.6	149	7.5	186
$p \to e^- \pi^+ K^+$	0.46	81	127.2	518	160.3	518
$p \to e^+ \eta^0$	0.28	0	0.2	315	0.2	315
$p \to e^+ e^+ e^-$	0.71	0	0.5	799	0.9	799
$p \to e^+ \gamma$	0.60	0	0.1	675	0.1	675
$p \to e^+ K^0$	0.12	23	25.2	55	26.4	68
$p \to e^+ K^{*0}$	0.39	38	52.0	433	61.1	433
$p \to e^+ \mu^+ \mu^-$	0.47	1	0.9	396	1.2	429
$p \to e^+ \nu \nu$	0.32	152	153.7	49	138.7	26
$p \to e^+ \omega^0$	0.21	7	10.8	236	13.5	236
$p \to e^+ \pi^0$	0.48	0	0.2	540	0.2	540
$p \to e^+ \pi^0 \pi^0$	0.26	2	0.8	147	0.7	143
$p \to e^+ \pi^+ \pi^-$	0.23	16	23.1	253	27.2	253
$p \to \mu^- \pi^+ K^+$	0.40	3	4.0	349	5.6	444
$p \to \mu^+ \eta^0$	0.23	3	2.8	144	2.8	145
$p \to \mu^+ e^+ e^-$	0.47	0	1.0	529	1.1	529
$p \to \mu^+ \gamma$	0.42	0	0.1	478	0.2	478
$p \to \mu^+ K^0$	0.19	4	7.2	214	9.5	214
$p \to \mu^+ \mu^+ \mu^-$	0.60	0	0.3	675	0.4	675
$p \to \mu^+ \omega^0$	0.33	11	12.1	171	21.5	371
$p \to \mu^+ \pi^0$	0.42	0	0.6	473	0.6	473
$p \to \mu^+ \pi^0 \pi^0$	0.20	3	1.6	103	1.7	103
$p \to \mu^+ \pi^+ \pi^-$	0.44	25	38.0	495	45.1	495
$p \to \nu K^+$	0.41	15	21.4	461	29.5	461
$p \to \nu K^{*+}$	0.11	7	9.1	92	11.3	124
$p \to \nu \pi^+$	0.03	15	20.3	34	23.5	34
$p \to \nu \rho^+$	0.54	18	21.7	385	25.1	608

Table 1: Nucleon decay limits determined using the IMB-3 Detector. The estimated background and limits are presented for two situations; the ratio of muon neutrinos to electron neutrinos at the detector is the measured value of $\frac{\nu_\mu}{\nu_e} = 0.54$, or the ratio is the predicted value of $\frac{\nu_\mu}{\nu_e} = 1.04$

Table 2: Summary of result of each nucleon decay mode from Kamiokande.

decay mode	# of ev	ν M.C.	$\epsilon \cdot B_m$	τ/B sub. ($10^{31}yr$)	τ/B unsub. ($10^{31}yr$)
$n \to e^-\pi^+$	0	<0.2	0.28	12	12
$n \to \mu^-\pi^+$	0	<0.2	0.18	7.5	7.5
$n \to e^+e^-\nu$	3	1.0	0.68	12	9.8
$n \to \mu^+\mu^-\nu$	0	1.8	0.50	20	20
$n \to e^-\mu^+\nu$	3	4.0	0.58	14	8.4
$p \to e^+\nu\nu$	81	76(46)$^{(1)}$	0.91	2.3	1.2
$p \to \mu^+\nu\nu$	39	82(49)$^{(1)}$	0.78	13	1.9
$p \to e^-\pi^+\pi^+$	1	1.4	0.27	11	8.4
$p \to \mu^-\pi^+\pi^+$	0	0.5	0.29	15	15
$n \to e^-\pi^+\pi^0$	0	<0.2	0.06	3.3	3.3
$n \to \mu^-\pi^+\pi^0$	0	0.5	0.17	7.9	7.9
$n \to e^-K^+(\mu^+\nu)$	0	0.5	0.34	15	15
$(\pi^+\pi^0)$	0	0.5	0.02		
$n \to \mu^-K^+(\mu^+\nu)$	0	0.2	0.35	17	17
$(\pi^+\pi^0)$	0	<0.2	0.05		

(1) A number in parenthesis is multiplied by a factor 0.6 for the conservative estimate of the neutrino background.

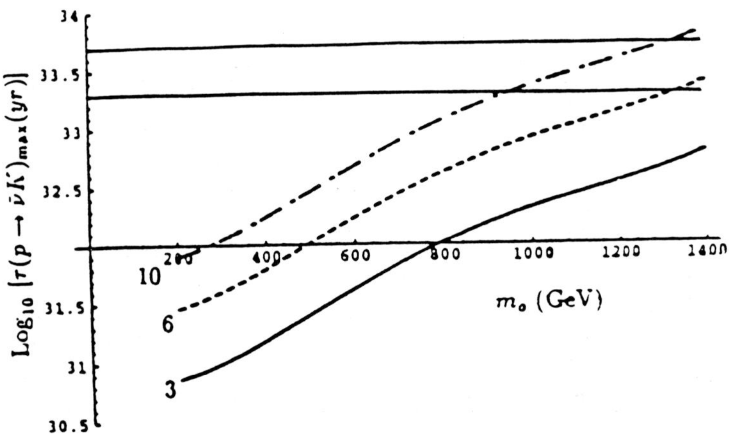

Figure 1: Dependence of the proton lifetime on the chargino mass in a SUSY-SU(5) model, for a top mass of 150 GeV. The lines represent the model predictions for different model parameters; the horizontal lines are the expected experimental sensitivities of SuperKamiokande and ICARUS.

Cosmology: Summary

Joshua A. Frieman

NASA/Fermilab Astrophysics Center, Fermi National Accelerator Laboratory, Batavia, IL 60510

Department of Astronomy and Astrophysics, The University of Chicago, Chicago, IL 60637

A brief summary of the cosmology contributions to Snowmass '94 is presented. The current status of the field is overviewed, and principal directions and goals for cosmology at the turn of the millennium are highlighted.

§1 Introduction

As the preceding chapters amply demonstrate, cosmology has made dramatic advances in recent decades, and has emerged as a vibrant, robust scientific field. This renaissance has been fueled by the symbiotic relationship between theoretical developments, particularly with regard to the early universe and the study of large-scale structure evolution, and observational advances that exploit a variety of novel experimental techniques. Cosmology is the paramount arena for the new interdisciplinary field of particle astrophysics, where particle physics, nuclear physics, and astronomy are increasingly interwoven.

At the same time, despite the significant progress in our understanding of the Universe, it is clear that basic issues at the very core of cosmology remain unresolved. We do not yet know the physical origin of the large-scale structure of the universe, the nature and identity of the dark matter, the precise values of the cosmological parameters, or the origin of the baryon asymmetry. The decade ahead holds the promise of real progress on these fundamental issues, for they are being attacked on a number of fronts. New galaxy surveys will map the large-scale structure of the universe over substantially larger scales than presently in hand, complemented by larger, more realistic N-body simulations of structure formation that include the physical complexities of galaxy formation; microwave background anisotropy experiments from the ground, balloons, and space will probe fluctuations in the early universe on finer scales with unprecedented accuracy; dark matter searches will identify or strongly constrain the dark matter, complemented by accelerator and non-accelerator particle experiments that probe the physics of elementary particles beyond the standard model; and the new generation of ground-based and space-borne telescopes will probe the evolution of the universe and determine the cosmological parameters with increasing accuracy.

For each of these endeavors, there are projects in place, under development, or proposed; if they are all vigorously pursued, the data base for cosmology a decade hence will be remarkably rich, of an altogether different scale and scope from what preceded it. Once it has been sifted through, we may emerge with a coherent world-picture, a more focused vision that fleshes out but essentially validates current ideas. On the other hand, we may end up in utter confusion, with the necessity for a new theoretical paradigm. Either way, we stand at the threshold of the era in which the bold new ideas about the early universe developed over the last fifteen years or so will be tested. The first steps through this door were taken with the detection of microwave background anisotropies by the COBE satellite, but it is the observations to come which will determine whether the early universe emperor wears any clothes.

While this prognosis for progress may appear overly optimistic, it gains credibility by virtue of the fact that the avenues being pursued are varied and ultimately mutually reinforcing. For example, the next round of galaxy redshift surveys will not only probe the large-scale power spectrum, but, through redshift distortions, will yield a potentially accurate determination of the cosmic mass density Ω. Similarly, accurate cosmic microwave background (CMB) measurements of the anisotropy power spectrum on small angular scales open up the possibility not only of testing inflation and defect models, but of providing independent determinations of cosmological parameters such as Ω, Ω_B, and Λ as well.

Over the last thirty years, the hot big bang model has been elevated to the 'standard model' of cosmology, because it provides a remarkably successful operating hypothesis describing the large-scale evolution of the Universe. In particular, it provides a consistent theoretical framework for three fundamental observations: the Hubble law of recession of galaxies, the pattern of light element abundances, and the thermal, isotropic microwave background radiation. The health and success of this paradigm provides a starting point for asking the fundamental questions about the universe above, such as: how did structure form? what is the dark matter? why are there baryons? how fast is the universe expanding, how long has it been expanding, and will it stop in the future? It is important to emphasize that the standard cosmology as currently conceived provides only the framework for studying these questions, not their solutions; consequently, our lack of definitive answers to these questions so far should not be misinterpreted as a failure of the big bang paradigm itself. As with any paradigm, the standard cosmology will presumably be subsumed eventually into an as yet unimagined theory with greater explanatory power, but the prospects for such a radical shift in the foundations of cosmology do not yet appear close at hand.

§2 Cosmological Parameters

The current status of measurements of the cosmological parameters H_0, t_0, Ω, and Λ is confused. While new techniques have appeared on the scene and heroic efforts have gone into measurements covering larger volumes with correspondingly reduced statistical errors, there remains a lack of consensus between different techniques, indicating that substantial systematic errors lurk in some of the measurements. Resolving these discrepancies, particularly in measurements of H_0 and Ω, will be a priority for observational astronomy over the next decade. In particular, we would like to know whether they are closer to the theoretically popular values of $H_0 = 50$ km/sec/Mpc and $\Omega = 1$ or to the values indicated by a number of astronomical observations, namely $H_0 \simeq 80$ km/sec/Mpc and $\Omega \simeq 0.2$. The answers impact directly on ideas about how structure forms in the universe. The answers will also determine whether the current age crisis is a true catastrophe for the standard cosmology or, as it was in Hubble's time, a temporary problem due to systematic errors. On the theoretical side, the smallness of the cosmological constant Λ compared to the Planck scale remains obscure; whether resolution of this puzzle will entail a major alteration in cosmic evolution is an open question.

• H_0: The myriad techniques being used and developed to determine the Hubble parameter, including distant Cepheid variables with HST, Type I and II supernovae, Tully-Fisher, surface brightness fluctuations, the Sunyaev-Zel'dovich effect in clusters, and gravitational lens time delays, raise the hope that H_0 can be determined to 10% accuracy within a decade. Additional Cepheids will place the calibration of secondary distance indicators on firmer footing. The acquisition of significant numbers of systems, e.g., measurements of time delays for many lensed quasars and of the Sunyaev-Zel'dovich effect in a substantial number of clusters, will help quantify the usefulness of these new techniques and provide statistically significant determinations.

• Ω: Deep spectroscopic surveys offer the hope of finally disentangling galaxy evolution from cosmology with sufficient accuracy to realize the classical global cosmological tests for determining Ω and q_0. Quasar lensing surveys should provide improved limits on Λ and Ω, while the detection of a significant number of Type Ia SN at high redshift should constrain q_0. Dynamical estimates of Ω on scales of clusters and larger have been in a state of confusion. X-ray and virial determinations of cluster masses suggest $\Omega \sim 0.1 - 0.2$, while strong and weak gravitational lens determinations indicate substantially higher cluster masses in at least some cases. Large-scale flows have so far been inconclusive; while some techniques have suggested high values of Ω, there are heated debates about the effects of biases; taken as a whole, the large-scale peculiar velocities indicate $\Omega \simeq 0.3 - 1$ (modulo bias). X-ray information about clusters is growing rapidly with ROSAT and ASCA and accurate X-ray temperature profiles for clusters should become available with AXAF. Weak lensing studies, now at an early stage, are opnening a new window on the large-scale distribution of dark matter. In this context, one goal is a large sample of clusters that have all been probed by virial, X-ray, Sunyaev-Zel'dovich, and lensing techniques. In this case, the internal structure of clusters themselves may provide a new test of Ω. New galaxy redshift surveys should enable much more accurate large-scale flow determinations of Ω via the anisotropy of galaxy clustering in redshift space. Finally, in the context of primordial adiabatic fluctuations, the positions of the Doppler peaks in multipole space in the microwave background anisotropy in principle provide an independent determination of Ω. Out of all this, optimistically one might hope for a consistent determination of Ω to an accuracy of 0.1 a decade from now.

§3 Nucleosynthesis

Primordial nucleosynthesis remains a key linchpin of the standard cosmology, and provides important constraints on the baryon density and the number of light neutrinos. Goals for the millennium include improved observations of the light element abundances, e.g., deuterium observations from Lyman alpha clouds, and additional theoretical investigations of chemical evolution and stellar processing. For stellar nucleosynthesis, important areas for development include 3-d hydrodynamics with increased resolution, studies of turbulence and convection, and measurement of key nuclear reaction rates.

§4 Baryogenesis

The basic ingredients for the generation of the baryon asymmetry were put forward by Sakharov in the late 1960s. With the advent of grand unified theories in the 1970s, a theoretical framework for producing the observed baryon to photon ratio emerged. Over the last few years, following the recognition of the importance of sphaleron-mediated baryon-violating processes, attention has focused on the electroweak phase transition as the possible origin for the baryon asymmetry. These models can be probed by improved experimental constraints on electron and neutron electric dipole moments. On the theoretical side, studies of the nature and dynamics of the electroweak phase transition will be critical for understanding the viability of this scenario. Accelerator probes of physics beyond the standard model will ultimately point to new directions.

§5 Dark Matter

Comparison of the baryon density inferred from primordial nucleosynthesis with the density of luminous matter suggests that some (perhaps most) of the baryons in the universe are dark. Comparison of Ω_B from nucleosynthesis with dynamical estimates of Ω from clusters and large-scale flows suggests that most of the mass in the universe is non-baryonic. While suggestive, the nature of the dark matter(s) must ultimately be decided by experiment. Active experimental programs to search for both baryonic and non-baryonic dark matter are underway.

5.1 Baryons

Theoretically, the most plausible candidates for baryonic dark matter are stellar remnants and sub-stellar objects (planets or brown dwarfs) which did not reach sufficiently high internal temperatures to initiate hydrogen burning. Three microlensing experiments are currently underway to detect dark stellar and sub-stellar objects (MACHOs) in the disk and the halo of our galaxy, and a number of others are planned. A number of things have been learned from this program to date:

• Gravitational microlensing has been observed.

• The microlensing event rate toward the LMC is lower than expected for a "standard halo" filled with MACHOs of substellar mass (the LMC rate is of order 20% of that for a standard halo, but with large errors).

• The event rate towards the galactic bulge is roughly 3 times higher than expected in standard models for the inner Milky Way.

Priorities for the next few years include:

• getting good statistics for the event rate towards the LMC,

• attempting to get the event rate towards the SMC (tougher since the number of stars is low, but the best handle on the shape of the halo)

• mapping the event rate as a function of galactic longitude and latitutde near the galactic center to disentangle galactic models.

Long-term priorities include:

• developing a photometric follow-up network to determine lightcurves more accurately (to measure higher order effects such as the finite angular size of the lensed star, which yields much more information per event; and to search for planetary companions)

• a parallax satellite to measure lightcurves simultaneously from earth and a spacecraft 1 AU away to resolve the lens; this will determine where the lensing objects are.

5.2 Non-baryonic Dark Matter

• *Wimps*: A prime candidate for (cold) dark matter is the lightest supersymmetric particle, which is generally predicted to have a mass in the GeV-TeV range to yield $\Omega \sim 1$. A number of technologies are being pursued to search for halo WIMPs through their scattering off detector nuclei. Devices with low thresholds, high sensitivity, and low radioactive backgrounds are necessary for these efforts to reach the theoretically predicted range. Expected event rates are of order 1 event per kg per day with an energy deposition in the keV or sub-keV range. The basic idea for detecting WIMPs was proposed ten years ago. It is hoped that a large-scale (kg or multi-kg) experiment in an ultra-low background environment will be mounted within the next decade.

• *Axions*: Axions, light pseudo-Nambu-Goldstone bosons which emerge from the Peccei-Quinn solution to the strong CP problem, are another prime theoretical candidate for cold dark matter. Cosmological and astrophysical arguments constrain the axion mass to the window $10^{-6} - 10^{-3}$ eV, and a second-generation Sikivie-type detector to probe this window is under construction at Livermore. A Japanese collaboration is building a complementary axion search experiment using Rydberg atoms.

As of this writing, the LSND collaboration at Los Alamos has reported evidence for neutrino oscillations; combined with limits from other experiments, it is possible that the implied neutrino mass is in the cosmologically interesting range of order eV. Future accelerator and non-accelerator neutrino experiments will help determine whether neutrinos can play a dynamically important role as (part of the) dark matter.

§6 Large-scale Structure

The advent of large-area redshift surveys in the early 1980s began the era of constructing three-dimensional catalogs of the large-scale structure of the universe. In recent years, deeper catalogs with large sky coverage have led to quantitative statistical information on the large-scale clustering of galaxies in space. However, the current surveys still probe a relatively small volume of the local cosmic neighborhood and typically contain of order 10,000 galaxies. Over the next several years, new surveys, in particular the Sloan Digital Sky Survey and the Anglo-Australian 2dF, with redshifts for hundreds of thousands to of order 10^6 galaxies and images for tens of millions of galaxies will substantially extend our picture of the galaxy distribution and allow qualitatively new features to be extracted. These surveys will probe the galaxy power spectrum and higher order statistics to significantly larger scales, comparable to those probed by the COBE satellite, and provide more accurate determinations of Ω and a finer probe of bias. They will also place limits on non-Gaussian fluctuations at some level and probe the evolution of clustering over time. These surveys of the galaxy distribution will be complemented by weak lensing studies, which will ultimately probe the dark matter distribution and power spectrum over large scales; deep spectroscopic surveys to probe cosmological parameters and galaxy evolution; and Sunyaev-Zel'dovich studies of clusters, which will provide a new probe of large-scale flows.

On the theoretical side, the modelling of large-scale structure formation is developing rapidly and along with it more careful methods for the comparison of models with the data on large-scale structure. N-body simulations on supercomputers are beginning to reach the dynamic range where useful predictions can be made, and the incorporation of gas dynamics and the complexities of non-gravitational phenomena on small scales is developing.

§7 Microwave Background

The first phase in our understanding of the cosmic microwave background (CMB) began some thirty years ago with the discovery of the 3° K radiation and the realization that it supported the hot big bang picture. The second phase began in 1992 with the discovery by the COBE satellite of anisotropies in the microwave background. Since then, more than a dozen other experiments have also observed anisotropies at the same level of $\Delta T/T \sim 10^{-5}$ on smaller angular scales. Gravitational instability theories predicted that such anisotropies should exist at roughly the level that has been observed. This concordance provides the best evidence that structure formed in the Universe via gravitational instability.

COBE provided substantial information about the CMB multipole moments $L = 2 - 10$. In the next phase of CMB exploration, a key goal is the measurement of the multipole moments up to $L = 500$ or even higher. On these smaller scales, the spectrum of anisotropies the C_L's depends sensitively on a number of cosmological parameters. Thus, we may exploit this to determine Ω (via the positions of the Doppler peaks), and to help pin down such unknowns as the Hubble constant; the baryon fraction; the nature of the dark matter; the cosmological constant; and the neutrino number density and temperature. As we get information out to $L \sim 250$, we can possibly uncover the source of the primordial perturbations, determining whether inflation or topological defects provide better descriptions of the data, and thus opening an observational window on GUT-scale physics. On very small scales ($L > 1000$), non-linear physics begins to become important. We might then learn something about the ionization history of the Universe; large structures such as clusters; and gravitational lensing of anisotropies.

To achieve these goals, a full or nearly full sky map with an angular resolution of order 10 – 30 arcminutes is needed. Currently, several groups in the U.S. and Europe have proposed a dedicated satellite instrument for this purpose.

§8 Conclusion

We are now at a most exciting juncture in the development of cosmology. For the past fifteen years, the infusion of particle physics ideas has led to the development of theories of the Universe of unprecedented detail, scope, and imagination. The ideas which have emerged from particle cosmology over this period such as inflation, topological defects, relic cold dark matter particles, the generation of density perturbations, the origin of the baryon asymmetry have had a profound impact throughout cosmology, astrophysics, and particle physics.

The wealth of data now starting to come in from space-borne and earth-based observatories, as well as experiments at accelerators, is beginning to put these theories to the test. This is a very healthy development which will spur further progress in the field. At the present time, there is still debate about the mutual consistency and interpretation of much of the data. This state of affairs requires cosmologists to become more 'phenomenological' in confronting model predictions with data to understand the astronomical complexity that acts as a filter between observations and the pristine world of theory. The interpretation of the mushrooming data base that will be accumulated in the near future will point the field in new theoretical and observational directions.

Parts of this summary were contributed by Charles Alcock and Scott Dodelson. Research supported by the DOE and by NASA grant NAG5-2788 at Fermilab.

List of Participants

Abashian, Al
Virginia Polytechnic Institute and State University

Abramovici, Alex
California Institute of Technology

Adams, James H.
Naval Research Laboratory

Adelberger, Eric G.
University of Washington

Ahluwalia, Dharam V.
Los Alamos National Laboratory

Akerib, Dan
CfPA, U. of California, Berkeley

Akerlof, Carl
University of Michigan

Albrecht, Andreas
Imperial College

Alcock, Charles
Lawrence Livermore National Laboratory

Allen, Bruce
University of Wisconsin

Allen, Glenn E.
University of Maryland

Amendola, Luca
Fermi National Accelerator Laboratory

Aoki, Shigeki
Kobe University

Applegate, James
Columbia University

Argyres, Philip C.
Institute for Advanced Study

Atac, Muzaffer
Fermi National Accelerator Laboratory

Atwood, William B.
Stanford Linear Accelerator Center

Aufderheide, Maurice
University of Pennsylvania

Avignone, Frank
University of South Carolina

Ayres, David
Argonne National Laboratory

Bahcall, John
Institute for Advanced Study

Baker, Howard
Butler University

Balantekin, A. Baha
U. of Wisconsin, Madison

Barnes, Peter
Center for Particle Astropysics

Barvinsky, Andrei
The University of Alberta

Barwick, Steven
U. of California, Irvine

Battye, Richard
University of Cambridge

Beatty, James J.
Washington University

Bender, Peter L.
JILA, University of Colorado

Bennett, David P.
LLNL/U. of California, Berkeley

Benson, Katherine M.
U. of California, San Diego

Bernstein, Robert
Fermi National Accelerator Laboratory

Biller, Steve
University of Leeds

Bird, David
University of Utah

Block, Martin
Northwestern University

Bloom, Elliott
Stanford Linear Accelerator Center

Bloom, Stewart D.
Lawrence Livermore National Laboratory

Blucher, Edward C.
EFI, University of Chicago

Bludman, Sidney
University of Pennsylvania

Bly, Charles Albert
Oak Ridge National Laboratory

Bock, Greg
Fermi National Accelerator Laboratory

Bolton, Tim
Kansas State University

Bonvicini, Giovanni
CERN

Boratav, Murat
University of Paris VI

Border, Peter
University of Minnesota

Bowles, Thomas J.
Los Alamos National Laboratory

Bradner, Hugh
U. of California, San Diego

Bucher, Martin
Institute for Advanced Study

Buckland, Kerry N.
U. of California, San Diego

Buckley, Jim
Harvard-Smithsonian Center for Astrophysics

Burdyuzha, V.
Astro Space Center of Lebedev Physical Institute

Cabrera, Blas
Stanford University

Calaprice, Frank P.
Princeton University

Caldwell, David O.
U. of California, Santa Barbara

Caldwell, Robert
Fermi National Accelerator Laboratory

Cei, Fabrizio
INFN, Pisa

Charlton, Gordon
U.S. Department of Energy

Chugg, Barron
Stanford University

Chumney, Pamela R.
U. of California, Irvine

Cline, David B.
U. of California, Los Angeles

Cohen, Andrew
Boston University

Cole, K. C.
Discover Magazine

Cole, Terry
Jet Propulsion Laboratory

Cominsky, Lynn
SLAC/Sonoma State University

Coppi, Paolo
University of Chicago

Covault, Corbin E.
EFI, University of Chicago

Cowan, John J.
University of Oklahoma

Cox, Bradley
University of Virginia

Crane, Daniel A.
Argonne National Laboratory

Cronin, James
EFI, University of Chicago

Cushman, Priscilla
University of Minnesota

DaSilva, Angela
CfPA, U. of California, Berkeley

Davis, Marc
U. of California, Berkeley

Daw, Edward
Massachusetts Institute of Technology

Dicus, Duane A.
University of Texas, Austin

Dingus, Brenda L.
NASA/Goddard Space Flight Center

Diwan, Milind
Brookhaven National Laboratory

Dodelson, Scott
Fermi National Accelerator Laboratory

Domokos, Gabor
The Johns Hopkins University

Dougherty, Brian
Stanford University

Dragovan, Mark
Princeton University

Drever, Ronald W. P.
California Institute of Technology

Duncan, Douglas
University of Chicago

Durrer, Ruth
University of Zurich

Engel, Jonathan
University of North Carolina

Evenson, Paul
University of Delaware

Fejer, Martin
Stanford University

Feldman, Hume A.
University of Michigan

Ficenec, David
Washington State University

Fick, Brian E.
EFI, University of Chicago

Filippenko, Alexei
U. of California, Berkeley

Finn, Lee Samuel
Northwestern University

Flaminio, Raffaele
INFN, Pisa

Fletcher, Robert S.
University of Arizona

Fleury, Patrick
Ecole Polytechnique

Fontaine, Gérard
Institut National de Physique Nucleaire France

Ford, Holland
Space Telescope Science Institute

Fortson, Lucy F.
EFI, University of Chicago

Frazer, William R.
Lawrence Berkeley Laboratory

Freese, Katherine
University of Michigan

Frichter, George M.
University of Kansas

Frieman, Joshua A.
Fermi National Accelerator Laboratory

Frolov, Valery
University of Alberta

Fuller, George M.
U. of California, San Diego

Gai, Moshe
University of Connecticut

Gaisser, Thomas K.
Bartol Research Institute

Gallagher, Hugh M.
Argonne National Laboratory

Gates, Evalyn
EFI, University of Chicago

Gelmini, Graciela
U. of California, Los Angeles

Gibbs, Richard L.
Louisiana Tech University

Giddings, Steve
U. of California, Santa Barbara

Godfrey, Gary
Stanford Linear Accelerator Center

Goldberg, Haim
Northeastern University

Golwala, Sunil
CfPA, U. of California, Berkeley

Gonzalez-Garcia, Concha
CERN

Goodman, Jordan
University of Maryland

Goodman, Maury
Argonne National Laboratory

Gottfried, Kurt
Cornell University

Gourlay, Steve
Fermi National Accelerator Laboratory

Gray, Lori Ann
Phenomenology Institute

Gray, Michael
U. of California, Santa Barbara

Graziani, Frank
Lawrence Livermore National Laboratory

Green, Kevin
University of Michigan

Gregorich, David T.
CSULA/Caltech

Griest, Kim
U. of California, San Diego

Grigorov, Naum L.
Skobeltsyn, Moscow State University

Grindlay, Jonathan
Harvard-Smithsonian Center for Astrophysics

Grishchuk, Leonid P.
Washington University

Gundersen, Joshua
U. of California, Santa Barbara

Gustafson, Eric
Stanford University

Gustafson, Richard
University of Michigan

Habib, Salman
Los Alamos National Laboratory

Hafen, Elizabeth S.
Massachusetts Institute of Technology

Haines, Todd
Los Alamos National Laboratory

Hallin, Aksel
Queen's University

Halpern, Leopold
Florida State University

Halzen, Francis
University of Wisconsin

Hanany, Shaul
CfPA, U. of California, Berkeley

Hanna, David
McGill University

Harding, Alice K.
NASA/Goddard Space Flight Center

Harmon, Alan
NASA/Marshall Space Flight Center

Hata, Naoya
University of Pennsylvania

Hauptman, John
Iowa State University

Haxton, Wick
University of Washington

Haymes, Robert C.
Rice University

Heinz, Richard
Indiana University

Heitowit, Ezra
Universities Research Association, Inc.

Heller, Kenneth
University of Minnesota

Hello, Patrice
Universite Paris-Sud

Helmer, Rich
TRIUMF

Hillebrandt, Wolfgang
Max-Planck-Institute

Hindi, Munther M.
Tennessee Technological University

Ho, Yunsil
Nevis Laboratories

Hoffman, Cyrus M.
Los Alamos National Laboratory

Hong, Woopyo
U. of California, Los Angeles

Hunter, Stanley D.
NASA/Goddard Space Flight Center

Isaacson, Richard
National Science Foundation

Jacobsen, John
University of Wisconsin

Jaffe, Andrew
University of Toronto

Janssen, Michael
Jet Propulsion Laboratory

Jedamzik, K.
Lawrence Livermore National Laboratory

Johnson, Robert P.
U. of California, Santa Cruz

Johnson, Warren
Louisiana State University

Jokipii, J. R.
University of Arizona

Jones, Frank C.
NASA/Goddard Space Flight Center

Jones, W. Vernon
NASA Headquarters

Jöstlein, Hans
Fermi National Accelerator Laboratory

Jungman, Gerard
Syracuse University

Kamionkowski, Marc
Institute for Advanced Study

Kaneyuki, Kenji
Tokyo Institute of Technology

Kauts, Vladimir L.
Astro Space Center of Lebedev Physical Institute

Kawabe, Keita
University of Tokyo

Kayser, Boris
National Science Foundation

Kearns, Edward
Boston University

Kennedy, Dallas C.
University of Florida

Kephart, Thomas W.
Vanderbilt University

Kim, Alex
U. of California, Berkeley

Kleinfeller, Jonny
Kernforschungszentrum Karlsruhe

Knapp, Bruce C.
Columbia University

Koizumi, Gordon
Fermi National Accelerator Laboratory

Kolb, Edward W.
Fermi National Accelerator Laboratory

Konoplich, Rostislav
Moscow Engineering Physics Institute

Kosik, Dan W.
Butler University

Kovalev, Yuri A.
Astro Space Center of Lebedev Physical Institute

Kovesi-Domokos, S.
Johns Hopkins University

Kozub, Raymond L.
Tennessee Technological University

Krastev, Plamen
University of North Carolina

Krauss, Lawrence M.
Case Western Reserve University

Kron, Richard G.
Fermilab/University of Chicago

Kycia, Thaddeus F.
Brookhaven National Laboratory

Lai, Kwan-Wu
Jet Propulsion Laboratory

Lalanne, Dominique
Laboratoire Accelerateur Lineaire

Lamb, Richard
Iowa State University

Lande, Kenneth
University of Pennsylvania

Lanou, Robert E.
Brown University

Laplanche, Francis
Laboratoire Accelerateur Lineaire

Lattimer, James M.
State Univeristy of New York

Leahy, Denis A.
University of Calgary

Learned, John
University of Hawaii

Lee, Andrew
Stanford Linear Accelerator Center

Lee, Chul M.
Hanyang University

Lee, Wonyong
Nevis Laboratories

Lesko, Kevin
Lawrence Berekely Laboratory

Levin, Daniel
University of Michigan

Levin, Janna
CITA, University of Toronto

Lin, Ying-chi
Stanford University

Ling, James
Jet Propulsion Laboratory

Linsky, Jeffrey
JILA, University of Colorado

Lipari, Paolo
INFN, Rome

Lloyd-Evans, Jeremy
University of Leeds

Loh, Eugene
University of Utah

Longley, Nathaniel P.
California Institute of Technology

Lowe, David
U. of California, Santa Barbara

Lu, Al
Stanford Linear Accelerator Center

Lubowich, Donald
AIP/Hofstra University

Lundberg, Byron
Fermi National Accelerator Laboratory

Magnussen, Norbert
University of Wuppertal

Malensek, Anthony J.
Fermi National Accelerator Laboratory

Mannel, Eric J.
Nevis Laboratories

Mantsch, Paul M.
Fermi National Accelerator Laboratory

Marshak, Marvin L.
University of Minnesota

Martello, Daniele
Bartol Research Institute

Mascarenhas, Nicholas
California Institute of Technology

Mather, John
NASA/Goddard Space Flight Center

Mathews, Grant J.
Lawrence Livermore National Laboratory

Matthews, James
University of Michigan

Mazur, Peter O.
Fermi National Accelerator Laboratory

McDonald, Art B.
Queens University

McGrew, Clark
U. of California, Irvine

McKay, Timothy
Fermi National Accelerator Laboratory

McKee, Shawn
University of Michigan

McLerran, Larry
University of Minnesota

McManus, Arthur
University of Virginia

Mewaldt, Richard
California Institute of Technology

Meyer, H.
University of Wuppertal

Meyer, Richard A.
U. S. Department of Energy

Meyer, Stephan
EFI, University of Chicago

Michael, Douglas
California Institute of Technology

Michelson, Peter F.
Stanford University

Miller, Tim
Bartol Research Institute

Mills, Geoffrey B.
Los Alamos National Laboratory

Minamitani, Takahisa
Hughes STX/Marshall Space Flight Center

Mock, Patrick C.
U. of California, Irvine

Mohapatra, R. N.
University of Maryland

Monteleoni, Bianca
INFN, Firenze

Morfín, Jorge G.
Fermi National Accelerator Laboratory

Morris, Duncan
U. of California, Los Angeles

Morris, Michael S.
Butler University

Moscoso, Luciano
DAPNIA/SPP, CE Saclay

Mours, Benoit
Lab. d'Annecy-le-Vieus de Phys. des Particules

Muller, Dietrich
University of Chicago

Müller, Volker
Astrophysikalisches Institut Potsdam

Myers, Steven T.
California Institute of Technology

Nagle, Darragh
Los Alamos National Laboratory

Narita, Tom
University of Wisconsin

Nash, Thomas
Fermi National Accelerator Laboratory

Newport, Brian
Univeristy of Chicago

Ng, Kin-Wang
Academia Sinica, Taipei

Nicholson, David
University of Wales

Nissen, Poul E.
University of Aarhus

Nitz, David F.
University of Michigan

Niwa, Kimio
Nagoya University

Nolan, Patrick
Stanford University

Novikov, Igor
Nordita, Nordisk Inst. for Teoretixk Fysik

O'Connor, Daniel
INFN, Pisa

Olinto, Angela A.
EFI, University of Chicago

Ong, Rene
EFI, University of Chicago

Oreglia, Mark
University of Chicago

Ormes, Jonathan
NASA/Goddard Space Flight Center

Page, Lyman
Princeton University

Pain, Reynald
Lawrence Berkeley Laboratory

Palmer, Bill
Ohio State University

Panasyuk, Mikhail I.
Skobeltsyn, Moscow State University

Paolone, Vittorio
U. of California, Davis

Parente, Gonzalo
U. of California, Irvine

Parker, Peter
Yale University

Pasachoff, Jay M.
Harvard-Smithsonian Center for Astrophysics

Pearce, G. F.
Rutherford Appleton Laboratory

Peccei, Roberto D.
U. of California, Los Angeles

Peoples, John
Fermi National Accelerator Laboratory

Perez, Patrice
DAPNIA, CE Saclay

Perlmutter, Saul
Lawrence Berkeley Laboratory

Peterson, Earl A.
University of Minnesota

Peterson, Jeffrey
Carnegie-Mellon University

Piepke, Andreas
California Institute of Technology

Playa, Rainer
Max-Planck-Institute

Postnov, K. A.
Sternberg Astronomical Inst., Moscow University

Primack, Joel
U. of California, Santa Cruz

Pritychenko, Boris
CfPA, U. of California, Berkeley

Pröbst, Franz
Max-Planck-Institute

Prosper, Harrison B.
Florida State University

Qian, Yong Zhong
Institute for Nuclear Theory

Rajagopal, Krishna
Harvard University

Rameika, Regina
Fermi National Accelerator Laboratory

Reay, Neville W.
Kansas State University

Ressell, Ted
Lawrence Livermore National Laboratory

Revol, Jean-Pierre
CERN

Rhie, Sun Hong
Lawrence Livermore National Laboratory

Rhoades, Jean
U. of California, Los Angeles

Robertson, Hamish
Los Alamos National Laboratory

Robertson, Norna
University of Glasgow

Rolli, Simona
University of Pavia

Rosen, S. Peter
University of Texas

Rosenfeld, Carl
University of South Carolina

Ross, Ronald
Lawrence Berkeley Laboratory

Rubakov, Valeri
Institute for Nuclear Research

Ruddick, Keith
University of Minnesota

Ruediger, Albrecht
Max-Planck-Institute

Sadoulet, Bernard
CfPA, U. of California, Berkeley

Salamon, Michael
University of Utah

Salopek, David
University of Alberta

Sandberg, Vern
Los Alamos National Laboratory

Sandeman, R. John
Emeritus Professor

Sanders, Gary H.
Los Alamos National Laboratory

Saulson, Peter
Syracuse University

Sawyer, Raymond F.
U. of California, Santa Barbara

Scargle, Jeffrey D.
NASA

Schmid, Christoph
Institute for Theoretical Physics, Zurich

Schmidt, David
Los Alamos National Laboratory

Schnee, Richard
Los Alamos National Laboratory

Schneps, Jack
Tufts University

Schramm, David
University of Chicago

Schub, Martin
University of Minnesota

Schubnell, Michael
University of Michigan

Seckel, David
University of Delaware

Segré, Gino
University of Pennsylvania

Seidel, Ed
University of Illinois

Seidlein, Rupert
Argonne National Laboratory

Seo, Eun-Suk
University of Maryland

Shaevitz, Michael H.
Columbia University

Shapero, Don
National Academy of Sciences

Shen, Benjamin
U. of California, Riverside

Shields, Gregory A.
University of Texas, Austin

Shoemaker, David H.
Massachusetts Institute of Technology

Shutt, Tom
CfPA, U. of California, Berkeley

Sigl, Günter
University of Chicago

Singh, Vinay M.
University of Minnesota

Sinuis, Gus
Los Alamos National Laboratory

Skillman, Evan
University of Minnesota

Smith, David A.
INFN, Pisa

Smith, Verne V.
University of Texas, Austin

Smoot, George
Lawrence Berkeley Laboratory

Sommers, Paul
University of Utah

Sonnenschein, Andrew
U. of California, Santa Barbara

Spooner, Neil
University of Oxford

Stanev, Todor
Bartol Research Institute

Stanton, Noel R.
Kansas State Universtiy

Starkman, Glenn
University of Toronto

Stecker, F. W.
NASA/Goddard Space Flight Center

Steinhardt, Paul Joseph
University of Pennsylvania

Stenger, Victor J.
University of Hawaii

Stephens, S. Alfred
Tata Institute of Fundamental Research

Stockwell, Walter
CfPA, U. of California, Berkeley

Stokstad, Robert G.
Lawrence Berkeley Laboratory

Stoughton, Chris
Fermi National Accelerator Laboratory

Streitmatter, Robert E.
NASA/Goddard Space Flight Center

Strominger, Andrew
U. of California, Santa Barbara

Stroynowski, Ryszard
Southern Methodist University

Stubbs, Christopher
U. of California, Santa Barbara

Suen, Wai-Mo
Washington University

Sullivan, Greg
University of Chicago

Sussman, Gerald Jay
Massachusetts Institute of Technology

Sutton, C. Sean
Mount Holyoke College

Swank, Jean H.
NASA/Goddard Space Flight Center

Swesty, F. Douglas
University of Illinois

Swordy, Simon
EFI, University of Chicago

Szalay, Alexander
Johns Hopkins University

Tang, Jonathan
University of Utah

Tao, Charling
College de France

Tarle, Gregory
University of Michigan

Taylor, Richard E.
Stanford Linear Accelerator Center

Thomas, David
Ohio State University

Thomas, Jennifer
Fermi National Accelerator Laboratory

Thorlacius, Larus
U. of California, Santa Barbara

Thorne, Kip
California Institute of Technology

Thun, Rudolf P.
University of Michigan

Tilav, Serap
U. of Wisconsin, Madison

Timbie, Peter T.
Brown University

Timmes, Frank
EFI, University of Chicago

Tinto, Massimo
Jet Propulsion Laboratory

Tkachev, Igor
Fermi National Accelerator Laboratory

Tobar, Mike
U. of Western Australia

Tolstaya, Ekaterina D.
Skobeltsyn, Moscow State University

Tomozawa, Yukio
University of Michigan

Toohig, Timothy E.
SSC Laboratory

Torii, Shoji
Kanagawa University

Travis, John
Science Magazine

Tsytovich, Vadim Nikolaevich
General Physics Institute

Tucker, Gregory
University of British Columbia

Tümer, Tümay
U. of California, Riverside

Tung, Michael M.
Universidad de Valencia

Turner, Michael
U. of Chicago/Fermilab

Turok, Neil
Princeton University

Tyson, Anthony
AT&T-Bell Labs

Tzanakos, George
University of Athens

Ulmer, Andrew
Princeton University

Vazquez, Ricardo
University of Wisconsin

Vilanova, Didier
DAPNIA/SPP, CE Saclay

Waga, Ioav
Fermi National Accelerator Laboratory

Walker, Terrance
Ohio State University

Wallerstein, George
University of Washington

Wang, Yun
Fermi National Accelerator Laboratory

Watson, Alan A.
University of Leeds

Weaver, Thomas
Lawrence Livermore National Laboratory

Weekes, Trevor C.
Harvard-Smithsonian Center for Astrophysics

Weiss, Rainer
Massachusetts Institute of Technology

Weller, Henry R.
Duke University

Wen, Han
Stanford Linear Accelerator Center

Wheeler, J. Craig
University of Texas, Austin

Whitcomb, Stanley E.
California Institute of Technology

White, Storn
CfPA, U. of California, Berkeley

Widrow, Larry
Queen's University

Wiencke, Lawrence R.
University of Utah

Wilkes, R. Jeffrey
University of Washington

Williams, David A.
U. of California, Santa Cruz

Wilson, James R.
Lawrence Livermore National Laboratory

Winn, David R.
Fairfield University

Wischnewski, Ralf
DESY

Wojcicki, Stanley G.
Stanford University

Wood, Kent S.
Naval Research Laboratory

Woosley, Stan E.
U. of California, Santa Cruz

Wright, Dennis
TRIUMF

Wright, Edward L.
U. of California, Los Angeles

Xu, Chun Xian
IHEP, Beijing

Yodh, Gaurang B.
U. of California, Irvine

Zas, Enrique
Universidad de Santiago

Zucker, Mike
California Institute of Technology